Discover Digital

Engaging and informing today's learner

> "Learners who used an integrated learning solution reported a stronger relationship to score improvements than those who used only a print textbook."
>
> (Source: The Book Industry Study Group Inc., May 2012)

Discover Digital takes you step-by-step toward solutions designed around learner engagement and better outcomes.

○ Pinpoint online resources for your disciplines.

○ Choose among course, assignment or textbook solutions.

○ Access walk-throughs that simulate your user experience.

Explore today at:
www.cengage.com/discoverdigital

EDITION 5

PREALGEBRA

ALAN S. TUSSY
CITRUS COLLEGE

DIANE R. KOENIG
ROCK VALLEY COLLEGE

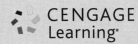

Australia • Brazil • Mexico • Singapore • United Kingdom • United States

To **Gordon Guillaume**, an excellent mathematics instructor and an outstanding department chairperson.

AST

To my daughters, **Ashley**, **Brianna**, and **Carly**, whom I love very much and who make me extremely proud.

DRK

Prealgebra, Fifth Edition
Alan S. Tussy, Diane R. Koenig

Product Manager: Marc Bove
Content Developer: Stefanie Beeck
Associate Content Developer: Cynthia Ashton
Product Assistant: Kathryn Clark
Media Developer: Guanglei Zhang, Bryon Spencer
Brand Manager: Gordon Lee
Market Development Manager: Mark Linton
Content Project Manager: Jennifer Risden
Art Director: Vernon Boes
Manufacturing Planner: Becky Cross
Rights Acquisitions Specialist: Don Schlotman
Production Service: Scratchgravel Publishing Services
Photo Researcher: PreMedia Global
Text Researcher: PreMedia Global
Copy Editor: Peggy Tropp
Illustrator: Lori Heckelman Illustration; Scratchgravel Publishing Services
Text Designer: Diane Beasley
Cover Designer: Terri Wright
Cover Image: Lizard/Shutterstock.com
Compositor: Graphic World, Inc.

© 2015, 2011 Cengage Learning

ALL RIGHTS RESERVED. No part of this work covered by the copyright herein may be reproduced, transmitted, stored, or used in any form or by any means graphic, electronic, or mechanical, including but not limited to photocopying, recording, scanning, digitizing, taping, Web distribution, information networks, or information storage and retrieval systems, except as permitted under Section 107 or 108 of the 1976 United States Copyright Act, without the prior written permission of the publisher.

> For product information and technology assistance, contact us at
> **Cengage Learning Customer & Sales Support, 1-800-354-9706**.
> For permission to use material from this text or product,
> submit all requests online at **www.cengage.com/permissions**.
> Further permissions questions can be e-mailed to
> **permissionrequest@cengage.com**.

Library of Congress Control Number: 2013944989

ISBN-13: 978-1-285-73725-6

ISBN-10: 1-285-73725-3

Cengage Learning
200 First Stamford Place, 4th Floor
Stamford, CT 06902
USA

Cengage Learning is a leading provider of customized learning solutions with office locations around the globe, including Singapore, the United Kingdom, Australia, Mexico, Brazil, and Japan. Locate your local office at **www.cengage.com/global**.

Cengage Learning products are represented in Canada by Nelson Education, Ltd.

To learn more about Cengage Learning Solutions, visit **www.cengage.com**. Purchase any of our products at your local college store or at our preferred online store **www.cengagebrain.com**.

Printed in the United States of America
1 2 3 4 5 6 7 17 16 15 14 13

CONTENTS

Study Skills Workshop S-1

CHAPTER 1
Whole Numbers 1

1.1 An Introduction to the Whole Numbers 2
THINK IT THROUGH *Re-entry Students* 9
1.2 Adding and Subtracting Whole Numbers 15
1.3 Multiplying Whole Numbers 34
1.4 Dividing Whole Numbers 48
1.5 Prime Factors and Exponents 63
1.6 The Least Common Multiple and the Greatest Common Factor 72
1.7 Order of Operations 84
THINK IT THROUGH *Education Pays* 91
1.8 Solving Equations Using Addition and Subtraction 96
1.9 Solving Equations Using Multiplication and Division 105
Chapter Summary and Review 114
Chapter Test 132

CHAPTER 2
The Integers 135

2.1 An Introduction to the Integers 136
THINK IT THROUGH *Credit Card Debt* 139
2.2 Adding Integers 148
THINK IT THROUGH *Cash Flow* 152
2.3 Subtracting Integers 160
2.4 Multiplying Integers 169
2.5 Dividing Integers 179
2.6 Order of Operations and Estimation 187
2.7 Solving Equations That Involve Integers 196
Chapter Summary and Review 208
Chapter Test 219
Cumulative Review 221

CHAPTER 3

The Language of Algebra 225

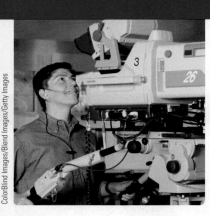

- **3.1** Algebraic Expressions 226
- **3.2** Evaluating Algebraic Expressions and Formulas 237
- **THINK IT THROUGH** *Study Time* 245
- **3.3** Simplifying Algebraic Expressions and the Distributive Property 251
- **3.4** Combining Like Terms 259
- **3.5** Simplifying Expressions to Solve Equations 268
- **3.6** Using Equations to Solve Application Problems 276

Chapter Summary and Review 289

Chapter Test 299

Cumulative Review 301

CHAPTER 4

Fractions and Mixed Numbers 303

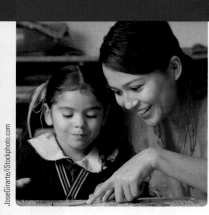

- **4.1** An Introduction to Fractions 304
- **4.2** Multiplying Fractions 318
- **4.3** Dividing Fractions 333
- **4.4** Adding and Subtracting Fractions 343
- **THINK IT THROUGH** *Budgets* 354
- **4.5** Multiplying and Dividing Mixed Numbers 360
- **4.6** Adding and Subtracting Mixed Numbers 374
- **THINK IT THROUGH** *381*
- **4.7** Order of Operations and Complex Fractions 387
- **4.8** Solving Equations That Involve Fractions 399

Chapter Summary and Review 416

Chapter Test 437

Cumulative Review 439

Contents   v

CHAPTER 5

Decimals   443

5.1 An Introduction to Decimals   444
5.2 Adding and Subtracting Decimals   458
5.3 Multiplying Decimals   472
THINK IT THROUGH   *Overtime*   474
5.4 Dividing Decimals   486
THINK IT THROUGH   *GPA*   496
5.5 Fractions and Decimals   500
5.6 Square Roots   514
5.7 Solving Equations That Involve Decimals   522
     Chapter Summary and Review   535
     Chapter Test   552
     Cumulative Review   555

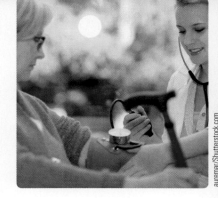

CHAPTER 6

Ratio, Proportion, and Measurement   557

6.1 Ratios   558
THINK IT THROUGH   *Student-to-Instructor Ratio*   561
6.2 Proportions   572
6.3 American Units of Measurement   587
6.4 Metric Units of Measurement; Medical Applications   600
6.5 Converting between American and Metric Units   614
THINK IT THROUGH   *Studying in Other Countries*   617
     Chapter Summary and Review   623
     Chapter Test   638
     Cumulative Review   640

CHAPTER 7

Percent 643

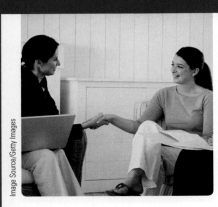

- **7.1** Percents, Decimals, and Fractions 644
- **7.2** Solving Percent Problems Using Percent Equations and Proportions 657
- THINK IT THROUGH *Community College Students* 673
- **7.3** Applications of Percent 679
- THINK IT THROUGH *Studying Mathematics* 687
- **7.4** Estimation with Percent 696
- **7.5** Interest 703
  - Chapter Summary and Review 714
  - Chapter Test 732
  - Cumulative Review 735

CHAPTER 8

Graphs and Statistics 739

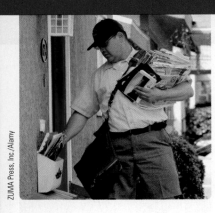

- **8.1** Reading Graphs and Tables 740
- **8.2** Mean, Median, and Mode 755
- THINK IT THROUGH *The Value of an Education* 762
- **8.3** Equations in Two Variables; The Rectangular Coordinate System 767
- THINK IT THROUGH *Population Shift* 774
- **8.4** Graphing Linear Equations 780
  - Chapter Summary and Review 797
  - Chapter Test 810
  - Cumulative Review 814

CHAPTER 9

An Introduction to Geometry 819

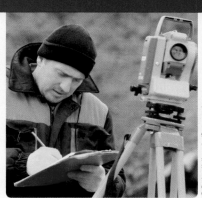

- **9.1** Basic Geometric Figures; Angles 820
- **9.2** Parallel and Perpendicular Lines 833
- **9.3** Triangles 844
- **9.4** The Pythagorean Theorem 855
- **9.5** Congruent Triangles and Similar Triangles 862
- **9.6** Quadrilaterals and Other Polygons 875
- **9.7** Perimeters and Areas of Polygons 885
- **THINK IT THROUGH** *Dorm Rooms* 890
- **9.8** Circles 900
- **9.9** Volume 909
  - Chapter Summary and Review 919
  - Chapter Test 942
  - Cumulative Review 946

CHAPTER 10

Exponents and Polynomials 949

- **10.1** Multiplication Rules for Exponents 950
- **10.2** Introduction to Polynomials 957
- **10.3** Adding and Subtracting Polynomials 961
- **10.4** Multiplying Polynomials 967
  - Chapter Summary and Review 975
  - Chapter Test 979
  - Cumulative Review 980

APPENDIXES

- **Appendix 1** Probability A-1
- **Appendix 2** Inductive and Deductive Reasoning A-7
- **Appendix 3** Roots and Powers A-15
- **Appendix 4** Answers to Selected Exercises (appears in Student Edition only) A-17

Index I-1

PREFACE

We are excited to present the Fifth Edition of *Prealgebra*. We believe the revision process has produced an even stronger instructional experience for both students and teachers. First, we have fine-tuned several of the popular features of our series, including the *Strategy* and *Why* example structure, the problem-solving strategy, and the real-life applications.

Second, we have expanded the quantity and types of problems available in Enhanced WebAssign® to include more vocabulary and application problems. In addition we have added a wider variety of *Guided Practice* and *Try It Yourself* problems.

We want to thank all of you across the country who provided suggestions and input about the previous edition. Your insights have proven invaluable. Throughout the revision process, our fundamental belief has remained the same: Algebra is a language in its own right. As always, the prime objective of this textbook (and its accompanying ancillaries) is to teach students how to read, write, speak, and think using the language of algebra.

NEW TO THIS EDITION

Extensively Updated Application Problems

Many real-life application problems have been updated to keep them current and relevant. Students can see how math relates to their everyday lives through these timely topics.

Additional Medical Applications

To better prepare students who plan to enter the health care field, more medical applications have been added to Section 6.4, *Metric Units of Measurement; Medical Applications*.

Appendix on Probability

Because of its increasing importance in daily life, many schools are adding the topic of probability to their Prealgebra curriculum. We have inserted a new appendix, Appendix 1, *Probability*, to address this need.

Additional Emphasis on Estimation

Estimation offers the students a quick way to determine the reasonableness of their answers. We have inserted more detailed estimation problems to several Study Sets to promote better student understanding of this topic.

Six-Step Problem-Solving Strategy

In an effort to better describe the problem-solving strategy used in this book, we have inserted a new second step in what was formerly a five-step problem process. This additional step (*Assign a variable to represent an unknown value in the problem*) better delineates the thought process students should use as they solve application problems. The six steps of the problem-solving strategy are now: *Analyze the problem*, *Assign a variable*, *Form an equation*, *Solve the equation*, *State the conclusion*, and *Check the result*.

Additional *Try It Yourself* Problems

In response to reviewers' suggestions, we have inserted additional problems in the *Try It Yourself* sections of selected Study Sets.

TRUSTED FEATURES

Emphasis on Study Skills

Prealgebra begins with a *Study Skills Workshop* module. Instead of simple, unrelated suggestions printed in the margins, this module contains one-page discussions of study skills topics, followed by a *Now Try This* section offering students actionable skills, assignments, and projects that will impact their study habits throughout the course.

Chapter Openers That Answer the Question: When Will I Use This?

Instructors are asked this question time and again by students. In response, we have written chapter openers called *From Campus to Careers*. This feature highlights vocations that require various algebraic skills. Designed to inspire career exploration, each includes job outlook, educational requirements, and annual earnings information. Careers presented in the openers are tied to an exercise found later in the *Study Sets*.

Examples That Tell Students Not Just How, But WHY

Why? That question is often asked by students as they watch their instructor solve problems in class and as they are working on problems at home. It's not enough to know *how* a problem is solved. Students gain a deeper understanding of the algebraic concepts if they know *why* a particular approach was taken. This instructional truth was the motivation for including a *Strategy* and a *Why* explanation in each worked example.

Examples That Offer Immediate Feedback

Each worked example includes a *Self Check*. These can be completed by students on their own or as classroom lecture examples, which is how Alan Tussy uses them. Alan asks selected students to read aloud the *Self Check* problems as he writes what the student says on the board. The other students, with their books open to that page, can quickly copy the *Self Check* problem to their notes. This speeds up the note-taking process and encourages student participation in his lectures. It also teaches students how to read mathematical symbols. Each *Self Check* answer is printed adjacent to the corresponding problem in the *Annotated Instructor's Edition* for easy reference. *Self Check* solutions can be found at the end of each section in the student edition before each *Study Set*.

Examples That Show the Behind-the-Scenes Calculations

Some steps to worked examples in *Prealgebra* involve arithmetic calculations that are too complicated to be performed mentally. In these instances, we have shown the actual computations that must be made to complete the formal solution. These computations appear directly to the right of the author notes and are separated from them by a thin, gray line. The necessary addition, subtraction, multiplication, or division (usually done on scratch paper) is placed at the appropriate stage of the solution where such a computation is required. Rather than simply list the steps of the solution horizontally, making no mention of how the numerical values within the solution are obtained, this unique feature will help answer the often heard question from a struggling student, "How did you get that answer?" It also serves as a model for the calculations that students must perform independently to solve the problems in the Study Sets.

Examples That Ask Students to Work Independently

Each worked example ends with a *Now Try* problem. These are the final steps in the learning process. Each one is linked to a similar problem found within the *Guided Practice* section of the *Study Sets*.

Emphasis on Problem Solving

New to *Prealgebra*, the six-step problem-solving strategy guides students through applied worked examples using the *Analyze, Assign, Form, Solve, State,* and *Check* process. This approach clarifies the thought process and mathematical skills necessary to solve a wide variety of problems. As a result, students' confidence is increased, and their problem-solving abilities are strengthened.

Detailed Author Notes

These notes guide students along in a step-by-step process and appear in the solution to every worked out example.

Integrated Focus on the Language of Algebra

Language of Algebra boxes draw connections between mathematical terms and everyday references to reinforce the language of algebra philosophy that runs throughout the text.

The Language of Algebra As you read the application problems, watch for the following words and phrases. They often indicate negative numbers.

behind	below	before	deficit	debt	drop
in the red	overdrawn	under	loss	B.C.	

Guidance When Students Need It Most

Appearing at key teaching moments, *Success Tips* and *Caution* boxes improve students' problem-solving abilities, warn students of potential pitfalls, and increase clarity.

Success Tip By the commutative property of multiplication, we can change the *order* of factors. By the associative property of multiplication, we can change the *grouping* of factors.

Caution! Be careful when translating division. As with subtraction, order is important. For example, *s* divided by *d* is *not* written $\frac{d}{s}$.

Useful Objectives Help Keep Students Focused

Each section begins with a set of numbered *Objectives* that focus students' attention on the skills that they will learn. As each objective is discussed in the section, the number and heading reappear to remind readers of the objective at hand.

Study Skills That Point Out Common Student Mistakes

In Chapter 1 we have included four *Study Skills Checklists* designed to actively show students how to effectively use the key features in this text. Subsequent chapters include one checklist just before the *Chapter Summary and Review* that provides another layer of preparation to promote student success. These *Study Skills Checklists* warn students of common errors, giving them time to consider these pitfalls before taking their exam.

Study Sets

Study Sets are found in each section and offer a multifaceted approach to practicing and reinforcing the concepts taught in each section. They are designed for students to methodically build their knowledge of the section concepts, from basic recall to increasingly complex problem solving, through reading, writing, and thinking mathematically.

Vocabulary—Each *Study Set* begins with the important *Vocabulary* discussed in that section. The fill-in-the-blank vocabulary problems emphasize the main concepts taught in the chapter and provide the foundation for learning and communicating the language of algebra.

Concepts—In *Concepts*, students are asked about the specific subskills and procedures necessary to successfully complete the *Guided Practice* and *Try It Yourself* problems that follow.

Notation—In *Notation*, the students review the new symbols introduced in a section. Often they are asked to fill in steps of a sample solution. This strengthens their ability to read and write mathematics and prepares them for *Guided Practice* problems by modeling solution formats.

Guided Practice—The problems in *Guided Practice* are linked to a corresponding worked example or objective from that section. This feature promotes student success by referring them to the proper examples if they encounter difficulties solving homework problems.

Try It Yourself—To promote problem recognition, the *Try It Yourself* problems are thoroughly mixed and are *not* linked to worked examples, giving students an opportunity to practice decision making and strategy selection as they would when taking a test or quiz.

Applications—The *Applications* provide students the opportunity to apply their newly acquired algebraic skills to relevant and interesting real-life situations.

Writing—The *Writing* problems help students build mathematical communication skills.

Review—The *Review* problems consist of randomly selected problems from previous chapters. These problems are designed to keep students' successfully mastered skills up to date before they move on to the next section.

Using Your Calculator

This is an optional feature that is designed for instructors who wish to use calculators as part of the instruction in this course. This feature introduces keystrokes and shows how scientific and graphing calculators can be used to solve problems. In the *Study Sets*, icons are used to denote problems that may be solved using a calculator.

Think It Through

The *Think It Through* feature makes the connection between mathematics and student life. These relevant topics often require algebra skills from the chapter to be applied to a real-life situation. Topics include tuition costs, student enrollment, job opportunities, credit cards, and many more.

Comprehensive End-of-Chapter Summary with Integrated Chapter Review

The end-of-chapter material has been designed to function as a complete study guide for students. Chapter summaries include definitions, concepts, and examples—organized by section. Review problems for each section immediately follow the summary for that section. Students will find the detailed summaries a very valuable study aid when preparing for exams.

Chapter Tests

Appearing at the end of every chapter, Chapter Tests can be used as preparation for the class exam.

Cumulative Reviews

The *Cumulative Review* follows the end-of-chapter material and keeps students' skills current before they move on to the next chapter. For reference, each problem is linked to the associated section from which the problem came. The final *Cumulative Review* is often used by instructors as a Final Exam Review.

ANCILLARIES

For the Student	For the Instructor
	Annotated Instructor's Edition (ISBN: 978-1-285-73728-7) This special version of the complete student text contains all answers printed next to all respective exercises.
Student Solutions Manual (ISBN: 978-1-285-73807-9) Author: Rhoda Oden, Gadsden State Community College The Student Solutions Manual provides worked-out solutions to all of the odd-numbered exercises in the text.	**Complete Solutions Manual** (ISBN: 978-1-285-73808-6) Author: Rhoda Oden, Gadsden State Community College The Complete Solutions Manual provides worked-out solutions to all of the problems in the text.
Student Workbook (ISBN: 978-1-285-85703-9) Author: Maria H. Andersen, former math faculty at Muskegon Community College and now working in the learning software industry The Student Workbook contains the entire student Assessments, Activities, and Worksheets from the Instructor's Resource Binder for classroom discussions, in-class activities, and group work.	**Instructor's Resource Binder** (ISBN: 978-0-538-73675-6) Author: Maria H. Andersen, former math faculty at Muskegon Community College and now working in the learning software industry Each section of the main text is discussed in uniquely designed Teaching Guides containing instruction tips, examples, activities, worksheets, overheads, assessments, and solutions to all worksheets and activities.
	Instructor's Companion Website Everything you need for your course in one place! This collection of book-specific lecture and class tools is available online via www.cengage.com/login. Formerly found on the PowerLecture, access and download PowerPoint presentations, images, and more.
Enhanced WebAssign Homework with LOE Access (Printed Access Card ISBN: 978-1-285-85770-1, Online Access Code ISBN: 978-1-285-85773-2) Enhanced WebAssign (assigned by the instructor) provides you with instant feedback on homework assignments. This online homework system is easy to use and includes helpful links to textbook sections, video examples, and problem-specific tutorials.	**Enhanced WebAssign Homework with LOE Access** (Printed Access Card ISBN: 978-1-285-85770-1, Online Access Code ISBN: 978-1-285-85773-2) Exclusively from Cengage Learning, Enhanced WebAssign combines the exceptional mathematics content that you know and love with immediate feedback, rich tutorial content, and interactive, fully customizable eBooks (YouBook), helping students to develop a deeper conceptual understanding of their subject matter. Online assignments can be built by selecting from thousands of text-specific problems or can be supplemented with problems from any Cengage Learning textbook.
	Solution Builder This online database offers complete worked solutions to all exercises in the text, allowing you to create customized, secure solutions printouts (in PDF format) matched exactly to the problems you assign in class. For more information, visit www.cengage.com/solutionbuilder.

For the Student	For the Instructor
Math Study Skills Workbook, 4e	**Math Study Skills Workbook, 4e**
(ISBN: 978-0-8400-5309-1)	(ISBN: 978-0-8400-5309-1)
Author: Paul D. Nolting	Author: Paul D. Nolting
This best-selling workbook helps traditionally unsuccessful students learn to effectively study mathematics. Typically used for a Math Study Skills course, Freshman Seminar, or as a supplement to class lectures, the Nolting workbook helps students identify their strengths, weaknesses, and personal learning styles in math. Nolting offers proven study tips, test-taking strategies, a homework system, and recommendations for reducing anxiety and improving grades.	This best-selling workbook helps traditionally unsuccessful students learn to effectively study mathematics. Typically used for a Math Study Skills course, Freshman Seminar, or as a supplement to class lectures, the Nolting workbook helps students identify their strengths, weaknesses, and personal learning styles in math. Nolting offers proven study tips, test-taking strategies, a homework system, and recommendations for reducing anxiety and improving grades.

ACKNOWLEDGMENTS

We want to express our gratitude to all those who helped with this project: Brenda Keller, Rhoda Oden, Steve Odrich, Mary Lou Wogan, Paul McCombs, Maria H. Andersen, Sheila Pisa, Laurie McManus, Alexander Lee, Ed Kavanaugh, Karl Hunsicker, Cathy Gong, Dave Ryba, Terry Damron, Marion Hammond, Lin Humphrey, Doug Keebaugh, Robin Carter, Tanja Rinkel, Jeff Cleveland, Jo Morrison, Sheila White, Jim McClain, Paul Swatzel, Matt Stevenson, Carole Carney, Joyce Low, Rob Everest, David Casey, Heddy Paek, Ralph Tippins, Mo Trad, Eagle Zhuang, Chris Scott, Victoria Dominguez, Esme Medrano, Sam M. Ditzion, Lisa Brown, Elaine Tucker, Laura McInerney, Solomon Willis, Tracy Nehnevaji, and the Citrus College library staff (including Barbara Rugeley) for their help with this project. Your encouragement, suggestions, and insight have been invaluable to us.

We would also like to express our thanks to the Cengage Learning editorial, marketing, production, and design staff for helping us craft this new edition: Charlie Van Wagner, Danielle Derbenti, Gordon Lee, Rita Lombard, Stefanie Beeck, Jennifer Cordoba, Heleny Wong, Jennifer Risden, Vernon Boes, Diane Beasley, Anne Draus, and Graphic World.

Additionally, we would like to say that authoring a textbook is a tremendous undertaking. A revision of this scale would not have been possible without the thoughtful feedback and support from the following colleagues listed below. Their contributions to this edition have shaped this revision in countless ways.

Alan S. Tussy
Diane R. Koenig

Reviewers

Tonya Adkins, *Johnson & Wales University*
Darla Aguilar, *Pima Community College*
Catherine Albright, *Brookdale Community College*
Patricia Barrientos, *El Paso Community College*
Gail Burkett, *Palm Beach State College*
Dorothy Carver, *Portland Community College*
Kristin Chatas, *Washtenaw Community College*
Seth Daugherty, *St. Louis Community College at Forest Park*
Mary Deas, *Johnson County Community College*
Steve Dostal, *College of the Desert*
Joe Estephan, *Victor Valley College*
Suzie Goss, *Lone Star College Kingwood*
Sheyleah Harris-Plant, *South Plains College*
Shelle Hartzel, *Lake Land College*
Steven Jackson, *Ivy Tech Community College*
Cynthia Johnson, *Arizona Western College*
Greta Johnson, *Amarillo College*
Jonathan Lee, *University of Oklahoma, Oklahoma City Community College*
Edith Lester, *Volunteer State Community College*
Susan McLoughlin, *Union County College*
Stacey Nichols, *Anne Arundel Community College*
Heather Niehoff, *Ivy Tech Community College*
Edward Pierce, *Ivy Tech Community College*
Jim Pierce, *Lincoln Land Community College*
Glenn Sandifer, *San Jacinto College, Central Campus*

Larry Small, *Pierce College*
Alexis Thurman, *County College of Morris*
Robyn Toman, *Anne Arundel Community College*
Diane Valade, *Piedmont Virginia Community College*
Betty Vix Weinberger, *Delgado Community College*
Lynn White, *Jones County Junior College*
Mary Wolyniak, *Broome Community College*

Reviewers of Previous Editions

Sheila Anderson, *Housatonic Community College*
Cedric E. Atkins, *Mott Community College*
William D. Barcus, *SUNY, Stony Brook*
David Behrman, *Somerset Community College*
Kathy Bernunzio, *Portland Community College*
Linda Bettie, *Western New Mexico University*
Michael Branstetter, *Hartnell College*
Joseph A. Bruno, Jr., *Community College of Allegheny County*
Girish Budhwar, *United Tribes Technical College*
Sharon Camner, *Pierce College–Fort Steilacoom*
Robin Carter, *Citrus College*
John Coburn, *Saint Louis Community College–Florissant Valley*
Eric Compton, *Brookdale Community College*
Joy Conner, *Tidewater Community College*
Sally Copeland, *Johnson County Community College*
Ann Corbeil, *Massasoit Community College*
Ben Cornelius, *Oregon Institute of Technology*
Ruth Dalrymple, *Saint Philip's College*
Nathalie Darden, *Brookdale Community College*
Carolyn Detmer, *Seminole Community College*
John D. Driscoll, *Middlesex Community College*
James Edmondson, *Santa Barbara Community College*
LaTonya Ellis, *Bishop State Community College*
David L. Fama, *Germanna Community College*
Steven Felzer, *Lenoir Community College*
Rhoderick Fleming, *Wake Technical Community College*
Maggie Flint, *Northeast State Technical Community College*
Charles Ford, *Shasta College*
J. Donato Fortin, *Johnson and Wales University*
Heather Gallacher, *Cleveland State University*
Barbara Gentry, *Parkland College*
Kathirave Giritharan, *John A. Logan College*
Marilyn Green, *Merritt College and Diablo Valley College*
Joseph Guiciardi, *Community College of Allegheny County*
Geoff Hagopian, *College of the Desert*
Deborah Hanus, *Brookhaven College*
A.T. Hayashi, *Oxnard College*
Michael Heeren, *Hamilton College*
Cheryl Hobneck, *Illinois Valley Community College*
Laurie Hoecherl, *Kishwaukee College*
Todd J. Hoff, *Wisconsin Indianhead Technical College*
Judith Jones, *Valencia Community College*
Therese Jones, *Amarillo College*
Joanne Juedes, *University of Wisconsin–Marathon County*
Susan Kautz, *Cy-Fair College*
Jack Keating, *Massasoit Community College*

Russ Alan Killingsworth, *Seattle Pacific University*
Dennis Kimzey, *Rogue Community College*
Monica C. Kurth, *Scott Community College*
Sally Leski, *Holyoke Community College*
Sandra Lofstock, *St. Petersberg College–Tarpon Springs Center*
Lynn Marecek, *Santa Ana College*
Lois Martin, *Massasoit Community College*
Mikal McDowell, *Cedar Valley College*
Chris Mirbaha, *The Community College of Baltimore County*
Elizabeth Morrison, *Valencia Community College*
Jan Alicia Nettler, *Holyoke Community College*
Marge Palaniuk, *United Tribes Technical College*
K. Maggie Pasqua, *Brookdale Community College*
Gregory Perkins, *Hartnell College*
Scott Perkins, *Lake-Sumter Community College*
Angela Peterson, *Portland Community College*
Euguenia Peterson, *City Colleges of Chicago–Richard Daley*
Jane Pinnow, *University of Wisconsin–Parkside*
Carol Ann Poore, *Hinds Community College*
Christopher Quarles, *Shoreline Community College*
George Reed, *Angelina College*
J. Doug Richey, *Northeast Texas Community College*
Patricia C. Rome, *Delgado Community College*
Patricia B. Roux, *Delgado Community College*
Rebecca Rozario, *Brookdale Community College*
Angelo Segalla, *Orange Coast College*
Eric Sims, *Art Institute of Dallas*
Lee Ann Spahr, *Durham Technical Community College*
Annette Squires, *Palomar College*
John Squires, *Cleveland State Community College*
John Strasser, *Scottsdale Community College*
June Strohm, *Pennsylvania State Community College–Dubois*
Rita Sturgeon, *San Bernardino Valley College*
Stuart Swain, *University of Maine at Machias*
Celeste M. Teluk, *D'Youville College*
Jo Anne Temple, *Texas Technical University*
Sharon Testone, *Onondaga Community College*
Bill Thompson, *Red Rocks Community College*
Barbara Tozzi, *Brookdale Community College*
Marilyn Treder, *Rochester Community College*
Sven Trenholm, *Herkeimer County Community College*
Donna Tupper, *Community College of Baltimore County–Essex*
Thomas Vanden Eynden, *Thomas More College*
Andreana Walker, *Calhoun Community College*
Jane Wampler, *Housatonic Community College*
Arminda Wey, *Brookdale Community College*
Stephen Whittle, *Augusta State University*
Mary Lou Wogan, *Klamath Community College*
Valerie Wright, *Central Piedmont Community College*
Kevin Yokoyama, *College of the Redwoods*
Mary Young, *Brookdale Community College*

ABOUT THE AUTHORS

Alan S. Tussy

Alan Tussy has taught all levels of developmental mathematics at Citrus College in Glendora, California. He has written nine math books—a paperback series and a hardcover series. A creative and visionary teacher who maintains a keen focus on his students' greatest challenges, Alan Tussy is an extraordinary author, dedicated to his students' success. Alan received his Bachelor of Science degree in Mathematics from the University of Redlands and his Master of Science degree in Applied Mathematics from California State University, Los Angeles. He has taught up and down the curriculum from Prealgebra to Differential Equations. He is currently focusing on the developmental math courses. Professor Tussy is a member of the American Mathematical Association of Two-Year Colleges.

Diane R. Koenig

Diane Koenig received a Bachelor of Science degree in Secondary Math Education from Illinois State University in 1980. She began her career at Rock Valley College in 1981, when she became the Math Supervisor for the newly formed Personalized Learning Center. Earning her Master's Degree in Applied Mathematics from Northern Illinois University, Ms. Koenig in 1984 had the distinction of becoming the first full-time woman mathematics faculty member at Rock Valley College. In addition to being nominated for AMATYC's Excellence in Teaching Award, Diane Koenig was chosen as the Rock Valley College Faculty of the Year by her peers in 2005, and, in 2006, she was awarded the NISOD Teaching Excellence Award as well as the Illinois Mathematics Association of Community Colleges Award for Teaching Excellence. In addition to her teaching, Ms. Koenig has been an active member of the Illinois Mathematics Association of Community Colleges (IMACC). As a member, she has served on the board of directors, on a state-level task force rewriting the course outlines for the developmental mathematics courses, and as the association's newsletter editor.

APPLICATIONS INDEX

Examples that are applications are shown with **boldface** page numbers.
Exercises that are applications are shown with lightface page numbers.

Agriculture. *See* **Farming**

Ancient history
Great Pyramid, 598
Great Sphinx, 599
history, 147, 167, 207
Roman Empire, 206

Animals
animal shelters, 113
aquariums, **78,** 642
birdbaths, 918
birds, **46**
cattle ranching, 413
cheetahs, 621
dogs, 586, 628
elephants, **26,** 599
farm animals, A-13
fishing, 248
frogs, **47**
Grand Canyon, 185
hippos, 599
horse racing, 144–145
horses, **26**
insects, **42**
koalas, **47**
life span, 113
lions, 621
marine science, 234
pet doors, 430
pet medication, 981
pets, 438, A-13
polar bears, 637
sharks, **20,** 440
speed of animals, **742**
spending on pets, 983
turtles, 247
weight of a baby, 599, 613
whales, 291, 621
wolves, endangered, **20–21**
zoos, A-13

Architecture
architecture, 317, **903**
blueprints, 585
bridge safety, 31
dimensions of a house, 31
drafting, 512, 585, 854
Greek architecture, 967
high-rise buildings, 637
interior decorating, 287
kitchen design, 331
new homes, 118
New York architecture, 235
reading blueprints, 966
retaining walls, 291
retrofits, 484
scale drawings, **580,** 628, 982
scale models, **580**

Sears Tower, 599
skyscrapers, 613

Arts and entertainment. *See*
Entertainment

Business and industry
accounting, 168, 248, 317, 553, **706,** 980
advertising, **43,** 179, 550
airline accidents, **20**
airlines, 159, 206
airports, **3,** 118
antique collecting, 283
appliance sales, **683**
ATMs, 83, **745**
attorney's fees, **277**
average years of experience, 292
awards, 534
bakery supplies, 483
bank takeovers, 220
banking, 119, 218, 220, 223, 470
bankruptcy, 569
banks, 439
bedding sales, 692
book sales, 692
bookstores, 207
bottling, 632
bouquets, **80**
budgets, **354,** 703
business, 570, 736
business accounts, 283, 285
business expenses, **529,** 816
business performance, 584
business takeovers, 194
butcher shops, 498
butchers, 290
buying a business, 223
buying fishing equipment, 693
buying paint, 599
capacity of a gym, **663**
capital gains tax, 691
car loans, 498
car sales, 290
car shows, 734
casting a movie, **409**
catering, 599
cell phones, **3**
checking accounts, 14, 168
civil service, **278**
clothes designers, 290
clothes shopping, 584
clothing labels, **614**
clothing sales, **228**
clothing stores, 247
code violations, 731
coffee, 13
coins, **479,** 540
commercials, 287, **686–687**

commissions, 285, 287, 694
commuting, **228**
comparison shopping, **495,** 622
computer companies, 236
corporate downsizing, 286
cost overruns, 112
coupons, 284
crude oil, **46**
customer satisfaction, 250
cutting budgets, 186
daily pay, **40**
declining sales, 692
delicatessens, **509,** 513
dining out, **689,** 702
dinnerware sales, 692
discount hotels, 692
discounts, **689–690**
draining pools, **61**
e-commerce, 484
employment agencies, 691
entertainment costs, **108**
entry-level jobs, **61**
farm loans, 712
federal taxes, **2**
fences, 32
films, **241**
financial statements, 218
financing, 129
fire escapes, 373
fitness clubs, **281**
floor space, 133
florists, 247
footwear trends, 206
franchises, 223
full-time jobs, **566**
fundraisers, 247
furnishings, **281**
gambling, 692
giveaways, 534
gold production, **744**
government grants, **367**
grand openings, 292
health care, 129
helium balloons, 534
help wanted, 104
home entertainment, **202**
home sales, 415, 722
Home Shopping network, 471
hourly wages, 795
infomercials, 113, 130, 695
inheritance tax, **681**
insurance claims, 982
Internet companies, 146–147
Internet sales, 570
investing, **280**
jewelry sales, **683**
job losses, 178
job training, 728
kitchen remodeling, 134

labor costs, 737
ladder sales, 692
landscape design, **61**
landscape designer, **1**
lawyer's fees, 286
layoffs, 692
legal fees, **46**
lift systems, **61**
the Lottery, 498
lowering prices, 185
lumber, 675
magazines, 21
managing a soup kitchen, **57**
markdowns, 186
market share, 206, 735
meeting payrolls, 712
meetings, 126
men's clothing sales, 692
mileage claims, 483
modeling, 233
money, 21
monthly payments, 534
mortgages, **101**
mover's pay scale, 287
movie tickets, **280**
night shift staffing, 751
office furnishings, 247
office supplies sales, 692
offshore drilling, 469
online shopping, 372
open houses, 113
ordering snacks, **61**
overdraft protection, 168
overtime, **474,** 693
packaging, **616**
parking, 694
part-time jobs, **566**
pay rate, 571
paychecks, 146, 483, 586, 981
personal financial adviser, **135**
petitions, 533
pharmaceutical sales, 691
pizza deliveries, **743**
pizza sales, 674
postage rates, 397
postal rates, **741**
price guarantees, 676
pricing, 247, **280, 465,** 469
product labeling, 371
product promotion, 677
production planning, 342
production time, 215
profits, 299
profits and losses, 205, 440
publishing, 286
quality control, 586, 727
raises, 691
reading blueprints, 33
real estate, **183,** 439, 812

xviii

rebates, 105, 676
recalls, 223
Red Cross, **654**
rentals, **46**, 286, 287, **663, 671**
retail price, 292
retailing, 469
ring sale, 695
room tax, 692
salaries, 483
salary schedules, 797
sale prices, 234, 292, 540
sales receipts, 692
sales tax, **680**, 691, 692, 703, 733
school lunches, 599
school supplies, **80**
scooter sale, 695
seafood, 547
self-employment tax, 691
selling boats, **183**
selling cars, 691
selling clocks, 691
selling condos, 317
selling electronics, **683**, 691
selling insurance, **683**
selling medical supplies, 725
selling shoes, 691
selling tires, 691
seminars, 436
service stations, 283
sheet metal, 32
shoe sales, **689**
shopping, 121, 373, **509**, 542, **722**
short-term business loans, **706**
small businesses, **100**, 247, 291
smartphones, 674
social networks, **648**
special offers, 702, 728
splitting the tip, 702
spreadsheets, 249
Sprint Nextel, 220
stadiums, 223
stock market, 186, **191,** 194
stock market records, 194
stocking shelves, **61,** 133
store sales, **755–756**
sunglasses sales, **689**
t-shirt sales, 725
takeovers, 194
tanning, 692
tax hikes, 693
tax refunds, 285
taxes, 570
telemarketing, 733
textbook sales, **61**
ticket sales, **61**
timeshares, **58**
tipping, 702, 722
tips, 677
total cost, **681**
towel sales, 734
traffic fines, **109**
unions, **646**
U.S. college costs, 72
U.S. national parks, **495**
Visa receipts, 702
watch sale, 695

websites, 31
weddings, 32
wishing wells, 981
withholding tax, **681**
working couples, 83
yard sales, 277

Careers
broadcasting technician, **225**
butchers, 290
chef, **557**
civil service, **278**
education, **91**
employment history, A-13
entry-level jobs, **61**
filmmaking, **409**
firefighting, **856,** 861
full-time jobs, **566**
health care, 179
home health aide, **443,** 470
landscape designer, **1,** 61
loan officer, **643**
mover, 287
occupations, A-12
personal financial adviser, **135,** 158
police officer, **949,** 961
postal service mail carrier, **739,** 749
school guidance counselor, **303,** 358
small business, **100**
sports, 438
surveyors, **819**
teacher, 33
telemarketing, 427, 733
working two jobs, A-14

Clothing and footwear
basketball shoes, 967
belts, 642, 736
clothes designers, 290
clothes shopping, 584, 816
clothing design, 372
clothing labels, **614**
clothing sales, **228,** 982
clothing stores, 247
footwear, 206, 284
letterman jackets, 534
men's clothing sales, 692
pants sale, 813
quality control, 586
sewing, 372, 398, 423, 438
shrinkage, 733
tuxedos, 286

Collectibles
antiques, 283, 286
autographs, 287
collectibles, 983
JFK, **685**
sports memorabilia, **228**
stamps, 385, 974

Communication
announcements, 14
broadcasting technician, **225**

bumper stickers, **366**
checking email, 799
constructing tables, 248
copyediting, 14
cryptography, 72
discussion groups, 134
displays, A-12
downloading, 675
Facebook, **720**
freeway signs, 385, 386
Gettysburg Address, 94
graph paper, 372
graphic arts, 414
infomercials, 113, 130, 695
Internet, 437
iPhones, 145
license plates, 372
logos, **656**
long-distance calls, **87**
magazine covers, 357
magazines, 33
mileage signs, 512
mobile phones, **761**
postage rates, 397
postal regulations, 622
poster boards, **47**
product labeling, 371
radio antennas, 521
radio stations, **27,** 286
rebates, 676
smartphones, **47**
social networks, **648**
sports pages, 470
Sprint Nextel, 220
stamps, 112, 332, 385
telemarketing, 427
telephone books, 414, 553, 737
television, 471
TV interviews, **367**
typing, 737
underground cables, 342
web traffic, **245**
word count, **46**
word processing, **47**
word processors, 542

Community life
city planning, 484
crime scenes, **279**
interstate speed limits, **242**
jury duty, A-14
meetings, 126
playgrounds, **279**
response time, **245**
speed checks, 222
speed limits, **242**

Community service
awards, 534
civil service, 815
disaster relief, 534
managing a soup kitchen, **57**
recycling, 233
Red Cross, 267, **654**
school supplies, **80**
service club, **276**

telethons, 534
volunteer service, 812

Construction. *See also* Engineering
accounting, 233
architecture, **903**
bridge repair, 248
building a pier, 178
building materials, 234, 642
buying paint, 599
capacity of a gym, **663**
carpentry, 521, 843, 860, 899, 900
cement mixers, 373
concrete, 775
concrete blocks, 918
constructing pyramids, 842
construction, 286, 533, **877**
construction delays, 130
cost overruns, 112
dance floors, 83
deck supports, 179
distance fallen, 292
drafting, 854
earth moving, 694
emergency exits, 372
engineering, **230**
fences, 898, 937
fire escapes, 373
flooring, 898
glass, 556
height of a building, 874
interior decorating, 287
kitchen remodeling, 134
lighthouses, 248
lighting design, 414
lumber, 441
masonry, 389
mobile homes, 267
new homes, 118, 483
offshore drilling, 469
painting, 569
parking, 694
pipe, 469
playground equipment, 246
playgrounds, **279**
plywood, 396
pool construction, 439
reading blueprints, 966
redevelopment, 713
remodeling, 816, 899
remodeling a bathroom, 396
scale drawings, 982
sinkholes, 317
stadiums, 223
storm damage, 485
structural engineering, **886–887**
subdivisions, 373
swimming pools, 485
tiles, 899
window replacements, 513

Demographics
Japan, 178
Native Americans, **408**

Applications Index

population increases, 733
population shift, **774**

Disasters
disaster relief, 534
earthquakes, 778
gasoline leaks, **175**
wind damage, 861

Education
accounting, 159
art classes, 84
buying pencils, 918
capacity of a gym, **663, 671**
child care, 290
class time, 287, 302, 815
college courses, 702
college employees, 733
college expenses, 722
college funds, 713
community college students, **673**
comparing grades, 765
construction, 286
counseling questionnaire, A-11
CPR class, 414
declining enrollment, 194
discussion groups, 134
dorm rooms, **890**
driving schools, 300
education costs, **705**
education pays, **91**
enrollments, **229,** 692
exam averages, 765
exam scores, 765
faculty-student ratios, 569
field trips, **529**
finding GPAs, **758–759**
fundraisers, 247
GPAs, 805, 812
grade distributions, **760**
grade point average (GPA), **496,** 534
grade summaries, 805
grades, 94, 133, 248, 291, 805
grading, 441
grading scales, 287
graduation, 121, 233
home schooling, **685**
honor roll, 702
libraries, 287
medical schools, 798–799
music education, **700**
no-shows, 702
observation hours, 440
open houses, 113
parking, 134
paying off loans, 983
P.E. classes, 133
physical education, 112
preschool enrollments, 287
purchasing textbooks, A-13
quiz results, **757**
re-entry students, **9**
reading, 571
reading programs, 168, 397
scheduling classes, A-9
scholarships, **229,** 283, 285
school supplies, **80**
semester grades, 764
seminars, 436
service club, **276**
service hours, **276**
speed reading, **109,** 112
staffing, 586
student drivers, **700**
student-to-instructor ratio, **561**
study time, **245**
studying, 233
studying in other countries, **617**
studying mathematics, **687**
surveys, 95
teacher's salaries, 33
teaching schedules, A-12
team GPA, 817
testing, 194, 702
textbook sales, **61**
textbooks, 436
tuition, 691, 712
tutoring, 285, 534
valedictorians, 538
value of an education, **762**

Elections
elections, **229,** 236, 677
politics, 213, 218
polls, 206, 640
presidential elections, **672**
voting, 703

Electricity
amperage, 167
power outages, 104
reading meters, 14, 455
utility bills, **452,** 456

Electronics and computers. See also Technology
computer companies, 236
computer printouts, 342
computers, 498
electronics, 498
pixels, **41**
quality control, 470
spreadsheets, 113
tax write-off, 178
word processing, **47**

Energy
air conditioning, 300
alternative fuels, 712
amperage, 167
auto travel, 571
batteries, 247
coastal drilling, 268
draining pools, **61**
draining tanks, 571
electric bills, 484
electricity rates, 571
electricity usage, **699**
energy, 677
energy conservation, **264**
energy savings, **47**
energy usage, 94, 249
fire hazards, 440
fuel economy, 46
gas mileage, 571
gas stations, 441
gas tanks, 317
gasoline, 301, 386
gasoline barrels, 358
gasoline leaks, **175**
gasoline prices, 457
gasoline storage, **100**
greenhouse gases, 678
holiday lights, 295
magnification, 178
meter readings, 299
mileage, 133, 586
mixing fuels, 586
nuclear power, 221
offshore drilling, 469
oil changes, 83, 301
oil wells, 499
power outages, 104
radiators, 599
rates of speed, 571
reading meters, 14, 371, 455
recycling, 233
reserves, 14
service stations, 283, 287
solar power, 676
sprinkler systems, 287
tanks, 918
U.S. gasoline prices, 551
utility bills, **452,** 456, 534
utility costs, 638
utility taxes, 484
waste, 677
water management, 164
water purity, 552
water storage, 186
water towers, **913**
water usage, 485, 678
weight of water, 553, 599

Engineering. See also Construction
bridge repair, 248
carousels, 585
distance fallen, 292
drafting, 585
engineering, **230**
Great Pyramid, 598
high-rise buildings, 637
Hoover Dam, 599
model railroads, 585
motors, 641
offshore drilling, 469
plumbing, 233
quality control, 628
ramps, 585
retaining walls, 291
scale drawings, **580**
scale models, **580**
the Sears Tower, 599
skyscrapers, 613, 632
structural engineering, **886–887**

Entertainment. See also Sports
amusement parks, 398
art design, 246
art galleries, 694
art history, 570, 956
balloon rides, **244**
billiards, 966
birthday presents, 130
bouquets, **80**
Broadway musicals, 414
cable television, 300
camcorder sale, 695
cameras, 427
camping, 223, 398, 599, 874
candy, 32
carousels, 585
casting a movie, **409**
catering, 373
classical music, **108**
concert parking, 694
concert seating, 484
concert tickets, **579,** 642
concerts, 298
crowd control, 626
dance floors, 83
decorating, 32
disc players, 695
draining pools, **61**
entertainment costs, **108**
figure drawing, 358
filmmaking, **409**
flags, 32
flowers, 126
gambling, 692
game shows, 12
gift wrapping, **43**
gin rummy, 167
Grand Canyon, 185
graphic arts, 414
guitar design, 434
guitar sale, 737
hearing protection, 134
high-ropes adventure courses, 929
hiking, 247, 359, 397, 498
hip hop, 104
hit records, 105
holiday lights, 295
home entertainment, **202**
horse racing, 144–145, **318,** 373, **380,** 513
hot-air balloons, 918
Indy 500, 499
infomercials, 113, 130, 695
interior decorating, 287
jewelry, 33, 386, 613, 957
lottery winners, **61**
lotto winners, 113
lottos, **229**
magazine covers, 357
magazines, 33
making jewelry, 424
marching bands, 72
medieval fair, 298
modeling, 233
movie tickets, **57, 280**
music, 317
musical instruments, 832
musical notes, 359
noise, 233

Applications Index

painting, **229**, 735
parties, 694
party invitations, 104
party preparations, 267
photography, 222, 429
picture framing, 300
piñatas, 967
playgrounds, **279**
pool, 854
racing programs, 676
radio stations, **27**, 286
rap music, **655**
rating movies, 812
ratings, 812
recreation, 300, 637
roller coasters, 206
shopping, 233
sight seeing, 247
sports memorabilia, **228**
stamps, 112, 332, 385
summer reading, 805
synthesizer, 832
television, **353**, 471, **650, 724,** 945
theater seating, 807
ticket sales, **61**
timeshares, **58**
tossing a coin, 133, **645**
touring, **58**
trampoline, 908
travel, 499
treats, 123
TV interviews, **367**
TV ratings, 95
water balloons, 976
water slides, 386
websites, 31
weddings, 32, 415
wishing wells, 981
word count, **46**
world records, 32
wrapping gifts, **47**, 94
YouTube, 95

Farming

crop damage, 693
crop loss, 178
farm loans, 712
farmer's markets, 32
farming, 941, 945
number of U.S. farms, 754
size of U.S. farms, 754
sod farms, 292

Finance

accounting, 159, 168, 233, 248, 553, **706,** 980
advertising, 179, 550
airlines, 159
annual income, **452**
appliance sales, **683**
art galleries, 694
ATMs, 83, **745**
attorney's fees, **277**
auctions, 691
auto care, 733
auto insurance, 735
auto repairs, 104, 129

autographs, 287
awards, 534
bank takeovers, 220
banking, 93, 119, 218, 220, 223, **228,** 470, 946
bankruptcy, 569
banks, 439
bedding sales, 692
blinds sale, 695
book sales, 692
bookstores, 207
bowling, 551
budget deficits, 186
budgets, **354,** 570, 703, 764
bus passes, 694
business, 570
business expenses, **529,** 816
business performance, 584
buying a business, 223
buying paint, 599
buying pencils, 918
camcorder sale, 695
candy bars, 764
capital gains tax, 691
car insurance, 694
car loans, 498, 712
car sales, 290
cash awards, 764
cash flow, **152**
cash gifts, 713
cash grants, 731
certificate of deposits, 713
checkbooks, 815
checking accounts, 14, 168, 206, 455
chocolate, 513
clothes shopping, 584, 816
clothing sales, **228,** 982
clothing stores, 247
coins, 285, **479,** 540
collectibles, 983
college funds, 713
commercials, 287
commissions, 285, 287, **723,** 725
comparison, **567**
comparison shopping, **495,** 567, 571, 622, 626, 638, 765, 899, 982
compound interest, **707–708,** 729
compounding annually, 713
compounding daily, **708–709,** 730
compounding semiannually, 713
concert parking, 694
concert seating, 484
concert tickets, **579**
congressional pay, **46**
cost-of-living, 733
cost-of-living increases, 693
coupons, 284, 754
credit, 734
credit card debt, **139**
cutting budgets, 186
daily pay, **40**
day care, 571
declining sales, 692
delicatessens, **509,** 513
depreciation, 796
dining out, **689,** 702

dinnerware sales, 692
disaster relief, 534
disc players, 695
discount hotels, 692
discount tickets, 692
discounts, **689–690,** 702
double coupons, 695
down payments, 728
e-commerce, 484
economic forecasts, **656**
education costs, **705**
electric bills, 484
electricity rates, 571
emergency loans, 731
entertainment costs, **108**
farm loans, 712
fax machines, 695
federal budget, 195, 209
federal debt, 484
field trips, **529**
filing a joint return, 752
filing a single return, 752
films, **241**
financial statements, 218
financing, 129
fines, 729
fire damage, 702
fitness clubs, **281**
florists, 247
foreign travel, 693
fuel economy, **46**
full-time jobs, **566**
fundraisers, 247
furnishings, **281**
furniture sales, **725**
gasoline cost, 571
gasoline prices, 457
gasoline tax, 693
giveaways, 534
government grants, **367**
government income, 678
government spending, 676
guitar sale, 737
helium balloons, 534
help wanted, 104
home repairs, 728
home sales, 722
Home Shopping network, 471
hourly pay, 499
hourly wages, 795
housing, 676
income tax, 456
inheritance, 713
inheritance tax, **681**
inheritances, 286
insurance, 676, 733
insurance claims, 982
interest charges, 734
interest rates, **655**
Internet sales, 570
investing, **280**
investment accounts, 731
investments, 712, 731, 734, 738, 795
jewelry sales, **683**
kitchen remodeling, 134
kitchen sinks, **465**

labor costs, 737
ladder sales, 692
landscaping, 571
legal fees, **46**
letterman jackets, 534
living on the interest, 713
loan applications, 712, 713
loan officer, **643**
loans, 286, 816
lottery, 713
the Lottery, 498
lotto winners, 113
lotto winnings, 752
low-interest loans, 713
lowering prices, 185
markdowns, 186
marriage penalty, 752
meeting payrolls, 712
men's clothing sales, 692
millionaires, 752
money, **21,** 455
monthly payments, 534, 731
mortgages, **101**
mover's pay scale, 287
movie tickets, **280**
moving expenses, 298
new homes, 483
office furnishings, 247
office supplies sales, 692
overdraft fees, 213
overdraft protection, 168, 735, 980
overtime, **474**
pants sale, 813
parking, 571
part-time jobs, **566**
pay rates, 571, 626
paychecks, 121, 146, **318,** 483, 586, 693, 981
paying off loans, 983
personal financial adviser, **135,** 158
petitions, 533
pharmaceutical sales, 691
piggy banks, 285
pizza sales, 674
plumbing bills, 485
postage rates, 397
postal rates, **741**
price guarantees, 676
pricing, 247, **280, 465,** 469
profits, 299
profits and losses, 440
purchasing, 123
raises, 691
real estate, **183,** 439, 812
rebates, 105, 676, 695
Red Cross, **654**
redevelopment, 713
remodeling, 712
rentals, **46,** 286, 287
rents, 726
retail price, 292
retailing, 469
retirement income, 712
ring sale, 695
room tax, 692

salaries, 299, 483
salary schedules, 797
sale prices, 234, 292, 540
sales receipts, 725
sales tax, **680,** 691, 692, 703, 733, 816, 982
sales tax rates, 725
savings accounts, 236, 691, 712, 731
scholarships, **229,** 283, 285
school lunches, 584
seafood, 547
segways, 695
self-employment tax, 691
selling a home, 702
selling boats, **183**
selling cars, 691
selling clocks, 691
selling electronics, **683,** 691, 816
selling insurance, **683**
selling medical supplies, 725
selling shoes, 691
selling tires, 691
shoe sales, **689**
shopping, 121, 233, **509,** 542, **579,** 638, **722,** 980
short-term business loans, **706**
short-term loans, **705,** 712, 734
small businesses, 247
smartphones, 674
smoke damage, 712
snacks, 247
Social Security, 717
special offers, 728
spending on pets, 983
splitting the tip, 702
sporting goods, 247
sports agents, 694
Sprint Nextel, 220
stock market, 112, 186, **191,** 194
stock market records, 194
student loans, **101**
sunglasses sales, **689**
t-shirt sales, 725
take-out food, 974
tax hikes, 693
tax refunds, 283, 285, 713, 946
tax-saving strategy, 752
tax write-off, 178
taxes, 570, **656**
telemarketing, **723–724,** 733
telephone books, 553
telethons, 534
term insurance, 694
textbook sales, **61**
Thanksgiving dinner, 544
ticket sales, **61**
tipping, **699,** 702, 722, 734, 738, 816, 983
tips, 677
tool sales, **724–725**
total cost, **681,** 733
towel sales, 734
traffic fines, **109**
tuition, 691, 712
tutoring, 285, 534
tuxedos, 286

TV shopping, 695
unit costs, 571
unit prices, 571
U.S. college costs, 72
U.S. economy, **742–743**
U.S. gasoline prices, 551
utility bills, **452,** 534
utility costs, 638
utility taxes, 484
value of a car, 795
videocassettes, 534
Visa receipts, 702
watch sale, 695
wedding costs, 554
weekly earnings, **480**
whole life insurance, 694
wishing wells, 285, 981
withdrawing only interest, 713
withholding tax, **681**
working couples, 83
yard sales, **277**

Food and beverages. *See also* **Health and nutrition**
bakery supplies, 483
baking, 222, **381, 581,** 981
bottled water, 637
bottled water dispenser, 809
bottling, 613
bowls of soup, 83
breakfast cereal, **46,** 373
butchers, 290
caffeine, 642
calories, 372
candy bars, 764
cereal boxes, 918
chili heat scale, 113
chocolate, 513
coffee, **565,** 619
coffee drinkers, 437, 981
cold storage, 298
comparison, **567**
comparison shopping, 571, 622, 626
containers, 613, 638
cooking, 331, 438, 584, **594,** 982
cooking meat, 622, 639
deli shops, 397
delicatessens, **509,** 513
desserts, 234, 569, 918
digging for clams, 457
dining out, **689,** 702
dinners, 357
drinking water, 544, **617,** 622
energy drinks, 811
fast food, 31, 95, 104, **202**
freeze drying, 168
French bread, **494**
fruit cakes, **494**
fruit storage, 736
hamburgers, 736
health food, 298
ice, 985
lunch meats, 642
making brownies, 586, 947
making cookies, 586
nutrition facts, 677

olives, 613
packaging, **616**
peanut butter, 627
pizza sales, 674
pretzel packaging, 805
reduced calories, 693
reducing fat intake, **686**
refreshments, 299
salads, **380**
salt intake, 484
school lunches, 584, 599
seafood, 547
serving size, 544
shopping, 373, **579,** 638
six packs, 613
snacks, **46,** 236, 247
soft drinks, **609**
splitting the tip, 702
sub sandwiches, 236
take-out food, 974
Thanksgiving dinner, 544
trail mix, 384
unit costs, 571
vegetarians, 983
waffle cones, 941
watermelons, **645**
world's coldest ice cream, 220

Games and toys. *See also* **Entertainment**
billiards, 795, 966
board games, **21, 645**
bouncing balls, 330
card games, 220, 302
cards, 569
dice, 779
games, 778
helium balloons, 534
kites, 556, 736
paper airplane, 861
piñatas, 967
Ping-Pong, 13, 267, **890,** 946
pool, 854
Scrabble, 94
sheets of stickers, 439
Sudoku, 555
toys, **366,** 974
trampoline, 908
video games, 104
water balloons, 976
Yahtzee, 248

Gardening and lawn care
gardening, 148, 222, 232, 832, 899, 966
hose repairs, 385
landscape design, 908
landscaping, 234, 267, 571, **879,** 899, 931
lawns, 937
planting trees, 542
sprinkler systems, 287
sprinklers, 641

Geography
Amazon River, 982
Centennial State, 287

city planning, 484
coastal drilling, 268
Dead Sea, 621
digging for clams, 457
elevations, 158
flags, 294, 555, 570, A-10–A-11, A-13
free fall, **244**
geography, 116, 148, 167, 220, 234, 326, 332, 542, 641, **650,** 875, 899, 918
globe, 778
Grand Canyon, 185
Hot Springs, 622
Japan, 178
lakes, 908
land area, 119
landmarks, 300
Middle East, 621
Mount Washington, 621
mountain elevations, **615**
New York City, 553
population, **61,** 439, 484, **649, 656**
regions of the country, **655**
sight seeing, 247
state flags, A-10–A-11
Suez Canal, 613
U.S. cities, **8**
Volunteer State, 414
weather maps, 146
Wyoming, **47**

Geometry
automobile jack, 853
baseball, 860
carpentry, 860
firefighting, 861
flags, 294
geometry, 233, 283, 285, 290, 302, 440, 814, 947, 973, 974
ladder sales, 692
length of guy wires, 860
paper airplane, 861
phrases, 833
picture frames, 860
polygons in nature, 853
pool construction, 439
quadrilaterals in everyday life, 884
roadside emergency, 547
wind damage, 861

Government
Bill of Rights, 717
budget deficits, 186
civil service, **278,** 815
congressional pay, **46**
elections, **229,** 236, 331, 677
federal budget, 195, 209
federal debt, 484
federal taxes, **2**
flags, 32
government grants, **367**
government income, 678
government spending, 676
how a bill becomes law, **325**
JFK, **685**

license plates, 372
military science, 159
petition drives, 642
political parties, 317
political polls, 159
politics, 213, 218
polls, 206, 427, 640
postal regulations, 622
presidential elections, **672**
presidents, 12
Senate rules, 330
Social Security, 717
taxes, **656,** 717
United Nations, 726
UN Security Council, **655**
U.S. House of Representatives, **8–9**
U.S. presidents, 637

Hardware
aluminum, **466**
automobile jack, 853
comparison shopping, 571
computer supplies, 764
dishwashers, **619**
drafting, 854
garage door openers, 357
gear ratios, 569
hardware, 289, 342, 357, 385, 641
hose repairs, 385
kitchen sinks, **465**
length of guy wires, 860
machinery, 809
machinist's tools, **761–762**
microwave ovens, 540
paper shredders, 641
pipe, 469
playground equipment, 246
quality control, 470
storage tanks, **911**
synthesizer, 832
tachometers, 483
tanks, 918
telescopes, 545
tool chests, 726
tool sales, **724–725**
tools, 359, 638, 842, 884
tuneups, 456
vehicle specifications, 471
vises, 332
water towers, **913**

Health and nutrition. *See also* **Medicine and health**
bakery supplies, 483
baking, 222, **381,** 569, **581,** 638
beverages, **61**
blood samples, 805
bottled water delivery, 286
breakfast cereal, **46,** 373
burning calories, 775
butchers, 290
caffeine, 642
calories, 372
cancer deaths, 798
cancer survival rates, 810
candy, 32

candy bars, 764
child care, 290
coffee, **565**
coffee drinkers, 437
cold storage, 298
comparison shopping, 571
conditioning programs, **466**
cooking, 331, **339,** 342, 438, 584, **594,** 779
cooking meat, 622, 639
CPR, 569
CPR class, 414
deli shops, 397
dentistry, 316, 754
dermatology, 434
desserts, 234, 569
dieting, 179
diets, 32
dinners, 357
drinking water, 544
energy drinks, 811
eye droppers, 613
eyesight, 168
farmer's markets, 32
fast food, 31, 95, 104, **202**
fevers, **619**
first aid, 854
fitness, 298
fitness clubs, **281**
freeze drying, 168
genetics, 414
health, 159
health care, 129, 179, 613
health clubs, 737
health food, 298
health statistics, 358
healthy diets, **27**
hearing protection, 104, 134
heartbeats, **47**
home health aide, **443**
injections, 455
jogging, 158
juice, **46**
lasers, 455
lunch time, **61**
making brownies, 586
making cookies, 586
managing a soup kitchen, **57**
medical dosage, **610**
medical schools, 798
medications, **607**
Nobel prize, 111
nursing, 83
nutrition, **46,** 438, **565,** 622, 766
nutrition facts, 677
octuplets, 385
ordering snacks, **61**
patient recovery, **77**
P.E. classes, 133
physical fitness, 397
physical therapy, 397
picnics, 83
prescriptions, **47**
the Red Cross, 267
reduced calories, 693
reducing fat intake, **686**
reflexes, 499

refreshments, 299
salads, **380**
salt intake, 484
school lunches, 584, 599
septuplets, 385
service hours, **276**
serving size, 544
shopping, 93, 373
skin creams, 397
sleep, 121, 398, 800
snacks, **46,** 236, 247
soft drinks, **609**
spinal cord injuries, 816
sub sandwiches, 236
surveys, 95
survival guide, 642
tooth development, 414
trail mix, 384
transplants, 33
vegetarians, 983
water purity, 552
world's coldest ice cream, 220
wrestling, **466**

History
airline accidents, **20**
art history, 570, 956
baseball history, **656**
basketball records, 640, **653**
history, 147, 167, 207, 944
missions to Mars, 13
NFL records, 599
record holders, 470
Roman Empire, 206
sunken ships, 158
U.S. presidents, 12

Home management
air conditioning, 622
anniversary gifts, 584
appliances, 286, **722**
blinds sale, 695
carpeting, **894,** 898, 908
carpeting a room, **894**
changing units, **46**
comparing rooms, **47**
decorating, 943, 982
dishwashers, **619**
down payments, 728
emergency exits, 372
fences, 898, 937
fire damage, 702
fire hazards, 440
floor space, 676
flooring, 898
frames, 898
gardening, **61,** 148, 222, 232, 513
hanging wallpaper, 842
heating, **243**
home sales, 415, 722
housing, 676
kitchen design, 331
kitchen sinks, **465**
landscaping, 234
lawns, 937
making a frame, 884
microwave ovens, 540

mobile homes, 267
new homes, 483
painting, 899
patio furniture, 341
pet doors, 430
picture frames, 300, 860
plumbing, 233, **237,** 485
quilts, **653**
real estate, 439, 694, 812
refrigerators, 918
remodeling, 396, 712, 816, 899
rentals, **663, 671,** 726
room dividers, 301
Scotch tape, 233
selling a home, 702
selling condos, 317
seminars, 436
sewing, 267, 372, 386, 874
shipping furniture, 221
shopping, 121, 233
solar covers, 899
subdivisions, 373
thread count, 221
tiles, 332, 899
waste disposal, 234
window replacements, 513

Law enforcement
crime scenes, **279**
fugitives, 733
jury duty, A-14
lie detector tests, 168, 223
NYPD, 811
police force, 693
police officer, **949,** 961

Marketing
advertising, **43,** 179, 550
basketball shoes, 967
cash awards, 764
cereal boxes, 918
clothing labels, **614**
clothing sales, 982
commercials, 287, **686–687**
giveaways, 534
Home Shopping network, 471
infomercials, 113, 130, 695
logos, **656**
magazine covers, 357
product labeling, 371
product promotion, 677
shaving, 441
soap, 656
social networks, **648**
special offers, 702, 728
telemarketing, 427
TV shopping, 695

Mathematics
algebra, 398
counting numbers, 94
divisibility, **653**
geometry, 233, 283, 285, 290, 302, 440, 814, 947, 973, 974
graph paper, 372
history, 167
number problems, 286

perfect numbers, 72
polygons in nature, 853
prime numbers, 94
quadrilaterals in everyday life, 884
quizzes, 284
studying mathematics, **687**

Measurement. *See* **Weights and measures**

Medicine and health. *See also* **Health and nutrition**
blood samples, 805
blood transfusions, 779
dosages, 585
eye droppers, 613
first aid, 854
ibuprofen, 613
injections, 613
lab work, 585
medical dosage, **610,** 613
medical schools, 798
medications, **607**
medicine, 613
prescriptions, 639
spinal cord injuries, 816
surgery, 634
Tylenol, 634

Miscellaneous
adjusting ladders, 860
batteries, 247
best-selling books, 290
birthdays, **656**
bottled water dispenser, 809
bubble wrap, **61**
cameras, 427
capacity of a gym, **663, 671**
carry-on luggage, **560**
cattle ranching, 413
changing units, **46**
cleaning supplies, **617**
clubs, 691
comparison shopping, 899
crowd control, 626
cryptography, 72
Dewey Decimal System, 456
displays, A-12
draining tanks, 571
driving directions, 469
easels, 854
electronic poll, A-14
engineering, **230**
enlargements, 676
estimation, 195
fences, 898
flowchart, 884
funnels, 357
genealogy, 737
glass, 556
Internet surveys, 702
ladders, 522
libraries, 287
lie detector tests, 168
lumber, 441
magazine covers, 357

mixing solutions, 236
mixtures, 677
moving, 810
moving expenses, 298
noise, 233
note cards, 342
packaging, 222, **230,** 233
painting, 735, 899
painting signs, 842
paper shredders, 641
passports, 431
photography, 429
poster boards, **47**
printing, 430
quality control, 628
quizzes, 284
reading meters, 371
reams of paper, 483
recycling, 233
room capacity, **47**
sails, 899
scheduling classes, A-9
Scotch tape, 233
sewing, 398, 423, 424, 438
siblings, A-13
signs, 702, 974
speed reading, **109,** 112
spray bottles, 498, 796
stamps, 974
Stars and Stripes, 331
supplies, 233
surveying, 521
surveys, 764
tape measures, 556
test tubes, 634
testing, 194
thread count, 221
time, **61**
time clocks, 316
tools, 842
tossing a coin, 133
UN Security Council, **655**
vises, 332
Washington, D.C., 874
waste disposal, 234
woodworking, 332
words of wisdom, 622

Natural resources
crude oil, **46,** 165, 637
energy reserves, 14
gold mining, 439
lake shorelines, 693
mining, 212
ocean exploration, 215
water distribution, 717
water pollution, 676

People
Amelia Earhart, 599
Beatles, 236
famous Bills, **230**
famous Toms, 235
genealogy, 737
JFK, **685**
Lewis and Clark, 599
Native Americans, **408**

Nobel prize, 111
octuplets, 765
the Red Cross, 267
the Wright brothers, 598

Personal care
bathing, **619**
beauty tips, 842
child care, 676
dorm rooms, **890**
haircuts, 385
luggage, 386
online shopping, 372
playpens, 974
shaving, 441
skin creams, 397
sleep habits, **380**
soap, **656**
taking a shower, 622

Reading and language
alphabet, 677
children's books, 235
diagramming sentences, 842
literature, 639
phrases, 833
reading, 571
reading programs, 168, 397
spelling, 736
summer reading, 805

Recordkeeping
business accounts, 283
checkbooks, 815
constructing tables, 248
flowchart, 884
graph paper, 372
spreadsheets, 113, 159, 194, 249
surveys, 292

Safety
bridge safety, 31
CPR, 569
drunk driving, **655**
emergency exits, 372
fire hazards, 440
firefighting, 415
fires, 811
roadside emergency, 547
safety inspections, 702
seat belts, 728

Science
alcohol, 223
Amazon River, 982
anatomy class, **408**
astronomy, 147, **595**
atoms, 158
avalanches, 816
average temperatures, 765
biology, 483, **655**
biorhythms, 83
birds, **46**
body weight, **616**
botany, 267, 331, 357
brain, 634

cell division, 72
chemistry, 158, 186, 223, **451,** 554, 853
chili heat scale, 113
climate, 94
clouds, 15
conservation, 484
digging for clams, 457
distance fallen, 292
distance traveled, 249
diving, 148
dosages, 585
drought, 209
earth, **645,** 974
earthquakes, 766
Earth's surface, 331, 722
eggs, 569
elevations, 158
energy reserves, 14
energy usage, 249
erosion, 178
falling objects, 249, 299
fingernails, 570
flooding, 146, 158
forestry, 342, 513
free fall, 145, **244,** 247
gardening, 148
Gateway City, 164
genetics, 330
geography, 116, 148, 167, 220, 234, 875
geology, 457, 513
giant sequoia, 908
gold mining, 439, 570
gravity, 113, 421
growth rates, **564**
hair growth, 621
heating, **243**
height of a tree, 874, 875, 931
human skin, **655**
human spine, **656**
hurricanes, 764
ice, 985
icebergs, 331
insects, **42**
lab work, 585
lakes, 908
life span, 113
light, 72, 178
lightning, 796
line graphs, 147
literature, 639
lumber, 675
making cologne, 585
marine science, 234
mercury, 207, 555
metric system, 456
microscope, 457
mining, 212
missions to Mars, 13
mixing perfumes, 585
noise, 233
ocean exploration, 186, 215
oceanography, **175**
octuplets, 765
oil wells, 499
pH scale, 469

planets, 178, 302, 814, 832
planting trees, 542
polygons in nature, 853
record temperatures, **155,** 158, 213
Saturn, **243**
seismology, 842
shadows, 945, 984
skin, 569
snowfall, **564**
snowy weather, 622
solar system, A-14
speed of light, 14
spreadsheets, 159
storm damage, 430, 485
the sun, **595**
sunny days, 119
survival guide, 642
temperature changes, **155,** 765
temperature conversion, 292
temperature drop, 185, 440
temperature extremes, 168, 209
thermometers, 249
waterfalls, **61**
weather, 168, 555, 735
weather forecasts, 206, 814
weather maps, 146
weather reports, 470
weight of water, 553
wind damage, 861
wind speeds, 801
windchill temperatures, 801
Windy City, 164
with/against the wind, 236

Sports
aerobics, 967
archery, 908
baseball, 13, 521, 694, 860, 884
baseball history, **656**
baseball teams, 233
baseball trades, 186
basketball, **18, 281,** 385
basketball records, 640, **653**
basketball shoes, 967
bicycle race, 811
bouncing balls, 330
bowling, 248, 551
boxing, **47,** 438, **656**
buying fishing equipment, 693
conditioning programs, **466**
diving, 93–94, 371
drag racing, 421
fishing, 248
football, 167, **590**
football statistics, 206
golf, 13, 146, 640
golf clubs, 104
hiking, 247, 359, 397, 498, 599
horse racing, 144–145, **380,** 513
Indy 500, 499
javelin throw, 984
jogging, 158, 908
Ladies Professional Golf Association, 209
letterman jackets, 534
marathons, 341, **590**

NASCAR, 145, 456
NFL offensive linemen, **90**
NFL records, 599
Olympic diving, A-13
Olympics (2012), 456
physical education, 112
Ping-Pong, 13, 267, **890**
racing, 722
racing programs, 676
record holders, 470
running, **61**
sailing, 973
scouting reports, 195
scuba diving, 167
sewing, 267
skateboarding, 800
snowboarding, 799
soccer, 484
speed skating, 612, 639
sport fishing, 766
sporting goods, 247
sports agents, 694
sports contracts, 438
sports equipment, 290
sports memorabilia, **228**
sports pages, 470
surfboard designs, **338–339**
swimming, **229,** 637
swimming workouts, 733
table tennis, 983
team rosters, 415, 815
tennis, 13, 287, 302, 815
track, 639
track and field, **615,** 621, 622
TV ratings, 95
volleyball, 13, **61**
walk-a-thons, 805
weightlifting, **42,** 485, 621, 702
windsurfing, 332
WNBA (Women's National Basketball Association), **18**
women's sports, **890**
won-lost records, **656**
wrestling, **466**

Technology
amperage, 167
automation, 778
broadcasting technician, **225**
computer companies, 236
computer printouts, 342
computer speed, 586
computer supplies, 764
computers, 498
copy machines, 676
data conversion, 534
downloading, 675
electronics, 498
Facebook, **720**
fax machines, 695
flatscreen televisions, 522
freeze drying, 168
Home Shopping network, 471
Internet, 437, **653**
Internet companies, 146–147
Internet sales, 570
inventions, 235

iPhones, 145
lasers, 342, 455
long-distance calls, **87**
microwave ovens, 540
mobile phones, **761**
photography, 222
pixels, **41**
printing, 430
quality control, 470
radio antennas, 521
radio stations, **27**
selling electronics, 816
smartphones, 674
Sprint Nextel, 220
submarines, 167, 177
tax write-off, 178
technology, 146
telescopes, 545
television, 471, 945
typing, 570
underground cables, 342
videocassettes, 534
web traffic, **245**
word processing, **47**
word processors, 542
world's coldest ice cream, 220

Travel and transportation
acceleration, 457
air traffic control, 796
air travel, 384
aircraft, 626
airline complaints, 570
airline seating, 284, 555
airlines, 159, 206, 247, 980
airports, 118
auto care, 733
auto insurance, 735
auto mechanics, 513
auto painting, 542
auto repairs, 104, 129
auto travel, 571
automobile jack, 853
automobiles, 638, 814
aviation, 832
brake inspections, 734
building a pier, 178
bumper stickers, **366**
bus passes, 694
camping, 599
car insurance, 694
car loans, 712
car repairs, 413, 947
car sales, 290
car shows, 734
carry-on language, **560, 747**
commuting, **228,** 267
commuting miles, 751
commuting time, 812
comparison shopping, 571, 622
construction, 533
crude oil, **46**
dashboards, 249
dealer markups, 249
delivery trucks, 357
discount hotels, 692
discount tickets, 692

distance, rate, and time, 795
distance traveled, 292, 299
driver's license, 676
driving, 484, **756–757**
driving directions, 469
driving schools, 300
drunk driving, **655**
dump trucks, **61**
elevators, **47,** 220
engines, 918
field trips, **529**
fleet mileage, 764
flight paths, 470, 875
foreign travel, 693
freeway signs, 385, 386
freeways, 983
fuel economy, **46**
fuel efficiency, 766
gas mileage, 571, 726
gas stations, 441
gas tanks, 317
gasoline cost, 571
gasoline prices, 457
gasoline tax, 693
gear ratios, 569
Grand Canyon, 185
hearing protection, 104
helicopters, 908
hiking, 247, 599
hot-air balloons, 918
hotel reservations, **757–758**
insurance, 676
insurance claims, 982
interstate speed limits, **242**
jogging, 908
license plates, 372
lift systems, **61**
magnification, 178
maps, 778
mileage, 32, 133, 484, 586, 909
mileage claims, 483
mileage signs, 512
model railroads, 585
motors, 641
ocean travel, 283, 286
oil changes, 83, 301
parking, 134, 571, 694, 974, A-12
parking design, 843
parking lots, 298
purchasing, 123
radiators, 599
rates of speed, 571
rebates, 105
recalls, 223
retrofits, 484
riding buses, 105
road signs, **655**
road trips, 234, 247, 290, 814
roadside emergency, 547
room tax, 692
safety inspections, 702
safety requirements, 414
scooter sale, 695
seat belts, 798
segways, 695
selling boats, **183**
shipping, 84, 384

shipping furniture, 221
sinkholes, 317
space travel, 599
speed checks, 222
speed limits, **242**
storm damage, 430
student drivers, **700**
submarines, 167, 177, 185, 555
Suez Canal, 613
tachometers, 483
timeshares, **58**
tire tread, 359
tires, 946
tourism, 693
traffic fines, **109**
traffic studies, 702
trains, **746**
trams, 220
travel time, 692
truck repair, 233
trucking, **381**, 632, **756**
trucks, 628
tuneups, 456
used cars, 980, A-10
value of a car, 795
vehicle specifications, 471
vehicle weights, 236, 247
with/against the wind, 236

U.S. history
airline accidents, **20**
Centennial State, 287
community college students, **673**
energy, 677
Gettysburg Address, 94
greenhouse gases, 678
historical documents, 385
hurricanes, 764
JFK, **685**
Lewis and Clark, 599
missions to Mars, 13
Native Americans, **408**
number of U.S. farms, 754
population, **61,** 439
population shift, **774**
presidential elections, **672**
presidents, 12
regions of the country, **655**
size of U.S. farms, 754
stamp collecting, 385
Stars and Stripes, 331
tourism, 693
unions, **646**
U.S. economy, **742–743**
U.S. gasoline prices, 551
U.S. national parks, **495**
U.S. presidents, 637
Volunteer State, 414
waste, 677
Western settlers, 637

Weather
allergy forecast, 538
average temperatures, 765
clouds, 15
drought, 209
forecasts, 206
Gateway City, 164
hurricane damage, 722
hurricanes, 764
lightning, 796
line graphs, 147
record temperatures, **155,** 158, 213
snowfall, **564**
snowy weather, 622
spreadsheets, 159
sunny days, 119
temperature changes, **155,** 765
temperature conversion, 292
temperature drop, 185, 440
temperature extremes, 168, 209
weather, 168, 555, 735
weather forecasts, 814
weather maps, 146
weather reports, 470
wind speeds, 801
windchill temperatures, 801
Windy City, 164

Weights and measures
batteries, 247
body weight, **616,** 639
cooking, 331
dashboards, 249
Dewey Decimal System, 456
drafting, 512
globe, 778
height of a building, 874
height of a tree, 874, 875, 931
height of an object, 959
leap year, 386
measurement, 455
metric system, 456
ounces and fluid ounces, 622
reading meters, 14
rulers, 316
surveying, 521
surveys, 95
tachometers, 483
tape measures, 556
thermometers, 249
time clocks, 316
vehicle weights, 236, 247
Vietnamese calendar, 301
weight of a baby, 599, 613
weight of water, 553, 599
weights and measures, 234–235, 358
windchill temperatures, 801

Study Skills Workshop

OBJECTIVES

1. Make the Commitment
2. Prepare to Learn
3. Manage Your Time
4. Listen and Take Notes
5. Build a Support System
6. Do Your Homework
7. Prepare for the Test

YURALAITS ALBERT/Shutterstock.com

SUCCESS IN YOUR COLLEGE COURSES requires more than just mastery of the content. The development of strong study skills and disciplined work habits plays a crucial role as well. Good note-taking, listening, test-taking, team-building, and time management skills are habits that can serve you well, not only in this course, but throughout your life and into your future career. Students often find that the approach to learning that they used for their high school classes no longer works when they reach college. In this Study Skills Workshop, we will discuss ways of improving and fine-tuning your study skills, providing you with the best chance for a successful college experience.

1 Make the Commitment

YURALAITS ALBERT/Shutterstock.com

Starting a new course is exciting, but it also may be a little frightening. Like any new opportunity, in order to be successful, it will require a commitment of both time and resources. You can decrease the anxiety of this commitment by having a plan to deal with these added responsibilities.

Set Your Goals for the Course. Explore the reasons why you are taking this course. What do you hope to gain upon completion? Is this course a prerequisite for further study in mathematics? Maybe you need to complete this course in order to begin taking coursework related to your field of study. No matter what your reasons, setting goals for yourself will increase your chances of success. Establish your ultimate goal and then break it down into a series of smaller goals; it is easier to achieve a series of short-term goals rather than focusing on one larger goal.

Keep a Positive Attitude. Since your level of effort is significantly influenced by your attitude, strive to maintain a positive mental outlook throughout the class. From time to time, remind yourself of the ways in which you will benefit from passing the course. Overcome feelings of stress or math anxiety with extra preparation, campus support services, and activities you enjoy. When you accomplish short-term goals such as studying for a specific period of time, learning a difficult concept, or completing a homework assignment, reward yourself by spending time with friends, listening to music, reading a novel, or playing a sport.

Attend Each Class. Many students don't realize that missing even one class can have a great effect on their grade. Arriving late takes its toll as well. If you are just a few minutes late, or miss an entire class, you risk getting behind. So, keep these tips in mind.

- Arrive on time, or a little early.
- If you must miss a class, get a set of notes, the homework assignments, and any handouts that the instructor may have provided for the day that you missed.
- Study the material you missed. Take advantage of the help that comes with this textbook, such as the video examples and problem-specific tutorials.

Now Try This

1. List six ways in which you will benefit from passing this course.
2. List six short-term goals that will help you achieve your larger goal of passing this course. For example, you could set a goal to read through the entire *Study Skills Workshop* within the first 2 weeks of class or attend class regularly and on time. (*Success Tip:* Revisit this action item once you have read through all seven *Study Skills Workshop* learning objectives.)
3. List some simple ways you can reward yourself when you complete one of your short-term class goals.
4. Plan ahead! List five possible situations that could cause you to be late for class or miss a class. (Some examples are parking/traffic delays, lack of a babysitter, oversleeping, or job responsibilities.) What can you do ahead of time so that these situations won't cause you to be late or absent?

2 Prepare to Learn

Many students believe that there are two types of people—those who are good at math and those who are not—and that this cannot be changed. This is not true! You can increase your chances for success in mathematics by taking time to prepare and taking inventory of your skills and resources.

Discover Your Learning Style. Are you a visual, verbal, or auditory learner? The answer to this question will help you determine how to study, how to complete your homework, and even where to sit in class. For example, visual-verbal learners learn best by reading and writing; a good study strategy for them is to rewrite notes and examples. However, auditory learners learn best by listening, so listening to the video examples of important concepts may be their best study strategy.

Get to Know Your Textbook and Its Resources. You have made a significant investment in your education by purchasing this book and the resources that accompany it. It has been designed with you in mind. Use as many of the features and resources as possible in ways that best fit your learning style.

Know What Is Expected. Your course syllabus maps out your instructor's expectations for the course. Read the syllabus completely and make sure you understand all that is required. If something is not clear, contact your instructor for clarification.

Organize Your Notebook. You will definitely appreciate a well-organized notebook when it comes time to study for the final exam. So let's start now! Refer to your syllabus and create a separate section in the notebook for each chapter (or unit of study) that your class will cover this term. Now, set a standard order within each section. One recommended order is to begin with your class notes, followed by your completed homework assignments, then any study sheets or handouts, and, finally, all graded quizzes and tests.

Now Try This

1. To determine what type of learner you are, take the *Index of Learning Styles Questionnaire* at http://www.engr.ncsu.edu/learningstyles/ilsweb.html, which will help you determine your learning type and offer study suggestions by type. List what you learned from taking this survey. How will you use this information to help you succeed in class?

2. Complete the *Study Skills Checklists* found at the end of sections 1–4 of Chapter 1 in order to become familiar with the many features that can enhance your learning experience using this book.

3. Read through the list of Student Resources found in the Preface of this book. Which ones will you use in this class?

4. Read through your syllabus and write down any questions that you would like to ask your instructor.

5. Organize your notebook using the guidelines given above. Place your syllabus at the very front of your notebook so that you can see the dates over which the material will be covered and for easy reference throughout the course.

3 Manage Your Time

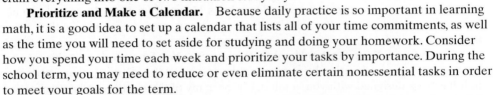

Now that you understand the importance of attending class, how will you make time to study what you have learned while attending? Much like learning to play the piano, math skills are best learned by practicing a little every day.

Make the Time. In general, 2 hours of independent study time is recommended for every hour in the classroom. If you are in class 3 hours per week, plan on 6 hours per week for reviewing your notes and completing your homework. It is best to schedule this time over the length of a week rather than to try to cram everything into one or two marathon study days.

Prioritize and Make a Calendar. Because daily practice is so important in learning math, it is a good idea to set up a calendar that lists all of your time commitments, as well as the time you will need to set aside for studying and doing your homework. Consider how you spend your time each week and prioritize your tasks by importance. During the school term, you may need to reduce or even eliminate certain nonessential tasks in order to meet your goals for the term.

Maximize Your Study Efforts. Using the information you learned from determining your learning style, set up your blocks of study time so that you get the most out of these sessions. Do you study best in groups or do you need to study alone to get anything done? Do you learn best when you schedule your study time in 30-minute time blocks or do you need at least an hour before the information kicks in? Consider your learning style to set up a schedule that truly suits your needs.

Avoid Distractions. Between texting and social networking, we have so many opportunities for distraction and procrastination. On top of these, there are the distractions of TV, video games, and friends stopping by to hang out. Once you have set your schedule, honor your study times by turning off any electronic devices and letting your voicemail take messages for you. After this time, you can reward yourself by returning phone calls and messages or spending time with friends after the pressure of studying has been lifted.

Now Try This

1. Keep track of how you spend your time for a week. Rate each activity on a scale from 1 (not important) to 5 (very important). Are there any activities that you need to reduce or eliminate in order to have enough time to study this term?

2. List three ways that you learn best according to your learning style. How can you use this information when setting up your study schedule?

3. Download the *Weekly Planner Form* from www.cengagebrain.com/shop/search/tussy and complete your schedule. If you prefer, you may set up a schedule in Google Calendar (calendar.google.com), www.rememberthemilk.com, your cell, or your email system. Many of these have the ability to set up useful reminders and to-do lists in addition to a weekly schedule.

4. List three ways in which you are most often distracted. What can you do to avoid these distractions during your scheduled study times?

4 Listen and Take Notes

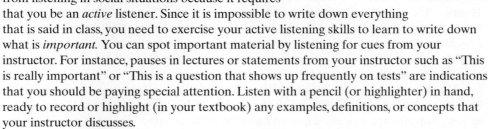
Monkey Business Images/Shutterstock.com

Make good use of your class time by listening and taking notes. Because your instructor will be giving explanations and examples that may not be found in your textbook, as well as other information about your course (test dates, homework assignments, and so on), it is important that you keep a written record of what was said in class.

Listen Actively. Listening in class is different from listening in social situations because it requires that you be an *active* listener. Since it is impossible to write down everything that is said in class, you need to exercise your active listening skills to learn to write down what is *important*. You can spot important material by listening for cues from your instructor. For instance, pauses in lectures or statements from your instructor such as "This is really important" or "This is a question that shows up frequently on tests" are indications that you should be paying special attention. Listen with a pencil (or highlighter) in hand, ready to record or highlight (in your textbook) any examples, definitions, or concepts that your instructor discusses.

Take Notes You Can Use. Don't worry about making your notes really neat. After class you can rework them into a format that is more useful to you. However, you should organize your notes as much as possible as you write them. Copy the examples your instructor uses in class. Circle or star any key concepts or definitions that your instructor mentions while explaining the example. Later, your homework problems will look a lot like the examples given in class, so be sure to copy each of the steps in detail.

Listen with an Open Mind. Even if there are concepts presented that you feel you already know, keep tuned in to the presentation of the material and look for a deeper understanding of the material. If the material being presented is something that has been difficult for you in the past, listen with an open mind; your new instructor may have a fresh presentation that works for you.

Avoid Classroom Distractions. Some of the same things that can distract you from your study time can distract you, and others, during class. Because of this, be sure to turn off your cell phone during class. If you take notes on a laptop, log out of your email and social networking sites during class. In addition to these distractions, avoid getting into side conversations with other students. Even if you feel you were only distracted for a few moments, you may have missed important verbal or body language cues about an upcoming exam or hints that will aid in your understanding of a concept.

Now Try This

1. Before your next class, refer to your syllabus and read the section(s) that will be covered. Make a list of the terms that you predict your instructor will think are most important.

2. During your next class, bring your textbook and keep it open to the sections being covered. If your instructor mentions a definition, concept, or example that is found in your text, highlight it.

3. Find at least one classmate with whom you can review notes. Make an appointment to compare your class notes as soon as possible after the class. Did you find differences in your notes?

4. Go to www.cengagebrain.com/shop/search/tussy and read the *Reworking Your Notes* handout. Complete the action items given in this document.

5 Build a Support System

Have you ever had the experience where you understand everything that your instructor is saying in class, only to go home and try a homework problem and be completely stumped? This is a common complaint among math students. The key to being a successful math student is to take care of these problems before you go on to tackle new material. That is why you should know what resources are available outside of class.

Make Good Use of Your Instructor's Office Hours. The purpose of your instructor's office hours is to be available to help students with questions. Usually these hours are listed in your syllabus and no appointment is needed. When you visit your instructor, have a list of questions and try to pinpoint exactly where in the process you are getting stuck. This will help your instructor answer your questions efficiently.

Use Your Campus Tutoring Services. Many colleges offer tutorial services for free. Sometimes tutorial assistance is available in a lab setting where you are able to drop in at your convenience. In some cases, you need to make an appointment to see a tutor in advance. Make sure to seek help as soon as you recognize the need, and come to see your tutor with a list of identified problems.

Form a Study Group. Study groups are groups of classmates who meet outside of class to discuss homework problems or study for tests. Get the most out of your study group by following these guidelines:

- Keep the group small—a maximum of four committed students. Set a regularly scheduled meeting day, time, and place.
- Find a place to meet where you can talk and spread out your work.
- Members should attempt all homework problems before meeting.
- All members should contribute to the discussion.
- When you meet, practice verbalizing and explaining problems and concepts to each other. The best way to really learn a topic is by teaching it to someone else.

Now Try This

1. Refer to your syllabus. Highlight your instructor's office hours and location. Next, pay a visit to your instructor during office hours this week and introduce yourself. (**Success Tip:** Program your instructor's office phone number and email address into your cell phone or email contact list.)

2. Locate your campus tutoring center or math lab. Write down the office hours, phone number, and location on your syllabus. Drop by or give them a call and find out how to go about making an appointment with a tutor.

3. Find two to three classmates who are available to meet at a time that fits your schedule. Plan to meet 2 days before your next homework assignment is due and follow the guidelines given above. After your group has met, evaluate how well it worked. Is there anything that the group can do to make it better next time you meet?

4. Download the *Support System Worksheet* at www.cengagebrain.com/shop/search/tussy. Complete the information and keep it at the front of your notebook following your syllabus.

6 Do Your Homework

Attending class and taking notes are important, but the only way that you are really going to learn mathematics is by completing your homework. Sitting in class and listening to lectures will help you to place concepts in short-term memory, but in order to do well on tests and in future math classes, you want to put these concepts in long-term memory. When completed regularly, homework assignments will help with this.

Give Yourself Enough Time. In Objective 3, you made a study schedule, setting aside 2 hours for study and homework for every hour that you spend in class. If you are not keeping this schedule, make changes to ensure that you can spend enough time outside of class to learn new material.

Review Your Notes and the Worked Examples from Your Text. In Objective 4, you learned how to take useful notes. Before you begin your homework, review or rework your notes. Then, read the sections in your textbook that relate to your homework problems, paying special attention to the worked examples. With a pencil in hand, work the *Self Check* and *Now Try* problems that are listed next to the examples in your text. Using the worked example as a guide, solve these problems and try to understand each step. As you read through your notes and your text, keep a list of anything that you don't understand.

Now Try Your Homework Problems. Once you have reviewed your notes and the textbook worked examples, you should be able to successfully manage the bulk of your homework assignment easily. When working on your homework, keep your textbook and notes close by for reference. If you have trouble with a homework question, look through your textbook and notes to see if you can identify an example that is similar to the homework question. See if you can apply the same steps to your homework problem. If there are places where you get stuck, add these to your list of questions.

Get Answers to Your Questions. At least one day before your assignment is due, seek help with the questions you have been listing. You can contact a classmate for assistance, make an appointment with a tutor, or visit your instructor during office hours.

Now Try This

1. Review your study schedule. Are you following it? If not, what changes can you make to adhere to the rule of 2 hours of homework and study for every hour of class?

2. Find five homework problems that are similar to the worked examples in your textbook. Were there any homework problems in your assignment that didn't have a worked example that was similar? (***Success Tip:*** Look for the *Now Try* and *Guided Practice* features for help linking problems to worked examples.)

3. As suggested in this Objective, make a list of questions while completing your homework. Visit your tutor or your instructor with your list of questions and ask one of them to work through these problems with you.

4. Go to www.cengagebrain.com/shop/search/tussy and read the *Study and Memory Techniques* handout. List the techniques that will be most helpful to you in your math course.

7 Prepare for the Test

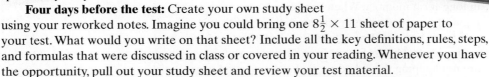

Taking a test does not need to be an unpleasant experience. Use your time management, organization, and these test-taking strategies to make this a learning experience and improve your score.

Make Time to Prepare. Schedule at least four daily 1-hour sessions to prepare specifically for your test.

Four days before the test: Create your own study sheet using your reworked notes. Imagine you could bring one $8\frac{1}{2} \times 11$ sheet of paper to your test. What would you write on that sheet? Include all the key definitions, rules, steps, and formulas that were discussed in class or covered in your reading. Whenever you have the opportunity, pull out your study sheet and review your test material.

Three days before the test: Create a sample test using the in-class examples from your notes and reading material. As you review and work these examples, make sure you understand how each example relates to the rules or definitions on your study sheet. While working through these examples, you may find that you forgot a concept that should be on your study sheet. Update your study sheet and continue to review it.

Two days before the test: Use the *Chapter Test* from your textbook or create one by matching problems from your text to the example types from your sample test. Now, with your book closed, take a timed trial test. When you are done, check your answers. Make a list of the topics that were difficult for you and review or add these to your study sheet.

One day before the test: Review your study sheet once more, paying special attention to the material that was difficult for you when you took your practice test the day before. Be sure you have all the materials that you will need for your test laid out ahead of time (two sharpened pencils, a good eraser, possibly a calculator or protractor, and so on). The most important thing you can do today is get a good night's rest.

Test day: Review your study sheet, if you have time. Focus on how well you have prepared and take a moment to relax. When taking your test, complete the problems that you are sure of first. Skip the problems that you don't understand right away, and return to them later. Bring a watch or make sure there will be some kind of time-keeping device in your test room so that you can keep track of your time. Try not to spend too much time on any one problem.

Now Try This

1. Create a study schedule using the guidelines given above.
2. Read the *Preparing for a Test* handout at www.cengagebrain.com/shop/search/tussy.
3. Read the *Taking the Test* handout at www.cengagebrain.com/shop/search/tussy.
4. After your test has been returned and scored, read the *Analyzing Your Test Results* handout at www.cengagebrain.com/shop/search/tussy.
5. Take time to reflect on your homework and study habits after you have received your test score. What actions are working well for you? What do you need to improve?
6. To prepare for your final exam, read the *Preparing for Your Final Exam* handout at www.cengagebrain.com/shop/search/tussy. Complete the action items given in this document.

1 Whole Numbers

1.1 An Introduction to the Whole Numbers
1.2 Adding and Subtracting Whole Numbers
1.3 Multiplying Whole Numbers
1.4 Dividing Whole Numbers
1.5 Prime Factors and Exponents
1.6 The Least Common Multiple and the Greatest Common Factor
1.7 Order of Operations
1.8 Solving Equations Using Addition and Subtraction
1.9 Solving Equations Using Multiplication and Division
Chapter Summary and Review
Chapter Test

from Campus to Careers
Landscape Designer

Landscape designers make outdoor places more beautiful and useful. They work on all types of projects. Some focus on yards and parks, others on land around buildings and highways. The training of a landscape designer should include botany classes to learn about plants; art classes to learn about color, line, and form; and mathematics classes to learn how to take measurements and keep business records.

In **Problem 104** of **Study Set 1.4,** you will see how a landscape designer uses division to determine the number of pine trees that are needed to form a windscreen for a flower garden. And in **Problem 112** of **Study Set 1.7**, the rule for the order of operations is used to calculate the available planting area.

JOB TITLE: Landscape designer
EDUCATION: A bachelor's degree in landscape design. Most states require a license.
JOB OUTLOOK: Expected to grow 24% from 2010 to 2020
ANNUAL EARNINGS: The average annual salary is $66,520.
FOR MORE INFORMATION
www.ashs.org/careers/profiles/landscape.lasso

Objectives

1. Identify the place value of a digit in a whole number.
2. Write whole numbers in words and in standard form.
3. Write a whole number in expanded form.
4. Compare whole numbers using inequality symbols.
5. Round whole numbers.
6. Read tables and graphs involving whole numbers.

SECTION 1.1
An Introduction to the Whole Numbers

The **whole numbers** are 0, 1, 2, 3, 4, 5, 6, 7, 8, 9, 10, 11, 12, and so on. They are used to answer questions such as How many?, How fast?, and How far?

- The movie *Titanic* won 11 Academy Awards.
- The average American adult reads at a rate of 250 to 300 words per minute.
- The driving distance from New York City to Los Angeles is 2,786 miles.

The *set of whole numbers* is written using **braces { }**, as shown below. The three dots indicate that the list continues forever—there is no largest whole number. The smallest whole number is 0.

The Set of Whole Numbers
{0, 1, 2, 3, 4, 5, 6, 7, 8, 9, 10, 11, 12, . . .}

1 Identify the place value of a digit in a whole number.

When a whole number is written using the **digits** 0, 1, 2, 3, 4, 5, 6, 7, 8, 9, it is said to be in **standard form** (also called **standard notation**). The position of a digit in a whole number determines its **place value**. In the number 325, the 5 is in the *ones column*, the 2 is in the *tens column*, and the 3 is in the *hundreds column*.

```
              Tens column
Hundreds column ↓   Ones column
        3 2 5
```

To make large whole numbers easier to read, we use commas to separate their digits into groups of three, called **periods**. Each period has a name, such as *ones*, *thousands*, *millions*, *billions*, and *trillions*. The following **place-value chart** shows the place value of each digit in the number 2,414,952,112,000, which is read as:

Two trillion, four hundred fourteen billion, nine hundred fifty-two million, one hundred twelve thousand

In 2011, the federal government collected a total of $2,414,952,112,000 in taxes.
Source: Internal Revenue Service

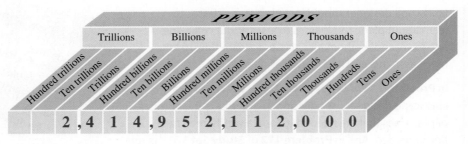

Each of the 2's in 2,414,952,112,000 has a different place value because of its position. The place value of the red 2 is 2 *trillions*. The place value of the blue 2 is 2 *millions*, and the place value of the green 2 is 2 *thousands*.

> **The Language of Algebra** As we move to the left in the chart, the place value of each column is 10 times the column directly to its right. This is why we call our number system the *base-10 number system*.

EXAMPLE 1

Airports Hartsfield-Jackson Atlanta International Airport is the busiest airport in the United States, handling 92,389,024 passengers in 2011. (Source: Airports Council International–North America)

a. What is the place value of the digit 3?
b. Which digit tells the number of millions?

Strategy We will begin in the ones column of 92,389,024. Then, moving to the left, we will name each column (ones, tens, hundreds, and so on) until we reach the digit 3.

WHY It's easier to remember the names of the columns if you begin with the smallest place value and move to the columns that have larger place values.

Solution

a. 92,389,024   Say, "Ones, tens, hundreds, thousands, ten thousands, hundred thousands" as you move from column to column.

3 hundred thousands is the place value of the digit 3.

b. 92,389,024

The digit 2 is in the millions column.

> **The Language of Algebra** Each of the worked examples in this textbook includes a *Strategy* and *Why* explanation. A **strategy** is a plan of action to follow to solve the given problem.

Self Check 1

CELL PHONES In 2012, there were 321,716,905 cellular telephone subscriber connections in the United States. (Source: CTIA The Wireless Association)

a. What is the place value of the digit 3? 3 hundred millions
b. Which digit tells the number of hundred thousands? 7

Now Try Problem 23

Teaching Example 1 CORN In 2011, there were 12,358,412,000 bushels of corn produced in the United States. (Source: *The World Almanac and Book of Facts*, 2013)
a. What is the place value of the digit 5?
b. Which digit tells the number of hundred thousands?
Answers:
a. 5 ten millions
b. 4

2 Write whole numbers in words and in standard form.

Since we use whole numbers so often in our daily lives, it is important to be able to read and write them.

Reading and Writing Whole Numbers

To write a whole number in words, start from the left. Write the number in each period followed by the name of the period (except for the *ones period*, which is not used). Use commas to separate the periods.

To read a whole number out loud, follow the same procedure. The commas are read as slight pauses.

> **The Language of Algebra** The word *and* should not be said when reading a whole number. It should only be used when reading a mixed number such as $5\frac{1}{2}$ (five *and* one-half) or a decimal such as 3.9 (three *and* nine-tenths).

EXAMPLE 2

Write each number in words:
a. 63   b. 499   c. 89,015   d. 6,070,534

Strategy For the larger numbers in parts c and d, we will name the periods from right to left to find the *greatest* period.

WHY To write a whole number in words, we must give the name of each period (except for the ones period). Finding the largest period helps to start the process.

Solution

a. 63 is written: *sixty-three*.   Use a hyphen to write whole numbers from 21 to 99 in words (except for 30, 40, 50, 60, 70, 80, and 90).

b. 499 is written: *four hundred ninety-nine*.

Self Check 2

Write each number in words:
a. 42
b. 798
c. 97,053
d. 23,000,017

Now Try Problems 31, 33, and 35
Self Check 2 Answers
a. forty-two
b. seven hundred ninety-eight
c. ninety-seven thousand, fifty-three
d. twenty-three million, seventeen

4 Chapter 1 Whole Numbers

Teaching Example 2 Write each number in words: a. 74 b. 851 c. 63,095 d. 7,320,116
Answers:
a. seventy-four
b. eight hundred fifty-one
c. sixty-three thousand, ninety-five
d. seven million, three hundred twenty thousand, one hundred sixteen

c. Thousands Ones   Say the names of the periods, working from right to left.

   89 , 015

Eighty-nine **thousand**, **fifteen**   We do not use a hyphen to write numbers between 1 and 20, such as 15. The ones period is not written.

d. Millions Thousands Ones   Say the names of the periods, working from right to left.

   6 , 0 7 0 , 5 3 4

Six **million**, seventy **thousand**, five **hundred** thirty-four.   The ones period is not written.

Caution! Two numbers, 40 and 90, are often misspelled: write *forty* (not *fourty*) and *ninety* (not *ninty*).

Self Check 3

Write each number in standard form:
a. *Two hundred three thousand, fifty-two* 203,052
b. *Nine hundred forty-six million, four hundred sixteen thousand, twenty-two* 946,416,022
c. *Three million, five hundred seventy-nine* 3,000,579

Now Try Problems 39 and 45

Teaching Example 3 Write each number in standard form: a. *Fourteen thousand, six hundred thirty-seven* b. *Eight hundred nine million, fifty-six thousand, two* c. *Sixty-five million, forty-eight*
Answers:
a. 14,637
b. 809,056,002
c. 65,000,048

EXAMPLE 3 Write each number in standard form:
a. *Twelve thousand, four hundred seventy-two*
b. *Seven hundred one million, thirty-six thousand, six*
c. *Forty-three million, sixty-eight*

Strategy We will locate the commas in the written-word form of each number.

WHY When a whole number is written in words, commas are used to separate periods.

Solution
a. Twelve thousand , four hundred seventy-two

   12, 472

b. Seven hundred one million , thirty-six thousand , six

   701,036,006

c. Forty-three million , sixty-eight   The written-word form does not mention the thousands period.

   43,000,068   If a period is not named, three zeros hold its place.

Success Tip Four-digit whole numbers are sometimes written without a comma. For example, we may write 3,911 or 3911 to represent three thousand, nine hundred eleven.

3 Write a whole number in expanded form.

In the number 6,352, the digit 6 is in the thousands column, 3 is in the hundreds column, 5 is in the tens column, and 2 is in the ones (or units) column. The meaning of 6,352 becomes clear when we write it in **expanded form** (also called **expanded notation**).

$$6{,}352 = 6 \text{ thousands} + 3 \text{ hundreds} + 5 \text{ tens} + 2 \text{ ones}$$

or

$$6{,}352 = 6{,}000 + 300 + 50 + 2$$

EXAMPLE 4 Write each number in expanded form:
a. 85,427 **b.** 1,251,609

Strategy Working from left to right, we will give the place value of each digit and combine them with + symbols.

WHY The term *expanded form* means to write the number as an addition of the place values of each of its digits.

Solution

a. The expanded form of 85,427 is:

8 ten thousands + 5 thousands + 4 hundreds + 2 tens + 7 ones

which can be written as:

80,000 + 5,000 + 400 + 20 + 7

b. The expanded form of 1,251,609 is:

1 million + 2 hundred thousands + 5 ten thousands + 1 thousand + 6 hundreds + 0 tens + 9 ones

Since 0 tens is zero, the expanded form can also be written as:

1 million + 2 hundred thousands + 5 ten thousands + 1 thousand + 6 hundreds + 9 ones

which can be written as:

1,000,000 + 200,000 + 50,000 + 1,000 + 600 + 9

Self Check 4
Write 708,413 in expanded form.

Now Try Problems 49, 53, and 57

Self Check 4 Answer
700,000 + 8,000 + 400 + 10 + 3

Teaching Example 4 Write each number in expanded form:
a. 62,189 **b.** 4,773,052
Answers:
a. 60,000 + 2,000 + 100 + 80 + 9
b. 4,000,000 + 700,000 + 70,000 + 3,000 + 50 + 2

4 Compare whole numbers using inequality symbols.

Whole numbers can be shown by drawing points on a **number line**. Like a ruler, a number line is straight and has uniform markings. To construct a number line, we begin on the left with a point on the line representing the number 0. This point is called the **origin**. We then move to the right, drawing equally spaced marks and labeling them with whole numbers that increase in value. The arrowhead at the right indicates that the number line continues forever.

Using a process known as **graphing**, we can represent a single number or a set of numbers on a number line. **The graph of a number** is the point on the number line that corresponds to that number. *To graph a number* means to locate its position on the number line and highlight it with a heavy dot. The graphs of 5 and 8 are shown on the number line below.

As we move to the right on the number line, the numbers increase in value. Because 8 lies to the right of 5, we say that 8 is greater than 5. The **inequality symbol** > ("is greater than") can be used to write this fact:

8 > 5   Read as "8 is greater than 5."

Since 8 > 5, it is also true that 5 < 8. We read this as "5 is less than 8."

Inequality Symbols

> means *is greater than*
< means *is less than*

Success Tip To tell the difference between these two inequality symbols, remember that they always point to the smaller of the two numbers involved.

$$8 > 5 \qquad 5 < 8$$
Points to the smaller number

EXAMPLE 5 Place an < or an > symbol in the box to make a true statement: **a.** 3 ☐ 7 **b.** 18 ☐ 16

Strategy To pick the correct inequality symbol to place between a pair of numbers, we need to determine the position of each number on the number line.

WHY For any two numbers on a number line, the number to the *left* is the smaller number and the number to the *right* is the larger number.

Solution
a. Since 3 is to the left of 7 on the number line, we have $3 < 7$.
b. Since 18 is to the right of 16 on the number line, we have $18 > 16$.

Self Check 5
Place an < or an > symbol in the box to make a true statement:
a. 12 > 4
b. 7 < 10
Now Try Problems 59 and 61

5 Round whole numbers.

When we don't need exact results, we often round numbers. For example, when a teacher with 36 students orders 40 textbooks, he has rounded the actual number to the *nearest ten*, because 36 is closer to 40 than it is to 30. We say 36, rounded to the nearest 10, is 40. This process is called **rounding up**.

Round up — 36 is closer to 40 than to 30.

When a geologist says that the height of Alaska's Mount McKinley is "about 20,300 feet," she has rounded to the *nearest hundred*, because its actual height of 20,320 feet is closer to 20,300 than it is to 20,400. We say that 20,320, rounded to the nearest hundred, is 20,300. This process is called **rounding down**.

20,320 is closer to 20,300 than 20,400.
Round down

The Language of Algebra When we round a whole number, we are finding an *approximation* of the number. An *approximation* is close to, but not the same as, the exact value.

To round a whole number, we follow an established set of rules. To round a number to the nearest ten, for example, we locate the **rounding digit** in the tens column. If the **test digit** to the right of that column (the digit in the ones column) is 5 or greater, we *round up* by increasing the tens digit by 1 and replacing the test digit with 0. If the test digit is less than 5, we *round down* by leaving the tens digit unchanged and replacing the test digit with 0.

EXAMPLE 6 Round each number to the nearest ten: **a.** 3,761 **b.** 12,087

Strategy We will find the digit in the tens column and the digit in the ones column.

WHY To round to the nearest ten, the digit in the tens column is the rounding digit and the digit in the ones column is the test digit.

Solution
a. We find the rounding digit in the tens column, which is 6. Then we look at the test digit to the right of 6, which is the 1 in the ones column. Since $1 < 5$, we round down by leaving the 6 unchanged and replacing the test digit with 0.

⌐Rounding digit: tens column   ⌐Keep the rounding digit: Do not add 1.
3,761   3,761
  └Test digit: 1 is less than 5.   └Replace with 0.

Thus, 3,761 rounded to the nearest ten is 3,760.

b. We find the rounding digit in the tens column, which is 8. Then we look at the test digit to the right of 8, which is the 7 in the ones column. Because 7 is 5 or greater, we round up by adding 1 to 8 and replacing the test digit with 0.

⌐Rounding digit: tens column   ⌐Add 1.
12,087   12,087
   └Test digit: 7 is 5 or greater.   └Replace with 0.

Thus, 12,087 rounded to the nearest ten is 12,090.

A similar method is used to round numbers to the nearest hundred, the nearest thousand, the nearest ten thousand, and so on.

Rounding a Whole Number

1. To round a number to a certain place value, locate the **rounding digit** in that place.
2. Look at the **test digit**, which is directly to the right of the rounding digit.
3. If the test digit is 5 or greater, round up by adding 1 to the rounding digit and replacing all of the digits to its right with 0.

   If the test digit is less than 5, replace it and all of the digits to its right with 0.

EXAMPLE 7 Round each number to the nearest hundred:
a. 18,349 **b.** 7,960

Strategy We will find the rounding digit in the hundreds column and the test digit in the tens column.

Self Check 6
Round each number to the nearest ten:
a. 35,642   35,640
b. 9,756   9,760

Now Try Problem 63

Teaching Example 6
Round each number to the nearest ten:
a. 4,183 **b.** 65,049
Answers:
a. 4,180
b. 65,050

Self Check 7
Round 365,283 to the nearest hundred.   365,300

Now Try Problems 69 and 71

Teaching Example 7
Round each number to the nearest hundred: **a.** 50,727 **b.** 1,953
Answers:
a. 50,700
b. 2,000

WHY To round to the nearest hundred, the digit in the hundreds column is the rounding digit and the digit in the tens column is the test digit.

Solution
a. First, we find the rounding digit in the hundreds column, which is 3. Then we look at the test digit 4 to the right of 3 in the tens column. Because $4 < 5$, we round down and leave the 3 in the hundreds column. We then replace the two rightmost digits with 0's.

⎡Rounding digit: hundreds column   ⎡Keep the rounding digit: Do not add 1.
18,349   18,349
 ⎣Test digit: 4 is less than 5.   ⎣Replace with 0's.

Thus, 18,349 rounded to the nearest hundred is 18,300.

b. First, we find the rounding digit in the hundreds column, which is 9. Then we look at the test digit 6 to the right of 9. Because 6 is 5 or greater, we round up and increase 9 in the hundreds column by 1. Since the 9 in the hundreds column represents 900, increasing 9 by 1 represents increasing 900 to 1,000. Thus, we replace the 9 with a 0 and add 1 to the 7 in the thousands column. Finally, we replace the two rightmost digits with 0's.

                                          Add 1. Since $9 + 1 = 10$, write 0 in this
⎡Rounding digit: hundreds column   ⎣column and carry 1 to the next column.
                                                 7+1 0
7,960   7,960
 ⎣Test digit: 6 is 5 or greater.   ⎣Replace with 0s.

Thus, 7,960 rounded to the nearest hundred is 8,000.

> *Caution!* To round a number, use *only* the test digit directly to the right of the rounding digit to determine whether to round up or round down.

Self Check 8

U.S. CITIES Round the elevation of Denver:
a. to the nearest hundred feet
b. to the nearest thousand feet

Now Try Problems 75 and 79
Self Check 8 Answers
a. 5,300 ft b. 5,000 ft

Teaching Example 8 U.S. CITIES
In 2011, San Antonio, Texas, with a population of 1,359,758, was the nation's 7th largest city. Round the population of San Antonio to
a. the nearest thousand and
b. the nearest million.
Answers:
a. 1,360,000
b. 1,000,000

EXAMPLE 8 **U.S. Cities** In 2011, Denver was the nation's 23rd largest city. Round the 2011 population of Denver shown on the sign to:
a. the nearest hundred thousand
b. the nearest thousand

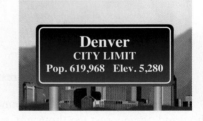

Strategy In each case, we will identify the rounding digit and the test digit.

WHY We need to know the value of the test digit to determine whether we round the population up or down.

Solution
a. The rounding digit in the hundred thousands column is 6. Since the test digit 1 is less than 5, we round down. To the nearest hundred thousand, Denver's population in 2011 was 600,000.
b. The rounding digit in the thousands column is 9. Since the test digit 9 is 5 or greater, we round up, by writing a 0 in the thousands column and carrying a 1 to the ten thousands column. To the nearest thousand, Denver's population in 2011 was 620,000.

6 Read tables and graphs involving whole numbers.

The following table is an example of the use of whole numbers. It shows the number of women members of the U.S. House of Representatives for the years 2003–2013.

Unless otherwise noted, all art on this page is © Cengage Learning.

Year	Number of women members
2003	63
2005	71
2007	75
2009	76
2011	75
2013	78

Source: Data from http://www.fas.org/sgp/crs/misc/RL30261.pdf

(a)

(b)

In figure (a), the information in the table is presented in a **bar graph**. The *horizontal* scale is labeled "Year" and units of 2 years are used. The *vertical* scale is labeled "Number of women members" and units of 10 are used. The bar directly over each year extends to a height that shows the number of women members of the House of Representatives that year.

> **The Language of Algebra** *Horizontal* is a form of the word *horizon*. Think of the sun setting over the *horizon*. *Vertical* means in an upright position. Pro basketball player LeBron James' *vertical* leap measures more than 49 inches.

Another way to present the information in the table is with a **line graph**. Instead of using a bar to represent the number of women members, we use a dot drawn at the correct height. After drawing data points for 2003, 2005, 2007, 2009, 2011, and 2013, we connect the points to create the line graph in figure (b).

THINK IT THROUGH   *Re-entry Students*

"A re-entry student is considered one who is the age of 25 or older, or those students that have had a break in their academic work for 5 years or more. Nationally, this group of students is growing at an astounding rate."
Student Life and Leadership Department, University Union, Cal Poly University, San Luis Obispo

Some common concerns expressed by adult students considering returning to school are listed below in Column I. Match each concern to an encouraging reply in Column II.

Column I
1. I'm too old to learn. c
2. I don't have the time. b
3. I didn't do well in school the first time around. I don't think a college would accept me. e
4. I'm afraid I won't fit in. d
5. I don't have the money to pay for college. a

Column II
a. Many students qualify for some type of financial aid.
b. Taking even a single class puts you one step closer to your educational goal.
c. There's no evidence that older students can't learn as well as younger ones.
d. More than 41% of the students in college are older than 25.
e. Typically, community colleges and career schools have an open admissions policy.

Source: Adapted from *Common Concerns for Adult Students*, Minnesota Higher Education Services Office

10 Chapter 1 Whole Numbers

ANSWERS TO SELF CHECKS

1. **a.** 3 hundred millions **b.** 7   2. **a.** forty-two **b.** seven hundred ninety-eight
   c. ninety-seven thousand, fifty-three **d.** twenty-three million, seventeen
3. **a.** 203,052 **b.** 946,416,022 **c.** 3,000,579   4. 700,000 + 8,000 + 400 + 10 + 3
5. **a.** > **b.** <   6. **a.** 35,640 **b.** 9,760   7. 365,300   8. **a.** 5,300 ft **b.** 5,000 ft

STUDY SKILLS CHECKLIST

Get to Know Your Textbook

Congratulations. You now own a state-of-the-art textbook that has been written especially for you. The following checklist will help you become familiar with the organization of this book. Place a check mark in each box after you answer the question.

☐ Turn to the **Table of Contents** on page iii. How many chapters does the book have?

☐ Each chapter of the book is divided into **sections**. How many sections are there in Chapter 1, which begins on page 1?

☐ Learning **Objectives** are listed at the start of each section. How many objectives are there for Section 1.2, which begins on page 15?

☐ Each section ends with a **Study Set**. How many problems are there in Study Set 1.2, which begins on page 29?

☐ Each chapter has a **Chapter Summary and Review**. Which column of the Chapter 1 Summary found on page 114 contains examples?

☐ How many review problems are there for Section 1.1 in the **Chapter 1 Summary and Review**, which begins on page 114?

☐ Each chapter has a **Chapter Test**. How many problems are there in the Chapter 1 Test, which begins on page 132?

☐ Each chapter (except Chapter 1) ends with a **Cumulative Review**. Which chapters are covered by the Cumulative Review that begins on page 221?

Answers: 10, 9, 10, 122, the right, 14, 48, 1–2

SECTION 1.1 STUDY SET

VOCABULARY

Fill in the blanks.

1. The numbers 0, 1, 2, 3, 4, 5, 6, 7, 8, and 9 are the digits .
2. The set of whole numbers is {0, 1, 2, 3, 4, 5, . . .}.
3. When we write five thousand eighty-nine as 5,089, we are writing the number in standard form.
4. To make large whole numbers easier to read, we use commas to separate their digits into groups of three, called periods .
5. When 297 is written as 200 + 90 + 7, we are writing 297 in expanded form.
6. Using a process called *graphing*, we can represent whole numbers as points on a number line.
7. The symbols > and < are inequality symbols.
8. If we round 627 to the nearest ten, we get 630.

CONCEPTS

9. Copy the following place-value chart. Then enter the whole number 1,342,587,200,946 and fill in the place value names and the periods.

1.1 An Introduction to the Whole Numbers

10. a. Insert commas in the proper positions for the following whole number written in standard form: 5467010 5,467,010

 b. Insert commas in the proper positions for the following whole number written in words: *seventy-two million four hundred twelve thousand six hundred thirty-five* seventy-two million, four hundred twelve thousand, six hundred thirty-five

11. Write each number in words.
 a. 40 forty
 b. 90 ninety
 c. 68 sixty-eight
 d. 15 fifteen

12. Write each number in standard form.
 a. 8 ten thousands + 1 thousand + 6 hundreds + 9 tens + 2 ones 81,692
 b. 900,000 + 60,000 + 5,000 + 300 + 40 + 7 965,347

Graph the following numbers on a number line.

13. 1, 3, 5, 7

14. 0, 2, 4, 6, 8

15. 2, 4, 5, 8

16. 2, 3, 5, 7, 9

17. the whole numbers less than 6

18. the whole numbers less than 9

19. the whole numbers between 2 and 8

20. the whole numbers between 0 and 6

NOTATION

Fill in the blanks.

21. The symbols { }, called braces , are used when writing a set.

22. The symbol > means is   greater   than , and the symbol < means is   less   than .

GUIDED PRACTICE

Find the place values. See Example 1.

23. Consider the number 57,634.
 a. What is the place value of the digit 3? 3 tens
 b. What digit is in the thousands column? 7
 c. What is the place value of the digit 6? 6 hundreds
 d. What digit is in the ten thousands column? 5

24. Consider the number 128,940.
 a. What is the place value of the digit 8? 8 thousands
 b. What digit is in the hundreds column? 9
 c. What is the place value of the digit 2? 2 ten thousands
 d. What digit is in the hundred thousands column? 1

25. WORLD HUNGER On the website Freerice.com, sponsors donate grains of rice to feed the hungry. From 2007 through 2012, there have been 97,736,292,140 grains of rice donated.
 a. What is the place value of the digit 6? 6 millions
 b. What digit is in the billions place? 7
 c. What is the place value of the 3? 3 ten millions
 d. What digit is in the ten billions place? 9

26. RECYCLING The website *Container Recycling Institute CRI* estimated that the number of beverage cans and bottles that were *not* recycled in the United States in 2012 was 128,901,701,453.
 a. What is the place value of the digit 7? 7 hundred thousands
 b. What digit is in the ten thousands place? 0
 c. What is the place value of the digit 8? 8 billions
 d. What digit is in the ten billions place? 2

Write each number in words. See Example 2.

27. 93 ninety-three
28. 48 forty-eight
29. 732 seven hundred thirty-two
30. 259 two hundred fifty-nine
31. 154,302 one hundred fifty-four thousand, three hundred two
32. 615,019 six hundred fifteen thousand, nineteen
33. 14,432,500 fourteen million, four hundred thirty-two thousand, five hundred
34. 104,052,005 one hundred four million, fifty-two thousand, five
35. 970,031,500,104 nine hundred seventy billion, thirty-one million, five hundred thousand, one hundred four
36. 5,800,010,700 five billion, eight hundred million, ten thousand, seven hundred
37. 82,000,415 eighty-two million, four hundred fifteen
38. 51,000,201,078 fifty-one billion, two hundred one thousand, seventy-eight

Write each number in standard form. See Example 3.

39. Three thousand, seven hundred thirty-seven 3,737
40. Fifteen thousand, four hundred ninety-two 15,492
41. Nine hundred thirty 930
42. Six hundred forty 640

43. Seven thousand, twenty-one 7,021
44. Four thousand, five hundred 4,500
45. Twenty-six million, four hundred thirty-two 26,000,432
46. Ninety-two billion, eighteen thousand, three hundred ninety-nine 92,000,018,399

Write each number in expanded form. See Example 4.

47. 245 200 + 40 + 5
48. 518 500 + 10 + 8
49. 3,609 3,000 + 600 + 9
50. 3,961 3,000 + 900 + 60 + 1
51. 72,533 70,000 + 2,000 + 500 + 30 + 3
52. 73,009 70,000 + 3,000 + 9
53. 104,401 100,000 + 4,000 + 400 + 1
54. 570,003 500,000 + 70,000 + 3
55. 8,403,613 8,000,000 + 400,000 + 3,000 + 600 + 10 + 3
56. 3,519,807 3,000,000 + 500,000 + 10,000 + 9,000 + 800 + 7
57. 26,000,156 20,000,000 + 6,000,000 + 100 + 50 + 6
58. 48,000,061 40,000,000 + 8,000,000 + 60 + 1

Place an < or an > symbol in the box to make a true statement. See Example 5.

59. a. 11 > 8   b. 29 < 54
60. a. 410 < 609   b. 3,206 < 3,231
61. a. 12,321 > 12,209   b. 23,223 < 23,231
62. a. 178,989 > 178,898   b. 850,234 < 850,342

Round to the nearest ten. See Example 6.

63. 98,154 98,150
64. 26,742 26,740
65. 512,967 512,970
66. 621,116 621,120

Round to the nearest hundred. See Example 7.

67. 8,352 8,400
68. 1,845 1,800
69. 32,439 32,400
70. 73,931 73,900
71. 65,981 66,000
72. 5,346,975 5,347,000
73. 2,580,952 2,581,000
74. 3,428,961 3,429,000

Round each number to the nearest thousand and then to the nearest ten thousand. See Example 8.

75. 52,867 53,000; 50,000
76. 85,432 85,000; 90,000
77. 76,804 77,000; 80,000
78. 34,209 34,000; 30,000
79. 816,492 816,000; 820,000
80. 535,600 536,000; 540,000
81. 296,500 297,000; 300,000
82. 498,903 499,000; 500,000

TRY IT YOURSELF

83. Round 79,593 to the nearest . . .
    a. ten 79,590
    b. hundred 79,600
    c. thousand 80,000
    d. ten thousand 80,000
84. Round 5,925,830 to the nearest . . .
    a. thousand 5,926,000
    b. ten thousand 5,930,000
    c. hundred thousand 5,900,000
    d. million 6,000,000

85. Round $419,161 to the nearest . . .
    a. $10 $419,160
    b. $100 $419,200
    c. $1,000 $419,000
    d. $10,000 $420,000
86. Round 5,436,483 ft to the nearest . . .
    a. 10 ft 5,436,480 ft
    b. 100 ft 5,436,500 ft
    c. 1,000 ft 5,436,000 ft
    d. 10,000 ft 5,440,000 ft

Write each number in standard notation.

87. 4 ten thousands + 2 tens + 5 ones 40,025
88. 7 millions + 7 tens + 7 ones 7,000,077
89. 200,000 + 2,000 + 30 + 6 202,036
90. 7,000,000,000 + 300 + 50 7,000,000,350
91. Twenty-seven thousand, five hundred ninety-eight 27,598
92. Seven million, four hundred fifty-two thousand, eight hundred sixty 7,452,860
93. Ten million, seven hundred thousand, five hundred six 10,700,506
94. Eighty-six thousand, four hundred twelve 86,412

APPLICATIONS

95. **GAME SHOWS** On *The Price Is Right* television show, the winning contestant is the person who comes closest to (without going over) the price of the item up for bid. Which contestant shown below will win if they are bidding on a bedroom set that has a suggested retail price of $4,745? Aisha

96. **PRESIDENTS** The following list shows the ten youngest U.S. presidents and their ages (in years/days) when they took office. Construct a two-column table that presents the data in order, beginning with the youngest president.

J. Polk 49 yr/122 days	U. Grant 46 yr/236 days
G. Cleveland 47 yr/351 days	J. Kennedy 43 yr/236 days
W. Clinton 46 yr/154 days	F. Pierce 48 yr/101 days
M. Filmore 50 yr/184 days	B. Obama 47 yr/169 days
J. Garfield 49 yr/105 days	T. Roosevelt 42 yr/322 days

97. **MISSIONS TO MARS** The United States, Russia, Europe, and Japan have launched Mars space probes. The graph shows the success rate of the missions, by decade.

   a. Which decade had the greatest number of successful or partially successful missions? How many? the 1970s, 7
   b. Which decade had the greatest number of unsuccessful missions? How many? the 1960s, 9
   c. Which decade had the greatest number of missions? How many? the 1960s, 12
   d. Which decade had no successful missions? the 1980s

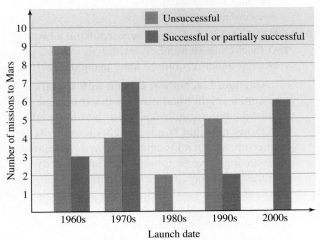

Source: The Planetary Society

98. **SPORTS** The graph shows the maximum recorded ball speeds for five sports.

   a. Which sport had the fastest recorded maximum ball speed? Estimate the speed. golf, 205 mph
   b. Which sport had the slowest maximum recorded ball speed? Estimate the speed. Ping-Pong, 70 mph
   c. Which sport had the second fastest maximum recorded ball speed? Estimate the speed. tennis, 155 mph

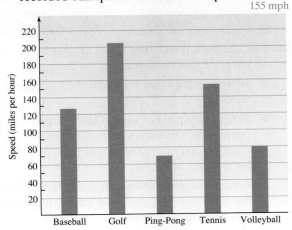

99. **COFFEE** Complete the bar graph and line graph using the data in the table.

Starbucks Locations

Year	Number
2000	3,501
2002	5,886
2004	8,569
2006	12,440
2008	16,680
2010	16,858
2012	17,651

Source: Starbucks Company

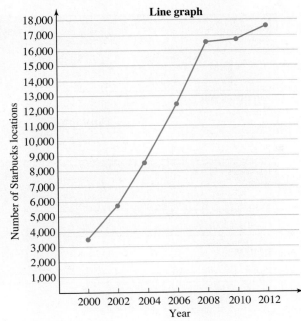

100. **ENERGY RESERVES** Complete the bar graph and line graph using the data in the table.

Natural Gas Reserves, 2010 Estimates (in Trillion Cubic Feet)

United States	300
Venezuela	195
Canada	70
Brazil	16
Mexico	13

Source: BP Statistical Review of World Energy, 2012

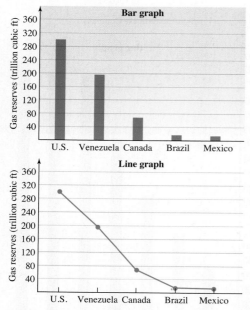

101. **CHECKING ACCOUNTS** Complete each check by writing the amount in words on the proper line.

a.

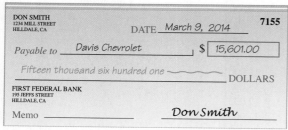

Fifteen thousand six hundred one

b.

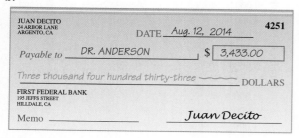

Three thousand four hundred thirty-three

102. **ANNOUNCEMENTS** One style used when printing formal invitations and announcements is to write all numbers in words. Use this style to write each of the following phrases.

   a. This diploma awarded this 27th day of June, 2014. This diploma awarded this twenty-seventh day of June, two thousand fourteen.

   b. The suggested contribution for the fundraiser is $850 a plate, or an entire table may be purchased for $5,250. The suggested contribution for the fundraiser is eight hundred fifty dollars a plate, or an entire table may be purchased for five thousand, two hundred fifty dollars.

103. **COPYEDITING** Edit this excerpt from a history text by circling all numbers written in words and rewriting them in standard form using digits.

   Abraham Lincoln was elected with a total of one million, eight hundred sixty-five thousand, five hundred ninety-three votes—four hundred eighty-two thousand, eight hundred eighty more than the runner-up, Stephen Douglas. He was assassinated after having served a total of one thousand, five hundred three days in office. Lincoln's Gettysburg Address, a mere two hundred sixty-nine words long, was delivered at the battle site where forty-three thousand, four hundred forty-nine casualties occurred. 1,865,593; 482,880; 1,503; 269; 43,449

104. **READING METERS** The amount of electricity used in a household is measured in kilowatt-hours (kwh). Determine the reading on the meter shown below. (When the pointer is between two numbers, read the *smaller* number.) 6,504 kwh

105. **SPEED OF LIGHT** The speed of light is 983,571,072 feet per second.

   a. In what place value column is the 5? hundred thousands

   b. Round the speed of light to the nearest ten million. Give your answer in standard notation and in expanded notation. 980,000,000; 900,000,000 + 80,000,000

   c. Round the speed of light to the nearest hundred million. Give your answer in standard notation and in written-word form. 1,000,000,000; one billion

106. **CLOUDS** Graph each cloud type given in the table at the proper altitude on the vertical number line below.

Cloud type	Altitude (ft)
Altocumulus	21,000
Cirrocumulus	37,000
Cirrus	38,000
Cumulonimbus	15,000
Cumulus	8,000
Stratocumulus	9,000
Stratus	4,000

WRITING

107. Explain how you would round 687 to the nearest ten.
108. The houses in a new subdivision are priced "in the low 130s." What does this mean?
109. A million is a thousand thousands. Explain why this is so.
110. Many television infomercials offer the viewer creative ways to make a six-figure income. What is a six-figure income? What is the smallest and what is the largest six-figure income?
111. What whole number is associated with each of the following words?

    duo   decade   zilch   a grand   four score
    dozen   trio   century   a pair   nil

112. Explain what is wrong by reading 20,003 as *twenty thousand and three*.
113. The words *two*, *to*, and *too* sound the same but are spelled differently and have different meanings. (Such words are called **homonyms**.) Write a sentence that contains all three words.
114. Write each statement in words.
    a. 2,016 < 2,106
    b. 7,080,008 > 7,008,800

SECTION 1.2
Adding and Subtracting Whole Numbers

Objectives
1. Add whole numbers.
2. Use properties of addition to add whole numbers.
3. Estimate sums of whole numbers.
4. Solve application problems by adding whole numbers.
5. Find the perimeter of a rectangle and a square.
6. Subtract whole numbers.
7. Check subtractions using addition.
8. Estimate differences of whole numbers.
9. Solve application problems by subtracting whole numbers.
10. Evaluate expressions involving addition and subtraction.

Everyone uses addition and subtraction of whole numbers. For example, to prepare an annual budget, an accountant adds separate line item costs. To determine the number of yearbooks to order, a principal adds the number of students in each grade level. To find the sale price of an item, a store clerk subtracts the discount from the regular price.

1 Add whole numbers.

Addition is the process of finding the total of two (or more) numbers. It can be illustrated using a number line, as shown below. For example, to compute 4 + 5, we begin at 0 and draw an arrow 4 units long, extending to the right. This represents 4. From the tip of that arrow, we draw another arrow 5 units long, also extending to the right. The second arrow points to 9. This result corresponds to the addition fact 4 + 5 = 9.

We can write this addition problem in **horizontal** or **vertical form** using an **addition symbol +**, which is read as "plus." The numbers that are being added are called **addends** and the answer is called the **sum** or **total**.

To add whole numbers that are less than 10, we rely on our understanding of basic addition facts. For example,

$$2 + 3 = 5, \quad 6 + 4 = 10, \quad \text{and} \quad 9 + 7 = 16$$

To add whole numbers that are greater than 10, we can use vertical form by stacking them with their corresponding place values lined up. Then we simply add the digits in each corresponding column. If an addition of the digits in any place value column produces a sum that is greater than 9, we must **carry**.

Self Check 1

Add: 675 + 1,497 + 1,527   3,699

Now Try Problems 27 and 31

Teaching Example 1
Add: 8,584 + 397 + 6,623
Answer:
15,604

EXAMPLE 1   Add: 9,835 + 692 + 7,275

Strategy We will write the numbers in vertical form so that corresponding place value columns are lined up. Then we will add the digits in each column, watching for any sums that are greater than 9.

WHY If the sum of the digits in any column is more than 9, we must carry.

Solution
We write the addition in vertical form, so that the corresponding digits are lined up. Each step of this addition is explained separately. Your solution need only look like the *last* step.

$$\begin{array}{r} \overset{1}{9{,}83}5 \\ 692 \\ +\ 7{,}275 \\ \hline 2 \end{array}$$ Add the digits in the ones column: 5 + 2 + 5 = 12. Write 2 in the ones column of the answer and carry 1 to the tens column.

$$\begin{array}{r} \overset{2\ 1}{9{,}83}5 \\ 692 \\ +\ 7{,}275 \\ \hline 0\ 2 \end{array}$$ Add the digits in the tens column: 1 + 3 + 9 + 7 = 20. Write 0 in the tens column of the answer and carry 2 to the hundreds column.

$$\begin{array}{r} \overset{1\ 2\ 1}{9{,}83}5 \\ 692 \\ +\ 7{,}275 \\ \hline 8\ 0\ 2 \end{array}$$ Add the digits in the hundreds column: 2 + 8 + 6 + 2 = 18. Write 8 in the hundreds column of the answer and carry 1 to the thousands column.

$$\begin{array}{r} \overset{1\ 2\ 1}{9{,}83}5 \\ 692 \\ +\ 7{,}275 \\ \hline 17{,}802 \end{array}$$ Add the digits in the thousands column: 1 + 9 + 7 = 17. Write 7 in the thousands column of the answer. Write 1 in the ten thousands column.

Your solution should look like this:
$$\begin{array}{r} \overset{1\ 2\ 1}{9{,}83}5 \\ 692 \\ +\ 7{,}275 \\ \hline 17{,}802 \end{array}$$

The sum is 17,802.

Success Tip In Example 1, the digits in each place value column were added from *top to bottom*. To check the answer, we can instead add from *bottom to top*. Adding down or adding up should give the same result. If it does not, an error has been made and you should re-add. You will learn why the two results should be the same in Objective 2, which follows.

```
              17,802
First add      9,835       To check,
top to           692       add
bottom.      + 7,275       bottom
             -------       to top.
              17,802
```

2 Use properties of addition to add whole numbers.

We have used a number line to find that 4 + 5 = 9. If we add 4 and 5 in the opposite order, we see on the number line below that the result is the same:

This example illustrates that the order in which we add two numbers does not affect the result. This property is called the **commutative property of addition**. To state the commutative property of addition in a compact form, we can use *variables*.

Variables

A **variable** is a letter (or a symbol) that is used to stand for a number.

We now use the variables a and b to state the communtative property of addition.

Commutative Property of Addition

The order in which whole numbers are added does not change their sum.
For any whole number a and b,
$$a + b = b + a$$

The Language of Algebra *Commutative* is a form of the word *commute*, meaning to go back and forth. *Commuter* trains take people to and from work.

To find the sum of three whole numbers, we add two of them and then add the sum to the third number. In the following examples, we add 3 + 4 + 7 in two ways. We will use the grouping symbols ( ), called **parentheses**, to show this. It is standard practice to perform the operations within the parentheses first. The steps of the solutions are written in horizontal form.

The Language of Algebra In the following example, read $(3 + 4) + 7$ as "The *quantity* of 3 plus 4," pause slightly, and then say "plus 7." We read $3 + (4 + 7)$ as, "3 plus the *quantity* of 4 plus 7." The word *quantity* alerts the reader to the parentheses that are used as grouping symbols.

Method 1: Group 3 and 4

$(3 + 4) + 7 = 7 + 7$ Because of the parentheses, add 3 and 4 first to get 7. Then add 7 and 7 to get 14.

$= 14$

Method 2: Group 4 and 7

$3 + (4 + 7) = 3 + 11$ Because of the parentheses, add 4 and 7 first to get 11. Then add 3 and 11 to get 14.

$= 14$

—Same result—

Either way, the answer is 14. This example illustrates that changing the grouping when adding numbers doesn't affect the result. This property is called the **associative property of addition**.

Associative Property of Addition

The way in which whole numbers are grouped does not change their sum.
    For any whole numbers a, b, and c,

$$(a + b) + c = a + (b + c)$$

The Language of Algebra *Associative* is a form of the word *associate*, meaning to join a group. The WNBA (Women's National Basketball *Association*) is a group of 12 professional basketball teams.

Sometimes, an application of the associative property can simplify a calculation.

Self Check 2
Find the sum:
$(139 + 25) + 75$ 239

Now Try Problem 35

Teaching Example 2
Find the sum: $91 + (9 + 27)$
Answer:
127

EXAMPLE 2   Find the sum: $98 + (2 + 17)$

Strategy We will use the associative property to group 2 with 98.

WHY It is helpful to regroup because 98 and 2 are a pair of numbers that are easily added.

Solution
We will write the steps of the solution in horizontal form.

$98 + (2 + 17) = (98 + 2) + 17$ Use the associative property of addition to regroup the addends.

$= 100 + 17$ Do the addition within the parentheses first.

$= 117$

Whenever we add 0 to a whole number, the number is unchanged. This property is called the **addition property of 0**.

Addition Property of 0

The sum of any whole number and 0 is that whole number.
    For any whole number a,

$$a + 0 = a \quad \text{and} \quad 0 + a = a$$

3 Estimate sums of whole numbers.

Estimation is used to find an approximate answer to a problem. Estimates are helpful in two ways. First, they serve as an accuracy check that can find errors. If an answer does not seem reasonable when compared to the estimate, the original problem should be reworked. Second, some situations call for only an approximate answer rather than the exact answer.

There are several ways to estimate, but the objective is the same: Simplify the numbers in the problem so that the calculations can be made easily and quickly. One popular method of estimation is called **front-end rounding**.

EXAMPLE 3 Use front-end rounding to estimate the sum: $3{,}714 + 2{,}489 + 781 + 5{,}500 + 303$

Strategy We will use front-end rounding to approximate each addend. Then we will find the sum of the approximations.

WHY Front-end rounding produces addends containing many 0's. Such numbers are easier to add.

Solution
Each of the addends is rounded to its *largest place value* so that all but its first digit is zero. Then we add the approximations using vertical form.

3,714	→	4,000	Round to the nearest thousand.
2,489	→	2,000	Round to the nearest thousand.
781	→	800	Round to the nearest hundred.
5,500	→	6,000	Round to the nearest thousand.
+ 303	→	+ 300	Round to the nearest hundred.
		13,100	

The estimate is 13,100.

If we calculate $3{,}714 + 2{,}489 + 781 + 5{,}500 + 303$, the sum is exactly 12,787. Note that the estimate is close: It's just 313 more than 12,787. This illustrates the tradeoff when using estimation: The calculations are easier to perform and they take less time, but the answers are not exact.

Self Check 3
Use front-end rounding to estimate the sum:

```
  6,780
  3,278
    566
  4,230
 +1,923
 16,600
```

Now Try Problem 37

Teaching Example 3
Use front-end rounding to estimate the sum:
$2{,}508 + 3{,}199 + 680 + 6{,}600 + 101$
Answer:
13,800

> **Success Tip** Estimates can be greater than or less than the exact answer. It depends on how often rounding up and rounding down occurs in the estimation.

4 Solve application problems by adding whole numbers.

Since application problems are almost always written in words, the ability to understand what you read is very important.

> **The Language of Algebra** Here are some key words and phrases that are often used to indicate addition:
>
> | gain | increase | up | forward | rise | more than |
> |---|---|---|---|---|---|
> | total | combined | in all | in the future | altogether | extra |

Self Check 4

AIRLINE ACCIDENTS The numbers of accidents involving U.S. airlines for the years 2004 through 2011 are listed in the table below. Find the total number of accidents for those years.

Year	Accidents
2004	30
2005	40
2006	33
2007	28
2008	28
2009	30
2010	29
2011	31

Source: National Transportation Safety Board

Now Try Problem 99

Self Check 4 Answer
The total number of accidents for 2004–2011 was 249.

Teaching Example 4
HURRICANES The number of hurricanes to strike the U.S. mainland for each decade from 1951 are listed in the table below. Find the total number of hurricanes for that time span. (Source: National Hurricane Center)

Decade	Number of hurricanes
1951–1960	18
1961–1970	14
1971–1980	12
1981–1990	15
1991–2000	14
2001–2010	19

Answer:
The total number of hurricanes from 1951 through 2010 was 92.

EXAMPLE 4

Sharks The graph on the right shows the number of shark attacks worldwide for the years 2004 through 2011. Find the total number of shark attacks for those years.

Strategy We will carefully read the problem looking for a key word or phrase.

WHY Key words and phrases indicate which arithmetic operation(s) should be used to solve the problem.

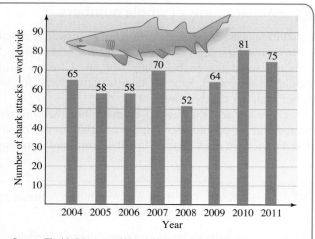

Source: Florida Museum of Natural History, Ichthyology Department

Solution
In the second sentence of the problem, the key word *total* indicates that we should add the number of shark attacks for the years 2004 through 2011. We can use vertical form to find the sum.

Add the digits, one column at a time, working from right to left. To simplify the calculations, we can look for groups of two or three numbers in each column whose sum is 10.

$$\begin{array}{r} \overset{5\,3}{65} \\ 58 \\ 58 \\ 70 \\ 52 \\ 64 \\ 81 \\ +\ 75 \\ \hline 523 \end{array}$$

The total number of shark attacks worldwide for the years 2004 through 2011 was 523.

> **The Language of Algebra** To solve the application problems, we must often *translate* the words of the problem to numbers and symbols. To *translate* means to change from one form to another, as in *translating* from Spanish to English.

EXAMPLE 5

Endangered Wolves In 1989, there were 1,814 gray wolves in the western Great Lakes region of Minnesota, Wisconsin, and Michigan. By 2011, the number had increased by 2,576. Find the number of gray wolves in 2011 in that region. (Source: U.S. Fish and Wildlife Service)

Strategy We will carefully read the problem looking for key words or phrases.

WHY Key words and phrases indicate which arithmetic operation(s) should be used to solve the problem.

Solution
The phrase *increased by* indicates addition. With that in mind, we translate the words of the problem to numbers and symbols.

The number of gray wolves in 2011 is equal to the number of gray wolves in 1989 increased by 2,576.

The number of gray wolves in 2011 = 1,814 + 2,576

Use vertical form to perform the addition:

$$\begin{array}{r} \overset{1\ \ 1}{1,814} \\ +\,2,576 \\ \hline 4,390 \end{array}$$

In 2011, the number of gray wolves in the western Great Lakes region was 4,390.

5 Find the perimeter of a rectangle and a square.

Figure (a) below is an example of a four-sided figure called a **rectangle**. Either of the longer sides of a rectangle is called its **length** and either of the shorter sides is called its **width**. Together, the length and width are called the **dimensions** of the rectangle. For any rectangle, opposite sides have the same measure.

When all four of the sides of a rectangle are the same length, we call the rectangle a **square**. An example of a square is shown in figure (b).

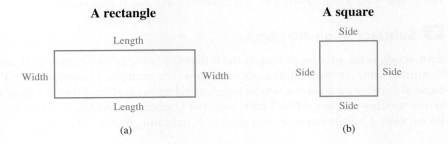

The distance around a rectangle or a square is called its **perimeter**. To find the perimeter of a rectangle, we add the lengths of its four sides.

The perimeter of a rectangle = length + length + width + width

To find the perimeter of a square, we add the lengths of its four sides.

The perimeter of a square = side + side + side + side

> **The Language of Algebra** When you hear the word *perimeter*, think of the distance around the "rim" of a flat figure.

EXAMPLE 6 **Money** Find the perimeter of the dollar bill shown below.

Width = 65 mm

mm stands for millimeters

Length = 156 mm

Self Check 5

MAGAZINES In 2011, the monthly circulation of *Game Informer* magazine was 5,954,884 copies. By 2012, the circulation had increased by 2,214,640 copies per month. What was the monthly circulation of *Game Informer* magazine in 2012? (Source: *The World Almanac and Book of Facts*, 2012, 2013) The monthly circulation in 2012 was 8,169,524.

Now Try Problem 97

Teaching Example 5
MOVIE SCREENS In 1987, the total number of movie screens (indoor and drive-in) in the United States was 22,679. By 2011 the number of screens had increased by 16,901. How many movie screens were there in 2011? (Source: National Association of Theater Owners)
Answer:
The number of movie screens in 2011 was 39,580.

Self Check 6

BOARD GAMES A Monopoly game board is a square with sides 19 inches long. Find the perimeter of the board. 76 in.

Now Try Problems 41 and 43

Unless otherwise noted, all art on this page is © Cengage Learning.

Teaching Example 6
FOOTBALL FIELDS A professional football field (including the end zones) is 160 feet wide and 360 feet long. Find the perimeter of the field.
Answer:
1,040 ft

Strategy We will add two lengths and two widths of the dollar bill.

WHY A dollar bill is rectangular-shaped, and this is how the perimeter of a rectangle is found.

Solution
We translate the words of the problem to numbers and symbols.

The perimeter of the dollar bill	is equal to	the length of the dollar bill	plus	the length of the dollar bill	plus	the width of the dollar bill	plus	the width of the dollar bill.
The perimeter of the dollar bill	=	156	+	156	+	65	+	65

Use vertical form to perform the addition:

$$\begin{array}{r} \overset{2\,2}{156} \\ 156 \\ 65 \\ +\ 65 \\ \hline 442 \end{array}$$

The perimeter of the dollar bill is 442 mm.
To see whether this result is reasonable, we estimate the answer. Because the rectangle is about 160 mm by 70 mm, its perimeter is approximately 160 + 160 + 70 + 70, or 460 mm. An answer of 442 mm is reasonable. ∎

6 Subtract whole numbers.

Subtraction is the process of finding the difference between two numbers. It can be illustrated using a number line, as shown below. For example, to compute 9 − 4, we begin at 0 and draw an arrow 9 units long, extending to the right. From the tip of that arrow, we draw another arrow 4 units long, but extending to the left. (This represents taking away 4.) The second arrow points to 5, indicating that 9 − 4 = 5.

We can write this subtraction problem in **horizontal** or **vertical form** using a **subtraction symbol** −, which is read as "minus." We call the number from which another number is subtracted the **minuend**. The number being subtracted is called the **subtrahend**, and the answer is called the **difference**.

> **The Language of Algebra** The prefix *sub* means *below*, as in *sub*marine or *sub*way. Notice that in vertical form, the *sub*trahend is written below the minuend.

To subtract two whole numbers that are less than 10, we rely on our understanding of basic subtraction facts. For example,

$$6 - 3 = 3, \quad 7 - 2 = 5, \quad \text{and} \quad 9 - 8 = 1$$

To subtract two whole numbers that are greater than 10, we can use vertical form by stacking them with their corresponding place values lined up. Then we simply subtract the digits in each corresponding column.

EXAMPLE 7   Subtract 235 from 6,496.

Strategy We will translate the sentence to mathematical symbols and then perform the subtraction. We must be careful when translating the instruction to subtract one number *from* another number.

WHY The order of the numbers in the sentence must be reversed when we translate to symbols.

Solution
Since 235 is the number to be subtracted, it is the subtrahend.

$$6{,}496 - 235$$

To find the difference, we write the subtraction in vertical form and subtract the digits in each column, working from right to left.

```
  6,496
-   235
  -----
  6,261
```
↳ Bring down the 6 in the thousands column.

When 235 is subtracted from 6,496, the difference is 6,261.

Self Check 7
Subtract 817 from 1,958.   1,141
Now Try Problem 49

Teaching Example 7
Subtract 372 from 4,689.
Answer:
4,317

Caution! When subtracting two numbers, it is important that we write them in the correct order, because subtraction is *not* commutative. For instance, in Example 7, if we had incorrectly translated "*Subtract 235 from 6,496*" as $235 - 6{,}496$, we see that the difference is not 6,261. In fact, the difference is not even a whole number.

If the subtraction of the digits in any place value column requires that we subtract a larger digit from a smaller digit, we must **borrow** or **regroup**. Some subtractions require borrowing from two (or more) place value columns.

EXAMPLE 8   Subtract: $9{,}927 - 568$

Strategy We will write the subtraction in vertical form and subtract as usual. In each column, we must watch for a digit in the subtrahend that is greater than the digit directly above it in the minuend.

WHY If a digit in the subtrahend is greater than the digit above it in the minuend, we need to borrow (regroup) to subtract in that column.

Self Check 8
Subtract: $6{,}734 - 356$ 6,378
Now Try Problem 53

Teaching Example 8
Subtract: $7{,}835 - 297$
Answer:
7,538

Solution

We write the subtraction in vertical form, so that the corresponding digits are lined up. Each step of this subtraction is explained separately. Your solution should look like the *last* step.

$$\begin{array}{r} 9,927 \\ -568 \\ \hline \end{array}$$

Since 8 in the ones column of 568 is greater than 7 in the ones column of 9,927, we cannot immediately subtract. To subtract in that column, we must regroup by borrowing 1 ten from 2 in the tens column. In this process, we use the fact that 1 ten = 10 ones.

$$\begin{array}{r} 1\,17 \\ 9,9\cancel{2}\cancel{7} \\ -568 \\ \hline 9 \end{array}$$ Borrow 1 ten from 2 in the tens column and change the 2 to 1. Add the borrowed 10 to the digit 7 in the ones column of the minuend to get 17. Then subtract in the ones column: $17 - 8 = 9$.

Since 6 in the tens column of 568 is greater than 1 in the tens column directly above it, we cannot immediately subtract. To subtract in that column, we must regroup by borrowing 1 hundred from 9 in the hundreds column. In this process, we use the fact that 1 hundred = 10 tens.

$$\begin{array}{r} 11 \\ 8\,\cancel{1}\,17 \\ 9,\cancel{9}\cancel{2}\cancel{7} \\ -568 \\ \hline 59 \end{array}$$ Borrow 1 hundred from 9 in the hundreds column and change the 9 to 8. Add the borrowed 10 to the digit 1 in the tens column of the minuend to get 11. Then subtract in the tens column: $11 - 6 = 5$.

Complete the solution by subtracting in the hundreds column ($8 - 5 = 3$) and bringing down the 9 in the thousands column.

$$\begin{array}{r} 11 \\ 8\,\cancel{1}\,17 \\ 9,\cancel{9}\cancel{2}\cancel{7} \\ -568 \\ \hline 9,359 \end{array}$$

Your solution should look like this:	$\begin{array}{r} 11 \\ 8\,\cancel{1}\,17 \\ 9,\cancel{9}\cancel{2}\cancel{7} \\ -568 \\ \hline 9,359 \end{array}$

The difference is 9,359.

The borrowing process is more difficult when the minuend contains one or more zeros.

Self Check 9

Subtract: $65,304 - 1,445$ 63,859

Now Try Problem 57

Teaching Example 9
Subtract: $93,601 - 2,859$
Answer:
90,742

EXAMPLE 9 Subtract: $42,403 - 1,675$

Strategy We will write the subtraction in vertical form. To subtract in the ones column, we will borrow from the hundreds column of the minuend 42,403.

WHY Since the digit in the tens column of 42,403 is 0, it is not possible to borrow from that column.

Solution

We write the subtraction in vertical form so that the corresponding digits are lined up. Each step of this subtraction is explained separately. Your solution should look like the *last* step.

$$\begin{array}{r} 42,403 \\ -1,675 \\ \hline \end{array}$$

Since 5 in the ones column of 1,675 is greater than 3 in the ones column of 42,403, we cannot immediately subtract. It is not possible to borrow from the digit 0 in the tens column of 42,403. We can, however, borrow from the hundreds column to regroup in the tens column, as shown on the next page. In this process, we use the fact that 1 hundred = 10 tens.

$$\begin{array}{r} {\scriptstyle 3\ 10}\\ 42{,}4\cancel{0}3\\ -\ 1{,}675\\ \hline \end{array}$$ Borrow 1 hundred from 4 in the hundreds column and change the 4 to 3. Add the borrowed 10 to the digit 0 in the tens column of the minuend to get 10.

Now we can borrow from the 10 in the tens column to subtract in the ones column.

$$\begin{array}{r} {\scriptstyle\ \ \ \ 9}\\ {\scriptstyle 3\ 10\ 13}\\ 42{,}\cancel{4}\cancel{0}\cancel{3}\\ -\ 1{,}675\\ \hline 8 \end{array}$$ Borrow 1 ten from 10 in the tens column and change the 10 to 9. Add the borrowed 10 to the digit 3 in the ones column of the minuend to get 13. Then subtract in the ones column: $13 - 5 = 8$.

Next, we perform the subtraction in the tens column: $9 - 7 = 2$.

$$\begin{array}{r} {\scriptstyle\ \ \ \ 9}\\ {\scriptstyle 3\ 10\ 13}\\ 42{,}\cancel{4}\cancel{0}\cancel{3}\\ -\ 1{,}675\\ \hline 28 \end{array}$$

To subtract in the hundreds column, we borrow from the 2 in the thousands column. In this process, we use the fact that 1 thousand = 10 hundreds.

$$\begin{array}{r} {\scriptstyle\ \ \ 13\ 9}\\ {\scriptstyle 1\ \ 3\ 10\ 13}\\ 4\cancel{2}{,}\cancel{4}\cancel{0}\cancel{3}\\ -\ 1{,}675\\ \hline 728 \end{array}$$ Borrow 1 thousand from 2 in the thousands column and change the 2 to 1. Add the borrowed 10 to the digit 3 in the hundreds column of the minuend to get 13. Then subtract in the hundreds column: $13 - 6 = 7$.

Complete the solution by subtracting in the thousands column ($1 - 1 = 0$) and bringing down the 4 in the ten thousands column.

$$\begin{array}{r} {\scriptstyle\ \ \ 13\ 9}\\ {\scriptstyle 1\ \ 3\ 10\ 13}\\ 4\cancel{2}{,}\cancel{4}\cancel{0}\cancel{3}\\ -\ 1{,}675\\ \hline 40{,}728 \end{array}$$

Your solution should look like this:
$$\begin{array}{r} {\scriptstyle\ \ \ 13\ 9}\\ {\scriptstyle 1\ \ 3\ 10\ 13}\\ 4\cancel{2}{,}\cancel{4}\cancel{0}\cancel{3}\\ -\ 1{,}675\\ \hline 40{,}728 \end{array}$$

The difference is 40,728.

7 Check subtractions using addition.

Every subtraction has a **related addition statement**. For example,

$9 - 4 = 5$	because	$5 + 4 = 9$
$25 - 15 = 10$	because	$10 + 15 = 25$
$100 - 1 = 99$	because	$99 + 1 = 100$

These examples illustrate how we can check subtractions. If a subtraction is done correctly, *the sum of the difference and the subtrahend will always equal the minuend*:

Difference + subtrahend = minuend

> **The Language of Algebra** To describe the special relationship between addition and subtraction, we say that they are **inverse operations**.

EXAMPLE 10
Check the following subtraction using addition:
$$\begin{array}{r} 3{,}682\\ -1{,}954\\ \hline 1{,}728 \end{array}$$

Strategy We will add the difference (1,728) and the subtrahend (1,954) and compare that result to the minuend (3,682).

Self Check 10
Check the following subtraction using addition:
$$\begin{array}{r} 9{,}784\\ -4{,}792\\ \hline 4{,}892 \end{array}$$
The subtraction is incorrect.

Now Try Problem 61

Teaching Example 10
Check the following subtraction using addition:

$$\begin{array}{r} 8{,}037 \\ -6{,}395 \\ \hline 1{,}642 \end{array}$$

Answer:
The subtraction is correct.

WHY If the sum of the difference and the subtrahend gives the minuend, the subtraction checks.

Solution

The subtraction to check		Its related addition statement
$\begin{array}{r}3{,}682 \\ -1{,}954 \\ \hline 1{,}728\end{array}$ — difference + subtrahend — minuend		$\begin{array}{r}\overset{1\ \ 1}{1{,}728} \\ +1{,}954 \\ \hline 3{,}682\end{array}$

Since the sum of the difference and the subtrahend is the minuend, the subtraction is correct.

8 Estimate differences of whole numbers.

Estimation is used to find an approximate answer to a problem.

Self Check 11
Estimate the difference:
64,259 − 7,604   52,000

Now Try Problem 65

Teaching Example 11
Estimate the difference: 57,208 − 3,499
Answer:
57,000

EXAMPLE 11   Estimate the difference: 89,070 − 5,431

Strategy We will use front-end rounding to approximate the 89,070 and 5,431. Then we will find the difference of the approximations.

WHY Front-end rounding produces whole numbers containing many 0's. Such numbers are easier to subtract.

Solution
Both the minuend and the subtrahend are rounded to their *largest place value* so that all but their first digit is zero. Then we subtract the approximations using vertical form.

$$\begin{array}{rcl} 89{,}070 & \to & 90{,}000 \\ -\ \ 5{,}431 & \to & -\ \ 5{,}000 \\ \hline & & 85{,}000 \end{array}$$

Round to the nearest ten thousand.
Round to the nearest thousand.

The estimate is 85,000. If we calculate 89,070 − 5,431, the difference is exactly 83,639. Note that the estimate is close: It's only 1,361 more than 83,639.

9 Solve application problems by subtracting whole numbers.

To answer questions about *how much more* or *how many more*, we use subtraction.

Self Check 12

ELEPHANTS An average male African elephant weighs 13,000 pounds. An average male Asian elephant weighs 11,900 pounds. How much more does an African elephant weigh than an Asian elephant?   1,100 lb

Now Try Problem 105

EXAMPLE 12   *Horses* Big Jake, the world's largest horse, weighs 2,600 pounds. Thumbelina, the world's smallest horse, weighs 57 pounds. How much more does Big Jake weigh than Thumbelina? (Source: *Guinness Book of World Records*, 2013)

Strategy We will carefully read the problem, looking for a key word or phrase.

WHY Key words and phrases indicate which arithmetic operation(s) should be used to solve the problem.

Solution

In the second sentence of the problem, the phrase *How much more* indicates that we should subtract the weights of the horses. We translate the words of the problem to numbers and symbols.

| The number of pounds more that Big Jake weighs | is equal to | the weight of Big Jake | minus | the weight of Thumbelina. |

$$\text{The number of pounds more that Big Jake weighs} = 2{,}600 - 57$$

Use vertical form to perform the subtraction:

$$\begin{array}{r} \overset{9}{5\,\cancel{10}\,10} \\ 2{,}\cancel{6}\cancel{0}\cancel{0} \\ -57 \\ \hline 2{,}543 \end{array}$$

Big Jake weighs 2,543 pounds more than Thumbelina.

> **The Language of Algebra** Here are some more key words and phrases that often indicate subtraction:
>
> loss   decrease   down   backward   fell   less than   fewer
> reduce   remove   debit   in the past   remains   declined   take away

EXAMPLE 13   Radio Stations

In 2005, there were 763 oldies radio stations in the United States. By 2012, there were 166 fewer. How many oldies radio stations were there in 2012? (Source: *The World Almanac and Book of Facts*, 2013)

Strategy We will carefully read the problem, looking for a key word or phrase.

WHY Key words and phrases indicate which arithmetic operations should be used to solve the problem.

Solution

The key phrase 166 *fewer* indicates subtraction. We translate the words of the problem to numbers and symbols.

| The number of oldies radio stations in 2012 | is | 166 | fewer than | the number of oldies radio stations in 2005. |

$$\text{The number of oldies radio stations in 2012} = 763 - 166$$

Use vertical form to perform the subtraction:

$$\begin{array}{r} \overset{15}{6\,\cancel{5}\,13} \\ \cancel{7}\cancel{6}\cancel{3} \\ -166 \\ \hline 597 \end{array}$$

In 2012, there were 597 oldies radio stations in the United States.

10 Evaluate expressions involving addition and subtraction.

In arithmetic, numbers are combined with the operations of addition, subtraction, multiplication, and division to create **expressions**. For example,

$$15 + 6, \quad 873 - 99, \quad 6{,}512 \times 24, \quad \text{and} \quad 42 \div 7$$

are expressions.

Teaching Example 12
SHIPS The supertanker *Seawise Giant*, at 1,504 feet, is the longest ship ever made. The *Allure of the Seas*, at 1,181 feet, is the longest cruise ship ever made. How much longer is the supertanker?
Source: marineinsight.com
Answer:
The supertanker is 323 ft longer than the cruise ship.

Self Check 13
HEALTHY DIETS When Jared Fogle began his reduced-calorie diet of Subway sandwiches, he weighed 425 pounds. With dieting and exercise, he eventually dropped 245 pounds. What was his weight then?

Now Try Problem 111

Self Check 13 Answer
After the dieting and exercise program, Jared weighed 180 lb.

Teaching Example 13
PLANETS It takes 687 Earth days for Mars to orbit the Sun. It takes Venus 462 fewer Earth days. How many Earth days does it take Venus to orbit the Sun? (Source: Enchantedlearning.com)
Answer:
It takes 225 Earth days for Venus to orbit the sun.

Expressions can contain more than one operation. That is the case for the expression 27 − 16 + 5, which contains addition *and* subtraction. To **evaluate** (find the value of) expressions written in horizontal form that involve addition and subtraction, we perform the operations as they occur *from left to right*.

Self Check 14

Evaluate: 75 − 29 + 8 54

Now Try Problem 71

Teaching Example 14
Evaluate: 39 − 27 + 8
Answer:
20

EXAMPLE 14

Evaluate: 27 − 16 + 5

Strategy We will perform the subtraction first and add 5 to that result.

WHY The operations of addition and subtraction must be performed as they occur from left to right.

Solution

We will write the steps of the solution in horizontal form.

27 − 16 + 5 = **11** + 5   Working left to right, do the subtraction first: 27 − 16 = 11.
            = 16   Now do the addition.

Caution! When making the calculation in Example 14, we must perform the subtraction first, because it occurs first when reading left to right. If the addition is done first, we get the incorrect answer 6.

27 − 16 + 5 = 27 − 21
            = 6

Using Your CALCULATOR   The Addition and Subtraction Keys

Calculators are useful for making lengthy calculations and checking results. They should not, however, be used until you have a solid understanding of the basic arithmetic facts. This textbook ***does not*** require you to have a calculator. Ask your instructor if you are allowed to use a calculator in the course.

The *Using Your Calculator* feature explains the keystrokes for an inexpensive scientific calculator. If you have any questions about your specific model, see your user's manual.

To check the result in Example 6 using a scientific calculator, we can use the addition key $\boxed{+}$.

156 $\boxed{+}$ 156 $\boxed{+}$ 65 $\boxed{+}$ 65 $\boxed{=}$ $\boxed{442}$

On some calculator models, the $\boxed{\text{Enter}}$ key is pressed instead of the $\boxed{=}$ for the result to be displayed.

We can use a scientific calculator to check the result in Example 9 using the subtraction key $\boxed{-}$.

42403 $\boxed{-}$ 1675 $\boxed{=}$ $\boxed{40728}$

ANSWERS TO SELF CHECKS

1. 3,699 **2.** 239 **3.** 16,600 **4.** The total number of accidents for 2004–2011 was 249.
5. The monthly circulation in 2012 was 8,169,524. **6.** 76 in. **7.** 1,141 **8.** 6,378
9. 63,859 **10.** The subtraction is incorrect. **11.** 52,000 **12.** An African elephant weighs 1,100 lb more than an Asian elephant. **13.** After the dieting and exercise program, Jared weighed 180 lb. **14.** 54

1.2 Adding and Subtracting Whole Numbers

STUDY SKILLS CHECKLIST

Learning from the Worked Examples

The following checklist will help you become familiar with the example structure in this book. Place a check mark in each box after you answer the question.

☐ Each section of the book contains worked **Examples** that are numbered. How many worked examples are there in Section 1.3, which begins on page 34?

☐ Each worked example contains a **Strategy**. Fill in the blanks to complete the following strategy for Example 3 on page 4: We will locate the commas in the written-word form of each number .

☐ Each Strategy statement is followed by an explanation of **Why** that approach is used. Fill in the blanks to complete the following Why for Example 3 on page 4: When a whole number is written in words, commas are used to separate periods .

☐ Each worked example has a **Solution**. How many lettered parts are there to the Solution in Example 3 on page 4?

☐ Each example uses red **Author notes** to explain the steps of the solution. Fill in the blanks to complete the first author note in the solution of Example 3 on page 19: Round to the nearest   thousand .

☐ After reading a worked example, you should work the Self Check problem. How many **Self Check** problems are there for Example 5 on page 6?

☐ At the end of each section, you will find the **Answers to Self Checks**. What is the answer to Self Check problem 2 on page 28?

☐ After completing a Self Check problem, you can **Now Try** similar problems in the Study Sets. For Example 7 on page 23, which Study Set problem is suggested?

Answers: 11, form of each number, used to separate periods, 3, nearest thousand, 2, 239, 49

SECTION 1.2   STUDY SET

VOCABULARY

Fill in the blanks.

1. In the addition problem shown below, label each *addend* and the *sum*.

   10 + 15 = 25
   ↑   ↑   ↑
   addend addend sum

2. When using the vertical form to add whole numbers, if the addition of the digits in any one column produces a sum greater than 9, we must carry .

3. The commutative property of addition states that the order in which whole numbers are added does not change their sum.

4. The associative property of addition states that the way in which whole numbers are grouped does not change their sum.

5. To see whether the result of an addition is reasonable, we can round the addends and estimate the sum.

6. The words *rise*, *gain*, *total*, and *increase* are often used to indicate the operation of addition . The words *fall*, *lose*, *reduce*, and *decrease* often indicate the operation of subtraction .

7. The figure below on the left is an example of a rectangle . The figure on the right is an example of a square .

8. Label the *length* and the *width* of the rectangle below. Together, the length and width of a rectangle are called its dimensions .

9. When all the sides of a rectangle are the same length, we call the rectangle a square .
10. The distance around a rectangle is called its perimeter .
11. In the subtraction problem shown below, label the *minuend*, *subtrahend*, and the *difference*.

    25 − 10 = 15
    ↑ minuend ↑ subtrahend ↑ difference

12. If the subtraction of the digits in any place value column requires that we subtract a larger digit from a smaller digit, we must borrow or *regroup*.
13. Every subtraction has a related addition statement. For example,

    $7 - 2 = 5$ because $5 + 2 = 7$

14. To *evaluate* an expression such as $58 - 33 + 9$ means to find its value .

CONCEPTS

15. Which property of addition is shown?
    a. $3 + 4 = 4 + 3$
       commutative property of addition
    b. $(3 + 4) + 5 = 3 + (4 + 5)$
       associative property of addition
    c. $(36 + 58) + 32 = 36 + (58 + 32)$
       associative property of addition
    d. $319 + 507 = 507 + 319$
       commutative property of addition

Fill in the blanks.

16. a. Use the commutative property of addition to complete the following:
       $19 + 33 =$ 33 + 19
    b. Use the associative property of addition to complete the following:
       $3 + (97 + 16) =$ (3 + 97) + 16
17. The subtraction $7 - 3 = 4$ is related to the addition statement 4 + 3 = 7 .
18. The operation of addition can be used to check the result of a subtraction: If a subtraction is done correctly, *the* sum *of the difference and the subtrahend will always equal the minuend.*
19. To *evaluate* (find the value of) an expression that contains both addition and subtraction, we perform the operations as they occur from left to right .
20. To answer questions about *how much more* or *how many more*, we can use subtraction .

NOTATION

21. Fill in the blanks. The symbols ( ) are called parentheses . It is standard practice to perform the operations within them first .

22. Which expression is the correct translation of the sentence: *Subtract 30 from 83*.

    $83 - 30$ or $30 - 83$ 83 − 30

23. Complete each step to find the sum.

    $12 + (15 + 2) = 12 +$ 17
    $=$ 29

24. Fill in the blanks to complete each step:

    $36 - 11 + 5 =$ 25 $+ 5$
    $=$ 30

GUIDED PRACTICE

Add. See Example 1.

25. $25 + 13$ 38
26. $47 + 12$ 59
27. 406
    $+283$
    $\overline{689}$
28. 213
    $+751$
    $\overline{964}$
29. $156 + 305$ 461
30. $647 + 138$ 785
31. $4,301 + 789 + 3,847$ 8,937
32. $5,576 + 649 + 1,922$ 8,147

Apply the associative property of addition to find the sum. See Example 2.

33. $(13 + 8) + 12$ 33
34. $(19 + 7) + 13$ 39
35. $94 + (6 + 37)$ 137
36. $92 + (8 + 88)$ 188

Use front-end rounding to estimate the sum. See Example 3.

37. $686 + 789 + 12,233 + 24,500 + 5,768$ 37,500
38. $404 + 389 + 11,802 + 36,902 + 7,777$ 58,800
39. $567,897 + 23,943 + 309,900 + 99,113$ 1,020,000
40. $822,365 + 15,444 + 302,417 + 99,010$ 1,220,000

Find the perimeter of each rectangle or square. See Example 6.

41.
    32 feet (ft); 12 ft; 88 ft

42.
    127 meters (m); 91 m; 436 m

43.
    17 inches (in.); 17 in.; 68 in.

44.
    5 yards (yd); 5 yd; 20 yd

45. 94 mi (miles); 94 mi; 376 mi

46.
    56 ft (feet); 56 ft; 224 ft

1.2 Adding and Subtracting Whole Numbers

47. 87 cm (centimeters)   6 cm   186 cm

48. 77 in. (inches)   306 in.   76 in.

Subtract. See Example 7.

49. 347 from 7,989   7,642
50. 283 from 9,799   9,516
51. 405 from 2,967   2,562
52. 304 from 1,736   1,432

Subtract. See Example 8.

53. 8,746 − 289   8,457
54. 7,531 − 276   7,255
55. 6,961 − 478   6,483
56. 4,823 − 667   4,156

Subtract. See Example 9.

57. 54,506 − 2,829   51,677
58. 69,403 − 4,635   64,768
59. 48,402 − 3,958   44,444
60. 39,506 − 1,729   37,777

Check each subtraction using addition. See Example 10.

61. 298 − 175 / 123   correct
62. 469 − 237 / 132   incorrect
63. 4,539 − 3,275 / 1,364   incorrect
64. 2,698 − 1,569 / 1,129   correct

Estimate each difference. See Example 11.

65. 67,219 − 4,076   66,000
66. 45,333 − 3,410   47,000
67. 83,872 − 27,281   50,000
68. 74,009 − 37,405   30,000

Evaluate each expression. See Example 14.

69. 35 − 12 + 6   29
70. 47 − 23 + 4   28
71. 574 + 47 − 13   608
72. 863 + 39 − 11   891

TRY IT YOURSELF

Perform the operations.

73. 8,539 + 7,368   15,907
74. 5,799 + 6,879   12,678
75. 3,430 − 529   2,901
76. 2,470 − 863   1,607
77. 51,246 + 578 + 37 + 4,599   56,460
78. 4,689 + 73,422 + 26 + 433   78,570
79. 633 − 598 + 30   65
80. 600 − 497 + 60   163

81. (45 + 16) + 4   65
82. 7 + (63 + 23)   93
83. 20,007 − 78   19,929
84. 70,006 − 48   69,958
85. 852 − 695 + 40   197
86. 397 − 348 + 65   114
87. 632 + 347   979
88. 423 + 570   993
89. 15,700 − 15,397   303
90. 35,600 − 34,799   801
91. 16,427 increased by 13,573   30,000
92. 13,567 more than 18,788   32,355
93. Subtract 1,249 from 50,009.   48,760
94. Subtract 2,198 from 20,020.   17,822

APPLICATIONS

95. DIMENSIONS OF A HOUSE Find the length of the house shown in the blueprint.   91 ft.

96. FAST FOOD Find the total number of calories in the following lunch from McDonald's: Big Mac (540 calories), small French fries (230 calories), Fruit 'n Yogurt Parfait (160 calories), medium Coca-Cola Classic (210 calories).   1,140 calories

97. WEBSITES The number of persons age 15 or older who visited the eBay website at least once in April through June of 2012 was 241,288,000. The number that visited the Apple iTunes website during those months was 57,253,000 greater than the eBay site. How many visitors did the Apple iTunes site have? (Source: *The World Almanac and Book of Facts,* 2012)   298,541,000 visitors

98. BRIDGE SAFETY The results of a 2010 report of the condition of U.S. highway bridges is shown below. Each bridge was classified as either *safe, in need of repair,* or *should be replaced.* Complete the table.

Number of safe bridges	Number of bridges that need repair	Number of outdated bridges that should be replaced	Total number of bridges
457,828	69,220	77,412	604,460

Source: Bureau of Transportation Statistics

99. WEDDINGS The average wedding costs for 2011 are listed in the table below. Find the total cost of a wedding. $27,021

Clothing/hair/makeup	$1,621
Ceremony/music/flowers	$4,052
Photography/video	$2,162
Favors/gifts	$540
Jewelry	$4,055
Reception	$11,619
Honeymoon	$2,972

Source: jennlanedesign.com, prettypracticalbride.com

100. CANDY The graph below shows U.S. candy sales in 2010 during four holiday periods. Find the sum of these seasonal candy sales. $6,550,000,000

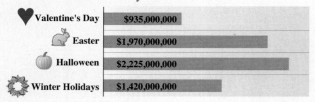

Valentine's Day	$935,000,000
Easter	$1,970,000,000
Halloween	$2,225,000,000
Winter Holidays	$1,420,000,000

Source: National Confectioners Association

101. FLAGS To decorate a city flag, yellow fringe is to be sewn around its outside edges, as shown. The fringe is sold by the inch. How many inches of fringe must be purchased to complete the project? 196 in.

102. DECORATING A child's bedroom is rectangular in shape with dimensions 15 feet by 11 feet. How many feet of wallpaper border are needed to wrap around the entire room? 52 ft

103. SHEET METAL A square region is to be cut from the center of a piece of sheet metal, as shown below. If a laser cutter follows the dashed lines, what is the total length of the cut? 104 in.

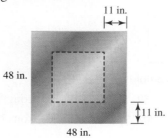

104. FENCES A square piece of land measuring 209 feet on all four sides is approximately one *acre*. How many feet of chain-link fencing are needed to enclose a piece of land this size? 836 ft

105. WORLD RECORDS The world's largest pumpkin weighed in at 2,009 pounds and the world's largest watermelon weighed in at 291 pounds. How much more did the pumpkin weigh? (Source: zimbio.com, pumpkinnook.com) 1,718 lb

106. FARMER'S MARKETS See the graph below. Between what two years was there the greatest increase in the number of farmer's markets in the U.S.? What was the increase? 2010–2011; 1,043 markets

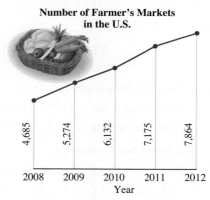

Number of Farmer's Markets in the U.S.

| 2008 | 2009 | 2010 | 2011 | 2012 |
| 4,685 | 5,274 | 6,132 | 7,175 | 7,864 |

Source: USDA-AMS-marketing services division

107. MILEAGE Find the distance (in miles) that a trucker drove on a trip from San Diego to Houston using the odometer readings shown below. 1,495 mi

70154 — Truck odometer reading leaving San Diego

71649 — Truck odometer reading arriving in Houston

108. DIETS Use the bathroom scale readings shown below to find the number of pounds that a dieter lost. 34 lb

January   October

Unless otherwise noted, all art on this page is © Cengage Learning.

109. MAGAZINES In 2012, *Reader's Digest* had a circulation of 5,577,717. By what amount did this exceed *TV Guide*'s circulation of 2,010,879? (Source: *The World Almanac and Book of Facts*, 2013) 3,566,838 magazines

110. TRANSPLANTS See the graph below. Find the decrease in the number of patients waiting for a liver transplant from:

   a. 2005 to 2006 94 patients **b.** 2007 to 2008 91 patients

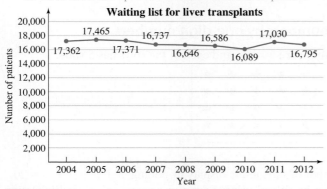

Source: Organ Procurement and Transplant Network; United Network for Organ Sharing

111. JEWELRY Gold melts at about 1,947°F. The melting point of silver is 183°F lower. What is the melting point of silver? 1,764°F

112. READING BLUEPRINTS Find the length of the motor on the machine shown in the blueprint. 34 cm

Refer to the teachers' salary schedule shown in the next column. To use this table, note that a fourth-year teacher (Step 4) in Column 2 makes $52,209 per year.

113. a. What is the salary of a teacher on Step 2/Column 2? $49,565

   b. How much more will that teacher make next year when she gains 1 year of teaching experience and moves down to Step 3 in that column? $1,322

114. a. What is the salary of a teacher on Step 4/Column 1? $50,751

   b. How much more will that teacher make next year when he gains 1 year of teaching experience and takes enough coursework to move over to Column 2? $2,780

Teachers' Salary Schedule
ABC Unified School District

Years teaching	Column 1	Column 2	Column 3
Step 1	$46,785	$48,243	$49,701
Step 2	$48,107	$49,565	$51,023
Step 3	$49,429	$50,887	$52,345
Step 4	$50,751	$52,209	$53,667
Step 5	$52,073	$53,531	$54,989

WRITING

115. In this section, it is said that estimation is a *tradeoff*. Give one benefit and one drawback of estimation.

116. A student added three whole numbers top to bottom and then bottom to top, as shown below. What does the result in red indicate? What should the student do next?

$$\begin{array}{r} 1{,}689 \\ 496 \\ 315 \\ +\ 788 \\ \hline 1{,}599 \end{array}$$

117. Explain why the operation of subtraction is not commutative.

118. Explain how addition can be used to check subtraction.

REVIEW

119. Write each number in expanded notation.
   a. 3,125 3,000 + 100 + 20 + 5
   b. 60,037 60,000 + 30 + 7

120. Round 6,354,784 to the nearest . . .
   a. ten 6,354,780
   b. hundred 6,354,800
   c. ten thousand 6,350,000
   d. hundred thousand 6,400,000

121. Round 5,370,645 to the nearest . . .
   a. ten 5,370,650
   b. ten thousand 5,370,000
   c. hundred thousand 5,400,000

122. Write 72,001,015
   a. in words seventy-two million, one thousand, fifteen
   b. in expanded notation
      70,000,000 + 2,000,000 + 1,000 + 10 + 5

Objectives

1. Multiply whole numbers by one-digit numbers.
2. Multiply whole numbers that end with zeros.
3. Multiply whole numbers by two- (or more) digit numbers.
4. Use properties of multiplication to multiply whole numbers.
5. Estimate products of whole numbers.
6. Solve application problems by multiplying whole numbers.
7. Find the area of a rectangle.

SECTION 1.3
Multiplying Whole Numbers

Everyone uses multiplication of whole numbers. For example, to double a recipe, a cook multiplies the amount of each ingredient by two. To determine the floor space of a dining room, a carpeting salesperson multiplies its length by its width. An accountant multiplies the number of hours worked by the hourly pay rate to calculate the weekly earnings of employees.

1 Multiply whole numbers by one-digit numbers.

In the following display, there are 4 rows, and each of the rows has 5 stars.

We can find the total number of stars in the display by adding: $5 + 5 + 5 + 5 = 20$.

This problem can also be solved using a simpler process called **multiplication**. Multiplication is repeated addition, and it is written using a **multiplication symbol ×**, which is read as "times." Instead of *adding* four 5's to get 20, we can multiply 4 and 5 to get 20.

Repeated addition	**Multiplication**	
$5 + 5 + 5 + 5$	$= \quad 4 \times 5 = 20$	Read as "4 times 5 equals (or is) 20."

We can write multiplication problems in **horizontal** or **vertical form**. The numbers that are being multiplied are called **factors** and the answer is called the **product**.

Horizontal form
$$4 \times 5 = 20$$
Factor Factor Product

Vertical form
$$\begin{array}{r} 5 \leftarrow \text{Factor} \\ \times\ 4 \leftarrow \text{Factor} \\ \hline 20 \leftarrow \text{Product} \end{array}$$

A **raised dot** · and **parentheses ( )** are also used to write multiplication in horizontal form.

Symbols Used for Multiplication

Symbol		Example
×	times symbol	4×5
·	raised dot	$4 \cdot 5$
( )	parentheses	$(4)(5)$ or $4(5)$ or $(4)5$

Recall that a variable is a letter that stands for a number. We often multiply a variable by another number or multiply a variable by another variable. When we do this, we don't need to use a symbol for multiplication.

$5a$ means $5 \cdot a$, ab means $a \cdot b$ and xyz means $x \cdot y \cdot z$

> **Caution!** In this book, we seldom use the × symbol, because it can be confused with the letter x.

To multiply whole numbers that are less than 10, we rely on our understanding of basic multiplication facts. For example,

$2 \cdot 3 = 6$, $8(4) = 32$,   and $9 \times 7 = 63$

To multiply larger whole numbers, we can use vertical form by stacking them with their corresponding place values lined up. Then we make repeated use of basic multiplication facts.

EXAMPLE 1   Multiply: 8×47

Strategy   We will write the multiplication in vertical form. Then, working right to left, we will multiply each digit of 47 by 8 and carry, if necessary.

WHY   This process is simpler than treating the problem as repeated addition and adding eight 47's.

Solution
To help you understand the process, each step of this multiplication is explained separately. Your solution need only look like the *last* step.

We begin by multiplying 7 by 8.

$$\begin{array}{r} \overset{5}{4}7 \\ \times\ \ 8 \\ \hline 6 \end{array}$$ Multiply 7 by 8. The product is 56.
Write 6 in the ones column of the answer,
and carry 5 to the tens column.

$$\begin{array}{r} \overset{5}{4}7 \\ \times\ \ 8 \\ \hline 376 \end{array}$$ Multiply 4 by 8. The product is 32.
To the 32, add the carried 5 to get 37.
Write 7 in the tens column and the
3 in the hundreds column of the answer.

Your solution should look like this: $\begin{array}{r} \overset{5}{4}7 \\ \times\ \ 8 \\ \hline 376 \end{array}$

The product is 376.

Self Check 1
Multiply: 6×54 324
Now Try Problem 19

Teaching Example 1
Multiply: 9×38
Answer: 342

2 Multiply whole numbers that end with zeros.

An interesting pattern develops when a whole number is multiplied by 10, 100, 1,000, and so on. Consider the following multiplications involving 8:

$8 \cdot 10 = 80$ There is one zero in 10. The product is 8 with one 0 attached.

$8 \cdot 100 = 800$ There are two zeros in 100. The product is 8 with two 0's attached.

$8 \cdot 1,000 = 8,000$ There are three zeros in 1,000. The product is 8 with three 0's attached.

$8 \cdot 10,000 = 80,000$ There are four zeros in 10,000. The product is 8 with four 0's attached.

These examples illustrate the following rule.

Multiplying a Whole Number by 10, 100, 1,000, and So On

To find the product of a whole number and 10, 100, 1,000, and so on, attach the number of zeros in that number to the right of the whole number.

Chapter 1 Whole Numbers

Self Check 2

Multiply:
a. 9 × 1,000   9,000
b. 25 · 100   2,500
c. 875(1,000)   875,000

Now Try Problems 23 and 25

Teaching Example 2
Multiply:
a. 3 × 1,000
b. 26 · 100
c. 871(10,000)
Answers:
a. 3,000 b. 2,600 c. 8,710,000

EXAMPLE 2 Multiply: a. 6 × 1,000   b. 45 · 100   c. 912(10,000)

Strategy For each multiplication, we will identify the factor that ends in zeros and count the number of zeros that it contains.

WHY Each product can then be found by attaching that number of zeros to the other factor.

Solution
a. 6 × 1,000 = 6,000   Since 1,000 has three zeros, attach three 0's after 6.
b. 45 · 100 = 4,500   Since 100 has two zeros, attach two 0's after 45.
c. 912(10,000) = 9,120,000   Since 10,000 has four zeros, attach four 0's after 912.

We can use an approach similar to that of Example 2 for multiplication involving any whole numbers that end in zeros. For example, to find 67 · 2,000, we have

67 · 2,000 = 67 · 2 · 1,000   Write 2,000 as 2 · 1,000.
= 134 · 1,000   Working left to right, multiply 67 and 2 to get 134.
= 134,000   Since 1,000 has three zeros, attach three 0's after 134.

This example suggests that to find 67 · 2,000 we simply multiply 67 and 2 and attach three zeros to that product. This method can be extended to find products of two factors that *both* end in zeros.

Self Check 3

Multiply:
a. 15 · 900   13,500
b. 3,100 · 7,000   21,700,000

Now Try Problems 29 and 33

Teaching Example 3
Multiply:
a. 18 · 40
b. 7,500 · 30,000
Answers:
a. 720
b. 225,000,000

EXAMPLE 3 Multiply: a. 14 · 300   b. 3,500 · 50,000

Strategy We will multiply the nonzero leading digits of each factor. To that product, we will attach the sum of the number of trailing zeros in the factors.

WHY This method is faster than the standard vertical form multiplication of factors that contain many zeros.

Solution
a. The factor **300** has two trailing zeros.
   14 · **300** = 4,2**00**   Attach two 0's after 42.
   └────┘↑
   Multiply 14 and 3
   to get 42.

   $\overset{1}{14}$
   $\underline{\times3}$
   42

b. The factors **3,500** and **50,000** have a total of six trailing zeros.
   3,500 · **50,000** = 175,**000,000**   Attach six 0's after 175.
   └────┘↑
   Multiply 35 and 5
   to get 175.

   $\overset{2}{35}$
   $\underline{\times5}$
   175

Success Tip Calculations that you cannot perform in your head should be shown outside the steps of your solution.

3 Multiply whole numbers by two- (or more) digit numbers.

Self Check 4

Multiply: 36 · 334   12,024

Now Try Problem 37

Teaching Example 4
Multiply: 37 · 234
Answer:
8,658

EXAMPLE 4 Multiply: 23 · 436

Strategy We will write the multiplication in vertical form. Then we will multiply 436 by 3 and by 20, and add those products.

WHY Since 23 = 3 + 20, we can multiply 436 by 3 and by 20, and add those products.

Solution

Each step of this multiplication is explained separately. Your solution need only look like the *last* step.

Vertical form

$$\begin{array}{r} 4\,3\,6 \\ \times\ \ 2\,3 \end{array}$$

Vertical form multiplication is often easier if the factor with the larger number of digits is written on top.

We begin by multiplying 436 by 3.

$$\begin{array}{r} \overset{1}{4}3\,6 \\ \times\ \ 2\,3 \\ \hline 8 \end{array}$$

Multiply 6 by 3. The product is 18. Write 8 in the ones column and carry 1 to the tens column.

$$\begin{array}{r} \overset{1}{4}\overset{1}{3}\,6 \\ \times\ \ 2\,3 \\ \hline 0\,8 \end{array}$$

Multiply 3 by 3. The product is 9. To the 9, add the carried 1 to get 10. Write the 0 in the tens column and carry the 1 to the hundreds column.

$$\begin{array}{r} \overset{1}{4}\overset{1}{3}\,6 \\ \times\ \ 2\,3 \\ \hline 1\,3\,0\,8 \end{array}$$

Multiply 4 by 3. The product is 12. Add the 12 to the carried 1 to get 13. Write 13.

We continue by multiplying 436 by 2 tens, or 20. If we think of 20 as $2 \cdot 10$, then we simply multiply 436 by 2 and attach one zero to the result.

$$\begin{array}{r} \overset{1}{\overset{1}{4}}\overset{1}{3}\,6 \\ \times\ \ 2\,3 \\ \hline 1\,3\,0\,8 \\ 2\,0 \end{array}$$

Write the 0 that is to be attached to the result of $20 \cdot 436$ in the ones column (shown in blue). Then multiply 6 by 2. The product is 12. Write 2 in the tens column and carry 1.

$$\begin{array}{r} \overset{1}{\overset{1}{4}}\overset{1}{3}\,6 \\ \times\ \ 2\,3 \\ \hline 1\,3\,0\,8 \\ 7\,2\,0 \end{array}$$

Multiply 3 by 2. The product is 6. Add 6 to the carried 1 to get 7. Write the 7 in the hundreds column. There is no carry.

$$\begin{array}{r} \overset{1}{\overset{1}{4}}\overset{1}{3}\,6 \\ \times\ \ 2\,3 \\ \hline 1\,3\,0\,8 \\ 8\,7\,2\,0 \end{array}$$

Multiply 4 by 2. The product is 8. There is no carried digit to add. Write the 8 in the thousands column.

$$\begin{array}{r} \overset{1}{\overset{1}{4}}\overset{1}{3}\,6 \\ \times\ \ 2\,3 \\ \hline 1\,3\,0\,8 \\ 8\,7\,2\,0 \\ \hline 10,0\,2\,8 \end{array}$$

Draw another line beneath the two completed rows. Add column by column, working right to left. This sum gives the product of 436 and 23.

The product is 10,028.

The Language of Algebra In Example 4, the numbers **1,308** and **8,720** are called **partial products**. We added the partial products to get the answer, 10,028. The word *partial* means *only a part*, as in a *partial* eclipse of the moon.

```
    4 3 6
  ×   2 3
  ─────────
    1 3 0 8
    8 7 2 0
  ─────────
  1 0, 0 2 8
```

When a factor in a multiplication contains one or more zeros, we must be careful to enter the correct number of zeros when writing the partial products.

Self Check 5

Multiply:
a. 706(351) 247,806
b. 4,004(2,008) 8,040,032

Now Try Problem 41

Teaching Example 5
Multiply:
a. 502 · 594
b. 1,007(6,009)
Answers:
a. 298,188
b. 6,051,063

EXAMPLE 5 Multiply: **a.** 406 · 253   **b.** 3,009(2,007)

Strategy We will think of 406 as 6 + 400 and 3,009 as 9 + 3,000.

WHY Thinking of the multipliers (406 and 3,009) in this way is helpful when determining the correct number of zeros to enter in the partial products.

Solution
We will use vertical form to perform each multiplication.

a. Since 406 = 6 + 400, we will multiply 253 by 6 and by 400, and add those partial products.

```
      253
  ×   406
  ─────────
    1 518    ← 6 · 253
  101 200    ← 400 · 253. Think of 400 as 4 · 100 and simply multiply
  ─────────     253 by 4 and attach two zeros (shown in blue) to the result.
  102,718
```

The product is 102,718.

b. Since 3,009 = 9 + 3,000, we will multiply 2,007 by 9 and by 3,000, and add those partial products.

```
        2,007
  ×     3,009
  ─────────────
       18 063   ← 9 · 2,007
    6 021 000   ← 3,000 · 2,007. Think of 3,000 as 3 · 1,000 and simply multiply
  ─────────────   2,007 by 3 and attach three zeros (shown in blue) to the result.
    6,039,063
```

The product is 6,039,063.

4 Use properties of multiplication to multiply whole numbers.

Have you ever noticed that two whole numbers can be multiplied in either order because the result is the same? For example,

$$4 \cdot 6 = 24 \quad \text{and} \quad 6 \cdot 4 = 24$$

This example illustrates the **commutative property of multiplication**.

Commutative Property of Multiplication

The order in which whole numbers are multiplied does not change their product.
   For any whole numbers *a* and *b*,

$$a \cdot b = b \cdot a \quad \text{or, more simply,} \quad ab = ba$$

Whenever we multiply a whole number by 0, the product is 0. For example,

$$0 \cdot 5 = 0, \quad 0 \cdot 8 = 0, \quad \text{and} \quad 9 \cdot 0 = 0$$

Whenever we multiply a whole number by 1, the number remains the same. For example,

$$3 \cdot 1 = 3, \quad 7 \cdot 1 = 7, \quad \text{and} \quad 1 \cdot 9 = 9$$

These examples illustrate the multiplication properties of 0 and 1.

> **Multiplication Properties of 0 and 1**
>
> The product of any whole number and 0 is 0.
> The product of any whole number and 1 is that whole number.
> For any whole number a,
>
> $$a \cdot 0 = 0 \quad \text{and} \quad 0 \cdot a = 0$$
> $$a \cdot 1 = a \quad \text{and} \quad 1 \cdot a = a$$

> **Success Tip** If one (or more) of the factors in a multiplication is 0, the product will be 0. For example,
>
> $$16(27)(0) = 0 \quad \text{and} \quad 109 \cdot 53 \cdot 0 \cdot 2 = 0$$

To multiply three numbers, we first multiply two of them and then multiply that result by the third number. In the following examples, we multiply $3 \cdot 2 \cdot 4$ in two ways. The parentheses show us which multiplication to perform first. The steps of the solutions are written in horizontal form.

> **The Language of Algebra** In the following example, read $(3 \cdot 2) \cdot 4$ as "The *quantity* of 3 times 2," pause slightly, and then say "times 4." Read $3 \cdot (2 \cdot 4)$ as "3 times the *quantity* of 2 times 4." The word *quantity* alerts the reader to the parentheses that are used as grouping symbols.

Method 1: Group $3 \cdot 2$

$(3 \cdot 2) \cdot 4 = 6 \cdot 4$ Multiply 3 and 2 to get 6.

$ = 24$ Multiply 6 and 4 to get 24.

Method 2: Group $2 \cdot 4$

$3 \cdot (2 \cdot 4) = 3 \cdot 8$ Then multiply 2 and 4 to get 8.

$ = 24$ Then multiply 3 and 8 to get 24.

Same result

Either way, the answer is 24. This example illustrates that changing the grouping when multiplying numbers doesn't affect the result. This property is called the **associative property of multiplication**.

> **Associative Property of Multiplication**
>
> The way in which whole numbers are grouped does not change their product.
> For any whole numbers a, b, and c,
>
> $$(a \cdot b) \cdot c = a \cdot (b \cdot c) \quad \text{or, more simply,} \quad (ab)c = a(bc)$$

Sometimes, an application of the associative property can simplify a calculation.

Chapter 1 Whole Numbers

Self Check 6

Find the product: $(23 \cdot 25) \cdot 4$

Now Try Problem 45

Self Check 6 Answer
2,300

Teaching Example 6
Find the product: $2 \cdot (50 \cdot 89)$
Answer:
8,900

EXAMPLE 6 Find the product: $(17 \cdot 50) \cdot 2$

Strategy We will use the associative property to group 50 with 2.

WHY It is helpful to regroup because 50 and 2 are a pair of numbers that are easily multiplied.

Solution
We will write the solution in horizontal form.

$(17 \cdot 50) \cdot 2 = 17 \cdot (50 \cdot 2)$ Use the associative property of multiplication to regroup the factors.
$= 17 \cdot 100$ Do the multiplication within the parentheses first.
$= 1,700$ Since 100 has two zeros, attach two 0's after 17.

5 Estimate products of whole numbers.

Estimation is used to find an approximate answer to a problem.

Self Check 7

Estimate the product: $74 \cdot 488$

Now Try Problem 51

Self Check 7 Answer
35,000

Teaching Example 7
Estimate the product: $56 \cdot 713$
Answer:
42,000

EXAMPLE 7 Estimate the product: $59 \cdot 334$

Strategy We will use front-end rounding to approximate the factors 59 and 334. Then we will find the product of the approximations.

WHY Front-end rounding produces whole numbers containing many 0's. Such numbers are easier to multiply.

Solution
Both of the factors are rounded to their *largest place value* so that all but their first digit is zero.

```
       ┌─Round to the nearest ten.─┐
       ↓                            ↓
   59 · 334                       60 · 300
       └─Round to the nearest hundred.─┘
```

To find the product of the approximations, $60 \cdot 300$, we simply multiply 6 by 3, to get 18, and attach 3 zeros. Thus, the estimate is **18,000**.

If we calculate $59 \cdot 334$, the product is exactly 19,706. Note that the estimate is close: It's only 1,706 less than 19,706.

6 Solve application problems by multiplying whole numbers.

Application problems that involve repeated addition are often more easily solved using multiplication.

Self Check 8

DAILY PAY In 2011, the average U.S. construction worker made $26 per hour. At that rate, how much money was earned in an 8-hour workday? (Source: Bureau of Labor Statistics) $208

Now Try Problem 86

EXAMPLE 8 *Daily Pay* In 2011, the average U.S. manufacturing worker made $24 per hour. At that rate, how much money was earned in an 8-hour workday? (Source: Bureau of Labor Statistics)

Strategy To find the amount earned in an 8-hour workday, we will multiply the hourly rate of $24 by 8.

WHY For each of the 8 hours, the average manufacturing worker earned $24. The amount earned for the day is the sum of eight 24's: $24 + 24 + 24 + 24 + 24 + 24 + 24 + 24$. This repeated addition can be calculated more simply by multiplication.

Solution
We translate the words of the problem to numbers and symbols.

| The amount earned in an 8-hr workday | is equal to | the rate per hour | times | 8 hours. |

| The amount earned in an 8-hr workday | = | 24 | · | 8 |

Use vertical form to perform the multiplication:

$$\begin{array}{r} \overset{3}{2}4 \\ \times\ 8 \\ \hline 192 \end{array}$$

In 2011, the average U.S. manufacturing worker earned $192 in an 8-hour workday.

We can use multiplication to count objects arranged in patterns of neatly arranged rows and columns called **rectangular arrays**.

> **The Language of Algebra** An *array* is an orderly arrangement. For example, a jewelry store might display a beautiful *array* of gemstones.

EXAMPLE 9

Pixels Refer to the illustration at the right. Small squares of color, called *pixels*, create the digital images seen on computer screens. If a 14-inch screen has 640 pixels from side to side and 480 pixels from top to bottom, how many pixels are displayed on the screen?

Strategy We will multiply 640 by 480 to determine the number of pixels that are displayed on the screen.

WHY The pixels form a rectangular array with 640 pixels in each row and 480 in each column. Multiplication can be used to count objects in a rectangular array.

Solution
We translate the words of the problem to numbers and symbols.

| The number of pixels on the screen | is equal to | the number of pixels in a row | times | the number of pixels in a column. |

| The number of pixels on the screen | = | 640 | · | 480 |

To find the product of 640 and 480, we use vertical form to multiply 64 and 48 and attach two zeros to that result.

$$\begin{array}{r} 48 \\ \times\ 64 \\ \hline 192 \\ 2\ 880 \\ \hline 3{,}072 \end{array}$$

Since the product of 64 and 48 is 3,072, the product of 640 and 480 is 307,200. The screen displays 307,200 pixels.

Teaching Example 8
DAILY PAY In 2011, the average U.S. miner made $29 per hour. At that rate, how much money was earned in an 8-hour workday? (Source: Bureau of Labor Statistics)
Answer: $232

Self Check 9
PIXELS If a 17-inch computer screen has 1,024 pixels from side to side and 768 from top to bottom, how many pixels are displayed on the screen? 786,432
Now Try Problem 93

Teaching Example 9
SEATING How many people can be seated in a rectangular arrangement of chairs, if there are 18 rows, and each row contains 25 chairs?
Answer: 450 people

> **The Language of Algebra** Here are some key words that are often used to indicate multiplication:
>
> double   triple   twice   of   times   tenfold

Self Check 10

INSECTS Leaf cutter ants can carry pieces of leaves that weigh 30 times their body weight. How much can an ant lift if it weighs 25 milligrams? 750 milligrams

Now Try Problem 99

Teaching Example 10 SHEEP New Zealand has about 7 times as many sheep as people. If the country's population in December 2012 was 4,450,270, how many sheep were there in New Zealand at that time? (Source: economist.com)
Answer:
31,151,890 sheep

EXAMPLE 10   *Weight Lifting*

In 1983, Stefan Topurov of Bulgaria was the first man to lift three times his body weight over his head. If he weighed 132 pounds at the time, how much weight did he lift over his head?

Strategy To find how much weight he lifted over his head, we will multiply his body weight by 3.

WHY We can use multiplication to determine the result when a quantity increases in size by 2 *times*, 3 *times*, 4 *times*, and so on.

Solution
We translate the words of the problem to numbers and symbols.

The amount he lifted over his head   was   3   times   his body weight.

The amount he lifted over his head = 3 · 132

Use vertical form to perform the multiplication:

$$\begin{array}{r} 132 \\ \times\ \ 3 \\ \hline 396 \end{array}$$

Stefan Topurov lifted 396 pounds over his head.

Using Your CALCULATOR   The Multiplication Key: Seconds in a Year

There are 60 seconds in 1 minute, 60 minutes in 1 hour, 24 hours in 1 day, and 365 days in 1 year. We can find the number of seconds in 1 year using the multiplication key $\times$ on a calculator.

60 $\times$ 60 $\times$ 24 $\times$ 365 $=$ 31536000

On some calculator models, the ENTER key is pressed instead of the $=$ for the result to be displayed.

There are 31,536,000 seconds in 1 year.

---

7 Find the area of a rectangle.

One important application of multiplication is finding the area of a rectangle. The **area of a rectangle** is the measure of the amount of surface it encloses. Area is measured in square units, such as square inches (written in.^{2}) or square centimeters (written cm^2), as shown below.

One square inch (1 in.^{2})   One square centimeter (1 cm^2)

Unless otherwise noted, all art on this page is © Cengage Learning.

The rectangle in the figure below has a length of 5 centimeters and a width of 3 centimeters. Since each small square region covers an area of one square centimeter, each small square region measures 1 cm². The small square regions form a rectangular pattern, with 3 rows of 5 squares.

Because there are 5 · 3, or 15, small square regions, the area of the rectangle is 15 cm². This suggests that the area of any rectangle is the product of its length and its width.

Area of a rectangle = length · width

By using the letter A to represent the area of the rectangle, the letter l to represent the length of the rectangle, and the letter w to represent its width, we can write this **formula** in simpler form. Letters (or symbols), such as A, l, and w, that are used to represent numbers are called **variables**.

Area of a Rectangle

The area, A, of a rectangle is the product of the rectangle's length, l, and its width, w.

Area = length · width   or $A = l \cdot w$ or $A = lw$

EXAMPLE 11 *Gift Wrapping* When completely unrolled, a long sheet of gift wrapping paper has the dimensions shown below. How many square feet of gift wrap are on the roll?

Strategy We will substitute 12 for the length and 3 for the width in the formula for the area of a rectangle.

WHY To find the number of square feet of paper, we need to find the area of the rectangle shown in the figure.

Solution

$A = lw$ This is the formula for the area of a rectangle.

$A = 12 \cdot 3$ Replace the length l with 12 and the width w with 3.

$A = 36$ Do the multiplication.

There are 36 square feet of wrapping paper on the roll. This can be written in more compact form as 36 ft².

Self Check 11

ADVERTISING The rectangular posters used on small billboards in the New York subway are 59 inches wide by 45 inches tall. Find the area of a subway poster. 2,655 in.²

Now Try Problems 53 and 55

Teaching Example 11 MIRRORS
Find the area of a rectangular bathroom mirror that is 36 inches high and 60 inches long.
Answer:
2,160 in.²

> **Caution!** Remember that the perimeter of a rectangle is the distance around it and is measured in units such as inches, feet, and miles. The area of a rectangle is the amount of surface it encloses and is measured in square units such as in.^{2}, ft^2, and mi^2.

ANSWERS TO SELF CHECKS

1. 324   **2. a.** 9,000   **b.** 2,500   **c.** 875,000   **3. a.** 13,500   **b.** 21,700,000
4. 12,024   **5. a.** 247,806   **b.** 8,040,032   **6.** 2,300   **7.** 35,000   **8.** $208
9. 786,432   **10.** 750 milligrams   **11.** 2,655 in.^{2}

STUDY SKILLS CHECKLIST
Getting the Most from the Study Sets

The following checklist will help you become familiar with the Study Sets in this book. Place a check mark in each box after you answer the question.

☐ Answers to the odd-numbered **Study Set** problems are located in the appendix on page A-17. On what page do the answers to Study Set 1.3 appear?

☐ Each Study Set begins with **Vocabulary** problems. How many Vocabulary problems appear in Study Set 1.3?

☐ Following the Vocabulary problems, you will see **Concepts** problems. How many Concepts problems appear in Study Set 1.3?

☐ Following the Concepts problems, you will see **Notation** problems. How many Notation problems appear in Study Set 1.3?

☐ After the Notation problems are **Guided Practice** problems, which are linked to similar examples within the section. How many Guided Practice problems appear in Study Set 1.3?

☐ After the Guided Practice problems are **Try It Yourself** problems, which can be used to help you prepare for quizzes. How many Try It Yourself problems appear in Study Set 1.3?

☐ Following the Try It Yourself problems, you will see **Applications** problems. How many Applications problems appear in Study Set 1.3?

☐ After the Applications problems in Study Set 1.3, how many **Writing** problems are given?

☐ Each Study Set ends with a few **Review** problems. How many Review problems appear in Study Set 1.3?

Answers: A-18, 6, 6, 4, 40, 24, 26, 2, 2

SECTION 1.3   STUDY SET

VOCABULARY

Fill in the blanks.

1. In the multiplication problem shown below, label each *factor* and the *product*.

2. Multiplication is <u>repeated</u> addition.

3. The <u>commutative</u> property of multiplication states that the order in which whole numbers are multiplied does not change their product. The <u>associative</u> property of multiplication states that the way in which whole numbers are grouped does not change their product.

4. Letters that are used to represent numbers are called <u>variables</u>.

5. If a square measures 1 inch on each side, its area is 1 <u>square</u> inch.

6. The <u>area</u> of a rectangle is a measure of the amount of surface it encloses.

CONCEPTS

7. **a.** Write the repeated addition $8 + 8 + 8 + 8$ as a multiplication. $4 \cdot 8$

   b. Write the multiplication $7 \cdot 15$ as a repeated addition. $15 + 15 + 15 + 15 + 15 + 15 + 15$

8. a. Fill in the blank: A rectangular <u>array</u> of red squares is shown below.
   b. Write a multiplication statement that will give the number of red squares shown below. $5 \cdot 12 = 60$

9. a. How many zeros do you attach to the right of 25 to find $25 \cdot 1,000$? 3
   b. How many zeros do you attach to the right of 8 to find $400 \cdot 2,000$? 5

10. a. Using the variables x and y, write a statement that illustrates the commutative property of multiplication. $xy = yx$
    b. Using the variables x, y, and z, write a statement that illustrates the associative property of multiplication. $(xy)z = x(yz)$

11. Determine whether the concept of *perimeter* or that of *area* should be applied to find each of the following.
    a. The amount of floor space to carpet   area
    b. The number of inches of lace needed to trim the sides of a handkerchief   perimeter
    c. The amount of clear glass to be tinted   area
    d. The number of feet of fencing needed to enclose a playground   perimeter

12. Perform each multiplication.
    a. $1 \cdot 25$ 25
    b. $62(1)$ 62
    c. $10 \cdot 0$ 0
    d. $0(4)$ 0

NOTATION

13. Write three symbols that are used for multiplication. $\times$, $\cdot$, ( )
14. What does ft^2 mean? square feet
15. Write the formula for the area of a rectangle using variables. $A = l \cdot w$ or $A = lw$
16. Write each multiplication in simpler form.
    a. $8 \cdot x$ $8x$
    b. $a \cdot b$ ab

GUIDED PRACTICE

Multiply. See Example 1.

17. 15×7 105
18. 19×9 171
19. 34×8 272
20. 37×6 222

Perform each multiplication without using pencil and paper or a calculator. See Example 2.

21. $37 \cdot 100$ 3,700
22. $63 \cdot 1,000$ 63,000
23. 75×10 750
24. $88 \times 10,000$ 880,000
25. $107(10,000)$ 1,070,000
26. $323(100)$ 32,300
27. $512(1,000)$ 512,000
28. $673(10)$ 6,730

Multiply. See Example 3.

29. $68 \cdot 40$ 2,720
30. $83 \cdot 30$ 2,490
31. $56 \cdot 200$ 11,200
32. $222 \cdot 500$ 111,000
33. $130(3,000)$ 390,000
34. $630(7,000)$ 4,410,000
35. $2,700(40,000)$ 108,000,000
36. $5,100(80,000)$ 408,000,000

Multiply. See Example 4.

37. $73 \cdot 128$ 9,344
38. $54 \cdot 173$ 9,342
39. $64(287)$ 18,368
40. $72(461)$ 33,192

Multiply. See Example 5.

41. $602 \cdot 679$ 408,758
42. $504 \cdot 729$ 367,416
43. $3,002(5,619)$ 16,868,238
44. $2,003(1,376)$ 2,756,128

Apply the associative property of multiplication to find the product. See Example 6.

45. $(18 \cdot 20) \cdot 5$ 1,800
46. $(29 \cdot 2) \cdot 50$ 2,900
47. $250 \cdot (4 \cdot 135)$ 135,000
48. $250 \cdot (4 \cdot 289)$ 289,000

Estimate each product. See Example 7.

49. $86 \cdot 249$ 18,000
50. $56 \cdot 631$ 36,000
51. $215 \cdot 1,908$ 400,000
52. $434 \cdot 3,789$ 1,600,000

Find the area of each rectangle or square. See Example 11.

53.
84 in.^{2}

54.
1,100 m^2

55.
12 in. × 12 in.
144 in.^{2}

56.
400 cm^2

TRY IT YOURSELF

Multiply.

57. 213
    $\times\ 7$
    1,491

58. 863
    $\times\ 9$
    7,767

59. $34,474 \cdot 2$ 68,948
60. $54,912 \cdot 4$ 219,648

61. 99 × 77 = 7,623
62. 73 × 59 = 4,307
63. 44(55)(0) = 0
64. 81 · 679 · 0 · 5 = 0
65. 53 · 30 = 1,590
66. 20 · 78 = 1,560
67. 754 × 59 = 44,486
68. 846 × 79 = 66,834
69. (2,978)(3,004) = 8,945,912
70. (2,003)(5,003) = 10,021,009
71. 916 × 409 = 374,644
72. 889 × 507 = 450,723
73. 25 · (4 · 99) = 9,900
74. (41 · 5) · 20 = 4,100
75. 4,800 × 500 = 2,400,000
76. 6,400 × 700 = 4,480,000
77. 2,779 × 128 = 355,712
78. 3,596 × 136 = 489,056
79. 370 · 450 = 166,500
80. 280 · 340 = 95,200

APPLICATIONS

81. **BREAKFAST CEREAL** A cereal maker advertises "Two cups of raisins in every box." Find the number of cups of raisins in a case of 36 boxes of cereal. 72 cups

82. **SNACKS** A candy warehouse sells large four-pound bags of M & M's. There are approximately 180 peanut M & M's per pound. How many peanut M & M's are there in one bag? 720 peanut M & M's

83. **NUTRITION** There are 17 grams of fat in one Krispy Kreme chocolate-iced, custard-filled donut. How many grams of fat are there in one dozen of those donuts? 204 grams

84. **JUICE** It takes 13 oranges to make one can of orange juice. Find the number of oranges used to make a case of 24 cans. 312 oranges

85. **BIRDS** How many times do a hummingbird's wings beat each minute? 3,900 times

65 wingbeats per second

86. **LEGAL FEES** The average hourly rate for consumer law attorneys in New York with 20 years' experience is $312. If such an attorney bills her client for 15 hours of legal work, what is the fee? $4,680

87. **CHANGING UNITS** There are 12 inches in 1 foot and 5,280 feet in 1 mile. How many inches are there in a mile? 63,360 in.

88. **FUEL ECONOMY** Mileage figures for a 2013 Ford Mustang GT convertible are shown in the table.

Fuel tank capacity	16 gal
Fuel economy (miles per gallon)	18 city/25 hwy

a. For city driving, how far can it travel on a tank of gas? 288 mi

b. For highway driving, how far can it travel on a tank of gas? 400 mi

89. **WORD COUNT** Generally, the number of words on a page for a published novel is 250. What would be the expected word count for the 308-page children's novel *Harry Potter and the Philosopher's Stone*? 77,000 words

90. **RENTALS** Mia owns an apartment building with 18 units. Each unit generates a monthly income of $450. Find her total monthly income. $8,100

91. **CONGRESSIONAL PAY** The annual salary of a U.S. House of Representatives member is $174,900. What does it cost per year to pay the salaries of all 36 members of the House from Texas? $6,296,400

92. **CRUDE OIL** In 2011, the United States used about 18,949,000 barrels of crude oil per day. One barrel contains 42 gallons of crude oil. How many gallons of crude oil did the United States use in one day? 795,858,000 gal

93. WORD PROCESSING A student used the *Insert Table* options shown when typing a report. How many entries will the table hold? 72 entries

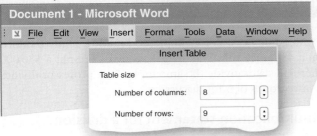

94. BOXING How many feet of padded rope are needed to make a square boxing ring, 24 feet on each side? 384 ft

95. ROOM CAPACITY A college lecture hall has 17 rows of 33 seats each. A sign on the wall reads, "Occupancy by more than 570 persons is prohibited." If all of the seats are taken, and there is one instructor in the room, is the college breaking the rule? no

96. ELEVATORS There are 14 people in an elevator with a capacity of 2,000 pounds. If the average weight of a person in the elevator is 150 pounds, is the elevator overloaded? yes

97. KOALAS In one 24-hour period, a koala sleeps 3 times as many hours as it is awake. If it is awake for 6 hours, how many hours does it sleep? 18 hr

98. FROGS Bullfrogs can jump as far as 10 times their body length. How far could an 8-inch-long bullfrog jump? 80 in.

99. SMARTPHONES According to a report from Trend Micro, from June 2012 to September 2012 the number of malware attacks on smartphones increased *sixfold*. If there were 28,000 attacks in June, how many were there by September? (Source: TrendMicro, nationalnotary.org) 168,000 attacks

100. ENERGY SAVINGS An ENERGY STAR light bulb lasts 8 times as long as a standard 60-watt light bulb. If a standard bulb normally lasts 11 months, how long will an ENERGY STAR bulb last? 88 mo

101. PRESCRIPTIONS How many tablets should a pharmacist put in the container shown in the illustration? 84 tablets

102. HEARTBEATS A normal pulse rate for a healthy adult, while resting, can range from 60 to 100 beats per minute.
a. How many beats is that in one day at the lower end of the range? 86,400 beats
b. How many beats is that in one day at the upper end of the range? 144,000 beats

103. WRAPPING PRESENTS When completely unrolled, a long sheet of wrapping paper has the dimensions shown. How many square feet of gift wrap are on the roll? 54 ft²

104. POSTER BOARDS A rectangular-shaped poster board has dimensions of 24 inches by 36 inches. Find its area. 864 in.²

105. WYOMING The state of Wyoming is approximately rectangular-shaped, with dimensions 360 miles long and 270 miles wide. Find its perimeter and its area. 1,260 mi, 97,200 mi²

106. COMPARING ROOMS Which has the greater area, a rectangular room that is 14 feet by 17 feet or a square room that is 16 feet on each side? Which has the greater perimeter? the square room, the square room

WRITING

107. Explain the difference between 1 foot and 1 square foot.

108. When two numbers are multiplied, the result is 0. What conclusion can be drawn about the numbers?

REVIEW

109. Find the sum of 10,357, 9,809, and 476. 20,642

110. DISCOUNTS A radio, originally priced at $367, has been marked down to $179. By how many dollars was the radio discounted? $188

Objectives

1. Write the related multiplication statement for a division.
2. Use properties of division to divide whole numbers.
3. Perform long division (no remainder).
4. Perform long division (with a remainder).
5. Use tests for divisibility.
6. Divide whole numbers that end with zeros.
7. Estimate quotients of whole numbers.
8. Solve application problems by dividing whole numbers.

SECTION 1.4
Dividing Whole Numbers

Everyone uses division of whole numbers. For example, to find how many 6-ounce servings a chef can get from a 48-ounce roast, he divides 48 by 6. To split a $36,000 inheritance equally, a brother and sister divide the amount by 2. A professor divides the 35 students in her class into groups of 5 for discussion.

1 Write the related multiplication statement for a division.

To divide whole numbers, think of separating a quantity into equal-sized groups. For example, if we start with a set of 12 stars and divide them into groups of 4 stars, we will obtain 3 groups.

A set of 12 stars.

There are 3 groups of 4 stars.

We can write this division problem using a **division symbol** ÷, a **long division symbol** $\overline{)}$, or a **fraction bar** —. We call the number being divided the **dividend**, and the number that we are dividing by is called the **divisor**. The answer is called the **quotient**.

We read each form as "12 divided by 4 equals (or is) 3."

Recall from Section 1.3 that multiplication is repeated addition. Likewise, division is repeated subtraction. To divide 12 by 4, we ask, "How many 4's can be subtracted from 12?"

$$\left.\begin{array}{r}12 \\ -\ 4\end{array}\right\}\text{Subtract 4 one time.}$$

$$\left.\begin{array}{r}8 \\ -\ 4\end{array}\right\}\text{Subtract 4 a second time.}$$

$$\left.\begin{array}{r}4 \\ -\ 4\end{array}\right\}\text{Subtract 4 a third time.}$$
$$0$$

Since exactly three 4's can be subtracted from 12 to get 0, we know that $12 \div 4 = 3$.

Another way to answer a division problem is to think in terms of multiplication. For example, the division $12 \div 4$ asks the question, "What must I multiply 4 by to get 12?" Since the answer is 3, we know that

$$12 \div 4 = 3 \text{ because } 3 \cdot 4 = 12$$

We call $3 \cdot 4 = 12$ the **related multiplication statement** for the division $12 \div 4 = 3$. In general, to write the related multiplication statement for a division, we use:

Quotient · divisor = dividend

EXAMPLE 1 Write the related multiplication statement for each division.

a. $10 \div 5 = 2$ b. $6\overline{)24}$ (quotient 4) c. $\dfrac{21}{3} = 7$

Strategy We will identify the quotient, the divisor, and the dividend in each division statement.

WHY A related multiplication statement has the following form:
Quotient · divisor = dividend.

Solution

a. $10 \div 5 = 2$ because $2 \cdot 5 = 10$. (Dividend, Quotient, Divisor labeled)

b. $6\overline{)24}$ because $4 \cdot 6 = 24$. 4 is the quotient, 6 is the divisor, and 24 is the dividend.

c. $\dfrac{21}{3} = 7$ because $7 \cdot 3 = 21$. 7 is the quotient, 3 is the divisor, and 21 is the dividend.

The Language of Algebra To describe the special relationship between multiplication and division, we say that they are **inverse operations**.

Self Check 1
Write the related multiplication statement for each division.
a. $8 \div 2 = 4$ $4 \cdot 2 = 8$
b. $7\overline{)56}$ (quotient 8) $8 \cdot 7 = 56$
c. $\dfrac{36}{4} = 9$ $9 \cdot 4 = 36$

Now Try Problems 19 and 23

Teaching Example 1
Write the related multiplication statement for each division.
a. $30 \div 5 = 6$
b. $4\overline{)28}$ (quotient 7)
c. $\dfrac{10}{2} = 5$
Answers:
a. $6 \cdot 5 = 30$
b. $7 \cdot 4 = 28$
c. $5 \cdot 2 = 10$

2 Use properties of division to divide whole numbers.

Recall from Section 1.3 that *the product of any whole number and 1 is that whole number*. We can use that fact to establish two important properties of division. Consider the following examples where a whole number is divided by 1:

$8 \div 1 = 8$ because $8 \cdot 1 = 8$.

$1\overline{)4}$ (quotient 4) because $4 \cdot 1 = 4$.

$\dfrac{20}{1} = 20$ because $20 \cdot 1 = 20$.

These examples illustrate that *any whole number divided by 1 is equal to the number itself*.

Consider the following examples where a whole number is divided by itself:

$6 \div 6 = 1$ because $1 \cdot 6 = 6$.

$9\overline{)9}$ (quotient 1) because $1 \cdot 9 = 9$.

$\dfrac{35}{35} = 1$ because $1 \cdot 35 = 35$.

These examples illustrate that *any nonzero whole number divided by itself is equal to 1*.

Properties of Division

Any whole number divided by 1 is equal to that number.
Any nonzero whole number divided by itself is equal to 1.
For any whole number a,

$\dfrac{a}{1} = a$ and $\dfrac{a}{a} = 1$ (where $a \neq 0$) Read $\neq$ as "is not equal to."

Recall from Section 1.3 that *the product of any whole number and 0 is 0*. We can use that fact to establish another property of division. Consider the following examples where 0 is divided by a whole number:

$0 \div 2 = 0$ because $0 \cdot 2 = 0$.

$7\overline{)0}$ with quotient 0 because $0 \cdot 7 = 0$.

$\dfrac{0}{42} = 0$ because $0 \cdot 42 = 0$.

These examples illustrate that *0 divided by any nonzero whole number is equal to 0*.

We cannot divide a whole number by 0. To illustrate why, we will attempt to find the quotient when 2 is divided by 0 using the related multiplication statement shown below.

Division statement	*Related multiplication statement*
$\dfrac{2}{0} = ?$	$? \cdot 0 = 2$

There is no number that gives 2 when multiplied by 0.

Since $\dfrac{2}{0}$ does not have a quotient, we say that division of 2 by 0 is *undefined*. Our observations about division of 0 and division by 0 are listed below.

Division with Zero

1. Zero divided by any nonzero number is equal to 0.
2. Division by 0 is undefined.

For any nonzero whole number a,

$$\dfrac{0}{a} = 0 \quad \text{and} \quad \dfrac{a}{0} \text{ is undefined}$$

---

3 Perform long division (no remainder).

A process called **long division** can be used to divide larger whole numbers.

EXAMPLE 2 Divide using long division: $2{,}514 \div 6$. Check the result.

Strategy We will write the problem in long division form and follow a four-step process: **estimate**, **multiply**, **subtract**, and **bring down**.

WHY The repeated subtraction process would take too long to perform and the related multiplication statement ($? \cdot 6 = 2{,}514$) is too difficult to solve.

Solution
To help you understand the process, each step of this division is explained separately. Your solution need only look like the *last* step.

We write the problem in the form $6\overline{)2514}$. The quotient will appear above the long division symbol. Since 6 will not divide 2,

$6\overline{)2514}$

we divide 25 by 6.

4
$6\overline{)2514}$ Ask: "How many times will 6 divide 25?" We estimate that $25 \div 6$ is about 4, and write the 4 in the hundreds column above the long division symbol.

Self Check 2
Divide using long division: $2{,}968 \div 4$. Check the result.
Now Try Problem 31
Self Check 2 Answer
742; $742 \cdot 4 = 2{,}968$

Teaching Example 2
Divide using long division: $2{,}205 \div 7$. Check the result.
Answer:
315; $315 \cdot 7 = 2{,}205$

Next, we multiply 4 and 6, and subtract their product, 24, from 25, to get 1.

```
     4
   ─────
 6)2514
   -24
   ───
     1
```

Now we bring down the next digit in the dividend, the 1, and again estimate, multiply, and subtract.

```
     41
   ─────
 6)2514
   -24↓
   ───
    11
    -6
    ──
     5
```

Ask: "How many times will 6 divide 11?" We estimate that 11 ÷ 6 is about 1, and write the 1 in the tens column above the long division symbol. Multiply 1 and 6, and subtract their product, 6, from 11, to get 5.

To complete the process, we bring down the last digit in the dividend, the 4, and estimate, multiply, and subtract one final time.

```
     419
   ─────
 6)2514
   -24
   ───
    11
    -6↓
    ──
    54
   -54
   ───
     0
```

Ask: "How many times will 6 divide 54?" We estimate that 54 ÷ 6 is 9, and we write the 9 in the ones column above the long division symbol. Multiply 9 and 6, and subtract their product, 54, from 54, to get 0.

Your solution should look like this:
```
     419
   ─────
 6)2514
   -24
   ───
    11
    -6
    ──
    54
   -54
   ───
     0
```

To check the result, we see if the product of the quotient and the divisor equals the dividend.

```
    1 5
    419  ← Quotient
  ×   6  ← Divisor        6)2514
  ─────
  2,514  ← Dividend
```

The check confirms that 2,514 ÷ 6 = 419.

> **The Language of Algebra** In Example 2, the long division process ended with a 0. In such cases, we say that the divisor divides the dividend *exactly*.

We can see how the long division process works if we write the names of the place-value columns above the quotient. The solution for Example 2 is shown in more detail on the next page.

52 Chapter 1 Whole Numbers

$$\begin{array}{r} \text{Hundreds Tens Ones} \\ 419 \\ 6\overline{)2514} \\ -2400 \\ \hline 114 \\ -60 \\ \hline 54 \\ -54 \\ \hline 0 \end{array}$$

Here, we are really subtracting 400 · 6, which is 2,400, from 2,514. That is why the 4 is written in the hundreds column of the quotient.

Here, we are really subtracting 10 · 6, which is 60, from 114. That is why the 1 is written in the tens column of the quotient.

Here, we are subtracting 9 · 6, which is 54, from 54. That is why the 9 is written in the ones column of the quotient.

The extra zeros (shown in the steps highlighted in red and blue) are often omitted.

We can use long division to perform divisions when the divisor has more than one digit. The estimation step is often made easier if we approximate the divisor.

Self Check 3
Divide using long division:

$57\overline{)45{,}885}$ 805

Now Try Problem 35

Teaching Example 3
Divide using long division: $39\overline{)31{,}278}$
Answer:
802

EXAMPLE 3
Divide using long division: $48\overline{)33{,}888}$

Strategy We will follow a four-step process: **estimate**, **multiply**, **subtract**, and **bring down**.

WHY This is how long division is performed.

Solution
To help you understand the process, each step of this division is explained separately. Your solution need only look like the *last* step.

Since 48 will not divide 3, nor will it divide 33, we divide 338 by 48.

$$\begin{array}{r} 6 \\ 48\overline{)33888} \end{array}$$

Ask: "How many times will 48 divide 338?" Since 48 is almost 50, we can estimate the answer to that question by thinking 33 ÷ 5 is about 6, and we write the 6 in the hundreds column of the quotient.

$$\begin{array}{r} 6 \\ 48\overline{)33888} \\ -288 \\ \hline 50 \end{array}$$

Multiply 6 and 48, and subtract their product, 288, from 338 to get 50. Since 50 is greater than the divisor, 48, the estimate of 6 for the hundreds column of the quotient is *too small*. We will erase the 6 and increase the estimate of the quotient by 1 and try again.

$$\begin{array}{r} 7 \\ 48\overline{)33888} \\ -336 \\ \hline 2 \end{array}$$

Change the estimate from 6 to 7 in the hundreds column of the quotient. Multiply 7 and 48, and subtract their product, 336, from 338 to get 2. Since 2 is less than the divisor, we can proceed with the long division.

$$\begin{array}{r} 70 \\ 48\overline{)33888} \\ -336\downarrow \\ \hline 28 \\ -0 \\ \hline 28 \end{array}$$

Bring down the 8 from the tens column of the dividend. Ask: "How many times will 48 divide 28?" Since 28 cannot be divided by 48, write a 0 in the tens column of the quotient. Multiply 0 and 48, and subtract their product, 0, from 28 to get 28.

Bring down the 8 from the ones column of the dividend. Ask: "How many times will 48 divide 288?" We can estimate the answer to that question by thinking 28 ÷ 5 is about 5, and we write the 5 in the ones column of the quotient. Multiply 5 and 48, and subtract their product, 240, from 288 to get 48. Since 48 is equal to the divisor, the estimate of 5 for the ones column of the quotient is *too small*. We will erase the 5 and increase the estimate of the quotient by 1 and try again.

Caution! If a difference at any time in the long division process is greater than or equal to the divisor, the estimate made at that point should be increased by 1, and you should try again.

```
         706
    48) 33888
       -336
          28
         - 0
         288    Change the estimate from 5 to 6 in the ones column of the quotient.
        -288    Multiply 6 and 48, and subtract their product, 288, from 288 to
           0    get 0. Your solution should look like this.
```

The quotient is 706. Check the result using multiplication.

4 Perform long division (with a remainder).

Sometimes, it is not possible to separate a group of objects into a whole number of equal-sized groups. For example, if we start with a set of 14 stars and divide them into groups of 4 stars, we will have 3 groups of 4 stars and 2 stars left over. We call the leftover part the **remainder**.

A set of 14 stars.

★★★★ ★★★★ ★★★★ ★★

There are 3 groups of 4 stars.   There are 2 stars left over.

In the next long division example, there is a remainder. To check such a problem, we add the remainder to the product of the quotient and divisor. The result should equal the dividend.

(Quotient · divisor) + remainder = dividend   Recall that the operation within the parentheses must be performed first.

EXAMPLE 4   Divide: 23)832. Check the result.

Strategy We will follow a four-step process: **estimate**, **multiply**, **subtract**, and **bring down**.

WHY This is how long division is performed.

Solution
Since 23 will not divide 8, we divide 83 by 23.

```
        4
   23) 832
```
Ask: "How many times will 23 divide 83?" Since 23 is about 20, we can estimate the answer to that question by thinking 8 ÷ 2 is 4, and we write the 4 in the tens column of the quotient.

```
        4
   23) 832
      - 92
```
Multiply 4 and 23, and write their product, 92, under the 83. Because 92 is greater than 83, the estimate of 4 for the tens column of the quotient is too large. We will erase the 4 and decrease the estimate of the quotient by 1 and try again.

Self Check 4

Divide: 34)792. Check the result.

Now Try Problem 39

Self Check 4 Answer
23 R 10; (23 · 34) + 10 = 792

Teaching Example 4 Divide: 19)899. Check the result.
Answer:
47 R 6; (47 · 19) + 6 = 899

Chapter 1 Whole Numbers

$$\begin{array}{r} 3 \\ 23\overline{)832} \\ -69 \\ \hline 14 \end{array}$$

Change the **estimate** from 4 to 3 in the tens column of the quotient. **Multiply** 3 and 23, and **subtract** their product, 69, from 83, to get 14.

$$\begin{array}{r} 3 \\ 23\overline{)832} \\ -69\downarrow \\ \hline 142 \end{array}$$

Bring down the 2 from the ones column of the dividend.

$$\begin{array}{r} 37 \\ 23\overline{)832} \\ -69 \\ \hline 142 \\ -161 \end{array}$$

Ask: "How many times will 23 divide 142?" We can **estimate** the answer to that question by thinking 14 ÷ 2 is 7, and we write the 7 in the ones column of the quotient. **Multiply** 7 and 23, and write their product, 161, under 142. Because 161 is greater than 142, the estimate of 7 for the ones column of the quotient is *too large*. We will erase the 7 and decrease the estimate of the quotient by 1 and try again.

$$\begin{array}{r} 36 \\ 23\overline{)832} \\ -69 \\ \hline 142 \\ -138 \\ \hline 4 \end{array}$$ ← The remainder

Change the **estimate** from 7 to 6 in the ones column of the quotient. **Multiply** 6 and 23, and **subtract** their product, 138, from 142, to get 4.

The quotient is 36, and the remainder is 4. We can write this result as 36 R 4.

To check the result, we multiply the divisor by the quotient and then add the remainder. The result should be the dividend.

Check: Quotient   Divisor   Remainder
$$(36 \cdot 23) + 4 = 828 + 4$$
$$= 832 \leftarrow \text{Dividend}$$

Since 832 is the dividend, the answer 36 R 4 is correct.

Self Check 5

Divide: $\dfrac{28{,}992}{629}$ 46 R 58

Now Try Problem 43

Teaching Example 5 Divide: $\dfrac{12{,}602}{431}$

Answer:
29 R 103

EXAMPLE 5

Divide: $\dfrac{13{,}011}{518}$

Strategy We will write the problem in long-division form and follow a four-step process: **estimate**, **multiply**, **subtract**, and **bring down**.

WHY This is how long division is performed.

Solution
We write the division in the form: $518\overline{)13011}$. Since 518 will not divide 1, nor 13, nor 130, we divide 1,301 by 518.

$$\begin{array}{r} 2 \\ 518\overline{)13011} \\ -1036 \\ \hline 265 \end{array}$$

Ask: "How many times will 518 divide 1,301?" Since 518 is about 500, we can **estimate** the answer to that question by thinking 13 ÷ 5 is about 2, and we write the 2 in the tens column of the quotient. **Multiply** 2 and 518, and **subtract** their product, 1,036, from 1,301, to get 265.

$$\begin{array}{r}25\\518{\overline{\smash{\big)}\,13011}}\\-1036\\\hline 2651\\-2590\\\hline 61\end{array}$$

Bring down the 1 from the ones column of the dividend. Ask: "How many times will 518 divide 2,651?" We can estimate the answer to that question by thinking $26 \div 5$ is about 5, and we write the 5 in the ones column of the quotient. Multiply 5 and 518, and subtract their product, 2,590, from 2,651, to get a remainder of 61.

The result is 25 R 61. To check, verify that $(25 \cdot 518) + 61$ is 13,011.

5 Use tests for divisibility.

We have seen that some divisions end with a 0 remainder and others do not. The word *divisible* is used to describe such situations.

> **Divisibility**
>
> One number is **divisible** by another if, when dividing them, we get a remainder of 0.

Since $27 \div 3 = 9$, with a 0 remainder, we say that *27 is divisible by 3*. Since $27 \div 5 = 5$ R 2, we say that *27 is not divisible by 5*.

There are tests to help us decide whether one number is divisible by another.

> **Tests for Divisibility**
>
> A number is divisible by
> - 2 if its last digit is divisible by 2.
> - 3 if the sum of its digits is divisible by 3.
> - 4 if the number formed by its last two digits is divisible by 4.
> - 5 if its last digit is 0 or 5.
> - 6 if it is divisible by 2 and 3.
> - 9 if the sum of its digits is divisible by 9.
> - 10 if its last digit is 0.

There are tests for divisibility by a number other than 2, 3, 4, 5, 6, 9, or 10, but they are more complicated. See problems 109 and 110 of Study Set 1.4 for some examples.

EXAMPLE 6 Is 534,840 divisible by:
a. 2 **b.** 3 **c.** 4 **d.** 5 **e.** 6 **f.** 9 **g.** 10

Strategy We will look at the last digit, the last two digits, and the sum of the digits of each number.

WHY The divisibility rules call for these types of examination.

Solution

a. 534,84**0** is divisible by 2, because its last digit **0** is divisible by 2.

b. 534,840 is divisible by 3, because the sum of its digits is divisible by 3.

$5 + 3 + 4 + 8 + 4 + 0 = 24$ and $24 \div 3 = 8$

Self Check 6

Is 73,311,435 divisible by:
a. 2 no **b.** 3 yes **c.** 5 yes
d. 6 no **e.** 9 yes **f.** 10 no

Now Try Problems 49 and 53

Teaching Example 6 Is 165,820 divisible by: **a.** 2 **b.** 3 **c.** 4 **d.** 5 **e.** 6 **f.** 9 **g.** 10
Answers:
a. yes **b.** no
c. yes **d.** yes
e. no **f.** no
g. yes

c. 534,840 is divisible by 4, because the number formed by its last two digits is divisible by 4.

$$40 \div 4 = 10$$

d. 534,840 is divisible by 5, because its last digit is 0 or 5.

e. 534,840 is divisible by 6, because it is divisible by 2 and 3. (See parts a and b.)

f. 534,840 is not divisible by 9, because the sum of its digits is not divisible by 9. There is a remainder.

$$24 \div 9 = 2 \text{ R } 6$$

g. 534,840 is divisible by 10, because its last digit is 0.

6 Divide whole numbers that end with zeros.

There is a shortcut for dividing a dividend by a divisor when both end with zeros. We simply *remove the ending zeros in the divisor and remove the same number of ending zeros in the dividend.*

Self Check 7
Divide: **a.** $50 \div 10$ 5
b. $62,000 \div 100$ 620
c. $12,000 \div 1,500$ 8
Now Try Problems 55 and 57

EXAMPLE 7 Divide: **a.** $80 \div 10$ **b.** $47,000 \div 100$ **c.** $350\overline{)9,800}$

Strategy We will look for ending zeros in each divisor.

WHY If a divisor has ending zeros, we can simplify the division by removing the same number of ending zeros in the divisor and dividend.

Solution

There is one zero in the divisor.

a. $80 \div 10 = 8 \div 1 = 8$

Remove one zero from the dividend and the divisor, and divide.

Teaching Example 7 Divide:
a. $60 \div 10$
b. $32,000 \div 100$
c. $580\overline{)9,280}$
Answers:
a. 6
b. 320
c. 16

There are two zeros in the divisor.

b. $47,000 \div 100 = 470 \div 1 = 470$

Remove two zeros from the dividend and the divisor, and divide.

c. To find

$$350\overline{)9,800}$$

we can drop *one zero* from the divisor and the dividend and perform the division $35\overline{)980}$.

$$\begin{array}{r} 28 \\ 35\overline{)980} \\ -70 \\ \hline 280 \\ -280 \\ \hline 0 \end{array}$$

Thus, $9,800 \div 350$ is 28.

7 Estimate quotients of whole numbers.

To estimate quotients, we use a method that approximates both the dividend and the divisor so that they divide easily. There is one rule of thumb for this method: If possible, round both numbers up or both numbers down.

EXAMPLE 8   Estimate the quotient: 170,715 ÷ 57

Strategy   We will round the dividend and the divisor up and find 180,000 ÷ 60.

WHY   The division can be made easier if the dividend and the divisor end with zeros. Also, 6 divides 18 exactly.

Solution

The dividend is approximately ↓
170,715 ÷ 57   180,000 ÷ 60 = 3,000   To divide, drop one zero from 180,000 and from 60 and find 18,000 ÷ 6.
The divisor is approximately ↑

The estimate is 3,000.

If we calculate 170,715 ÷ 57, the quotient is exactly 2,995. Note that the estimate is close: It's just 5 more than 2,995.

Self Check 8

Estimate the quotient:
33,642 ÷ 42   800

Now Try Problem 59

Teaching Example 8 Estimate the quotient: 482,901 ÷ 67
Answer:
7,000

8 Solve application problems by dividing whole numbers.

Application problems that involve forming equal-sized groups can be solved by division.

EXAMPLE 9   *Managing a Soup Kitchen*   A soup kitchen plans to feed 1,990 people. Because of space limitations, only 144 people can be served at one time. How many group seatings will be necessary to feed everyone? How many will be served at the last seating?

Strategy   We will divide 1,990 by 144.

WHY   Separating 1,990 people into equal-sized groups of 144 indicates division.

Solution
We translate the words of the problem to numbers and symbols.

The number of group seatings	is equal to	the number of people to be fed	divided by	the number of people at each seating.
The number of group seatings	=	1,990	÷	144

Use long division to find 1,990 ÷ 144.

```
      13
144)1,990
    −144
     550
    −432
     118
```

The quotient is 13, and the remainder is 118. This indicates that 14 group seatings are needed: 13 full-capacity seatings and one partial seating to serve the remaining 118 people.

Self Check 9

MOVIE TICKETS   On a Saturday, 3,924 movie tickets were purchased at an IMAX theater. Each showing of the movie was sold out, except for the last. If the theater seats 346 people, how many times was the movie shown on Saturday? How many people were at the last showing?   12; 118

Now Try Problem 91

Teaching Example 9 SCHEDULING A college counselor needs to meet with 231 students to check their graduation requirements. Because of other duties, he can only meet with 8 each day. How many days will be necessary to see all of the students? How many will be seen the last day?
Answer:
29 days; 7

The Language of Algebra Here are some key words and phrases that are often used to indicate division:

split equally	distributed equally	how many does each
goes into	per	how much extra (remainder)
shared equally	among	how many left (remainder)

Self Check 10

TOURING A rock band will take a 275-day world tour and spend the same number of days in each of 25 cities. How long will they stay in each city? 11 days

Now Try Problem 97

Teaching Example 10 SALARIES
The starting yearly salary for 2012 college graduates majoring in computer science was approximately $60,036. How much would such a graduate make each month? (Source: National Association of Colleges and Employers)
Answer:
$5,003

EXAMPLE 10

Timeshares Every year, the 73 part-owners of a timeshare resort condominium get use of it for an equal number of days. How many days does each part-owner get to stay at the condo? (Use a 365-day year.)

Strategy We will divide 365 by 73.

WHY Since the part-owners get use of the condo for an equal number of days, the phrase "How many days does each" indicates division.

Solution
We translate the words of the problem to numbers and symbols.

The number of days each part-owner gets to stay at the condo **is equal to** the number of days in a year **divided by** the number of part-owners.

The number of days each part-owner gets to stay at the condo = 365 ÷ 73

Use long division to find 365 ÷ 73.

$$\begin{array}{r} 5 \\ 73\overline{)365} \\ -365 \\ \hline 0 \end{array}$$

Each part-owner gets to stay at the condo for 5 days during the year.

Using Your CALCULATOR
The Division Key

A beverage company production run of 604,800 bottles of mountain spring water will be shipped to stores on pallets that hold 1,728 bottles each. We can find the number of full pallets to be shipped using the division key ÷ on a calculator.

604800 ÷ 1728 =   350

On some calculator models, the ENTER key is pressed instead of = for the result to be displayed.

The beverage company will ship 350 full pallets of bottled water.

ANSWERS TO SELF CHECKS

1. a. $4 \cdot 2 = 8$ **b.** $8 \cdot 7 = 56$ **c.** $9 \cdot 4 = 36$ **2.** 742 **3.** 805 **4.** 23 R 10 **5.** 46 R 58
6. a. no **b.** yes **c.** yes **d.** no **e.** yes **f.** no **7. a.** 5 **b.** 620 **c.** 8 **8.** 800
9. 12 showings, 118 **10.** 11 days

STUDY SKILLS CHECKLIST

Get the Most from Your Textbook

The following checklist will help you become familiar with some useful features in this book. Place a check mark in each box after you answer the question.

☐ Locate the **Definition** for divisibility on page 55 and the order of operations rules on page 85. What color are these boxes?

☐ Find the **Caution** box on page 34 and the **Language of Algebra** box on page 39. What color is used to identify these boxes?

☐ Each chapter begins with **From Campus to Careers**. Chapter 3 gives information on how to become a broadcast technician (see page 225). On what page does a related problem appear in Study Set 3.6?

☐ Locate the **Study Skills Workshop** at the beginning of your text beginning on page S-1. How many Objectives appear in the Study Skills Workshop?

Answers: Green, Red, 287, 7

SECTION 1.4 STUDY SET

VOCABULARY

Fill in the blanks.

1. In the three division problems shown below, label the *dividend*, *divisor*, and the *quotient*.

2. We call $5 \cdot 8 = 40$ the related __multiplication__ statement for the division $40 \div 8 = 5$.

3. The problem $6\overline{)246}$ is written in __long__ division form.

4. If a division is not exact, the leftover part is called the __remainder__.

5. One number is __divisible__ by another number if, when we divide them, the remainder is 0.

6. Phrases such as *split equally* and *how many does each* indicate the operation of __division__.

CONCEPTS

7. **a.** Divide the objects below into groups of 3. How many groups of 3 are there? 7

• •

b. Divide the objects below into groups of 4. How many groups of 4 are there? How many objects are left over? 5, 2

✻ ✻

8. Tell whether each statement is true or false.

a. Any whole number divided by 1 is equal to that number. true

b. Any nonzero whole number divided by itself is equal to 1. true

c. Zero divided by any nonzero number is undefined. false

d. Division of a number by 0 is equal to 0. false

Fill in the blanks.

9. Divide, if possible.

a. $\dfrac{25}{25} = $ __1__   **b.** $\dfrac{6}{1} = $ __6__

c. $\dfrac{100}{0}$ is __undefined__   **d.** $\dfrac{0}{12} = $ __0__

10. To perform long division, we follow a four-step process: _estimate_, _multiply_, _subtract_, and _bring_ _down_.

11. Find the *first* digit of each quotient.

a. $5\overline{)1147}$ → 2

b. $9\overline{)587}$ → 6

c. $23\overline{)7501}$ → 3

d. $16\overline{)892}$ → 5

12. a. Quotient · divisor = _dividend_

b. (Quotient · divisor) + _remainder_ = dividend

13. To check whether the division $9\overline{)333}$ is correct, we use multiplication:

$$\begin{array}{r} 37 \\ \times\ 9 \\ \hline 333 \end{array}$$

14. a. A number is divisible by _2_ if its last digit is divisible by 2.

b. A number is divisible by 3 if the _sum_ of its digits is divisible by 3.

c. A number is divisible by 4 if the number formed by its last _two_ digits is divisible by 4.

15. a. A number is divisible by 5 if its last digit is _0_ or _5_.

b. A number is divisible by 6 if it is divisible by _2_ and _3_.

c. A number is divisible by 9 if the _sum_ of its digits is divisible by 9.

d. A number is divisible by _10_ if its last digit is 0.

16. We can simplify the division 43,800 ÷ 200 by removing two _zeros_ from the dividend and the divisor.

NOTATION

17. Write three symbols that can be used for division.

÷, $\overline{)\ }$, —

18. In a division, 35 R 4 means "a quotient of 35 and a _remainder_ of 4."

GUIDED PRACTICE

Fill in the blanks. See Example 1.

19. $9\overline{)45}$ → 5 because _5_ · _9_ = _45_.

20. $\dfrac{54}{6} = 9$ because _9_ · _6_ = _54_.

21. 44 ÷ 11 = 4 because _4_ · _11_ = _44_.

22. 120 ÷ 12 = 10 because _10_ · _12_ = _120_.

Write the related multiplication statement for each division. See Example 1.

23. 21 ÷ 3 = 7   7 · 3 = 21

24. 32 ÷ 4 = 8   8 · 4 = 32

25. $\dfrac{72}{12} = 6$ 6 · 12 = 72

26. $15\overline{)75}$ → 5   5 · 15 = 75

Divide using long division. Check the result. See Example 2.

27. 96 ÷ 6   16

28. 72 ÷ 4   18

29. $\dfrac{87}{3}$ 29

30. $\dfrac{98}{7}$ 14

31. 2,275 ÷ 7   325

32. 1,728 ÷ 8   216

33. $9\overline{)1,962}$ 218

34. $5\overline{)1,635}$ 327

Divide using long division. Check the result. See Example 3.

35. $62\overline{)31,248}$ 504

36. $71\overline{)28,613}$ 403

37. $37\overline{)22,274}$ 602

38. $28\overline{)19,712}$ 704

Divide using long division. Check the result. See Example 4.

39. $24\overline{)951}$ 39 R 15

40. $33\overline{)943}$ 28 R 19

41. 999 ÷ 46   21 R 33

42. 979 ÷ 49   19 R 48

Divide using long division. Check the result. See Example 5.

43. $\dfrac{24,714}{524}$ 47 R 86

44. $\dfrac{29,773}{531}$ 56 R 37

45. $178\overline{)3,514}$ 19 R 132

46. $164\overline{)2,929}$ 17 R 141

If the given number is divisible by 2, 3, 4, 5, 6, 9, *or* 10, *enter a checkmark* ✓ *in the box. See Example 6.*

	Divisible by →	2	3	4	5	6	9	10
47.	2,940	✓	✓	✓	✓	✓		✓
48.	5,850	✓	✓		✓	✓	✓	✓
49.	43,785		✓		✓		✓	
50.	72,954	✓	✓			✓	✓	
51.	181,223							
52.	379,157							
53.	9,499,200	✓	✓	✓	✓	✓		✓
54.	6,653,100	✓	✓	✓	✓	✓		✓

Use a division shortcut to find each quotient. See Example 7.

55. 700 ÷ 10   70

56. 900 ÷ 10   90

57. $450\overline{)9,900}$ 22

58. $260\overline{)9,100}$ 35

Estimate each quotient. See Example 8.

59. 353,922 ÷ 38   9,000

60. 237,621 ÷ 55   4,000

61. 46,080 ÷ 933   50

62. 81,097 ÷ 419   200

TRY IT YOURSELF

Divide.

63. $\dfrac{25{,}950}{6}$ 4,325
64. $\dfrac{23{,}541}{7}$ 3,363
65. 54 ÷ 9 6
66. 72 ÷ 8 9
67. 273 ÷ 31 8 R 25
68. 295 ÷ 35 8 R 15
69. $\dfrac{64{,}000}{400}$ 160
70. $\dfrac{125{,}000}{5{,}000}$ 25
71. 745 divided by 7 106 R 3
72. 931 divided by 9 103 R 4
73. $29\overline{)14{,}761}$ 509
74. $27\overline{)10{,}989}$ 407
75. 539,000 ÷ 175 3,080
76. 749,250 ÷ 185 4,050
77. 75 ÷ 15 5
78. 96 ÷ 16 6
79. $212\overline{)5{,}087}$ 23 R 211
80. $214\overline{)5{,}777}$ 26 R 213
81. $42\overline{)1{,}273}$ 30 R 13
82. $83\overline{)3{,}363}$ 40 R 43
83. 89,000 ÷ 1,000 89
84. 930,000 ÷ 1,000 930
85. $\dfrac{57}{8}$ 7 R 1
86. $\dfrac{82}{9}$ 9 R 1

APPLICATIONS

87. **TICKET SALES** A movie theater makes a $4 profit on each ticket sold. How many tickets must be sold to make a profit of $2,500? 625 tickets

88. **RUNNING** Brian runs 7 miles each day. In how many days will Brian run 371 miles? 53 days

89. **DUMP TRUCKS** A 15-cubic-yard dump truck must haul 405 cubic yards of dirt to a construction site. How many trips must the truck make? 27 trips

90. **STOCKING SHELVES** After receiving a delivery of 288 bags of potato chips, a store clerk stocked each shelf of an empty display with 36 bags. How many shelves of the display did he stock with potato chips? 8 shelves

91. **LUNCH TIME** A fifth grade teacher received 50 half-pint cartons of milk to distribute evenly to his class of 23 students. How many cartons did each child get? How many cartons were left over? 2 cartons, 4 cartons

92. **BUBBLE WRAP** A furniture manufacturer uses an 11-foot-long strip of bubble wrap to protect a lamp when it is boxed and shipped to a customer. How many lamps can be packaged in this way from a 200-foot-long roll of bubble wrap? How many feet will be left on the roll? 18 lamps, 2 ft

93. **GARDENING** A metal can holds 640 fluid ounces of gasoline. How many times can the 68-ounce tank of a lawnmower be filled from the can? How many ounces of gasoline will be left in the can? 9 times, 28 ounces

94. **BEVERAGES** A plastic container holds 896 ounces of punch. How many 6-ounce cups of punch can be served from the container? How many ounces will be left over? 149 servings, 2 ounces

95. **LIFT SYSTEMS** If the bus shown below weighs 58,000 pounds, how much weight is on each jack? 14,500 lb

96. **LOTTERY WINNERS** In 2008, a group of 22 postal workers, who had been buying Pennsylvania Lotto tickets for years, won a $10,282,800 jackpot. If they split the prize evenly, how much money did each person win? $467,400

97. **TEXTBOOK SALES** A store received $25,200 on the sale of 240 algebra textbooks. What was the cost of each book? $105

98. **DRAINING POOLS** A 950,000-gallon pool is emptied in 20 hours. How many gallons of water are drained each hour? 47,500 gal

99. **MILEAGE** A tour bus has a range of 700 miles on one tank (140 gallons) of gasoline. How far does the bus travel on one gallon of gas? 5 mi

100. **WATERFALLS** Every 45 seconds, about 33,750,000 gallons of water flow over Niagara Falls. How many gallons flow over the falls in one second? (Source: water.epa.gov) 750,000 gal

101. **ORDERING SNACKS** How many *dozen* doughnuts must be ordered for a meeting if 156 people are expected to attend, and each person will be served one doughnut? 13 dozen

102. **TIME** A *millennium* is a period of time equal to one thousand years. How many decades are in a millennium? 100 decades

103. **VOLLEYBALL** A total of 216 girls are going to play in a city volleyball league. How many girls should be put on each team if the following requirements must be met? 9 girls

    - All the teams are to have the same number of players.
    - A reasonable number of players on a team is 7 to 10.
    - For scheduling purposes, there must be an even number of teams (2, 4, 6, 8, and so on).

104. A landscape designer intends to plant pine trees 12 feet apart to form a windscreen along one side of a flower garden, as shown below. How many trees are needed if the length of the flower garden is 744 feet? 62 + 1 = 63 trees

from Campus to Careers
Landscape Designer

105. ENTRY-LEVEL JOBS The typical starting salaries for 2012 college graduates majoring in health sciences, business, and social sciences are shown below. Complete the last column of the table.

College major	Yearly salary	Monthly salary
Health Sciences	$46,560	$3,880
Business	$51,540	$4,295
Social Sciences	$36,828	$3,069

Source: National Association of Colleges and Employers

106. POPULATION To find the **population density** of a state, divide its population by its land area (in square miles). The result is the number of people per square mile. Use the data in the table to approximate the population density for each state.
AZ: 57, OK: 55, RI: 1,100, SC: 155 (all amounts are given in persons per square mile)

State	2011 Population*	Land area* (square miles)
Arizona	6,498,000	114,000
Oklahoma	3,795,000	69,000
Rhode Island	1,100,000	1,000
South Carolina	4,650,000	30,000

Source: Worldpopulationreview.com   *approximation

WRITING

107. Explain how 24 ÷ 6 can be calculated by repeated subtraction.

108. Explain why division of 0 is possible, but division by 0 is impossible.

109. DIVISIBILTY TEST FOR 7 Use the following rule to show that 308 is divisible by 7. Show each of the steps of your solution in writing.

> Subtract twice the units digit from the number formed by the remaining digits. If that result is divisible by 7, then the original number is divisible by 7.

110. DIVISIBILTY TEST FOR 11 Use the following rule to show that 1,848 is divisible by 11. Show each of the steps of your solution in writing.

> Start with the digit in the one's place. From it, subtract the digit in the ten's place. To that result, add the digit in the hundred's place. From that result, subtract the digit in the thousands place, and so on. If the final result is a number divisible by 11, the original number is divisible by 11.

REVIEW

111. Add: 2,903 + 378   3,281
112. Subtract: 2,903 − 378   2,525
113. Multiply: 2,903 × 378   1,097,334
114. DISCOUNTS A car, originally priced at $17,550, is being sold for $13,970. By how many dollars has the price been decreased? $3,580

SECTION 1.5
Prime Factors and Exponents

In this section, we will discuss how to express whole numbers in factored form. The procedures used to find the factored form of a whole number involve multiplication and division.

1 Factor whole numbers.

The statement $3 \cdot 2 = 6$ has two parts: the numbers that are being multiplied and the answer. The numbers that are being multiplied are called *factors*, and the answer is the *product*. We say that 3 and 2 are factors of 6.

Factors

Numbers that are multiplied together are called **factors**.

EXAMPLE 1 Find the factors of 12.

Strategy We will find all the pairs of whole numbers whose product is 12.

WHY Each of the numbers in those pairs is a factor of 12.

Solution
The pairs of whole numbers whose product is 12 are:

$1 \cdot 12 = 12, \quad 2 \cdot 6 = 12, \quad$ and $\quad 3 \cdot 4 = 12$

In order, from least to greatest, the factors of 12 are 1, 2, 3, 4, 6, and 12.

Success Tip In Example 1, once we determine the pair 1 and 12 are factors of 12, any remaining factors must be *between* 1 and 12. Once we determine that the pair 2 and 6 are factors of 12, any remaining factors must be *between* 2 and 6. Once we determine that the pair 3 and 4 are factors of 12, any remaining factors of 12 must be *between* 3 and 4. Since there are no whole numbers between 3 and 4, we know that all the possible factors of 12 have been found.

In Example 1, we found that 1, 2, 3, 4, 6, and 12 are the factors of 12. Notice that each of the factors divides 12 exactly, leaving a remainder of 0.

$\frac{12}{1} = 12 \quad \frac{12}{2} = 6 \quad \frac{12}{3} = 4 \quad \frac{12}{4} = 3 \quad \frac{12}{6} = 2 \quad \frac{12}{12} = 1$

In general, if a whole number is a factor of a given number, it also divides the given number exactly.

When we say that 3 is a factor of 6, we are using the word *factor* as a noun. The word *factor* is also used as a verb.

Factoring a Whole Number

To **factor** a whole number means to express it as the product of other whole numbers.

Objectives

1. Factor whole numbers.
2. Identify even and odd whole numbers, prime numbers, and composite numbers.
3. Find prime factorizations using a factor tree.
4. Find prime factorizations using a division ladder.
5. Use exponential notation.
6. Evaluate exponential expressions.

Self Check 1

Find the factors of 20.

Now Try Problems 21 and 27

Self Check 1 Answer
1, 2, 4, 5, 10, and 20

Teaching Example 1 Find the factors of 24.
Answer:
1, 2, 3, 4, 6, 8, 12, 24

Self Check 2

Factor 18 using **a.** two factors **b.** three factors

Now Try Problems 39 and 45

Self Check 2 Answers
a. $1 \cdot 18, 2 \cdot 9,$ or $3 \cdot 6$
b. Two possibilities are $2 \cdot 3 \cdot 3$ and $1 \cdot 2 \cdot 9$

Teaching Example 2 Factor 36 using
a. two factors
b. three factors
Answers:
a. $1 \cdot 36, 2 \cdot 18, 3 \cdot 12, 4 \cdot 9,$ or $6 \cdot 6$
b. Two possibilities are $2 \cdot 2 \cdot 9$ and $3 \cdot 3 \cdot 4$

Self Check 3

Find the factors of 23. 1 and 23

Now Try Problem 49

Teaching Example 3 Find the factors of 31.
Answer:
1 and 31

EXAMPLE 2
Factor 40 using **a.** two factors **b.** three factors

Strategy We will find a pair of whole numbers whose product is 40 and three whole numbers whose product is 40.

WHY To *factor* a number means to express it as the product of two (or more) numbers.

Solution

a. To factor 40 using two factors, there are several possibilities.

$40 = 1 \cdot 40, \quad 40 = 2 \cdot 20, \quad 40 = 4 \cdot 10, \quad$ and $\quad 40 = 5 \cdot 8$

b. To factor 40 using three factors, there are several possibilities. Two of them are:

$40 = 5 \cdot 4 \cdot 2 \quad$ and $\quad 40 = 2 \cdot 2 \cdot 10$

EXAMPLE 3
Find the factors of 17.

Strategy We will find all the pairs of whole numbers whose product is 17.

WHY Each of the numbers in those pairs is a factor of 17.

Solution
The only pair of whole numbers whose product is 17 is:

$1 \cdot 17 = 17$

Therefore, the only factors of 17 are 1 and 17.

2 Identify even and odd whole numbers, prime numbers, and composite numbers.

A whole number is either *even* or *odd*.

Even and Odd Whole Numbers

If a whole number is divisible by 2, it is called an **even** number.

If a whole number is not divisible by 2, it is called an **odd** number.

The even whole numbers are the numbers

0, 2, 4, 6, 8, 10, 12, 14, 16, 18, …

The odd whole numbers are the numbers

1, 3, 5, 7, 9, 11, 13, 15, 17, 19, …

The three dots at the end of each list shown above indicate that there are infinitely many even and infinitely many odd whole numbers.

> **The Language of Algebra** The word *infinitely* is a form of the word *infinite*, meaning *unlimited*.

In Example 3, we saw that the only factors of 17 are 1 and 17. Numbers that have only two factors, 1 and the number itself, are called **prime numbers**.

Prime Numbers

A **prime number** is a whole number greater than 1 that has only 1 and itself as factors.

The prime numbers are the numbers:

2, 3, 5, 7, 11, 13, 17, 19, 23, 29, 31, 37, 41, 43, 47, 53, 59, 61, 67, 71, 73, 79, 83, 89, 97, 101, …

There are infinitely many prime numbers.

Note that the only even prime number is 2. Any other even whole number is divisible by 2, and thus has 2 as a factor, in addition to 1 and itself. Also note that not all odd whole numbers are prime numbers. For example, since 15 has factors of 1, 3, 5, and 15, it is not a prime number.

The set of whole numbers contains many prime numbers. It also contains many numbers that are not prime.

Composite Numbers

The **composite numbers** are whole numbers greater than 1 that are *not* prime.

The composite numbers are the numbers

4, 6, 8, 9, 10, 12, 14, 15, 16, 18, …

There are infinitely many composite numbers.

Caution! The numbers 0 and 1 are neither prime nor composite, because neither is a whole number greater than 1.

EXAMPLE 4   a. Is 37 a prime number?   b. Is 45 a prime number?

Strategy We will determine whether the given number has only 1 and itself as factors.

WHY If that is the case, it is a prime number.

Solution
a. Since 37 is a whole number greater than 1 and its only factors are 1 and 37, it is prime. Since 37 is not divisible by 2, we say it is an odd prime number.
b. The factors of 45 are 1, 3, 5, 9, 15, and 45. Since it has factors other than 1 and 45, 45 is *not* prime. It is an odd composite number.

Self Check 4
a. Is 39 a prime number? no
b. Is 57 a prime number? no
Now Try Problems 53 and 57

Teaching Example 4 a. Is 31 a prime number?
b. Is 55 a prime number?
Answers:
a. yes b. no

3 Find prime factorizations using a factor tree.

Every composite number can be formed by multiplying a specific combination of prime numbers. The process of finding that combination is called **prime factorization**.

Prime Factorization

To find the **prime factorization** of a whole number means to write it as the product of only prime numbers.

One method for finding the prime factorization of a number is called a **factor tree**. The factor trees shown below are used to find the prime factorization of 90 in two ways.

1. Factor 90 as 9 · 10.
2. Neither 9 nor 10 is prime, so we factor each of them.
3. The process is complete when only prime numbers appear at the bottom of all branches.

1. Factor 90 as 6 · 15.
2. Neither 6 nor 15 is prime, so we factor each of them.
3. The process is complete when only prime numbers appear at the bottom of all branches.

Either way, the prime factorization of 90 contains one factor of 2, two factors of 3, and one factor of 5. Writing the factors in order, from least to greatest, the **prime-factored form** of 90 is 2 · 3 · 3 · 5. It is true that no other combination of prime factors will produce 90. This example illustrates an important fact about composite numbers.

Fundamental Theorem of Arithmetic

Any composite number has exactly one set of prime factors.

Self Check 5

Use a factor tree to find the prime factorization of 126.

Now Try Problems 61 and 71

Self Check 5 Answer
2 · 3 · 3 · 7

Teaching Example 5 Use a factor tree to find the prime factorization of 140.
Answer:
2 · 2 · 5 · 7

EXAMPLE 5 Use a factor tree to find the prime factorization of 210.

Strategy We will factor each number that we encounter as a product of two whole numbers (other than 1 and itself) until *all the factors involved are prime*.

WHY The prime factorization of a whole number contains only prime numbers.

Solution

Factor 210 as 7 · 30. (The resulting prime factorization will be the same no matter which two factors of 210 you begin with.) Since 7 is prime, circle it. That branch of the tree is completed.

Since 30 is not prime, factor it as 5 · 6. (The resulting prime factorization will be the same no matter which two factors of 30 you use.) Since 5 is prime, circle it. That branch of the tree is completed.

Since 6 is not prime, factor it as 2 · 3. Since 2 and 3 are prime, circle them. All the branches of the tree are now completed.

The prime factorization of 210 is 7 · 5 · 2 · 3. Writing the prime factors in order, from least to greatest, we have 210 = 2 · 3 · 5 · 7.

Check: Multiply the prime factors. The product should be 210.

$$2 \cdot 3 \cdot 5 \cdot 7 = 6 \cdot 5 \cdot 7$$ Write the multiplication in horizontal form. Working left to right, multiply 2 and 3.

$$= 30 \cdot 7$$ Working left to right, multiply 6 and 5.

$$= 210$$ Multiply 30 and 7. The result checks.

Caution! Remember that there is a difference between the *factors* and the *prime factors* of a number. For example:

The factors of 15 are 1, 3, 5, and 15.

The prime factors of 15 are 3 and 5.

4 Find prime factorizations using a division ladder.

We can also find the prime factorization of a whole number using an inverted division process called a **division ladder**. It is called that because of the vertical "steps" that it produces.

Success Tip The divisibility rules found in Section 1.4 are helpful when using the division ladder method. You may want to review them at this time.

EXAMPLE 6 Use a division ladder to find the prime factorization of 280.

Strategy We will perform repeated divisions by prime numbers until the final quotient is itself a prime number.

WHY If a prime number is a factor of 280, it will divide 280 exactly.

Solution
It is helpful to begin with the *smallest prime*, 2, as the first trial divisor. Then, if necessary, try the primes 3, 5, 7, 11, 13, ... in that order.

Step 1 The prime number 2 divides 280 exactly.

$$2\underline{)280}$$
$$140$$

The result is 140, which is not prime. Continue the division process.

Step 2 Since 140 is even, divide by 2 again.

$$2\underline{)280}$$
$$2\underline{)140}$$
$$70$$

The result is 70, which is not prime. Continue the division process.

Step 3 Since 70 is even, divide by 2 a third time. The result is 35, which is not prime.

$$2\underline{)280}$$
$$2\underline{)140}$$
$$2\underline{)70}$$
$$35$$

Continue the division process.

Step 4 Since neither the prime number 2 nor the next greatest prime number 3 divide 35 exactly, we try 5. The result is 7, which is prime. We are done.

$$2\underline{)280}$$
$$2\underline{)140}$$
$$2\underline{)70}$$
$$5\underline{)35}$$
$$7 \leftarrow \text{Prime}$$

The prime factorization of 280 appears in the left column of the division ladder: $2 \cdot 2 \cdot 2 \cdot 5 \cdot 7$. Check this result using multiplication.

Self Check 6
Use a division ladder to find the prime factorization of 108.

Now Try Problems 63 and 73

Self Check 6 Answer
$2 \cdot 2 \cdot 3 \cdot 3 \cdot 3$

Teaching Example 6 Use a division ladder to find the prime factorization of 550.
Answer:
$2 \cdot 5 \cdot 5 \cdot 11$

Caution! In Example 6, it would be incorrect to begin the division process with

$$4\underline{)280}$$
$$70$$

because 4 is not a prime number.

5 Use exponential notation.

In Example 6, we saw that the prime factorization of 280 is $2 \cdot 2 \cdot 2 \cdot 5 \cdot 7$. Because this factorization has three factors of 2, we call 2 a *repeated factor*. We can use **exponential notation** to write $2 \cdot 2 \cdot 2$ in a more compact form.

> ### Exponent and Base
>
> An **exponent** is used to indicate repeated multiplication. It tells how many times the **base** is used as a factor.

The exponent is 3.

$$\underbrace{2 \cdot 2 \cdot 2}_{\text{Repeated factors}} = 2^3 \qquad \text{Read } 2^3 \text{ as "2 to the third power" or "2 cubed."}$$

The base is 2.

The prime factorization of 280 can be written using exponents: $2 \cdot 2 \cdot 2 \cdot 5 \cdot 7 = 2^3 \cdot 5 \cdot 7$.

In the **exponential expression** 2^3, the number 2 is the base and 3 is the exponent. The expression itself is called a **power of 2**.

Self Check 7

Write each product using exponents:

a. $3 \cdot 3 \cdot 7$ $3^2 \cdot 7$

b. $5(5)(7)(7)$ $5^2(7^2)$

c. $2 \cdot 2 \cdot 2 \cdot 3 \cdot 3 \cdot 5$ $2^3 \cdot 3^2 \cdot 5$

Now Try Problems 77 and 81

Teaching Example 7 Use exponents to write each prime factorization:
a. $7 \cdot 7$
b. $2 \cdot 2 \cdot 2 \cdot 19$
c. $3(3)(3)(3)(3)(5)(5)$
Answers:
a. 7^2
b. $2^3 \cdot 19$
c. $3^5(5^2)$

EXAMPLE 7

Write each product using exponents:

a. $5 \cdot 5 \cdot 5 \cdot 5$ **b.** $7 \cdot 7 \cdot 11$ **c.** $2(2)(2)(2)(3)(3)(3)$

Strategy We will determine the number of repeated factors in each expression.

WHY An exponent can be used to represent repeated multiplication.

Solution

a. The factor 5 is repeated 4 times. We can represent this repeated multiplication with an exponential expression having a base of 5 and an exponent of 4:

$$5 \cdot 5 \cdot 5 \cdot 5 = 5^4$$

b. $7 \cdot 7 \cdot 11 = 7^2 \cdot 11$ 7 is used as a factor 2 times.

c. $2(2)(2)(2)(3)(3)(3) = 2^4(3^3)$ 2 is used as a factor 4 times, and 3 is used as a factor 3 times.

6 Evaluate exponential expressions.

We can use the definition of exponent to **evaluate** (find the value of) exponential expressions.

Self Check 8

Evaluate each expression:

a. 9^2 81   **b.** 6^3 216

c. 3^4 81   **d.** 12^1 12

Now Try Problem 89

EXAMPLE 8

Evaluate each expression:

a. 7^2 **b.** 2^5 **c.** 10^4 **d.** 6^1

Strategy We will rewrite each exponential expression as a product of repeated factors, and then perform the multiplication. This requires that we identify the base and the exponent.

WHY The exponent tells the number of times the base is to be written as a factor.

Solution

We can write the steps of the solutions in horizontal form.

a. $7^2 = 7 \cdot 7$ Read 7^2 as "7 to the second power" or "7 squared." The base is 7 and the exponent is 2. Write the base as a factor 2 times.

$ = 49$ Multiply.

b. $2^5 = 2 \cdot 2 \cdot 2 \cdot 2 \cdot 2$ Read 2^5 as "2 to the 5th power." The base is 2 and the exponent is 5. Write the base as a factor 5 times.

$ = 4 \cdot 2 \cdot 2 \cdot 2$ Multiply, working left to right.

$ = 8 \cdot 2 \cdot 2$

$ = 16 \cdot 2$

$ = 32$

c. $10^4 = 10 \cdot 10 \cdot 10 \cdot 10$ Read 10^4 as "10 to the 4th power." The base is 10 and the exponent is 4. Write the base as a factor 4 times.

$ = 100 \cdot 10 \cdot 10$ Multiply, working left to right.

$ = 1{,}000 \cdot 10$

$ = 10{,}000$

d. $6^1 = 6$ Read 6^1 as "6 to the first power." Write the base 6 once.

Teaching Example 8 Evaluate each expression:
a. 3^5 b. 4^4
c. 5^3 d. 10^1
Answers:
a. 243   b. 256
c. 125   d. 10

Caution! Note that 2^5 means $2 \cdot 2 \cdot 2 \cdot 2 \cdot 2$. It does not mean $2 \cdot 5$. That is, $2^5 = 32$ and $2 \cdot 5 = 10$.

EXAMPLE 9

The prime factorization of a number is $2^3 \cdot 3^4 \cdot 5$. What is the number?

Strategy To find the number, we will evaluate each exponential expression and then do the multiplication.

WHY The exponential expressions must be evaluated first.

Solution
We can write the steps of the solutions in horizontal form.

$2^3 \cdot 3^4 \cdot 5 = 8 \cdot 81 \cdot 5$ Evaluate the exponential expressions: $2^3 = 8$ and $3^4 = 81$.

$ = 648 \cdot 5$ Multiply, working left to right.

$ = 3{,}240$ Multiply.

$2^2 \cdot 3^4 \cdot 5$ is the prime factorization of 3,240.

81
$\underline{\times8}$
648

$^{2\,4}$
648
$\underline{\times5}$
$3{,}240$

Self Check 9
The prime factorization of a number is $2 \cdot 3^3 \cdot 5^2$. What is the number? 1,350

Now Try Problems 93 and 97

Teaching Example 9 The prime factorization of a number is $2^2 \cdot 3^3 \cdot 7$. What is the number?
Answer:
756

Success Tip Calculations that you cannot perform in your head should be shown outside the steps of your solution.

Using Your CALCULATOR
The Exponential Key

At the end of 1 hour, a culture contains two bacteria. Suppose the number of bacteria doubles every hour thereafter. Use exponents to determine how many bacteria the culture will contain after 24 hours.

We can use a table to help model the situation. From the table, we see a pattern developing: The number of bacteria in the culture after 24 hours will be 2^{24}.

Time	Number of bacteria
1 hr	$2 = 2^1$
2 hr	$4 = 2^2$
3 hr	$8 = 2^3$
4 hr	$16 = 2^4$
24 hr	$? = 2^{24}$

Unless otherwise noted, all art on this page is © Cengage Learning.

We can evaluate this exponential expression using the exponential key y^x on a scientific calculator (x^y on some models).

2 y^x 24 =   16777216

On a graphing calculator, we use the carat key ^ to raise a number to a power.

2 ^ 24 ENTER   16777216

Since $2^{24} = 16,777,216$, there will be 16,777,216 bacteria after 24 hours.

ANSWERS TO SELF CHECKS

1. 1, 2, 4, 5, 10, and 20   **2. a.** $1 \cdot 18$, $2 \cdot 9$, or $3 \cdot 6$ **b.** Two possibilities are $2 \cdot 3 \cdot 3$ and $1 \cdot 2 \cdot 9$ **3.** 1 and 23   **4. a.** no   **b.** no   **5.** $2 \cdot 3 \cdot 3 \cdot 7$ **6.** $2 \cdot 2 \cdot 3 \cdot 3 \cdot 3$ **7. a.** $3^2 \cdot 7$ **b.** $5^2(7^2)$ **c.** $2^3 \cdot 3^2 \cdot 5$ **8. a.** 81   **b.** 216   **c.** 81   **d.** 12   **9.** 1,350

SECTION 1.5   STUDY SET

VOCABULARY

Fill in the blanks.

1. Numbers that are multiplied together are called __factors__.

2. To __factor__ a whole number means to express it as the product of other whole numbers.

3. A __prime__ number is a whole number greater than 1 that has only 1 and itself as factors.

4. Whole numbers greater than 1 that are not prime numbers are called __composite__ numbers.

5. To prime factor a number means to write it as a product of only __prime__ numbers.

6. An exponent is used to represent __repeated__ multiplication. It tells how many times the __base__ is used as a factor.

7. In the exponential expression 6^4, the number 6 is the __base__, and 4 is the __exponent__.

8. We can read 5^2 as "5 to the second power" or as "5 __squared__." We can read 7^3 as "7 to the third power" or as "7 __cubed__."

CONCEPTS

9. Fill in the blanks to find the pairs of whole numbers whose product is 45.

     $1 \cdot$ __45__ $= 45$ $3 \cdot$ __15__ $= 45$ $5 \cdot$ __9__ $= 45$

The factors of 45, in order from least to greatest, are: __1__, __3__, __5__, __9__, __15__, __45__

10. Fill in the blanks to find the pairs of whole numbers whose product is 28.

     $1 \cdot$ __28__ $= 28$ $2 \cdot$ __14__ $= 28$ $4 \cdot$ __7__ $= 28$

The factors of 28, in order from least to greatest, are: __1__, __2__, __4__, __7__, __14__, __28__

11. If 4 is a factor of a whole number, will 4 divide the number exactly? yes

12. Suppose a number is divisible by 10. Is 10 a factor of the number? yes

13. **a.** Fill in the blanks: If a whole number is divisible by 2, it is an __even__ number. If it is not divisible by 2, it is an __odd__ number.

     b. List the first 10 even whole numbers.
         0, 2, 4, 6, 8, 10, 12, 14, 16, 18

     c. List the first 10 odd whole numbers.
         1, 3, 5, 7, 9, 11, 13, 15, 17, 19

14. **a.** List the first 10 prime numbers.
         2, 3, 5, 7, 11, 13, 17, 19, 23, 29

     b. List the first 10 composite numbers.
         4, 6, 8, 9, 10, 12, 14, 15, 16, 18

15. Fill in the blanks to prime factor 150 using a factor tree.

The prime factorization of 150 is __2__ · __3__ · __5__ · __5__.

16. Which of the whole numbers, 1, 2, 3, 4, 5, 6, 7, 8, 9, and 10, could be at the top of this factor tree? 4, 6, 9, 10

17. Fill in the blanks to prime factor 150 using a division ladder.

```
2 | 150
  3 | 75
    5 | 25
       5
```

The prime factorization of 150 is 2 · 3 · 5 · 5.

18. a. When using the division ladder method to find the prime factorization of a number, what is the first divisor to try? 2

b. If 2 does not divide the given number exactly, what other divisors should be tried?
3, 5, 7, 11, 13, …

NOTATION

19. For each exponential expression, what is the base and the exponent?
  a. 7^6 **b.** 15^1
  base: 7, exponent: 6   base: 15, exponent: 1

20. Consider the expression 2 · 2 · 2 · 3 · 3.
  a. How many repeated factors of 2 are there? 3
  b. How many repeated factors of 3 are there? 2

GUIDED PRACTICE

Find the factors of each whole number. List them from least to greatest. See Example 1.

21. 10 1, 2, 5, 10
22. 6 1, 2, 3, 6
23. 40 1, 2, 4, 5, 8, 10, 20, 40
24. 75 1, 3, 5, 15, 25, 75
25. 18 1, 2, 3, 6, 9, 18
26. 32 1, 2, 4, 8, 16, 32
27. 44 1, 2, 4, 11, 22, 44
28. 65 1, 5, 13, 65
29. 77 1, 7, 11, 77
30. 81 1, 3, 9, 27, 81
31. 100 1, 2, 4, 5, 10, 20, 25, 50, 100
32. 441 1, 3, 7, 9, 21, 49, 63, 147, 441

Factor each of the following whole numbers using two factors. Do not use the factor 1 in your answer. See Example 2.

33. 8 2 · 4
34. 9 3 · 3
35. 27 3 · 9
36. 35 5 · 7
37. 49 7 · 7
38. 25 5 · 5
39. 20 2 · 10 or 4 · 5
40. 16 2 · 8 or 4 · 4

Factor each of the following whole numbers using three factors. Do not use the factor 1 in your answer. See Example 2.

41. 30 2 · 3 · 5
42. 28 2 · 2 · 7
43. 63 3 · 3 · 7
44. 50 2 · 5 · 5
45. 54 2 · 3 · 9 or 3 · 3 · 6
46. 56 2 · 4 · 7 or 2 · 2 · 14
47. 60 2 · 3 · 10 or 2 · 2 · 15 or 2 · 5 · 6 or 3 · 4 · 5
48. 64 2 · 4 · 8 or 2 · 2 · 16 or 4 · 4 · 4

Find the factors of each whole number. See Example 3.

49. 11 1 and 11
50. 29 1 and 29
51. 37 1 and 37
52. 41 1 and 41

Determine whether each of the following numbers is a prime number. See Example 4.

53. 17 yes
54. 59 yes
55. 99 no, (9 · 11)
56. 27 no, (3 · 9)
57. 51 no, (3 · 17)
58. 91 no, (7 · 13)
59. 43 yes
60. 83 yes

Find the prime factorization of each number. Use exponents in your answer, when it is helpful. See Examples 5 and 6.

61. 30 2 · 3 · 5
62. 20 2 · 2 · 5 or 2^2 · 5
63. 39 3 · 13
64. 105 3 · 5 · 7
65. 99 3 · 3 · 11 or 3^2 · 11
66. 400 2 · 2 · 2 · 2 · 5 · 5 or 2^4 · 5^2
67. 162 2 · 3 · 3 · 3 · 3 or 2 · 3^4
68. 98 2 · 7 · 7 or 2 · 7^2
69. 64 2 · 2 · 2 · 2 · 2 · 2 or 2^6
70. 243 3 · 3 · 3 · 3 · 3 or 3^5
71. 147 3 · 7 · 7 or 3 · 7^2
72. 140 2 · 2 · 5 · 7 or 2^2 · 5 · 7
73. 220 2 · 2 · 5 · 11 or 2^2 · 5 · 11
74. 385 5 · 7 · 11
75. 102 2 · 3 · 17
76. 114 2 · 3 · 19

Write each product using exponents. See Example 7.

77. 2 · 2 · 2 · 2 · 2 2^5
78. 3 · 3 · 3 · 3 · 3 · 3 3^6
79. 5 · 5 · 5 · 5 5^4
80. 9 · 9 · 9 9^3
81. 4(4)(8)(8)(8) $4^2(8^3)$
82. 12(12)(12)(16) $12^3(16)$
83. 7 · 7 · 7 · 9 · 9 · 7 · 7 · 7 · 7 $7^7 · 9^2$
84. 6 · 6 · 6 · 5 · 5 · 6 · 6 · 6 $6^6 · 5^2$

Evaluate each exponential expression. See Example 8.

85. a. 3^4 81   **b.** 4^3 64
86. a. 5^3 125   **b.** 3^5 243
87. a. 2^5 32   **b.** 5^2 25
88. a. 4^5 1,024   **b.** 5^4 625
89. a. 7^3 343   **b.** 3^7 2,187
90. a. 8^2 64   **b.** 2^8 256
91. a. 9^1 9   **b.** 1^9 1
92. a. 20^1 20   **b.** 1^{20} 1

The prime factorization of a number is given. What is the number? See Example 9.

93. 2 · 3 · 3 · 5 90
94. 2 · 2 · 2 · 7 56
95. 7 · 11^2 847
96. 2 · 3^4 162

97. $3^2 \cdot 5^2$ 225   **98.** $3^3 \cdot 5^3$ 3,375
99. $2^3 \cdot 3^3 \cdot 13$ 2,808   **100.** $2^3 \cdot 3^2 \cdot 11$ 792

APPLICATIONS

101. PERFECT NUMBERS A whole number is called a **perfect number** when the sum of its factors that are less than the number equals the number. For example, 6 is a perfect number, because $1 + 2 + 3 = 6$. Find the factors of 28. Then use addition to show that 28 is also a perfect number.
1, 2, 4, 7, 14, 28, $1 + 2 + 4 + 7 + 14 = 28$

102. CRYPTOGRAPHY Information is often transmitted in code. Many codes involve writing products of large primes, because they are difficult to factor. To see how difficult, try finding two prime factors of 7,663. (*Hint:* Both primes are greater than 70.) 79, 97

103. LIGHT The illustration shows that the light energy that passes through the first unit of area, 1 yard away from the bulb, spreads out as it travels away from the source. How much area does that light energy cover 2 yards, 3 yards, and 4 yards from the bulb? Express each answer using exponents.
2^2 square units, 3^2 square units, 4^2 square units

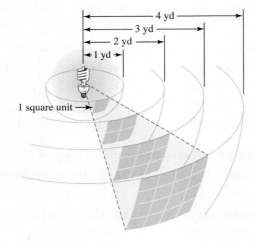

104. CELL DIVISION After 1 hour, a cell has divided to form another cell. In another hour, these two cells have divided so that four cells exist. In another hour, these four cells divide so that eight exist.
  a. How many cells exist at the end of the fourth hour? 16
  b. The number of cells that exist after each division can be found using an exponential expression. What is the base? 2
  c. Find the number of cells after 12 hours. 4,096

WRITING

105. Explain how to check a prime factorization.

106. Explain the difference between the *factors* of a number and the *prime factors* of a number. Give an example.

107. Find 1^2, 1^3, and 1^4. From the results, what can be said about any power of 1?

108. Use the phrase *infinitely many* in a sentence.

REVIEW

109. MARCHING BANDS When a university band lines up in 8 rows of 15 musicians, there are 5 musicians left over. How many band members are there? 125 band members

110. U.S. COLLEGE COSTS In 2013, the average yearly cost for tuition and fees at a private four-year college was $29,056. The average yearly cost for tuition and fees at a public four-year college for in-state students was $8,655. At these rates, how much less are the tuition costs and fees at a public college over four years? (Source: The College Board) $81,604

Objectives

1. Find the LCM by listing multiples.
2. Find the LCM using prime factorization.
3. Find the GCF by listing factors.
4. Find the GCF using prime factorization.

SECTION 1.6
The Least Common Multiple and the Greatest Common Factor

As a child, you probably learned how to count by 2's and 5's and 10's. Counting in that way is an example of an important concept in mathematics called *multiples*.

1 Find the LCM by listing multiples.

The **multiples** of a number are the products of that number and 1, 2, 3, 4, 5, and so on.

EXAMPLE 1   Find the first eight multiples of 6.

Strategy We will multiply 6 by 1, 2, 3, 4, 5, 6, 7, and 8.

WHY The *multiples of a number* are the products of that number and 1, 2, 3, 4, 5, and so on.

Solution
To find the multiples, we proceed as follows:

$6 \cdot 1 = 6$ This is the first multiple of 6.
$6 \cdot 2 = 12$
$6 \cdot 3 = 18$
$6 \cdot 4 = 24$
$6 \cdot 5 = 30$
$6 \cdot 6 = 36$
$6 \cdot 7 = 42$
$6 \cdot 8 = 48$ This is the eighth multiple of 6.

The first eight multiples of 6 are 6, 12, 18, 24, 30, 36, 42, and 48.

Self Check 1
Find the first eight multiples of 9.
Now Try Problems 17 and 85
Self Check 1 Answer
9, 18, 27, 36, 45, 54, 63, 72

Teaching Example 1 Find the first eight multiples of 7.
Answer:
7, 14, 21, 28, 35, 42, 49, 56

The first eight multiples of 3 and the first eight multiples of 4 are shown below. The numbers highlighted in red are *common multiples* of 3 and 4.

$3 \cdot 1 = 3$ $4 \cdot 1 = 4$
$3 \cdot 2 = 6$ $4 \cdot 2 = 8$
$3 \cdot 3 = 9$ $4 \cdot 3 = \mathbf{12}$
$3 \cdot 4 = \mathbf{12}$ $4 \cdot 4 = 16$
$3 \cdot 5 = 15$ $4 \cdot 5 = 20$
$3 \cdot 6 = 18$ $4 \cdot 6 = \mathbf{24}$
$3 \cdot 7 = 21$ $4 \cdot 7 = 28$
$3 \cdot 8 = \mathbf{24}$ $4 \cdot 8 = 32$

If we extend each list, it soon becomes apparent that 3 and 4 have infinitely many common multiples.

The common multiples of 3 and 4 are: **12, 24, 36, 48, 60, 72,** …

Because 12 is the smallest number that is a multiple of both 3 and 4, it is called the **least common multiple (LCM)** of 3 and 4. We can write this in compact form as:

LCM (3, 4) = 12   Read as "The least common multiple of 3 and 4 is 12."

The Least Common Multiple (LCM)

The **least common multiple** of two whole numbers is the smallest common multiple of the numbers.

We have seen that the LCM of 3 and 4 is 12. It is important to note that 12 is divisible by both 3 and 4.

$$\frac{12}{3} = 4 \quad \text{and} \quad \frac{12}{4} = 3$$

This observation illustrates an important relationship between divisibility and the least common multiple.

The Least Common Multiple (LCM)

The **least common multiple (LCM)** of two whole numbers is the smallest whole number that is divisible by both of those numbers.

When finding the LCM of two numbers, writing both lists of multiples can be tiresome. From the previous definition of LCM, it follows that we need only list the multiples of the larger number. The LCM is simply *the first multiple of the larger number that is divisible by the smaller number.* For example, to find the LCM of 3 and 4, we observe that

The multiples of 4 are: 4, 8, 12, 16, 20, 24, …

4 is not divisible by 3. 8 is not divisible by 3. 12 is divisible by 3.

Recall that one number is *divisible* by another if, when dividing them, we get a remainder of 0.

Since 12 is the first multiple of 4 that is divisible by 3, the LCM of 3 and 4 is 12. As expected, this is the same result that we obtained using the two-list method.

Finding the LCM by Listing the Multiples of the Largest Number

To find the least common multiple of two (or more) whole numbers:

1. Write multiples of the largest number by multiplying it by 1, 2, 3, 4, 5, and so on.
2. Continue this process until you find the first multiple of the larger number that is divisible by each of the smaller numbers. That multiple is their LCM.

Self Check 2

Find the LCM of 8 and 10. 40

Now Try Problem 25

Teaching Example 2 Find the LCM of 6 and 10.
Answer:
30

EXAMPLE 2 Find the LCM of 6 and 8.

Strategy We will write the multiples of the larger number, 8, until we find one that is divisible by the smaller number, 6.

WHY The LCM of 6 and 8 is the smallest multiple of 8 that is divisible by 6.

Solution

The 1st multiple of 8: $8 \cdot 1 = 8$ ← 8 is not divisible by 6. (When we divide, we get a remainder of 2.) Since 8 is not divisible by 6, find the next multiple.

The 2nd multiple of 8: $8 \cdot 2 = 16$ ← 16 is not divisible by 6. Find the next multiple.

The 3rd multiple of 8: $8 \cdot 3 = 24$ ← 24 is divisible by 6. This is the LCM.

The first multiple of 8 that is divisible by 6 is 24. Thus,

LCM (6, 8) = 24   Read as "The least common multiple of 6 and 8 is 24."

We can extend this method to find the LCM of three whole numbers.

Self Check 3

Find the LCM of 3, 4, and 8. 24

Now Try Problem 35

EXAMPLE 3 Find the LCM of 2, 3, and 10.

Strategy We will write the multiples of the largest number, 10, until we find one that is divisible by both of the smaller numbers, 2 and 3.

WHY The LCM of 2, 3, and 10 is the smallest multiple of 10 that is divisible by 2 and 3.

Solution

The 1st multiple of 10: $10 \cdot 1 = 10$ ← 10 is divisible by 2, but not by 3. Find the next multiple.

The 2nd multiple of 10: $10 \cdot 2 = 20$ ← 20 is divisible by 2, but not by 3. Find the next multiple.

The 3rd multiple of 10: $10 \cdot 3 = 30$ ← 30 is divisible by 2 and by 3. It is the LCM.

The first multiple of 10 that is divisible by 2 and 3 is 30. Thus,

LCM $(2, 3, 10) = 30$ Read as "The least common multiple of 2, 3, and 10 is 30."

Teaching Example 3 Find the LCM of 3, 5, and 20.
Answer:
60

2 Find the LCM using prime factorization.

Another method for finding the LCM of two (or more) whole numbers uses prime factorization. This method is especially helpful when working with larger numbers. As an example, we will find the LCM of 36 and 54. First, we find their prime factorizations:

$36 = 2 \cdot 2 \cdot 3 \cdot 3$ Factor trees (or division ladders) can be used to find the prime factorizations.

$54 = 2 \cdot 3 \cdot 3 \cdot 3$

The LCM of 36 and 54 must be divisible by 36 and 54. If the LCM is divisible by 36, it must have the prime factors of 36, which are $2 \cdot 2 \cdot 3 \cdot 3$. If the LCM is divisible by 54, it must have the prime factors of 54, which are $2 \cdot 3 \cdot 3 \cdot 3$. The smallest number that meets both requirements is

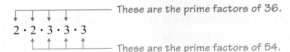

$2 \cdot 2 \cdot 3 \cdot 3 \cdot 3$

These are the prime factors of 36.

These are the prime factors of 54.

To find the LCM, we perform the indicated multiplication:

LCM $(36, 54) = 2 \cdot 2 \cdot 3 \cdot 3 \cdot 3 = 108$

> **Caution!** The LCM (36, 54) is not the product of the prime factorization of 36 and the prime factorization of 54. That gives an incorrect answer of 1,944.
>
> LCM $(36, 54) = 2 \cdot 2 \cdot 3 \cdot 3 \cdot 2 \cdot 3 \cdot 3 \cdot 3 = 1{,}944$
>
> The LCM should contain all the prime factors of 36 and all the prime factors of 54, but the prime factors that 36 and 54 have in common are not repeated.

The prime factorizations of 36 and 54 contain the numbers 2 and 3.

$36 = 2 \cdot 2 \cdot 3 \cdot 3 \qquad 54 = 2 \cdot 3 \cdot 3 \cdot 3$

We see that

- The greatest number of times the factor 2 appears in any one of the prime factorizations is twice and the LCM of 36 and 54 has 2 as a factor twice.
- The greatest number of times that 3 appears in any one of the prime factorizations is three times and the LCM of 36 and 54 has 3 as a factor three times.

These observations suggest a procedure to use to find the LCM of two (or more) numbers using prime factorization.

> ### Finding the LCM Using Prime Factorization
>
> To find the least common multiple of two (or more) whole numbers:
>
> 1. Prime factor each number.
> 2. The LCM is a product of prime factors, where each factor is used the greatest number of times it appears in any one factorization.

Self Check 4

Find the LCM of 18 and 32. 288

Now Try Problem 37

Teaching Example 4 Find the LCM of 21 and 45.
Answer:
315

EXAMPLE 4   Find the LCM of 24 and 60.

Strategy We will begin by finding the prime factorizations of 24 and 60.

WHY To find the LCM, we need to determine the greatest number of times each prime factor appears in any one factorization.

Solution

Step 1 Prime factor 24 and 60.

$24 = 2 \cdot 2 \cdot 2 \cdot 3$ Division ladders (or factor trees) can be used to find the prime factorizations.

$60 = 2 \cdot 2 \cdot 3 \cdot 5$

Step 2 The prime factorizations of 24 and 60 contain the prime factors 2, 3, and 5. To find the LCM, we use each of these factors the greatest number of times it appears in any one factorization.

- We will use the factor 2 three times, because 2 appears three times in the factorization of 24. Circle $2 \cdot 2 \cdot 2$, as shown below.
- We will use the factor 3 once, because it appears one time in the factorization of 24 and one time in the factorization of 60. When the number of times a factor appears are equal, circle either one, but not both, as shown below.
- We will use the factor 5 once, because it appears one time in the factorization of 60. Circle the 5, as shown below.

$24 = \boxed{2 \cdot 2 \cdot 2} \cdot \boxed{3}$
$60 = 2 \cdot 2 \cdot 3 \cdot \boxed{5}$

Since there are no other prime factors in either prime factorization, we have

LCM (24, 60) = $2 \cdot 2 \cdot 2 \cdot 3 \cdot 5 = 120$

(Use 2 three times. Use 3 one time. Use 5 one time.)

Note that 120 is the smallest number that is divisible by both 24 and 60:

$\dfrac{120}{24} = 5$ and $\dfrac{120}{60} = 2$

In Example 4, we can express the prime factorizations of 24 and 60 using exponents. To determine the greatest number of times each factor appears in any one factorization, we circle the factor with the greatest exponent.

$24 = \boxed{2^3} \cdot \boxed{3^1}$ The greatest exponent on the factor 2 is 3.
The greatest exponent on the factor 3 is 1.

$60 = 2^2 \cdot 3^1 \cdot \boxed{5^1}$ The greatest exponent on the factor 5 is 1.

The LCM of 24 and 60 is

$2^3 \cdot 3^1 \cdot 5^1 = 8 \cdot 3 \cdot 5 = 120$ Evaluate: $2^3 = 8$.

EXAMPLE 5 Find the LCM of 28, 42, and 45.

Strategy We will begin by finding the prime factorizations of 28, 42, and 45.

WHY To find the LCM, we need to determine the greatest number of times each prime factor appears in any one factorization.

Solution

Step 1 Prime factor 28, 42, and 45.

$28 = \boxed{2 \cdot 2} \cdot 7$ This can be written as $\boxed{2^2} \cdot 7^1$.
$42 = 2 \cdot 3 \cdot \boxed{7}$ This can be written as $2^1 \cdot 3^1 \cdot \boxed{7^1}$.
$45 = \boxed{3 \cdot 3} \cdot \boxed{5}$ This can be written as $\boxed{3^2} \cdot \boxed{5}$.

Step 2 The prime factorizations of 28, 42, and 45 contain the prime factors 2, 3, 5, and 7. To find the LCM (28, 42, 45), we use each of these factors the greatest number of times it appears in any one factorization.

- We will use the factor 2 two times, because 2 appears two times in the factorization of 28. Circle 2 · 2, as shown above.
- We will use the factor 3 twice, because it appears two times in the factorization of 45. Circle 3 · 3, as shown above.
- We will use the factor 5 once, because it appears one time in the factorization of 45. Circle the 5, as shown above.
- We will use the factor 7 once, because it appears one time in the factorization of 28 and one time in the factorization of 42. You may circle either 7, but only circle one of them.

Since there are no other prime factors in either prime factorization, we have

Use the factor 2 two times.
Use the factor 3 two times.
Use the factor 5 one time.
Use the factor 7 one time.

LCM (28, 42, 45) = $2 \cdot 2 \cdot 3 \cdot 3 \cdot 5 \cdot 7 = 1{,}260$

If we use exponents, we have

LCM (28, 42, 45) = $2^2 \cdot 3^2 \cdot 5 \cdot 7 = 1{,}260$

Either way, we have found that the LCM (28, 42, 45) = 1,260. Note that 1,260 is the smallest number that is divisible by 28, 42, and 45:

$\dfrac{1{,}260}{28} = 45$ $\dfrac{1{,}260}{42} = 30$ $\dfrac{1{,}260}{45} = 28$

Self Check 5
Find the LCM of 45, 60, and 75. 900

Now Try Problem 45

Teaching Example 5 Find the LCM of 20, 70, and 98.
Answer:
980

EXAMPLE 6 *Patient Recovery* Two patients recovering from heart surgery exercise daily by walking around a track. One patient can complete a lap in 4 minutes. The other can complete a lap in 6 minutes. If they begin at the same time and at the same place on the track, in how many minutes will they arrive together at the starting point of their workout?

Self Check 6

AQUARIUMS A pet store owner changes the water in a fish aquarium every 45 days and he changes the pump filter every 20 days. If the water and filter are changed on the same day, in how many days will they be changed again together? 180 days

Now Try Problem 87

Teaching Example 6 CLOCKS In a clock store, a grandfather clock chimes every 15 minutes and a cuckoo clock chirps every 20 minutes. If both clocks sound at the same time, in how many minutes will they sound again together?
Answer:
60 min

Strategy We will find the LCM of 4 and 6.

WHY Since one patient reaches the starting point of the workout every 4 minutes, and the other is there every 6 minutes, we want to find the least common multiple of those numbers. At that time, they will both be at the starting point of the workout.

Solution
To find the LCM, we prime factor 4 and 6, and circle each prime factor the greatest number of times it appears in any one factorization.

$4 = \boxed{2 \cdot 2}$ Use the factor 2 two times, because 2 appears two times in the factorization of 4.

$6 = 2 \cdot \boxed{3}$ Use the factor 3 once, because it appears one time in the factorization of 6.

Since there are no other prime factors in either prime factorization, we have

$$\text{LCM}(4, 6) = 2 \cdot 2 \cdot 3 = 12$$

The patients will arrive together at the starting point 12 minutes after beginning their workout.

3 Find the GCF by listing factors.

We have seen that two whole numbers can have common multiples. They can also have *common factors*. To explore this concept, let's find the factors of 26 and 39 and see what factors they have in common.

To find the factors of 26, we find all the pairs of whole numbers whose product is 26. There are two possibilities:

$$1 \cdot 26 = 26 \qquad 2 \cdot 13 = 26$$

Each of the numbers in the pairs is a factor of 26. From least to greatest, the factors of 26 are 1, 2, 13, and 26.

To find the factors of 39, we find all the pairs of whole numbers whose product is 39. There are two possibilities:

$$1 \cdot 39 = 39 \qquad 3 \cdot 13 = 39$$

Each of the numbers in the pairs is a factor of 39. From least to greatest, the factors of 39 are 1, 3, 13, and 39. As shown below, the *common factors* of 26 and 39 are 1 and 13.

$\boxed{1}, 2, \boxed{13}, 26$ These are the factors of 26.
$\boxed{1}, 3, \boxed{13}, 39$ These are the factors of 39.

Because 13 is the largest number that is a factor of both 26 and 39, it is called the **greatest common factor (GCF)** of 26 and 39. We can write this in compact form as:

GCF (26, 39) = 13 Read as "The greatest common factor of 26 and 39 is 13."

The Greatest Common Factor (GCF)

The **greatest common factor** of two whole numbers is the largest common factor of the numbers.

Self Check 7

Find the GCF of 30 and 42. 6

Now Try Problem 49

EXAMPLE 7 Find the GCF of 18 and 45.

Strategy We will find the factors of 18 and 45.

WHY Then we can identify the largest factor that 18 and 45 have in common.

Solution

To find the factors of 18, we find all the pairs of whole numbers whose product is 18. There are three possibilities:

$1 \cdot 18 = 18 \qquad 2 \cdot 9 = 18 \qquad 3 \cdot 6 = 18$

To find the factors of 45, we find all the pairs of whole numbers whose product is 45. There are three possibilities:

$1 \cdot 45 = 45 \qquad 3 \cdot 15 = 45 \qquad 5 \cdot 9 = 45$

The factors of 18 and 45 are listed below. Their common factors are circled.

Factors of 18: 1, 2, 3, 6, 9, 18
Factors of 45: 1, 3, 5, 9, 15, 45

The common factors of 18 and 45 are 1, 3, and 9. Since 9 is their largest common factor,

GCF (18, 45) = 9   *Read as "The greatest common factor of 18 and 45 is 9."*

In Example 7, we found that the GCF of 18 and 45 is 9. Note that 9 is the greatest number that divides 18 and 45.

$$\frac{18}{9} = 2 \qquad \frac{45}{9} = 5$$

In general, the greatest common factor of two (or more) numbers is the largest number that divides them exactly. For this reason, the greatest common factor is also known as the **greatest common divisor (GCD)** and we can write GCD (18, 45) = 9.

Teaching Example 7 Find the GCF of 42 and 70.
Answer:
14

4 Find the GCF using prime factorization.

We can find the GCF of two (or more) numbers by listing the factors of each number. However, this method can be lengthy. Another way to find the GCF uses the prime factorization of each number.

Finding the GCF Using Prime Factorization

To find the greatest common factor of two (or more) whole numbers:

1. Prime factor each number.
2. Identify the common prime factors.
3. The GCF is a product of all the common prime factors found in Step 2. If there are no common prime factors, the GCF is 1.

EXAMPLE 8   Find the GCF of 48 and 72.

Strategy We will begin by finding the prime factorizations of 48 and 72.

WHY Then we can identify any prime factors that they have in common.

Solution

Step 1 Prime factor 48 and 72.

$48 = 2 \cdot 2 \cdot 2 \cdot 2 \cdot 3$
$72 = 2 \cdot 2 \cdot 2 \cdot 3 \cdot 3$

Self Check 8
Find the GCF of 36 and 60. 12
Now Try Problem 57

Teaching Example 8 Find the GCF of 45 and 105.
Answer:
15

Step 2 The circling on the previous page shows that 48 and 72 have four common prime factors: Three common factors of 2 and one common factor of 3.

Step 3 The GCF is the product of the circled prime factors.

$$\text{GCF}(48, 72) = 2 \cdot 2 \cdot 2 \cdot 3 = 24$$

Self Check 9
Find the GCF of 8 and 25. 1

Now Try Problem 61

Teaching Example 9 Find the GCF of 9 and 16.
Answer:
1

EXAMPLE 9 Find the GCF of 8 and 15.

Strategy We will begin by finding the prime factorizations of 8 and 15.

WHY Then we can identify any prime factors that they have in common.

Solution
The prime factorizations of 8 and 15 are shown below.

$$8 = 2 \cdot 2 \cdot 2$$
$$15 = 3 \cdot 5$$

Since there are no common factors, the GCF of 8 and 15 is 1. Thus,

$$\text{GCF}(8, 15) = 1 \quad \text{Read as "The greatest common factor of 8 and 15 is 1."}$$

Self Check 10
Find the GCF of 45, 60, and 75. 15

Now Try Problem 67

Teaching Example 10 Find the GCF of 42, 63, and 84.
Answer:
21

EXAMPLE 10 Find the GCF of 20, 60, and 140.

Strategy We will begin by finding the prime factorizations of 20, 60, and 140.

WHY Then we can identify any prime factors that they have in common.

Solution
The prime factorizations of 20, 60, and 140 are shown below.

$$20 = 2 \cdot 2 \cdot 5$$
$$60 = 2 \cdot 2 \cdot 3 \cdot 5$$
$$140 = 2 \cdot 2 \cdot 5 \cdot 7$$

The circling above shows that 20, 60, and 140 have three common factors: two common factors of 2 and one common factor of 5. The GCF is the product of the circled prime factors.

$$\text{GCF}(20, 60, 140) = 2 \cdot 2 \cdot 5 = 20 \quad \text{Read as "The greatest common factor of 20, 60, and 140 is 20."}$$

Note that 20 is the greatest number that divides 20, 60, and 140 exactly.

$$\frac{20}{20} = 1 \qquad \frac{60}{20} = 3 \qquad \frac{140}{20} = 7$$

Self Check 11
SCHOOL SUPPLIES A bookstore manager wants to use some leftover items (36 markers, 54 pencils, and 108 pens) to make identical gift packs to donate to an elementary school.

a. What is the greatest number of gift packs that can be made? *(continued)*

EXAMPLE 11 *Bouquets* A florist wants to use 12 white tulips, 30 pink tulips, and 42 purple tulips to make as many identical arrangements as possible. Each bouquet is to have the same number of each color tulip.

a. What is the greatest number of arrangements that she can make?
b. How many of each type of tulip can she use in each bouquet?

Strategy We will find the GCF of 12, 30, and 42.

WHY Since an equal number of tulips of each color will be used to create the identical arrangements, division is indicated. The greatest common factor of three numbers is the largest number that divides them exactly.

Solution

a. To find the GCF, we prime factor 12, 30, and 42, and circle the prime factors that they have in common.

$$12 = \boxed{2} \cdot 2 \cdot \boxed{3}$$
$$30 = \boxed{2} \cdot \boxed{3} \cdot 5$$
$$42 = \boxed{2} \cdot \boxed{3} \cdot 7$$

The GCF is the product of the circled numbers.

$$\text{GCF } (12, 30, 42) = 2 \cdot 3 = 6$$

The florist can make 6 identical arrangements from the tulips.

b. To find the number of white, pink, and purple tulips in each of the 6 arrangements, we divide the number of tulips of each color by 6.

White tulips: Pink tulips: Purple tulips:

$$\frac{12}{6} = 2 \qquad \frac{30}{6} = 5 \qquad \frac{42}{6} = 7$$

Each of the 6 identical arrangements will contain 2 white tulips, 5 pink tulips, and 7 purple tulips.

b. How many of each type of item will be in each gift pack?

Now Try Problem 93

Self Check 11 Answer
a. 18 gift packs
b. 2 markers, 3 pencils, 6 pens

Teaching Example 11 FANTASY CAMPS Equally balanced rock bands are to be formed from the 42 guitarists, 14 drummers, and 28 singers attending a fantasy rock-n-roll camp.
a. What is the greatest number of bands that can be formed?
b. How many guitarists, drummers, and singers will be in each band?
Answers:
a. 14 bands
b. 3 guitarists, 1 drummer, 2 singers

ANSWERS TO SELF CHECKS
1. 9, 18, 27, 36, 45, 54, 63, 72 **2.** 40 **3.** 24 **4.** 288 **5.** 900 **6.** 180 days **7.** 6
8. 12 **9.** 1 **10.** 15 **11. a.** 18 gift packs **b.** 2 markers, 3 pencils, 6 pens

---

SECTION 1.6 STUDY SET

VOCABULARY

Fill in the blanks.

1. The __multiples__ of a number are the products of that number and 1, 2, 3, 4, 5, and so on.

2. Because 12 is the smallest number that is a multiple of both 3 and 4, it is the __least__ __common__ __multiple__ of 3 and 4.

3. One number is __divisible__ by another if, when dividing them, we get a remainder of 0.

4. Because 6 is the largest number that is a factor of both 18 and 24, it is the __greatest__ __common__ __factor__ of 18 and 24.

CONCEPTS

5. **a.** The LCM of 4 and 6 is 12. What is the smallest whole number divisible by 4 and 6? 12

   b. Fill in the blank: In general, the LCM of two whole numbers is the __smallest__ whole number that is divisible by both numbers.

6. **a.** What are the common multiples of 2 and 3 that appear in the list of multiples shown in the next column? 6, 12

   b. What is the LCM of 2 and 3? 6

Multiples of 2	Multiples of 3
$2 \cdot 1 = 2$	$3 \cdot 1 = 3$
$2 \cdot 2 = 4$	$3 \cdot 2 = 6$
$2 \cdot 3 = 6$	$3 \cdot 3 = 9$
$2 \cdot 4 = 8$	$3 \cdot 4 = 12$
$2 \cdot 5 = 10$	$3 \cdot 5 = 15$
$2 \cdot 6 = 12$	$3 \cdot 6 = 18$

7. **a.** The first six multiples of 5 are 5, 10, 15, 20, 25, and 30. What is the first multiple of 5 that is divisible by 4? 20

   b. What is the LCM of 4 and 5? 20

8. Fill in the blanks to complete the prime factorization of 24.

9. The prime factorizations of 36 and 90 are:

    $36 = 2 \cdot 2 \cdot 3 \cdot 3$
    $90 = 2 \cdot 3 \cdot 3 \cdot 5$

    What is the greatest number of times
    a. 2 appears in any one factorization? two
    b. 3 appears in any one factorization? two
    c. 5 appears in any one factorization? one
    d. Fill in the blanks to find the LCM of 36 and 90:
       LCM = 2 · 2 · 3 · 3 · 5 = 180

10. The prime factorizations of 14, 70, and 140 are:

    $14 = 2 \cdot 7$
    $70 = 2 \cdot 5 \cdot 7$
    $140 = 2 \cdot 2 \cdot 5 \cdot 7$

    What is the greatest number of times
    a. 2 appears in any one factorization? two
    b. 5 appears in any one factorization? one
    c. 7 appears in any one factorization? one
    d. Fill in the blanks to find the LCM of 14, 70, and 140:
       LCM = 2 · 2 · 5 · 7 = 140

11. The prime factorizations of 12 and 54 are:

    $12 = 2^2 \cdot 3^1$
    $54 = 2^1 \cdot 3^3$

    What is the greatest number of times
    a. 2 appears in any one factorization? two
    b. 3 appears in any one factorization? three
    c. Fill in the blanks to find the LCM of 12 and 54:
       LCM = $2^2 \cdot 3^3$ = 108

12. The factors of 18 and 45 are shown below.

    Factors of 18:   1, 2, 3, 6, 9, 18
    Factors of 45:   1, 3, 5, 9, 15, 45

    a. Circle the common factors of 18 and 45. 1, 3, 9
    b. What is the GCF of 18 and 45? 9

13. The prime factorizations of 60 and 90 are:

    $60 = 2 \cdot 2 \cdot 3 \cdot 5$
    $90 = 2 \cdot 3 \cdot 3 \cdot 5$

    a. Circle the common prime factors of 60 and 90. 2, 3, 5
    b. What is the GCF of 60 and 90? 30

14. The prime factorizations of 36, 84, and 132 are:

    $36 = 2 \cdot 2 \cdot 3 \cdot 3$
    $84 = 2 \cdot 2 \cdot 3 \cdot 7$
    $132 = 2 \cdot 2 \cdot 3 \cdot 11$

    a. Circle the common factors of 36, 84, and 132. 2, 2, 3
    b. What is the GCF of 36, 84, and 132? 12

NOTATION

Fill in the blanks.

15. a. The abbreviation for the greatest common factor is GCF .
    b. The abbreviation for the least common multiple is LCM .

16. a. We read LCM (2, 15) = 30 as "The least   commmon multiple of 2 and 15 is 30."
    b. We read GCF (18, 24) = 6 as "The greatest   common factor of 18 and 24 is 6."

GUIDED PRACTICE

Find the first eight multiples of each number. See Example 1.

17. 4
    4, 8, 12, 16, 20, 24, 28, 32
18. 2
    2, 4, 6, 8, 10, 12, 14, 16
19. 11
    11, 22, 33, 44, 55, 66, 77, 88
20. 10
    10, 20, 30, 40, 50, 60, 70, 80
21. 8
    8, 16, 24, 32, 40, 48, 56, 64
22. 9
    9, 18, 27, 36, 45, 54, 63, 72
23. 20
    20, 40, 60, 80, 100, 120, 140, 160
24. 30
    30, 60, 90, 120, 150, 180, 210, 240

Find the LCM of the given numbers. See Example 2.

25. 3, 5 15
26. 6, 9 18
27. 8, 12 24
28. 10, 25 50
29. 5, 11 55
30. 7, 11 77
31. 4, 7 28
32. 5, 8 40

Find the LCM of the given numbers. See Example 3.

33. 3, 4, 6 12
34. 2, 3, 8 24
35. 2, 3, 10 30
36. 3, 6, 15 30

Find the LCM of the given numbers. See Example 4.

37. 16, 20 80
38. 14, 21 42
39. 30, 50 150
40. 21, 27 189
41. 35, 45 315
42. 36, 48 144
43. 100, 120 600
44. 120, 180 360

Find the LCM of the given numbers. See Example 5.

45. 6, 24, 36 72
46. 6, 10, 18 90
47. 5, 12, 15 60
48. 8, 12, 16 48

Find the GCF of the given numbers. See Example 7.

49. 4, 6 2
50. 6, 15 3
51. 9, 12 3
52. 10, 12 2

Find the GCF of the given numbers. See Example 8.

53. 22, 33 11
54. 14, 21 7
55. 15, 30 15
56. 15, 75 15
57. 18, 96 6
58. 30, 48 6
59. 28, 42 14
60. 63, 84 21

Find the GCF of the given numbers. See Example 9.

61. 16, 51 1
62. 27, 64 1
63. 81, 125 1
64. 57, 125 1

Find the GCF of the given numbers. See Example 10.

65. 12, 68, 92 4
66. 24, 36, 40 4
67. 72, 108, 144 36
68. 81, 108, 162 27

TRY IT YOURSELF

Find the LCM and the GCF of the given numbers.

69. 100, 120 600; 20
70. 120, 180 360; 60
71. 14, 140 140; 14
72. 15, 300 300; 15
73. 66, 198, 242 2,178; 22
74. 52, 78, 130 780; 26
75. 8, 9, 49 3,528; 1
76. 9, 16, 25 3,600; 1
77. 120, 125 3,000; 5
78. 98, 102 4,998; 2
79. 34, 68, 102 204; 34
80. 26, 39, 65 390; 13
81. 46, 69 138; 23
82. 38, 57 114; 19
83. 50, 81 4,050; 1
84. 65, 81 5,265; 1

APPLICATIONS

85. OIL CHANGES Ford has officially extended the oil change interval for 2007 and newer cars to every 7,500 miles. (It used to be every 5,000 miles.) Complete the table below that shows Ford's new recommended oil change mileages.

1st oil change	2nd oil change	3rd oil change	4th oil change	5th oil change	6th oil change
7,500 mi	15,000 mi	22,500 mi	30,000 mi	37,500 mi	45,000 mi

86. ATMs An ATM offers the customer cash withdrawal choices in multiples of $20. The minimum withdrawal is $20 and the maximum is $200. List the dollar amounts of cash that can be withdrawn from the ATM. $20, $40, $60, $80, $100, $120, $140, $160, $180, $200

87. NURSING A nurse is instructed to check a patient's blood pressure every 45 minutes and another is instructed to take the same patient's temperature every 60 minutes. If both nurses are in the patient's room now, how long will it be until the nurses are together in the room once again? 180 min or 3 hr

88. BIORHYTHMS Some scientists believe that there are natural rhythms of the body, called *biorhythms*, that affect our physical, emotional, and mental cycles. Our physical biorhythm cycle lasts 23 days, the emotional biorhythm cycle lasts 28 days, and our mental biorhythm cycle lasts 33 days. Each biorhythm cycle has a high, low, and critical zone. If your three cycles are together one day, all at their lowest point, in how many more days will they be together again, all at their lowest point? 21,252 days

89. PICNICS A package of hot dogs usually contains 10 hot dogs and a package of buns usually contains 12 buns. How many packages of hot dogs and buns should a person buy to be sure that there are equal numbers of each? 6 packages of hot dogs and 5 packages of buns

90. WORKING COUPLES A husband works for 6 straight days and then has a day off. His wife works for 7 straight days and then has a day off. If the husband and wife are both off from work on the same day, in how many days will they both be off from work again? 42 days

91. DANCE FLOORS A dance floor is to be made from rectangular sheets of plywood that are 6 feet by 8 feet. What is the minimum number of full sheets of plywood that are needed to make a square dance floor, as shown below? 12 pieces

92. BOWLS OF SOUP Each of the bowls shown below holds an exact number of *full ladles* of soup.

a. If there is no spillage, what is the greatest-size ladle (in ounces) that a chef can use to fill all three bowls? 3 ounces

b. How many ladles will it take to fill each bowl?
1st bowl: 4 ladles, 2nd bowl: 7 ladles, 3rd bowl: 6 ladles

93. **ART CLASSES** Students in a painting class must pay an art supplies fee. On the first day of class, the instructor collected a total of $28 in fees from several students. On the second day she collected $21 more from some other students, and on the third day she collected another $63 from some more students.

   a. What is the most the art supplies fee could cost per student? $7

   b. Use your answer in part a to determine how many students paid the art supplies fee each day.
   1st day: 4 students, 2nd day: 3 students, 3rd day: 9 students

94. **SHIPPING** A toy manufacturer needs to ship 135 brown teddy bears, 105 black teddy bears, and 30 white teddy bears. They can pack only one type of teddy bear in each box, and they must pack the same number of teddy bears in each box. What is the greatest number of teddy bears they can pack in each box? 15 bears

WRITING

95. Explain how to find the LCM of 8 and 28 using prime factorization.

96. Explain how to find the GCF of 8 and 28 using prime factorization.

97. The prime factorization of 12 is 2 · 2 · 3 and the prime factorization of 15 is 3 · 5. Explain why the LCM of 12 and 15 is *not* 2 · 2 · 3 · 3 · 5.

98. How can you tell by looking at the prime factorizations of two whole numbers that their GCF is 1?

REVIEW

Perform each operation.

99. 9,999 + 1,111   11,110   100. 10,000 − 7,989   2,011
101. 305 · 50   15,250   102. 2,100 ÷ 105   20

Objectives

1. Use the order of operations rule.
2. Evaluate expressions containing grouping symbols.
3. Find the mean (average) of a set of values.

SECTION 1.7
Order of Operations

Recall that numbers are combined with the operations of addition, subtraction, multiplication, and division to create **expressions**. We often have to **evaluate** (find the value of) expressions that involve more than one operation. In this section, we introduce an order of operations rule to follow in such cases.

1 Use the order of operations rule.

Suppose you are asked to contact a friend if you see a Rolex watch for sale while you are traveling in Europe. While in Switzerland, you find the watch and send the following text message, shown on the left. The next day, you get the response shown on the right from your friend.

You sent this message.

You get this response.

Unless otherwise noted, all art on this page is © Cengage Learning.

Something is wrong. The first part of the response (No price too high!) says to buy the watch at any price. The second part (No! Price too high.) says not to buy it, because it's too expensive. The placement of the exclamation point makes us read the two parts of the response differently, resulting in different meanings. When reading a mathematical statement, the same kind of confusion is possible. For example, consider the expression

$2 + 3 \cdot 6$

We can evaluate this expression in two ways. We can add first, and then multiply. Or we can multiply first, and then add. However, the results are different.

$2 + 3 \cdot 6 = 5 \cdot 6$ Add 2 and 3 first. $2 + 3 \cdot 6 = 2 + 18$ Multiply 3 and 6 first.
$ = 30$ Multiply 5 and 6. $ = 20$ Add 2 and 18.

— Different results —

Every numerical expression has only one correct value. If we don't establish a uniform order of operations, the expression $2 + 3 \cdot 6$ has two different values. To avoid this possibility, always use the following order of operations rule.

Order of Operations

1. Perform all calculations within parentheses and other grouping symbols following the order listed in Steps 2–4 below, working from the innermost pair of grouping symbols to the outermost pair.
2. Evaluate all exponential expressions.
3. Perform all multiplications and divisions as they occur from left to right.
4. Perform all additions and subtractions as they occur from left to right.

When grouping symbols have been removed, repeat Steps 2–4 to complete the calculation.

If a fraction bar is present, evaluate the expression above the bar (called the **numerator**) and the expression below the bar (called the **denominator**) separately. Then perform the division indicated by the fraction bar, if possible.

It isn't necessary to apply all of these steps in every problem. For example, the expression $2 + 3 \cdot 6$ does not contain any parentheses, and there are no exponential expressions. So we look for multiplications and divisions to perform and proceed as follows:

$2 + 3 \cdot 6 = 2 + 18$ Do the multiplication first.
$ = 20$ Do the addition.

EXAMPLE 1   Evaluate: $2 \cdot 4^2 - 8$

Strategy We will scan the expression to determine what operations need to be performed. Then we will perform those operations, one at a time, following the order of operations rule.

WHY Every numerical expression has only one correct value. If we don't follow the correct order of operations, the expression can have more than one value.

Solution
Since the expression does not contain any parentheses, we begin with Step 2 of the order of operations rule: Evaluate all exponential expressions. We will write the steps of the solution in horizontal form.

Self Check 1
Evaluate: $4 \cdot 3^3 - 6$ 102

Now Try Problem 19

Teaching Example 1 Evaluate: $3 \cdot 5^2 - 11$
Answer:
64

$$2 \cdot 4^2 - 8 = 2 \cdot 16 - 8 \quad \text{Evaluate the exponential expression: } 4^2 = 16.$$
$$= 32 - 8 \quad \text{Do the multiplication: } 2 \cdot 16 = 32.$$
$$= 24 \quad \text{Do the subtraction.}$$

$$\begin{array}{r} \overset{1}{16} \\ \times\ 2 \\ \hline 32 \end{array}$$

$$\begin{array}{r} \overset{2\ 12}{\cancel{3}\cancel{2}} \\ -\ 8 \\ \hline 24 \end{array}$$

Success Tip Calculations that you cannot perform in your head should be shown outside the steps of your solution.

Self Check 2

Evaluate: $60 - 2 \cdot 3 + 22$ 76

Now Try Problem 23

Teaching Example 2 Evaluate:
$50 - 7 \cdot 3 + 9$
Answer:
38

EXAMPLE 2 Evaluate: $80 - 3 \cdot 2 + 16$

Strategy We will perform the multiplication first.

WHY The expression does not contain any parentheses, nor are there any exponents.

Solution
We will write the steps of the solution in horizontal form.

$$80 - 3 \cdot 2 + 16 = 80 - 6 + 16 \quad \text{Do the multiplication: } 3 \cdot 2 = 6.$$
$$= 74 + 16 \quad \text{Working from left to right, do the subtraction: } 80 - 6 = 74.$$
$$= 90 \quad \text{Do the addition.}$$

$$\begin{array}{r} \overset{1}{74} \\ +\ 16 \\ \hline 90 \end{array}$$

Caution! In Example 2, a common mistake is to forget to work from left to right and *incorrectly* perform the addition before the subtraction. This error produces the wrong answer, 58.

$$80 - 3 \cdot 2 + 16 = 80 - 6 + 16$$
$$= 80 - 22$$
$$= 58$$

Remember to perform additions and subtractions *in the order in which they occur*. The same is true for multiplications and divisions.

Self Check 3

Evaluate: $144 \div 9 + 4(2)3$ 40

Now Try Problem 27

Teaching Example 3 Evaluate:
$156 \div 6 + 5(7)2$
Answer:
96

EXAMPLE 3 Evaluate: $192 \div 6 - 5(3)2$

Strategy We will perform the division first.

WHY Although the expression contains parentheses, there are no calculations to perform *within* them. Since there are no exponents, we perform multiplications and divisions as they occur from left to right.

Solution
We will write the steps of the solution in horizontal form.

$$192 \div 6 - 5(3)2 = 32 - 5(3)2 \quad \text{Working from left to right, do the division: } 192 \div 6 = 32.$$
$$= 32 - 15(2) \quad \text{Working from left to right, do the multiplication: } 5(3) = 15.$$
$$= 32 - 30 \quad \text{Complete the multiplication: } 15(2) = 30.$$
$$= 2 \quad \text{Do the subtraction.}$$

$$\begin{array}{r} 32 \\ 6\overline{)192} \\ -\ 18 \\ \hline 12 \\ -\ 12 \\ \hline 0 \end{array}$$

EXAMPLE 4   Long-Distance Calls

The rates that Skype charges for overseas landline calls from the United States are shown on the right. A newspaper editor in Washington, D.C., made a 60-minute call to Iraq, a 45-minute call to Panama, and a 30-minute call to Vietnam. What was the total cost of the calls?

Landline calls
All rates are per minute.

	Afghanistan	37¢
	Iraq	39¢
	Haiti	27¢
	Panama	11¢
	Russia	5¢
	Vietnam	34¢

Includes tax

Solution It is helpful to list the given facts and what you are to find.

- The 60-minute call to Iraq costs 39 cents per minute.   Given
- The 45-minute call to Panama costs 11 cents per minute.   Given
- The 30-minute call to Vietnam costs 34 cents per minute. Given
- What is the total cost of the calls?   Find

Now we translate the words of the problem to numbers and symbols. We can find the cost of each call by multiplying the length of the call (in minutes) by the rate charged per minute (in cents). Since the word *total* indicates addition, we will add to find the total cost of the calls.

The total cost of the calls	is equal to	the cost of the call to Iraq	plus	the cost of the call to Panama	plus	the cost of the call to Vietnam.
The total cost of the calls	=	60(39)	+	45(11)	+	30(34)

To evaluate this expression, we apply the order of operations rule.

$$\begin{aligned} \text{The total cost of the calls} &= 60(39) + 45(11) + 30(34) &&\text{The units are cents.} \\ &= 2{,}340 + 495 + 1{,}020 &&\text{Do the multiplication first.} \\ &= 3{,}855 &&\text{Do the addition.} \end{aligned}$$

$$\begin{array}{r} \overset{1}{2{,}340} \\ 495 \\ +\ 1{,}020 \\ \hline 3{,}855 \end{array}$$

The total cost of the overseas calls is 3,855¢, or $38.55.

Check: We can check the result by finding an estimate using front-end rounding. The total cost of the calls is approximately 60(40¢) + 50(10¢) + 30(30¢) = 2,400¢ + 500¢ + 900¢ or 3,800¢. The result of 3,855¢ seems reasonable.

Self Check 4

LONG-DISTANCE CALLS A newspaper reporter in Chicago made a 90-minute call to Afghanistan, a 25-minute call to Haiti, and a 55-minute call to Russia. What was the total cost of the calls? 4,280¢ = $42.80

Now Try Problem 105

Teaching Example 4
LONG-DISTANCE CALLS A South American tourist official, with an office in the United States, made a 35-minute call to Bolivia (at 13¢ per minute), a 15-minute call to Colombia (at 6¢ per minute), and a 45-minute call to Brazil (at 3¢ per minute). What was the total cost of the calls?
Answer:
680¢ = $6.80

2 Evaluate expressions containing grouping symbols.

Grouping symbols determine the order in which an expression is evaluated. Examples of grouping symbols are parentheses ( ), brackets [ ], braces { }, and the fraction bar —.

EXAMPLE 5   Evaluate each expression: **a.** 12 − 3 + 5   **b.** 12 − (3 + 5)

Strategy To evaluate the expression in part a, we will perform the subtraction first. To evaluate the expression in part b, we will perform the addition first.

WHY The similar-looking expression in part b is evaluated in a different order because it contains parentheses. Any operations within parentheses must be performed first.

Self Check 5

Evaluate each expression:

a. 20 − 7 + 6   19

b. 20 − (7 + 6)   7

Now Try Problem 33

Unless otherwise noted, all art on this page is © Cengage Learning.

Teaching Example 5 Evaluate:
a. $39 - 15 + 1$
b. $39 - (15 + 1)$
Answers:
a. 25
b. 23

Solution
a. The expression does not contain any parentheses, nor are there any exponents, nor any multiplication or division. We perform the additions and subtractions as they occur, from left to right.

$$12 - 3 + 5 = 9 + 5 \quad \text{Do the subtraction: } 12 - 3 = 9.$$
$$= 14 \quad \text{Do the addition.}$$

b. By the order of operations rule, we must perform the operation within the parentheses first.

$$12 - (3 + 5) = 12 - 8 \quad \text{Do the addition: } 3 + 5 = 8. \text{ Read as "12 minus the quantity of 3 plus 5."}$$
$$= 4 \quad \text{Do the subtraction.}$$

> **The Language of Algebra** When we read the expression $12 - (3 + 5)$ as "12 minus the *quantity* of 3 plus 5," the word *quantity* alerts the reader to the parentheses that are used as grouping symbols.

Self Check 6

Evaluate: $(1 + 3)^4$ 256

Now Try Problem 35

Teaching Example 6
Evaluate: $(2 + 3)^3$
Answer:
125

EXAMPLE 6 Evaluate: $(2 + 6)^3$

Strategy We will perform the operation within the parentheses first.

WHY This is the first step of the order of operations rule.

Solution
$$(2 + 6)^3 = 8^3 \quad \text{Read as "The cube of the quantity of 2 plus 6." Do the addition.}$$
$$= 512 \quad \text{Evaluate the exponential expression: } 8^3 = 8 \cdot 8 \cdot 8 = 512.$$

$$\begin{array}{r} \overset{3}{6}4 \\ \times\ 8 \\ \hline 512 \end{array}$$

Self Check 7

Evaluate: $50 - 4(12 - 5 \cdot 2)$ 42

Now Try Problem 39

Teaching Example 7 Evaluate:
$8 + 3(27 - 4 \cdot 4)$
Answer:
41

EXAMPLE 7 Evaluate: $5 + 2(13 - 5 \cdot 2)$

Strategy We will perform the multiplication within the parentheses first.

WHY When there is more than one operation to perform within parentheses, we follow the order of operations rule. Multiplication is to be performed before subtraction.

Solution
We apply the order of operations rule within the parentheses to evaluate $13 - 5 \cdot 2$.

$$5 + 2(13 - 5 \cdot 2) = 5 + 2(13 - 10) \quad \text{Do the multiplication within the parentheses.}$$
$$= 5 + 2(3) \quad \text{Do the subtraction within the parentheses.}$$
$$= 5 + 6 \quad \text{Do the multiplication: } 2(3) = 6.$$
$$= 11 \quad \text{Do the addition.}$$

Some expressions contain two or more sets of grouping symbols. Since it can be confusing to read an expression such as $16 + 6(4^2 - 3(5 - 2))$, we use a pair of **brackets** in place of the outer pair of parentheses.

$$16 + 6[4^2 - 3(5 - 2)]$$

If an expression contains more than one pair of grouping symbols, we always begin by working within the **innermost pair** and then work to the **outermost pair**.

Innermost parentheses
$$16 + 6[4^2 - 3(5 - 2)]$$
Outermost brackets

> **The Language of Algebra** Multiplication is indicated when a number is next to a parenthesis or a bracket. For example,
> $$16 + 6[4^2 - 3(5 - 2)]$$
> Multiplication   Multiplication

EXAMPLE 8
Evaluate: $16 + 6[4^2 - 3(5 - 2)]$

Strategy We will work within the parentheses first and then within the brackets. Within each set of grouping symbols, we will follow the order of operations rule.

WHY By the order of operations, we must work from the *innermost* pair of grouping symbols to the *outermost*.

Solution

$$16 + 6[4^2 - 3(5 - 2)] = 16 + 6[4^2 - 3(3)]$$ Do the subtraction within the parentheses.
$$= 16 + 6[16 - 3(3)]$$ Evaluate the exponential expression: $4^2 = 16$.
$$= 16 + 6[16 - 9]$$ Do the multiplication within the brackets.
$$= 16 + 6[7]$$ Do the subtraction within the brackets.
$$= 16 + 42$$ Do the multiplication: $6[7] = 42$.
$$= 58$$ Do the addition.

Self Check 8
Evaluate:
$130 - 7[2^2 + 3(6 - 2)]$ 18
Now Try Problem 43

Teaching Example 8 Evaluate:
$40 - 3[5^2 - 7(9 - 6)]$
Answer:
28

> **Caution!** In Example 8, a common mistake is to *incorrectly* add 16 and 6 instead of *correctly* multiplying 6 and 7 first. This error produces a wrong answer, 154.
>
> $16 + 6[4^2 - 3(5 - 2)] = 16 + 6[4^2 - 3(3)]$
> $= 16 + 6[16 - 3(3)]$
> $= 16 + 6[16 - 9]$
> $= 16 + 6[7]$
> $= \cancel{22[7]}$
> $= \cancel{154}$

EXAMPLE 9
Evaluate: $\dfrac{2(13) - 2}{3(2^3)}$

Strategy We will evaluate the expression above and the expression below the fraction bar separately. Then we will do the indicated division, if possible.

WHY Fraction bars are grouping symbols. They group the numerator and denominator. The expression could be written $[2(13) - 2] \div [3(2^3)]$.

Self Check 9
Evaluate: $\dfrac{3(14) - 6}{2(3^2)}$ 2
Now Try Problem 47

Teaching Example 9 Evaluate:
$$\frac{10(15) - 42}{2(3^3)}$$
Answer:
2

Solution

$$\frac{2(13) - 2}{3(2^3)} = \frac{26 - 2}{3(8)}$$ In the numerator, do the multiplication. In the denominator, evaluate the exponential expression within the parentheses.

$$= \frac{24}{24}$$ In the numerator, do the subtraction. In the denominator, do the multiplication.

$$= 1$$ Do the division indicated by the fraction bar: $24 \div 24 = 1$.

3 Find the mean (average) of a set of values.

The **mean** (sometimes called the **arithmetic mean** or **average**) of a set of numbers is a value around which the values of the numbers are grouped. It gives you an indication of the "center" of the set of numbers. To find the mean of a set of numbers, we must apply the order of operations rule.

> **Finding the Mean**
>
> To find the mean (average) of a set of values, divide the sum of the values by the number of values.

Self Check 10

NFL DEFENSIVE LINEMEN The weights of the 2011–2012 New York Giants starting defensive linemen were 268 lb, 324 lb, 318 lb, and 278 lb. What was their mean (average) weight? (Source: nfl.com) 297 lb

Now Try Problems 51 and 113

Teaching Example 10 PRO BASKETBALL The starting lineup for the 2011–2012 NBA champion Miami Heat was: LeBron James (250 lb), Dwayne Wade (220 lb), Chris Bosh (235 lb), Shane Battier (225 lb), and Mario Chalmers (190 lb). What was their mean (average) weight? (Source: nba.com)
Answer:
224 lb

EXAMPLE 10 *NFL Offensive Linemen* The weights of the 2011–2012 Super Bowl champion New York Giants starting offensive linemen are shown below. What was their mean (average) weight?

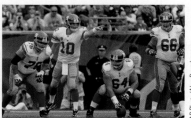

| Left tackle #66 D. Diehl 305 lb | Left guard #77 K. Boothe 320 lb | Center #64 D. Baas 312 lb | Right guard #76 C. Snee 306 lb | Right tackle #67 K. McKenzie 327 lb |

(Source: nfl.com/New York Giants depth chart)

Strategy We will add 305, 320, 312, 306, and 327 and divide the sum by 5.

WHY To find the mean (average) of a set of values, we divide the sum of the values by the number of values.

Solution
Since there are 5 weights, divide the sum by 5.

$$\text{Mean} = \frac{305 + 320 + 312 + 306 + 327}{5}$$

$$\text{Mean} = \frac{1{,}570}{5}$$ In the numerator, do the addition.

$$\text{Mean} = 314$$ Do the indicated division: $1{,}570 \div 5$.

```
    2
   305
   320
   312
   306
 + 327
 1,570
```

```
     314
 5)1,570
    -15
      7
     -5
     20
    -20
      0
```

In 2011–2012, the mean (average) weight of the starting offensive linemen on the New York Giants was 314 pounds.

Using Your CALCULATOR   Order of Operations and Parentheses

Calculators have the rules for order of operations built in. A left parenthesis key (and a right parenthesis key) should be used when grouping symbols, including a fraction bar, are needed. For example, to evaluate $\frac{240}{20-5}$, the parentheses keys must be used, as shown below.

240 ÷ (20 − 5) =   16

On some calculator models, the ENTER key is pressed instead of = for the result to be displayed.

If the parentheses are not entered, the calculator will find 240 ÷ 20 and then subtract 5 from that result, to produce the wrong answer, 7.

THINK IT THROUGH   Education Pays

"Education does pay. It has a high rate of return for students from all racial/ethnic groups, for men and for women, and for those from all family backgrounds. It also has a high rate of return for society."

The College Board, Trends in Higher Education *Series*

Attending school requires an investment of time, effort, and sacrifice. Is it all worth it? The graph below shows how average weekly earnings in the U.S. increase as the level of education increases. Begin at the bottom of the graph and work upward. Use the given clues to determine each of the missing weekly earnings amounts.

Average earnings per week in 2011

Doctoral degree	$114 decrease ?
Professional degree	$402 increase ?
Master's degree	$210 increase ?
Bachelor's degree	$265 increase ?
Associate degree	$49 increase ?
Some college, no degree	$81 increase ?
High-school graduate	$187 increase ?
Less than a high school diploma	**$451 per week**

Source: Bureau of Labor Statistics $638, $719, $768, $1,033, $1,243, $1,645, $1,531

ANSWERS TO SELF CHECKS

1. 102   **2.** 76   **3.** 40   **4.** 4,280¢ = $42.80   **5. a.** 19   **b.** 7   **6.** 256   **7.** 42   **8.** 18
9. 2   **10.** 297 lb

Unless otherwise noted, all art on this page is © Cengage Learning.

SECTION 1.7 STUDY SET

VOCABULARY

Fill in the blanks.

1. Numbers are combined with the operations of addition, subtraction, multiplication, and division to create expressions .

2. To *evaluate* the expression $2 + 5 \cdot 4$ means to find its value .

3. The grouping symbols ( ) are called parentheses , and the symbols [ ] are called brackets .

4. The rule for the order of operations guarantees that an evaluation of a numerical expression will result in a single answer.

5. In the expression $9 + 6[8 + 6(4 - 1)]$, the parentheses are the inner most grouping symbols and the brackets are the outer most grouping symbols.

6. To find the mean (average) of a set of values, we add the values and divide by the number of values.

CONCEPTS

7. List the operations in the order in which they should be performed to evaluate each expression. *You do not have to evaluate the expression.*

   a. $5(2)^2 - 1$ square, multiply, subtract
   b. $15 + 90 - (2 \cdot 2)^3$ multiply, cube, add, subtract
   c. $7 \cdot 4^2$ square, multiply
   d. $(7 \cdot 4)^2$ multiply, square

8. List the operations in the order in which they should be performed to evaluate each expression. *You do not have to evaluate the expression.*

   a. $50 + 8 - 40$ add, subtract
   b. $50 - 40 + 8$ subtract, add
   c. $16 \cdot 2 \div 4$ multiply, divide
   d. $16 \div 4 \cdot 2$ divide, multiply

9. Consider the expression $\dfrac{5 + 5(7)}{(5 \cdot 20 - 8^2) - 28}$. In the numerator, what operation should be performed first? In the denominator, what operation should be performed first? multiply, square

10. To find the mean (average) of 15, 33, 45, 12, 6, 19, and 3, we add the values and divide by what number? 7

NOTATION

11. In the expression $\dfrac{60 - 5 \cdot 2}{5 \cdot 2 + 40}$, what symbol serves as a grouping symbol? What does it group?
    the fraction bar, the numerator and the denominator

12. Use brackets to write $2(12 - (5 + 4))$ in clearer form. $2[12 - (5 + 4)]$

Fill in the blanks.

13. We read the expression $16 - (4 + 9)$ as "16 minus the quantity of 4 plus 9."

14. We read the expression $(8 - 3)^3$ as "The cube of the quantity of 8 minus 3."

Complete the steps to evaluate each expression.

15. $7 \cdot 4 - 5(2)^2 = 7 \cdot 4 - 5(\,4\,)$
    $= 28 - 20$
    $= 8$

16. $2 + (5 + 6 \cdot 2) = 2 + (5 + 12)$
    $= 2 + 17$
    $= 19$

17. $[4(2 + 7)] - 4^2 = [4(\,9\,)] - 4^2$
    $= 36 - 4^2$
    $= 36 - 16$
    $= 20$

18. $\dfrac{12 + 5 \cdot 3}{3^2 - 2 \cdot 3} = \dfrac{12 + 15}{9 - 6}$
    $= \dfrac{27}{3}$
    $= 9$

GUIDED PRACTICE

Evaluate each expression. See Example 1.

19. $3 \cdot 5^2 - 28$ 47
20. $4 \cdot 2^2 - 11$ 5
21. $6 \cdot 3^2 - 41$ 13
22. $5 \cdot 4^2 - 32$ 48

Evaluate each expression. See Example 2.

23. $52 - 6 \cdot 3 + 4$ 38
24. $66 - 8 \cdot 7 + 16$ 26
25. $32 - 9 \cdot 3 + 31$ 36
26. $62 - 5 \cdot 8 + 27$ 49

Evaluate each expression. See Example 3.

27. $192 \div 4 - 4(2)3$ 24
28. $455 \div 7 - 3(4)5$ 5
29. $252 \div 3 - 6(2)6$ 12
30. $264 \div 4 - 7(4)2$ 10

Evaluate each expression. See Example 5.

31. a. $26 - 2 + 9$ 33
    b. $26 - (2 + 9)$ 15
32. a. $37 - 4 + 11$ 44
    b. $37 - (4 + 11)$ 22
33. a. $51 - 16 + 8$ 43
    b. $51 - (16 + 8)$ 27
34. a. $73 - 35 + 9$ 47
    b. $73 - (35 + 9)$ 29

Evaluate each expression. See Example 6.

35. $(4 + 6)^2$ 100
36. $(3 + 4)^2$ 49
37. $(3 + 5)^3$ 512
38. $(5 + 2)^3$ 343

Evaluate each expression. See Example 7.

39. $8 + 4(29 - 5 \cdot 3)$ 64
40. $33 + 6(56 - 9 \cdot 6)$ 45
41. $77 + 9(38 - 4 \cdot 6)$ 203
42. $162 + 7(47 - 6 \cdot 7)$ 197

Evaluate each expression. See Example 8.

43. $46 + 3[5^2 - 4(9 - 5)]$ 73
44. $53 + 5[6^2 - 5(8 - 1)]$ 58
45. $81 + 9[7^2 - 7(11 - 4)]$ 81
46. $81 + 3[8^2 - 7(13 - 5)]$ 105

Evaluate each expression. See Example 9.

47. $\dfrac{2(50) - 4}{2(4^2)}$ 3
48. $\dfrac{4(34) - 1}{5(3^2)}$ 3
49. $\dfrac{25(8) - 8}{6(2^3)}$ 4
50. $\dfrac{6(31) - 26}{4(2^3)}$ 5

Find the mean (average) of each list of numbers. See Example 10.

51. 6, 9, 4, 3, 8   6
52. 7, 1, 8, 2, 2   4
53. 3, 5, 9, 1, 7, 5   5
54. 8, 7, 7, 2, 4, 8   6
55. 19, 15, 17, 13   16
56. 11, 14, 12, 11   12
57. 5, 8, 7, 0, 3, 1   4
58. 9, 3, 4, 11, 14, 1   7

TRY IT YOURSELF

Evaluate each expression, if possible.

59. $(8 - 6)^2 + (4 - 3)^2$ 5
60. $(2 + 1)^2 + (3 + 2)^2$ 34
61. $2 \cdot 3^4$ 162
62. $3^3 \cdot 5$ 135
63. $7 + 4 \cdot 5$ 27
64. $6 \cdot 8 - 3$ 45
65. $(7 - 4)^2 + 1$ 10
66. $(9 - 5)^3 + 8$ 72
67. $\dfrac{10 + 5}{52 - 47}$ 3
68. $\dfrac{18 + 12}{61 - 55}$ 5
69. $5 \cdot 10^3 + 2 \cdot 10^2 + 3 \cdot 10^1 + 9$ 5,239
70. $8 \cdot 10^3 + 0 \cdot 10^2 + 7 \cdot 10^1 + 4$ 8,074
71. $20 - 10 + 5$ 15
72. $80 - 5 + 4$ 79
73. $25 \div 5 \cdot 5$ 25
74. $6 \div 2 \cdot 3$ 9
75. $150 - 2(2 \cdot 6 - 4)^2$ 22
76. $760 - 2(2 \cdot 3 - 4)^2$ 752
77. $190 - 2[10^2 - (5 + 2^2)] + 45$ 53
78. $161 - 8[6(6) - 6^2] + 2^2(5)$ 181
79. $2 + 3(0)$ 2
80. $5(0) + 8$ 8
81. $\dfrac{(5 - 3)^2 + 2}{4^2 - (8 + 2)}$ 1
82. $\dfrac{(4^3 - 2) + 7}{5(2 + 4) - 7}$ 3
83. $4^2 + 3^2$ 25
84. $12^2 + 5^2$ 169
85. $3 + 2 \cdot 3^4 \cdot 5$ 813
86. $3 \cdot 2^3 \cdot 4 - 12$ 84
87. $60 - \left(6 + \dfrac{40}{2^3}\right)$ 49
88. $7 + \left(5^3 - \dfrac{200}{2}\right)$ 32
89. $\dfrac{(3 + 5)^2 + 2}{2(8 - 5)}$ 11
90. $\dfrac{25 - (2 \cdot 3 - 1)}{2 \cdot 9 - 8}$ 2
91. $(18 - 12)^3 - 5^2$ 191
92. $(9 - 2)^2 - 3^3$ 22
93. $30(1)^2 - 4(2) + 12$ 34
94. $5(1)^3 + (1)^2 + 2(1) - 6$ 2
95. $16^2 - \dfrac{25}{5} + 6(3)4$ 323
96. $15^2 - \dfrac{24}{6} + 8(2)(3)$ 269
97. $\dfrac{3^2 - 2^2}{(3 - 3)^2}$ undefined
98. $\dfrac{5^2 + 17}{4 - 2^2}$ undefined
99. $3\left(\dfrac{18}{3}\right) - 2(2)$ 14
100. $2\left(\dfrac{12}{3}\right) + 3(5)$ 23
101. $4[50 - (3^3 - 5^2)]$ 192
102. $6[15 + (5 \cdot 2^2)]$ 210
103. $80 - 2[12 - (5 + 4)]$ 74
104. $15 + 5[12 - (2^2 + 4)]$ 35

APPLICATIONS

Write an expression to solve each problem and evaluate it.

105. SHOPPING At the supermarket, Carlos is buying 3 cases of soda, 4 bags of tortilla chips, and 2 bottles of salsa. Each case of soda costs $7, each bag of chips costs $4, and each bottle of salsa costs $3. Find the total cost of the snacks. $3(7) + 4(4) + 2(3)$, $43

106. BANKING When a customer deposits cash, a teller must complete a currency count on the back of the deposit slip. In the illustration, a teller has written the number of each type of bill to be deposited. What is the total amount of cash being deposited? $24 + 6(5) + 10(10) + 12(20) + 2(50) + 100$, $594

Currency count, for financial use only	
24	x 1's
—	x 2's
6	x 5's
10	x 10's
12	x 20's
2	x 50's
1	x 100's
	TOTAL $

107. DIVING The scores awarded to a diver by seven judges as well as the degree of difficulty of his dive are shown on the next page. Use the two-step process shown on the next page to calculate the diver's overall score. $3(8 + 7 + 8 + 8 + 7)$, 114

Step 1 Throw out the lowest score and the highest score.

Step 2 Add the sum of the remaining scores and multiply by the degree of difficulty.

108. **WRAPPING GIFTS** How much ribbon is needed to wrap the package shown if 15 inches of ribbon are needed to make the bow?
$2(9) + 2(16) + 4(4) + 15$, 81 in.

109. **SCRABBLE** Illustration (a) shows part of the game board before and illustration (b) shows it after the words *brick* and *aphid* were played. Determine the scoring for each word. (*Hint:* The number on each tile gives the point value of the letter.)
brick: $3(3) + 1 + 1 + 3 + 3(5)$, 29; aphid: $3[1 + 2(3) + 4 + 1 + 2]$, 42

110. **THE GETTYSBURG ADDRESS** Here is an excerpt from Abraham Lincoln's Gettysburg Address:

> Fourscore and seven years ago, our fathers brought forth on this continent a new nation, conceived in liberty, and dedicated to the proposition that all men are created equal.

Lincoln's comments refer to the year 1776, when the United States declared its independence. If a score is 20 years, in what year did Lincoln deliver the Gettysburg Address? $1776 + 4(20) + 7$, 1863

111. **PRIME NUMBERS** Show that 87 is the sum of the squares of the first four prime numbers.
$2^2 + 3^2 + 5^2 + 7^2 = 4 + 9 + 25 + 49 = 87$

112. A 27-foot-long by 19-foot-wide rectangular garden is one feature of a landscape design for a community park. A concrete walkway is to run through the garden and will occupy 125 square feet of space. How many square feet are left for planting in the garden? 388 ft²

from Campus to Careers
Landscape Designer

113. **CLIMATE** One December week, the high temperatures in Honolulu, Hawaii, were 75°, 80°, 83°, 80°, 77°, 72°, and 86°. Find the week's mean (average) high temperature. 79°

114. **GRADES** In a science class, a student had test scores of 94, 85, 81, 77, and 89. He also overslept, missed the final exam, and received a 0 on it. What was his test average (mean) in the class? 71

115. **ENERGY USAGE** See the graph below. Find the mean (average) number of therms of natural gas used per month for the year 2013. 31 therms

116. **COUNTING NUMBERS** What is the average (mean) of the first nine counting numbers: 1, 2, 3, 4, 5, 6, 7, 8, and 9? 5

117. FAST FOODS The table shows the sandwiches Subway advertises on its 6 grams of fat or less menu. What is the mean (average) number of calories for the group of sandwiches? 300 calories

6-inch subs	Calories
Veggie Delite	230
Turkey Breast	280
Turkey Breast & Ham	295
Ham	290
Roast Beef	290
Subway Club	330
Roasted Chicken Breast	310
Chicken Teriyaki	375

Source: Subway.com/NutritionInfo

118. TV RATINGS The table below shows the number of viewers* of the 2012 Major League Baseball World Series between the San Francisco Giants and the Detroit Tigers. How large was the average (mean) audience? 12,625,000

Game 1	Wednesday, Oct. 24	12,200,000
Game 2	Thursday, Oct. 25	12,300,000
Game 3	Saturday, Oct. 27	10,500,000
Game 4	Sunday, Oct. 28	15,500,000

* Rounded to the nearest hundred thousand
Source: Nielsen Media Research

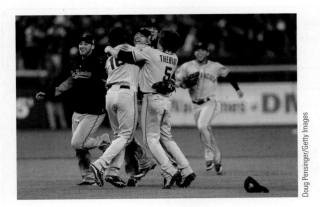

119. YOUTUBE A YouTube video contest is to be part of a kickoff for a new sports drink. The cash prizes to be awarded are shown below.

a. How many prizes will be awarded? 125
b. What is the total amount of money that will be awarded? $11,875
c. What is the average (mean) cash prize? $95

> **YouTube Video Contest**
> **Grand prize: Disney World vacation plus $2,500**
> Four 1st place prizes of $500
> Thirty-five 2nd place prizes of $150
> Eighty-five 3rd place prizes of $25

120. SURVEYS Some students were asked to rate their college cafeteria food on a scale from 1 to 5. The responses are shown on the tally sheet.

a. How many students took the survey? 15
b. Find the mean (average) rating. 3

Poor		Fair		Excellent
1	2	3	4	5
IIII	I	THL	I	IIII

WRITING

121. Explain why the order of operations rule is necessary.

122. What does it mean when we say to do all additions and subtractions *as they occur from left to right*? Give an example.

123. Explain the error in the following work:

Evaluate:
$$8 + 2[6 - 3(9 - 8)] = 8 + 2[6 - 3(1)]$$
$$= 8 + 2[6 - 3]$$
$$= 8 + 2(3)$$
$$= 10(3)$$
$$= 30$$

124. Explain the error in the following work:

Evaluate:
$$24 - 4 + 16 = 24 - 20$$
$$= 4$$

REVIEW

Write each number in words.

125. 254,309
two hundred fifty-four thousand, three hundred nine

126. 504,052,040
five hundred four million, fifty-two thousand, forty

Objectives

1. Determine whether a number is a solution.
2. Use the addition property of equality.
3. Use the subtraction property of equality.
4. Use equations to solve application problems.

SECTION 1.8
Solving Equations Using Addition and Subtraction

The first seven sections of this textbook have been devoted to an in-depth study of whole-number arithmetic. It's now time to begin the move toward *algebra*. **Algebra** is the language of mathematics. It is the result of contributions from many cultures over thousands of years. The word *algebra* comes from the title of the book *Ihm Al-jabr wa'l muqābalah*, written by an Arabian mathematician around A.D. 800. In this section, we will introduce one of the most powerful concepts in algebra, the *equation*.

1 Determine whether a number is a solution.

An **equation** is a statement that two expressions are equal. All equations contain an = symbol. An example is $x + 5 = 15$. The equal symbol = separates the equation into two parts: The expression $x + 5$ is the **left side** and 15 is the **right side**. The letter x is the **variable** (or the **unknown**). The sides of an equation can be reversed, so we can write $x + 5 = 15$ or $15 = x + 5$.

- An equation can be true: $6 + 3 = 9$
- An equation can be false: $2 + 4 = 7$
- An equation can be neither true nor false. For example, $x + 5 = 15$ is neither true nor false because we don't know what number x represents.

An equation that contains a variable is made true or false by substituting a number for the variable. If we substitute 10 for x in $x + 5 = 15$, the resulting equation is true: $10 + 5 = 15$. If we substitute 1 for x, the resulting equation is false: $1 + 5 = 15$. A number that makes an equation true when substituted for the variable is called a **solution** and it is said to **satisfy** the equation. Therefore, 10 is a solution of $x + 5 = 15$, and 1 is not.

> **The Language of Algebra**   To *substitute* means to put or use in place of another, as with a *substitute* teacher. In the previous example, we *substituted* 10 for x in $x + 5 = 15$.

Self Check 1

Is 8 a solution of $x + 17 = 25$? yes

Now Try Problem 17

Teaching Example 1
Is 9 a solution of $x - 3 = 6$?
Answer:
yes

EXAMPLE 1   Is 18 a solution of $x - 3 = 15$?

Strategy We will substitute 18 for x in the equation and evaluate the left side.

WHY If a true statement results, 18 is a solution of the equation. If we obtain a false statement, 18 is not a solution.

Solution

$$x - 3 = 15 \quad \text{This is the given equation.}$$
$$18 - 3 \stackrel{?}{=} 15 \quad \text{Substitute 18 for } x. \text{ Read } \stackrel{?}{=} \text{ as "is possibly equal to."}$$
$$15 = 15 \quad \text{On the left side, do the subtraction.}$$

$$\begin{array}{r} 18 \\ -\ 3 \\ \hline 15 \end{array}$$

Since $15 = 15$ is a true statement, 18 is a solution of $x - 3 = 15$.

> **The Language of Algebra**   It is important to know the difference between an *equation* and *expression*. An equation contains an = symbol and an expression does not.

EXAMPLE 2   Is 23 a solution of $32 = y + 10$?

Strategy We will substitute 23 for y in the equation and evaluate the right side.

WHY If a true statement results, 23 is a solution of the equation. If we obtain a false statement, 23 is not a solution.

Solution

$32 = y + 10$ This is the given equation.

$32 \stackrel{?}{=} 23 + 10$ Substitute 23 for y.

$32 = 33$ On the right side, do the addition.

$\begin{array}{r} 23 \\ +10 \\ \hline 33 \end{array}$

Since $32 = 33$ is a false statement, 23 is not a solution of $32 = y + 10$.

Self Check 2

Is 35 a solution of $20 = y - 17$?
no

Now Try Problem 21

Teaching Example 2
Is 16 a solution of $30 = y + 15$?
Answer:
no

2 Use the addition property of equality.

Since the solution of an equation is usually not given, we must develop a process to find it. This process is called *solving the equation*.

To **solve an equation** means to find all values of the variable that make the equation true. To develop an understanding of how to solve equations, refer to the scales shown on the right.

The first scale represents the equation $x - 2 = 3$. The scale is in balance because the weights on the left side and right side are equal. To find x, we must add 2 to the left side. To keep the scale in balance, we must also add 2 to the right side. After doing this, we see from the second scale that x is balanced by 5. Therefore, x must be 5. We say that we have solved the equation $x - 2 = 3$ and that the solution is 5.

In this example, we solved $x - 2 = 3$ by transforming it to a simpler *equivalent equation*, $x = 5$.

Equivalent Equations

Equations with the same solutions are called **equivalent equations**.

The Language of Algebra We *solve* equations. An expression can be *evaluated* (or *simplified*), but never *solved*.

The procedure that we used with the scales suggests the following property of equality.

Addition Property of Equality

Adding the same number to both sides of an equation does not change its solution.

For any numbers a, b, and c,

if $a = b$, then $a + c = b + c$

When we use this property, the resulting equation is *equivalent to the original one*. We will now show how it is used to solve $x - 2 = 3$ algebraically.

Chapter 1 Whole Numbers

Self Check 3

Solve $x - 10 = 33$ and check the result. 43

Now Try Problem 25

Teaching Example 3
Solve $x - 8 = 2$ and check the result.
Answer:
10

EXAMPLE 3 Solve: $x - 2 = 3$

Strategy We will use the addition property of equality to isolate the variable x on the left side of the equation.

WHY To solve the original equation, we want to find a simpler equivalent equation of the form $x =$ **a number**, whose solution is obvious.

Solution

$$x - 2 = 3 \quad \text{This is the equation to solve.}$$
$$x - 2 + 2 = 3 + 2 \quad \text{To isolate x, undo the subtraction of 2 by adding 2 to both sides.}$$
$$x = 5 \quad \text{On the left side, adding 2 undoes the subtraction of 2 and leaves x. On the right side, do the addition: } 3 + 2 = 5.$$

Since 5 is obviously the solution of the equivalent equation $x = 5$, the solution of the original equation, $x - 2 = 3$, is also 5.

To check this result, we substitute 5 for x in the original equation and simplify.

Check: $x - 2 = 3$ This is the original equation.
$5 - 2 \stackrel{?}{=} 3$ Substitute 5 for x.
$3 = 3$ On the left side, do the subtraction.

Since $3 = 3$ is a true statement, 5 is the solution of $x - 2 = 3$.

> **The Language of Algebra** We solve equations by writing a series of steps that result in an equivalent equation of the form
>
> $x =$ *a number*   or   *a number* $= x$
>
> We say the variable is *isolated* on one side of the equation. *Isolated* means alone or by itself.

Self Check 4

Solve $75 = b - 38$ and check the result. 113

Now Try Problem 29

Teaching Example 4
Solve $32 = y - 4$ and check the result.
Answer:
36

EXAMPLE 4 Solve: $19 = y - 7$

Strategy We will use the addition property of equality to isolate the variable y on the right side of the equation.

WHY To solve the original equation, we want to find a simpler equivalent equation of the form **a number** $= y$, whose solution is obvious.

Solution

$$19 = y - 7 \quad \text{This is the equation to solve.}$$
$$19 + 7 = y - 7 + 7 \quad \text{To isolate y, undo the subtraction of 7 by adding 7 to both sides.}$$
$$26 = y \quad \text{On the left side, do the addition: } 19 + 7 = 26. \text{ On the right side, adding 7 undoes the subtraction of 7 and leaves y.}$$

$$\begin{array}{r} \overset{1}{1}9 \\ + 7 \\ \hline 26 \end{array}$$

Success Tip Calculations that you cannot perform in your head should be shown outside the steps of your solution.

Since 26 is obviously the solution of the equivalent equation $26 = y$, the solution of the original equation, $19 = y - 7$, is also 26.

To check this result, we substitute 26 for y in the original equation and simplify.

Check: $19 = y - 7$ This is the original equation.
$19 \stackrel{?}{=} 26 - 7$ Substitute 26 for y.
$19 = 19$ On the right side, do the subtraction.

Since $19 = 19$ is a true statement, 26 is the solution of $19 = y - 7$.

Success Tip Perhaps you are more comfortable by first reversing the sides of equations like that of Example 4 before attempting to solve them:

$19 = y - 7$ can be rewritten as $y - 7 = 19$

That step is fine; however, when solving equations, it is not necessary that the variable be isolated on the left side of the equation.

3 Use the subtraction property of equality.

To introduce another property of equality, consider the first scale shown on the right, which represents the equation $x + 3 = 5$. The scale is in balance because the weights on the left and right sides are equal. To find x, we need to remove 3 from the left side. To keep the scale in balance, we must also remove 3 from the right side. After doing this, we see from the second scale that x is balanced by 2. Therefore, x must be 2. We say that we have solved the equation $x + 3 = 5$ and that the solution is 2. This example illustrates the following property of equality.

Subtraction Property of Equality

Subtracting the same number from both sides of an equation does not change its solution.

For any numbers a, b, and c,

if $a = b$, then $a - c = b - c$

When we use this property, the resulting equation is equivalent to the original one.

EXAMPLE 5   Solve: $x + 3 = 5$

Strategy We will use the subtraction property of equality to isolate the variable x on the left side of the equation.

WHY To solve the original equation, we want to find a simpler equivalent equation of the form $x = $ **a number**, whose solution is obvious.

Solution

$x + 3 = 5$	This is the equation to solve.
$x + 3 - 3 = 5 - 3$	To isolate x, undo the addition of 3 by subtracting 3 from both sides.
$x = 2$	On the left side, subtracting 3 undoes the addition of 3 and leaves x. On the right side, do the subtraction: $5 - 3 = 2$.

We check by substituting 2 for x in the original equation and simplifying. If 2 is the solution, we will obtain a true statement.

Check: $x + 3 = 5$ This is the original equation.
  $2 + 3 \stackrel{?}{=} 5$ Substitute 2 for x.
  $5 = 5$ On the left side, do the addition.

Since the resulting equation $5 = 5$ is true, 2 is the solution of $x + 3 = 5$.

Self Check 5

Solve $m + 7 = 14$ and check the result. 7

Now Try Problem 33

Teaching Example 5
Solve $x + 4 = 9$ and check the result.
Answer:
5

4 Use equations to solve application problems.

The key to problem solving is to understand the problem and then to develop a plan for solving it. The following list of steps provides a good strategy to follow.

> **The Language of Algebra** A **strategy** is a careful plan or method. For example, a businessman might develop a new advertising *strategy* to increase sales or a long-distance runner might have a *strategy* to win a marathon.

Strategy for Problem Solving

1. **Analyze the problem** by reading it carefully to understand the given facts. What information is given? What are you asked to find? What vocabulary is given? Often, a diagram will help you understand the facts of the problem.
2. **Assign a variable** to represent an unknown value in the problem. This means, in most cases, to let $x = $ what you are asked to find. If there are other unknown values, represent each of them using an algebraic expression that involves the variable.
3. **Form an equation** by translating the words of the problem into mathematical symbols.
4. **Solve the equation** formed in Step 3.
5. **State the conclusion** clearly. Be sure to include the units (such as feet, seconds, or pounds) in your answer.
6. **Check the result** using the original wording of the problem, not the equation that was formed in Step 3 from the words.

The purpose of the following examples is to help you learn the six-step strategy, even though you can probably solve the problems without it. If you learn how to use the strategy now, you will gain valuable problem-solving experience that will pay off later in the course when you are asked to solve more difficult problems.

Self Check 6

GASOLINE STORAGE A tank currently contains 1,325 gallons of gasoline. If 450 gallons were pumped from the tank earlier, how many gallons did it originally contain? The tank originally contained 1,775 gallons of gasoline.

Now Try Problems 38 and 75

Teaching Example 6
LAYOFFS Last year, a company had to lay off 24 full-time employees and put a "hiring freeze" into effect. If the company now has 367 full-time employees, and no one retired, how many did the company have originally?
Answer:
The company originally had 391 full-time employees.

EXAMPLE 6   *Small Businesses*   Last year a hairstylist lost 17 customers who moved away. If she now has 73 customers, how many did she have originally?

Analyze
- She lost 17 customers.   Given
- She now has 73 customers.   Given
- How many customers did she originally have?   Find

Caution! Unlike an arithmetic approach, you do *not* have to determine whether to add, subtract, multiply, or divide at this stage. Simply translate the words of the problem to mathematical symbols to form an equation that describes the situation. Then solve the equation.

Assign
We can let $c = $ the original number of customers.

Form
To form an equation involving c, we look for a key word or phrase in the problem.

     Key phrase: *moved away*   **Translation:** *subtraction*

Now we translate the words of the problem into an equation.

This is called the verbal model.

The original number of customers	minus	17	is equal to	the number of customers she now has.
c	$-$	17	$=$	73

Solve

$c - 17 = 73$ We need to isolate c on the left side.

$c - 17 + 17 = 73 + 17$ To isolate c, add 17 to both sides to undo the subtraction of 17.

$c = 90$ Do the addition.

$$\begin{array}{r} \overset{1}{}73 \\ +\ 17 \\ \hline 90 \end{array}$$

State

She originally had 90 customers.

Check

If the hairstylist originally had 90 customers, and we decrease that number by the 17 that moved away, we should obtain the number of customers she now has.

$$\begin{array}{r} \overset{8\ 10}{9\cancel{0}} \\ -\ 17 \\ \hline 73 \end{array}$$ This is the number of customers the hairstylist now has.

The result, 90, checks.

> **Caution!** Check the result using the original wording of the problem, not by substituting it into the equation. Why? The equation may have been solved correctly, but the danger is that you may have formed it incorrectly.

EXAMPLE 7   Mortgages

Sue wants to buy a condo that costs $87,000. Since she has only $15,000 for a down payment, she will have to borrow some money by taking a mortgage. How much will she have to borrow?

Analyze
- The condo costs $87,000.   Given
- Sue has $15,000 for a down payment.   Given
- How much money does she need to borrow?   Find

Assign

We can let x = the amount of money that she needs to borrow.

Form

To form an equation involving x, we look for a key word or phrase in the problem.

Key phrase: borrow some additional money   **Translation:** addition

Now we translate the words of the problem into an equation.

The amount Sue now has	plus	the amount she borrows	is equal to	the total cost of the condo.
15,000	+	x	=	87,000

Solve

$15,000 + x = 87,000$ We need to isolate x on the left side.

$15,000 + x - 15,000 = 87,000 - 15,000$ To isolate x, subtract 15,000 from both sides to undo the addition of 15,000.

$x = 72,000$ Do the subtraction.

$$\begin{array}{r} 87,000 \\ -\ 15,000 \\ \hline 72,000 \end{array}$$

Self Check 7

STUDENT LOANS A student has saved $1,500 to pay for his first year of college. How much money will he have to borrow if books, tuition, and expenses for first-year students are estimated to total $3,750?

Now Try Problem 73

Self Check 7 Answer

The student needs to borrow $2,250.

Teaching Example 7
MORTGAGES Matt wants to buy a condo that costs $135,000. Since he only has $22,000 for a down payment, he will have to borrow some money by taking a mortgage. How much will he have to borrow?
Answer:
Matt will have to borrow $113,000.

State
Sue must borrow $72,000.

Check
If Sue has $15,000 and we add the amount of money she needs to borrow, we should obtain the cost of the condo.

$$\begin{array}{r}\$15,000\\+\$72,000\\\hline\$87,000\end{array}$$ This is the cost of the condo.

The result, $72,000, checks.

ANSWERS TO SELF CHECKS

1. yes **2.** no **3.** 43 **4.** 113 **5.** 7 **6.** The tank originally contained 1,775 gallons of gasoline. **7.** The student needs to borrow $2,250.

SECTION 1.8 STUDY SET

VOCABULARY

Fill in the blanks.

1. An equation is a statement indicating that two expressions are equal. All equations contain an = symbol.
2. A number that makes an equation true when substituted for the variable is called a solution of the equation. Such numbers are said to satisfy the equation.
3. To solve an equation means to find all values of the variable that make the equation true.
4. To solve an equation, we isolate the variable on one side of the equal symbol.
5. Equations with the same solutions are called equivalent equations.
6. To check the solution of an equation, we substitute the value for the variable in the original equation and determine whether the result is a true statement.

CONCEPTS

7. Given: $x + 6 = 12$
   a. What is the left side of the equation? $x + 6$
   b. Is this equation true or false? neither
   c. Is 5 a solution of this equation? no
   d. Does 6 satisfy the equation? yes
8. Tell whether each of the following is an equation.
   a. $x - 3$ no
   b. $m + 12 = 40$ yes
   c. $7 < 8$ no
   d. $18 > 0$ no

9. Fill in the blanks.
   a. The addition property of equality: Adding the same number to both sides of an equation does not change its solution.
   b. If $a = b$, then $a + c = b +$ c .
10. Fill in the blanks.
    a. The subtraction property of equality: Subtracting the same number from both sides of an equation does not change its solution.
    b. If $a = b$, then $a - c = b -$ c .
11. Fill in the blanks.
    a. To solve $x - 8 = 24$, we add 8 to both sides of the equation.
    b. To solve $x + 4 = 11$, we subtract 4 from both sides of the equation.
12. Simplify each expression.
    a. $x + 7 - 7$ x
    b. $y - 2 + 2$ y

NOTATION

Complete the steps to solve the equation. Check the result.

13. $x - 5 = 45$
     $x - 5 + \boxed{5} = 45 + \boxed{5}$
     $x = \boxed{50}$

Check: $x - 5 = 45$
        $\boxed{50} - 5 \stackrel{?}{=} 45$
        $\boxed{45} = 45$ True

$\boxed{50}$ is the solution.

14.
$$y + 11 = 12$$
$$y + 11 - \boxed{11} = 12 - \boxed{11}$$
$$y = \boxed{1}$$

Check: $y + 11 = 12$
$$\boxed{1} + 11 \stackrel{?}{=} 12$$
$$\boxed{12} = 12 \text{ True}$$

$\boxed{1}$ is the solution.

15. What does the symbol $\stackrel{?}{=}$ mean? is possibly equal to

16. If you solve an equation and obtain $50 = x$, can you write $x = 50$? yes

GUIDED PRACTICE

Check to determine whether the given number is a solution of the equation. See Example 1.

17. Is 1 a solution of $x + 2 = 3$? yes
18. Is 4 a solution of $x + 2 = 6$? yes
19. Is 7 a solution of $a - 7 = 0$? yes
20. Is 16 a solution of $x - 8 = 8$? yes

Check to determine whether the given number is a solution of the equation. See Example 2.

21. Is 40 a solution of $50 = y - 8$? no
22. Is 5 a solution of $16 = 10 + c$? no
23. Is 2 a solution of $1 = x + 2$? no
24. Is 4 a solution of $8 = x + 1$? no

Solve each equation and check the result. See Example 3.

25. $x - 7 = 3$ 10
26. $y - 11 = 7$ 18
27. $a - 20 = 50$ 70
28. $z - 31 = 60$ 91

Solve each equation and check the result. See Example 4.

29. $1 = b - 2$ 3
30. $0 = t - 1$ 1
31. $19 = n - 42$ 61
32. $17 = m - 16$ 33

Solve each equation and check the result. See Example 5.

33. $x + 9 = 12$ 3
34. $x + 3 = 9$ 6
35. $y + 7 = 12$ 5
36. $c + 11 = 22$ 11

In Exercises 37 and 38, complete the steps to answer the question.

37. HISTORY A 1,700-year-old scroll is 425 years older than the clay jar in which it was found. How old is the jar? *See Example 6.*

Analyze
- The scroll is $\underline{1,700}$ years old. *Given*
- The scroll is $\underline{425}$ years older than the jar. *Given*
- How old is the $\underline{\text{jar}}$? *Find*

Assign Let $x =$ the age of the $\underline{\text{jar}}$.

Form Now we look for a key word or phrase in the problem.

Key phrase: *older than* **Translation:** addition

Now we translate the words of the problem into an equation.

The age of the scroll	is	425 years	plus	the age of the jar.
1,700	=	425	+	x

Solve
$$\boxed{1{,}700} = 425 + x$$
$$1{,}700 - \boxed{425} = 425 + x - \boxed{425}$$
$$\boxed{1{,}275} = x$$

State The jar is $\underline{1{,}275}$ years old.

Check If the jar is 1,275 years old, and if we add 425 years to its age, we should get the age of the scroll.

$$\begin{array}{r} \overset{1\,1}{1{,}275} \\ + 425 \\ \hline \boxed{1{,}700} \end{array}$$ This is the age of the scroll.

The result checks.

38. BANKING After a student wrote a $1,500 check to pay for a car, he had a new balance of $750 in his account. What was the account balance before he wrote the check?

Analyze
- The student wrote a $\underline{\$1{,}500}$ check. *Given*
- The new balance in the account was $\underline{\$750}$. *Given*
- What was the $\underline{\text{balance}}$ before he wrote the check? *Find*

Assign Let $x =$ the account $\underline{\text{balance}}$ before he wrote the check.

Form Now we look for a key word or phrase in the problem.

Key phrase: *wrote a check* **Translation:** subtraction

Now we translate the words of the problem into an equation.

The account balance before the check	minus	the amount of the check	is equal to	the new balance.
x	−	1,500	=	750

Solve

$$x - 1{,}500 = 750$$
$$x - 1{,}500 + 1{,}500 = 750 + 1{,}500$$
$$x = 2{,}250$$

State The account balance before he wrote the check was $2,250.

Check If the old balance was $2,250, and if we subtract the $1,500 check from it, we should get the new balance.

$$\begin{array}{r} \$2{,}250 \\ -\$1{,}500 \\ \hline \$\ 750 \end{array}$$ This is the new balance in the account.

The result checks.

TRY IT YOURSELF

Solve each equation and check the result.

39. $s + 55 = 100$ 45
40. $n + 37 = 200$ 163
41. $x - 4 = 0$ 4
42. $c - 3 = 0$ 3
43. $y - 7 = 6$ 13
44. $a - 2 = 4$ 6
45. $70 = x - 5$ 75
46. $66 = b - 6$ 72
47. $312 = x - 428$ 740
48. $113 = x - 307$ 420
49. $x - 117 = 222$ 339
50. $y - 27 = 317$ 344
51. $t + 19 = 28$ 9
52. $s + 45 = 84$ 39
53. $23 + x = 33$ 10
54. $34 + y = 34$ 0
55. $5 = 4 + c$ 1
56. $41 = 23 + x$ 18
57. $99 = r + 43$ 56
58. $92 = r + 37$ 55
59. $512 = 428 + x$ 84
60. $513 = 307 + x$ 206
61. $x + 117 = 222$ 105
62. $y + 38 = 321$ 283
63. $3 + x = 7$ 4
64. $4 + b = 8$ 4
65. $y - 5 = 7$ 12
66. $z - 9 = 23$ 32
67. $4 + a = 12$ 8
68. $5 + x = 13$ 8
69. $x - 13 = 34$ 47
70. $x - 23 = 19$ 42

APPLICATIONS

Let a variable represent the unknown quantity. Then write and solve an equation to answer the question.

71. **FAST FOOD** The franchise fee and start-up costs for a Pizza Hut restaurant are $316,500. If a woman has $68,500 to invest, how much money will she need to borrow to open her own Pizza Hut restaurant? (Source: yumfranchises.com) She will need to borrow $248,000.

72. **PARTY INVITATIONS** Three of Mia's party invitations were lost in the mail, but 59 were delivered. How many invitations did she send? Mia sent 62 invitations.

73. **HIP HOP** *Forbes* magazine estimates that in 2011, Shawn "Jay-Z" Carter earned $38 million. If this was $72 million less than Andre "Dr. Dre" Young's earnings, how much did Dr. Dre earn in 2011? (Source: *Forbes* Magazine) Dr. Dre earned $110 million in 2011.

74. **GOLF CLUBS** A man wants to buy a new set of golf clubs for $345. How much more money does he need if he now has $317? The man needs $28 more to buy the golf clubs.

75. **HEARING PROTECTION** The sound intensity of a jet engine is 110 decibels. If an airplane mechanic wears earplugs when working near a jet, she only experiences 81 decibels of sound intensity. By how many decibels do the earplugs reduce the noise level? The earplugs reduce the noise level by 29 decibels.

76. **HELP WANTED** From the following ad from the classifed section of a newspaper, determine the value of the benefit package. ($45 K means $45,000.) The value of the benefit package is $7,000.

> ★ACCOUNTS PAYABLE★
> 2-3 yrs exp as supervisor. Degree a +. High vol company. Good pay, $45K & xlnt benefits; total compensation worth $52K. Fax resume.

77. **POWER OUTAGES** The electrical system in a building automatically shuts down when the meter shown below reads 85. By how much must the current reading increase to cause the system to shut down? The reading must increase by 25 units to cause the system to shut down.

78. **VIDEO GAMES** After a week of playing *Angry Birds*, a boy scored 105,880 points in one game—an improvement of 77,420 points over the very first time he played. What was the score for his first game? His score for the first game was 28,460 points.

79. **AUTO REPAIRS** A woman paid $29 less to have her car repaired at a muffler shop than she would have paid at a gas station. If she paid $190 at the muffler shop, what was the gas station going to charge her? The gas station was going to charge her $219.

80. **RIDING BUSES** A man had to wait 20 minutes for a bus today. Three days ago, he had to wait 15 minutes longer than he did today, because several buses passed by without stopping. How long did he wait three days ago?
Three days ago he waited for 35 minutes.

81. **HIT RECORDS** The oldest artist to have a number 1 single was 67-year-old Louis Armstrong, with his version of *Hello Dolly*. He was 55 years older than the youngest artist to have a number 1 single, Jimmy Boyd, who sang *I Saw Mommy Kissing Santa Claus*. How old was Jimmy Boyd when he had the number 1 song? (Source: *The Top 10 of Everything*, 2000)
Jimmy Boyd was 12 years old when he had the number 1 song.

82. **REBATES** The price of a new Honda Civic was advertised in a newspaper as $16,727*. A note at the bottom of the ad read, "*Reflects $1,550 factory rebate." What was the car's original sticker price?
The car's original sticker price was $18,277.

WRITING

83. Explain what it means for a number to *satisfy* an equation.

84. Explain how to tell whether a number is a solution of an equation.

85. Explain what the pair of figures on page 97 are trying to show.

86. Think of a number. Add 8 to it. Now subtract 8 from that result. Explain why we will always obtain the original number.

87. When solving equations, we *isolate* the variable. Write a sentence in which the word *isolate* is used in a different context.

88. What do you find to be the most difficult step of the six-step problem solving strategy? Explain why it is.

89. Unlike an arithmetic approach, you do *not* have to determine whether to add, subtract, multiply, or divide to solve the application problems in this section. That decision is made for you when you solve the equation that mathematically describes the situation. Explain.

90. What does the word *translate* mean?

REVIEW

91. Round 325,784 to the nearest ten. 325,780
92. Evaluate: 1^5 1
93. Evaluate: $2 \cdot 3^2 \cdot 5$ 90
94. a. Represent $4 + 4 + 4$ as a multiplication. $3 \cdot 4$
    b. Represent $4 \cdot 4 \cdot 4$ using an exponential expression. 4^3
95. Evaluate: $8 - 2(2^2 - 1) + 1^3$ 3
96. Write 1,055 in words. one thousand, fifty-five

SECTION 1.9
Solving Equations Using Multiplication and Division

Objectives
1. Use the multiplication property of equality.
2. Use the division property of equality.
3. Use equations to solve application problems.

In the previous section, we solved simple equations such as

$$x - 2 = 3 \quad \text{and} \quad x + 8 = 11$$

by using the addition and subtraction properties of equality. In this section, we will learn how to solve equations such as

$$\frac{x}{3} = 25 \quad \text{and} \quad 2x = 6$$

by using the multiplication and division properties of equality.

1 Use the multiplication property of equality.

To introduce a third property of equality, consider the first scale shown on the right, which represents the equation $\frac{x}{3} = 25$. The scale is in balance because the weights on the left side and right side are equal. To find x, we must triple (multiply by 3) the weight on the left side. To keep the scale in balance, we must also triple the weight on the right side. After doing this, we see in the second scale that x is balanced by 75. Therefore, x must be 75.

The procedure that we just used suggests the following property of equality.

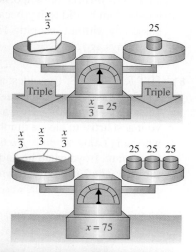

Multiplication Property of Equality

Multiplying both sides of an equation by the same nonzero number does not change its solution.

For any numbers a, b, and c, where c is not 0,

if $a = b$, then $ca = cb$

When we use this property, the resulting equation is equivalent to the original one. We will now show how it is used to solve $\frac{x}{3} = 25$ algebraically.

Self Check 1

Solve $\frac{x}{12} = 24$ and check the result. 288

Now Try Problem 13

Teaching Example 1

Solve $\frac{x}{6} = 16$ and check the result.
Answer:
96

EXAMPLE 1 Solve: $\frac{x}{3} = 25$

Strategy We will use the multiplication property of equality to isolate the variable x on the left side of the equation.

WHY To solve the original equation, we want to find a simpler equivalent equation of the form $x = $ a number, whose solution is obvious.

Solution

$\frac{x}{3} = 25$ This is the equation to solve.

$3 \cdot \frac{x}{3} = 3 \cdot 25$ To isolate x, undo the division by 3 by multiplying both sides by 3.

$x = 75$ On the left side, when x is divided by 3 and that quotient is then multiplied by 3, the result is x. Multiplication by 3 undoes division by 3. On the right side, do the multiplication: $3 \cdot 25 = 75$.

$$\begin{array}{r} \overset{1}{25} \\ \times\ 3 \\ \hline 75 \end{array}$$

Since 75 is obviously the solution of $x = 75$, the solution of the original equation, $\frac{x}{3} = 25$, is also 75.

Check: $\frac{x}{3} = 25$ This is the original equation.

$\frac{75}{3} \stackrel{?}{=} 25$ Substitute 75 for x.

$25 = 25$ On the left side, do the division: $75 \div 3 = 25$.

$$\begin{array}{r} 25 \\ 3\overline{)75} \\ \underline{-6} \\ 15 \\ \underline{-15} \\ 0 \end{array}$$

Since $25 = 25$ is a true statement, 75 is the solution of $\frac{x}{3} = 25$.

EXAMPLE 2   Solve: $84 = \dfrac{n}{16}$

Strategy We will use the multiplication property of equality to isolate the variable n on the right side of the equation.

WHY To solve the original equation, we want to find a simpler equivalent equation of the form **a number** $= n$, whose solution is obvious.

Solution

$84 = \dfrac{n}{16}$ This is the equation to solve.

$16 \cdot 84 = 16 \cdot \dfrac{n}{16}$ To isolate n, undo the division by 16 by multiplying both sides by 16.

$1{,}344 = n$ On the left side, do the multiplication: $16 \cdot 84 = 1{,}344$. On the right side, when n is divided by 16 and that quotient is then multiplied by 16, the result is n.

```
    84
  × 16
   504
   840
 1,344
```

To check this result, we substitute 1,344 for n in the original equation

Check: $84 = \dfrac{n}{16}$ This is the original equation.

$84 \stackrel{?}{=} \dfrac{1{,}344}{16}$ Substitute 1,344 for n.

$84 = 84$ On the right side, do the division.

Since $84 = 84$ is a true statement, 1,344 is the solution of $84 = \dfrac{n}{16}$.

```
       84
  16)1,344
    −128
      64
     −64
       0
```

Self Check 2

Solve $30 = \dfrac{b}{34}$ and check the result.   1,020

Now Try Problem 17

Teaching Example 2
Solve $27 = \dfrac{s}{43}$ and check the result.
Answer:
1,161

Success Tip Calculations that you cannot perform in your head should be shown outside the steps of your solution.

2 Use the division property of equality.

To introduce a fourth property of equality, consider the first scale shown on the right, which represents the equation $2x = 6$. The scale is in balance because the weights on the left and right sides are equal. To find x, we need to split the amount of weight on the left side in half (divide by 2). To keep the scale in balance, we must split the amount of weight in half on the right side. After doing this, we see in the second scale that x is balanced by 3. Therefore, x must be 3. We say that we have solved the equation $2x = 6$ and that the solution is 3. This example illustrates the following property of equality.

Chapter 1 Whole Numbers

> **Division Property of Equality**
>
> Dividing both sides of an equation by the same nonzero number does not change its solution.
>
> For any numbers a, b, and c, where c is not 0,
>
> $$\text{if } a = b, \text{ then } \frac{a}{c} = \frac{b}{c}$$

When we use this property, the resulting equation is equivalent to the original one. We will now show how it is used to solve $2x = 6$ algebraically.

Self Check 3

Solve $17x = 153$ and check the result. 9

Now Try Problem 23

Teaching Example 3
Solve $8x = 32$ and check the result.
Answer:
4

EXAMPLE 3   Solve: $2x = 6$

Strategy We will use the division property of equality to isolate the variable x on the left side of the equation.

WHY To solve the original equation, we want to find a simpler equivalent equation of the form $x = $ **a number**, whose solution is obvious.

Solution   Recall that $2x = 6$ means $2 \cdot x = 6$. To isolate x on the left side of the equation, we undo the multiplication by 2 by dividing both sides by 2.

$2x = 6$ This is the equation to solve.

$\dfrac{2x}{2} = \dfrac{6}{2}$ Divide both sides by 2.

$x = 3$ When x is multiplied by 2 and that product is then divided by 2, the result is x. On the right side, do the division: 6 ÷ 2 = 3.

To check this result, we substitute 3 for x in $2x = 6$.

Check: $2x = 6$ This is the original equation.

$2 \cdot 3 \stackrel{?}{=} 6$ Substitute 3 for x.

$6 = 6$ On the left side, do the multiplication: 2 · 3 = 6.

Since $6 = 6$ is a true statement, 3 is the solution of $2x = 6$.

3 Use equations to solve application problems.

As before, we can use equations to solve application problems. Remember that the purpose of these early examples is to help you learn the strategy, even though you can probably solve the problems without it.

Self Check 4

CLASSICAL MUSIC A woodwind quartet (four musicians) was hired to play at an art exhibit. If each musician made $85 for the performance, what fee did the quartet charge?

Now Try Problem 25

Self Check 4 Answer
The quartet charged $340 for the performance.

EXAMPLE 4   Entertainment Costs   A five-piece band worked on New Year's Eve. If each player earned $120, what fee did the band charge?

Analyze
- There were 5 players in the band.   Given
- Each player made $120.   Given
- What fee did the band charge?   Find

Assign
We can let $f = $ the band's fee.

Form
To form an equation, we look for a key word or phrase. In this case, we find it in the analysis of the problem. If each player earned the same amount ($120), the band's fee must have been divided into 5 equal parts.

Key phrase: *divided into 5 equal parts*   **Translation:** *division*

Now we translate the words of the problem into an equation.

The band's fee	divided by	the number of players in the band	is	each person's share.
f	$\div$	5	$=$	120

Solve

$\dfrac{f}{5} = 120$ We need to isolate f on the left side.

$5 \cdot \dfrac{f}{5} = 5 \cdot 120$ To isolate f, multiply both sides by 5 to undo the division by 5.

$f = 600$ Do the multiplication.

$$\begin{array}{r} \overset{1}{1}20 \\ \times\ \ 5 \\ \hline 600 \end{array}$$

State
The band's fee was $600.

Check
If the band's fee was $600, and we divide it into 5 equal parts, we should get the amount that each player earned.

$$\begin{array}{r} 120 \\ 5\overline{)600} \\ \underline{-5} \\ 10 \\ \underline{-10} \\ 00 \\ \underline{-0} \\ 0 \end{array}$$ ← This is the amount each band member earned.

The result, $600, checks.

EXAMPLE 5   **Traffic Fines**   For speeding in a construction zone, a motorist had to pay a fine of $592. The violation occurred on a highway posted with signs like the one shown here. What would the fine have been if such signs were not posted?

TRAFFIC FINES DOUBLED IN CONSTRUCTION ZONE

Analyze
- For speeding, the motorist was fined $592.   Given
- The fine was double what it would normally have been.   Given
- What would the fine have been, had the sign not been posted?   Find

Assign
We can let f = the amount that the fine would normally have been.

Form
To form an equation, we look for a key word or phrase in the problem or analysis.

Key word: *double*   **Translation:** *multiply by 2*

Now we translate the words of the problem into an equation.

Two	times	the normal speeding fine	is	the new fine.
2	$\cdot$	f	$=$	592

Teaching Example 4
ENTERTAINMENT COSTS A seven-piece band played at a high school prom. If each player earned $175 that evening, what fee did the band charge?
Answer:
The band charged a fee of $1,225 to play at the prom.

Self Check 5

SPEED READING A speed reading course claims it can teach a person to read four times faster. After taking the course, a student can now read 700 words per minute. If the company's claims are true, what was the student's reading rate before taking the course?

Now Try Problem 26

Self Check 5 Answer
The student used to read 175 words per minute.

Teaching Example 5
READ-A-THONS The number of students that are currently part of a "Read for Dollars" fund-raiser is triple that of last year. If 384 students are now participating, how many took part last year?
Answer:
The number of students that took part in the fund-raiser last year was 128.

Solve

$2f = 592$ We need to isolate f on the left side.

$\dfrac{2f}{2} = \dfrac{592}{2}$ To isolate f, divide both sides by 2 to undo the multiplication by 2.

$f = 296$ Do the division.

```
    296
2)592
   −4
   19
   −18
    12
   −12
     0
```

State

The fine would normally have been $296.

Check

If the normal fine was $296, and we double it, we should get the new fine.

$$\begin{array}{r} \overset{1\;1}{296} \\ \times\; 2 \\ \hline 592 \end{array}$$ This is the new fine.

The result, $296, checks.

ANSWERS TO SELF CHECKS

1. 288   **2.** 1,020   **3.** 9   **4.** The quartet charged $340 for the performance.   **5.** The student used to read 175 words per minute.

SECTION 1.9   STUDY SET

VOCABULARY

Fill in the blanks.

1. To _solve_ an equation means to find all values of the variable that make the equation true.
2. A number that makes an equation true when substituted for the variable is called a _solution_ of the equation.
3. To solve an equation, we _isolate_ the variable on one side of the equal symbol.
4. In this section, we used the multiplication and division properties of _equality_ to solve equations.

CONCEPTS

Fill in the blanks.

5. **a.** The multiplication property of equality: Multiplying both sides of an equation by the _same_ nonzero number does not change its solution.

   b. If $a = b$, then $ca = \boxed{cb}$.   (provided c is not 0)

6. **a.** The division property of equality: Dividing both sides of an equation by the _same_ nonzero number does not change its solution.

   b. If $a = b$, then $\dfrac{a}{c} = \dfrac{\boxed{b}}{c}$.   (provided c is not 0)

7. **a.** If we multiply x by 6 and then divide that product by 6, the result is $\boxed{x}$.

   b. If we divide x by 8 and then multiply that quotient by 8, the result is $\boxed{x}$.

8. Simplify each expression.

   a. $9 \cdot \dfrac{x}{9}$ x **b.** $\dfrac{6y}{6}$ y

9. Fill in the blanks.

   a. To solve $\dfrac{x}{5} = 10$, we _multiply_ both sides of the equation by 5.

   b. To solve $5x = 10$, we _divide_ both sides of the equation by 5.

   c. To solve $x - 5 = 10$, we _add_ 5 to both sides of the equation.

   d. To solve $x + 5 = 10$, we _subtract_ 5 from both sides of the equation.

10. Use a check to determine whether the given number is a solution of the equation.

    a. Is 8 a solution of $16 = 8t$?   no

    b. Is 2 a solution of $16 = \dfrac{t}{8}$?   no

NOTATION

Complete the steps to solve the equation. Check the result.

11. $\dfrac{x}{5} = 9$

   $\boxed{5} \cdot \dfrac{x}{5} = \boxed{5} \cdot 9$

   $x = \boxed{45}$

   Check: $\dfrac{x}{5} = 9$

   $\dfrac{\boxed{45}}{5} \stackrel{?}{=} 9$

   $\boxed{9} = 9$ True

   $\boxed{45}$ is the solution.

12. $3x = 12$

   $\dfrac{3x}{\boxed{3}} = \dfrac{12}{\boxed{3}}$

   $x = \boxed{4}$

   Check: $3x = 12$

   $3 \cdot \boxed{4} \stackrel{?}{=} 12$

   $\boxed{12} = 12$ True

   $\boxed{4}$ is the solution.

GUIDED PRACTICE

Solve each equation and check the result. See Example 1.

13. $\dfrac{x}{7} = 2$ 14
14. $\dfrac{x}{12} = 4$ 48
15. $\dfrac{y}{14} = 3$ 42
16. $\dfrac{y}{13} = 5$ 65

Solve each equation and check the result. See Example 2.

17. $16 = \dfrac{x}{24}$ 384
18. $22 = \dfrac{x}{18}$ 396
19. $31 = \dfrac{t}{11}$ 341
20. $33 = \dfrac{m}{19}$ 627

Solve each equation and check the result. See Example 3.

21. $3x = 3$ 1
22. $5x = 5$ 1
23. $9z = 90$ 10
24. $3z = 60$ 20

In Exercises 25 and 26, complete the steps to answer the question.

25. **THE NOBEL PRIZE** In 1998, three Americans, Louis Ignarro, Robert Furchgott, and Fred Murad, were awarded the Nobel Prize for Medicine. They shared the prize money equally. If each person received $318,500, what was the amount of the Nobel Prize cash award? **See Example 4.**

Analyze
- $\boxed{3}$ people shared the cash award equally.   Given
- Each person received $\boxed{\$318{,}500}$.   Given
- What was the $\boxed{\text{amount}}$ of the Nobel Prize cash award?   Find

Assign
Let $x =$ the $\boxed{\text{amount}}$ of the Nobel Prize cash award.

Form
Now we look for a key word or phrase in the problem.

   Key phrase: *shared the prize money equally*
   Translation: division

Now we translate the words of the problem into an equation.

The Nobel Prize cash award	divided by	the number of prize winners	is equal to	each person's share.
x	÷	3	=	318,500

Solve

$\dfrac{x}{3} = 318{,}500$

$\boxed{3} \cdot \dfrac{x}{3} = \boxed{3} \cdot 318{,}500$

$x = \boxed{955{,}500}$

State
The amount of the Nobel Prize cash award was $\boxed{\$955{,}500}$.

Check
If we divide the Nobel Prize cash award by 3, we should get the amount each winner received.

$\dfrac{\$955{,}500}{3} = \boxed{\$318{,}500}$ This is the amount each winner received.

The result checks.

26. **THE STOCK MARKET** An investor has seen the value of his stock double in the last 12 months. If the current value of his stock is $274,552, what was its value one year ago?

Analyze

- The value of the stock <u>doubled</u> in 12 months. *Given*
- The current value of the stock is <u>$274,552</u>. *Given*
- What was the <u>value</u> of the stock one year ago? *Find*

Assign
We can let x = the <u>value</u> of the stock one year ago (in dollars).

Form
We now look for a key phrase in the problem.

Key phrase: *double* **Translation:** <u>multiply</u> by 2

Now we translate the words of the problem into an equation.

2	times	the value of the stock one year ago	is equal to	the current value of the stock.
2	·	x	=	274,552

Solve

$2x = 274{,}552$

$\dfrac{2x}{2} = \dfrac{274{,}552}{2}$

$x = 137{,}276$

State
The value of the stock one year ago was <u>$137,276</u>.

Check If we multiply the value that the stock had one year ago by 2, we should get its current value.

$137{,}276 \times 2 = \$274{,}552$ This is the current value of the stock.

The result checks.

TRY IT YOURSELF

Solve each equation and check the result.

27. $100 = 100x$ 1
28. $35 = 35y$ 1
29. $\dfrac{a}{15} = 5$ 75
30. $\dfrac{b}{25} = 5$ 125
31. $16 = 8r$ 2
32. $44 = 11m$ 4
33. $21s = 210$ 10
34. $155 = 31x$ 5
35. $\dfrac{c}{1{,}000} = 3$ 3,000
36. $\dfrac{d}{100} = 11$ 1,100
37. $1 = \dfrac{x}{50}$ 50
38. $1 = \dfrac{x}{25}$ 25
39. $7 = \dfrac{t}{7}$ 49
40. $4 = \dfrac{m}{4}$ 16
41. $7x = 21$ 3
42. $13x = 52$ 4
43. $172 = 43t$ 4
44. $288 = 96t$ 3
45. $\dfrac{d}{20} = 201$ 4,020
46. $\dfrac{x}{60} = 106$ 6,360
47. $417 = \dfrac{t}{3}$ 1,251
48. $259 = \dfrac{y}{7}$ 1,813
49. $170y = 5{,}100$ 30
50. $190y = 7{,}600$ 40
51. $\dfrac{t}{3} = 47$ 141
52. $\dfrac{d}{9} = 83$ 747
53. $34y = 204$ 6
54. $18y = 162$ 9

APPLICATIONS

Let a variable represent the unknown quantity. Then write and solve an equation to answer the question.

55. **SPEED READING** An advertisement for a speed reading program claimed that successful completion of the course could triple a person's reading rate. After taking the course, Alicia can now read 399 words per minute. If the company's claims are true, what was her reading rate before taking the course?
Before the course, Alicia could read 133 words per minute.

56. **PHYSICAL EDUCATION** A high school PE teacher had the students in her class form three-person teams for a basketball tournament. Thirty-two teams participated in the tournament. How many students were in the PE class?
There were 96 students in the class.

57. **COST OVERRUNS** Lengthy delays and skyrocketing costs caused a rapid-transit construction project to go over budget by a factor of 10. The final audit showed the project costing $540 million. What was the initial cost estimate?
The initial cost estimate was $54 million.

58. **STAMPS** Large sheets of commemorative stamps honoring Marilyn Monroe are to be printed. On each sheet, there are 112 stamps, with 8 stamps per row. How many rows of stamps are on a sheet? There are 14 rows of stamps on the sheet.

Carolina K. Smith MD/Shutterstock.com

59. SPREADSHEETS The grid shown below is a computerized spreadsheet. The rows are labeled with numbers, and the columns are labeled with letters. Each empty box of the grid is called a *cell*. Suppose a certain project calls for a spreadsheet with 294 cells, using columns A through F. How many rows will need to be used? 49 rows will need to be used.

60. LOTTO WINNERS The grocery store employees listed below pooled their money to buy $120 worth of lottery tickets each week, with the understanding they would split the prize equally if they happened to win. One week they did have the winning ticket and won $480,000. What was each employee's share of the winnings?

Each employee's share of the winnings was $40,000.

Sam M. Adler	Ronda Pellman	Manny Fernando
Lorrie Jenkins	Tom Sato	Sam Lin
Kiem Nguyen	H. R. Kinsella	Tejal Neeraj
Virginia Ortiz	Libby Sellez	Alicia Wen

61. ANIMAL SHELTERS The number of phone calls to an animal shelter quadrupled after the evening news showed a segment explaining the services the shelter offered. Before the publicity, the shelter received 8 calls a day. How many calls did the shelter receive each day after being featured on the news? The shelter received 32 calls each day after being featured on the news.

62. OPEN HOUSES The attendance at an elementary school open house was only half of what the principal had expected. If 120 people visited the school that evening, how many had she expected to attend? The principal expected 240 people to attend.

63. GRAVITY The weight of an object on Earth is 6 times greater than what it is on the moon. The situation shown to the right took place on Earth. If it took place on the moon, what weight would the scale register? The scale would register 55 pounds.

On Earth

64. INFOMERCIALS The number of orders received each week by a company selling skin care products increased fivefold after a Hollywood celebrity was added to the company's infomercial. After adding the celebrity, the company received about 175 orders each week. How many orders were received each week before the celebrity took part? Before the celebrity took part, 35 orders were received each week.

65. LIFE SPAN The average life span of an Amazon parrot is 104 years. That is thirteen times longer than the average life span of a Guinea pig. Find the average life span of a Guinea pig. (Source: petdoc.com) The average life span of a Guinea pig is 8 years.

66. CHILI HEAT SCALE In 1912, a chemist by the name of Wilbur Scoville developed a method to measure the heat level of chili peppers. For example, the heat rating on the Scoville scale for a habanero chili is 320,000 units. That is forty times the heat rating of a jalapeño chili. What is the Scoville rating for a jalapeño chili? (Source: ushotstuff.com) The Scoville rating for a jalapeño chili is 8,000 units.

WRITING

67. Explain what the pair of figures on page 106 are trying to show.

68. Draw a pair of figures like those on page 107. Explain what the figures illustrate.

69. What does it mean to *solve an equation*?

70. Think of a number. Double it. Now divide it by 2. Explain why you always obtain the original number.

REVIEW

71. Find the perimeter of a rectangle with sides measuring 8 cm and 16 cm. 48 cm

72. Find the area of a rectangle with sides measuring 23 inches and 37 inches. 851 in.^{2}

73. Find the prime factorization of 120. $2^3 \cdot 3 \cdot 5$

74. Find the prime factorization of 150. $2 \cdot 3 \cdot 5^2$

75. Evaluate: $3^2 \cdot 2^3$ 72

76. Evaluate: $5 + 6 \cdot 3$ 23

77. Divide, if possible: $\dfrac{0}{12}$ 0

78. Divide, if possible: $\dfrac{50}{0}$ undefined

CHAPTER 1 SUMMARY AND REVIEW

SECTION 1.1 An Introduction to the Whole Numbers

DEFINITIONS AND CONCEPTS	EXAMPLES
The set of **whole numbers** is {0, 1, 2, 3, 4, 5, …}. When a whole number is written using the **digits** 0, 1, 2, 3, 4, 5, 6, 7, 8, 9, it is said to be in **standard form**.	Some examples of whole numbers written in standard form are: 2, 16, 530, 7,894, and 3,201,954
The position of a digit in a whole number determines its **place value**. A **place-value chart** shows the place value of each digit in the number. To make large whole numbers easier to read, we use commas to separate their digits into groups of three, called **periods**.	*PERIODS*: Trillions, Billions, Millions, Thousands, Ones (Hundred trillions, Ten trillions, Trillions, Hundred billions, Ten billions, Billions, Hundred millions, Ten millions, Millions, Hundred thousands, Ten thousands, Thousands, Hundreds, Tens, Ones) 5,206,379,814,256 The place value of the digit 7 is 7 ten millions. The digit 4 tells the number of thousands.
To **write a whole number in words**, start from the left. Write the number in each period followed by the name of the period (except for the *ones period*, which is not used). Use commas to separate the periods. To **read a whole number out loud**, follow the same procedure. The commas are read as slight pauses.	Millions Thousands Ones 2,568,019 Two **million**, five hundred sixty-eight **thousand**, nineteen
To change from the **written-word form of a number to standard form**, look for the commas. Commas are used to separate periods.	Six billion , forty-one million , two hundred eight thousand , thirty-six 6,041,208,036
To write a number in **expanded form** (**expanded notation**) means to write it as an addition of the place values of each of its digits.	The expanded form of 32,159 is: 30,000 + 2,000 + 100 + 50 + 9
Whole numbers can be shown by drawing points on a **number line**.	The graphs of 3 and 7 are shown on the number line below. 0 1 2 3 4 5 6 7 8
Inequality symbols are used to compare whole numbers: > means *is greater than* < means *is less than*	9 > 8 and 2,343 > 762 1 < 2 and 9,000 < 12,453

When we don't need exact results, we often **round** numbers.	Round 9,842 to the nearest ten.
Rounding a Whole Number	Rounding digit: tens column → 9,842 ← Test digit: Since 2 is less than 5, leave the rounding digit unchanged and replace the test digit with 0.
1. To round a number to a certain place value, locate the **rounding digit** in that place.	Thus, 9,842 rounded to the nearest ten is 9,840.
2. Look at the **test digit**, which is directly to the right of the rounding digit.	Round 63,179 to the nearest hundred.
3. If the test digit is 5 or greater, round up by adding 1 to the rounding digit and replacing all of the digits to its right with 0. If the test digit is less than 5, replace it and all of the digits to its right with 0.	Rounding digit: hundreds column → 63,179 ← Test digit: Since 7 is 5 or greater, add 1 to the rounding digit and replace all the digits to its right with 0. Thus, 63,179 rounded to the nearest hundred is 63,200.
Whole numbers are often used in **tables**, **bar graphs**, and **line graphs**.	See page 9 for an example of a table, a bar graph, and a line graph.

REVIEW EXERCISES

Consider the number 41,948,365,720.

1. a. Which digit is in the ten thousands column? 6
   b. Which digit is in the hundreds column? 7
   c. What is the place value of the digit 1? 1 billion
   d. Which digit tells the number of millions? 8

2. Write each number in words.
   a. 97,283 ninety-seven thousand, two hundred eighty-three
   b. 5,444,060,017
      five billion, four hundred forty-four million, sixty thousand, seventeen

3. Write each number in standard form.
   a. Three thousand, two hundred seven 3,207
   b. Twenty-three million, two hundred fifty-three thousand, four hundred twelve 23,253,412

4. Write 60,000 + 1,000 + 200 + 4 in standard form.
   61,204

Write each number in expanded form.

5. 570,302 500,000 + 70,000 + 300 + 2

6. 37,309,154
   30,000,000 + 7,000,000 + 300,000 + 9,000 + 100 + 50 + 4

Graph the following numbers on a number line.

7. 0, 2, 8, 10

8. The whole numbers between 3 and 7.

Place an < or an > symbol in the box to make a true statement.

9. 9 > 7

10. 301 < 310

11. Round 2,507,348
    a. to the nearest hundred 2,507,300
    b. to the nearest ten thousand 2,510,000
    c. to the nearest ten 2,507,350
    d. to the nearest million 3,000,000

12. Round 969,501
    a. to the nearest thousand 970,000
    b. to the nearest hundred thousand 1,000,000

13. CONSTRUCTION The following table lists the number of building permits issued in the city of Springsville for the period 2006–2013.

Year	2006	2007	2008	2009	2010	2011	2012	2013
Building permits	12	13	10	7	9	14	6	5

a. Construct a bar graph of the data.

b. Construct a line graph of the data.

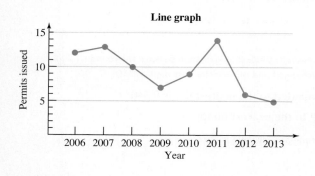

14. GEOGRAPHY The names and lengths of the five longest rivers in the world are listed below. Write them in order, beginning with the longest.

Nile, Amazon, Yangtze, Mississippi-Missouri, Ob-Irtysh

Amazon (South America)	4,049 mi
Mississippi-Missouri (North America)	3,709 mi
Nile (Africa)	4,160 mi
Ob-Irtysh (Russia)	3,459 mi
Yangtze (China)	3,964 mi

Source: geography.about.com

SECTION 1.2 Adding and Subtracting Whole Numbers

DEFINITIONS AND CONCEPTS	EXAMPLES
To **add whole numbers**, think of combining sets of similar objects. **Vertical form:** Stack the addends. Add the digits in the ones column, the tens column, the hundreds column, and so on. **Carry** when necessary.	Add: $10{,}892 + 5{,}467 + 499$ Carrying $\quad 1\ 2\ 1$ $\ \ 10{,}892\ \leftarrow$ Addend $\quad\ 5{,}467\ \leftarrow$ Addend $+\ \ \ \ 499\ \leftarrow$ Addend $\ \ 16{,}858\ \leftarrow$ Sum To check, add bottom to top.
A **variable** is a letter (or symbol) that stands for a number.	Variables: x, a, and y
Commutative property of addition: The order in which whole numbers are added does not change their sum. For any whole numbers a and b, $a + b = b + a$	$6 + 5 = 5 + 6$ By the commutative property, the sum is the same.
Associative property of addition: The way in which whole numbers are grouped does not change their sum. For any whole numbers a and b, $(a + b) + c = a + (b + c)$	$(17 + 5) + 25 = 17 + (5 + 25)$ By the associative property, the sum is the same.
To estimate a sum, use **front-end rounding** to approximate the addends. Then add.	Estimate the sum: $\quad 7{,}219 \rightarrow \ \ 7{,}000$ Round to the nearest thousand. $\quad\ \ \ 592 \rightarrow \ \ \ \ \ 600$ Round to the nearest hundred. $+3{,}425 \rightarrow +3{,}000$ Round to the nearest thousand. $\qquad\qquad\ \ \ \ 10{,}600$ The estimate is 10,600.

To solve the application problems, we must often *translate* the **key words** and **phrases** of the problem to numbers and symbols. Some key words and phrases that are often used to indicate addition are: gain　　increase　　up　　forward rise　　more than　　total　　combined in all　　in the future　　extra　　altogether	Translate the words to numbers and symbols: VACATIONS There were 3,901,408 visitors to Yosemite National Park in 2010. The following year, attendance increased by 49,985. How many people visited the park in 2011? The phrase *increased by* indicates addition: The number of visitors to the park in 2011 = 3,901,408 + 49,985
The distance around a rectangle or a square is called its **perimeter**. Perimeter of a rectangle = length + length + width + width Perimeter of a square = side + side + side + side	Find the perimeter of the rectangle shown below. 15 ft 10 ft Perimeter = 15 + 15 + 10 + 10　　Add the two lengths and the two widths. Perimeter = 50 The perimeter of the rectangle is 50 feet.
To **subtract whole numbers**, think of taking away objects from a set. **Vertical form:** Stack the numbers. Subtract the digits in the ones column, the tens column, the hundreds column, and so on. **Borrow** when necessary. To **check:** Difference + subtrahend = minuend	Subtract: 4,957 − 869
Be careful when translating the instruction to subtract one number *from* another number. The order of the numbers in the sentence must be reversed when we translate to symbols.	Translate the words to numbers and symbols: Subtract 41 from 97.　　Since 41 is the number to be subtracted, it is the subtrahend. 97 − 41
Every subtraction has a **related addition statement**.	10 − 3 = 7　　because　　7 + 3 = 10
To estimate a difference, use **front-end rounding** to approximate the minuend and subtrahend. Then subtract.	Estimate the difference: 59,033 → 60,000　　Round to the nearest ten thousand. − 4,124 → − 4,000　　Round to the nearest thousand. 　　　　　　56,000 The estimate is 56,000.

Some of the **key words** and **phrases** that are often used to indicate subtraction are:

loss	decrease	down	backward
fell	less than	fewer	reduce
remove	debit	in the past	remains
declined			take away

To answer questions about *how much more* or *how many more*, we use subtraction.

WEIGHTS OF CARS A Chevy Suburban weighs 5,607 pounds and a Smart car weighs 1,852 pounds. How much heavier is the Suburban?

The phrase *how much heavier* indicates subtraction:

$$\begin{array}{r} 5{,}607 \\ -1{,}852 \\ \hline 3{,}755 \end{array}$$

Weight of the Suburban
Weight of the Smart car

The Suburban weighs 3,755 pounds more than the Smart car.

To **evaluate** (find the value of) expressions that involve addition and subtraction written in **horizontal form**, we perform the operations as they occur *from left to right*.

Evaluate: $75 - 23 + 9$

$75 - 23 + 9 = 52 + 9$ Working left to right, do the subtraction first.
$ = 61$ Now do the addition.

REVIEW EXERCISES

Add.

15. $27 + 436$ 463

16. $4 + (36 + 19)$ 59

17. $\begin{array}{r} 5{,}345 \\ + 655 \\ \hline 6{,}000 \end{array}$

18. $2 + 1 + 38 + 3 + 6$ 50

19. $4{,}447 + 7{,}478 + 676$ 12,601

20. $\begin{array}{r} 32{,}812 \\ 65{,}034 \\ +54{,}323 \\ \hline 152{,}169 \end{array}$

21. Use front-end rounding to estimate the sum.

$615 + 789 + 14{,}802 + 39{,}902 + 8{,}098$ 59,400

22. a. Use the commutative property of addition to complete the following:

$24 + 61 = \boxed{61 + 24}$

b. Use the associative property of addition to complete the following:

$9 + (91 + 29) = \boxed{(9 + 91) + 29}$

23. AIRPORTS The nation's three busiest airports in 2011 are listed below. Find the total number of passengers passing through those airports.
220,952,316 passengers

Airport	Total passengers
Hartsfield-Jackson Atlanta	92,389,023
Chicago O'Hare	66,701,241
Los Angeles International	61,862,052

Source: Airports Council International–North America

24. Add from bottom to top to check the sum. Is it correct? no

$\begin{array}{r} 1{,}291 \\ 859 \\ 345 \\ + 226 \\ \hline 1{,}821 \end{array}$

25. What is the sum of three thousand, seven hundred six and ten thousand, nine hundred fifty-five? 14,661

26. What is 451,775 more than 327,891? 779,666

27. NEW HOMES In 1974, the average size of a newly built home in the United States was 1,695 square feet. By 2011, the average size had increased by 785 square feet. What was the average size of a new home in 2011? (Source: U.S. Census Bureau) 2,480 square feet

28. Find the perimeter of the rectangle shown below.
2,746 ft

Subtract.

29. $148 - 87$ 61

30. Subtract 10,218 from 10,435. 217

31. $750 - 259 + 14$ 505

32. $\begin{array}{r} 7{,}800 \\ -5{,}725 \\ \hline 2{,}075 \end{array}$

33. Check the subtraction using addition.

$\begin{array}{r} 8{,}017 \\ -6{,}949 \\ \hline 1{,}168 \end{array}$ incorrect

34. Fill in the blank: 20 − 8 = 12 because 12 + 8 = 20 .

35. Estimate the difference: 181,232 − 44,810   160,000

36. LAND AREA Use the data in the table below to determine how much larger the land area of Russia is compared to that of Canada. 2,772,459 square miles

Country	Land area (square miles)
Russia	6,323,482
Canada	3,551,023

Source: *The World Almanac and Book of Facts*, 2013

37. BANKING A savings account contains $12,975. If the owner makes a withdrawal of $3,800 and later deposits $4,270, what is the new account balance? $13,445

38. SUNNY DAYS In the United States, the city of Yuma, Arizona, typically has the most sunny days per year—about 242. The city of Buffalo, New York, typically has 188 days fewer than that. How many sunny days per year does Buffalo have? 54 days

SECTION 1.3 Multiplying Whole Numbers

DEFINITIONS AND CONCEPTS	EXAMPLES
Multiplication of whole numbers is repeated addition but with different notation.	Repeated addition: The sum of four 6's Multiplication 6 + 6 + 6 + 6 = 4 × 6 = 24
To write multiplication, we use a times symbol ×, a raised dot ·, or parentheses ( ).	4 × 6 4 · 6 4(6) or (4)(6) or (4)6
When multiplying a variable by a number or a variable by a variable, we can omit the symbol for multiplication.	$3x$ means $3 \cdot x$ and ab means $a \cdot b$

Vertical form: Stack the factors. If the bottom factor has more than one digit, multiply in steps to find the partial products. Then add them to find the product.	Multiply: 24 · 163
	
	     163 ← Factor
	  × 24 ← Factor
	     652 ← Partial product: 4 · 163
	   3260 ← Partial product: 20 · 163
	  3,912 ← Product

To find the **product of a whole number and 10, 100, 1,000, and so on**, attach the number of zeros in that number to the right of the whole number. This rule can be extended to multiply any two whole numbers that end in zeros.	Multiply: 8 · 1,000 = 8,000 Since 1,000 has three zeros, attach three 0's after 8. 43(10,000) = 430,000 Since 10,000 has four zeros, attach four 0's after 43. 160 · 20,000 = 3,200,000 160 and 20,000 have a total of five trailing zeros. Attach five 0's after 32. Multiply 16 and 2 to get 32.

Multiplication Properties of 0 and 1
The product of any whole number and 0 is 0.
   For any whole number a,
      $a \cdot 0 = 0$ and $0 \cdot a = 0$
The product of any whole number and 1 is that whole number.
   For any whole number a,
      $a \cdot 1 = a$ and $1 \cdot a = a$
	$0 \cdot 9 = 0$ and $3(0) = 0$
	
	$15 \cdot 1 = 15$ and $1(6) = 6$

Commutative property of multiplication: The order in which whole numbers are multiplied does not change their product. For any whole numbers a and b, $ab = ba$ **Associative property of multiplication:** The way in which whole numbers are grouped does not change their product. For any whole numbers a and b, $(ab)c = a(bc)$	$5 \cdot 9 = 9 \cdot 5$ By the commutative property, the product is the same. $(3 \cdot 7) \cdot 10 = 3 \cdot (7 \cdot 10)$ By the associative property, the product is the same.
To **estimate** a product, use **front-end rounding** to approximate the factors. Then multiply.	To estimate the product for $74 \cdot 873$, find $70 \cdot 900$. ┌─ Round to the nearest ten ─┐ $74 \cdot 873$ $70 \cdot 900$ └─ Round to the nearest hundred ─┘
Application problems that involve **repeated addition** are often more easily solved using multiplication.	HEALTH CARE A doctor's office is open 210 days a year. Each day the doctor sees 25 patients. How many patients does the doctor see in 1 year? This **repeated addition** can be calculated by multiplication: The number of patients seen each year $= 25 \cdot 210$

We can use multiplication to count objects arranged in rectangular patterns of neatly arranged rows and columns called **rectangular arrays**.
 	
Some **key words** that are often used to indicate multiplication are:
 	
    double   triple   twice
    of   times   tenfold
	CLASSROOMS A large lecture hall has 16 rows of desks and there are 12 desks in each row. How many desks are in the lecture hall?
	
	The **rectangular array** of desks indicates multiplication:
	
	The number of desks in the lecture hall $= 16 \cdot 12$

The **area of a rectangle** is the measure of the amount of surface it encloses. Area is measured in square units, such as square inches (written in.^{2}) or square centimeters (written cm^2).
 	
  Area of a rectangle $=$ length $\cdot$ width
                      or
                        $A = lw$
 	
Letters (or symbols) that are used to represent numbers are called **variables**.
	Find the area of the rectangle shown below.
	
	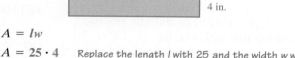
	
	$A = lw$
	$A = 25 \cdot 4$ Replace the length l with 25 and the width w with 4.
	$A = 100$ Multiply.
	
	The area of the rectangle is 100 square inches, which can be written in more compact form as 100 in.^{2}.

REVIEW EXERCISES

Multiply.

39. 47×9 423

40. $5 \cdot (7 \cdot 6)$ 210

41. $72 \cdot 10,000$ 720,000

42. $157 \cdot 59$ 9,263

43. $\begin{array}{r} 5,624 \\ \times\ 281 \\ \hline 1,580,344 \end{array}$

44. $502 \cdot 459$ 230,418

45. Estimate the product: $6,891 \cdot 438$ 2,800,000

46. a. Write the repeated addition $7 + 7 + 7 + 7 + 7$ as a multiplication. $5 \cdot 7$

  b. Write $2 \cdot t$ in simpler form. $2t$

  c. Write $m \cdot n$ in simpler form. mn

47. Find each product:

  a. $8 \cdot 0$ 0

  b. $7 \cdot 1$ 7

48. What property of multiplication is shown?

  a. $2 \cdot (5 \cdot 7) = (2 \cdot 5) \cdot 7$ associative property of multiplication

  b. $100(50) = 50(100)$ commutative property of multiplication

Find the area of the rectangle and the square.

49.

8 cm, 4 cm, 32 cm²

50.

78 in., 78 in., 6,084 in.²

51. SLEEP The National Sleep Foundation recommends that adults get from 7 to 9 hours of sleep each night.

  a. How many hours of sleep is that in one year using the smaller number? (Use a 365-day year.) 2,555 hr

  b. How many hours of sleep is that in one year using the larger number? 3,285 hr

52. GRADUATION For a graduation ceremony, the graduates were assembled in a rectangular 22-row and 15-column formation. How many members are in the graduating class? 330 members

53. PAYCHECKS Sarah worked 12 hours at $9 per hour, and Santiago worked 14 hours at $8 per hour. Who earned more money? Santiago

54. SHOPPING There are 12 eggs in one dozen, and 12 dozen in one gross. How many eggs are in a shipment of 100 gross? 14,400 eggs

SECTION 1.4 Dividing Whole Numbers

DEFINITIONS AND CONCEPTS	EXAMPLES
To **divide whole numbers**, think of separating a quantity into equal-sized groups. To write division, we can use a division symbol $\div$, a long division symbol $\overline{)\ }$, or a fraction bar —.	Dividend $\downarrow$ Divisor $\downarrow$ $8 \div 2 = 4$ $\uparrow$ Quotient $2\overline{)8}^{\,4}$ $\dfrac{8}{2} = 4$
Another way to answer a division problem is to think in terms of multiplication and write a **related multiplication statement**.	$8 \div 2 = 4$ because $4 \cdot 2 = 8$
A process called **long division** can be used to divide whole numbers. Follow a four-step process: • Estimate • Multiply • Subtract • Bring down	Divide: $8,317 \div 23$ Quotient: 361 R 14; Divisor → 23)8,317 ← Dividend; −69; 141; −138; 37; −23; 14 — Remainder

Unless otherwise noted, all art on this page is © Cengage Learning.

To **check** the result of a division, we multiply the divisor by the quotient and add the remainder. The result should be the dividend.	For the division on the previous page, the result checks. Quotient · divisor remainder $(361 \cdot 23) + 14 = 8{,}303 + 14$ $ = 8{,}317 \leftarrow$ Dividend
Properties of Division Any whole number divided by 1 is equal to that number. For any whole number a, $$\frac{a}{1} = a.$$ Any nonzero whole number divided by itself is equal to 1. For any nonzero whole number a, $$\frac{a}{a} = 1$$	$\dfrac{4}{1} = 4$ and $\dfrac{58}{1} = 58$ $\dfrac{9}{9} = 1$ and $\dfrac{103}{103} = 1$
Division with Zero Zero divided by any nonzero number is equal to 0. Division by 0 is undefined. For any nonzero whole number a, $$\frac{0}{a} = 0 \quad \text{and} \quad \frac{a}{0} \text{ is undefined}$$	$\dfrac{0}{7} = 0$ and $\dfrac{0}{23} = 0$ $\dfrac{7}{0}$ is undefined and $\dfrac{2{,}190}{0}$ is undefined
There are **divisibility tests** to help us decide whether one number is divisible by another. They are listed on page 55.	Is 21,507 divisible by 3? 21,507 is divisible by 3, because the sum of its digits is divisible by 3. $2 + 1 + 5 + 0 + 7 = 15$ and $15 \div 3 = 5$
There is a shortcut for **dividing a dividend by a divisor when both end with zeros**. We simply *remove the ending zeros in the divisor and remove the same number of ending zeros in the dividend*.	Divide: $64{,}000 \div 1{,}600 = 640 \div 16$ Remove two zeros from the dividend and the divisor, and divide.
To **estimate quotients**, we use a method that approximates both the dividend and the divisor so that they divide easily.	Estimate the quotient for $154{,}908 \div 51$ by finding $150{,}000 \div 50$. ⎡— The dividend is approximately —⎤ $154{,}908 \div 51 \quad\quad 150{,}000 \div 50$ ⎣— The divisor is approximately —⎦
Application problems that involve **forming equal-sized groups** can be solved by division. Some **key words** and **phrases** that are often used to indicate division: *split equally distributed equally shared equally how many does each how many left (remainder) per how much extra (remainder) among*	BRACES An orthodontist offers his patients a plan to pay the $5,400 cost of braces in 36 equal payments. What is the amount of each payment? The phrase *36 equal payments* indicates division: The amount of each payment $= 5{,}400 \div 36$

REVIEW EXERCISES

Divide, if possible.

55. $\dfrac{72}{4}$ 18

56. $1{,}443 \div 39$ 37

57. $68\overline{)20{,}876}$ 307

58. $21\overline{)405}$ 19 R 6

59. $\dfrac{0}{10}$ 0

60. $\dfrac{165}{0}$ undefined

61. $127\overline{)5{,}347}$ 42 R 13

62. $1{,}482{,}000 \div 3{,}900$ 380

63. Write the related multiplication statement for $160 \div 4 = 40$. $40 \cdot 4 = 160$

64. Use a check to determine whether the following division is correct. It is not correct.

$$\begin{array}{r} 45 \text{ R } 6 \\ 7\overline{)320} \end{array}$$

65. Is 364,545 divisible by 2, 3, 4, 5, 6, 9, or 10? It is divisible by 3, 5, and 9.

66. Estimate the quotient: $210{,}999 \div 53$ 4,000

67. TREATS If 745 candies are distributed equally among 45 children, how many will each child receive? How many candies will be left over? 16; 25

68. PURCHASING A county received an $850,000 grant to purchase some new police patrol cars. If a fully equipped patrol car costs $25,000, how many can the county purchase with the grant money? 34 cars

SECTION 1.5 Prime Factors and Exponents

DEFINITIONS AND CONCEPTS	EXAMPLES
Numbers that are multiplied together are called **factors**. To **factor** a whole number means to express it as the product of other whole numbers. If a whole number is a factor of a given number, it also *divides the given number exactly.*	The pairs of whole numbers whose product is 6 are: $1 \cdot 6 = 6$ and $2 \cdot 3 = 6$ From least to greatest, the factors of 6 are 1, 2, 3, and 6. Each of the factors of 6 divides 6 exactly (no remainder): $\dfrac{6}{1} = 6 \quad \dfrac{6}{2} = 3 \quad \dfrac{6}{3} = 2 \quad \dfrac{6}{6} = 1$
If a whole number is divisible by 2, it is called an **even** number. If a whole number is not divisible by 2, it is called an **odd** number.	Even whole numbers: 0, 2, 4, 6, 8, 10, 12, 14, 16, 18, … Odd whole numbers: 1, 3, 5, 7, 9, 11, 13, 15, 17, 19, …
A **prime number** is a whole number greater than 1 that has only 1 and itself as factors. There are infinitely many prime numbers.	Prime numbers: 2, 3, 5, 7, 11, 13, 17, 19, 23, 29, 31, …
The **composite numbers** are whole numbers greater than 1 that are *not* prime. There are infinitely many composite numbers.	Composite numbers: 4, 6, 8, 9, 10, 12, 14, 15, 16, 18, …

To find the **prime factorization** of a whole number means to write it as the product of only prime numbers.

A **factor tree** and a **division ladder** can be used to find prime factorizations.

Use a *factor tree* to find the prime factorization of 30.

Factor each number that is encountered as a product of two whole numbers (other than 1 and itself) until all the factors involved are prime.

The prime factorization of 30 is $2 \cdot 3 \cdot 5$.

Use a *division ladder* to find the prime factorization of 70.

$$\begin{array}{r} 2\,\underline{|70} \\ 5\,\underline{|35} \\ 7 \end{array}$$

Perform repeated divisions by prime numbers until the final quotient is itself a prime number.

The prime factorization of 70 is $2 \cdot 5 \cdot 7$.

An **exponent** is used to indicate repeated multiplication. It tells how many times the **base** is used as a factor.

$$\underbrace{2 \cdot 2 \cdot 2 \cdot 2}_{\text{Repeated factors}} = 2^{\overset{\text{Exponent}}{4}}$$

Base

2^4 is called an exponential expression.

We can use the definition of exponent to **evaluate** (find the value of) exponential expressions.

Evaluate: 7^3

$\begin{aligned} 7^3 &= 7 \cdot 7 \cdot 7 & \text{Write the base 7 as a factor 3 times.} \\ &= 49 \cdot 7 & \text{Multiply, working left to right.} \\ &= 343 & \text{Multiply.} \end{aligned}$

Evaluate: $2^2 \cdot 3^3$

$\begin{aligned} 2^2 \cdot 3^3 &= 4 \cdot 27 & \text{Evaluate the exponential expressions first.} \\ &= 108 & \text{Multiply.} \end{aligned}$

REVIEW EXERCISES

Find all of the factors of each number. List them from least to greatest.

69. 18   1, 2, 3, 6, 9, 18

70. 75   1, 3, 5, 15, 25, 75

71. Factor 20 using two factors. *Do not use the factor 1 in your answer.* $2 \cdot 10$ or $4 \cdot 5$

72. Factor 54 using three factors. *Do not use the factor 1 in your answer.* $2 \cdot 3 \cdot 9$ or $3 \cdot 3 \cdot 6$

Tell whether each number is a prime number, a composite number, or neither.

73. a. 31   prime   **b.** 100   composite

   c. 1   neither   **d.** 0   neither

   e. 125   composite   **f.** 47   prime

Tell whether each number is an even or an odd number.

74. a. 171   odd   **b.** 214   even

   c. 0   even   **d.** 1   odd

Find the prime factorization of each number. Use exponents in your answer, when helpful.

75. 42 $2 \cdot 3 \cdot 7$ **76.** 75 $3 \cdot 5^2$

77. 220 $2^2 \cdot 5 \cdot 11$ **78.** 140 $2^2 \cdot 5 \cdot 7$

Write each expression using exponents.

79. $6 \cdot 6 \cdot 6 \cdot 6$ 6^4 **80.** $5(5)(5)(13)(13)$ $5^3 \cdot 13^2$

Evaluate each expression.

81. 5^3 125   **82.** 11^2 121

83. $2^4 \cdot 7^2$ 784   **84.** $2^2 \cdot 3^3 \cdot 5^2$ 2,700

SECTION 1.6 The Least Common Multiple and the Greatest Common Factor

DEFINITIONS AND CONCEPTS	EXAMPLES
The **multiples** of a number are the products of that number and 1, 2, 3, 4, 5, and so on.	Multiples of 2: 2, 4, **6**, 8, 10, **12**, 14, 16, **18**, 20, 22, **24**, … Multiples of 3: 3, **6**, 9, **12**, 15, **18**, 21, **24**, 27, … The common multiples of 2 and 3 are: 6, 12, 18, 24, 30, …
The **least common multiple (LCM)** of two whole numbers is the smallest common multiple of the numbers. The **LCM** of two whole numbers is the *smallest* whole number that is divisible by both of those numbers.	The least common multiple of 2 and 3 is 6, which is written as: LCM (2, 3) = 6. $\frac{6}{2} = 3$ and $\frac{6}{3} = 2$

To **find the LCM** of two (or more) whole numbers by **listing**:
 1. Write multiples of the largest number by multiplying it by 1, 2, 3, 4, 5, and so on.
 2. Continue this process until you find the *first* multiple of the larger number that is divisible by each of the smaller numbers. That multiple is their LCM.
	Find the LCM of 3 and 5.
	 Multiples of 5: 5, 10, 15, 20, 25, …
	                 Not divisible Not divisible Divisible by 3.
	                 by 3. by 3.
	 Since 15 is the first multiple of 5 that is divisible by 3, the LCM (3, 5) = 15.

To **find the LCM** of two (or more) whole numbers using **prime factorization**:
 1. Prime factor each number.
 2. The LCM is a product of prime factors, where each factor is used the greatest number of times it appears in any one factorization.
	Find the LCM of 6 and 20.
	 $6 = 2 \cdot \textcircled{3}$ The greatest number of times 3 appears is once.
	 $20 = \textcircled{2 \cdot 2} \cdot \textcircled{5}$ The greatest number of times 2 appears is twice.
	               The greatest number of times 5 appears is once.
	          Use the factor 2 two times.
	          Use the factor 3 one time.
	          Use the factor 5 one time.
	 LCM (6, 20) = $2 \cdot 2 \cdot 3 \cdot 5 = 60$

The **greatest common factor (GCF)** of two (or more) whole numbers is the largest common factor of the numbers. The greatest common factor of two (or more) numbers is the *largest* whole number that divides them exactly.	The factors of 18: ①, ②, ③, ⑥, 9, 18 The factors of 30: ①, ②, ③, 5, ⑥, 10, 15, 30 The common factors of 18 and 30 are 1, 2, 3, and 6. The greatest common factor of 18 and 30 is 6, which is written as: GCF (18, 30) = 6. $\frac{18}{6} = 3$ and $\frac{30}{6} = 5$
To **find the GCF** of two (or more) whole numbers using **prime factorization**: 1. Prime factor each number. 2. Identify the common prime factors. 3. The GCF is a product of all the common prime factors found in Step 2. If there are no common prime factors, the GCF is 1.	Find the GCF of 36 and 60. $36 = \textcircled{2} \cdot \textcircled{2} \cdot \textcircled{3} \cdot 3$ 36 and 60 have two common factors $60 = \textcircled{2} \cdot \textcircled{2} \cdot \textcircled{3} \cdot 5$ of 2 and one common factor of 3. The GCF is the product of the circled prime factors. GCF (36, 60) = $2 \cdot 2 \cdot 3 = 12$

REVIEW EXERCISES

85. Find the first ten multiples of 9.
    9, 18, 27, 36, 45, 54, 63, 72, 81, 90

86. a. Find the common multiples of 6 and 8 in the lists below.

    Multiples of 6: 6, 12, 18, 24, 30, 36, 42, 48, 54 …
    Multiples of 8: 8, 16, 24, 32, 40, 48, 56, 64, 72 …
    24, 48

    b. Find the common factors of 6 and 8 in the lists below.

    Factors of 6: 1, 2, 3, 6   1, 2
    Factors of 8: 1, 2, 4, 8

Find the LCM of the given numbers.

87. 4, 6   12
88. 3, 4   12
89. 9, 15 45
90. 12, 18 36
91. 18, 21 126
92. 24, 45 360
93. 4, 14, 20 140
94. 21, 28, 42 84

Find the GCF of the given numbers.

95. 8, 12   4
96. 9, 12   3
97. 30, 40 10
98. 30, 45 15
99. 63, 84 21
100. 112, 196 28
101. 48, 72, 120 24
102. 88, 132, 176 44

103. MEETINGS The Rotary Club meets every 14 days and the Kiwanis Club meets every 21 days. If both clubs have a meeting on the same day, in how many more days will they again meet on the same day? 42 days

104. FLOWERS A florist is making flower arrangements for a 4th of July party. He has 32 red carnations, 24 white carnations, and 16 blue carnations. He wants each arrangement to be identical.

    a. What is the greatest number of arrangements that he can make if every carnation is used?
    8 arrangements

    b. How many of each type of carnation will be used in each arrangement?
    4 red carnations, 3 white carnations, 2 blue carnations

SECTION 1.7 Order of Operations

DEFINITIONS AND CONCEPTS

To **evaluate** (find the value of) expressions that involve more than one operation, use the order of operations rule.

Order of Operations

1. Perform all calculations within parentheses and other grouping symbols following the order listed in Steps 2–4 below, working from the innermost pair of grouping symbols to the outermost pair.
2. Evaluate all exponential expressions.
3. Perform all multiplications and divisions as they occur from left to right.
4. Perform all additions and subtractions as they occur from left to right.

When grouping symbols have been removed, repeat Steps 2–4 to complete the calculation.

If a fraction bar is present, evaluate the expression above the bar (called the **numerator**) and the expression below the bar (called the **denominator**) separately. Then perform the division indicated by the fraction bar, if possible.

EXAMPLES

Evaluate: $10 + 3[2^4 - 3(5 - 2)]$

Work within the *innermost* parentheses first and then within the *outermost* brackets.

$10 + 3[2^4 - 3(5 - 2)] = 10 + 3[2^4 - 3(3)]$ Subtract within the parentheses.

$= 10 + 3[16 - 3(3)]$ Evaluate the exponential expression within the brackets: $2^4 = 16$.

$= 10 + 3[16 - 9]$ Multiply within the brackets.

$= 10 + 3[7]$ Subtract within the brackets.

$= 10 + 21$ Multiply: $3[7] = 21$.

$= 31$ Do the addition.

Caution! A common error is to incorrectly add 10 and 3 in the fifth step of the solution.

$= 10 + 3[7]$
$= 13[7]$ Multiply before adding.
$= 91$

Evaluate: $\dfrac{3^3 + 8}{7(15 - 14)}$

Evaluate the expressions above and below the fraction bar separately.

$\dfrac{3^3 + 8}{7(15 - 14)} = \dfrac{27 + 8}{7(1)}$ In the numerator, evaluate the exponential expression. In the denominator, subtract.

$= \dfrac{35}{7}$ In the numerator, add. In the denominator, multiply.

$= 5$ Divide.

The **mean**, or **average**, of a set of numbers is a value around which the values of the numbers are grouped. To **find the mean (average)** of a set of values, divide the sum of the values by the number of values.	Find the mean (average) of the test scores 74, 83, 79, 91, and 73. Mean $= \dfrac{74 + 83 + 79 + 91 + 73}{5}$ Since there are 5 scores, divide by 5. Mean $= \dfrac{400}{5}$ Do the addition in the numerator. Mean $= 80$ Divide. The mean (average) test score is 80.

REVIEW EXERCISES

Evaluate each expression.

105. $3^2 + 12 \cdot 3$ 45

106. $35 - 5 \cdot 3 + 3$ 23

107. $(6 \div 2 \cdot 3)^2 \cdot 3$ 243

108. $(35 - 5 \cdot 3) \div 5$ 4

109. $2^3 \cdot 5 - 4 \div 2 \cdot 4$ 32

110. $8 \cdot (5 - 4 \div 2)^2$ 72

111. $2 + 3\left(\dfrac{100}{10} - 2^2 \cdot 2\right)$ 8

112. $4(4^2 - 5 \cdot 3 + 2) - 4$ 8

113. $\dfrac{4(6) - 6}{2(3^2)}$ 1

114. $\dfrac{6 \cdot 2 + 3 \cdot 7}{5^2 - 2(7)}$ 3

115. $7 + 3[3^3 - 10(4 - 2)]$ 28

116. $5 + 2\left[\left(2^4 - 3 \cdot \dfrac{8}{2}\right) - 2\right]$ 9

Find the arithmetic mean (average) of each set of test scores.

117. 77

Test	1	2	3	4
Score	80	74	66	88

118. 60

Test	1	2	3	4	5
Score	73	77	81	0	69

Unless otherwise noted, all art on this page is © Cengage Learning.

SECTION 1.8 Solving Equations Using Addition and Subtraction

DEFINITIONS AND CONCEPTS	EXAMPLES
An **equation** is a statement indicating that two expressions are equal. All equations contain an = symbol. The equal symbol separates an equation into two parts: the left side and the right side.	Equations: $$x + 4 = 10 \quad y - 7 = 15 \quad \frac{x}{3} = 9 \quad 6x = 42$$
A number that makes an equation a true statement when substituted for the variable is called a **solution** of the equation.	Use a check to determine whether 6 is a solution of $x + 4 = 10$. **Check:** $x + 4 = 10$ $\quad\quad 6 + 4 \stackrel{?}{=} 10 \quad$ Substitute 6 for x. $\quad\quad\quad\; 10 = 10 \quad$ On the left side, do the addition. Since the resulting statement $10 = 10$ is true, 6 is the solution.
Equivalent equations have the same solutions.	$x - 2 = 6$ and $x = 8$ are equivalent equations because they have the same solution, 8.
To **solve an equation**, isolate the variable on one side of the equation by undoing the operation performed on it using a property of equality. The **addition property of equality**: Adding the same number to both sides of an equation does not change its solution. $\quad$ If $a = b$, then $a + c = b + c$	Solve: $x - 8 = 12$ We can use the addition property of equality to isolate x on the left side of the equation. $\quad x - 8 = 12 \quad$ This is the equation to solve. $\quad x - 8 + 8 = 12 + 8 \quad$ Undo the subtraction of 8 by adding 8 to both sides. $\quad x = 20 \quad$ On the left side, adding 8 undoes the subtraction of 8 and leaves x. On the right side add: $12 + 8 = 20$. The solution is 20. Check this result by substituting 20 for x in the original equation.
The **subtraction property of equality**: Subtracting the same number from both sides of an equation does not change its solution. $\quad$ If $a = b$, then $a - c = b - c$	Solve: $59 = y + 31$ We can use the subtraction property of equality to isolate y on the right side of the equation. $\quad 59 = y + 31 \quad$ This is the equation to solve. $\quad 59 - 31 = y + 31 - 31 \quad$ Undo the addition of 31 by subtracting 31 from both sides. $\quad 28 = y \quad$ On the left side, do the subtraction: $59 - 31 = 28$. On the right side, subtracting 31 undoes the addition of 31 and leaves y. The solution is 28. Check this result by substituting 28 for y in the original equation.

To solve application problems, use the six-step problem-solving strategy.

1. **Analyze the problem** by reading it carefully to understand the given facts. What information is given? What are you asked to find? What vocabulary is given? Often, a diagram will help you understand the facts of the problem.
2. **Assign a variable** to represent an unknown value in the problem. This means, in most cases, to let $x = $ what you are asked to find. If there are other unknown values, represent each of them using an algebraic expression that involves the variable.
3. **Form an equation** by translating the words of the problem into mathematical symbols.
4. **Solve the equation** formed in Step 3.
5. **State the conclusion** clearly. Be sure to include the units (such as feet, seconds, or pounds) in your answer.
6. **Check the result** using the original wording of the problem, not the equation that was formed in Step 3 from the words.

AUTO REPAIRS A man paid $34 less for a new set of tires at a gas station than he would have paid for the same tires at a car dealer. If he paid $356 at the gas station, what was the car dealer going to charge him for the tires?

Analyze

- He paid $356 for the tires at the gas station. Given
- The gas station charged $34 less than what the car dealer would have charged. Given
- What would the car dealer have charged for the tires? Find

Assign Let $x = $ the amount the car dealer would have charged for the tires.

Form

Key phrase: *$34 less* **Translation:** subtraction

Now we translate the words of the problem into an equation.

The amount the car dealer would have charged	minus	$34	is equal to	the amount the gas station charged.
x	$-$	34	$=$	356

Solve

$x - 34 = 356$ We need to isolate x on the left side.

$x - 34 + 34 = 356 + 34$ To isolate x, undo the subtraction of 34 by adding 34 to both sides.

$x = 390$ Do the addition.

State The car dealer would have charged $390 for the tires.

Check If the car dealer was going to charge $390 for the tires, and if we subtract the $34 from that cost, we should get the amount the gas station charged.

$$\begin{array}{r} \overset{810}{\$39\!\!\!/0} \\ -\ \$\ 34 \\ \hline \$\ 356 \end{array}$$ This is what the gas station charged for the tires.

The result checks.

REVIEW EXERCISES

Use a check to determine whether the given number is a solution of the equation.

119. Is 5 a solution of $x + 2 = 13$? no
120. Is 4 a solution of $x - 3 = 1$? yes

Solve each equation and check the result.

121. $x - 7 = 2$ 9
122. $x - 11 = 20$ 31
123. $225 = y - 115$ 340
124. $101 = p - 32$ 133
125. $x + 9 = 18$ 9
126. $b + 12 = 26$ 14
127. $175 = p + 55$ 120
128. $212 = m + 207$ 5

Let a variable represent the unknown quantity. Then write and solve an equation to answer the question.

129. FINANCING A newly married couple made a $25,500 down payment on a $122,750 house. How much did they need to borrow? The couple needed to borrow $97,250.

130. HEALTH CARE After moving his office, a doctor lost 13 patients. If he had 172 patients left, how many did he have originally? The doctor originally had 185 patients.

SECTION 1.9 Solving Equations Using Multiplication and Division

DEFINITIONS AND CONCEPTS	EXAMPLES
The **multiplication property of equality**: Multiplying both sides of an equation by the same nonzero number does not change its solution. If $a = b$, then $ca = cb$ (provided $c \neq 0$)	Solve: $\dfrac{m}{5} = 32$ We can use the multiplication property of equality to isolate m on the left side. $\dfrac{m}{5} = 32$ This is the equation to solve. $5 \cdot \dfrac{m}{5} = 5 \cdot 32$ Undo the division by 5 by multiplying both sides by 5. $m = 160$ On the left side, multiplying both sides by 5 undoes the division by 5 and leaves m. On the right side, multiply: $5 \cdot 32 = 160$. The solution is 160. Verify this by substituting 160 into the original equation.
The **division property of equality**: Dividing both sides of an equation by the same nonzero number does not change its solution. If $a = b$, then $\dfrac{a}{c} = \dfrac{b}{c}$ (provided $c \neq 0$)	Solve: $17 = 17c$ We can use the division property of equality to isolate c on the right side. $17 = 17c$ This is the equation to solve. $\dfrac{17}{17} = \dfrac{17c}{17}$ Undo the multiplication by 17 by dividing both sides by 17. $1 = c$ On the left side, divide: $17 \div 17 = 1$. On the right side, dividing both sides by 17 undoes the multiplication by 17 and leaves c. The solution is 1. Verify this by substituting 1 into the original equation.
To solve application problems, use the **six-step problem-solving strategy**.	CONSTRUCTION DELAYS Because of bad weather and a labor stoppage, the final cost of a construction project was three times greater than the original estimate. Upon completion, the project cost $126 million. What was the original cost estimate? *Analyze* • The final cost of the construction project was 3 times greater than the estimate. Given • The completed project cost $126 million. Given • What was the original cost estimate? Find *Assign* Let x = the original cost estimate (in millions of dollars). *Form* We now look for a key phrase in the problem. **Key phrase:** *three times* **Translation:** multiplication

Now we translate the words of the problem into an equation. The units are millions of dollars.

3	times	the original cost	is equal to	the final cost.
3	·	x	=	126

Solve

$3x = 126$ We need to isolate x on the left side.

$\dfrac{3x}{3} = \dfrac{126}{3}$ To isolate x, undo the multiplication by 3 by dividing both sides by 3.

$x = 42$ Do the division.

State

The original cost estimate for the project was $42 million.

Check

If we multiply the original estimate $42 million by 3, we should get the final cost.

$$\begin{array}{r} \$42 \\ \times\ 3 \\ \hline \$126 \end{array}$$ This is the final cost in millions of dollars.

The result checks.

REVIEW EXERCISES

Solve each equation and check the result.

131. $3x = 12$ 4

132. $15y = 45$ 3

133. $105 = 5r$ 21

134. $224 = 16q$ 14

135. $\dfrac{x}{7} = 3$ 21

136. $\dfrac{a}{3} = 12$ 36

137. $15 = \dfrac{s}{21}$ 315

138. $25 = \dfrac{d}{17}$ 425

Let a variable represent the unknown quantity. Then write and solve an equation to answer the question.

139. INFOMERCIALS The number of orders received by a company selling juicers doubled the week after a sports celebrity was added to the company's infomercial. If the company received 364 orders that week, how many did they receive the week before? The week before, the comany received 182 orders.

140. BIRTHDAY PRESENTS Four sisters split the cost of a gold chain that they were giving to their mother as a birthday present. How much did the chain cost if each sister's share was $32? The chain cost $128.

CHAPTER 1 TEST

1. a. The set of __whole__ numbers is {0, 1, 2, 3, 4, 5, … }.
    b. The symbols > and < are __inequality__ symbols.
    c. The __area__ of a rectangle is a measure of the amount of surface it encloses.
    d. The grouping symbols ( ) are called __parentheses__, and the symbols [ ] are called __brackets__.
    e. A __prime__ number is a whole number greater than 1 that has only 1 and itself as factors.
    f. An __equation__ is a statement indicating that two expressions are equal.
    g. A number that makes an equation true when substituted for the variable is called a __solution__ of the equation.
    h. In this chapter, we used the addition, subtraction, multiplication, and division properties of __equality__ to solve equations.

2. Graph the whole numbers less than 7 on a number line.

3. Consider the whole number 402,198.
   a. What is the place value of the digit 1? 1 hundred
   b. What digit is in the ten thousands column? 0

4. a. Write 7,018,641 in words.
      seven million, eighteen thousand, six hundred forty-one
   b. Write "one million, three hundred eighty-five thousand, two hundred sixty-six" in standard form.
      1,385,266
   c. Write 92,561 in expanded form.
      90,000 + 2,000 + 500 + 60 + 1

5. Place an < or an > symbol in the box to make a true statement.
   a. 15 > 10
   b. 1,247 < 1,427

6. Round 34,759,841 to the …
   a. nearest million 35,000,000
   b. nearest hundred thousand 34,800,000
   c. nearest thousand 34,760,000

7. THE NHL The table below shows the number of teams in the National Hockey League at various times during its history. Use the data to complete the bar graph.

Year	1960	1970	1980	1990	2000	2010
Number of teams	6	14	21	21	28	30

   Source: www.rauzulusstreet.com

8. Subtract 287 from 535. Show a check of your result.
   248, 248 + 287 = 535

9. Add: 136,231
          82,574
        + 6,359

        225,164

10. Subtract: 4,521
              −3,579

                942

11. Multiply: 53
              × 8

              424

12. Multiply: 74 · 562
    41,588

13. Divide: 6)432   72

14. Divide: 8,379 ÷ 73. Show a check of your result.
    114 R 57, (73 · 114) + 57 = 8,379

15. Find the product of 23,000 and 600. 13,800,000

16. Find the quotient of 125,000 and 500. 250

17. Use front-end rounding to estimate the difference:
    49,213 − 7,198   43,000

18. A rectangle is 327 inches wide and 757 inches long. Find its perimeter. 2,168 in.

19. Find the area of the square shown below. 529 cm²

20. a. Find the factors of 12. 1, 2, 3, 4, 6, 12
    b. Find the first six multiples of 4. 4, 8, 12, 16, 20, 24
    c. Write $5 + 5 + 5 + 5 + 5 + 5 + 5 + 5$ as a multiplication. $8 \cdot 5$

21. Find the prime factorization of 1,260. $2^2 \cdot 3^2 \cdot 5 \cdot 7$

22. TOSSING A COIN During World War II, John Kerrich, a prisoner of war, tossed a coin 10,000 times and wrote down the results. If he recorded 5,067 heads, how many tails occurred? (Source: *Figure This!*) 4,933 tails

23. P.E. CLASSES In a physical education class, the students stand in a rectangular formation of 8 rows and 12 columns when the instructor takes attendance. How many students are in the class? 96 students

24. FLOOR SPACE The men's, women's, and children's departments in a clothing store occupy a total of 12,255 square feet. Find the square footage of each department if they each occupy the same amount of floor space. 4,085 ft²

25. MILEAGE The fuel tank of a Hummer H3 holds 23 gallons of gasoline. How far can a Hummer travel on one tank of gas if it gets 18 miles per gallon on the highway? 414 mi

26. What property is illustrated by each statement?
    a. $18 \cdot (9 \cdot 40) = (18 \cdot 9) \cdot 40$
       associative property of multiplication
    b. $23{,}999 + 1 = 1 + 23{,}999$
       commutative property of addition

27. Perform each operation, if possible.
    a. $15 \cdot 0$ 0
    b. $\frac{0}{15}$ 0
    c. $\frac{8}{8}$ 1
    d. $\frac{8}{0}$ undefined

28. Find the LCM of 15 and 18. 90

29. Find the LCM of 8, 9, and 12. 72

30. Find the GCF of 30 and 54. 6

31. Find the GCF of 24, 28, and 36. 4

32. STOCKING SHELVES Boxes of rice are being stacked next to boxes of instant mashed potatoes on the same bottom shelf in a supermarket display. The boxes of rice are 8 inches tall and the boxes of instant potatoes are 10 inches high.
    a. What is the shortest height at which the two stacks will be the same height? 40 in.
    b. How many boxes of rice and how many boxes of potatoes will be used in each stack?
       rice: 5 boxes, potatoes: 4 boxes

33. Is 521,340 divisible by 2, 3, 4, 5, 6, 9, or 10?
    It is divisible by 2, 3, 4, 5, 6, and 10.

34. GRADES A student scored 73, 52, 95, and 70 on four exams and received 0 on one missed exam. Find his mean (average) exam score. 58

Evaluate each expression.

35. $9 + 4 \cdot 5$ 29

36. $3^4 \cdot 10 - 2(6)(4)$ 762

37. $20 + 2[4^2 - 2(6 - 2^2)]$ 44

38. $\dfrac{3^3 - 2(15 - 14)^2}{33 - 9 + 1}$ 1

39. Use a check to determine whether 3 is a solution of the equation $x + 13 = 16$. yes

40. Explain what it means to *solve* an equation.
    To solve an equation means to find all the values of the variable that, when substituted into the equation, make a true statement.

Solve each equation and check the result.

41. $100 = x + 1$ 99

42. $y - 12 = 18$ 30

43. $5m = 55$ 11

44. $\dfrac{q}{3} = 27$ 81

Let a variable represent the unknown quantity. Then write and solve an equation to answer the question.

45. PARKING After many student complaints, a college decided to triple the number of parking spaces on campus by constructing a parking structure. That increase will bring the total number of spaces up to 6,240. How many parking spaces does the college have at this time? At this time, the college has 2,080 parking spaces.

46. HEARING PROTECTION When a sound technician at a rock concert wears ear plugs, the sound intensity that he experiences from a heavy metal band is only 73 decibels. If the ear plugs reduce the sound intensity by 41 decibels, what is the actual sound intensity of the band? The sound intensity of the band is 114 decibels.

47. DISCUSSION GROUPS A sociology professor had the students in her class split up into six-person discussion groups. If there were exactly twelve discussion groups of that size, how many students were in the class? There were 72 students in the class.

48. KITCHEN REMODELING A woman wants to have her kitchen remodeled. If she has saved $12,500, and the project costs $27,250, how much money does she need to borrow? She needs to borrow $14,750.

2 The Integers

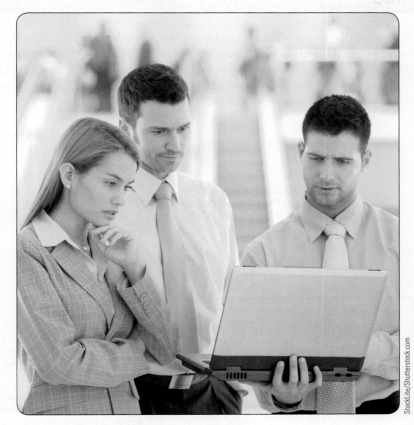

2.1 An Introduction to the Integers
2.2 Adding Integers
2.3 Subtracting Integers
2.4 Multiplying Integers
2.5 Dividing Integers
2.6 Order of Operations and Estimation
2.7 Solving Equations That Involve Integers
Chapter Summary and Review
Chapter Test
Cumulative Review

from Campus to Careers
Personal Financial Adviser

Personal financial advisers help people manage their money and teach them how to make their money grow. They offer advice on how to budget for monthly expenses, as well as how to save for retirement. A bachelor's degree in business, accounting, finance, economics, or statistics provides good preparation for the occupation. Strong communication and problem-solving skills are equally important to achieve success in this field.

In **Problem 90** of **Study Set 2.2,** you will see how a personal financial adviser uses integers to determine whether a duplex rental unit would be a moneymaking investment for a client.

JOB TITLE: Personal Financial Adviser
EDUCATION: Must have at least a bachelor's degree. Some states require a certificate or license.
JOB OUTLOOK: Excellent—Jobs are projected to grow by 32% over the next decade.
ANNUAL EARNINGS: In 2011, the median annual salary was $66,580.
FOR MORE INFORMATION:
http://money.usnews.com/careers/best-jobs/financial_adviser

Objectives

1. Define the set of integers.
2. Graph integers on a number line.
3. Use inequality symbols to compare integers.
4. Find the absolute value of an integer.
5. Find the opposite of an integer.

SECTION 2.1
An Introduction to the Integers

We have seen that whole numbers can be used to describe many situations that arise in everyday life. However, we cannot use whole numbers to express temperatures below zero, the balance in a checking account that is overdrawn, or how far an object is below sea level. In this section, we will see how negative numbers can be used to describe these three situations as well as many others.

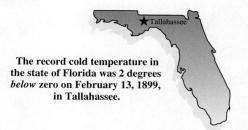

The record cold temperature in the state of Florida was 2 degrees *below* zero on February 13, 1899, in Tallahassee.

A check for $500 was written when there was only $450 in the account. The checking account is *overdrawn*.

The American lobster is found off the East Coast of North America at depths as much as 600 feet *below* sea level.

1 Define the set of integers.

To describe a temperature of 2 degrees above zero, a balance of $50, or 600 feet above sea level, we can use numbers called **positive numbers**. All positive numbers are greater than 0, and we can write them with or without a **positive sign** +.

In words	In symbols	Read as
2 degrees above zero	+2 or 2	positive two
A balance of $50	+50 or 50	positive fifty
600 feet above sea level	+600 or 600	positive six hundred

To describe a temperature of 2 degrees below zero, $50 overdrawn, or 600 feet below sea level, we need to use negative numbers. **Negative numbers** are numbers less than 0, and they are written using a **negative sign** −.

In words	In symbols	Read as
2 degrees below zero	−2	negative two
$50 overdrawn	−50	negative fifty
600 feet below sea level	−600	negative six hundred

Together, positive and negative numbers are called **signed numbers**.

Unless otherwise noted, all art on this page is © Cengage Learning.

Positive and Negative Numbers

Positive numbers are greater than 0. **Negative numbers** are less than 0.

Caution! Zero is neither positive nor negative.

The collection of positive whole numbers, the negatives of the whole numbers, and 0 is called the set of **integers** (read as "in-ti-jers").

The Set of Integers

$$\{\ldots, -5, -4, -3, -2, -1, 0, 1, 2, 3, 4, 5, \ldots\}$$

The three dots on the right indicate that the list continues forever—there is no largest integer. The three dots on the left indicate that the list continues forever—there is no smallest integer. The set of **positive integers** is $\{1, 2, 3, 4, 5, \ldots\}$ and the set of **negative integers** is $\{\ldots, -5, -4, -3, -2, -1\}$.

The Language of Algebra Since every whole number is an integer, we say that the set of whole numbers is a **subset** of the integers.

The set of integers → $\{\ldots, -5, -4, -3, -2, -1, \underbrace{0, 1, 2, 3, 4, 5, \ldots}_{\text{The set of whole numbers}}\}$

2 Graph integers on a number line.

In Section 1.1, we introduced the number line. We can use an extension of the number line to learn about negative numbers.

Negative numbers can be represented on a number line by extending the line to the left and drawing an arrowhead. Beginning at the origin (the 0 point), we move to the left, marking equally spaced points as shown below. As we move to the right on the number line, the values of the numbers increase. As we move to the left, the values of the numbers decrease.

The thermometer shown on the next page is an example of a vertical number line. It is scaled in degrees and shows a temperature of $-10°$. The time line is an example of a horizontal number line. It is scaled in units of 500 years.

138 Chapter 2 The Integers

A vertical number line

A horizontal number line

Self Check 1

Graph −4, −2, 1, and 3 on a number line.

Now Try Problem 23

Teaching Example 1 Graph −4, −2, 0, and 3 on a number line.
Answer:

EXAMPLE 1 Graph −3, 2, −1, and 4 on a number line.

Strategy We will locate the position of each integer on the number line and draw a bold dot.

WHY To *graph a number* means to make a drawing that represents the number.

Solution
The position of each negative integer is to the left of 0. The position of each positive integer is to the right of 0.

By extending the number line to include negative numbers, we can represent more situations using bar graphs and line graphs. For example, the following bar graph shows the net income of the Kodak Company for the years 2000 through 2011. Since the net income in 2004 was positive $556 million, the company made a *profit*. Since the net income in 2005 was −$1,362 million, the company had a *loss*.

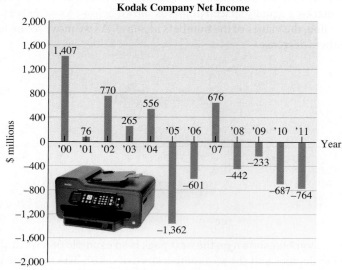

Source: morningstar.com; photo, Hugh Threlfall/Alamy

The Language of Algebra *Net* refers to what remains after all the deductions (losses) have been accounted for. **Net income** is a term used in business that often is referred to as the *bottom line*. Net income indicates what a company has earned (or lost) in a given period of time (usually 1 year).

THINK IT THROUGH Credit Card Debt

"The most dangerous pitfall for many college students is the overuse of credit cards. Many banks do their best to entice new card holders with low or zero-interest cards."

Gary Schatsky, certified financial planner

Which numbers on the credit card statement below are actually debts and, therefore, could be represented using negative numbers? $4,621, $1,073, $3,325

		Account Summary	
Previous Balance	New Purchases	Payments & Credits	New Balance
$4,621	$1,073	$2,369	$3,325
04/21/14 Billing Date	05/16/14 Date Payment Due		$67 Minimum payment

BANK STAR — Periodic rates may vary. See reverse for explanation and important information. Please allow sufficient time for mail to reach Bank Star.

3 Use inequality symbols to compare integers.

Recall that the symbol < means "is less than" and that > means "is greater than." The figure below shows the graph of the integers -2 and 1. Since -2 is to the left of 1 on the number line, $-2 < 1$. Since $-2 < 1$, it is also true that $1 > -2$.

EXAMPLE 2
Place an < or an > symbol in the box to make a true statement.   **a.** 4 ☐ -5 **b.** -8 ☐ -7

Strategy To pick the correct inequality symbol to place between the pair of numbers, we will determine the position of each number on the number line.

WHY For any two numbers on a number line, the number to the *left* is the smaller number and the number on the *right* is the larger number.

Solution

a. Since 4 is to the right of -5 on the number line, $4 > -5$.

b. Since -8 is to the left of -7 on the number line, $-8 < -7$.

Self Check 2

Place an < or an > symbol in the box to make a true statement.

a. 6 ☐ -6 >

b. -11 ☐ -10 <

Now Try Problems 31 and 35

Teaching Example 2 Place an < or an > symbol in the box to make a true statement.
a. -15 ☐ 51   **b.** -3 ☐ -4
Answers:
a. <   **b.** >

The Language of Algebra Because the symbol < requires one number to be strictly less than another number and the symbol > requires one number to be strictly greater than another number, mathematical statements involving the symbols < and > are called *strict inequalities*.

There are three other commonly used inequality symbols.

Inequality Symbols

≠   means *is not equal to*
≥   means *is greater than or equal to*
≤   means *is less than or equal to*

−5 ≠ −2   Read as "−5 is not equal to −2."

−6 ≤ 10   Read as "−6 is less than or equal to 10."
           This statement is true, because −6 < 10.

12 ≤ 12   Read as "12 is less than or equal to 12."
           This statement is true, because 12 = 12.

−15 ≥ −17 Read as "−15 is greater than or equal to −17."
           This statement is true, because −15 > −17.

−20 ≥ −20 Read as "−20 is greater than or equal to −20."
           This statement is true, because −20 = −20.

Self Check 3

Tell whether each statement is true or false.

a. −17 ≥ −15   false
b. −35 ≤ −35   true
c. −2 ≥ −2   true
d. −61 ≤ −62   false

Now Try Problems 41 and 45

Teaching Example 3 Tell whether each statement is true or false.
a. −56 ≤ −57
b. −3 ≤ −3
c. −14 ≥ −19
d. −8 ≤ −8
Answers:
a. false
b. true
c. true
d. true

EXAMPLE 3   Tell whether each statement is true or false.

a. −9 ≥ −9   b. −1 ≤ −5   c. −27 ≥ 6   d. −32 ≤ −31

Strategy We will determine if either the strict inequality or the equality that the symbols ≤ and ≥ allow is true.

WHY If either is true, then the given statement is true.

Solution

a. −9 ≥ −9   This statement is true, because −9 = −9.
b. −1 ≤ −5   This statement is false, because neither −1 < −5 nor −1 = −5 is true.
c. −27 ≥ 6   This statement is false, because neither −27 > 6 nor −27 = 6 is true.
d. −32 ≤ −31   This statement is true, because −32 < −31.

4 Find the absolute value of an integer.

Using a number line, we can see that the numbers 3 and −3 are both a distance of 3 units away from 0, as shown below.

The **absolute value** of a number gives the distance between the number and 0 on the number line. To indicate absolute value, the number is inserted between two vertical bars, called the **absolute value symbol**. For example, we can write |−3| = 3. This is read as "The absolute value of negative 3 is 3," and it tells us that the distance between −3 and 0 on the number line is 3 units. From the figure, we also see that |3| = 3.

Absolute Value

The **absolute value** of a number is the distance on the number line between the number and 0.

Caution! Absolute value expresses distance. The absolute value of a number is always positive or 0. It is never negative.

EXAMPLE 4 Find each absolute value: **a.** $|8|$ **b.** $|-5|$ **c.** $|0|$

Strategy We need to determine the distance that the number within the vertical absolute value bars is from 0 on a number line.

WHY The absolute value of a number is the distance between 0 and the number on a number line.

Solution
a. On the number line, the distance between 8 and 0 is 8. Therefore,

$$|8| = 8$$

b. On the number line, the distance between −5 and 0 is 5. Therefore,

$$|-5| = 5$$

c. On the number line, the distance between 0 and 0 is 0. Therefore,

$$|0| = 0$$

Self Check 4
Find each absolute value:
a. $|-9|$ 9
b. $|4|$ 4
Now Try Problems 47 and 49

Teaching Example 4 Find each absolute value:
a. $|16|$
b. $|-7|$
c. $|0|$
Answers:
a. 16
b. 7
c. 0

5 Find the opposite of an integer.

Opposites

Two numbers that are the same distance from 0 on the number line, but on opposite sides of it, are called **opposites**.

The figure below shows that for each integer on the number line, there is a corresponding integer, called its *opposite*, on the opposite side of 0. For example, we see that 3 and −3 are opposites, as are −5 and 5. Note that 0 is its own opposite.

Opposites

To write the opposite of a number, a − symbol is used. For example, the opposite of 5 is −5 (read as "negative 5"). Parentheses are needed to express the opposite of a negative number. The opposite of −5 is written as −(−5). Since 5 and −5 are the same distance from 0, the opposite of −5 is 5. Therefore, −(−5) = 5. This illustrates the following rule.

The Opposite of the Opposite Rule

The opposite of the opposite of a number is that number.
   For any number a,

$$-(-a) = a \quad \text{Read as "the opposite of the opposite of } a\text{."}$$

Unless otherwise noted, all art on this page is © Cengage Learning.

Number	Opposite	
57	−57	Read as "negative fifty-seven."
−8	−(−8) = 8	Read as "the opposite of negative eight is eight."
0	−0 = 0	Read as "the opposite of 0 is 0."

The concept of opposite can also be applied to an absolute value. For example, the opposite of the absolute value of −8 can be written as −|−8|. Think of this as a two-step process, where the absolute value symbol serves as a grouping symbol. Find the absolute value first, and then attach a − sign to that result.

First, find the absolute value.

$$-|-8| = -8$$

Read as "the opposite of the absolute value of negative eight is negative eight."

Then attach a − sign.

Self Check 5

Simplify each expression:

a. −(−1) 1
b. −|4| −4
c. −|−99| −99

Now Try Problems 55, 65, and 67

Teaching Example 5 Simplify each expression:
a. −(−23)
b. −|200|
c. −|−81|
Answers:
a. 23
b. −200
c. −81

EXAMPLE 5

Simplify each expression: **a.** −(−44) **b.** −|11| **c.** −|−225|

Strategy We will find the opposite of each number.

WHY In each case, the − symbol written outside the grouping symbols means "the opposite of."

Solution

a. −(−44) means the opposite of −44. Since the opposite of −44 is 44, we write

$$-(-44) = 44$$

b. −|11| means the opposite of the absolute value of 11. Since |11| = 11, and the opposite of 11 is −11, we write

$$-|11| = -11$$

c. −|−225| means the opposite of the absolute value of −225. Since |−225| = 225, and the opposite of 225 is −225, we write

$$-|-225| = -225$$

The − symbol is used to indicate a negative number, the opposite of a number, and the operation of subtraction. The key to reading the − symbol correctly is to examine the context in which it is used.

Reading the − Symbol

−12	Negative twelve	A − symbol directly in front of a number is read as "negative."
−(−12)	The opposite of negative twelve	The first − symbol is read as "the opposite of" and the second as "negative."
12 − 5	Twelve minus five	Notice the space used before and after the − symbol. This indicates subtraction and is read as "minus."

ANSWERS TO SELF CHECKS

1.
2. **a.** > **b.** <
3. **a.** false **b.** true **c.** true **d.** false
4. **a.** 9 **b.** 4
5. **a.** 1 **b.** −4 **c.** −99

SECTION 2.1 STUDY SET

VOCABULARY

Fill in the blanks.

1. _Positive_ numbers are greater than 0 and _negative_ numbers are less than 0.
2. $\{\ldots, -5, -4, -3, -2, -1, 0, 1, 2, 3, 4, 5, \ldots\}$ is called the set of _integers_.
3. To _graph_ an integer means to locate it on the number line and highlight it with a dot.
4. The symbols $>$ and $<$ are called _inequality_ symbols.
5. The _absolute_ _value_ of a number is the distance between the number and 0 on the number line.
6. Two numbers that are the same distance from 0 on the number line, but on opposite sides of it, are called _opposites_.

CONCEPTS

7. Represent each of these situations using a signed number.
   a. $225 overdrawn −$225
   b. 10 seconds before liftoff −10 sec
   c. 3 degrees below normal −3°
   d. A deficit of $12,000 −$12,000
   e. A 1-mile retreat by an army −1 mi

8. Represent each of these situations using a signed number, and then describe its opposite in words.
   a. A trade surplus of $3 million
      +3 million, $3 million deficit
   b. A bacteria count 70 more than the standard
      +70, 70 less than the standard
   c. A profit of $67
      +67, a loss of $67
   d. A business $1 million in the "black"
      +1 million, $1 million in the red
   e. 20 units over their quota
      +20, 20 units under quota

9. Determine what is wrong with each number line.
   a. The spacing is not uniform.
   b. The numbering is not uniform.
   c. Zero is missing.
   d. The arrowheads are not drawn.

10. a. If a number is less than 0, what type of number must it be? It is negative.
    b. If a number is greater than 0, what type of number must it be? It is positive.

11. On the number line, what number is
    a. 3 units to the right of -7? -4
    b. 4 units to the left of 2? -2

12. Name two numbers on the number line that are a distance of
    a. 5 away from -3. -8 and 2
    b. 4 away from 3. 7 and -1

13. a. Which number is closer to -3 on the number line: 2 or -7? -7
    b. Which number is farther from 1 on the number line: -5 or 8? 8

14. Is there a number that is both greater than 10 and less than 10 at the same time? no

15. a. Express the fact $-12 < 15$ using an $>$ symbol.
       $15 > -12$
    b. Express the fact $-4 > -5$ using an $<$ symbol.
       $-5 < -4$

16. Fill in the blank: The opposite of the _opposite_ of a number is that number. For any number a, $-(-a) = $ _a_.

17. Complete the table by finding the opposite and the absolute value of the given numbers.

Number	Opposite	Absolute value
-25	25	25
39	-39	39
0	0	0

18. Is the absolute value of a number always positive? no, $|0| = 0$

NOTATION

19. Translate each phrase to mathematical symbols.
    a. The opposite of negative eight $-(-8)$
    b. The absolute value of negative eight $|-8|$
    c. Eight minus eight $8 - 8$
    d. The opposite of the absolute value of negative eight $-|-8|$

20. a. Write the set of integers.
{..., −5, −4, −3, −2, −1, 0, 1, 2, 3, 4, 5, ...}
b. Write the set of positive integers.
{1, 2, 3, 4, 5, ...}
c. Write the set of negative integers.
{..., −5, −4, −3, −2, −1}

21. Fill in the blanks.
a. We read ≥ as "is _greater_ than or _equal_ to."
b. We read ≤ as "is _less_ than or _equal_ to."

22. Which of the following expressions contains a minus sign?
15 − 8   −(−15)   −15   15 − 8

GUIDED PRACTICE

Graph the following numbers on a number line. See Example 1.

23. −3, 4, 3, 0, −1

24. 2, −4, 5, 1, −1

25. The integers that are less than 3 but greater than −5

26. The integers that are less than 4 but greater than −3

27. The opposite of −3, the opposite of 5, and the absolute value of −2

28. The absolute value of 3, the opposite of 3, and the number that is 1 less than −3

29. 2 more than 0, 4 less than 0, 2 more than negative 5, and 5 less than 4

30. 4 less than 0, 1 more than 0, 3 less than −2, and 6 more than −4

Place an < or an > symbol in the box to make a true statement. See Example 2.

31. −5 < 5
32. 0 > −1
33. −12 < −6
34. −7 < −6
35. −10 > −17
36. −11 > −20
37. −325 > −532
38. −401 < −104

Tell whether each statement is true or false. See Example 3.

39. −15 ≤ −14 true
40. −77 ≤ −76 true
41. 210 ≥ 210 true
42. 37 ≥ 37 true
43. −1,255 ≥ −1,254 false
44. −6,546 ≥ −6,465 false
45. 0 ≤ −8 false
46. −6 ≤ −6 true

Find each absolute value. See Example 4.

47. |9| 9
48. |12| 12
49. |−8| 8
50. |−1| 1
51. |−14| 14
52. |−85| 85
53. |180| 180
54. |371| 371

Simplify each expression. See Example 5.

55. −(−11) 11
56. −(−1) 1
57. −(−4) 4
58. −(−9) 9
59. −(−102) 102
60. −(−295) 295
61. −(−561) 561
62. −(−703) 703
63. −|20| −20
64. −|143| −143
65. −|6| −6
66. −|0| 0
67. −|−253| −253
68. −|−11| −11
69. −|−0| 0
70. −|97| −97

TRY IT YOURSELF

Place an < or an > symbol in the box to make a true statement.

71. |−12| > −(−7)
72. |−50| > −(−40)
73. −|−71| < −|−65|
74. −|−163| < −|−150|
75. −(−343) > −(−161)
76. −(−999) > −(−998)
77. −|−30| < −|−(−8)|
78. −|−100| < −|−(−88)|

Write the integers in order, from least to greatest.

79. 82, −52, 52, −22, 12, −12   −52, −22, −12, 12, 52, 82
80. 49, −9, 19, −39, 89, −49   −49, −39, −9, 19, 49, 89

Fill in the blanks to continue each pattern.

81. 5, 3, 1, −1, _−3_, _−5_, _−7_, ...
82. 4, 2, 0, −2, _−4_, _−6_, _−8_, ...

APPLICATIONS

83. HORSE RACING In the 1973 Belmont Stakes, Secretariat won by 31 lengths over second place finisher Twice a Prince. Some experts call it the greatest performance by a thoroughbred in the history of racing. Express the position of Twice a Prince compared to Secretariat as a signed number. (Source: ezinearticles.com) −31 lengths

84. NASCAR In the NASCAR driver standings, negative numbers are used to tell how many points behind the leader a given driver is. Brad Keselowski was the leading driver in 2012. The other drivers in the top ten, listed alphabetically, were Clint Bowyer (−39), Greg Biffle (−68), Jeff Gordon (−97), Denny Hamlin (−71), Kevin Harvick (−79), Jimmie Johnson (−40), Kasey Kahne (−55), Matt Kenseth (−76), and Tony Stewart (−89). Use this information to rank the drivers in the table below. (Source: NASCAR.com)

2012 NASCAR Final Driver Standings

Rank	Driver	Points behind leader
1	Brad Keselowski	Leader
2	Clint Bowyer	−39
3	Jimmie Johnson	−40
4	Kasey Kahne	−55
5	Greg Biffle	−68
6	Denny Hamlin	−71
7	Matt Kenseth	−76
8	Kevin Harvick	−79
9	Tony Stewart	−89
10	Jeff Gordon	−97

Source: NASCAR.com

85. FREE FALL A boy launches a water balloon from the top of a building, as shown below. At that instant, his friend starts a stopwatch and keeps track of the time as the balloon sails above the building and then falls to the ground. Use the number line to estimate the position of the balloon at each time listed in the table below.

Time	Position of balloon
0 sec	0
1 sec	20
2 sec	5
3 sec	−40
4 sec	−120

86. IPHONES You can get a more accurate reading of an iPhone's signal strength by dialing *3001#12345#*. Field test mode is then activated and the standard signal strength bars (in the upper left corner of the display) are replaced by a negative number. The closer the negative number is to zero, the stronger the signal. Which iPhone shown below is receiving the strongest signal? c

(a) (b) (c) (d)

87. **TECHNOLOGY** The readout from a testing device is shown below. Use the number line to find the height of each of the peaks and the depth of each of the valleys. peaks: 2, 4, 0; valleys: −3, −5, −2

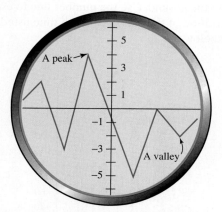

88. **FLOODING** A week of daily reports listing the height of a river in comparison to flood stage is given in the table. Complete the bar graph shown below.

Flood Stage Report

Sun.	2 ft below
Mon.	3 ft over
Tue.	4 ft over
Wed.	2 ft over
Thu.	1 ft below
Fri.	3 ft below
Sat.	4 ft below

89. **GOLF** In golf, *par* is the standard number of strokes considered necessary on a given hole. A score of −2 indicates that a golfer used 2 strokes less than par. A score of +2 means 2 more strokes than par were used. In the graph in the next column, each golf ball represents the score of a professional golfer on the 16th hole of a certain course.

   a. What score was shot most often on this hole?
      −1 (1 below par)
   b. What was the best score on this hole?
      −3 (3 below par)
   c. Explain why this hole appears to be too easy for a professional golfer.
      Most of the scores are below par.

90. **PAYCHECKS** Examine the items listed on the following paycheck stub. Then write two columns on your paper—one headed "positive" and the other "negative." List each item under the proper heading.
positives: gross pay, overtime, Christmas bonus; negatives: all others

Tom Dryden Dec. 14	Christmas bonus	$100
Gross pay $2,000	**Reductions**	
Overtime $300	Retirement	$200
Deductions	**Taxes**	
Union dues $30	Federal withholding	$160
U.S. Bonds $100	State withholding	$35

91. **WEATHER MAPS** The illustration shows the predicted Fahrenheit temperatures for a day in mid-January.

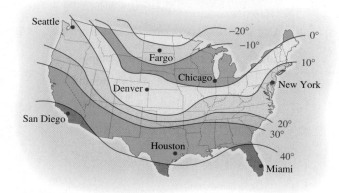

   a. What is the temperature range for the region including Fargo, North Dakota? −20° to −10°
   b. According to the prediction, what is the warmest it should get in Houston? 40°
   c. According to this prediction, what is the coldest it should get in Seattle? 10°

92. **INTERNET COMPANIES** The graph on the next page shows the net income of Amazon.com for the years 1998–2011. (Source: Morningstar)

a. In what years did Amazon suffer a loss? Estimate each loss. 1998: −$120 million, 1999: −$720 million, 2000: −$1,410 million, 2001: −$570 million, 2002: −$150 million

b. In what year did Amazon first turn a profit? Estimate it. 2003: $30 million

c. In what year did Amazon have the greatest profit? Estimate it. 2010: $1,050 million

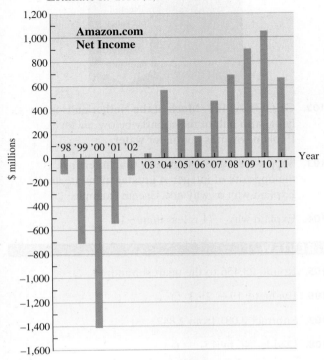

93. HISTORY Number lines can be used to display historical data. Some important world events are shown on the time line below.

a. What basic unit is used to scale this time line? 200 yr

b. What can be thought of as positive numbers? A.D.

c. What can be thought of as negative numbers? B.C.

d. What important event distinguishes the positive from the negative numbers? the birth of Jesus

94. ASTRONOMY Astronomers use an inverted vertical number line called the *apparent magnitude scale* to denote the brightness of objects in the sky. The brighter an object appears to an observer on Earth, the more negative is its apparent magnitude. Graph each of the following on the scale to the right.

- Visual limit of binoculars +10
- Visual limit of large telescope +20
- Visual limit of naked eye +6
- Full moon −12
- Pluto +15
- Sirius (a bright star) −2
- Sun −26
- Venus −4

95. LINE GRAPHS Each thermometer in the illustration gives the daily high temperature in degrees Fahrenheit. Use the data to complete the line graph below.

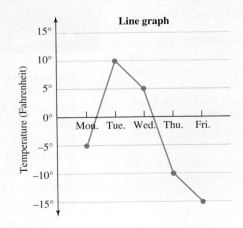

96. GARDENING The illustration shows the depths at which the bottoms of various types of flower bulbs should be planted. (The symbol ″ represents inches.)

a. At what depth should a tulip bulb be planted? 8″ below the surface (−8″)

b. How much deeper are hyacinth bulbs planted than gladiolus bulbs? 3″

c. Which bulb must be planted the deepest? How deep? daffodil, 11″ below the surface (−11″)

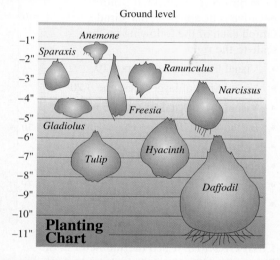

WRITING

97. Explain the concept of *the opposite of a number*.

98. What real-life situation do you think gave rise to the concept of a negative number?

99. Explain why the absolute value of a number is never negative.

100. Give an example of the use of the number line that you have seen in another course.

101. DIVING Divers use the terms *positive buoyancy*, *neutral buoyancy*, and *negative buoyancy* as shown. What do you think each of these terms means?

102. GEOGRAPHY Much of the Netherlands is low-lying, with half of the country below sea level. Explain why it is not under water.

103. Suppose integer A is greater than integer B. Is the opposite of integer A greater than integer B? Explain why or why not. Use an example.

104. Explain why −11 is less than −10.

REVIEW

105. Round 23,456 to the nearest hundred. 23,500

106. Evaluate: $19 - 2 \cdot 3$ 13

107. Subtract 2,081 from 2,842. 761

108. Divide 346 by 15. 23 R 1

109. Give the name of the property shown below:
$(13 \cdot 2) \cdot 5 = 13 \cdot (2 \cdot 5)$ associative property of multiplication

110. Write *four times five* using three different symbols. $4 \times 5, 4 \cdot 5, 4(5)$

Objectives

1. Add two integers that have the same sign.
2. Add two integers that have different signs.
3. Perform several additions to evaluate expressions.
4. Identify opposites (additive inverses) when adding integers.
5. Solve application problems by adding integers.

SECTION 2.2
Adding Integers

An amazing change in temperature occurred in 1943 in Spearfish, South Dakota. On January 22, at 7:30 A.M., the temperature was −4 degrees Fahrenheit. Strong warming winds suddenly kicked up and, in just 2 minutes, the temperature rose 49 degrees! To calculate the temperature at 7:32 A.M., we need to add 49 to −4.

$$-4 + 49$$

Unless otherwise noted, all art on this page is © Cengage Learning.

To perform this addition, we must know how to add positive and negative integers. In this section, we develop rules to help us make such calculations.

> **The Language of Algebra** In 1724, Daniel Gabriel *Fahrenheit*, a German scientist, introduced the temperature scale that bears his name. The United States is one of the few countries that still uses this scale. The temperature -4 degrees Fahrenheit can be written in more compact form as $-4°F$.

1 Add two integers that have the same sign.

We can use the number line to explain addition of integers. For example, to find $4 + 3$, we begin at 0 and draw an arrow 4 units long that points to the right. It represents positive 4. From the tip of that arrow, we draw a second arrow, 3 units long, that points to the right. It represents positive 3. Since we end up at 7, it follows that $4 + 3 = 7$.

To check our work, let's think of the problem in terms of money. If you had $4 and earned $3 more, you would have a total of $7.

To find $-4 + (-3)$ on a number line, we begin at 0 and draw an arrow 4 units long that points to the left. It represents -4. From the tip of that arrow, we draw a second arrow, 3 units long, that points to the left. It represents -3. Since we end up at -7, it follows that $-4 + (-3) = -7$.

Let's think of this problem in terms of money. If you lost $4 ($-4$) and then lost another $3 ($-3$), overall, you would have lost a total of $7 ($-7$).

Here are some observations about the process of adding two numbers that have the same sign on a number line.

- The arrows representing the integers point in the same direction and they build upon each other.
- The answer has the same sign as the integers that we added.

These observations illustrate the following rules.

> **Adding Two Integers That Have the Same (Like) Signs**
>
> 1. To add two positive integers, add them as usual. The final answer is positive.
> 2. To add two negative integers, add their absolute values and make the final answer negative.

Chapter 2 The Integers

> **The Language of Algebra** When writing additions that involve integers, write negative integers within parentheses to separate the negative sign − from the plus symbol +.
>
> $9 + (-4)$ ~~$9 + -4$~~   and $-9 + (-4)$ ~~$-9 + -4$~~

Self Check 1

Add:

a. $-7 + (-2)$ -9
b. $-25 + (-48)$ -73
c. $-325 + (-169)$ -494

Now Try Problems 19, 23, and 27

Teaching Example 1 Add:
a. $-4 + (-3)$
b. $-38 + (-25)$
c. $-502 + (-189)$
Answers:
a. -7 b. -63 c. -691

EXAMPLE 1 Add: **a.** $-3 + (-5)$ **b.** $-26 + (-65)$ **c.** $-456 + (-177)$

Strategy We will use the rule for adding two integers that have the *same sign*.

WHY In each case, we are asked to add two negative integers.

Solution

a. To add two negative integers, we add the absolute values of the integers and make the final answer negative. Since $|-3| = 3$ and $|-5| = 5$, we have

$$-3 + (-5) = -8$$ Add their absolute values, 3 and 5, to get 8.
Then make the final answer negative.

b. Find the absolute values: $|-26| = 26$ and $|-65| = 65$

$$-26 + (-65) = -91$$ Add their absolute values, 26 and 65, to get 91. Then make the final answer negative.

$$\begin{array}{r} \overset{1}{26} \\ +65 \\ \hline 91 \end{array}$$

c. Find the absolute values: $|-456| = 456$ and $|-177| = 177$

$$-456 + (-177) = -633$$ Add their absolute values, 456 and 177, to get 633. Then make the final answer negative.

$$\begin{array}{r} \overset{11}{456} \\ +177 \\ \hline 633 \end{array}$$

> **Success Tip** Calculations that you cannot perform in your head should be shown outside the steps of your solution.

> **The Language of Algebra** Two negative integers, as well as two positive integers, are said to have *like* signs.

2 Add two integers that have different signs.

To find $4 + (-3)$ on a number line, we begin at 0 and draw an arrow 4 units long that points to the right. This represents positive 4. From the tip of that arrow, we draw a second arrow, 3 units long, that points to the left. It represents -3. Since we end up at 1, it follows that $4 + (-3) = 1$.

In terms of money, if you won $4 and then lost $3 ($-3$), overall, you would have $1 left.

To find $-4 + 3$ on a number line, we begin at 0 and draw an arrow 4 units long that points to the left. It represents -4. From the tip of that arrow, we draw a second arrow, 3 units long, that points to the right. It represents positive 3. Since we end up at −1, it follows that −4 + 3 = −1.

In terms of money, if you lost $4 (−4) and then won $3, overall, you have lost $1 (−1).

Here are some observations about the process of adding two integers that have different signs on a number line.

- The arrows representing the integers point in opposite directions.
- The longer of the two arrows determines the sign of the answer. If the longer arrow represents a positive integer, the sum is positive. If it represents a negative integer, the sum is negative.

These observations suggest the following rules.

Adding Two Integers That Have Different (Unlike) Signs

To add a positive integer and a negative integer, subtract the smaller absolute value from the larger.

1. If the positive integer has the larger absolute value, the final answer is positive.
2. If the negative integer has the larger absolute value, make the final answer negative.

EXAMPLE 2   Add: 5 + (−7)

Strategy We will use the rule for adding two integers that have different signs.

WHY The addend 5 is positive and the addend −7 is negative.

Solution

Step 1 To add two integers with different signs, we first subtract the smaller absolute value from the larger absolute value. Since $|5|$, which is 5, is smaller than $|-7|$, which is 7, we begin by subtracting 5 from 7.

$$7 - 5 = 2$$

Step 2 Since the negative number, −7, has the larger absolute value, we attach a negative sign − to the result from step 1. Therefore,

$$5 + (-7) = -2$$

— Make the final answer negative.

The Language of Algebra A positive integer and a negative integer are said to have *unlike* signs.

Self Check 2

Add: 6 + (−9)   −3

Now Try Problem 31

Teaching Example 2 Add: 3 + (−8)
Answer:
−5

Self Check 3

Add:

a. $7 + (-2)$ 5

b. $-53 + 39$ -14

c. $-506 + 888$ 382

Now Try Problems 33, 35, and 39

Teaching Example 3 Add:
a. $6 + (-3)$
b. $-43 + 18$
c. $-301 + 566$
Answers:
a. 3
b. -25
c. 265

EXAMPLE 3

Add: **a.** $8 + (-4)$ **b.** $-41 + 17$ **c.** $-206 + 568$

Strategy We will use the rule for adding two integers that have different signs.

WHY In each case, we are asked to add a positive integer and a negative integer.

Solution

a. Find the absolute values: $|8| = 8$ and $|-4| = 4$

$$8 + (-4) = 4$$

Subtract the smaller absolute value from the larger: $8 - 4 = 4$. Since the positive number, 8, has the larger absolute value, the final answer is positive.

b. Find the absolute values: $|-41| = 41$ and $|17| = 17$

$$-41 + 17 = -24$$

Subtract the smaller absolute value from the larger: $41 - 17 = 24$. Since the negative number, -41, has the larger absolute value, make the final answer negative.

$$\begin{array}{r} \overset{3\,11}{\cancel{4}\cancel{1}} \\ -17 \\ \hline 24 \end{array}$$

c. Find the absolute values: $|-206| = 206$ and $|568| = 568$

$$-206 + 568 = 362$$

Subtract the smaller absolute value from the larger: $568 - 206 = 362$. Since the positive number, 568, has the larger absolute value, the answer is positive.

$$\begin{array}{r} 568 \\ -206 \\ \hline 362 \end{array}$$

Caution! Did you notice that the answers to the addition problems in Examples 2 and 3 were found using subtraction? This is the case when the addition involves two integers that have *different signs*.

THINK IT THROUGH Cash Flow

"College can be trial by fire — a test of how to cope with pressure, freedom, distractions, and a flood of credit card offers. It's easy to get into a cycle of overspending and unnecessary debt as a student."

Planning for College, Wells Fargo Bank

If your income is less than your expenses, you have a *negative* cash flow. A negative cash flow can be a red flag that you should increase your income and/or reduce your expenses. Which of the following activities can increase income and which can decrease expenses?

- Buy generic or store-brand items. decrease expenses
- Get training and/or more education. increase income
- Use your student ID to get discounts at stores, events, etc. decrease expenses
- Work more hours. increase income
- Turn a hobby or skill into a money-making business. increase income
- Tutor young students. increase income
- Stop expensive habits, like smoking, buying snacks every day, etc. decrease expenses
- Attend free activities and free or discounted days at local attractions. decrease expenses
- Sell rarely used items, like an old CD player. increase income
- Compare the prices of at least three products or at three stores before buying. decrease expenses

Based on the *Building Financial Skills* by National Endowment for Financial Education.

3 Perform several additions to evaluate expressions.

To evaluate expressions that contain several additions, we make repeated use of the rules for adding two integers.

EXAMPLE 4 Evaluate: $-3 + 5 + (-12) + 2$

Strategy Since there are no calculations within parentheses, no exponential expressions, and no multiplication or division, we will perform the additions, working from the left to the right.

WHY This is step 4 of the order of operations rule that was introduced in Section 1.7.

Solution

$-3 + 5 + (-12) + 2 = 2 + (-12) + 2$ Use the rule for adding two integers that have different signs: $-3 + 5 = 2$.

$\qquad\qquad\qquad\qquad = -10 + 2$ Use the rule for adding two integers that have different signs: $2 + (-12) = -10$.

$\qquad\qquad\qquad\qquad = -8$ Use the rule for adding two integers that have different signs.

Self Check 4

Evaluate:
$-12 + 8 + (-6) + 1$ -9

Now Try Problem 43

Teaching Example 4 Evaluate:
$-7 + 3 + (-15) + 6$
Answer:
-13

The properties of addition that were introduced in Section 1.2, *Adding and Subtracting Whole Numbers*, are also true for integers.

Commutative Property of Addition

The order in which integers are added does not change their sum.

Associative Property of Addition

The way in which integers are grouped does not change their sum.

Another way to evaluate an expression like that in Example 4 is to use these properties to reorder and regroup the integers in a helpful way.

EXAMPLE 5 Use the commutative and/or associative properties of addition to help evaluate the expression: $-3 + 5 + (-12) + 2$

Strategy We will use the commutative and/or associative properties of addition so that we can add the positives and add the negatives separately. Then we will add those results to obtain the final answer.

WHY It is easier to add integers that have the same sign than integers that have different signs. This approach lessens the possibility of an error, because we only have to add integers that have different signs once.

Solution

$-3 + 5 + (-12) + 2$

$= -3 + (-12) + 5 + 2$ Use the commutative property of addition to reorder the integers.

$\qquad\quad$ Negatives $\quad$ Positives

$= [-3 + (-12)] + (5 + 2)$ Use the associative property of addition to group the negatives and group the positives.

Self Check 5

Use the commutative and/or associative properties of addition to help evaluate the expression:
$-12 + 8 + (-6) + 1$ -9

Now Try Problem 45

Teaching Example 5 Use the commutative and/or associative properties of addition to help evaluate the expression from Teaching Example 4 in a different way: $-7 + 3 + (-15) + 6$
Answer:
-13

$= -15 + 7$ Use the rule for adding two integers that have the same sign twice. Add the negatives within the brackets. Add the positives within the parentheses.

$= -8$ Use the rule for adding two integers that have different signs. This is the same result as in Example 4.

Self Check 6
Evaluate:
$(-6 + 8) + [10 + (-17)]$ -5
Now Try Problem 47

Teaching Example 6 Evaluate:
$[-8 + (-4)] + (-9 + 3)$
Answer:
-18

EXAMPLE 6 Evaluate: $[-21 + (-5)] + (-17 + 6)$

Strategy We will perform the addition within the brackets and the addition within the parentheses first. Then we will add those results.

WHY By the order of operations rule, we must perform the calculations within the grouping symbols first.

Solution Use the rule for adding two integers that have the same sign to do the addition within the brackets and the rule for adding two integers that have different signs to do the addition within the parentheses.

$[-21 + (-5)] + (-17 + 6) = -26 + (-11)$ Add within each pair of grouping symbols.

$ = -37$ Use the rule for adding two integers that have the same sign.

4 Identify opposites (additive inverses) when adding integers.

Recall from Section 1.2 that when 0 is added to a whole number, the whole number remains the same. This is true for any number. For example, $-5 + 0 = -5$ and $0 + (-43) = -43$. Because of this, we call 0 the **additive identity**.

> **The Language of Algebra** *Identity* is a form of the word *identical*, meaning the same. You have probably seen *identical* twins.

Addition Property of 0

The sum of any number and 0 is that number.
  For any number a
    $a + 0 = a$ and $0 + a = a$

There is another important fact about the operation of addition and 0. To illustrate it, we use the number line below to add 6 and its opposite, -6. Notice that $6 + (-6) = 0$.

If the sum of two numbers is 0, the numbers are said to be **additive inverses** of each other. Since $6 + (-6) = 0$, we say that 6 and -6 are additive inverses. Likewise, -7 is the additive inverse of 7, and 51 is the additive inverse of -51.

We can now classify a pair of integers such as 6 and -6 in two ways: as opposites or as additive inverses.

Addition Property of Opposites

The sum of a number and its opposite (additive inverse) is 0.

For any number a,

$$a + (-a) = 0 \quad \text{and} \quad -a + a = 0$$

At certain times, the addition property of opposites can be used to make addition of several integers easier.

EXAMPLE 7
Evaluate: $12 + (-5) + 6 + 5 + (-12)$

Strategy Instead of working from left to right, we will use the commutative and associative properties of addition to add *pairs of opposites*.

WHY Since the sum of an integer and its opposite is 0, it is helpful to identify such pairs in an addition.

Solution

$$12 + (-5) + 6 + 5 + (-12) = 0 + 0 + 6$$ Locate pairs of opposites and add them to get 0.

$$= 6$$ The sum of any integer and 0 is that integer.

Self Check 7
Evaluate:
$8 + (-1) + 6 + (-8) + 1$ 6

Now Try Problem 51

Teaching Example 7 Evaluate:
$16 + (-3) + 9 + 3 + (-16)$
Answer:
9

5 Solve application problems by adding integers.

Since application problems are almost always written in words, the ability to understand what you read is very important. Recall from Chapter 1 that words and phrases such as *gained*, *increased by*, and *rise* indicate addition.

EXAMPLE 8
Record Temperature Change At the beginning of this section, we learned that at 7:30 A.M. on January 22, 1943, in Spearfish, South Dakota, the temperature was −4°F. The temperature then rose 49 degrees in just 2 minutes. What was the temperature at 7:32 A.M.?

Strategy We will carefully read the problem looking for a key word or phrase.

WHY Key words and phrases indicate what arithmetic operations should be used to solve the problem.

Solution The phrase *rose 49 degrees* indicates addition. With that in mind, we translate the words of the problem to numbers and symbols.

The temperature at 7:32 A.M.	was	the temperature at 7:30 A.M.	plus	49 degrees.
The temperature at 7:32 A.M.	=	−4	+	49

To find the sum, we will use the rule for adding two integers that have different signs. First, we find the absolute values: $|-4| = 4$ and $|49| = 49$.

$-4 + 49 = 45$ Subtract the smaller absolute value from the larger absolute value: $49 - 4 = 45$. Since the positive number, 49, has the larger absolute value, the final answer is positive.

At 7:32 A.M., the temperature was 45°F.

Self Check 8
TEMPERATURE CHANGE On the morning of February 21, 1918, in Granville, North Dakota, the morning low temperature was −33°F. By the afternoon, the temperature had risen a record 83 degrees. What was the afternoon high temperature in Granville? (Source: *Extreme Weather* by Christopher C. Burt)

Now Try Problem 83

Self Check 8 Answer
50°F

Teaching Example 8
RECORD TEMPERATURE CHANGE The greatest temperature change ever recorded in a 24-hour period occurred in Loma, Montana, in 1972. On January 14, the low temperature there was −54°F. By the next day, the temperature had risen 103°F. What was the high temperature in Loma on January 15? (Source: *Extreme Weather* by Christopher C. Burt)
Answer:
49°F

Using Your CALCULATOR Entering Negative Numbers

Canada is the largest U.S. trading partner. To calculate the 2011 U.S. trade balance with Canada, we add the $281 billion worth of U.S. exports *to* Canada (considered positive) to the $315 billion worth of U.S. imports *from* Canada (considered negative). We can use a calculator to perform the addition: $281 + (-315)$.

We do not have to do anything special to enter a positive number. Negative numbers are entered using either **direct** or **reverse entry**, depending on the type of calculator you have.

To enter -315 using reverse entry, press the change-of-sign key $\boxed{+/-}$ *after* entering 315. To enter -315 using direct entry, press the negative key $\boxed{(-)}$ *before* entering 315. In either case, note that $\boxed{+/-}$ and the $\boxed{(-)}$ keys are different from the subtraction key $\boxed{-}$.

Reverse entry: 281 $\boxed{+}$ 315 $\boxed{+/-}$ $\boxed{=}$

Direct entry: 281 $\boxed{+}$ $\boxed{(-)}$ 315 $\boxed{\text{ENTER}}$ $\boxed{-34}$

In 2011, the United States had a trade balance of −$34 billion with Canada. Because the result is negative, it is called a trade *deficit*.

ANSWERS TO SELF CHECKS

1. a. −9 b. −73 c. −494   2. −3   3. a. 5 b. −14 c. 382   4. −9   5. −9
6. −5   7. 6   8. 50°F

SECTION 2.2   STUDY SET

VOCABULARY

Fill in the blanks.

1. Two negative integers, as well as two positive integers, are said to have the same or __like__ signs.

2. A positive integer and a negative integer are said to have different or __unlike__ signs.

3. When 0 is added to a number, the number remains the same. We call 0 the additive __identity__ .

4. Since $-5 + 5 = 0$, we say that 5 is the additive __inverse__ of −5. We can also say that 5 and −5 are __opposites__ .

5. __Commutative__ property of addition: The order in which integers are added does not change their sum.

6. __Associative__ property of addition: The way in which integers are grouped does not change their sum.

CONCEPTS

7. a. What is the absolute value of 10? What is the absolute value of −12? $|10| = 10, |-12| = 12$

   b. Which number has the larger absolute value, 10 or −12? −12

   c. Using your answers to part a, subtract the smaller absolute value from the larger absolute value. What is the result? 2

8. a. If you lost $6 and then lost $8, overall, what amount of money was lost? $14 was lost.

   b. If you lost $6 and then won $8, overall, what amount of money have you won? $2 was won.

Fill in the blanks.

9. To add two integers with unlike signs, __subtract__ their absolute values, the smaller from the larger. Then attach to that result the sign of the number with the __larger__ absolute value.

10. To add two integers with like signs, add their __absolute__ values and attach their common __sign__ to the sum.

11.
  a. Is the sum of two positive integers always positive? yes
  b. Is the sum of two negative integers always negative? yes
  c. Is the sum of a positive integer and a negative integer always positive? no
  d. Is the sum of a positive integer and a negative integer always negative? no

12. Complete the table by finding the additive inverse, opposite, and absolute value of the given numbers.

Number	Additive inverse	Opposite	Absolute value
19	−19	−19	19
−2	2	2	2
0	0	0	0

13.
  a. What is the sum of an integer and its additive inverse? 0
  b. What is the sum of an integer and its opposite? 0

14.
  a. What number must be added to −5 to obtain 0? 5
  b. What number must be added to 8 to obtain 0? −8

NOTATION

Complete each step to evaluate the expression.

15. $-16 + (-2) + (-1) = \boxed{-18} + (-1)$
$= \boxed{-19}$

16. $-8 + (-2) + 6 = \boxed{-10} + 6$
$= \boxed{-4}$

17. $(-3 + 8) + (-3) = \boxed{5} + (-3)$
$= \boxed{2}$

18. $-5 + [2 + (-9)] = -5 + (\boxed{-7})$
$= \boxed{-12}$

GUIDED PRACTICE

Add. See Example 1.

19. $-6 + (-3)$ −9
20. $-2 + (-3)$ −5
21. $-5 + (-5)$ −10
22. $-8 + (-8)$ −16
23. $-51 + (-11)$ −62
24. $-43 + (-12)$ −55
25. $-69 + (-27)$ −96
26. $-55 + (-36)$ −91
27. $-248 + (-131)$ −379
28. $-423 + (-164)$ −587
29. $-565 + (-309)$ −874
30. $-709 + (-187)$ −896

Add. See Examples 2 and 3.

31. $-8 + 5$ −3
32. $-9 + 3$ −6
33. $7 + (-6)$ 1
34. $4 + (-2)$ 2
35. $20 + (-42)$ −22
36. $-18 + 10$ −8
37. $71 + (-23)$ 48
38. $75 + (-56)$ 19
39. $479 + (-122)$ 357
40. $589 + (-242)$ 347
41. $-339 + 279$ −60
42. $-704 + 649$ −55

Evaluate each expression. See Examples 4 and 5.

43. $9 + (-3) + 5 + (-4)$ 7
44. $-3 + 7 + (-4) + 1$ 1
45. $6 + (-4) + (-13) + 7$ −4
46. $59 + (-24) + (-31) + (-7)$ −3

Evaluate each expression. See Example 6.

47. $[-3 + (-4)] + (-5 + 2)$ −10
48. $[9 + (-10)] + (-7 + 9)$ 1
49. $(-1 + 34) + [16 + (-8)]$ 41
50. $(-32 + 13) + [5 + (-14)]$ −28

Evaluate each expression. See Example 7.

51. $23 + (-5) + 3 + 5 + (-23)$ 3
52. $41 + (-1) + 9 + 1 + (-41)$ 9
53. $-10 + (-1) + 10 + (-6) + 1$ −6
54. $-14 + (-30) + 14 + (-9) + 9$ −30

TRY IT YOURSELF

Add.

55. $-2 + 6 + (-1)$ 3
56. $4 + (-3) + (-2)$ −1
57. $-7 + 0$ −7
58. $0 + (-15)$ −15
59. $24 + (-15)$ 9
60. $-4 + 14$ 10
61. $-435 + (-127)$ −562
62. $-346 + (-273)$ −619
63. $-7 + 9$ 2
64. $-3 + 6$ 3
65. $2 + (-2)$ 0
66. $-10 + 10$ 0
67. $-5 + (-9 + 8) + (-2)$ −8
68. $(-9 + 12) + (-4)$ −1
69. $-9 + 1 + (-2) + (-1) + 9$ −2
70. $5 + 4 + (-6) + (-4) + (-5)$ −6
71. $[6 + (-4)] + [8 + (-11)]$ −1
72. $[5 + (-8)] + [9 + (-15)]$ −9
73. $(-4 + 8) + (-11 + 4)$ −3
74. $(-12 + 6) + (-6 + 8)$ −4
75. $-675 + (-456) + 99$ −1,032
76. $-9,750 + (-780) + 2,345$ −8,185
77. Find the sum of $-6, -7$, and -8. −21
78. Find the sum of $-11, -12$, and -13. −36
79. $-2 + [789 + (-9,135)]$ −8,348
80. $-8 + [2,701 + (-4,089)]$ −1,396
81. What is 25 more than -45? −20
82. What is 31 more than -65? −34

APPLICATIONS

Use signed numbers to solve each problem.

83. **RECORD TEMPERATURES** The lowest recorded temperatures for Michigan and Minnesota are shown below. Use the given information to find the highest recorded temperature for each state. 112°F, 115°F

State	Lowest temperature	Highest temperature
Michigan	Feb. 9, 1934: −51°F	July 13, 1936: 163°F warmer than the record low
Minnesota	Feb. 2, 1996: −60°F	July 29, 1917: 175°F warmer than the record low

Source: *The World Almanac and Book of Facts*, 2013

84. **ELEVATIONS** The lowest point in the United States is Death Valley, California, with an elevation of −282 feet (282 feet below sea level). Mt. McKinley (Alaska) is the highest point in the United States. Its elevation is 20,602 feet higher than Death Valley. What is the elevation of Mt. McKinley? (Source: *The World Almanac and Book of Facts*, 2013) 20,320 ft

85. **SUNKEN SHIPS** Refer to the map below.
   a. The German battleship *Bismarck*, one of the most feared warships of World War II, was sunk by the British in 1941. It lies on the ocean floor 15,720 feet below sea level off the west coast of France. Represent that depth using a signed number. −15,720 ft
   b. In 1912, the luxury liner *Titanic* sank after striking an iceberg. It lies on the North Atlantic ocean floor, 3,220 feet higher than the *Bismarck*. At what depth is the *Titanic* resting? −12,500 ft

86. **JOGGING** A businessman's lunchtime workout includes jogging up ten stories of stairs in his high-rise office building. He starts the workout on the fourth level below ground in the underground parking garage.
   a. Represent that level using a signed number. −4
   b. On what story of the building will he finish his workout? 6th story

87. **FLOODING** After a heavy rainstorm, a river that had been 9 feet under flood stage rose 11 feet in a 48-hour period.
   a. Represent the level of the river before the storm using a signed number. −9 ft
   b. Find the height of the river after the storm in comparison to flood stage. 2 ft above flood stage

88. **ATOMS** An atom is composed of protons, neutrons, and electrons. A proton has a positive charge (represented by +1), a neutron has no charge, and an electron has a negative charge (−1). Two simple models of atoms are shown below.
   a. How many protons does the atom in figure (a) have? How many electrons? 8, 10
   b. What is the net charge of the atom in figure (a)? −2
   c. How many protons does the atom in figure (b) have? How many electrons? 4, 4
   d. What is the net charge of the atom in figure (b)? 0

(a)   (b)

89. **CHEMISTRY** The three steps of a chemistry lab experiment are listed here. The experiment begins with a compound that is stored at −40°F.
   Step 1 Raise the temperature of the compound 200°.
   Step 2 Add sulfur and then raise the temperature 10°.
   Step 3 Add 10 milliliters of water, stir, and raise the temperature 25°.

   What is the resulting temperature of the mixture after step 3? 195°

90. Suppose you are a personal financial adviser and your clients are considering purchasing income property. You find a duplex apartment unit that is for sale and learn that the maintenance costs, utilities, and taxes on it total $900 per month. If the current owner receives monthly rental payments of $450 and $380 from the tenants, does the duplex produce a positive cash flow each month? no, $70 shortfall each month (−$70)

from Campus to Careers
Personal Financial Adviser

91. HEALTH Find the point total for the six risk factors (shown with blue headings) on the medical questionnaire below. Then use the table at the bottom of the form (under the red heading) to determine the risk of contracting heart disease for the man whose responses are shown. 5, 4% risk

Age		Total Cholesterol	
Age	Points	Reading	Points
35	−4	280	3

Cholesterol		Blood Pressure	
HDL	Points	Systolic/Diastolic	Points
62	−3	124/100	3

Diabetic		Smoker	
	Points		Points
Yes	4	Yes	2

10-Year Heart Disease Risk			
Total Points	Risk	Total Points	Risk
−2 or less	1%	5	4%
−1 to 1	2%	6	6%
2 to 3	3%	7	6%
4	4%	8	7%

Source: National Heart, Lung, and Blood Institute

92. POLITICAL POLLS Six months before a general election, the incumbent senator found himself trailing the challenger by 18 points. To overtake his opponent, the campaign staff decided to use a four-part strategy. Each part of this plan is shown below, with the anticipated point gain.

Part 1 Intense TV ad blitz: gain 10 points

Part 2 Ask for union endorsement: gain 2 points

Part 3 Voter mailing: gain 3 points

Part 4 Get-out-the-vote campaign: gain 1 point

With these gains, will the incumbent overtake the challenger on election day?
No, he will still be behind by 2 points.

93. MILITARY SCIENCE During a battle, an army retreated 1,500 meters, regrouped, and advanced 3,500 meters. The next day, it advanced 1,250 meters. Find the army's net gain. 3,250 m

94. AIRLINES The graph in the next column shows the annual net income for Delta Air Lines during the years 2004–2011.

 a. Estimate the company's total net income over this span of eight years in millions of dollars. Use front-end rounding. about −$21,000 million

 b. About how many billions of dollars is your answer to part a? about −$21 billion

Source: *The Wall Street Journal* and marketwatch.com

95. ACCOUNTING On a financial balance sheet, debts (considered negative numbers) are written within parentheses. Assets (considered positive numbers) are written without parentheses. What is the 2014 fund balance for the preschool whose financial records are shown below? ($967)

ABC Preschool Balance Sheet, June 2014

Fund	Balance $
Classroom supplies	$5,889
Emergency needs	$927
Holiday program	($2,928)
Insurance	$1,645
Janitorial	($894)
Licensing	$715
Maintenance	($6,321)
BALANCE	?

96. SPREADSHEETS Monthly rain totals for four counties are listed in the spreadsheet below. The −1 entered in cell B1 means that the rain total for Suffolk County for a certain month was 1 inch *below* average. We can analyze these data by asking the computer to perform various operations.

Book 1

	A	B	C	D	E	F
1	Suffolk	−1	−1	0	+1	+1
2	Marin	0	−2	+1	+1	−1
3	Logan	−1	+1	+2	+1	+1
4	Tipton	−2	−2	+1	−1	−3
5						

 a. To ask the computer to add the numbers in cells B1, B2, B3, and B4, we type SUM(B1:B4). Find this sum. −4

 b. Find SUM(F1:F4). −2

WRITING

97. Is the sum of a positive and a negative number always positive? Explain why or why not.

98. How do you explain the fact that when asked to *add* −4 and 8, we must actually *subtract* to obtain the result?

99. Explain why the sum of two negative numbers is a negative number.

100. Write an application problem that will require adding −50 and −60.

101. If the sum of two integers is 0, what can be said about the integers? Give an example.

102. Explain why the expression −6 + −5 is not written correctly. How should it be written?

103. Explain the commutative property of addition in your own words. Give an example using a positive integer and a negative integer.

104. Explain the associative property of addition in your own words. Give an example using two positive integers and a negative integer.

REVIEW

105. **a.** Find the perimeter of the rectangle shown below. 16 ft

  b. Find the area of the rectangle shown below. 15 ft²

106. What property is illustrated by the statement 5 · 15 = 15 · 5? commutative property of multiplication

107. Prime factor 250. Use exponents to express the result. $2 \cdot 5^3$

108. Divide: $\dfrac{144}{12}$ 12

Objectives

1 Use the subtraction rule.

2 Evaluate expressions involving subtraction and addition.

3 Solve application problems by subtracting integers.

SECTION 2.3
Subtracting Integers

In this section, we will discuss a rule that is helpful when subtracting signed numbers.

1 Use the subtraction rule.

The subtraction problem 6 − 4 can be thought of as taking away 4 from 6. We can use a number line to illustrate this. Beginning at 0, we draw an arrow of length 6 units long that points to the right. It represents positive 6. From the tip of that arrow, we draw a second arrow, 4 units long, that points to the left. It represents taking away 4. Since we end up at 2, it follows that 6 − 4 = 2.

Note that the illustration above also represents the *addition* 6 + (−4) = 2. We see that

Subtracting 4 from 6 . . .   is the same as . . .   adding the opposite of 4 to 6.

6 − 4 = 2   6 + (−4) = 2

The results are the same.

This observation suggests the following rule.

Rule for Subtraction

To subtract two numbers, add the first number to the opposite (additive inverse) of the number to be subtracted.

For any numbers a and b,

$$a - b = a + (-b)$$

Put more simply, this rule says that *subtraction is the same as adding the opposite*.

After rewriting a subtraction as addition of the opposite, we then use one of the rules for the addition of signed numbers discussed in Section 2.2 to find the result.

You won't need to use this rule for every subtraction problem. For example, $6 - 4$ is obviously 2; it does not need to be rewritten as adding the opposite. But for more complicated problems such as $-6 - 4$ or $3 - (-5)$, where the result is not obvious, the subtraction rule will be quite helpful.

EXAMPLE 1 Subtract and check the result:

a. $-6 - 4$ **b.** $3 - (-5)$ **c.** $7 - 23$

Strategy To find each difference, we will apply the rule for subtraction: Add the first integer to the opposite of the integer to be subtracted.

WHY It is easy to make an error when subtracting signed numbers. We will probably be more accurate if we write each subtraction as addition of the opposite.

Solution

a. We read $-6 - 4$ as "negative six *minus* four." Thus, the number to be subtracted is 4. Subtracting 4 is the same as adding its opposite, -4.

$-6 - 4 = -6 + (-4) = -10$ Use the rule for adding two integers with the same sign.

To check, we add the *difference*, -10, and the *subtrahend*, 4. We should get the *minuend*, -6.

Check: $-10 + 4 = -6$ The result checks.

Caution! Don't forget to write the opposite of the number to be subtracted within parentheses if it is negative.

$-6 - 4 = -6 + \boxed{(-4)}$

b. We read $3 - (-5)$ as "three *minus* negative five." Thus, the number to be subtracted is -5. Subtracting -5 is the same as adding its opposite, 5.

$3 - (-5) = 3 + 5 = 8$

Check: $8 + (-5) = 3$ The result checks.

Self Check 1

Subtract and check the result:

a. $-2 - 3$ -5
b. $4 - (-8)$ 12
c. $6 - 85$ -79

Now Try Problems 21, 25, and 29

Teaching Example 1 Subtract and check the result:
a. $-5 - 3$
b. $9 - (-2)$
c. $4 - 67$
Answers:
a. -8
b. 11
c. -63

c. We read $7 - 23$ as "seven *minus* twenty-three." Thus, the number to be subtracted is 23. Subtracting 23 is the same as adding its opposite, -23.

$$7 - 23 = 7 + (-23) = -16$$ Use the rule for adding two integers with different signs.

Check: $-16 + 23 = 7$ The result checks.

Caution! When applying the subtraction rule, *do not change* the first number.

$$-6 - 4 = -6 + (-4) \qquad 3 - (-5) = 3 + 5$$

Self Check 2
a. Subtract -10 from -7. 3
b. Subtract -7 from -10. -3
Now Try Problem 33

Teaching Example 2
a. Subtract -14 from -2.
b. Subtract -2 from -14.
Answers:
a. 12
b. -12

EXAMPLE 2 **a.** Subtract -12 from -8. **b.** Subtract -8 from -12.

Strategy We will translate each phrase to mathematical symbols and then perform the subtraction. We must be careful when translating the instruction to subtract one number *from* another number.

WHY The order of the numbers in each word phrase must be reversed when we translate it to mathematical symbols.

Solution

a. Since -12 is the number to be subtracted, we reverse the order in which -12 and -8 appear in the sentence when translating to symbols.

Subtract -12 from -8

$$-8 - (-12) \quad \text{Write } -12 \text{ within parentheses.}$$

To find this difference, we write the subtraction as addition of the opposite:

$$-8 - (-12) = -8 + 12 = 4 \quad \text{Use the rule for adding two integers with different signs.}$$

b. Since -8 is the number to be subtracted, we reverse the order in which -8 and -12 appear in the sentence when translating to symbols.

Subtract -8 from -12

$$-12 - (-8) \quad \text{Write } -8 \text{ within parentheses.}$$

To find this difference, we write the subtraction as addition of the opposite:

$$-12 - (-8) = -12 + 8 = -4 \quad \text{Use the rule for adding two integers with different signs.}$$

The Language of Algebra When we change a number to its opposite, we say we have *changed* (or *reversed*) its sign.

Remember that any subtraction problem can be rewritten as an equivalent addition. We just add the opposite of the number that is to be subtracted. Here are four examples:

- $4 - 8 = 4 + (-8) = -4$
- $4 - (-8) = 4 + 8 = 12$
- $-4 - 8 = -4 + (-8) = -12$
- $-4 - (-8) = -4 + 8 = 4$

Any subtraction can be written as addition of the opposite of the number to be subtracted.

2 Evaluate expressions involving subtraction and addition.

Expressions can involve repeated subtraction or combinations of subtraction and addition. To evaluate them, we use the order of operations rule discussed in Section 1.7.

EXAMPLE 3 Evaluate: $-1 - (-2) - 10$

Strategy This expression involves two subtractions. We will write each subtraction as addition of the opposite and then evaluate the expression using the order of operations rule.

WHY It is easy to make an error when subtracting signed numbers. We will probably be more accurate if we write each subtraction as addition of the opposite.

Solution We apply the rule for subtraction twice and then perform the additions, working from left to right. (We could also add the positives and the negatives separately, and then add those results.)

$$-1 - (-2) - 10 = -1 + 2 + (-10)$$ Add the opposite of -2, which is 2. Add the opposite of 10, which is -10.

$$= 1 + (-10)$$ Work from left to right. Add $-1 + 2$ using the rule for adding integers that have different signs.

$$= -9$$ Use the rule for adding integers that have different signs.

Self Check 3
Evaluate: $-3 - 5 - (-1)$ -7
Now Try Problem 37

Teaching Example 3 Evaluate:
$-4 - (-9) - 12$
Answer:
-7

EXAMPLE 4 Evaluate: $-80 - (-2 - 24)$

Strategy We will consider the subtraction within the parentheses first and rewrite it as addition of the opposite.

WHY By the order of operations rule, we must perform all calculations within parentheses first.

Solution
$$-80 - (-2 - 24) = -80 - [-2 + (-24)]$$ Add the opposite of 24, which is -24. Since -24 must be written within parentheses, we write $-2 + (-24)$ within brackets.

$$= -80 - (-26)$$ Within the brackets, add -2 and -24. Since only one set of grouping symbols is now needed, we can write the answer, -26, within parentheses.

$$= -80 + 26$$ Add the opposite of -26, which is 26.
$$= -54$$ Use the rule for adding integers that have different signs.

$$\begin{array}{r} 7\,10 \\ 8\cancel{0} \\ -26 \\ \hline 54 \end{array}$$

Self Check 4
Evaluate: $-72 - (-6 - 51)$ -15
Now Try Problem 49

Teaching Example 4 Evaluate:
$-61 - (-1 - 22)$
Answer:
-38

EXAMPLE 5 Evaluate: $-(-6) + (-18) - 4 - (-51)$

Strategy This expression involves one addition and two subtractions. We will write each subtraction as addition of the opposite and then evaluate the expression.

Self Check 5
Evaluate:
$-(-3) + (-16) - 9 - (-28)$ 6
Now Try Problem 55

Teaching Example 5 Evaluate:
−(−6) + (−7) − 14 − (−39)
Answer:
24

WHY It is easy to make an error when subtracting signed numbers. We will probably be more accurate if we write each subtraction as addition of the opposite.

Solution We apply the rule for subtraction twice. Then we will add the positives and the negatives separately, and add those results. (By the commutative and associative properties of addition, we can add the integers in any order.)

$$-(-6) + (-18) - 4 - (-51)$$
$$= 6 + (-18) + (-4) + 51 \quad \text{Simplify: } -(-6) = 6. \text{ Add the opposite of 4,}$$
$$ \text{which is } -4, \text{ and add the opposite of } -51,$$
$$ \text{which is 51.}$$
$$= (6 + 51) + [(-18) + (-4)] \quad \text{Reorder the integers. Then group the positives}$$
$$ \text{together and group the negatives together.}$$
$$= 57 + (-22) \quad \text{Add the positives and add the negatives.}$$
$$= 35 \quad \text{Use the rule for adding integers that have different signs.}$$

3 Solve application problems by subtracting integers.

Subtraction finds the *difference* between two numbers. When we find the difference between the maximum value and the minimum value of a collection of measurements, we are finding the **range** of the values.

$$\text{Range} = \text{maximum value} - \text{minimum value}$$

Self Check 6

THE GATEWAY CITY The record high temperature for St. Louis, Missouri, is 105°F. The record low temperature is −17°F. Find the temperature range for these extremes.
(Source: weather.com) 122°F

Now Try Problem 101

Teaching Example 6
THE STEEL CITY The record high temperature for Pittsburgh, Pennsylvania, is 101°F. The record low is −19°F. Find the temperature range for these extremes. (Source: weather.com)
Answer:
120°F

EXAMPLE 6 *The Windy City* The record high temperature for Chicago, Illinois, is 107°F. The record low is −25°F. Find the temperature range for these extremes. (Source: weather.com)

Strategy We will subtract the minimum temperature (−25°F) from the maximum temperature (107°F).

WHY The *range* of a collection of data indicates the spread of the data. It is the difference between the maximum and minimum values.

Solution We apply the rule for subtraction and add the opposite of −25.

$$107 - (-25) = 107 + 25 \quad \text{107° is the highest temperature and } -25° \text{ is the lowest.}$$
$$= 132$$

The temperature range for these extremes is 132°F.

Things are constantly changing in our daily lives. The amount of money we have in the bank, the price of gasoline, and our ages are examples. In mathematics, the operation of subtraction is used to measure change. To find the **change** in a quantity, we subtract the earlier value from the later value.

$$\text{Change} = \text{later value} - \text{earlier value}$$

EXAMPLE 7

Water Management On Monday, the water level in a city storage tank was 16 feet above normal. By Friday, the level had fallen to a mark 14 feet below normal. Find the change in the water level from Monday to Friday.

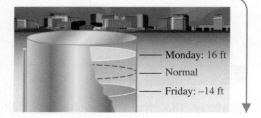

Solution It is helpful to list the given facts and what you are to find.

- On Monday, the water level was 16 feet above normal.   Given
- On Friday, the water level was 14 feet below normal.   Given
- What was the change in the water level?   Find

To find the change in the water level, we *subtract the earlier value from the later value*. The water levels of 16 feet above normal (the earlier value) and 14 feet below normal (the later value) can be represented by 16 and −14.

We translate the words of the problem to numbers and symbols.

The change in the water level	is equal to	the later water level (Friday)	minus	the earlier water level (Monday).
The change in the water level	=	−14	−	16

We can use the rule for subtraction to find the difference.

$-14 - 16 = -14 + (-16)$ Add the opposite of 16, which is −16.

$ = -30$ Use the rule for adding integers with the same sign.

The negative result means the water level *fell* 30 feet from Monday to Friday.

Check: If we represent the change in water level on a horizontal number line, we see that the water level fell $16 + 14 = 30$ units. The result checks.

Self Check 7

CRUDE OIL On Wednesday, the level of crude oil in a storage tank was 5 feet above standard capacity. Thursday, after a large refining session, the level fell to a mark 76 feet below standard capacity. Find the change in the crude oil level from Wednesday to Thursday. The crude oil level fell 81 ft.

Now Try Problem 103

Teaching Example 7
WATER MANAGEMENT On Monday, the water level in a city storage tank was 19 feet above normal. By the following Monday, the level had fallen to a mark 15 feet below normal. Find the change in the water level that week.
Answer:
The water level fell 34 ft.

Using Your CALCULATOR   Subtraction with Negative Numbers

The world's highest peak is Mount Everest in the Himalayas. The greatest ocean depth yet measured lies in the Mariana Trench near the island of Guam in the western Pacific. To find the range between the highest peak and the greatest depth, we must subtract:

$29{,}035 - (-36{,}025)$

To perform this subtraction on a calculator, we enter the following:

Reverse entry: 29035 [−] 36025 [+/−] [=]

Direct entry: 29035 [−] [(−)] 36025 [ENTER]   65060

The range is 65,060 feet between the highest peak and the lowest depth. (We could also write $29{,}035 - (-36{,}025)$ as $29{,}035 + 36{,}025$ and then use the addition key [+] to find the answer.)

ANSWERS TO SELF CHECKS

1. a. −5 **b.** 12 **c.** −79   **2. a.** 3 **b.** −3   **3.** −7   **4.** −15   **5.** 6   **6.** 122°F
7. The crude oil level fell 81 ft.

SECTION 2.3 STUDY SET

VOCABULARY

Fill in the blanks.

1. −8 is the **opposite** (or **additive** inverse) of 8.
2. When we change a number to its opposite, we say we have *changed* (or *reversed*) its **sign**.
3. To evaluate an expression means to find its **value**.
4. The difference between the maximum and the minimum value of a collection of measurements is called the **range** of the values.

CONCEPTS

Fill in the blanks.

5. To subtract two integers, add the first integer to the **opposite** (additive inverse) of the integer to be subtracted.
6. Subtracting is the same as **adding** the opposite.
7. Subtracting 3 is the same as adding **−3**. Subtracting −6 is the same as adding **6**.
8. For any numbers a and b, $a - b = a +$ **$(-b)$**.
9. We can find the **change** in a quantity by subtracting the earlier value from the later value.
10. After rewriting a subtraction as addition of the opposite, we then use one of the rules for the **addition** of signed numbers discussed in the previous section to find the result.
11. In each case, determine what number is being subtracted.
    a. $-7 - 3$ **3**
    b. $1 - (-12)$ **−12**
12. Fill in the blanks to rewrite each subtraction as addition of the opposite of the number being subtracted.
    a. $2 - 7 = 2 +$ **(-7)**
    b. $2 - (-7) = 2 +$ **7**
    c. $-2 - 7 = -2 +$ **(-7)**
    d. $-2 - (-7) = -2 +$ **7**
13. Apply the rule for subtraction and fill in the three blanks.
    $3 - (-6) = 3 +$ **+** **6** $=$ **9**
14. Use addition to check this subtraction: $14 - (-2) = 12$. Is the result correct? **no, $12 + (-2) \neq 14$**

NOTATION

15. Write each phrase using symbols.
    a. negative eight minus negative four **$-8 - (-4)$**
    b. negative eight subtracted from negative four **$-4 - (-8)$**

16. Write each phrase in words.
    a. $7 - (-2)$ seven minus negative two
    b. $-2 - (-7)$ negative two minus negative seven

Complete each step to evaluate each expression.

17. $1 - 3 - (-2) = 1 + ($ **−3** $) + 2$
    $\qquad = -2 +$ **2**
    $\qquad =$ **0**

18. $-6 + 5 - (-5) = -6 + 5 +$ **5**
    $\qquad =$ **−1** $+ 5$
    $\qquad =$ **4**

19. $(-8 - 2) - (-6) = [-8 + ($ **−2** $)] - (-6)$
    $\qquad =$ **−10** $- (-6)$
    $\qquad = -10 +$ **6**
    $\qquad =$ **−4**

20. $-(-5) - (-1 - 4) =$ **5** $- [-1 + ($ **−4** $)]$
    $\qquad = 5 - ($ **−5** $)$
    $\qquad = 5 +$ **5**
    $\qquad =$ **10**

GUIDED PRACTICE

Subtract. See Example 1.

21. $-4 - 3$ −7
22. $-4 - 1$ −5
23. $-5 - 5$ −10
24. $-7 - 7$ −14
25. $8 - (-1)$ 9
26. $3 - (-8)$ 11
27. $11 - (-7)$ 18
28. $18 - (-25)$ 43
29. $3 - 21$ −18
30. $8 - 32$ −24
31. $15 - 65$ −50
32. $12 - 82$ −70

Perform the indicated operation. See Example 2.

33. a. Subtract -1 from -11. −10
    b. Subtract -11 from -1. 10
34. a. Subtract -2 from -19. −17
    b. Subtract -19 from -2. 17
35. a. Subtract -41 from -16. 25
    b. Subtract -16 from -41. −25
36. a. Subtract -57 from -15. 42
    b. Subtract -15 from -57. −42

Evaluate each expression. See Example 3.

37. $-4 - (-4) - 15$ −15
38. $-3 - (-3) - 10$ −10
39. $10 - 9 - (-8)$ 9
40. $16 - 14 - (-9)$ 11

41. $-1-(-3)-4$ -2 42. $-2-4-(-1)$ -5
43. $-5-8-(-3)$ -10 44. $-6-5-(-1)$ -10

Evaluate each expression. See Example 4.

45. $-1-(-4-6)$ 9   46. $-7-(-2-14)$ 9
47. $-42-(-16-14)$ -12 48. $-45-(-8-32)$ -5
49. $-9-(6-7)$ -8 50. $-13-(6-12)$ -7
51. $-8-(4-12)$ 0   52. $-9-(1-10)$ 0

Evaluate each expression. See Example 5.

53. $-(-5)+(-15)-6-(-48)$ 32
54. $-(-2)+(-30)-3-(-66)$ 35
55. $-(-3)+(-41)-7-(-19)$ -26
56. $-(-1)+(-52)-4-(-21)$ -34

Use a calculator to perform each subtraction. See Using Your CALCULATOR.

57. $-1{,}557-890$ $-2{,}447$ 58. $20{,}007-(-496)$ 20,503
59. $-979-(-44{,}879)$ 43,900   60. $-787-1{,}654-(-232)$ $-2{,}209$

TRY IT YOURSELF

Evaluate each expression.

61. $-5+7-(-1)-8$ -5 62. $-5+9-(-4)-16$ -8
63. Subtract -3 from 7. 10   64. Subtract 8 from -2. -10
65. $-2-(-10)$ 8   66. $-6-(-12)$ 6
67. $0-(-5)$ 5   68. $0-8$ -8
69. $(6-4)-(1-2)$ 3   70. $(5-3)-(4-6)$ 4
71. $-5-(-4)$ -1 72. $-12-(-7)$ -5
73. $-3-3-3$ -9 74. $-1-1-1$ -3
75. $-(-9)+(-20)-14-(-3)$ -22
76. $-(-8)+(-33)-7-(-21)$ -11
77. $[-4+(-8)]-(-6)+15$ 9
78. $[-5+(-4)]-(-2)+22$ 15
79. Subtract -6 from -10. -4
80. Subtract -4 from -9. -5
81. $-3-(-3)$ 0   82. $-5-(-5)$ 0
83. $-8-[4-(-6)]$ -18 84. $-1-[5-(-2)]$ -8
85. $4-(-4)$ 8   86. $-3-3$ -6
87. $(-6-5)-3+(-11)$ -25 88. $(-2-1)-5+(-19)$ -27

APPLICATIONS

Use signed numbers to solve each problem.

89. SUBMARINES A submarine was traveling 2,000 feet below the ocean's surface when the sonar system warned of a possible collision with another sub. The captain ordered the navigator to dive an additional 200 feet and then level off. Use a signed number to represent the depth of the submarine after the dive. $-2{,}200$ ft

90. SCUBA DIVING A diver jumps from his boat into the water and descends to a depth of 50 feet. He pauses to check his equipment and then descends an additional 70 feet. Use a signed number to represent the diver's final depth. -120 ft

91. GEOGRAPHY Death Valley, California, is the lowest land point in the United States, at 282 feet below sea level. The lowest land point on Earth is the Dead Sea, which is 1,348 feet below sea level. How much lower is the Dead Sea than Death Valley? 1,066 ft

92. HISTORY Two of the greatest Greek mathematicians were Archimedes (287–212 B.C.) and Pythagoras (569–500 B.C.).
   a. Express the year of Archimedes' birth as a negative number. -287
   b. Express the year of Pythagoras' birth as a negative number. -569
   c. How many years apart were they born? 282 yr

93. AMPERAGE During normal operation, the ammeter on a car reads $+5$. If the headlights are turned on, they lower the ammeter reading 7 amps. If the radio is turned on, it lowers the reading 6 amps. What number will the ammeter register if they are both turned on? -8

94. GIN RUMMY After a losing round, a card player must deduct the value of each of the cards left in his hand from his previous point total of 21. If face cards are counted as 10 points, what is his new score? -8

95. FOOTBALL A college football team records the outcome of each of its plays during a game on a stat sheet. Find the net gain (or loss) after the third play. -4 yd

Down	Play	Result
1st	Run	Lost 1 yd
2nd	Pass—sack!	Lost 6 yd
Penalty	Delay of game	Lost 5 yd
3rd	Pass	Gained 8 yd

96. ACCOUNTING Complete the balance sheet below. Then determine the overall financial condition of the company by subtracting the total debts from the total assets. $86,514; $99,218; −$12,704

Walker Corporation Balance Sheet 2014	
Assets	
Cash	$11,109
Supplies	7,862
Land	67,543
Total assets	$
Debts	
Accounts payable	$79,037
Income taxes	20,181
Total debts	$

97. OVERDRAFT PROTECTION A student forgot that she had only $15 in her bank account and wrote a check for $25, used an ATM to get $40 cash, and used her debit card to buy $30 worth of groceries. On each of the three transactions, the bank charged her a $20 overdraft protection fee. Find the new account balance. −$140

98. CHECKING ACCOUNTS Michael has $1,303 in his checking account. Can he pay his car insurance premium of $676, his utility bills of $121, and his rent of $750 without having to make another deposit? Explain. No, he will be $244 overdrawn (−$244).

99. TEMPERATURE EXTREMES The highest and lowest temperatures ever recorded in several cities are shown below. List the cities in order, from the largest to smallest range in temperature extremes. Portland, Barrow, Kansas City, Norfolk, Atlantic City

Extreme Temperatures

City	Highest	Lowest
Atlantic City, NJ	102	−3
Barrow, AK	79	−56
Kansas City, MO	112	−19
Norfolk, VA	105	−3
Portland, ME	103	−39

Source: weather.com

100. EYESIGHT *Nearsightedness*, the condition in which near objects are clear and far objects are blurry, is measured using negative numbers. *Farsightedness*, the condition in which far objects are clear and near objects are blurry, is measured using positive numbers. Find the range in the measurements shown in the next column. 6

Nearsighted −2   Farsighted +4

101. FREEZE DRYING To make freeze-dried coffee, the coffee beans are roasted at a temperature of 360°F and then the ground coffee bean mixture is frozen at a temperature of −110°F. What is the temperature range of the freeze-drying process? 470°F

102. WEATHER Rashawn flew from his New York home to Hawaii for a week of vacation. He left blizzard conditions and a temperature of −6°F, and stepped off the airplane into 85°F weather. What temperature change did he experience? 91°F increase

103. READING PROGRAMS In a state reading test given at the start of a school year, an elementary school's performance was 23 points below the county average. The principal immediately began a special tutorial program. At the end of the school year, retesting showed the students to be only 7 points below the average. How did the school's reading score change over the year? 16-point increase

104. LIE DETECTOR TESTS On one lie detector test, a burglar scored −18, which indicates deception. However, on a second test, he scored −1, which is inconclusive. Find the change in his scores. 17

WRITING

105. Explain what is meant when we say that subtraction is the same as addition of the opposite.

106. Give an example showing that it is possible to subtract something from nothing.

107. Explain how to check the result: $-7 - 4 = -11$

108. Explain why students don't need to change every subtraction they encounter to an addition of the opposite. Give some examples.

REVIEW

109. a. Round 24,085 to the nearest ten. 24,090
    b. Round 5,999 to the nearest hundred. 6,000

110. List the factors of 20 from least to greatest. 1, 2, 4, 5, 10, 20

111. It takes 13 oranges to make one can of orange juice. Find the number of oranges used to make 12 cans. 156

112. a. Find the LCM of 15 and 18. 90
    b. Find the GCF of 15 and 18. 3

SECTION 2.4
Multiplying Integers

Objectives
1. Multiply two integers that have different signs.
2. Multiply two integers that have the same sign.
3. Perform several multiplications to evaluate expressions.
4. Evaluate exponential expressions that have negative bases.
5. Solve application problems by multiplying integers.

Multiplication of integers is very much like multiplication of whole numbers. The only difference is that we must determine whether the answer is positive or negative.

When we multiply two nonzero integers, they either have different signs or they have the same sign. This means that there are two possibilities to consider.

1 Multiply two integers that have different signs.

To develop a rule for multiplying two integers that have different signs, we will find $4(-3)$, which is the product of a positive integer and negative integer. We say that the signs of the factors are *unlike*. By the definition of multiplication, $4(-3)$ means that we are to add -3 four times.

$4(-3) = (-3) + (-3) + (-3) + (-3)$ Write -3 as an addend four times.
$ = -12$ Use the rule for adding two integers that have the same sign.

The result is negative. As a check, think in terms of money. If you lose $3 four times, you have lost a total of $12, which is written $-$12. This example illustrates the following rule.

> **Multiplying Two Integers That Have Different (Unlike) Signs**
>
> To multiply a positive integer and a negative integer, multiply their absolute values. Then make the final answer negative.

EXAMPLE 1 Multiply:
a. $7(-5)$ **b.** $20(-8)$ **c.** $-93 \cdot 16$ **d.** $-34(1,000)$

Strategy We will use the rule for multiplying two integers that have different (unlike) signs.

WHY In each case, we are asked to multiply a positive integer and a negative integer.

Solution

a. Find the absolute values: $|7| = 7$ and $|-5| = 5$.

$7(-5) = -35$ Multiply the absolute values, 7 and 5, to get 35.
Then make the final answer negative.

b. Find the absolute values: $|20| = 20$ and $|-8| = 8$.

$20(-8) = -160$ Multiply the absolute values, 20 and 8, to get 160.
Then make the final answer negative.

c. Find the absolute values: $|-93| = 93$ and $|16| = 16$.

$-93 \cdot 16 = -1,488$ Multiply the absolute values, 93 and 16, to get 1,488.
Then make the final answer negative.

$$\begin{array}{r} 93 \\ \times\ 16 \\ \hline 558 \\ 930 \\ \hline 1,488 \end{array}$$

d. Recall from Section 1.4, to find the product of a whole number and 10, 100, 1,000, and so on, *attach the number of zeros in that number to the right of the whole number*. This rule can be extended to products of integers and 10, 100, 1,000, and so on.

$-34(1,000) = -34,000$ Since 1,000 has three zeros, attach three 0's after -34.

Self Check 1
Multiply:
a. $2(-6)$ -12
b. $30(-4)$ -120
c. $-75 \cdot 17$ $-1,275$
d. $-98(1,000)$ $-98,000$

Now Try Problems 21, 25, 29, and 31

Teaching Example 1 Multiply:
a. $9(-4)$
b. $30(-6)$
c. $-57 \cdot 18$
d. $-29(1,000)$
Answers:
a. -36
b. -180
c. $-1,026$
d. $-29,000$

Caution! When writing multiplication involving signed numbers, do not write a negative sign − next to a raised dot · (the multiplication symbol). Instead, use parentheses to show the multiplication.

$6(-2)$ ~~$6 \cdot -2$~~ and $-6(-2)$ ~~$-6 \cdot -2$~~

2 Multiply two integers that have the same sign.

To develop a rule for multiplying two integers that have the same sign, we will first consider $4(3)$, which is the product of two positive integers. We say that the signs of the factors are *like*. By the definition of multiplication, $4(3)$ means that we are to add 3 four times.

$4(3) = 3 + 3 + 3 + 3$ Write 3 as an addend four times.
$ = 12$ The result is 12, which is a positive number.

As expected, the result is positive.

To develop a rule for multiplying two negative integers, consider the following list, where we multiply -4 by factors that decrease by 1. We know how to find the first four products. Graphing those results on a number line is helpful in determining the last three products.

From the pattern, we see that the product increases by 4 each time. Thus,

$-4(-1) = 4,$ $-4(-2) = 8,$ and $-4(-3) = 12$

These results illustrate that *the product of two negative integers is positive.* As a check, think of it as losing four debts of $3. This is equivalent to gaining $12. Therefore, $-4(-\$3) = \$12$.

We have seen that the product of two positive integers is positive, and the product of two negative integers is also positive. Those results illustrate the following rule.

Multiplying Two Integers That Have the Same (Like) Signs

To multiply two integers that have the same sign, multiply their absolute values. The final answer is positive.

EXAMPLE 2   Multiply:
a. $-5(-9)$ **b.** $-8(-10)$ **c.** $-23(-42)$ **d.** $-2,500(-30,000)$

Strategy We will use the rule for multiplying two integers that have the same (like) signs.

WHY In each case, we are asked to multiply two negative integers.

Solution

a. Find the absolute values: $|-5| = 5$ and $|-9| = 9$.

$-5(-9) = 45$ Multiply the absolute values, 5 and 9, to get 45. The final answer is positive.

b. Find the absolute values: $|-8| = 8$ and $|-10| = 10$.

$-8(-10) = 80$ Multiply the absolute values, 8 and 10, to get 80. The final answer is positive.

c. Find the absolute values: $|-23| = 23$ and $|-42| = 42$.

$-23(-42) = 966$ Multiply the absolute values, 23 and 42, to get 966. The final answer is positive.

$$\begin{array}{r} 42 \\ \times\ 23 \\ \hline 126 \\ 840 \\ \hline 966 \end{array}$$

d. We can extend the method discussed in Section 1.4 for multiplying whole-number factors with trailing zeros to products of integers with trailing zeros.

$-2,500(-30,000) = 75,000,000$ Attach six 0's after 75.

Multiply -25 and -3 to get 75.

We now summarize the multiplication rules for two integers.

Self Check 2
Multiply:

a. $-9(-7)$ 63

b. $-12(-2)$ 24

c. $-34(-15)$ 510

d. $-4,100(-20,000)$ 82,000,000

Now Try Problems 33, 37, 41, and 43

Teaching Example 2 Multiply:
a. $-3(-7)$
b. $-9(-9)$
c. $-46(-24)$
d. $-1,200(-40,000)$
Answers:
a. 21
b. 81
c. 1,104
d. 48,000,000

Multiplying Two Integers

To multiply two nonzero integers, multiply their absolute values.

1. The product of two integers that have the same (*like*) signs is positive.
2. The product of two integers that have different (*unlike*) signs is negative.

Using Your CALCULATOR   Multiplication with Negative Numbers

At Thanksgiving time, a large supermarket chain offered customers a free turkey with every grocery purchase of $200 or more. Each turkey cost the store $8, and 10,976 people took advantage of the offer. Since each of the 10,976 turkeys given away represented a loss of $8 (which can be expressed as $-\$8$), the company lost a total of $10,976(-\$8)$. To perform this multiplication using a calculator, we enter the following:

Reverse entry: 10976 [×] 8 [+/−] [=] $\boxed{-87808}$

Direct entry: 10976 [×] [(−)] 8 [ENTER] $\boxed{-87808}$

The negative result indicates that with the turkey giveaway promotion, the supermarket chain lost $87,808.

---

3 Perform several multiplications to evaluate expressions.

To evaluate expressions that contain several multiplications, we make repeated use of the rules for multiplying two integers.

Self Check 3

Evaluate each expression:
a. $3(-12)(-2)$ 72
b. $-1(9)(-6)$ 54
c. $-4(-5)(8)(-3)$ −480

Now Try Problems 45, 47, and 49

Teaching Example 3 Evaluate each expression:
a. $4(-9)(-3)$
b. $-5(12)(-1)$
c. $-6(-3)(2)(-7)$
Answers:
a. 108
b. 60
c. −252

EXAMPLE 3
Evaluate each expression:
a. $6(-2)(-7)$ b. $-9(8)(-1)$ c. $-3(-5)(2)(-4)$

Strategy Since there are no calculations within parentheses and no exponential expressions, we will perform the multiplications, working from the left to the right.

WHY This is step 3 of the order of operations rule that was introduced in Section 1.9.

Solution

a. $6(-2)(-7) = -12(-7)$ Use the rule for multiplying two integers that have different signs: $6(-2) = -12$.

$= 84$ Use the rule for multiplying two integers that have the same sign.

$$\begin{array}{r} \overset{1}{12} \\ \times 7 \\ \hline 84 \end{array}$$

b. $-9(8)(-1) = -72(-1)$ Use the rule for multiplying two integers that have different signs: $-9(8) = -72$.

$= 72$ Use the rule for multiplying two integers that have the same sign.

c. $-3(-5)(2)(-4) = 15(2)(-4)$ Use the rule for multiplying two integers that have the same sign: $-3(-5) = 15$.

$= 30(-4)$ Use the rule for multiplying two integers that have the same sign: $15(2) = 30$.

$= -120$ Use the rule for multiplying two integers that have different signs.

The properties of multiplication that were introduced in Section 1.3, *Multiplying Whole Numbers*, are also true for integers.

Properties of Multiplication

Commutative property of multiplication: The order in which integers are multiplied does not change their product.

Associative property of multiplication: The way in which integers are grouped does not change their product.

Multiplication property of 0: The product of any integer and 0 is 0.

Multiplication property of 1: The product of any integer and 1 is that integer.

Another approach to evaluate expressions like those in Example 3 is to use the properties of multiplication to reorder and regroup the factors in a helpful way.

Self Check 4

Use the commutative and/or associative properties of multiplication to evaluate each expression from Self Check 3 in a different way:

a. $3(-12)(-2)$ 72
b. $-1(9)(-6)$ 54
c. $-4(-5)(8)(-3)$ −480

Now Try Problems 45, 47, and 49

EXAMPLE 4
Use the commutative and/or associative properties of multiplication to evaluate each expression from Example 3 in a different way:
a. $6(-2)(-7)$ b. $-9(8)(-1)$ c. $-3(-5)(2)(-4)$

Strategy When possible, we will use the commutative and/or associative properties of multiplication to multiply pairs of negative factors.

WHY The product of two negative factors is positive. With this approach, we work with fewer negative numbers, and that lessens the possibility of an error.

Solution

a. $6(-2)(-7) = 6(14)$ Multiply the last two negative factors to produce a positive product: $-7(-2) = 14$.

$= 84$

$$\begin{array}{r} \overset{2}{14} \\ \times 6 \\ \hline 84 \end{array}$$

b. $-9(8)(-1) = 9(8)$ Multiply the negative factors to produce a positive product: $-9(-1) = 9$.
$= 72$

c. $-3(-5)(2)(-4) = 15(-8)$ Multiply the first two negative factors to produce a positive product. Multiply the last two factors.
$= -120$ Use the rule for multiplying two integers that have different signs.

$$\begin{array}{r} \overset{4}{15} \\ \times 8 \\ \hline 120 \end{array}$$

EXAMPLE 5   Evaluate: **a.** $-2(-4)(-5)$ **b.** $-3(-2)(-6)(-5)$

Strategy When possible, we will use the commutative and/or associative properties of multiplication to multiply pairs of negative factors.

WHY The product of two negative factors is positive. With this approach, we work with fewer negative numbers, and that lessens the possibility of an error.

Solution
a. Note that this expression is the product of three (an odd number) negative integers.

$-2(-4)(-5) = 8(-5)$ Multiply the first two negative factors to produce a positive product.
$= -40$ The product is negative.

b. Note that this expression is the product of four (an even number) negative integers.

$-3(-2)(-6)(-5) = 6(30)$ Multiply the first two negative factors and the last two negative factors to produce positive products.
$= 180$ The product is positive.

Example 5, part a, illustrates that a product is negative when there is an odd number of negative factors. Example 5, part b, illustrates that a product is positive when there is an even number of negative factors.

Multiplying an Even and an Odd Number of Negative Integers

The product of an even number of negative integers is positive.
The product of an odd number of negative integers is negative.

4 Evaluate exponential expressions that have negative bases.

Recall that exponential expressions are used to represent repeated multiplication. For example, 2 to the third power, or 2^3, is a shorthand way of writing $2 \cdot 2 \cdot 2$. In this expression, the *exponent* is 3 and the base is *positive* 2. In the next example, we evaluate exponential expressions with bases that are negative numbers.

EXAMPLE 6   Evaluate each expression: **a.** $(-2)^4$ **b.** $(-5)^3$ **c.** $(-1)^5$

Strategy We will write each exponential expression as a product of repeated factors and then perform the multiplication. This requires that we identify the base and the exponent.

WHY The exponent tells the number of times the base is to be written as a factor.

Teaching Example 4 Use the commutative and/or associative properties of multiplication to evaluate each expression from Teaching Example 3 in a different way:
a. $4(-9)(-3)$
b. $-5(12)(-1)$
c. $-6(-3)(2)(-7)$
Answers:
a. 108   b. 60   c. -252

Self Check 5
Evaluate each expression:
a. $-1(-2)(-5)$ -10
b. $-2(-7)(-1)(-2)$ 28

Now Try Problems 53 and 57

Teaching Example 5 Evaluate each expression:
a. $-3(-6)(-5)$
b. $-2(-1)(-3)(-8)$
Answers:
a. -90
b. 48

Self Check 6
Evaluate each expression:
a. $(-3)^4$ 81
b. $(-4)^3$ -64
c. $(-1)^7$ -1

Now Try Problems 61, 65, and 67

Teaching Example 6 Evaluate each expression:
a. $(-4)^4$
b. $(-6)^3$
c. $(-1)^9$
Answers:
a. 256
b. −216
c. −1

Solution

a. We read $(-2)^4$ as "negative two raised to the fourth power" or as "the fourth power of negative two." Note that the exponent is even.

$(-2)^4 = (-2)(-2)(-2)(-2)$ Write the base, −2, as a factor 4 times.
$ = 4(4)$ Multiply the first two negative factors and the last two negative factors to produce positive products.
$ = 16$ The result is positive.

b. We read $(-5)^3$ as "negative five raised to the third power" or as "the third power of negative five," or as "negative five, cubed." Note that the exponent is odd.

$(-5)^3 = (-5)(-5)(-5)$ Write the base, −5, as a factor 3 times.
$ = 25(-5)$ Multiply the first two negative factors to produce a positive product.
$ = -125$ The result is negative.

$$\begin{array}{r} \overset{2}{2}5 \\ \times 5 \\ \hline 125 \end{array}$$

c. We read $(-1)^5$ as "negative one raised to the fifth power" or as "the fifth power of negative one." Note that the exponent is odd.

$(-1)^5 = (-1)(-1)(-1)(-1)(-1)$ Write the base, −1, as a factor 5 times.
$ = 1(1)(-1)$ Multiply the first and second negative factors and multiply the third and fourth negative factors to produce positive products.
$ = -1$ The result is negative.

In Example 6, part a, −2 was raised to an even power, and the answer was positive. In parts b and c, −5 and −1 were raised to odd powers, and, in each case, the answer was negative. These results suggest a general rule.

Even and Odd Powers of a Negative Integer

When a negative integer is raised to an even power, the result is positive.
When a negative integer is raised to an odd power, the result is negative.

Although the exponential expressions $(-3)^2$ and -3^2 look similar, they are not the same. We read $(-3)^2$ as "negative 3 squared" and -3^2 as "the opposite of the square of three." When we evaluate them, it becomes clear that they are not equivalent.

$(-3)^2 = (-3)(-3)$ Because of the parentheses, the base is −3. The exponent is 2.
$ = 9$

$-3^2 = -(3 \cdot 3)$ Since there are no parentheses around −3, the base is 3. The exponent is 2.
$ = -9$

Different results

Caution! The base of an exponential expression *does not include* the negative sign unless parentheses are used.

-7^3 $(-7)^3$

Positive base: 7   Negative base: −7

EXAMPLE 7   Evaluate: -2^2

Strategy We will rewrite the expression as a product of repeated factors, and then perform the multiplication. We must be careful when identifying the base. It is 2, not -2.

WHY Since there are no parentheses around -2, the base is 2.

Solution

$-2^2 = -(2 \cdot 2)$ Read as "the opposite of the square of two."

$\quad\quad = -4$ Do the multiplication within the parentheses to get 4. Then write the opposite of that result.

Self Check 7
Evaluate: -4^2 -16
Now Try Problem 71

Teaching Example 7 Evaluate: -5^2
Answer:
-25

Using Your CALCULATOR   Raising a Negative Number to a Power

We can find powers of negative integers, such as $(-5)^6$, using a calculator. The keystrokes that are used to evaluate such expressions vary from model to model, as shown below. You will need to determine which keystrokes produce the positive result that we would expect when raising a negative number to an even power.

5 [+/−] [y^x] 6 [=]   Some calculators don't require the parentheses to be entered.

[(] 5 [+/−] [)] [y^x] 6 [=]   Other calculators require the parentheses to be entered.

[(] [(−)] 5 [)] [^] 6 [ENTER] | 15625 |

From the calculator display, we see that $(-5)^6 = 15,625$.

---

5 Solve application problems by multiplying integers.

Problems that involve repeated addition are often more easily solved using multiplication.

EXAMPLE 8   Oceanography

Scientists lowered an underwater vessel called a *submersible* into the Pacific Ocean to record the water temperature. The first measurement was made 75 feet below sea level, and more were made every 75 feet until it reached the ocean floor. Find the depth of the submersible when the 25th measurement was made.

Emory Kristof/National Geographic/Getty Images

Solution
It is helpful to list the given facts and what you are to find.

- The first measurement was made 75 feet below sea level.   Given
- More measurements were made every 75 feet.   Given
- What was the depth of the submersible when it made the 25th measurement?   Find

If we use negative numbers to represent the depths at which the measurements were made, then the first was at -75 feet. The depth (in feet) of the submersible when the 25th measurement was made can be found by adding -75 twenty-five times. This repeated addition can be calculated more simply by multiplication.

Self Check 8
GASOLINE LEAKS To determine how badly a gasoline tank was leaking, inspectors used a drilling process to take soil samples nearby. The first sample was taken 6 feet below ground level, and more were taken every 6 feet after that. The 14th sample was the first one that did not show signs of gasoline. How far below ground level was that?

Now Try Problem 97
Self Check 8 Answer
It was 84 ft below ground level (-84 ft).

Teaching Example 8
FORESTRY To determine the moisture content of the ground in a forest, rangers took soil samples at various depths. The first sample was taken 18 inches below ground, and more were taken every 18 inches after that. Find the depth of the 12th sample.
Answer:
216 in. below ground level (−216 in.)

We translate the words of the problem to numbers and symbols.

The depth of the submersible when it made the 25th measurement	is equal to	the number of measurements made	times	the amount it was lowered each time.
The depth of the submersible when it made the 25th measurement	=	25	·	(−75)

To find the product, we use the rule for multiplying two integers that have different signs. First, we find the absolute values: $|25| = 25$ and $|-75| = 75$.

$25(-75) = -1,875$ Multiply the absolute values, 25 and 75, to get 1,875. Since the integers have different signs, make the final answer negative.

$$\begin{array}{r} 75 \\ \times\ 25 \\ \hline 375 \\ 1\ 500 \\ \hline 1,875 \end{array}$$

The depth of the submersible was 1,875 feet below sea level (−1,875 feet) when the 25th temperature measurement was taken.

Check: We can use estimation or simply perform the actual multiplication again to see if the result seems reasonable.

ANSWERS TO SELF CHECKS
1. a. −12 **b.** −120 **c.** −1,275 **d.** −98,000 **2. a.** 63 **b.** 24 **c.** 510 **d.** 82,000,000
3. a. 72 **b.** 54 **c.** −480 **4. a.** 72 **b.** 54 **c.** −480 **5. a.** −10 **b.** 28 **6. a.** 81
b. −64 **c.** −1 **7.** −16 **8.** 84 ft below ground level (−84 ft)

SECTION 2.4   STUDY SET

VOCABULARY

Fill in the blanks.

1. In the multiplication problem shown below, label each *factor* and the *product*.

    −5 · 10 = −50
    ↑   ↑   ↑
    factor factor product

2. Two negative integers, as well as two positive integers, are said to have the same signs or __like__ signs.

3. A positive integer and a negative integer are said to have different signs or __unlike__ signs.

4. __Commutative__ property of multiplication: The order in which integers are multiplied does not change their product.

5. __Associative__ property of multiplication: The way in which integers are grouped does not change their product.

6. In the expression $(-3)^5$, the __base__ is −3, and 5 is the __exponent__.

CONCEPTS

Fill in the blanks.

7. Multiplication of integers is very much like multiplication of whole numbers. The only difference is that we must determine whether the answer is __positive__ or __negative__.

8. When we multiply two nonzero integers, they either have __unlike/different__ signs or __like/the same__ signs.

9. To multiply a positive integer and a negative integer, multiply their absolute values. Then make the final answer __negative__.

10. To multiply two integers that have the same sign, multiply their absolute values. The final answer is __positive__.

11. The product of two integers with __unlike/different__ signs is negative.

12. The product of two integers with __like/the same__ signs is positive.

13. The product of any integer and 0 is __0__.

14. The product of an even number of negative integers is __positive__ and the product of an odd number of negative integers is __negative__.

15. Find each absolute value.
    a. $|-3|$ 3
    b. $|12|$ 12

16. If each of the following expressions were evaluated, what would be the *sign* of the result?
    a. $(-5)^{13}$ negative
    b. $(-3)^{20}$ positive

NOTATION

17. For each expression, identify the base and the exponent.
    a. -8^4 base: 8, exponent: 4
    b. $(-7)^9$ base: -7, exponent: 9

18. Translate to mathematical symbols.
    a. negative three times negative two $-3(-2)$
    b. negative five squared $(-5)^2$
    c. the opposite of the square of five -5^2

Complete each step to evaluate the expression.

19. $-3(-2)(-4) = \boxed{6}\,(-4)$
    $= \boxed{-24}$

20. $(-3)^4 = (-3)(-3)(-3)\boxed{(-3)}$
    $= \boxed{9}\,(9)$
    $= \boxed{81}$

GUIDED PRACTICE

Multiply. See Example 1.

21. $5(-3)$ -15
22. $4(-6)$ -24
23. $9(-2)$ -18
24. $5(-7)$ -35
25. $18(-4)$ -72
26. $17(-8)$ -136
27. $21(-6)$ -126
28. $39(-3)$ -117
29. $-45 \cdot 37$ $-1{,}665$
30. $-42 \cdot 24$ $-1{,}008$
31. $-94 \cdot 1{,}000$ $-94{,}000$
32. $-76 \cdot 1{,}000$ $-76{,}000$

Multiply. See Example 2.

33. $(-8)(-7)$ 56
34. $(-9)(-3)$ 27
35. $-7(-1)$ 7
36. $-5(-1)$ 5
37. $-3(-52)$ 156
38. $-4(-73)$ 292
39. $-6(-46)$ 276
40. $-8(-48)$ 384
41. $-59(-33)$ 1,947
42. $-61(-29)$ 1,769
43. $-60{,}000(-1{,}200)$ 72,000,000
44. $-20{,}000(-3{,}200)$ 64,000,000

Evaluate each expression. See Examples 3 and 4.

45. $6(-3)(-5)$ 90
46. $9(-3)(-4)$ 108
47. $-5(10)(-3)$ 150
48. $-8(7)(-2)$ 112
49. $-2(-4)(6)(-8)$ -384
50. $-3(-5)(2)(-9)$ -270
51. $-8(-3)(7)(-2)$ -336
52. $-9(-3)(4)(-2)$ -216

Evaluate each expression. See Example 5.

53. $-4(-2)(-6)$ -48
54. $-4(-6)(-3)$ -72
55. $-3(-9)(-3)$ -81
56. $-5(-2)(-5)$ -50
57. $-1(-3)(-2)(-6)$ 36
58. $-1(-4)(-2)(-4)$ 32
59. $-9(-4)(-1)(-4)$ 144
60. $-6(-3)(-6)(-1)$ 108

Evaluate each expression. See Example 6.

61. $(-3)^3$ -27
62. $(-6)^3$ -216
63. $(-2)^5$ -32
64. $(-3)^5$ -243
65. $(-5)^4$ 625
66. $(-7)^4$ 2,401
67. $(-1)^8$ 1
68. $(-1)^{10}$ 1

Evaluate each expression. See Example 7.

69. $(-7)^2$ and -7^2 49, -49
70. $(-5)^2$ and -5^2 25, -25
71. $(-12)^2$ and -12^2 144, -144
72. $(-11)^2$ and -11^2 121, -121

TRY IT YOURSELF

Evaluate each expression.

73. $6(-5)(2)$ -60
74. $4(-2)(2)$ -16
75. $-8(0)$ 0
76. $0(-27)$ 0
77. $(-4)^3$ -64
78. $(-8)^3$ -512
79. $(-2)10$ -20
80. $(-3)8$ -24
81. $-2(-3)(3)(-1)$ -18
82. $5(-2)(3)(-1)$ 30
83. Find the product of -6 and the opposite of 10. 60
84. Find the product of the opposite of 9 and the opposite of 8. 72
85. $-6(-4)(-2)$ -48
86. $-3(-2)(-3)$ -18
87. $-42 \cdot 200{,}000$ $-8{,}400{,}000$
88. $-56 \cdot 10{,}000$ $-560{,}000$
89. -5^4 -625
90. -2^4 -16
91. $-12(-12)$ 144
92. $-5(-5)$ 25
93. $(-1)^6$ 1
94. $(-1)^5$ -1
95. $(-1)(-2)(-3)(-4)(-5)$ -120
96. $(-10)(-8)(-6)(-4)(-2)$ $-3{,}840$

APPLICATIONS

Use signed numbers to solve each problem.

97. SUBMARINES As part of a training exercise, the captain of a submarine ordered it to descend 250 feet, level off for 5 minutes, and then repeat the process several times. If the sub was on the ocean's surface at the beginning of the exercise, what signed number represents its depth after the 8th dive? $-2{,}000$ ft

98. **BUILDING A PIER** A *pile driver* uses a heavy weight to pound tall poles into the ocean floor. If each strike of a pile driver on the top of a pole sends it 6 inches deeper, what signed number represents the depth of the pole after 20 strikes? −120 in.

99. **MAGNIFICATION** A mechanic used an electronic testing device to check the smog emissions of a car. The results of the test are displayed on a screen.

   a. Find the high and low values for this test as shown on the screen. high: 2, low: −3

   b. By switching a setting, the picture on the screen can be magnified. What would be the new high and new low if every value were doubled?
   high: 4, low: −6

100. **LIGHT** Sunlight is a mixture of all colors. When sunlight passes through water, the water absorbs different colors at different rates, as shown.

   a. Use a signed number to represent the depth to which red light penetrates water. −15 ft

   b. Green light penetrates four times as deep as red light. How deep is this? −60 ft

   c. Blue light penetrates three times as deep as orange light. How deep is this? −75 ft

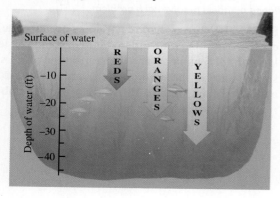

101. **JOB LOSSES** The recession of 2008 was a major worldwide economic downturn that caused tremendous job losses in the United States. See the bar graph below. Find the number of jobs lost in . . .

   a. September 2008 if it was about six times the number lost in April. −402,000 jobs

   b. October 2008 if it was about nine times the number lost in May. −423,000 jobs

   c. November 2008 if it was about seven times the number lost in February. −581,000 jobs

   d. December if it was about six times the number lost in March. −528,000 jobs

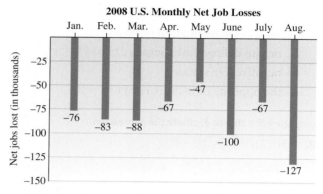

Source: Bureau of Labor Statistics

102. **JAPAN** The population of Japan is decreasing by about 870,000 per year because of low birth rates. If this pattern continues, what will be the total decline in Japan's population over the next 45 years? (Source: huffingtonpost.com) −39,150,000

103. **PLANETS** The average surface temperature of Mars is −81°F. Find the average surface temperature of Uranus if it is four times colder than Mars. (Source: *The World Almanac and Book of Facts*, 2013) −324°F

104. **CROP LOSS** A farmer, worried about his fruit trees suffering frost damage, calls the weather service for temperature information. He is told that temperatures will be decreasing approximately 5 degrees every hour for the next five hours. What signed number represents the total change in temperature expected over the next five hours? −25°

105. **TAX WRITE-OFF** For each of the last six years, a businesswoman has filed a $200 depreciation allowance on her income tax return for an office computer system. What signed number represents the total amount of depreciation written off over the six-year period? −$1,200

106. **EROSION** A levee protects a town in a low-lying area from flooding. According to geologists, the banks of the levee are eroding at a rate of 2 feet per year. If something isn't done to correct the problem, what signed number indicates how much of the levee will erode during the next decade? −20 ft

Unless otherwise noted, all art on this page is © Cengage Learning.

107. DECK SUPPORTS After a winter storm, a homeowner has an engineering firm inspect his damaged deck. Their report concludes that the original foundation poles were not sunk deep enough, by a factor of 3. What signed number represents the depth to which the poles should have been sunk? −18 ft

108. DIETING After giving a patient a physical exam, a physician felt that the patient should begin a diet. The two options that were discussed are shown in the following table.

	Plan #1	Plan #2
Length	10 weeks	14 weeks
Daily exercise	1 hr	30 min
Weight loss per week	3 lb	2 lb

a. Find the expected weight loss from Plan 1. Express the answer as a signed number. −30 lb

b. Find the expected weight loss from Plan 2. Express the answer as a signed number. −28 lb

c. With which plan should the patient expect to lose more weight? Explain why the patient might not choose it. plan #1, the workout time is double that of plan #2

109. ADVERTISING On Thursday, March 15, 2012, Rodeo Houston had an all-time record paid attendance: 74,905. Suppose a local radio station gave a sports bag, worth $3, to all paying customers. What signed number expresses the radio station's financial loss from this giveaway? −$224,715

110. HEALTH CARE A health care provider for a company estimates that 75 hours per week are lost by employees suffering from stress-related or preventable illness. In a 52-week year, how many hours are lost? Use a signed number to answer. −3,900 hr

WRITING

111. Explain why the product of a positive number and a negative number is negative, using 5(−3) as an example.

112. Explain the multiplication rule for integers that is shown in the pattern of signs below.

$$(-)(-) = +$$
$$(-)(-)(-) = -$$
$$(-)(-)(-)(-) = +$$
$$(-)(-)(-)(-)(-) = -$$
$$\vdots$$

113. When a number is multiplied by −1, the result is the opposite of the original number. Explain why.

114. A student claimed, "A positive and a negative is negative." What is wrong with this statement?

REVIEW

115. List the first ten prime numbers. 2, 3, 5, 7, 11, 13, 17, 19, 23, 29

116. ENROLLMENT The number of students attending a college went from 10,250 to 12,300 in one year. What was the increase in enrollment? 2,050

117. Divide: 978 ÷ 49   19 R 47

118. What does the symbol < mean?   is less than

SECTION 2.5
Dividing Integers

Objectives
1. Divide two integers.
2. Identify *division of 0* and *division by 0*.
3. Solve application problems by dividing integers.

In this section, we will develop rules for division of integers, just as we did earlier for multiplication of integers.

1 Divide two integers.

Recall from Section 1.4 that every division has a related multiplication statement. For example,

$$\frac{6}{3} = 2 \quad \text{because} \quad 2(3) = 6$$

and

$$\frac{20}{5} = 4 \quad \text{because} \quad 4(5) = 20$$

We can use the relationship between multiplication and division to help develop rules for dividing integers. There are four cases to consider.

Case 1: A positive integer divided by a positive integer
From years of experience, we already know that the result is positive. Therefore, *the quotient of two positive integers is positive.*

Case 2: A negative integer divided by a negative integer
As an example, consider the division $\frac{-12}{-2} = ?$. We can find ? by examining the related multiplication statement.

Related multiplication statement

$?(-2) = -12$

↑ This must be positive 6 if the product is to be negative 12.

Division statement

$\frac{-12}{-2} = ?$

↑ So the quotient is positive 6.

Therefore, $\frac{-12}{-2} = 6$. This example illustrates that *the quotient of two negative integers is positive.*

Case 3: A positive integer divided by a negative integer
Let's consider $\frac{12}{-2} = ?$. We can find ? by examining the related multiplication statement.

Related multiplication statement

$?(-2) = 12$

↑ This must be −6 if the product is to be positive 12.

Division statement

$\frac{12}{-2} = ?$

↑ So the quotient is −6.

Therefore, $\frac{12}{-2} = -6$. This example illustrates that *the quotient of a positive integer and a negative integer is negative.*

Case 4: A negative integer divided by a positive integer
Let's consider $\frac{-12}{2} = ?$. We can find ? by examining the related multiplication statement.

Related multiplication statement

$?(2) = -12$

↑ This must be −6 if the product is to be −12.

Division statement

$\frac{-12}{2} = ?$

↑ So the quotient is −6.

Therefore, $\frac{-12}{2} = -6$. This example illustrates that *the quotient of a negative integer and a positive integer is negative.*

We now summarize the results from the previous examples and note that they are similar to the rules for multiplication.

Dividing Two Integers

To divide two integers, divide their absolute values.

1. The quotient of two integers that have the same (*like*) signs is positive.
2. The quotient of two integers that have different (*unlike*) signs is negative.

2.5 Dividing Integers

EXAMPLE 1
Divide and check the result:

a. $\dfrac{-14}{7}$ **b.** $30 \div (-5)$ **c.** $\dfrac{176}{-11}$ **d.** $-24{,}000 \div 600$

Strategy We will use the rule for dividing two integers that have different (unlike) signs.

WHY Each division involves a positive and a negative integer.

Solution

a. Find the absolute values: $|-14| = 14$ and $|7| = 7$.

$\dfrac{-14}{7} = -2$ Divide the absolute values, 14 by 7, to get 2. Then make the final answer negative.

To check, we multiply the *quotient*, -2, and the *divisor*, 7. We should get the *dividend*, -14.

Check: $-2(7) = -14$ The result checks.

b. Find the absolute values: $|30| = 30$ and $|-5| = 5$.

$30 \div (-5) = -6$ Divide the absolute values, 30 by 5, to get 6. Then make the final answer negative.

Check: $-6(-5) = 30$ The result checks.

c. Find the absolute values: $|176| = 176$ and $|-11| = 11$.

$\dfrac{176}{-11} = -16$ Divide the absolute values, 176 by 11, to get 16. Then make the final answer negative.

```
    16
11)176
   -11
    66
   -66
     0
```

Check: $-16(-11) = 176$ The result checks.

d. Recall from Section 1.5, that if a divisor has ending zeros, we can simplify the division by removing the same number of ending zeros in the divisor and dividend.

There are two zeros in the divisor.

$-24{,}000 \div 600 = -240 \div 6 = -40$ Divide the absolute values, 240 by 6, to get 40. Then make the final answer negative.

Remove two zeros from the dividend and the divisor, and divide.

Check: $-40(600) = -24{,}000$ Use the original divisor and dividend in the check.

Self Check 1
Divide and check the result:

a. $\dfrac{-45}{5}$ -9

b. $28 \div (-4)$ -7

c. $\dfrac{336}{-14}$ -24

d. $-18{,}000 \div 300$ -60

Now Try Problems 13, 17, 21, and 27

Teaching Example 1 Divide and check the result:

a. $\dfrac{-40}{5}$

b. $18 \div (-2)$

c. $\dfrac{468}{-18}$

d. $-56{,}000 \div 800$

Answers:
a. -8
b. -9
c. -26
d. -70

EXAMPLE 2
Divide and check the result:

a. $\dfrac{-12}{-3}$ **b.** $-48 \div (-6)$ **c.** $\dfrac{-315}{-9}$ **d.** $-200 \div (-40)$

Strategy We will use the rule for dividing two integers that have the same (like) signs.

WHY In each case, we are asked to find the quotient of two negative integers.

Solution

a. Find the absolute values: $|-12| = 12$ and $|-3| = 3$.

$\dfrac{-12}{-3} = 4$ Divide the absolute values, 12 by 3, to get 4. The final answer is positive.

Check: $4(-3) = -12$ The result checks.

Self Check 2
Divide and check the result:

a. $\dfrac{-27}{-3}$ 9

b. $-24 \div (-4)$ 6

c. $\dfrac{-301}{-7}$ 43

d. $-400 \div (-20)$ 20

Now Try Problems 31, 35, 39, and 43

Teaching Example 2 Divide and check the result:
a. $\dfrac{-36}{-6}$
b. $-63 \div (-9)$
c. $\dfrac{-624}{-16}$
d. $-600 \div (-30)$

Answers:
a. 6
b. 7
c. 39
d. 20

b. Find the absolute values: $|-48| = 48$ and $|-6| = 6$.

$-48 \div (-6) = 8$ Divide the absolute values, 48 by 6, to get 8. The final answer is positive.

Check: $8(-6) = -48$ The result checks.

c. Find the absolute values: $|-315| = 315$ and $|-9| = 9$.

$\dfrac{-315}{-9} = 35$ Divide the absolute values, 315 by 9, to get 35. The final answer is positive.

$$\begin{array}{r} 35 \\ 9\overline{)315} \\ -27 \\ \hline 45 \\ -45 \\ \hline 0 \end{array}$$

Check: $35(-9) = -315$ The result checks.

d. We can simplify the division by removing the same number of ending zeros in the divisor and dividend.

There is one zero in the divisor.

$-200 \div (-40) \;=\; -20 \div (-4) \;=\; 5$ Divide the absolute values, 20 by 4, to get 5. The final answer is positive.

Remove one zero from the dividend and the divisor, and divide.

Check: $5(-40) = -200$ The result checks.

2 Identify *division of 0* and *division by 0*.

To review the concept of division of 0, we consider $\dfrac{0}{-2} = ?$. We can attempt to find ? by examining the related multiplication statement.

Related multiplication statement

$(?)(-2) = 0$

— *This must be 0 if the product is to be 0.*

Division statement

$\dfrac{0}{-2} = ?$

— *So the quotient is 0.*

Therefore, $\dfrac{0}{-2} = 0$. This example illustrates that *the quotient of 0 divided by any nonzero number is 0.*

To review division by 0, let's consider $\dfrac{-2}{0} = ?$. We can attempt to find ? by examining the related multiplication statement.

Related multiplication statement

$(?)0 = -2$

— *There is no number that gives -2 when multiplied by 0.*

Division statement

$\dfrac{-2}{0} = ?$

— *There is no quotient.*

Therefore, $\dfrac{-2}{0}$ does not have an answer and we say that $\dfrac{-2}{0}$ is undefined. This example illustrates that *the quotient of any nonzero number divided by 0 is undefined.*

> **Division with 0**
>
> 1. If 0 is divided by any nonzero number, the quotient is 0.
>   For any nonzero number a,
> $$\dfrac{0}{a} = 0$$
>
> 2. Division of any nonzero number by 0 is undefined.
>   For any nonzero number a,
> $$\dfrac{a}{0} \text{ is undefined}$$

EXAMPLE 3
Divide, if possible: **a.** $\dfrac{-4}{0}$ **b.** $0 \div (-8)$

Strategy In each case, we need to determine if we have division *of* 0 or division *by* 0.

WHY Division of 0 by a nonzero integer is defined, and the answer is 0. However, division of a nonzero integer by 0 is undefined; there is no answer.

Solution

a. $\dfrac{-4}{0}$ is undefined. This is division by 0.

b. $0 \div (-8) = 0$ because $0(-8) = 0$. This is division of 0.

Self Check 3
Divide, if possible:

a. $\dfrac{-12}{0}$ undefined **b.** $0 \div (-6)$ 0

Now Try Problems 45 and 47

Teaching Example 3 Divide, if possible:
a. $\dfrac{-15}{0}$
b. $0 \div (-9)$
Answers:
a. undefined
b. 0

3 Solve application problems by dividing integers.

Problems that involve forming equal-sized groups can be solved by division.

EXAMPLE 4
Real Estate Over the course of a year, a homeowner reduced the price of his house by an equal amount each month because it was not selling. By the end of the year, the price was $11,400 less than at the beginning of the year. By how much was the price of the house reduced each month?

Solution

It is helpful to list the given facts and what you are to find.

- The homeowner dropped the price $11,400 in 1 year.   Given
- The price was reduced by an equal amount each month.   Given
- By how much was the price of the house reduced each month?   Find

We can express the drop in the price of the house for the year as −$11,400. The phrase *reduced by an equal amount each month* indicates division.

We translate the words of the problem to numbers and symbols.

The amount the price was reduced each month	is equal to	the drop in the price of the house for the year	divided by	the number of months in 1 year.
The amount the price was reduced each month	=	−11,400	÷	12

To find the quotient, we use the rule for dividing two integers that have different signs. First, we find the absolute values: $|-11{,}400| = 11{,}400$ and $|12| = 12$.

$-11{,}400 \div 12 = -950$ Divide the absolute values, 11,400 and 12, to get 950. Then make the final answer negative.

$$\begin{array}{r} 950 \\ 12\overline{)11{,}400} \\ -10\ 8 \\ \hline 60 \\ -60 \\ \hline 00 \\ -00 \\ \hline 0 \end{array}$$

The negative result indicates that the price of the house was *reduced* by $950 each month.

Check: We can use estimation to check the result. A reduction of $1,000 each month would cause the price to drop $12,000 in 1 year. It seems reasonable that a reduction of $950 each month would cause the price to drop $11,400 in a year.

Self Check 4
SELLING BOATS The owner of a sailboat reduced the price of the boat by an equal amount each month because there were no interested buyers. After 8 months, and a $960 reduction in price, the boat sold. By how much was the price of the boat reduced each month?

Now Try Problem 81
Self Check 4 Answer
The price was reduced by $120 each month.

Teaching Example 4
PRICE REDUCTIONS For 1 year, an electronics store reduced the price of a television set by an equal amount each month because it was not selling. By the end of the year, the price was $132 less than at the beginning of the year. By how much was the price reduced each month?
Answer:
The price was reduced by $11 each month.

Using Your CALCULATOR   Division with Negative Numbers

According to the Bureau of Labor Statistics, the United States lost 162,000 auto manufacturing jobs (motor vehicles and parts) during the recession year 2008. Because the jobs were lost, we write this as $-162,000$. To find the average number of manufacturing jobs lost each month, we divide: $\frac{-162,000}{12}$. We can use a calculator to perform the division.

Reverse entry: 162000 $\boxed{+/-}$ $\boxed{\div}$ 12 $\boxed{=}$

Direct entry: $\boxed{(-)}$ 162000 $\boxed{\div}$ 12 $\boxed{\text{ENTER}}$ $\boxed{-13500}$

The average number of auto manufacturing jobs lost each month in 2008 was 13,500.

ANSWERS TO SELF CHECKS

1. a. -9 **b.** -7 **c.** -24 **d.** -60 **2. a.** 9 **b.** 6 **c.** 43 **d.** 20 **3. a.** undefined **b.** 0 **4.** The price was reduced by $120 each month.

SECTION 2.5   STUDY SET

VOCABULARY

Fill in the blanks.

1. In the division problems shown below, label the *dividend*, *divisor*, and *quotient*.

$$12 \div (-4) = -3$$
  ↑   ↑   ↑
dividend divisor quotient

$$\frac{12}{-4} = -3$$
dividend ⟶ (12), quotient ⟶ (−3), divisor ⟶ (−4)

2. The related _multiplication_ statement for $\frac{-6}{3} = -2$ is $-2(3) = -6$.

3. $\frac{-3}{0}$ is division _by_ 0 and $\frac{0}{-3} = 0$ is division _of_ 0.

4. Division of a nonzero integer by 0, such as $\frac{-3}{0}$, is _undefined_.

CONCEPTS

5. Write the related multiplication statement for each division.

a. $\frac{-25}{5} = -5$ **b.** $-36 \div (-6) = 6$ **c.** $\frac{0}{-15} = 0$

  $-5(5) = -25$ $6(-6) = -36$ $0(-15) = 0$

6. Using multiplication, check to determine whether $-720 \div 45 = -12$. The answer is incorrect: $-12(45) \neq -720$

7. Fill in the blanks.
To divide two integers, divide their absolute values.

a. The quotient of two integers that have the same (*like*) signs is _positive_.

b. The quotient of two integers that have different (*unlike*) signs is _negative_.

8. If a divisor has ending zeros, we can simplify the division by removing the same number of ending zeros in the divisor and dividend. Fill in the blank: $-2,400 \div 60 = -240 \div$ _6_

9. Fill in the blanks.

a. If 0 is divided by any nonzero integer, the quotient is _0_.

b. Division of any nonzero integer by 0 is _undefined_.

10. What operation can be used to solve problems that involve forming equal-sized groups? division

11. Determine whether each statement is always true, sometimes true, or never true.

a. The product of a positive integer and a negative integer is negative. always true

b. The sum of a positive integer and a negative integer is negative. sometimes true

c. The quotient of a positive integer and a negative integer is negative. always true

12. Determine whether each statement is always true, sometimes true, or never true.

a. The product of two negative integers is positive. always true

b. The sum of two negative integers is negative. always true

c. The quotient of two negative integers is negative. never true

2.5 Dividing Integers

GUIDED PRACTICE

Divide and check the result. See Example 1.

13. $\dfrac{-14}{2}$ -7
14. $\dfrac{-10}{5}$ -2
15. $\dfrac{-20}{5}$ -4
16. $\dfrac{-24}{3}$ -8
17. $36 \div (-6)$ -6
18. $36 \div (-9)$ -4
19. $24 \div (-3)$ -8
20. $42 \div (-6)$ -7
21. $\dfrac{264}{-12}$ -22
22. $\dfrac{364}{-14}$ -26
23. $\dfrac{702}{-18}$ -39
24. $\dfrac{396}{-12}$ -33
25. $-9{,}000 \div 300$ -30
26. $-12{,}000 \div 600$ -20
27. $-250{,}000 \div 5{,}000$ -50
28. $-420{,}000 \div 7{,}000$ -60

Divide and check the result. See Example 2.

29. $\dfrac{-8}{-4}$ 2
30. $\dfrac{-12}{-4}$ 3
31. $\dfrac{-45}{-9}$ 5
32. $\dfrac{-81}{-9}$ 9
33. $-63 \div (-7)$ 9
34. $-21 \div (-3)$ 7
35. $-32 \div (-8)$ 4
36. $-56 \div (-7)$ 8
37. $\dfrac{-400}{-25}$ 16
38. $\dfrac{-490}{-35}$ 14
39. $\dfrac{-651}{-31}$ 21
40. $\dfrac{-736}{-32}$ 23
41. $-800 \div (-20)$ 40
42. $-800 \div (-40)$ 20
43. $-15{,}000 \div (-30)$ 500
44. $-36{,}000 \div (-60)$ 600

Divide, if possible. See Example 3.

45. a. $\dfrac{-3}{0}$ undefined b. $\dfrac{0}{-3}$ 0
46. a. $\dfrac{-5}{0}$ undefined b. $\dfrac{0}{-5}$ 0
47. a. $\dfrac{0}{-24}$ 0 b. $\dfrac{-24}{0}$ undefined
48. a. $\dfrac{0}{-32}$ 0 b. $\dfrac{-32}{0}$ undefined

TRY IT YOURSELF

Divide, if possible.

49. $-36 \div (-12)$ 3
50. $-45 \div (-15)$ 3
51. $\dfrac{425}{-25}$ -17
52. $\dfrac{462}{-42}$ -11
53. $\dfrac{0}{120}$ 0
54. $\dfrac{0}{-11}$ 0
55. Find the quotient of -45 and 9. -5
56. Find the quotient of -36 and -4. 9
57. $-2{,}500 \div 500$ -5
58. $-52{,}000 \div 4{,}000$ -13
59. $\dfrac{-6}{0}$ undefined
60. $\dfrac{-8}{0}$ undefined
61. $\dfrac{-19}{1}$ -19
62. $\dfrac{-9}{1}$ -9
63. $-23 \div (-23)$ 1
64. $-11 \div (-11)$ 1
65. $\dfrac{40}{-2}$ -20
66. $\dfrac{35}{-7}$ -5
67. $9 \div (-9)$ -1
68. $15 \div (-15)$ -1
69. $\dfrac{-10}{-1}$ 10
70. $\dfrac{-12}{-1}$ 12
71. $\dfrac{-888}{37}$ -24
72. $\dfrac{-456}{24}$ -19
73. $\dfrac{3{,}000}{-100}$ -30
74. $\dfrac{-60{,}000}{-1{,}000}$ 60
75. Divide 8 by -2. -4
76. Divide -16 by -8. 2

Use a calculator to perform each division.

77. $\dfrac{-13{,}550}{25}$ -542
78. $\dfrac{3{,}876}{-19}$ -204
79. $\dfrac{27{,}778}{-17}$ $-1{,}634$
80. $\dfrac{-168{,}476}{-77}$ $2{,}188$

APPLICATIONS

Use signed numbers to solve each problem.

81. **LOWERING PRICES** A furniture store owner reduced the price of an oak table an equal amount each week because it was not selling. After six weeks and a $210 reduction in price, the table was purchased. By how much did the price of the table change each week? $-\$35$ per week

82. **TEMPERATURE DROP** During a five-hour period, the temperature steadily dropped 20°F. By how many degrees did the temperature change each hour? $-4°$ per hour

83. **SUBMARINES** In a series of three equal dives, a submarine is programmed to reach a depth of 3,030 feet below the ocean surface. What signed number describes how deep each of the dives will be? $-1{,}010$ ft

84. **GRAND CANYON** A mule train is to travel from a stable on the rim of the Grand Canyon to a camp on the canyon floor, approximately 5,500 feet below the rim. If the guide wants the mules to be rested after every 500 feet of descent, how many stops will be made on the trip? 11

85. **CHEMISTRY** During an experiment, a solution was steadily chilled and the times and temperatures were recorded, as shown in the illustration below. What signed number represents the change in temperature each minute? −7° per min

Beginning of experiment 8:00 A.M.   End of experiment 8:06 A.M.

86. **OCEAN EXPLORATION** The Mariana Trench is the deepest part of the world's oceans. It is located in the North Pacific Ocean near the Philippines and has a maximum depth of 36,201 feet. If a remote-controlled vessel is sent to the bottom of the trench in a series of 11 equal descents, what signed number describes how far the vessel will descend on each dive? (Source: marianatrench.com) −3,291 ft

87. **BASEBALL TRADES** At the midway point of the season, a baseball team finds itself 12 games behind the league leader. Team management decides to trade for a talented hitter, in hopes of making up at least half of the deficit in the standings by the end of the year. What signed number describes where in the league standings management expects to finish at season's end? −6 (6 games behind)

88. **BUDGET DEFICITS** A politician proposed a two-year plan for cutting a county's $20-million budget deficit, as shown. If this plan is put into effect, how will the deficit change in two years?
The deficit will be reduced by $15 million.

	Plan	Prediction
1st year	Raise taxes, drop failing programs	Will cut deficit in half
2nd year	Search out waste and fraud	Will cut remaining deficit in half

89. **MARKDOWNS** The owner of a clothing store decides to reduce the price on a line of jeans that are not selling. She feels she can afford to lose $300 of projected income on these pants. What signed number represents the amount that she can mark down each of the 20 pairs of jeans? −$15

90. **WATER STORAGE** Over a week's time, engineers at a city water reservoir steadily released enough water to lower the water level 105 feet. What signed number represents the daily change in the water level? −15 ft

91. **THE STOCK MARKET** On Monday, the value of Maria's 255 shares of stock was at an all-time high. By Friday, the value had fallen $4,335. What signed number represents her per-share loss that week? −$17

92. **CUTTING BUDGETS** In a cost-cutting effort, a company decides to cut $5,840,000 from its annual budget. To do this, all of the company's 160 departments will have their budgets reduced by an equal amount. What signed number represents each department's budget reduction? −$36,500

WRITING

93. Explain why the quotient of two negative integers is positive.

94. How do the rules for multiplying integers compare with the rules for dividing integers?

95. Use a specific example to explain how multiplication can be used as a check for division.

96. Explain what it means when we say that division by 0 is undefined.

97. Explain the division rules for integers that are shown below using symbols.

$$\frac{+}{+} = + \qquad \frac{-}{-} = + \qquad \frac{-}{+} = - \qquad \frac{+}{-} = -$$

98. Explain the difference between *division of 0* and *division by 0*.

REVIEW

99. Evaluate: $\dfrac{6(32) - 72}{3(2^3)}$ 5

100. Find the prime factorization of 210. $2 \cdot 3 \cdot 5 \cdot 7$

101. The statement $(4 + 8) + 10 = 4 + (8 + 10)$ illustrates what property? associative property of addition

102. Is $17 \geq 17$ a true statement? yes

103. Does $8 - 2 = 2 - 8$? no

104. Sharif has scores of 55, 70, 80, and 75 on four mathematics tests. What is his mean (average) score? 70

SECTION 2.6
Order of Operations and Estimation

In this chapter, we have discussed the rules for adding, subtracting, multiplying, and dividing integers. Now we will use those rules in combination with the order of operations rule from Section 1.7 to evaluate expressions involving more than one operation.

Objectives

1. Use the order of operations rule.
2. Evaluate expressions containing grouping symbols.
3. Evaluate expressions containing absolute values.
4. Estimate the value of an expression.

1 Use the order of operations rule.

Recall that every numerical expression has only one correct value. If we don't establish a uniform order of operations, an expression such as $2 + 3 \cdot 6$ can have more than one value. To avoid this possibility, always use the following rule for the order of operations.

Order of Operations

1. Perform all calculations within parentheses and other grouping symbols in the following order listed in Steps 2–4 below, working from the innermost pair of grouping symbols to the outermost pair.
2. Evaluate all the exponential expressions.
3. Perform all multiplications and divisions as they occur from left to right.
4. Perform all additions and subtractions as they occur from left to right.

When grouping symbols have been removed, repeat Steps 2–4 to complete the calculation.

If a fraction bar is present, evaluate the expression above the bar (called the **numerator**) and the expression below the bar (the **denominator**) separately. Then perform the division indicated by the fraction bar, if possible.

We can use this rule to evaluate expressions involving integers.

EXAMPLE 1   Evaluate: $-4(-3)^2 - (-2)$

Strategy We will scan the expression to determine what operations need to be performed. Then we will perform those operations, one at a time, following the order of operations rule.

WHY Every numerical expression has only one correct value. If we don't follow the correct order of operations, the expression can have more than one value.

Solution Although the expression contains parentheses, there are no calculations to perform *within* them. We begin with Step 2 of the order of operations rule: Evaluate all exponential expressions.

$$-4(-3)^2 - (-2) = -4(9) - (-2) \quad \text{Evaluate the exponential expression: } (-3)^2 = 9.$$
$$= -36 - (-2) \quad \text{Do the multiplication: } -4(9) = -36.$$
$$= -36 + 2 \quad \text{If it is helpful, use the subtraction rule: Add the opposite of } -2, \text{ which is } 2.$$
$$= -34 \quad \text{Do the addition.}$$

Self Check 1
Evaluate: $-5(-2)^2 - (-6)$ -14

Now Try Problem 13

Teaching Example 1 Evaluate: $-2(-5)^2 - (-9)$
Answer:
-41

Self Check 2

Evaluate:
$4(9) + (-4)(-3)(-2)$ 12

Now Try Problem 17

Teaching Example 2 Evaluate:
$9(5) + (-4)(-2)(-5)$
Answer:
5

EXAMPLE 2 Evaluate: $12(3) + (-5)(-3)(-2)$

Strategy We will perform the multiplication first.

WHY There are no operations to perform within parentheses, nor are there any exponents.

Solution

$12(3) + (-5)(-3)(-2) = 36 + (-30)$ Working from left to right, do the multiplications.

$= 6$ Do the addition.

Self Check 3

Evaluate: $45 \div (-5)3$ -27

Now Try Problem 21

Teaching Example 3 Evaluate:
$20 \div (-5)2$
Answer:
-8

EXAMPLE 3 Evaluate: $40 \div (-4)5$

Strategy This expression contains the operations of division and multiplication. We will perform the divisions and multiplications as they occur from left to right.

WHY There are no operations to perform within parentheses, nor are there any exponents.

Solution

$40 \div (-4)5 = -10 \cdot 5$ Do the division first: $40 \div (-4) = -10$.

$= -50$ Do the multiplication.

Caution! In Example 3, a common mistake is to forget to work from left to right and incorrectly perform the multiplication first. This produces the wrong answer, -2.

$40 \div (-4)5 = 40 \div (-20)$
$= -2$

Self Check 4

Evaluate: $-3^2 - (-3)^2$ -18

Now Try Problem 25

Teaching Example 4 Evaluate:
$-4^2 - (-4)^2$
Answer:
-32

EXAMPLE 4 Evaluate: $-2^2 - (-2)^2$

Strategy There are two exponential expressions to evaluate and a subtraction to perform. We will begin with the exponential expressions.

WHY Since there are no operations to perform within parentheses, we begin with Step 2 of the order of operations rule: Evaluate all exponential expressions.

Solution Recall from Section 2.4 that the values of -2^2 and $(-2)^2$ are not the same.

$-2^2 - (-2)^2 = -4 - 4$ Evaluate the exponential expressions: $-2^2 = -(2 \cdot 2) = -4$ and $(-2)^2 = -2(-2) = 4$.

$= -4 + (-4)$ If it is helpful, use the subtraction rule: Add the opposite of 4, which is -4.

$= -8$ Do the addition.

2 Evaluate expressions containing grouping symbols.

Recall that **parentheses** ( ), **brackets** [ ], **absolute value symbols** | |, and the **fraction bar** — are called **grouping symbols**. When evaluating expressions, we must perform all calculations within parentheses and other grouping symbols first.

EXAMPLE 5 Evaluate: $-15 + 3(-4 + 7 \cdot 2)$

Strategy We will begin by evaluating the expression $-4 + 7 \cdot 2$ that is within the parentheses. Since it contains more than one operation, we will use the order of operations rule to evaluate it. We will perform the multiplication first and then the addition.

WHY By the order of operations rule, we must perform all calculations within the parentheses first following the order listed in Steps 2–4 of the rule.

Solution

$-15 + 3(-4 + \mathbf{7 \cdot 2}) = -15 + 3(-4 + \mathbf{14})$ Do the multiplication within the parentheses: $7 \cdot 2 = 14$.

$= -15 + 3(10)$ Do the addition within the parentheses: $-4 + 14 = 10$.

$= -15 + 30$ Do the multiplication: $3(10) = 30$.

$= 15$ Do the addition.

Expressions can contain two or more pairs of grouping symbols. To evaluate the following expression, we begin within the innermost pair of grouping symbols, the parentheses. Then we work within the outermost pair, the brackets.

$$67 - 5[-1 + (2 - 8)^2]$$

Innermost pair: $(2-8)$
Outermost pair: $[\ldots]$

EXAMPLE 6 Evaluate: $67 - 5[-1 + (2 - 8)^2]$

Strategy We will work within the parentheses first and then within the brackets. Within each pair of grouping symbols, we will follow the order of operations rule.

WHY We must work from the *innermost* pair of grouping symbols to the *outermost*.

Solution

$67 - 5[-1 + (\mathbf{2 - 8})^2]$

$= 67 - 5[-1 + (\mathbf{-6})^2]$ Do the subtraction within the parentheses: $2 - 8 = -6$.

$= 67 - 5[-1 + 36]$ Evaluate the exponential expression within the brackets.

$= 67 - 5[35]$ Do the addition within the brackets: $-1 + 36 = 35$.

$= 67 - 175$ Do the multiplication: $5(35) = 175$.

$= 67 + (-175)$ If it is helpful, use the subtraction rule: Add the opposite of 175, which is -175.

$= -108$ Do the addition.

$$\begin{array}{r} \overset{2}{35} \\ \times\ 5 \\ \hline 175 \end{array}$$

$$\begin{array}{r} \overset{6\ 15}{17\not{5}} \\ -\ 67 \\ \hline 108 \end{array}$$

Success Tip Any arithmetic steps that you cannot perform in your head should be shown outside of the horizontal steps of your solution.

Self Check 5

Evaluate: $-18 + 6(-7 + 9 \cdot 2)$

Now Try Problem 29

Self Check 5 Answer
48

Teaching Example 5 Evaluate:
$-19 + 7(-6 + 5 \cdot 2)$
Answer:
9

Self Check 6

Evaluate:
$81 - 4[-2 + (5 - 9)^2]$ 25

Now Try Problem 33

Teaching Example 6 Evaluate:
$53 - 3[-2 + (7 - 11)^2]$
Answer:
11

Self Check 7

Evaluate: $-\left[8 - \left(3^3 + \dfrac{90}{-9}\right)\right]$ 9

Now Try Problem 37

Teaching Example 7 Evaluate:
$-\left[37 - \left(4^3 + \dfrac{20}{-5}\right)\right]$
Answer:
23

EXAMPLE 7

Evaluate: $-\left[1 - \left(2^4 + \dfrac{66}{-6}\right)\right]$

Strategy We will work within the parentheses first and then within the brackets. Within each pair of grouping symbols, we will follow the order of operations rule.

WHY We must work from the *innermost* pair of grouping symbols to the *outermost*.

Solution

$$-\left[1 - \left(2^4 + \dfrac{66}{-6}\right)\right] = -\left[1 - \left(16 + \dfrac{66}{-6}\right)\right]$$ Evaluate the exponential expression within the parentheses: $2^4 = 16$.

$$= -[1 - (16 + (-11))]$$ Do the division within the parentheses: $66 \div (-6) = -11$.

$$= -[1 - 5]$$ Do the addition within the parentheses: $16 + (-11) = 5$.

$$= -[-4]$$ Do the subtraction within the brackets: $1 - 5 = -4$.

$$= 4$$ The opposite of -4 is 4.

Self Check 8

Evaluate: $\dfrac{-9 + 6(-4)}{28 - (-5)^2}$ -11

Now Try Problem 41

Teaching Example 8 Evaluate:
$\dfrac{-22 + 5(-10)}{40 - (-6)^2}$
Answer:
-18

EXAMPLE 8

Evaluate: $\dfrac{-20 + 3(-5)}{21 - (-4)^2}$

Strategy We will evaluate the expression above and the expression below the fraction bar separately. Then we will do the indicated division, if possible.

WHY Fraction bars are grouping symbols that group the numerator and the denominator. The expression could be written $[-20 + 3(-5)] \div [21 - (-4)^2]$.

Solution

$$\dfrac{-20 + 3(-5)}{21 - (-4)^2} = \dfrac{-20 + (-15)}{21 - 16}$$ In the numerator, do the multiplication: $3(-5) = -15$. In the denominator, evaluate the exponential expression: $(-4)^2 = 16$.

$$= \dfrac{-35}{5}$$ In the numerator, add: $-20 + (-15) = -35$. In the denominator, subtract: $21 - 16 = 5$.

$$= -7$$ Do the division indicated by the fraction bar.

3 Evaluate expressions containing absolute values.

Earlier in this chapter, we found the absolute values of integers. For example, recall that $|-3| = 3$ and $|10| = 10$. We use the order of operations rule to evaluate more complicated expressions that contain absolute values.

Self Check 9

Evaluate each expression:

a. $|(-6)(5)|$ 30

b. $|-3 + 96|$ 93

Now Try Problem 45

Teaching Example 9 Evaluate each expression:
a. $|-6(9)|$
b. $|-14 + 8|$
Answers:
a. 54
b. 6

EXAMPLE 9

Evaluate each expression: **a.** $|-4(3)|$ **b.** $|-6 + 1|$

Strategy We will perform the calculation within the absolute value symbols first. Then we will find the absolute value of the result.

WHY Absolute value symbols are grouping symbols, and by the order of operations rule, all calculations within grouping symbols must be performed first.

Solution

a. $|-4(3)| = |-12|$ Do the multiplication within the absolute value symbol: $-4(3) = -12$.

$= 12$ Find the absolute value of -12.

b. $|-6 + 1| = |-5|$ Do the addition within the absolute value symbol: $-6 + 1 = -5$.

$= 5$ Find the absolute value of -5.

The Language of Algebra Multiplication is indicated when a number is outside and next to an absolute value symbol. For example,

$$8 - 4|-6 - 2| \quad \text{means} \quad 8 - 4 \cdot |-6 - 2|$$

EXAMPLE 10   Evaluate: $8 - 4|-6 - 2|$

Strategy The absolute value bars are grouping symbols. We will perform the subtraction within them first.

WHY By the order of operations rule, we must perform all calculations within parentheses and other grouping symbols (such as absolute value bars) first.

Solution

$$\begin{aligned}
8 - 4|-6 - 2| &= 8 - 4|-6 + (-2)| && \text{If it is helpful, use the subtraction rule within the absolute value symbol: Add the opposite of 2, which is } -2. \\
&= 8 - 4|-8| && \text{Do the addition within the absolute value symbol: } -6 + (-2) = -8. \\
&= 8 - 4(8) && \text{Find the absolute value: } |-8| = 8. \\
&= 8 - 32 && \text{Do the multiplication: } 4(8) = 32. \\
&= 8 + (-32) && \text{If it is helpful, use the subtraction rule: Add the opposite of 32, which is } -32. \\
&= -24 && \text{Do the addition.}
\end{aligned}$$

Self Check 10
Evaluate: $7 - 5|-1 - 6|$ -28
Now Try Problem 49

Teaching Example 10 Evaluate: $30 - 3|-7 - 4|$
Answer:
-3

4 Estimate the value of an expression.

Recall that the idea behind estimation is to simplify calculations by using rounded numbers that are close to the actual values in the problem. When an exact answer is not necessary and a quick approximation will do, we can use estimation.

EXAMPLE 11   *The Stock Market*

The change in the Dow Jones Industrial Average is announced at the end of each trading day to give a general picture of how the stock market is performing. A positive change means a good performance, while a negative change indicates a poor performance. The week of October 13–17, 2008, had some record changes, as shown below. Round each number to the nearest ten and estimate the net gain or loss of points in the Dow that week.

Strategy To estimate the net gain or loss, we will round each number to the nearest ten and *add* the approximations.

+936	-78	-733	+402	-123
Monday Oct. 13, 2008 (largest 1-day increase)	Tuesday Oct. 14, 2008	Wednesday Oct. 15, 2008 (second-largest 1-day decline)	Thursday Oct. 16, 2008 (tenth-largest 1-day increase)	Friday Oct. 17, 2008

Source: finance.yahoo.com

Self Check 11
THE STOCK MARKET For the week of December 15–19, 2008, the Dow Jones Industrial Average performance was as follows, Monday: −63, Tuesday: +358, Wednesday: −98, Thursday: −219, Friday: −27. Round each number to the nearest ten and estimate the net gain or loss of points in the Dow for that week. (Source: finance.yahoo.com)

Now Try Problems 53 and 97
Self Check 11 Answer
There was a net loss that week of approximately 50 points.

Teaching Example 11
THE STOCK MARKET For the week of November 17–21, 2008, the Dow Jones Industrial Average performance was as follows: Monday: −221, Tuesday: +151, Wednesday: −423, Thursday: −443, Friday: +494. Round each number to the nearest ten and estimate the net gain or loss of points in the Dow for that week. (Source: finance.yahoo.com)
Answer:
There was a net loss that week of approximately 440 points.

WHY The phrase *net gain or loss* refers to what remains after all of the losses and gains have been combined (added).

Solution To nearest ten:

936 rounds to 940   −78 rounds to −80   −733 rounds to −730
402 rounds to 400   −123 rounds to −120

To estimate the net gain or loss for the week, we add the rounded numbers.

$$940 + (-80) + (-730) + 400 + (-120)$$
$$= 1,340 + (-930) \quad \text{Add the positives and the negatives separately.}$$
$$= 410 \quad \text{Do the addition.}$$

$$\begin{array}{r} \overset{1}{1},\overset{3}{3}40 \\ -930 \\ \hline 410 \end{array}$$

The positive result means there was a net gain that week of approximately 410 points in the Dow.

ANSWERS TO SELF CHECKS
1. −14 **2.** 12 **3.** −27 **4.** −18 **5.** 48 **6.** 25 **7.** 9 **8.** −11 **9. a.** 30 **b.** 93
10. −28 **11.** There was a net loss that week of approximately 50 points.

SECTION 2.6   STUDY SET

VOCABULARY

Fill in the blanks.

1. To evaluate expressions that contain more than one operation, we use the order of operations rule.
2. Absolute value symbols, parentheses, and brackets are types of grouping symbols.
3. In the expression $-9 + 2[-5 - 6(-3 - 1)]$, the parentheses are the inner most grouping symbols and the brackets are the outer most grouping symbols.
4. In situations where an exact answer is not needed, an approximation or estimate is a quick way of obtaining a rough idea of the size of the actual answer.

CONCEPTS

5. List the operations in the order in which they should be performed to evaluate each expression. *You do not have to evaluate the expression.*
   a. $5(-2)^2 - 1$ square, multiplication, subtraction
   b. $15 - 3 + (-5 \cdot 2)^3$ multiplication, cube, subtraction, addition
   c. $4 + 2(-7 - 3)$ subtraction, multiplication, addition
   d. $-2 \cdot 3^2$ square, multiplication
6. Consider the expression $\dfrac{5 + 5(7)}{2 + (4 - 8)}$. In the numerator, what operation should be performed first? In the denominator, what operation should be performed first? multiplication, subtraction

NOTATION

7. Give the name of each grouping symbol:

   ( )   [ ] | | —   parentheses, brackets, absolute value symbols, fraction bar

8. What operation is indicated?

   $-2 + 9\underset{\uparrow}{|}8 - (-2 + 4)|$ multiplication

Complete each step to evaluate the expression.

9. $-8 - 5(-2)^2 = -8 - 5(\boxed{4})$
   $= -8 - \boxed{20}$
   $= -8 + (\boxed{-20})$
   $= \boxed{-28}$

10. $2 + (5 - 6 \cdot 2) = 2 + (5 - \boxed{12})$
    $= 2 + [5 + (\boxed{-12})]$
    $= 2 + (\boxed{-7})$
    $= \boxed{-5}$

11. $-9 + 5(-4 \cdot 2 + 7) = -9 + 5(\boxed{-8} + 7)$
    $= -9 + 5(\boxed{-1})$
    $= -9 + (\boxed{-5})$
    $= \boxed{-14}$

12. $\dfrac{|-9 + (-3)|}{9 - 6} = \dfrac{|\boxed{-12}|}{3}$
    $= \dfrac{\boxed{12}}{3}$
    $= \boxed{4}$

2.6 Order of Operations and Estimation

GUIDED PRACTICE

Evaluate each expression. See Example 1.

13. $-2(-3)^2 - (-8)$ -10
14. $-6(-2)^2 - (-9)$ -15
15. $-5(-4)^2 - (-18)$ -62
16. $-3(-5)^2 - (-24)$ -51

Evaluate each expression. See Example 2.

17. $9(7) + (-6)(-2)(-4)$ 15
18. $9(8) + (-2)(-5)(-7)$ 2
19. $8(6) + (-2)(-9)(-2)$ 12
20. $7(8) + (-3)(-6)(-2)$ 20

Evaluate each expression. See Example 3.

21. $30 \div (-5)2$ -12
22. $50 \div (-2)5$ -125
23. $60 \div (-3)4$ -80
24. $120 \div (-4)3$ -90

Evaluate each expression. See Example 4.

25. $-6^2 - (-6)^2$ -72
26. $-7^2 - (-7)^2$ -98
27. $-10^2 - (-10)^2$ -200
28. $-8^2 - (-8)^2$ -128

Evaluate each expression. See Example 5.

29. $-14 + 2(-9 + 6 \cdot 3)$ 4
30. $-18 + 3(-10 + 3 \cdot 7)$ 15
31. $-23 + 3(-15 + 8 \cdot 4)$ 28
32. $-31 + 6(-12 + 5 \cdot 4)$ 17

Evaluate each expression. See Example 6.

33. $77 - 2[-6 + (3 - 9)^2]$ 17
34. $84 - 3[-7 + (5 - 8)^2]$ 78
35. $99 - 4[-9 + (6 - 10)^2]$ 71
36. $67 - 5[-6 + (4 - 7)^2]$ 52

Evaluate each expression. See Example 7.

37. $-\left[4 - \left(3^3 + \dfrac{22}{-11}\right)\right]$ 21
38. $-\left[1 - \left(2^3 + \dfrac{40}{-20}\right)\right]$ 5
39. $-\left[50 - \left(5^3 + \dfrac{50}{-2}\right)\right]$ 50
40. $-\left[12 - \left(2^5 + \dfrac{40}{-4}\right)\right]$ 10

Evaluate each expression. See Example 8.

41. $\dfrac{-24 + 3(-4)}{42 - (-6)^2}$ -6
42. $\dfrac{-18 + 6(-2)}{52 - (-7)^2}$ -10
43. $\dfrac{-38 + 11(-2)}{69 - (-8)^2}$ -12
44. $\dfrac{-36 + 8(-2)}{85 - (-9)^2}$ -13

Evaluate each expression. See Example 9.

45. a. $|-6(2)|$ 12 b. $|-12 + 7|$ 5
46. a. $|-4(9)|$ 36 b. $|-15 + 6|$ 9
47. a. $|15(-4)|$ 60 b. $|16 + (-30)|$ 14
48. a. $|12(-5)|$ 60 b. $|47 + (-70)|$ 23

Evaluate each expression. See Example 10.

49. $16 - 6|-2 - 1|$ -2
50. $15 - 6|-3 - 1|$ -9
51. $17 - 2|-6 - 4|$ -3
52. $21 - 9|-3 - 1|$ -15

Estimate the value of each expression by rounding each number to the nearest ten. See Example 11.

53. $-379 + (-13) + 287 + (-671)$ -770
54. $-363 + (-781) + 594 + (-42)$ -590

Estimate the value of each expression by rounding each number to the nearest hundred. See Example 11.

55. $-3,887 + (-5,806) + 4,701$ $-5,000$
56. $-5,684 + (-2,270) + 3,404 + 2,689$ $-1,900$

TRY IT YOURSELF

Evaluate each expression.

57. $(-3)^2 - 4^2$ -7
58. $9 - 9 \cdot 4$ -27
59. $3^2 - 4(-2)(-1)$ 1
60. $2^3 - 3^3$ -19
61. $|-3 \cdot 4 + (-5)|$ 17
62. $|6 - 4(1 - 5)|$ 22
63. $(2 - 5)(5 + 2)$ -21
64. $-3(2)^2 4$ -48
65. $6 + \dfrac{25}{-5} + 6 \cdot 3$ 19
66. $-(5 \cdot 7 - 4 \cdot 8)^4$ -81
67. $\dfrac{-6 - 2^3}{-2 - (-4)}$ -7
68. $\dfrac{-6 - 6}{-2 - 2}$ 3
69. $-12 \div (-2)2$ 12
70. $-60(-2) \div 3$ 40
71. $-16 - 4 \div (-2)$ -14
72. $80 - 40 \div 10 + 5$ 81
73. $-|2 \cdot 7 - (-5)^2|$ -11
74. $-|8 \div (-2) - 5|$ -9
75. $-3|4 - 7|$ -9
76. $|-2 + 6 - 5|$ 1
77. $(7 - 5)^2 - (1 - 4)^2$ -5
78. $5^2 - (-9 - 3)$ 37
79. $-1(2^2 - 2 + 1^2)$ -3
80. $(-7 - 4)^2 - (-1)$ 122
81. $\dfrac{-5 - 5}{1^4 + 1^5}$ -5
82. $\dfrac{-7 - (-3)}{2 - 2^2}$ 2
83. $-50 - 2(-3)^3(4)$ 166
84. $8 - 3[5^2 - (7 - 3)^2]$ -19

85. $-6^2 + 6^2$ 0
86. $-9^2 + 9^2$ 0
87. $3\left(\dfrac{-18}{3}\right) - 2(-2)$ -14
88. $2\left(\dfrac{-12}{3}\right) + 3(-5)$ -23
89. $2|1-8| \cdot |-8|$ 112
90. $2(5) - 6(|-3|)^2$ -44
91. $\dfrac{2 + 3[5 - (1 - 10)]}{|2(-8+2) + 10|}$ 22
92. $\dfrac{11 + (-2 \cdot 2 + 3)}{|15 + (-3 \cdot 4 - 8)|}$ 2
93. $-2 + |6 - 4^2|$ 8
94. $-3 - 4|6 - 7|$ -7
95. $\dfrac{-4(-5) - 2}{3 - 3^2}$ -3
96. $\dfrac{(-6)^2 - 1}{-(2^2 - 3)}$ -35

APPLICATIONS

97. **STOCK MARKET RECORDS** For the week of January 5–9, 2009, the Dow Jones Industrial Average performance was as follows, Monday: -74, Tuesday: $+61$, Wednesday: -227, Thursday: -27, Friday: -129. Round each number to the nearest ten and estimate the net gain or loss of points in the Dow for that week. (Source: finance.yahoo.com) -400 points

98. **STOCK MARKET RECORDS** Refer to the tables below. Round each of the record Dow Jones point gains and losses to the nearest hundred and then add all ten of them. There is an interesting result. What is it? 0

5 Greatest Dow Jones Daily Point Gains

Rank	Date	Gain
1	10/13/2008	+936
2	10/28/2008	+889
3	11/13/2008	+553
4	11/21/2008	+494
5	9/30/2008	+485

5 Greatest Dow Jones Daily Point Losses

Rank	Date	Loss
1	9/29/2008	-778
2	10/15/2008	-733
3	12/1/2008	-680
4	10/9/2008	-679
5	10/22/2008	-514

Source: Dow Jones Indexes

99. **TESTING** In an effort to discourage her students from guessing on multiple-choice tests, a professor uses the grading scale shown in the table in the next column. If unsure of an answer, a student does best to skip the question, because incorrect responses are penalized very heavily. Find the test score of a student who gets 12 correct and 3 wrong and leaves 5 questions blank. 19

Response	Value
Correct	$+3$
Incorrect	-4
Left blank	-1

100. **SPREADSHEETS** The table shows the data from a chemistry experiment in spreadsheet form. To obtain a result, the chemist needs to add the values in row 1, double that sum, and then divide that number by the smallest value in column C. What is the final result of these calculations? -11

	A	B	C	D
1	12	-5	6	-2
2	15	4	5	-4
3	6	4	-2	8

101. **BUSINESS TAKEOVERS** Six investors are taking over a poorly managed company, but first they must repay the debt that the company built up over the past four quarters. (See the graph below.) If the investors plan equal ownership, how much of the company's total debt is each investor responsible for? $-\$8$ million

102. **DECLINING ENROLLMENT** Find the drop in enrollment for each high school shown in the table below. Express each drop as a negative number. Then find the mean (average) drop in enrollment for these four schools. -85 students

High school	2013 enrollment	2014 enrollment	Drop
Lincoln	2,683	2,573	-110
Camble	2,754	2,662	-92
Munson	1,948	1,875	-73
Hardy	2,257	2,192	-65

103. **THE FEDERAL BUDGET** See the graph below. Suppose you were hired to write a speech for a politician who wanted to highlight the improvement in the federal government's finances during the 1990s. Would it be better for the politician to talk about the mean (average) budget deficit/surplus for the last half of the decade, or for the last four years of that decade? Explain your reasoning. It's better to refer to the last four years, because there was an average budget surplus of $16 billion.

U.S. Budget Deficit/Surplus
($ billions)

Deficit	Year	Surplus
−164	1995	
−107	1996	
−22	1997	
	1998	+70
	1999	+123

104. **SCOUTING REPORTS** The illustration below shows a football coach how successful his opponent was running a "28 pitch" the last time the two teams met. What was the opponent's mean (average) gain with this play? 11 yd

Play: 28 pitch

| Gain 16 yd | Gain 10 yd | Loss 2 yd | No gain |
| Gain 4 yd | Loss 4 yd | TD Gain 66 yd | Loss 2 yd |

105. **ESTIMATION** Quickly determine a reasonable estimate of the exact answer in each of the following situations. Use front-end rounding.

   a. A scuba diver, swimming at a depth of 34 feet below sea level, spots a sunken ship beneath him. He dives down another 57 feet to reach it. What is the depth of the sunken ship? 90 ft below sea level (−90)

   b. A dental hygiene company offers a money-back guarantee on its tooth whitener kit. When the kit is returned by a dissatisfied customer, the company loses the $11 it cost to produce it, because it cannot be resold. How much money has the company lost because of this return policy if 56 kits have been mailed back by customers? $600 lost (−600)

   c. A tram line makes a 7,891-foot descent from a mountaintop in 18 equal stages. How much does it descend in each stage? −400 ft

106. **ESTIMATION** Quickly determine a reasonable estimate of the exact answer in each of the following situations. Use front-end rounding.

   a. A submarine, cruising at a depth of −175 feet, descends another 605 feet. What is the depth of the submarine? 800 ft below sea level (−800)

   b. A married couple has assets that total $840,756 and debts that total $265,789. What is their net worth? $500,000

   c. According to pokerlistings.com, the top five online poker losses for 2012 were −$3,552,742, −$3,109,619, −$2,673,899, −$1,791,000, and −$1,377,402. Find the total amount lost. $13,000,000 lost (−13,000,000)

WRITING

107. When evaluating expressions, why is the order of operations rule necessary?

108. In the rules for the order of operations, what does the phrase *as they occur from left to right* mean?

109. Explain the error in each evaluation below.

   a. $80 \div (-2)4 = 80 \div (-8)$
   $= -10$

   b. $-1 + 8|4 - 9| = -1 + 8|-5|$
   $= 7|-5|$
   $= 35$

110. Describe a situation in daily life in which you use estimation.

REVIEW

111. Solve: $34 + y = 34$ 0

112. Solve: $\dfrac{a}{12} = 3$ 36

113. Is 834,540 divisible by: **a.** 2 yes **b.** 3 yes **c.** 4 yes **d.** 5 yes **e.** 6 yes **f.** 9 no **g.** 10 yes

114. **a.** Find the LCM of 12 and 44. 132

   b. Find the GCF of 12 and 44. 4

Objectives

1. Use one property of equality to solve equations.
2. Solve equations involving $-x$.
3. Use more than one property of equality to solve equations.
4. Use equations to solve application problems involving integers.

SECTION 2.7
Solving Equations That Involve Integers

In this section, we revisit the topic of solving equations. The equations that we will solve involve negative numbers, and some of the solutions are negative numbers as well.

1 Use one property of equality to solve equations.

Recall that **to solve an equation** means to find all the values of the variable that make the equation true. In Chapter 1, we used the following properties of equality to solve equations involving whole numbers.

> **Properties of Equality**
>
> **Addition Property of Equality:** Adding the same number to both sides of an equation does not change its solution.
>
> **Subtraction Property of Equality:** Subtracting the same number from both sides of an equation does not change its solution.
>
> **Multiplication Property of Equality:** Multiplying both sides of an equation by the same nonzero number does not change its solution.
>
> **Division Property of Equality:** Dividing both sides of an equation by the same nonzero number does not change its solution.

These properties are also used to solve equations involving integers.

Self Check 1
Solve $x + (-4) = -12$ and check the result. -8

Now Try Problem 17

Teaching Example 1
Solve $x + (-5) = -11$ and check the result.
Answer:
-6

EXAMPLE 1 Solve: $x + (-8) = -10$

Strategy We will use a property of equality to isolate the variable on one side of the equation.

WHY To solve the original equation, we want to find a simpler equivalent equation of the form $x = $ **a number**, whose solution is obvious.

Solution We will use the addition property of equality to isolate x on the left side of the equation. We can undo the addition of -8 by adding 8 to both sides.

$$x + (-8) = -10 \qquad \text{This is the equation to solve.}$$
$$x + (-8) + 8 = -10 + 8 \qquad \text{Add 8 to both sides.}$$
$$x + 0 = -2 \qquad \text{On the left side, the sum of a number and its opposite is zero: } (-8) + 8 = 0. \text{ On the right side add: } -10 + 8 = -2.$$
$$x = -2 \qquad \text{On the left side, the sum of any number and 0 is that number: } x + 0 = x.$$

To check, we substitute -2 for x in the original equation and simplify.

$$x + (-8) = -10 \qquad \text{This is the original equation.}$$
$$-2 + (-8) \stackrel{?}{=} -10 \qquad \text{Substitute } -2 \text{ for } x.$$
$$-10 = -10 \qquad \text{On the left side, do the addition.}$$

Since the resulting statement $-10 = -10$ is true, -2 is the solution of $x + (-8) = -10$.

2.7 Solving Equations That Involve Integers

> **Success Tip** From Example 1, we see that to undo addition, we can *add the opposite* of the number that is added to the variable.

EXAMPLE 2 Solve: $t + 16 = -8$

Strategy We will use a property of equality to isolate the variable on one side of the equation.

WHY To solve the original equation, we want to find a simpler equation of the form $t = $ a number, whose solution is obvious.

Solution We will use the subtraction property of equality to isolate t on the left side of the equation. We can undo the addition of 16 by subtracting 16 from both sides.

$$t + 16 = -8 \quad \text{This is the equation to solve.}$$
$$t + 16 - 16 = -8 - 16 \quad \text{Subtract 16 from both sides.}$$
$$t + 0 = -8 + (-16) \quad \text{On the left side, } 16 - 16 = 0. \text{ On the right side, write the subtraction as addition of the opposite.}$$
$$t = -24 \quad \text{On the left side, the sum of any number and 0 is that number: } t + 0 = t. \text{ On the right side, do the addition.}$$

$$\begin{array}{r} \overset{1}{1}6 \\ +\ 8 \\ \hline 24 \end{array}$$

Check:

$$t + 16 = -8 \quad \text{This is the original equation.}$$
$$-24 + 16 \overset{?}{=} -8 \quad \text{Substitute } -24 \text{ for } t.$$
$$-8 = -8 \quad \text{On the left side, do the addition.}$$

$$\begin{array}{r} \overset{1\ 1\ 4}{2}\cancel{4} \\ -16 \\ \hline 8 \end{array}$$

Since the resulting statement $-8 = -8$ is true, -24 is the solution of $t + 16 = -8$.

Self Check 2
Solve $c + 4 = -3$ and check the result. -7

Now Try Problem 21

Teaching Example 2
Solve $m + 9 = -5$ and check the result.
Answer:
-14

EXAMPLE 3 Solve: $-3 + 7 = h + 11(-2)$

Strategy We will begin by performing the addition on the left side of the equation and the multiplication on the right side.

WHY The expressions on each side of the equation should be simplified before we use any properties of equality.

Solution

$$-3 + 7 = h + 11(-2) \quad \text{This is the equation to solve.}$$
$$4 = h + (-22) \quad \text{On the left side, do the addition: } -3 + 7 = 4. \text{ On the right side, do the multiplication: } 11(-2) = -22.$$

Now we use the addition property of equality to isolate h on the right side of the equation.

$$4 + 22 = h + (-22) + 22 \quad \text{To isolate } h, \text{ undo the addition of } -22 \text{ by adding 22 to both sides.}$$
$$26 = h \quad \text{Simplify each side: } 4 + 22 = 26 \text{ and } (-22) + 22 = 0.$$

Check:

$$-3 + 7 = h + 11(-2) \quad \text{This is the original equation.}$$
$$-3 + 7 \overset{?}{=} 26 + 11(-2) \quad \text{Substitute 26 for } h.$$
$$4 \overset{?}{=} 26 + (-22) \quad \text{On the left side, add. On the right side, multiply.}$$
$$4 = 4 \quad \text{On the right side, do the addition.}$$

Since the resulting statement $4 = 4$ is true, 26 is the solution.

Self Check 3
Solve $-2 + 8 = y + 3(-4)$ and check the result. 18

Now Try Problem 25

Teaching Example 3
Solve $-1 + 7 = y + 2(-8)$ and check the result.
Answer:
22

Chapter 2 The Integers

Self Check 4

Solve each equation and check the result:

a. $-7k = 28$ -4

b. $-40 = -8k$ 5

Now Try Problem 29

Teaching Example 4
Solve each equation and check the result:
a. $-6t = 30$
b. $-72 = -9t$
Answers:
a. -5
b. 8

EXAMPLE 4 Solve: a. $-3y = 15$ b. $-16 = -4y$

Strategy We will use a property of equality to isolate the variable on one side of the equation.

WHY To solve each of the original equations, we want to find a simpler equivalent equation of the form $y = $ **a number** or **a number** $= y$, whose solution is obvious.

Solution

a. Recall that $-3y$ indicates multiplication: $-3 \cdot y$. We must undo the multiplication of y by -3. To do this, we use the division property of equality and divide both sides of the equation by -3.

$$-3y = 15 \qquad \text{This is the equation to solve.}$$

$$\frac{-3y}{-3} = \frac{15}{-3} \qquad \text{Divide both sides by } -3.$$

$$y = -5 \qquad \text{On the left side, } -3 \text{ times } y, \text{ divided by } -3, \text{ is } y. \text{ On the right side, do the division: } 15 \div (-3) = -5.$$

Check:

$$-3y = 15 \qquad \text{This is the original equation.}$$

$$-3(-5) \stackrel{?}{=} 15 \qquad \text{Substitute } -5 \text{ for } y.$$

$$15 = 15 \qquad \text{On the left side, do the multiplication: } -3(-5) = 15.$$

Since the resulting statement $15 = 15$ is true, -5 is the solution of $-3y = 15$.

b. $-16 = -4y$ This is the equation to solve.

$$\frac{-16}{-4} = \frac{-4y}{-4} \qquad \text{To isolate } y, \text{ undo the multiplication by } -4, \text{ by dividing both sides by } -4.$$

$$4 = y \qquad \text{On the left side, do the division: } -16 \div (-4) = 4. \text{ On the right side, } -4 \text{ times } y, \text{ divided by } -4 \text{ is } y.$$

Check the result to verify that 4 is the solution.

Self Check 5

Solve $\dfrac{t}{-3} = 4$ and check the result. -12

Now Try Problem 33

Teaching Example 5
Solve $\dfrac{a}{-10} = 2$ and check the result.
Answer:
-20

EXAMPLE 5 Solve: $\dfrac{x}{-5} = -10$

Strategy We will use a property of equality to isolate the variable on one side of the equation.

WHY To solve the original equation, we want to find a simpler equivalent equation of the form $x = $ **a number**, whose solution is obvious.

Solution In this equation, x is being divided by -5. To undo this division, we use the multiplication property of equality and multiply both sides of the equation by -5.

$$\frac{x}{-5} = -10 \qquad \text{This is the equation to solve.}$$

$$-5\left(\frac{x}{-5}\right) = -5(-10) \qquad \text{Multiply both sides by } -5.$$

$$x = 50 \qquad \text{On the left side, when } x \text{ is divided by } -5 \text{ and then multiplied by } -5, \text{ the result is } x. \text{ On the right side, do the multiplication: } -5(-10) = 50.$$

Check:

$$\frac{x}{-5} = -10 \quad \text{This is the original equation.}$$

$$\frac{50}{-5} \stackrel{?}{=} -10 \quad \text{Substitute 50 for } x.$$

$$-10 = -10 \quad \text{On the left side, do the division: } 50 \div (-5) = -10.$$

Since the resulting statement $-10 = -10$ is true, 50 is the solution of $\frac{x}{-5} = -10$.

2 Solve equations involving −x.

Recall from Chapter 1 that we don't need to write a multiplication symbol when multiplying a variable by a number. For example,

$5a$ means $5 \cdot a$, $-9m$ means $-9 \cdot m$,   and $-1x$ means $-1 \cdot x$

A simpler way to write the last expression, $-1x$, is $-x$. When we examine what each notation means, it becomes clear why this is true.

$$-1x = -x$$

This means multiply the value of x by −1.   This means find the opposite of the value of x.

We can use the fact that $-1x = -x$ to solve equations that involve the expression $-x$.

EXAMPLE 6   Solve: $-x = 3$

Strategy The variable x is not isolated, because there is a $-$ sign in front of it. Since the term $-x$ has an understood coefficient of -1, the equation can be written as $-1x = 3$. We need to select a property of equality and use it to isolate the variable on one side of the equation.

WHY To find the solution of the original equation, we want to find a simpler equivalent equation of the form $x = $ **a number**, whose solution is obvious.

Solution To isolate x, we can either multiply or divide both sides by -1.

Multiply both sides by −1:

$-x = 3$	The equation to solve
$-1x = 3$	Write $-x$ as $-1x$.
$(-1)(-1x) = (-1)3$	
$1x = -3$	On the left, $(-1)(-1) = 1$.
$x = -3$	$1x = x$

Divide both sides by −1:

$-x = 3$	The equation to solve
$-1x = 3$	Write $-x$ as $-1x$.
$\dfrac{-1x}{-1} = \dfrac{3}{-1}$	
$1x = -3$	On the left side, $\frac{-1}{-1} = 1$.
$x = -3$	$1x = x$

Check: $-x = 3$ This is the original equation.
   $-(-3) \stackrel{?}{=} 3$ Substitute -3 for x.
   $3 = 3$ On the left side, the opposite of -3 is 3.

Since the statement $3 = 3$ is true, -3 is the solution of $-x = 3$.

Self Check 6

Solve $-h = -17$ and check the result.   17

Now Try Problem 37

Teaching Example 6
Solve $-a = 2$ and check the result.
Answer:
-2

3 Use more than one property of equality to solve equations.

In the previous examples, each equation was solved by using a single property of equality. Sometimes we must use two (or more) properties of equality to solve more complicated equations. For example, on the left side of $2x + 6 = 10$, the variable x is multiplied by 2, and then 6 is added to that product. To solve the equation, we use the order of operations rule in reverse. First, we isolate the variable term $2x$ by undoing the addition of 6. Then isolate the variable x by undoing the multiplication by 2.

$$2x + 6 = 10 \qquad \text{This is the equation to solve.}$$
$$2x + 6 - 6 = 10 - 6 \qquad \text{To undo the addition of 6, subtract 6 from both sides.}$$
$$2x = 4 \qquad \text{Do the subtractions.}$$
$$\frac{2x}{2} = \frac{4}{2} \qquad \text{To undo the multiplication by 2, divide both sides by 2.}$$
$$x = 2 \qquad \text{Do the division.}$$

The solution is 2.

> **The Language of Algebra** In the example above, we subtracted 6 from both sides to *isolate the variable term*, $2x$. Then we divided both sides by 2 to *isolate the variable*, x.
>
> $$2x + 6 = 10$$
> ↑
> The variable term

Self Check 7
Solve $-6b - 1 = 11$ and check the result. -2

Now Try Problem 41

Teaching Example 7
Solve $-9a - 3 = 78$ and check the result.
Answer:
-9

EXAMPLE 7 Solve: $-4x - 5 = 15$

Strategy First we will use a property of equality to isolate the *variable term* on one side of the equation. Then we will use a second property of equality to isolate the *variable* itself.

WHY To solve the original equation, we want to find a simpler equivalent equation of the form $x = $ **a number**, whose solution is obvious.

Solution On the left side of the equation, x is multiplied by -4, and then 5 is subtracted from that product. To solve the equation, we undo the operations in the opposite order.

- To isolate the variable term, $-4x$, we add 5 to both sides to undo the subtraction of 5.
- To isolate the variable, x, we divide both sides by -4 to undo the multiplication by -4.

$$-4x - 5 = 15 \qquad \text{This is the equation to solve.}$$
$$-4x - 5 + 5 = 15 + 5 \qquad \text{Use the addition property of equality: Add 5 to both sides to isolate } -4x.$$
$$-4x = 20 \qquad \text{Do the additions: } -5 + 5 = 0 \text{ and } 15 + 5 = 20. \text{ Now we want to isolate } x.$$
$$\frac{-4x}{-4} = \frac{20}{-4} \qquad \text{Use the division property of equality: Divide both sides by } -4 \text{ to isolate } x.$$
$$x = -5 \qquad \text{Do the division.}$$

Check:

$-4x - 5 = 15$ This is the original equation.

$-4(-5) - 5 \stackrel{?}{=} 15$ Substitute -5 for x.

$20 - 5 \stackrel{?}{=} 15$ On the left side, do the multiplication: $-4(-5) = 20$.

$15 = 15$ On the left side, do the subtraction.

Since the resulting statement $15 = 15$ is true, -5 is the solution of $-4x - 5 = 15$.

EXAMPLE 8   Solve: $-1 = 2 - 3p$

Strategy First we will use a property of equality to isolate the *variable term* on one side of the equation. Then we will use a second property of equality to isolate the *variable* itself.

WHY To solve the original equation, we want to find a simpler equivalent equation of the form **a number = p**, whose solution is obvious.

Solution On the right side of the equation, p is multiplied by -3, and then 2 is added to that product. Think of $2 - 3p$ as $2 + (-3p)$.

- To isolate the variable term, $-3p$, we subtract 2 from both sides to undo the addition of 2.
- To isolate the variable, p, we divide both sides by -3 to undo the multiplication by -3.

$-1 = 2 - 3p$ This is the equation to solve.

$-1 - 2 = 2 - 3p - 2$ Use the subtraction property of equality: Subtract 2 from both sides to isolate $-3p$.

$-3 = -3p$ On the right side, do the subtraction: $2 - 2 = 0$.
On the left side do the subtraction: $-1 - 2 = -3$.

$\dfrac{-3}{-3} = \dfrac{-3p}{-3}$ Use the division property of equality: Divide both sides by -3 to isolate p.

$1 = p$ Do the division.

Check this result in the *original* equation to verify that 1 is the solution.

Caution! In Example 8, a common error is to forget to write the $-$ symbol in front of $3p$ after subtracting 2 from both sides of the equation.

$-1 - 2 = 2 - 3p - 2$

$-3 = -3p$
↑
Don't forget to write the $-$ symbol.

Self Check 8

Solve $-34 = 6 - 8k$ and check the result.   5

Now Try Problem 45

Teaching Example 8
Solve $-46 = 9 - 5d$ and check the result.
Answer:
11

EXAMPLE 9   Solve: $\dfrac{y}{-2} - 6 = -43$

Strategy First we will use a property of equality to isolate the *variable term* on one side of the equation. Then we will use a second property of equality to isolate the *variable* itself.

WHY To solve the original equation, we want to find a simpler equivalent equation of the form **y = a number**, whose solution is obvious.

Self Check 9

Solve $\dfrac{m}{-8} - 10 = -74$ and check the result.   512

Now Try Problem 49

Teaching Example 9

Solve $\dfrac{s}{-2} - 18 = -52$ and check the result.

Answer: 68

Solution On the left side of the equation, y is divided by -2, and 6 is subtracted from the quotient.

- To isolate the variable term, $\dfrac{y}{-2}$, we add 6 to both sides to undo the subtraction of 6.
- To isolate the variable, y, we multiply both sides by -2 to undo the division by -2.

$$\dfrac{y}{-2} - 6 = -43 \quad \text{This is the equation to solve.}$$

$$\dfrac{y}{-2} - 6 + 6 = -43 + 6 \quad \text{Use the addition property of equality: Add 6 to both sides to isolate } \dfrac{y}{-2}.$$

$$\dfrac{y}{-2} = -37 \quad \text{Do the addition: } -6 + 6 = 0 \text{ and } -43 + 6 = -37.$$

$$-2\left(\dfrac{y}{-2}\right) = -2(-37) \quad \text{Use the multiplication property of equality: Multiply both sides by } -2 \text{ to isolate } y.$$

$$y = 74 \quad \text{Do the multiplication.}$$

Check:

$$\dfrac{y}{-2} - 6 = -43 \quad \text{This is the original equation.}$$

$$\dfrac{74}{-2} - 6 \stackrel{?}{=} -43 \quad \text{Substitute 74 for } y.$$

$$-37 - 6 \stackrel{?}{=} -43 \quad \text{On the left side, do the division: } 74 \div (-2) = -37.$$

$$-43 = -43 \quad \text{On the left side, do the subtraction.}$$

Since the resulting statement $-43 = -43$ is true, 74 is the solution. ∎

4 Use equations to solve application problems involving integers.

In Chapter 1, we used the concepts of variable and equation to solve application problems involving whole numbers. We will now use a similar approach to solve problems involving integers. As in Chapter 1, we will follow the six-step problem-solving strategy of analyze, assign, form, solve, state, and check.

> **The Language of Algebra** As you read the application problems, watch for the following words and phrases. They often indicate negative numbers.
>
> | behind | below | before | deficit | debt | drop |
> |---|---|---|---|---|---|
> | in the red | overdrawn | under | loss | B.C. | |

Self Check 10

FAST FOOD In 2010, Wendy's (the restaurant chain) lost $4,320,000. The year after, the company made a profit. If the company made a total of $5,560,000 over this two-year span, how much profit did Wendy's make in 2011? (Source: wikinvest.com)

Now Try Problem 91

EXAMPLE 10 **Home Entertainment** In 2011, TiVo, Inc., suffered a loss due to large operating expenses and ended the year $85 million in the red. In 2012, the company did much better and made a large profit. If the company made a total of $17 million over this two-year span, how much profit did TiVo make in 2012? (Source: wikinvest.com)

Analyze

- In 2011, TiVo lost $85 million. *Given*
- The company made a total of $17 million in 2011 and 2012. *Given*
- How much profit did TiVo make in 2012? *Find*

Assign We will let x = the profit (in millions) that TiVo made in 2012.

Form If we work in terms of millions of dollars, we can represent the loss in 2011 using the negative number $-\$85$, and the total amount made in 2011 and 2012 can be represented by the positive number $17.

The key word *total* suggests addition. Now we translate the words of the problem to numbers and symbols.

The loss in 2011	plus	the profit in 2012	equals	the total amount made in 2011 and 2012.
-85	$+$	x	$=$	17

Solve
$$-85 + x = 17$$
$$-85 + x + 85 = 17 + 85 \quad \text{To isolate } x \text{ on the left side, add 85 to both sides.}$$
$$x = 102 \quad \text{Do the addition. The units are millions of dollars.}$$

$$\begin{array}{r} \overset{1}{8}5 \\ +17 \\ \hline 102 \end{array}$$

Success Tip It is a good idea to work in terms of millions of dollars. If we don't, the equation we must solve involves many zeros:
$$-32{,}000{,}000 + x = 72{,}000{,}000$$

State In 2012, TiVo made a profit of $102 million.

Check We can check the result by going back to the original problem. If TiVo had a loss of $85 million in 2011 and a profit of $102 million in 2012, total profit for the two years would be $17 million. The solution is correct.

Self Check 10 Answer
Wendy's made a profit of $9,880,000 in 2011.

Teaching Example 10
PROFITS AND LOSSES In 2011, Sears made a profit of $150 million. The next year the company suffered a large loss. If the company lost a total of $3,050 million in this two-year span, how much did Sears lose in 2012? (Source: wikinvest.com)
Answer:
Sears lost $3,200 million in 2012.

ANSWERS TO SELF CHECKS
1. -8 **2.** -7 **3.** 18 **4. a.** -4 **b.** 5 **5.** -12 **6.** 17 **7.** -2 **8.** 5 **9.** 512
10. Wendy's made a profit of $9,880,000 in 2011.

SECTION 2.7 STUDY SET

VOCABULARY

Fill in the blanks.

1. To **solve** an equation means to find all the values of the variable that make the equation true.
2. In the equation $3x + 1 = 10$, we call $3x$ the **variable** term.
3. To **check** the solution of an equation, we substitute the value for the variable in the original equation and determine whether the result is a true statement.
4. Words such as *debt*, *overdrawn*, and *loss* are often used to indicate a **negative** number.

CONCEPTS

5. What operation is performed on the variable x?
   a. $-2x = -100$ multiplication by -2
   b. $-6 + x = -9$ addition of -6
   c. $\dfrac{x}{-5} = 2$ division by -5
   d. $-20 = x - 4$ subtraction of 4

6. What operations are performed on the variable x?
   a. $-4x - 1 = 11$ multiplication by -4, subtraction of 1
   b. $-1 = -28 + 9x$ multiplication by 9, addition of -28
   c. $\dfrac{x}{-6} - 3 = 9$ division by -6, subtraction of 3

7. What step should be used to isolate the *variable* on one side of the equation?

   a. $x + (-9) = 14$ add 9 to both sides

   b. $-32 = -8x$ divide both sides by -8

8. What step should be used to isolate the *variable term* on one side of the equation?

   a. $-11x + 3 = -19$ subtract 3 from both sides

   b. $-6 + \dfrac{h}{-3} = -14$ add 6 to both sides

Fill in the blanks.

9. The addition property of equality: Adding the same number to both sides of an equation does not change its solution.

10. The multiplication property of equality : Multiplying both sides of an equation by the same nonzero number does not change its solution.

11. It takes two steps to solve the equation
    $$4x + 10 = -6$$
    - To isolate the variable term $4x$, we undo the addition of 10 by subtracting 10 from both sides.
    - To isolate the variable x, we undo the multiplication by 4 by dividing both sides by 4.

12. To solve $-x = 6$, we can multiply or divide both sides of the equation by -1 .

NOTATION

Complete each step to solve the equation. Then check the result.

13. $$y + (-7) = -16 + 3$$
    $$y + (-7) = \boxed{-13}$$
    $$y + (-7) + \boxed{7} = -13 + \boxed{7}$$
    $$y = \boxed{-6}$$
    Check: $y + (-7) = -16 + 3$
    $$\boxed{-6} + (-7) \stackrel{?}{=} -13$$
    $$\boxed{-13} = -13 \quad \text{True}$$
    The solution is -6 .

14. $$-31 = -4y + 1$$
    $$-31 - \boxed{1} = -4y + 1 - \boxed{1}$$
    $$\boxed{-32} = -4y$$
    $$\dfrac{-32}{-4} = \dfrac{-4y}{-4}$$
    $$\boxed{8} = y$$
    Check: $-31 = -4y + 1$
    $$-31 \stackrel{?}{=} -4(\boxed{8}) + 1$$
    $$-31 \stackrel{?}{=} \boxed{-32} + 1$$
    $$-31 = \boxed{-31} \quad \text{True}$$
    The solution is 8 .

15. a. What does $-10x$ mean? $-10 \cdot x$

    b. What does $\dfrac{x}{-8}$ mean? $x \div (-8)$

16. Fill in the blank: $-x = \boxed{-1}\,x$

GUIDED PRACTICE

Solve each equation and check the result. See Example 1.

17. $x + (-3) = -12$ -9
18. $y + (-1) = -4$ -3
19. $m + (-6) = -1$ 5
20. $r + (-12) = -2$ 10

Solve each equation and check the result. See Example 2.

21. $y + 20 = -4$ -24
22. $s + 18 = -10$ -28
23. $t + 19 = -33$ -52
24. $x + 17 = -32$ -49

Solve each equation and check the result. See Example 3.

25. $-7 + 9 = x + 5(-3)$ 17
26. $-1 + 7 = x + 2(-9)$ 24
27. $-6 + 3 = f + 2(-4)$ 5
28. $-10 + 4 = t + 3(-3)$ 3

Solve each equation and check the result. See Example 4.

29. $-2s = 16$ -8
30. $-3t = 9$ -3
31. $-150 = -6g$ 25
32. $-60 = -6m$ 10

Solve each equation and check the result. See Example 5.

33. $\dfrac{t}{-3} = -9$ 27
34. $\dfrac{w}{-4} = -5$ 20
35. $\dfrac{x}{-7} = 11$ -77
36. $\dfrac{s}{-9} = 9$ -81

Solve each equation and check the result. See Example 6.

37. $-x = 14$ -14
38. $-m = 32$ -32
39. $-y = -58$ 58
40. $-x = -73$ 73

Solve each equation and check the result. See Example 7.

41. $-5x - 9 = 11$ -4
42. $-6x - 4 = 44$ -8
43. $-11y - 1 = 87$ -8
44. $-12y - 9 = 39$ -4

Solve each equation and check the result. See Example 8.

45. $-22 = 8 - 3x$ 10
46. $-60 = 3 - 7x$ 9
47. $49 = 4 - 5t$ -9
48. $21 = 15 - 6n$ -1

Solve each equation and check the result. See Example 9.

49. $\dfrac{x}{-2} - 6 = -9$ 6
50. $\dfrac{a}{-5} - 7 = -16$ 45
51. $\dfrac{y}{4} + 5 = -8$ -52
52. $\dfrac{r}{2} + 5 = -13$ -36

TRY IT YOURSELF

Solve each equation and check the result.

53. $-21 = 4h - 5$ -4
54. $-22 = 7l - 8$ -2
55. $-9h = -3(-3)$ -1
56. $-6k = -2(-3)$ -1
57. $0 = \dfrac{y}{8}$ 0
58. $0 = \dfrac{h}{7}$ 0
59. $-5 - 6 - 5x = 4$ -3
60. $-7 - 5 - 7x = 16$ -4
61. $-15 = -k$ 15
62. $-4 = -p$ 4
63. $\dfrac{h}{-6} + 4 = 5$ -6
64. $\dfrac{p}{-3} + 3 = 8$ -15
65. $r - (-7) = -1 - 6$ -14
66. $x - (-1) = -4 - 3$ -8
67. $-8 = t - 6$ -2
68. $x - 1 = -7$ -6
69. $2x + 3(0) = -6$ -3
70. $3x - 4(0) = -12$ -4
71. $-x = 8$ -8
72. $-y = 12$ -12
73. $0 = y + 9$ -9
74. $0 = t + 5$ -5
75. $-1 = -8 + \dfrac{h}{-2}$ -14
76. $-5 = 4 + \dfrac{g}{-4}$ 36
77. $34 = 4 - 5x$ -6
78. $21 = 15 - 6x$ -1
79. $t - 4 = -8 - (-2)$ -2
80. $r - 1 = -3 - (-4)$ 2
81. $5 - t = 500$ -495
82. $4 - r = 300$ -296
83. $4 = -3x + (-2)$ -2
84. $15 = -2x + (-11)$ -13
85. $-2(4) = \dfrac{t}{-6} + 1$ 54
86. $-2(5) = \dfrac{y}{-3} + 3$ 39
87. $2y + 8 = -6$ -7
88. $5y + 1 = -9$ -2

APPLICATIONS

Complete each solution.

89. SHARKS During a research project, a diver inside a shark cage made the first observations at a depth of 120 feet below sea level. For a second set of observations, the cage was raised to a depth of 75 feet below sea level. How many feet was the cage raised between observations?

Analyze

We can represent depths below sea level using negative numbers.

- The first observations were at a depth of -120 ft. *Given*
- The second set of observations were at a depth of -75 ft. *Given*
- How many feet was the cage raised? *Find*

Assign

Let x = the number of feet the cage was raised between observations.

Form

The key word *raised* suggests addition . We now translate the words of the problem into an equation.

The first depth of the cage	plus	the amount the cage was raised	is equal to	the second depth of the cage.
-120	+	x	=	-75

Solve

$$-120 + x = -75$$
$$-120 + x + 120 = -75 + 120$$
$$x = 45$$

State

The shark cage was raised 45 feet.

Check

If we add the number of feet that the cage was raised to the first depth, we should get the second depth.

$$-120 \text{ ft} + 45 \text{ ft} = -75 \text{ ft}$$

The result checks.

90. PROFITS AND LOSSES In its first year of business, a plant nursery suffered a loss due to frost damage, ending the year $11,500 in the red. In the second year, it made a sizable profit. If the nursery made a total of $32,000 the first two years in business, how much profit was made the second year?

Analyze

We can represent a loss using a negative number and a profit using a positive number.

- The first year loss was − $11,500 . *Given*
- The total amount made the first two years in business was $32,000 . *Given*
- How much profit was made the second year? *Find*

Assign

Let x = the profit made the second year.

Form

The key word *total* suggests addition . We now translate the words of the problem into an equation.

The first-year loss	plus	the second-year profit	is equal to	the total amount made in two years.
$-11,500$	+	x	=	$32,000$

Solve

$$-11{,}500 + \boxed{x} = 32{,}000$$
$$-11{,}500 + x + \boxed{11{,}500} = 32{,}000 + \boxed{11{,}500}$$
$$x = \boxed{43{,}500}$$

State

The business made a profit of $\boxed{\$43{,}500}$ the second year.

Check

If we add the second-year profit to the first-year loss, we should get the total amount made in two years.

$$-\$11{,}500 + \$43{,}500 = \boxed{\$32{,}000}$$

The result checks.

In each of the following problems, let a variable represent the unknown quantity. Then write and solve an equation to answer the question.

91. **FOOTWEAR TRENDS** Because of tough economic times and cheap knock-offs from competitors, Crocs, Inc. (a shoe manufacturer), lost $42 million in 2009. Just one year later, the company made a large profit. If the company made a total of $26 million in this two-year span, how much profit did Crocs have in 2010? (Source: google.com/finance) In 2010, Crocs had a profit of $68 million.

92. **AIRLINES** In 2008, Jet Blue Airways lost $84 million. In 2009, the company made a profit. If the company lost a total of $23 million in this two-year span, how much profit did Jet Blue make in 2009? (Source: google.com/finance) In 2009, Jet Blue made $61 million in profit.

93. **MARKET SHARE** After its first year of business, a manufacturer of smoke detectors found its market share 43 points behind the industry leader. Five years later, it trailed the leader by only 9 points. How many points of market share did the company gain over this five-year span? The company gained 34 points of market share in five years.

94. **POLLS** Six months before an election, a political candidate was 31 points behind in the polls. Two days before the election, polls showed that his support had skyrocketed; he was now only 2 points behind. How much support had he gained over the six-month period? The candidate gained 29 points in six months.

95. **CHECKING ACCOUNTS** After he made deposits of $95 and $65, a student's account was still $15 overdrawn. What was his checking account balance before the deposit? His checking account balance before the deposit was −$175.

96. **WEATHER FORECASTS** The weather forecast for Fairbanks, Alaska, warned listeners that the daytime high temperature of 2° below zero would drop to a nighttime low of 28° below. By how many degrees did the temperature fall overnight? The temperature fell 26° overnight.

97. **FOOTBALL STATISTICS** Most football teams keep track of how many yards their offense gains or loses by rushing (running) and by passing the ball during a game. Then they combine those two numbers to find the total yards gained (or lost). The chart below shows the statistics for a game in 1943 between the Detroit Lions and the Chicago Cardinals in which Detroit set the NFL record for *fewest rushing yards in a game*. Incredibly, the Lions still won the game 7-0. Find the number of yards Detroit had rushing that day. (Source: pro-football-reference.com) Detroit had −53 yards rushing that day.

Detroit Lions	7	vs.	Chicago Cardinals	0
		October 10, 1943		
	Team Stats: Detroit Lions			
Passing yards		Rushing yards		Total yards
189		?		136

98. **ROLLER COASTERS** The end of a roller-coaster ride consists of a steep plunge from a peak 145 feet above ground level. The car then comes to a screeching halt in a cave that is 25 feet below ground level. How many feet does the roller coaster drop at the end of the ride shown in the illustration below? The roller coaster drops 170 feet at the end of the ride.

99. **THE ROMAN EMPIRE** Historians usually date the beginning of the Roman Empire as 27 B.C. The date given for the fall of the Roman Empire is 476 A.D. For how many years did the Roman Empire last? The Roman Empire lasted for 503 years.

100. **HISTORY** The Roman–Persian wars were a series of conflicts between the Greco-Roman world and two Iranian empires that began in 92 B.C. and finally concluded in 627 A.D. For how many years did the Roman–Persian wars last? The Roman–Persian wars lasted for 719 years.

101. **BOOKSTORES** Refer to the graph below. Find the missing net income for Barnes & Noble for the year 2011. In 2011, Barnes & Noble lost $74 million.

Source: wikinvest.com

102. **MERCURY** The freezing point of mercury is −38°F. By how many degrees must it be heated to reach its boiling point, which is 674°F? It must be heated 712°F.

WRITING

103. Explain why the variable is not isolated in the equation $-x = 10$.

104. Explain the two-step process to solve the equation $-3x + 6 = -9$. What properties of equality are used?

REVIEW

105. Write the product using exponents.
$$(-4)(-4)(-4)(-4)(-4) \quad (-4)^5$$

106. How can the addition $2 + 2 + 2 + 2 + 2$ be represented using multiplication? $5 \cdot 2$

107. Perform the division, if possible: $\dfrac{0}{-17}$ 0

108. Find the prime factorization of 126. $2 \cdot 3^2 \cdot 7$

109. Subtract: $10,000 - 782$ 9,218

110. Divide: $54\overline{)2,303}$ 42 R 35

111. Add: $23 + 234 + 2,345 + 23,456$ 26,058

112. Evaluate the expression: $26 - 15 \div 5$ 23

STUDY SKILLS CHECKLIST

Do You Know the Basics?

The key to mastering the material in Chapter 2 is to know the basics. Put a checkmark in the box if you can answer "yes" to the statement.

☐ I understand order on the number line:
$$-4 < -3 \quad \text{and} \quad -15 > -20$$

☐ I know how to add two integers that have the same sign.
- The sum of two positive numbers is *positive*.
$$4 + 5 = 9$$
- The sum of two negative numbers is *negative*.
$$-4 + (-5) = -9$$

☐ I know how to add two integers that have different signs.
- If the positive integer has the larger absolute value, the sum is positive.
$$-7 + 11 = 4$$
- If the negative integer has the larger absolute value, the sum is negative.
$$12 + (-20) = -8$$

☐ I know how to use the subtraction rule: *Subtraction is the same as addition of the opposite.*
$$-2 - (-7) = -2 + 7 = 5$$
and
$$-9 - 3 = -9 + (-3) = -12$$

☐ I know that the rules for multiplying and dividing two integers are the same.
- Like signs: positive result
$$(-2)(-3) = 6 \quad \text{and} \quad \dfrac{-15}{-3} = 5$$
- Unlike signs: negative result
$$2(-3) = -6 \quad \text{and} \quad \dfrac{-15}{3} = -5$$

☐ I know the meaning of a − symbol:
$$-(-6) = 6 \quad -|-6| = -6$$

Teaching Guide: Refer to the Instructor's Resource Binder to find activities, worksheets on key concepts, more examples, instruction tips, overheads, and assessments.

Unless otherwise noted, all art on this page is © Cengage Learning.

CHAPTER 2 SUMMARY AND REVIEW

SECTION 2.1 An Introduction to the Integers

DEFINITIONS AND CONCEPTS	EXAMPLES
The collection of positive whole numbers, the negatives of the whole numbers, and 0 is called the set of **integers**.	The set of integers: $\{\ldots, -5, -4, -3, -2, -1, 0, 1, 2, 3, 4, 5, \ldots\}$
Positive numbers are greater than 0 and **negative numbers** are less than 0.	The set of positive integers: $\{1, 2, 3, 4, 5, \ldots\}$ The set of negative integers: $\{\ldots, -5, -4, -3, -2, -1\}$
Negative numbers can be represented on a **number line** by extending the line to the left and drawing an arrowhead. As we move to the right on the number line, the values of the numbers increase. As we move to the left, the values of the numbers decrease.	Graph $-1, 6, 0, -4,$ and 3 on a number line.
Inequality symbols: $\neq$ means *is not equal to* $\geq$ means *is greater than or equal to* $\leq$ means *is less than or equal to*	Each of the following statements is true: $5 \neq -3$ Read as "5 is not equal to -3." $4 \geq -6$ Read as "4 is greater than or equal to -6." $-2 \leq -2$ Read as "-2 is less than or equal to -2."
The **absolute value** of a number is the distance on a number line between the number and 0.	Find each absolute value: $\|12\| = 12$ $\|-9\| = 9$ $\|0\| = 0$
Two numbers that are the same distance from 0 on the number line, but on opposite sides of it, are called **opposites**.	The opposite of 4 is -4. The opposite of -77 is 77. The opposite of 0 is 0.
The opposite of the opposite rule The opposite of the opposite of a number is that number. For any number a, $-(-a) = a$ Read as "the opposite of the opposite of a."	Simplify each expression: $-(-6) = 6$ $-\|8\| = -8$ $-\|-26\| = -26$
The $-$ symbol is used to indicate a negative number, the opposite of a number, and the operation of subtraction.	-2 — negative 2 $-(-4)$ — the opposite of negative four $6 - 1$ — six minus one

REVIEW EXERCISES

1. Write the set of integers.
   $\{\ldots, -5, -4, -3, -2, -1, 0, 1, 2, 3, 4, 5, \ldots\}$

2. Represent each of the following situations using a signed number.
   a. a deficit of $1,200 $-\$1,200$
   b. 10 seconds before going on the air -10 sec

3. **WATER PRESSURE** Salt water exerts a pressure of approximately 29 pounds per square inch at a depth of 33 feet. Express the depth using a signed number. -33 ft

4. Graph the following integers on a number line.
   a. $-3, 0, 4, -1$

   b. the integers greater than -3 but less than 4

5. Place an $<$ or an $>$ symbol in the box to make a true statement.
   a. $0 \; > \; -7$ b. $-20 \; < \; -19$

6. Tell whether each statement is true or false.
   a. $-17 \geq -16$ false   b. $-56 \leq -56$ true

7. Find each absolute value.
   a. $|5|$ 5   b. $|-43|$ 43   c. $|0|$ 0

8. a. What is the opposite of 8? -8
   b. What is the opposite of -8? 8
   c. What is the opposite of 0? 0

9. Simplify each expression.
   a. $-|12|$ -12
   b. $-(-12)$ 12
   c. -0 0

10. Explain the meaning of each red $-$ symbol.
    a. -5 negative   b. $-(-5)$ the opposite of
    c. $-(-5)$ negative   d. $5 - (-5)$ minus

11. **LADIES PROFESSIONAL GOLF ASSOCIATION** The scores of the top six finishers of the 2012 CME Group Titleholders Tournament and their final scores related to par are listed alphabetically: Na Yeon Choi (-14), Karine Icher (-8), Brittany Lincicome (-11), Ai Miyazato (-9), So Yeon Ryu (-12), and Karrie Webb (-10). Complete the table below. Remember, in golf, the lowest score wins.

Position	Player	Score to Par
1	Na Yeon Choi	-14
2	So Yeon Ryu	-12
3	Brittany Lincicome	-11
4	Karrie Webb	-10
5	Ai Miyazato	-9
6	Karine Icher	-8

Source: lpga.com

12. **FEDERAL BUDGET** The graph shows the U.S. government's deficit/surplus budget data for the years 1980–2011.
    a. When did the first budget surplus occur? Estimate it. 1998, $60 billion
    b. In what year was there the largest surplus? Estimate it. 2000, $230 billion
    c. In what year was there the greatest deficit? Estimate it. 2009, $-\$1,420$ billion

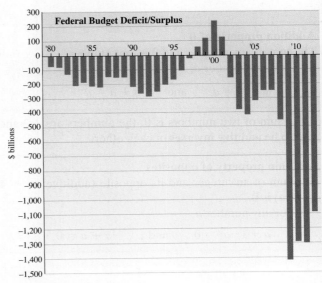

Source: U.S. Department of the Treasury

SECTION 2.2 Adding Integers

DEFINITIONS AND CONCEPTS	EXAMPLES
Adding two integers that have the same (like) signs 1. To add two positive integers, add them as usual. The final answer is positive. 2. To add two negative integers, add their absolute values and make the final answer negative.	Add: $-5 + (-10)$ Find the absolute values: $\lvert -5 \rvert = 5$ and $\lvert -10 \rvert = 10$. $-5 + (-10) = -15$ Add their absolute values, 5 and 10, to get 15. Then make the final answer negative.
Adding two integers that have different (unlike) signs To add a positive integer and a negative integer, subtract the smaller absolute value from the larger. 1. If the positive integer has the larger absolute value, the final answer is positive. 2. If the negative integer has the larger absolute value, make the final answer negative.	Add: $-7 + 12$ Find the absolute values: $\lvert -7 \rvert = 7$ and $\lvert 12 \rvert = 12$. $-7 + 12 = 5$ Subtract the smaller absolute value from the larger: $12 - 7 = 5$. Since the positive number, 12, has the larger absolute value, the final answer is positive. Add: $-8 + 3$ Find the absolute values: $\lvert -8 \rvert = 8$ and $\lvert 3 \rvert = 3$. $-8 + 3 = -5$ Subtract the smaller absolute value from the larger: $8 - 3 = 5$. Since the negative number, -8, has the larger absolute value, make the final answer negative.
To **evaluate expressions** that contain several additions, we make repeated use of the rules for adding two integers.	Evaluate: $-7 + 1 + (-20) + 1$ Perform the additions working left to right. $\begin{aligned}-7 + 1 + (-20) + 1 &= -6 + (-20) + 1\\ &= -26 + 1\\ &= -25\end{aligned}$
We can use the **commutative** and **associative properties of addition** to *reorder* and *regroup* addends.	Another way to evaluate this expression is to add the negatives and add the positives separately. Then add those results. $\quad\quad\quad\quad\quad\quad\quad\quad\quad\text{Negatives}\quad\text{Positives}$ $-7 + 1 + (-20) + 1 = [-7 + (-20)] + (1 + 1)$ $\quad\quad\quad\quad\quad\quad\quad\quad = -27 + 2$ $\quad\quad\quad\quad\quad\quad\quad\quad = -25$
Addition property of 0 The sum of any number and 0 is that number. For any number a, $\quad a + 0 = a \quad \text{and} \quad 0 + a = a$	$-2 + 0 = -2 \quad \text{and} \quad 0 + (-25) = -25$
If the sum of two numbers is 0, the numbers are said to be **additive inverses** of each other.	3 and -3 are *additive inverses* because $3 + (-3) = 0$.
Addition property of opposites The sum of an integer and its opposite (additive inverse) is 0. For any number a, $\quad a + (-a) = 0 \quad \text{and} \quad -a + a = 0$	$4 + (-4) = 0 \quad \text{and} \quad -712 + 712 = 0$

At certain times, the **addition property of opposites** can be used to make addition of several integers easier.

Evaluate: $14 + (-9) + 8 + 9 + (-14)$

Locate pairs of opposites and add them to get 0.

$$14 + (-9) + 8 + 9 + (-14) = 0 + 0 + 8$$
$$= 8$$

The sum of any integer and 0 is that integer.

REVIEW EXERCISES

Add.

13. $-6 + (-4)$ -10

14. $-3 + (-6)$ -9

15. $-28 + 60$ 32

16. $93 + (-20)$ 73

17. $-8 + 8$ 0

18. $73 + (-73)$ 0

19. $-1 + (-4) + (-3)$ -8

20. $3 + (-2) + (-4)$ -3

21. $[7 + (-9)] + (-4 + 16)$ 10

22. $(-2 + 11) + [(-5) + 4]$ 8

23. $-4 + 0$ -4

24. $0 + (-20)$ -20

25. $-2 + (-1) + (-76) + 1 + 2$ -76

26. $-5 + (-31) + 9 + (-9) + 5$ -31

27. Find the sum of -102, 73, and -345. -374

28. What is $3{,}187$ more than -59? $3{,}128$

29. What is the additive inverse of each number?
   a. -11 11
   b. 4 -4

30. a. Is the sum of two positive integers always positive? yes

   b. Is the sum of two negative integers always negative? yes

   c. Is the sum of a positive integer and a negative integer always positive? no

   d. Is the sum of a positive integer and a negative integer always negative? no

31. DROUGHT During a drought, the water level in a reservoir fell to a point 100 feet below normal. After a lot of rain in April it rose 16 feet, and after even more rain in May it rose another 18 feet.

   a. Express the water level of the reservoir before the rainy months as a signed number. -100 ft

   b. What was the water level after the rain? -66 ft

32. TEMPERATURE EXTREMES The world record for lowest temperature is $-129°$F. It was set on July 21, 1983, in Antarctica. The world record for highest temperature is an amazing $263°$F warmer. It was set on July, 1913, in Death Valley, California. Find the record high temperature. (Source: *The World Almanac and Book of Facts*, 2013) $134°$F

SECTION 2.3 Subtracting Integers

DEFINITIONS AND CONCEPTS	EXAMPLES
The **rule for subtraction** is helpful when subtracting signed numbers. To subtract two integers, add the first integer to the opposite of the integer to be subtracted. *Subtracting is the same as adding the opposite.* For any numbers a and b, $$a - b = a + (-b)$$	Subtract: $3 - (-5)$ $$3 - (-5) = 3 + 5 = 8$$ Use the rule for adding two integers with the same sign. Check using addition: $8 + (-5) = 3$
After rewriting a subtraction as addition of the opposite, use one of the rules for the addition of signed numbers discussed in Section 2.2 to find the result.	Subtract: $-3 - 5 = -3 + (-5) = -8$ Add the opposite of 5, which is -5. $-4 - (-7) = -4 + 7 = 3$ Add the opposite of -7, which is 7.
Be careful when translating the instruction to subtract one number *from* another number.	Subtract -6 from -9. $-9 - (-6)$ The number to be subtracted is -6.

Expressions can involve repeated subtraction or combinations of subtraction and addition. To evaluate them, we use the **order of operations rule** discussed in Section 1.7.	Evaluate: $-43 - (-6 - 15)$ $-43 - (-6 - 15) = -43 - [-6 + (-15)]$ Within the parentheses, add the opposite of 15, which is -15. $\qquad\qquad\qquad\quad = -43 - [-21]$ Within the brackets, add -6 and -15. $\qquad\qquad\qquad\quad = -43 + 21$ Add the opposite of -21, which is 21. $\qquad\qquad\qquad\quad = -22$ Use the rule for adding integers that have different signs.
When we find the difference between the maximum value and the minimum value of a collection of measurements, we are finding the **range** of the values. Range = maximum value − minimum value	GEOGRAPHY The highest point in the United States is Mt. McKinley at 20,230 feet. The lowest point is −282 feet at Death Valley, California. Find the range between the highest and lowest points. Range = 20,320 − (−282) Range = 20,320 + 282 Add the opposite of −282, which is 282. Range = 20,602 Do the addition. The range between the highest point and lowest point in the United States is 20,602 feet.
To find the **change** in a quantity, we subtract the *earlier value* from the *later value*. Change = later value − earlier value	SUBMARINES A submarine was traveling at a depth of 165 feet below sea level. The captain ordered it to a new position of only 8 feet below the surface. Find the change in the depth of the submarine. We can represent 165 feet below sea level as −165 feet and 8 feet below the surface as −8 feet. Change of depth = −8 − (−165) Subtract the earlier depth from the later depth. Change of depth = −8 + 165 Add the opposite of −165, which is 165. Change of depth = 157 Use the rule for adding integers that have different signs. The change in the depth of the submarine was 157 feet.

REVIEW EXERCISES

33. Fill in the blank: Subtracting an integer is the same as adding the ___opposite___ of that integer.

34. Write each phrase using symbols.
   a. negative nine minus negative one. $-9 - (-1)$
   b. negative ten subtracted from negative six
   $\qquad -6 - (-10)$

Subtract.

35. $5 - 8$ -3
36. $-9 - 12$ -21
37. $-4 - (-8)$ 4
38. $-8 - (-2)$ -6
39. $-6 - 106$ -112
40. $-7 - 1$ -8
41. $0 - 37$ -37
42. $0 - (-30)$ 30

Evaluate each expression.

43. $12 - 2 - (-6)$ 16
44. $-16 - 9 - (-1)$ -24
45. $-9 - 7 + 12$ -4
46. $-5 - 6 + 33$ 22

47. $1 - (2 - 7)$ 6
48. $-12 - (6 - 10)$ -8
49. $-70 - [(-6) - 2]$ -62
50. $89 - [(-2) - 12]$ 103
51. $-(-5) + (-28) - 2 - (-100)$ 75
52. a. Subtract 27 from -50. -77
   b. Subtract -50 from 27. 77

Use signed numbers to solve each problem.

53. MINING Some miners discovered a small vein of gold at a depth of 150 feet. This encouraged them to continue their exploration. After descending another 75 feet, they came upon a much larger find. Use a signed number to represent the depth of the second discovery. -225 ft

54. RECORD TEMPERATURES The lowest and highest recorded temperatures for Alaska and Virginia are shown. For each state, find the range between the record high and low temperatures.

Alaska		Virginia	
Low:	−80° Jan. 23, 1971	Low:	−30° Jan. 22, 1985
High:	100° June 27, 1915	High:	110° July 15, 1954

180° 140°

55. POLITICS On July 20, 2007, a CNN/Opinion Research poll had Barack Obama trailing Hillary Clinton in the South Carolina Democratic presidential primary race by 16 points. On January 26, 2008, Obama finished 28 points ahead of Clinton in the actual primary. Find the point change in Barack Obama's support. 44 points

56. OVERDRAFT FEES A student had a balance of $255 in her checking account. She wrote a check for rent for $300, and when it arrived at the bank she was charged an overdraft fee of $35. What is the new balance in her account? −$80

SECTION 2.4 Multiplying Integers

DEFINITIONS AND CONCEPTS	EXAMPLES
Multiplying two integers that have different (unlike) signs To multiply a positive integer and a negative integer, multiply their absolute values. Then make the final answer negative. **Multiplying two integers that have the same (like) signs** To multiply two integers that have the same sign, multiply their absolute values. The final answer is positive.	Multiply: $6(-8)$ Find the absolute values: $\|6\| = 6$ and $\|-8\| = 8$. $6(-8) = -48$ Multiply the absolute values, 6 and 8, to get 48. Then make the final answer negative. Multiply: $-2(-7)$ Find the absolute values: $\|-2\| = 2$ and $\|-7\| = 7$. $-2(-7) = 14$ Multiply the absolute values, 2 and 7, to get 14. The final answer is positive.
To **evaluate expressions** that contain several multiplications, we make repeated use of the rules for multiplying two integers. Another approach to evaluate expressions is to use the **commutative** and/or **associative properties of multiplication** to reorder and regroup the factors in a helpful way.	Evaluate $-5(3)(-6)$ in two ways. Perform the multiplications, working left to right. $-5(3)(-6) = -15(-6)$ $= 90$ First, multiply the pair of negative factors. $-5(3)(-6) = 30(3)$ Multiply the negative factors to produce a positive product. $= 90$
Multiplying an even and an odd number of negative integers The product of an even number of negative integers is positive. The product of an odd number of negative integers is negative.	Four negative factors: $-5(-1)(-6)(-2) = 60$ (positive) Five negative factors: $-2(-4)(-3)(-1)(-5) = -120$ (negative)
Even and odd powers of a negative integer When a negative integer is raised to an even power, the result is positive. When a negative integer is raised to an odd power, the result is negative.	Evaluate: $(-3)^4 = (-3)(-3)(-3)(-3)$ The exponent is even. $= 9(9)$ Multiply pairs of integers. $= 81$ The answer is positive. Evaluate: $(-2)^3 = (-2)(-2)(-2)$ The exponent is odd. $= -8$ The answer is negative.

Although the exponential expressions $(-6)^2$ and -6^2 look similar, they are not the same. The bases are different.

Evaluate: $(-6)^2$ and -6^2

Because of the parentheses, the base is -6. The exponent is 2.

$$(-6)^2 = (-6)(-6)$$
$$= 36$$

Since there are no parentheses around -6, the base is 6. The exponent is 2.

$$-6^2 = -(6 \cdot 6)$$
$$= -36$$

Application problems that involve **repeated addition** are often more easily solved using multiplication.

CHEMISTRY A chemical compound that is normally stored at 0°F had its temperature lowered 8°F each hour for 6 hours. What signed number represents the change in temperature of the compound after 6 hours?

$-8 \cdot 6 = -48$ Multiply the change in temperature each hour by the number of hours.

The change in temperature of the compound is $-48°$F.

REVIEW EXERCISES

Multiply.

57. $7(-2)$ -14
58. $(-8)(47)$ -376
59. $-23(-14)$ 322
60. $-5(-5)$ 25
61. $-1 \cdot 25$ -25
62. $(6)(-34)$ -204
63. $-4,000(17,000)$ $-68,000,000$
64. $-100,000(-300)$ $30,000,000$
65. $(-6)(-2)(-3)$ -36
66. $-4(-3)(-3)$ -36
67. $(-3)(4)(2)(-5)$ 120
68. $(-1)(10)(10)(-1)$ 100

69. Find the product of -15 and the opposite of 30. 450

70. Find the product of the opposite of 16 and the opposite of 3. 48

71. DEFICITS A state treasurer's prediction of a tax shortfall was two times worse than the actual deficit of $130 million. The governor's prediction of the same shortfall was even worse—three times the amount of the actual deficit. Complete the labeling of the vertical axis of the graph in the next column to show the two incorrect predictions. $-260, -390$

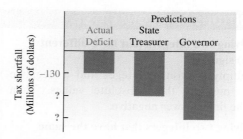

72. MINING An elevator is used to lower coal miners from the ground level entrance to various depths in the mine. The elevator stops every 45 vertical feet to let off miners. At what depth do the miners work who get off the elevator at the 12th stop? -540 ft

Evaluate each expression.

73. $(-5)^3$ -125
74. $(-2)^5$ -32
75. $(-8)^4$ $4,096$
76. $(-4)^4$ 256

77. When $(-17)^9$ is evaluated, will the result be positive or negative? negative

78. Explain the difference between -9^2 and $(-9)^2$ and then evaluate each expression. In the first expression, the base is 9. In the second expression, the base is -9. $-81, 81$

SECTION 2.5 Dividing Integers

DEFINITIONS AND CONCEPTS

Dividing two integers
To divide two integers, divide their absolute values.

1. The quotient of two integers that have the same (*like*) signs is positive.
2. The quotient of two integers that have different (*unlike*) signs is negative.

To **check division** of integers, multiply the *quotient* and the *divisor*. You should get the *dividend*.

EXAMPLES

Divide: $\dfrac{-21}{-7}$

Find the absolute values: $|-21| = 21$ and $|-7| = 7$.

$\dfrac{-21}{-7} = 3$ Divide the absolute values, 21 by 7, to get 3. The final answer is positive.

Check: $3(-7) = -21$ The result checks.

	Divide: $-54 \div 9$ Find the absolute values: $\|-54\| = 54$ and $\|9\| = 9$. $-54 \div 9 = -6$ Divide the absolute values, 54 by 9, to get 6. Then make the final answer negative. **Check:** $-6(9) = -54$ The result checks.
Division with 0 If 0 is divided *by* any nonzero number, the quotient is 0. For any nonzero number a, $$\frac{0}{a} = 0$$ Division *of* any nonzero number by 0 is undefined. For any nonzero number a, $$\frac{a}{0} \text{ is undefined.}$$	$\frac{0}{-8} = 0 \qquad 0 \div (-20) = 0$ $\frac{-2}{0}$ is undefined. $-6 \div 0$ is undefined.
Problems that involve forming **equal-sized groups** can be solved by division.	USED CAR SALES The price of a used car was reduced each day by an equal amount because it was not selling. After 7 days, and a $1,050 reduction in price, the car was finally purchased. By how much was the price of the car reduced each day? $$\frac{-1,050}{7} = -150 \quad \text{Divide the change in the price of the car by the number of days the price was reduced.}$$ The negative result indicates that the price of the car was *reduced* by $150 each day.

REVIEW EXERCISES

79. Fill in the blanks: We know that $\frac{-15}{5} = -3$ because $-3\ (\ 5\ ) = -15$.

80. Check using multiplication to determine whether $-152 \div (-8) = 18$. The answer is incorrect: $18(-8) \neq -152$

Divide, if possible.

81. $\frac{25}{-5}$ -5

82. $\frac{-14}{7}$ -2

83. $-64 \div (-8)$ 8

84. $72 \div (-9)$ -8

85. $\frac{-10}{-1}$ 10

86. $\frac{-673}{-673}$ 1

87. $-150,000 \div 3,000$ -50

88. $-24,000 \div (-60)$ 400

89. $\frac{-1,058}{-46}$ 23

90. $-272 \div 16$ -17

91. $\frac{0}{-5}$ 0

92. $\frac{-4}{0}$ undefined

93. Divide -96 by 3. -32

94. Find the quotient of -125 and -25. 5

95. PRODUCTION TIME Because of improved production procedures, the time needed to produce an electronic component dropped by 12 minutes over the past 6 months. If the drop in production time was uniform, how much did it change each month over this period of time? -2 min

96. OCEAN EXPLORATION The Puerto Rico Trench is the deepest part of the Atlantic Ocean. It has a maximum depth of 28,374 feet. If a remote-controlled unmanned submarine is sent to the bottom of the trench in a series of 6 equal dives, how far will the vessel descend on each dive? (Source: marianatrench.com) $-4,729$ ft

SECTION 2.6 Order of Operations and Estimation

DEFINITIONS AND CONCEPTS

Order of operations

1. Perform all calculations within parentheses and other grouping symbols following the order listed in Steps 2–4 below, working from the innermost pair of grouping symbols to the outermost pair.
2. Evaluate all exponential expressions.
3. Perform all multiplications and divisions as they occur from left to right.
4. Perform all additions and subtractions as they occur from left to right.

When grouping symbols have been removed, repeat Steps 2–4 to complete the calculation.

If a fraction bar is present, evaluate the expression above the bar (called the **numerator**) and the expression below the bar (called the **denominator**) separately. Then perform the division indicated by the fraction bar, if possible.

EXAMPLES

Evaluate: $-3(-5)^2 - (-40)$

$-3(-5)^2 - (-40) = -3(25) - (-40)$ Evaluate the exponential expression.
$= -75 - (-40)$ Do the multiplication.
$= -75 + 40$ Use the subtraction rule: Add the opposite of -40.
$= -35$ Do the addition.

Evaluate: $\dfrac{-6 + 4(-2)}{16 - (-3)^2}$

$\dfrac{-6 + 4(-2)}{16 - (-3)^2} = \dfrac{-6 + (-8)}{16 - 9}$ In the numerator, do the multiplication. In the denominator, evaluate the exponential expression.
$= \dfrac{-14}{7}$ In the numerator, do the addition. In the denominator, do the subtraction.
$= -2$ Do the division.

Absolute value symbols are grouping symbols, and by the order of operations rule, all calculations within grouping symbols must be performed first.

Evaluate: $10 - 2|-8 + 1|$

$10 - 2|-8 + 1| = 10 - 2|-7|$ Do the addition within the absolute value symbol.
$= 10 - 2(7)$ Find the absolute value of -7.
$= 10 - 14$ Do the multiplication.
$= -4$ Do the subtraction.

When an exact answer is not necessary and a quick approximation will do, we can use **estimation**.

Estimate the value of $-56 + (-67) + 89 + (-41) + 14$ by rounding each number to the nearest ten.

$-60 + (-70) + 90 + (-40) + 10$
$= -170 + 100$ Add the positives and the negatives separately.
$= -70$ Do the addition.

REVIEW EXERCISES

Evaluate each expression.

97. $7 - 8 \cdot 6$ -41
98. $7 - (-2)^2 + 1$ 4
99. $65 - 8(9) - (-47)$ 40
100. $-3(-2)^3 - 16$ 8
101. $-2(5)(-4) + \dfrac{|-9|}{3^2}$ 41
102. $-4^2 + (-4)^2$ 0
103. $-12 - (8 - 9)^2$ -13
104. $7|-8| - 2(3)(4)$ 32
105. $-4\left(\dfrac{15}{-3}\right) - 2^3$ 12
106. $-20 + 2(12 - 5 \cdot 2)$ -16
107. $-20 + 2[12 - (-7 + 5)^2]$ -4
108. $8 - 6|-3 \cdot 4 + 5|$ -34
109. $\dfrac{2 \cdot 5 + (-6)}{-3 - 1^5}$ -1
110. $\dfrac{3(-6) - 11 + 1}{4^2 - 3^2}$ -4
111. $-\dfrac{1}{5}\left[1 - \left(2^3 + \dfrac{100}{-50}\right)\right]$
112. $-\dfrac{1}{55}\left[45 - \left(5^3 + \dfrac{100}{-4}\right)\right]$
113. Round each number to the nearest hundred to estimate the value of the following expression:
$-4,471 + 7,935 + 2,094 + (-3,188)$ $2,300$
114. Find the mean (average) of $-8, 4, 7, -11, 2, 0, -6,$ and -4. -2

SECTION 2.7 Solving Equations That Involve Integers

DEFINITIONS AND CONCEPTS	EXAMPLES
To **solve an equation** means to find all the values of the variable that make the equation true. To isolate the variable on one side of the equation, we use: 1. Addition property of equality 2. Subtraction property of equality 3. Multiplication property of equality 4. Division property of equality	Solve: $x + (-3) = 8$ $x + (-3) + 3 = 8 + 3$ To isolate x, undo the addition of −3 by adding 3 to both sides. $x + 0 = 11$ On the left side, the sum of a number and its opposite is 0. $x = 11$ On the left side, the sum of any number and 0 is that number. **Check:** $x + (-3) = 8$ This is the original equation. $11 + (-3) \stackrel{?}{=} 8$ Substitute 11 for x. $8 = 8$ True Since the resulting statement $8 = 8$ is true, the solution is 11.
The expressions on each side of an equation should be simplified before using any properties of equality to isolate the variable.	Solve: $-19 + 4 = y + 2(3)$ $-15 = y + 6$ On the left side, do the addition. On the right side, do the multiplication. $-15 - 6 = y + 6 - 6$ To isolate y, undo the addition of 6 by subtracting 6 from both sides. $-21 = y$ Do the subtraction. Check the result in the *original* equation to verify that -21 is the solution.
The notation $-x$ means $-1x$.	Solve: $-x = 14$ We can multiply or divide both sides by -1 to isolate x. $-1x = 14$ Write $-x$ as $-1x$. $\dfrac{-1x}{-1} = \dfrac{14}{-1}$ To isolate x, undo the multiplication by −1 by dividing both sides by −1. $x = -14$ Do the division. Check the result in the *original* equation to verify that -14 is the solution.
Sometimes we must use two (or more) properties of equality to solve more complicated equations.	Solve: $-6x + 2 = -10$ To solve the equation, we use the order of operations rule in reverse. • To isolate the variable term $-6x$, subtract 2 from both sides to undo the addition of 2. • To isolate the variable x, divide both sides by -6 to undo the multiplication by -6. $-6x + 2 - 2 = -10 - 2$ Subtract 2 from both sides to isolate −6x. $-6x = -12$ Do the subtraction. $\dfrac{-6x}{-6} = \dfrac{-12}{-6}$ Divide both sides by −6 to isolate x. $x = 2$ Do the division. Check the result in the *original* equation to verify that 2 is the solution.

We can use the concepts of variable and equation to solve application problems involving integers.

The following words are often used to indicate negative numbers.

behind	below	before	deficit
debt	drop	under	loss
in the red	B.C.	overdrawn	

BANKING After a student made a deposit of $165, his checking account was still $38 overdrawn. What was his checking account balance before the deposit?

Analyze

An *overdrawn* account balance can be represented by a negative number.
- A deposit of $165 was made.   Given
- After the deposit, the account balance was $-\$38$.   Given
- What was the account balance before the deposit?   Find

Assign Let $x =$ the account balance before the deposit.

Form The word *deposit* indicates addition. Now we translate the words of the problem to numbers and symbols.

The balance before the deposit	plus	the deposit	is equal to	the new balance.
x	$+$	165	$=$	-38

Solve
$$x + 165 = -38$$
$$x + 165 - 165 = -38 - 165$$
$$x = -203$$

State His checking account balance before the deposit was $-\$203$.

Check If we add the deposit to the original balance, we should get the new balance.
$$-\$203 + \$165 = -\$38$$

The result checks.

REVIEW EXERCISES

Solve each equation and check the result.

115. $x + (-16) = -6$ 10   **116.** $-y = -32$ 32

117. $9x - 27 = 0$ 3   **118.** $-20 + 4 = a + 3(-7)$ 5

119. $-84 = -2t$ 42   **120.** $\dfrac{n}{5} + (-2) = -7$ -25

121. $\dfrac{s}{3} = -1$ -3 **122.** $-11y - 10 = 67$ -7

123. $16 - 6n = 22$ -1 **124.** $\dfrac{n}{-2} - 27 = -27$ 0

125. $-9 - 3 - 20x = -52$ 2

126. $15 = -13b + (-11)$ -2

In Exercises 127 and 128, let a variable represent the unknown quantity. Then write and solve an equation to answer the question.

127. FINANCIAL STATEMENTS In 2009, Dole Food Company made a profit of $74 million. In 2010, they suffered a loss, and in 2011, they made a profit of $33 million. If the company made a total of $65 million in this three-year span, how much did Dole lose in 2010? (Source: wikinvest.com)   In 2010, Dole lost $42 million.

128. POLITICS Eight weeks before an election, a political candidate was 32 points behind in the polls. On election day, she narrowly lost the race by 3 points. How much support had she gained over the eight-week period?   The candidate gained 29 points in eight weeks.

CHAPTER 2 TEST

1. Fill in the blanks.
   a. $\{\ldots, -5, -4, -3, -2, -1, 0, 1, 2, 3, 4, 5, \ldots\}$ is called the set of __integers__.
   b. The symbols $>$ and $<$ are called __inequality__ symbols.
   c. The __absolute__ __value__ of a number is the distance between the number and 0 on the number line.
   d. Two numbers that are the same distance from 0 on the number line, but on opposite sides of it, are called __opposites__.
   e. In the expression $(-3)^5$, the __base__ is -3 and 5 is the __exponent__.
   f. To __solve__ an equation means to find all the values of the variable that make the equation true.
   g. To __check__ the solution of an equation, we substitute the value for the variable in the original equation and determine whether the result is a true statement.

2. Insert one of the symbols $>$ or $<$ in the blank to make the statement true.
   a. $-8 \;>\; -9$ b. $-213 \;<\; 123$ c. $-5 \;<\; 0$

3. Tell whether each statement is true or false.
   a. $19 \geq 19$ true
   b. $-(-8) = 8$ true
   c. $-|-2| > |6|$ false
   d. $-7 + 0 = 0$ false
   e. $-5(0) = 0$ true

4. SCHOOL ENROLLMENT According to the projections in the table, which high school will face the greatest shortage of classroom seats in the year 2020? Poly

High Schools with Shortage of Classroom Seats by 2020	
Lyons	-669
Tolbert	$-1,630$
Poly	$-2,488$
Cleveland	-350
Samuels	-586
South	$-2,379$
Van Owen	$-1,690$
Twin Park	-462
Heywood	$-1,004$
Hampton	-774

5. Graph the following numbers on a number line: $-3, 4, -1,$ and 3

6. Add.
   a. $-16 + 14$ -2
   b. $-72 + (-73)$ -145
   c. $8 + (-6) + (-9) + 5 + 1$ -1
   d. $(-31 + 12) + [3 + (-16)]$ -32
   e. $-24 + (-3) + 24 + (-5) + 5$ -3

7. Subtract.
   a. $-7 - 6$ -13
   b. $-7 - (-6)$ -1
   c. $82 - (-109)$ 191
   d. $0 - 15$ -15
   e. $-29 - (-60) - 71$ -40

8. Multiply.
   a. $-10 \cdot 7$ -70
   b. $(-4)(-8)$ 32
   c. $-4(2)(-6)$ 48
   d. $-9(-3)(-1)(-2)$ 54
   e. $-20{,}000(1{,}300)$ $-26{,}000{,}000$

9. Write the related multiplication statement for $\dfrac{-20}{-4} = 5$. $5(-4) = -20$

10. Divide and check the result.
    a. $\dfrac{-32}{4}$ -8
    b. $24 \div (-3)$ -8
    c. $-54 \div (-6)$ 9
    d. $\dfrac{408}{-12}$ -34
    e. $-560{,}000 \div 7{,}000$ -80

11. a. What is 15 more than -27? -12
    b. Subtract -19 from -1. 18
    c. Divide -28 by -7. 4
    d. Find the product of 10 and the opposite of 8. -80

12. a. What property is shown: $-3 + 5 = 5 + (-3)$
       commutative property of addition
    b. What property is shown: $-4(-10) = -10(-4)$
       commutative property of multiplication
    c. Fill in the blank: Subtracting is the same as __adding__ the opposite.

13. Divide, if possible.
    a. $\dfrac{-21}{0}$ undefined
    b. $\dfrac{-5}{1}$ -5
    c. $\dfrac{0}{-6}$ 0
    d. $\dfrac{-18}{-18}$ 1

14. Evaluate each expression:
    a. $(-4)^2$ 16
    b. -4^2 -16

Evaluate each expression.

15. $4 - (-3)^2 - (-6)$ 1   **16.** $-18 \div 2 \cdot 3$ -27

17. $-3 + \left(\dfrac{-16}{4}\right) - 3^3$ -34 **18.** $94 - 3[-7 + (5 - 8)^2]$ 88

19. $\dfrac{4(-6) - 4^2 + (-2)}{-3 - 4 \cdot 1^5}$ 6   **20.** $6(-2 \cdot 6 + 5 \cdot 4)$ 48

21. $21 - 9|-3 - 4 + 2|$ -24 **22.** $-\left[2 - \left(4^3 + \dfrac{20}{-5}\right)\right]$ 58

23. CHEMISTRY In a lab, the temperature of a fluid was reduced 6°F per hour for 12 hours. What signed number represents the change in temperature? $-72°F$

24. GAMBLING On the first hand of draw poker, a player won the chips shown on the left. On the second hand, he lost the chips shown on the right. Determine his net gain or loss for the first two hands. The dollar value of each colored poker chip is shown. $203 lost ($-203$)

25. GEOGRAPHY The lowest point on the African continent is the Qattarah Depression in the Sahara Desert, 436 feet below sea level. The lowest point on the North American continent is Death Valley, California, 282 feet below sea level. Find the difference in these elevations. 154 ft

26. TRAMS A tram line makes a 5,250-foot descent from a mountaintop to the base of the mountain in 15 equal stages. How much does it descend in each stage? -350 ft

27. CARD GAMES After the first round of a card game, Tommy had a score of 8. When he lost the second round, he had to deduct the value of the cards left in his hand from his first-round score. (See the illustration.) What was his score after two rounds of the game? For scoring, face cards (Kings, Queens, and Jacks) are counted as 10 points and aces as 1 point. -15

28. BANK TAKEOVERS Before three investors can take over a failing bank, they must repay the losses that the bank had over the past three quarters, as shown in the next column. If the investors plan equal ownership, how much of the bank's total losses is each investor responsible for? $-$60$ million

Bank Losses

Solve each equation and check the result.

29. $\dfrac{x}{-4} = 10$ -40 **30.** $-10 = 6 - x$ 16

31. $c - (-7) = -8$ -15 **32.** $-6x = 0$ 0

33. $3x + (-7) = -11 + (-11)$ -5

34. $-a = 38$ -38

35. $-5 = -6s + 7$ 2   **36.** $\dfrac{x}{-2} + 3 = (-2)(-6)$ -18

Let a variable represent the unknown quantity. Then write and solve an equation to answer each question.

37. BANKING After making deposits of $125 and $100, a student's account was still $19 overdrawn. What was her account balance before the deposits? Her account balance before the deposits was $-$244$.

38. ELEVATORS The weight of the passengers on board an elevator as it traveled from the first to the second floor was 165 pounds *under* capacity. When the doors opened at the second floor, no one exited, and several people entered. The weight of the passengers in the elevator was then 85 pounds *over* capacity. What was the weight of the people who boarded the elevator on the second floor? The weight of the people who boarded the elevator on the second floor was 250 pounds.

39. WORLD'S COLDEST ICE CREAM Dippin'Dots is an ice cream snack that is produced by flash freezing an ice cream mixture in liquid nitrogen. The temperature at which the multicolored dots are produced is about nine times as cold as the temperature at which they are stored for sale. If Dippin'Dots are produced at $-360°F$, at what temperature are they stored? Dippin'Dots are stored at $-40°F$

40. SPRINT NEXTEL If you divide the 2007 net income of Sprint Nextel by ten, the result is approximately the net income for the company in 2011. If the company lost $3 billion in 2011, what was its net income in 2007? (Source: wikinvest.com) The net income for Sprint Nextel in 2007 was $-$30$ billion.

CHAPTERS 1–2 CUMULATIVE REVIEW

1. Consider the number 7,326,549. [Section 1.1]
   a. What is the place value of the digit 7? 7 millions
   b. Which digit is in the hundred thousands column? 3
   c. Round to the nearest hundred. 7,326,500
   d. Round to the nearest ten thousand. 7,330,000

2. BIDS A school district received the bids shown in the table for electrical work. If the lowest bidder wins, which company should be awarded the contract? [Section 1.1] CRF Cable

   Citrus Unified School District Bid 02-9899
   Cabling and Conduit Installation

Datatel	$2,189,413
Walton Electric	$2,201,999
Advanced Telecorp	$2,175,081
CRF Cable	$2,174,999
Clark & Sons	$2,175,801

3. NUCLEAR POWER The table gives the number of nuclear power plants operating in the United States for selected years. Complete the bar graph using the given data. [Section 1.1]

Year	1979	1984	1989	1994	1999	2004	2009
Plants	69	87	111	109	104	104	104

   Source: allcountries.org and *The World Almanac and Book of Facts*, 2013

4. THREAD COUNT The thread count of a fabric is the sum of the number of horizontal and vertical threads woven in one square inch of fabric. One square inch of a bed sheet is shown below. Find the thread count. [Section 1.2] 360

Add. [Section 1.2]

5. 1,237 + 68 + 549 1,854

6. 8,907
   2,345
   7,899
   + 5,237
   ———
   24,388

Subtract. [Section 1.2]

7. 6,375 − 2,569 3,806

8. 5,369
   − 685
   ———
   4,684

9. 39,506
   − 1,729
   ———
   37,777

10. Subtract 304 from 1,736. [Section 1.2] 1,432

11. Check the subtraction below using addition. Is it correct? [Section 1.2] no

    469
    − 237
    ———
    132

12. SHIPPING FURNITURE In a shipment of 147 pieces of furniture, 27 pieces were sofas, 55 were leather chairs, and the rest were wooden chairs. Find the number of wooden chairs. [Section 1.2] 65 wooden chairs

Unless otherwise noted, all art on this page is © Cengage Learning.

Multiply. [Section 1.3]

13. 435 · 27   11,745

14. 9,183
    × 602
    5,528,166

15. 3,100 · 7,000   21,700,000

16. PACKAGING There are 3 tennis balls in one can, 24 cans in one case, and 12 cases in one box. How many tennis balls are there in one box? [Section 1.3] 864 tennis balls

17. GARDENING Find the perimeter and the area of the rectangular garden shown below. [Section 1.3] 104 ft, 595 ft²

17 ft
35 ft

18. PHOTOGRAPHY The photographs below are the same except that different numbers of *pixels* (squares of color) are used to display them. The number of pixels in each row and each column of the photographs are given. Find the total number of pixels in each photograph. [Section 1.3] 25; 144; 10,000

5 pixels

12 pixels

100 pixels

Divide. [Section 1.4]

19. $\frac{701}{8}$ 87 R 5

20. 1,261 ÷ 97   13

21. 38)17,746   467

22. 350)9,800   28

23. Check the division below using multiplication. Is it correct? [Section 1.4] yes

    9)1,962 = 218

24. GARDENING A metal can holds 320 fluid ounces of gasoline. How many times can the 30-ounce tank of a lawnmower be filled from the can? How many ounces of gasoline will be left in the can? [Section 1.4] 10 times, 20 ounces

25. BAKING A baker uses 4-ounce pieces of bread dough to make dinner rolls. How many dinner rolls can he make from 15 pounds of dough? (*Hint:* There are 16 ounces in one pound.) [Section 1.4] 60 rolls

26. List the factors of 18, from least to greatest. [Section 1.5] 1, 2, 3, 6, 9, 18

27. Identify each number as a prime number, a composite number, or neither. Then identify it as an even number or an odd number. [Section 1.5]

    a. 17   prime number, odd number
    b. 18   composite number, even number
    c. 0   neither, even number
    d. 1   neither, odd number

28. Find the prime factorization of 504. Use exponents to express your answer. [Section 1.5] $2^3 \cdot 3^2 \cdot 7$

29. Write the expression 11 · 11 · 11 · 11 using an exponent. [Section 1.5] 11^4

30. Evaluate: $5^2 \cdot 7$ [Section 1.5] 175

31. Find the LCM of 8 and 12. [Section 1.6] 24

32. Find the LCM of 3, 6, and 15. [Section 1.6] 30

33. Find the GCF of 30 and 48. [Section 1.6] 6

34. Find the GCF of 81, 108, and 162. [Section 1.6] 27

Evaluate each expression. [Section 1.7]

35. 4 − 2[26 + 2(5 − 3)]   −56

36. 264 ÷ 4 − 7(4)2   10

37. $\frac{4^2 - 2 \cdot 3}{2 + (3^2 - 3 \cdot 2)}$ 2

38. SPEED CHECKS A traffic officer used a radar gun and found that the speeds of several cars traveling on Main Street were:

    38 mph, 42 mph, 36 mph, 38 mph, 48 mph, 44 mph

    What was the mean (average) speed of the cars traveling on Main Street? [Section 1.7] 41 mph

39. Use a check to determine whether 6 is a solution of the equation $x - 2 = 4$. [Section 1.8] yes

40. Tell whether each of the following is an equation. [Section 1.8]

    a. $d + 4$ no
    b. $a - 11 = 19$ yes
    c. $4 < 5$ no
    d. $\frac{x}{6} + 12$ no

Solve each equation and check the result. [Sections 1.8 and 1.9]

41. $50 = x + 37$ 13
42. $a - 12 = 41$ 53
43. $5p = 135$ 27
44. $\dfrac{y}{8} = 3$ 24

Let a variable represent the unknown quantity. Then write and solve an equation to answer the question.

45. FRANCHISES As of January 2013, Dunkin' Donuts would have had to open up 26,999 more shops to match the number of Subway stores. If there were 37,199 Subway stores, how many Dunkin' Donuts shops were there? (Source: entreprenur.com) [Section 1.8] There were 10,200 Dunkin' Donut shops.

46. STADIUMS The May Day Stadium in Pyongyang, North Korea, has the largest nonracing stadium capacity in the world: 150,000 people. This is twice the capacity for a football game at Sun Life Stadium in Miami, Florida. What is the capacity of Sun Life Stadium? (Source: www.worldstadiums.com) [Section 1.8] The capacity of Sun Life Stadium is 75,000 people.

47. Graph the following integers on a number line. [Section 2.1]

a. $-2, -1, 0, 2$

b. The integers greater than -4 but less than 2

48. Find the sum of $-11, 20, -13,$ and 1. [Section 2.2] -3

Use signed numbers to solve each problem.

49. LIE DETECTOR TESTS A burglar scored -18 on a polygraph test, a score that indicates deception. However, on a second test, he scored $+3$, a score that is uncertain. Find the change in the scores. [Section 2.3]   21

50. BANKING A student has $48 in his checking account. He then writes a check for $105 to purchase books. The bank honors the check, but charges the student an additional $22 service fee for being overdrawn. What is the student's new checking account balance? [Section 2.3] The new account balance will be $-$79.

51. CHEMISTRY The *melting point* of a solid is the temperature range at which it changes state from solid to liquid. The melting point of helium is seven times colder than the melting point of mercury. If the melting point of mercury is $-39°$ Celsius, what is the melting point of helium? (Source: chemicalelements.com) [Section 2.4] The melting point of helium is $-273°$ Celsius.

52. BUYING A BUSINESS When 12 investors decided to buy a bankrupt company, they agreed to assume equal shares of the company's debt of $660,000. How much debt was each investor responsible for? [Section 2.5] Each investor was responsible for $55,000 of debt.

Evaluate each expression. [Section 2.6]

53. $5 + (-3)(-7)(-2)$ -37
54. $-2[-6(5 \cdot 1^3) - 5]$ 70
55. $\dfrac{10 - (-5)}{1 - 2 \cdot 3}$ -3
56. $\dfrac{3(-6) - 10}{3^2 - 4^2}$ 4
57. $3^4 + 6(-12 + 5 \cdot 4)$ 129
58. $15 - 2|-3 - 4|$ 1
59. $2\left(\dfrac{-12}{3}\right) + 3(-5)$ -23
60. $-9^2 + (-9)^2$ 0
61. $-\left|\dfrac{45}{-9} - (-9)\right|$ -4
62. $\dfrac{-4(-5) - 2}{3 - 3^2}$ -3

For Exercises 63 and 64, quickly determine a reasonable estimate of the exact answer. [Section 2.6]

63. CAMPING Hikers make a 1,150-foot descent into a canyon in 12 stages. What signed number represents how far they descend in each stage? -100 ft

64. RECALLS An automobile maker has to recall 19,250 cars because they have a faulty engine mount. If it costs $195 to repair each car, what signed number represents the loss the company will suffer because of the recall? $-$4,000,000

Solve each equation and check the result. [Section 2.7]

65. $-1 - h = -8$ 7
66. $8q - 7 = 25$ 4
67. $-1 + 7 = x + 2(-9)$ 24
68. $\dfrac{r}{2} + 5 = -13$ -36

Let a variable represent the unknown quantity. Then write and solve an equation to answer the question. [Section 2.7]

69. BANKING After she made deposits of $255 and $395, a business owner's account was still $85 overdrawn. What was the account balance before the deposits? The acount balance before the deposits was $-$735.

70. ALCOHOL The freezing point of ethanol alcohol is $-173°$F. By how many degrees must it be heated to reach its boiling point, which is $173°$F? (Source: about.com) It must be heated $346°$F.

3 The Language of Algebra

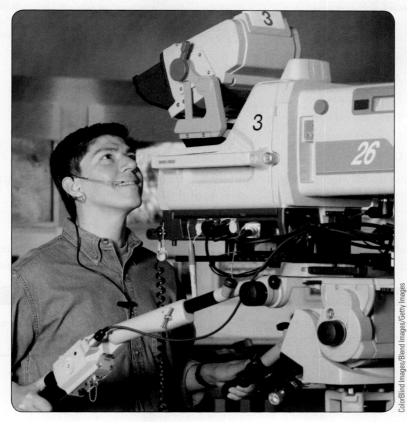

3.1 Algebraic Expressions
3.2 Evaluating Algebraic Expressions and Formulas
3.3 Simplifying Algebraic Expressions and the Distributive Property
3.4 Combining Like Terms
3.5 Simplifying Expressions to Solve Equations
3.6 Using Equations to Solve Application Problems
Chapter Summary and Review
Chapter Test
Cumulative Review

from Campus to Careers
Broadcasting Technician

It takes many people behind the scenes at television stations to make what we see and hear over the airwaves possible. Broadcast technicians set up, operate, and maintain transmitter and sound equipment. They also use computers to edit audio and video recordings. This job requires skills in programming and operating electronic equipment, and the mathematical ability to analyze data.

In **Problem 49** of **Study Set 3.6,** you will see how a television technician determines the amount of commercial time and program time he should schedule for a 30-minute time slot.

JOB TITLE: Broadcasting Technician
EDUCATION: Broadcasting jobs in large markets are usually offered to individuals who have an associate's degree or vocational certificate.
JOB OUTLOOK: Employment in broadcasting is expected to increase about 10 percent over the 2010–2020 period.
ANNUAL EARNINGS: Ranges from a low of $18,000 for an entry-level position to $75,000 for top positions.
FOR MORE INFORMATION:
http://www.bls.gov/ooh/media-and-communication

Objectives

1. Translate word phrases to algebraic expressions.
2. Write algebraic expressions to represent unknown quantities.

SECTION 3.1
Algebraic Expressions

In Chapter 1, we introduced the following strategy for solving application problems.

1. Analyze the problem.
2. Assign a variable.
3. Form an equation.
4. Solve the equation.
5. State the conclusion.
6. Check the result.

To successfully form an equation in step 3 of the strategy, we must be able to translate English words and phrases into mathematical symbols.

1 Translate word phrases to algebraic expressions.

Recall that a **variable** is a letter (or symbol) that stands for a number. When we combine variables and numbers using arithmetic operations, the result is an *algebraic expression*.

> **Algebraic Expressions**
>
> Variables and/or numbers can be combined with the operations of addition, subtraction, multiplication, and division to create **algebraic expressions**.

Here are some examples of algebraic expressions.

$4a + 7$ This expression is a combination of the numbers 4 and 7, the variable a, and the operations of multiplication and addition.

$\dfrac{10 - y}{3}$ This expression is a combination of the numbers 10 and 3, the variable y, and the operations of subtraction and division.

Algebraic expressions can contain two (or more) variables.

$15mn(2m)$ This expression is a combination of the numbers 15 and 2, the variables m and n, and the operation of multiplication.

> **The Language of Algebra**   We often refer to *algebraic expressions* as simply *expressions*.

    In order to solve application problems, which are almost always given in words, we must translate those words into mathematical symbols. The following tables show how key words and phrases can be translated into algebraic expressions.

Addition	
the sum of a and 8	$a + 8$
4 plus c	$4 + c$
16 added to m	$m + 16$
4 more than t	$t + 4$
20 greater than F	$F + 20$
T increased by r	$T + r$
exceeds y by 35	$y + 35$

Subtraction	
the difference of 23 and P	$23 - P$
550 minus h	$550 - h$
18 less than w	$w - 18$
7 decreased by j	$7 - j$
M reduced by x	$M - x$
12 subtracted from L	$L - 12$
5 less f	$5 - f$

Unless otherwise noted, all art on this page is © Cengage Learning.

Caution! Be careful when translating subtraction. Order is important. For example, when a translation involves the phrase *less than*, note how the terms are reversed.

18 less than w

w − 18

Multiplication	
the product of 4 and *x*	4*x*
20 times *B*	20*B*
twice *r*	2*r*
double the amount *a*	2*a*
triple the profit *P*	3*P*
three-fourths of *m**	$\frac{3}{4}m$

Division	
the quotient of *R* and 19	$\frac{R}{19}$
s divided by *d*	$\frac{s}{d}$
k split into 4 equal parts	$\frac{k}{4}$
the ratio of *c* to *d**	$\frac{c}{d}$

* This translation is discussed in more detail in Chapter 4.

* This translation is discussed in more detail in Chapter 6.

Caution! Be careful when translating division. As with subtraction, order is important. For example, *s* divided by *d* is *not* written $\frac{d}{s}$.

EXAMPLE 1
Write each phrase as an algebraic expression:
a. twice the profit *P*
b. 5 less than the capacity *c*
c. the product of the weight *w* and 2,000, increased by 300

Strategy We will begin by identifying any key words or phrases.

WHY Key words or phrases can be translated to mathematical symbols.

Solution
a. Key word: twice   **Translation:** multiplication by 2
The algebraic expression is: 2*P*.

b. Key phrase: *less than*   **Translation:** subtraction
Sometimes thinking in terms of specific numbers makes translating easier. Suppose the capacity was 100. Then 5 *less than* 100 would be 100 − 5. If the capacity is *c*, then we need to make *c* 5 less. The algebraic expression is: *c* − 5.

Caution! 5 < *c* is the translation of the statement 5 *is less than* the capacity *c* and not 5 *less than* the capacity *c*.

c. Key phrase: *product of*   **Translation:** multiplication
   Key phrase: *increased by*   **Translation:** addition
In the given wording, the comma after 2,000 means *w* is first multiplied by 2,000; then 300 is added to that product. The algebraic expression is: 2,000*w* + 300.

Self Check 1
Write each phrase as an algebraic expression:
a. 80 less than the total, *t*   *t* − 80
b. $\frac{2}{3}$ of the time *T* $\frac{2}{3}T$
c. the difference of twice *a* and 15, squared $(2a - 15)^2$

Now Try Problems 15, 17, and 23

Teaching Example 1 Write each phrase as an algebraic expression:
a. 15 less than the population *p*
b. twice the weight *w*
c. the product of the volume *V* and 3, increased by 60
Answers:
a. *p* − 15
b. 2*w*
c. 3*V* + 60

Unless otherwise noted, all art on this page is © Cengage Learning.

2 Write algebraic expressions to represent unknown quantities.

To solve application problems, we let a variable stand for an unknown quantity. We can use the translation skills just discussed to describe any other unknown quantities in the problem by using algebraic expressions.

Self Check 2

COMMUTING It takes Val m minutes to get to work if she drives her car. If she takes the bus, her travel time exceeds this by 15 minutes. Write an algebraic expression that represents the time (in minutes) that it takes her to get to work by bus. $m + 15$

Now Try Problem 43

Teaching Example 2
PACKAGING A regular-size package of Oreo cookies weighs x ounces. A family-size pack of Oreos weighs 16 ounces more. Write an algebraic expression that represents the weight (in ounces) of the family-size pack.
Answer:
$x + 16$

EXAMPLE 2

Banking Javier deposited d dollars in his checking account. He deposited $500 more than that in his savings account. Write an algebraic expression that represents the amount that he deposited in the savings account.

Strategy We will carefully read the problem, looking for a key word or key phrase.

WHY Then we can translate the key word (or phrase) to mathematical symbols to represent the unknown amount that Javier deposited in the savings account.

Solution The deposit that Javier made to the savings account was $500 *more than* the d dollars he deposited in his checking account.

Key phrase: *more than*   **Translation:** add

The number of dollars he deposited in the savings account was $d + 500$.

When solving application problems, we are rarely told which variable to use. We must decide what the unknown quantities are and how to represent them using variables.

Self Check 3

CLOTHING SALES The sale price of a sweater is $20 less than the regular price. Choose a variable to represent one price. Then write an algebraic expression that represents the other price.

Now Try Problem 55

Self Check 3 Answer
p = the regular price of the sweater (in dollars); $p - 20$ = the sale price of the sweater (in dollars)

Teaching Example 3
JEWELRY The value of a wedding ring is three times that of an engagement ring. Choose a variable to represent the value of one of the rings. Then write an algebraic expression that represents the value of the other ring.
Answer:
v = the value of the engagement ring; $3v$ = the value of the wedding ring

EXAMPLE 3

Sports Memorabilia The value of the baseball card shown on the right is four times that of the football card. Choose a variable to represent the value of one card. Then write an algebraic expression that represents the value of the other card.

FOOTBALL

BASEBALL

Strategy There are two unknowns—the value of the baseball card and the value of the football card. We will let v = the value of the football card.

WHY The words of the problem tell us that the value of the baseball card is related to (based on) the value of the football card.

Solution The baseball card's value is *four times* that of the football card.

Key phrase: *four times*   **Translation:** multiply by 4

Therefore, $4v$ is the value of the baseball card.

Caution! A variable is used to represent an unknown number. Therefore, in the previous example, it would be incorrect to write, "Let v = football card," because the football card is not a number. We need to write, "Let v = the *value* of the football card."

3.1 Algebraic Expressions 229

EXAMPLE 4 Swimming

A pool is to be sectioned into eight equally wide swimming lanes. Write an algebraic expression that represents the width of each lane.

Strategy There are two unknowns—the width of the pool and the width of each lane. We will begin by letting w = the width of the pool (in feet), as shown in the illustration.

WHY The width of each lane is related to (based on) the width of the pool.

Solution The width of the pool is sectioned into *eight equally wide lanes*.

Key phrase: *eight equally wide lanes*   **Translation:** division by 8

Therefore, the width of each lane is $\frac{w}{8}$ feet.

Self Check 4

LOTTOS The payoff for a winning lottery ticket is to be split equally among fifteen friends. Write an algebraic expression that represents each person's share of the prize (in dollars).

Now Try Problem 57

Self Check 4 Answer

x = lottery payoff (in dollars); $\frac{x}{15}$ = each person's share (in dollars)

Teaching Example 4 CAMPING A rope is to be cut into six equally long pieces. Write an algebraic expression that represents the length of each piece.

Answer:

r = length of the rope (in feet); $\frac{r}{6}$ = length of each piece (in feet)

EXAMPLE 5 Enrollments

Second-semester enrollment in a nursing program was 32 more than twice that of the first semester. Choose a variable to represent the enrollment for one of the semesters. Then write an algebraic expression that represents the enrollment for the other semester.

Strategy There are two unknowns—the enrollment for the first semester and the enrollment for the second semester. We will begin by letting x = the enrollment for the first semester.

WHY The second-semester enrollment is related to (based on) the first-semester enrollment.

Solution

Key phrase: *more than*   **Translation:** addition

Key phrase: *twice that*   **Translation:** multiplication by 2

The second-semester enrollment was $2x + 32$.

Self Check 5

ELECTIONS In an election, the incumbent received 55 fewer votes than three times the challenger's votes. Choose a variable to represent the number of votes received by one candidate. Then write an algebraic expression that represents the number of votes received by the other.

Now Try Problem 59

Self Check 5 Answer

x = the number of votes received by the challenger; $3x - 55$ = the number of votes received by the incumbent

Teaching Example 5 VEHICLE WEIGHTS The weight of a Cadillac Escalade is 900 pounds less than four times a Smart car. Choose a variable to represent the weight of one of the cars. Write an algebraic expression that represents the weight of the other car. *Answer:* x = the weight (in pounds) of the Smart car; $4x - 900$ = the weight (in pounds) of the Escalade

EXAMPLE 6 Painting

A 10-inch-long paintbrush has two parts: a handle and bristles. Choose a variable to represent the length of one of the parts. Then write an algebraic expression to represent the length of the other part.

Strategy There are two approaches. We can let h = the length of the handle or we can let b = the length of the bristles.

WHY Both the length of the handle and the length of the bristles are unknown.

Self Check 6

SCHOLARSHIPS Part of a $900 donation to a college went to the scholarship fund, the rest to the building fund. Choose a variable to represent the amount donated to one of the funds. Then write an expression that represents the amount donated to the other fund.

Now Try Problem 63

Self Check 6 Answer
s = the amount donated to the scholarship fund (in dollars); $900 - s$ = the amount donated to the building fund (in dollars)

Teaching Example 6 POLICE FORCE
Part of a 500-officer police force works in the field. The others have desk jobs. Choose a variable to represent the number of officers that work in the field. Then write an algebraic expression that represents the number that work at a desk.
Answer:
f = the number of officers in the field; $500 - f$ = the number of officers that work at a desk

Self Check 7
FAMOUS BILLS Bill Cosby was born 9 years before Bill Clinton. Bill Gates was born 9 years after Bill Clinton. Write algebraic expressions to represent the ages of each of these famous men. (Source: celebritybirthdaylist.com)

Now Try Problem 67

Self Check 7 Answer
x = the age of Bill Clinton; $x + 9$ = the age of Bill Cosby; $x - 9$ = the age of Bill Gates

Teaching Example 7
LEADERS Hillary Clinton was born 7 years before Oprah Winfrey. Michelle Obama was born 10 years after Oprah. Write algebraic expressions to represent the ages of each of these famous women. (Source: celebritybirthdaylist.com)
Answer:
x = the age of Oprah Winfrey; $x + 7$ = the age of Hillary Clinton; $x - 10$ = the age of Michelle Obama

Solution Refer to the first drawing on the right. If we let h = the length of the handle (in inches), then the length of the bristles is $10 - h$.

Now refer to the second drawing. If we let b = the length of the bristles (in inches), then the length of the handle is $10 - b$.

Sometimes we must analyze the wording of a problem carefully to detect hidden operations.

EXAMPLE 7 *Engineering* The Golden Gate Bridge was completed 28 years before the Houston Astrodome was opened. The CN Tower in Toronto was built 10 years after the Astrodome. Write algebraic expressions to represent the ages (in years) of each of these engineering wonders. (Source: Data from www.fhwa.dot.gov; www.astrosdaily.com; www.great-towers.com)

Strategy There are three unknowns—the ages of the Golden Gate Bridge, the Astrodome, and the CN Tower. We will begin by letting x = the age of the Astrodome (in years).

WHY The ages of the Golden Gate Bridge and the CN Tower are both related to (based on) the age of the Astrodome.

Solution Reading the problem carefully, we find that the Golden Gate Bridge was built 28 years before the dome, so its age is more than that of the Astrodome.

Key phrase: *more than* **Translation:** add

In years, the age of the Golden Gate Bridge is $x + 28$.

The CN Tower was built 10 years after the dome, so its age is less than that of the Astrodome.

Key phrase: *less than* **Translation:** subtract

In years, the age of the CN Tower is $x - 10$.
The results are summarized in the table at the right.

Engineering feat	Age
Astrodome	x
Golden Gate Bridge	$x + 28$
CN Tower	$x - 10$

EXAMPLE 8 *Packaging* Write an algebraic expression that represents the number of eggs in d dozen.

Strategy First, we will determine how many eggs are in 1 dozen, 2 dozen, and 3 dozen.

WHY There are no key words or phrases in the problem. It will be helpful to consider some specific cases to determine which operation (addition, subtraction, multiplication, or division) is called for.

Solution If we calculate the number of eggs in 1 dozen, 2 dozen, and 3 dozen (as shown in the table below), a pattern becomes apparent.

Number of dozen	Number of eggs
1	$12 \cdot 1 = 12$
2	$12 \cdot 2 = 24$
3	$12 \cdot 3 = 36$
d	$12 \cdot d = 12d$

We multiply the number of dozen by 12 to find the number of eggs.

If d = the number of dozen eggs, the number of eggs is $12 \cdot d$, or, more simply, $12d$.

Self Check 8
Complete the table. Then use that information to write an algebraic expression that represents the number of yards in f feet.

Number of feet	Number of yards
3	1
6	2
9	3
f	$\frac{f}{3}$

Now Try Problems 71 and 75

Teaching Example 8
Write an algebraic expression that represents the number of hours in d days.
Answer:
$24d$ hours

ANSWERS TO SELF CHECKS

1. **a.** $t - 80$ **b.** $\frac{2}{3}T$ **c.** $(2a - 15)^2$ 2. $m + 15$ 3. p = the regular price of the sweater (in dollars); $p - 20$ = the sale price of the sweater (in dollars)
4. x = the lottery payoff (in dollars); $\frac{x}{15}$ = each person's share (in dollars)
5. x = the number of votes received by the challenger; $3x - 55$ = the number of votes received by the incumbent   6. s = the amount donated to the scholarship fund (in dollars); $900 - s$ = the amount donated to the building fund (in dollars)   7. x = the age of Bill Clinton; $x + 9$ = the age of Bill Cosby; $x - 9$ = the age of Bill Gates   8. $1, 2, 3, \frac{f}{3}$

SECTION 3.1   STUDY SET

VOCABULARY

Fill in the blanks.

1. A __variable__ is a letter (or symbol) that stands for a number.
2. Variables and/or numbers can be combined with the operations of addition, subtraction, multiplication, and division to create algebraic __expressions__.
3. Phrases such as *increased by* and *more than* indicate the operation of __addition__. Phrases such as *decreased by* and *less than* indicate the operation of __subtraction__.
4. The word *product* indicates the operation of __multiplication__. The word *quotient* indicates the operation of __division__.

CONCEPTS

5. **a.** Write an algebraic expression that is a combination of the number 10, the variable x, and the operation of addition. $10 + x$
   b. Write an algebraic expression that is a combination of the numbers 3 and 2, the variable t, and the operations of multiplication and subtraction. $3t - 2$ (answers may vary)
6. The illustration below shows the commute to work (in miles) for two men, Mr. Lamb and Mr. Lopez, who work in the same office.
   a. Who lives farther from the office? Mr. Lamb
   b. How much farther? 15 mi

7. Match each algebraic expression to the correct phrase.
   a. $c + 2$ ii.
   b. $2 - c$ iii.
   c. $c - 2$ iv.
   d. $2c$ i.

   i. twice c
   ii. c increased by 2
   iii. c less than 2
   iv. 2 less than c

8. Fill in each blank to complete the translation.
   a. 16 less than m

      $m - 16$
   b. 16 is less than m
      $16 < m$

9. CUTLERY The knife shown below is 12 inches long. Write an algebraic expression that represents the length of the blade (in inches). $12 - h$

10. The following table shows the ages of three family members.
    a. Who is the youngest person shown in the table? Sarah
    b. Who is the oldest person listed in the table? Joshua
    c. On whose age are the ages in the table based? Matthew's

	Age (years)
Matthew	x
Sarah	$x - 8$
Joshua	$x + 2$

11. Complete the table. Then fill in the blank.

Number of decades	Number of years
1	10
2	20
3	30
d	$10d$

We __multiply__ the number of decades by 10 to find the number of years.

12. Complete the table. Then fill in the blank.

Number of inches	Number of feet
12	1
24	2
36	3
i	$\frac{i}{12}$

We __divide__ the number of inches by 12 to find the number of feet.

NOTATION

13. Write each algebraic expression in simpler form.
    a. $x \cdot 8$ $8x$ b. $5(t)$ $5t$ c. $10 \div g$ $\frac{10}{g}$

14. Consider the phrase:
    the product of 5 and w increased by 30
    Insert a comma in the phrase so that it translates to $5w + 30$.
    the product of 5 and w, increased by 30

GUIDED PRACTICE

Translate each phrase to an algebraic expression. If no variable is given, use x as the variable. See Example 1.

15. The sum of the length l and 15 $l + 15$
16. The difference of a number and 10 $x - 10$
17. The product of a number and 50 $50x$
18. Three-fourths of the population p $\frac{3}{4}p$
19. The ratio of the amount won w and lost l $\frac{w}{l}$
20. The tax t added to c $c + t$
21. P increased by two-thirds of p $P + \frac{2}{3}p$
22. 21 less than the total height h $h - 21$
23. The square of k, minus 2,005 $k^2 - 2{,}005$
24. s subtracted from S $S - s$
25. 1 less than twice the attendance a $2a - 1$
26. J reduced by 500 $J - 500$
27. 1,000 split n equal ways $\frac{1{,}000}{n}$
28. Exceeds the cost c by 25,000 $c + 25{,}000$
29. 90 more than twice the current price p $2p + 90$
30. 64 divided by the cube of y $\frac{64}{y^3}$
31. Three times the total of 35, h, and 300 $3(35 + h + 300)$
32. Decrease x by -17 $x - (-17)$
33. 680 fewer than the entire population p $p - 680$
34. Triple the number of expected participants $3x$
35. The product of d and 4, decreased by 15 $4d - 15$
36. The quotient of y and 6, cubed $\left(\frac{y}{6}\right)^3$
37. Twice the sum of 200 and t $2(200 + t)$
38. The square of the quantity 14 less than x $(x - 14)^2$
39. The absolute value of the difference of a and 2 $|a - 2|$
40. The absolute value of a, decreased by 2 $|a| - 2$
41. One-tenth of the distance d $0.1d$ or $\frac{1}{10}d$
42. Double the difference of x and 18 $2(x - 18)$

Write an algebraic expression that represents the unknown quantity. See Example 2.

43. GARDENING The height of a hedge was f feet before a gardener cut 2 feet off the top. Write an algebraic expression that represents the height of the trimmed hedge (in feet). $f - 2$

44. **SHOPPING** A married couple needed to purchase 21 presents for friends and relatives on their holiday gift list. If the husband purchased g presents, write an algebraic expression that represents the number of presents that the wife needs to buy. $21 - g$

45. **PACKAGING** A restaurant owner purchased s six-packs of cola. Write an algebraic expression that represents the number of cans that this would be. $6s$

46. **NOISE** The highest decibel reading during a rock concert was only 5 decibels shy of that of a jet engine. If a jet engine is normally j decibels, write an algebraic expression that represents the decibel reading for the concert. $j - 5$

47. **SUPPLIES** A pad of yellow legal paper contains p pages. If a lawyer uses 15 pages every day, write an algebraic expression that represents the number of days that one pad will last. $\dfrac{p}{15}$

48. **ACCOUNTING** The projected cost c (in dollars) of a freeway was too low by a factor of 10. Write an algebraic expression that represents the actual cost of the freeway (in dollars). $10c$

49. **RECYCLING** A campus ecology club collected t tons of newspaper. A Boy Scout troop then contributed an additional 2 tons. Write an algebraic expression that represents the number of tons of newspaper that were collected by the two groups. $t + 2$

50. **GRADUATION** A graduating class of x people took buses that held 40 students each to an all-night graduation party. Write an algebraic expression that represents the number of buses that were needed to transport the class. $\dfrac{x}{40}$

51. **STUDYING** A student will devote h hours to study for a government final exam. She wants to spread the studying evenly over a four-day period. Write an algebraic expression that represents the number of hours that she should study each day. $\dfrac{h}{4}$

52. **BASEBALL TEAMS** After all c children complete a Little League tryout, the league officials decide that they have enough players for 8 teams of equal size. Write an algebraic expression that represents the number of players that will be on each team. $\dfrac{c}{8}$

53. **SCOTCH TAPE** Suppose x inches of tape have been used off the roll shown below. Write an algebraic expression that represents the number of inches of tape that are left on the roll. $450 - x$

54. **MODELING** A model's skirt is x inches long. The designer then lets the hem down 2 inches. Write an algebraic expression that represents the length of the altered skirt (in inches). $x + 2$

In Problems 55–58, there are two unknowns. See Examples 3 and 4.

55. **GEOMETRY** The length of a rectangle is 6 inches longer than its width. Choose a variable to represent one of the unknown dimensions of the rectangle. Then write an algebraic expression that represents the other dimension. $w = $ the width of the rectangle (in inches); $w + 6 = $ the length of the rectangle (in inches)

56. **PLUMBING** The smaller pipe shown below takes three times longer to fill the tank than does the larger pipe. Choose a variable to represent one of the unknown times it takes to fill the tank. Then write an algebraic expression that represents the other time. $t = $ the time for the larger pipe to fill the tank; $3t = $ the time for the smaller pipe to fill the tank

57. **TRUCK REPAIR** The truck radiator shown below was full of coolant. Then three quarts of coolant were drained from it. Choose a variable to represent one of the unknown amounts of coolant in the radiator. Then write an algebraic expression that represents the other amount. $q = $ the number of quarts of coolant originally in the radiator; $q - 3 = $ the number of quarts of coolant that are left in the radiator

58. SALE PRICES During a sale, the regular price of a CD was reduced by $2. Choose a variable to represent one of the unknown prices of the CD. Then write an algebraic expression that represents the other price. p = the regular price of the CD (in dollars); $p - 2$ = the sale price of the CD (in dollars)

In Problems 59–62, there are two unknowns. **See Example 5.**

59. GEOGRAPHY Alaska is much larger than Vermont. To be exact, the area of Alaska is 380 square miles more than 50 times that of Vermont. Choose a variable to represent one area. Then write an algebraic expression that represents the other area. v = the area of Vermont (in square miles); $50v + 380$ = the area of Alaska (in square miles)

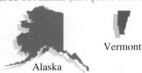

60. ROAD TRIPS On the second part of her trip, Tamiko drove 20 miles less than three times as far as the first part. Choose a variable to represent the number of miles driven on one part of her trip. Then write an algebraic expression that represents the number of miles driven on the other part. f = the number of miles driven on the first part of her trip; $3f - 20$ = the number of miles driven on the second part

61. DESSERTS The number of calories in a slice of pie is 100 more than twice the calories in a scoop of ice cream. Choose a variable to represent the number of calories in one type of dessert. Then write an algebraic expression that represents the number of calories in the other type. s = the number of calories in a scoop of ice cream; $2s + 100$ = the number of calories in a slice of pie

62. WASTE DISPOSAL A waste disposal tank buried in the ground holds 15 gallons less than four times what a tank mounted on a truck holds. Choose a variable to represent the number of gallons that one type of tank holds. Then write an algebraic expression that represents the number of gallons the other tank holds. t = the number of gallons the tank mounted on the truck holds; $4t - 15$ = the number of gallons the tank buried in the ground holds

In Problems 63–66, two approaches are used to represent the unknowns. **See Example 6.**

63. LANDSCAPING

a. Let b represent the height of the birch tree (in feet) that is shown in the next column. Write an algebraic expression that represents the height of the elm tree (in feet). $b + 30$

b. Let e stand for the height of the elm tree (in feet). Write an algebraic expression that represents the height of the birch tree (in feet). $e - 30$

64. BUILDING MATERIALS

a. Let b = the length of the beam shown below (in feet). Write an algebraic expression that represents the length of the pipe. $b - 15$

b. Let p = the length of the pipe (in feet). Write an algebraic expression that represents the length of the beam. $p + 15$

65. MARINE SCIENCE

a. Let s represent the length (in feet) of the great white shark shown below. Write an algebraic expression that represents the length (in feet) of the orca (killer whale). $s + 11$

b. Let w represent the length (in feet) of the orca (killer whale) shown below. Write an algebraic expression that represents the length (in feet) of the great white shark. $w - 11$

66. WEIGHTS AND MEASURES

a. Refer to the scale shown on the next page. Which mixture is heavier, A or B? How much heavier is it? mixture A, 3 ounces

b. Let *a* represent the weight (in ounces) of mixture A. Write an algebraic expression that represents the weight (in ounces) of mixture B. $a - 3$

c. Let *b* represent the weight (in ounces) of mixture B. Write an algebraic expression that represents the weight (in ounces) of mixture A. $b + 3$

In Problems 67–70, there are three unknowns. **See Example 7.**

67. INVENTIONS The digital clock was invented 11 years before the automatic teller machine (ATM). The camcorder was invented 15 years after the ATM. Write algebraic expressions to represent the ages (in years) of each of these inventions. (Source: Data from http://blog.onlineclock.net; http://patft.uspto.gov; www.screenonline.org.uk) x = the age of the ATM; $x + 11$ = the age of the digital clock; $x - 15$ = the age of the camcorder

68. FAMOUS TOMS Tom Petty was born 6 years before Tom Hanks. Tom Cruise was born 6 years after Tom Hanks. Write algebraic expressions to represent the ages of each of these celebrities. (Source: celebritybirthdaylist.com) x = the age of Tom Hanks; $x + 6$ = the age of Tom Petty; $x - 6$ = the age of Tom Cruise

69. NEW YORK ARCHITECTURE The Woolworth Building was completed 18 years before the Empire State Building. The United Nations Building was completed 21 years after the Empire State Building. Write algebraic expressions to represent the ages of each of these buildings. (Source: emporis.com) x = the age of the Empire State Building; $x + 18$ = the age of the Woolworth Building; $x - 21$ = the age of the United Nations Building

70. CHILDREN'S BOOKS *The Tale of Peter Rabbit* was first published 24 years before *Winnie-the-Pooh*. *The Cat in the Hat* was first published 31 years after *Winnie-the-Pooh*. Write algebraic expressions to represent the ages of each of these books. (Source: Data from Lear, Linda [2007], *Beatrix Potter: A Life in Nature*, New York: St. Martin's Press; Thwaite, Ann [2004], *Oxford Dictionary of National Biography: Alan Alexander Milne*, Oxford: Oxford University Press; www.worldcat.org) x = the age of *Winnie-the-Pooh*; $x + 24$ = the age of *The Tale of Peter Rabbit*; $x - 31$ = the age of *The Cat in the Hat*

Use a table to help answer Problems 71–78. **See Example 8.**

71. Write an algebraic expression that represents the number of seconds in *m* minutes. $60m$ sec

72. Write an algebraic expression that represents the number of minutes in *h* hours. $60h$ min

73. Write an algebraic expression that represents the number of inches in *f* feet. $12f$ in.

74. Write an algebraic expression that represents the number of feet in *y* yards. $3y$ ft

75. Write an algebraic expression that represents the number of centuries in *y* years. $\frac{y}{100}$ centuries

76. Write an algebraic expression that represents the number of decades in *y* years. $\frac{y}{10}$ decades

77. Write an algebraic expression that represents the number of dozen eggs in *e* eggs. $\frac{e}{12}$ dozen

78. Write an algebraic expression that represents the number of days in *h* hours. $\frac{h}{24}$ days

TRY IT YOURSELF

Translate each algebraic expression into words. (Answers may vary.)

79. $\frac{3}{4}r$ three-fourths of *r*   **80.** $\frac{2}{3}d$ two-thirds of *d*

81. $t - 50$ 50 less than *t*   **82.** $c + 19$ 19 more than *c*

83. xyz the product of *x*, *y*, and *z*

84. $10ab$ the product of 10, *a*, and *b*

85. $2m + 5$ twice *m*, increased by 5

86. $2s - 8$ twice *s*, decreased by 8

87. A man sleeps x hours per day. Write an algebraic expression that represents

  a. the number of hours that he sleeps in a week. $7x$ hours

  b. the number of hours that he sleeps in a year (non–leap year). $365x$ hours

88. A store manager earns d dollars an hour. Write an algebraic expression that represents

  a. the amount of money he will earn in an 8-hour day. $8d$ dollars

  b. the amount of money he will earn in a 40-hour week. $40d$ dollars

89. A secretary earns an annual salary of s dollars. Write an algebraic expression that represents

  a. her salary per month. $\frac{s}{12}$ dollars

  b. her salary per week. $\frac{s}{52}$ dollars

90. Write an algebraic expression that represents the number of miles in f feet. (Hint: There are 5,280 feet in one mile.) $\frac{f}{5,280}$ mi

APPLICATIONS

91. ELECTIONS In 1960, John F. Kennedy was elected President of the United States with a popular vote only 118,550 votes more than that of Richard M. Nixon. Choose a variable to represent the number of votes received by one candidate. Then write an algebraic expression that represents the number of votes received by the other candidate. $x =$ the number of votes received by Nixon; $x + 118,550 =$ the number of votes received by Kennedy

92. THE BEATLES According to music historians, sales of the Beatles' second most popular single, *Hey Jude*, trail the sales of their most popular single, *I Want to Hold Your Hand*, by 2,000,000 copies. Choose a variable to represent the number of copies sold of one song. Then write an algebraic expression that represents the number of copies sold of the other song. $c =$ the number of copies sold of *I Want to Hold Your Hand*; $c - 2,000,000 =$ the number of copies sold of *Hey Jude*

93. COMPUTER COMPANIES IBM was founded 80 years before Apple Computer. Dell Computer Corporation was founded 9 years after Apple. Let x represent the age (in years) of one of the companies. Write algebraic expressions to represent the ages (in years) of the other two companies. let $x =$ the age of Apple; $x + 80 =$ the age of IBM; $x - 9 =$ the age of Dell

94. VEHICLE WEIGHTS Refer to the illustration below. The car is 1,000 pounds lighter than the van. Choose a variable to represent the weight (in pounds) of one of the vehicles. Then write an algebraic expression that represents the weight (in pounds) of the other vehicle. $v =$ weight of the van, $v - 1,000 =$ weight of the car

95. WITH/AGAINST THE WIND On a flight from Dallas to Miami, a jet airliner, which can fly 500 mph in still air, has a tail wind of x mph. The tail wind increases the speed of the jet. On the return flight to Dallas, the airliner flies into a head wind of the same strength. The head wind decreases the speed of the jet. Use this information to complete the table.

Wind conditions	Speed of jet (mph)
In still air	500
With the tail wind	$500 + x$
Against the head wind	$500 - x$

96. SUB SANDWICHES Refer to the illustration below. Write an algebraic expression that represents the length (in inches) of the second piece of the sandwich. $72 - x$

97. SAVINGS ACCOUNTS A student inherited $5,000 and deposits x dollars in American Savings. Write an algebraic expression that represents the amount of money (in dollars) left to deposit in a City Mutual account. $5,000 - x$

98. a. MIXING SOLUTIONS Solution 1 is poured into solution 2. Write an algebraic expression that represents the number of ounces in the mixture. $x + 20$

b. **SNACKS** Cashews were mixed with p pounds of peanuts to make 100 pounds of a mixture. Write an algebraic expression that represents the number of pounds of cashews that were used.
$100 - p$

WRITING

99. Explain how variables are used in this section.
100. Explain the difference between the phrases *greater than* and *is greater than*.
101. Suppose in an application problem you were asked to find the unknown height of a building. Explain what is wrong with the following start.
    Let x = building

102. What is an algebraic expression?

REVIEW

103. Find the sum: $-5 + (-6) + 1$ -10
104. Evaluate: $2 + 3(-3)$ -7
105. Solve: $-x = 4$ -4
106. Write the set of integers.
    $\{\ldots, -3, -2, -1, 0, 1, 2, 3, \ldots\}$
107. Evaluate the expression:
    $(-1 - 2^2 \cdot 5)3^2$ -189
108. Evaluate the expression:
    $-3(2) - 7(-6)(-4)$ -174
109. Subtract: $-3 - 2$ -5
110. Evaluate: $(-5)^3$ -125

SECTION 3.2
Evaluating Algebraic Expressions and Formulas

Objectives
1. Evaluate algebraic expressions.
2. Use formulas from business to solve application problems.
3. Use formulas from science to solve application problems.
4. Find the mean (average) of a set of values.

Recall that an **algebraic expression** is a combination of variables and numbers with the operation symbols of addition, subtraction, multiplication, and division. In this section, we will be replacing the variables in algebraic expressions with numbers. Then, using the rule for the order of operations, we will *evaluate* each expression. We will also study formulas. Like algebraic expressions, formulas involve variables.

1 Evaluate algebraic expressions.

EXAMPLE 1 *Plumbing* The manufacturer's instructions for installing a kitchen garbage disposal are shown below.
a. Choose a variable to represent the length of one of the pieces of pipe (A, B, or C). Then write algebraic expressions to represent the lengths of the other two pieces.
b. Suppose model #201 is being installed. Find the length of each piece of pipe that is needed to connect the disposal to the drain line.

Model	Length of piece A
#101	2 inches
#201	3 inches
#301	4 inches

Self Check 1
Refer to Example 1. Suppose model #101 is being installed. Find the length of each piece of pipe that is needed to connect the disposal to the drain line.

Now Try Problem 11

Self Check 1 Answer
piece B: 4 in.; piece C: 1 in.

Teaching Example 1
Refer to Example 1.
a. Choose a variable to represent the length of one of the pieces of pipe. Then write algebraic expressions to represent the lengths of the other two pieces.
b. Suppose model #301 is being installed. Find the length of each piece of pipe that is needed to connect the disposal to the drain line.
Answer:
piece B: 6 in.; piece C: 3 in.

Unless otherwise noted, all art on this page is © Cengage Learning.

Strategy There are three lengths of pipe to represent. We will begin by letting x = the length (in inches) of piece A.

WHY The lengths of the other pieces are related to (based on) the length of piece A.

Solution

a. Since the instructions call for piece B to be *2 inches longer* than piece A, and the length of piece A is represented by x, we have:

$x + 2$ = the length of piece B (in inches)

Since piece C is to be *1 inch shorter* than piece A,

$x - 1$ = the length of piece C (in inches)

The illustration on the right shows the algebraic expressions that can be used to represent the length of each piece of pipe.

b. If model #201 is being installed, the table tells us that piece A should be 3 inches long. We can find the lengths of the other two pieces of pipe by replacing x with 3 in each of the algebraic expressions.

To find the length of piece B:	To find the length of piece C:
Replace x with 3.	Replace x with 3.
$x + 2 = 3 + 2$	$x - 1 = 3 - 1$
$= 5$	$= 2$
Piece B should be 5 inches long.	Piece C should be 2 inches long.

When we substitute given numbers for each of the variables in an algebraic expression and apply the order of operations rule, we are **evaluating the expression**. In the previous example, we say that we *substituted* 3 for x to evaluate the algebraic expressions $x + 2$ and $x - 1$.

> **Caution!** When replacing a variable with its numerical value, we must often write the replacement number within parentheses to convey the proper meaning.

Self Check 2

Evaluate each expression for $y = 5$:

a. $5y - 4$ 21

b. $\dfrac{-y - 15}{2}$ -10

Now Try Problems 15 and 17

Teaching Example 2 Evaluate each expression for $x = 4$:

a. $6x - 2$

b. $\dfrac{-x - 8}{3}$

Answers:
a. 22 b. -4

EXAMPLE 2 Evaluate each expression for $x = 3$:

a. $2x - 1$ b. $\dfrac{-x - 15}{6}$

Strategy We will replace x with the given value of the variable and evaluate the expression using the order of operations rule.

WHY To *evaluate an algebraic expression* means to find its numerical value, once we know the value of its variable.

Solution

a. $2x - 1 = 2(3) - 1$ Substitute 3 for x. Use parentheses.

$ = 6 - 1$ Do the multiplication first: 2(3) = 6.

$ = 5$ Do the subtraction.

b. $\dfrac{-x - 15}{6} = \dfrac{-(3) - 15}{6}$ Substitute 3 for x. Use parentheses. Don't forget to write the − sign in front of (3).

$= \dfrac{-3 - 15}{6}$ Simplify: −(3) = −3.

$= \dfrac{-3 + (-15)}{6}$ If it is helpful, write the subtraction of 15 as addition of the opposite of 15.

$\quad\quad\quad\quad\quad\quad$ $\begin{array}{r} 15 \\ +\ 3 \\ \hline 18 \end{array}$

$= \dfrac{-18}{6}$ Do the addition: −3 + (−15) = −18.

$= -3$ Do the division.

EXAMPLE 3
Evaluate each expression for $a = -2$:
a. $4a^2 - 3a$ **b.** $-a + 3(1 + a)$ **c.** $a^3 - 5$

Strategy We will replace each a in the expression with the given value of the variable and evaluate the expression using the order of operations rule.

WHY To *evaluate an algebraic expression* means to find its numerical value, once we know the value of its variable.

Solution

a. $4a^2 - 3a = 4(-2)^2 - 3(-2)$ Substitute −2 for each a. Use parentheses.

$= 4(4) - 3(-2)$ Evaluate the exponential expression: $(-2)^2 = 4$.

$= 16 - (-6)$ Do each multiplication.

$= 16 + 6$ If it is helpful, write the subtraction of −6 as addition of the opposite of −6.

$\quad\quad\quad\quad\quad\quad$ $\begin{array}{r} \overset{1}{1}6 \\ +\ 6 \\ \hline 22 \end{array}$

$= 22$ Do the addition.

b. $-a + 3(1 + a) = -(-2) + 3[1 + (-2)]$ Substitute −2 for each a. Use parentheses. Don't forget to write the − sign in front of (−2). Since another pair of grouping symbols are now needed, write brackets around 1 + (−2).

$= -(-2) + 3(-1)$ Do the addition within the brackets.

$= 2 + (-3)$ Simplify: −(−2) = 2. Do the multiplication: 3(−1) = −3.

$= -1$ Do the addition.

c. $a^3 - 5 = (-2)^3 - 5$ Substitute −2 for a. Use parentheses.

$= -8 - 5$ Evaluate the exponential expression: $(-2)^3 = -8$.

$= -8 + (-5)$ If it is helpful, write the subtraction of 5 as addition of the opposite of 5.

$= -13$ Do the addition.

Self Check 3
Evaluate each expression for $t = -3$:

a. $4t^2 - 2t$ 42

b. $-t + 2(t + 1)$ −1

c. $t^3 + 16$ −11

Now Try Problems 19, 21, and 23

Teaching Example 3 Evaluate each expression for $x = -4$:
a. $2x^2 - 3x$
b. $-x + 7(1 + x)$
c. $x^3 - 25$
Answers:
a. 44 **b.** −17 **c.** −89

To evaluate algebraic expressions containing two or more variables, we need to know the value of each variable.

Self Check 4

Evaluate each expression for $r = -1$ and $s = 5$:

a. $(5rs + 4s)^2$ 25

b. $|-8s - 2r|$ 38

Now Try Problems 27 and 29

Teaching Example 4 Evaluate each expression for $c = -3$ and $d = 9$:

a. $(2cd + 5d)^2$

b. $|-8d - 3c|$

Answers:
a. 81 b. 63

EXAMPLE 4 Evaluate each expression for $h = -1$ and $g = 5$:

a. $(8hg + 6g)^2$ b. $|-5g - 7h|$

Strategy We will replace each h and g in the expression with the given value of the variable and evaluate the expression using the order of operations rule.

WHY To *evaluate an expression* means to find its numerical value, once we know the values of its variables.

Solution

a. $(8hg + 6g)^2 = [8(-1)(5) + 6(5)]^2$ Substitute -1 for h and 5 for g. Use parentheses. Since another pair of grouping symbols are now needed, write brackets around $8(-1)(5) + 6(5)$.

$= (-40 + 30)^2$ Do the multiplication within the brackets.

$= (-10)^2$ Do the addition within the parentheses.

$= 100$ Evaluate the exponential expression: $(-10)^2 = 100$.

b. $|-5g - 7h| = |-5(5) - 7(-1)|$ Substitute -1 for h and 5 for g. Use parentheses.

$= |-25 - (-7)|$ Do the multiplication within the absolute value symbols.

$= |-25 + 7|$ If it is helpful, write the subtraction of -7 as the addition of the opposite of -7.

$= |-18|$ Do the addition.

$= 18$ Find the absolute value of -18.

$$\begin{array}{r} \overset{1\,1\,5}{2\,\cancel{5}} \\ -7 \\ \hline 18 \end{array}$$

2 Use formulas from business to solve application problems.

A **formula** is an equation that is used to state a relationship between two or more variables. Formulas are used in many fields: economics, physical education, biology, automotive repair, and nursing, just to name a few. In this section, we will consider several formulas from business, science, and mathematics.

A formula to find the sale price

If a car that usually sells for $22,850 is discounted $1,500, you can find the **sale price** using the formula

Sale price = original price − discount

Using the variables s to represent the sale price, p the original price, and d the discount, we can write this formula as

$s = p - d$

To find the sale price of the car, we substitute 22,850 for p, 1,500 for d, and evaluate the right side of the equation.

$s = p - d$ This is the sale price formula.
$s = 22{,}850 - 1{,}500$ Substitute 22,850 for p and 1,500 for d.
$s = 21{,}350$ Do the subtraction.

$\quad\quad 22{,}850$
$\quad - \ 1{,}500$
$\quad\quad 21{,}350$

The sale price of the car is $21,350.

A formula to find the retail price

To make a profit, a merchant must sell a product for more than he paid for it. The price at which he sells the product, called the **retail price**, is the *sum* of what the item cost him and the markup.

| Retail price | = | cost | + | markup |

Using the variables r to represent the retail price, c the cost, and m the markup, we can write this formula as

$$r = c + m$$

As an example, suppose that a store owner buys a lamp for $35 and then marks up the cost $20 before selling it. We can find the retail price of the lamp using this formula.

$r = c + m$ This is the retail price formula.
$r = 35 + 20$ Substitute 35 for c and 20 for m.
$r = 55$ Do the addition.

The retail price of the lamp is $55.

A formula to find profit

The **profit** a business makes is the *difference* of the revenue (the money it takes in) and the costs.

| Profit | = | revenue | − | costs |

Using the variables p to represent the profit, r the revenue, and c the costs, we have the formula

$$p = r - c$$

EXAMPLE 5 *Films* It cost Universal Studios about $525 million to make and distribute the film *Jurassic Park*. If the studio has received approximately $924 million to date in worldwide revenue from the film, find the profit the studio has made on this movie. (Source: the-numbers.com/movies)

Strategy To find the profit, we will substitute the given values in the formula $p = r - c$ and evaluate the expression on the right side of the equation.

WHY The variable p represents the unknown profit.

Self Check 5
FILMS It cost Buena Vista Pictures about $397 million to make and distribute the film *Lion King*. If the studio has received approximately $953 million to date in worldwide revenue from the film, find the profit the studio has made on this movie. (Source: the-numbers.com/movies)

Now Try Problem 35

Self Check 5 Answer
Buena Vista Pictures has made $556 million in profit on the movie *Lion King*.

Teaching Example 5 FILMS It cost DreamWorks SKG about $530 million to make and distribute the film *Shrek 2*. If the studio has received approximately $920 million to date in worldwide revenue from the film, find the profit the studio has made on this movie. (Source: the-numbers.com/movies)
Answer:
DreamWorks SKG has made $390 million in profit on the movie *Shrek 2*.

Solution The studio has received $924 million in revenue r, and the cost c to make and distribute the movie was $525 million. To find the profit p, we proceed as follows.

$p = r - c$ This is the formula for profit.

$p = 924 - 525$ Substitute 924 for r and 525 for c. The units are millions of dollars.

$p = 399$ Do the subtraction.

$$\begin{array}{r} \overset{11}{8}\overset{1}{4}\overset{1}{4} \\ 9\,2\,4 \\ -5\,2\,5 \\ \hline 3\,9\,9 \end{array}$$

Universal Studios has made $399 million in profit on the film *Jurassic Park*.

3 Use formulas from science to solve application problems.

A formula to find the distance traveled

If we know the rate (speed) at which we are traveling and the time we will be moving at that rate, we can find the distance traveled using the formula

$$\boxed{\text{Distance}} = \boxed{\text{rate}} \cdot \boxed{\text{time}}$$

Using the variables d to represent the distance, r the rate, and t the time, we have the formula

$$\boxed{d = rt}$$

Self Check 6

SPEED LIMITS Missouri's speed limit for trucks on rural interstate highways is 70 mph. How far would a truck travel in 3 hours at that speed? 210 mi

Now Try Problem 39

Teaching Example 6 SPEED LIMITS Texas's daytime speed limit for cars on most rural interstate highways is 75 mph. How far would a car travel in 4 hours at this speed?
Answer:
300 mi

EXAMPLE 6 *Interstate Speed Limits* Three state speed limits for trucks are shown below. At each of these speeds, how far would a truck travel in 3 hours?

Strategy To find the distance traveled, we will substitute the given values in the formula $d = rt$ and evaluate the expression on the right side of the equation.

WHY The variable d represents the unknown distance traveled.

Solution To find the distance traveled by a truck in Oregon, we write

$d = rt$ This is the formula for distance traveled.

$d = 55(3)$ Substitute: 55 mph is the rate r and 3 hours is the time t.

$d = 165$ Do the multiplication. The units of the answer are miles.

$$\begin{array}{r} \overset{1}{5}5 \\ \times\ 3 \\ \hline 165 \end{array}$$

At 55 mph, a truck would travel 165 miles in 3 hours. We can use a table to display the calculations for each state.

	r	$\cdot$	t	$=$	d
Oregon	55		3		165
Washington	60		3		180
Kansas	75		3		225

↑ This column gives the distance traveled, in miles.

$$\begin{array}{r} 60 \\ \times\ 3 \\ \hline 180 \end{array} \qquad \begin{array}{r} \overset{1}{7}5 \\ \times\ 3 \\ \hline 225 \end{array}$$

Caution! When using $d = rt$ to find distance, make sure that the units are similar. For example, if the rate is given in miles per hour, the time must be expressed in hours.

A formula for converting degrees Fahrenheit to degrees Celsius

Electronic message boards in front of some banks flash two temperature readings. This is because temperature can be measured using the Fahrenheit or the Celsius scales. The **Fahrenheit scale** is used in the American system of measurement, and temperatures are measured in degrees Fahrenheit, written °F. The **Celsius scale** is used in the metric system, and temperatures are measured in degrees Celsius, written °C. The two scales are shown on the thermometers to the right. This should help you to see how the two scales are related. There is a formula to convert a Fahrenheit reading F to a Celsius reading C.

$$C = \frac{5(F - 32)}{9} \ *$$

Later we will see that there is a formula to convert a Celsius reading to a Fahrenheit reading.

*An alternate form of this formula is $C = \frac{5}{9}(F - 32)$.

The Language of Algebra In 1724, Daniel Gabriel *Fahrenheit*, a German scientist, introduced the temperature scale that bears his name. The Celsius scale was invented in 1742 by Swedish astronomer Anders *Celsius*.

EXAMPLE 7 **Heating** The thermostat in an office building was set at 77°F. Convert this setting to degrees Celsius.

Strategy To find the temperature in degrees Celsius, we will substitute the given Fahrenheit temperature in the formula $C = \frac{5(F - 32)}{9}$ and evaluate the expression on the right side of the equation.

WHY The variable C represents the temperature in degrees Celsius.

Self Check 7

SATURN Change −283°F, the temperature on Saturn, to degrees Celsius. −175°C

Now Try Problem 43

Teaching Example 7 LAND OF ENCHANTMENT The record high temperature for New Mexico is 122°F, on June 27, 1994. Convert this temperature to degrees Celsius.
Answer:
50°C

Solution

$$C = \frac{5(F - 32)}{9}$$ This is the formula for temperature conversion.

$$C = \frac{5(77 - 32)}{9}$$ Substitute the Fahrenheit temperature, 77, for F.

$$C = \frac{5(45)}{9}$$ Do the subtraction within the parentheses first: 77 − 32 = 45.

$$C = \frac{225}{9}$$ Do the multiplication: 5(45) = 225.

$$C = 25$$ Do the division.

The thermostat is set at 25°C.

A formula to find the distance an object falls

The distance an object falls (in feet) when it is dropped from a height is related to the time (in seconds) that it has been falling by the formula

$$\text{Distance fallen} = 16 \cdot (\text{time})^2$$

Using the variables d to represent the distance and t the time, we have

$$d = 16t^2$$

Self Check 8

FREEFALL Find the distance a rock fell in 3 seconds if it was dropped over the edge of the Grand Canyon. 144 ft

Now Try Problem 47

Teaching Example 8 ELEVATORS An elevator mechanic dropped his wrench down an elevator shaft. How far will the wrench fall in 5 seconds?
Answer:
400 ft

EXAMPLE 8 **Balloon Rides** Find the distance a camera fell in 6 seconds if it was dropped overboard by a vacationer taking a hot-air balloon ride.

Strategy To find the distance the camera fell, we will substitute the given time in the formula $d = 16t^2$ and evaluate the expression on the right side of the equation.

WHY The variable d represents the distance fallen.

Solution

$d = 16t^2$ This is the formula for distance fallen.
$d = 16(6)^2$ The camera fell for 6 seconds. Substitute 6 for t.
$d = 16(36)$ Evaluate the exponential expression: $6^2 = 36$.
$d = 576$ Do the multiplication.

The camera fell 576 feet.

4 Find the mean (average) of a set of values.

A formula to find the mean (average)

The **mean**, or **average**, of a set of numbers is a value around which the numbers are grouped. To find the mean, we divide the *sum* of all the values by the *number* of values. Writing this as a formula, we get

$$\text{Mean} = \frac{\text{sum of the values}}{\text{number of values}}$$

Using the variables S to represent the sum and n the number of values, we have

$$\text{Mean} = \frac{S}{n}$$

EXAMPLE 9

Response Time A police department recorded the length of time between incoming 911 calls and the arrival of a police unit at the scene. The response times (in minutes) for one 24-hour period are listed below. Find the mean (average) response time.

Response times

| 5 min | 3 min | 6 min | 2 min | 7 min | 4 min | 3 min | 2 min |

Strategy We will count the number of response times and calculate their sum.

WHY To find the mean of a set of values, we divide the sum of the values by the number of values.

Solution There are eight response times. To find their sum, it is helpful to look for groups of numbers that add to 10.

$5 + 3 + 6 + 2 + 7 + 4 + 3 + 2 = 32$ Add: $5 + 3 + 2 = 10$
　　　　　　　　　　　　　　　　　　Add: $6 + 4 = 10$
　　　　　　　　　　　　　　　　　　Add: $7 + 3 = 10$

Now we use the formula to find the mean.

Mean $= \dfrac{S}{n}$　 This is the formula to find the mean (average).

Mean $= \dfrac{32}{8}$　 Substitute 32 for S, the sum of the response times.
　　　　　　 Substitute 8 for n, the number of response times.

Mean $= 4$　 Do the division.

The mean response time was 4 minutes.

Self Check 9

WEB TRAFFIC The number of hits a website received each day for one week are listed below.
Mon: 392, Tues: 931,
Wed: 842, Thurs: 566,
Fri: 301, Sat: 103, Sun: 43
Find the mean (average) number of hits each day.　454 hits

Now Try Problem 51

Teaching Example 9 STUDENT PARKING As part of a traffic survey, the number of cars in a college parking lot at noon were counted. The results are listed below.
Mon: 121, Tues: 204,
Wed: 166, Thurs: 152,
Fri: 87
Find the mean (average) number of cars in the parking lot at that time.
Answer:
146 cars

THINK IT THROUGH   *Study Time*

"Your success in school is dependent on your ability to study effectively and efficiently. The results of poor study skills are wasted time, frustration, and low or failing grades."
Effective Study Skills, Dr. Bob Kizlik, 2004

For a course that meets for h hours each week, the formula $H = 2h$ gives the suggested number of hours H that a student should study the course outside of class each week. If a student expects difficulty in a course, the formula can be adjusted upward to $H = 3h$. Use the formulas to complete the table on the right.

If a course meets for:	Suggested study time (hours per week)	Expanded study time (hours per week)
2 hours per week	4	6
3 hours per week	6	9
4 hours per week	8	12
5 hours per week	10	15

ANSWERS TO SELF CHECKS

1. piece B: 4 in.; piece C: 1 in.　**2. a.** 21　**b.** -10　**3. a.** 42　**b.** -1　**c.** -11
4. a. 25　**b.** 38　**5.** Buena Vista Pictures has made $556 million in profit on the movie *Lion King*.　**6.** 210 mi　**7.** $-175°C$　**8.** 144 ft　**9.** 454 hits

SECTION 3.2 STUDY SET

VOCABULARY

Fill in the blanks.

1. An algebraic _expression_ is a combination of variables, numbers, and the operation symbols for addition, subtraction, multiplication, and division.

2. When we substitute 5 for x in the algebraic expression $7x + 10$ and apply the order of operations rule, we are _evaluating_ the expression.

3. To evaluate $a^2 + 10a + 1$ for $a = -3$, we _substitute_ -3 for a and apply the order of operations rule.

4. A _formula_ is an equation that states a relationship between two or more variables.

5. Temperature can be measured using the Fahrenheit or _Celsius_ scale.

6. To find the _mean_ (or average) of a set of values, we divide the sum of the values by the number of values.

CONCEPTS

7. Use variables to write the formula that relates each of the quantities listed below.
   a. Sale price, original price, discount $s = p - d$
   b. Profit, revenue, costs $p = r - c$
   c. Retail price, cost, markup $r = c + m$

8. Use variables to write the formula that relates each of the quantities listed below.
   a. Distance, rate, time $d = rt$
   b. Celsius temperature, Fahrenheit temperature
      $C = \dfrac{5(F - 32)}{9}$
   c. The distance an object falls when dropped, time
      $d = 16t^2$
   d. Mean, number of values, sum of values   mean $= \dfrac{s}{n}$

NOTATION

9. Complete each step. Evaluate the expression for $a = 5$.
$$9a - a^2 = 9(\,5\,) - (5)^2$$
$$= 9(5) - \,25\,$$
$$= \,45\, - 25$$
$$= 20$$

10. Fill in the blanks. The symbol °F stands for degrees _Fahrenheit_ and the symbol °C stands for degrees _Celsius_.

GUIDED PRACTICE

In Problems 11–14, write algebraic expressions to represent the three unknowns and then evaluate each of them for the given value of the variable. See Example 1.

11. **PLAYGROUND EQUIPMENT** The plans for building a children's swing set are shown below.
    a. Choose a variable to represent the length (in inches) of one part of the swing set. Then write algebraic expressions that represent the lengths (in inches) of the other two parts. x = the length of part 1; $x - 40$ = the length of part 2; $x + 16$ = the length of part 3 (answers may vary)
    b. If the builder chooses to have part 1 be 60 inches long, how long should parts 2 and 3 be?   part 2: 20 in.; part 3: 76 in.

Part 3: crossbar. This is to be 16 inches longer than part 1.
Part 2: brace. This is to be 40 inches less than part 1.
Part 1: leg

12. **ART DESIGN** A television studio art department plans to construct two sets of decorations out of plywood, using the plan shown below.
    a. Choose a variable to represent the height (in inches) of one piece of plywood. Then write algebraic expressions that represent the heights (in inches) of the other two pieces. x = height of piece A; $2x$ = height of piece B; $3x$ = height of piece C (answers may vary)
    b. Designers will make the first set of three pieces for the foreground. Piece A will be 15 inches high. How high should pieces B and C be?   piece B: 30 in.; piece C: 45 in.
    c. Designers will make another set of three pieces for the background. Piece A will be 30 inches high. How high should pieces B and C be?   piece B: 60 in.; piece C: 90 in.

Piece C–three times as high as piece A
Piece B–twice as high as piece A
Piece A

13. VEHICLE WEIGHTS An H2 Hummer weighs 340 pounds less than twice a Nissan Leaf. A Smart Fortwo car weighs 1,720 pounds less than a Nissan Leaf.
  a. Choose a variable to represent the weight (in pounds) of one car. Then write algebraic expressions that represent the weights (in pounds) of the other two cars. x = the weight of a Nissan Leaf; $2x - 340$ = the weight of an H2 Hummer; $x - 1,720$ = the weight of a Smart Fortwo car (answers may vary)
  b. If the weight of the Nissan Leaf is 3,370 pounds, find the weights of the other two cars. H2 Hummer: 6,400 lb; Smart Fortwo car: 1,650 lb

14. BATTERIES An AAA-size battery weighs 53 grams less than a C-size battery. A D-size battery weighs 5 grams more than twice a C-size battery.
  a. Choose a variable to represent the weight (in grams) of one size battery. Then write algebraic expressions that represent the weights (in grams) of the other two batteries. x = the weight of a C-size battery; $x - 53$ = the weight of an AAA-size battery; $2x + 5$ = the weight of a D-size battery (answers may vary)
  b. If the weight of a C-size battery is 65 grams, find the weights of the other two batteries. AAA-size battery: 12 grams; D-size battery: 135 grams

Evaluate each expression for the given value of the variable. See Example 2.

15. $10x - 3$ for $x = 3$ 27
16. $4a - 2$ for $a = 9$ 34
17. $\dfrac{-n-1}{3}$ for $n = 11$ -4
18. $\dfrac{-b-2}{7}$ for $b = 5$ -1

Evaluate each expression for the given value of the variable. See Example 3.

19. $3x^2 - 2x$ for $x = -2$ 16
20. $4n^2 - 5n$ for $n = -3$ 51
21. $-y + 3(1 + y)$ for $y = -10$ -17
22. $-b + 6(2 + b)$ for $b = -8$ -28
23. $h^3 - 24$ for $h = -3$ -51
24. $t^3 - 30$ for $t = -4$ -94
25. $n^4 + n^2$ for $n = -1$ 2
26. $d^4 + d^3$ for $d = -2$ 8

Evaluate each expression for the given values of the variables. See Example 4.

27. $(2ab + 4b)^2$ for $a = -5$ and $b = 2$ 144
28. $(3xy + 2y)^2$ for $x = -4$ and $y = 3$ 900
29. $|-6r - 8s|$ for $r = -11$ and $s = 9$ 6
30. $|-7t - 10x|$ for $t = -12$ and $x = 15$ 66

Use the correct formula to solve each problem. See Objective 2 and Example 5.

31. SPORTING GOODS Find the sale price of a pair of skis that usually sells for $200 but is discounted $35. $165

32. OFFICE FURNISHINGS If a desk chair that usually sells for $199 is discounted $38, what is the sale price of the chair? $161

33. CLOTHING STORES A store owner buys a pair of pants for $125 and marks them up $65 for sale. What is the retail price of the pants? $190

34. SNACKS It costs a snack bar owner 20 cents to make a snow cone. If the markup is 50 cents, what is the retail price of a snow cone? 70 cents

35. SMALL BUSINESSES On its first night of business, a pizza parlor brought in $445. The owner estimated his costs that night to be $295. What was the profit? $150

36. FLORISTS For the month of June, a florist's cost of doing business was $3,795. If June revenues totaled $5,115, what was her profit for the month? $1,320

37. FUNDRAISERS A school carnival brought in revenues of $13,500 and had costs of $5,300. What was the profit? $8,200

38. PRICING A shopkeeper marks up the cost of every item she carries by the amount she paid for the item. If a fan costs her $27, what does she charge for the fan? $54

Use the correct formula to solve each problem. See Example 6.

39. AIRLINES Find the distance covered by a jet if it travels for 3 hours at 550 mph. 1,650 mi

40. ROAD TRIPS Find the distance covered by a car traveling 60 miles per hour for 5 hours. 300 mi

41. HIKING A hiker can cover 12 miles per day. At that rate, how far will the hiker travel in 8 days? 96 mi

42. TURTLES A turtle can walk 250 feet per minute. At that rate, how far can a turtle walk in 5 minutes? 1,250 ft

Use the correct formula to convert each Fahrenheit temperature to a Celsius temperature. See Example 7.

43. 59°F 15°C
44. 113°F 45°C
45. -4°F -20°C
46. -22°F -30°C

Use the correct formula to solve each problem. See Example 8.

47. FREE FALL Find the distance a ball has fallen 2 seconds after being dropped from a tall building. 64 ft

48. SIGHTSEEING A visitor to the Grand Canyon accidentally dropped her sunglasses over the edge. It took 9 seconds for the sunglasses to fall directly to the bottom of the canyon. How far above the canyon bottom was she standing? 1,296 ft

49. BRIDGE REPAIR A steel worker dropped his wrench while tightening a cable on the top of a bridge. It took 4 seconds for the wrench to fall straight to the ground. How far above ground level was the man working? 256 ft

50. LIGHTHOUSES An object was dropped from the top of the Tybee Island Lighthouse (located near Savannah, Georgia). It took 3 seconds for the object to hit the ground. How tall is the lighthouse? 144 ft

Use the correct formula to find each mean (average). See Example 9.

51. BOWLING Find the mean score for a bowler who rolled scores of 254, 225, and 238. 239

52. YAHTZEE A player had scores of 288, 192, 264, and 124 at a Yahtzee tournament. What was his mean score? 217

53. FISHING The weights of each of the fish caught by those on a deep-sea fishing trip are listed below. What was the mean weight? 25 lb

23 lb   18 lb   37 lb   11 lb   18 lb   26 lb   42 lb   25 lb

54. GRADES Find the mean score of the following test scores: 76, 83, 79, 91, 0, 73. 67

TRY IT YOURSELF

Evaluate each expression for the given value(s) of the variable(s).

55. $\dfrac{x-8}{2}$ for $x = -4$ −6

56. $\dfrac{-10+y}{-4}$ for $y = -6$ 4

57. $-p$ for $p = -4$ 4

58. $-j$ for $j = -9$ 9

59. $2(p+9) + 2p$ for $p = -12$ −30

60. $3(r-20) + 2r$ for $r = 15$ 15

61. $x^2 - x - 7$ for $x = -5$ 23

62. $a^2 + 3a - 9$ for $a = -3$ −9

63. $\dfrac{x-y}{a-b}$ for $x = -1, y = 8, a = 6,$ and $b = 3$ −3

64. $\dfrac{m-n}{c-d}$ for $m = -20, n = -40, c = -5,$ and $d = -10$ 4

65. $\dfrac{-b^2+3b}{2b+1}$ for $b = -4$ 4

66. $\dfrac{-a^2+5a}{2a+12}$ for $a = -3$ −4

67. $\dfrac{24+k}{3k}$ for $k = 3$ 3

68. $\dfrac{4-h}{h-4}$ for $h = -1$ −1

69. $(x-a)^2 + (y-b)^2$ for $x = -2, y = 1, a = 5,$ and $b = -3$ 65

70. $2a^2 + 2ab + b^2$ for $a = -5$ and $b = -1$ 61

71. $|6 - x|$ for $x = 50$ 44

72. $|3c - 1|$ for $c = -1$ 4

73. $-2|x| - 7$ for $x = -7$ −21

74. $|x^2 - 7^2|$ for $x = 7$ 0

75. $2 - [10 - x(5h-1)]$ for $x = -2$ and $h = 2$ −26

76. $1 - [8 - c(2k-7)]$ for $c = -3$ and $k = 4$ −10

77. $b^2 - 4ac$ for $b = -3, a = 4,$ and $c = -1$ 25

78. $\dfrac{2y+1}{x} - x$ for $x = -3$ and $y = -5$ 6

79. $\dfrac{x}{y+10}$ for $x = 30$ and $y = -10$ undefined

80. $\dfrac{e}{3f+24}$ for $e = 24$ and $f = -8$ undefined

81. $\dfrac{50-6s}{-t}$ for $s = 5$ and $t = 4$ −5

82. $\dfrac{7v-5r}{-r}$ for $v = 8$ and $r = 4$ −9

83. $5rs^2t$ for $r = 2, s = -3,$ and $t = -3$ −270

84. $-3bk^2t$ for $b = -5, k = -2,$ and $t = -3$ −180

85. $\dfrac{|a^2 - b^2|}{2a - b}$ for $a = -2$ and $b = -5$ 21

86. $\dfrac{-|2x - 3y + 10|}{-3 - y}$ for $x = 0$ and $y = -4$ −22

APPLICATIONS

87. ACCOUNTING Refer to the financial statement for Avon Products, Inc., shown below. Find the operating profit for the year ending December 2010 and the year ending December 2011.

Annual Financials: Income Statement
(All dollar amounts in millions)

	Year ending Dec. '10	Year ending Dec. '11
Total revenues	10,900	11,300
Cost of goods sold	3,850	3,910
Operating profit	7,050	7,390

Source: *wikinvest.com*

88. CONSTRUCTING TABLES Complete the table below by finding the distance traveled in each instance.

	Rate (mph)	· time (hr)	= distance (mi)
Bike	12	4	48
Walking	3	2	6
Car	3	x	$3x$

89. DASHBOARDS The illustration below shows part of a dashboard. Explain what each of the three instruments measures. What is the formula that mathematically relates these measurements?
speedometer: rate; odometer: distance; clock: time; $d = rt$

90. SPREADSHEETS A store manager wants to use a spreadsheet to post the prices of items on sale. If column B in the following table lists the regular price and column C lists the discount, write a formula using column names to have a computer find the sale price to print in column D. Then fill in column D with the correct sale price. $D = B - C$

	A	B	C	D
1	Bath towel set	$25	$5	$20
2	Pillows	$15	$3	$12
3	Comforter	$53	$11	$42

91. THERMOMETERS A thermometer manufacturer wishes to scale a thermometer in both degrees Celsius and degrees Fahrenheit. Find the missing Celsius degree measures in the illustration. $30°, 15°, -5°$

92. DEALER MARKUPS A car dealer marks up the cars he sells $500 above factory invoice (that is, $500 over what it costs him to purchase the car from the factory).

a. Complete the following table.

Model	Factory invoice ($)	Markup ($)	Price ($)
Minivan	25,600	500	26,100
Pickup	23,200	500	23,700
Convertible	x	500	$x + 500$

b. Write a formula that represents the price p of a car if the factory invoice is f dollars. $p = f + 500$

93. FALLING OBJECTS See the table below. First, find the distance in feet traveled by a falling object in 1, 2, 3, and 4 seconds. Enter the results in the middle column. Then find the distance the object traveled over each time interval and enter it in the right column.

Time falling	Distance traveled (ft)	Time intervals
1 sec	16	Distance traveled from 0 sec to 1 sec 16 ft
2 sec	64	Distance traveled from 1 sec to 2 sec 48 ft
3 sec	144	Distance traveled from 2 sec to 3 sec 80 ft
4 sec	256	Distance traveled from 3 sec to 4 sec 112 ft

94. DISTANCE TRAVELED

a. When in orbit, the space shuttle travels at a rate of approximately 17,250 miles per hour. How far does it travel in one day?
414,000 mi

b. The speed of light is approximately 186,000 miles per second. How far will light travel in 1 minute?
11,160,000 mi

c. The speed of a sound wave in air is about 1,100 feet per second at normal temperatures. How far does it travel in half a minute? 33,000 ft

95. ENERGY USAGE The number of therms of natural gas that were used each month by a household are listed below. Find the mean number of therms the household used per month that year.
32 therms

January: 39   May: 22   September: 33
February: 41   June: 23   October: 41
March: 37   July: 16   November: 35
April: 34   August: 16   December: 47

96. CUSTOMER SATISFACTION As customers were leaving a restaurant, they were asked to rate the service they had received. Good service was rated with a 5, fair service with a 3, and poor service with a 1. The tally sheet compiled by the questioner is shown below. What was the restaurant's average score on this survey? 4

Type of service	Point value	Number
Good	5	卌 卌 卌 卌 卌 卌 卌 卌 卌
Fair	3	卌 卌 卌 III
Poor	1	卌 IIII

WRITING

97. Explain the error in the student's work shown below.

Evaluate $-a + 3a$ for $a = -6$.

$$-a + 3a = -6 + 3(-6)$$
$$= -6 + (-18)$$
$$= -24$$

98. Explain how we can use a stopwatch to find the distance traveled by a falling object.

99. Write a definition for each of these business words: *revenue*, *markup*, and *profit*.

100. What is a formula?

101. In this section we *substituted* a number for a variable. List some other uses of the word *substitute* that you encounter in everyday life.

102. Temperature can be measured using the Fahrenheit or the Celsius scale. How do the scales differ?

103. Show the misunderstanding that occurs if we don't write parentheses around -8 when evaluating the expression $2x + 10$ for $x = -8$.

$$2x + 10 = 2 - 8 + 10$$
$$= -6 + 10$$
$$= 4$$

104. Explain why the following instruction is incomplete. Evaluate the algebraic expression $3a^2 - 4$.

105. What occupation might use a formula that finds:
  a. target heart rate after a workout
  b. gas mileage of a car
  c. age of a fossil
  d. equity in a home
  e. dose to administer
  f. cost-of-living index

106. A car travels at a rate of 65 mph for 15 minutes. What is wrong with the following thinking?

$$d = rt$$
$$= 65(15)$$
$$= 975$$

The car travels 975 miles in 15 minutes.

REVIEW

107. Which of these are prime numbers? 9, 15, 17, 33, 37, 41 17, 37, 41

108. How can this repeated multiplication be rewritten in simpler form? $2 \cdot 2 \cdot 2 \cdot 2 \cdot 2$ 2^5

109. Evaluate: $|-2 + (-5)|$ 7

110. Multiply: $-3(-2)(4)$ 24

111. In the equation $\frac{x}{3} = -4$, what operation is performed on the variable? division by 3

112. Is -6 a solution of $2t - 3 = 15$? Explain. No; when -6 is substituted for t, we obtain a false statement.

113. Subtract: $-3 - (-6)$ 3

114. Which is undefined: division of 0 or division by 0? division by 0

SECTION 3.3
Simplifying Algebraic Expressions and the Distributive Property

Objectives
1. Simplify products.
2. Use the distributive property.
3. Distribute a factor of -1.

In algebra, we frequently replace one algebraic expression with another that is equivalent and simpler in form. That process, called *simplifying an algebraic expression*, often involves the use of one or more properties of real numbers.

1 Simplify products.

The commutative and associative properties of multiplication can be used to simplify certain products. For example, let's simplify $8(4x)$.

$$
\begin{aligned}
8(4x) &= 8 \cdot (4 \cdot x) && \text{Rewrite } 4x \text{ as } 4 \cdot x. \\
&= (8 \cdot 4) \cdot x && \text{Use the associative property of multiplication to group 4 with 8.} \\
&= 32x && \text{Do the multiplication within the parentheses.}
\end{aligned}
$$

We have found that $8(4x) = 32x$. We say that $8(4x)$ and $32x$ are **equivalent expressions** because for each value of x, they represent the same number. For example, if $x = 10$, both expressions have a value of 320. If $x = -3$, both expressions have a value of -96.

If x = 10

$$
\begin{aligned}
8(4x) &= 8[4(10)] & 32x &= 32(10) \\
&= 8(40) & &= 320 \\
&= 320
\end{aligned}
$$

same result

If x = −3

$$
\begin{aligned}
8(4x) &= 8[4(-3)] & 32x &= 32(-3) \\
&= 8(-12) & &= -96 \\
&= -96
\end{aligned}
$$

same result

> **Success Tip** By the commutative property of multiplication, we can change the *order* of factors. By the associative property of multiplication, we can change the *grouping* of factors.

EXAMPLE 1 Simplify: **a.** $2 \cdot 7x$ **b.** $-12t(-6)$

Strategy We will use the commutative and associative properties of multiplication to reorder and regroup the factors in each expression.

WHY We want to group all of the numerical factors of an expression together so that we can find their product.

Solution

a. $2 \cdot 7x = (2 \cdot 7)x$ Use the associative property of multiplication to group the numerical factors together.

$\qquad\quad\ = 14x$ Do the multiplication within the parentheses: $2 \cdot 7 = 14$.

b. $-12t(-6) = -12(-6)t$ Use the commutative property of multiplication to change the order of the factors.

$\qquad\qquad\ = [-12(-6)]t$ Use the associative property of multiplication to group the numbers together. Use brackets to show this.

$\qquad\qquad\ = 72t$ Do the multiplication within the brackets: $-12(-6) = 72$.

$$
\begin{array}{r}
\overset{1}{1}2 \\
\times\ 6 \\
\hline
72
\end{array}
$$

Self Check 1
Simplify:

a. $4 \cdot 8r$ $32r$

b. $-3y(-5)$ $15y$

Now Try Problems 21 and 25

Teaching Example 1
Simplify:
a. $6 \cdot 9m$
b. $-11a(-10)$
Answers:
a. $54m$ **b.** $110a$

Chapter 3 The Language of Algebra

Self Check 2
Simplify:
a. $-7k \cdot 5t$ $-35kt$
b. $2(-3d)(4a)$ $-24ad$

Now Try Problems 29 and 33

Teaching Example 2
Simplify:
a. $-6x \cdot 3y$
b. $3(-5b)(2a)$
Answers:
a. $-18xy$ b. $-30ab$

EXAMPLE 2 Simplify: a. $-4m \cdot 5n$ b. $2(-4z)(6y)$

Strategy We will use the commutative and associative properties of multiplication to reorder and regroup the factors in each expression.

WHY We want to group all of the numerical factors of an expression together so that we can find their product.

Solution

a. $-4m \cdot 5n = (-4 \cdot 5)(m \cdot n)$ Group the numbers and variables separately, using the commutative and associative properties of multiplication.

$= -20mn$ Do the multiplication within the parentheses: $-4 \cdot 5 = -20$. Write $m \cdot n$ as mn.

b. $2(-4z)(6y) = [2(-4)(6)](z \cdot y)$ Use the commutative and associative properties to reorder and regroup the factors. Use brackets to show this.

$= -48zy$ Do the multiplication within the brackets: $2(-4)(6) = -48$. Write $z \cdot y$ as zy.

$= -48yz$ Standard practice is to write variable factors in alphabetical order: $zy = yz$.

The Language of Algebra Be careful when using the words *simplify* and *solve*. In mathematics, we *simplify expressions* and we *solve equations*.

2 Use the distributive property.

Another property that is often used to simplify algebraic expressions is the **distributive property**. To introduce it, we will evaluate $4(5 + 3)$ in two ways.

Method 1	Method 2
Use the order of operations:	Distribute the multiplication:
$4(5 + 3) = 4(8)$	$4(5 + 3) = 4(5) + 4(3)$
$= 32$	$= 20 + 12$
	$= 32$

Each method gives a result of 32. This observation suggests the following property.

The Distributive Property

For any numbers a, b, and c,

$a(b + c) = ab + ac$

The Language of Algebra To *distribute* means to give from one to several. You have probably *distributed* candy to children coming to your door on Halloween.

To illustrate one use of the distributive property, let's consider the expression $5(x + 3)$. Since we are not given the value of x, we cannot add x and 3 within the parentheses. However, we can distribute the multiplication by the factor of 5 that is outside the parentheses to x and to 3 and add those products.

$5(x + 3) = 5(x) + 5(3)$ Distribute the multiplication by 5.
$ = 5x + 15$ Do the multiplication.

In the expression $5(x + 3)$, we say that there are two *terms* within the parentheses, x and 3. In general, a **term** is a product or quotient of numbers and/or variables. A single number or variable is also a term. Some examples of terms are:

$$4, \quad -22, \quad y, \quad -6r, \quad x^2, \quad \text{and} \quad 15ab$$

We will discuss terms in more detail in the next section.

Since subtraction is the same as adding the opposite, the distributive property also holds for subtraction.

The Distributive Property

For any numbers a, b, and c,

$a(b - c) = ab - ac$

EXAMPLE 3
Multiply: **a.** $3(x + 7)$ **b.** $6(5x - 1)$

Strategy In each case, we will distribute the multiplication by the factor *outside* the parentheses over each term *within* the parentheses.

WHY In each case, we cannot simplify the expression within the parentheses. To multiply, we must use the distributive property.

Solution

a. We read $3(x + 7)$ as "three times the *quantity* of x plus seven." The word *quantity* alerts us to the grouping symbols in the expression.

$3(x + 7) = 3 \cdot x + 3 \cdot 7$ Distribute the multiplication by 3.
$ = 3x + 21$ Do the multiplication. Try to go to this step immediately.

Caution! A common mistake is to forget to distribute the multiplication over each of the terms within the parentheses.

$3(x + 7) = 3x + 7$

b. $6(5x - 1) = 6 \cdot 5x - 6 \cdot 1$ Distribute the multiplication by 6.
$ = 30x - 6$ Do the multiplication. Try to go to this step immediately.

Self Check 3
Multiply:

a. $5(h + 4)$ $5h + 20$

b. $9(2a - 3)$ $18a - 27$

Now Try Problems 37 and 41

Teaching Example 3
Multiply:
a. $10(d + 7)$
b. $4(3y - 8)$
Answers:
a. $10d + 70$ **b.** $12y - 32$

The Language of Algebra Formally, it is called the *distributive property of multiplication over addition*. When we use it to write a product, such as $3(x + 7)$, as a sum, $3x + 21$, we say that we have *removed* or *cleared* the parentheses.

Self Check 4

Multiply:

a. $-4(6y + 8)$ $-24y - 32$

b. $-7(2 - 8m)$ $-14 + 56m$

c. $-10(-9r - 5)$ $90r + 50$

d. $-1(x - 3)$ $-x + 3$

Now Try Problems 45, 49, 53, and 57

Teaching Example 4
Multiply:
a. $-2(21x + 3)$
b. $-8(4 - 10n)$
c. $-4(-3t - 16)$
d. $-1(a - 7)$
Answers:
a. $-42x - 6$ b. $-32 + 80n$
c. $12t + 64$ d. $-a + 7$

EXAMPLE 4   Multiply: a. $-3(4x + 2)$ b. $-9(3 - 2t)$
c. $-6(-3y - 8)$ d. $-1(t - 9)$

Strategy In each case, we will distribute the multiplication by the factor *outside* the parentheses over each term *within* the parentheses.

WHY In each case, we cannot simplify the expression within the parentheses. To multiply, we must use the distributive property.

Solution

a. $-3(4x + 2) = -3(4x) + (-3)(2)$ Distribute the multiplication by -3.

$ = -12x + (-6)$ Do the multiplication.

$ = -12x - 6$ Write the answer in simpler form. Adding -6 is the same as subtracting 6. Try to go to this step immediately.

b. $-9(3 - 2t) = -9(3) - (-9)(2t)$ Distribute the multiplication by -9.

$ = -27 - (-18t)$ Do the multiplication.

$ = -27 + 18t$ Write the answer in simpler form. Add the opposite of $-18t$. Try to go to this step immediately.

c. $-6(-3y - 8) = -6(-3y) - (-6)(8)$ Distribute the multiplication by -6.

$ = 18y - (-48)$ Do the multiplication.

$ = 18y + 48$ Write the result in simpler form. Add the opposite of -48. Try to go to this step immediately.

Another approach is to write the subtraction within the parentheses as addition of the opposite. Then we distribute the multiplication by -6 over the addition.

$-6(-3y - 8) = -6[-3y + (-8)]$ Add the opposite of 8.

$ = -6(-3y) + (-6)(-8)$ Distribute the multiplication by -6.

$ = 18y + 48$ Do the multiplication.

d. $-1(t - 9) = -1(t) - (-1)(9)$ Distribute the multiplication by -1.

$ = -t - (-9)$ Do the multiplication.

$ = -t + 9$ Write the result in simpler form. Add the opposite of -9. Try to go to this step immediately.

Notice that distributing the multiplication by -1 *changes the sign* of each term within the parentheses.

Success Tip It is common practice to write answers in simplified form. For instance, the answer to Example 4, part a, is expressed as $-12x - 6$ because it involves fewer symbols than $-12x + (-6)$. For the same reason, the answer to Example 4, part b, is given as $-27 + 18t$ instead of $-27 - (-18t)$.

3.3 Simplifying Algebraic Expressions and the Distributive Property

> **Caution!** The distributive property does not apply to every expression that contains parentheses—only those where multiplication is distributed over addition (or subtraction). For example, to simplify $6(5x)$, we do not use the distributive property.
>
> | Correct | Incorrect |
> |---|---|
> | $6(5x) = (6 \cdot 5)x = 30x$ | $6(5x) = 30 \cdot 6x = 180x$ |

The distributive property can be extended to several other useful forms. Since multiplication is commutative, we have:

$$(b + c)a = ba + ca \qquad (b - c)a = ba - ca$$

For situations in which there are more than two terms within parentheses, we have:

$$a(b + c + d) = ab + ac + ad \qquad a(b - c - d) = ab - ac - ad$$

EXAMPLE 5 Multiply: **a.** $(5 + 3r)7$ **b.** $(4 - x)2$ **c.** $2(a - 3b)8$ **d.** $-6(-3x - 6y + 8)$

Strategy We will multiply each term within the parentheses by the factor (or factors) outside the parentheses.

WHY In each case, we cannot simplify the expression within the parentheses. To multiply, we use the distributive property.

Solution

a. $(5 + 3r)7 = (5)7 + (3r)7$ Distribute the multiplication by 7.
$ = 35 + 21r$ Do the multiplication. Try to go to this step immediately.

b. $(4 - x)2 = (4)2 - (x)2$ Distribute the multiplication by 2.
$ = 8 - 2x$ Do the multiplication.

c. This expression contains 3 factors.
$2(a - 3b)8 = 2 \cdot 8(a - 3b)$ Use the commutative property of multiplication to reorder the factors.
$ = 16(a - 3b)$ Multiply 2 and 8 to get 16.
$ = 16(a) - 16(3b)$ Distribute the multiplication by 16.
$ = 16a - 48b$ Do the multiplication.

d. There are three terms within the parentheses.
$-6(-3x - 6y + 8)$
$= -6(-3x) - (-6)(6y) + (-6)(8)$ Distribute the multiplication by -6.
$= 18x - (-36y) + (-48)$ Do the multiplication.
$= 18x + 36y - 48$ Write the answer in simplest form. Try to go to this step immediately.

Self Check 5
Multiply:
a. $(8 + 7x)5$ $40 + 35x$
b. $(5 - c)3$ $15 - 3c$
c. $4(m - 6n)2$ $8m - 48n$
d. $-2(-7c - 4d + 1)$ $14c + 8d - 2$

Now Try Problems 61, 67, 69, and 75

Teaching Example 5
Multiply:
a. $(14x + 21)3$
b. $(32 - n)4$
c. $3(x - 6y)2$
d. $-7(-2a - 3b + 1)$
Answers:
a. $42x + 63$ **b.** $128 - 4n$
c. $6x - 36y$ **d.** $14a + 21b - 7$

3 Distribute a factor of −1.

We can use the distributive property to find the opposite of a sum. For example, to find $-(x + 10)$, we interpret the $-$ symbol as a factor of -1, and proceed as follows:

$$-(x + 10) = -1(x + 10) \quad \text{Replace the } - \text{ symbol with } -1.$$
$$= -1(x) + (-1)(10) \quad \text{Distribute the multiplication by } -1.$$
$$= -x + (-10) \quad \text{Do the multiplication.}$$
$$= -x - 10 \quad \text{Write the answer in simplest form.}$$

In general, we have the following property.

The Opposite of a Sum

The opposite of a sum is the sum of the opposites.
For any numbers a and b,
$$-(a + b) = -a + (-b)$$

Self Check 6

Simplify: $-(-5x + 18)$ $5x - 18$

Now Try Problem 77

Teaching Example 6
Simplify: $-(16r - 8)$
Answer:
$-16r + 8$

EXAMPLE 6   Simplify: $-(-9s - 3)$

Strategy We will multiply each term within the parentheses by -1.

WHY The $-$ outside the parentheses represents a factor of -1 that is to be distributed.

Solution

$$-(-9s - 3) = -1(-9s - 3) \quad \text{Replace the } - \text{ symbol in front of the parentheses with } -1.$$
$$= -1(-9s) - (-1)(3) \quad \text{Distribute the multiplication by } -1.$$
$$= 9s - (-3) \quad \text{Do the multiplication.}$$
$$= 9s + 3 \quad \text{Write the answer in simplest form. Try to go to this step immediately}$$

Success Tip After working several problems like Example 6, you will notice that it is not necessary to show each of the steps. The result can be obtained very quickly by *changing the sign of each term within the parentheses and dropping the parentheses.*

ANSWERS TO SELF CHECKS

1. a. $32r$ **b.** $15y$ **2. a.** $-35kt$ **b.** $-24ad$ **3. a.** $5h + 20$ **b.** $18a - 27$
4. a. $-24y - 32$ **b.** $-14 + 56m$ **c.** $90r + 50$ **d.** $-x + 3$ **5. a.** $40 + 35x$
b. $15 - 3c$ **c.** $8m - 48n$ **d.** $14c + 8d - 2$ **6.** $5x - 18$

SECTION 3.3 STUDY SET

VOCABULARY

Fill in the blanks.

1. To <u>simplify</u> the expression $5(6x)$ means to write it in simpler form: $5(6x) = 30x$.
2. $5(6x)$ and $30x$ are <u>equivalent</u> expressions because for each value of x, they represent the same number.
3. In the expression $2(x + 8)$, there are two <u>terms</u> within the parentheses, x and 8.
4. To perform the multiplication $2(x + 8)$, we use the <u>distributive</u> property.
5. When we use the distributive property to write a product, such as $7(4y + 3)$, as the sum $28y + 21$, we say we have <u>removed</u> or *cleared* the parentheses.
6. We call $-(c + 9)$ the <u>opposite</u> of a sum.

CONCEPTS

7. a. Fill in the blanks to simplify the expression.
   $$4(9t) = (\,\underline{4}\, \cdot 9)t = \underline{36}\, t$$
   b. What property did you use in part a?
   associative property of multiplication

8. a. Fill in the blanks to simplify the expression.
   $$-6y \cdot 2 = \underline{-6} \cdot 2 \cdot y = \underline{-12}\, y$$
   b. What property did you use in part a?
   commutative property of multiplication

9. State the distributive property using the variables x, y, and z. $x(y + z) = xy + xz$

10. Fill in the blanks.
    a. $2(x + 4) = 2x\, \underline{+}\, 8$
    b. $2(x - 4) = 2x\, \underline{-}\, 8$
    c. $-2(x + 4) = -2x\, \underline{-}\, 8$
    d. $-2(-x - 4) = 2x\, \underline{+}\, 8$

11. Fill in the blanks: Distributing multiplication by -1 changes the <u>sign</u> of each term within the parentheses.
    $$-(x + 10) = \underline{-1}\,(x + 10) = -x\, \underline{-}\, 10$$

12. For each of the following expressions, determine whether the distributive property applies. Write *yes* or *no*.
    a. $3(5t)$ no
    b. $3(t + 5)$ yes
    c. $5(3 \cdot t)$ no
    d. $(3t)5$ no
    e. $(3)(-t)5$ no
    f. $(5 - t)3$ yes

13. a. Simplify: $6(4x)$ $24x$
    b. Remove parentheses: $6(4 + x)$ $24 + 6x$

14. Explain what the arrows are illustrating.
    $-9(y - 7)$ distributing the multiplication by -9

NOTATION

Complete each step.

15. $-5 \cdot 7n = (\,\underline{-5}\, \cdot 7)n$
    $ = \underline{-35}\, n$

16. $6y(-9) = 6(\,\underline{-9}\,)y$
    $ = [6(\,\underline{-9}\,)]y$
    $ = -54y$

17. $9(5y - 4) = \underline{9}\,(5y) - \underline{9}\,(4)$
    $ = \underline{45y}\, - 36$

18. $4(2a + b - 1) = 4(\,\underline{2a}\,) + 4(\,\underline{b}\,) - \underline{4}\,(1)$
    $ = \underline{8a}\, + 4b - 4$

19. Write each expression in simpler form, using fewer mathematical symbols.
    a. $-(-x)$ x
    b. $x - (-5)$ $x + 5$
    c. $10y + (-15)$ $10y - 15$
    d. $5 \cdot x$ $5x$

20. In each expression, determine what number is to be distributed.
    a. $6(x - 2)$ 6
    b. $(t + 1)(-5)$ -5
    c. $(a + 24)8$ 8
    d. $-(z - 16)$ -1

GUIDED PRACTICE

Simplify. See Example 1.

21. $2 \cdot 6x$ $12x$
22. $4 \cdot 7b$ $28b$
23. $5 \cdot 8y$ $40y$
24. $12 \cdot 6t$ $72t$
25. $-10t(-10)$ $100t$
26. $-8k(-6)$ $48k$
27. $-15a(-3)$ $45a$
28. $-11n(-9)$ $99n$

Simplify. See Example 2.

29. $-7x \cdot 9y$ $-63xy$
30. $-13a \cdot 2b$ $-26ab$
31. $-4r \cdot 4s$ $-16rs$
32. $-7x \cdot 7y$ $-49xy$
33. $2(-5x)(3y)$ $-30xy$
34. $4(-3y)(4z)$ $-48yz$
35. $5r(2)(-3b)$ $-30br$
36. $4d(5)(-3e)$ $-60de$

Multiply. See Example 3.

37. $4(x + 1)$ $4x + 4$
38. $5(y + 3)$ $5y + 15$
39. $7(b + 2)$ $7b + 14$
40. $8(k + 7)$ $8k + 56$
41. $9(3e - 3)$ $27e - 27$
42. $10(7t - 2)$ $70t - 20$
43. $3(2q - 7)$ $6q - 21$
44. $6(3p - 1)$ $18p - 6$

Multiply. See Example 4.

45. $-2(3h + 5)$ $-6h - 10$
46. $-5(7t + 3)$ $-35t - 15$
47. $-10(4y + 6)$ $-40y - 60$
48. $-9(2t + 9)$ $-18t - 81$
49. $-8(2q - 4)$ $-16q + 32$
50. $-2(22x - 1)$ $-44x + 2$
51. $-5(7g - 1)$ $-35g + 5$
52. $-7(3p - 8)$ $-21p + 56$
53. $-4(-5s - 3)$ $20s + 12$
54. $-6(-3d - 1)$ $18d + 6$
55. $-6(-15t - 9)$ $90t + 54$
56. $-4(-5d - 6)$ $20d + 24$
57. $-1(x - 5)$ $-x + 5$
58. $-1(y - 1)$ $-y + 1$
59. $-1(5d + 8)$ $-5d - 8$
60. $-1(6w + 2)$ $-6w - 2$

Multiply. See Example 5.

61. $(4d + 7)6$ $24d + 42$
62. $(8r + 2)7$ $56r + 14$
63. $(3q + 20)7$ $21q + 140$
64. $(30x + 12)3$ $90x + 36$
65. $(-4 - d)6$ $-24 - 6d$
66. $(-9 - j)5$ $-45 - 5j$
67. $(t - 12)9$ $9t - 108$
68. $(x - 25)6$ $6x - 150$
69. $2(4t - 3)3$ $24t - 18$
70. $3(9m - 2)2$ $54m - 12$
71. $4(3h + 1)5$ $60h + 20$
72. $4(2w + 1)6$ $48w + 24$
73. $-3(-3z - 3x + 5)$
    $9z + 9x - 15$
74. $-10(-5e - 4a + 6)$
    $50e + 40a - 60$
75. $-8(2a + 4b - 6)$
    $-16a - 32b + 48$
76. $-9(3r + 6s - 9)$
    $-27r - 54s + 81$

Simplify. See Example 6.

77. $-(-3w - 4)$ $3w + 4$
78. $-(-4y - 6)$ $4y + 6$
79. $-(-18x - 19)$ $18x + 19$
80. $-(-50n - 100)$ $50n + 100$
81. $-(x + 3)$ $-x - 3$
82. $-(5 + y)$ $-5 - y$
83. $-(4t + 5)$ $-4t - 5$
84. $-(8x + 4)$ $-8x - 4$

TRY IT YOURSELF

Perform the indicated operations.

85. $-(x - 7)$
    $-x + 7$
86. $-(y + 1)$
    $-y - 1$
87. $(4s)3$ $12s$
88. $(9j)7$ $63j$
89. $5(-7q)$ $-35q$
90. $-7(5t)$ $-35t$
91. $6(-6c + 7)$ $-36c + 42$
92. $9(-9d + 3)$ $-81d + 27$
93. $5(9s - 12t - 3)$
    $45s - 60t - 15$
94. $5(4 - 5r + 8s)$
    $20 - 25r + 40s$
95. $-5 \cdot 8h$ $-40h$
96. $-8 \cdot 4d$ $-32d$
97. $5 \cdot 8c \cdot 2$ $80c$
98. $3 \cdot 6j \cdot 2$ $36j$
99. $2(3t + 2)8$ $48t + 32$
100. $3(2q + 1)9$ $54q + 27$
101. $(-1)(-2e)(-4)$ $-8e$
102. $(-1)(-5t)(-1)$ $-5t$
103. $-(5x - 4y + 1)$
     $-5x + 4y - 1$
104. $-(6r - 5f + 1)$
     $-6r + 5f - 1$

Each expression is the result of an application of the distributive property. What was the original algebraic expression?

105. $2(4x) + 2(5)$ $2(4x + 5)$
106. $3(3y) + 3(7)$ $3(3y + 7)$
107. $-3(4y) + (-3)(2)$
     $-3(4y + 2)$
108. $-5(11s) + (-5)(7t)$
     $-5(11s + 7t)$
109. $3(4) - 3(7t) - 3(5s)$
     $3(4 - 7t - 5s)$
110. $2(7y) + 2(8x) - 2(4)$
     $2(7y + 8x - 4)$
111. $-4(5) - 3x(5)$
     $(-4 - 3x)5$
112. $-8(7) - (4s)(7)$
     $(-8 - 4s)7$

WRITING

113. Explain what it means to simplify an algebraic expression. Give an example.
114. Explain how to apply the distributive property. Give an example.
115. Use the word *distribute* in a sentence that describes a situation from everyday life.
116. Explain why the distributive property applies to $2(3 + x)$ but does not apply to $2(3x)$.
117. Explain the mistake: $5(6x + 2) = 30x + 2$
118. The distributive property can be demonstrated using the following illustration.
    a. Fill in the blanks: Two groups of 6 plus three groups of 6 is 5 groups of 6.
       Therefore,
       $6 \cdot 2 + 6 \cdot 3 = 6(2 + 3)$

    b. Draw a diagram that illustrates $5 \cdot 4 + 5 \cdot 6 = 5(4 + 6)$.

REVIEW

119. Evaluate: $|-6 + 1|$ 5
120. Subtract: $-1 - (-4)$ 3
121. Identify the operation associated with each word: *product, quotient, difference, sum.* multiplication, division, subtraction, addition
122. What steps are used to find the mean (average) of a set of values? Add the values and divide by the number of values.
123. Insert the proper inequality symbol: $-6 \; > \; -7$
124. Fill in the blank: To factor a number means to express it as the product of other whole numbers.
125. Which of the following involve area: carpeting a room, fencing a yard, walking around a lake, painting a wall? carpeting, painting
126. Write seven squared and seven cubed. 7^2, 7^3

SECTION 3.4
Combining Like Terms

Objectives
1. Identify terms and coefficients of terms.
2. Identify like terms.
3. Combine like terms.
4. Find the perimeter of a rectangle and square.

In this section, we will show how the distributive property can be used to simplify algebraic expressions that involve addition and subtraction. We will also review the concept of perimeter and write the formulas for the perimeter of a rectangle and a square using variables.

1 Identify terms and coefficients of terms.

Addition symbols separate expressions into parts called *terms*. For example, the expression $x + 8$ has two terms.

$$\underbrace{x}_{\text{First term}} + \underbrace{8}_{\text{Second term}}$$

Since subtraction can be written as addition of the opposite, the expression $a^2 - 3a - 9$ has three terms.

$$a^2 - 3a - 9 = \underbrace{a^2}_{\text{First term}} + \underbrace{(-3a)}_{\text{Second term}} + \underbrace{(-9)}_{\text{Third term}}$$

In general, a **term** is a product or quotient of numbers and/or variables. A single number or variable is also a term. Examples of terms are:

$$4, \quad y, \quad 6r, \quad -w^3, \quad 7x^5, \quad \frac{3}{n}, \quad -15ab^2$$

> **Caution!** By the commutative property of multiplication, $r6 = 6r$ and $-15b^2a = -15ab^2$. However, when writing terms, we usually write the numerical factor first and the variable factors in alphabetical order.

EXAMPLE 1 Identify the terms of each expression:
a. $3x^2 + 5x + 8$ **b.** $-24rs$ **c.** $a - 5 - 3a + 10$

Strategy We will locate the addition symbols in each expression.

WHY Addition symbols separate expressions into terms.

Solution

a. After locating the addition symbols in the given expression, we see that it has three terms: $3x^2$, $5x$, and 8.

$$\underbrace{3x^2}_{\text{First term}} + \underbrace{5x}_{\text{Second term}} + \underbrace{8}_{\text{Third term}}$$

b. Since the given expression does not contain any addition or subtraction symbols, it has only one term, $-24rs$.

c. When we write each subtraction in the expression $a - 5 - 3a + 10$ as addition of the opposite, we see that it has four terms: a, -5, $-3a$, and 10.

$$a + (-5) + (-3a) + 10$$

Self Check 1
Identify the terms of each expression:

a. $12y^2 + y + 10$ $\quad 12y^2, y, 10$

b. $-4ab$ $\quad -4ab$

c. $9 - m - 6m + 12$ $\quad 9, -m, -6m, 12$

Now Try Problems 23 and 27

Teaching Example 1
Identify the terms of each expression:
a. $2y^2 + 7y + 15$
b. $-6ab$
c. $t - 4 - 9t + 8$
Answers:
a. $2y^2, 7y, 15$ **b.** $-6ab$ **c.** $t, -4, -9t, 8$

Chapter 3 The Language of Algebra

It is important to be able to distinguish between the *terms* of an expression and the *factors* of a term.

Self Check 2

Is b used as a *factor* or a *term* in each expression?
a. $-27b$ factor
b. $5a + b$ term

Now Try Problems 31 and 33

Teaching Example 2 Is x used as a *factor* or a *term* in each expression?
a. $x + 2$
b. $2x$
Answers:
a. term
b. factor

EXAMPLE 2 Is m used as a *factor* or a *term* in each expression?
a. $m + 6$ b. $8m$

Strategy We will begin by determining whether m is involved in an addition or a multiplication.

WHY Addition symbols separate expressions into *terms*. A factor is a number being multiplied.

Solution

a. Since m is added to 6, m is a term of $m + 6$.

b. Since m is multiplied by 8, m is a factor of $8m$.

The numerical factor of a term is called the **coefficient** of the term. For instance, the term $6r$ has a coefficient of 6 because $6r = 6 \cdot r$. The coefficient of $-15ab^2$ is -15 because $-15ab^2 = -15 \cdot ab^2$. More examples are shown below.

A term such as 4, that consists of a single number, is called a **constant term**.

Term	Coefficient	
$8y^2$	8	
$9pq$	9	
$-78m$	-78	
$-2b^2$	-2	
x	1	Because $x = 1x$
$-y$	-1	Because $-y = -1y$
27	27	The coefficient of a constant term is that constant.

Notice that when there is no number in front of a variable, the coefficient is understood to be 1. For example, the coefficient of the term x is 1. If there is only a negative (or opposite) sign in front of the variable, the coefficient is understood to be -1. Therefore, $-y$ can be thought of as $-1y$.

> **The Language of Algebra** Terms such as x and y have *implied* coefficients of 1. *Implied* means suggested without being directly expressed.

Self Check 3

Identify the coefficient of each term in the expression:
$p^3 - 12p^2 + 3p - 4$ $1, -12, 3, -4$

Now Try Problems 35 and 41

Teaching Example 3 Identify the coefficient of each term in the expression: $8x^2 - x + 9$
Answer:
$8, -1, 9$

EXAMPLE 3 Identify the coefficient of each term in the expression: $7x^2 - x + 6$

Strategy We will begin by writing the subtraction as addition of the opposite. Then we will determine the numerical factor of each term.

WHY Addition symbols separate expressions into terms.

Solution If we write $7x^2 - x + 6$ as $7x^2 + (-x) + 6$, we see that it has three terms: $7x^2$, $-x$, and 6. The numerical factor of each term is its coefficient.

The coefficient of $7x^2$ is 7 because $7x^2$ means $7 \cdot x^2$.

The coefficient of $-x$ is -1 because $-x$ means $-1 \cdot x$.

The coefficient of the constant 6 is 6.

2 Identify like terms.

Before we can discuss methods for simplifying algebraic expressions involving addition and subtraction, we need to introduce some new vocabulary.

> **Like Terms**
>
> **Like terms** are terms containing exactly the same variables raised to exactly the same powers. Any constant terms in an expression are considered to be like terms. Terms that are not like terms are called **unlike terms**.

Here are several examples.

Like terms	*Unlike terms*	
$4x$ and $7x$	$4x$ and $7y$	The variables are not the same.
$-10p^2$ and $25p^2$	$-10p$ and $25p^2$	Same variable, but different powers.
$8c^3d$ and c^3d	$8c^3d$ and c^3	The variables are not the same.

> **Success Tip** When looking for like terms, don't look at the coefficients of the terms. Consider only the variable factors of each term. If two terms are like terms, only their coefficients may differ.

EXAMPLE 4 Identify the like terms in each expression:
a. $7r + 5 + 3r$ **b.** $6x^4 - 6x^2 - 6x$ **c.** $-17m^3 + 3 - 2 + m^3$

Strategy First, we will identify the terms of the expression. Then we will look for terms that contain the same variables raised to exactly the same powers.

WHY If two terms contain the same variables raised to the same powers, they are like terms.

Solution

a. $7r + 5 + 3r$ contains the like terms $7r$ and $3r$.

b. Since the exponents on x are different, $6x^4 - 6x^2 - 6x$ contains no like terms.

c. $-17m^3 + 3 - 2 + m^3$ contains two pairs of like terms: $-17m^3$ and m^3 are like terms, and the constant terms, 3 and -2, are like terms.

Self Check 4
Identify the like terms in each expression:
a. $2x - 2y + 7y$
b. $5p^2 - 12 + 17p^2 + 2$

Now Try Problems 43 and 47

Self Check 4 Answers
a. $-2y$ and $7y$
b. $5p^2$ and $17p^2$; -12 and 2

Teaching Example 4 Identify the like terms in each expression:
a. $6x + 7 + 2x$
b. $8a^4 - 8a^3 - 8a^2$
c. $-4t^3 + 1 - 9 + t^3$
Answers:
a. $6x$ and $2x$
b. no like terms
c. $-4t^3$ and t^3; 1 and -9

3 Combine like terms.

To add or subtract objects, they must be similar. For example, we can add apples to apples, but we cannot add apples to oranges. The same is true when working with terms of an algebraic expression. They can be added or subtracted only if they are *like terms*.

This expression can be simplified because it contains like terms.

$$3x + 4x$$

This expression cannot be simplified because its terms are not like terms.

$$3x + 4y$$

Recall that the distributive property can be written in the following forms:

$$(b + c)a = ba + ca \qquad (b - c)a = ba - ca$$

We can use these forms of the distributive property in reverse to simplify a sum or difference of like terms. For example, we can simplify $3x + 4x$ as follows:

$3x + 4x = (3 + 4)x$ Use the form: $ba + ca = (b + c)a$.
$ = 7x$ Do the addition within the parentheses.

Success Tip Just as 3 apples plus 4 apples is 7 apples,
$$3x + 4x = 7x$$

We can simplify $15m^2 - 9m^2$ in a similar way:

$15m^2 - 9m^2 = (15 - 9)m^2$ Use the form: $ba - ca = (b - c)a$.
$ = 6m^2$ Do the subtraction within the parentheses.

The Language of Algebra Simplifying a sum or difference of like terms is called *combining like terms*. In each example above, we say that we combined like terms.

These examples suggest the following general rule.

Combining Like Terms

Like terms can be combined by adding or subtracting the coefficients of the terms and keeping the same variables with the same exponents.

Self Check 5
Simplify, if possible:
a. $3x + 5x$ $8x$
b. $-6y + (-6y) + 9y$ $-3y$
c. $4s^4 - 2s^4$ $2s^4$
d. $4a - 2$ does not simplify
e. $10r + 6r - 9r$ $7r$

Now Try Problems 51, 55, 59, 63, and 67

Teaching Example 5 Simplify by combining like terms, if possible:
a. $6x + 2x$
b. $-7m + (-6m) + 3m$
c. $7t^3 - 4t^3$
d. $9r + 11$
e. $6t + 9t - 3t$
Answers:
a. $8x$
b. $-10m$
c. $3t^3$
d. does not simplify
e. $12t$

EXAMPLE 5   Simplify by combining like terms, if possible:
a. $2x + 9x$ b. $-8p + (-2p) + 4p$ c. $5s^3 - 3s^3$ d. $4w + 6$ e. $8a + 2a - 3a$

Strategy We will use the distributive property in reverse to add (or subtract) the coefficients of the like terms. We will keep the same variables raised to the same powers.

WHY To *combine like terms* means to add or subtract the like terms in an expression.

Solution

a. Since $2x$ and $9x$ are like terms with the common variable x, we can combine them.

$$2x + 9x = 11x \quad \text{Think: } (2 + 9)x = 11x.$$

b. $-8p + (-2p) + 4p = -6p$ Think: $[-8 + (-2) + 4]p = -6p$.

c. $5s^3 - 3s^3 = 2s^3$ Think: $(5 - 3)s^3 = 2s^3$.

d. Since $4w$ and 6 are not like terms, they cannot be combined. The expression $4w + 6$ does not simplify.

e. $8a + 2a - 3a = 7a$ Think: $(8 + 2 - 3)a = 7a$.

3.4 Combining Like Terms

EXAMPLE 6 Simplify by combining like terms:
a. $16t - 15t$ b. $16t^2 - t^2$ c. $15t - 16t$ d. $16t + t$

Strategy As we combine like terms, we must be careful when working with terms such as t and $-t$.

WHY Coefficients of 1 and -1 are usually not written.

Solution
a. $16t - 15t = t$ Think: $(16 - 15)t = 1t = t$.
b. $16t^2 - t^2 = 15t^2$ Think: $16t^2 - 1t^2 = (16 - 1)t^2 = 15t^2$.
c. $15t - 16t = -t$ Think: $(15 - 16)t = -1t = -t$.
d. $16t + t = 17t$ Think: $16t + 1t = (16 + 1)t = 17t$.

Self Check 6
Simplify:
a. $9h - h$ $8h$ b. $9h + h$ $10h$
c. $9h - 8h$ h d. $8h - 9h$ $-h$

Now Try Problems 71 and 77

Teaching Example 6 Simplify by combining like terms:
a. $12y - 11y$ b. $12y - y$
c. $11y - 12y$ d. $12y + y$
Answers:
a. y b. $11y$ c. $-y$ d. $13y$

EXAMPLE 7 Simplify: $6a^2 + 54a - 4a - 36$

Strategy First, we will identify any like terms in the expression. Then we will use the distributive property in reverse to combine them.

WHY To *simplify* an expression we use properties of real numbers to write an equivalent expression in simpler form.

Solution We can combine the like terms that involve the variable a.

$6a^2 + 54a - 4a - 36 = 6a^2 + 50a - 36$ Think: $(54 - 4)a = 50a$.

Self Check 7
Simplify: $7y^2 + 21y - 2y - 6$

Now Try Problem 79
Self Check 7 Answer
$7y^2 + 19y - 6$

Teaching Example 7 Simplify:
$5x^2 + 13x - 3x - 4$
Answer:
$5x^2 + 10x - 4$

EXAMPLE 8 Simplify: $4(x + 5) - 5 - (2x - 4)$

Strategy First, we will remove the parentheses. Then we will identify any like terms and combine them.

WHY To *simplify* an expression we use properties of real numbers, such as the distributive property, to write an equivalent expression in simpler form.

Solution Here, the distributive property is used both *forward* (to remove parentheses) and in *reverse* (to combine like terms).

$4(x + 5) - 5 - (2x - 4) = 4(x + 5) - 5 - 1(2x - 4)$ Replace the $-$ symbol in front of $(2x - 4)$ with -1.

$ = 4x + 20 - 5 - 2x + 4$ Distribute the multiplication by 4 and -1.

$ = 2x + 19$ Think: $(4 - 2)x = 2x$.
Think: $(20 - 5 + 4) = 19$.

Self Check 8
Simplify:
$6(3y - 1) + 2 - (-3y + 4)$

Now Try Problem 83
Self Check 8 Answer
$21y - 8$

Teaching Example 8 Simplify:
$7(b + 6) - 1 - (3b - 9)$
Answer:
$4b + 50$

---

4 Find the perimeter of a rectangle and square.

To develop the formula for the perimeter of a rectangle, we let $l =$ the length of the rectangle and $w =$ the width of the rectangle, as shown on the right. Then

$P = l + w + l + w$ The perimeter is the distance around the rectangle.

$P = 2l + 2w$ Combine like terms: $l + l = 2l$ and $w + w = 2w$.

Unless otherwise noted, all art on this page is © Cengage Learning.

264 Chapter 3 The Language of Algebra

The Formula for the Perimeter of a Rectangle

The **perimeter P of a rectangle** with length l and width w is given by

$$P = 2l + 2w$$

To develop the formula for the perimeter of a square, we let s = the length of a side of the square, as shown on the right. Then

$P = s + s + s + s$ Add the lengths of the four sides.

$P = 4s$ Combine like terms. Recall that $s = 1s$.

The Formula for the Perimeter of a Square

The **perimeter of a square** with sides of length s is given by

$$P = 4s$$

Self Check 9

ENERGY CONSERVATION Refer to the figure in Example 9. Find the cost to weatherstrip around the door and window if the door is 8 feet tall and 3 feet wide and the window is 5 feet long and 3 feet high. 760¢ or $7.60

Now Try Problem 119

Teaching Example 9 WALLS Find the cost to construct a wall around a rectangular-shaped lot that is 116 feet long and 97 feet wide if the wall costs $27 a running foot.
Answer:
$11,502

EXAMPLE 9 Energy Conservation

Refer to the figure to the right. Find the cost to weatherstrip around the front door and the window of the house if the material costs 20¢ a foot.

Analyze

- The door is in the shape of a rectangle. Given
- The window is in the shape of a square. Given
- The weatherstripping material costs 20¢ a foot. Given
- What will it cost to weatherstrip around the front door and window? Find

Assign

Let P = the total perimeter.

Form

Now we translate the words of the problem into an equation. We will use the formulas for the perimeter of a rectangle and a square.

The total perimeter	is	the perimeter of the door	plus	the perimeter of the window.
P	=	$2l + 2w$	+	$4s$

Solve

$P = 2l + 2w + 4s$ This is a formula for the combined perimeter.

$P = 2(7) + 2(3) + 4(3)$ Substitute 7 for l, 3 for w, and 3 for s.

$P = 14 + 6 + 12$ Do the multiplication.

$P = 32$ Do the addition.

The total perimeter is 32 feet. At 20¢ a foot, the total cost will be $(32 \cdot 20)$¢.

```
   32
 × 20
  640
```

State

It will cost 640¢ or $6.40 to weatherstrip around the front door and window.

Check

We can check the result by estimation. The perimeter is approximately 30 feet, and $30 \cdot 20 = 600$¢, which is $6. The answer, $6.40, seems reasonable.

ANSWERS TO SELF CHECKS

1. **a.** $12y^2, y, 10$ **b.** $-4ab$ **c.** $9, -m, -6m, 12$ 2. **a.** factor   **b.** term
3. $1, -12, 3, -4$ 4. **a.** $-2y$ and $7y$ **b.** $5p^2$ and $17p^2$; -12 and 2 5. **a.** $8x$ **b.** $-3y$
   c. $2s^4$ **d.** does not simplify   **e.** $7r$ 6. **a.** $8h$ **b.** $10h$ **c.** h **d.** $-h$
7. $7y^2 + 19y - 6$ 8. $21y - 8$ 9. 760¢ or $7.60

SECTION 3.4   STUDY SET

VOCABULARY

Fill in the blanks.

1. A __term__ is a number or a product of a number and one or more variables. A single number or variable is also a __term__.
2. A __factor__ is a number being multiplied.
3. In the term $5t$, 5 is called the __coefficient__.
4. A term that consists of a single number is called a __constant__ term.
5. Terms such as x and y have an __implied__ coefficient of 1.
6. Terms with exactly the same variables raised to exactly the same powers are called __like__ terms.
7. When we write $9x + x$ as $10x$, we say that we have __combined__ like terms.
8. The __perimeter__ of a rectangle is the distance around it.

CONCEPTS

Fill in the blanks.

9. The expression $5x - 10 + 8x$ has __3__ terms. The second term is __-10__. The coefficient of the third term is __8__.
10. The expression $2a - 12 + 5a + 15$ has __4__ terms. The third term is __$5a$__. The coefficient of the first term is __2__.
11. The term $8m$ has a coefficient of 8 because $8m =$ __8__ $\cdot$ __m__.
12. Just as 5 pencils plus 6 pencils is 11 __pencils__, $5x + 6x = 11$ __x__.

13. Are the given pair of terms *like* or *unlike* terms?
    a. $6a$ and $6b$ unlike
    b. $5x^2$ and $5x^3$ unlike
    c. $3mn$ and $3m^2n$ unlike
    d. 15 and 16   like

14. What exponent must appear in each box to have like terms?
    a. $6x^2, 3x^{\boxed{2}}$ **b.** $-8h^5, -5h^{\boxed{5}}$
    c. $7a^3, 21a^{\boxed{3}}$ **d.** $25n^4, -15n^{\boxed{4}}$

NOTATION

Complete each step to simplify the expression.

15. $2x + 3x = (\boxed{2} + \boxed{3})x$
    $ = \boxed{5}x$

16. $16w^2 - 12w^2 = (\boxed{16} - \boxed{12})w^2$
    $ = \boxed{4}w^2$

17. $2(x - 1) + 3x = 2x - \boxed{2} + 3x$
    $ = \boxed{5x} - 2$

18. $-3(1 - b) - b = -3 + \boxed{3b} - b$
    $ = \boxed{2b} - 3$

19. In the formula $P = 2l + 2w$,
    a. what does P represent?   the perimeter of a rectangle
    b. what does $2l$ mean?   2 times the length
    c. what does $2w$ mean?   2 times the width

20. In the formula $P = 4s$,
    a. what does P represent?   the perimeter of a square
    b. what does $4s$ mean?   4 times the length of a side

21. Determine whether each statement is true or false.
    a. $x = 1x$ true
    b. $-y = -1y$ true
    c. $100yx = 100xy$ true
    d. $7x - x = 7$ false
22. Fill in the blanks:
    $y^2 - 7y - 3 = y^2 + (\,-7y\,) + (\,-3\,)$

GUIDED PRACTICE

Identify the terms of each expression. See Example 1.

23. $3x^2 - 9x + 4$
    $3x^2, -9x, 4$
24. $y^2 - 12y + 6$
    $y^2, -12y, 6$
25. $5 + 5t - 8t - 1$
    $5, 5t, -8t, -1$
26. $3x - y - 5x + y$
    $3x, -y, -5x, y$
27. $-35a$
    $-35a$
28. $-7t$
    $-7t$
29. $9mn - 6n$
    $9mn, -6n$
30. $3rs - 2r$
    $3rs, -2r$

Determine whether x is used as a factor or as a term.
See Example 2.

31. a. $x + 12$ term   b. $7x$ factor
32. a. $12x + 12y - 6$ factor b. $x - 36y$ term
33. a. $5x(-10)$ factor b. $8 + x - z$ term
34. a. $100 + x + z$ term b. $-xz$ factor

Identify the coefficient of each term in the expression.
See Example 3.

35. $5x^2 + x - 12$ $5, 1, -12$
36. $9y^2 + y - 8$ $9, 1, -8$
37. $a^3 - 27$ $1, -27$
38. $b^3 - 64$ $1, -64$
39. $xy - x + y + 10$ $1, -1, 1, 10$
40. $mn + m - n - 4$ $1, 1, -1, -4$
41. $-a^2 + 6b^2 - a + 5$ $-1, 6, -1, 5$
42. $-8x^3 - 4x^2 + 3x + 1$ $-8, -4, 3, 1$

Identify the like terms in each expression. See Example 4.

43. $8x + 7 + 2x$ $8x, 2x$
44. $9y + 12 + 11y$ $9y, 11y$
45. $6x^3 + 3x^2 + 6x$ none
46. $2m^2 - 2m + 2$ none
47. $-3k^3 + 6k + k^3 - 3k$ $-3k^3, k^3; 6k, -3k$
48. $-12m^4 - 3m^3 + 2m^2 - m^3$ $-3m^3, -m^3$
49. $12a - 8 + 15a + 1$ $12a, 15a; -8, 1$
50. $33t + 4 + 18t - 9$ $33t, 18t; 4, -9$

Simplify by combining like terms, if possible. See Example 5.

51. $6t + 9t$ $15t$
52. $7r + 5r$ $12r$
53. $20b + 30b$ $50b$
54. $18c + 12c$ $30c$
55. $-5x + (-6x) + 2x$ $-9x$
56. $-8m + (-6m) + 7m$ $-7m$
57. $-5d + (-9d) + 10d$ $-4d$
58. $-4a + (-12a) + 11a$ $-5a$
59. $5s^2 - 3s^2$ $2s^2$
60. $8y^2 - 5y^2$ $3y^2$
61. $3e^3 - 17e^3$ $-14e^3$
62. $2s^3 - 14s^3$ $-12s^3$

63. $h - 7$ does not simplify
64. $j - 8$ does not simplify
65. $14z - 8z + 2z$ $8z$
66. $9w - 3w + 8w$ $14w$
67. $53a + 6a - 21a$ $38a$
68. $72n + 8n - 35n$ $45n$
69. $2x + 2y$ does not simplify
70. $5a - 5b$ does not simplify

Simplify by combining like terms. See Example 6.

71. $10s - 9s$ s
72. $7q - 6q$ q
73. $40a^2 - a^2$ $39a^2$
74. $13z^2 - z^2$ $12z^2$
75. $43s^3 - 44s^3$ $-s^3$
76. $4h - 5h$ $-h$
77. $14r + r$ $15r$
78. $21w + w$ $22w$

Simplify. See Example 7.

79. $5x^2 + 19x - 3x + 6$
    $5x^2 + 16x + 6$
80. $2b^2 - 6b + 12b + 1$
    $2b^2 + 6b + 1$
81. $y^2 - 8y - 2y - 4$
    $y^2 - 10y - 4$
82. $n^2 - 4n - 7n - 3$
    $n^2 - 11n - 3$

Simplify. See Example 8.

83. $5(m + 2) - 8 - (3m - 1)$ $2m + 3$
84. $7(r + 1) - 9 - (2r - 4)$ $5r + 2$
85. $4(x - 1) - 2 - (x + 5)$ $3x - 11$
86. $10(x - 1) - 6 - (x + 8)$ $9x - 24$

Use the formulas from this section to find the perimeter of each figure. See Example 9.

87. A rectangle with length 16 feet and width 7 feet 46 ft
88. A rectangle with length 24 inches and width 11 inches 70 in.
89. A square with a side 37 yards long 148 yd
90. A square with a side 98 miles long 392 mi

TRY IT YOURSELF

Simplify each expression, if possible.

91. $-7b^2 + 27b^2$ $20b^2$
92. $-2c^3 + 12c^3$ $10c^3$
93. $-4(-4y + 5) + 4 - 6(y + 2)$ $10y - 28$
94. $-3(-6y - 8) - 15 - 4(5 - y)$ $22y - 11$
95. $6t + 9 + 5t + 3$
    $11t + 12$
96. $4b + 9 - 9b + 9$
    $-5b + 18$
97. $x + x + x + x$ $4x$
98. $s - s - s$ $-s$
99. $3t - (t - 8)$
    $2t + 8$
100. $-(z + 2) + 5(6 - z)$
    $-6z + 28$
101. $5(2x)(-5)$ $-50x$
102. $2(-3x)(3)$ $-18x$
103. $2x + 2y + 2z$
    does not simplify
104. $9y - 9$
    does not simplify
105. $4x^2 - 3x - 7 + 4x^2 - 2 - x$ $8x^2 - 4x - 9$
106. $2a^2 + 8 - a - 5 + 5a^2 - 9a$ $7a^2 - 10a + 3$
107. $-6s + 6s$ 0
108. $19c + (-19c)$ 0
109. $-4r + 8R + 2R - 3r + R$ $-7r + 11R$

110. $12a - A - a - 8A - a$ $10a - 9A$
111. $0 - 2y^3$ $-2y^3$
112. $0 - 7x^4$ $-7x^4$
113. $5(3 - 2s) + 4(2 - 3s) + 19s$ $-3s + 23$
114. $7(y + 1) + 8(2y + 3) + 12y$ $35y + 31$

APPLICATIONS

115. **a.** COMMUTING The illustration below shows the distances (in miles) that two men live from the office where they both work. Write an algebraic expression that represents the *total distance* the men must drive from their homes to the office. $(2d + 15)$ mi

    b. BOTANY Write an algebraic expression that represents the sum of the heights (in feet) of the two trees shown below. $2b + 30$

116. THE RED CROSS In 1891, Clara Barton founded the Red Cross. Its symbol is a white flag bearing a red cross. If each side of the cross has length x, write an algebraic expression that represents the perimeter of the cross. $12x$

117. PING-PONG Write an algebraic expression that represents the perimeter of the Ping-Pong table in feet. $(4x + 8)$ ft

118. SEWING Write an algebraic expression that represents the length (in cm) of the blue trim needed to outline a pennant with the given side lengths. $(5x - 30)$ cm

119. MOBILE HOMES The design of a mobile home calls for a thin strip of stained pine around the outside of all four exterior sides, as shown in brown below. If the strip costs 80¢ a running foot, how much will be spent on the pine used for the trim? (*Hint:* the left and right sides have the same design, as do the front and back of the mobile home.) $288

120. LANDSCAPING A landscape architect has designed a planter surrounding two birch trees, as shown below. The planter is to be outlined with redwood edging in the shape of a rectangle and two squares. If the material costs 17¢ a running foot, how much will the redwood cost for this project? $17

121. PARTY PREPARATIONS The appropriate size of a dance floor for a given number of dancers can be determined from the table shown. Find the perimeter of each of the dance floors listed. 36 ft, 48 ft, 60 ft, 72 ft, 84 ft

Slow dancers	Fast dancers	Size of floor (in feet)
8	5	9 × 9
14	9	12 × 12
22	15	15 × 15
32	20	18 × 18
50	30	21 × 21

122. COASTAL DRILLING The map shows an area of the California coast where oil drilling is planned. Use the scale to estimate the lengths of the sides of the area highlighted on the map. Then find its perimeter. 108 mi, 432 mi (answers may vary)

WRITING

123. Explain what it means for two terms to be like terms.

124. Explain what it means to say that the coefficient of x is an *implied* (or *understood*) 1.

125. Explain the difference between a term and a factor. Give some examples.

126. When simplifying an algebraic expression, some students use underlining, as shown below. What purpose does the underlining serve?

$$\underline{3y} + \underline{\underline{4}} + \underline{5y} + \underline{\underline{8}}$$

REVIEW

127. Solve: $-12y - 8 = 40$ -4

128. Find the prime factorization of 100. Use exponents in your answer. $2^2 \cdot 5^2$

129. Fill in the blank: To __check__ the solution of an equation, we substitute the value of the variable in the original equation and determine whether the result is a true statement.

130. A store manager earns d dollars an hour. Write an algebraic expression that represents

  a. the amount of money he will earn in an 8-hour day. $8d$ dollars

  b. the amount of money he will earn in a 40-hour week. $40d$ dollars

---

Objectives

1. Determine whether a number is a solution.
2. Combine like terms to solve equations.
3. Solve equations that have variable terms on both sides.
4. Use the distributive property to solve equations.
5. Apply a strategy to solve equations.

SECTION 3.5
Simplifying Expressions to Solve Equations

We must often simplify algebraic expressions to solve equations. Sometimes it will be necessary to combine like terms in order to isolate the variable on one side of the equation. At other times, it will be necessary to apply the distributive property to write an equation in a form that can be solved. In this section, we will discuss both of these situations.

1 Determine whether a number is a solution.

Recall that a number that makes an equation true when substituted for the variable is called a **solution** and is said to **satisfy** the equation.

Self Check 1

Is 25 a solution of $10 - x = 35 - 2x$? yes

Now Try Problem 19

Teaching Example 1 Is 8 a solution of $2x - 4 = x + 3$?
Answer: no

EXAMPLE 1 Is 9 a solution of $3y - 1 = 2y + 7$?

Strategy We will substitute 9 for each y in the equation and evaluate the expression on the left side and the expression on the right side separately.

WHY If a true statement results, 9 is a solution of the equation. If we obtain a false statement, 9 is not a solution.

Solution

Evaluate the expression on the left side.

$$3y - 1 = 2y + 7$$
$$3(9) - 1 \stackrel{?}{=} 2(9) + 7 \quad \text{Read } \stackrel{?}{=} \text{ as "is possibly equal to."}$$
$$27 - 1 \stackrel{?}{=} 18 + 7$$
$$26 = 25$$

Evaluate the expression on the right side.

Since $26 = 25$ is false, 9 is *not* a solution of $3y - 1 = 2y + 7$.

2 Combine like terms to solve equations.

Recall that **like terms** are terms containing exactly the same variables raised to exactly the same powers. Like terms can appear on the left side of an equation, on the right side of an equation, or on both sides. When asked to solve such equations, we should combine the like terms first before using a property of equality.

EXAMPLE 2 Solve: $7x - 4x = 15$

Strategy We will begin by combining the like terms on the left side of the equation.

WHY It is best to simplify the algebraic expressions on each side of an equation before using a property of equality.

Solution

$7x - 4x = 15$ This is the equation to solve.
$3x = 15$ Combine like terms: $7x - 4x = 3x$.
$\dfrac{3x}{3} = \dfrac{15}{3}$ To isolate x, undo the multiplication by 3 by dividing both sides by 3.
$x = 5$ Do the division.

To check, we substitute 5 for x in the original equation and evaluate the left side.

$7x - 4x = 15$ This is the original equation.
$7(5) - 4(5) \stackrel{?}{=} 15$ Substitute 5 for x.
$35 - 20 \stackrel{?}{=} 15$ Do the multiplication.
$15 = 15$ Do the subtraction.

Since the statement $15 = 15$ is true, 5 is the solution of $7x - 4x = 15$.

Self Check 2
Solve: $8r - 6r = 16$. Check the result. 8
Now Try Problem 23

Teaching Example 2
Solve: $10m - 6m = 24$. Check the result.
Answer:
6

EXAMPLE 3 Solve: $100 + 248 = t - 20 + t$

Strategy We will begin by combining the like terms on the left side and on the right side of the equation.

WHY It is best to simplify the algebraic expressions on each side of an equation before using a property of equality.

Self Check 3
Solve: $150 + 5 = d - 1 + 3d$. Check the result. 39
Now Try Problem 27

Teaching Example 3
Solve: $100 + 27 = 8a - 35 + a$. Check the result.
Answer:
18

Solution

$$100 + 248 = t - 20 + t \qquad \text{This is the equation to solve.}$$
$$348 = 2t - 20 \qquad \text{Combine like terms on each side of the equation: } 100 + 248 = 348 \text{ and } t + t = 2t.$$
$$348 + 20 = 2t - 20 + 20 \qquad \text{To isolate } 2t, \text{ undo the subtraction of 20 by adding 20 to both sides.}$$
$$368 = 2t \qquad \text{Do the addition.}$$
$$\frac{368}{2} = \frac{2t}{2} \qquad \text{To isolate } t, \text{ undo the multiplication by 2 by dividing both sides by 2.}$$
$$184 = t \qquad \text{Do the division: } 368 \div 2 = 184.$$

Verify that 184 is the solution by substituting it into the original equation.

3 Solve equations that have variable terms on both sides.

When solving an equation, if variables appear on both sides, we can use the addition (or subtraction) property of equality to get all variable terms on one side and all constant terms on the other.

Self Check 4
Solve: $30 + 6n = 4n - 2 \quad -16$
Now Try Problem 31

Teaching Example 4
Solve: $7x - 22 = 8x + 4$
Answer:
-26

EXAMPLE 4 Solve: $3x - 15 = 4x + 36$

Strategy There are variable terms ($3x$ and $4x$) on both sides of the equation. We will eliminate $3x$ from the left side by subtracting $3x$ from both sides.

WHY To solve for x, all the terms containing x must be on the same side of the equation.

Solution

$$3x - 15 = 4x + 36 \qquad \text{This is the equation to solve. There are variable terms (highlighted in blue) on both sides of the equation.}$$
$$3x - 15 - 3x = 4x + 36 - 3x \qquad \text{Subtract } 3x \text{ from both sides to isolate the variable term on the right side.}$$
$$-15 = x + 36 \qquad \text{Combine like terms: } 3x - 3x = 0 \text{ and } 4x - 3x = x. \text{ Now we want to isolate the variable, } x.$$
$$-15 - 36 = x + 36 - 36 \qquad \text{To undo the addition of 36, subtract 36 from both sides. This isolates } x.$$
$$-51 = x \qquad \text{Do the subtraction.}$$

Check:
$$3x - 15 = 4x + 36 \qquad \text{This is the original equation.}$$
$$3(-51) - 15 \stackrel{?}{=} 4(-51) + 36 \qquad \text{Substitute } -51 \text{ for } x.$$
$$-153 - 15 \stackrel{?}{=} -204 + 36 \qquad \text{Do the multiplications.}$$
$$-168 = -168 \qquad \text{True}$$

The solution is -51.

> **Success Tip** In Example 4, we could have eliminated $4x$ from the right side by subtracting $4x$ from both sides:
> $$3x - 15 - 4x = 4x + 36 - 4x$$
> $$-x - 15 = 36 \qquad \text{Note that the coefficient of } x \text{ is negative.}$$
> However, it is usually easier to isolate the variable term on the side that will result in a *positive* coefficient.

3.5 Simplifying Expressions to Solve Equations

EXAMPLE 5 Solve: $-9 - 2a - 4a = 5a + 2$

Strategy We will begin by combining the like terms on the left side of the equation.

WHY It is best to simplify the algebraic expressions on each side of an equation before using a property of equality.

Solution

$-9 - 2a - 4a = 5a + 2$ This is the equation to solve.

$-9 - 6a = 5a + 2$ Combine like terms: $-2a - 4a = -6a$.

There are variable terms (highlighted in blue) on both sides of the equation. Either we can subtract $5a$ from both sides to isolate a on the left side, or we can add $6a$ to both sides to isolate a on the right side. We will add $6a$ to both sides. That way, the coefficient of the resulting variable term on the right side will be positive.

$-9 - 6a + 6a = 5a + 2 + 6a$ Add $6a$ to both sides to isolate the variable term on the right side.

$-9 = 11a + 2$ Combine like terms: $-6a + 6a = 0$ and $5a + 6a = 11a$. Now we want to isolate the variable term, $11a$.

$-9 - 2 = 11a + 2 - 2$ To isolate $11a$, undo the addition of 2 by subtracting 2 from both sides.

$-11 = 11a$ Do the subtraction. Now we want to isolate the variable, a.

$\dfrac{-11}{11} = \dfrac{11a}{11}$ To isolate a, undo the multiplication by 11 by dividing both sides by 11.

$-1 = a$ Do the division.

Verify that the solution is -1 by substituting it into the original equation.

Self Check 5
Solve: $72 - 8d - 5d = 12d - 3$. Check the result.   3

Now Try Problem 39

Teaching Example 5
Solve: $-38 - 2b - b = 4b + 11$. Check the result.
Answer:
-7

4 Use the distributive property to solve equations.

At times, we must use the distributive property to remove parentheses when solving an equation.

EXAMPLE 6 Solve: $3(6x + 15) = 45$

Strategy We will use the distributive property on the left side of the equation.

WHY This will remove the parentheses and make it easier to see which properties of equality should be used to isolate x on the left side.

Solution

$3(6x + 15) = 45$ This is the equation to solve.

$3(6x) + 3(15) = 45$ Distribute the multiplication by 3.

$18x + 45 = 45$ Do the multiplication.

$18x + 45 - 45 = 45 - 45$ To isolate the variable term, $18x$, undo the addition of 45 by subtracting 45 from both sides.

$18x = 0$ Do the subtraction.

$\dfrac{18x}{18} = \dfrac{0}{18}$ To isolate x, undo the multiplication by 18 by dividing both sides by 18.

$x = 0$ Do the division.

$\begin{array}{r} \overset{1}{15} \\ \times\ 3 \\ \hline 45 \end{array}$

Verify that the solution is 0 by substituting it into the original equation.

Self Check 6
Solve: $7(5b + 5) = -70$. Check the result. -3

Now Try Problem 47

Teaching Example 6
Solve: $4(8n + 7) = 28$. Check the result.
Answer:
0

5 Apply a strategy to solve equations.

The previous examples suggest the following strategy for solving equations. You won't always have to use all four steps to solve a given equation. If a step doesn't apply, skip it and go to the next step.

> **Strategy for Solving Equations**
>
> 1. **Simplify each side of the equation:** Use the distributive property to remove parentheses, and then combine like terms on each side.
> 2. **Isolate the variable term on one side:** Add (or subtract) to get the variable term on one side of the equation and a number on the other using the addition (or subtraction) property of equality.
> 3. **Isolate the variable:** Multiply (or divide) to isolate the variable using the multiplication (or division) property of equality.
> 4. **Check the result:** Substitute the possible solution for the variable in the *original* equation to see if a true statement results.

Self Check 7
Solve:
$6(5x - 30) - 2x = 8(x + 50)$ 29
Now Try Problem 55

Teaching Example 7
Solve: $3(5x - 40) + 9x = 6(x + 70)$
Answer:
30

EXAMPLE 7   Solve: $3(4x - 80) + 6x = 2(x + 40)$

Strategy We will follow the steps of the equation-solving strategy to solve the equation.

WHY This is the most efficient way to solve an equation.

Solution

$3(4x - 80) + 6x = 2(x + 40)$ This is the equation to solve.

$12x - 240 + 6x = 2x + 80$ Distribute the multiplication by 3 and by 2.

$18x - 240 = 2x + 80$ On the left side, combine like terms: $12x + 6x = 18x$. There are variable terms on both sides.

$18x - 240 - 2x = 2x + 80 - 2x$ To eliminate the term $2x$ on the right side, subtract $2x$ from both sides.

$16x - 240 = 80$ Combine like terms on each side: $18x - 2x = 16x$ and $2x - 2x = 0$.

$16x - 240 + 240 = 80 + 240$ To isolate the variable term, $16x$, on the left side, add 240 to both sides to undo the subtraction of 240.

$$\begin{array}{r} \overset{1}{2}40 \\ +\ 80 \\ \hline 320 \end{array}$$

$16x = 320$ Do the addition on each side: $-240 + 240 = 0$ and $80 + 240 = 320$. Now we want to isolate the variable, x.

$\dfrac{16x}{16} = \dfrac{320}{16}$ To isolate x on the left side, divide both sides by 16 to undo the multiplication by 16.

$$\begin{array}{r} 20 \\ 16\overline{)320} \\ -\ 32 \\ \hline 00 \\ -\ 0 \\ \hline 0 \end{array}$$

$x = 20$ Do the division.

To check, we substitute 20 for x in the original equation and evaluate each side.

$3(4x - 80) + 6x = 2(x + 40)$ This is the original equation.

$3[4(20) - 80] + 6(20) \overset{?}{=} 2(20 + 40)$ Substitute 20 for each x.

$3[80 - 80] + 120 \overset{?}{=} 2(60)$

$3[0] + 120 \overset{?}{=} 120$

$120 = 120$ True

Since the statement $120 = 120$ is true, 20 is the solution of $3(4x - 80) + 6x = 2(x + 40)$.

ANSWERS TO SELF CHECKS
1. yes **2.** 8 **3.** 39 **4.** -16 **5.** 3 **6.** -3 **7.** 29

SECTION 3.5 STUDY SET

VOCABULARY

Fill in the blanks.

1. To __solve__ an equation means to find all values of the variable that make the equation a true statement.
2. A number that makes an equation true when substituted for the variable is called a __solution__ and is said to *satisfy* the equation.
3. To __check__ a solution means to substitute that value into the original equation to see whether a true statement results.
4. The equation $6x + 1 = 2x + 7$ has variable __terms__ on both sides.
5. When solving equations, __simplify__ the expressions that make up the left and right sides of the equation before using properties of equality to isolate the variable.
6. When we write the expression $2y + 3 + 6y$ as $8y + 3$, we say we have __combined__ __like__ terms.

CONCEPTS

7. a. Circle the variable terms.
   $(5x) + (3x) = 8$ $(5t) = (3t) + 8$ $7 = (5h) + (3h) - 1$
   b. Which equation has variable terms on both sides? $5t = 3t + 8$
8. To solve $6k = 5k - 18$, we need to eliminate $5k$ from the right side. To do this, what should we subtract from both sides? $5k$
9. Perform only the first step in solving each equation. **You do not have to solve the equation.**
   a. $2x + 4x = 36$ combine like terms: $6x = 36$
   b. $5(x + 1) = 15$ distribute the multiplication by 5: $5x + 5 = 15$
   c. $7x - 5 + 4x = x + 4 + x$ combine like terms: $11x - 5 = 2x + 4$
   d. $-3(x - 4) = 2(x + 1)$ distribute the multiplication by -3 and 2: $-3x + 12 = 2x + 2$

10. Consider the equation $2x - 8 = -4x - 14$.
    a. To solve this equation by isolating x on the left side, what should we add to both sides? $4x$
    b. To solve this equation by isolating x on the right side, what should we subtract from both sides? $2x$
11. Fill in the blanks.
    $6 - (d - 4) = 8$
    $6 - \underline{1}(d - 4) = 8$
    $6 - \underline{d} + \underline{4} = 8$
12. Fill in the blanks to complete the strategy for solving equations.
    Step 1: __Simplify__ each side of the equation.
    Step 2: Isolate the variable __term__ on one side.
    Step 3: Isolate the __variable__.
    Step 4: __Check__ the result.
13. a. Simplify: $3t - t - 8$ $2t - 8$
    b. Solve: $3t - t = -8$ -4
    c. Evaluate: $3t - t - 8$ for $t = -2$ -12
    d. Check: Is 5 a solution of $3t - t = -8$? no
14. a. Simplify: $2(x + 1) - 4$ $2x - 2$
    b. Solve: $2(x + 1) = -4$ -3
    c. Evaluate: $2(x + 1)$ for $x = -1$ 0
    d. Check: Is 3 a solution of $2(x + 1) = -4$? no

NOTATION

Complete each step to solve the equation and check the result.

15. $5x - 2x = -27$
    $\underline{3x} = -27$
    $\dfrac{3x}{\underline{3}} = \dfrac{-27}{\underline{3}}$
    $x = \underline{-9}$

    Check:
    $5x - 2x = -27$
    $5(\underline{-9}) - 2(\underline{-9}) \stackrel{?}{=} -27$
    $\underline{-45} - (-18) \stackrel{?}{=} -27$
    $-45 + \underline{18} \stackrel{?}{=} -27$
    $\underline{-27} = -27$ True

    The solution is $\underline{-9}$.

16.
$$8y - 6 = -2 + 10y$$
$$8y - 6 - \boxed{8y} = -2 + 10y - \boxed{8y}$$
$$-6 = -2 + \boxed{2y}$$
$$-6 + \boxed{2} = -2 + 2y + \boxed{2}$$
$$-4 = \boxed{2y}$$
$$\frac{-4}{2} = \frac{2y}{2}$$
$$\boxed{-2} = y$$

Check:
$$8y - 6 = -2 + 10y$$
$$8(\boxed{-2}) - 6 \stackrel{?}{=} -2 + 10(\boxed{-2})$$
$$-16 - 6 \stackrel{?}{=} -2 + (\boxed{-20})$$
$$\boxed{-22} = -22 \quad \text{True}$$

The solution is $\boxed{-2}$.

17.
$$5(x - 9) = 5$$
$$5x - 5(\boxed{9}) = 5$$
$$5x - \boxed{45} = 5$$
$$5x - 45 + \boxed{45} = 5 + \boxed{45}$$
$$\boxed{5x} = 50$$
$$\frac{5x}{5} = \frac{50}{5}$$
$$x = \boxed{10}$$

Check:
$$5(x - 9) = 5$$
$$5(\boxed{10} - 9) \stackrel{?}{=} 5$$
$$5(\boxed{1}) \stackrel{?}{=} 5$$
$$\boxed{5} = 5 \quad \text{True}$$

The solution is $\boxed{10}$.

18.
$$-4(-1 - x) = 16$$
$$4 + \boxed{4x} = 16$$
$$4 + 4x - \boxed{4} = 16 - \boxed{4}$$
$$4x = \boxed{12}$$
$$\frac{4x}{4} = \frac{12}{4}$$
$$x = \boxed{3}$$

Check:
$$-4(-1 - x) = 16$$
$$-4(-1 - \boxed{3}) \stackrel{?}{=} 16$$
$$-4(\boxed{-4}) \stackrel{?}{=} 16$$
$$\boxed{16} = 16 \quad \text{True}$$

The solution is $\boxed{3}$.

GUIDED PRACTICE

Use a check to determine whether the given number is a solution of the equation. See Example 1.

19. Is 3 a solution of $5f + 8 = 4f + 11$? yes
20. Is 5 a solution of $3r + 8 = 5r - 2$? yes
21. Is 12 a solution of $2(x - 1) = 33$? no
22. Is 8 a solution of $-6(x + 4) = -40$? no

Solve each equation. Check the result. See Example 2.

23. $3x + 6x = 54$ 6
24. $4c + 4c = 16$ 2
25. $6x - 3x = 9$ 3
26. $12b - 10b = 6$ 3

Solve each equation. Check the result. See Example 3.

27. $250 + 350 = m - 12 + m$ 306
28. $213 + 190 = x - 3 + x$ 203
29. $255 + 275 = a + 16 + a$ 257
30. $170 + 180 = m + 26 + m$ 162

Solve each equation. Check the result. See Example 4.

31. $3s - 1 = 4s + 7$ -8
32. $6v - 2 = 7v + 13$ -15
33. $x - 14 = 2x - 10$ -4
34. $2x - 27 = 3x - 20$ -7
35. $10m - 14 = 11m + 13$ -27
36. $5r - 10 = 6r + 2$ -12
37. $n + 5 = 2n - 3$ 8
38. $9y + 3 = 10y - 4$ 7

Solve each equation. Check the result. See Example 5.

39. $-16 - 3r - 5r = 2r + 4$ -2
40. $-29 - x - 6x = 2x + 7$ -4
41. $60 - a - 6a = 2a - 3$ 7
42. $19 - 4t - 7t = t - 5$ 2
43. $-3x - 7 + x = 6x + 20 - 5x$ -9
44. $-2x - 5 + x = 4x - 3 - 3x$ -1
45. $7 + 5y + 3y = 4y - 13$ -5
46. $36 + 9w + 2w = 6w + 1$ -7

Solve each equation. Check the result. See Example 6.

47. $2(6x + 7) = 14$ 0
48. $9(3y + 2) = 18$ 0
49. $8(9b + 5) = -32$ -1
50. $5(11n + 5) = -30$ -1
51. $-(6 - t) = -14$ -8
52. $-(-3 - x) = -11$ -14
53. $-2(3x + 5) = 20$ -5
54. $-3(5x + 1) = 87$ -6

Solve each equation. Check the result. See Example 7.

55. $3(x - 4) + 3x = 2(x + 10)$ 8
56. $9(w - 1) + 7w = 5(w + 7)$ 4
57. $6(2j + 6) + 4j = 4(j - 30)$ -13
58. $4(9h + 2) + 8h = 4(h - 18)$ -2

TRY IT YOURSELF

Solve each equation. Check the result.

59. $-16 = 2(t + 2)$ -10
60. $-10 = 5(y - 7)$ 5
61. $-7 + 5r = 83 - 10r$ 6
62. $-20 + t = 44 - 7t$ 8
63. $T + T - 17 = 57$ 37
64. $r + r - 15 = 95$ 55
65. $-15 = 5 + 5(2x + 10)$ -7
66. $-1 = 2 + 3(4x + 7)$ -2
67. $60 = 3v - 5v$ -30
68. $28 = x - 3x$ -14
69. $9q + 3(q - 7) = 18 - q$ 3
70. $q + 6(q - 4) = 24 - q$ 6
71. $60r - 50 = 15r - 5$ 1
72. $100f - 75 = 50f + 75$ 3
73. $3 - 5x - 2x = x + 27$ -3
74. $25 + 3x - x = x + 4$ -21
75. $38 + 6x - 5x = 2x + 7$ 31
76. $15 - 11y + 2y = -21$ 4
77. $5(13x + 3) = 210$ 3
78. $-2(19r - 5) = -28$ 1
79. $5 - (7 - y) = -5$ -3
80. $10 - (5 - x) = 40$ 35
81. $50a - 1 = 60a - 101$ 10
82. $25y - 2 = 75y - 202$ 4
83. $-(4 - c) = -3$ 1
84. $-(6 - 2x) = -8$ -1
85. $-20 - 8 = -m + 2m$ -28
86. $-100 - 20 = -p + 4p$ -40
87. $x + x + 6 = 90$ 42
88. $c + c + 1 = 51$ 25
89. $8 + 4(2x - 2) = -16 + 4x$ -4
90. $3 + 3(2x - 1) = -18 + 3x$ -6
91. $1,500 = b + 30 + b$ 735
92. $8,000 = h + 100 + h$ $3,950$
93. $7x = 3x + 8$ 2
94. $4x = 2x + 14$ 7
95. $100 - y = 100 + y$ 0
96. $-60 + z = -60 - z$ 0
97. $2(4y + 8) + 3y = -3(2 - 3y)$ -11
98. $3(7 - y) = 3(2y + 1)$ 2
99. $t + 5t + 3t - 40 = 14t$ -8
100. $5r - 24 = r + 5r + 2r$ -8
101. $-2(9 - 3s) - (5s + 2) = -25$ -5
102. $-(19 - 3s) - (8s + 1) = 35$ -11
103. $25 + 4j = 9j$ 5
104. $36 + 5j = 9j$ 9
105. $-3(3 - 2w) = -9$ 0
106. $-4(5t + 2) = -8$ 0
107. $-4(12) + 12t = 16t$ -12
108. $-4(7) + 7t = 21t$ -2
109. $-5g - 40 = -15g$ 4
110. $-20s - 20 = -40s$ 1
111. $4(p - 2) = 0$ 2
112. $10(4s - 4) = 0$ 1
113. $16 - (x + 3) = -13$ 26
114. $10 - (w + 4) = -12$ 18

WRITING

115. Explain the error in the work shown below.

    Solve: $2x = 4x - x$
    $2x = 4$
    $\dfrac{2x}{2} = \dfrac{4}{2}$
    $x = 2$

116. To solve $3x - 4 = 5x + 1$, one student began by subtracting $3x$ from both sides. Another student solved the same equation by first subtracting $5x$ from both sides. Will the students get the same solution? Explain why or why not.

117. Explain the error in the following solution.

    Solve: $2x + 4 = 30$
    $\dfrac{2x}{2} + 4 = \dfrac{30}{2}$
    $x + 4 = 15$
    $x + 4 - 4 = 15 - 4$
    $x = 11$

118. Consider $3x = 2x + 9$. Why is it necessary to eliminate one of the variable terms in order to solve for x?

119. What does it mean to *solve an equation*?

120. Explain how to determine whether a number is a solution of an equation.

REVIEW

121. Subtract: $-7 - 9$ -16

122. Which of the following numbers are *not* factors of 28? 4, 6, 7, 8 $6, 8$

123. Translate to mathematical symbols: *8 less than twice a number x* $2x - 8$

124. Translate to mathematical symbols: *4 less than x* $x - 4$

125. Simplify: $-(-5)$ 5

126. Using x and y, illustrate the commutative property of addition. $x + y = y + x$

127. What is the sign of the product of two negative integers?   positive

128. Complete the table.

Term	Coefficient
$6m$	6
$-75t$	-75
w	1
$4bh$	4

Objectives

1. Solve application problems to find one unknown.
2. Solve application problems to find two unknowns.
3. Solve number-value problems.

SECTION 3.6
Using Equations to Solve Application Problems

The skills that we have studied in this chapter can now be used to solve more complicated application problems. Once again, we will use the six-step problem-solving strategy as an outline for each solution.

1 Solve application problems to find one unknown.

EXAMPLE 1 *Service Hours* To receive a degree in child development, students at one college must complete 135 hours of volunteer service by working 3-hour shifts at a local preschool. If a student has already volunteered 87 hours, how many more 3-hour shifts must she work to meet the service requirement for her degree?

Analyze
- Students must complete 135 hours of volunteer service.   Given
- Students work 3-hour shifts.   Given
- A student has already completed 87 hours of service.   Given
- How many more 3-hour shifts must she work?   Find

Assign
Let x = the number of shifts needed to complete the service requirement.

Form
Since each shift is 3 hours long, multiplying 3 by the number of shifts will give the number of additional hours the student needs to volunteer.

The number of hours she has already completed	plus	3 times	the number of shifts yet to be completed	is	the number of hours required.
87	+	3 ·	x	=	135

Solve

$$87 + 3x = 135$$ We need to isolate x on the left side.

$$87 + 3x - 87 = 135 - 87$$ To isolate the variable term $3x$, subtract 87 from both sides to undo the addition of 87.

$$3x = 48$$ Do the subtraction.

$$\frac{3x}{3} = \frac{48}{3}$$ To isolate x, divide both sides by 3 to undo the multiplication by 3.

$$x = 16$$ Do the division.

$$\begin{array}{r} \overset{12}{\cancel{1}}\overset{15}{\cancel{3}\cancel{5}} \\ -\ 87 \\ \hline 48 \end{array}$$

$$\begin{array}{r} 16 \\ 3{\overline{\smash{\big)}\,48}} \\ \underline{-\ 3} \\ 18 \\ \underline{-18} \\ 0 \end{array}$$

State
The student needs to complete 16 more 3-hour shifts of volunteer service.

Check
The student has already completed 87 hours. If she works 16 more shifts, each 3 hours long, she will have $16 \cdot 3 = 48$ more hours. Adding the two sets of hours, we get:

$$\begin{array}{r} 87 \\ +\ 48 \\ \hline 135 \end{array}$$ ← This is the total number of hours needed.

The result, 16, checks.

Self Check 1

SERVICE CLUBS To become a member of a service club, students at one college must complete 72 hours of volunteer service by working 4-hour shifts at the tutoring center. If a student has already volunteered 48 hours, how many more 4-hour shifts must she work to meet the service requirement for membership in the club?

Now Try Problem 21

Self Check 1 Answer
The student needs to complete 6 more 4-hour shifts of volunteer service.

Teaching Example 1 COLLECTORS
A poster collector purchases 6 classic film posters each year. If he now owns 36 posters, in how many years will he have a collection of 120 posters?
Answer:
It will take 14 more years for him to have 120 posters.

EXAMPLE 2 Attorney's Fees
In return for her services, an attorney and her client split the jury's cash award equally. After paying her assistant $1,000, the attorney ended up making $10,000 from the case. What was the amount of the award?

Analyze
- The attorney and client split the award equally.   Given
- The attorney's assistant was paid $1,000.   Given
- The attorney made $10,000.   Given
- What was the amount of the award?   Find

Assign
Let x = the amount of the award.

Form
Two key phrases in the problem help us form an equation.

| Key phrase: | *split the award equally* | Translation: | divide by 2 |
| Key phrase: | *paying her assistant $1,000* | Translation: | subtract $1,000 |

Now we translate the words of the problem into an equation.

The award split in half	minus	the amount paid to the assistant	is	the amount the attorney makes.
$\frac{x}{2}$	−	1,000	=	10,000

Solve

$\frac{x}{2} - 1,000 = 10,000$ We need to isolate x on the left side.

$\frac{x}{2} - 1,000 + \mathbf{1,000} = 10,000 + \mathbf{1,000}$ To isolate the variable term $\frac{x}{2}$, add 1,000 to both sides to undo the subtraction of 1,000.

$\frac{x}{2} = 11,000$ Do the addition.

$2 \cdot \frac{x}{2} = 2 \cdot 11,000$ To isolate the variable x, multiply both sides by 2 to undo the division by 2.

$\phantom{2 \cdot \frac{x}{2} = }x = 22,000$ Do the multiplication.

$$\begin{array}{r} 11,000 \\ \times\ \ \ \ \ 2 \\ \hline 22,000 \end{array}$$

State
The amount of the award was $22,000.

Check
If the award of $22,000 is split in half, the attorney's share is $11,000. If $1,000 is paid to her assistant, we subtract to get:

$$\begin{array}{r} \$11,000 \\ -\ 1,000 \\ \hline \$10,000 \end{array}$$ ←This is what the attorney made.

The result, $22,000, checks.

2 Solve application problems to find two unknowns.

When solving application problems, we usually let the variable stand for the quantity we are asked to find. In the next two examples, each problem contains a second unknown quantity. We will look for a key word or phrase in the problem to help us describe it using an algebraic expression.

Self Check 2
YARD SALES A husband and wife split the money equally that they made on a yard sale. The husband gave $75 of his share to charity, leaving him with $210. How much money did the couple make at their yard sale?
Now Try Problem 22
Self Check 2 Answer
The couple made $570 at the yard sale.

Teaching Example 2 TAX REFUNDS
After receiving their tax refund check, a husband and wife split the refunded money equally. The wife then gave $110 of her share to her son, leaving her with $344. What was the amount of the tax refund check?
Answer:
The amount of the tax refund check was $908.

Self Check 3

CIVIL SERVICE A candidate for a position with the IRS scored 15 points higher on the written part of the civil service exam than he did on his interview. If his combined score was 155, what were his scores on the interview and on the written part?

Now Try Problem 23

Self Check 3 Answer
His score on the interview was 70 and his score on the written part was 85.

Teaching Example 3 MARCHING BAND At a field competition, a marching band scored 6 points more on the music portion of the judging than they did on the visual portion. If their combined score was 84, what were their scores on the music portion and on the visual portion?
Answer:
Their score on the visual portion was 39 and their score on the music portion was 45.

EXAMPLE 3 Civil Service
A candidate for a position with the FBI scored 12 points higher on the written part of the civil service exam than she did on her interview. If her combined score was 92, what were her scores on the interview and on the written part of the exam?

Analyze
- She scored 12 points higher on the written part than on the interview. **Given**
- Her combined score was 92. **Given**
- What were her scores on the interview and on the written part? **Find**

Assign
Since we are told that her score on the written part was related to her score on the interview, we let $x =$ her score on the interview.

There is a second unknown quantity—her score on the written part of the exam. We look for a key phrase to help us decide how to represent that score using an algebraic expression.

Key phrase: 12 points *higher* on the written part than on the interview

Translation: add 12 points to the interview score

So $x + 12 =$ her score on the written part of the exam.

Form
Now we translate the words of the problem into an equation.

The score on the interview	plus	the score on the written part	is	the overall score.
x	$+$	$x + 12$	$=$	92

Solve

$$x + x + 12 = 92$$ We need to isolate x on the left side.

$$2x + 12 = 92$$ On the left side, combine like terms: $x + x = 2x$.

$$2x + 12 - 12 = 92 - 12$$ To isolate the variable term, 2x, subtract 12 from both sides to undo the addition of 12.

$$2x = 80$$ Do the subtraction.

$$\frac{2x}{2} = \frac{80}{2}$$ To isolate the variable x, divide both sides by 2 to undo the multiplication by 2.

$$x = 40$$ Do the division. This is her score on the interview.

To find the second unknown, we substitute 40 for x in the expression that represents her score on the written part.

$$x + 12 = 40 + 12$$
$$= 52$$ This is her score on the written part.

State
Her score on the interview was 40 and her score on the written part was 52.

Check
Her score of 52 on the written part was 12 points higher than her score of 40 on the interview. Also, if we add the two scores, we get:

$$\begin{array}{r} 40 \\ + 52 \\ \hline 92 \end{array}$$ ← This is her combined score.

The results, 40 and 52, check.

EXAMPLE 4   Playgrounds

After receiving a donation of 400 feet of chain link fencing, the staff of a preschool decided to use it to enclose a playground that is rectangular. Find the length and the width of the playground if the length is three times the width.

The perimeter is 400 ft.   Width

The length is three times as long as the width.

Analyze
- The perimeter is 400 ft.   Given
- The length is three times as long as the width.   Given
- What is the length and what is the width of the rectangle? Find

Assign
Since we are told that the length is related to the width, we will let w = the width of the playground. There is a second unknown quantity: the length of the playground. We look for a key phrase to help us decide how to represent it using an algebraic expression.

Key phrase: length is *three times* the width   **Translation:** multiply width by 3

So $3w$ = the length of the playground.

Form
The formula for the perimeter of a rectangle is $P = 2l + 2w$. In words, we can write

2 · the length of the playground plus 2 · the width of the playground is the perimeter.

2 · 3w + 2 · w = 400

Solve

$2 \cdot 3w + 2w = 400$	We need to isolate w on the left side.
$6w + 2w = 400$	Do the multiplication: $2 \cdot 3w = 6w$.
$8w = 400$	On the left side, combine like terms: $6w + 2w = 8w$.
$\dfrac{8w}{8} = \dfrac{400}{8}$	To isolate w, divide both sides by 8 to undo the multiplication by 8.
$w = 50$	Do the division. This is the width.

$$\begin{array}{r} 50 \\ 8\overline{)400} \\ -40 \\ \hline 00 \\ -0 \\ \hline 0 \end{array}$$

To find the second unknown, we substitute 50 for w in the expression that represents the length of the playground.

$3w = 3(50)$ Substitute 50 for w.
$ = 150$ This is the length of the playground.

State
The width of the playground is 50 feet and the length is 150 feet.

Check
If we add two lengths and two widths, we get $2(150) + 2(50) = 300 + 100 = 400$. Also, the length (150 ft) is three times the width (50 ft). The results check.

Self Check 4

CRIME SCENES Police used 800 feet of yellow tape to fence off a rectangular lot for an investigation. Fifty less feet of tape was used for each width than for each length. Find the length and the width of the lot.

Now Try Problem 24

Self Check 4 Answer
The length of the lot is 225 feet and the width of the lot is 175 feet.

Teaching Example 4 GEOMETRY The perimeter of a rectangle is 100 inches. Find the length and the width if the length is four times the width.
Answer:
The length is 40 inches and the width is 10 inches.

---

3   Solve number-value problems.

Some problems deal with quantities that have a value. In these problems, we must distinguish between the *number of* and the *value of* the unknown quantity. For example, to find the value of 3 quarters, we multiply the number of quarters by the value (in cents) of one quarter. Therefore, the value of 3 quarters is $3 \cdot 25$ cents = 75 cents.

The same approach must be taken if the number is unknown. For example, the value of d dimes is not d cents. The value of d dimes is $d \cdot 10$ *cents* $= 10d$ cents. For problems of this type, we will use the relationship

> Number $\cdot$ value = total value   This is the number-value formula.

Self Check 5

INVESTING A T-bill (Treasury bill) is worth $10,000. Find the value of:

a. two T-bills $20,000
b. x T-bills $10,000x$ dollars
c. $(x + 3)$ T-bills $10,000(x + 3)$ dollars

Now Try Problem 27

Teaching Example 5 Bananas sell for 99 cents a pound. Find the cost of:
a. 6 pounds of bananas
b. p pounds of bananas
c. $(p + 3)$ pounds of bananas
Answers:
a. 594 cents, or $5.94
b. $99p$ cents
c. $99(p + 3)$ cents

EXAMPLE 5 **Pricing** Delicious apples sell for 89 cents a pound. Find the cost of:

a. 5 pounds of apples   b. p pounds of apples   c. $(p - 2)$ pounds of apples

Strategy In each case, we will multiply the number of pounds of apples by their value (89 cents a pound) to find the total cost.

WHY Since *value* and *cost* are similar concepts, we can use the number-value formula in this situation.

Solution

a. Total value = **number $\cdot$ value**   This is the number-value formula.
   Total value = 5 $\cdot$ 89   Substitute 5 for the number of pounds and 89 for the value (cost per pound) of the apples.
   Total value = 445   Do the multiplication.

$$\begin{array}{r} \overset{4}{8}9 \\ \times\ 5 \\ \hline 445 \end{array}$$

The cost of 5 pounds of apples is 445 cents, or $4.45.

b. Since the number of pounds of apples is unknown (p pounds), we cannot calculate the total cost as in part a. We can only represent it using an algebraic expression.

   Total value = **number $\cdot$ value**   This is the number-value formula.
   Total value = p $\cdot$ 89   Substitute p for the number of pounds and 89 for the value (cost per pound) of the apples.
   Total value = $89p$ It is standard practice to write the numerical factor, 89, in front of the variable factor.

The cost of p pounds of apples can be represented by the algebraic expression $89p$ cents.

c. Since the number of pounds of apples is unknown, $(p - 2)$ pounds, we cannot calculate the total cost, as in part a. We can only represent it using an algebraic expression.

   Total value = **number $\cdot$ value**   This is the number-value formula.
   Total value = $(p - 2) \cdot 89$ Substitute $p - 2$ for the number of pounds and 89 for the value (cost per pound) of the apples.
   Total value = $89(p - 2)$ It is standard practice to write the numerical factor, 89, in front of the quantity $(p - 2)$.

The cost of $(p - 2)$ pounds of apples can be represented by the algebraic expression $89(p - 2)$ cents.

EXAMPLE 6 **Movie Tickets** Ninety-five people attended a movie matinee. Ticket prices were $6 for adults and $4 for children. Write algebraic expressions that represent the income received from the sale of children's tickets and from the sale of adult tickets.

Strategy In each case, we will multiply the number of tickets sold by their value ($6 for adults and $4 for children) to find the income received.

WHY Since *total value* and *income received* are similar concepts, we can use the number-value formula in this situation.

Solution We will let c = the unknown number of children's tickets sold. If we subtract the number of children's tickets sold from the total number of tickets sold, we obtain an expression for the number of adult tickets sold:

$$95 - c = \text{number of adult tickets sold}$$

The value of a children's ticket is \$4. To find the income from the sale of c children's tickets, we multiply:

Total value = **number · value**
Total value = c · 4
Total value = $4c$

The income from the sale of the children's tickets is represented by the algebraic expression $4c$ dollars.

The value of an adult ticket is \$6. To find the income from the sale of $(95 - c)$ adult tickets we multiply:

Total value = **number · value**
Total value = $(95 - c)$ · 6
Total value = $6(95 - c)$

The income from the sale of the adult tickets is represented by the algebraic expression $6(95 - c)$ dollars.

These results can be presented in a number-value table, as shown below.

Type of ticket	Number ·	Value (\$)	=	Total value (\$)
Child	c	4		$4c$
Adult	$95 - c$	6		$6(95 - c)$

Multiply to obtain each of these expressions.

Enter this information first.

When solving number-value problems, it is often helpful to construct a table to organize all of the facts.

EXAMPLE 7   Basketball
On a night when they scored 110 points, a basketball team made only 5 free throws (worth 1 point each). The remainder of their points came from two- and three-point baskets. If the number of two- and three-point baskets totaled 45, how many two-point and how many three-point baskets did they make?

Analyze

- The team scored 110 points.   *Given*
- They made 5 free throws (1 point each).   *Given*
- They made a total of 45 two- and three-point baskets.   *Given*
- How many two-point and three-point baskets were made?   *Find*

Self Check 6

FITNESS CLUBS A fitness club has 150 members. Monthly membership fees are \$25 for nonseniors and \$15 for senior citizens. Find the income the club receives from nonseniors and from seniors each month. Use the table below to present your results. Let m represent the number of nonseniors.

Member's age	Number	· Fee	=	Total income
nonsenior	m	25		$25m$
senior	$150 - m$	15		$15(150 - m)$

Now Try Problem 29

Teaching Example 6 TICKET SALES
A total of 950 tickets were sold for a concert. Ticket prices were \$55 for general admission and \$105 for reserved. Write algebraic expressions that represent the income received from the sale of general admission tickets and the sale of reserved tickets. Use a table to present the results.
Answer:
$55g$ dollars from general admission ticket sales, $105(950 - g)$ dollars from reserved ticket sales.

Self Check 7

FURNISHINGS A restaurant owner purchased \$2,720 worth of tables and chairs for the dining area of her cafe. Each table cost \$200 and each chair cost \$60. If she purchased a total of 36 pieces of furniture, how many tables and how many chairs did she buy?

Now Try Problem 25
Self Check 7 Answer
She bought 4 tables and 32 chairs.

Teaching Example 7 BASKETBALL
On a night when they scored 104 points, a basketball team made 10 free throws (1 point each). The remainder of their points came from two- and three-point baskets. If the number of two- and three-point baskets totaled 44, how many two-point and how many three-point baskets did they make?
Answer:
They made 38 two-point baskets and 6 three-point baskets.

Assign

The number of two- and three-point baskets totaled 45. If we let x = the number of three-point baskets made, then $45 - x$ = the number of two-point baskets made. We can organize the data in a table. For each type of basket, multiply the *number* of baskets made by the point *value* to find an expression to represent the total value.

Type of basket	Number	· Value	= Total value
Three-point	x	3	$3x$
Two-point	$45 - x$	2	$2(45 - x)$
Free throw	5	1	5

Multiply to obtain each of these expressions.

Enter this information first. Total: 110

Use the information in this column to form an equation.

Form

$3 \cdot$ the number of three-point baskets $\text{plus } 2 \cdot$ the number of two-point baskets $\text{plus } 1 \cdot$ the number of free throws is the total points scored.

$3 \cdot x \quad + \quad 2 \cdot (45 - x) \quad + \quad 1 \cdot 5 \quad = \quad 110$

Solve

$3x + 2(45 - x) + 5 = 110$ This is the equation to solve.
$3x + 90 - 2x + 5 = 110$ Distribute the multiplication by 2.
$x + 95 = 110$ Combine like terms: $3x - 2x = x$.
$x + 95 - 95 = 110 - 95$ To isolate x, undo the addition of 95 by subtracting 95 from both sides.
$x = 15$ Do the subtraction. This is the number of three-point baskets.

$$\begin{array}{r} \overset{10}{\cancel{\overset{\cancel{1}}{1}}} \overset{10}{\cancel{0}} \\ -95 \\ \hline 15 \end{array}$$

We can substitute 15 for x in $45 - x$ to find the number of two-point baskets made.

$45 - x = 45 - 15$
$ = 30$ This is the number of two-point baskets.

State

The basketball team made 15 three-point baskets and 30 two-point baskets.

Check

If we multiply the number of three-point baskets by their value, we get $15 \cdot 3 = 45$ points. If we multiply the number of two-point baskets by their value, we get $30 \cdot 2 = 60$ points. If we add the number of made free throws to these two subtotals, we get $45 + 60 + 5 = 110$ points. The results check.

$$\begin{array}{r} \overset{1}{4}5 \\ 60 \\ +5 \\ \hline 110 \end{array}$$

ANSWERS TO SELF CHECKS

1. The student needs to complete 6 more 4-hour shifts of volunteer service.
2. The couple made $570 at the yard sale.
3. His score on the interview was 70 and his score on the written part was 85.
4. The length of the lot is 225 feet and the width of the lot is 175 feet.
5. **a.** $20,000 **b.** $10,000x$ dollars **c.** $10,000(x + 3)$ dollars
6. m, 25, $25m$; $150 - m$, 15, $15(150 - m)$
7. She bought 4 tables and 32 chairs.

Unless otherwise noted, all art on this page is © Cengage Learning.

SECTION 3.6 STUDY SET

VOCABULARY

Fill in the blanks.

1. The six-step problem-solving strategy is:
   - **Analyze** the problem.
   - **Assign** a variable.
   - Form an **equation**.
   - **Solve** the equation.
   - State the **conclusion**.
   - **Check** the result.
2. Words such as *doubled* and *tripled* indicate the operation of **multiplication**.
3. Phrases such as *distributed equally* and *sectioned off uniformly* indicate the operation of **division**.
4. Words such as *trimmed*, *removed*, and *melted* indicate the operation of **subtraction**.
5. Words such as *extended* and *reclaimed* indicate the operation of **addition**.
6. A letter (or symbol) that is used to represent a number is called a **variable**.

Fill in the blanks to complete each formula.

7. **Number** · value = total value
8. $P = 2l + 2w$

9. **BUSINESS ACCOUNTS** Every month, a salesperson adds five new accounts. Write an algebraic expression that represents the number of new accounts that he will add in x months. $5x$

10. **ANTIQUE COLLECTING** Every year, a woman purchases four antique spoons to add to her collection. Write an algebraic expression that represents the number of spoons that she will purchase in x years. $4x$

11. **SERVICE STATIONS** See the illustration below. Write an algebraic expression that represents the number of gallons that the smaller tank holds. $g - 100$

This tank holds g gallons.

This tank holds 100 gallons less than the premium tank.

12. **SCHOLARSHIPS** See the illustration below. Write an algebraic expression that represents the number of scholarships that were awarded this year. $s + 6$

Last year, s scholarships were awarded.

Six more scholarships were awarded this year than last year.

13. **OCEAN TRAVEL** See the illustration below. Write an algebraic expression that represents the number of miles that the passenger ship traveled. $3m$

The freighter traveled m miles.

The passenger ship traveled 3 times farther than the freighter.

14. **TAX REFUNDS** See the illustration below. Write an algebraic expression that represents the amount of the tax refund that the husband gets. $\dfrac{d}{2}$ dollars

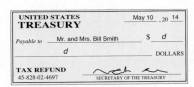

A husband and wife received a tax refund of $d.

The couple split the refund equally.

15. **GEOMETRY** See the illustration below. The length of a rectangle is twice its width. Write an algebraic expression that represents the length of the rectangle. $2w$

16. **GEOMETRY** Fill in the blanks to complete the equation that describes the perimeter of the rectangle shown below.

$$2 \cdot 5w + 2 \cdot w = 240$$

The perimeter is 240 ft.

17. **FOOTWEAR** The illustration below shows a rack that contains both dress shoes and athletic shoes.
    a. How many pairs of shoes are stored in the rack? 9
    b. Suppose there are d pairs of dress shoes in the rack. Write an algebraic expression that represents the number of pairs of athletic shoes in the rack. $9 - d$

18. **QUIZZES** The answers to a Prealgebra quiz are shown below.
    a. How many questions were on the quiz? 10
    b. Suppose the student answered c questions correctly. Write an algebraic expression that represents the number of questions she answered incorrectly. $10 - c$

> **PREALGEBRA QUIZ**
> **CHAPTER 3**
> 1. 44   6. 250 ft
> 2. 376   7. 165 mi
> 3. equal 8. no
> 4. 9 − x 9. yes
> 5. 4x   10. simplify

In Problems 19 and 20, complete each step.

19. **AIRLINE SEATING** An 88-seat passenger plane has ten times as many economy seats as first-class seats. Find the number of first-class seats.

    Analyze
    - There are 88 seats on the plane.
    - There are 10 times as many economy seats as first-class seats.
    - How many first–class seats are there?

    Assign Since the number of economy seats is related to the number of first-class seats, we let $x =$ the number of first–class seats.

    To write an algebraic expression to represent the number of economy seats, we look for a key phrase in the problem.

    Key phrase: ten times as many economy seats

    Translation: multiply by 10

    So $10 \cdot x =$ the number of economy seats.

    Form Now we translate the words of the problem into an equation.

The number of first-class seats	plus	the number of economy seats	is	88.
x	$+$	$10x$	$=$	88

    Solve
    $$x + 10x = 88$$
    $$11x = 88$$
    $$\frac{11x}{11} = \frac{88}{11}$$
    $$x = 8$$

    State There are 8 first-class seats

    Check If there are 8 first-class seats, there are $10 \cdot 8 = 80$ economy seats. Adding 8 and 80, we get 88. The result checks.

20. **COUPONS** A shopper redeemed some 20-cents-off and some 40-cents-off coupons at the supermarket to get $2.60 off her grocery bill. If she used a total of eight coupons, how many 20¢ and how many 40¢ coupons did she redeem?

    Analyze
    - 20 ¢ and 40 ¢ coupons were redeemed.
    - She got $2.60, which is 260 ¢, off her grocery bill.
    - She used a total of 8 coupons.
    - How many 20 ¢ and how many 40 ¢ coupons did she redeem?

    Assign The total number of coupons redeemed was 8. If we let $x =$ the number of 20¢ coupons she redeemed, then $8 - x =$ the number of 40¢ coupons she redeemed.

    Form

	the number of 20¢ coupons redeemed	plus		the number of 40¢ coupons redeemed	is	260.
20 ·	x	$+$	40 ·	$(8 - x)$	$=$	260

    Solve
    $$20x + 40(8 - x) = 260$$
    $$20x + 320 - 40x = 260$$
    $$-20x + 320 = 260$$
    $$-20x + 320 - 320 = 260 - 320$$
    $$-20x = -60$$
    $$\frac{-20x}{-20} = \frac{-60}{-20}$$
    $$x = 3$$

If 3 of the 20¢ coupons were redeemed, then 8 − 3 = 5 of the 40¢ coupons were redeemed.

State She redeemed 3 of the 20¢ coupons and 5 of the 40¢ coupons.

Check The value of 3 of the 20¢ coupons is 3 · 20 = 60 ¢. The value of 5 of the 40¢ coupons is 5 · 40 = 200 ¢. Adding these two subtotals, we get 260¢, which is $2.60. The results check.

GUIDED PRACTICE

Form an equation and solve it to answer the following question. See Example 1.

21. BUSINESS After beginning a new position with 15 established accounts, a salesman made it his objective to add 5 new accounts every month. His goal was to reach 100 accounts. At this rate, how many months would it take to reach his goal?
    It will take 17 months for him to reach his goal.

Form an equation and solve it to answer the following question. See Example 2.

22. TAX REFUNDS After receiving their tax refund, a husband and wife split the refunded money equally. The husband then gave $50 of his money to charity, leaving him with $70. What was the amount of the tax refund check?
    The amount of the tax refund was $240.

Form an equation and solve it to answer the following question. See Example 3.

23. SCHOLARSHIPS Because of increased giving, a college scholarship program awarded six more scholarships this year than last year. If a total of 20 scholarships were awarded over the last two years, how many were awarded last year and how many were awarded this year? Last year, 7 scholarships were awarded. This year 13 scholarships were awarded.

Form an equation and solve it to answer the following question. See Example 4.

24. GEOMETRY The perimeter of a rectangle is 150 inches. Find the length and the width if the length is four times the width.
    The length is 60 inches and the width is 15 inches.

In Problems 25 and 26, form an equation and solve it to answer the question. See Example 7.

25. PIGGY BANKS When a child emptied her coin bank, she had a collection of pennies, nickels, and dimes. There were a total of 20 pennies, and a combined total of 25 nickels and dimes. If the coins had a total value of 220 cents, how many nickels and dimes were in the bank? There were 10 nickels and 15 dimes in the piggy bank.

26. WISHING WELLS A city park employee collected 650 cents in nickels, dimes, and quarters at the bottom of a wishing well. There were 20 nickels, and a combined total of 40 dimes and quarters. How many dimes and quarters were at the bottom of the wishing well? There were 30 dimes and 10 quarters at the bottom of the wishing well.

In Problems 27–30, complete the number-value table. See Examples 5 and 6.

27. COMMISSIONS A shoe salesman receives a commission for every pair of shoes he sells. Complete the table.

Type of shoe	Number sold	·	Commission per pair ($)	=	Total commission ($)
Dress	10		3		30
Athletic	12		2		24
Child's	x		5		$5x$
Sandal	$9 - x$		4		$4(9 - x)$

28. COINS Complete the table.

Type of coin	Number	·	Value (¢)	=	Total value (¢)
Nickel	12		5		60
Dime	d		10		$10d$
Quarter	$q + 2$		25		$25(q + 2)$

29. TUTORING A tutoring center charges $18 an hour for English tutoring and $20 an hour for mathematics tutoring. One week, forty students were tutored one hour per week in these subjects, and no student took both types of tutoring. Write algebraic expressions that represent the weekly income received by the center from the English tutoring and from the mathematics tutoring. Present your results in the following table. (answers may vary)

Type of tutoring	Number of hours	·	Fee ($)	=	Total income ($)
English	h		18		$18h$
Mathematics	$40 - h$		20		$20(40 - h)$

30. **TUXEDOS** A formal wear shop rents prom tuxedos for $55 and wedding tuxedos for $75. One weekend, 80 tuxedos were rented. Write algebraic expressions that represent the income received that weekend from the rental of prom tuxedos and of wedding tuxedos. Present your results in the following table. (answers may vary)

Type of tuxedo	Number	·	Rental fee ($)	=	Total income ($)
Prom	p		55		$55p$
Wedding	$80 - p$		75		$75(80 - p)$

TRY IT YOURSELF

Form an equation and solve it to answer each question.

31. **NUMBER PROBLEMS** Eight more than a number is the same as twice the number. What is the number? The number is 8.

32. **NUMBER PROBLEMS** Twenty more than a number is the same as three times the number. What is the number? The number is 10.

33. **NUMBER PROBLEMS** Ten less than five times a number is the same as the number increased by six. What is the number? The number is 4.

34. **NUMBER PROBLEMS** Four less than seven times a number is the same as the number increased by eight. What is the number? The number is 2.

APPLICATIONS

Form an equation and solve it to answer each question.

35. **CONSTRUCTION** To get a heavy-equipment operator's certificate, 48 hours of on-the-job training are required. If a woman has completed 24 hours, and the training sessions last for 6 hours, how many more sessions must she take to get the certificate? She must take 4 more sessions.

36. **PUBLISHING** An editor needs to read a 600-page manuscript. Her goal is to proofread 24 pages each day. If she has already read 96 pages, how many more days will it take her to complete the proofreading? It will take her 21 more days to complete the proofreading.

37. **LOANS** A student plans to pay back a $600 loan with monthly payments of $30. How many payments has she made if the debt has been reduced to $420? She has made 6 payments.

38. **ANTIQUES** A woman purchases 4 antique spoons each year. She now owns 56 spoons. In how many years will she have 100 spoons in her collection? She will have 100 spoons in 11 years.

39. **CORPORATE DOWNSIZING** In an effort to cut costs, a corporation has decided to lay off 5 employees every month until the number of employees totals 465. If 510 people are now employed, how many months will it take to reach the employment goal? It will take 9 months to reach the goal.

40. **BOTTLED WATER DELIVERY** A truck driver left the plant carrying 300 bottles of drinking water. His delivery route consisted of office buildings, each of which was to receive 3 bottles of water. The driver returned to the plant at the end of the day with 117 bottles of water on the truck. To how many office buildings did he deliver? He delivered to 61 office buildings.

41. **OCEAN TRAVEL** At noon, a passenger ship and a freighter left a port traveling in opposite directions. By midnight, the passenger ship was 3 times farther from port than the freighter was. How far was the freighter from port if the distance between the ships was 84 miles? The freighter was 21 miles from port.

42. **RADIO STATIONS** The daily listening audience of an AM radio station is four times as large as that of its FM sister station. If 100,000 people listen to these two radio stations, how many listeners does the FM station have? The FM station has 20,000 listeners.

43. **INHERITANCES** Five brothers split an inheritance from their father equally. One of the brothers used part of his share to pay off a $5,575 balance on a credit card. That left him with $78,525. Find the total amount of the inheritance that the father left to his sons. The father left a total of $420,500 to his sons.

44. **APPLIANCES** A couple split the rebate check that they received after purchasing a new energy-efficient refrigerator. The wife then spent $25 of her share on a new energy-saving iron. If she was left with $35, what was the amount of the rebate check? The amount of the rebate check was $120.

45. **RENTALS** In renting an apartment with two other friends, Enrique agreed to pay the security deposit of $100 himself. The three of them agreed to contribute equally toward the monthly rent. Enrique's first check to the apartment owner was for $425. What was the monthly rent for the apartment? (*Hint:* First determine how many people are splitting the rent.) The monthly rent for the apartment was $975.

46. **LAWYER'S FEES** A lawyer and his client split the money that a jury awarded the client in a personal injury lawsuit. From his share, the lawyer paid his two assistants $15,000 each and ended up making $50,000 from the case. What was the amount of the jury award? The amount of the jury award was $160,000.

47. SERVICE STATIONS At a service station, the underground tank storing regular gas holds 100 gallons less than the tank storing premium gas. If the total storage capacity of the tanks is 700 gallons, how much does the premium gas tank hold? The premium gas tank holds 400 gallons.

48. LIBRARIES According to a 2010 survey, the state of New York had the most public libraries of the fifty states. Illinois was in second place with 122 fewer. Together, the two states had a total of 1,390 public libraries. How many public libraries did New York have in 2010? (Source: *The Institute of Museum and Library Service Public Survey*, 2012) In 2010, New York had 756 public libraries.

49. COMMERCIALS *from Campus to Careers — Broadcasting Technician* During a 30-minute television show, a viewer found that the actual program aired a total of 18 minutes more than the time devoted to commercials. How many minutes of commercials were there? There were 6 minutes of commercials.

50. CLASS TIME In a biology course, students spend a total of 250 minutes in lab and lecture each week. The lab time is 50 minutes shorter than the lecture time. How many minutes do the students spend in lecture per week? The students spend 150 minutes in lecture each week.

51. INTERIOR DECORATING As part of redecorating, crown molding was installed around the ceiling of a room. Sixty feet of molding was needed for the project. Find the width of the room if its length is twice the width. The width of the room is 10 feet.

52. SPRINKLER SYSTEMS A landscaper buried a water line around a rectangular lawn to serve as a supply line for a sprinkler system. The length of the lawn is five times its width. If 240 feet of pipe was used to do the job, what is the width of the lawn? The width of the lawn is 20 feet.

53. TENNIS The perimeter of a regulation singles tennis court is 210 feet and the length is 51 feet more than the width. Find the length and width of the court. The width of the court is 27 feet and the length is 78 feet.

54. THE CENTENNIAL STATE The border of the state of Colorado is approximately a rectangle with a perimeter of 1,320 miles. Find the length (east to west) and width (north to south), if the length is 100 miles longer than the width. The length of the border of the state of Colorado is approximately 380 miles and the width is approximately 280 miles.

Form an equation and then solve it to answer each question. Make a table to organize the data.

55. COMMISSIONS A salesman receives a commission of $3 for every pair of dress shoes he sells. He is paid $2 for every pair of athletic shoes he sells. After selling 9 pairs of shoes in a day, his commission was $24. How many pairs of each kind of shoe did he sell that day? He sold 6 pairs of dress shoes and 3 pairs of athletic shoes.

56. GRADING SCALES For every problem answered correctly on an exam, 3 points are awarded. For every incorrect answer, 4 points are deducted. In a 10-question test, a student scored 16 points. How many correct and incorrect answers did he have on the exam? He had 8 correct answers and 2 incorrect answers.

57. MOVER'S PAY SCALE A part-time mover's regular pay rate is $60 an hour. If the work involves going up and down stairs, his rate increases to $90 an hour. In one week, he earned $1,380 and worked 20 hours. How many hours did he work at each rate? He worked 14 hours at the regular rate and 6 hours going up and down stairs.

58. PRESCHOOL ENROLLMENTS A preschool charges $8 for a child to attend its morning session or $10 to attend the afternoon session. No child can attend both. Thirty children are enrolled in the preschool. If the daily receipts are $264, how many children attend each session? There are 18 children in the morning session and 12 children in the afternoon session.

59. AUTOGRAPHS Martin has collected the autographs of six more movie stars than he has television celebrities. Each movie star autograph is worth $200 and each television celebrity autograph is worth $75. If his collection is valued at $4,500, how many of each type of autograph does he have? He has 18 movie star autographs and 12 television celebrity autographs.

60. RENTALS In an apartment building, seven more 1-bedroom units are rented than 2-bedroom units. The monthly rent for a 1-bedroom is $500 and a 2-bedroom is $700. If the total monthly income from these units is $15,500, how many of each type of unit are there? There are 17 1-bedroom units and 10 2-bedroom units.

WRITING

61. Explain what should be accomplished in each of the six steps of the problem-solving strategy studied in this section.

62. Use an example to explain the difference between the number of quarters a person has and the value of those quarters.

63. Write a problem that could be represented by the following equation.

Age of father	plus	age of son	is	50.
x	$+$	$x - 20$	$=$	50

64. Write a problem that could be represented by the following equation.

2 ·	length of a field	plus	2 ·	width of a field	is	600 ft.
2 ·	$4x$	$+$	2 ·	x	$=$	600

REVIEW

65. What property is illustrated? associative property of addition
$$(2 + 9) + 1 = 2 + (9 + 1)$$

66. Solve: $4 - x = -8$ 12

67. Evaluate: -10^2 -100

68. List the factors of 18. 1, 2, 3, 6, 9, 18

69. Fill in the blank: Subtraction of a number is the same as __addition__ of the opposite of that number.

70. Round 123,808 to the nearest ten thousand. 120,000

71. Write this prime factorization using exponents:
$2 \cdot 2 \cdot 2 \cdot 5 \cdot 5$ $2^3 \cdot 5^2$

72. The value of a stock dropped $3 a day for 6 consecutive days. What was the change in the value of the stock over this period? $-$18$

STUDY SKILLS CHECKLIST

Solving Equations

The first step to solve an equation is often the most difficult for students to determine. Before taking the test on Chapter 3, make sure that you know what to do first when solving the following equations. Put a checkmark in the box if you can answer "yes" to the statement.

☐ I know that the first step to solve
$$3(3x + 8) = 51$$
is to use the distributive property on the left side.

☐ I know that the first step to solve
$$45 = 7a + 8a$$
is to combine like terms on the right side.

☐ I know that the first step to solve
$$9n + 12 = 6n - 9$$
is to subtract $6n$ from both sides.

☐ I know that the first step to solve
$$2(y + 40) - 6y = 3(4y - 80)$$
is to use the distributive property on both sides.

☐ I know that the first step to solve
$$100 + 25 = 4h - 15 + 3h$$
is to combine like terms on both sides.

☐ I know that the first step to solve
$$7(r - 3) = r + 4 + r$$
is to use the distributive property on the left side and combine like terms on the right side.

Teaching Guide: Refer to the Instructor's Resource Binder to find activities, worksheets on key concepts, more examples, instruction tips, overheads, and assessments.

CHAPTER 3 SUMMARY AND REVIEW

SECTION 3.1 Algebraic Expressions

DEFINITIONS AND CONCEPTS | EXAMPLES

A **variable** is a letter (or symbol) that stands for a number. Since numbers do not change value, they are called **constants**.

Variables: x, a, and y

Constants: 8, -10, 2, and 314

Variables and/or numbers can be combined with the operations of addition, subtraction, multiplication, and division to create **algebraic expressions**.

We often refer to *algebraic expressions* as simply **expressions**.

Expressions:

$5y + 7$ $\dfrac{12 - x}{5}$ $8a(b - 3)$

Key words and **key phrases** can be translated into algebraic expressions.

Review the tables on pages 226 and 227.

5 more than x can be expressed as $x + 5$.

25 less than twice y can be expressed as $2y - 25$.

One-half of the cost c can be expressed as $\dfrac{1}{2}c$.

REVIEW EXERCISES

1. The illustration below shows the distances from two towns to an airport. Which town is closer to the airport? How much closer is it? Brandon is closer by 250 mi.

   Brandon —— $(x - 250)$ mi —— Airport —— x mi —— Mill City

2. See the illustration below. Let h represent the height of the ladder, and write an algebraic expression for the height of the ceiling in feet. $h + 7$

7 ft, h

Translate each of the following phrases to an algebraic expression.

3. Five less than n $n - 5$
4. The product of 7 and x $7x$
5. The quotient of six and p $\dfrac{6}{p}$

6. The sum of s and -15 $s + (-15)$
7. Twice the length l $2l$
8. D reduced by 100 $D - 100$
9. Two more than r $r + 2$
10. 45 divided by x $\dfrac{45}{x}$
11. 100 reduced by twice the cutoff score s $100 - 2s$
12. The absolute value of the difference of 2 and the square of a $|2 - a^2|$
13. Translate the expression $m - 500$ into words. five hundred less than m (answers may vary)
14. HARDWARE Refer to the illustration below.
    a. Let n represent the length of the nail (in inches). Write an algebraic expression that represents the length of the bolt (in inches). $(n + 4)$ in.
    b. Let b represent the length of the bolt (in inches). Write an algebraic expression that represents the length of the nail (in inches). $(b - 4)$ in.

4 in.

Unless otherwise noted, all art on this page is © Cengage Learning.

15. **CHILD CARE** A child care center has six rooms, and the same number of children are in each room. If c children attend the center, write an algebraic expression that represents the number of children in each room. $\dfrac{c}{6}$

16. **CAR SALES** A used car, originally advertised for $1,000, did not sell. The owner decided to drop the price $$x. Write an algebraic expression that represents the new price of the car (in dollars). $1,000 - x$

17. **CLOTHES DESIGNERS** The legs on a pair of pants are x inches long. The designer then lets the hem down 1 inch. Write an algebraic expression that represents the new length (in inches) of the pants legs. $x + 1$

18. **BUTCHERS** A roast weighs p pounds. A butcher cut the roast into 8 equal-sized servings. Write an algebraic expression that represents the weight (in pounds) of one serving. $\dfrac{p}{8}$

19. **ROAD TRIPS** On a cross-country vacation, a husband drove for twice as many hours as his wife. Choose a variable to represent the hours driven by one of them. Then write an algebraic expression to represent the hours driven by the other. x = the number of hours driven by the wife, $2x$ = the number of hours driven by the husband

20. **GEOMETRY** The length of a rectangle is 3 units more than its width. Choose a variable to represent one of the dimensions. Then write an algebraic expression that represents the other dimension. w = the width, $w + 3$ = the length

21. **SPORTS EQUIPMENT** An NBA basketball weighs 2 ounces more than twice the weight of a volleyball. Let a variable represent the weight of one of the sports balls. Then write an algebraic expression that represents the weight of the other ball. x = the weight of the volleyball (in ounces), $2x + 2$ = the weight of the NBA basketball (in ounces)

22. **BEST-SELLING BOOKS** *The Lord of the Rings* was first published 6 years before *To Kill a Mockingbird*. *The Godfather* was first published 9 years after *To Kill a Mockingbird*. Write algebraic expressions to represent the ages of each of those books. x = the age of *To Kill a Mockingbird*, $x + 6$ = the age of *The Lord of the Rings*, $x - 9$ = the age of *The Godfather*

Use a table to help answer Problems 23 and 24.

23. How many eggs are in x dozen? $12x$

24. d days is how many weeks? $\dfrac{d}{7}$

SECTION 3.2 Evaluating Algebraic Expressions and Formulas

DEFINITIONS AND CONCEPTS	EXAMPLES
To **evaluate algebraic expressions**, we substitute the values of its variables and apply the order of operations rule.	Evaluate $\dfrac{x^2 - y^2}{x + y}$ for $x = 2$ and $y = -3$. $\dfrac{x^2 - y^2}{x + y} = \dfrac{2^2 - (-3)^2}{2 + (-3)}$ Substitute 2 for x and −3 for y. $= \dfrac{4 - 9}{-1}$ In the numerator, evaluate the exponential expressions. In the denominator, add. $= \dfrac{-5}{-1}$ In the numerator, subtract. $= 5$ Do the division.

A **formula** is an equation that states a relationship between two or more variables.

Formulas from business:

Sale price = original price − discount

Retail price = cost + markup

Profit = revenue − costs

SMALL BUSINESSES For the month of December, a nail salon's cost of doing business was $6,050. If December revenues totaled $18,295, what was the salon's profit for the month?

$P = r - c$	This is the formula for profit.
$P = 18{,}295 - 6{,}050$	Substitute 18,295 for the revenue r and 6,050 for the costs c.
$P = 12{,}245$	Do the subtraction.

The nail salon made a profit of $12,245 in December.

Formulas from science:

Distance = rate · time

Fahrenheit to Celsius temperature:

$$C = \frac{5(F - 32)}{9}$$

Distance fallen = 16 · (time)^2

WHALES As they migrate from the Bering Sea to Baja California, gray whales swim at an average rate of 3 mph. If they swim for 20 hours a day, find the distance they travel each day.

$d = rt$	This is the formula for distance traveled.
$d = 3(20)$	Substitute 3 for the rate r and 20 for the time t.
$d = 60$	Do the multiplication.

Gray whales travel a distance of 60 miles each day.

The **mean** (or **average**) of a set of numbers is a value around which the numbers are grouped.

$$\text{Mean} = \frac{\text{sum of values}}{\text{number of values}}$$

GRADES Find the mean of the test scores of 74, 83, 79, 91, and 73.

$$\text{Mean} = \frac{74 + 83 + 79 + 91 + 73}{5}$$

$$\text{Mean} = \frac{400}{5}$$

$$\text{Mean} = 80$$

The mean test score is 80.

REVIEW EXERCISES

25. **RETAINING WALLS** The illustration to the right shows the design for a retaining wall. The relationships between the lengths of its important parts are given in words.

    a. Choose a variable to represent one unknown dimension of the wall. Then write algebraic expressions to represent the lengths of the other two parts.
       h = the height of the wall, $h - 5$ = the length of the upper base, $2h - 3$ = the length of the lower base

    b. Suppose engineers determine that a 10-foot-high wall is needed. Find the lengths of the upper and lower bases. upper base: 5 ft, lower base: 17 ft

The length of the upper base is 5 ft less than the height.

Height

The length of the lower base is 3 ft less than twice the height.

26. **SOD FARMS** The expression $20{,}000 - 3s$ gives the number of square feet of sod that are left in a field after s strips have been removed. Suppose a city orders 7,000 strips of sod. Evaluate the expression and explain the result. $-1{,}000$ means the sod farm is short 1,000 ft² to fill the city's order

Strips of sod, cut and ready to be loaded on a truck for delivery

Evaluate each algebraic expression.

27. $-2x + 6$ for $x = -3$ 12
28. $\dfrac{6-a}{1+a}$ for $a = -2$ -8
29. $(-y - 40)^2$ for $y = -50$ 100
30. $|x^3 - 8x^2|$ for $x = 4$ 64
31. $b^2 - 4ac$ for $a = 4, b = 6,$ and $c = -4$ 100
32. $(x + y)^2 - |z + y|$ for $x = 3, y = 2,$ and $z = -4$ 23

Use the correct formula to solve each problem.

33. **SALE PRICE** Find the sale price of a trampoline that usually sells for $315 if a $37 discount is being offered. The sale price is $278.

34. **RETAIL PRICE** Find the retail price of a car if the dealer pays $14,505 and the markup is $725. The retail price is $15,230.

35. **GRAND OPENINGS** On its first month of business, a bookstore brought in $52,895. The costs for the month were $47,980. Find the profit the store made its first month. The profit was $4,915.

36. **DISTANCE TRAVELED** Complete the table by finding the distance traveled for a given time at a given rate.

	Rate (mph)	Time (hr)	Distance traveled (mi)
Monorail	65	2	130
Subway	38	3	114
Train	x	6	$6x$
Bus	55	$t + 1$	$55(t + 1)$

37. **TEMPERATURE CONVERSION** At a summer resort, visitors can relax by taking a dip in a swimming pool or a lake. The pool water is kept at a constant temperature of 77°F. The water in the lake is 23°C. Which water is warmer, and by how many degrees Celsius? The pool is 2°C warmer.

38. **DISTANCE FALLEN** A steelworker accidentally dropped a wrench while working atop a new high-rise building. How far will the wrench fall in 3 seconds? The wrench will fall 144 ft.

39. **AVERAGE YEARS OF EXPERIENCE** Three generations of Smiths now operate a family-owned real estate office. The two grandparents, who started the business, have been realtors for 40 years. Their son and daughter-in-law joined the company as realtors 18 years ago. Their grandson has worked as a realtor for 4 years. What is the average number of years a member of the Smith family has worked at Smith Realty? 24 yr

40. **SURVEYS** Some students were asked to rate their college cafeteria food on a scale from 1 to 5. The responses are shown on the tally sheet. Find the mean rating. 4

Poor		Fair		Excellent
1	2	3	4	5
	|||	|||	⟊⟊	⟊⟊ ||||

SECTION 3.3 Simplifying Algebraic Expressions and the Distributive Property

DEFINITIONS AND CONCEPTS	EXAMPLES
We often use the *commutative property of multiplication* to reorder factors and the *associative property of multiplication* to regroup factors when **simplifying expressions**.	Simplify: $-5 \cdot 3y = (-5 \cdot 3)y = -15y$ Simplify: $-45b\left(\dfrac{5}{9}\right) = \left(-45 \cdot \dfrac{5}{9}\right)b = -\dfrac{5 \cdot \overset{1}{9} \cdot 5}{\underset{1}{9}}b = -25b$

Unless otherwise noted, all art on this page is © Cengage Learning.

The **distributive property** can be used to remove parentheses: $a(b + c) = ab + ac$ $a(b - c) = ab - ac$	Multiply: $7(x + 3) = 7 \cdot x + 7 \cdot 3$ $= 7x + 21$
The distributive property can be extended to several other useful forms. $a(b + c + d) = ab + ac + ad$ $(b + c)a = ba + ca$ $(b - c)a = ba - ca$	Multiply: $-2(4m - 5n - 7) = -2(4m) - (-2)(5n) - (-2)(7)$ $= -8m + 10n + 14$ Multiply: $(6y - 10)5 = 6y \cdot 5 - 10 \cdot 5$ $= 30y - 50$
The **opposite of a sum** is the sum of the opposites. $-(a + b) = -a + (-b)$ The result can be obtained very quickly by changing the sign of each term within the parentheses and dropping the parentheses.	Simplify: $-(-3r + 14) = -1(-3r + 14)$ Replace the $-$ symbol with -1. $= (-1)(-3r) + (-1)(14)$ Distribute -1. $= 3r - 14$

REVIEW EXERCISES

Simplify each expression.

41. $-2 \cdot 5x$ $-10x$

42. $-7x(-6y)$ $42xy$

43. $4d \cdot 3e \cdot 5$ $60de$

44. $(4s)8$ $32s$

45. $-1(-e)(2)$ $2e$

46. $7x \cdot 7y$ $49xy$

47. $4 \cdot 3k \cdot 7$ $84k$

48. $(-10t)(-10)$ $100t$

Multiply.

49. $4(y + 5)$ $4y + 20$

50. $-5(6t + 9)$ $-30t - 45$

51. $(-3 - 3x)7$ $-21 - 21x$

52. $5(9s - 12t - 2)$ $45s - 60t - 10$

53. $4(6w - 3)2$ $48w - 24$

54. $-9(x + 1)4$ $-36x - 36$

Simplify.

55. $-(6t - 4)$ $-6t + 4$

56. $-(5 + x)$ $-5 - x$

57. $-(6t - 3s + 1)$ $-6t + 3s - 1$

58. $-(-5a - 3)$ $5a + 3$

SECTION 3.4 Combining Like Terms

DEFINITIONS AND CONCEPTS	EXAMPLES
A **term** is a product or quotient of numbers and/or variables. A single number or variable is also a term. A term such as 4, that consists of a single number, is called a **constant term**.	Terms: 4, y, $6r$, $-w^3$, $7x^5$, $-15ab^2$
Addition symbols separate expressions into parts called **terms**. The numerical factor of a term is called the **coefficient** of the term.	Since $6a^2 + a - 5$ can be written as $6a^2 + a + (-5)$, it has three terms. | Term | Coefficient | |---|---| | $6a^2$ | 6 | | a | 1 | | -5 | -5 |
It is important to be able to distinguish between the **terms of an expression** and the **factors of a term**.	$x + 6$ $6x$ ↑ ↑ x is a term. x is a factor.
Like terms are terms with exactly the same variables raised to exactly the same powers.	$3x$ and $-5x$ are like terms. $-4t^3$ and $3t^2$ are unlike terms because the variable t has different exponents. $0.5xyz$ and $3.7xy$ are unlike terms because they have different variables.
Simplifying the sum or difference of like terms is called **combining like terms**. Like terms can be combined by adding or subtracting the coefficients of the terms and keeping the same variables with the same exponents.	Simplify: $4a + 2a = 6a$ Think: $(4 + 2)a = 6a$. Simplify: $5p^2 + p - p^2 - 9p = 4p^2 - 8p$ Think: $(5 - 1)p^2 = 4p^2$ and $(1 - 9)p = -8p$. Simplify: $2(k - 1) - 3(k + 2) = 2k - 2 - 3k - 6$ $= -k - 8$
The perimeter of a rectangle is given by $P = 2l + 2w$ The perimeter of a square is given by $P = 4s$	**FLAGS** Find the perimeter of the flag of Eritrea, a country in east Africa, that is shown to the right. 32 in. 48 in. $P = 2l + 2w$ This is the formula for the perimeter of a rectangle. $P = 2(48) + 2(32)$ Substitute 48 for the length l and 32 for the width w. $P = 96 + 64$ Do the multiplication. $P = 160$ Do the addition. The perimeter of the flag is 160 in.

REVIEW EXERCISES

Identify the terms in each expression.

59. $8x^2 - 7x + 9$ $8x^2, -7x, 9$
60. $-15y$ $-15y$

61. $16ab - 6b$ $16ab, -6b$
62. $4x - 3 + 5x - 7$ $4x, -3, 5x, -7$

Identify the coefficient of each term in the expression.

63. $5x^2 - 4x + 8$ $5, -4, 8$
64. $7y + 3y + x - y$ $7, 3, 1, -1$

65. $t + r - t + 6$ $1, 1, -1, 6$
66. $-5y^2 + 125$ $-5, 125$

Determine whether x is used as a factor or a term.

67. $5x - 6y^2$ factor
68. $x + 6$ term

69. $-36 - x + b$ term
70. $6xy$ factor

Determine whether the following are like terms. Write yes or no.

71. $4x, -5x$ yes
72. $4x, 4x^2$ no

73. $3xy, xy$ yes
74. $-5b^2c, -5bc^2$ no

Simplify by combining like terms, if possible.

75. $3x + 4x$ $7x$
76. $-3t^3 - 6t^2$ does not simplify

77. $2z + (-5z)$ $-3z$
78. $6x - x$ $5x$

79. $-6y - 7y - (-y)$ $-12y$
80. $5w^2 - 8 - 4w^2 + 3$ $w^2 - 5$

81. $-45d - 2a + 4a - d$ $-46d + 2a$

82. $5y + 8h - 3 + 7h + 5y + 2$ $10y + 15h - 1$

83. $10a^2 + 6a - 17a + 6$ $10a^2 - 11a + 6$
84. $28w + w$ $29w$

Simplify each expression.

85. $7(y + 6) + 3(2y + 2)$ $13y + 48$

86. $-4(t - 7) - (t + 6)$ $-5t + 22$

87. $5x - 4 - 2(x - 6)$ $3x + 8$

88. $6f + (-11) + 7(12 - 8f)$ $-50f + 73$

89. ROBOTS Find an algebraic expression that represents the total length (in feet) of the robotic arm shown below. $3x + 3$

90. HOLIDAY LIGHTS To decorate a house, lights will be hung around the entire home, as shown. They will also be placed around the two 5-foot-by-5-foot windows in the front. How many feet of lights will be needed?   194 ft

SECTION 3.5 Simplifying Expressions to Solve Equations

DEFINITIONS AND CONCEPTS	EXAMPLES
A number that makes an equation a true statement when substituted for the variable is called a **solution** of the equation. We say such a number **satisfies** the equation.	Use a check to determine whether -51 is a solution of $3x - 15 = 4x + 36$. $3x - 15 = 4x + 36$ The original equation. $3(-51) - 15 \stackrel{?}{=} 4(-51) + 36$ Substitute -51 for x. $-153 - 15 \stackrel{?}{=} -204 + 36$ Do the multiplication. $-168 = -168$ True Since the resulting statement, $-168 = -168$, is true, -51 is a solution of $3x - 15 = 4x + 36$.

When solving equations, we should simplify the expressions that make up the left and right sides before applying any properties of equality.

A strategy for solving equations:

1. *Simplify* each side. Use the distributive property and combine like terms when necessary.
2. *Isolate the variable term.* Use the addition and subtraction properties of equality.
3. *Isolate the variable.* Use the multiplication and division properties of equality.
4. *Check* the result in the original equation.

Solve: $2(y + 2) + 4y = 11 - y$

$2y + 4 + 4y = 11 - y$ Distribute the multiplication by 2.
$6y + 4 = 11 - y$ Combine like terms: $2y + 4y = 6y$.
$6y + 4 + y = 11 - y + y$ To eliminate $-y$ on the right, add y to both sides.
$7y + 4 = 11$ Combine like terms.
$7y + 4 - 4 = 11 - 4$ To isolate the variable term $7y$, subtract 4 from both sides.
$7y = 7$ Simplify each side of the equation.
$\frac{7y}{7} = \frac{7}{7}$ To isolate y, divide both sides by 7.
$y = 1$

The solution is 1. Check by substituting it into the original equation.

REVIEW EXERCISES

91. Use a check to determine whether -1 is a solution of $6a + (-7) = 5a - 9$.   not a solution

92. Use a check to determine whether 4 is a solution of $4(8 - 3t) = 32 - 8(t + 2)$.   it is a solution

Solve each equation. Check the result.

93. $5a - 3a = -36$ -18
94. $3x - 4x = -8$ 8
95. $250 + 350 = x - 10 + x$ 305
96. $7x + 1 = 3x - 11$ -3

97. $5(y - 15) = 0$ 15
98. $3a - (2a - 1) = -2$ -3
99. $15 + b = 5b + 1 + 3b$ 2
100. $-6(2x + 3) = -(5x - 10)$ -4
101. $5(2x - 4) - 5x = 0$ 4
102. $4(9d + 2) = 4(d - 18) - 8d$ -2

SECTION 3.6   Using Equations to Solve Application Problems

DEFINITIONS AND CONCEPTS	EXAMPLES
To solve application problems, use the six-step problem-solving strategy. 1. **Analyze the problem:** What information is given? What are you asked to find? 2. **Assign a variable:** Let a variable represent an unknown value in the problem. 3. **Form an equation:** Translate the words of the problem into an equation. 4. **Solve the equation.**	SOUND SYSTEMS A 45-foot-long speaker wire is cut into two pieces. One piece is 9 feet longer than the other. Find the length of each piece of wire. *Analyze* • A 45-foot long wire is cut into two pieces. Given • One piece is 9 feet longer than the other. Given • What is the length of the shorter piece and the length of the longer piece of wire? Find

5. **State the conclusion clearly:** Be sure to include the units (such as feet, seconds, or pounds) in your answer.
6. **Check the result:** Use the original wording of the problem, not the equation that was formed in Step 3 from the words.

The six-step problem-solving strategy can also be used to solve application problems to find **two unknowns**.

Assign

Since we are told that the length of the longer piece of wire is related to the length of the shorter piece,

   Let x = the length of the shorter piece of wire

There is a second unknown quantity. Look for a key phrase to help represent the length of the longer piece of wire using an algebraic expression.

   Key phrase: 9 feet longer   **Translation:** addition

So $x + 9$ = the length of the longer piece of wire.

Form

Now, translate the words of the problem to an equation.

The length of the shorter piece	plus	the length of the longer piece	is	45 feet.
x	$+$	$x + 9$	$=$	45

Solve

$x + x + 9 = 45$	We need to isolate x on the left side.
$2x + 9 = 45$	Combine like terms: x + x = 2x.
$2x + 9 - 9 = 45 - 9$	To isolate 2x, subtract 9 from both sides.
$2x = 36$	Do the subtraction.
$\dfrac{2x}{2} = \dfrac{36}{2}$	To isolate x, undo the multiplication by 2 by dividing both sides by 2.
$x = 18$	Do the division. This is the length of the shorter piece.

To find the second unknown, we substitute 18 for x in the expression that represents the length of the longer piece of wire.

$$x + 9 = 18 + 9 = 27$$

State

The length of the shorter piece of wire is 18 feet and the length of the longer piece is 27 feet.

Check

The length of the longer piece of wire, 27 feet, is 9 feet longer than the length of the shorter piece, 18 feet. Adding the two lengths, we get

```
  18
+ 27
  45  ← This is the original length of the wire,
         before it was cut into two pieces.
```

The results, 18 ft and 27 ft, check.

Be careful to distinguish between the *number* and the *value* of a set of objects.

   Total value = number · value

Determine the total value of x \$20 bills.

   Total value = **number · value**

   Total value = x · 20

   Total value = $20x$

The total value is $20x$ dollars.

REVIEW EXERCISES

103. CONCERTS The fee to rent a concert hall is $2,250 plus $150 per hour to pay for the support staff. For how many hours can an orchestra rent the hall and stay within a budget of $3,300? They can rent the hall for 7 hours.

104. CHEMISTRY As part of a stress test, chemists are lowering the temperature of a chemical compound 7° Fahrenheit every hour. How long will it take for the compound to go from a room temperature of 71°F to 29°F? It will take 6 hr to lower the temperature to 29°F.

105. MOVING EXPENSES Tom and his friend split the cost of renting a U-Haul trailer equally. Tom also agreed to pay the $4 to rent a refrigerator dolly. In all, Tom paid $20. What did it cost to rent the trailer? It cost $32 to rent the trailer.

106. FITNESS The midweek workout for a fitness instructor consists of walking and running. She walks 3 fewer miles than she runs. If her workout covers a total of 15 miles, how many miles does she run and how many miles does she walk? She runs 9 miles and she walks 6 miles.

107. MEDIEVAL FAIR Attendance during the first day of a two-day medieval fair was low. On the second day, attendance doubled. If a total of 6,600 people attended the fair, what was the attendance on the first day and what was the attendance on the second day? The attendance on the first day was 2,200 people. The attendance on the second day was 4,400 people.

108. PARKING LOTS A rectangular parking lot is four times as long as it is wide. If the perimeter of the parking lot is 250 feet, what is its length and width? The width of the parking lot is 25 feet and the length is 100 feet.

109. Complete the table.

Type of coin	Number	Value (¢)	Total value (¢)
Dime	6	10	60
Quarter	7	25	175
Penny	x	1	x
Nickel	$n + 25$	5	$5(n + 25)$

110. HEALTH FOOD A fruit juice bar sells two types of drinks, one priced at $3 and the other at $4. One day at lunchtime, business was very brisk. If a total of 50 drinks were sold and the receipts were $185, how many $3 drinks and how many $4 drinks were purchased? There were 15 $3 drinks and 35 $4 drinks sold.

CHAPTER 3 TEST

Fill in the blanks.

1.
   a. _Variables_ are letters (or symbols) that stand for numbers.
   b. To perform the multiplication $3(x + 4)$, we use the _distributive_ property.
   c. Terms such as $7x^2$ and $5x^2$, which have the same variables raised to exactly the same power, are called _like_ terms.
   d. To _solve_ an equation means to find all values of the variable that make the equation true.
   e. The _coefficient_ of the term $9y$ is 9.
   f. Variables and/or numbers can be combined with the operations of addition, subtraction, multiplication, and division to create algebraic _expressions_.
   g. To evaluate $y^2 + 9y - 3$ for $y = -5$, we _substitute_ -5 for y and apply the order of operations rule.
   h. An _equation_ is a statement indicating that two expressions are equal.
   i. When we write $4x + x$ as $5x$, we say we have _combined_ like terms.
   j. To _check_ the solution of an equation, we substitute the value for the variable in the original equation and determine whether the result is a true statement.

2. SALARIES A wife's monthly salary is $1,000 less than twice her husband's monthly salary.
   a. If her husband's monthly salary is h dollars, write an algebraic expression that represents the wife's monthly salary (in dollars). $2h - 1,000$
   b. Suppose the husband's monthly salary is $2,350. Find the wife's monthly salary. $3,700

3. REFRESHMENTS How many cups of coffee are left in the coffeemaker shown if c cups have already been poured from it? $56 - c$

4. Let a variable represent the length of one of the fish shown. Then write an expression that represents the length (in inches) of the other fish. Give two possible sets of answers. t = the length of the trout, $t + 10$ = the length of the salmon; or s = the length of the salmon, $s - 10$ = the length of the trout

5. Translate each phrase to mathematical symbols.
   a. 2 less than r $r - 2$
   b. The product of 3, x, and y $3xy$
   c. x increased by 100 $x + 100$
   d. The absolute value of the quotient of x and -9 $\left|\frac{x}{-9}\right|$

6. Write an algebraic expression that represents the number of years in d decades. $10d$

7. Evaluate each expression.
   a. $x - 16$ for $x = 4$ -12
   b. $2t^2 - 3(t - s)$ for $t = -2$ and $s = 4$ 26
   c. $-a^2 + 10$ for $a = -3$ 1
   d. $\left|\dfrac{-10d + f^3}{-f}\right|$ for $d = -1$ and $f = -5$ 23

8. DISTANCE TRAVELED Find the distance traveled by a motorist who departed from home at 9:00 A.M. and arrived at his destination at noon, traveling at a rate of 55 miles per hour. The distance traveled is 165 mi.

9. PROFITS A craft show promoter had revenues and costs as shown. Find the profit. The profit is $37,000.

Revenues	Costs
Ticket sales: $40,000	Supplies: $13,000
Booth rental: $15,000	Facility rental fee: $5,000

10. FALLING OBJECTS If a tennis ball were dropped from the top of a 200-foot-tall building, would it hit the ground after falling for 3 seconds? If not, how far short of the ground would it be? It would be 56 ft short of hitting the ground.

11. METER READINGS Every hour between 8 A.M. and 5 P.M., a technician noted the value registered by a meter in a power plant and recorded that number on a line graph. Find the mean meter value reading for this period. The mean meter reading is 1.

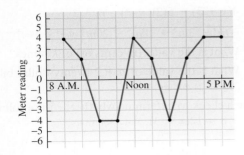

12. **LANDMARKS** Overlund College is going to construct a gigantic block letter O on a foothill slope near campus. The outline of the letter is to be done using redwood edging. How many feet of edging will be needed? The project requires 250 ft of edging.

13. **AIR CONDITIONING** After the air conditioner in a classroom was accidentally left on all night, the room's temperature in the morning was a cool 59°F. What was the temperature in degrees Celsius? The temperature was 15°C.

14. Multiply.
    a. $5(5x + 1)$ $25x + 5$
    b. $-6(7 - x)$ $-42 + 6x$
    c. $-(6y + 4)$ $-6y - 4$
    d. $3(2a + 3b - 7)$ $6a + 9b - 21$
    e. $(a - 15)8$ $8a - 120$
    f. $2(6r + 9)3$ $36r + 54$

15. Determine whether x is used as a factor or as a term.
    a. $5xy$ factor
    b. $8y + x + 6$ term

16. Simplify each expression, if possible.
    a. $7x + 4x$ $11x$
    b. $3 \cdot 4e$ $12e$
    c. $6x^2 - x^2$ $5x^2$
    d. $-5y(-6)$ $30y$
    e. $0 - 7x$ $-7x$
    f. $0 + 9y$ $9y$
    g. $-8a \cdot 9b$ $-72ab$
    h. $8(-7m)5$ $-280m$

17. a. Identify each term in this algebraic expression: $8x^2 - x - 6$ $8x^2, -x, -6$
    b. What is the coefficient of each term? $8, -1, -6$

18. Simplify.
    a. $-20y + 6 - 8y + 4$ $-28y + 10$
    b. $-t - t - t$ $-3t$
    c. $4(y + 3) - 5(2y + 3)$ $-6y - 3$
    d. $-m^4 + 3m^3 + 10m^4 + 20m^3$ $9m^4 + 23m^3$

19. a. What is the value (in cents of) of k dimes? $10k¢$
    b. What is the value of $p + 2$ twenty-dollar bills? $20(p + 2)$ dollars

20. Use a check to determine whether -5 is a solution of $6x - 8 = 12(x - 3)$.   not a solution

Solve each equation. Check the result.

21. $5x - 3x = -18$ -9
22. $6r = r - 12 + r$ -3
23. $-55 + 10 = 3(1 - 4t)$ 4
24. $6 - (y - 3) = 19$ -10
25. $8 + 2(3x - 4) = -60$ -10
26. $-23 - n - 6n = 13 + 2n$ -4
27. $5(x + 7) - 7x = 9(x - 1)$ 4
28. $-80 + y = -80 - y$ 0

Form an equation and then solve it to answer each question.

29. **DRIVING SCHOOLS** A driver's training program requires students to attend six equally long classroom sessions. Then the students take a 2-hour final exam at the end of the training. If the entire program requires 20 hours of a student's time, how long is each classroom session? Each classroom session is 3 hr long.

30. **CABLE TELEVISION** In order to receive its broadcasting license, a cable television station was required to broadcast locally produced shows in addition to its national programming. During a typical 24-hour period, the national shows aired for 8 hours more than the local shows. How many hours of local shows and how many hours of national shows were broadcast each day? Each day, there were 8 hr of local shows and 16 hr of national shows.

31. **RECREATION** A developer donated a large plot of land to a city for a park. Half of the acres will be used for sports fields. From the other half, 4 acres will be used for parking. This will leave 18 acres for a nature habitat. How many acres of land did the developer donate to the city? The developer donated 44 acres of land to the city.

32. **PICTURE FRAMING** A rectangular picture frame is twice as long as it is wide. If 144 inches of framing material were used to make it, what is the width and what is the length of the frame? The width of the frame is 24 inches and the length is 48 inches.

33. Do the instructions *simplify* and *solve* mean the same thing? Explain. We simplify expressions and we solve equations.

34. Explain why we can simplify $5x \cdot 2$ but we cannot simplify $5x + 2$. $5x \cdot 2 = 10x$; $5x$ and 2 are not like terms and therefore cannot be combined.

CHAPTERS 1–3 CUMULATIVE REVIEW

1. **GASOLINE** In 2011, the United States produced three billion, one hundred ninety-four million, seven hundred fifty-four thousand barrels of finished motor gasoline. Write this number in standard notation. (Source: U.S Energy Information Administration). [Section 1.1]
   3,194,754,000 barrels

2. Round 49,999 to the nearest thousand. [Section 1.1]
   50,000

Perform each operation.

3. 38,908 [Section 1.2]
   +15,696
   ──────
    54,604

4. 9,700 [Section 1.2]
   −5,491
   ──────
    4,209

5. 345 [Section 1.3]
   × 67
   ──────
   23,115

6. $23\overline{)2{,}001}$ [Section 1.4] 87

7. a. Explain how to check the following result using addition. [Section 1.2] 683 + 459 = 1,142

   1,142
   − 459
   ─────
     683

   b. Write an expression showing division by 0 and an expression showing division of 0. Which is undefined? [Section 1.4] $\frac{5}{0}, \frac{0}{5}$ (answers may vary); division by 0

8. **VIETNAMESE CALENDAR** An animal represents each Vietnamese lunar year. Recent Years of the Tiger are listed below. If the cycle continues, what year will be the next Year of the Tiger? [Section 1.2] 2022

   1962 1974 1986 1998 2010

9. Consider the multiplication statement 4 · 5 = 20. Show that multiplication is repeated addition. [Section 1.3] 4 · 5 = 5 + 5 + 5 + 5 = 20

10. **ROOM DIVIDERS** Four pieces of plywood, each 22 inches wide and 62 inches high, are to be covered with fabric, front and back, to make the room divider shown. How many square inches of fabric will be used? [Section 1.3] 10,912 in.²

11. **OIL CHANGES** In July 2009, the 1964 Mercury Comet that Rachel Veitch of Orlando, Florida, drives notched its 558,000th mile. The 90-year-old retired nurse has changed the oil every 3,000 miles since she bought the car new. How many oil changes did the car have to that point? (Source: foxnews.com) [Section 1.4] 186 oil changes

12. a. Find the factors of 18. [Section 1.5] 1, 2, 3, 6, 9, 18
    b. Why isn't 27 a prime number? [Section 1.5] It has factors other than 1 and itself: 27 = 3 · 9.
    c. Find the prime factorization of 18. [Section 1.5] $2 \cdot 3^2$

13. Write the first ten prime numbers. [Section 1.5]
    2, 3, 5, 7, 11, 13, 17, 19, 23, 29

14. a. Find the LCM of 35 and 45. [Section 1.6] 315
    b. Find the GCF of 12, 68, and 92. [Section 1.6] 4

15. Evaluate each expression. [Section 1.7]
    a. $(9 - 2)^2 - 3^3$ 22
    b. $\dfrac{80 - 2[12 - (5 + 4)]}{8 \div 8 \cdot 2}$ 37

16. What property was used to solve the equation shown below? [Section 1.8] the addition property of equality

    $x - 3 = 47$
    $x - 3 + 3 = 47 + 3$
    $x = 50$

17. Solve $250 = \dfrac{y}{2}$ and check the result. [Section 1.9] 500

18. a. Simplify: $-(-6)$ [Section 2.1] 6
    b. Find the absolute value: $|-5|$ 5
    c. Is the statement $-12 > -10$ true or false? false

19. Graph the integers greater than −3 but less than 4. [Section 2.1]

20. Translate the following phrase to mathematical symbols: *Negative twenty-one minus negative seventy-three* [Section 2.1] $-21 - (-73)$

21. Perform the indicated operations.
    a. $-25 + 5$ [Section 2.2] -20
    b. $25 - (-5)$ [Section 2.3] 30
    c. $-25(5)(-1)$ [Section 2.4] 125
    d. $\dfrac{-25}{-5}$ [Section 2.5] 5

22. CARD GAMES Canasta is a card game commonly played by four players in two teams. It is possible for a team to have a negative score for a hand. The canasta scores for two teams are shown to the right. Find the total for each team. [Section 2.2]

	Team 1	Team 2
	305	295
	−75	−120
	600	300
	500	0
	200	100
	100	0
Total:	1,630	575

23. PLANETS Mercury orbits closer to the sun than any other planet. Temperatures on Mercury can get as high as 810°F and as low as −290°F. What is the temperature range? [Section 2.3] 1,100°F

24. a. Explain how to evaluate -3^2 and $(-3)^2$.
[Section 2.4] $-3^2 = -(3 \cdot 3) = -9$; $(-3)^2 = (-3)(-3) = 9$

b. What property allows us to rewrite $x \cdot 5$ as $5x$?
[Section 2.4] the commutative property of multiplication

Evaluate each expression. [Section 2.6]

25. $\dfrac{(-6)^2 - 1^5}{-4 - 3}$ −5

26. $-3 + 3(-4 - 4 \cdot 2)^2$ 429

Solve each equation and check the result. [Section 2.7]

27. $-4x + 4 = -24$ 7   **28.** $-y = 10$ −10

29. $\dfrac{m}{-2} - 6 = -9$ 6   **30.** $-7 + 9 = a + 5(-3)$ 17

31. $-90 = x + (-3)$ −87   **32.** $16 = -7 - 5 - 7x$ −4

33. Translate each phrase to mathematical symbols. [Section 3.1]

a. h increased by 12 $h + 12$

b. 4 less than the width w $w - 4$

c. 1,000 split x equal ways $\dfrac{1{,}000}{x}$

34. a. TENNIS Write an algebraic expression that represents the length of the handle of the tennis racket in inches. [Section 3.1] $(26 - x)$ in.

b. What is the value (in cents) of q quarters?
[Section 3.6] $25q$¢

35. Write an algebraic expression that represents the number of inches in f feet. [Section 3.1] $12f$ in.

36. Evaluate $x^2 - 2x + 1$ for $x = -5$. [Section 3.2] 36

37. Complete the table. [Section 3.2]

	Rate (mph)	Time (hr)	Distance traveled (mi)
Truck	55	4	220

38. Multiply. [Section 3.3]

a. $5(2x - 7)$ $10x - 35$ **b.** $-(5t - 7)$ $-5t + 7$

39. Simplify. [Section 3.3]

a. $-6(-4t)$ $24t$ **b.** $4(-3y)(4z)$ $-48yz$

40. Complete the table. [Section 3.4]

Term	Coefficient
$4a$	4
$-2y^2$	−2
x	1
$-m$	−1

41. Write an expression in which x is used as a term. Then write an expression in which x is used as a factor. [Section 3.4] $x + 3$; $3x$ (answers may vary)

42. Simplify: [Section 3.4]

a. $5b + 8 - 6b - 7$ $-b + 1$

b. $4(x + 5) - 5 - (2x - 4)$ $2x + 19$

Solve each equation and check the result. [Section 3.5]

43. $8p + 2p - 1 = -11$ −1

44. $7 + 2x = 2 - (4x + 7)$ −2

45. $5(2x - 4) - 5x = 0$ 4

46. $4(m - 30) - 4m = 6(2m + 6)$ −13

Form an equation and then solve it to answer the following questions. [Section 3.6]

47. CLASS TIME In a chemistry course, students spend a total of 300 minutes in lab and lecture each week. The time spent in lab is 50 minutes less than the time spent in lecture. How many minutes do the students spend in lecture and lab each week? The students spend 175 min in lecture and 125 min in lab each week.

48. GEOMETRY The perimeter of a rectangle is 120 feet, and the length is five times as long as the width. Find the length and the width. The width is 10 ft and the length is 50 ft.

4

Fractions and Mixed Numbers

- **4.1** An Introduction to Fractions
- **4.2** Multiplying Fractions
- **4.3** Dividing Fractions
- **4.4** Adding and Subtracting Fractions
- **4.5** Multiplying and Dividing Mixed Numbers
- **4.6** Adding and Subtracting Mixed Numbers
- **4.7** Order of Operations and Complex Fractions
- **4.8** Solving Equations That Involve Fractions

Chapter Summary and Review
Chapter Test
Cumulative Review

from Campus to Careers
School Guidance Counselor

School guidance counselors plan academic programs and help students choose the best courses to take to achieve their educational goals. Counselors often meet with students to discuss the life skills needed for personal and social growth. To prepare for this career, guidance counselors take classes in an area of mathematics called *statistics*, where they learn how to collect, analyze, explain, and present data.

In **Problem 117** of **Study Set 4.4**, you will see how a counselor must be able to add fractions to better understand a graph that shows students' study habits.

JOB TITLE: School Guidance Counselor
EDUCATION: A master's degree is usually required to be licensed as a counselor. However, some schools accept a bachelor's degree with the appropriate counseling courses.
JOB OUTLOOK: The number of openings will increase by about 19% from 2010 through 2020.
ANNUAL EARNINGS: The average (median) salary in 2011 was $54,130.
FOR MORE INFORMATION:
www.bls.gov/ooh/community-and-social-service

Objectives

1. Identify the numerator and denominator of a fraction.
2. Simplify special fraction forms.
3. Define equivalent fractions.
4. Build equivalent fractions.
5. Simplify fractions.
6. Build and simplify algebraic fractions.

SECTION 4.1
An Introduction to Fractions

Whole numbers are used to count objects, such as CDs, stamps, eggs, and magazines. When we need to describe a part of a whole, such as one-half of a pie, three-quarters of an hour, or a one-third-pound burger, we can use *fractions*.

One-half of a cherry pie
$$\frac{1}{2}$$

Three-quarters of an hour
$$\frac{3}{4}$$

A one-third-pound burger
$$\frac{1}{3}$$

1 Identify the numerator and denominator of a fraction.

A **fraction** describes the number of equal parts of a whole. For example, consider the figure below with 5 of the 6 equal parts colored red. We say that $\frac{5}{6}$ (five-sixths) of the figure is shaded.

In a fraction, the number above the **fraction bar** is called the **numerator**, and the number below is called the **denominator**.

Fraction bar ⟶ $\frac{5}{6}$ ⟵ numerator
⟵ denominator

> **The Language of Algebra** The word *fraction* comes from the Latin word *fractio* meaning "breaking in pieces."

Self Check 1
Identify the numerator and denominator of each fraction:

a. $\frac{7}{9}$ numerator: 7; denominator: 9

b. $\frac{21}{20}$ numerator: 21; denominator: 20

Now Try Problem 21

EXAMPLE 1
Identify the numerator and denominator of each fraction:

a. $\frac{11}{12}$ b. $\frac{8}{3}$

Strategy We will find the number above the fraction bar and the number below it.

WHY The number above the fraction bar is the numerator, and the number below is the denominator.

Solution

a. $\frac{11}{12}$ ⟵ numerator
⟵ denominator

b. $\frac{8}{3}$ ⟵ numerator
⟵ denominator

If the numerator of a fraction is less than its denominator, the fraction is called a **proper fraction**. A proper fraction is less than 1. If the numerator of a fraction is greater than or equal to its denominator, the fraction is called an **improper fraction**. An improper fraction is greater than or equal to 1.

Proper fractions

$\dfrac{1}{4}, \dfrac{2}{3},$ and $\dfrac{98}{99}$

Improper fractions

$\dfrac{7}{2}, \dfrac{98}{97}, \dfrac{16}{16},$ and $\dfrac{5}{1}$

Teaching Example 1 Identify the numerator and denominator of each fraction:
a. $\dfrac{31}{32}$ b. $\dfrac{12}{5}$

Answers:
a. numerator: 31, denominator: 32
b. numerator: 12, denominator: 5

> ***The Language of Algebra*** The phrase *improper fraction* is somewhat misleading. In algebra and other mathematics courses, we often use such fractions "properly" to solve many types of problems.

EXAMPLE 2 Write fractions that represent the shaded and unshaded portions of the figure below.

Strategy We will determine the number of equal parts into which the figure is divided. Then we will determine how many of those parts are shaded.

WHY The denominator of a fraction shows the number of equal parts in the whole. The numerator shows how many of those parts are being considered.

Solution
Since the figure is divided into 3 equal parts, the denominator of the fraction is 3. Since 2 of those parts are shaded, the numerator is 2, and we say that

$\dfrac{2}{3}$ of the figure is shaded.   Write: $\dfrac{\text{number of parts shaded}}{\text{number of equal parts}}$

Since 1 of the 3 equal parts of the figure is not shaded, the numerator is 1, and we say that

$\dfrac{1}{3}$ of the figure is not shaded.   Write: $\dfrac{\text{number of parts not shaded}}{\text{number of equal parts}}$

Self Check 2
Write fractions that represent the portion of the month that has passed and the portion that remains. $\dfrac{11}{31}, \dfrac{20}{31}$

DECEMBER

X	X	X	X	X	X	X
X	X	X	X	12	13	14
15	16	17	18	19	20	21
22	23	24	25	26	27	28
29	30	31				

Now Try Problems 25 and 113

Teaching Example 2 Write fractions to represent the shaded and unshaded portions of the figure.

Answer:
$\dfrac{3}{5}, \dfrac{2}{5}$

There are times when a negative fraction is needed to describe a quantity. For example, if an earthquake causes a road to sink seven-eighths of an inch, the amount of downward movement can be represented by $-\dfrac{7}{8}$. Negative fractions can be written in three ways. The negative sign can appear in the numerator, in the denominator, or in front of the fraction.

$$\dfrac{-7}{8} = \dfrac{7}{-8} = -\dfrac{7}{8} \qquad \dfrac{-15}{4} = \dfrac{15}{-4} = -\dfrac{15}{4}$$

Notice that the examples above agree with the rule from Chapter 2 for dividing integers with different (unlike) signs: *the quotient of a negative integer and a positive integer is negative.*

iStockphoto.com/Jamie VanBuskirk

2 Simplify special fraction forms.

Recall from Section 1.4 that a fraction bar indicates division. This fact helps us simplify four special fraction forms.

- **Fractions that have the same numerator and denominator:** In this case, we have a number divided by itself. The result is 1 (provided the numerator and denominator are not 0). We call each of the following fractions a **form of 1**.

$$1 = \frac{1}{1} = \frac{2}{2} = \frac{3}{3} = \frac{4}{4} = \frac{5}{5} = \frac{6}{6} = \frac{7}{7} = \frac{8}{8} = \frac{9}{9} = \cdots$$

- **Fractions that have a denominator of 1:** In this case, we have a number divided by 1. The result is simply the numerator.

$$\frac{5}{1} = 5 \qquad \frac{24}{1} = 24 \qquad \frac{-7}{1} = -7$$

- **Fractions that have a numerator of 0:** In this case, we have division of 0. The result is 0 (provided the denominator is not 0).

$$\frac{0}{8} = 0 \qquad \frac{0}{56} = 0 \qquad \frac{0}{-11} = 0$$

- **Fractions that have a denominator of 0:** In this case, we have division by 0. The division is undefined.

$$\frac{7}{0} \text{ is undefined} \qquad \frac{-18}{0} \text{ is undefined}$$

> **The Language of Algebra** Perhaps you are wondering about the fraction form $\frac{0}{0}$. It is said to be *undetermined*. This form is important in advanced mathematics courses.

Self Check 3
Simplify, if possible:

a. $\frac{4}{4}$ b. $\frac{51}{1}$ c. $\frac{45}{0}$ d. $\frac{0}{6}$

Now Try Problem 33

Self Check 3 Answers
a. 1 b. 51 c. undefined d. 0

Teaching Example 3 Simplify, if possible:
a. $\frac{10}{1}$ b. $\frac{9}{9}$ c. $\frac{0}{17}$ d. $\frac{21}{0}$
Answers:
a. 10 b. 1 c. 0 d. undefined

EXAMPLE 3
Simplify, if possible: a. $\frac{12}{12}$ b. $\frac{0}{24}$ c. $\frac{18}{0}$ d. $\frac{9}{1}$

Strategy To simplify each fraction, we will divide the numerator by the denominator, if possible.

WHY A fraction bar indicates division.

Solution

a. $\frac{12}{12} = 1$ This corresponds to dividing a quantity into 12 equal parts, and then considering all 12 of them. We would get 1 whole quantity.

b. $\frac{0}{24} = 0$ This corresponds to dividing a quantity into 24 equal parts, and then considering 0 (none) of them. We would get 0.

c. $\frac{18}{0}$ is undefined   This corresponds to dividing a quantity into 0 equal parts, and then considering 18 of them. That is not possible.

d. $\frac{9}{1} = 9$ This corresponds to "dividing" a quantity into 1 equal part, and then considering 9 of them. We would get 9 of those quantities.

The Language of Algebra Fractions are often referred to as **rational numbers**. All integers are rational numbers, because every integer can be written as a fraction with a denominator of 1. For example,

$$2 = \frac{2}{1}, \quad -5 = \frac{-5}{1}, \quad \text{and} \quad 0 = \frac{0}{1}$$

3 Define equivalent fractions.

Fractions can look different but still represent the same part of a whole. To illustrate this, consider the identical rectangular regions on the right. The first one is divided into 10 equal parts. Since 6 of those parts are red, $\frac{6}{10}$ of the figure is shaded.

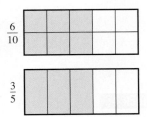

The second figure is divided into 5 equal parts. Since 3 of those parts are red, $\frac{3}{5}$ of the figure is shaded. We can conclude that $\frac{6}{10} = \frac{3}{5}$ because $\frac{6}{10}$ and $\frac{3}{5}$ represent the same shaded portion of the figure. We say that $\frac{6}{10}$ and $\frac{3}{5}$ are *equivalent fractions*.

Equivalent Fractions

Two fractions are **equivalent** if they represent the same number. **Equivalent fractions** represent the same portion of a whole.

4 Build equivalent fractions.

Writing a fraction as an equivalent fraction with a *larger* denominator is called **building the fraction**. To build a fraction, we use a familiar property from Chapter 1 that is also true for fractions:

Multiplication Property of 1

The product of any fraction and 1 is that fraction.

We also use the following rule for multiplying fractions. (It will be discussed in greater detail in the next section.)

Multiplying Fractions

To multiply two fractions, multiply the numerators and multiply the denominators.

To build an equivalent fraction for $\frac{1}{2}$ with a denominator of 8, we first ask, "What number times 2 equals 8?" To answer that question we *divide* 8 by 2 to get 4. Since we need to multiply the denominator of $\frac{1}{2}$ by 4 to obtain a denominator of 8, it follows that $\frac{4}{4}$ should be the form of 1 that is used to build an equivalent fraction for $\frac{1}{2}$.

$$\frac{1}{2} = \frac{1}{2} \cdot \frac{4}{4}$$ Multiply $\frac{1}{2}$ by 1 in the form of $\frac{4}{4}$. Note the form of 1 highlighted in red.

$$= \frac{1 \cdot 4}{2 \cdot 4}$$ Use the rule for multiplying two fractions. Multiply the numerators. Multiply the denominators.

$$= \frac{4}{8}$$

We have found that $\frac{4}{8}$ is equivalent to $\frac{1}{2}$. To build an equivalent fraction for $\frac{1}{2}$ with a denominator of 8, we *multiplied by a factor equal to 1* in the form of $\frac{4}{4}$. Multiplying $\frac{1}{2}$ by $\frac{4}{4}$ changes its appearance but does not change its value, because we are multiplying it by 1.

> **Building Fractions**
>
> To build a fraction, *multiply it by a factor equal to 1* in the form of $\frac{2}{2}, \frac{3}{3}, \frac{4}{4}, \frac{5}{5}$, and so on.

> ***The Language of Algebra*** Building an equivalent fraction with a larger denominator is also called *expressing a fraction in higher terms*.

Self Check 4
Write $\frac{5}{8}$ as an equivalent fraction with a denominator of 24. $\frac{15}{24}$

Now Try Problem 37

Teaching Example 4 Write $\frac{5}{6}$ as an equivalent fraction with a denominator of 48.
Answer:
$\frac{40}{48}$

EXAMPLE 4 Write $\frac{3}{5}$ as an equivalent fraction with a denominator of 35.

Strategy We will compare the given denominator to the required denominator and ask, "What number times 5 equals 35?"

WHY The answer to that question helps us determine the form of 1 to use to build an equivalent fraction.

Solution
To answer the question "What number times 5 equals 35?" we *divide* 35 by 5 to get 7. Since we need to multiply the denominator of $\frac{3}{5}$ by 7 to obtain a denominator of 35, it follows that $\frac{7}{7}$ should be the form of 1 that is used to build an equivalent fraction for $\frac{3}{5}$.

$$\frac{3}{5} = \frac{3}{5} \cdot \boxed{\frac{7}{7}} \quad \text{Multiply } \tfrac{3}{5} \text{ by a form of 1: } \tfrac{7}{7} = 1.$$

$$= \frac{3 \cdot 7}{5 \cdot 7} \quad \begin{array}{l}\text{Multiply the numerators.}\\\text{Multiply the denominators.}\end{array}$$

$$= \frac{21}{35}$$

We have found that $\frac{21}{35}$ is equivalent to $\frac{3}{5}$.

> ***Success Tip*** To build an equivalent fraction in Example 4, we multiplied $\frac{3}{5}$ by 1 in the form of $\frac{7}{7}$. As a result of that step, the numerator and the denominator of $\frac{3}{5}$ were multiplied by 7:
>
> $$\frac{3 \cdot 7}{5 \cdot 7} \quad \begin{array}{l}\leftarrow \text{ The numerator is multiplied by 7.}\\\leftarrow \text{ The denominator is multiplied by 7.}\end{array}$$
>
> This process illustrates the following property of fractions.

> **The Fundamental Property of Fractions**
>
> If the numerator and denominator of a fraction are multiplied by the same nonzero number, the resulting fraction is equivalent to the original fraction.

Since multiplying the numerator and denominator of a fraction by the same nonzero number produces an equivalent fraction, your instructor may allow you to begin your solution to problems like Example 4 as shown in the Success Tip above.

EXAMPLE 5
Write 4 as an equivalent fraction with a denominator of 6.

Strategy We will express 4 as the fraction $\frac{4}{1}$ and build an equivalent fraction by multiplying it by $\frac{6}{6}$.

WHY Since we need to multiply the denominator of $\frac{4}{1}$ by 6 to obtain a denominator of 6, it follows that $\frac{6}{6}$ should be the form of 1 that is used to build an equivalent fraction for $\frac{4}{1}$.

Solution

$$4 = \frac{4}{1} \qquad \text{Write 4 as a fraction: } 4 = \frac{4}{1}.$$

$$= \frac{4}{1} \cdot \boxed{\frac{6}{6}} \qquad \text{Build an equivalent fraction by multiplying } \frac{4}{1} \text{ by a form of 1: } \frac{6}{6} = 1.$$

$$= \frac{4 \cdot 6}{1 \cdot 6} \qquad \text{Multiply the numerators.}$$
$$\qquad\qquad\quad \text{Multiply the denominators.}$$

$$= \frac{24}{6}$$

Self Check 5
Write 10 as an equivalent fraction with a denominator of 3. $\frac{30}{3}$

Now Try Problem 49

Teaching Example 5 Write 7 as an equivalent fraction with a denominator of 4.
Answer:
$\frac{28}{4}$

5 Simplify fractions.

Every fraction can be written in infinitely many equivalent forms. For example, some equivalent forms of $\frac{10}{15}$ are:

$$\frac{2}{3} = \frac{4}{6} = \frac{6}{9} = \frac{8}{12} = \frac{\mathbf{10}}{\mathbf{15}} = \frac{12}{18} = \frac{14}{21} = \frac{16}{24} = \frac{18}{27} = \frac{20}{30} = \ldots$$

Of all of the equivalent forms in which we can write a fraction, we often need to determine the one that is in *simplest form*.

> **Simplest Form of a Fraction**
>
> A fraction is in **simplest form**, or **lowest terms**, when the numerator and denominator have no common factors other than 1.

EXAMPLE 6
Are the following fractions in simplest form? **a.** $\frac{12}{27}$ **b.** $\frac{5}{8}$

Strategy We will determine whether the numerator and denominator have any common factors other than 1.

WHY If the numerator and denominator have no common factors other than 1, the fraction is in simplest form.

Solution
a. The factors of the numerator, 12, are: **1**, 2, **3**, 4, 6, 12
   The factors of the denominator, 27, are: **1**, **3**, 9, 27

Since the numerator and denominator have a common factor of 3, the fraction $\frac{12}{27}$ is *not* in simplest form.

b. The factors of the numerator, 5, are: **1**, 5
   The factors of the denominator, 8, are: **1**, 2, 4, 8

Since the only common factor of the numerator and denominator is 1, the fraction $\frac{5}{8}$ is in simplest form.

Self Check 6
Are the following fractions in simplest form?

a. $\frac{4}{21}$ yes

b. $\frac{6}{20}$ no

Now Try Problem 53

Teaching Example 6 Are the following fractions in simplest form?
a. $\frac{14}{18}$ **b.** $\frac{4}{9}$
Answers:
a. no **b.** yes

Chapter 4 Fractions and Mixed Numbers

To **simplify a fraction**, we write it in simplest form by *removing a factor equal to 1*. For example, to simplify $\frac{10}{15}$, we note that the greatest factor common to the numerator and denominator is 5 and proceed as follows:

$$\frac{10}{15} = \frac{2 \cdot 5}{3 \cdot 5} \quad \text{Factor 10 and 15. Note the form of 1 highlighted in red.}$$

$$= \frac{2}{3} \cdot \frac{5}{5} \quad \text{Use the rule for multiplying fractions in reverse: write } \frac{2 \cdot 5}{3 \cdot 5} \text{ as the product of two fractions, } \frac{2}{3} \text{ and } \frac{5}{5}.$$

$$= \frac{2}{3} \cdot 1 \quad \text{A number divided by itself is equal to 1: } \frac{5}{5} = 1.$$

$$= \frac{2}{3} \quad \text{Use the multiplication property of 1: The product of any fraction and 1 is that fraction.}$$

We have found that the simplified form of $\frac{10}{15}$ is $\frac{2}{3}$. To simplify $\frac{10}{15}$, we *removed a factor equal to 1* in the form of $\frac{5}{5}$. The result, $\frac{2}{3}$, is equivalent to $\frac{10}{15}$.

To streamline the simplifying process, we can replace pairs of factors common to the numerator and denominator with the equivalent fraction $\frac{1}{1}$.

Self Check 7

Simplify each fraction:

a. $\frac{10}{25}$ $\frac{2}{5}$

b. $\frac{3}{9}$ $\frac{1}{3}$

Now Try Problems 57 and 61

Teaching Example 7 Simplify each fraction:
a. $\frac{9}{15}$ b. $\frac{2}{18}$
Answers:
a. $\frac{3}{5}$ b. $\frac{1}{9}$

EXAMPLE 7

Simplify each fraction: **a.** $\frac{6}{10}$ **b.** $\frac{7}{21}$

Strategy We will factor the numerator and denominator. Then we will look for any factors common to the numerator and denominator and remove them.

WHY We need to make sure that the numerator and denominator have no common factors other than 1. If that is the case, then the fraction is in *simplest form*.

Solution

a. $\frac{6}{10} = \frac{2 \cdot 3}{2 \cdot 5}$ To prepare to simplify, factor 6 and 10. Note the form of 1 highlighted in red.

$= \frac{\overset{1}{\cancel{2}} \cdot 3}{\underset{1}{\cancel{2}} \cdot 5}$ Simplify by removing the common factor of 2 from the numerator and denominator. A slash / and the 1's are used to show that $\frac{2}{2}$ is replaced by the equivalent fraction $\frac{1}{1}$. A factor equal to 1 in the form of $\frac{2}{2}$ was removed.

$= \frac{3}{5}$ Multiply the remaining factors in the numerator: $1 \cdot 3 = 3$. Multiply the remaining factors in the denominator: $1 \cdot 5 = 5$.

Since 3 and 5 have no common factors (other than 1), $\frac{3}{5}$ is in simplest form.

b. $\frac{7}{21} = \frac{7}{3 \cdot 7}$ To prepare to simplify, factor 21.

$= \frac{\overset{1}{\cancel{7}}}{3 \cdot \underset{1}{\cancel{7}}}$ Simplify by removing the common factor of 7 from the numerator and denominator.

$= \frac{1}{3}$ Multiply the remaining factors in the denominator: $1 \cdot 3 = 3$.

Caution! Don't forget to write the 1's when removing common factors of the numerator and the denominator. Failure to do so can lead to the common mistake shown below.

$$\cancel{\frac{7}{21} = \frac{7}{3 \cdot 7} = \frac{0}{3}}$$

We can easily identify common factors of the numerator and the denominator of a fraction if we write them in prime-factored form.

EXAMPLE 8

Simplify each fraction, if possible: **a.** $\dfrac{90}{105}$ **b.** $\dfrac{25}{27}$

Strategy We begin by prime factoring the numerator, 90, and denominator, 105. Then we look for any factors common to the numerator and denominator and remove them.

WHY When the numerator and/or denominator of a fraction are large numbers, such as 90 and 105, writing their prime factorizations is helpful in identifying any common factors.

Solution

a. $\dfrac{90}{105} = \dfrac{2 \cdot 3 \cdot 3 \cdot 5}{3 \cdot 5 \cdot 7}$ To prepare to simplify, write 90 and 105 in prime-factored form.

$= \dfrac{2 \cdot \overset{1}{\cancel{3}} \cdot 3 \cdot \overset{1}{\cancel{5}}}{\underset{1}{\cancel{3}} \cdot \underset{1}{\cancel{5}} \cdot 7}$ Remove the common factors of 3 and 5 from the numerator and denominator. Slashes and 1's are used to show that $\tfrac{3}{3}$ and $\tfrac{5}{5}$ are replaced by the equivalent fraction $\tfrac{1}{1}$. A factor equal to 1 in the form of $\tfrac{3 \cdot 5}{3 \cdot 5} = \tfrac{15}{15}$ was removed.

$= \dfrac{6}{7}$ Multiply the remaining factors in the numerator: $2 \cdot 1 \cdot 3 \cdot 1 = 6$.
Multiply the remaining factors in the denominator: $1 \cdot 1 \cdot 7 = 7$.

Since 6 and 7 have no common factors (other than 1), $\dfrac{6}{7}$ is in simplest form.

b. $\dfrac{25}{27} = \dfrac{5 \cdot 5}{3 \cdot 3 \cdot 3}$ Write 25 and 27 in prime-factored form.

Since 25 and 27 have no common factors, other than 1, the fraction $\dfrac{25}{27}$ is in simplest form.

Self Check 8

Simplify each fraction, if possible:

a. $\dfrac{70}{126}$ $\dfrac{5}{9}$

b. $\dfrac{16}{81}$ in simplest form

Now Try Problems 67 and 69

Teaching Example 8 Simplify each fraction, if possible:

a. $\dfrac{75}{315}$ **b.** $\dfrac{8}{45}$

Answers:

a. $\dfrac{5}{21}$ **b.** in simplest form

EXAMPLE 9

Simplify: $\dfrac{63}{36}$

Strategy We will prime factor the numerator and denominator. Then we will look for any factors common to the numerator and denominator and remove them.

WHY We need to make sure that the numerator and denominator have no common factors other than 1. If that is the case, then the fraction is in *simplest form*.

Solution

$\dfrac{63}{36} = \dfrac{3 \cdot 3 \cdot 7}{2 \cdot 2 \cdot 3 \cdot 3}$ To prepare to simplify, write 63 and 36 in prime-factored form.

$\begin{array}{r|l} 3 & 63 \\ 3 & 21 \\ & 7 \end{array}$ $\begin{array}{r|l} 2 & 36 \\ 2 & 18 \\ 3 & 9 \\ & 3 \end{array}$

$= \dfrac{\overset{1}{\cancel{3}} \cdot \overset{1}{\cancel{3}} \cdot 7}{2 \cdot 2 \cdot \underset{1}{\cancel{3}} \cdot \underset{1}{\cancel{3}}}$ Simplify by removing the common factors of 3 from the numerator and denominator.

$= \dfrac{7}{4}$ Multiply the remaining factors in the numerator: $1 \cdot 1 \cdot 7 = 7$.
Multiply the remaining factors in the denominator: $2 \cdot 2 \cdot 1 \cdot 1 = 4$.

Self Check 9

Simplify: $\dfrac{162}{72}$ $\dfrac{9}{4}$

Now Try Problem 81

Teaching Example 9 Simplify: $\dfrac{132}{48}$
Answer:
$\dfrac{11}{4}$

Success Tip If you recognized that 63 and 36 have a common factor of 9, you may remove that common factor from the numerator and denominator without writing the prime factorizations. However, make sure that the numerator and denominator of the resulting fraction do not have any common factors. If they do, continue to simplify.

$\dfrac{63}{36} = \dfrac{7 \cdot \overset{1}{\cancel{9}}}{4 \cdot \underset{1}{\cancel{9}}} = \dfrac{7}{4}$ Factor 63 as $7 \cdot 9$ and 36 as $4 \cdot 9$, and then remove the common factor of 9 from the numerator and denominator.

Use the following steps to simplify a fraction.

> **Simplifying Fractions**
>
> To simplify a fraction, *remove factors equal to 1* of the form $\frac{2}{2}, \frac{3}{3}, \frac{4}{4}, \frac{5}{5}$, and so on, using the following procedure:
>
> 1. Factor (or prime factor) the numerator and denominator to determine their common factors.
> 2. Remove factors equal to 1 by replacing each pair of factors common to the numerator and denominator with the equivalent fraction $\frac{1}{1}$.
> 3. Multiply the remaining factors in the numerator and in the denominator.

Negative fractions are simplified in the same way as positive fractions. Just remember to write a negative sign − in front of each step of the solution. For example, to simplify $-\frac{15}{33}$ we proceed as follows:

$$-\frac{15}{33} = -\frac{\overset{1}{\cancel{3}} \cdot 5}{\underset{1}{\cancel{3}} \cdot 11}$$

$$= -\frac{5}{11}$$

6 Build and simplify algebraic fractions.

Since a variable is a letter that stands for a number, variables can appear in fractions. Fractions that contain a variable (or variables) in the numerator, the denominator, or both are called **algebraic fractions**. Here are some examples of algebraic fractions.

$$\frac{x}{2}, \quad \frac{10}{y}, \quad \frac{m}{25n}, \quad \frac{4a^2 b}{6ab^3}, \quad \frac{x+3}{x-5}$$

Algebraic fractions are built up and simplified just like numerical fractions.

Self Check 10

Write $\frac{2}{9}$ as an equivalent fraction with a denominator of 72y. $\frac{16y}{72y}$

Now Try Problem 91

Teaching Example 10
Write $\frac{3}{8}$ as an equivalent fraction with a denominator of 32x.
Answer:
$\frac{12x}{32x}$

EXAMPLE 10 Write $\frac{5}{7}$ as an equivalent fraction with a denominator of 42a.

Strategy We will compare the given denominator to the required denominator and ask, "What expression times 7 equals 42a?"

WHY The answer to that question helps us determine the form of 1 to use to build an equivalent fraction.

Solution We need to multiply the denominator of $\frac{5}{7}$ by 6a to obtain 42a. It follows that $\frac{6a}{6a}$ should be the form of 1 that is used to build $\frac{5}{7}$.

$$\frac{5}{7} = \frac{5}{7} \cdot \boxed{\frac{6a}{6a}} \quad \text{Multiply } \tfrac{5}{7} \text{ by a form of 1: } \tfrac{6a}{6a} = 1.$$

$$= \frac{5 \cdot 6a}{7 \cdot 6a} \quad \begin{array}{l}\text{Multiply the numerators.}\\ \text{Multiply the denominators.}\end{array}$$

$$= \frac{30a}{42a}$$

4.1 An Introduction to Fractions

EXAMPLE 11 Simplify each fraction: **a.** $\dfrac{3}{15x}$ **b.** $\dfrac{9y^3}{10y^2}$ **c.** $\dfrac{24ab^2}{64ab^4}$

Strategy We will factor the numerator and denominator of each algebraic fraction. Then we will look for any factors common to the numerator and denominator and remove them.

WHY We need to make sure that the numerator and denominator have no common factors other than 1. If that is the case, then the fraction is in *simplest form*.

Solution

a. $\dfrac{3}{15x} = \dfrac{3}{3 \cdot 5 \cdot x}$ Recall that $15x$ means $15 \cdot x$. To prepare to simplify, factor $15x$ as $3 \cdot 5 \cdot x$.

$= \dfrac{\cancel{3}}{\cancel{3} \cdot 5 \cdot x}$ Simplify by removing the common factor of 3 from the numerator and denominator.

$= \dfrac{1}{5x}$ Multiply the remaining factors in the denominator: $1 \cdot 5 \cdot x = 5x$.

b. $\dfrac{9y^3}{10y^2} = \dfrac{9 \cdot y \cdot y \cdot y}{10 \cdot y \cdot y}$ To prepare to simplify, use the definition of exponent to write y^3 and y^2 in factored form.

$= \dfrac{9 \cdot \cancel{y} \cdot \cancel{y} \cdot y}{10 \cdot \cancel{y} \cdot \cancel{y}}$ Simplify by removing the common factors of y from the numerator and denominator. Slashes and 1's are used to show that $\dfrac{y}{y}$ is replaced by the equivalent fraction $\dfrac{1}{1}$. A factor equal to 1 in the form of $\dfrac{y \cdot y}{y \cdot y} = \dfrac{y^2}{y^2}$ was removed.

$= \dfrac{9y}{10}$ Multiply the remaining factors in the numerator: $9 \cdot 1 \cdot 1 \cdot y = 9y$. Multiply the remaining factors in the denominator: $10 \cdot 1 \cdot 1 = 10$.

c. $\dfrac{24ab^2}{64ab^4} = \dfrac{3 \cdot 8 \cdot a \cdot b \cdot b}{8 \cdot 8 \cdot a \cdot b \cdot b \cdot b \cdot b}$ To prepare to simplify, factor 24, b^2, 64, and b^4.

$= \dfrac{3 \cdot \cancel{8} \cdot \cancel{a} \cdot \cancel{b} \cdot \cancel{b}}{8 \cdot \cancel{8} \cdot \cancel{a} \cdot \cancel{b} \cdot \cancel{b} \cdot b \cdot b}$ Simplify by removing the common factors of 8, a, and b from the numerator and denominator. A factor equal to 1 in the form of $\dfrac{8 \cdot a \cdot b \cdot b}{8 \cdot a \cdot b \cdot b} = \dfrac{8ab^2}{8ab^2}$ was removed.

$= \dfrac{3}{8b^2}$ Multiply the remaining factors in the numerator. Multiply the remaining factors in the denominator.

Self Check 11

Simplify each fraction:

a. $\dfrac{2}{16d}$ $\dfrac{1}{8d}$

b. $\dfrac{14x^5}{15x^2}$ $\dfrac{14x^3}{15}$

c. $\dfrac{35mn^2}{21mn^5}$ $\dfrac{5}{3n^3}$

Now Try Problems 97, 101, and 105

Teaching Example 11
Simplify each fraction:
a. $\dfrac{7}{21m}$ **b.** $\dfrac{4r^4}{11r^3}$ **c.** $\dfrac{30c^3d}{12c^5d}$
Answers:
a. $\dfrac{1}{3m}$ **b.** $\dfrac{4r}{11}$ **c.** $\dfrac{5}{2c^2}$

ANSWERS TO SELF CHECKS

1. a. numerator: 7; denominator: 9 **b.** numerator: 21; denominator: 20 **2.** $\dfrac{11}{31}, \dfrac{20}{31}$

3. a. 1 **b.** 51 **c.** undefined **d.** 0 **4.** $\dfrac{15}{24}$ **5.** $\dfrac{30}{3}$ **6. a.** yes **b.** no **7. a.** $\dfrac{2}{5}$ **b.** $\dfrac{1}{3}$

8. a. $\dfrac{5}{9}$ **b.** in simplest form **9.** $\dfrac{9}{4}$ **10.** $\dfrac{16y}{72y}$ **11. a.** $\dfrac{1}{8d}$ **b.** $\dfrac{14x^3}{15}$ **c.** $\dfrac{5}{3n^3}$

SECTION 4.1 STUDY SET

VOCABULARY

Fill in the blanks.

1. A fraction describes the number of equal parts of a whole.
2. For the fraction $\frac{7}{8}$, the numerator is 7 and the denominator is 8.
3. If the numerator of a fraction is less than its denominator, the fraction is called a proper fraction. If the numerator of a fraction is greater than or equal to its denominator it is called an improper fraction.
4. Each of the following fractions is a form of 1 .

$$\frac{1}{1} = \frac{2}{2} = \frac{3}{3} = \frac{4}{4} = \frac{5}{5} = \frac{6}{6} = \frac{7}{7} = \frac{8}{8} = \frac{9}{9} = \ldots$$

5. Two fractions are equivalent if they represent the same number.
6. Writing a fraction as an equivalent fraction with a larger denominator is called building the fraction.
7. A fraction is in simplest form, or lowest terms, when the numerator and denominator have no common factors other than 1.
8. Algebraic fractions, such as $\frac{x}{2}$ and $\frac{4a^2b}{6ab^3}$, are fractions that contain variables in the numerator, the denominator, or both.

CONCEPTS

9. What concept studied in this section is shown on the right?

   equivalent fractions: $\frac{2}{6} = \frac{1}{3}$

10. What concept studied in this section does the following statement illustrate?

    $$\frac{1}{2} = \frac{2}{4} = \frac{3}{6} = \frac{4}{8} = \frac{5}{10} = \ldots \text{ equivalent fractions}$$

11. Classify each fraction as a proper fraction or an improper fraction.

    a. $\frac{37}{24}$ improper fraction
    b. $\frac{1}{3}$ proper fraction
    c. $\frac{71}{100}$ proper fraction
    d. $\frac{9}{9}$ improper fraction

12. Remove the common factors of the numerator and denominator to simplify the fraction:

    $\frac{2 \cdot 3 \cdot 3 \cdot 5}{2 \cdot 3 \cdot 5 \cdot 7}$ $\frac{3}{7}$

13. What common factor (other than 1) do the numerator and the denominator of the fraction $\frac{10}{15}$ have? 5

Fill in the blanks.

14. Multiplication property of 1: The product of any fraction and 1 is that fraction .
15. Multiplying fractions: To multiply two fractions, multiply the numerators and multiply the denominators.
16. a. Consider the following work: $\frac{2}{3} = \frac{2}{3} \cdot \frac{4}{4}$

    $$= \frac{8}{12}$$

    To build an equivalent fraction for $\frac{2}{3}$ with a denominator of 12, multiply it by a factor equal to 1 in the form of $\frac{4}{4}$.

    b. Consider the following work: $\frac{15}{27} = \frac{\overset{1}{\cancel{3}} \cdot 5}{\underset{1}{\cancel{3}} \cdot 9}$

    $$= \frac{5}{9}$$

    To simplify the fraction $\frac{15}{27}$, remove a factor equal to 1 of the form $\frac{3}{3}$.

NOTATION

17. Write the fraction $\frac{7}{-8}$ in two other ways. $\frac{-7}{8}, -\frac{7}{8}$
18. Write each integer as a fraction.

    a. 8 $\frac{8}{1}$
    b. −25 $-\frac{25}{1}$

Fill in the blanks to complete each step.

19. Build an equivalent fraction for $\frac{1}{6}$ with a denominator of 18.

    $$\frac{1}{6} = \frac{1}{6} \cdot \frac{3}{3}$$

    $$= \frac{1 \cdot 3}{6 \cdot 3}$$

    $$= \frac{3}{18}$$

20. Simplify:

    $$\frac{18}{24} = \frac{2 \cdot 3 \cdot 3}{2 \cdot 2 \cdot 2 \cdot 3}$$

    $$= \frac{\overset{1}{\cancel{2}} \cdot 3 \cdot \overset{1}{\cancel{3}}}{\underset{1}{\cancel{2}} \cdot 2 \cdot 2 \cdot \underset{1}{\cancel{3}}}$$

    $$= \frac{3}{4}$$

4.1 An Introduction to Fractions 315

GUIDED PRACTICE

Identify the numerator and denominator of each fraction. See Example 1.

21. $\dfrac{4}{5}$ numerator: 4; denominator: 5

22. $\dfrac{7}{8}$ numerator: 7; denominator: 8

23. $\dfrac{17}{10}$ numerator: 17; denominator: 10

24. $\dfrac{29}{21}$ numerator: 29; denominator: 21

Write a fraction to describe what part of the figure is shaded. Write a fraction to describe what part of the figure is not shaded. See Example 2.

25. $\dfrac{3}{4}, \dfrac{1}{4}$

26. $\dfrac{2}{3}, \dfrac{1}{3}$

27. $\dfrac{5}{8}, \dfrac{3}{8}$

28. $\dfrac{7}{12}, \dfrac{5}{12}$

29. $\dfrac{1}{4}, \dfrac{3}{4}$

30. $\dfrac{3}{5}, \dfrac{2}{5}$

31. $\dfrac{7}{12}, \dfrac{5}{12}$

32. $\dfrac{9}{14}, \dfrac{5}{14}$

Simplify, if possible. See Example 3.

33. a. $\dfrac{4}{1}$ 4 b. $\dfrac{8}{8}$ 1 c. $\dfrac{0}{12}$ 0 d. $\dfrac{1}{0}$ undefined

34. a. $\dfrac{25}{1}$ 25 b. $\dfrac{14}{14}$ 1 c. $\dfrac{0}{1}$ 0 d. $\dfrac{83}{0}$ undefined

35. a. $\dfrac{5}{0}$ undefined b. $\dfrac{0}{50}$ 0 c. $\dfrac{33}{33}$ 1 d. $\dfrac{75}{1}$ 75

36. a. $\dfrac{0}{64}$ 0 b. $\dfrac{27}{0}$ undefined c. $\dfrac{125}{125}$ 1 d. $\dfrac{98}{1}$ 98

Write each fraction as an equivalent fraction with the indicated denominator. See Example 4.

37. $\dfrac{7}{8}$, denominator 40 $\dfrac{35}{40}$

38. $\dfrac{3}{4}$, denominator 24 $\dfrac{18}{24}$

39. $\dfrac{4}{9}$, denominator 27 $\dfrac{12}{27}$

40. $\dfrac{5}{7}$, denominator 49 $\dfrac{35}{49}$

41. $\dfrac{5}{6}$, denominator 54 $\dfrac{45}{54}$

42. $\dfrac{2}{3}$, denominator 27 $\dfrac{18}{27}$

43. $\dfrac{2}{7}$, denominator 14 $\dfrac{4}{14}$

44. $\dfrac{3}{10}$, denominator 50 $\dfrac{15}{50}$

Write each whole number as an equivalent fraction with the indicated denominator. See Example 5.

45. 4, denominator 9 $\dfrac{36}{9}$

46. 4, denominator 3 $\dfrac{12}{3}$

47. 6, denominator 8 $\dfrac{48}{8}$

48. 3, denominator 6 $\dfrac{18}{6}$

49. 3, denominator 5 $\dfrac{15}{5}$

50. 7, denominator 4 $\dfrac{28}{4}$

51. 14, denominator 2 $\dfrac{28}{2}$

52. 10, denominator 9 $\dfrac{90}{9}$

Are the following fractions in simplest form? See Example 6.

53. a. $\dfrac{12}{16}$ no b. $\dfrac{3}{25}$ yes

54. a. $\dfrac{9}{24}$ no b. $\dfrac{7}{36}$ yes

55. a. $\dfrac{35}{36}$ yes b. $\dfrac{18}{21}$ no

56. a. $\dfrac{22}{45}$ yes b. $\dfrac{21}{56}$ no

Simplify each fraction, if possible. See Example 7.

57. $\dfrac{6}{9}$ $\dfrac{2}{3}$

58. $\dfrac{15}{20}$ $\dfrac{3}{4}$

59. $\dfrac{16}{20}$ $\dfrac{4}{5}$

60. $\dfrac{25}{35}$ $\dfrac{5}{7}$

61. $\dfrac{5}{15}$ $\dfrac{1}{3}$

62. $\dfrac{6}{30}$ $\dfrac{1}{5}$

63. $\dfrac{2}{48}$ $\dfrac{1}{24}$

64. $\dfrac{2}{42}$ $\dfrac{1}{21}$

Simplify each fraction, if possible. See Example 8.

65. $\dfrac{16}{17}$ in simplest form

66. $\dfrac{14}{25}$ in simplest form

67. $\dfrac{36}{96}$ $\dfrac{3}{8}$

68. $\dfrac{48}{120}$ $\dfrac{2}{5}$

69. $\dfrac{55}{62}$ in simplest form

70. $\dfrac{41}{51}$ in simplest form

71. $\dfrac{50}{55}$ $\dfrac{10}{11}$

72. $\dfrac{22}{88}$ $\dfrac{1}{4}$

73. $\dfrac{60}{108}$ $\dfrac{5}{9}$

74. $\dfrac{75}{275}$ $\dfrac{3}{11}$

75. $\dfrac{180}{210}$ $\dfrac{6}{7}$

76. $\dfrac{90}{120}$ $\dfrac{3}{4}$

Simplify each fraction, if possible. See Example 9.

77. $\dfrac{306}{234}$ $\dfrac{17}{13}$

78. $\dfrac{208}{117}$ $\dfrac{16}{9}$

79. $\dfrac{105}{42}$ $\dfrac{5}{2}$

80. $\dfrac{120}{80}$ $\dfrac{3}{2}$

81. $\dfrac{420}{144}$ $\dfrac{35}{12}$

82. $\dfrac{216}{189}$ $\dfrac{8}{7}$

83. $-\dfrac{4}{68}$ $-\dfrac{1}{17}$

84. $-\dfrac{3}{42}$ $-\dfrac{1}{14}$

Unless otherwise noted, all art on this page is © Cengage Learning.

85. $-\dfrac{90}{105}$ $-\dfrac{6}{7}$

86. $-\dfrac{98}{126}$ $-\dfrac{7}{9}$

87. $-\dfrac{16}{26}$ $-\dfrac{8}{13}$

88. $-\dfrac{81}{132}$ $-\dfrac{27}{44}$

Write each fraction as an equivalent fraction with the indicated denominator. See Example 10.

89. $\dfrac{1}{2}$, denominator $6a$

$\dfrac{3a}{6a}$

90. $\dfrac{1}{3}$, denominator $12b$

$\dfrac{4b}{12b}$

91. $\dfrac{9}{10}$, denominator $50c$

$\dfrac{45c}{50c}$

92. $\dfrac{11}{16}$, denominator $32m$

$\dfrac{22m}{32m}$

93. $\dfrac{5}{4n}$, denominator $44n$

$\dfrac{55}{44n}$

94. $\dfrac{9}{7n}$, denominator $63n$

$\dfrac{81}{63n}$

95. $\dfrac{14}{15x}$, denominator $45x$

$\dfrac{42}{45x}$

96. $\dfrac{12}{13r}$, denominator $39r$

$\dfrac{36}{39r}$

Simplify each fraction, if possible. See Example 11.

97. $\dfrac{7}{14a}$ $\dfrac{1}{2a}$

98. $\dfrac{5}{25y}$ $\dfrac{1}{5y}$

99. $\dfrac{3x}{12}$ $\dfrac{x}{4}$

100. $\dfrac{7x}{35}$ $\dfrac{x}{5}$

101. $\dfrac{4m^5}{25m^4}$ $\dfrac{4m}{25}$

102. $\dfrac{2n^4}{13n^2}$ $\dfrac{2n^2}{13}$

103. $\dfrac{6b^4}{9b}$ $\dfrac{2b^3}{3}$

104. $\dfrac{4c^4}{10c^3}$ $\dfrac{2c}{5}$

105. $\dfrac{35a^3b^2}{25a^2b^3}$ $\dfrac{7a}{5b}$

106. $\dfrac{56n^2p^4}{28np^5}$ $\dfrac{2n}{p}$

107. $-\dfrac{16n^5p}{24np}$ $-\dfrac{2n^4}{3}$

108. $-\dfrac{36cd^6}{54cd^4}$ $-\dfrac{2d^2}{3}$

TRY IT YOURSELF

Tell whether each pair of fractions are equivalent by simplifying each fraction.

109. a. $\dfrac{2}{14}$ and $\dfrac{6}{36}$ not equivalent

b. $\dfrac{4}{30}$ and $\dfrac{12}{90}$ equivalent

110. a. $\dfrac{22}{34}$ and $\dfrac{33}{51}$ equivalent

b. $\dfrac{3}{12}$ and $\dfrac{4}{24}$ not equivalent

111. Shade $\dfrac{1}{6}$ of the figure. As a result of the shading, what fraction equivalent to $\dfrac{1}{6}$ is shown? $\dfrac{2}{12}$

112. Shade the *same portion* of each of the figures below so that a fraction equivalent to $\dfrac{1}{4}$ is showing. Then, for each figure, give the equivalent fraction that is illustrated. $\dfrac{4}{16}, \dfrac{25}{100}$

APPLICATIONS

113. **DENTISTRY** Refer to the dental chart.

a. How many teeth are shown on the chart? 32

b. What fraction of this set of teeth have fillings? $\dfrac{5}{32}$

114. **TIME CLOCKS** For each clock, what fraction of the hour has passed? Write your answers in simplified form. (*Hint:* There are 60 minutes in an hour.)

a. $\dfrac{15}{60} = \dfrac{1}{4}$ b. $\dfrac{45}{60} = \dfrac{3}{4}$

c. $\dfrac{25}{60} = \dfrac{5}{12}$ d. $\dfrac{40}{60} = \dfrac{2}{3}$

115. **RULERS** The illustration to the right shows a ruler.

a. How many spaces are there between the numbers 0 and 1? 16

b. To what fraction is the arrow pointing? Write your answer in simplified form. $\dfrac{5}{8}$

116. **SINKHOLES** The illustration below shows a side view of a drop in the sidewalk near a sinkhole. Describe the movement of the sidewalk using a signed fraction. $-\frac{15}{16}$ in.

117. **POLITICAL PARTIES** The graph shows the number of Democratic, Republican, and Independent governors of the 50 states, as of January 2013.

    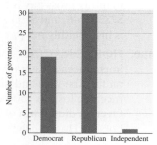

    Source: thegreenpapers.com

    a. How many Democratic governors are there? How many Republican governors are there? How many Independent governors are there? 19, 30, 1
    b. What fraction of the governors are Democrats? $\frac{19}{50}$
    c. What fraction of the governors are Republicans? Write your answer in simplified form. $\frac{30}{50} = \frac{3}{5}$
    d. What fraction of the governors are Independents? $\frac{1}{50}$

118. **GAS TANKS** Write fractions to describe the amount of gas left in the tank and the amount of gas that has been used. $\frac{5}{8}, \frac{3}{8}$

119. **SELLING CONDOS** The model below shows a new condominium development. The condos that have been sold are shaded.

    a. How many units are there in the development? 20
    b. What fraction of the units in the development have been sold? What fraction have not been sold? Write your answers in simplified form. $\frac{2}{5}, \frac{3}{5}$

120. **ARCHITECTURE** The floor plan of a movie theater complex is shown below. It has three Screen Rooms, a Lobby (that *includes* the Men's and Women's Restrooms), and an Entrance.

    a. What fraction of the complex floor space is used for Screen Room 1? $\frac{18}{96} = \frac{3}{16}$
    b. What fraction of the complex floor space is used for the Lobby? $\frac{36}{96} = \frac{3}{8}$
    c. What fraction of the complex floor space is used for the Lobby and Entrance (combined)? $\frac{42}{96} = \frac{7}{16}$
    d. What fraction of the Lobby floor space is used for the Men's and Women's Restrooms (combined)? $\frac{4}{36} = \frac{1}{9}$

121. **ACCOUNTING** Businesses often divide a calendar year into four quarters for accounting purposes. On what day would the second quarter begin and on what day would it end? Begin: April 1; end June 30

122. **MUSIC** The illustration shows a side view of the finger position needed to produce a length of string (from the *bridge* to the *fingertip*) that gives low C on a violin. To play other notes, fractions of that length are used. Draw arrows to show these finger positions on the string.

    a. $\frac{1}{2}$ of the length gives middle C.
    b. $\frac{3}{4}$ of the length gives F above low C.
    c. $\frac{2}{3}$ of the length gives G.

WRITING

123. Explain the concept of equivalent fractions. Give an example.

124. What does it mean for a fraction to be in simplest form? Give an example.

125. Why can't we say that $\frac{2}{5}$ of the figure is shaded?

126. Perhaps you have heard the following joke. Explain what is wrong with the customer's thinking.

    A pizza parlor waitress asks a customer if he wants the pizza cut into four pieces or six pieces or eight pieces. The customer then declares that he wants it cut into either four or six pieces "because I can't eat eight."

127. What type of problem is shown below? Explain the steps used to find the answer.

    a. $\dfrac{1}{2} = \dfrac{1}{2} \cdot \dfrac{4}{4} = \dfrac{4}{8}$

    b. $\dfrac{15}{35} = \dfrac{3 \cdot \cancel{5}}{\cancel{5} \cdot 7} = \dfrac{3}{7}$

128. Explain the difference in the two approaches used to simplify $\frac{20}{28}$. Are the results the same?

$$\dfrac{\cancel{4} \cdot 5}{\cancel{4} \cdot 7} \quad \text{and} \quad \dfrac{\cancel{2} \cdot \cancel{2} \cdot 5}{\cancel{2} \cdot \cancel{2} \cdot 7}$$

REVIEW

129. **PAYCHECKS** *Gross pay* is what a worker makes before deductions and *net pay* is what is left after taxes, health benefits, union dues, and other deductions are taken out. Suppose a worker's monthly gross pay is $3,575. If deductions of $235, $782, $148, and $103 are taken out of his check, what is his monthly net pay? $2,307

130. **HORSE RACING** One day, a man bet on all eight horse races at Santa Anita Racetrack. He won $168 on the first race and he won $105 on the fourth race. He lost his $50-bets on each of the other races. Overall, did he win or lose money betting on the horses? How much? He lost $27 (−$27).

SECTION 4.2
Multiplying Fractions

Objectives

1. Multiply fractions.
2. Simplify answers when multiplying fractions.
3. Multiply algebraic fractions.
4. Evaluate exponential expressions that have fractional bases.
5. Solve application problems by multiplying fractions.
6. Find the area of a triangle.

In the next three sections, we discuss how to add, subtract, multiply, and divide fractions. We begin with the operation of multiplication.

1 Multiply fractions.

To develop a rule for multiplying fractions, let's consider a real-life application.

Suppose $\frac{3}{5}$ of the last page of a school newspaper is devoted to campus sports coverage. To show this, we can divide the page into fifths, and shade three of them red.

Sports coverage: $\frac{3}{5}$ of the page

Unless otherwise noted, all art on this page is © Cengage Learning.

Furthermore, suppose that $\frac{1}{2}$ of the sports coverage is about women's teams. We can show that portion of the page by dividing the already colored region into two halves, and shading one of them in purple.

Women's teams coverage: $\frac{1}{2}$ of $\frac{3}{5}$ of the page

To find the fraction represented by the purple-shaded region, the page needs to be divided into equal-size parts. If we extend the dashed line downward, we see there are 10 equal-sized parts. The purple-shaded parts are 3 out of 10, or $\frac{3}{10}$, of the page. Thus, $\frac{3}{10}$ of the last page of the school newspaper is devoted to women's sports.

Women's teams coverage: $\frac{3}{10}$ of the page

In this example, we have found that

$$\frac{1}{2} \text{ of } \frac{3}{5} \text{ is } \frac{3}{10}$$
$$\frac{1}{2} \cdot \frac{3}{5} = \frac{3}{10}$$

Since the key word *of* indicates multiplication, and the key word *is* means equals, we can translate this statement to symbols.

Two observations can be made from this result.

- The numerator of the answer is the product of the numerators of the original fractions.

$$\overset{1 \cdot 3 = 3}{\frac{1}{2} \cdot \frac{3}{5} = \frac{3}{10}} \quad \text{Answer}$$
$$2 \cdot 5 = 10$$

- The denominator of the answer is the product of the denominators of the original fractions.

These observations illustrate the following rule for multiplying two fractions.

Multiplying Fractions

To multiply two fractions, multiply the numerators and multiply the denominators. Simplify the result, if possible.

Success Tip In the newspaper example, we found a *part of a part* of a page. Multiplying proper fractions can be thought of in this way. When taking a *part of a part* of something, the result is always smaller than the original part that you began with.

Self Check 1

Multiply:

a. $\dfrac{1}{2} \cdot \dfrac{1}{8}$ $\dfrac{1}{16}$

b. $\dfrac{5}{9} \cdot \dfrac{2}{3}$ $\dfrac{10}{27}$

Now Try Problems 17 and 21

Teaching Example 1

Multiply:

a. $\dfrac{1}{3} \cdot \dfrac{1}{4}$ b. $\dfrac{4}{7} \cdot \dfrac{2}{5}$

Answers:

a. $\dfrac{1}{12}$ b. $\dfrac{8}{35}$

EXAMPLE 1

Multiply: a. $\dfrac{1}{6} \cdot \dfrac{1}{4}$ b. $\dfrac{7}{8} \cdot \dfrac{3}{5}$

Strategy We will multiply the numerators and denominators, and make sure that the result is in simplest form.

WHY This is the rule for multiplying two fractions.

a. $\dfrac{1}{6} \cdot \dfrac{1}{4} = \dfrac{1 \cdot 1}{6 \cdot 4}$ Multiply the numerators.
Multiply the denominators.

$= \dfrac{1}{24}$ Since 1 and 24 have no common factors other than 1, the result is in simplest form.

Solution

b. $\dfrac{7}{8} \cdot \dfrac{3}{5} = \dfrac{7 \cdot 3}{8 \cdot 5}$ Multiply the numerators.
Multiply the denominators.

$= \dfrac{21}{40}$ Since 21 and 40 have no common factors other than 1, the result is in simplest form.

The sign rules for multiplying integers also hold for multiplying fractions. When we multiply two fractions with *like* signs, the product is positive. When we multiply two fractions with *unlike* signs, the product is negative.

Self Check 2

Multiply: $\dfrac{5}{6}\left(-\dfrac{1}{3}\right)$ $-\dfrac{5}{18}$

Now Try Problem 25

Teaching Example 2

Multiply: $-\dfrac{5}{8}\left(\dfrac{1}{9}\right)$

Answer: $-\dfrac{5}{72}$

EXAMPLE 2

Multiply: $-\dfrac{3}{4}\left(\dfrac{1}{8}\right)$

Strategy We will use the rule for multiplying two fractions that have different (unlike) signs.

WHY One fraction is positive and one is negative.

Solution

$-\dfrac{3}{4}\left(\dfrac{1}{8}\right) = -\dfrac{3 \cdot 1}{4 \cdot 8}$ Multiply the numerators.
Multiply the denominators.
Since the fractions have unlike signs, make the answer negative.

$= -\dfrac{3}{32}$ Since 3 and 32 have no common factors other than 1, the result is in simplest form.

Self Check 3

Multiply: $\dfrac{1}{3} \cdot 7$ $\dfrac{7}{3}$

Now Try Problem 29

EXAMPLE 3

Multiply: $\dfrac{1}{2} \cdot 3$

Strategy We will begin by writing the integer 3 as a fraction.

WHY Then we can use the rule for multiplying two fractions to find the product.

Solution

$$\frac{1}{2} \cdot 3 = \frac{1}{2} \cdot \frac{3}{1}$$ Write 3 as a fraction: $3 = \frac{3}{1}$.

$$= \frac{1 \cdot 3}{2 \cdot 1}$$ Multiply the numerators. Multiply the denominators.

$$= \frac{3}{2}$$ Since 3 and 2 have no common factors other than 1, the result is in simplest form.

Teaching Example 3 Multiply: $\frac{1}{4} \cdot 9$

Answer: $\frac{9}{4}$

2 Simplify answers when multiplying fractions.

After multiplying two fractions, we need to simplify the result, if possible. To do that, we can use the procedure discussed in Section 4.1 by removing pairs of common factors of the numerator and denominator.

EXAMPLE 4 Multiply and simplify: $\frac{5}{8} \cdot \frac{4}{5}$

Strategy We will multiply the numerators and denominators, and make sure that the result is in simplest form.

WHY This is the rule for multiplying two fractions.

Solution

$$\frac{5}{8} \cdot \frac{4}{5} = \frac{5 \cdot 4}{8 \cdot 5}$$ Multiply the numerators. Multiply the denominators.

$$= \frac{5 \cdot 2 \cdot 2}{2 \cdot 2 \cdot 2 \cdot 5}$$ To prepare to simplify, write 4 and 8 in prime-factored form.

$$= \frac{\overset{1}{\cancel{5}} \cdot \overset{1}{\cancel{2}} \cdot \overset{1}{\cancel{2}}}{\underset{1}{\cancel{2}} \cdot \underset{1}{\cancel{2}} \cdot 2 \cdot \underset{1}{\cancel{5}}}$$ To simplify, remove the common factors of 2 and 5 from the numerator and denominator.

$$= \frac{1}{2}$$ Multiply the remaining factors in the numerator: $1 \cdot 1 \cdot 1 = 1$. Multiply the remaining factors in the denominator: $1 \cdot 1 \cdot 2 \cdot 1 = 2$.

Self Check 4 Multiply and simplify: $\frac{11}{25} \cdot \frac{10}{11}$ $\frac{2}{5}$

Now Try Problem 33

Teaching Example 4 Multiply and simplify: $\frac{7}{12} \cdot \frac{4}{7}$

Answer: $\frac{1}{3}$

Success Tip If you recognized that 4 and 8 have a common factor of 4, you may remove that common factor from the numerator and denominator of the product without writing the prime factorizations. However, make sure that the numerator and denominator of the resulting fraction do not have any common factors. If they do, continue to simplify.

$$\frac{5}{8} \cdot \frac{4}{5} = \frac{5 \cdot 4}{8 \cdot 5} = \frac{\overset{1}{\cancel{5}} \cdot \overset{1}{\cancel{4}}}{2 \cdot \underset{1}{\cancel{4}} \cdot \underset{1}{\cancel{5}}} = \frac{1}{2}$$ Factor 8 as $2 \cdot 4$, and then remove the common factors of 4 and 5 in the numerator and denominator.

The rule for multiplying two fractions can be extended to find the product of three or more fractions.

EXAMPLE 5 Multiply and simplify: $\frac{2}{3}\left(-\frac{9}{14}\right)\left(-\frac{7}{10}\right)$

Strategy We will multiply the numerators and denominators, and make sure that the result is in simplest form.

WHY This is the rule for multiplying three (or more) fractions.

Self Check 5 Multiply and simplify:

$\frac{2}{5}\left(-\frac{15}{22}\right)\left(-\frac{11}{26}\right)$ $\frac{3}{26}$

Now Try Problem 37

Teaching Example 5
Multiply and simplify:
$$\frac{3}{5}\left(-\frac{25}{22}\right)\left(-\frac{11}{12}\right)$$
Answer:
$$\frac{5}{8}$$

Solution Recall from Section 2.4 that a product is positive when there are an even number of negative factors. Since $\frac{2}{3}\left(-\frac{9}{14}\right)\left(-\frac{7}{10}\right)$ has *two* negative factors, the product is positive.

$$\frac{2}{3}\left(-\frac{9}{14}\right)\left(-\frac{7}{10}\right) = \frac{2}{3}\left(\frac{9}{14}\right)\left(\frac{7}{10}\right)$$ Since the answer is positive, drop both − signs and continue.

$$= \frac{2 \cdot 9 \cdot 7}{3 \cdot 14 \cdot 10}$$ Multiply the numerators. Multiply the denominators.

$$= \frac{2 \cdot 3 \cdot 3 \cdot 7}{3 \cdot 2 \cdot 7 \cdot 2 \cdot 5}$$ To prepare to simplify, write 9, 14, and 10 in prime-factored form.

$$= \frac{\overset{1}{2} \cdot \overset{1}{3} \cdot 3 \cdot \overset{1}{7}}{\underset{1}{3} \cdot 2 \cdot \underset{1}{7} \cdot 2 \cdot 5}$$ To simplify, remove the common factors of 2, 3, and 7 from the numerator and denominator.

$$= \frac{3}{10}$$ Multiply the remaining factors in the numerator. Multiply the remaining factors in the denominator.

Caution! In Example 5, it was very helpful to prime factor and simplify when we did (the third step of the solution). If, instead, you find the product of the numerators and the product of the denominators, the resulting fraction is difficult to simplify because the numerator, 126, and the denominator, 420, are large.

$$\frac{2}{3} \cdot \frac{9}{14} \cdot \frac{7}{10} \quad = \quad \frac{2 \cdot 9 \cdot 7}{3 \cdot 14 \cdot 10} \quad = \quad \cancel{\frac{126}{420}}$$

↑ Factor and simplify at this stage, before multiplying in the numerator and denominator.

↑ Don't multiply in the numerator and denominator and then try to simplify the result. You will get the same answer, but it takes much more work.

3 Multiply algebraic fractions.

To multiply two algebraic fractions, we use the same approach as with numerical fractions: *Multiply the numerators and multiply the denominators. Then simplify the result, if possible.*

Self Check 6
Multiply and simplify:
a. $\dfrac{5}{12y} \cdot \dfrac{3y}{8} \cdot \dfrac{5}{32}$
b. $-\dfrac{2a^3}{15b} \cdot \dfrac{35ab}{8a^5} \cdot \dfrac{7}{12a}$

Now Try Problems 41 and 45

Teaching Example 6
Multiply and simplify:
a. $\dfrac{9}{16b} \cdot \dfrac{4b}{5}$
b. $-\dfrac{3c}{35d} \cdot \dfrac{28cd}{9c^5}$
Answers:
a. $\dfrac{9}{20}$
b. $-\dfrac{4}{15c^3}$

EXAMPLE 6 Multiply and simplify: a. $\dfrac{5b}{2} \cdot \dfrac{4}{7b}$ b. $-\dfrac{t^2}{21r} \cdot \dfrac{14r}{t^4}$

Strategy To find each product, we will use the rule for multiplying fractions. In the process, we must be prepared to factor the numerators and denominators so that any common factors can be removed.

WHY We want to give each result in simplified form.

Solution
a. $\dfrac{5b}{2} \cdot \dfrac{4}{7b} = \dfrac{5b \cdot 4}{2 \cdot 7b}$ Multiply the numerators. Multiply the denominators.

It's obvious that the numerator and denominator of $\frac{5b \cdot 4}{2 \cdot 7b}$ have two common factors, 2 and b. These common factors become more apparent when we factor the numerator and denominator completely.

$$= \frac{5 \cdot b \cdot 2 \cdot 2}{2 \cdot 7 \cdot b}$$ To prepare to simplify, factor 4 as 2 · 2. Write 5b as 5 · b and 7b as 7 · b.

$$= \frac{5 \cdot \overset{1}{b} \cdot \overset{1}{2} \cdot 2}{\underset{1}{2} \cdot 7 \cdot \underset{1}{b}}$$ To simplify, remove the common factors of 2 and b from the numerator and denominator

$$= \frac{10}{7}$$ Multiply the remaining factors in the numerator: 5 · 1 · 1 · 2 = 10. Multiply the remaining factors in the denominator: 1 · 7 · 1 = 7.

b. $-\dfrac{t^2}{21r} \cdot \dfrac{14r}{t^4} = -\dfrac{t^2 \cdot 14r}{21r \cdot t^4}$ Multiply the numerators. Multiply the denominators.

The product of two fractions with unlike signs is negative.

$= -\dfrac{t \cdot t \cdot 7 \cdot 2 \cdot r}{3 \cdot 7 \cdot r \cdot t \cdot t \cdot t \cdot t}$ To prepare to simplify, factor t^2, 14, 21, and t^4.

$= -\dfrac{\overset{1}{t} \cdot \overset{1}{t} \cdot \overset{1}{7} \cdot 2 \cdot \overset{1}{r}}{3 \cdot \underset{1}{7} \cdot \underset{1}{r} \cdot \underset{1}{t} \cdot \underset{1}{t} \cdot t \cdot t}$ To simplify, remove the common factors of 7, r, and t from the numerator and denominator.

$= -\dfrac{2}{3t^2}$ Multiply the remaining factors in the numerator: $1 \cdot 1 \cdot 1 \cdot 2 \cdot 1 = 2$. Multiply the remaining factors in the denominator: $3 \cdot 1 \cdot 1 \cdot 1 \cdot 1 \cdot t \cdot t = 3t^2$.

EXAMPLE 7 Multiply and simplify: **a.** $\dfrac{1}{4}(4y)$ **b.** $9a \cdot \dfrac{1}{3a}$

Strategy We will write the expressions $4y$ and $9a$ as fractions.

WHY Then we can use the rule for multiplying two fractions to find each product.

Solution

a. $\dfrac{1}{4}(4y) = \dfrac{1}{4} \cdot \left(\dfrac{4y}{1}\right)$ Write $4y$ as a fraction: $4y = \dfrac{4y}{1}$.

$= \dfrac{1 \cdot 4y}{4 \cdot 1}$ Multiply the numerators. Multiply the denominators.

$= \dfrac{1 \cdot \overset{1}{4} \cdot y}{\underset{1}{4} \cdot 1}$ To simplify, write $4y$ as $4 \cdot y$ and remove the common factor of 4 from the numerator and denominator.

$= \dfrac{y}{1}$ Multiply the remaining factors in the numerator: $1 \cdot 1 \cdot y = y$. Multiply the remaining factors in the denominator: $1 \cdot 1 = 1$.

$= y$ Any number divided by 1 is equal to that number.

b. $9a \cdot \dfrac{1}{3a} = \dfrac{9a}{1} \cdot \dfrac{1}{3a}$ Write $9a$ as a fraction: $9a = \dfrac{9a}{1}$.

$= \dfrac{9a \cdot 1}{1 \cdot 3a}$ Multiply the numerators. Multiply the denominators.

$= \dfrac{\overset{1}{3} \cdot 3 \cdot \overset{1}{a} \cdot 1}{1 \cdot \underset{1}{3} \cdot \underset{1}{a}}$ To simplify, factor 9 as $3 \cdot 3$. Then remove the common factors of 3 and a from the numerator and denominator.

$= \dfrac{3}{1}$ Multiply the remaining factors in the numerator: $1 \cdot 3 \cdot 1 \cdot 1 = 3$. Multiply the remaining factors in the denominator: $1 \cdot 1 \cdot 1 = 1$.

$= 3$ Any number divided by 1 is equal to that number.

Self Check 7
Multiply and simplify:

a. $\dfrac{1}{5} \cdot 5m$ m

b. $12m\left(\dfrac{1}{6m}\right)$ 2

Now Try Problems 49 and 53

Teaching Example 7
Multiply and simplify:

a. $\dfrac{1}{8}(8x)$

b. $15d \cdot \dfrac{1}{5d}$

Answers:
a. x
b. 3

To multiply $\dfrac{1}{2}$ and x, we can express the product as $\dfrac{1}{2}x$, or we can use the concept of multiplying fractions to write it in a different form.

$\dfrac{1}{2} \cdot x = \dfrac{1}{2} \cdot \dfrac{x}{1}$ Write x as a fraction: $x = \dfrac{x}{1}$.

$= \dfrac{1 \cdot x}{2 \cdot 1}$ Multiply the numerators. Multiply the denominators.

$= \dfrac{x}{2}$

The product of $\dfrac{1}{2}$ and x can be expressed as $\dfrac{1}{2}x$ or $\dfrac{x}{2}$. Similarly,

$\dfrac{3}{4}t = \dfrac{3t}{4}$ and $-\dfrac{5}{16}y = -\dfrac{5y}{16}$

4 Evaluate exponential expressions that have fractional bases.

We have evaluated exponential expressions that have whole-number bases and integer bases. If the base of an exponential expression is a fraction, the exponent tells us how many times to write that fraction as a factor. For example,

$$\left(\frac{2}{3}\right)^2 = \frac{2}{3} \cdot \frac{2}{3} = \frac{2 \cdot 2}{3 \cdot 3} = \frac{4}{9}$$ Since the exponent is 2, write the base, $\frac{2}{3}$, as a factor 2 times.

Self Check 8
Evaluate each expression.
a. $\left(\frac{2}{5}\right)^3$ $\frac{8}{125}$
b. $\left(-\frac{3}{4}\right)^2$ $\frac{9}{16}$
c. $-\left(\frac{3}{4}\right)^2$ $-\frac{9}{16}$

Now Try Problems 57 and 59

Teaching Example 8 Evaluate each expression:
a. $\left(\frac{4}{5}\right)^3$ b. $\left(-\frac{3}{7}\right)^2$ c. $-\left(\frac{3}{7}\right)^2$

Answers:
a. $\frac{64}{125}$ b. $\frac{9}{49}$ c. $-\frac{9}{49}$

EXAMPLE 8
Evaluate each expression: a. $\left(\frac{1}{4}\right)^3$ b. $\left(-\frac{2}{3}\right)^2$ c. $-\left(\frac{2}{3}\right)^2$

Strategy We will write each exponential expression as a product of repeated factors, and then perform the multiplication. This requires that we identify the base and the exponent.

WHY The exponent tells the number of times the base is to be written as a factor.

Solution
Recall that exponents are used to represent repeated multiplication.

a. We read $\left(\frac{1}{4}\right)^3$ as "one-fourth raised to the third power," or as "one-fourth, cubed."

$$\left(\frac{1}{4}\right)^3 = \frac{1}{4} \cdot \frac{1}{4} \cdot \frac{1}{4}$$ Since the exponent is 3, write the base, $\frac{1}{4}$, as a factor 3 times.

$$= \frac{1 \cdot 1 \cdot 1}{4 \cdot 4 \cdot 4}$$ Multiply the numerators. Multiply the denominators.

$$= \frac{1}{64}$$

b. We read $\left(-\frac{2}{3}\right)^2$ as "negative two-thirds raised to the second power," or as "negative two-thirds, squared."

$$\left(-\frac{2}{3}\right)^2 = \left(-\frac{2}{3}\right)\left(-\frac{2}{3}\right)$$ Since the exponent is 2, write the base, $-\frac{2}{3}$, as a factor 2 times.

$$= \frac{2 \cdot 2}{3 \cdot 3}$$ The product of two fractions with like signs is positive: Drop the − signs. Multiply the numerators. Multiply the denominators.

$$= \frac{4}{9}$$

c. We read $-\left(\frac{2}{3}\right)^2$ as "the opposite of two-thirds squared." Recall that if the − symbol is not within the parentheses, it is not part of the base.

$$-\left(\frac{2}{3}\right)^2 = -\frac{2}{3} \cdot \frac{2}{3}$$ Since the exponent is 2, write the base, $\frac{2}{3}$, as a factor 2 times.

$$= -\frac{2 \cdot 2}{3 \cdot 3}$$ Multiply the numerators. Multiply the denominators.

$$= -\frac{4}{9}$$

We can use the rule for multiplying fractions to find powers of algebraic fractions.

Self Check 9
Find the power: $\left(-\frac{3t}{4}\right)^3$ $-\frac{27t^3}{64}$

Now Try Problem 61

EXAMPLE 9
Find the power: $\left(-\frac{4x}{5}\right)^2$

Strategy We will write the exponential expression as a product of repeated factors, and then perform the multiplication. This requires that we identify the base and the exponent.

WHY The exponent tells the number of times the base is to be written as a factor.

Solution

$$\left(-\frac{4x}{5}\right)^2 = \left(-\frac{4x}{5}\right)\left(-\frac{4x}{5}\right)$$ Since the exponent is 2, write the base, $-\frac{4x}{5}$, as a factor 2 times.

$$= \frac{4x \cdot 4x}{5 \cdot 5}$$ Since the product of two fractions with like signs is positive, we can drop the − symbols. Multiply the numerators. Multiply the denominators.

$$= \frac{16x^2}{25}$$ Since 16 and 25 have no common factors other than 1, the result is in simplest form.

Teaching Example 9
Find the power:
$$\left(-\frac{2x}{3}\right)^2$$
Answer:
$\frac{4x^2}{9}$

5 Solve application problems by multiplying fractions.

The key word *of* often appears in application problems involving fractions. When a fraction is followed by the word *of*, such as $\frac{1}{2}$ *of* or $\frac{3}{4}$ *of*, it indicates that we are to find a part of some quantity using multiplication.

EXAMPLE 10 *How a Bill Becomes Law* If the President vetoes (refuses to sign) a bill, it takes $\frac{2}{3}$ of those voting in the House of Representatives (and the Senate) to override the veto for it to become law. If all 435 members of the House cast a vote, how many of their votes does it take to override a presidential veto?

Solution

- It takes $\frac{2}{3}$ *of* those voting to override a veto.   Given
- All 435 members of the House cast a vote.   Given
- How many votes does it take to override a presidential veto?   Find

The key phrase $\frac{2}{3}$ *of* suggests that we are to find a part of the 435 possible votes using multiplication.
We translate the words of the problem to numbers and symbols.

| The number of votes needed in the House to override a veto | is equal to | $\frac{2}{3}$ | of | the number of House members that vote. |

$$\text{The number of votes needed in the House to override a veto} = \frac{2}{3} \cdot 435$$

Self Check 10

HOW A BILL BECOMES LAW If only 96 senators are present and cast a vote, how many of their votes does it take to override a presidential veto?   64 votes

Now Try Problems 65 and 103

Teaching Example 10 COLLEGE LIFE It takes $\frac{3}{4}$ of the active members of Alpha Gamma Sigma to vote to accept a pledge into their fraternity. How many votes are needed if the fraternity has 84 active members?
Answer:
63 votes

To find the product, we will express 435 as a fraction and then use the rule for multiplying two fractions.

$$\frac{2}{3} \cdot 435 = \frac{2}{3} \cdot \frac{435}{1}$$ Write 435 as a fraction: $435 = \frac{435}{1}$.

$$= \frac{2 \cdot 435}{3 \cdot 1}$$ Multiply the numerators. Multiply the denominators.

$$= \frac{2 \cdot 3 \cdot 5 \cdot 29}{3 \cdot 1}$$ To prepare to simplify, write 435 in prime-factored form: $3 \cdot 5 \cdot 29$.

$$= \frac{2 \cdot \overset{1}{\cancel{3}} \cdot 5 \cdot 29}{\underset{1}{\cancel{3}} \cdot 1}$$ Remove the common factor of 3 from the numerator and denominator.

$$= \frac{290}{1}$$ Multiply the remaining factors in the numerator: $2 \cdot 1 \cdot 5 \cdot 29 = 290$.
Multiply the remaining factors in the denominator: $1 \cdot 1 = 1$.

$$= 290$$ Any number divided by 1 is equal to that number.

435
③ 145
⑤ ㉙

It would take 290 votes in the House to override a veto.

Check: We can estimate to check the result. We will use 440 to approximate the number of House members voting. Since $\frac{1}{2}$ of 440 is 220, and since $\frac{2}{3}$ is a greater part than $\frac{1}{2}$, we would expect the number of votes needed to be *more than* 220. The result of 290 seems reasonable.

6 Find the area of a triangle.

As the figures below show, a triangle has three sides. The length of the base of the triangle can be represented by the letter b and the height by the letter h. The height of a triangle is always perpendicular (makes a square corner) to the base. This is shown by using the symbol ⌐.

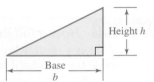

Recall that the area of a figure is the amount of surface that it encloses. The area of a triangle can be found by using the following formula.

Area of a Triangle

The area A of a triangle is one-half the product of its base b and its height h.

$$\text{Area} = \frac{1}{2}(\text{base})(\text{height}) \quad \text{or} \quad A = \frac{1}{2} \cdot b \cdot h$$

The Language of Algebra The formula $A = \frac{1}{2} \cdot b \cdot h$ can be written more simply as $A = \frac{1}{2}bh$. The formula for the area of a triangle can also be written as $A = \frac{bh}{2}$.

Self Check 11

Find the area of the triangle shown below. 216 in.²

Now Try Problems 69 and 115

EXAMPLE 11 *Geography* Approximate the area of the state of Virginia (in square miles) using the triangle shown below.

Strategy We will find the product of $\frac{1}{2}$, 405, and 200.

WHY The formula for the area of a triangle is $A = \frac{1}{2}(\text{base})(\text{height})$.

Solution

$$A = \frac{1}{2}bh$$ This is the formula for the area of a triangle.

$$A = \frac{1}{2} \cdot 405 \cdot 200$$ $\frac{1}{2}bh$ means $\frac{1}{2} \cdot b \cdot h$. Substitute 405 for b and 200 for h.

$$A = \frac{1}{2} \cdot \frac{405}{1} \cdot \frac{200}{1}$$ Write 405 and 200 as fractions.

$$A = \frac{1 \cdot 405 \cdot 200}{2 \cdot 1 \cdot 1}$$ Multiply the numerators. Multiply the denominators.

$$A = \frac{1 \cdot 405 \cdot \overset{1}{\cancel{2}} \cdot 100}{\underset{1}{\cancel{2}} \cdot 1 \cdot 1}$$ Factor 200 as $2 \cdot 100$. Then remove the common factor of 2 from the numerator and denominator.

$$A = 40{,}500$$ In the numerator, multiply: $405 \cdot 100 = 40{,}500$.

The area of the state of Virginia is approximately 40,500 square miles. This can be written as 40,500 mi².

Teaching Example 11 Find the area of the triangle shown below.

Answer:
33 ft²

> **Caution!** Remember that area is measured in square units, such as in.², ft², and cm². Don't forget to write the units in your answer when finding the area of a figure.

ANSWERS TO SELF CHECKS

1. a. $\frac{1}{16}$ **b.** $\frac{10}{27}$ **2.** $-\frac{5}{18}$ **3.** $\frac{7}{3}$ **4.** $\frac{2}{5}$ **5.** $\frac{3}{26}$ **6. a.** $\frac{5}{32}$ **b.** $-\frac{7}{12a}$ **7. a.** m **b.** 2
8. a. $\frac{8}{125}$ **b.** $\frac{9}{16}$ **c.** $-\frac{9}{16}$ **9.** $-\frac{27t^3}{64}$ **10.** 64 votes **11.** 216 in.²

SECTION 4.2 STUDY SET

VOCABULARY

Fill in the blanks.

1. When a fraction is followed by the word *of*, such as $\frac{1}{3}$ *of*, it indicates that we are to find a part of some quantity using __multiplication__.
2. The answer to a multiplication is called the __product__.
3. To __simplify__ a fraction, we remove common factors of the numerator and denominator.
4. In the expression $\left(\frac{1}{4}\right)^3$, the __base__ is $\frac{1}{4}$ and the __exponent__ is 3.
5. The __area__ of a triangle is the amount of surface that it encloses.

6. Label the *base* and the *height* of the triangle shown below.

CONCEPTS

7. Fill in the blanks: To multiply two fractions, multiply the __numerators__ and multiply the __denominators__. Then __simplify__, if possible.

8. Use the following rectangle to find $\frac{1}{3} \cdot \frac{1}{4}$.

   a. Draw three vertical lines that divide the given rectangle into four equal parts and lightly shade one part. What fractional part of the rectangle did you shade? $\frac{1}{4}$
   b. To find $\frac{1}{3}$ of the shaded portion, draw two horizontal lines to divide the given rectangle into three equal parts and lightly shade one part. Into how many equal parts is the rectangle now divided? How many parts have been shaded twice? 12, 1
   c. What is $\frac{1}{3} \cdot \frac{1}{4}$? $\frac{1}{12}$

9. Determine whether each product is positive or negative. *You do not have to find the answer.*

   a. $-\frac{1}{8} \cdot \frac{3}{5}$ negative
   b. $-\frac{7}{16}\left(-\frac{2}{21}\right)$ positive
   c. $-\frac{4}{5}\left(\frac{1}{3}\right)\left(-\frac{1}{8}\right)$ positive
   d. $-\frac{3}{4}\left(-\frac{8}{9}\right)\left(-\frac{1}{2}\right)$ negative

10. Translate each phrase to symbols. *You do not have to find the answer.*

    a. $\frac{7}{10}$ of $\frac{4}{9}$ $\frac{7}{10} \cdot \frac{4}{9}$
    b. $\frac{1}{5}$ of 40 $\frac{1}{5} \cdot 40$

11. Fill in the blanks.

    a. Area of a triangle = $\frac{1}{2}($ base $)($ height $)$
       or
       $A = \frac{1}{2}bh$
    b. Area is measured in   square   units, such as in.² and ft².

12. Determine whether each statement is true or false.

    a. $\frac{1}{2}x = \frac{x}{2}$ true
    b. $\frac{2t}{3} = \frac{2}{3}t$ true
    c. $-\frac{3}{8}a = -\frac{3}{8a}$ false
    d. $\frac{-4e}{7} = -\frac{4e}{7}$ true

NOTATION

13. Write each of the following as a fraction.
    a. 4 $\frac{4}{1}$
    b. -3 $-\frac{3}{1}$
    c. x $\frac{x}{1}$

14. Fill in the blanks: $\left(\frac{1}{2}\right)^2$ represents the repeated multiplication $\frac{1}{2} \cdot \frac{1}{2}$.

Fill in the blanks to complete each step.

15. Multiply and simplify:

$$\frac{5}{8} \cdot \frac{7}{15} = \frac{5 \cdot 7}{8 \cdot 15}$$
$$= \frac{5 \cdot 7}{2 \cdot 2 \cdot 2 \cdot 3 \cdot 5}$$
$$= \frac{\cancel{5} \cdot 7}{2 \cdot 2 \cdot 2 \cdot 3 \cdot \cancel{5}}$$
$$= \frac{7}{24}$$

16. Multiply and simplify:

$$\frac{7a}{12} \cdot \frac{4}{21a} = \frac{7a \cdot 4}{12 \cdot 21a}$$
$$= \frac{7 \cdot a \cdot 4}{3 \cdot 4 \cdot 3 \cdot 7 \cdot a}$$
$$= \frac{\cancel{7} \cdot \cancel{a} \cdot \cancel{4}}{3 \cdot \cancel{4} \cdot 3 \cdot \cancel{7} \cdot \cancel{a}}$$
$$= \frac{1}{9}$$

GUIDED PRACTICE

Multiply. See Example 1.

17. $\frac{1}{4} \cdot \frac{1}{2}$ $\frac{1}{8}$
18. $\frac{1}{3} \cdot \frac{1}{5}$ $\frac{1}{15}$
19. $\frac{1}{9} \cdot \frac{1}{5}$ $\frac{1}{45}$
20. $\frac{1}{2} \cdot \frac{1}{8}$ $\frac{1}{16}$
21. $\frac{2}{3} \cdot \frac{7}{9}$ $\frac{14}{27}$
22. $\frac{3}{4} \cdot \frac{5}{7}$ $\frac{15}{28}$
23. $\frac{8}{11} \cdot \frac{3}{7}$ $\frac{24}{77}$
24. $\frac{11}{13} \cdot \frac{2}{3}$ $\frac{22}{39}$

Multiply. See Example 2.

25. $-\frac{4}{5} \cdot \frac{1}{3}$ $-\frac{4}{15}$
26. $-\frac{7}{9} \cdot \frac{1}{4}$ $-\frac{7}{36}$
27. $\frac{5}{6}\left(-\frac{7}{12}\right)$ $-\frac{35}{72}$
28. $\frac{2}{15}\left(-\frac{4}{3}\right)$ $-\frac{8}{45}$

Multiply. See Example 3.

29. $\frac{1}{8} \cdot 9$ $\frac{9}{8}$
30. $\frac{1}{6} \cdot 11$ $\frac{11}{6}$
31. $\frac{1}{2} \cdot 5$ $\frac{5}{2}$
32. $\frac{1}{2} \cdot 21$ $\frac{21}{2}$

4.2 Multiplying Fractions

Multiply. Write the product in simplest form. See Example 4.

33. $\dfrac{11}{10} \cdot \dfrac{5}{11}$ $\dfrac{1}{2}$

34. $\dfrac{5}{4} \cdot \dfrac{2}{5}$ $\dfrac{1}{2}$

35. $\dfrac{6}{49} \cdot \dfrac{7}{6}$ $\dfrac{1}{7}$

36. $\dfrac{13}{4} \cdot \dfrac{4}{39}$ $\dfrac{1}{3}$

Multiply. Write the product in simplest form. See Example 5.

37. $\dfrac{3}{4}\left(-\dfrac{8}{35}\right)\left(-\dfrac{7}{12}\right)$ $\dfrac{1}{10}$

38. $\dfrac{9}{10}\left(-\dfrac{4}{15}\right)\left(-\dfrac{5}{18}\right)$ $\dfrac{1}{15}$

39. $-\dfrac{5}{8}\left(\dfrac{16}{27}\right)\left(-\dfrac{9}{25}\right)$ $\dfrac{2}{15}$

40. $-\dfrac{15}{28}\left(\dfrac{7}{9}\right)\left(-\dfrac{18}{35}\right)$ $\dfrac{3}{14}$

Multiply. Write the product in simplest form. See Example 6.

41. $\dfrac{3}{4x} \cdot \dfrac{2x}{5}$ $\dfrac{3}{10}$

42. $\dfrac{4}{15m} \cdot \dfrac{3m}{5}$ $\dfrac{4}{25}$

43. $\dfrac{6a}{49} \cdot \dfrac{14}{3a}$ $\dfrac{4}{7}$

44. $\dfrac{9x}{56} \cdot \dfrac{8}{3x}$ $\dfrac{3}{7}$

45. $-\dfrac{2m^3}{21n} \cdot \dfrac{27mn}{16m^5}$ $-\dfrac{9}{56m}$

46. $-\dfrac{2s^2}{27r} \cdot \dfrac{33rs}{16s^5}$ $-\dfrac{11}{72s^2}$

47. $\left(-\dfrac{25xy}{8y}\right)\left(-\dfrac{8x^2}{45x^2y^2}\right)$ $\dfrac{5x}{9y^2}$

48. $\left(-\dfrac{20cd}{21d^2}\right)\left(-\dfrac{14c^2d}{55c^2d^2}\right)$ $\dfrac{8c}{33d^2}$

Multiply. Write the product in simplest form. See Example 7.

49. $\dfrac{1}{8}(8w)$ w

50. $\dfrac{1}{9}(9e)$ e

51. $7x \cdot \dfrac{1}{7}$ x

52. $8r \cdot \dfrac{1}{8}$ r

53. $12a \cdot \dfrac{1}{6a}$ 2

54. $18b \cdot \dfrac{1}{3b}$ 6

55. $\dfrac{1}{9}(-36y)$ $-4y$

56. $\dfrac{1}{12}(-24d)$ $-2d$

Evaluate each expression. See Example 8.

57. a. $\left(\dfrac{3}{5}\right)^2$ $\dfrac{9}{25}$ b. $\left(-\dfrac{3}{5}\right)^2$ $\dfrac{9}{25}$

58. a. $\left(\dfrac{4}{9}\right)^2$ $\dfrac{16}{81}$ b. $\left(-\dfrac{4}{9}\right)^2$ $\dfrac{16}{81}$

59. a. $-\left(-\dfrac{1}{6}\right)^2$ $-\dfrac{1}{36}$ b. $\left(-\dfrac{1}{6}\right)^3$ $-\dfrac{1}{216}$

60. a. $-\left(-\dfrac{2}{5}\right)^2$ $-\dfrac{4}{25}$ b. $\left(-\dfrac{2}{5}\right)^3$ $-\dfrac{8}{125}$

Find each power. See Example 9.

61. $\left(-\dfrac{6t}{7}\right)^2$ $\dfrac{36t^2}{49}$

62. $\left(-\dfrac{3s}{8}\right)^2$ $\dfrac{9s^2}{64}$

63. $\left(-\dfrac{2a}{5}\right)^3$ $-\dfrac{8a^3}{125}$

64. $\left(\dfrac{4x}{5}\right)^3$ $\dfrac{64x^3}{125}$

Unless otherwise noted, all art on this page is © Cengage Learning.

Find each product. Write your answer in simplest form. See Example 10.

65. $\dfrac{3}{4}$ of $\dfrac{5}{8}$ $\dfrac{15}{32}$

66. $\dfrac{4}{5}$ of $\dfrac{3}{7}$ $\dfrac{12}{35}$

67. $\dfrac{1}{6}$ of 54 9

68. $\dfrac{1}{9}$ of 36 4

Find the area of each triangle. See Example 11.

69. 15 ft² (3 ft, 10 ft)

70. 10 yd² (5 yd, 4 yd)

71. 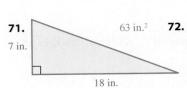 63 in.² (7 in., 18 in.)

72. 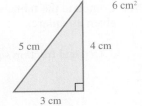 6 cm² (5 cm, 4 cm, 3 cm)

73. 6 m² (3 m, 4 m)

74. 102 in.² (12 in., 17 in.)

75. 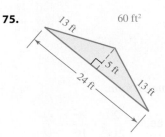 60 ft² (13 ft, 5 ft, 24 ft, 13 ft)

76. 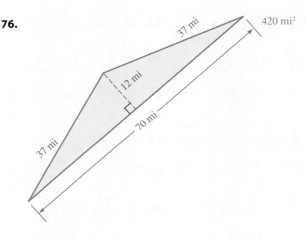 420 mi² (37 mi, 12 mi, 70 mi, 37 mi)

TRY IT YOURSELF

77. Complete the multiplication table of fractions.

·	$\frac{1}{2}$	$\frac{1}{3}$	$\frac{1}{4}$	$\frac{1}{5}$	$\frac{1}{6}$
$\frac{1}{2}$	$\frac{1}{4}$	$\frac{1}{6}$	$\frac{1}{8}$	$\frac{1}{10}$	$\frac{1}{12}$
$\frac{1}{3}$	$\frac{1}{6}$	$\frac{1}{9}$	$\frac{1}{12}$	$\frac{1}{15}$	$\frac{1}{18}$
$\frac{1}{4}$	$\frac{1}{8}$	$\frac{1}{12}$	$\frac{1}{16}$	$\frac{1}{20}$	$\frac{1}{24}$
$\frac{1}{5}$	$\frac{1}{10}$	$\frac{1}{15}$	$\frac{1}{20}$	$\frac{1}{25}$	$\frac{1}{30}$
$\frac{1}{6}$	$\frac{1}{12}$	$\frac{1}{18}$	$\frac{1}{24}$	$\frac{1}{30}$	$\frac{1}{36}$

78. Complete the table by finding the original fraction, given its square.

Original fraction squared	Original fraction
$\frac{1}{9}$	$\frac{1}{3}$
$\frac{1}{100}$	$\frac{1}{10}$
$\frac{4}{25}$	$\frac{2}{5}$
$\frac{16}{49}$	$\frac{4}{7}$
$\frac{81}{36}$	$\frac{9}{6}$
$\frac{9}{121}$	$\frac{3}{11}$

Multiply. Write the product in simplest form.

79. $-\frac{15x^3}{24} \cdot \frac{8}{25x^2}$ $\;-\frac{x}{5}$

80. $-\frac{20}{21x^4} \cdot \frac{7x^5}{16}$ $\;-\frac{5x}{12}$

81. $\frac{3}{8} \cdot \frac{7}{16}$ $\;\frac{21}{128}$

82. $18 \cdot \frac{2}{9}$ $\;4$

83. $\frac{1}{2}\left(-\frac{1}{3}\right)\left(-\frac{1}{4}\right)$ $\;\frac{1}{24}$

84. $\left(\frac{3}{8}\right)\left(-\frac{2}{3}\right)\left(-\frac{12}{27}\right)$ $\;\frac{1}{9}$

85. $-\frac{5}{6} \cdot 18x$ $\;-15x$

86. $6\left(-\frac{2a}{3}\right)$ $\;-4a$

87. $\left(-\frac{3}{4n}\right)^3$ $\;-\frac{27}{64n^3}$

88. $\left(-\frac{2}{5m}\right)^3$ $\;-\frac{8}{125m^3}$

89. $\frac{3a^3}{4} \cdot \frac{4}{3a^2}$ $\;a$

90. $\frac{4m^2}{5} \cdot \frac{5}{4m^3}$ $\;\frac{1}{m}$

91. $\frac{5}{3}\left(-\frac{6}{15}\right)(-4)$ $\;\frac{8}{3}$

92. $\frac{5}{6}\left(-\frac{2}{3}\right)(-12)$ $\;\frac{20}{3}$

93. $-\frac{11}{12} \cdot \frac{18}{55} \cdot 5$ $\;-\frac{3}{2}$

94. $-\frac{24}{5} \cdot \frac{7}{12} \cdot \frac{1}{14}$ $\;-\frac{1}{5}$

95. $\left(-\frac{11}{21}\right)\left(-\frac{14}{33}\right)$ $\;\frac{2}{9}$

96. $\left(-\frac{16}{35}\right)\left(-\frac{25}{48}\right)$ $\;\frac{5}{21}$

97. $-\left(-\frac{5}{9}\right)^2$ $\;-\frac{25}{81}$

98. $-\left(-\frac{5}{6}\right)^2$ $\;-\frac{25}{36}$

99. $\frac{7x^4}{10xy^3}\left(\frac{20y^2}{21x^2}\right)$ $\;\frac{2x}{3y}$

100. $\frac{7r^4}{6rt^3}\left(\frac{9rt^2}{49r^2}\right)$ $\;\frac{3r^2}{14t}$

101. $\frac{3}{4}\left(\frac{5}{7}\right)\left(\frac{2}{3}\right)\left(\frac{7}{3}\right)$ $\;\frac{5}{6}$

102. $-\frac{5}{4}\left(\frac{8}{15}\right)\left(\frac{2}{3}\right)\left(\frac{7}{2}\right)$ $\;-\frac{14}{9}$

APPLICATIONS

103. SENATE RULES A *filibuster* is a method U.S. senators sometimes use to block passage of a bill or appointment. It takes $\frac{3}{5}$ of those voting in the Senate to break a filibuster. If all 100 senators cast a vote, how many of their votes does it take to break a filibuster? 60 votes

104. GENETICS Gregor Mendel (1822–1884), an Augustinian monk, is credited with developing a model that became the foundation of modern genetics. In his experiments, he crossed purple-flowered plants with white-flowered plants and found that $\frac{3}{4}$ of the offspring plants had purple flowers and $\frac{1}{4}$ of them had white flowers. Refer to the illustration below, which shows a group of offspring plants. According to this concept, when the plants begin to flower, how many will have purple flowers? 9

105. BOUNCING BALLS A tennis ball is dropped from a height of 54 inches. Each time it hits the ground, it rebounds one-third of the previous height that it fell. Find the three missing rebound heights in the illustration. 18 in., 6 in., and 2 in.

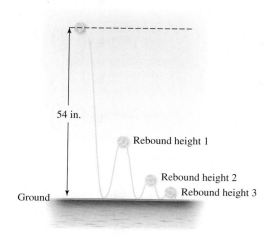

Unless otherwise noted, all art on this page is © Cengage Learning.

106. ELECTIONS The final election returns for a city bond measure are shown below.
   a. Find the total number of votes cast. 188,400
   b. Find two-thirds of the total number of votes cast. 125,600
   c. Did the bond measure pass? It failed to pass by 1 vote.

MEASURE 1	
100% of the precincts reporting	
Fire–Police–Paramedics General Obligation Bonds	
(Requires two-thirds vote to pass)	
YES	No
125,599	62,801

107. COOKING Use the recipe below, along with the concept of multiplication of fractions, to find how much sugar and how much molasses are needed to make *one dozen* cookies. (*Hint:* this recipe is for *two dozen* cookies.) $\frac{3}{8}$ cup sugar, $\frac{1}{6}$ cup molasses

Gingerbread Cookies

$\frac{3}{4}$ cup sugar $\frac{1}{2}$ cup water
2 cups flour $\frac{2}{3}$ cup shortening
$\frac{1}{8}$ teaspoon allspice $\frac{1}{4}$ teaspoon salt
$\frac{1}{3}$ cup dark molasses $\frac{3}{4}$ teaspoon ginger

Makes two dozen gingerbread cookies.

108. THE EARTH'S SURFACE The surface of Earth covers an area of approximately 196,800,000 square miles. About $\frac{3}{4}$ of that area is covered by water. Find the number of square miles of the surface covered by water. 147,600,000 mi^2

109. BOTANY In an experiment, monthly growth rates of three types of plants doubled when nitrogen was added to the soil. Complete the graph by drawing the improved growth rate bar next to each normal growth rate bar.

110. ICEBERGS About $\frac{9}{10}$ of the volume of an iceberg is below the water line.
   a. What fraction of the volume of an iceberg is *above* the water line? $\frac{1}{10}$
   b. Suppose an iceberg has a total volume of 18,700 cubic meters. What is the volume of the part of the iceberg that is above the water line? 1,870 cubic meters

111. KITCHEN DESIGN Find the area of the *kitchen work triangle* formed by the paths between the refrigerator, the range, and the sink shown below. 27 ft^2

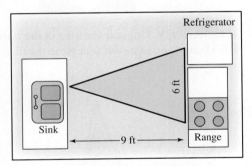

112. STARS AND STRIPES The illustration shows a folded U.S. flag. When it is placed on a table as part of an exhibit, how much area will it occupy? 121 in.^{2}

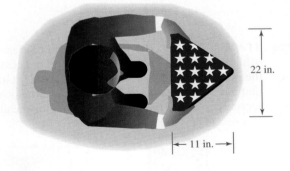

113. WINDSURFING Estimate the area of the sail on the windsurfing board. 42 ft²

114. TILE DESIGN A design for bathroom tile is shown. Find the amount of area on a tile that is blue. 18 in.²

115. GEOGRAPHY Estimate the area of the state of New Hampshire, using the triangle in the illustration. 9,646 mi²

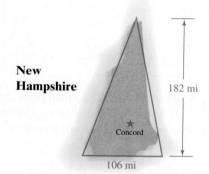

116. STAMPS The best designs in a contest to create a wildlife stamp are shown. To save on paper costs, the postal service has decided to choose the stamp that has the smaller area. Which one did the postal service choose? (*Hint:* use the formula for the area of a rectangle.) the natural beauty stamp

117. VISES Each complete turn of the handle of the bench vise shown below tightens its jaws exactly $\frac{1}{16}$ of an inch. How much tighter will the jaws of the vise get if the handle is turned 12 complete times? $\frac{3}{4}$ in.

118. WOODWORKING Each time a board is passed through a power sander, the machine removes $\frac{1}{64}$ of an inch of thickness. If a rough pine board is passed through the sander 6 times, by how much will its thickness change? It will lose $\frac{3}{32}$ of an inch of thickness $\left(-\frac{3}{32}\right)$

WRITING

119. In a word problem, when a fraction is followed by the word *of*, multiplication is usually indicated. Give three real-life examples of this type of use of the word *of*.

120. Can you multiply the number 5 and another number and obtain an answer that is less than 5? Explain why or why not.

121. In the following solution, what step did the student forget to use that caused him to have to work with such large numbers?

Multiply. Simplify the product, if possible.

$$\frac{44}{63} \cdot \frac{27}{55} = \frac{44 \cdot 27}{63 \cdot 55}$$

$$= \frac{1{,}188}{3{,}465}$$

122. Is the product of two proper fractions always smaller than either of those fractions? Explain why or why not.

REVIEW

Divide and check each result using multiplication.

123. $\frac{-8}{4}$ −2

124. $21 \div (-3)$ −7

125. $-36 \div (-9)$ 4

126. $\frac{-200}{-200}$ 1

SECTION 4.3
Dividing Fractions

We will now discuss how to divide fractions. The fraction multiplication skills that you learned in Section 4.2 will also be useful in this section.

1 Find the reciprocal of a fraction.

Division with fractions involves working with *reciprocals*. To present the concept of reciprocal, we consider the problem $\frac{7}{8} \cdot \frac{8}{7}$.

$$\frac{7}{8} \cdot \frac{8}{7} = \frac{7 \cdot 8}{8 \cdot 7} \quad \text{Multiply the numerators.}$$
$$\text{Multiply the denominators.}$$

$$= \frac{\overset{1}{7} \cdot \overset{1}{8}}{\underset{1}{8} \cdot \underset{1}{7}} \quad \text{To simplify, remove the common factors of 7 and 8 from the numerator and denominator.}$$

$$= \frac{1}{1} \quad \text{Multiply the remaining factors in the numerator.}$$
$$\text{Multiply the remaining factors in the denominator.}$$

$$= 1 \quad \text{Any whole number divided by 1 is equal to that number.}$$

The product of $\frac{7}{8}$ and $\frac{8}{7}$ is 1.

Whenever the product of two numbers is 1, we say that those numbers are *reciprocals*. Therefore, $\frac{7}{8}$ and $\frac{8}{7}$ are reciprocals. To find the reciprocal of a fraction, *we invert the numerator and the denominator*.

Reciprocals

Two numbers are called **reciprocals** if their product is 1.

Caution! Zero does not have a reciprocal, because the product of 0 and a number can never be 1.

Objectives

1. Find the reciprocal of a fraction.
2. Divide fractions.
3. Divide algebraic fractions.
4. Solve application problems by dividing fractions.

EXAMPLE 1
For each number, find its reciprocal and show that their product is 1: **a.** $\frac{2}{3}$ **b.** $-\frac{3}{4}$ **c.** 5

Strategy To find each reciprocal, we will invert the numerator and denominator.

WHY This procedure will produce a new fraction that, when multiplied by the original fraction, gives a result of 1.

Solution

a. Fraction   Reciprocal

$$\frac{2}{3} \longrightarrow \frac{3}{2}$$
invert

The reciprocal of $\frac{2}{3}$ is $\frac{3}{2}$.

Check: $\frac{2}{3} \cdot \frac{3}{2} = \frac{\overset{1}{2} \cdot \overset{1}{3}}{\underset{1}{3} \cdot \underset{1}{2}} = 1$

Self Check 1
For each number, find its reciprocal and show that their product is 1.
a. $\frac{3}{5} \quad \frac{5}{3}$ **b.** $-\frac{5}{6} \quad -\frac{6}{5}$ **c.** $8 \quad \frac{1}{8}$

Now Try Problem 13

Teaching Example 1 For each number, find its reciprocal and show that their product is 1.
a. $\frac{11}{12}$ **b.** $-\frac{4}{5}$ **c.** 2
Answers:
a. $\frac{12}{11}$ **b.** $-\frac{5}{4}$ **c.** $\frac{1}{2}$

b. Fraction   Reciprocal

$$-\frac{3}{4} \quad \quad -\frac{4}{3}$$
$$\text{invert}$$

The reciprocal of $-\dfrac{3}{4}$ is $-\dfrac{4}{3}$.

Check: $-\dfrac{3}{4}\left(-\dfrac{4}{3}\right) = \dfrac{\overset{1}{\cancel{3}} \cdot \overset{1}{\cancel{4}}}{\underset{1}{\cancel{4}} \cdot \underset{1}{\cancel{3}}} = 1$ The product of two fractions with like signs is positive.

c. Since $5 = \dfrac{5}{1}$, the reciprocal of 5 is $\dfrac{1}{5}$.

Check: $5 \cdot \dfrac{1}{5} = \dfrac{5}{1} \cdot \dfrac{1}{5} = \dfrac{\overset{1}{\cancel{5}} \cdot 1}{1 \cdot \underset{1}{\cancel{5}}} = 1$

Caution! Don't confuse the concepts of the *opposite* of a negative number and the *reciprocal* of a negative number. For example:

The reciprocal of $-\dfrac{9}{16}$ is $-\dfrac{16}{9}$.

The opposite of $-\dfrac{9}{16}$ is $\dfrac{9}{16}$.

---

2   Divide fractions.

To develop a rule for dividing fractions, let's consider a real-life application.

Suppose that the manager of a candy store buys large bars of chocolate and divides each one into four equal parts to sell. How many fourths can be obtained from 5 bars?

We are asking, "How many $\frac{1}{4}$'s are there in 5?" To answer the question, we need to use the operation of division. We can represent this division as $5 \div \frac{1}{4}$.

5 bars of chocolate   Total number of fourths = 5 · 4 = 20

There are 20 fourths in the 5 bars of chocolate. Two observations can be made from this result.

- This division problem involves a fraction: $5 \div \frac{1}{4}$.
- Although we were asked to find $5 \div \frac{1}{4}$, we solved the problem using *multiplication* instead of *division*: $5 \cdot 4 = 20$. That is, division by $\frac{1}{4}$ (a fraction) is the same as multiplication by 4 (its reciprocal).

$$5 \div \dfrac{1}{4} = 5 \cdot 4$$

These observations suggest the following rule for dividing two fractions.

> **Dividing Fractions**
>
> To divide two fractions, multiply the first fraction by the reciprocal of the second fraction. Simplify the result, if possible.

For example, to find $\frac{5}{7} \div \frac{3}{4}$, we multiply $\frac{5}{7}$ by the reciprocal of $\frac{3}{4}$.

$$\frac{5}{7} \div \frac{3}{4} = \frac{5}{7} \cdot \frac{4}{3}$$

Change the division to multiplication. The reciprocal of $\frac{3}{4}$ is $\frac{4}{3}$.

$$= \frac{5 \cdot 4}{7 \cdot 3} \quad \text{Multiply the numerators.}$$
$$\phantom{=\frac{5 \cdot 4}{7 \cdot 3}} \quad \text{Multiply the denominators.}$$
$$= \frac{20}{21}$$

Thus, $\frac{5}{7} \div \frac{3}{4} = \frac{20}{21}$. We say that the *quotient* of $\frac{5}{7}$ and $\frac{3}{4}$ is $\frac{20}{21}$.

EXAMPLE 2

Divide: $\frac{1}{3} \div \frac{4}{5}$

Strategy We will multiply the first fraction, $\frac{1}{3}$, by the reciprocal of the second fraction, $\frac{4}{5}$. Then, if possible, we will simplify the result.

WHY This is the rule for dividing two fractions.

Solution

$$\frac{1}{3} \div \frac{4}{5} = \frac{1}{3} \cdot \frac{5}{4} \quad \text{Multiply } \frac{1}{3} \text{ by the reciprocal of } \frac{4}{5}, \text{ which is } \frac{5}{4}.$$
$$= \frac{1 \cdot 5}{3 \cdot 4} \quad \text{Multiply the numerators.}$$
$$\phantom{=\frac{1 \cdot 5}{3 \cdot 4}} \quad \text{Multiply the denominators.}$$
$$= \frac{5}{12}$$

Since 5 and 12 have no common factors other than 1, the result is in simplest form.

Self Check 2

Divide: $\frac{2}{3} \div \frac{7}{8}$ $\frac{16}{21}$

Now Try Problem 17

Teaching Example 2 Divide: $\frac{1}{6} \div \frac{2}{5}$
Answer:
$\frac{5}{12}$

EXAMPLE 3

Divide and simplify: $\frac{9}{16} \div \frac{3}{20}$

Strategy We will multiply the first fraction, $\frac{9}{16}$, by the reciprocal of the second fraction, $\frac{3}{20}$. Then, if possible, we will simplify the result.

WHY This is the rule for dividing two fractions.

Self Check 3

Divide and simplify: $\frac{4}{5} \div \frac{8}{25}$ $\frac{5}{2}$

Now Try Problem 21

Chapter 4 Fractions and Mixed Numbers

Teaching Example 3 Divide and simplify: $\dfrac{3}{7} \div \dfrac{9}{28}$

Answer: $\dfrac{4}{3}$

Solution

$$\dfrac{9}{16} \div \dfrac{3}{20} = \dfrac{9}{16} \cdot \dfrac{20}{3}$$ Multiply $\dfrac{9}{16}$ by the reciprocal of $\dfrac{3}{20}$, which is $\dfrac{20}{3}$.

$$= \dfrac{9 \cdot 20}{16 \cdot 3}$$ Multiply the numerators. Multiply the denominators.

$$= \dfrac{\overset{1}{\cancel{3}} \cdot 3 \cdot \overset{1}{\cancel{4}} \cdot 5}{\underset{1}{\cancel{4}} \cdot 4 \cdot \underset{1}{\cancel{3}}}$$ To simplify, factor 9 as $3 \cdot 3$, factor 20 as $4 \cdot 5$, and factor 16 as $4 \cdot 4$. Then remove the common factors of 3 and 4 from the numerator and denominator.

$$= \dfrac{15}{4}$$ Multiply the remaining factors in the numerator: $1 \cdot 3 \cdot 1 \cdot 5 = 15$
Multiply the remaining factors in the denominator: $1 \cdot 4 \cdot 1 = 4$.

Self Check 4
Divide and simplify:
$80 \div \dfrac{20}{11}$ 44

Now Try Problem 27

Teaching Example 4 Divide and simplify: $90 \div \dfrac{30}{17}$

Answer: 51

EXAMPLE 4
Divide and simplify: $120 \div \dfrac{10}{7}$

Strategy We will write 120 as a fraction and then multiply the first fraction by the reciprocal of the second fraction.

WHY This is the rule for dividing two fractions.

Solution

$$120 \div \dfrac{10}{7} = \dfrac{120}{1} \div \dfrac{10}{7}$$ Write 120 as a fraction: $120 = \dfrac{120}{1}$.

$$= \dfrac{120}{1} \cdot \dfrac{7}{10}$$ Multiply $\dfrac{120}{1}$ by the reciprocal of $\dfrac{10}{7}$, which is $\dfrac{7}{10}$.

$$= \dfrac{120 \cdot 7}{1 \cdot 10}$$ Multiply the numerators. Multiply the denominators.

$$= \dfrac{\overset{1}{\cancel{10}} \cdot 12 \cdot 7}{1 \cdot \underset{1}{\cancel{10}}}$$ To simplify, factor 120 as $10 \cdot 12$, then remove the common factor of 10 from the numerator and denominator.

$$= \dfrac{84}{1}$$ Multiply the remaining factors in the numerator: $1 \cdot 12 \cdot 7 = 84$.
Multiply the remaining factors in the denominator: $1 \cdot 1 = 1$.

$$= 84$$ Any whole number divided by 1 is the same number.

Because of the relationship between multiplication and division, the sign rules for *dividing* fractions are the same as those for *multiplying* fractions.

Self Check 5
Divide and simplify:
$\dfrac{2}{3} \div \left(-\dfrac{7}{6}\right)$ $-\dfrac{4}{7}$

Now Try Problem 29

EXAMPLE 5
Divide and simplify: $\dfrac{1}{6} \div \left(-\dfrac{1}{18}\right)$

Strategy We will multiply the first fraction, $\dfrac{1}{6}$, by the reciprocal of the second fraction, $-\dfrac{1}{18}$. To determine the sign of the result, we will use the rule for multiplying two fractions that have different (unlike) signs.

WHY One fraction is positive and one is negative.

Solution

$$\frac{1}{6} \div \left(-\frac{1}{18}\right) = \frac{1}{6}\left(-\frac{18}{1}\right)$$ Multiply $\frac{1}{6}$ by the reciprocal of $-\frac{1}{18}$, which is $-\frac{18}{1}$.

$$= -\frac{1 \cdot 18}{6 \cdot 1}$$ Multiply the numerators. Multiply the denominators. Since the fractions have unlike signs, make the answer negative.

$$= -\frac{1 \cdot 3 \cdot \overset{1}{\cancel{6}}}{\underset{1}{\cancel{6}} \cdot 1}$$ To simplify, factor 18 as $3 \cdot 6$. Then remove the common factor of 6 from the numerator and denominator.

$$= -\frac{3}{1}$$ Multiply the remaining factors in the numerator. Multiply the remaining factors in the denominator.

$$= -3$$

EXAMPLE 6
Divide and simplify: $-\frac{21}{36} \div (-3)$

Strategy We will multiply the first fraction, $-\frac{21}{36}$, by the reciprocal of -3. To determine the sign of the result, we will use the rule for multiplying two fractions that have the same (like) signs.

WHY Both fractions are negative.

Solution

$$-\frac{21}{36} \div (-3) = -\frac{21}{36}\left(-\frac{1}{3}\right)$$ Multiply $-\frac{21}{36}$ by the reciprocal of -3, which is $-\frac{1}{3}$.

$$= \frac{21}{36}\left(\frac{1}{3}\right)$$ Since the product of two negative fractions is positive, drop both $-$ signs and continue.

$$= \frac{21 \cdot 1}{36 \cdot 3}$$ Multiply the numerators. Multiply the denominators.

$$= \frac{\overset{1}{\cancel{3}} \cdot 7 \cdot 1}{36 \cdot \underset{1}{\cancel{3}}}$$ To simplify, factor 21 as $3 \cdot 7$. Then remove the common factor of 3 from the numerator and denominator.

$$= \frac{7}{36}$$ Multiply the remaining factors in the numerator: $1 \cdot 7 \cdot 1 = 7$. Multiply the remaining factors in the denominator: $36 \cdot 1 = 36$.

3 Divide algebraic fractions.

To work problems involving division of algebraic fractions, we must find the reciprocal of an algebraic fraction. We learned earlier that the reciprocal of a numerical fraction is found by inverting its numerator and denominator. The same is true for algebraic fractions. Here are some examples.

Algebraic fraction   **Reciprocal**   **Algebraic fraction**   **Reciprocal**
$\frac{2a}{3}$ ⟶ $\frac{3}{2a}$ (Invert) $-\frac{m^2}{n}$ ⟶ $\frac{n}{m^2}$ (Invert)

Since $x = \frac{x}{1}$, the reciprocal of x is $\frac{1}{x}$.

To divide algebraic functions, we use the same approach as with numerical fractions.

Teaching Example 5 Divide and simplify:
$\frac{2}{9} \div \left(-\frac{25}{18}\right)$
Answer: $-\frac{4}{25}$

Self Check 6
Divide and simplify:
$-\frac{35}{16} \div (-7)$ $\frac{5}{16}$

Now Try Problem 33

Teaching Example 6
Divide and simplify: $-\frac{64}{35} \div (-8)$
Answer: $\frac{8}{35}$

Self Check 7

Divide: $\dfrac{7}{4} \div \dfrac{3}{b}$ $\dfrac{7b}{12}$

Now Try Problem 37

Teaching Example 7
Divide: $\dfrac{9}{5} \div \dfrac{4}{x}$
Answer:
$\dfrac{9x}{20}$

EXAMPLE 7
Divide: $\dfrac{2}{3} \div \dfrac{a}{5}$

Strategy We will multiply the first fraction, $\dfrac{2}{3}$, by the reciprocal of the second fraction, $\dfrac{a}{5}$.

WHY This is the rule for dividing two fractions.

Solution

$$\dfrac{2}{3} \div \dfrac{a}{5} = \dfrac{2}{3} \cdot \dfrac{5}{a}$$ Multiply $\dfrac{2}{3}$ by the reciprocal of $\dfrac{a}{5}$, which is $\dfrac{5}{a}$.

$$= \dfrac{2 \cdot 5}{3 \cdot a}$$ Multiply the numerators. Multiply the denominators.

$$= \dfrac{10}{3a}$$ Since 10 and 3 have no common factors other than 1, the result is in simplest form.

Self Check 8

Divide and simplify: $\dfrac{9y^8}{10x} \div \dfrac{18y}{5x}$ $\dfrac{y^7}{4}$

Now Try Problem 41

Teaching Example 8
Divide and simplify: $\dfrac{6a^3}{11b} \div \dfrac{15a}{22b^2}$
Answer:
$\dfrac{4a^2 b}{5}$

EXAMPLE 8
Divide and simplify: $\dfrac{15x^2}{8y} \div \dfrac{10x}{y^3}$

Strategy We will multiply the first algebraic fraction, $\dfrac{15x^2}{8y}$, by the reciprocal of the second algebraic fraction, $\dfrac{10x}{y^3}$.

WHY This is the rule for dividing fractions.

Solution

$$\dfrac{15x^2}{8y} \div \dfrac{10x}{y^3} = \dfrac{15x^2}{8y} \cdot \dfrac{y^3}{10x}$$ Multiply $\dfrac{15x^2}{8y}$ by the reciprocal of $\dfrac{10x}{y^3}$, which is $\dfrac{y^3}{10x}$.

$$= \dfrac{15x^2 \cdot y^3}{8y \cdot 10x}$$ Multiply the numerators. Multiply the denominators.

$$= \dfrac{3 \cdot 5 \cdot x \cdot x \cdot y \cdot y \cdot y}{8 \cdot y \cdot 2 \cdot 5 \cdot x}$$ To prepare to simplify, factor 15, x^2, y^3, and 10.

$$= \dfrac{3 \cdot \overset{1}{\cancel{5}} \cdot \cancel{x} \cdot x \cdot \overset{1}{\cancel{y}} \cdot y \cdot y}{8 \cdot \underset{1}{\cancel{y}} \cdot 2 \cdot \underset{1}{\cancel{5}} \cdot \underset{1}{\cancel{x}}}$$ Remove the common factors of 5, x, and y from the numerator and denominator.

$$= \dfrac{3xy^2}{16}$$ Multiply the remaining factors in the numerator: $3 \cdot 1 \cdot 1 \cdot x \cdot 1 \cdot y \cdot y = 3xy^2$. Multiply the remaining factors in the denominator: $8 \cdot 1 \cdot 2 \cdot 1 \cdot 1 = 16$.

4 Solve application problems by dividing fractions.

Problems that involve forming equal-sized groups can be solved by division.

Finish: $\dfrac{3}{8}$ in. thick

Foam core

EXAMPLE 9
Surfboard Designs   Most surfboards are made of a foam core covered with several layers of fiberglass to keep them watertight. How many layers are needed to build up a finish $\dfrac{3}{8}$ of an inch thick if each layer of fiberglass has a thickness of $\dfrac{1}{16}$ of an inch?

Solution

- The surfboard is to have a $\dfrac{3}{8}$-inch-thick fiberglass finish.   Given
- Each layer of fiberglass is $\dfrac{1}{16}$ of an inch thick.   Given
- How many layers of fiberglass need to be applied?   Find

Unless otherwise noted, all art on this page is © Cengage Learning.

Think of the $\frac{3}{8}$-inch-thick finish separated into an unknown number of equally thick layers of fiberglass. This indicates division.

We translate the words of the problem to numbers and symbols.

The number of layers of fiberglass that are needed	is equal to	the thickness of the finish	divided by	the thickness of 1 layer of fiberglass.
The number of layers of fiberglass that are needed	=	$\frac{3}{8}$	÷	$\frac{1}{16}$

To find the quotient, we will use the rule for dividing two fractions.

$$\frac{3}{8} \div \frac{1}{16} = \frac{3}{8} \cdot \frac{16}{1} \quad \text{Multiply } \frac{3}{8} \text{ by the reciprocal of } \frac{1}{16}, \text{ which is } \frac{16}{1}.$$

$$= \frac{3 \cdot 16}{8 \cdot 1} \quad \begin{array}{l}\text{Multiply the numerators.} \\ \text{Multiply the denominators.}\end{array}$$

$$= \frac{3 \cdot 2 \cdot \overset{1}{8}}{\underset{1}{8} \cdot 1} \quad \begin{array}{l}\text{To simplify, factor 16 as } 2 \cdot 8. \text{ Then remove the common} \\ \text{factor of 8 from the numerator and denominator.}\end{array}$$

$$= \frac{6}{1} \quad \begin{array}{l}\text{Multiply the remaining factors in the numerator: } 3 \cdot 2 \cdot 1 = 6. \\ \text{Multiply the remaining factors in the denominator: } 1 \cdot 1 = 1.\end{array}$$

$$= 6 \quad \text{Any number divided by 1 is the same number.}$$

The number of layers of fiberglass needed is 6.

Check: If 6 layers of fiberglass, each $\frac{1}{16}$ of an inch thick, are used, the finished thickness will be $\frac{6}{16}$ of an inch. If we simplify $\frac{6}{16}$, we see that it is equivalent to the desired finish thickness:

$$\frac{6}{16} = \frac{\overset{1}{2} \cdot 3}{\underset{1}{2} \cdot 8} = \frac{3}{8}$$

The result checks.

> **Self Check 9**
>
> **COOKING** A recipe calls for 4 cups of sugar, and the only measuring container you have holds $\frac{1}{3}$ cup. How many $\frac{1}{3}$ cups of sugar would you need to add to follow the recipe? 12
>
> **Now Try** Problem 89
>
> **Teaching Example 9 GOLD COINS**
> How many $\frac{1}{16}$-ounce coins can be cast from a $\frac{7}{8}$-ounce bar of gold?
> Answer:
> 14 coins

ANSWERS TO SELF CHECKS

1. a. $\frac{5}{3}$ **b.** $-\frac{6}{5}$ **c.** $\frac{1}{8}$ **2.** $\frac{16}{21}$ **3.** $\frac{5}{2}$ **4.** 44 **5.** $-\frac{4}{7}$ **6.** $\frac{5}{16}$ **7.** $\frac{7b}{12}$ **8.** $\frac{y^7}{4}$ **9.** 12

SECTION 4.3 STUDY SET

VOCABULARY

Fill in the blanks.

1. The __reciprocal__ of $\frac{5}{12}$ is $\frac{12}{5}$.
2. To find the reciprocal of a fraction, __invert__ the numerator and denominator.
3. The answer to a division is called the __quotient__.
4. To simplify $\frac{2 \cdot 2 \cdot 3}{2 \cdot 3 \cdot 5 \cdot 7}$, we __remove__ common factors of the numerator and denominator.

CONCEPTS

5. Fill in the blanks.
   a. To divide two fractions, __multiply__ the first fraction by the __reciprocal__ of the second fraction.
   b. $\frac{1}{2} \div \frac{2}{3} = \frac{1}{2} \cdot \frac{3}{2}$

6. a. What division problem is illustrated below? $4 \div \frac{1}{3}$
   b. What is the answer? 12

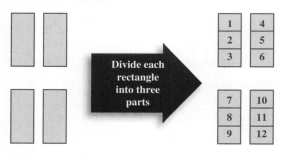

7. Determine whether each quotient is positive or negative. *You do not have to find the answer.*
   a. $-\frac{1}{4} \div \frac{3}{4}$ negative
   b. $-\frac{7}{8} \div \left(-\frac{21}{32}\right)$ positive

8. Complete the table.

Number	Opposite	Reciprocal
$\frac{3}{10}$	$-\frac{3}{10}$	$\frac{10}{3}$
$-\frac{7}{11}$	$\frac{7}{11}$	$-\frac{11}{7}$
6	-6	$\frac{1}{6}$

9. a. Multiply $\frac{4}{5}$ and its reciprocal. What is the result? 1
   b. Multiply $-\frac{3}{5}$ and its reciprocal. What is the result? 1

10. a. Find: $15 \div 3$ 5
    b. Rewrite $15 \div 3$ as multiplication by the reciprocal of 3, and find the result. $15 \cdot \frac{1}{3} = 5$
    c. Complete this statement: Division by 3 is the same as multiplication by $\frac{1}{3}$.

NOTATION

Fill in the blanks to complete each step.

11. $\frac{4}{9} \div \frac{8}{27} = \frac{4}{9} \cdot \frac{27}{8}$

    $= \frac{4 \cdot 27}{9 \cdot 8}$

    $= \frac{4 \cdot 3 \cdot 9}{9 \cdot 2 \cdot 4}$

    $= \frac{\cancel{4} \cdot 3 \cdot \cancel{9}}{\cancel{9} \cdot 2 \cdot \cancel{4}}$

    $= \frac{3}{2}$

12. $\frac{4a}{9b} \div \frac{8a}{27b} = \frac{4a}{9b} \cdot \frac{27b}{8a}$

    $= \frac{4a \cdot 27b}{9b \cdot 8a}$

    $= \frac{4 \cdot a \cdot 3 \cdot 9 \cdot b}{9 \cdot b \cdot 2 \cdot 4 \cdot a}$

    $= \frac{\cancel{4} \cdot \cancel{a} \cdot 3 \cdot \cancel{9} \cdot \cancel{b}}{\cancel{9} \cdot \cancel{b} \cdot 2 \cdot \cancel{4} \cdot \cancel{a}}$

    $= \frac{3}{2}$

GUIDED PRACTICE

Find the reciprocal of each number or algebraic fraction. See Example 1 and Objective 3.

13. a. $\frac{6}{7}$ $\frac{7}{6}$ b. $-\frac{15}{8}$ $-\frac{8}{15}$ c. 10 $\frac{1}{10}$

14. a. $\frac{2}{9}$ $\frac{9}{2}$ b. $-\frac{9}{4}$ $-\frac{4}{9}$ c. 7 $\frac{1}{7}$

15. a. $\frac{11a}{8}$ $\frac{8}{11a}$ b. $-\frac{1}{14b}$ $-14b$ c. $-63x$ $-\frac{1}{63x}$

16. a. $\frac{13}{2n}$ $\frac{2n}{13}$ b. $-\frac{b}{5}$ $-\frac{5}{b}$ c. $-21y$ $-\frac{1}{21y}$

Divide. Simplify each quotient, if possible. See Example 2.

17. $\frac{1}{8} \div \frac{2}{3}$ $\frac{3}{16}$

18. $\frac{1}{2} \div \frac{8}{9}$ $\frac{9}{16}$

19. $\frac{2}{23} \div \frac{1}{7}$ $\frac{14}{23}$

20. $\frac{4}{21} \div \frac{1}{5}$ $\frac{20}{21}$

Divide. Simplify each quotient, if possible. See Example 3.

21. $\dfrac{25}{32} \div \dfrac{5}{28}$ $\dfrac{35}{8}$
22. $\dfrac{4}{25} \div \dfrac{2}{35}$ $\dfrac{14}{5}$
23. $\dfrac{27}{32} \div \dfrac{9}{8}$ $\dfrac{3}{4}$
24. $\dfrac{16}{27} \div \dfrac{20}{21}$ $\dfrac{28}{45}$

Divide. Simplify each quotient, if possible. See Example 4.

25. $50 \div \dfrac{10}{9}$ 45
26. $60 \div \dfrac{10}{3}$ 18
27. $150 \div \dfrac{15}{32}$ 320
28. $170 \div \dfrac{17}{6}$ 60

Divide. Simplify each quotient, if possible. See Example 5.

29. $\dfrac{1}{8} \div \left(-\dfrac{1}{32}\right)$ -4
30. $\dfrac{1}{9} \div \left(-\dfrac{1}{27}\right)$ -3
31. $\dfrac{2}{5} \div \left(-\dfrac{4}{35}\right)$ $-\dfrac{7}{2}$
32. $\dfrac{4}{9} \div \left(-\dfrac{16}{27}\right)$ $-\dfrac{3}{4}$

Divide. Simplify each quotient, if possible. See Example 6.

33. $-\dfrac{28}{55} \div (-7)$ $\dfrac{4}{55}$
34. $-\dfrac{32}{45} \div (-8)$ $\dfrac{4}{45}$
35. $-\dfrac{33}{23} \div (-11)$ $\dfrac{3}{23}$
36. $-\dfrac{21}{31} \div (-7)$ $\dfrac{3}{31}$

Divide. See Example 7.

37. $\dfrac{4a}{5} \div \dfrac{3}{7}$ $\dfrac{28a}{15}$
38. $\dfrac{2x}{3} \div \dfrac{3}{2}$ $\dfrac{4x}{9}$
39. $\dfrac{6x}{7} \div \dfrac{5}{11}$ $\dfrac{66x}{35}$
40. $\dfrac{9b}{4} \div \dfrac{4}{3}$ $\dfrac{27b}{16}$

Divide. Simplify each quotient, if possible. See Example 8.

41. $\dfrac{6b^2}{7a} \div \dfrac{9b}{14a^3}$ $\dfrac{4a^2b}{3}$
42. $\dfrac{8c^2}{27d} \div \dfrac{4c^3}{9d^3}$ $\dfrac{2d^2}{3c}$
43. $\dfrac{20x}{33y} \div \dfrac{30x^4}{77y}$ $\dfrac{14}{9x^3}$
44. $\dfrac{16r^4}{27s^3} \div \dfrac{32r^4}{21s}$ $\dfrac{7}{18s^2}$
45. $-\dfrac{ab^2}{15} \div \dfrac{a^2b^2}{25}$ $-\dfrac{5}{3a}$
46. $-\dfrac{xy^3}{12} \div \dfrac{x^2y^2}{10}$ $-\dfrac{5y}{6x}$
47. $-\dfrac{48}{m^{11}} \div \left(-\dfrac{16}{m^8}\right)$ $\dfrac{3}{m^3}$
48. $-\dfrac{64}{q^{12}} \div \left(-\dfrac{24}{q^6}\right)$ $\dfrac{8}{3q^6}$

TRY IT YOURSELF

Divide. Simplify each quotient, if possible.

49. $120 \div \dfrac{12}{5x}$ $50x$
50. $360 \div \dfrac{36}{5y}$ $50y$
51. $-8x \div \left(-\dfrac{4x^3}{9}\right)$ $\dfrac{18}{x^2}$
52. $-12y \div \left(-\dfrac{3y^7}{10}\right)$ $\dfrac{40}{y^6}$
53. $\left(-\dfrac{7}{4}\right) \div \left(-\dfrac{21}{8}\right)$ $\dfrac{2}{3}$
54. $\left(-\dfrac{15}{16}\right) \div \left(-\dfrac{5}{8}\right)$ $\dfrac{3}{2}$
55. $\dfrac{4}{5} \div \dfrac{4}{5}$ 1
56. $\dfrac{2}{3} \div \dfrac{2}{3}$ 1
57. Divide $-\dfrac{15}{32}$ by $\dfrac{3}{4}$ $-\dfrac{5}{8}$
58. Divide $-\dfrac{7}{10}$ by $\dfrac{4}{5}$ $-\dfrac{7}{8}$
59. $3a \div \dfrac{1}{12a}$ $36a^2$
60. $9x \div \dfrac{3}{4x}$ $12x^2$
61. $-\dfrac{4}{5} \div (-6)$ $\dfrac{2}{15}$
62. $-\dfrac{7}{8} \div (-14)$ $\dfrac{1}{16}$
63. $\dfrac{15}{16} \div 180$ $\dfrac{1}{192}$
64. $\dfrac{7}{8} \div 210$ $\dfrac{1}{240}$
65. $-\dfrac{9n^5}{10} \div \dfrac{4n^4}{15}$ $-\dfrac{27n}{8}$
66. $-\dfrac{3m^3}{4} \div \dfrac{3m^2}{2}$ $-\dfrac{m}{2}$
67. $\dfrac{9a^3}{10b} \div \left(-\dfrac{3a^4}{25b^3}\right)$ $-\dfrac{15b^2}{2a}$
68. $\dfrac{11d^4}{16n} \div \left(-\dfrac{9d^5}{16n^3}\right)$ $-\dfrac{11n^2}{9d}$
69. $-\dfrac{x^2}{y^3} \div \dfrac{x}{y}$ $-\dfrac{x}{y^2}$
70. $-\dfrac{k}{t^8} \div \dfrac{k}{t^2}$ $-\dfrac{1}{t^6}$
71. $-\dfrac{1}{8} \div 8$ $-\dfrac{1}{64}$
72. $-\dfrac{1}{15} \div 15$ $-\dfrac{1}{225}$

The following problems involve multiplication and division. Perform each operation. Simplify the result, if possible.

73. $-\dfrac{7m^3}{6}\left(-\dfrac{9}{49m^2}\right)$ $\dfrac{3m}{14}$
74. $-\dfrac{7n^4}{10}\left(-\dfrac{20}{21n^3}\right)$ $\dfrac{2n}{3}$
75. $-\dfrac{1}{3} \div \dfrac{4}{5}$ $-\dfrac{5}{12}$
76. $-\dfrac{2}{3} \div \left(-\dfrac{3}{2}\right)$ $\dfrac{4}{9}$
77. $\dfrac{13}{16} \div x$ $\dfrac{13}{16x}$
78. $\dfrac{7}{8} \div x$ $\dfrac{7}{8x}$
79. $\left(-\dfrac{11}{21}\right)\left(-\dfrac{14}{33}\right)$ $\dfrac{2}{9}$
80. $\left(-\dfrac{16}{35}\right)\left(-\dfrac{25}{48}\right)$ $\dfrac{5}{21}$
81. $-\dfrac{15x}{32y} \div \dfrac{5x}{64y^4}$ $-6y^3$
82. $-\dfrac{28r}{15s} \div \dfrac{21r^4}{10s}$ $-\dfrac{8}{9r^3}$
83. $11 \cdot \dfrac{1}{6}$ $\dfrac{11}{6}$
84. $9 \cdot \dfrac{1}{8}$ $\dfrac{9}{8}$
85. $\dfrac{3x}{4} \cdot \dfrac{5}{7}$ $\dfrac{15x}{28}$
86. $\dfrac{2c}{3} \cdot \dfrac{7}{9}$ $\dfrac{14c}{27}$
87. $\dfrac{25}{7} \div \left(-\dfrac{30}{21}\right)$ $-\dfrac{5}{2}$
88. $\dfrac{39}{25} \div \left(-\dfrac{13}{10}\right)$ $-\dfrac{6}{5}$

APPLICATIONS

89. **PATIO FURNITURE** A production process applies several layers of a clear plastic coat to outdoor furniture to help protect it from the weather. If each protective coat is $\dfrac{3}{32}$-inch thick, how many applications will be needed to build up $\dfrac{3}{8}$ inch of clear finish?
    4 applications

90. **MARATHONS** Each lap around a stadium track is $\dfrac{1}{4}$ mile. How many laps would a runner have to complete to get a 26-mile workout? 104 laps

91. COOKING A recipe calls for $\frac{3}{4}$ cup of flour, and the only measuring container you have holds $\frac{1}{8}$ cup. How many $\frac{1}{8}$ cups of flour would you need to add to follow the recipe? 6 one-eighth cups

92. LASERS A technician uses a laser to slice thin pieces of aluminum off the end of a rod that is $\frac{7}{8}$-inch long. How many $\frac{1}{64}$-inch-wide slices can be cut from this rod? (Assume that there is no waste in the process.) 56 slices

93. UNDERGROUND CABLES Refer to the illustration and table below.

a. How many days will it take to install underground TV cable from the broadcasting station to the new homes using route 1? 30 days

b. How long is route 2? 15 mi

c. How many days will it take to install the cable using route 2? 25 days

d. Which route will require the fewer number of days to install the cable? route 2

Proposal	Amount of cable installed per day	Comments
Route 1	$\frac{2}{5}$ of a mile	Ground very rocky
Route 2	$\frac{3}{5}$ of a mile	Longer than Route 1

94. PRODUCTION PLANNING The materials used to make a pillow are shown. Examine the inventory list to decide how many pillows can be manufactured in one production run with the materials in stock. 147 pillows

Factory Inventory List

Materials	Amount in stock
Lace trim	135 yd
Corduroy fabric	154 yd
Cotton filling	98 lb

95. NOTE CARDS Ninety 3 × 5 cards are stacked next to a ruler, as shown below.

a. Into how many parts is 1 inch divided on the ruler? 16

b. How thick is the stack of cards? $\frac{3}{4}$ in.

c. How thick is one 3 × 5 card? $\frac{1}{120}$ in.

96. COMPUTER PRINTERS The illustration shows how the letter E is formed by a dot matrix printer. What is the height of one dot? $\frac{3}{224}$ in.

97. FORESTRY A set of forestry maps divides the 6,284 acres of an old-growth forest into $\frac{4}{5}$-acre sections. How many sections do the maps contain? 7,855 sections

98. HARDWARE A hardware chain purchases large amounts of nails and packages them in $\frac{9}{16}$-pound bags for sale. How many of these bags of nails can be obtained from 2,871 pounds of nails? 5,104 bags

WRITING

99. Explain how to divide two fractions.

100. Why do you need to know how to multiply fractions to be able to divide fractions?

101. Explain why 0 does not have a reciprocal.

102. What number is its own reciprocal? Explain why this is so.

103. Write an application problem that could be solved by finding $10 \div \frac{1}{5}$.

104. Explain why dividing a fraction by 2 is the same as finding $\frac{1}{2}$ of it. Give an example.

REVIEW

Fill in the blanks.

105. The symbol < means is   less   than .
106. The statement $9 \cdot 8 = 8 \cdot 9$ illustrates the   commutative   property of multiplication.
107.   Zero   is neither positive nor negative.
108. The sum of two negative numbers is   negative .
109. Find the LCM of 6 and 9. 18
110. Evaluate each expression.
   a. 3^5 243
   b. $(-2)^5$ –32

SECTION 4.4
Adding and Subtracting Fractions

Objectives
1. Add and subtract fractions that have the same denominator.
2. Add and subtract fractions that have different denominators.
3. Find the LCD to add and subtract fractions.
4. Add and subtract algebraic fractions.
5. Identify the greater of two fractions.
6. Solve application problems by adding and subtracting fractions.

In mathematics and everyday life, we can only add (or subtract) objects that are similar. For example, we can add dollars to dollars, but we cannot add dollars to oranges. This concept is important when adding or subtracting fractions.

1 Add and subtract fractions that have the same denominator.

Consider the problem $\frac{3}{5} + \frac{1}{5}$. When we write it in words, it is apparent that we are adding similar objects.

Because the denominators of $\frac{3}{5}$ and $\frac{1}{5}$ are the same, we say that they have a **common denominator**. Since the fractions have a common denominator, we can add them. The following figure explains the addition process.

We can make some observations about the addition shown in the figure.

The sum of the numerators is the numerator of the answer.

$$\frac{3}{5} + \frac{1}{5} = \frac{4}{5}$$

The answer is a fraction that has the same denominator as the two fractions that were added.

These observations illustrate the following rule.

Adding and Subtracting Fractions That Have the Same Denominator

To add (or subtract) fractions that have the same denominator, add (or subtract) their numerators and write the sum (or difference) over the common denominator. Simplify the result, if possible.

Chapter 4 Fractions and Mixed Numbers

> **Caution!** We **do not** add fractions by adding the numerators and adding the denominators!
>
> $$\frac{3}{5} + \frac{1}{5} = \frac{3+1}{5+5} = \frac{4}{10}$$ ~~(crossed out)~~
>
> The same caution applies when subtracting fractions.

Self Check 1

Perform each operation and simplify the result, if possible.

a. Add: $\dfrac{5}{12} + \dfrac{1}{12}$

b. Subtract: $\dfrac{8}{9} - \dfrac{1}{9}$

Now Try Problems 11 and 15

Teaching Example 1 Perform each operation and simplify the result, if possible.

a. Add: $\dfrac{3}{10} + \dfrac{1}{10}$

b. Subtract: $\dfrac{6}{7} - \dfrac{2}{7}$

Answers:

a. $\dfrac{2}{5}$ b. $\dfrac{4}{7}$

EXAMPLE 1

Perform each operation and simplify the result, if possible.

a. Add: $\dfrac{1}{8} + \dfrac{5}{8}$ b. Subtract: $\dfrac{11}{15} - \dfrac{4}{15}$

Strategy We will use the rule for adding and subtracting fractions that have *the same* denominator.

WHY In part a, the fractions have the same denominator, 8. In part b, the fractions have the same denominator, 15.

Solution

a. $\dfrac{1}{8} + \dfrac{5}{8} = \dfrac{1+5}{8}$ Add the numerators and write the sum over the common denominator 8.

$= \dfrac{6}{8}$ This fraction can be simplified.

$= \dfrac{\overset{1}{\cancel{2}} \cdot 3}{\underset{1}{\cancel{2}} \cdot 4}$ To simplify, factor 6 as 2 · 3 and 8 as 2 · 4. Then remove the common factor of 2 from the numerator and denominator.

$= \dfrac{3}{4}$ Multiply the remaining factors in the numerator: 1 · 3 = 3. Multiply the remaining factors in the denominator: 1 · 4 = 4.

b. $\dfrac{11}{15} - \dfrac{4}{15} = \dfrac{11-4}{15}$ Subtract the numerators and write the difference over the common denominator 15.

$= \dfrac{7}{15}$

Since 7 and 15 have no common factors other than 1, the result is in simplest form.

The rule for subtraction from Section 2.3 can be extended to subtraction involving signed fractions:

> *To subtract two fractions, add the first to the opposite of the fraction to be subtracted.*

Self Check 2

Subtract: $-\dfrac{9}{11} - \left(-\dfrac{3}{11}\right)$

Now Try Problem 19

EXAMPLE 2

Subtract: $-\dfrac{7}{3} - \left(-\dfrac{2}{3}\right)$

Strategy To find the difference, we will apply the rule for subtraction.

WHY It is easy to make an error when subtracting signed fractions. We will probably be more accurate if we write the subtraction as addition of the opposite.

Solution

We read $-\frac{7}{3} - \left(-\frac{2}{3}\right)$ as "negative seven-thirds *minus* negative two-thirds." Thus, the number to be subtracted is $-\frac{2}{3}$. Subtracting $-\frac{2}{3}$ is the same as adding its opposite, $\frac{2}{3}$.

$$-\frac{7}{3} - \left(-\frac{2}{3}\right) = -\frac{7}{3} + \frac{2}{3} \quad \text{Add the opposite of } -\frac{2}{3}, \text{ which is } \frac{2}{3}.$$

$$= \frac{-7}{3} + \frac{2}{3} \quad \text{Write } -\frac{7}{3} \text{ as } \frac{-7}{3}.$$

$$= \frac{-7 + 2}{3} \quad \text{Add the numerators and write the sum over the common denominator 3.}$$

$$= \frac{-5}{3} \quad \text{Use the rule for adding two integers with different signs: } -7 + 2 = -5.$$

$$= -\frac{5}{3} \quad \text{Rewrite the result with the } - \text{ sign in front: } \frac{-5}{3} = -\frac{5}{3}.$$

This fraction is in simplest form.

Teaching Example 2
Subtract: $-\frac{9}{5} - \left(-\frac{2}{5}\right)$
Answer: $-\frac{7}{5}$

EXAMPLE 3
Perform the operations and simplify: $\frac{18}{25} - \frac{2}{25} - \frac{1}{25}$

Strategy We will use the rule for subtracting fractions that have *the same* denominator.

WHY All three fractions have the same denominator, 25.

Solution

$$\frac{18}{25} - \frac{2}{25} - \frac{1}{25} = \frac{18 - 2 - 1}{25} \quad \text{Subtract the numerators and write the difference over the common denominator 25.}$$

$$= \frac{15}{25} \quad \text{This fraction can be simplified.}$$

$$= \frac{3 \cdot \overset{1}{\cancel{5}}}{\underset{1}{\cancel{5}} \cdot 5} \quad \text{To simplify, factor 15 as } 3 \cdot 5 \text{ and 25 as } 5 \cdot 5. \text{ Then remove the common factor of 5 from the numerator and denominator.}$$

$$= \frac{3}{5} \quad \text{Multiply the remaining factors in the numerator: } 3 \cdot 1 = 3.$$
$$\text{Multiply the remaining factors in the denominator: } 1 \cdot 5 = 5.$$

Self Check 3
Perform the operations and simplify:

$$\frac{2}{9} + \frac{2}{9} + \frac{2}{9} - \frac{2}{3}$$

Now Try Problem 23

Teaching Example 3 Perform the operations and simplify:
$\frac{23}{30} - \frac{11}{30} - \frac{7}{30}$
Answer: $\frac{1}{6}$

2 Add and subtract fractions that have different denominators.

Now we consider the problem $\frac{3}{5} + \frac{1}{3}$. Since the denominators are different, we cannot add these fractions in their present form.

three-**fifths** + one-**third**
Not similar objects

To add (or subtract) fractions with different denominators, we express them as equivalent fractions that have a common denominator. The smallest common denominator, called the **least** or **lowest common denominator**, is usually the easiest common denominator to use.

Least Common Denominator

The **least common denominator (LCD)** for a set of fractions is the smallest number each denominator will divide exactly (divide with no remainder).

The denominators of $\frac{3}{5}$ and $\frac{1}{3}$ are 5 and 3. The numbers 5 and 3 divide many numbers exactly (30, 45, and 60, to name a few), but the smallest number that they divide exactly is 15. Thus, 15 is the LCD for $\frac{3}{5}$ and $\frac{1}{3}$.

To find $\frac{3}{5} + \frac{1}{3}$, we *build* equivalent fractions that have denominators of 15. (This procedure was introduced in Section 4.1.) Then we use the rule for adding fractions that have the same denominator.

$$\frac{3}{5} + \frac{1}{3} = \frac{3}{5} \cdot \frac{3}{3} + \frac{1}{3} \cdot \frac{5}{5}$$

We need to multiply this denominator by 5 to obtain 15. It follows that $\frac{5}{5}$ should be the form of 1 used to build $\frac{1}{3}$.

We need to multiply this denominator by 3 to obtain 15. It follows that $\frac{3}{3}$ should be the form of 1 that is used to build $\frac{3}{5}$.

$$= \frac{9}{15} + \frac{5}{15}$$ Multiply the numerators. Multiply the denominators. Note that the denominators are now the same.

$$= \frac{9 + 5}{15}$$ Add the numerators and write the sum over the common denominator 15.

$$= \frac{14}{15}$$ Since 14 and 15 have no common factors other than 1, this fraction is in simplest form.

The figure below shows $\frac{3}{5}$ and $\frac{1}{3}$ expressed as equivalent fractions with a denominator of 15. Once the denominators are the same, the fractions are similar objects and can be added easily.

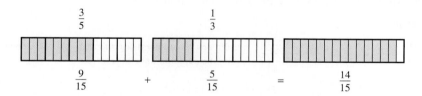

We can use the following steps to add or subtract fractions with different denominators.

Adding and Subtracting Fractions That Have Different Denominators

1. Find the LCD.
2. Rewrite each fraction as an equivalent fraction with the LCD as the denominator. To do so, build each fraction using a form of 1 that involves any factors needed to obtain the LCD.
3. Add or subtract the numerators and write the sum or difference over the LCD.
4. Simplify the result, if possible.

Self Check 4

Add: $\frac{1}{2} + \frac{2}{5} \quad \frac{9}{10}$

Now Try Problem 29

EXAMPLE 4 Add: $\frac{1}{7} + \frac{2}{3}$

Strategy We will express each fraction as an equivalent fraction that has the LCD as its denominator. Then we will use the rule for adding fractions that have the same denominator.

WHY To add (or subtract) fractions, the fractions must have *like* denominators.

Solution
Since the smallest number the denominators 7 and 3 divide exactly is 21, the LCD is 21.

$$\frac{1}{7} + \frac{2}{3} = \frac{1}{7} \cdot \frac{3}{3} + \frac{2}{3} \cdot \frac{7}{7}$$ To build $\frac{1}{7}$ and $\frac{2}{3}$ so that their denominators are 21, multiply each by a form of 1.

$$= \frac{3}{21} + \frac{14}{21}$$ Multiply the numerators. Multiply the denominators. The denominators are now the same.

$$= \frac{3 + 14}{21}$$ Add the numerators and write the sum over the common denominator 21.

$$= \frac{17}{21}$$ Since 17 and 21 have no common factors other than 1, this fraction is in simplest form.

EXAMPLE 5
Subtract: $\frac{5}{2} - \frac{7}{3}$

Strategy We will express each fraction as an equivalent fraction that has the LCD as its denominator. Then we will use the rule for subtracting fractions that have the same denominator.

WHY To add (or subtract) fractions, the fractions must have *like* denominators.

Solution
Since the smallest number the denominators 2 and 3 divide exactly is 6, the LCD is 6.

$$\frac{5}{2} - \frac{7}{3} = \frac{5}{2} \cdot \frac{3}{3} - \frac{7}{3} \cdot \frac{2}{2}$$ To build $\frac{5}{2}$ and $\frac{7}{3}$ so that their denominators are 6, multiply each by a form of 1.

$$= \frac{15}{6} - \frac{14}{6}$$ Multiply the numerators. Multiply the denominators. The denominators are now the same.

$$= \frac{15 - 14}{6}$$ Subtract the numerators and write the difference over the common denominator 6.

$$= \frac{1}{6}$$ This fraction is in simplest form.

EXAMPLE 6
Subtract: $\frac{2}{5} - \frac{11}{15}$

Strategy Since the smallest number the denominators 5 and 15 divide exactly is 15, the LCD is 15. We will only need to build an equivalent fraction for $\frac{2}{5}$.

WHY We do not have to build the fraction $\frac{11}{15}$ because it already has a denominator of 15.

Solution

$$\frac{2}{5} - \frac{11}{15} = \frac{2}{5} \cdot \frac{3}{3} - \frac{11}{15}$$ To build $\frac{2}{5}$ so that its denominator is 15, multiply it by a form of 1.

$$= \frac{6}{15} - \frac{11}{15}$$ Multiply the numerators. Multiply the denominators. The denominators are now the same.

$$= \frac{6 - 11}{15}$$ Subtract the numerators and write the difference over the common denominator 15.

$$= -\frac{5}{15}$$ If it is helpful, use the subtraction rule and add the opposite in the numerator: $6 + (-11) = -5$. Write the − sign in front of the fraction.

$$= -\frac{\overset{1}{\cancel{5}}}{3 \cdot \underset{1}{\cancel{5}}}$$ To simplify, factor 15 as $3 \cdot 5$. Then remove the common factor of 5 from the numerator and denominator.

$$= -\frac{1}{3}$$ Multiply the remaining factors in the denominator: $3 \cdot 1 = 3$.

Teaching Example 4
Add: $\frac{1}{2} + \frac{3}{7}$
Answer:
$\frac{13}{14}$

Self Check 5
Subtract: $\frac{6}{7} - \frac{3}{5}$ $\frac{9}{35}$
Now Try Problem 31

Teaching Example 5
Subtract: $\frac{4}{5} - \frac{2}{3}$
Answer:
$\frac{2}{15}$

Self Check 6
Subtract: $\frac{2}{3} - \frac{13}{6}$ $-\frac{3}{2}$
Now Try Problem 35

Teaching Example 6
Subtract: $\frac{1}{3} - \frac{17}{6}$
Answer:
$-\frac{5}{2}$

Success Tip In Example 6, did you notice that the denominator 5 is a factor of the denominator 15, and that the LCD is 15? In general, when adding (or subtracting) two fractions with different denominators, *if the smaller denominator is a factor of the larger denominator, the larger denominator is the LCD.*

Caution! You might not have to build each fraction when adding or subtracting fractions with different denominators. For instance, the step in blue shown below is unnecessary when solving Example 6.

$$\frac{2}{5} - \frac{11}{15} = \frac{2}{5} \cdot \frac{3}{3} - \frac{11}{15} \cdot \cancel{\frac{1}{1}}$$

Self Check 7

Add: $-6 + \frac{3}{8} - \frac{45}{8}$

Now Try Problem 39

Teaching Example 7

Add: $-3 + \frac{11}{12}$

Answer:

$-\frac{25}{12}$

EXAMPLE 7 Add: $-5 + \frac{3}{4}$

Strategy We will write -5 as the fraction $\frac{-5}{1}$. Then we will follow the steps for adding fractions that have different denominators.

WHY The fractions $\frac{-5}{1}$ and $\frac{3}{4}$ have different denominators.

Solution

Since the smallest number the denominators 1 and 4 divide exactly is 4, the LCD is 4.

$$-5 + \frac{3}{4} = \frac{-5}{1} + \frac{3}{4} \quad \text{Write } -5 \text{ as } \frac{-5}{1}.$$

$$= \frac{-5}{1} \cdot \frac{4}{4} + \frac{3}{4} \quad \text{To build } \frac{-5}{1} \text{ so that its denominator is 4, multiply it by a form of 1.}$$

$$= \frac{-20}{4} + \frac{3}{4} \quad \text{Multiply the numerators. Multiply the denominators. The denominators are now the same.}$$

$$= \frac{-20 + 3}{4} \quad \text{Add the numerators and write the sum over the common denominator 4.}$$

$$= \frac{-17}{4} \quad \text{Use the rule for adding two integers with different signs: } -20 + 3 = -17.$$

$$= -\frac{17}{4} \quad \text{Write the result with the } - \text{sign in front: } \frac{-17}{4} = -\frac{17}{4}. \text{ This fraction is in simplest form.}$$

3 Find the LCD to add and subtract fractions.

When we add or subtract fractions that have different denominators, the least common denominator is not always obvious. We can use a concept studied earlier to determine the LCD for more difficult problems that involve larger denominators. To illustrate this, let's find the least common denominator of $\frac{3}{8}$ and $\frac{1}{10}$. (Note, the LCD is *not* 80.)

We have learned that both 8 and 10 must divide the LCD exactly. This divisibility requirement should sound familiar. Recall the following fact from Section 1.6.

The Least Common Multiple (LCM)

The **least common multiple (LCM)** of two whole numbers is the smallest whole number that is divisible by both of those numbers.

Thus, the least common denominator of $\frac{3}{8}$ and $\frac{1}{10}$ is simply the *least common multiple* of 8 and 10.

We can find the LCM of 8 and 10 by listing multiples of the larger number, 10, until we find one that is divisible by the smaller number, 8. (This method is explained in Example 2 of Section 1.6.)

Multiples of 10:   10,   20,   30,   **40**,   50,   60,   …

This is the first multiple of 10 that is divisible by 8 (no remainder).

Since the LCM of 8 and 10 is 40, it follows that the LCD of $\frac{3}{8}$ and $\frac{1}{10}$ is 40.

We can also find the LCM of 8 and 10 using prime factorization. We begin by prime factoring 8 and 10. (This method is explained in Example 4 of Section 1.6.)

$$8 = \boxed{2 \cdot 2 \cdot 2}$$
$$10 = 2 \cdot \boxed{5}$$

The LCM of 8 and 10 is a product of prime factors, where each factor is used the greatest number of times it appears in any one factorization.

- We will use the factor 2 three times, because 2 appears three times in the factorization of 8. Circle $2 \cdot 2 \cdot 2$, as shown above.
- We will use the factor 5 once, because it appears one time in the factorization of 10. Circle 5 as shown above.

Since there are no other prime factors in either prime factorization, we have

Use 2 three times.
Use 5 one time.

$$\text{LCM}(8, 10) = 2 \cdot 2 \cdot 2 \cdot 5 = 40$$

Finding the LCD

The least common denominator (LCD) of a set of fractions is the least common multiple (LCM) of the denominators of the fractions. Two ways to find the LCM of the denominators are as follows:

- Write the multiples of the largest denominator in increasing order, until one is found that is divisible by the other denominators.
- Prime factor each denominator. The LCM is a product of prime factors, where each factor is used the greatest number of times it appears in any one factorization.

EXAMPLE 8   Add: $\dfrac{7}{15} + \dfrac{3}{10}$

Strategy We begin by expressing each fraction as an equivalent fraction that has the LCD for its denominator. Then we use the rule for adding fractions that have the same denominator.

WHY To add (or subtract) fractions, the fractions must have *like* denominators.

Solution

To find the LCD, we find the prime factorization of both denominators and use each prime factor the *greatest* number of times it appears in any one factorization:

$$\left.\begin{array}{l} 15 = \boxed{3} \cdot \boxed{5} \\ 10 = \boxed{2} \cdot 5 \end{array}\right\} \text{LCD} = 2 \cdot 3 \cdot 5 = 30$$

2 appears once in the factorization of 10.
3 appears once in the factorization of 15.
5 appears once in the factorizations of 15 and 10.

Self Check 8

Add: $\dfrac{1}{8} + \dfrac{5}{6}$ $\dfrac{23}{24}$

Now Try Problem 43

Teaching Example 8

Add: $\dfrac{3}{8} + \dfrac{5}{12}$

Answer:
$\dfrac{19}{24}$

The LCD for $\frac{7}{15}$ and $\frac{3}{10}$ is 30.

$$\frac{7}{15} + \frac{3}{10} = \frac{7}{15} \cdot \frac{2}{2} + \frac{3}{10} \cdot \frac{3}{3}$$ To build $\frac{7}{15}$ and $\frac{3}{10}$ so that their denominators are 30, multiply each by a form of 1.

$$= \frac{14}{30} + \frac{9}{30}$$ Multiply the numerators. Multiply the denominators. The denominators are now the same.

$$= \frac{14 + 9}{30}$$ Add the numerators and write the sum over the common denominator 30.

$$= \frac{23}{30}$$ Since 23 and 30 have no common factors other than 1, this fraction is in simplest form.

Self Check 9

Subtract and simplify:

$$\frac{21}{56} - \frac{9}{40} \cdot \frac{3}{20}$$

Now Try Problem 47

Teaching Example 9

Subtract and simplify: $\frac{13}{24} - \frac{3}{40}$

Answer: $\frac{7}{15}$

EXAMPLE 9

Subtract and simplify: $\frac{13}{28} - \frac{1}{21}$

Strategy We begin by expressing each fraction as an equivalent fraction that has the LCD for its denominator. Then we use the rule for subtracting fractions with *like* denominators.

WHY To add (or subtract) fractions, the fractions must have like denominators.

Solution

To find the LCD, we find the prime factorization of both denominators and use each prime factor the *greatest* number of times it appears in any one factorization:

$$\left.\begin{array}{l} 28 = \boxed{2 \cdot 2} \cdot \boxed{7} \\ 21 = \boxed{3} \cdot 7 \end{array}\right\} \text{LCD} = 2 \cdot 2 \cdot 3 \cdot 7 = 84$$

2 appears twice in the factorization of 28.
3 appears once in the factorization of 21.
7 appears once in the factorizations of 28 and 21.

The LCD for $\frac{13}{28}$ and $\frac{1}{21}$ is 84.

We will compare the prime factorizations of 28, 21, and the prime factorization of the LCD, 84, to determine what forms of 1 to use to build equivalent fractions for $\frac{13}{28}$ and $\frac{1}{21}$ with a denominator of 84.

LCD = 2 · 2 · **3** · 7 | LCD = **2 · 2** · 3 · 7

Cover the prime factorization of 28. Since 3 is left uncovered, use $\frac{3}{3}$ to build $\frac{13}{28}$.

Cover the prime factorization of 21. Since $2 \cdot 2 = 4$ is left uncovered, use $\frac{4}{4}$ to build $\frac{1}{21}$.

$$\frac{13}{28} - \frac{1}{21} = \frac{13}{28} \cdot \frac{3}{3} - \frac{1}{21} \cdot \frac{4}{4}$$ To build $\frac{13}{28}$ and $\frac{1}{21}$ so that their denominators are 84, multiply each by a form of 1.

$$= \frac{39}{84} - \frac{4}{84}$$ Multiply the numerators. Multiply the denominators. The denominators are now the same.

$$= \frac{39 - 4}{84}$$ Subtract the numerators and write the difference over the common denominator.

$$= \frac{35}{84}$$ This fraction is not in simplest form.

$$= \frac{\overset{1}{5 \cdot 7}}{2 \cdot 2 \cdot 3 \cdot \underset{1}{7}}$$ To simplify, factor 35 and 84. Then remove the common factor of 7 from the numerator and denominator.

$$= \frac{5}{12}$$ Multiply the remaining factors in the numerator: $5 \cdot 1 = 5$. Multiply the remaining factors in the denominator: $2 \cdot 2 \cdot 3 \cdot 1 = 12$.

4 Add and subtract algebraic fractions.

To add or subtract algebraic fractions, we use the same approach as with numerical fractions.

EXAMPLE 10 Perform each operation and simplify the result, if possible:

a. $\dfrac{x}{16} + \dfrac{3x}{16}$ b. $\dfrac{9}{25m} - \dfrac{2}{25m}$

Strategy We will add (or subtract) the numerators and write the sum (or difference) over the common denominator. Then, if possible, we will simplify the result.

WHY This is the rule for adding (or subtracting) fractions that have like denominators.

Solution

a. $\dfrac{x}{16} + \dfrac{3x}{16} = \dfrac{x + 3x}{16}$ Add the numerators and write the sum over the common denominator, 16.

$= \dfrac{4x}{16}$ Combine like terms in the numerator: $x + 3x = 4x$. This result can be simplified.

$= \dfrac{\overset{1}{\cancel{4}} \cdot x}{\underset{1}{\cancel{4}} \cdot 4}$ To simplify, factor 16 as $4 \cdot 4$ and then remove the common factor of 4 from the numerator and denominator.

$= \dfrac{x}{4}$ Multiply the remaining factors in the numerator: $1 \cdot x = x$. Multiply the remaining factors in the denominator: $1 \cdot 4 = 4$.

b. $\dfrac{9}{25m} - \dfrac{2}{25m} = \dfrac{9 - 2}{25m}$ Subtract the numerators and write the difference over the common denominator, 25m.

$= \dfrac{7}{25m}$ Since 7 and 25 have no common factors other than 1, the result is in simplified form.

Self Check 10
Perform each operation and simplify the result, if possible:

a. $\dfrac{2x}{15} + \dfrac{4x}{15}$ $\dfrac{2x}{5}$

b. $\dfrac{11}{3n} - \dfrac{7}{3n}$ $\dfrac{4}{3n}$

Now Try Problems 51 and 55

Teaching Example 10
Perform each operation and simplify the result, if possible:

a. $\dfrac{4a}{27} + \dfrac{5a}{27}$

b. $\dfrac{14}{9d} - \dfrac{7}{9d}$

Answers:

a. $\dfrac{a}{3}$ b. $\dfrac{7}{9d}$

EXAMPLE 11 Perform each operation: a. $\dfrac{r}{3} + \dfrac{1}{4}$ b. $\dfrac{2}{y} - \dfrac{5}{18}$

Strategy We will use the procedure for adding and subtracting fractions that have unlike denominators. The first step is to determine the LCD.

WHY If we are to add (or subtract) fractions, their denominators must be the same. Since the denominators of these fractions are different, we cannot add (nor subtract) them in their present form.

Solution

a. The denominators of $\dfrac{r}{3}$ and $\dfrac{1}{4}$ are 3 and 4. By inspection, we see that the LCD is $3 \cdot 4 = 12$.

$\dfrac{r}{3} + \dfrac{1}{4} = \dfrac{r}{3} \cdot \dfrac{4}{4} + \dfrac{1}{4} \cdot \dfrac{3}{3}$ Build each fraction so that each has a denominator of 12.

$= \dfrac{4r}{12} + \dfrac{3}{12}$ Multiply the numerators. Multiply the denominators.

$= \dfrac{4r + 3}{12}$ Add the numerators and write the sum over the common denominator.

Self Check 11
Perform each operation:

a. $\dfrac{y}{5} + \dfrac{7}{9}$ $\dfrac{9y + 35}{45}$

b. $\dfrac{6}{a} - \dfrac{1}{8}$ $\dfrac{48 - a}{8a}$

Now Try Problems 59 and 63

Teaching Example 11
Perform each operation:

a. $\dfrac{x}{4} + \dfrac{1}{5}$

b. $\dfrac{6}{n} - \dfrac{7}{3}$

Answers:

a. $\dfrac{5x + 4}{20}$ b. $\dfrac{18 - 7n}{3n}$

Caution! The result, $\frac{4r+3}{12}$, does not simplify. Do not make either of the two common mistakes shown below.

$$\frac{4r+3}{12} \;\;\cancel{=}\;\; \frac{7r}{12} \quad \text{4r and 3 are not like terms. Don't add them.} \qquad \frac{4r+3}{12} \;\;\cancel{=}\;\; \frac{4r+\overset{1}{3}}{\underset{1}{4\cdot 3}} \quad \text{3 is a term of the numerator, not a factor, and therefore cannot be removed.}$$

b. The denominators of $\frac{2}{y}$ and $\frac{5}{18}$ are y and 18. By inspection, we see that the LCD is $y \cdot 18 = 18y$.

$$\frac{2}{y} - \frac{5}{18} = \frac{2}{y} \cdot \frac{18}{18} - \frac{5}{18} \cdot \frac{y}{y} \qquad \text{Build each fraction so that each has a denominator of 18y.}$$

$$= \frac{36}{18y} - \frac{5y}{18y} \qquad \text{Multiply the numerators.}$$
$$\text{Multiply the denominators.}$$

$$= \frac{36 - 5y}{18y} \qquad \text{Subtract the numerators and write the difference over the common denominator.}$$

Caution! The result, $\frac{36-5y}{18y}$, is in simplest form. Do not make the following common mistake to try to "simplify" it.

$$\frac{36-5y}{18y} \;\;\cancel{=}\;\; \frac{36-5\overset{1}{\cancel{y}}}{18\underset{1}{\cancel{y}}} = \frac{31}{18} \quad \text{We cannot remove a common factor of } y, \text{ because } y \text{ is not a factor of the entire numerator. It is a factor of one term of the numerator.}$$

5 Identify the greater of two fractions.

If two fractions have the same denominator, the fraction with the greater numerator is the greater fraction.

For example,

$$\frac{7}{8} > \frac{3}{8} \quad \text{because } 7 > 3 \qquad\qquad -\frac{1}{3} > -\frac{2}{3} \quad \text{because } -1 > -2$$

If the denominators of two fractions are different, we need to write the fractions with a common denominator (preferably the LCD) before we can make a comparison.

Self Check 12

Which fraction is larger:
$\frac{7}{12}$ or $\frac{3}{5}$? $\frac{3}{5}$

Now Try Problem 67

Teaching Example 12

Which fraction is larger: $\frac{3}{7}$ or $\frac{1}{2}$?

Answer:
$\frac{1}{2}$

EXAMPLE 12 Which fraction is larger: $\frac{5}{6}$ or $\frac{7}{8}$?

Strategy We will express each fraction as an equivalent fraction that has the LCD for its denominator. Then we will compare their numerators.

WHY We cannot compare the fractions as given. They are not similar objects.

five-**sixths**   seven-**eighths**

Solution

Since the smallest number the denominators will divide exactly is 24, the LCD for $\frac{5}{6}$ and $\frac{7}{8}$ is 24.

$$\frac{5}{6} = \frac{5}{6} \cdot \frac{4}{4} \qquad\qquad \frac{7}{8} = \frac{7}{8} \cdot \frac{3}{3} \qquad \text{To build } \tfrac{5}{6} \text{ and } \tfrac{7}{8} \text{ so that their denominators are 24, multiply each by a form of 1.}$$

$$= \frac{20}{24} \qquad\qquad\qquad = \frac{21}{24} \qquad \text{Multiply the numerators.}$$
$$\text{Multiply the denominators.}$$

Next, we compare the numerators. Since $21 > 20$, it follows that $\frac{21}{24}$ is greater than $\frac{20}{24}$. Thus, $\frac{7}{8} > \frac{5}{6}$.

6 Solve application problems by adding and subtracting fractions.

EXAMPLE 13 *Television Viewing Habits* Students on a college campus were asked to estimate to the nearest hour how much television they watched each day. The results are given in the **circle graph** (also called a **pie chart**) to the right. For example, the chart tells us that $\frac{1}{4}$ of those responding watched 1 hour per day. What fraction of the student body watches from 0 to 2 hours daily?

Solution

- $\frac{1}{6}$ of the student body watches no TV daily.   Given
- $\frac{1}{4}$ of the student body watches 1 hour of TV daily.   Given
- $\frac{7}{15}$ of the student body watches 2 hours of TV daily.   Given
- What fraction of the student body watches 0 to 2 hours of TV daily?   Find

We translate the words of the problem to numbers and symbols.

The fraction of the student body that watches from 0 to 2 hours of TV daily	is equal to	the fraction that watches no TV daily	plus	the fraction that watches 1 hour of TV daily	plus	the fraction that watches 2 hours of TV daily.

The fraction of the student body that watches from 0 to 2 hours of TV daily $= \frac{1}{6} + \frac{1}{4} + \frac{7}{15}$

We must find the sum of three fractions with different denominators. To find the LCD, we prime factor the denominators and use each prime factor the *greatest* number of times it appears in any one factorization:

$$\left.\begin{array}{l} 6 = 2 \cdot \circled{3} \\ 4 = \circled{2 \cdot 2} \\ 15 = 3 \cdot \circled{5} \end{array}\right\} \text{LCD} = 2 \cdot 2 \cdot 3 \cdot 5 = 60$$

2 appears twice in the factorization of 4.
3 appears once in the factorization of 6 and 15.
5 appears once in the factorization of 15.

The LCD for $\frac{1}{6}$, $\frac{1}{4}$, and $\frac{7}{15}$ is 60.

$\frac{1}{6} + \frac{1}{4} + \frac{7}{15} = \frac{1}{6} \cdot \frac{10}{10} + \frac{1}{4} \cdot \frac{15}{15} + \frac{7}{15} \cdot \frac{4}{4}$ Build each fraction so that its denominator is 60.

$= \frac{10}{60} + \frac{15}{60} + \frac{28}{60}$ Multiply the numerators. Multiply the denominators. The denominators are now the same.

$= \frac{10 + 15 + 28}{60}$ Add the numerators and write the sum over the common denominator 60.

$= \frac{53}{60}$ This fraction is in simplest form.

$\begin{array}{r} 10 \\ 15 \\ + 28 \\ \hline 53 \end{array}$

The fraction of the student body that watches 0 to 2 hours of TV daily is $\frac{53}{60}$.

Self Check 13
Refer to the circle graph below. Find the fraction of the student body that watches 2 or more hours of television daily. $\frac{7}{12}$

Now Try Problems 75 and 118

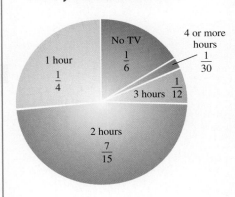

Teaching Example 13 Refer to the circle graph above. Find the fraction of the student body that watches from 1 to 3 hours of television daily.
Answer:
$\frac{4}{5}$

Unless otherwise noted, all art on this page is © Cengage Learning.

Check: We can check by estimation. The result, $\frac{53}{60}$, is approximately $\frac{50}{60}$, which simplifies to $\frac{5}{6}$. The red, yellow, and blue shaded areas appear to shade about $\frac{5}{6}$ of the pie chart. The result seems reasonable.

ANSWERS TO SELF CHECKS

1. a. $\frac{1}{2}$ b. $\frac{7}{9}$ 2. $-\frac{6}{11}$ 3. $\frac{2}{3}$ 4. $\frac{9}{10}$ 5. $\frac{9}{35}$ 6. $-\frac{3}{2}$ 7. $-\frac{45}{8}$ 8. $\frac{23}{24}$ 9. $\frac{3}{20}$
10. a. $\frac{2x}{5}$ b. $\frac{4}{3n}$ 11. a. $\frac{9y+35}{45}$ b. $\frac{48-a}{8a}$ 12. $\frac{3}{5}$ 13. $\frac{7}{12}$

THINK IT THROUGH Budgets

"Putting together a budget is crucial if you don't want to spend your way into serious problems. You're also developing a habit that can serve you well throughout your life."

Liz Pulliam Weston, MSN Money

The circle graph below shows a suggested budget for new college graduates as recommended by Springboard, a nonprofit consumer credit counseling service. What fraction of net take-home pay should be spent on housing? $\frac{7}{20}$

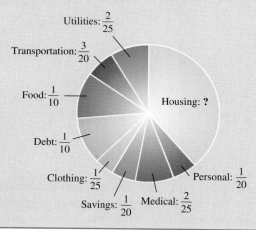

Utilities: $\frac{2}{25}$
Transportation: $\frac{3}{20}$
Food: $\frac{1}{10}$
Debt: $\frac{1}{10}$
Clothing: $\frac{1}{25}$
Savings: $\frac{1}{20}$
Medical: $\frac{2}{25}$
Personal: $\frac{1}{20}$
Housing: ?

SECTION 4.4 STUDY SET

VOCABULARY

Fill in the blanks.

1. Because the denominators of $\frac{3}{8}$ and $\frac{7}{8}$ are the same number, we say that they have a __common__ denominator.

2. Consider the solution below. To __build__ an equivalent fraction with a denominator of 18, we multiply $\frac{4}{9}$ by a 1 in the form of $\frac{2}{2}$.

$$\frac{4}{9} = \frac{4}{9} \cdot \frac{2}{2}$$
$$= \frac{8}{18}$$

CONCEPTS

Fill in the blanks.

3. **a.** To add (or subtract) fractions that have the same denominator, add (or subtract) their __numerators__ and write the sum (or difference) over the __common__ denominator. __Simplify__ the result, if possible.

   b. To add (or subtract) fractions that have different denominators, we express each fraction as an equivalent fraction that has the __LCD__ for its denominator. Then we use the rule for adding (subtracting) fractions that have the __same__ denominator.

4. The least common denominator for a set of fractions is the __smallest__ number each denominator will divide exactly (no remainder).

5. Consider $\frac{3}{4}$. By what form of 1 should we multiply the numerator and denominator to express it as an equivalent fraction with a denominator of 36? $\frac{9}{9}$

6. The *denominators* of two fractions are given. Find the least common denominator.

   a. 2 and 3 6   **b.** 3 and 5 15

   c. Fill in the blank: When adding (or subtracting) two fractions with different denominators, if the smaller denominator is a factor of the larger denominator, the __larger__ denominator is the LCD.

7. Consider the following prime factorizations:

   $$24 = 2 \cdot 2 \cdot 2 \cdot 3$$
   $$90 = 2 \cdot 3 \cdot 3 \cdot 5$$

   For any one factorization, what is the greatest number of times

   a. a 5 appears? once
   b. a 3 appears? twice
   c. a 2 appears? three times

8. **a.** The *denominators* of two fractions have their prime-factored forms shown below. Fill in the blanks to find the LCD for the fractions.

   $$\left.\begin{array}{l}20 = 2 \cdot 2 \cdot 5\\30 = 2 \cdot 3 \cdot 5\end{array}\right\} \text{LCD} = 2 \cdot 2 \cdot 3 \cdot 5 = 60$$

   b. The *denominators* of three fractions have their prime-factored forms shown below. Fill in the blanks to find the LCD for the fractions.

   $$\left.\begin{array}{l}20 = 2 \cdot 2 \cdot 5\\30 = 2 \cdot 3 \cdot 5\\90 = 2 \cdot 3 \cdot 3 \cdot 5\end{array}\right\} \text{LCD} = 2 \cdot 2 \cdot 3 \cdot 3 \cdot 5 = 180$$

NOTATION

Fill in the blanks to complete each step.

9. $\frac{2}{5} + \frac{1}{7} = \frac{2}{5} \cdot \frac{7}{7} + \frac{1}{7} \cdot \frac{5}{5}$

   $= \frac{14}{35} + \frac{5}{35}$

   $= \frac{14 + 5}{35}$

   $= \frac{19}{35}$

10. $\frac{7x}{8} - \frac{2}{9} = \frac{7x}{8} \cdot \frac{9}{9} - \frac{2}{9} \cdot \frac{8}{8}$

    $= \frac{63x}{72} - \frac{16}{72}$

    $= \frac{63x - 16}{72}$

GUIDED PRACTICE

Perform each operation and simplify, if possible. See Example 1.

11. $\frac{4}{9} + \frac{1}{9}$ $\frac{5}{9}$

12. $\frac{3}{7} + \frac{1}{7}$ $\frac{4}{7}$

13. $\frac{3}{8} + \frac{1}{8}$ $\frac{1}{2}$

14. $\frac{7}{12} + \frac{1}{12}$ $\frac{2}{3}$

15. $\frac{11}{15} - \frac{7}{15}$ $\frac{4}{15}$

16. $\frac{10}{21} - \frac{5}{21}$ $\frac{5}{21}$

17. $\frac{11}{20} - \frac{3}{20}$ $\frac{2}{5}$

18. $\frac{7}{18} - \frac{5}{18}$ $\frac{1}{9}$

Subtract and simplify, if possible. See Example 2.

19. $-\frac{11}{5} - \left(-\frac{8}{5}\right)$ $-\frac{3}{5}$

20. $-\frac{15}{9} - \left(-\frac{11}{9}\right)$ $-\frac{4}{9}$

21. $-\frac{7}{21} - \left(-\frac{2}{21}\right)$ $-\frac{5}{21}$

22. $-\frac{21}{25} - \left(-\frac{9}{25}\right)$ $-\frac{12}{25}$

Perform the operations and simplify, if possible. See Example 3.

23. $\frac{19}{40} - \frac{3}{40} - \frac{1}{40}$ $\frac{3}{8}$

24. $\frac{11}{24} - \frac{1}{24} - \frac{7}{24}$ $\frac{1}{8}$

25. $\frac{13}{33} + \frac{1}{33} + \frac{7}{33}$ $\frac{7}{11}$

26. $\frac{21}{50} + \frac{1}{50} + \frac{13}{50}$ $\frac{7}{10}$

Add and simplify, if possible. See Example 4.

27. $\frac{1}{3} + \frac{1}{7}$ $\frac{10}{21}$

28. $\frac{1}{4} + \frac{1}{5}$ $\frac{9}{20}$

29. $\frac{2}{5} + \frac{1}{9}$ $\frac{23}{45}$

30. $\frac{2}{7} + \frac{1}{2}$ $\frac{11}{14}$

Chapter 4 Fractions and Mixed Numbers

Subtract and simplify, if possible. See Example 5.

31. $\dfrac{4}{5} - \dfrac{3}{4}$ $\dfrac{1}{20}$
32. $\dfrac{2}{3} - \dfrac{3}{5}$ $\dfrac{1}{15}$
33. $\dfrac{3}{4} - \dfrac{2}{7}$ $\dfrac{13}{28}$
34. $\dfrac{6}{7} - \dfrac{2}{3}$ $\dfrac{4}{21}$

Subtract and simplify, if possible. See Example 6.

35. $\dfrac{11}{12} - \dfrac{2}{3}$ $\dfrac{1}{4}$
36. $\dfrac{11}{18} - \dfrac{1}{6}$ $\dfrac{4}{9}$
37. $\dfrac{9}{14} - \dfrac{1}{7}$ $\dfrac{1}{2}$
38. $\dfrac{13}{15} - \dfrac{2}{3}$ $\dfrac{1}{5}$

Add and simplify, if possible. See Example 7.

39. $-2 + \dfrac{5}{9}$ $-\dfrac{13}{9}$
40. $-3 + \dfrac{5}{8}$ $-\dfrac{19}{8}$
41. $-3 + \dfrac{9}{4}$ $-\dfrac{3}{4}$
42. $-1 + \dfrac{7}{10}$ $-\dfrac{3}{10}$

Add and simplify, if possible. See Example 8.

43. $\dfrac{1}{6} + \dfrac{5}{8}$ $\dfrac{19}{24}$
44. $\dfrac{7}{12} + \dfrac{3}{8}$ $\dfrac{23}{24}$
45. $\dfrac{4}{9} + \dfrac{5}{12}$ $\dfrac{31}{36}$
46. $\dfrac{1}{9} + \dfrac{5}{6}$ $\dfrac{17}{18}$

Subtract and simplify, if possible. See Example 9.

47. $\dfrac{9}{10} - \dfrac{3}{14}$ $\dfrac{24}{35}$
48. $\dfrac{11}{12} - \dfrac{11}{30}$ $\dfrac{11}{20}$
49. $\dfrac{11}{12} - \dfrac{7}{15}$ $\dfrac{9}{20}$
50. $\dfrac{7}{15} - \dfrac{5}{12}$ $\dfrac{1}{20}$

Perform each operation and simplify, if possible. See Example 10.

51. $\dfrac{x}{12} + \dfrac{5x}{12}$ $\dfrac{x}{2}$
52. $\dfrac{d}{16} + \dfrac{7d}{16}$ $\dfrac{d}{2}$
53. $\dfrac{2c}{7} + \dfrac{3c}{7}$ $\dfrac{5c}{7}$
54. $\dfrac{4n}{19} + \dfrac{8n}{19}$ $\dfrac{12n}{19}$
55. $\dfrac{16}{21m} - \dfrac{11}{21m}$ $\dfrac{5}{21m}$
56. $\dfrac{9}{17a} - \dfrac{5}{17a}$ $\dfrac{4}{17a}$
57. $\dfrac{16}{15y} - \dfrac{7}{15y}$ $\dfrac{3}{5y}$
58. $\dfrac{11}{12r} - \dfrac{1}{12r}$ $\dfrac{5}{6r}$

Perform each operation and simplify, if possible. See Example 11.

59. $\dfrac{a}{5} + \dfrac{2}{3}$ $\dfrac{3a+10}{15}$
60. $\dfrac{d}{4} + \dfrac{2}{7}$ $\dfrac{7d+8}{28}$
61. $\dfrac{1}{8} + \dfrac{x}{3}$ $\dfrac{3+8x}{24}$
62. $\dfrac{1}{9} + \dfrac{m}{4}$ $\dfrac{4+9m}{36}$
63. $\dfrac{3}{n} - \dfrac{5}{12}$ $\dfrac{36-5n}{12n}$
64. $\dfrac{6}{f} - \dfrac{4}{5}$ $\dfrac{30-4f}{5f}$
65. $\dfrac{8}{9} - \dfrac{11}{d}$ $\dfrac{8d-99}{9d}$
66. $\dfrac{12}{13} - \dfrac{6}{r}$ $\dfrac{12r-78}{13r}$

Determine which fraction is larger. See Example 12.

67. $\dfrac{3}{8}$ or $\dfrac{5}{16}$ $\dfrac{3}{8}$
68. $\dfrac{5}{6}$ or $\dfrac{7}{12}$ $\dfrac{5}{6}$
69. $\dfrac{4}{5}$ or $\dfrac{2}{3}$ $\dfrac{4}{5}$
70. $\dfrac{7}{9}$ or $\dfrac{4}{5}$ $\dfrac{4}{5}$
71. $\dfrac{7}{9}$ or $\dfrac{11}{12}$ $\dfrac{11}{12}$
72. $\dfrac{3}{8}$ or $\dfrac{5}{12}$ $\dfrac{5}{12}$
73. $\dfrac{23}{20}$ or $\dfrac{7}{6}$ $\dfrac{7}{6}$
74. $\dfrac{19}{15}$ or $\dfrac{5}{4}$ $\dfrac{19}{15}$

Add and simplify, if possible. See Example 13.

75. $\dfrac{1}{6} + \dfrac{5}{18} + \dfrac{2}{9}$ $\dfrac{2}{3}$
76. $\dfrac{1}{10} + \dfrac{1}{8} + \dfrac{1}{5}$ $\dfrac{17}{40}$
77. $\dfrac{4}{15} + \dfrac{2}{3} + \dfrac{1}{6}$ $\dfrac{11}{10}$
78. $\dfrac{1}{2} + \dfrac{3}{5} + \dfrac{3}{20}$ $\dfrac{5}{4}$

TRY IT YOURSELF

Perform each operation and simplify, if possible.

79. $-\dfrac{1}{12} - \left(-\dfrac{5}{12}\right)$ $\dfrac{1}{3}$
80. $-\dfrac{1}{16} - \left(-\dfrac{15}{16}\right)$ $\dfrac{7}{8}$
81. $\dfrac{3n}{4} + \dfrac{2}{3}$ $\dfrac{9n+8}{12}$
82. $\dfrac{4b}{5} + \dfrac{2}{3}$ $\dfrac{12b+10}{15}$
83. $\dfrac{12}{25} - \dfrac{1}{25} - \dfrac{1}{25}$ $\dfrac{2}{5}$
84. $\dfrac{7}{9} + \dfrac{1}{9} + \dfrac{1}{9}$ 1
85. $-\dfrac{7}{20} - \dfrac{1}{5}$ $-\dfrac{11}{20}$
86. $-\dfrac{5}{8} - \dfrac{1}{3}$ $-\dfrac{23}{24}$
87. $-\dfrac{7}{16} + \dfrac{1}{4}$ $-\dfrac{3}{16}$
88. $-\dfrac{17}{20} + \dfrac{4}{5}$ $-\dfrac{1}{20}$
89. $\dfrac{x}{9} + \dfrac{2x}{9}$ $\dfrac{x}{3}$
90. $\dfrac{a}{8} + \dfrac{5a}{8}$ $\dfrac{3a}{4}$
91. $\dfrac{2}{3} + \dfrac{4}{5} + \dfrac{5}{6}$ $\dfrac{23}{10}$
92. $\dfrac{3}{4} + \dfrac{2}{5} + \dfrac{3}{10}$ $\dfrac{29}{20}$
93. $\dfrac{9}{20} - \dfrac{1}{30}$ $\dfrac{5}{12}$
94. $\dfrac{5}{6} - \dfrac{3}{10}$ $\dfrac{8}{15}$
95. $\dfrac{27}{50} + \dfrac{5}{16}$ $\dfrac{341}{400}$
96. $\dfrac{49}{50} - \dfrac{15}{16}$ $\dfrac{17}{400}$

97. $\dfrac{13}{20} - \dfrac{1}{5}$ $\dfrac{9}{20}$

98. $\dfrac{71}{100} - \dfrac{1}{10}$ $\dfrac{61}{100}$

99. $-\dfrac{3}{4} - 5$ $-\dfrac{23}{4}$

100. $-2 - \dfrac{7}{8}$ $-\dfrac{23}{8}$

101. $\dfrac{7}{30} - \dfrac{19}{75}$ $-\dfrac{1}{50}$

102. $\dfrac{73}{75} - \dfrac{31}{30}$ $-\dfrac{3}{50}$

103. Find the difference of $\dfrac{13}{d}$ and $\dfrac{3}{2}$. $\dfrac{26 - 3d}{2d}$

104. Find the sum of $\dfrac{25}{x}$ and $\dfrac{4}{3}$. $\dfrac{75 + 4x}{3x}$

105. Subtract $\dfrac{5}{12}$ from $\dfrac{2}{15}$. $-\dfrac{17}{60}$

106. What is the sum of $\dfrac{11}{24}$ and $\dfrac{7}{36}$ increased by $\dfrac{5}{48}$? $\dfrac{109}{144}$

APPLICATIONS

107. **BOTANY** To determine the effects of smog on tree development, a scientist cut down a pine tree and measured the width of the growth rings for two five-year time periods.

    a. What was the total growth over the ten-year period? $\dfrac{31}{32}$ in.

    b. What is the difference in growth for the two five-year time periods? $\dfrac{11}{32}$ in.

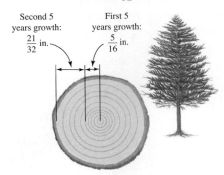

108. **GARAGE DOOR OPENERS** What is the difference in strength between a $\dfrac{1}{3}$-hp and a $\dfrac{1}{2}$-hp garage door opener? $\dfrac{1}{6}$ hp

109. **FUNNELS** Find an algebraic expression that represents the difference between the diameter of the opening at the top of the funnel and the diameter of its spout. $\dfrac{16 - x}{2x}$ cm

110. **MAGAZINE COVERS** The page design for the magazine cover shown below includes a blank strip at the top, called a *header*, and a blank strip at the bottom of the page, called a *footer*. How much page length is lost because of the header and footer? $\dfrac{11}{16}$ in.

111. **DELIVERY TRUCKS** A truck can safely carry a one-ton load. Should it be used to deliver one-half ton of sand, one-third ton of gravel, and one-fifth ton of cement in one trip to a job site? no

112. **DINNERS** A family bought two large pizzas for dinner. Some pieces of each pizza were not eaten, as shown.

    a. What fraction of the first pizza was not eaten? $\dfrac{3}{8}$

    b. What fraction of the second pizza was not eaten? $\dfrac{2}{6} = \dfrac{1}{3}$

    c. What fraction of a pizza was left? $\dfrac{17}{24}$ of a pizza was left

    d. Could the family have been fed with just one pizza? no

113. **HARDWARE** Find the inside diameter of the circular washer shown below. $\dfrac{13}{32}$ in.

114. GASOLINE BARRELS Three identical-sized barrels are shown below. If the contents of two of the barrels are poured into the empty third barrel, what fraction of the third barrel will be filled? $\frac{17}{20}$

115. WEIGHTS AND MEASURES A consumer protection agency determines the accuracy of butcher shop scales by placing a known three-quarter-pound weight on the scale and then comparing that to the scale's readout. According to the illustration, by how much is this scale off? Does it result in undercharging or overcharging customers on their meat purchases? $\frac{1}{16}$ lb, undercharge

116. FIGURE DRAWING As an aid in drawing the human body, artists divide the body into three parts. Each part is then expressed as a fraction of the total body height. For example, the torso is $\frac{4}{15}$ of the body height. What fraction of body height is the head? $\frac{2}{15}$

117. Suppose you work as a school guidance counselor at a community college and your department has conducted a survey of the full-time students to learn more about their study habits. As part of a Power Point presentation of the survey results to the school board, you show the following circle graph. At that time, you are asked, "What fraction of the full-time students study 2 hours or more daily?" What would you answer? $\frac{7}{10}$ of the full-time students study 2 or more hours a day.

from Campus to Careers
School Guidance Counselor

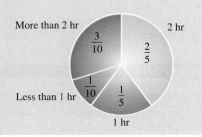

118. HEALTH STATISTICS The circle graph below shows the leading causes of death in the United States for 2010. For example, $\frac{6}{25}$ of all of the deaths that year were caused by heart disease. What fraction of all the deaths were caused by heart disease, cancer, respiratory diseases, or stroke, combined? $\frac{29}{50}$

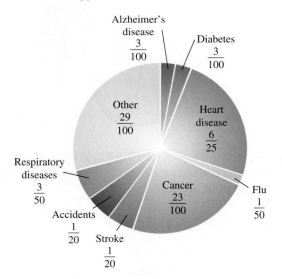

Source: National Center for Health Statistics

119. **MUSICAL NOTES** The notes used in music have fractional values. Their names and the symbols used to represent them are shown in illustration (a). In common time, the values of the notes in each measure must add to 1. Is the measure in illustration (b) complete? no

120. **TOOLS** A mechanic likes to hang his wrenches above his tool bench in order of narrowest to widest. What is the proper order of the wrenches in the illustration? $\frac{5}{32}, \frac{3}{16}, \frac{1}{4}, \frac{3}{8}$

121. **TIRE TREAD** A mechanic measured the tire tread depth on each of the tires on a car and recorded them on the form shown below. (The letters LF stand for *left front*, RR stands for *right rear*, and so on.)
   a. Which tire has the most tread? RR: right rear
   b. Which tire has the least tread? LR: left rear

122. **HIKING** The illustration below shows the length of each part of a three-part hike. Rank the lengths of the parts from longest to shortest. $\frac{4}{5}, \frac{3}{4}, \frac{5}{8}$

WRITING

123. Explain why we cannot add or subtract the fractions $\frac{2}{9}$ and $\frac{2}{5}$ as they are written.

124. To multiply fractions, must they have the same denominators? Explain why or why not. Give an example.

125. Explain the error in the following work.

$$\frac{5x + 2}{6} = \frac{5x + 2}{2 \cdot 3}$$

$$= \frac{5x + \overset{1}{\cancel{2}}}{\underset{1}{\cancel{2}} \cdot 3}$$

$$= \frac{5x + 1}{3}$$

126. How do we compare the sizes of two fractions with different denominators?

REVIEW

Perform each operation and simplify, if possible.

127. a. $\frac{1}{4} + \frac{1}{8} \quad \frac{3}{8}$ b. $\frac{1}{4} - \frac{1}{8} \quad \frac{1}{8}$

     c. $\frac{1}{4} \cdot \frac{1}{8} \quad \frac{1}{32}$ d. $\frac{1}{4} \div \frac{1}{8} \quad 2$

128. a. $\frac{5}{21} + \frac{3}{14} \quad \frac{19}{42}$ b. $\frac{5}{21} - \frac{3}{14} \quad \frac{1}{42}$

     c. $\frac{5}{21} \cdot \frac{3}{14} \quad \frac{5}{98}$ d. $\frac{5}{21} \div \frac{3}{14} \quad \frac{10}{9}$

Chapter 4 Fractions and Mixed Numbers

Objectives

1. Identify the whole-number and fractional parts of a mixed number.
2. Write mixed numbers as improper fractions.
3. Write improper fractions as mixed numbers.
4. Graph fractions and mixed numbers on a number line.
5. Multiply and divide mixed numbers.
6. Solve application problems by multiplying and dividing mixed numbers.

SECTION 4.5
Multiplying and Dividing Mixed Numbers

In the next two sections, we show how to add, subtract, multiply, and divide *mixed numbers*. These numbers are widely used in daily life.

The recipe calls for $2\frac{1}{3}$ cups of flour.
(Read as "two and one-third.")

It took $3\frac{3}{4}$ hours to paint the living room.
(Read as "three and three-fourths.")

The entrance to the park is $1\frac{1}{2}$ miles away.
(Read as "one and one-half.")

1 Identify the whole-number and fractional parts of a mixed number.

A **mixed number** is the *sum* of a whole number and a proper fraction. For example, $3\frac{3}{4}$ is a mixed number.

$$3\frac{3}{4} \quad = \quad 3 \quad + \quad \frac{3}{4}$$

Mixed number ——— Whole-number part ↑ ——— Fractional part ↑

Mixed numbers can be represented by shaded regions. In the illustration below, each rectangular region outlined in black represents one whole. To represent $3\frac{3}{4}$, we shade 3 *whole* rectangular regions and 3 out of 4 *parts* of another.

Caution! Note that $3\frac{3}{4}$ means $3 + \frac{3}{4}$, even though the + symbol is not written. Do not confuse $3\frac{3}{4}$ with $3 \cdot \frac{3}{4}$ or $3\left(\frac{3}{4}\right)$, which indicate the multiplication of 3 by $\frac{3}{4}$.

Self Check 1

In the illustration below, each oval region represents one whole. Write an improper fraction and a mixed number to represent the shaded portion. $\frac{9}{2}$, $4\frac{1}{2}$

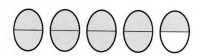

Now Try Problem 19

EXAMPLE 1
In the illustration below, each disk represents one whole. Write an improper fraction and a mixed number to represent the shaded portion.

Strategy We will determine the number of equal parts into which a disk is divided. Then we will determine how many of those *parts* are shaded and how many of the *whole* disks are shaded.

Unless otherwise noted, all art on this page is © Cengage Learning.

WHY To write an improper fraction, we need to find its numerator and its denominator. To write a mixed number, we need to find its whole number part and its fractional part.

Solution
Since each disk is divided into 5 equal parts, the denominator of the improper fraction is 5. Since a total of 11 of those parts are shaded, the numerator is 11, and we say that

$\frac{11}{5}$ is shaded. Write: $\frac{\text{total number of parts shaded}}{\text{number of equal parts in one disk}}$

Since 2 whole disks are shaded, the whole number part of the mixed number is 2. Since 1 out of 5 of the parts of the last disk is shaded, the fractional part of the mixed number is $\frac{1}{5}$, and we say that

$2\frac{1}{5}$ is shaded.

2 wholes $\quad\quad \frac{1}{5}$

In this section, we will work with negative as well as positive mixed numbers. For example, the negative mixed number $-3\frac{3}{4}$ could be used to represent $3\frac{3}{4}$ feet below sea level. Think of $-3\frac{3}{4}$ as $-3 - \frac{3}{4}$ or as $-3 + \left(-\frac{3}{4}\right)$.

2 Write mixed numbers as improper fractions.

In Example 1, we saw that the shaded portion of the illustration can be represented by the mixed number $2\frac{1}{5}$ and by the improper fraction $\frac{11}{5}$. To develop a procedure to write any mixed number as an improper fraction, consider the following steps that show how to do this for $2\frac{1}{5}$. The objective is to find how many *fifths* that the mixed number $2\frac{1}{5}$ represents.

$2\frac{1}{5} = 2 + \frac{1}{5}$ Write the mixed number $2\frac{1}{5}$ as a sum.

$\quad = \frac{2}{1} + \frac{1}{5}$ Write 2 as a fraction: $2 = \frac{2}{1}$.

$\quad = \frac{2}{1} \cdot \frac{5}{5} + \frac{1}{5}$ To build $\frac{2}{1}$ so that its denominator is 5, multiply it by a form of 1.

$\quad = \frac{10}{5} + \frac{1}{5}$ Multiply the numerators.
Multiply the denominators.

$\quad = \frac{11}{5}$ Add the numerators and write the sum over the common denominator 5.

Thus, $2\frac{1}{5} = \frac{11}{5}$.

Unless otherwise noted, all art on this page is © Cengage Learning.

Teaching Example 1 In the illustration below, each disk represents one whole. Write an improper fraction and a mixed number to represent the shaded portion.

Answer:
$\frac{23}{4}, 5\frac{3}{4}$

We can obtain the same result with far less work. To change $2\frac{1}{5}$ to an improper fraction, we simply multiply 5 by 2 and add 1 to get the numerator, and keep the denominator of 5.

$$2\frac{1}{5} = \frac{5 \cdot 2 + 1}{5} = \frac{10 + 1}{5} = \frac{11}{5}$$

This example illustrates the following procedure.

> **Writing a Mixed Number as an Improper Fraction**
>
> To write a mixed number as an improper fraction:
>
> 1. Multiply the denominator of the fraction by the whole-number part.
> 2. Add the numerator of the fraction to the result from Step 1.
> 3. Write the sum from Step 2 over the original denominator.

Self Check 2

Write the mixed number $3\frac{3}{8}$ as an improper fraction. $\frac{27}{8}$

Now Try Problems 23 and 27

Teaching Example 2
Write the mixed number $4\frac{3}{9}$ as an improper fraction.
Answer:
$\frac{39}{9}$

EXAMPLE 2 Write the mixed number $7\frac{5}{6}$ as an improper fraction.

Strategy We will use the three-step procedure to find the improper fraction.

WHY It's faster than writing $7\frac{5}{6}$ as $7 + \frac{5}{6}$, building to get an LCD, and adding.

Solution
To find the numerator of the improper fraction, multiply 6 by 7, and add 5 to that result. The denominator of the improper fraction is the same as the denominator of the fractional part of the mixed number.

Step 2: Add.

$$7\frac{5}{6} = \frac{6 \cdot 7 + 5}{6} = \frac{42 + 5}{6} = \frac{47}{6}$$

By the order of operations rule, multiply first and then add.

Step 1: Multiply. Step 3: Use the same denominator.

To write a *negative mixed number* in fractional form, ignore the − sign and use the method shown in Example 2 for a positive mixed number. Once that procedure is completed, write a − sign in front of the result. For example,

$$-6\frac{1}{4} = -\frac{25}{4} \qquad -1\frac{9}{10} = -\frac{19}{10} \qquad -12\frac{3}{8} = -\frac{99}{8}$$

3 Write improper fractions as mixed numbers.

To write an improper fraction as a mixed number, we must find two things: the *whole-number part* and the *fractional part* of the mixed number. To develop a procedure to do this, let's consider the improper fraction $\frac{7}{3}$. To find the number of groups of 3 in 7, we can divide 7 by 3. This will find the whole-number part of the mixed number. The remainder is the numerator of the fractional part of the mixed number.

This example suggests the following procedure.

> **Writing an Improper Fraction as a Mixed Number**
>
> To write an improper fraction as a mixed number:
> 1. Divide the numerator by the denominator to obtain the whole-number part.
> 2. The remainder over the divisor is the fractional part.

EXAMPLE 3 Write each improper fraction as a mixed number or a whole number: **a.** $\frac{29}{6}$ **b.** $\frac{40}{16}$ **c.** $\frac{84}{3}$ **d.** $-\frac{9}{5}$

Strategy We will divide the numerator by the denominator and write the remainder over the divisor.

WHY A fraction bar indicates division.

Solution

a. To write $\frac{29}{6}$ as a mixed number, divide 29 by 6:

$$\begin{array}{r} 4 \\ 6\overline{)29} \\ -24 \\ \hline 5 \end{array}$$ ← The whole-number part is 4.

← Write the remainder 5 over the divisor 6 to get the fractional part.

Thus, $\frac{29}{6} = 4\frac{5}{6}$.

b. To write $\frac{40}{16}$ as a mixed number, divide 40 by 16:

$$\begin{array}{r} 2 \\ 16\overline{)40} \\ -32 \\ \hline 8 \end{array}$$

Thus, $\frac{40}{16} = 2\frac{8}{16} = 2\frac{1}{2}$. Simplify the fractional part: $\frac{8}{16} = \frac{\cancel{8}}{2 \cdot \cancel{8}} = \frac{1}{2}$.

c. For $\frac{84}{3}$, divide 84 by 3:

$$\begin{array}{r} 28 \\ 3\overline{)84} \\ -6 \\ \hline 24 \\ -24 \\ \hline 0 \end{array}$$

Thus, $\frac{84}{3} = 28$.

← Since the remainder is 0, the improper fraction represents a whole number.

d. To write $-\frac{9}{5}$ as a mixed number, ignore the − sign, and use the method for the positive improper fraction $\frac{9}{5}$. Once that procedure is completed, write a − sign in front of the result.

$$\begin{array}{r} 1 \\ 5\overline{)9} \\ -5 \\ \hline 4 \end{array}$$

Thus, $-\frac{9}{5} = -1\frac{4}{5}$.

Self Check 3

Write each improper fraction as a mixed number or a whole number:

a. $\frac{31}{7}$ $4\frac{3}{7}$ **b.** $\frac{50}{26}$ $1\frac{12}{13}$

c. $\frac{51}{3}$ 17 **d.** $-\frac{10}{3}$ $-3\frac{1}{3}$

Now Try Problems 31, 35, 39, and 43

Teaching Example 3
Write each improper fraction as a mixed number or a whole number:

a. $\frac{31}{9}$ **b.** $\frac{33}{6}$ **c.** $\frac{81}{3}$ **d.** $-\frac{9}{4}$

Answers:

a. $3\frac{4}{9}$ **b.** $5\frac{1}{2}$ **c.** 27 **d.** $-2\frac{1}{4}$

4 Graph fractions and mixed numbers on a number line.

In Chapters 1 and 2, we graphed whole numbers and integers on a number line. Fractions and mixed numbers can also be graphed on a number line.

Chapter 4 Fractions and Mixed Numbers

Self Check 4
Graph $-1\frac{7}{8}$, $-\frac{2}{3}$, $\frac{3}{5}$, and $\frac{9}{4}$ on a number line.

Now Try Problem 47

Teaching Example 4
Graph $-2\frac{1}{8}$, $\frac{9}{10}$, $-\frac{3}{4}$, and $\frac{8}{3}$ on a number line.
Answer:

EXAMPLE 4
Graph $-2\frac{3}{4}$, $-1\frac{1}{2}$, $-\frac{1}{8}$, and $\frac{13}{5}$ on a number line.

Strategy We will locate the position of each fraction and mixed number on the number line and draw a bold dot.

WHY To *graph a number* means to make a drawing that represents the number.

Solution
- Since $-2\frac{3}{4} < -2$, the graph of $-2\frac{3}{4}$ is to the left of -2 on the number line.
- The number $-1\frac{1}{2}$ is between -1 and -2.
- The number $-\frac{1}{8}$ is less than 0.
- Expressed as a mixed number, $\frac{13}{5} = 2\frac{3}{5}$.

5 Multiply and divide mixed numbers.

We will use the same procedures for multiplying and dividing mixed numbers as we used in Sections 4.2 and 4.3 to multiply and divide fractions. However, we must write the mixed numbers as improper fractions before we actually multiply or divide.

> **Multiplying and Dividing Mixed Numbers**
>
> To multiply or divide mixed numbers, first change the mixed numbers to improper fractions. Then perform the multiplication or division of the fractions. Write the result as a mixed number or a whole number in simplest form.

The sign rules for multiplying and dividing integers also hold for multiplying and dividing mixed numbers.

Self Check 5
Multiply and simplify, if possible.

a. $3\frac{1}{3} \cdot 2\frac{1}{3}$ $7\frac{7}{9}$ **b.** $9\frac{3}{5} \cdot \left(3\frac{3}{4}\right)$ 36

c. $-4\frac{5}{6}(2)$ $-9\frac{2}{3}$

Now Try Problems 51, 55, and 57

Teaching Example 5
Multiply and simplify, if possible.
a. $2\frac{1}{3} \cdot 1\frac{1}{4}$ b. $3\frac{1}{2}\left(5\frac{3}{7}\right)$ c. $-4\frac{5}{6}(3)$
Answers:
a. $2\frac{11}{12}$ b. 19 c. $-14\frac{1}{2}$

EXAMPLE 5
Multiply and simplify, if possible.

a. $1\frac{3}{4} \cdot 2\frac{1}{3}$ **b.** $5\frac{1}{5} \cdot \left(1\frac{2}{13}\right)$ **c.** $-4\frac{1}{9}(3)$

Strategy We will write the mixed numbers and whole numbers as improper fractions.

WHY Then we can use the rule for multiplying two fractions from Section 4.2.

Solution

a. $1\frac{3}{4} \cdot 2\frac{1}{3} = \frac{7}{4} \cdot \frac{7}{3}$ Write $1\frac{3}{4}$ and $2\frac{1}{3}$ as improper fractions.

$= \frac{7 \cdot 7}{4 \cdot 3}$ Use the rule for multiplying two fractions. Multiply the numerators and the denominators.

$= \frac{49}{12}$ Since there are no common factors to remove, perform the multiplication in the numerator and in the denominator. The result is an improper fraction.

$= 4\frac{1}{12}$ Write the improper fraction $\frac{49}{12}$ as a mixed number.

$\begin{array}{r} 4 \\ 12\overline{)49} \\ -48 \\ \hline 1 \end{array}$

Unless otherwise noted, all art on this page is © Cengage Learning.

b. $5\dfrac{1}{5}\left(1\dfrac{2}{13}\right) = \dfrac{26}{5} \cdot \dfrac{15}{13}$ Write $5\dfrac{1}{5}$ and $1\dfrac{2}{13}$ as improper fractions.

$= \dfrac{26 \cdot 15}{5 \cdot 13}$ Multiply the numerators. Multiply the denominators.

$= \dfrac{2 \cdot 13 \cdot 3 \cdot 5}{5 \cdot 13}$ To prepare to simplify, factor 26 as $2 \cdot 13$ and 15 as $3 \cdot 5$.

$= \dfrac{2 \cdot \cancel{13} \cdot 3 \cdot \cancel{5}}{\cancel{5} \cdot \cancel{13}}$ Remove the common factors of 13 and 5 from the numerator and denominator.

$= \dfrac{6}{1}$ Multiply the remaining factors in the numerator: $2 \cdot 1 \cdot 3 \cdot 1 = 6$. Multiply the remaining factors in the denominator: $1 \cdot 1 = 1$.

$= 6$ Any whole number divided by 1 remains the same.

c. $-4\dfrac{1}{9} \cdot 3 = -\dfrac{37}{9} \cdot \dfrac{3}{1}$ Write $-4\dfrac{1}{9}$ as an improper fraction and write 3 as a fraction.

$= -\dfrac{37 \cdot 3}{9 \cdot 1}$ Multiply the numerators and multiply the denominators. Since the fractions have unlike signs, make the answer negative.

$= -\dfrac{37 \cdot \cancel{3}}{3 \cdot \cancel{3} \cdot 1}$ To simplify, factor 9 as $3 \cdot 3$, and then remove the common factor of 3 from the numerator and denominator.

$= -\dfrac{37}{3}$ Multiply the remaining factors in the numerator and in the denominator. The result is an improper fraction.

$= -12\dfrac{1}{3}$ Write the negative improper fraction $-\dfrac{37}{3}$ as a negative mixed number.

$\begin{array}{r}12\\3\overline{)37}\\-3\\\hline 7\\-6\\\hline 1\end{array}$

> **Success Tip** We can use estimation by rounding to check the results when multiplying mixed numbers. If the fractional part of the mixed number is $\dfrac{1}{2}$ or greater, round up by adding 1 to the whole-number part and dropping the fraction. If the fractional part of the mixed number is less than $\dfrac{1}{2}$, round down by dropping the fraction and using only the whole-number part. To check the answer $4\dfrac{1}{12}$ from Example 5a, we proceed as follows:
>
> $1\dfrac{3}{4} \cdot 2\dfrac{1}{3} \approx 2 \cdot 2 = 4$ Since $\dfrac{3}{4}$ is greater than $\dfrac{1}{2}$, round $1\dfrac{3}{4}$ up to 2. Since $\dfrac{1}{3}$ is less than $\dfrac{1}{2}$, round $2\dfrac{1}{3}$ down to 2.
>
> Since $4\dfrac{1}{12}$ is close to 4, it is a reasonable answer.

EXAMPLE 6   Divide and simplify, if possible:

a. $-3\dfrac{3}{8} \div \left(-2\dfrac{1}{4}\right)$ **b.** $1\dfrac{11}{16} \div \dfrac{3}{4}$

Strategy We will write the mixed numbers as improper fractions.

WHY Then we can use the rule for dividing two fractions from Section 4.3.

Solution

a. $-3\dfrac{3}{8} \div \left(-2\dfrac{1}{4}\right) = -\dfrac{27}{8} \div \left(-\dfrac{9}{4}\right)$ Write $-3\dfrac{3}{8}$ and $-2\dfrac{1}{4}$ as improper fractions.

$= -\dfrac{27}{8}\left(-\dfrac{4}{9}\right)$ Use the rule for dividing two fractions: Multiply $-\dfrac{27}{8}$ by the reciprocal of $-\dfrac{9}{4}$, which is $-\dfrac{4}{9}$.

Self Check 6

Divide and simplify, if possible:

a. $-3\dfrac{4}{15} \div \left(-2\dfrac{1}{10}\right)$ $1\dfrac{5}{9}$

b. $5\dfrac{3}{5} \div \dfrac{7}{8}$ $6\dfrac{2}{5}$

Now Try Problems 59 and 65

Teaching Example 6
Divide and simplify, if possible:
a. $-8\frac{1}{3} \div \left(-4\frac{4}{9}\right)$ b. $2\frac{11}{12} \div \frac{5}{6}$

Answers:
a. $1\frac{7}{8}$ b. $3\frac{1}{2}$

$$= \frac{27}{8}\left(\frac{4}{9}\right)$$ Since the product of two negative fractions is positive, drop both − signs and continue.

$$= \frac{27 \cdot 4}{8 \cdot 9}$$ Multiply the numerators. Multiply the denominators.

$$= \frac{3 \cdot \overset{1}{\cancel{9}} \cdot \overset{1}{\cancel{4}}}{2 \cdot \underset{1}{\cancel{4}} \cdot \underset{1}{\cancel{9}}}$$ To simplify, factor 27 as 3 · 9 and 8 as 2 · 4. Then remove the common factors of 9 and 4 from the numerator and denominator.

$$= \frac{3}{2}$$ Multiply the remaining factors in the numerator: $3 \cdot 1 \cdot 1 = 3$. Multiply the remaining factors in the denominator: $2 \cdot 1 \cdot 1 = 2$.

$$= 1\frac{1}{2}$$ Write the improper fraction $\frac{3}{2}$ as a mixed number by dividing 3 by 2.

b. $1\frac{11}{16} \div \frac{3}{4} = \frac{27}{16} \div \frac{3}{4}$ Write $1\frac{11}{16}$ as an improper fraction.

$$= \frac{27}{16} \cdot \frac{4}{3}$$ Multiply $\frac{27}{16}$ by the reciprocal of $\frac{3}{4}$, which is $\frac{4}{3}$.

$$= \frac{27 \cdot 4}{16 \cdot 3}$$ Multiply the numerators. Multiply the denominators.

$$= \frac{\overset{1}{\cancel{3}} \cdot 9 \cdot \overset{1}{\cancel{4}}}{\underset{1}{\cancel{4}} \cdot 4 \cdot \underset{1}{\cancel{3}}}$$ To simplify, factor 27 as 3 · 9 and 16 as 4 · 4. Then remove the common factors of 3 and 4 from the numerator and denominator.

$$= \frac{9}{4}$$ Multiply the remaining factors in the numerator and in the denominator. The result is an improper fraction.

$$= 2\frac{1}{4}$$ Write the improper fraction $\frac{9}{4}$ as a mixed number by dividing 9 by 4.

6 Solve application problems by multiplying and dividing mixed numbers.

Self Check 7

BUMPER STICKERS A rectangular-shaped bumper sticker is $8\frac{1}{4}$ inches long by $3\frac{1}{4}$ inches wide. Find its area. $26\frac{13}{16}$ in.²

Now Try Problem 99

Teaching Example 7 KLEENEX
When unfolded and laid flat, a standard sheet of Kleenex brand tissue is $8\frac{1}{5}$ inches wide by $8\frac{2}{5}$ inches long. Find the area of a sheet of Kleenex.
Answer:
$68\frac{22}{25}$ in.²

EXAMPLE 7 **Toys** The dimensions of the rectangular-shaped screen of an Etch-a-Sketch are shown in the illustration below. Find the area of the screen.

Strategy To find the area, we will multiply $6\frac{1}{4}$ by $4\frac{1}{2}$.

WHY The formula for the area of a rectangle is Area = length · width.

Solution

$A = lw$ This is the formula for the area of a rectangle.

$A = 6\frac{1}{4} \cdot 4\frac{1}{2}$ Substitute $6\frac{1}{4}$ for l and $4\frac{1}{2}$ for w.

$A = \frac{25}{4} \cdot \frac{9}{2}$ Write $6\frac{1}{4}$ and $4\frac{1}{2}$ as improper fractions.

$A = \frac{25 \cdot 9}{4 \cdot 2}$ Multiply the numerators.
Multiply the denominators.

$A = \frac{225}{8}$ Since there are no common factors to remove, perform the multiplication in the numerator and in the denominator. The result is an improper fraction.

$A = 28\frac{1}{8}$ Write the improper fraction $\frac{225}{8}$ as a mixed number.

```
    28
 8)225
  -16
   65
  -64
    1
```

The area of the screen of an Etch-a-Sketch is $28\frac{1}{8}$ in.²

EXAMPLE 8   Government Grants
If $\$12\frac{1}{2}$ million is to be split equally among five cities to fund recreation programs, how much will each city receive?

Solution

- There is $\$12\frac{1}{2}$ million in grant money.   *Given*
- Five cities will split the money equally.   *Given*
- How much grant money will each city receive?   *Find*

The key phrase *split equally* suggests division. We translate the words of the problem to numbers and symbols.

The amount of money that each city will receive (in millions of dollars)	is equal to	the total amount of grant money (in millions of dollars)	divided by	the number of cities receiving money.
The amount of money that each city will receive (in millions of dollars)	=	$12\frac{1}{2}$	÷	5

To find the quotient, we will express $12\frac{1}{2}$ and 5 as fractions and then use the rule for dividing two fractions.

$12\frac{1}{2} \div 5 = \frac{25}{2} \div \frac{5}{1}$ Write $12\frac{1}{2}$ as an improper fraction, and write 5 as a fraction.

$= \frac{25}{2} \cdot \frac{1}{5}$ Multiply by the reciprocal of $\frac{5}{1}$, which is $\frac{1}{5}$.

$= \frac{25 \cdot 1}{2 \cdot 5}$ Multiply the numerators.
Multiply the denominators.

$= \frac{\overset{1}{\cancel{5}} \cdot 5 \cdot 1}{2 \cdot \underset{1}{\cancel{5}}}$ To simplify, factor 25 as 5 · 5. Then remove the common factor of 5 from the numerator and denominator.

$= \frac{5}{2}$ Multiply the remaining factors in the numerator.
Multiply the remaining factors in the denominator.

$= 2\frac{1}{2}$ Write the improper fraction $\frac{5}{2}$ as a mixed number by dividing 5 by 2. The units are millions of dollars.

Self Check 8

TV INTERVIEWS An $18\frac{3}{4}$-minute taped interview with an actor was played in equally long segments over five consecutive nights on a celebrity news program. How long was each interview segment? $3\frac{3}{4}$ min

Now Try Problem 107

Teaching Example 8 HIGHWAYS
It took workers three days to paint the yellow centerline on a new $17\frac{1}{2}$-mile stretch of highway. If they covered the same distance each day, how much did they do in one day?
Answer:
$5\frac{5}{6}$ mi

Each city will receive $2\frac{1}{2}$ million in grant money.

Check: We can estimate to check the result. If there was $10 million in grant money, each city would receive $\frac{\$10 \text{ million}}{5}$, or $2 million. Since there is actually $12\frac{1}{2}$ million in grant money, the answer that each city would receive $2\frac{1}{2}$ million seems reasonable.

ANSWERS TO SELF CHECKS

1. $\frac{9}{2}$, $4\frac{1}{2}$ 2. $\frac{27}{8}$ 3. a. $4\frac{3}{7}$ b. $1\frac{12}{13}$ c. 17 d. $-3\frac{1}{3}$

4. (number line showing $-1\frac{7}{8}$, $-\frac{2}{3}$, $\frac{3}{5}$, $\frac{9}{4} = 2\frac{1}{4}$ graphed between -3 and 3)

5. a. $7\frac{7}{9}$ b. 36 c. $-9\frac{2}{3}$ 6. a. $1\frac{5}{9}$ b. $6\frac{2}{5}$

7. $26\frac{13}{16}$ in.² 8. $3\frac{3}{4}$ min

SECTION 4.5 STUDY SET

VOCABULARY

Fill in the blanks.

1. A __mixed__ number, such as $8\frac{4}{5}$, is the sum of a whole number and a proper fraction.

2. In the mixed number $8\frac{4}{5}$, the __whole__-number part is 8 and the __fractional__ part is $\frac{4}{5}$.

3. The numerator of an __improper__ fraction is greater than or equal to its denominator.

4. To __graph__ a number means to locate its position on a number line and highlight it using a dot.

CONCEPTS

5. What signed mixed number could be used to describe each situation?

   a. A temperature of five and one-third degrees above zero $5\frac{1}{3}°$

   b. The depth of a sprinkler pipe that is six and seven-eighths inches below the sidewalk $-6\frac{7}{8}$ in.

6. What signed mixed number could be used to describe each situation?

   a. A rain total two and three-tenths of an inch lower than the average $-2\frac{3}{10}$ in.

   b. Three and one-half minutes after the liftoff of a rocket $3\frac{1}{2}$ min

Fill in the blanks.

7. To write a mixed number as an improper fraction:
   1. __Multiply__ the denominator of the fraction by the whole-number part.
   2. __Add__ the numerator of the fraction to the result from Step 1.
   3. Write the sum from Step 2 over the original __denominator__.

8. To write an improper fraction as a mixed number:
   1. __Divide__ the numerator by the denominator to obtain the whole-number part.
   2. The __remainder__ over the divisor is the fractional part.

9. What fractions have been graphed on the number line? $-\frac{4}{5}, -\frac{2}{5}, \frac{1}{5}$

10. What mixed numbers have been graphed on the number line? $-1\frac{3}{4}, 1\frac{1}{4}$

11. Fill in the blank: To multiply or divide mixed numbers, first change the mixed numbers to __improper__ fractions. Then perform the multiplication or division of the fractions as usual.

12. Simplify the fractional part of each mixed number.

    a. $11\frac{2}{4}$ $11\frac{1}{2}$

    b. $1\frac{3}{9}$ $1\frac{1}{3}$

    c. $7\frac{15}{27}$ $7\frac{5}{9}$

13. Use *estimation* to determine whether the following answer seems reasonable:

    $4\frac{1}{5} \cdot 2\frac{5}{7} = 7\frac{2}{35}$ not reasonable: $4\frac{1}{5} \cdot 2\frac{5}{7} \approx 4 \cdot 3 = 12$

14. What is the formula for the

    a. area of a rectangle? Area = (length)(width)

    b. area of a triangle? Area = $\frac{1}{2}$(base)(height)

NOTATION

15. Fill in the blanks.

   a. We read $5\frac{11}{16}$ as "five __and__ eleven-__sixteenths__."

   b. We read $-4\frac{2}{3}$ as "__negative__ four and __two__ -thirds."

16. Determine the sign of the result. *You do not have to find the answer.*

   a. $1\frac{1}{9}\left(-7\frac{3}{14}\right)$ negative

   b. $-3\frac{4}{15} \div \left(-1\frac{5}{6}\right)$ positive

Fill in the blanks to complete each step.

17. Multiply:

$$5\frac{1}{4} \cdot 1\frac{1}{7} = \frac{21}{4} \cdot \frac{8}{7}$$

$$= \frac{21 \cdot 8}{4 \cdot 7}$$

$$= \frac{3 \cdot \overset{1}{\cancel{7}} \cdot 2 \cdot \overset{1}{\cancel{4}}}{\underset{1}{\cancel{4}} \cdot \underset{1}{\cancel{7}}}$$

$$= \frac{6}{1}$$

$$= 6$$

18. Divide:

$$-5\frac{5}{6} \div 2\frac{1}{12} = -\frac{35}{6} \div \frac{25}{12}$$

$$= -\frac{35}{6} \cdot \frac{12}{25}$$

$$= -\frac{35 \cdot 12}{6 \cdot 25}$$

$$= -\frac{\overset{1}{\cancel{5}} \cdot 7 \cdot 2 \cdot \overset{1}{\cancel{6}}}{\underset{1}{\cancel{6}} \cdot \underset{1}{\cancel{5}} \cdot 5}$$

$$= -\frac{14}{5}$$

$$= -2\frac{4}{5}$$

GUIDED PRACTICE

Each region outlined in black represents one whole. Write an improper fraction and a mixed number to represent the shaded portion. See Example 1.

19. $\frac{19}{8}, 2\frac{3}{8}$

20. $\frac{29}{6}, 4\frac{5}{6}$

21. $\frac{34}{25}, 1\frac{9}{25}$

22. $\frac{51}{16}, 3\frac{3}{16}$

Chapter 4 Fractions and Mixed Numbers

Write each mixed number as an improper fraction. See Example 2.

23. $6\frac{1}{2}$ $\frac{13}{2}$

24. $8\frac{2}{3}$ $\frac{26}{3}$

25. $20\frac{4}{5}$ $\frac{104}{5}$

26. $15\frac{3}{8}$ $\frac{123}{8}$

27. $-7\frac{5}{9}$ $-\frac{68}{9}$

28. $-7\frac{1}{12}$ $-\frac{85}{12}$

29. $-8\frac{2}{3}$ $-\frac{26}{3}$

30. $-9\frac{3}{4}$ $-\frac{39}{4}$

Write each improper fraction as a mixed number or a whole number. Simplify the result, if possible. See Example 3.

31. $\frac{13}{4}$ $3\frac{1}{4}$

32. $\frac{41}{6}$ $6\frac{5}{6}$

33. $\frac{28}{5}$ $5\frac{3}{5}$

34. $\frac{28}{3}$ $9\frac{1}{3}$

35. $\frac{42}{9}$ $4\frac{2}{3}$

36. $\frac{62}{8}$ $7\frac{3}{4}$

37. $\frac{84}{8}$ $10\frac{1}{2}$

38. $\frac{93}{9}$ $10\frac{1}{3}$

39. $\frac{52}{13}$ 4

40. $\frac{80}{16}$ 5

41. $\frac{34}{17}$ 2

42. $\frac{38}{19}$ 2

43. $-\frac{58}{7}$ $-8\frac{2}{7}$

44. $-\frac{33}{7}$ $-4\frac{5}{7}$

45. $-\frac{20}{6}$ $-3\frac{1}{3}$

46. $-\frac{28}{8}$ $-3\frac{1}{2}$

Graph the given numbers on a number line. See Example 4.

47. $-2\frac{8}{9}, 1\frac{2}{3}, \frac{16}{5}, -\frac{1}{2}$

48. $-\frac{3}{4}, -3\frac{1}{4}, \frac{5}{2}, 4\frac{3}{4}$

49. $3\frac{1}{7}, -\frac{98}{99}, -\frac{10}{3}, \frac{3}{2}$

50. $-2\frac{1}{5}, \frac{4}{5}, -\frac{11}{3}, \frac{17}{4}$

Multiply and simplify, if possible. See Example 5.

51. $3\frac{1}{2} \cdot 2\frac{1}{3}$ $8\frac{1}{6}$

52. $1\frac{5}{6} \cdot 1\frac{1}{2}$ $2\frac{3}{4}$

53. $2\frac{2}{5}\left(3\frac{1}{12}\right)$ $7\frac{2}{5}$

54. $\frac{40}{13}\left(\frac{26}{5}\right)$ 16

55. $6\frac{1}{2} \cdot 1\frac{3}{13}$ 8

56. $12\frac{3}{5} \cdot 1\frac{3}{7}$ 18

57. $-2\frac{1}{2}(4)$ -10

58. $-3\frac{3}{4}(8)$ -30

Divide and simplify, if possible. See Example 6.

59. $-1\frac{13}{15} \div \left(-4\frac{1}{5}\right)$ $\frac{4}{9}$

60. $-2\frac{5}{6} \div \left(-8\frac{1}{2}\right)$ $\frac{1}{3}$

61. $15\frac{1}{3} \div 2\frac{2}{9}$ $6\frac{9}{10}$

62. $6\frac{1}{4} \div 3\frac{3}{4}$ $1\frac{2}{3}$

63. $1\frac{3}{4} \div \frac{3}{4}$ $2\frac{1}{3}$

64. $5\frac{3}{5} \div \frac{9}{10}$ $6\frac{2}{9}$

65. $1\frac{7}{24} \div \frac{7}{8}$ $1\frac{10}{21}$

66. $4\frac{1}{2} \div \frac{3}{17}$ $25\frac{1}{2}$

TRY IT YOURSELF

Perform each operation and simplify, if possible.

67. $-6 \cdot 2\frac{7}{24}$ $-13\frac{3}{4}$

68. $-7 \cdot 1\frac{3}{28}$ $-7\frac{3}{4}$

69. $-6\frac{3}{5} \div 7\frac{1}{3}$ $-\frac{9}{10}$

70. $-4\frac{1}{4} \div 4\frac{1}{2}$ $-\frac{17}{18}$

71. $\left(1\frac{2}{3}\right)^2$ $\frac{25}{9} = 2\frac{7}{9}$

72. $\left(3\frac{1}{2}\right)^2$ $\frac{49}{4} = 12\frac{1}{4}$

73. $8 \div 3\frac{1}{5}$ $2\frac{1}{2}$

74. $15 \div 3\frac{1}{3}$ $4\frac{1}{2}$

75. $-20\frac{1}{4} \div \left(-1\frac{11}{16}\right)$ 12

76. $-2\frac{7}{10} \div \left(-1\frac{1}{14}\right)$ $2\frac{13}{25}$

77. $3\frac{1}{16} \cdot 4\frac{4}{7}$ 14

78. $5\frac{3}{5} \cdot 1\frac{11}{14}$ 10

79. Find the quotient of $-4\frac{1}{2}$ and $2\frac{1}{4}$. -2

80. Find the quotient of 25 and $-10\frac{5}{7}$. $-2\frac{1}{3}$

81. $2\frac{1}{2}\left(-3\frac{1}{3}\right)$ $-8\frac{1}{3}$ 82. $\left(-3\frac{1}{4}\right)\left(1\frac{1}{5}\right)$ $-3\frac{9}{10}$

83. $7 \cdot 1\frac{3}{28}$ $7\frac{3}{4}$ 84. $4\frac{2}{3} \cdot 7$ $32\frac{2}{3}$

85. $8 \div 3\frac{1}{5}$ $2\frac{1}{2}$ 86. $4\frac{2}{5} \div 11$ $\frac{2}{5}$

87. Find the product of $1\frac{2}{3}$, 6, and $-\frac{1}{8}$. $-1\frac{1}{4}$

88. Find the product of $-\frac{5}{6}$, -8, and $-2\frac{1}{10}$. -14

89. $\left(-1\frac{1}{3}\right)^3$ $-\frac{64}{27} = -2\frac{10}{27}$

90. $\left(-1\frac{1}{5}\right)^3$ $-\frac{216}{125} = -1\frac{91}{125}$

APPLICATIONS

91. In the illustration below, each barrel represents one whole.

    a. Write a mixed number to represent the shaded portion. $3\frac{2}{3}$

    b. Write an improper fraction to represent the shaded portion. $\frac{11}{3}$

92. Draw $\frac{17}{8}$ pizzas.

93. DIVING Fill in the blank with a mixed number to describe the dive shown below: forward $2\frac{1}{2}$ somersaults

94. PRODUCT LABELING Several mixed numbers appear on the label shown below. Write each mixed number as an improper fraction. $1\frac{3}{4} = \frac{7}{4}$, $23\frac{1}{4} = \frac{93}{4}$, $18\frac{7}{8} = \frac{151}{8}$, $10\frac{1}{2} = \frac{21}{2}$

95. READING METERS

    a. Use a mixed number to describe the value to which the arrow is currently pointing. $2\frac{2}{3}$

    b. If the arrow moves twelve tick marks to the left, to what value will it be pointing? $-1\frac{1}{3}$

96. READING METERS

    a. Use a mixed number to describe the value to which the arrow is currently pointing. $-2\frac{1}{2}$

    b. If the arrow moves up six tick marks, to what value will it be pointing? $\frac{1}{2}$

97. ONLINE SHOPPING A mother is ordering a pair of jeans for her daughter from the screen shown below. If the daughter's height is $60\frac{3}{4}$ in. and her waist is $24\frac{1}{2}$ in., on what size and what cut (regular or slim) should the mother point and click? size 14, slim cut

98. SEWING Use the following table to determine the number of yards of fabric needed . . .

  a. to make a size 16 top if the fabric to be used is 60 inches wide. $2\frac{1}{8}$ yd

  b. to make size 18 pants if the fabric to be used is 45 inches wide. $2\frac{5}{8}$ yd

99. LICENSE PLATES Find the area of the license plate shown below. $76\frac{9}{16}$ in.²

100. GRAPH PAPER Mathematicians use specially marked paper, called graph paper, when drawing figures. It is made up of squares that are $\frac{1}{4}$-inch long by $\frac{1}{4}$-inch high.

  a. Find the length of the piece of graph paper shown below. $2\frac{3}{4}$ in.

  b. Find its height. $1\frac{1}{4}$ in.

  c. What is the area of the piece of graph paper? $3\frac{7}{16}$ in.²

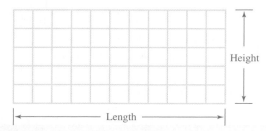

101. EMERGENCY EXITS The following sign marks the emergency exit on a school bus. Find the area of the sign. $42\frac{5}{8}$ in.²

102. CLOTHING DESIGN Find the number of square yards of material needed to make the triangular-shaped shawl shown in the illustration. $1\frac{1}{9}$ yd²

103. CALORIES A company advertises that its mints contain only $3\frac{1}{5}$ calories a piece. What is the calorie intake if you eat an entire package of 20 mints? 64 calories

104. **CEMENT MIXERS** A cement mixer can carry $9\frac{1}{2}$ cubic yards of concrete. If it makes 8 trips to a job site, how much concrete will be delivered to the site? 76 yd³

105. **SHOPPING** In the illustration, what is the cost of buying the fruit in the scale? Give your answer in cents and in dollars. 357¢ = $3.57

106. **PICTURE FRAMES** How many inches of molding is needed to make the square picture frame below? $40\frac{1}{2}$ in.

107. **BREAKFAST CEREAL** A box of cereal contains about $13\frac{3}{4}$ cups. Refer to the nutrition label shown below and determine the recommended size of one serving. $1\frac{1}{4}$ cups

108. **BREAKFAST CEREAL** A box of cereal contains about $14\frac{1}{4}$ cups. Refer to the nutrition label shown below. Determine how many servings there are for children under 4 in one box. 19 servings

109. **CATERING** How many people can be served $\frac{1}{3}$-pound hamburgers if a caterer purchases 200 pounds of ground beef? 600 people

110. **SUBDIVISIONS** A developer donated to the county 100 of the 1,000 acres of land she owned. She divided the remaining acreage into $1\frac{1}{3}$-acre lots. How many lots were created? 675 lots

111. **HORSE RACING** The race tracks on which thoroughbred horses run are marked off in $\frac{1}{8}$-mile-long segments called *furlongs*. How many furlongs are there in a $1\frac{1}{16}$-mile race? $8\frac{1}{2}$ furlongs

112. **FIRE ESCAPES** Part of the fire escape stairway for one story of an office building is shown below. Each riser is $7\frac{1}{2}$ inches high and each story of the building is 105 inches high.

   a. How many stairs are there in one story of the fire escape stairway? 14 stairs

   b. If the building has 43 stories, how many stairs are there in the entire fire escape stairway? 602 stairs

WRITING

113. Explain the difference between $2\frac{3}{4}$ and $2(\frac{3}{4})$.

114. Give three examples of how you use mixed numbers in daily life.

REVIEW

Find the LCM of the given numbers.

115. 5, 12, 15   60

116. 8, 12, 16   48

Find the GCF of the given numbers.

117. 12, 68, 92   4

118. 24, 36, 40   4

Objectives

1. Add mixed numbers.
2. Add mixed numbers in vertical form.
3. Subtract mixed numbers.
4. Solve application problems by adding and subtracting mixed numbers.

SECTION 4.6
Adding and Subtracting Mixed Numbers

In this section, we discuss several methods for adding and subtracting mixed numbers.

1 Add mixed numbers.

We can add mixed numbers by writing them as improper fractions. To do so, we follow these steps.

> **Adding Mixed Numbers: Method 1**
>
> 1. Write each mixed number as an improper fraction.
> 2. Write each improper fraction as an equivalent fraction with a denominator that is the LCD.
> 3. Add the fractions.
> 4. Write the result as a mixed number, if desired.

Method 1 works well when the whole-number parts of the mixed numbers are small.

Self Check 1

Add: $3\frac{2}{3} + 1\frac{1}{5}$ $4\frac{13}{15}$

Now Try Problem 13

Teaching Example 1
Add: $1\frac{5}{6} + 2\frac{1}{4}$
Answer:
$4\frac{1}{12}$

EXAMPLE 1   Add: $4\frac{1}{6} + 2\frac{3}{4}$

Strategy We will write each mixed number as an improper fraction, and then use the rule for adding two fractions that have different denominators.

WHY We cannot add the mixed numbers as they are; their fractional parts are not similar objects.

$$4\frac{1}{6} + 2\frac{3}{4}$$

Four and one-sixth ↑   ↑ Two and three-fourths

Solution

$$4\frac{1}{6} + 2\frac{3}{4} = \frac{25}{6} + \frac{11}{4} \quad \text{Write } 4\frac{1}{6} \text{ and } 2\frac{3}{4} \text{ as improper fractions.}$$

By inspection, we see that the lowest common denominator is 12.

$$= \frac{25}{6} \cdot \frac{2}{2} + \frac{11}{4} \cdot \frac{3}{3} \quad \text{To build } \tfrac{25}{6} \text{ and } \tfrac{11}{4} \text{ so that their denominators are 12, multiply each by a form of 1.}$$

$$= \frac{50}{12} + \frac{33}{12} \quad \text{Multiply the numerators. Multiply the denominators.}$$

$$= \frac{83}{12} \quad \text{Add the numerators and write the sum over the common denominator 12. The result is an improper fraction.}$$

$$= 6\frac{11}{12} \quad \text{Write the improper fraction } \tfrac{83}{12} \text{ as a mixed number.}$$

$$\begin{array}{r} 6 \\ 12\overline{)83} \\ -72 \\ \hline 11 \end{array}$$

Success Tip We can use estimation by rounding to check the results when adding (or subtracting) mixed numbers. To check the answer $6\frac{11}{12}$ from Example 1, we proceed as follows:

$$4\frac{1}{6} + 2\frac{3}{4} \approx 4 + 3 = 7 \quad \begin{array}{l} \text{Since } \frac{1}{6} \text{ is less than } \frac{1}{2}, \text{ round } 4\frac{1}{6} \text{ down to 4.} \\ \text{Since } \frac{3}{4} \text{ is greater than } \frac{1}{2}, \text{ round } 2\frac{3}{4} \text{ up to 3.} \end{array}$$

Since $6\frac{11}{12}$ is close to 7, it is a reasonable answer.

EXAMPLE 2 Add: $-3\frac{1}{8} + 1\frac{1}{2}$

Strategy We will write each mixed number as an improper fraction, and then use the rule for adding two fractions that have different denominators.

WHY We cannot add the mixed numbers as they are; their fractional parts are not similar objects.

$$-3\frac{1}{8} + 1\frac{1}{2}$$

Negative three and one-eighth ⎯⎯⎯↑   ↑⎯⎯One and one-half

Solution

$$-3\frac{1}{8} + 1\frac{1}{2} = -\frac{25}{8} + \frac{3}{2} \quad \text{Write } -3\frac{1}{8} \text{ and } 1\frac{1}{2} \text{ as improper fractions.}$$

Since the smallest number the denominators 8 and 2 divide exactly is 8, the LCD is 8. We will only need to build an equivalent fraction for $\frac{3}{2}$.

$$= -\frac{25}{8} + \frac{3}{2} \cdot \frac{4}{4} \quad \begin{array}{l} \text{To build } \frac{3}{2} \text{ so that its denominator is 8,} \\ \text{multiply it by a form of 1.} \end{array}$$

$$= -\frac{25}{8} + \frac{12}{8} \quad \begin{array}{l} \text{Multiply the numerators.} \\ \text{Multiply the denominators.} \end{array}$$

$$= \frac{-25 + 12}{8} \quad \begin{array}{l} \text{Add the numerators and write the sum} \\ \text{over the common denominator 8.} \end{array}$$

$$= \frac{-13}{8} \quad \begin{array}{l} \text{Use the rule for adding integers that have} \\ \text{different signs: } -25 + 12 = -13. \end{array}$$

$$= -1\frac{5}{8} \quad \text{Write } \frac{-13}{8} \text{ as a negative mixed number by dividing 13 by 8.}$$

Self Check 2
Add: $-4\frac{1}{12} + 2\frac{1}{4} - 1\frac{5}{6}$

Now Try Problem 17

Teaching Example 2
Add: $-3\frac{1}{10} + 1\frac{2}{5}$
Answer:
$-1\frac{7}{10}$

We can also add mixed numbers by adding their whole-number parts and their fractional parts. To do so, we follow these steps.

> ### Adding Mixed Numbers: Method 2
>
> 1. Write each mixed number as the sum of a whole number and a fraction.
> 2. Use the commutative property of addition to write the whole numbers together and the fractions together.
> 3. Add the whole numbers and the fractions separately.
> 4. Write the result as a mixed number, if necessary.

Method 2 works well when the whole number parts of the mixed numbers are large.

Self Check 3

Add: $275\dfrac{1}{6} + 81\dfrac{3}{5}$ $356\dfrac{23}{30}$

Now Try Problem 21

Teaching Example 3 Add:
$353\dfrac{3}{7} + 71\dfrac{1}{9}$
Answer:
$424\dfrac{34}{63}$

EXAMPLE 3   Add: $168\dfrac{3}{7} + 85\dfrac{2}{9}$

Strategy We will write each mixed number as the sum of a whole number and a fraction. Then we will add the whole numbers and the fractions separately.

WHY If we change each mixed number to an improper fraction, build equivalent fractions, and add, the resulting numerators will be very large and difficult to work with.

Solution
We will write the solution in *horizontal* form.

$168\dfrac{3}{7} + 85\dfrac{2}{9} = 168 + \dfrac{3}{7} + 85 + \dfrac{2}{9}$ Write each mixed number as the sum of a whole number and a fraction.

$= 168 + 85 + \dfrac{3}{7} + \dfrac{2}{9}$ Use the commutative property of addition to change the order of the addition so that the whole numbers are together and the fractions are together.

$= 253 + \dfrac{3}{7} + \dfrac{2}{9}$ Add the whole numbers.

$\begin{array}{r} 1\,1 \\ 168 \\ +\ 85 \\ \hline 253 \end{array}$

$= 253 + \dfrac{3}{7} \cdot \dfrac{9}{9} + \dfrac{2}{9} \cdot \dfrac{7}{7}$ Prepare to add the fractions. To build $\dfrac{3}{7}$ and $\dfrac{2}{9}$ so that their denominators are 63, multiply each by a form of 1.

$= 253 + \dfrac{27}{63} + \dfrac{14}{63}$ Multiply the numerators. Multiply the denominators.

$= 253 + \dfrac{41}{63}$ Add the numerators and write the sum over the common denominator 63.

$\begin{array}{r} 1 \\ 27 \\ +\ 14 \\ \hline 41 \end{array}$

$= 253\dfrac{41}{63}$ Write the sum as a mixed number.

Caution! If we use method 1 to add the mixed numbers in Example 3, the numbers we encounter are very large. As expected, the result is the same: $253\dfrac{41}{63}$.

$168\dfrac{3}{7} + 85\dfrac{2}{9} = \dfrac{1{,}179}{7} + \dfrac{767}{9}$ Write $168\dfrac{3}{7}$ and $85\dfrac{2}{9}$ as improper fractions.

$= \dfrac{1{,}179}{7} \cdot \dfrac{9}{9} + \dfrac{767}{9} \cdot \dfrac{7}{7}$ The LCD is 63.

$= \dfrac{10{,}611}{63} + \dfrac{5{,}369}{63}$ Note how large the numerators are.

$= \dfrac{15{,}980}{63}$ Add the numerators and write the sum over the common denominator 63.

$= 253\dfrac{41}{63}$ To write the improper fraction as a mixed number, divide 15,980 by 63.

Generally speaking, the larger the whole-number parts of the mixed numbers, the more difficult it becomes to add those mixed numbers using method 1.

2 Add mixed numbers in vertical form.

We can add mixed numbers quickly when they are written in **vertical form** by working in columns. The strategy is the same as in Example 3: Add whole numbers to whole numbers and fractions to fractions.

EXAMPLE 4

Add: $25\frac{3}{4} + 31\frac{1}{5}$

Strategy We will perform the addition in *vertical form* with the fractions in a column and the whole numbers lined up in columns. Then we will add the fractional parts and the whole-number parts separately.

WHY It is often easier to add the fractional parts and the whole-number parts of mixed numbers vertically—especially if the whole-number parts contain two or more digits, such as 25 and 31.

Solution

The sum is $56\frac{19}{20}$.

Self Check 4

Add: $71\frac{5}{8} + 23\frac{1}{3}$ $94\frac{23}{24}$

Now Try Problem 25

Teaching Example 4
Add: $62\frac{2}{5} + 15\frac{1}{3}$
Answer:
$77\frac{11}{15}$

EXAMPLE 5

Add and simplify, if possible: $75\frac{1}{12} + 43\frac{1}{4} + 54\frac{1}{6}$

Strategy We will write the problem in *vertical form*. We will make sure that the fractional part of the answer is in simplest form.

WHY When adding, subtracting, multiplying, or dividing fractions or mixed numbers, the answer should always be written in simplest form.

Solution

The LCD for $\frac{1}{12}, \frac{1}{4},$ and $\frac{1}{6}$ is 12.

The sum is $172\frac{1}{2}$.

Self Check 5

Add and simplify, if possible:
$68\frac{1}{6} + 37\frac{5}{18} + 52\frac{1}{9}$ $157\frac{5}{9}$

Now Try Problem 29

Teaching Example 5
Add and simplify, if possible:
$44\frac{1}{3} + 91\frac{1}{5} + 27\frac{2}{15}$
Answer:
$162\frac{2}{3}$

When we add mixed numbers, sometimes the sum of the fractions is an improper fraction.

Self Check 6

Add: $76\frac{11}{12} + 49\frac{5}{8}$ $126\frac{13}{24}$

Now Try Problem 33

Teaching Example 6
Add: $83\frac{7}{8} + 39\frac{3}{5}$
Answer:
$123\frac{19}{40}$

EXAMPLE 6 Add: $45\frac{2}{3} + 96\frac{4}{5}$

Strategy We will write the problem in *vertical form*. We will make sure that the fractional part of the answer is in simplest form.

WHY When adding, subtracting, multiplying, or dividing fractions or mixed numbers, the answer should always be written in simplest form.

Solution

The LCD for $\frac{2}{3}$ and $\frac{4}{5}$ is 15.

Since we don't want an improper fraction in the answer, we write $\frac{22}{15}$ as a mixed number. Then we *carry* 1 from the fraction column to the whole-number column.

$$141\frac{22}{15} = 141 + \frac{22}{15} \quad \text{Write the mixed number as the sum of a whole number and a fraction.}$$

$$= 141 + 1\frac{7}{15} \quad \text{To write the improper fraction as a mixed number divide 22 by 15.}$$

$$= 142\frac{7}{15} \quad \text{Carry the 1 and add it to 141 to get 142.}$$

$$\begin{array}{r}1\\15\overline{)22}\\-15\\\hline 7\end{array}$$

3 Subtract mixed numbers.

Subtracting mixed numbers is similar to adding mixed numbers.

Self Check 7

Subtract and simplify, if possible:

$12\frac{9}{20} - 8\frac{1}{30}$ $4\frac{5}{12}$

Now Try Problem 37

EXAMPLE 7 Subtract and simplify, if possible: $16\frac{7}{10} - 9\frac{8}{15}$

Strategy We will perform the subtraction in *vertical form* with the fractions in a column and the whole numbers lined up in columns. Then we will subtract the fractional parts and the whole-number parts separately.

WHY It is often easier to subtract the fractional parts and the whole-number parts of mixed numbers vertically.

Solution

The LCD for $\frac{7}{10}$ and $\frac{8}{15}$ is 30.

$$16\frac{7}{10} = 16\frac{7}{10} \cdot \frac{3}{3} = 16\frac{21}{30} = 16\frac{21}{30}$$
$$-9\frac{8}{15} = -9\frac{8}{15} \cdot \frac{2}{2} = -9\frac{16}{30} = -9\frac{16}{30}$$
$$\overline{\hspace{2cm}} \quad \overline{\hspace{2cm}} \quad \overline{\hspace{2cm}} \quad \overline{\hspace{2cm}}$$
$$\frac{5}{30} \quad 7\frac{5}{30} = 7\frac{1}{6}$$

- Write the mixed numbers in vertical form.
- Build $\frac{7}{10}$ and $\frac{8}{15}$ so that their denominators are 30.
- Subtract the fractions separately.
- Subtract the whole numbers separately.

Simplify: $\frac{5}{30} = \frac{\cancel{5}}{\cancel{5} \cdot 6} = \frac{1}{6}$.

The difference is $7\frac{1}{6}$.

Subtraction of mixed numbers (like subtraction of whole numbers) sometimes involves borrowing. When the fraction we are subtracting is greater than the fraction we are subtracting it from, it is necessary to borrow.

EXAMPLE 8

Subtract: $34\frac{1}{8} - 11\frac{2}{3}$

Strategy We will perform the subtraction in *vertical form* with the fractions in a column and the whole numbers lined up in columns. Then we will subtract the fractional parts and the whole-number parts separately.

WHY It is often easier to subtract the fractional parts and the whole-number parts of mixed numbers vertically.

Solution

The LCD for $\frac{1}{8}$ and $\frac{2}{3}$ is 24.

$$34\frac{1}{8} = 34\frac{1}{8} \cdot \frac{3}{3} = 34\frac{3}{24}$$
$$-11\frac{2}{3} = -11\frac{2}{3} \cdot \frac{8}{8} = -11\frac{16}{24}$$

- Write the mixed numbers in vertical form.
- Build $\frac{1}{8}$ and $\frac{2}{3}$ so that their denominators are 24.
- Note that $\frac{16}{24}$ is greater than $\frac{3}{24}$.

Since $\frac{16}{24}$ is greater than $\frac{3}{24}$, borrow 1 (in the form of $\frac{24}{24}$) from 34 and add it to $\frac{3}{24}$ to get $\frac{27}{24}$.

$$34\overset{33}{\frac{3}{24}} + \frac{24}{24} = 33\frac{27}{24} = 33\frac{27}{24}$$
$$-11\frac{16}{24} = -11\frac{16}{24} = -11\frac{16}{24}$$
$$\overline{\hspace{2cm}} \quad \overline{\hspace{2cm}} \quad \overline{\hspace{2cm}}$$
$$\frac{11}{24} \quad 22\frac{11}{24}$$

- Subtract the fractions separately.
- Subtract the whole numbers separately.

The difference is $22\frac{11}{24}$.

Teaching Example 7
Subtract and simplify, if possible:
$15\frac{2}{3} - 7\frac{5}{12}$
Answer:
$8\frac{1}{4}$

Self Check 8
Subtract: $258\frac{3}{4} - 175\frac{15}{16}$ $82\frac{13}{16}$

Now Try Problem 41

Teaching Example 8
Subtract: $219\frac{1}{4} - 63\frac{10}{11}$
Answer:
$155\frac{15}{44}$

Success Tip We can use estimation by rounding to check the results when subtracting mixed numbers. To check the answer $22\frac{11}{24}$ from Example 8, we proceed as follows:

$$34\frac{1}{8} - 11\frac{2}{3} \approx 34 - 12 = 22$$

Since $\frac{1}{8}$ is less than $\frac{1}{2}$, round $34\frac{1}{8}$ down to 34.
Since $\frac{2}{3}$ is greater than $\frac{1}{2}$, round $11\frac{2}{3}$ up to 12.

Since $22\frac{11}{24}$ is close to 22, it is a reasonable answer.

Self Check 9

Subtract: $2{,}300 - 129\frac{31}{32}$ $2{,}170\frac{1}{32}$

Now Try Problem 45

Teaching Example 9

Subtract: $643 - 81\frac{3}{28}$

Answer:
$561\frac{25}{28}$

EXAMPLE 9

Subtract: $419 - 53\frac{11}{16}$

Strategy We will write the numbers in vertical form and borrow 1 (in the form of $\frac{16}{16}$) from 419.

WHY In the fraction column, we need to have a fraction from which to subtract $\frac{11}{16}$.

Solution

Write the numbers in vertical form.
Borrow 1 (in the form of $\frac{16}{16}$) from 419.
Then subtract the fractions separately.
Subtract the whole numbers separately.
This also requires borrowing.

$$\begin{array}{r} 419 \\ -\ 53\frac{11}{16} \end{array} = \begin{array}{r} 418\frac{16}{16} \\ -\ 53\frac{11}{16} \\ \hline 365\frac{5}{16} \end{array} = \begin{array}{r} \overset{3\ 11}{4\cancel{1}8}\frac{16}{16} \\ -\ 53\frac{11}{16} \\ \hline 365\frac{5}{16} \end{array}$$

The difference is $365\frac{5}{16}$.

4 Solve application problems by adding and subtracting mixed numbers.

Self Check 10

SALADS A three-bean salad calls for one can of green beans ($14\frac{1}{2}$ ounces), one can of garbanzo beans ($10\frac{3}{4}$ ounces), and one can of kidney beans ($15\frac{7}{8}$ ounces). How many ounces of beans are called for in the recipe? $41\frac{1}{8}$ oz

Now Try Problem 93

EXAMPLE 10 Horse Racing

In order to become the Triple Crown Champion, a thoroughbred horse must win three races: the Kentucky Derby ($1\frac{1}{4}$ miles long), the Preakness Stakes ($1\frac{3}{16}$ miles long), and the Belmont Stakes ($1\frac{1}{2}$ miles long). What is the combined length of the three races of the Triple Crown?

Affirmed, in 1978, was the last of only 11 horses in history to win the Triple Crown.

Solution

- The Kentucky Derby is $1\frac{1}{4}$ miles long.   *Given*
- The Preakness Stakes is $1\frac{3}{16}$ miles long.   *Given*
- The Belmont Stakes is $1\frac{1}{2}$ miles long.   *Given*
- What is the combined length of the three races?   *Find*

The key phrase *combined length* indicates addition.
We translate the words of the problem to numbers and symbols.

The combined length of the three races	is equal to	the length of the Kentucky Derby	plus	the length of the Preakness Stakes	plus	the length of the Belmont Stakes.
The combined length of the three races	=	$1\frac{1}{4}$	+	$1\frac{3}{16}$	+	$1\frac{1}{2}$

To find the sum, we will write the mixed numbers in vertical form. To add in the fraction column, the LCD for $\frac{1}{4}$, $\frac{3}{16}$, and $\frac{1}{2}$ is 16.

— Build $\frac{1}{4}$ and $\frac{1}{2}$ so that their denominators are 16.
— Add the fractions separately.
— Add the whole numbers separately.

$$1\frac{1}{4} = 1\frac{1}{4} \cdot \frac{4}{4} = 1\frac{4}{16} = 1\frac{4}{16}$$
$$1\frac{3}{16} = 1\frac{3}{16} = 1\frac{3}{16} = 1\frac{3}{16}$$
$$+ 1\frac{1}{2} = + 1\frac{1}{2} \cdot \frac{8}{8} = + 1\frac{8}{16} = + 1\frac{8}{16}$$
$$\overline{ \frac{15}{16} 3\frac{15}{16}}$$

The combined length of the three races of the Triple Crown is $3\frac{15}{16}$ miles.

Check: We can estimate to check the result. If we round $1\frac{1}{4}$ down to 1, round $1\frac{3}{16}$ down to 1, and round $1\frac{1}{2}$ up to 2, the approximate combined length of the three races is $1 + 1 + 2 = 4$ miles. Since $3\frac{15}{16}$ is close to 4, the result seems reasonable.

Teaching Example 10 BAKING A baker wants to make a cookie recipe that calls for $1\frac{1}{8}$ cups of flour, a bread recipe that calls for $3\frac{3}{4}$ cups of flour, and a cake recipe that calls for $2\frac{1}{2}$ cups. How many cups of flour does she need to make all three recipes?
Answer:
$7\frac{3}{8}$ cups

THINK IT THROUGH

"You must schedule sleep like any other daily activity, so put it on your 'to-do' list and cross it off every night."
National Sleep Foundation

The 1,500 people who took part in the 2011 Sleep in America poll were asked how long they sleep on both workdays and non-workdays. The results are shown on the right. Write the average hours slept on a workday and on a non-workday as mixed numbers. How much longer does the average person sleep on a non-workday? workday: $6\frac{1}{2}$ hr; non-workday: $7\frac{1}{5}$ hr; $\frac{7}{10}$ hr

Typical Workday and Non-workday Sleep Habits

Average hours slept on workdays
6 hours 30 minutes

Average hours slept on non-workdays
7 hours 12 minutes

Source: National Sleep Foundation, 2011

Self Check 11

TRUCKING The mixing barrel of a cement truck holds 9 cubic yards of concrete. How much concrete is left in the barrel if $6\frac{3}{4}$ cubic yards have already been unloaded? $2\frac{1}{4}$ yd³

Now Try Problem 99

Teaching Example 11
REFRESHMENTS How much punch is left in a 20-gallon container if $9\frac{1}{2}$ gallons have been used?
Answer:
$10\frac{1}{2}$ gal

EXAMPLE 11

Baking How much butter is left in a 10-pound tub after $2\frac{2}{3}$ pounds are used for a wedding cake?

Solution

- The tub contained 10 pounds of butter.   Given
- $2\frac{2}{3}$ pounds of butter are used for a cake.   Given
- How much butter is left in the tub?   Find

The key phrase *how much butter is left* indicates subtraction. We translate the words of the problem to numbers and symbols.

The amount of butter left in the tub	is equal to	the amount of butter in one tub	minus	the amount of butter used for the cake.
The amount of butter left in the tub	=	10	−	$2\frac{2}{3}$

To find the difference, we will write the numbers in vertical form and borrow 1 (in the form of $\frac{3}{3}$) from 10.

In the fraction column, we need to have a fraction from which to subtract $\frac{2}{3}$.
Subtract the fractions separately.
Subtract the whole numbers separately.

$$10 = 10\frac{3}{3} \overset{9}{=} 10\frac{3}{3}\overset{9}{}$$
$$-2\frac{2}{3} = -2\frac{2}{3} = -2\frac{2}{3}$$
$$\frac{1}{3} \quad 7\frac{1}{3}$$

There are $7\frac{1}{3}$ pounds of butter left in the tub.

Check: We can check using addition. If $2\frac{2}{3}$ pounds of butter were used and $7\frac{1}{3}$ pounds of butter are left in the tub, then the tub originally contained $2\frac{2}{3} + 7\frac{1}{3} = 9\frac{3}{3} = 10$ pounds of butter. The result checks.

ANSWER TO SELF CHECKS

1. $4\frac{13}{15}$ **2.** $-1\frac{5}{6}$ **3.** $356\frac{23}{30}$ **4.** $94\frac{23}{24}$ **5.** $157\frac{5}{9}$ **6.** $126\frac{13}{24}$ **7.** $4\frac{5}{12}$ **8.** $82\frac{13}{16}$
9. $2{,}170\frac{1}{32}$ **10.** $41\frac{1}{8}$ oz   **11.** $2\frac{1}{4}$ yd³

SECTION 4.6   STUDY SET

VOCABULARY

Fill in the blanks.

1. A __mixed__ number, such as $1\frac{7}{8}$, contains a whole-number part and a fractional part.

2. We can add (or subtract) mixed numbers quickly when they are written in __vertical__ form by working in columns.

3. To add (or subtract) mixed numbers written in vertical form, we add (or subtract) the __fractions__ separately and the __whole__ numbers separately.

4. A fraction such as $\frac{11}{8}$, whose numerator is greater than or equal to the denominator, is called an __improper__ fraction.

5. Consider the following problem:

$$36\frac{5}{7}$$
$$+ 42\frac{4}{7}$$
$$\overline{78\frac{9}{7}} = 78 + 1\frac{2}{7} = 79\frac{2}{7}$$

Since we don't want an improper fraction in the answer, we write $\frac{9}{7}$ as $1\frac{2}{7}$, __carry__ the 1, and add it to 78 to get 79.

6. Consider the following problem:

$$86\frac{1}{3} = \quad 86\frac{\overset{5}{1}}{3} + \frac{3}{3}$$
$$-24\frac{2}{3} = -24\frac{2}{3}$$

To subtract in the fraction column, we __borrow__ 1 from 86 in the form of $\frac{3}{3}$.

CONCEPTS

7. **a.** For $76\frac{3}{4}$, list the whole-number part and the fractional part. $76, \frac{3}{4}$

   b. Write $76\frac{3}{4}$ as a sum. $76 + \frac{3}{4}$

8. Use the commutative property of addition to rewrite the following expression with the whole numbers together and the fractions together. *You do not have to find the answer.*

$$14 + \frac{5}{8} + 53 + \frac{1}{6} \quad 14 + 53 + \frac{5}{8} + \frac{1}{6}$$

9. The *denominators* of two fractions are given. Find the least common denominator.

   a. 3 and 4 12   **b.** 5 and 6 30
   c. 6 and 9 18   **d.** 8 and 12 24

10. Simplify.

    a. $9\frac{17}{16}$ $10\frac{1}{16}$ **b.** $1{,}288\frac{7}{3}$ $1{,}290\frac{1}{3}$
    c. $16\frac{12}{8}$ $17\frac{1}{2}$ **d.** $45\frac{24}{20}$ $46\frac{1}{5}$

NOTATION

Fill in the blanks to complete each step.

11. $6\frac{3}{5} = \quad 6\frac{3}{5} \cdot \frac{7}{7} = \quad 6\frac{21}{35}$
     $+3\frac{2}{7} = +3\frac{2}{7} \cdot \frac{5}{5} = +3\frac{10}{35}$
     $\overline{ 9\frac{31}{35}}$

12. $67\frac{3}{8} = \quad 67\frac{3}{8} \cdot \frac{3}{3} = \quad 67\frac{9}{24} = \quad 67\frac{\overset{6}{9}}{24} + \frac{24}{24} = \quad 66\frac{33}{24}$
    $-23\frac{2}{3} = -23\frac{2}{3} \cdot \frac{8}{8} = -23\frac{16}{24} = -23\frac{16}{24} \quad\quad = -23\frac{16}{24}$
    $\overline{ 43\frac{17}{24}}$

GUIDED PRACTICE

Add. See Example 1.

13. $1\frac{1}{4} + 2\frac{1}{3}$ $3\frac{7}{12}$ 14. $2\frac{2}{5} + 3\frac{1}{4}$ $5\frac{13}{20}$
15. $2\frac{1}{3} + 4\frac{2}{5}$ $6\frac{11}{15}$ 16. $4\frac{1}{3} + 1\frac{1}{7}$ $5\frac{10}{21}$

Add. See Example 2.

17. $-4\frac{1}{8} + 1\frac{3}{4}$ $-2\frac{3}{8}$ 18. $-3\frac{11}{15} + 2\frac{1}{5}$ $-1\frac{8}{15}$
19. $-6\frac{5}{6} + 3\frac{2}{3}$ $-3\frac{1}{6}$ 20. $-6\frac{3}{14} + 1\frac{2}{7}$ $-4\frac{13}{14}$

Add. See Example 3.

21. $334\frac{1}{7} + 42\frac{2}{3}$ $376\frac{17}{21}$ 22. $259\frac{3}{8} + 40\frac{1}{3}$ $299\frac{17}{24}$
23. $667\frac{1}{5} + 47\frac{3}{4}$ $714\frac{19}{20}$ 24. $568\frac{1}{6} + 52\frac{3}{4}$ $620\frac{11}{12}$

Add. See Example 4.

25. $41\frac{2}{9} + 18\frac{2}{5}$ $59\frac{28}{45}$ 26. $60\frac{3}{11} + 24\frac{2}{3}$ $84\frac{31}{33}$
27. $89\frac{6}{11} + 43\frac{1}{3}$ $132\frac{29}{33}$ 28. $77\frac{5}{8} + 55\frac{1}{7}$ $132\frac{43}{56}$

Add and simplify, if possible. See Example 5.

29. $14\frac{1}{4} + 29\frac{1}{20} + 78\frac{3}{5}$ $121\frac{9}{10}$ 30. $11\frac{1}{12} + 59\frac{1}{4} + 82\frac{1}{6}$ $152\frac{1}{2}$
31. $106\frac{5}{18} + 22\frac{1}{2} + 19\frac{1}{9}$ $147\frac{8}{9}$ 32. $75\frac{2}{5} + 43\frac{7}{30} + 54\frac{1}{3}$
$172\frac{29}{30}$

Add and simplify, if possible. See Example 6.

33. $39\frac{5}{8} + 62\frac{11}{12}$ $102\frac{13}{24}$ 34. $53\frac{5}{6} + 47\frac{3}{8}$ $101\frac{5}{24}$
35. $82\frac{8}{9} + 46\frac{11}{15}$ $129\frac{28}{45}$ 36. $44\frac{2}{9} + 76\frac{20}{21}$ $121\frac{11}{63}$

Subtract and simplify, if possible. See Example 7.

37. $19\frac{11}{12} - 9\frac{2}{3}$ $10\frac{1}{4}$ 38. $32\frac{2}{3} - 7\frac{1}{6}$ $25\frac{1}{2}$
39. $21\frac{5}{6} - 8\frac{3}{10}$ $13\frac{8}{15}$ 40. $41\frac{2}{5} - 6\frac{3}{20}$ $35\frac{1}{4}$

Subtract. See Example 8.

41. $47\frac{1}{11} - 15\frac{2}{3}$ $31\frac{14}{33}$
42. $58\frac{4}{11} - 15\frac{1}{2}$ $42\frac{19}{22}$
43. $84\frac{5}{8} - 12\frac{6}{7}$ $71\frac{43}{56}$
44. $95\frac{4}{7} - 23\frac{5}{6}$ $71\frac{31}{42}$

Subtract. See Example 9.

45. $674 - 94\frac{11}{15}$ $579\frac{4}{15}$
46. $437 - 63\frac{6}{23}$ $373\frac{17}{23}$
47. $112 - 49\frac{9}{32}$ $62\frac{23}{32}$
48. $221 - 88\frac{35}{64}$ $132\frac{29}{64}$

TRY IT YOURSELF

Add or subtract and simplify, if possible.

49. $140\frac{5}{6} - 129\frac{4}{5}$ $11\frac{1}{30}$
50. $291\frac{1}{4} - 289\frac{1}{12}$ $2\frac{1}{6}$
51. $4\frac{1}{6} + 1\frac{1}{5}$ $5\frac{11}{30}$
52. $2\frac{2}{5} + 3\frac{1}{4}$ $5\frac{13}{20}$
53. $5\frac{1}{2} + 3\frac{4}{5}$ $9\frac{3}{10}$
54. $6\frac{1}{2} + 2\frac{2}{3}$ $9\frac{1}{6}$
55. $2 + 1\frac{7}{8}$ $3\frac{7}{8}$
56. $3\frac{3}{4} + 5$ $8\frac{3}{4}$
57. $8\frac{7}{9} - 3\frac{1}{9}$ $5\frac{2}{3}$
58. $9\frac{9}{10} - 6\frac{3}{10}$ $3\frac{3}{5}$
59. $140\frac{3}{16} - 129\frac{3}{4}$ $10\frac{7}{16}$
60. $442\frac{1}{8} - 429\frac{2}{3}$ $12\frac{11}{24}$
61. $380\frac{1}{6} + 17\frac{1}{4}$ $397\frac{5}{12}$
62. $103\frac{1}{2} + 210\frac{2}{5}$ $313\frac{9}{10}$
63. $-2\frac{5}{6} + 1\frac{3}{8}$ $-1\frac{11}{24}$
64. $-4\frac{5}{9} + 2\frac{1}{6}$ $-2\frac{7}{18}$
65. $3\frac{1}{4} + 4\frac{1}{4}$ $7\frac{1}{2}$
66. $2\frac{1}{8} + 3\frac{3}{8}$ $5\frac{1}{2}$
67. $-3\frac{3}{4} + \left(-1\frac{1}{2}\right)$ $-5\frac{1}{4}$
68. $-3\frac{2}{3} + \left(-1\frac{4}{5}\right)$ $-5\frac{7}{15}$
69. $7 - \frac{2}{3}$ $6\frac{1}{3}$
70. $6 - \frac{1}{8}$ $5\frac{7}{8}$
71. $12\frac{1}{2} + 5\frac{3}{4} + 35\frac{1}{6}$ $53\frac{5}{12}$
72. $31\frac{1}{3} + 20\frac{2}{5} + 10\frac{1}{15}$ $61\frac{4}{5}$
73. $16\frac{1}{4} - 13\frac{3}{4}$ $2\frac{1}{2}$
74. $40\frac{1}{7} - 19\frac{6}{7}$ $20\frac{2}{7}$
75. $-4\frac{5}{8} - 1\frac{1}{4}$ $-5\frac{7}{8}$
76. $-2\frac{1}{16} - 3\frac{7}{8}$ $-5\frac{15}{16}$
77. $6\frac{5}{8} - 3$ $3\frac{5}{8}$
78. $10\frac{1}{2} - 6$ $4\frac{1}{2}$
79. $\frac{7}{3} + 2$ $4\frac{1}{3}$
80. $\frac{9}{7} + 3$ $4\frac{2}{7}$
81. $58\frac{7}{8} + 340\frac{1}{2} + 61\frac{3}{4}$ $461\frac{1}{8}$
82. $191\frac{1}{2} + 233\frac{1}{16} + 16\frac{5}{8}$ $441\frac{3}{16}$
83. $9 - 8\frac{3}{4}$ $\frac{1}{4}$
84. $11 - 10\frac{4}{5}$ $\frac{1}{5}$

For each problem, fill in the blanks of the estimate. Then tell whether the given result seems reasonable.

85.
Exact	Estimate
$167\frac{3}{8}$	167
$+ 44\frac{3}{4}$	$+ 45$
$202\frac{5}{8}$	212

not reasonable

86.
Exact	Estimate
$293\frac{5}{9}$	294
$+ 458\frac{7}{18}$	$+ 458$
$751\frac{17}{18}$	752

reasonable

87.
Exact	Estimate
$616\frac{2}{5}$	616
$- 488\frac{2}{3}$	$- 489$
$127\frac{11}{15}$	127

reasonable

88.
Exact	Estimate
$207\frac{1}{2}$	208
$- 189\frac{7}{11}$	$- 190$
$27\frac{19}{22}$	18

not reasonable

APPLICATIONS

89. **AIR TRAVEL** A businesswoman's flight left Los Angeles and in $3\frac{3}{4}$ hours she landed in Minneapolis. She then boarded a commuter plane in Minneapolis and arrived at her final destination in $1\frac{1}{2}$ hours. Find the total time she spent on the flights. $5\frac{1}{4}$ hr

90. **SHIPPING** A passenger ship and a cargo ship left San Diego harbor at midnight. During the first hour, the passenger ship traveled south at $16\frac{1}{2}$ miles per hour, while the cargo ship traveled north at a rate of $5\frac{1}{5}$ miles per hour. How far apart were they at 1:00 A.M.? $21\frac{7}{10}$ mi

91. **TRAIL MIX** How many cups of trail mix will the recipe shown below make? $7\frac{1}{6}$ cups

Trail Mix
A healthy snack—great for camping trips

$2\frac{3}{4}$ cups peanuts
$\frac{1}{2}$ cup sunflower seeds
$\frac{2}{3}$ cup raisins
$\frac{1}{3}$ cup coconut
$2\frac{2}{3}$ cups oat flakes
$\frac{1}{4}$ cup pretzels

92. HARDWARE Refer to the illustration below. How long should the threaded part of the bolt be? $7\frac{9}{16}$ in.

93. OCTUPLETS On January 26, 2009, at Kaiser Permanente Bellflower Medical Center in California, Nadya Suleman gave birth to eight babies. (The United States' first live octuplets were born in Houston in 1998 to Nkem Chukwu and Iyke Louis Udobi). Find the combined birthweights of the babies from the information shown below. (Source: The Nadya Suleman family website) $20\frac{1}{16}$ lb

No. 1: Noah, male, $2\frac{11}{16}$ pounds

No. 2: Maliah, female, $2\frac{3}{4}$ pounds

No. 3: Isaiah, male, $3\frac{1}{4}$ pounds

No. 4: Nariah, female, $2\frac{1}{2}$ pounds

No. 5: Makai, male, $1\frac{1}{2}$ pounds

No. 6: Josiah, male, $2\frac{3}{4}$ pounds

No. 7: Jeremiah, male, $1\frac{15}{16}$ pounds

No. 8: Jonah, male, $2\frac{11}{16}$ pounds

94. SEPTUPLETS On November 19, 1997, at Iowa Methodist Medical Center, Bobbie McCaughey gave birth to seven babies. Find the combined birthweights of the babies from the following information. (Source: *Los Angeles Times*, Nov. 20, 1997) $19\frac{7}{8}$ lb

95. HISTORICAL DOCUMENTS The Declaration of Independence on display at the National Archives in Washington, D.C., is $24\frac{1}{2}$ inches wide by $29\frac{3}{4}$ inches high. How many inches of molding would be needed to frame it? $108\frac{1}{2}$ in.

96. STAMP COLLECTING The Pony Express Stamp, shown below, was issued in 1940. It is a favorite of collectors all over the world. A Postal Service document describes its size in an unusual way:

"The dimensions of the stamp are $\frac{84}{100}$ by $1\frac{44}{100}$ inches, arranged horizontally."

To display the stamp, a collector wants to frame it with gold braid. How many inches of braid are needed? $4\frac{14}{25}$ in.

97. FREEWAY SIGNS A freeway exit sign is shown. How far apart are the Citrus Ave. and Grand Ave. exits? $2\frac{3}{4}$ mi

98. BASKETBALL See the graph below. What is the difference in height between the tallest and the shortest of the starting players? $9\frac{3}{8}$ in.

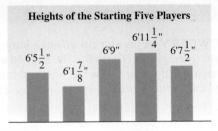

99. HOSE REPAIRS To repair a bad connector, a gardener removes $1\frac{1}{2}$ feet from the end of a 50-foot hose. How long is the hose after the repair? $48\frac{1}{2}$ ft

100. HAIRCUTS A mother makes her child get a haircut when his hair measures 3 inches in length. His barber uses clippers with attachment #2 that leaves $\frac{3}{8}$-inch of hair. How many inches does the child's hair grow between haircuts? $2\frac{5}{8}$ in.

101. **GASOLINE** Use the service station sign below to answer the following questions.
    a. What is the difference in price per gallon between the least and most expensive types of gasoline at the self-serve pump? 20¢
    b. For each type of gasoline, how much more is the cost per gallon for full service compared to self-serve? 20¢

102. **WATER SLIDES** An amusement park added a new section to a water slide to create a slide $311\frac{5}{12}$ feet long. How long was the slide before the addition? $191\frac{2}{3}$ ft

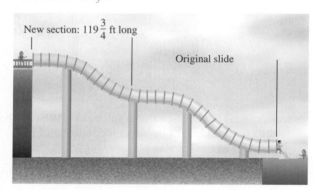

103. **JEWELRY** A jeweler cut a 7-inch-long silver wire into three pieces. To do this, he aligned a 6-inch-long ruler directly below the wire and made the indicated cuts. Find the length of piece 2 of the wire. $3\frac{1}{4}$ in.

104. **SEWING** To make some draperies, an interior decorator needs $12\frac{1}{4}$ yards of material for the den and $8\frac{1}{2}$ yards for the living room. If the material comes only in 21-yard bolts, how much will be left over after completing both sets of draperies? $\frac{1}{4}$ yd

105. **LUGGAGE** On many flights, airlines do not accept luggage whose total dimensions (length + width + height) exceed 62 inches. Estimate the total dimensions of the piece of luggage shown below. about 60 in.

106. **FREEWAY SIGNS** Estimate the distance between the Downtown San Diego off ramp and the Sea World off ramp on the 5 Freeway. about 5 mi.

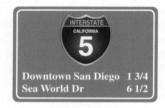

WRITING

107. Of the methods studied to add mixed numbers, which do you like better, and why?
108. **LEAP YEAR** It actually takes Earth $365\frac{1}{4}$ days, give or take a few minutes, to make one revolution around the sun. Explain why every four years we add a day to the calendar to account for this fact.
109. Explain the process of simplifying $12\frac{7}{5}$.
110. Consider the following problem:

$$108\frac{1}{3}$$
$$-99\frac{2}{3}$$

   a. Explain why borrowing is necessary.
   b. Explain how the borrowing is done.
111. Explain why $18\frac{13}{25}$ rounds to 19.
112. Explain why $18\frac{12}{25}$ rounds to 18.

REVIEW

Perform each operation and simplify, if possible.

113. a. $3\frac{1}{2} + 1\frac{1}{4}$ $4\frac{3}{4}$ b. $3\frac{1}{2} - 1\frac{1}{4}$ $2\frac{1}{4}$
     c. $3\frac{1}{2} \cdot 1\frac{1}{4}$ $4\frac{3}{8}$ d. $3\frac{1}{2} \div 1\frac{1}{4}$ $2\frac{4}{5}$

114. a. $4\frac{1}{8} + 1\frac{5}{6}$ $5\frac{23}{24}$ b. $4\frac{1}{8} - 1\frac{5}{6}$ $2\frac{7}{24}$
     c. $4\frac{1}{8} \cdot 1\frac{5}{6}$ $7\frac{9}{16}$ d. $4\frac{1}{8} \div 1\frac{5}{6}$ $2\frac{1}{4}$

SECTION 4.7
Order of Operations and Complex Fractions

Objectives
1. Use the order of operations rule.
2. Solve application problems by using the order of operations rule.
3. Evaluate formulas.
4. Simplify complex fractions.

We have seen that the order of operations rule is used to evaluate expressions that contain more than one operation. In Chapter 1, we used it to evaluate expressions involving whole numbers, and in Chapter 2, we used it to evaluate expressions involving integers. We will now use it to evaluate expressions involving fractions and mixed numbers.

1 Use the order of operations rule.

Recall from Section 1.7 that every numerical expression has only one correct value. If we don't establish a uniform order of operations, an expression can have more than one value. To avoid this possibility, always use the following rule for the order of operations.

Order of Operations

1. Perform all calculations within parentheses and other grouping symbols following the order listed in Steps 2–4 below, working from the innermost pair of grouping symbols to the outermost pair.
2. Evaluate all exponential expressions.
3. Perform all multiplications and divisions as they occur from left to right.
4. Perform all additions and subtractions as they occur from left to right.

When grouping symbols have been removed, repeat Steps 2–4 to complete the calculation.

If a fraction bar is present, evaluate the expression above the bar (called the **numerator**) and the expression below the bar (called the **denominator**) separately. Then perform the division indicated by the fraction bar, if possible.

EXAMPLE 1
Evaluate: $\dfrac{3}{4} + \dfrac{5}{3}\left(-\dfrac{1}{2}\right)^3$

Strategy We will scan the expression to determine what operations need to be performed. Then we will perform those operations, one at a time, following the order of operations rule.

WHY Every numerical expression has only one correct value. If we don't follow the correct order of operations, the expression can have more than one value.

Solution

Although the expression contains parentheses, there are no calculations to perform *within* them. We will begin with Step 2 of the rule: Evaluate all exponential expressions. We will write the steps of the solution in horizontal form.

$$\dfrac{3}{4} + \dfrac{5}{3}\left(-\dfrac{1}{2}\right)^3 = \dfrac{3}{4} + \dfrac{5}{3}\left(-\dfrac{1}{8}\right) \quad \text{Evaluate: } \left(-\tfrac{1}{2}\right)^3 = \left(-\tfrac{1}{2}\right)\left(-\tfrac{1}{2}\right)\left(-\tfrac{1}{2}\right) = -\tfrac{1}{8}.$$

$$= \dfrac{3}{4} + \left(-\dfrac{5}{24}\right) \quad \text{Multiply: } \tfrac{5}{3}\left(-\tfrac{1}{8}\right) = -\tfrac{5 \cdot 1}{3 \cdot 8} = -\tfrac{5}{24}.$$

$$= \dfrac{3}{4} \cdot \dfrac{6}{6} + \left(-\dfrac{5}{24}\right) \quad \text{Prepare to add the fractions: Their LCD is 24. To build the first fraction so that its denominator is 24, multiply it by a form of 1.}$$

Self Check 1
Evaluate: $\dfrac{7}{8} + \dfrac{3}{2}\left(-\dfrac{1}{4}\right)^2 \quad \dfrac{31}{32}$

Now Try Problem 15

Teaching Example 1
Evaluate: $\dfrac{1}{6} + \dfrac{5}{4}\left(-\dfrac{1}{3}\right)^2$
Answer: $\dfrac{11}{36}$

Chapter 4 Fractions and Mixed Numbers

$$= \frac{18}{24} + \left(-\frac{5}{24}\right)$$

Multiply the numerators: $3 \cdot 6 = 18$.
Multiply the denominators: $4 \cdot 6 = 24$.

$$= \frac{13}{24}$$

Add the numerators: $18 + (-5) = 13$. Write the sum over the common denominator 24.

If an expression contains grouping symbols, we perform the operations within the grouping symbols first.

Self Check 2

Evaluate: $\left(\frac{19}{21} - \frac{2}{3}\right) \div \left(-2\frac{1}{7}\right)$ $-\frac{1}{9}$

Now Try Problem 19

Teaching Example 2

Evaluate: $\left(\frac{29}{32} - \frac{1}{8}\right) \div \left(-1\frac{7}{8}\right)$

Answer: $-\frac{5}{12}$

EXAMPLE 2 Evaluate: $\left(\frac{7}{8} - \frac{1}{4}\right) \div \left(-2\frac{3}{16}\right)$

Strategy We will perform any operations within parentheses first.

WHY This is the first step of the order of operations rule.

Solution

We will begin by performing the subtraction within the first set of parentheses. The second set of parentheses does not contain an operation to perform.

$$\left(\frac{7}{8} - \frac{1}{4}\right) \div \left(-2\frac{3}{16}\right)$$

$$= \left(\frac{7}{8} - \frac{1}{4} \cdot \frac{2}{2}\right) \div \left(-2\frac{3}{16}\right)$$

Within the first set of parentheses, prepare to subtract the fractions: Their LCD is 8. Build $\frac{1}{4}$ so that its denominator is 8.

$$= \left(\frac{7}{8} - \frac{2}{8}\right) \div \left(-2\frac{3}{16}\right)$$

Multiply the numerators: $1 \cdot 2 = 2$.
Multiply the denominators: $4 \cdot 2 = 8$.

$$= \frac{5}{8} \div \left(-2\frac{3}{16}\right)$$

Subtract the numerators: $7 - 2 = 5$.
Write the difference over the common denominator 8.

$$= \frac{5}{8} \div \left(-\frac{35}{16}\right)$$

Write the mixed number as an improper fraction.

$$= \frac{5}{8}\left(-\frac{16}{35}\right)$$

Use the rule for division of fractions: Multiply the first fraction by the reciprocal of $-\frac{35}{16}$.

$$= -\frac{5 \cdot 16}{8 \cdot 35}$$

Multiply the numerators and multiply the denominators.
The product of two fractions with unlike signs is negative.

$$= -\frac{\overset{1}{\cancel{5}} \cdot 2 \cdot \overset{1}{\cancel{8}}}{\underset{1}{\cancel{8}} \cdot \underset{1}{\cancel{5}} \cdot 7}$$

To simplify, factor 16 as $2 \cdot 8$ and factor 35 as $5 \cdot 7$. Remove the common factors of 5 and 8 from the numerator and denominator.

$$= -\frac{2}{7}$$

Multiply the remaining factors in the numerator.
Multiply the remaining factors in the denominator.

Self Check 3

Add $2\frac{1}{4}$ to the difference of $\frac{7}{8}$ and $\frac{2}{3}$. $2\frac{11}{24}$

Now Try Problem 23

EXAMPLE 3 Add $7\frac{1}{3}$ to the difference of $\frac{5}{6}$ and $\frac{1}{4}$.

Strategy We will translate the words of the problem to numbers and symbols. Then we will use the order of operations rule to evaluate the resulting expression.

WHY Since the expression involves two operations, addition and subtraction, we need to perform them in the proper order.

Solution

The key word *difference* indicates subtraction. Since we are to add $7\frac{1}{3}$ to the difference, the difference should be written first within parentheses, followed by the addition.

Add $7\frac{1}{3}$ to the difference of $\frac{5}{6}$ and $\frac{1}{4}$.

$$\left(\frac{5}{6} - \frac{1}{4}\right) + 7\frac{1}{3} \quad \text{Translate from words to numbers and mathematical symbols.}$$

$$\left(\frac{5}{6} - \frac{1}{4}\right) + 7\frac{1}{3} = \left(\frac{5}{6} \cdot \frac{2}{2} - \frac{1}{4} \cdot \frac{3}{3}\right) + 7\frac{1}{3} \quad \begin{array}{l}\text{Prepare to subtract the fractions}\\\text{within the parentheses. Build the}\\\text{fractions so that their denominators}\\\text{are the LCD 12.}\end{array}$$

$$= \left(\frac{10}{12} - \frac{3}{12}\right) + 7\frac{1}{3} \quad \begin{array}{l}\text{Multiply the numerators.}\\\text{Multiply the denominators.}\end{array}$$

$$= \frac{7}{12} + 7\frac{1}{3} \quad \begin{array}{l}\text{Subtract the numerators: } 10 - 3 = 7.\\\text{Write the difference over the common}\\\text{denominator 12.}\end{array}$$

$$= \frac{7}{12} + 7\frac{4}{12} \quad \begin{array}{l}\text{Prepare to add the fractions. Build } \frac{1}{3} \text{ so that its}\\\text{denominator is 12: } \frac{1}{3} \cdot \frac{4}{4} = \frac{4}{12}.\end{array}$$

$$= 7\frac{11}{12} \quad \begin{array}{l}\text{Add the numerators of the fractions: } 7 + 4 = 11.\\\text{Write the sum over the common denominator 12.}\end{array}$$

Teaching Example 3 Add $4\frac{1}{6}$ to the difference of $\frac{7}{9}$ and $\frac{3}{4}$.
Answer:
$4\frac{7}{36}$

2 Solve application problems by using the order of operations rule.

Sometimes more than one operation is needed to solve a problem.

EXAMPLE 4
Masonry To build a wall, a mason will use blocks that are $5\frac{3}{4}$ inches high, held together with $\frac{3}{8}$-inch-thick layers of mortar. If the plans call for 8 layers, called *courses*, of blocks, what will be the height of the wall when completed?

Blocks $5\frac{3}{4}$ in. high
Mortar $\frac{3}{8}$ in. thick

Solution
- The blocks are $5\frac{3}{4}$ inches high.   Given
- A layer of mortar is $\frac{3}{8}$ inch thick.   Given
- There are 8 layers (courses) of blocks.   Given
- What is the height of the wall when completed?   Find

To find the height of the wall when it is completed, we could add the heights of 8 blocks and 8 layers of mortar. However, it will be simpler if we find the height of one block and one layer of mortar, and multiply that result by 8.

The height of the wall when completed	is equal to	8	times	(the height of one block	plus	the thickness of one layer of mortar.)
The height of the wall when completed	=	8		($5\frac{3}{4}$	+	$\frac{3}{8}$)

Self Check 4

MASONRY Find the height of a wall if 8 layers (called *courses*) of $7\frac{3}{8}$-inch-high blocks are held together by $\frac{1}{4}$-inch-thick layers of mortar.   61 in.

Now Try Problem 77

Teaching Example 4 MASONRY Find the height of a planter if 8 layers (called *courses*) of $2\frac{1}{2}$-inch-high bricks are held together by $\frac{3}{8}$-inch-thick layers of mortar.
Answer:
23 in.

Unless otherwise noted, all art on this page is © Cengage Learning.

To evaluate the expression, we use the order of operations rule.

$$8\left(5\frac{3}{4}+\frac{3}{8}\right)=8\left(5\frac{6}{8}+\frac{3}{8}\right)$$ Prepare to add the fractions within the parentheses: Their LCD is 8. Build $\frac{3}{4}$ so that its denominator is 8: $\frac{3}{4}\cdot\frac{2}{2}=\frac{6}{8}$.

$$=8\left(5\frac{9}{8}\right)$$ Add the numerators of the fractions: $6+3=9$. Write the sum over the common denominator 8.

$$=\frac{8}{1}\left(\frac{49}{8}\right)$$ Prepare to multiply the fractions. Write $5\frac{9}{8}$ as an improper fraction.

$$=\frac{\overset{1}{8}\cdot 49}{1\cdot \underset{1}{8}}$$ Multiply the numerators and multiply the denominators. To simplify, remove the common factor of 8 from the numerator and denominator.

$$=49$$ Simplify: $\frac{49}{1}=49$.

The completed wall will be 49 inches high.

Check: We can estimate to check the result. Since one block and one layer of mortar is about 6 inches high, eight layers of blocks and mortar would be $8 \cdot 6$ inches, or 48 inches high. The result of 49 inches seems reasonable.

3 Evaluate formulas.

To evaluate a formula, we replace its letters, called **variables**, with specific numbers and evaluate the right side using the order of operations rule.

Self Check 5

The formula for the area of a triangle is $A=\frac{1}{2}bh$. Find the area of a triangle whose base is $12\frac{1}{2}$ meters long and whose height is $15\frac{1}{3}$ meters. $95\frac{5}{6}$ m²

Now Try Problems 27 and 87

A trapezoid

Teaching Example 5 The formula for the area of a trapezoid is $A=\frac{1}{2}h(a+b)$. Find A when $h=2\frac{1}{8}$ ft, $a=1\frac{3}{4}$ ft, and $b=4\frac{1}{4}$ ft.
Answer:
$6\frac{3}{8}$ ft²

EXAMPLE 5 The formula for the area of a trapezoid is $A=\frac{1}{2}h(a+b)$, where A is the area, h is the height, and a and b are the lengths of its bases. Find A when $h=1\frac{2}{3}$ in., $a=2\frac{1}{2}$ in., and $b=5\frac{1}{2}$ in.

Strategy In the formula, we will replace the letter h with $1\frac{2}{3}$, the letter a with $2\frac{1}{2}$, and the letter b with $5\frac{1}{2}$.

WHY Then we can use the order of operations rule to find the value of the expression on the right side of the $=$ symbol.

Solution

$$A=\frac{1}{2}h(a+b)$$ This is the formula for the area of a trapezoid.

$$A=\frac{1}{2}\left(1\frac{2}{3}\right)\left(2\frac{1}{2}+5\frac{1}{2}\right)$$ Replace h, a, and b with the given values.

$$A=\frac{1}{2}\left(1\frac{2}{3}\right)(8)$$ Do the addition within the parentheses: $2\frac{1}{2}+5\frac{1}{2}=8$.

$$A=\frac{1}{2}\left(\frac{5}{3}\right)\left(\frac{8}{1}\right)$$ To prepare to multiply fractions, write $1\frac{2}{3}$ as an improper fraction and 8 as $\frac{8}{1}$.

$$A=\frac{1\cdot 5\cdot 8}{2\cdot 3\cdot 1}$$ Multiply the numerators. Multiply the denominators.

$$A=\frac{1\cdot 5\cdot \overset{1}{2}\cdot 4}{\underset{1}{2}\cdot 3\cdot 1}$$ To simplify, factor 8 as $2\cdot 4$. Then remove the common factor of 2 from the numerator and denominator.

$$A=\frac{20}{3}$$ Multiply the remaining factors in the numerator. Multiply the remaining factors in the denominator.

$$A=6\frac{2}{3}$$ Write the improper fraction $\frac{20}{3}$ as a mixed number by dividing 20 by 3.

The area of the trapezoid is $6\frac{2}{3}$ in.².

4 Simplify complex fractions.

Fractions whose numerators and/or denominators contain fractions are called *complex fractions*. Here is an example of a complex fraction:

A fraction in the numerator ⟶ $\dfrac{\frac{3}{4}}{\frac{7}{8}}$ ⟵ The main fraction bar

A fraction in the denominator ⟶

Complex Fraction

A **complex fraction** is a fraction whose numerator or denominator, or both, contain one or more fractions or mixed numbers.

Here are more examples of complex fractions:

$$\dfrac{-\frac{1}{4} - \frac{4}{5}}{2\frac{4}{5}} \quad\quad \dfrac{\frac{1}{3} + \frac{1}{4}}{\frac{1}{3} - \frac{1}{4}}$$

⟵ Numerator ⟶
⟵ Main fraction bar ⟶
⟵ Denominator ⟶

To *simplify* a complex fraction means to express it as a fraction in simplified form.

The following method for simplifying complex fractions is based on the fact that the main fraction bar indicates division.

$$\dfrac{\frac{1}{4}}{\frac{2}{5}}$$ ⟵ The main fraction bar means "divide the fraction in the numerator by the fraction in the denominator." ⟶ $\dfrac{1}{4} \div \dfrac{2}{5}$

Simplifying a Complex Fraction

To simplify a complex fraction:

1. Add or subtract in the numerator and/or denominator so that the numerator is a single fraction and the denominator is a single fraction.
2. Perform the indicated division by multiplying the numerator of the complex fraction by the reciprocal of the denominator.
3. Simplify the result, if possible.

After simplifying a complex fraction, the result is often an improper fraction. Check with your instructor to see if he or she will accept an answer in that form or prefers that you express the final result as a mixed number.

EXAMPLE 6

Simplify: $\dfrac{\frac{1}{4}}{\frac{2}{5}}$

Strategy We will perform the division indicated by the main fraction bar using the rule for dividing fractions from Section 4.3.

WHY We can skip step 1 and immediately divide because the numerator and the denominator of the complex fraction are already single fractions.

Self Check 6

Simplify: $\dfrac{\frac{1}{6}}{\frac{3}{8}} \quad \dfrac{4}{9}$

Now Try Problem 31

Teaching Example 6

Simplify: $\dfrac{\frac{1}{8}}{\frac{2}{7}}$

Answer: $\dfrac{7}{16}$

Solution

$$\dfrac{\frac{1}{4}}{\frac{2}{5}} = \dfrac{1}{4} \div \dfrac{2}{5}$$ Write the division indicated by the main fraction bar using a ÷ symbol.

$$= \dfrac{1}{4} \cdot \dfrac{5}{2}$$ Use the rule for dividing fractions: Multiply the first fraction by the reciprocal of $\frac{2}{5}$, which is $\frac{5}{2}$.

$$= \dfrac{1 \cdot 5}{4 \cdot 2}$$ Multiply the numerators. Multiply the denominators.

$$= \dfrac{5}{8}$$

Self Check 7

Simplify: $\dfrac{-\frac{5}{8}+\frac{1}{3}}{\frac{3}{4}-\frac{1}{3}} - \dfrac{7}{10}$

Now Try Problem 35

Teaching Example 7

Simplify: $\dfrac{-\frac{8}{9}+\frac{3}{4}}{\frac{5}{6}-\frac{1}{9}}$

Answer: $-\dfrac{5}{26}$

EXAMPLE 7

Simplify: $\dfrac{-\frac{1}{4}+\frac{2}{5}}{\frac{1}{2}-\frac{4}{5}}$

Strategy Recall that a fraction bar is a type of grouping symbol. We will work above and below the main fraction bar separately to write $-\frac{1}{4} + \frac{2}{5}$ and $\frac{1}{2} - \frac{4}{5}$ as single fractions.

WHY The numerator and the denominator of the complex fraction must be written as single fractions before dividing.

Solution To write the numerator as a single fraction, we build $-\frac{1}{4}$ and $\frac{2}{5}$ to have an LCD of 20, and then add. To write the denominator as a single fraction, we build $\frac{1}{2}$ and $\frac{4}{5}$ to have an LCD of 10, and subtract.

$$\dfrac{-\frac{1}{4}+\frac{2}{5}}{\frac{1}{2}-\frac{4}{5}} = \dfrac{-\frac{1}{4}\cdot\frac{5}{5}+\frac{2}{5}\cdot\frac{4}{4}}{\frac{1}{2}\cdot\frac{5}{5}-\frac{4}{5}\cdot\frac{2}{2}}$$
The LCD for the numerator is 20. Build each fraction so that each has a denominator of 20.
The LCD for the denominator is 10. Build each fraction so that each has a denominator of 10.

$$= \dfrac{-\frac{5}{20}+\frac{8}{20}}{\frac{5}{10}-\frac{8}{10}}$$ Multiply in the numerator. Multiply in the denominator.

$$= \dfrac{\frac{3}{20}}{-\frac{3}{10}}$$ In the numerator of the complex fraction, add the fractions. In the denominator, subtract the fractions.

$$= \dfrac{3}{20} \div \left(-\dfrac{3}{10}\right)$$ Write the division indicated by the main fraction bar using a ÷ symbol.

$$= \dfrac{3}{20}\left(-\dfrac{10}{3}\right)$$ Multiply the first fraction by the reciprocal of $-\frac{3}{10}$, which is $-\frac{10}{3}$.

$$= -\dfrac{3 \cdot 10}{20 \cdot 3}$$ The product of two fractions with unlike signs is negative. Multiply the numerators. Multiply the denominators.

$$= -\dfrac{\overset{1}{\cancel{3}} \cdot \overset{1}{\cancel{10}}}{2 \cdot \underset{1}{\cancel{10}} \cdot \underset{1}{\cancel{3}}}$$ To simplify, factor 20 as 2 · 10. Then remove the common factors of 3 and 10 from the numerator and denominator.

$$= -\dfrac{1}{2}$$ Multiply the remaining factors in the numerator. Multiply the remaining factors in the denominator.

EXAMPLE 8

Simplify: $\dfrac{7 - \dfrac{2}{3}}{4\dfrac{5}{6}}$

Strategy Recall that a fraction bar is a type of grouping symbol. We will work above and below the main fraction bar separately to write $7 - \dfrac{2}{3}$ as a single fraction and $4\dfrac{5}{6}$ as an improper fraction.

WHY The numerator and the denominator of the complex fraction must be written as single fractions before dividing.

Solution

$\dfrac{7 - \dfrac{2}{3}}{4\dfrac{5}{6}} = \dfrac{\dfrac{7}{1} \cdot \dfrac{3}{3} - \dfrac{2}{3}}{\dfrac{29}{6}}$ In the numerator, write 7 as $\dfrac{7}{1}$. The LCD for the numerator is 3. Build $\dfrac{7}{1}$ so that it has a denominator of 3. In the denominator, write $4\dfrac{5}{6}$ as the improper fraction $\dfrac{29}{6}$.

$= \dfrac{\dfrac{21}{3} - \dfrac{2}{3}}{\dfrac{29}{6}}$ Multiply in the numerator.

$= \dfrac{\dfrac{19}{3}}{\dfrac{29}{6}}$ In the numerator of the complex fraction, subtract the numerators: $21 - 2 = 19$. Then write the difference over the common denominator 3.

$= \dfrac{19}{3} \div \dfrac{29}{6}$ Write the division indicated by the main fraction bar using a $\div$ symbol.

$= \dfrac{19}{3} \cdot \dfrac{6}{29}$ Multiply the first fraction by the reciprocal of $\dfrac{29}{6}$, which is $\dfrac{6}{29}$.

$= \dfrac{19 \cdot 6}{3 \cdot 29}$ Multiply the numerators. Multiply the denominators.

$= \dfrac{19 \cdot 2 \cdot \overset{1}{\cancel{3}}}{\underset{1}{\cancel{3}} \cdot 29}$ To simplify, factor 6 as $2 \cdot 3$. Then remove the common factor of 3 from the numerator and denominator.

$= \dfrac{38}{29}$ Multiply the remaining factors in the numerator. Multiply the remaining factors in the denominator.

Self Check 8

Simplify: $\dfrac{5 - \dfrac{3}{4}}{1\dfrac{7}{8}}$ $\dfrac{34}{15}$

Now Try Problem 39

Teaching Example 8

Simplify: $\dfrac{8 - \dfrac{1}{4}}{1\dfrac{7}{8}}$

Answer: $\dfrac{62}{15}$

ANSWERS TO SELF CHECKS

1. $\dfrac{31}{32}$ 2. $-\dfrac{1}{9}$ 3. $2\dfrac{11}{24}$ 4. 61 in.   5. $95\dfrac{5}{6}$ m²   6. $\dfrac{4}{9}$ 7. $-\dfrac{7}{10}$ 8. $\dfrac{34}{15}$

SECTION 4.7   STUDY SET

VOCABULARY

Fill in the blanks.

1. We use the order of __operations__ rule to evaluate expressions that contain more than one operation.

2. To evaluate a formula such as $A = \frac{1}{2}h(a + b)$, we substitute specific numbers for the letters, called __variables__, in the formula and find the value of the right side.

Chapter 4 Fractions and Mixed Numbers

3. $\dfrac{\frac{1}{2}}{\frac{3}{4}}$ and $\dfrac{\frac{7}{8}+\frac{2}{5}}{\frac{1}{2}-\frac{1}{3}}$ are examples of __complex__ fractions.

4. In the complex fraction $\dfrac{\frac{2}{5}+\frac{1}{4}}{\frac{2}{5}-\frac{1}{4}}$, the __numerator__ is $\frac{2}{5}+\frac{1}{4}$ and the __denominator__ is $\frac{2}{5}-\frac{1}{4}$.

CONCEPTS

5. What operations are involved in this expression?

   $5\left(6\frac{1}{3}\right)+\left(-\frac{1}{4}\right)^3$ raising to a power (exponent), multiplication, and addition

6. a. To evaluate $\frac{7}{8}+\left(\frac{1}{3}\right)\left(\frac{1}{4}\right)$, what operation should be performed first? multiplication
   b. To evaluate $\frac{7}{8}+\left(\frac{1}{3}-\frac{1}{4}\right)^2$, what operation should be performed first? subtraction

7. Translate the following to numbers and symbols. *You do not have to find the answer.*

   Add $1\frac{2}{15}$ to the difference of $\frac{2}{3}$ and $\frac{1}{10}$. $\left(\frac{2}{3}-\frac{1}{10}\right)+1\frac{2}{15}$

8. Refer to the trapezoid shown below. Label the length of the upper base $3\frac{1}{2}$ inches, the length of the lower base $5\frac{1}{2}$ inches, and the height $2\frac{2}{3}$ inches.

9. What division is represented by this complex fraction?

   $\dfrac{\frac{2}{3}}{\frac{1}{5}}$ $\frac{2}{3} \div \frac{1}{5}$

10. Consider: $\dfrac{\frac{2}{3}-\frac{1}{5}}{\frac{1}{2}+\frac{4}{5}}$
    a. What is the LCD for the fractions in the numerator of this complex fraction? 15
    b. What is the LCD for the fractions in the denominator of this complex fraction? 10

11. Write the denominator of the following complex fraction as an improper fraction.

    $\dfrac{\frac{1}{8}-\frac{3}{16}}{5\frac{3}{4}}$ $\dfrac{\frac{1}{8}-\frac{3}{16}}{\frac{23}{4}}$

12. When this complex fraction is simplified, will the result be positive or negative?

    $\dfrac{-\frac{2}{3}}{\frac{3}{4}}$ negative

NOTATION

Fill in the blanks to complete each step.

13. $\dfrac{7}{12}-\dfrac{1}{2}\cdot\dfrac{1}{3}=\dfrac{7}{12}-\dfrac{1\cdot 1}{2\cdot 3}$

    $=\dfrac{7}{12}-\dfrac{1}{6}$

    $=\dfrac{7}{12}-\dfrac{1}{6}\cdot\dfrac{2}{2}$

    $=\dfrac{7}{12}-\dfrac{2}{12}$

    $=\dfrac{5}{12}$

14. $\dfrac{\frac{1}{8}}{\frac{3}{4}}=\dfrac{1}{8}\div\dfrac{3}{4}$

    $=\dfrac{1}{8}\cdot\dfrac{4}{3}$

    $=\dfrac{1\cdot 4}{8\cdot 3}$

    $=\dfrac{1\cdot \cancel{4}}{2\cdot \cancel{4}\cdot 3}$

    $=\dfrac{1}{6}$

GUIDED PRACTICE

Evaluate each expression. See Example 1.

15. $\dfrac{3}{4}+\dfrac{2}{5}\left(-\dfrac{1}{2}\right)^2$ $\dfrac{17}{20}$

16. $\dfrac{1}{4}+\dfrac{8}{27}\left(-\dfrac{3}{2}\right)^2$ $\dfrac{11}{12}$

17. $\dfrac{1}{6}+\dfrac{9}{8}\left(-\dfrac{2}{3}\right)^3$ $-\dfrac{1}{6}$

18. $\dfrac{1}{5}+\dfrac{1}{9}\left(-\dfrac{3}{2}\right)^3$ $-\dfrac{7}{40}$

Evaluate each expression. See Example 2.

19. $\left(\dfrac{3}{4} - \dfrac{1}{6}\right) \div \left(-2\dfrac{1}{6}\right)$ $-\dfrac{7}{26}$

20. $\left(\dfrac{7}{8} - \dfrac{3}{7}\right) \div \left(-1\dfrac{3}{7}\right)$ $-\dfrac{5}{16}$

21. $\left(\dfrac{15}{16} - \dfrac{1}{8}\right) \div \left(-9\dfrac{3}{4}\right)$ $-\dfrac{1}{12}$

22. $\left(\dfrac{19}{36} - \dfrac{1}{6}\right) \div \left(-8\dfrac{2}{3}\right)$ $-\dfrac{1}{24}$

Evaluate each expression. See Example 3.

23. Add $5\dfrac{4}{15}$ to the difference of $\dfrac{5}{6}$ and $\dfrac{2}{3}$. $5\dfrac{13}{30}$

24. Add $8\dfrac{5}{24}$ to the difference of $\dfrac{3}{4}$ and $\dfrac{1}{6}$. $8\dfrac{19}{24}$

25. Add $2\dfrac{7}{18}$ to the difference of $\dfrac{7}{9}$ and $\dfrac{1}{2}$. $2\dfrac{2}{3}$

26. Add $1\dfrac{19}{30}$ to the difference of $\dfrac{4}{5}$ and $\dfrac{1}{2}$. $1\dfrac{14}{15}$

Evaluate the formula $A = \tfrac{1}{2}h(a+b)$ *for the given values. See Example 5.*

27. $a = 2\dfrac{1}{2}, b = 7\dfrac{1}{2}, h = 5\dfrac{1}{4}$ $26\dfrac{1}{4}$

28. $a = 4\dfrac{1}{2}, b = 5\dfrac{1}{2}, h = 2\dfrac{1}{8}$ $10\dfrac{5}{8}$

29. $a = 1\dfrac{1}{4}, b = 6\dfrac{3}{4}, h = 4\dfrac{1}{2}$ 18

30. $a = 1\dfrac{1}{3}, b = 4\dfrac{2}{3}, h = 2\dfrac{2}{5}$ $7\dfrac{1}{5}$

Simplify each complex fraction. See Example 6.

31. $\dfrac{\tfrac{1}{16}}{\tfrac{2}{5}}$ $\dfrac{5}{32}$

32. $\dfrac{\tfrac{2}{11}}{\tfrac{3}{4}}$ $\dfrac{8}{33}$

33. $\dfrac{\tfrac{5}{8}}{\tfrac{3}{4}}$ $\dfrac{5}{6}$

34. $\dfrac{\tfrac{1}{5}}{\tfrac{8}{15}}$ $\dfrac{3}{8}$

Simplify each complex fraction. See Example 7.

35. $\dfrac{-\tfrac{1}{4} + \tfrac{2}{3}}{\tfrac{5}{6} + \tfrac{2}{3}}$ $\dfrac{5}{18}$

36. $\dfrac{-\tfrac{1}{2} + \tfrac{7}{8}}{\tfrac{3}{4} - \tfrac{1}{2}}$ $\dfrac{3}{2}$

37. $\dfrac{\tfrac{1}{3} - \tfrac{3}{4}}{\tfrac{1}{6} + \tfrac{2}{3}}$ $-\dfrac{1}{2}$

38. $\dfrac{\tfrac{1}{3} - \tfrac{3}{4}}{\tfrac{1}{6} + \tfrac{1}{3}}$ $-\dfrac{5}{6}$

Simplify each complex fraction. See Example 8.

39. $\dfrac{5 - \tfrac{5}{6}}{1\tfrac{1}{12}}$ $\dfrac{50}{13}$

40. $\dfrac{4 - \tfrac{3}{4}}{1\tfrac{7}{8}}$ $\dfrac{26}{15}$

41. $\dfrac{4 - \tfrac{7}{8}}{3\tfrac{1}{4}}$ $\dfrac{25}{26}$

42. $\dfrac{6 - \tfrac{2}{7}}{6\tfrac{2}{3}}$ $\dfrac{6}{7}$

TRY IT YOURSELF

Evaluate each expression or formula. Simplify each complex fraction.

43. $\dfrac{7}{8} - \left(\dfrac{4}{5} + 1\dfrac{3}{4}\right)$ $-1\dfrac{27}{40}$

44. $\left(\dfrac{5}{4}\right)^2 + \left(\dfrac{2}{3} - 2\dfrac{1}{6}\right)$ $\dfrac{1}{16}$

45. $\dfrac{-\tfrac{14}{15}}{\tfrac{7}{10}}$ $-1\dfrac{1}{3}$

46. $\dfrac{\tfrac{5}{27}}{-\tfrac{5}{9}}$ $-\dfrac{1}{3}$

47. $A = \tfrac{1}{2}bh$ for $b = 10$ and $h = 7\dfrac{1}{5}$ 36

48. $V = lwh$ for $l = 12$, $w = 8\dfrac{1}{2}$, and $h = 3\dfrac{1}{3}$ 340

49. $\dfrac{2}{3}\left(-\dfrac{1}{4}\right) + \dfrac{1}{2}$ $\dfrac{1}{3}$

50. $-\dfrac{7}{8} - \left(\dfrac{1}{8}\right)\left(\dfrac{2}{3}\right)$ $-\dfrac{23}{24}$

51. $\dfrac{1}{2}\left(\dfrac{1}{8}\right) + \left(-\dfrac{1}{4}\right)^2$ $\dfrac{1}{8}$

52. $-\dfrac{3}{16} - \left(-\dfrac{1}{2}\right)^3$ $-\dfrac{1}{16}$

53. $\dfrac{\tfrac{3}{8} + \tfrac{1}{4}}{\tfrac{3}{8} - \tfrac{1}{4}}$ 5

54. $\dfrac{\tfrac{2}{3} - \tfrac{5}{2}}{\tfrac{2}{3} - \tfrac{3}{2}}$ $\dfrac{11}{5}$

55. Add $12\dfrac{11}{12}$ to the difference of $5\dfrac{1}{6}$ and $3\dfrac{7}{8}$. $14\dfrac{5}{24}$

56. Add $18\dfrac{1}{3}$ to the difference of $11\dfrac{3}{5}$ and $9\dfrac{11}{15}$. $20\dfrac{1}{5}$

57. $\dfrac{5\frac{1}{2}}{-\frac{1}{4}+\frac{3}{4}}$ 11

58. $\dfrac{4\frac{1}{4}}{\frac{2}{3}+\left(-\frac{1}{6}\right)}$ $\dfrac{17}{2}$

59. $\left|\dfrac{2}{3}-\dfrac{9}{10}\right|\div\left(-\dfrac{1}{5}\right)$ $-1\frac{1}{6}$

60. $\left|-\dfrac{3}{16}\div 2\dfrac{1}{4}\right|+\left(-2\dfrac{1}{8}\right)$ $-2\dfrac{1}{24}$

61. $\dfrac{\frac{1}{5}-\left(-\frac{1}{4}\right)}{\frac{1}{4}+\frac{4}{5}}$ $\dfrac{3}{7}$

62. $\dfrac{\frac{1}{8}-\left(-\frac{1}{2}\right)}{\frac{1}{4}+\frac{3}{8}}$ 1

63. $1\dfrac{3}{5}\left(\dfrac{1}{2}\right)^2\left(\dfrac{3}{4}\right)$ $\dfrac{3}{10}$

64. $2\dfrac{3}{5}\left(-\dfrac{1}{3}\right)^2\left(\dfrac{1}{2}\right)$ $\dfrac{13}{90}$

65. $A = lw$ for $l = 5\dfrac{5}{6}$ and $w = 7\dfrac{3}{5}$ $44\dfrac{1}{3}$

66. $P = 2l + 2w$ for $l = \dfrac{7}{8}$ and $w = \dfrac{3}{5}$ $2\dfrac{19}{20}$

67. $\left(2-\dfrac{1}{2}\right)^2+\left(2+\dfrac{1}{2}\right)^2$ $8\dfrac{1}{2}$

68. $\left(\dfrac{9}{20}\div 2\dfrac{2}{5}\right)+\left(\dfrac{3}{4}\right)^2$ $\dfrac{3}{4}$

69. $\dfrac{-\frac{5}{6}}{-1\frac{7}{8}}$ $\dfrac{4}{9}$

70. $\dfrac{-\frac{4}{3}}{-2\frac{5}{6}}$ $\dfrac{8}{17}$

71. Subtract $9\dfrac{1}{10}$ from the sum of $7\dfrac{3}{7}$ and $3\dfrac{1}{5}$. $1\dfrac{37}{70}$

72. Subtract $3\dfrac{2}{3}$ from the sum of $2\dfrac{5}{12}$ and $1\dfrac{5}{8}$. $\dfrac{3}{8}$

73. $\dfrac{\frac{3}{5}-2}{\frac{2}{5}-2}$ $\dfrac{7}{8}$

74. $\dfrac{\frac{1}{3}+\frac{1}{4}}{\frac{1}{3}-\frac{1}{4}}$ 7

75. $\left(\dfrac{8}{5}-1\dfrac{1}{3}\right)-\left(-\dfrac{4}{5}\cdot 10\right)$ $8\dfrac{4}{15}$

76. $\left(1-\dfrac{3}{4}\right)\left(1+\dfrac{3}{4}\right)$ $\dfrac{7}{16}$

APPLICATIONS

77. **REMODELING A BATHROOM** A handyman installed 20 rows of grout and tile on a bathroom wall using the pattern shown below. How high above floor level does the tile work reach? (*Hint:* There is no grout line above the last row of tiles.) $91\dfrac{1}{4}$ in.

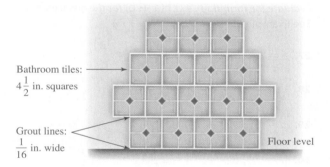

Bathroom tiles: $4\dfrac{1}{2}$ in. squares

Grout lines: $\dfrac{1}{16}$ in. wide

Floor level

78. **PLYWOOD** To manufacture a sheet of plywood, several thin layers of wood are glued together, as shown. Then an exterior finish is attached to the top and the bottom, as shown below. How thick is the final product? 1 in.

Exterior finish pieces: $\dfrac{1}{8}$ in. each

Inner layers: $\dfrac{3}{16}$ in. each

Unless otherwise noted, all art on this page is © Cengage Learning.

79. POSTAGE RATES Can the advertising package shown below be mailed for the 1-ounce rate? yes

80. PHYSICAL THERAPY After back surgery, a patient followed a walking program shown in the table below to strengthen her muscles. What was the total distance she walked over this three-week period? $10\frac{1}{2}$ mi

Week	Distance per day
#1	$\frac{1}{4}$ mile
#2	$\frac{1}{2}$ mile
#3	$\frac{3}{4}$ mile

81. READING PROGRAMS To improve reading skills, elementary school children read silently at the end of the school day for $\frac{1}{4}$ hour on Mondays and for $\frac{1}{2}$ hour on Fridays. For the month of January, how many total hours did the children read silently in class? $3\frac{1}{4}$ hr

S	M	T	W	T	F	S
	1	2	3	4	5	6
7	8	9	10	11	12	13
14	15	16	17	18	19	20
21	22	23	24	25	26	27
28	29	30	31			

82. PHYSICAL FITNESS Two people begin their workouts from the same point on a bike path and travel in opposite directions, as shown below. How far apart are they in $1\frac{1}{2}$ hours? Use the table to help organize your work. $14\frac{11}{20}$ mi

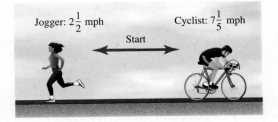

83. HIKING A scout troop plans to hike from the campground to Glenn Peak, as shown below. Since the terrain is steep, they plan to stop and rest after every $\frac{2}{3}$ mile. With this plan, how many parts will there be to this hike? 9 parts

84. DELI SHOPS A sandwich shop sells a $\frac{1}{2}$-pound club sandwich made of turkey and ham. The owner buys the turkey in $1\frac{3}{4}$-pound packages and the ham in $2\frac{1}{2}$-pound packages. If he mixes two packages of turkey and one package of ham together, how many sandwiches can he make from the mixture? 12 sandwiches

85. SKIN CREAMS Using a formula of $\frac{1}{2}$ ounce of sun block, $\frac{2}{3}$ ounce of moisturizing cream, and $\frac{3}{4}$ ounce of lanolin, a beautician mixes her own brand of skin cream. She packages it in $\frac{1}{4}$-ounce tubes. How many full tubes can be produced using this formula? How much skin cream is left over? 7 full tubes, $\frac{2}{3}$ of a tube is left over

86. **SLEEP** The graph below compares the amount of sleep a 1-month-old baby got to the $15\frac{1}{2}$-hour daily requirement recommended by Children's Hospital of Orange County, California. For the week, how far below the baseline was the baby's daily average? $\frac{3}{4}$ hr under baseline

87. **CAMPING** The four sides of a tent are all the same trapezoid-shape. (See the illustration below.) How many square yards of canvas are used to make one of the sides of the tent? 7 yd²

88. **SEWING** A seamstress begins with a trapezoid-shaped piece of denim to make the back pocket on a pair of jeans. (See the illustration below.) How many square inches of denim are used to make the pocket? $43\frac{1}{2}$ in.²

89. **AMUSEMENT PARKS** At the end of a ride at an amusement park, a boat splashes into a pool of water. The time (in seconds) that it takes two pipes to refill the pool is given by

$$\frac{1}{\frac{1}{10} + \frac{1}{15}}$$

Simplify the complex fraction to find the time. 6 sec

90. **ALGEBRA** Complex fractions, like the one shown below, are seen in an algebra class when the topic of *slope of a line* is studied. Simplify this complex fraction and, as is done in algebra, write the answer as an improper fraction. $\frac{10}{3}$

$$\frac{\frac{1}{2} - \frac{1}{3}}{\frac{1}{4} - \frac{1}{5}}$$

WRITING

91. Why is an order of operations rule necessary?
92. What does it mean to evaluate a formula?
93. What is a complex fraction?
94. In the complex fraction $\dfrac{\frac{3}{8} + \frac{1}{4}}{\frac{3}{8} - \frac{1}{4}}$, the fraction bar serves as a grouping symbol. Explain why this is so.

REVIEW

95. Solve: $a - 5 = 66$ 71
96. Solve: $9 = p - 9$ 18
97. Solve: $2x + 5 = 17$ 6
98. Solve: $-1 - h = -9$ 8
99. List the factors of 24. 1, 2, 3, 4, 6, 8, 12, 24
100. Find the prime factorization of 24. $2^3 \cdot 3$

SECTION 4.8
Solving Equations That Involve Fractions

In this section, we will discuss how to solve equations that involve fractions and equations whose solutions are fractions. We will make use of several concepts from this chapter, including the reciprocal and the LCD.

Objectives

1. Use the addition and subtraction properties of equality to solve equations that involve fractions.
2. Use reciprocals to solve equations.
3. Clear equations of fractions.
4. Use equations to solve application problems that involve fractions.

1 Use the addition and subtraction properties of equality to solve equations that involve fractions.

Recall that to **solve an equation**, we find all the values of the variable that make the equation true. The properties of equality that we used to solve equations involving whole numbers and integers are also used to solve equations involving fractions.

EXAMPLE 1 Solve: $y - \dfrac{15}{32} = \dfrac{5}{32}$

Strategy We will use the addition property of equality to isolate the variable y on the left side of the equation.

WHY To solve the original equation, we want to find a simpler equivalent equation of the form $y =$ **a number**, whose solution is obvious.

Solution

$$y - \dfrac{15}{32} = \dfrac{5}{32}$$ This is the equation to solve.

$$y - \dfrac{15}{32} + \dfrac{15}{32} = \dfrac{5}{32} + \dfrac{15}{32}$$ To isolate y, undo the subtraction of $\dfrac{15}{32}$ by adding $\dfrac{15}{32}$ to both sides.

$$y = \dfrac{20}{32}$$ On the left side, $-\dfrac{15}{32} + \dfrac{15}{32} = 0$. On the right side, $\dfrac{5}{32} + \dfrac{15}{32} = \dfrac{20}{32}$.

$$y = \dfrac{\overset{1}{\cancel{4}} \cdot 5}{\underset{1}{\cancel{4}} \cdot 8}$$ To simplify the fraction, factor 20 as $4 \cdot 5$ and 32 as $4 \cdot 8$. Then remove the common factor of 4 from the numerator and denominator.

$$y = \dfrac{5}{8}$$ Multiply the remaining factors in the numerator: $1 \cdot 5 = 5$. Multiply the remaining factors in the denominator: $1 \cdot 8 = 8$.

To check this result, substitute $\dfrac{5}{8}$ for y in the original equation and simplify.

Check:

$$y - \dfrac{15}{32} = \dfrac{5}{32}$$ This is the original equation.

$$\dfrac{5}{8} - \dfrac{15}{32} \overset{?}{=} \dfrac{5}{32}$$ Substitute $\dfrac{5}{8}$ for y.

$$\dfrac{5}{8} \cdot \dfrac{4}{4} - \dfrac{15}{32} \overset{?}{=} \dfrac{5}{32}$$ To prepare to subtract the fractions on the left side, build $\dfrac{5}{8}$ so that its denominator is 32.

$$\dfrac{20}{32} - \dfrac{15}{32} \overset{?}{=} \dfrac{5}{32}$$ On the left side, multiply the numerators and multiply the denominators.

$$\dfrac{5}{32} = \dfrac{5}{32}$$ On the left side, subtract the numerators and write the difference over the common denominator, 32.

Since $\dfrac{5}{32} = \dfrac{5}{32}$ is a true statement, $\dfrac{5}{8}$ is the solution of $y - \dfrac{15}{32} = \dfrac{5}{32}$.

Self Check 1

Solve: $a - \dfrac{1}{16} = \dfrac{11}{16} \dfrac{3}{4}$

Now Try Problem 17

Teaching Example 1

Solve: $x - \dfrac{13}{18} = \dfrac{1}{18}$

Answer: $\dfrac{7}{9}$

Chapter 4 Fractions and Mixed Numbers

Self Check 2
Solve $\frac{2}{3} = y + \frac{1}{5}$ and check the result. $\frac{7}{15}$

Now Try Problem 21

Teaching Example 2
Solve $\frac{7}{8} = m + \frac{1}{3}$ and check the result.
Answer: $\frac{13}{24}$

EXAMPLE 2
Solve: $\frac{3}{4} = x + \frac{1}{6}$

Strategy We will use the subtraction property of equality to isolate the variable x on the right side of the equation.

WHY To solve the original equation, we want to find a simpler equivalent equation of the form **a number = x**, whose solution is obvious.

Solution

$\frac{3}{4} = x + \frac{1}{6}$ This is the equation to solve.

$\frac{3}{4} - \frac{1}{6} = x + \frac{1}{6} - \frac{1}{6}$ To isolate x, undo the addition of $\frac{1}{6}$ by subtracting $\frac{1}{6}$ from both sides.

$\frac{3}{4} - \frac{1}{6} = x$ On the right side, do the addition: $\frac{1}{6} - \frac{1}{6} = 0$.

$\frac{3}{4} \cdot \frac{3}{3} - \frac{1}{6} \cdot \frac{2}{2} = x$ To build $\frac{3}{4}$ and $\frac{1}{6}$ so that their denominators are 12, multiply each by a form of 1.

$\frac{9}{12} - \frac{2}{12} = x$ Multiply the numerators. Multiply the denominators.

$\frac{7}{12} = x$ On the left side, subtract the numerators and write the difference over the common denominator, 12.

Since 7 and 12 have no common factors other than 1, the result is in simplest form. The solution is $\frac{7}{12}$. Verify this result by substituting it into the original equation.

2 Use reciprocals to solve equations.

Recall that the product of a number and its reciprocal is 1.

$$\frac{2}{3} \cdot \frac{3}{2} = 1 \qquad -\frac{4}{5}\left(-\frac{5}{4}\right) = 1 \qquad 9 \cdot \frac{1}{9} = 1$$

We can use this fact to solve equations such as $\frac{2}{3}x = 6$ and $-\frac{5}{4}x = 3$, where the coefficient of the variable term is a fraction.

$\frac{2}{3}x = 6$ $-\frac{5}{4}x = 3$

↑   ↑
The coefficient of x is   The coefficient of x is
a fraction.   a negative fraction.

Self Check 3
Solve and check the result.
a. $\frac{7}{2}b = 21$ 6
b. $-\frac{3}{8}b = 2$ $-\frac{16}{3}$

Now Try Problems 25 and 29

EXAMPLE 3
Solve: **a.** $\frac{2}{3}x = 6$ **b.** $-\frac{5}{4}x = 3$

Strategy To isolate the variable x, we will multiply both sides of the equation by the reciprocal of the coefficient of the variable term.

WHY To isolate the variable means that we want its coefficient to be 1. The product of a number and its reciprocal is 1.

Solution

a. Recall that $\frac{2}{3}x = 6$ means $\frac{2}{3} \cdot x = 6$. Since the coefficient of x is $\frac{2}{3}$, we can isolate x by multiplying both sides of the equation by the reciprocal of $\frac{2}{3}$.

4.8 Solving Equations That Involve Fractions

$$\frac{2}{3}x = 6$$ This is the equation to solve.

$$\frac{3}{2} \cdot \frac{2}{3}x = \frac{3}{2} \cdot 6$$ To isolate x, undo the multiplication by $\frac{2}{3}$, multiplying both sides by the reciprocal of $\frac{2}{3}$, which is $\frac{3}{2}$.

$$\left(\frac{3}{2} \cdot \frac{2}{3}\right)x = \frac{3}{2} \cdot \frac{6}{1}$$ On the left side, use the associative property of multiplication to group $\frac{3}{2}$ and $\frac{2}{3}$. On the right side, write 6 as $\frac{6}{1}$.

$$1x = \frac{3 \cdot 6}{2 \cdot 1}$$ On the left side, the product of a number and its reciprocal is 1: $\frac{3}{2} \cdot \frac{2}{3} = 1$. On the right side, multiply the numerators and multiply the denominators.

$$x = \frac{3 \cdot \overset{1}{\cancel{2}} \cdot 3}{\underset{1}{\cancel{2}} \cdot 1}$$ On the left side, the coefficient of 1 need not be written since 1x = x. To simplify the right side, factor 6 as 2 · 3 and remove the common factor of 2.

$$x = \frac{9}{1}$$ On the right side, multiply the remaining factors in the numerator (3 · 1 · 3 = 9) and multiply the remaining factors in the denominator (1 · 1 = 1.).

$$x = 9$$ Any number divided by 1 is equal to that number.

Teaching Example 3
Solve and check the result.
a. $\frac{3}{5}y = 15$
b. $-\frac{5}{8}y = 3$

Answers:
a. 25   b. $-\frac{24}{5}$

To check this result, substitute 9 for x in the original equation and simplify.

Check: $\frac{2}{3}x = 6$ This is the original equation.

$\frac{2}{3}(9) \overset{?}{=} 6$ Substitute 9 for x.

$6 = 6$ On the left side, $\frac{2}{3}(9) = \frac{18}{3} = 6$.

Since the statement $6 = 6$ is true, 9 is the solution of $\frac{2}{3}x = 6$.

b. Recall that $-\frac{5}{4}x = 3$ means $-\frac{5}{4} \cdot x = 3$. Since the coefficient of x is $-\frac{5}{4}$, we can isolate x by multiplying both sides of the equation by the reciprocal of $-\frac{5}{4}$.

$$-\frac{5}{4}x = 3$$ This is the equation to solve.

$$-\frac{4}{5}\left(-\frac{5}{4}x\right) = -\frac{4}{5}(3)$$ To isolate x, undo the multiplication by $-\frac{5}{4}$ by multiplying both sides by the reciprocal of $-\frac{5}{4}$, which is $-\frac{4}{5}$.

$$\left[-\frac{4}{5}\left(-\frac{5}{4}\right)\right]x = -\frac{4}{5}(3)$$ On the left side, use the associative property of multiplication to group $-\frac{4}{5}$ and $-\frac{5}{4}$.

$$1x = -\frac{4}{5}\left(\frac{3}{1}\right)$$ On the left side, the product of a number and its reciprocal is 1: $-\frac{4}{5}\left(-\frac{5}{4}\right) = 1$. On the right side, write 3 as $\frac{3}{1}$.

$$x = -\frac{12}{5}$$ On the left side, the coefficient 1 need not be written since 1x = x. On the right side, multiply the numerators and multiply the denominators.
The product of two numbers with unlike signs is negative.

Since 12 and 5 have no common factors other than 1, the result is in simplest form. The solution is $-\frac{12}{5}$. Verify that this is correct by checking.

Caution! In algebra, we usually leave a solution to an equation as an improper fraction (in simplified form) rather than converting it to a mixed number. The one exception is when solving application problems, where presenting the solution in mixed-number form is often more informative.

Success Tip Variable terms with fractional coefficients can be written in two ways. For example,

$$\frac{2}{3}x = \frac{2}{3} \cdot \frac{x}{1} = \frac{2x}{3} \quad \text{Thus,} \quad \frac{2}{3}x = \frac{2x}{3}.$$

Similarly,

$$-\frac{5}{4}x = -\frac{5}{4} \cdot \frac{x}{1} = -\frac{5x}{4} \quad \text{Thus,} \quad -\frac{5}{4}x = -\frac{5x}{4}.$$

Another method for solving equations such as $\frac{2}{3}x = 6$ uses *two steps* to isolate the variable.

Self Check 4

Solve $\frac{7}{2}b = 21$ using a two-step process. 6

Now Try Problem 33

Teaching Example 4
Solve $\frac{3}{5}y = 15$ using a two-step process.

Answer:
25

EXAMPLE 4

Solve $\frac{2}{3}x = 6$ using a two-step process.

Strategy We will use two properties of equality to isolate the variable x on one side of the equation.

WHY In the expression $\frac{2}{3}x$, we will consider the variable x to be multiplied by 2 and that the product divided by 3.

Solution We will undo the multiplication and division performed on the variable in reverse order.

$\frac{2}{3}x = 6$ This is the equation to solve.

$3 \cdot \frac{2}{3}x = 3 \cdot 6$ To isolate 2x on the left side, undo the division by 3 by multiplying both sides by 3.

$\left(3 \cdot \frac{2}{3}\right)x = 3 \cdot 6$ On the left side, use the associative property of multiplication to regroup the factors.

$\left(\frac{\overset{1}{\cancel{3}} \cdot 2}{1 \cdot \underset{1}{\cancel{3}}}\right)x = 18$ On the left side, write 3 as $\frac{3}{1}$, multiply the numerators and multiply the denominators, and then remove the common factor of 3. On the right side, do the multiplication.

$2x = 18$ On the left side, simplify the expression within the parentheses.

$\frac{2x}{2} = \frac{18}{2}$ To isolate x, undo the multiplication by 2 by dividing both sides by 2.

$x = 9$ Do the division.

The solution is 9. (As expected, this is the same as the solution obtained using the reciprocal method in Example 3, part a.)

Self Check 5

Solve $\frac{1}{11}q = \frac{19}{55}$ and check the result. $\frac{19}{5}$

Now Try Problem 37

EXAMPLE 5

Solve: $\frac{1}{9}r = \frac{17}{27}$

Strategy To isolate the variable r, we will multiply both sides of the equation by the reciprocal of the coefficient of the variable term $\frac{1}{9}r$.

WHY To isolate the variable means that we want the coefficient of r to be 1. The product of $\frac{1}{9}$ and its reciprocal will produce a coefficient of 1.

Solution Recall that $\frac{1}{9}r = \frac{1}{9} \cdot r$. Since the coefficient of r is $\frac{1}{9}$, we multiply both sides of the equation by the reciprocal of $\frac{1}{9}$.

$$\frac{1}{9}r = \frac{17}{27}$$ This is the equation to solve.

$$9 \cdot \frac{1}{9}r = 9 \cdot \frac{17}{27}$$ To isolate r, undo the multiplication by $\frac{1}{9}$ by multiplying both sides by the reciprocal of $\frac{1}{9}$, which is 9.

$$\left(9 \cdot \frac{1}{9}\right)r = \frac{9}{1} \cdot \frac{17}{27}$$ On the left side, use the associative property of multiplication to group 9 and $\frac{1}{9}$. Write 9 as $\frac{9}{1}$.

$$1r = \frac{9 \cdot 17}{1 \cdot 27}$$ On the left side, the product of a number and its reciprocal is 1: $9 \cdot \frac{1}{9} = 1$. On the right side, multiply the numerators and multiply the denominators.

$$r = \left(\frac{\overset{1}{\cancel{9}} \cdot 17}{1 \cdot 3 \cdot \underset{1}{\cancel{9}}}\right)$$ One the left side, the coefficient of 1 need not be written since $1r = r$. To simplify the right side, factor 27 as $3 \cdot 9$ and remove the common factor of 9.

$$r = \frac{17}{3}$$ Multiply the remaining factors in the numerator: $1 \cdot 17 = 17$. Multiply the remaining factors in the denominator: $1 \cdot 3 \cdot 1 = 3$.

To check this result, we substitute $\frac{17}{3}$ for r in the original equation.

Check: $\frac{1}{9}r = \frac{17}{27}$ This is the original equation.

$\frac{1}{9} \cdot \frac{17}{3} \overset{?}{=} \frac{17}{27}$ Substitute $\frac{17}{3}$ for r.

$\frac{17}{27} = \frac{17}{27}$ On the left side, multiply the numerators and multiply the denominators.

Since the statement $\frac{17}{27} = \frac{17}{27}$ is true, $\frac{17}{3}$ is the solution of $\frac{1}{9}r = \frac{17}{27}$.

EXAMPLE 6   Solve: $24z = -\frac{11}{3}$

Strategy To isolate the variable z, we will multiply both sides of the equation by the reciprocal of the coefficient of the variable term $24z$.

WHY To isolate z, we can either divide both sides by 24 or multiply both sides by the reciprocal of 24, which is $\frac{1}{24}$. Since it is easier to find $\frac{1}{24}\left(-\frac{11}{3}\right)$ than $\frac{-\frac{11}{3}}{24}$, we will use the reciprocal approach.

Solution

$$24z = -\frac{11}{3}$$ This is the equation to solve.

$$\frac{1}{24}(24z) = \frac{1}{24}\left(-\frac{11}{3}\right)$$ To isolate z, undo the multiplication by 24 by multiplying both sides by the reciprocal of 24, which is $\frac{1}{24}$.

$$\left(\frac{1}{24} \cdot 24\right)z = -\frac{1 \cdot 11}{24 \cdot 3}$$ On the left side, use the associative property of multiplication to group $\frac{1}{24}$ and 24. On the right side, multiply the numerators and multiply the denominators. The product of two numbers with unlike signs is negative.

$$1z = -\frac{11}{72}$$ On the left side, the product of a number and its reciprocal is 1: $\frac{1}{24} \cdot 24 = 1$. On the right, do the multiplication in the numerator and the denominator.

$$z = -\frac{11}{72}$$ On the left side, the coefficient of 1 need not be written since $1z = z$.

$$\begin{array}{r} 24 \\ \times\ 3 \\ \hline 72 \end{array}$$

Since 11 and 72 have no common factors other than 1, the result is in simplest form. The solution is $-\frac{11}{72}$. Verify that this is correct by checking.

Teaching Example 5
Solve $\frac{1}{6}a = \frac{23}{36}$ and check the result.
Answer:
$\frac{23}{6}$

Self Check 6
Solve $42n = -\frac{13}{2}$ and check the result. $-\frac{13}{84}$

Now Try Problem 41

Teaching Example 6
Solve $36d = -\frac{23}{3}$ and check the result.
Answer:
$-\frac{23}{108}$

Sometimes several properties of equality must be used to solve an equation.

Self Check 7

Solve $\frac{7}{12}a - 6 = -27$ and check the result. -36

Now Try Problem 45

Teaching Example 7
Solve $\frac{8}{9}n - 7 = -63$ and check the result.
Answer:
-63

EXAMPLE 7 Solve: $\frac{5}{8}m - 2 = -12$

Strategy We will use two properties of equality to isolate the variable m on left side of the equation.

WHY To solve the original equation, we want to find a simpler equivalent equation of the form $m = $ **a number**, whose solution is obvious.

Solution We note that the coefficient of m is $\frac{5}{8}$ and proceed as follows.

- To isolate the variable term $\frac{5}{8}m$, we add 2 to both sides to undo the subtraction of 2.
- To isolate the variable m, we multiply both sides by $\frac{8}{5}$ to undo the multiplication by $\frac{5}{8}$.

$$\frac{5}{8}m - 2 = -12 \qquad \text{This is the equation to solve.}$$

$$\frac{5}{8}m - 2 + 2 = -12 + 2 \qquad \text{To isolate the variable term } \tfrac{5}{8}m, \text{ undo the subtraction of 2 by adding 2 to both sides.}$$

$$\frac{5}{8}m = -10 \qquad \text{Do the addition: } -2 + 2 = 0 \text{ and } -12 + 2 = -10.$$

$$\frac{8}{5}\left(\frac{5}{8}m\right) = \frac{8}{5}(-10) \qquad \text{To isolate } m, \text{ undo the multiplication by } \tfrac{5}{8} \text{ by multiplying both sides by the reciprocal of } \tfrac{5}{8}, \text{ which is } \tfrac{8}{5}.$$

$$\left(\frac{8}{5} \cdot \frac{5}{8}\right)m = \frac{8}{5}(-10) \qquad \text{On the left side, use the associative property of multiplication to group } \tfrac{8}{5} \text{ and } \tfrac{5}{8}.$$

$$1m = \frac{8}{5}\left(-\frac{10}{1}\right) \qquad \text{On the left side, the product of a number and its reciprocal is 1: } \tfrac{8}{5} \cdot \tfrac{5}{8} = 1. \text{ On the right side, write } -10 \text{ as } \tfrac{-10}{1}.$$

$$m = -\frac{8 \cdot 10}{5 \cdot 1} \qquad \text{On the left side, the coefficient of 1 need not be written: } 1m = m. \text{ On the right side, multiply the numerators and multiply the denominators.}$$
The product of two numbers with unlike signs is negative.

$$m = -\frac{8 \cdot 2 \cdot \overset{1}{\cancel{5}}}{\underset{1}{\cancel{5}} \cdot 1} \qquad \text{To simplify the right side, factor 10 as } 2 \cdot 5 \text{ and then remove the common factor of 5.}$$

$$m = -\frac{16}{1} \qquad \text{Multiply the remaining factors in the numerator: } 8 \cdot 2 \cdot 1 = 16.$$
Multiply the remaining factors in the denominator: $1 \cdot 1 = 1$.

$$m = -16 \qquad \text{Any number divided by 1 is the same number.}$$

To check this result, we substitute -16 for m in the original equation and evaluate the left side.

Check: $\frac{5}{8}m - 2 = -12$ This is the original equation.

$\frac{5}{8}(-16) - 2 \stackrel{?}{=} -12$ Substitute –16 for m.

$-\frac{5 \cdot 16}{8 \cdot 1} - 2 \stackrel{?}{=} -12$ On the left side, write 16 as $\frac{16}{1}$. Then multiply the numerators and multiply the denominators. The product of two numbers with unlike signs is negative.

$-\frac{5 \cdot 2 \cdot \overset{1}{\cancel{8}}}{\underset{1}{\cancel{8}} \cdot 1} - 2 \stackrel{?}{=} -12$ To simplify the fraction, factor 16 as 2 · 8 and remove the common factor of 8.

$-10 - 2 \stackrel{?}{=} -12$ Multiply the remaining factors in the numerator and the denominator. Then simplify: $-\frac{10}{1} = -10$.

$-12 = -12$ On the left side, do the subtraction.

Since the statement $-12 = -12$ is true, -16 is the solution of $\frac{5}{8}m - 2 = -12$.

3 Clear equations of fractions.

To solve the equation $\frac{3}{4} = x + \frac{1}{6}$ in Example 2, we had to find an LCD and build equivalent fractions for $\frac{3}{4}$ and $\frac{1}{6}$ to subtract in the third step of the solution. We will now discuss a method in which we *clear* such an equation of fractions.

EXAMPLE 8 Solve $\frac{3}{4} = x + \frac{1}{6}$ by first clearing the equation of fractions.

Strategy We will use the multiplication property of equality to clear this equation of fractions by multiplying both sides by the LCD.

WHY Equations that involve only integers are usually easier to solve than equations that involve fractions.

Solution Since the denominators of the fractions in the equation are 4 and 6, we multiply both sides of the equation by the LCD, 12.

$\frac{3}{4} = x + \frac{1}{6}$ This is the equation to solve.

$12\left(\frac{3}{4}\right) = 12\left(x + \frac{1}{6}\right)$ Multiply both sides of the equation by the LCD of $\frac{3}{4}$ and $\frac{1}{6}$, which is 12. Don't forget to write the parentheses on each side.

$12\left(\frac{3}{4}\right) = 12(x) + 12\left(\frac{1}{6}\right)$ On the right side, distribute the multiplication by 12.

$\frac{12}{1}\left(\frac{3}{4}\right) = 12(x) + \frac{12}{1}\left(\frac{1}{6}\right)$ Write 12 as $\frac{12}{1}$. This makes the numerators and denominators in the fraction multiplication process clearer.

$\frac{3 \cdot \overset{1}{\cancel{4}}}{1}\left(\frac{3}{\underset{1}{\cancel{4}}}\right) = 12(x) + \frac{2 \cdot \overset{1}{\cancel{6}}}{1}\left(\frac{1}{\underset{1}{\cancel{6}}}\right)$ On the left side, factor 12 as 3 · 4 and remove the common factor of 4 from the numerator and denominator. On the right side, factor 12 as 2 · 6 and remove the common factor of 6. Try to do these steps in your head.

Self Check 8

Solve $\frac{2}{3} = y + \frac{1}{5}$ by first clearing the equation of fractions. $\frac{7}{15}$

Now Try Problem 49

Teaching Example 8

Solve $\frac{7}{8} = m + \frac{1}{3}$ by first clearing the equation of fractions.

Answer: $\frac{13}{24}$

Success Tip Here is an alternate way to show how the common factors of the numerator and denominator are removed in the multiplication process:

$$\frac{\overset{3}{\cancel{12}}}{1}\left(\frac{3}{\underset{1}{\cancel{4}}}\right) = 12(x) + \frac{\overset{2}{\cancel{12}}}{1}\left(\frac{1}{\underset{1}{\cancel{6}}}\right)$$

$9 = 12x + 2$ Complete each multiplication. Note that the fractions have been cleared from the equation.

$9 - 2 = 12x + 2 - 2$ To isolate the variable term 12x, undo the addition of 2 by subtracting 2 from both sides.

$7 = 12x$ Do the subtraction.

$\dfrac{7}{12} = \dfrac{12x}{12}$ To isolate the variable x, undo the multiplication by 12 by dividing both sides by 12.

$\dfrac{7}{12} = x$ Since the only common factor of 7 and 12 is 1, the fraction is in simplest form.

The solution is $\frac{7}{12}$. Note that this is the same result that we obtained in Example 2.

Self Check 9

Solve $\dfrac{m}{2} - \dfrac{m}{5} = -6$ and check the result. -20

Now Try Problem 53

Teaching Example 9
Solve $\dfrac{x}{5} - \dfrac{x}{7} = -4$ and check the result.
Answer:
-70

EXAMPLE 9 Solve: $\dfrac{n}{3} - \dfrac{n}{5} = -4$

Strategy We will use the multiplication property of equality to clear the equation of fractions by multiplying both sides by the LCD.

WHY Equations that involve only integers are usually easier to solve than equations that involve fractions.

Solution Since the denominators of the fractions in the equation are 3 and 5, we multiply both sides of the equation by the LCD, 15.

$\dfrac{n}{3} - \dfrac{n}{5} = -4$ This is the equation to solve.

$15\left(\dfrac{n}{3} - \dfrac{n}{5}\right) = 15(-4)$ Multiply both sides of the equation by the LCD of $\frac{n}{3}$ and $\frac{n}{5}$, which is 15. Don't forget to write the parentheses on each side.

$15\left(\dfrac{n}{3}\right) - 15\left(\dfrac{n}{5}\right) = 15(-4)$ On the left side, distribute the multiplication by 15.

$\dfrac{15}{1}\left(\dfrac{n}{3}\right) - \dfrac{15}{1}\left(\dfrac{n}{5}\right) = 15(-4)$ Write 15 as $\frac{15}{1}$. This makes the numerators and denominators in the fraction multiplication process clearer.

$\dfrac{\overset{1}{\cancel{3}} \cdot 5}{1}\left(\dfrac{n}{\underset{1}{\cancel{3}}}\right) - \dfrac{3 \cdot \overset{1}{\cancel{5}}}{1}\left(\dfrac{n}{\underset{1}{\cancel{5}}}\right) = 15(-4)$ On the left side, factor 15 as 3 · 5 and remove the common factor of 3 and the common factor of 5. Try to do these steps in your head.

Success Tip Here is an alternate way to show how the common factors of the numerator and denominator are removed in the multiplication process:

$$\frac{\overset{5}{\cancel{15}}}{1}\left(\frac{n}{\underset{1}{\cancel{3}}}\right) - \frac{\overset{3}{\cancel{15}}}{1}\left(\frac{n}{\underset{1}{\cancel{5}}}\right) = 15(-4)$$

$5n - 3n = -60$ Complete each multiplication. Note that the fractions have been cleared from the equation.

$2n = -60$ On the left side, combine like terms: $5n - 3n = 2n$.

$\dfrac{2n}{2} = \dfrac{-60}{2}$ To isolate the variable n, undo the multiplication by 2 by dividing both sides by 2.

$n = -30$ Do the division.

Check: $\frac{n}{3} - \frac{n}{5} = -4$ This is the original equation.

$\frac{-30}{3} - \frac{-30}{5} \stackrel{?}{=} -4$ Substitute -30 for each n.

$-10 - (-6) \stackrel{?}{=} -4$ On the left side, do each division. Recall that the quotient of two numbers with unlike signs is negative.

$-4 \stackrel{?}{=} -4$ On the left side, write the subtraction as addition of the opposite: $-10 - (-6) = -10 + 6 = -4$.

Since the statement $-4 = -4$ is true, -30 is the solution of $\frac{n}{3} - \frac{n}{5} = -4$.

EXAMPLE 10

Solve: $\frac{3}{4}h - \frac{1}{2} = \frac{5}{8}h$

Strategy We will use the multiplication property of equality to clear the equation of fractions by multiplying both sides by the LCD.

WHY Equations that involve only integers are usually easier to solve than equations that involve fractions.

Solution Since the denominators of the fractions in the equation are 4, 2, and 8, we multiply both sides of the equation by the LCD, 8.

$\frac{3}{4}h - \frac{1}{2} = \frac{5}{8}h$ This is the equation to solve.

$8\left(\frac{3}{4}h - \frac{1}{2}\right) = 8\left(\frac{5}{8}h\right)$ Multiply both sides of the equation by the LCD of $\frac{3}{4}, \frac{1}{2}$, and $\frac{5}{8}$, which is 8. Don't forget to write the parentheses on both sides.

$8\left(\frac{3}{4}h\right) - 8\left(\frac{1}{2}\right) = 8\left(\frac{5}{8}h\right)$ On the left side, distribute the multiplication by 8.

$\frac{8}{1}\left(\frac{3}{4}h\right) - \frac{8}{1}\left(\frac{1}{2}\right) = \frac{8}{1}\left(\frac{5}{8}h\right)$ Write 8 as $\frac{8}{1}$. This makes the numerators and denominators in the fraction multiplication process clearer.

$\frac{2 \cdot \overset{1}{4}}{1}\left(\frac{3}{\underset{1}{4}}h\right) - \frac{\overset{1}{2} \cdot 4}{1}\left(\frac{1}{\underset{1}{2}}\right) = \frac{\overset{1}{8}}{1}\left(\frac{5}{\underset{1}{8}}h\right)$ On the left side, factor 8 as $2 \cdot 4$ and remove the common factor of 4 and the common factor of 2. On the right side, remove the common factor of 8. Try to do this step in your head.

Success Tip Here is an alternate way to show how the common factors of the numerator and denominator are removed in the multiplication process.

$\frac{\overset{2}{8}}{1}\left(\frac{3}{\underset{1}{4}}h\right) - \frac{\overset{4}{8}}{1}\left(\frac{1}{\underset{1}{2}}\right) = \frac{\overset{1}{8}}{1}\left(\frac{5}{\underset{1}{8}}h\right)$

$2(3h) - 4(1) = 5h$ Simplify each term. Note that the fractions have been cleared from the equation.

$6h - 4 = 5h$ Complete the multiplication.

$6h - 4 - 5h = 5h - 5h$ To eliminate the term $5h$ from the right side, subtract $5h$ from both sides.

$h - 4 = 0$ Combine like terms: $6h - 5h = h$ and $5h - 5h = 0$.

$h - 4 + 4 = 0 + 4$ To isolate the variable h on the left side, undo the subtraction of 4 by adding 4 to both sides.

$h = 4$ Do the addition.

The solution is 4. Verify this result by substituting it into the original equation.

Self Check 10

Solve $\frac{4}{5}w - \frac{1}{2} = \frac{3}{4}w$ and check the result. 10

Now Try Problem 57

Teaching Example 10
Solve $\frac{3}{2}y - \frac{8}{5} = \frac{7}{10}y$ and check the result.
Answer: 2

Success Tip After multiplying both sides by the LCD and simplifying, the equation should not contain any fractions. If it does, check for an algebraic error, or perhaps your LCD is incorrect.

We can now complete the strategy for solving equations discussed in Chapter 3. You won't always have to use all five steps to solve a given equation. If a step doesn't apply, skip it and move to the next step.

Strategy for Solving Equations

1. **Clear the equation of fractions:** Multiply both sides by the LCD to clear fractions.
2. **Simplify each side of the equation:** Use the distributive property to remove parentheses, and then combine like terms on each side.
3. **Isolate the variable term on one side:** Add (or subtract) to get the variable term on one side of the = symbol and a number on the other using the addition (or subtraction) property of equality.
4. **Isolate the variable:** Multiply (or divide) to isolate the variable using the multiplication (or division) property of equality.
5. **Check the result:** Substitute the possible solution for the variable in the *original* equation to see if a true statement results.

4 Use equations to solve application problems that involve fractions.

We can use the concepts of variable and equation to solve application problems involving fractions. Once again, we will follow the strategy of analyze, assign, form, solve, state, and check.

Self Check 11

ANATOMY CLASS A premed student has to memorize the name of each bone in the human hand. So far, he has learned 18 of them, which is two-thirds of the total. How many bones are in the human hand? 27 bones

Now Try Problems 97 and 99

Teaching Example 11
AMERICAN LITERATURE A student has read the first 768 pages of the paperback version of the novel *Gone with the Wind*. If that is three-fourths of the book, how many pages are there in the paperback version?
Answer:
1,024 pages

EXAMPLE 11 *Native Americans* The U.S. Constitution requires a population count, called a *census*, to be taken every 10 years. In the 2010 census, the population of the Sioux tribe was approximately 130,000. This was about two-fifths of the population of the largest Native American tribe, the Navajo. What was the population of the Navajo tribe in 2010? (Source: infoplease.com)

Analyze
- In 2010, the population of the Sioux tribe was about 130,000. *Given*
- In 2010, the population of the Sioux tribe was about $\frac{2}{5}$ of the population of the Navajo tribe. *Given*
- What was the population of the Navajo tribe in 2010? *Find*

Assign
Let $x =$ the population of the Navajo tribe in 2010. Next, we look for a key word or phrase in the problem.

Key phrase: *two-fifths of*   **Translation:** multiply by $\frac{2}{5}$

Form
Now we translate the words of the problem into an equation.

In 2010, the population of the Sioux tribe	was	$\frac{2}{5}$	of	the population of the Navajo tribe.
130,000	=	$\frac{2}{5}$	·	x

Solve

$$130,000 = \frac{2}{5}x$$

$$\frac{5}{2}(130,000) = \frac{5}{2}\left(\frac{2}{5}x\right)$$ To isolate x, undo the multiplication by $\frac{2}{5}$ by multiplying both sides by the reciprocal of $\frac{2}{5}$, which is $\frac{5}{2}$.

$$\frac{5}{2}\left(\frac{130,000}{1}\right) = \left(\frac{5}{2} \cdot \frac{2}{5}\right)x$$ On the left side, write 130,000 as $\frac{130,000}{1}$. On the right side, use the associative property of multiplication to group $\frac{5}{2}$ and $\frac{2}{5}$.

$$\frac{5 \cdot 130,000}{2 \cdot 1} = 1x$$ On the left side, multiply the numerators and multiply the denominators. On the right side, the product of a number and its reciprocal is 1: $\frac{5}{2} \cdot \frac{2}{5} = 1$.

$$\frac{5 \cdot \overset{1}{\cancel{2}} \cdot 65,000}{\underset{1}{\cancel{2}} \cdot 1} = x$$ To simplify the left side, factor 130,000 as $2 \cdot 65,000$ and remove the common factor of 2. On the right side, the coefficient of 1 need not be written since $1x = x$.

$$325,000 = x$$ On the left side, multiply the remaining factors in the numerator: $5 \cdot 1 \cdot 65,000 = 325,000$. Then simplify the fraction.

$$\begin{array}{r} \overset{2}{}65,000 \\ \times 5 \\ \hline 325,000 \end{array}$$

State
In 2010, the population of the Navajo tribe was about 325,000.

Check
If we use a fraction to compare the two populations, we should get $\frac{2}{5}$.

$$\frac{130,000}{325,000} = \frac{130}{325} = \frac{2 \cdot \overset{1}{\cancel{65}}}{5 \cdot \underset{1}{\cancel{65}}} = \frac{2}{5}$$ Write $\frac{\text{the population of the Sioux tribe}}{\text{the population of the Navajo tribe}}$ and simplify.

The result checks.

EXAMPLE 12 *Filmmaking* A movie director has sketched out a "storyboard" for a film that is in the planning stages. He estimates the amount of time in the film that will be devoted to scenes involving dialogue, action scenes, and scenes that make a transition between the two. From the information on the storyboard shown below, how long will this film be, in minutes?

Storyboard	Film: "Terminating Force"	
Dialogue One-half of film	Action scenes One-third of film	Transition scenes 20 minutes

Self Check 12

CASTING A MOVIE Two-thirds of the cast of a movie are male adults, one-fourth are female adults, and there are 6 children in the movie. Find the number of people in the cast. 72 people

Now Try Problem 107

Teaching Example 12
TRAFFIC TICKET DISMISSAL A traffic school class has three parts. In the first part, a film is shown that takes one-sixth of the class time. In the second part, the instructor lectures for 105 minutes. In the final part, a test is given that takes one-fourth of the class time. How many minutes long is the traffic school class?
Answer:
180 minutes

Analyze

- $\frac{1}{2}$ of the film is dialogue.   Given
- $\frac{1}{3}$ of the film is action scenes.   Given
- There are 20 minutes of transition scenes.   Given
- How long is the film?   Find

Assign

Let x = the length of the film in minutes. To represent the number of minutes for the dialogue and the action scenes, look for a key word or phrase in the storyboard.

    Key phrases: one-half of, one-third of   **Translation:** multiply

Thus, $\frac{1}{2}x$ = the number of minutes for dialogue scenes and $\frac{1}{3}x$ = the number of minutes for action scenes.

Form

The time for dialogue scenes	plus	the time for action scenes	plus	the time for transition scenes	is	the total length of the film.
$\frac{1}{2}x$	+	$\frac{1}{3}x$	+	20	=	x

Solve

$$\frac{1}{2}x + \frac{1}{3}x + 20 = x$$

$$6\left(\frac{1}{2}x + \frac{1}{3}x + 20\right) = 6(x)$$ To clear the equation of fractions, multiply both sides by the LCD of $\frac{1}{2}x$ and $\frac{1}{3}x$, which is 6. Don't forget to write the parentheses.

$$6\left(\frac{1}{2}x\right) + 6\left(\frac{1}{3}x\right) + 6(20) = 6(x)$$ On the left side, distribute the multiplication by 6.

$$\frac{6}{1}\left(\frac{1}{2}x\right) + \frac{6}{1}\left(\frac{1}{3}x\right) + 6(20) = 6x$$ Write 6 as $\frac{6}{1}$. This makes the numerators and denominators in the fraction multiplication process clearer.

$$\frac{\overset{1}{2}\cdot 3}{1}\left(\frac{1}{\underset{1}{2}}x\right) + \frac{2\cdot \overset{1}{3}}{1}\left(\frac{1}{\underset{1}{3}}x\right) + 6(20) = 6(x)$$ On the left side, factor 6 as 2 · 3 and remove the common factors in the numerator and denominator of the first two terms.

$$3x + 2x + 120 = 6x$$ Complete the multiplication on both sides of the equation.

$$5x + 120 = 6x$$ On the left side, combine like terms: $3x + 2x = 5x$.

$$5x + 120 - 5x = 6x - 5x$$ To eliminate the term $5x$ from the left side, subtract $5x$ from both sides.

$$120 = x$$ Combine like terms: $5x - 5x = 0$ and $6x - 5x = x$.

State
The length of the film will be 120 minutes.

Check
If x is **120**, the time for dialogue scenes is $\frac{1}{2}x = \frac{1}{2} \cdot 120 = 60$ minutes. The time for action scenes is $\frac{1}{3}x = \frac{1}{3} \cdot 120 = 40$ minutes. The time for transition scenes is 20 minutes. Adding the three times, we get $60 + 40 + 20 = 120$ minutes. The result checks. ∎

4.8 Solving Equations That Involve Fractions   411

ANSWERS TO SELF CHECKS
1. $\dfrac{3}{4}$ 2. $\dfrac{7}{15}$ 3. a. 6   b. $-\dfrac{16}{3}$ 4. 6   5. $\dfrac{19}{5}$ 6. $-\dfrac{13}{84}$ 7. -36 8. $\dfrac{7}{15}$ 9. -20
10. 10   11. 27 bones   12. 72 people

SECTION 4.8   STUDY SET

VOCABULARY

Fill in the blanks.

1. To _solve_ an equation, we find all the values of the variable that make the equation true.
2. In the term $\tfrac{5}{12}x$, the _coefficient_ of x is $\tfrac{5}{12}$.
3. To find the _reciprocal_ of a fraction, we invert the numerator and the denominator.
4. To _clear_ the equation $\tfrac{1}{4}y - \tfrac{2}{3} = \tfrac{1}{2}y$ of fractions, we multiply both sides by the LCD of $\tfrac{1}{4}$, $\tfrac{2}{3}$, and $\tfrac{1}{2}$, which is 12.

CONCEPTS

5. Use a check to determine whether 40 is the solution of $\tfrac{5}{8}x = 25$. Since $25 = 25$ is a true statement, 40 is the solution of $\tfrac{5}{8}x = 25$.
6. Find the reciprocal of the coefficient of x.
   a. $\tfrac{7}{9}x$ $\tfrac{9}{7}$
   b. $-\tfrac{1}{2}x$ -2
7. What is the result when a fraction is multiplied by its reciprocal? 1
8. Multiply.
   a. $\tfrac{3}{2}\left(\tfrac{2}{3}x\right)$ x
   b. $-\tfrac{16}{15}\left(-\tfrac{15}{16}t\right)$ t
9. Translate to mathematical symbols.
   a. Four-fifths of the population p $\tfrac{4}{5}p$
   b. One-quarter of the time t $\tfrac{1}{4}t$
10. On the left side of the equation, what property is illustrated by the arrows? the distributive property

    $12\left(\tfrac{1}{6}y - \tfrac{1}{4}\right) = 12\left(\tfrac{1}{2}\right)$

11. By what should both sides of the equation be multiplied to clear it of fractions?
    a. $\tfrac{1}{3}x = \tfrac{5}{2}x - 8$ 6
    b. $\tfrac{m}{8} = \tfrac{m}{3} + \tfrac{1}{2}$ 24
12. Fill in the blanks.
    $8\left(\tfrac{1}{4}x\right) = 8\left(\tfrac{1}{2}x\right) + 8\left(\tfrac{3}{8}\right)$
    $\underbrace{2x}\ =\ \underbrace{4x}\ +\ \underbrace{3}$

NOTATION

Complete each step to solve the equation.

13. $\tfrac{7}{8}x = 21$

    $\boxed{\tfrac{8}{7}}\left(\tfrac{7}{8}x\right) = \boxed{\tfrac{8}{7}}(21)$

    $x = \boxed{24}$

    The solution is $\boxed{24}$.

14. $h + \tfrac{1}{2} = \tfrac{2}{3}$

    $6\left(h + \tfrac{1}{2}\right) = 6\left(\tfrac{2}{3}\right)$

    $6h + 6\left(\tfrac{1}{2}\right) = 6\left(\tfrac{2}{3}\right)$

    $6h + \boxed{3} = 4$

    $6h + 3 - \boxed{3} = 4 - \boxed{3}$

    $6h = \boxed{1}$

    $\tfrac{6h}{6} = \tfrac{1}{6}$

    $h = \tfrac{1}{6}$

    The solution is $\boxed{\tfrac{1}{6}}$.

15. Determine whether each statement is true or false.
    a. $\tfrac{1}{2}x = \tfrac{x}{2}$ true
    b. $\tfrac{1}{8}y = 8y$ false
    c. $-\tfrac{1}{2}x = \tfrac{-x}{2} = \tfrac{x}{-2}$ true
    d. $\tfrac{7p}{8} = \tfrac{7}{8}p$ true
16. Write the product of $\tfrac{4}{7}$ and x in two ways. $\tfrac{4x}{7},\ \tfrac{4}{7}x$

GUIDED PRACTICE

Solve each equation and check the result. See Example 1.

17. $y - \tfrac{1}{20} = \tfrac{13}{20}$ $\tfrac{7}{10}$
18. $a - \tfrac{1}{24} = \tfrac{17}{24}$ $\tfrac{3}{4}$
19. $x - \tfrac{1}{15} = \tfrac{11}{15}$ $\tfrac{4}{5}$
20. $y - \tfrac{1}{16} = \tfrac{9}{16}$ $\tfrac{5}{8}$

Solve each equation and check the result. See Example 2.

21. $\dfrac{5}{6} = x + \dfrac{1}{9}$ $\dfrac{13}{18}$
22. $\dfrac{1}{6} = x + \dfrac{1}{8}$ $\dfrac{1}{24}$
23. $\dfrac{7}{10} = a + \dfrac{1}{4}$ $\dfrac{9}{20}$
24. $\dfrac{7}{15} = c + \dfrac{1}{10}$ $\dfrac{11}{30}$

Solve each equation and check the result. See Example 3.

25. $\dfrac{4}{9}x = 12$ 27
26. $\dfrac{5}{6}x = 10$ 12
27. $\dfrac{3}{7}w = 30$ 70
28. $\dfrac{2}{15}y = 6$ 45
29. $-\dfrac{8}{3}w = 7$ $-\dfrac{21}{8}$
30. $-\dfrac{9}{5}w = 4$ $-\dfrac{20}{9}$
31. $-\dfrac{16}{9}w = 5$ $-\dfrac{45}{16}$
32. $-\dfrac{11}{6}w = 7$ $-\dfrac{42}{11}$

Solve each equation using a two-step process. See Example 4.

33. $\dfrac{4}{9}x = 12$ 27
34. $\dfrac{5}{6}x = 10$ 12
35. $\dfrac{3}{7}w = 30$ 70
36. $\dfrac{2}{15}y = 6$ 45

Solve each equation and check the result. See Example 5.

37. $\dfrac{1}{8}x = \dfrac{25}{72}$ $\dfrac{25}{9}$
38. $\dfrac{1}{6}x = \dfrac{29}{54}$ $\dfrac{29}{9}$
39. $\dfrac{1}{5}x = \dfrac{43}{55}$ $\dfrac{43}{11}$
40. $\dfrac{1}{9}x = \dfrac{20}{27}$ $\dfrac{20}{3}$

Solve each equation and check the result. See Example 6.

41. $17x = -\dfrac{15}{4}$ $-\dfrac{15}{68}$
42. $21x = -\dfrac{31}{2}$ $-\dfrac{31}{42}$
43. $29x = -\dfrac{13}{3}$ $-\dfrac{13}{87}$
44. $41x = -\dfrac{27}{5}$ $-\dfrac{27}{205}$

Solve each equation and check the result. See Example 7.

45. $\dfrac{5}{6}k - 5 = -15$ -12
46. $\dfrac{2}{5}c - 12 = -32$ -50
47. $\dfrac{7}{16}h + 28 = 21$ -16
48. $\dfrac{5}{8}a + 25 = 15$ -16

Solve each equation by first clearing it of fractions. Check the result. See Example 8.

49. $\dfrac{2}{9} = r - \dfrac{1}{6}$ $\dfrac{7}{18}$
50. $\dfrac{5}{8} = x - \dfrac{1}{6}$ $\dfrac{19}{24}$
51. $\dfrac{1}{4} = x - \dfrac{2}{9}$ $\dfrac{17}{36}$
52. $\dfrac{1}{7} = y - \dfrac{1}{2}$ $\dfrac{9}{14}$

Solve each equation by first clearing it of fractions. Check the result. See Example 9.

53. $\dfrac{a}{4} - \dfrac{a}{7} = -6$ -56
54. $\dfrac{f}{3} - \dfrac{f}{8} = -10$ -48
55. $\dfrac{k}{7} - \dfrac{k}{3} = -24$ 126
56. $\dfrac{a}{11} - \dfrac{a}{5} = -12$ 110

Solve each equation by first clearing it of fractions. Check the result. See Example 10.

57. $\dfrac{3}{5}d - \dfrac{4}{5} = \dfrac{1}{3}d$ 3
58. $\dfrac{11}{10}x - \dfrac{5}{2} = \dfrac{3}{5}x$ 5
59. $\dfrac{7}{8}w + \dfrac{1}{4} = \dfrac{3}{4}w$ -2
60. $\dfrac{1}{2}d + \dfrac{17}{2} = \dfrac{12}{7}d$ 7

TRY IT YOURSELF

Solve each equation and check the result.

61. $\dfrac{1}{2}x - \dfrac{1}{9} = \dfrac{1}{3}$ $\dfrac{8}{9}$
62. $\dfrac{1}{4}y - \dfrac{2}{3} = \dfrac{1}{2}$ $\dfrac{14}{3}$
63. $\dfrac{a}{2} - \dfrac{a}{12} = -\dfrac{5a}{4}$ 0
64. $\dfrac{n}{6} - \dfrac{n}{18} = -\dfrac{4n}{3}$ 0
65. $\dfrac{7}{8}t = -28$ -32
66. $\dfrac{5}{6}c = -25$ -30
67. $x + \dfrac{1}{9} = \dfrac{4}{9}$ $\dfrac{1}{3}$
68. $x - \dfrac{1}{12} = \dfrac{7}{12}$ $\dfrac{2}{3}$
69. $\dfrac{5f}{7} = -2$ $-\dfrac{14}{5}$
70. $\dfrac{3h}{5} = -35$ $-\dfrac{175}{3}$
71. $6x = 2x - 11$ $-\dfrac{11}{4}$
72. $5t = t - 7$ $-\dfrac{7}{4}$
73. $2(y - 3) = 7$ $\dfrac{13}{2}$
74. $3(r + 2) = 10$ $\dfrac{4}{3}$
75. $\dfrac{1}{3} = 2x - \dfrac{1}{2}$ $\dfrac{5}{12}$
76. $\dfrac{1}{8} = 3y - \dfrac{2}{5}$ $\dfrac{7}{40}$
77. $4 = -\dfrac{3}{5}h$ $-\dfrac{20}{3}$
78. $-2 = -\dfrac{5}{6}f$ $\dfrac{12}{5}$
79. $5 + \dfrac{1}{3}x = \dfrac{1}{2} + \dfrac{1}{6}x$ -27
80. $\dfrac{x}{4} = \dfrac{1}{2} - \dfrac{3x}{20}$ $\dfrac{5}{4}$
81. $\dfrac{2}{15}x = \dfrac{4}{5}$ 6
82. $\dfrac{5}{66}y = \dfrac{10}{11}$ 12
83. $\dfrac{2}{5}x + 1 = \dfrac{1}{3} + x$ $\dfrac{10}{9}$
84. $\dfrac{2}{3}y + 2 = \dfrac{1}{5} + y$ $\dfrac{27}{5}$
85. $\dfrac{5h}{6} - 8 = 12$ 24
86. $\dfrac{6a}{7} - 1 = 11$ 14
87. $\dfrac{1}{2}y - 2 = \dfrac{1}{5}y + 1$ 10
88. $\dfrac{1}{5}m - 5 = \dfrac{1}{3}m - 3$ -15
89. $\dfrac{x}{21} + \dfrac{3x}{7} = 1 - \dfrac{5x}{3}$ $\dfrac{7}{15}$
90. $\dfrac{3x}{26} - \dfrac{5x}{2} - 4 = \dfrac{9x}{13}$ $-\dfrac{13}{10}$
91. $\dfrac{x}{2} + \dfrac{x}{3} + \dfrac{x}{4} = 0$ 0
92. $\dfrac{x}{2} + \dfrac{x}{3} + \dfrac{x}{4} = \dfrac{x}{5}$ 0

In part a of Exercises 93–96, use the methods of this section to solve the equation. In part b, use the methods of Section 4.4 to perform the addition or subtraction.

93. a. $\dfrac{12}{5} + \dfrac{n}{2} = \dfrac{n}{10} - 6$ b. $\dfrac{12}{5} + \dfrac{n}{2} \quad \dfrac{24 + 5n}{10}$

94. a. $\dfrac{9}{2} + \dfrac{h}{3} = \dfrac{h}{6} - 27$ b. $\dfrac{9}{2} + \dfrac{h}{3} \quad \dfrac{27 + 2h}{6}$

95. a. $\dfrac{x}{4} - \dfrac{1}{3} = \dfrac{x}{2} - \dfrac{4}{3}$ b. $\dfrac{x}{4} - \dfrac{1}{3} \quad \dfrac{3x - 4}{12}$

96. a. $\dfrac{a}{5} - \dfrac{5}{6} = \dfrac{a}{3} - \dfrac{25}{4}$ b. $\dfrac{a}{5} - \dfrac{5}{6} \quad \dfrac{6a - 25}{30}$

APPLICATIONS

Complete each solution.

97. CAR REPAIRS One-eighth of the cars that an automobile repair shop serviced last year had transmission problems. If the shop repaired 32 transmissions, how many cars did they service that year?

Analyze

- $\tfrac{1}{8}$ of the cars serviced last year had transmission problems.   Given
- The shop repaired 32 transmissions last year.   Given
- How many cars did the shop service last year?   Find

Assign

Let x = the number of cars that the repair shop serviced last year.

Form

The key phrase *one-eighth of* suggests multiplication .

$\tfrac{1}{8}$	of	the number of cars repaired last year	had	transmission problems.
$\tfrac{1}{8}$	·	x	=	32

Solve

$$\dfrac{1}{8}x = 32$$

$$8\left(\dfrac{1}{8}x\right) = 8\,(32)$$

$$x = 256$$

State The shop repaired 256 cars last year.

Check If we can use a fraction to compare the number of cars with transmission problems to the number of cars serviced, we should get $\tfrac{1}{8}$.

$$\dfrac{32}{256} = \dfrac{\overset{1}{\cancel{32}}}{8 \cdot \underset{1}{\cancel{32}}} = \dfrac{1}{8}$$

The result checks.

98. CATTLE RANCHING A rancher is going to fence in a rectangular grazing area next to a $\tfrac{3}{4}$-mile-long stretch of lakeshore. He has determined that $1\tfrac{1}{2}$ square miles of land are needed to make sure that overgrazing does not occur. How wide should this grazing area be?

Analyze

- The rectangular grazing area is $1\tfrac{1}{2}$ square miles.   Given
- The length of the rectangle is $\tfrac{3}{4}$ mile.   Given
- How wide should the grazing area be?   Find

Assign

Let w = width of grazing area.

Form

The key word *area* suggests we multiply the length of the rectangle by the width.

The area of the rectangle	is equal to	the length	times	the width.
$\tfrac{3}{2}$	=	$\tfrac{3}{4}$	·	w

Solve

$$\dfrac{3}{2} = \dfrac{3}{4}w$$

$$\tfrac{4}{3}\left(\dfrac{3}{2}\right) = \tfrac{4}{3}\left(\dfrac{3}{4}w\right)$$

$$2 = w$$

State

The width of the grazing area must be 2 miles.

Check If we multiply the length and the width of the rectangular area, we should get $1\frac{1}{2}$ square miles.

$$\frac{3}{4} \cdot 2 = \frac{6}{4} = 1\frac{1}{2}$$

The result checks.

In Exercises 99–112, let a variable represent the unknown quantity. Then write and solve an equation to answer the question.

99. TOOTH DEVELOPMENT During a checkup, a pediatrician found that only four-fifths of a child's baby teeth had emerged. The mother counted 16 teeth in the child's mouth. How many baby teeth will the child eventually have? 20 teeth

100. GENETICS Bean plants with inflated pods were cross-bred with bean plants with constricted pods. Of the offspring plants, three-fourths had inflated pods and only one-fourth had constricted pods. If 244 offspring plants had constricted pods, how many offspring plants resulted from the cross-breeding experiment? 976 offspring plants

Inflated pod   Constricted pod

101. TELEPHONE BOOKS A telephone book consists of the white pages and the yellow pages. Two-thirds of the book consists of the white pages; the white pages number 300. Find the total number of pages in the telephone book. 450 pages

102. BROADWAY MUSICALS A theater usher at a Broadway musical finds that seven-eighths of the patrons attending a performance, which is 350 people, are in their seats by show time. If the show is always a complete sellout, how many seats does the theater have? 400 seats

103. LIGHTING DESIGN In the warehouse shown below, the distance between light fixtures is two-thirds of the floor-to-ceiling height. What is the height of the warehouse ceiling? 27 ft

104. THE VOLUNTEER STATE The dimensions of the flag of the State of Tennessee were adopted into law in 1905. The official description requires that "the flag is to be a banner whose length is one and two-thirds times its width." If the flag shown below meets this requirement, what is its width? 18 in.

Length = 30 in.

105. GRAPHIC ARTS A design for a yearbook is shown. The page is divided into 12 equal-size parts. Five of the twelve parts (those shaded in red) will contain photographs. If the photographs are to cover an area of 100 square inches, how many square inches are there on the page? 240 in.²

106. SAFETY REQUIREMENTS In developing taillights for an automobile, designers must be aware of a safety standard that requires an area of 30 square inches to be visible from behind the vehicle. If the designers want the taillights to be $3\frac{3}{4}$ inches high, how wide must they be to meet safety standards? (*Hint:* Write $3\frac{3}{4}$ as an improper fraction.) 8 in.

107. CPR CLASS The instructor for a course in CPR (cardiopulmonary resuscitation) has three segments in her lesson plan, as shown below. How many minutes long is the CPR course? 360 min

Lecture on subject	Practicing CPR techniques	Legal responsibilities
One-fourth of class	Two-thirds of class	30 min

108. FIREFIGHTING A firefighting crew is composed of three elements, as shown below. How many firefighters are in the crew? 300 firefighters

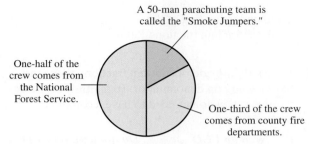

109. TEAM ROSTERS At the end of the season, major league baseball teams are allowed to add players to the roster. Suppose one-half of the players on a team's expanded roster are pitchers, one-fourth are infielders, three-twentieths are outfielders, and there are 4 catchers. How many players are on the expanded roster? 40 players

110. TEAM ROSTERS One-third of the players on a basketball team are forwards and one-fifth are centers. The remaining 7 players are guards. How many players are on the team? 15 players

111. HOME SALES In less than a month, three-quarters of the homes in a new subdivision were purchased. This left only 9 homes to be sold. How many homes are there in the subdivision? (*Hint:* First determine what fractional part of the homes in the subdivision were *not* yet sold.) 36 homes

112. WEDDING GUESTS Of those invited to a wedding, three-tenths were friends of the bride. The friends of the groom numbered 84. How many people were invited to the wedding? (*Hint:* First determine what fractional part of the people invited to the wedding were friends of the groom.) 120 people

WRITING

113. What is wrong with the following solution?

Solve: $\dfrac{x}{6} - \dfrac{3}{5} = 2$

$30\left(\dfrac{x}{6} - \dfrac{3}{5}\right) = 2$

$30\left(\dfrac{x}{6}\right) - 30\left(\dfrac{3}{5}\right) = 2$

$5x - 18 = 2$

$5x = 20$

$x = 4$

114. Explain two ways in which the variable x can be isolated to solve the equation $\dfrac{2}{3}x = -4$.

115. What does it mean to clear an equation of fractions before solving the equation?

116. Explain how the method used to add $\dfrac{a}{3} + \dfrac{1}{5}$ differs from the method used to solve the equation $\dfrac{a}{3} + \dfrac{1}{5} = \dfrac{7}{15}$.

REVIEW

117. Round 12,599,767:
   a. to the nearest million 13,000,000
   b. to the nearest ten thousand 12,600,00
   c. to the nearest hundred 12,599,800

118. Round 247,349,501:
   a. to the nearest ten million 250,000,000
   b. to the nearest hundred thousand 247,300,000
   c. to the nearest hundred 247,349,500

STUDY SKILLS CHECKLIST

Working with Fractions

Before taking the test on Chapter 4, make sure that you have a solid understanding of the following methods for simplifying, multiplying, dividing, adding, and subtracting fractions. Put a checkmark in the box if you can answer "yes" to the statement.

☐ I know how to simplify fractions by factoring the numerator and denominator and then removing the common factors.

$$\frac{42}{50} = \frac{2 \cdot 3 \cdot 7}{2 \cdot 5 \cdot 5}$$
$$= \frac{\overset{1}{\cancel{2}} \cdot 3 \cdot 7}{\underset{1}{\cancel{2}} \cdot 5 \cdot 5}$$
$$= \frac{21}{25}$$

☐ When multiplying fractions, I know that it is important to factor and simplify first, before multiplying.

Factor and simplify first

$$\frac{15}{16} \cdot \frac{24}{35} = \frac{15 \cdot 24}{16 \cdot 35}$$
$$= \frac{3 \cdot \overset{1}{\cancel{5}} \cdot 3 \cdot \overset{1}{\cancel{8}}}{2 \cdot \underset{1}{\cancel{8}} \cdot \underset{1}{\cancel{5}} \cdot 7}$$

Don't multiply first

$$\frac{15}{16} \cdot \frac{24}{35} \ne \frac{15 \cdot 24}{16 \cdot 35}$$
$$= \frac{360}{560}$$

☐ To divide fractions, I know to multiply the first fraction by the reciprocal of the second fraction.

$$\frac{7}{8} \div \frac{23}{24} = \frac{7}{8} \cdot \frac{24}{23}$$

☐ I know that to add or subtract fractions, they must have a common denominator. To multiply or divide fractions, they **do not** need to have a common denominator.

Need an LCD

$$\frac{2}{3} + \frac{1}{5} \qquad \frac{9}{20} - \frac{7}{12}$$

Do not need an LCD

$$\frac{4}{7} \cdot \frac{2}{9} \qquad \frac{11}{40} \div \frac{5}{8}$$

☐ I know how to find the LCD of a set of fractions using one of the following methods.
- Write the multiples of the largest denominator in increasing order, until one is found that is divisible by the other denominators.
- Prime factor each denominator. The LCM is a product of prime factors, where each factor is used the greatest number of times it appears in any one factorization.

☐ I know how to build equivalent fractions by multiplying the given fraction by a form of 1.

$$\frac{2}{3} = \frac{2}{3} \cdot \frac{5}{5}$$
$$= \frac{2 \cdot 5}{3 \cdot 5}$$
$$= \frac{10}{15}$$

Teaching Guide: Refer to the Instructor's Resource Binder to find activities, worksheets on key concepts, more examples, instruction tips, overheads, and assessments.

CHAPTER 4 SUMMARY AND REVIEW

SECTION 4.1 An Introduction to Fractions

DEFINITIONS AND CONCEPTS	EXAMPLES
A **fraction** describes the number of equal parts of a whole. In a fraction, the number above the **fraction bar** is called the **numerator**, and the number below is called the **denominator**.	Since 3 of 8 equal parts are colored red, $\frac{3}{8}$ (three-eighths) of the figure is shaded. Fraction bar ⟶ $\frac{3}{8}$ ⟵ numerator / denominator

Unless otherwise noted, all art on this page is © Cengage Learning.

Chapter 4 Summary and Review

If the numerator of a fraction is less than its denominator, the fraction is called a **proper fraction**. If the numerator of a fraction is greater than or equal to its denominator, the fraction is called an **improper fraction**.	Proper fractions: $\frac{1}{5}, \frac{7}{8}$, and $\frac{999}{1,000}$ — Proper fractions are less than 1. Improper fractions: $\frac{3}{2}, \frac{41}{16}$, and $\frac{15}{15}$ — Improper fractions are greater than or equal to 1.
There are four **special fraction forms** that involve 0 and 1. Each of these fractions is a **form of 1**: $1 = \frac{1}{1} = \frac{2}{2} = \frac{3}{3} = \frac{4}{4} = \frac{5}{5} = \frac{6}{6} = \frac{7}{7} = \frac{8}{8} = \frac{9}{9} = \ldots$	Simplify each fraction: $\frac{0}{8} = 0 \qquad \frac{7}{0}$ is undefined $\qquad \frac{5}{1} = 5 \qquad \frac{20}{20} = 1$
Two fractions are **equivalent** if they represent the same number. **Equivalent fractions** represent the same portion of a whole.	$\frac{2}{3}, \frac{4}{6}$, and $\frac{8}{12}$ are equivalent fractions. They represent the same shaded portion of the figure. $\frac{2}{3} \quad = \quad \frac{4}{6} \quad = \quad \frac{8}{12}$
To **build a fraction**, we multiply it by a factor equal to 1 in the form of $\frac{2}{2}, \frac{3}{3}, \frac{4}{4}, \frac{5}{5}$, and so on.	Write $\frac{3}{4}$ as an equivalent fraction with a denominator of 36. $\frac{3}{4} = \frac{3}{4} \cdot \frac{9}{9}$ — We must multiply the denominator of $\frac{3}{4}$ by 9 to obtain a denominator of 36. It follows that $\frac{9}{9}$ should be the form of 1 that is used to build $\frac{3}{4}$. $= \frac{3 \cdot 9}{4 \cdot 9}$ — Multiply the numerators. Multiply the denominators. $= \frac{27}{36}$ $\frac{27}{36}$ is equivalent to $\frac{3}{4}$.
A fraction is **in simplest form**, or **lowest terms**, when the numerator and denominator have no common factors other than 1.	Is $\frac{6}{14}$ in simplest form? The factors of the numerator, 6, are: **1, 2**, 3, 6. The factors of the denominator, 14, are: **1, 2**, 7, 14. Since the numerator and denominator have a common factor of 2, the fraction $\frac{6}{14}$ is *not* in simplest form.
To **simplify a fraction**, we write it in simplest form by removing a factor equal to 1: 1. Factor (or prime factor) the numerator and denominator to determine their common factors. 2. Remove factors equal to 1 by replacing each pair of factors common to the numerator and denominator with the equivalent fraction $\frac{1}{1}$. 3. Multiply the remaining factors in the numerator and in the denominator.	Simplify: $\frac{12}{30}$ $\frac{12}{30} = \frac{2 \cdot 2 \cdot 3}{2 \cdot 3 \cdot 5}$ — Prime factor 12 and 30. $= \frac{\overset{1}{\cancel{2}} \cdot 2 \cdot \overset{1}{\cancel{3}}}{\underset{1}{\cancel{2}} \cdot \underset{1}{\cancel{3}} \cdot 5}$ — Remove the common factors of 2 and 3 from the numerator and denominator. $= \frac{2}{5}$ — Multiply the remaining factors in the numerator: $1 \cdot 2 \cdot 1 = 2$. Multiply the remaining factors in the denominator: $1 \cdot 1 \cdot 5 = 5$. Since 2 and 5 have no common factors other than 1, we say that $\frac{2}{5}$ is in simplest form.

Unless otherwise noted, all art on this page is © Cengage Learning.

Fractions that contain a variable (or variables) in the numerator, the denominator, or both are called **algebraic fractions**.	Algebraic fractions: $$\frac{x}{5}, \frac{12}{a}, \frac{r}{25s}, \frac{4c^2d}{16cd^3}, \frac{x+1}{x-9}$$
Algebraic fractions are built up just like numerical fractions.	Write $\frac{4}{9}$ as an equivalent fraction with a denominator of $27x$. We need to multiply the denominator of $\frac{4}{9}$ by $3x$ to obtain $27x$. It follows that $\frac{3x}{3x}$ should be the form of 1 that is used to build $\frac{4}{9}$. $\frac{4}{9} = \frac{4}{9} \cdot \frac{3x}{3x}$ Multiply $\frac{4}{9}$ by a form of 1: $\frac{3x}{3x} = 1$. $= \frac{4 \cdot 3x}{9 \cdot 3x}$ Multiply the numerators. Multiply the denominators. $= \frac{12x}{27x}$
Algebraic fractions are simplified just like numerical fractions.	Simplify: $\frac{18x^2y^3}{42x^4y}$ $\frac{18x^2y^3}{42x^4y} = \frac{2 \cdot 3 \cdot 3 \cdot x \cdot x \cdot y \cdot y \cdot y}{2 \cdot 3 \cdot 7 \cdot x \cdot x \cdot x \cdot x \cdot y}$ To prepare to simplify, factor 18, x^2, y^3, 42, and x^4. $= \frac{\overset{1}{2} \cdot \overset{1}{3} \cdot 3 \cdot \overset{1}{x} \cdot \overset{1}{x} \cdot \overset{1}{y} \cdot y \cdot y}{\underset{1}{2} \cdot \underset{1}{3} \cdot 7 \cdot \underset{1}{x} \cdot \underset{1}{x} \cdot x \cdot x \cdot \underset{1}{y}}$ Simplify by removing the common factors of 2, 3, x, and y from the numerator and denominator. $= \frac{3y^2}{7x^2}$ Multiply the remaining factors in the numerator. Multiply the remaining factors in the denominator.

REVIEW EXERCISES

1. Identify the numerator and denominator of the fraction $\frac{11}{16}$. Is it a proper or an improper fraction? numerator: 11, denominator: 16; proper fraction

2. Write fractions that represent the shaded and unshaded portions of the figure to the right. $\frac{4}{7}, \frac{3}{7}$

3. In the illustration below, why can't we say that $\frac{3}{4}$ of the figure is shaded? The figure is not divided into equal parts.

4. Write the fraction $\frac{2}{-3}$ in two other ways. $-\frac{2}{3}, \frac{-2}{3}$

5. Simplify, if possible:

   a. $\frac{5}{5}$ 1

   b. $\frac{0}{10}$ 0

   c. $\frac{18}{1}$ 18

   d. $\frac{7}{0}$ undefined

6. What concept about fractions is illustrated below? equivalent fractions: $\frac{6}{8} = \frac{3}{4}$

Write each fraction as an equivalent fraction with the indicated denominator.

7. $\frac{2}{3}$, denominator 18 $\frac{12}{18}$

8. $\frac{3}{8}$, denominator 16 $\frac{6}{16}$

9. $\frac{7}{15}$, denominator $45a$ $\frac{21a}{45a}$

10. $\frac{13}{12x}$, denominator $60x$ $\frac{65}{60x}$

11. Write 5 as an equivalent fraction with denominator 9. $\frac{45}{9}$

12. Are the following fractions in simplest form?

    a. $\frac{6}{9}$ no

    b. $\frac{10}{81}$ yes

Simplify each fraction, if possible.

13. $\dfrac{15}{45}$ $\dfrac{1}{3}$

14. $\dfrac{20x^3}{48x^2}$ $\dfrac{5x}{12}$

15. $\dfrac{66}{108}$ $\dfrac{11}{18}$

16. $\dfrac{117a^2b^6}{208a^5b}$ $\dfrac{9b^5}{16a^3}$

17. $\dfrac{81}{64}$ in simplest form

18. Tell whether $\dfrac{8}{12}$ and $\dfrac{176}{264}$ are equivalent by simplifying each fraction. **equivalent**

19. SLEEP If a woman gets seven hours of sleep each night, write a fraction to describe the part of a whole day that she spends sleeping and another to describe the part of a whole day that she is not sleeping. $\dfrac{7}{24}, \dfrac{17}{24}$

20. a. What type of problem is shown below? Explain the step shown in red.

$$\dfrac{5}{8} = \dfrac{5}{8} \cdot \dfrac{2}{2} = \dfrac{10}{16}$$

The fraction $\dfrac{5}{8}$ is being expressed as an equivalent fraction with a denominator of 16. To build the fraction, multiply $\dfrac{5}{8}$ by 1 in the form of $\dfrac{2}{2}$.

b. What type of problem is shown below? Explain the step shown in red.

$$\dfrac{4}{6} = \dfrac{\overset{1}{2} \cdot 2}{2 \cdot 3} = \dfrac{2}{3}$$

The fraction $\dfrac{4}{6}$ is being simplified. To simplify the fraction, remove the common factors of 2 from the numerator and denominator. This removes a factor equal to 1: $\dfrac{2}{2} = 1$.

SECTION 4.2   Multiplying Fractions

DEFINITIONS AND CONCEPTS	EXAMPLES
To **multiply two fractions**, multiply the numerators and multiply the denominators. Simplify the result, if possible.	Multiply and simplify, if possible: $\dfrac{4}{5} \cdot \dfrac{2}{3}$ $\dfrac{4}{5} \cdot \dfrac{2}{3} = \dfrac{4 \cdot 2}{5 \cdot 3}$ Multiply the numerators. Multiply the denominators. $= \dfrac{8}{15}$ Since 8 and 15 have no common factors other than 1, the result is in simplest form.
Multiplying signed fractions The product of two fractions with the same (like) signs is positive. The product of two fractions with different (unlike) signs is negative.	Multiply and simplify, if possible: $-\dfrac{3}{4} \cdot \dfrac{2}{27}$ $-\dfrac{3}{4} \cdot \dfrac{2}{27} = -\dfrac{3 \cdot 2}{4 \cdot 27}$ Multiply the numerators. Multiply the denominators. Since the fractions have unlike signs, make the answer negative. $= -\dfrac{\overset{1}{3} \cdot \overset{1}{2}}{\underset{1}{2} \cdot 2 \cdot \underset{1}{3} \cdot 3 \cdot 3}$ Prime factor 4 and 27. Then simplify, by removing the common factors of 2 and 3 from the numerator and denominator. $= -\dfrac{1}{18}$ Multiply the remaining factors in the numerator: $1 \cdot 1 = 1$. Multiply the remaining factors in the denominator: $1 \cdot 2 \cdot 1 \cdot 3 \cdot 3 = 18$.

To **multiply two algebraic fractions**, we use the same approach as with numerical fractions.	Multiply and simplify, if possible: $\dfrac{35x^3}{11} \cdot \dfrac{2}{25x}$ $\dfrac{35x^3}{11} \cdot \dfrac{2}{25x} = \dfrac{35x^3 \cdot 2}{11 \cdot 25x}$ Multiply the numerators. Multiply the denominators. $= \dfrac{5 \cdot 7 \cdot x \cdot x \cdot x \cdot 2}{11 \cdot 5 \cdot 5 \cdot x}$ To prepare to simplify, factor 35, x^3, and 25. $= \dfrac{\overset{1}{\cancel{5}} \cdot 7 \cdot \overset{1}{\cancel{x}} \cdot x \cdot x \cdot 2}{11 \cdot \underset{1}{\cancel{5}} \cdot 5 \cdot \underset{1}{\cancel{x}}}$ Simplify by removing the common factors of 5 and x from the numerator and denominator. $= \dfrac{14x^2}{55}$ Multiply the remaining factors in the numerator. Multiply the remaining factors in the denominator.
The base of an **exponential expression** can be a positive or a negative fraction. The rule for multiplying two fractions can be extended to find the product of three or more fractions.	Evaluate: $\left(\dfrac{2}{3}\right)^3$ $\left(\dfrac{2}{3}\right)^3 = \dfrac{2}{3} \cdot \dfrac{2}{3} \cdot \dfrac{2}{3}$ Write the base, $\tfrac{2}{3}$, as a factor 3 times. $= \dfrac{2 \cdot 2 \cdot 2}{3 \cdot 3 \cdot 3}$ Multiply the numerators. Multiply the denominators. $= \dfrac{8}{27}$ This fraction is in simplest form.
When a **fraction is followed by the word *of***, it indicates that we are to find a part of some quantity using multiplication.	To find $\dfrac{2}{5}$ of 35, we multiply: $\dfrac{2}{5} \text{ of } 35 = \dfrac{2}{5} \cdot 35$ The word *of* indicates multiplication. $= \dfrac{2}{5} \cdot \dfrac{35}{1}$ Write 35 as a fraction: $35 = \tfrac{35}{1}$. $= \dfrac{2 \cdot 35}{5 \cdot 1}$ Multiply the numerators. Multiply the denominators. $= \dfrac{2 \cdot \overset{1}{\cancel{5}} \cdot 7}{\underset{1}{\cancel{5}} \cdot 1}$ Prime factor 35. Then simplify by removing the common factor of 5 from the numerator and denominator. $= \dfrac{14}{1}$ Multiply the remaining factors in the numerator and in the denominator. $= 14$ Any number divided by 1 is equal to that number.

The formula for the area of a triangle

Area of a triangle = $\frac{1}{2}$ (base)(height)

or

$A = \frac{1}{2}bh$

Find the area of the triangle shown on the right.

$A = \frac{1}{2}$ (base)(height)

$A = \frac{1}{2}(8)(5)$ Substitute 8 for the base and 5 for the height.

$A = \frac{1}{2}\left(\frac{8}{1}\right)\left(\frac{5}{1}\right)$ Write 8 and 5 as fractions.

$A = \frac{1 \cdot 8 \cdot 5}{2 \cdot 1 \cdot 1}$ Multiply the numerators. Multiply the denominators.

$A = \dfrac{1 \cdot \overset{1}{2} \cdot 2 \cdot 2 \cdot 5}{\underset{1}{2} \cdot 1 \cdot 1}$ Prime factor 8. Then simplify by removing the common factor of 2 from the numerator and denominator.

$A = 20$

The area of the triangle is 20 ft².

REVIEW EXERCISES

21. Fill in the blanks: To multiply two fractions, multiply the __numerators__ and multiply the __denominators__. Then __simplify__, if possible.

22. Translate the following phrase to symbols. *You do not have to find the answer.*

    $\frac{5}{6}$ of $\frac{2}{3}$ $\frac{5}{6} \cdot \frac{2}{3}$

Multiply. Simplify the product, if possible.

23. $\frac{1}{2} \cdot \frac{1}{3}$ $\frac{1}{6}$

24. $\frac{2}{5}\left(-\frac{7}{9}\right)$ $-\frac{14}{45}$

25. $\frac{9c^3}{16d} \cdot \frac{20d}{27c}$ $\frac{5c^2}{12}$

26. $-\frac{5}{6}\left(-\frac{1}{15}\right)\left(-\frac{18}{25}\right)$ $-\frac{1}{25}$

27. $\frac{3}{5} \cdot 7$ $\frac{21}{5}$

28. $-4\left(-\frac{9m}{16}\right)$ $\frac{9m}{4}$

29. $-3x\left(\frac{1}{3}\right)$ $-x$

30. $-\frac{6}{7}\left(-\frac{7}{6}\right)$ 1

Evaluate each expression.

31. $-\left(\frac{3}{4}\right)^2$ $-\frac{9}{16}$

32. $\left(-\frac{5a}{2}\right)^3$ $-\frac{125a^3}{8}$

33. $\left(-\frac{2}{5}\right)^3$ $-\frac{8}{125}$

34. $\left(\frac{2}{3}\right)^2$ $\frac{4}{9}$

35. **DRAG RACING** A top-fuel dragster had to make 8 trial runs on a quarter-mile track before it was ready for competition. Find the total distance it covered on the trial runs. 2 mi

36. **GRAVITY** Objects on the moon weigh only one-sixth of their weight on Earth. How much will an astronaut weigh on the moon if he weighs 180 pounds on Earth? 30 lb

37. Find the area of the triangular sign. 60 in.²

38. Find the area of the triangle shown below. 165 ft²

SECTION 4.3 Dividing Fractions

DEFINITIONS AND CONCEPTS	EXAMPLES
One number is the **reciprocal** of another if their product is 1. To find the **reciprocal of a fraction**, invert the numerator and denominator.	The reciprocal of $\dfrac{4}{5}$ is $\dfrac{5}{4}$ because $\dfrac{4}{5} \cdot \dfrac{5}{4} = 1$. Fraction $\dfrac{4}{5}$ ⟶ Reciprocal $\dfrac{5}{4}$ (Invert)
To **divide two fractions**, multiply the first fraction by the reciprocal of the second fraction. Simplify the result, if possible.	Divide and simplify, if possible: $\dfrac{4}{35} \div \dfrac{2}{21}$ $\dfrac{4}{35} \div \dfrac{2}{21} = \dfrac{4}{35} \cdot \dfrac{21}{2}$ Multiply $\dfrac{4}{35}$ by the reciprocal of $\dfrac{2}{21}$, which is $\dfrac{21}{2}$. $= \dfrac{4 \cdot 21}{35 \cdot 2}$ Multiply the numerators. Multiply the denominators. $= \dfrac{2 \cdot 2 \cdot 3 \cdot 7}{5 \cdot 7 \cdot 2}$ To prepare to simplify, write 4, 21, and 35 in prime-factored form. $= \dfrac{\overset{1}{2} \cdot 2 \cdot 3 \cdot \overset{1}{7}}{5 \cdot \underset{1}{7} \cdot \underset{1}{2}}$ To simplify, remove the common factors of 2 and 7 from the numerator and denominator. $= \dfrac{6}{5}$ Multiply the remaining factors in the numerator: $1 \cdot 2 \cdot 3 \cdot 1 = 6$. Multiply the remaining factors in the denominator: $5 \cdot 1 \cdot 1 = 5$.
The **sign rules for dividing fractions** are the same as those for multiplying fractions.	Divide and simplify, if possible: $\dfrac{9}{16} \div (-3)$ $\dfrac{9}{16} \div (-3) = \dfrac{9}{16} \cdot \left(-\dfrac{1}{3}\right)$ Multiply $\dfrac{9}{16}$ by the reciprocal of -3, which is $-\dfrac{1}{3}$. $= -\dfrac{9 \cdot 1}{16 \cdot 3}$ Multiply the numerators. Multiply the denominators. Since the fractions have unlike signs, make the answer negative. $= -\dfrac{\overset{1}{3} \cdot 3 \cdot 1}{16 \cdot \underset{1}{3}}$ To simplify, factor 9 as $3 \cdot 3$. Then remove the common factor of 3 from the numerator and denominator. $= -\dfrac{3}{16}$ Multiply the remaining factors in the numerator: $1 \cdot 3 \cdot 1 = 3$. Multiply the remaining factors in the denominator: $16 \cdot 1 = 16$.

To **divide two algebraic fractions**, we use the same approach as with numerical fractions. The **reciprocal** of an algebraic fraction is found by inverting the numerator and denominator. Algebraic fraction → Reciprocal $\dfrac{6x}{23}$ ⟶ $\dfrac{23}{6x}$ Invert	Divide and simplify: $-\dfrac{9x}{35y} \div \left(-\dfrac{15x^2}{14y}\right)$ $-\dfrac{9x}{35y} \div \left(-\dfrac{15x^2}{14y}\right) = -\dfrac{9x}{35y}\left(-\dfrac{14y}{15x^2}\right)$ Multiply by the reciprocal of $-\dfrac{15x^2}{14y}$. $= \dfrac{9x \cdot 14y}{35y \cdot 15x^2}$ Multiply the numerators. Multiply the denominators. Since the product of two fractions with like signs is positive, drop the negative signs and continue. $= \dfrac{3 \cdot 3 \cdot x \cdot 2 \cdot 7 \cdot y}{5 \cdot 7 \cdot y \cdot 3 \cdot 5 \cdot x \cdot x}$ To prepare to simplify, factor 9, 14, 35, and $15x^2$. $= \dfrac{3 \cdot \overset{1}{\cancel{3}} \cdot \overset{1}{\cancel{x}} \cdot 2 \cdot \overset{1}{\cancel{7}} \cdot \overset{1}{\cancel{y}}}{5 \cdot \underset{1}{\cancel{7}} \cdot \underset{1}{\cancel{y}} \cdot \underset{1}{\cancel{3}} \cdot 5 \cdot \underset{1}{\cancel{x}} \cdot x}$ Simplify by removing the common factors of 3, 7, x, and y from the numerator and denominator. $= \dfrac{6}{25x}$ Multiply the remaining factors in the numerator. Multiply the remaining factors in the denominator.
Problems that involve forming **equal-sized groups** can be solved by division.	**SEWING** How many Halloween costumes, which require $\frac{3}{4}$ yard of material, can be made from 6 yards of material? Since 6 yards of material is to be separated into an unknown number of equal-sized $\frac{3}{4}$-yard pieces, division is indicated. $6 \div \dfrac{3}{4} = \dfrac{6}{1} \cdot \dfrac{4}{3}$ Write 6 as a fraction: $6 = \frac{6}{1}$. Multiply $\frac{6}{1}$ by the reciprocal of $\frac{3}{4}$, which is $\frac{4}{3}$. $= \dfrac{6 \cdot 4}{1 \cdot 3}$ Multiply the numerators. Multiply the denominators. $= \dfrac{2 \cdot \overset{1}{\cancel{3}} \cdot 4}{1 \cdot \underset{1}{\cancel{3}}}$ To simplify, factor 6 as $2 \cdot 3$. Then remove the common factor of 3 from the numerator and denominator. $= \dfrac{8}{1}$ Multiply the remaining factors in the numerator. Multiply the remaining factors in the denominator. $= 8$ Any number divided by 1 is the same number. The number of Halloween costumes that can be made from 6 yards of material is 8.

REVIEW EXERCISES

39. Find the reciprocal of each number or algebraic fraction.

a. $\dfrac{1}{8}$ 8

b. $-\dfrac{11}{12}$ $-\dfrac{12}{11}$

c. 5 $\dfrac{1}{5}$

d. $\dfrac{8a}{7}$ $\dfrac{7}{8a}$

40. Fill in the blanks: To divide two fractions, __multiply__ the first fraction by the __reciprocal__ of the second fraction.

Divide. Simplify the quotient, if possible.

41. $\dfrac{1}{6} \div \dfrac{11}{25}$ $\dfrac{25}{66}$

42. $-\dfrac{7}{32} \div \dfrac{1}{4}$ $-\dfrac{7}{8}$

43. $-\dfrac{39m^4}{25n} \div \left(-\dfrac{13m^3}{10n}\right)$ $\dfrac{6m}{5}$

44. $54d \div \dfrac{63}{5}$ $\dfrac{30d}{7}$

45. $-\dfrac{3}{8} \div \dfrac{1}{4}$ $-\dfrac{3}{2}$

46. $\dfrac{4}{5} \div \dfrac{1}{2}$ $\dfrac{8}{5}$

47. $\dfrac{2}{3} \div (-120)$ $-\dfrac{1}{180}$

48. $\dfrac{7}{15} \div \dfrac{7}{15}$ 1

49. MAKING JEWELRY How many $\frac{1}{16}$-ounce silver angel pins can be made from a $\frac{3}{4}$-ounce bar of silver? 12 pins

50. SEWING How many pillow cases, which require $\frac{2}{3}$ yard of material, can be made from 20 yards of cotton cloth? 30 pillow cases

SECTION 4.4 Adding and Subtracting Fractions

DEFINITIONS AND CONCEPTS	EXAMPLES
To **add (or subtract) fractions that have the same denominator**, add (or subtract) the numerators and write the sum (or difference) over the common denominator. Simplify the result, if possible.	Add: $\dfrac{3}{16} + \dfrac{5}{16}$ $\dfrac{3}{16} + \dfrac{5}{16} = \dfrac{3+5}{16}$ Add the numerators and write the sum over the common denominator 16. $= \dfrac{8}{16}$ The resulting fraction can be simplified. $= \dfrac{\overset{1}{\cancel{8}}}{2 \cdot \underset{1}{\cancel{8}}}$ To simplify, factor 16 as $2 \cdot 8$. Then remove the common factor of 8 from the numerator and denominator. $= \dfrac{1}{2}$ Multiply the remaining factors in the denominator: $2 \cdot 1 = 2$.
Adding and subtracting fractions that have different denominators 1. Find the LCD. 2. Rewrite each fraction as an equivalent fraction with the LCD as the denominator. To do so, build each fraction using a form of 1 that involves any factors needed to obtain the LCD. 3. Add or subtract the numerators and write the sum or difference over the LCD. 4. Simplify the result, if possible.	Subtract: $\dfrac{4}{7} - \dfrac{1}{3}$ Since the smallest number the denominators 7 and 3 divide exactly is 21, the LCD is 21. $\dfrac{4}{7} - \dfrac{1}{3} = \dfrac{4}{7} \cdot \dfrac{3}{3} - \dfrac{1}{3} \cdot \dfrac{7}{7}$ To build $\frac{4}{7}$ and $\frac{1}{3}$ so that their denominators are 21, multiply each by a form of 1. $= \dfrac{12}{21} - \dfrac{7}{21}$ Multiply the numerators. Multiply the denominators. The denominators are now the same. $= \dfrac{12-7}{21}$ Subtract the numerators and write the difference over the common denominator 21. $= \dfrac{5}{21}$ This fraction is in simplest form.

The **least common denominator (LCD)** of a set of fractions is the **least common multiple (LCM)** of the denominators of the fractions. Two ways to find the LCM of the denominators are as follows: • Write the multiples of the largest denominator in increasing order, until one is found that is divisible by the other denominators. • Prime factor each denominator. The LCM is a product of prime factors, where each factor is used the greatest number of times it appears in any one factorization.	Add and simplify: $\dfrac{9}{20} + \dfrac{7}{15}$ To find the LCD, find the prime factorization of both denominators and use each prime factor the *greatest* number of times it appears in any one factorization: $\left. \begin{array}{l} 20 = \boxed{2 \cdot 2} \cdot \boxed{5} \\ 15 = \boxed{3} \cdot 5 \end{array} \right\} \text{LCD} = 2 \cdot 2 \cdot 3 \cdot 5 = 60$ $\dfrac{9}{20} + \dfrac{7}{15} = \dfrac{9}{20} \cdot \dfrac{3}{3} + \dfrac{7}{15} \cdot \dfrac{4}{4}$ To build $\dfrac{9}{20}$ and $\dfrac{7}{15}$ so that their denominators are 60, multiply each by a form of 1. $= \dfrac{27}{60} + \dfrac{28}{60}$ Multiply the numerators. Multiply the denominators. The denominators are now the same. $= \dfrac{27 + 28}{60}$ Add the numerators and write the sum over the common denominator 60. $= \dfrac{55}{60}$ This fraction is not in simplest form. $= \dfrac{\overset{1}{\cancel{5}} \cdot 11}{2 \cdot 2 \cdot 3 \cdot \underset{1}{\cancel{5}}}$ To simplify, prime factor 55 and 60. Then remove the common factor of 5 from the numerator and denominator. $= \dfrac{11}{12}$ Multiply the remaining factors in the numerator and in the denominator.
To **add or subtract algebraic fractions**, we use the same approach as with numerical fractions.	Add and simplify: $\dfrac{5w}{24} + \dfrac{11w}{24}$ $\dfrac{5w}{24} + \dfrac{11w}{24} = \dfrac{5w + 11w}{24}$ Since the fractions have like denominators, add the numerators and write the sum over the common denominator, 24. $= \dfrac{16w}{24}$ Combine like terms in the numerator: $5w + 11w = 16w$. $= \dfrac{2 \cdot \overset{1}{\cancel{8}} \cdot w}{3 \cdot \underset{1}{\cancel{8}}}$ To simplify, factor 16 as $2 \cdot 8$ and 24 as $3 \cdot 8$ and then remove the common factor of 8 from the numerator and denominator. $= \dfrac{2w}{3}$ Multiply the remaining factors in the numerator. Multiply the remaining factors in the denominator.

If we are to **add or subtract algebraic fractions**, their denominators must be the same.

Subtract and simplify, if possible: $\dfrac{6}{m} - \dfrac{3}{4}$

The denominators of $\dfrac{6}{m}$ and $\dfrac{3}{4}$ are m and 4. By inspection, we see that the LCD is $m \cdot 4 = 4m$.

$$\dfrac{6}{m} - \dfrac{3}{4} = \dfrac{6}{m} \cdot \dfrac{4}{4} - \dfrac{3}{4} \cdot \dfrac{m}{m}$$
Build each fraction so that each has a denominator of $4m$.

$$= \dfrac{24}{4m} - \dfrac{3m}{4m}$$
Multiply the numerators. Multiply the denominators. The denominators are now the same.

$$= \dfrac{24 - 3m}{4m}$$
Subtract the numerators and write the difference over the common denominator, $4m$.

Caution! The result, $\dfrac{24 - 3m}{4m}$, is in simplest form. We cannot remove a common factor of m, because m is not a factor of the entire numerator.

$$\dfrac{24 - 3m}{4m} = \dfrac{24 - 3\overset{1}{\cancel{m}}}{4\underset{1}{\cancel{m}}} = \dfrac{21}{4}$$

Comparing fractions
If two fractions have the **same denominator**, the fraction with the greater numerator is the greater fraction.

Which fraction is larger: $\dfrac{11}{18}$ or $\dfrac{7}{18}$?

$\dfrac{11}{18} > \dfrac{7}{18}$ because $11 > 7$

If two fractions have **different denominators**, express each of them as an equivalent fraction that has the LCD for its denominator. Then compare numerators.

Which fraction is larger: $\dfrac{2}{3}$ or $\dfrac{3}{4}$?

Build each fraction to have a denominator that is the LCD, 12.

$$\dfrac{2}{3} = \dfrac{2}{3} \cdot \dfrac{4}{4} = \dfrac{8}{12} \qquad \dfrac{3}{4} = \dfrac{3}{4} \cdot \dfrac{3}{3} = \dfrac{9}{12}$$

Since $9 > 8$, it follows that $\dfrac{9}{12} > \dfrac{8}{12}$ and therefore, $\dfrac{3}{4} > \dfrac{2}{3}$.

REVIEW EXERCISES

Add or subtract and simplify, if possible.

51. $\dfrac{2}{7} + \dfrac{3}{7} \quad \dfrac{5}{7}$

52. $\dfrac{3}{4} - \dfrac{1}{4} \quad \dfrac{1}{2}$

53. $\dfrac{7x}{8} + \dfrac{3x}{8} \quad \dfrac{5x}{4}$

54. $-\dfrac{3}{5} - \dfrac{3}{5} \quad -\dfrac{6}{5}$

55. **a.** Add the fractions represented by the figures below. $\dfrac{5}{8}$

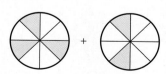

b. Subtract the fractions represented by the figures below. $\dfrac{1}{5}$

56. Fill in the blanks. Use the prime factorizations below to find the least common denominator for fractions with denominators of 45 and 30.

$$\left.\begin{array}{l}45 = 3 \cdot 3 \cdot 5\\ 30 = 2 \cdot 3 \cdot 5\end{array}\right\} \text{LCD} = 2 \cdot 3 \cdot 3 \cdot 5 = 90$$

Add or subtract and simplify, if possible.

57. $\dfrac{1}{6} + \dfrac{2}{3}$ $\dfrac{5}{6}$

58. $-\dfrac{2}{5} - \dfrac{3}{8}$ $-\dfrac{31}{40}$

59. $\dfrac{5}{24} + \dfrac{3}{16}$ $\dfrac{19}{48}$

60. $3 - \dfrac{1}{7}$ $\dfrac{20}{7}$

61. $-\dfrac{19}{18} + \dfrac{5}{12}$ $-\dfrac{23}{36}$

62. $\dfrac{17}{20} - \dfrac{4}{15}$ $\dfrac{7}{12}$

63. $-6 + \dfrac{13}{6}$ $-\dfrac{23}{6}$

64. $\dfrac{1}{3} + \dfrac{1}{4} + \dfrac{1}{5}$ $\dfrac{47}{60}$

65. $\dfrac{8}{n} + \dfrac{1}{2}$ $\dfrac{16 + n}{2n}$

66. $\dfrac{11}{9} - \dfrac{4}{x}$ $\dfrac{11x - 36}{9x}$

67. $\dfrac{a}{3} - \dfrac{4}{11}$ $\dfrac{11a - 12}{33}$

68. $\dfrac{7}{8} + \dfrac{r}{7}$ $\dfrac{49 + 8r}{56}$

69. MACHINE SHOPS How much must be milled off the $\dfrac{3}{4}$-inch-thick steel rod below so that the collar will slip over the end of it? $\dfrac{7}{32}$ in.

Steel rod

70. POLLS A group of adults were asked to rate the transportation system in their community. The results are shown below in a circle graph. What fraction of the group responded by saying either excellent, good, or fair? $\dfrac{3}{4}$

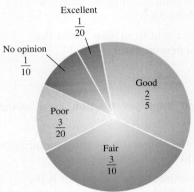

71. TELEMARKETING In the first hour of work, a telemarketer made 2 sales out of 9 telephone calls. In the second hour, she made 3 sales out of 11 calls. During which hour was the rate of sales to calls better? the second hour: $\dfrac{3}{11} > \dfrac{2}{9}$

72. CAMERAS When the shutter of a camera stays open longer than $\dfrac{1}{125}$ second, any movement of the camera will probably blur the picture. With this in mind, if a photographer is taking a picture of a fast-moving object, should she select a shutter speed of $\dfrac{1}{60}$ or $\dfrac{1}{250}$? $\dfrac{1}{250}$

SECTION 4.5 Multiplying and Dividing Mixed Numbers

DEFINITIONS AND CONCEPTS	EXAMPLES
A **mixed number** is the sum of a whole number and a proper fraction.	$2\dfrac{3}{4}$ (Mixed number) $=$ 2 (Whole-number part) $+$ $\dfrac{3}{4}$ (Fractional part)
There is a relationship between **mixed numbers and improper fractions** that can be seen using shaded regions.	Each disk represents one whole. $2\dfrac{3}{4} = \dfrac{11}{4}$

Unless otherwise noted, all art on this page is © Cengage Learning.

To write a mixed number as an improper fraction: 1. Multiply the denominator of the fraction by the whole-number part. 2. Add the numerator of the fraction to the result from Step 1. 3. Write the sum from Step 2 over the original denominator.	Write $3\frac{4}{5}$ as an improper fraction. $$3\frac{4}{5} = \frac{5 \cdot 3 + 4}{5} = \frac{15 + 4}{5} = \frac{19}{5}$$ Step 1: Multiply Step 2: Add Step 3: Use the same denominator Thus, $3\frac{4}{5} = \frac{19}{5}$.
To write an improper fraction as a mixed number: 1. Divide the numerator by the denominator to obtain the whole-number part. 2. The remainder over the divisor is the fractional part.	Write $\frac{47}{6}$ as a mixed number. $\begin{array}{r} 7 \\ 6{\overline{\smash{)}47}} \\ -42 \\ \hline 5 \end{array}$ ← The whole-number part is 7. ← Write the remainder 5 over the divisor 6 to get the fractional part. Thus, $\frac{47}{6} = 7\frac{5}{6}$.
Fractions and mixed numbers can be **graphed** on a number line.	Graph $-3\frac{1}{3}, 1\frac{1}{4}, \frac{18}{5},$ and $-\frac{7}{8}$ on a number line. $-3\frac{1}{3}$ $-\frac{7}{8}$ $1\frac{1}{4}$ $\frac{18}{5} = 3\frac{3}{5}$ ―4 ―3 ―2 ―1 0 1 2 3 4
To **multiply mixed numbers**, first change the mixed numbers to improper fractions. Then perform the multiplication of the fractions. Write the result as a mixed number or whole number in simplest form.	Multiply and simplify: $10\frac{1}{2} \cdot 1\frac{1}{6}$ $10\frac{1}{2} \cdot 1\frac{1}{6} = \frac{21}{2} \cdot \frac{7}{6}$ Write $10\frac{1}{2}$ and $1\frac{1}{6}$ as improper fractions. $= \frac{21 \cdot 7}{2 \cdot 6}$ Use the rule for multiplying two fractions. Multiply the numerators. Multiply the denominators. $= \frac{\overset{1}{\cancel{3}} \cdot 7 \cdot 7}{2 \cdot 2 \cdot \underset{1}{\cancel{3}}}$ To simplify, factor 21 as $3 \cdot 7$ and 6 as $2 \cdot 3$. Then remove the common factor of 3 from the numerator and denominator. $= \frac{49}{4}$ Multiply the remaining factors in the numerator and in the denominator. The result is an improper fraction. $= 12\frac{1}{4}$ Write the improper fraction $\frac{49}{4}$ as a mixed number. $\begin{array}{r} 12 \\ 4{\overline{\smash{)}49}} \\ -4 \\ \hline 09 \\ -8 \\ \hline 1 \end{array}$

To **divide mixed numbers**, first change the mixed numbers to improper fractions. Then perform the division of the fractions. Write the result as a mixed number or whole number in simplest form.

Divide and simplify: $5\frac{2}{3} \div \left(-3\frac{7}{9}\right)$

$5\frac{2}{3} \div \left(-3\frac{7}{9}\right) = \frac{17}{3} \div \left(-\frac{34}{9}\right)$ Write $5\frac{2}{3}$ and $3\frac{7}{9}$ as improper fractions.

$= \frac{17}{3}\left(-\frac{9}{34}\right)$ Multiply $\frac{17}{3}$ by the reciprocal of $-\frac{34}{9}$, which is $-\frac{9}{34}$.

$= -\frac{17 \cdot 9}{3 \cdot 34}$ Multiply the numerators. Multiply the denominators. Since the fractions have unlike signs, make the answer negative.

$= -\frac{\overset{1}{\cancel{17}} \cdot \overset{1}{\cancel{3}} \cdot 3}{\underset{1}{\cancel{3}} \cdot 2 \cdot \underset{1}{\cancel{17}}}$ To simplify, factor 9 as $3 \cdot 3$ and 34 as $2 \cdot 17$. Then remove the common factors of 3 and 17 from the numerator and denominator.

$= -\frac{3}{2}$ Multiply the remaining factors in the numerator and in the denominator. The result is a negative improper fraction.

$= -1\frac{1}{2}$ Write the negative improper fraction $-\frac{3}{2}$ as a negative mixed number.

REVIEW EXERCISES

73. In the illustration below, each triangular region outlined in black represents one whole. Write a mixed number and an improper fraction to represent what is shaded. $4\frac{1}{4} = \frac{17}{4}$

74. Graph $-2\frac{2}{3}, \frac{8}{9}, -\frac{3}{4}$, and $\frac{59}{24}$ on a number line.

$-2\frac{2}{3}$ $-\frac{3}{4}$ $\frac{8}{9}$ $\frac{59}{24} = 2\frac{11}{24}$

Write each improper fraction as a mixed number or a whole number.

75. $\frac{16}{5}$ $3\frac{1}{5}$

76. $-\frac{47}{12}$ $-3\frac{11}{12}$

77. $\frac{51}{3}$ 17

78. $\frac{14}{6}$ $2\frac{1}{3}$

Write each mixed number as an improper fraction.

79. $9\frac{3}{8}$ $\frac{75}{8}$

80. $-2\frac{1}{5}$ $-\frac{11}{5}$

81. $3\frac{11}{14}$ $\frac{53}{14}$

82. $1\frac{99}{100}$ $\frac{199}{100}$

Multiply or divide and simplify, if possible.

83. $1\frac{2}{5} \cdot 1\frac{1}{2}$ $2\frac{1}{10}$

84. $-3\frac{1}{2} \div 3\frac{2}{3}$ $-\frac{21}{22}$

85. $-6\left(-6\frac{2}{3}\right)$ 40

86. $8 \div 3\frac{1}{5}$ $2\frac{1}{2}$

87. $-11\frac{1}{5} \div \left(-\frac{7}{10}\right)$ 16

88. $5\frac{2}{3}\left(-7\frac{1}{5}\right)$ $-40\frac{4}{5}$

89. $\left(-2\frac{3}{4}\right)^2$ $7\frac{9}{16}$

90. $1\frac{5}{16} \cdot 1\frac{7}{9} \cdot 2\frac{2}{3}$ $6\frac{2}{9}$

91. PHOTOGRAPHY Each leg of a camera tripod can be extended to become $5\frac{1}{2}$ times its original length. If a leg is originally $8\frac{3}{4}$ inches long, how long will it become when it is completely extended? $48\frac{1}{8}$ in.

92. PET DOORS Find the area of the opening provided by the rectangular-shaped pet door shown below. 87 in.²

93. PRINTING It takes a color copier $2\frac{1}{4}$ minutes to print a movie poster. How many posters can be printed in 90 minutes? 40 posters

94. STORM DAMAGE A truck can haul $7\frac{1}{2}$ tons of trash in one load. How many loads would it take to haul away $67\frac{1}{2}$ tons from a hurricane cleanup site? 9 loads

SECTION 4.6 Adding and Subtracting Mixed Numbers

DEFINITIONS AND CONCEPTS	EXAMPLES
To add (or subtract) mixed numbers, we can change each to an improper fraction and use the method of Section 4.4.	Add: $3\frac{1}{2} + 1\frac{3}{5}$ $3\frac{1}{2} + 1\frac{3}{5} = \frac{7}{2} + \frac{8}{5}$ Write $3\frac{1}{2}$ and $1\frac{3}{5}$ as mixed numbers. $= \frac{7}{2} \cdot \frac{5}{5} + \frac{8}{5} \cdot \frac{2}{2}$ To build $\frac{7}{2}$ and $\frac{8}{5}$ so that their denominators are 10, multiply both by a form of 1. $= \frac{35}{10} + \frac{16}{10}$ Multiply the numerators. Multiply the denominators. $= \frac{51}{10}$ Add the numerators and write the sum over the common denominator 10. $= 5\frac{1}{10}$ To write the improper fraction $\frac{51}{10}$ as a mixed number, divide 51 by 10.
To add (or subtract) mixed numbers, we can also write them in **vertical form** and add (or subtract) the whole-number parts and the fractional parts separately.	Add: $42\frac{1}{3} + 89\frac{6}{7}$ Build to get the LCD, 21. Add the fractions. Add the whole numbers. $42\frac{1}{3} = \quad 42\frac{1}{3} \cdot \frac{7}{7} = \quad 42\frac{7}{21} = \quad 42\frac{7}{21}$ $+\ 89\frac{6}{7} = +\ 89\frac{6}{7} \cdot \frac{3}{3} = +\ 89\frac{18}{21} = +\ 89\frac{18}{21}$ $\qquad\qquad\qquad\qquad\qquad\qquad\qquad\qquad\quad \frac{25}{21} \qquad 131\frac{25}{21}$
When we add mixed numbers, sometimes the sum of the fractions is an improper fraction. If that is the case, write the improper fraction as a mixed number and **carry** its whole-number part to the whole-number column.	We don't want an improper fraction in the answer. Write $\frac{25}{21}$ as $1\frac{4}{21}$, carry the 1 to the whole-number column, and add it to 131 to get 132: $131\frac{25}{21} = 131 + 1\frac{4}{21} = 132\frac{4}{21}$

Unless otherwise noted, all art on this page is © Cengage Learning.

Subtraction of mixed numbers in vertical form sometimes involves **borrowing**. When the fraction we are subtracting is greater than the fraction we are subtracting it from, borrowing is necessary.

Subtract: $23\frac{1}{4} - 17\frac{5}{9}$

Build to get the LCD, 36.

Since $\frac{20}{36}$ is greater than $\frac{9}{36}$, we must borrow from 28.

$$28\frac{1}{4} = 28\frac{1}{4} \cdot \frac{9}{9} = 28\frac{9}{36} = 28\frac{9}{36} + \frac{36}{36} = 28\frac{7}{36}\frac{45}{36}$$
$$-17\frac{5}{9} = -17\frac{5}{9} \cdot \frac{4}{4} = -17\frac{20}{36} = -17\frac{20}{36} = -17\frac{20}{36}$$
$$\overline{10\frac{25}{36}}$$

REVIEW EXERCISES

Add or subtract and simplify, if possible.

95. $1\frac{3}{8} + 2\frac{1}{5}$ $3\frac{23}{40}$

96. $3\frac{1}{2} + 2\frac{2}{3}$ $6\frac{1}{6}$

97. $157\frac{11}{30} + 98\frac{7}{12}$ $255\frac{19}{20}$

98. $6\frac{3}{14} + 17\frac{7}{10}$ $23\frac{32}{35}$

99. $33\frac{8}{9} + 49\frac{1}{6}$ $83\frac{1}{18}$

100. $98\frac{11}{20} + 14\frac{4}{5}$ $113\frac{7}{20}$

101. $23\frac{1}{3} - 2\frac{5}{6}$ $20\frac{1}{2}$

102. $39 - 4\frac{5}{8}$ $34\frac{3}{8}$

103. PAINTING SUPPLIES In a project to restore a house, painters used $10\frac{3}{4}$ gallons of primer, $21\frac{1}{2}$ gallons of latex paint, and $7\frac{2}{3}$ gallons of enamel. Find the total number of gallons of paint used. $39\frac{11}{12}$ gal

104. PASSPORTS The required dimensions for a passport photograph are shown below. What is the distance from the subject's eyes to the top of the photograph? $\frac{5}{8}$ in.

SECTION 4.7 Order of Operations and Complex Fractions

DEFINITIONS AND CONCEPTS

Order of Operations

1. Perform all calculations within parentheses and other grouping symbols following the order listed in Steps 2–4 below, working from the innermost pair of grouping symbols to the outermost pair.
2. Evaluate all exponential expressions.
3. Perform all multiplications and divisions as they occur from left to right.
4. Perform all additions and subtractions as they occur from left to right.

When grouping symbols have been removed, repeat Steps 2–4 to complete the calculation.

If a fraction bar is present, evaluate the expression above the bar and the expression below the bar separately. Then perform the division indicated by the fraction bar, if possible.

EXAMPLES

Evaluate: $\left(\frac{1}{3}\right)^2 \div \left(\frac{3}{4} - \frac{1}{3}\right)$

First, we perform the subtraction within the second set of parentheses. (There is no operation to perform within the first set.)

$\left(\frac{1}{3}\right)^2 \div \left(\frac{3}{4} - \frac{1}{3}\right)$

$= \left(\frac{1}{3}\right)^2 \div \left(\frac{3}{4} \cdot \frac{3}{3} - \frac{1}{3} \cdot \frac{4}{4}\right)$ Within the parentheses, build each fraction so that its denominator is the LCD 12.

$= \left(\frac{1}{3}\right)^2 \div \left(\frac{9}{12} - \frac{4}{12}\right)$ Multiply the numerators. Multiply the denominators.

$= \left(\frac{1}{3}\right)^2 \div \frac{5}{12}$ Subtract the numerators: $9 - 4 = 5$. Write the difference over the common denominator 12.

$= \frac{1}{9} \div \frac{5}{12}$ Evaluate the exponential expression: $\left(\frac{1}{3}\right)^2 = \frac{1}{3} \cdot \frac{1}{3} = \frac{1}{9}$.

$= \frac{1}{9} \cdot \frac{12}{5}$ Use the rule for dividing fractions: Multiply the first fraction by the reciprocal of $\frac{5}{12}$, which is $\frac{12}{5}$.

$= \frac{1 \cdot 12}{9 \cdot 5}$ Multiply the numerators. Multiply the denominators.

$= \frac{1 \cdot \overset{1}{\cancel{3}} \cdot 4}{\underset{1}{\cancel{3}} \cdot 3 \cdot 5}$ To simplify, factor 12 as $3 \cdot 4$ and 9 as $3 \cdot 3$. Then remove the common factor of 3 from the numerator and denominator.

$= \frac{4}{15}$ Multiply the remaining factors in the numerator. Multiply the remaining factors in the denominator.

To **evaluate a formula**, we replace its variables (letters) with specific numbers and evaluate the right side using the order of operations rule.

Evaluate: $A = \frac{1}{2}h(a + b)$ for $a = 1\frac{1}{3}$, $b = 2\frac{2}{3}$, and $h = 2\frac{4}{5}$.

$A = \frac{1}{2}h(a + b)$ This is the given formula.

$A = \frac{1}{2}\left(2\frac{4}{5}\right)\left(1\frac{1}{3} + 2\frac{2}{3}\right)$ Replace h, a, and b with the given values.

$A = \frac{1}{2}\left(2\frac{4}{5}\right)(4)$ Do the addition within the parentheses.

$A = \frac{1}{2}\left(\frac{14}{5}\right)\left(\frac{4}{1}\right)$ To prepare to multiply fractions, write $2\frac{4}{5}$ as an improper fraction and 4 as $\frac{4}{1}$.

$A = \frac{1 \cdot 14 \cdot 4}{2 \cdot 5 \cdot 1}$ Multiply the numerators. Multiply the denominators.

$A = \frac{1 \cdot 14 \cdot \overset{1}{\cancel{2}} \cdot 2}{\underset{1}{\cancel{2}} \cdot 5 \cdot 1}$ To simplify, factor 4 as $2 \cdot 2$. Then remove the common factor of 2 from the numerator and denominator.

$A = \frac{28}{5}$ Multiply the remaining factors in the numerator. Multiply the remaining factors in the denominator.

$A = 5\frac{3}{5}$ Write the improper fraction $\frac{28}{5}$ as a mixed number by dividing 28 by 5.

Chapter 4 Summary and Review

A **complex fraction** is a fraction whose numerator or denominator, or both, contain one or more fractions or mixed numbers.	Complex fractions: $$\dfrac{\dfrac{9}{10}}{\dfrac{27}{5}} \qquad \dfrac{\dfrac{2}{5}-\dfrac{1}{3}}{\dfrac{3}{7}+\dfrac{1}{5}} \qquad \dfrac{-7\dfrac{1}{4}}{2-\dfrac{1}{9}}$$
The method for **simplifying complex fractions** is based on the fact that the main fraction bar indicates division.	Simplify: $\dfrac{\dfrac{9}{10}}{\dfrac{27}{5}}$ $\dfrac{\dfrac{9}{10}}{\dfrac{27}{5}} = \dfrac{9}{10} \div \dfrac{27}{5}$ Write the division indicated by the main fraction bar using a $\div$ symbol. $= \dfrac{9}{10} \cdot \dfrac{5}{27}$ Use the rule for dividing fractions: Multiply the first fraction by the reciprocal of $\dfrac{27}{5}$, which is $\dfrac{5}{27}$. $= \dfrac{9 \cdot 5}{10 \cdot 27}$ Multiply the numerators. Multiply the denominators. $= \dfrac{\overset{1}{\cancel{9}} \cdot \overset{1}{\cancel{5}}}{2 \cdot \underset{1}{\cancel{5}} \cdot 3 \cdot \underset{1}{\cancel{9}}}$ To simplify, factor 10 as $2 \cdot 5$ and 27 as $3 \cdot 9$. Then remove the common factors of 9 and 5 from the numerator and denominator. $= \dfrac{1}{6}$ Multiply the remaining factors in the numerator. Multiply the remaining factors in the denominator.
To **simplify a complex fraction:** 1. Add or subtract in the numerator and/or denominator so that the numerator is a single fraction and the denominator is a single fraction. 2. Perform the indicated division by multiplying the numerator of the complex fraction by the reciprocal of the denominator. 3. Simplify the result, if possible.	Simplify: $\dfrac{\dfrac{2}{5}-\dfrac{1}{3}}{\dfrac{3}{7}+\dfrac{1}{5}}$ $\dfrac{\dfrac{2}{5}-\dfrac{1}{3}}{\dfrac{3}{7}+\dfrac{1}{5}} = \dfrac{\dfrac{2}{5}\cdot\dfrac{3}{3}-\dfrac{1}{3}\cdot\dfrac{5}{5}}{\dfrac{3}{7}\cdot\dfrac{5}{5}+\dfrac{1}{5}\cdot\dfrac{7}{7}}$ In the numerator, build each fraction so that each has a denominator of 15. In the denominator, build each fraction so that each has a denominator of 35. $= \dfrac{\dfrac{6}{15}-\dfrac{5}{15}}{\dfrac{15}{35}+\dfrac{7}{35}}$ Multiply the numerators. Multiply the denominators. $= \dfrac{\dfrac{1}{15}}{\dfrac{22}{35}}$ Subtract the numerators and write the difference over the common denominator 15. Add the numerators and write the sum over the common denominator 35. $= \dfrac{1}{15} \div \dfrac{22}{35}$ Write the division indicated by the main fraction bar using a $\div$ symbol. $= \dfrac{1}{15} \cdot \dfrac{35}{22}$ Use the rule for dividing fractions: Multiply the first fraction by the reciprocal of $\dfrac{22}{35}$, which is $\dfrac{35}{22}$. $= \dfrac{1 \cdot 35}{15 \cdot 22}$ Multiply the numerators. Multiply the denominators. $= \dfrac{1 \cdot \overset{1}{\cancel{5}} \cdot 7}{3 \cdot \underset{1}{\cancel{5}} \cdot 22}$ To simplify, factor 35 as $5 \cdot 7$ and 15 as $3 \cdot 5$. Then remove the common factor of 5 from the numerator and denominator. $= \dfrac{7}{66}$ Multiply the remaining factors in the numerator. Multiply the remaining factors in the denominator.

REVIEW EXERCISES

Evaluate each expression.

105. $\frac{3}{4} + \left(-\frac{1}{3}\right)^2 \left(\frac{5}{4}\right)$ $\frac{8}{9}$

106. $\left(\frac{2}{3} \div \frac{16}{9}\right) - \left(1\frac{2}{3} \cdot \frac{1}{15}\right)$ $\frac{19}{72}$

107. $\left(\frac{11}{5} - 1\frac{2}{3}\right) - \left(-\frac{4}{9} \cdot 18\right)$ $8\frac{8}{15}$

108. $\left|-\frac{9}{16} \div 2\frac{1}{4}\right| + \left(-3\frac{7}{8}\right)$ $-3\frac{5}{8}$

Simplify each complex fraction.

109. $\dfrac{\frac{3}{5}}{-\frac{17}{20}}$ $-\frac{12}{17}$

110. $\dfrac{4 - \frac{2}{7}}{4\frac{1}{7}}$ $\frac{26}{29}$

111. $\dfrac{\frac{2}{3} - \frac{1}{6}}{\frac{3}{4} - \frac{1}{2}}$ $-\frac{2}{5}$

112. $\dfrac{5\frac{1}{4}}{\frac{7}{4} + \left(-\frac{1}{3}\right)}$ $\frac{63}{17}$

113. Subtract $4\frac{1}{8}$ from the sum of $5\frac{1}{5}$ and $1\frac{1}{2}$. $2\frac{23}{40}$

114. Add $12\frac{11}{16}$ to the difference of $4\frac{5}{8}$ and $3\frac{1}{4}$. $14\frac{1}{16}$

115. Evaluate the formula $A = \frac{1}{2}h(a + b)$ for $a = 1\frac{1}{8}$, $b = 4\frac{7}{8}$, and $h = 2\frac{7}{9}$. $8\frac{1}{3}$

116. Evaluate the formula $P = 2\ell + 2w$ for $\ell = 2\frac{1}{3}$ and $w = 3\frac{1}{4}$. $11\frac{1}{6}$

117. DERMATOLOGY A dermatologist mixes $1\frac{1}{2}$ ounces of cucumber extract, $2\frac{2}{3}$ ounces of aloe vera cream, and $\frac{3}{4}$ ounce of vegetable glycerin to make his own brand of anti-wrinkle cream. He packages it in $\frac{5}{6}$-ounce tubes. How many full tubes can be produced using this formula? How much cream is left over? 5 full tubes, $\frac{9}{10}$ of a tube is left over

118. GUITAR DESIGN Find the missing dimension on the vintage 1962 Stratocaster body shown below. 8 in.

SECTION 4.8 Solving Equations That Involve Fractions

DEFINITIONS AND CONCEPTS

The properties of equality that we used to solve equations involving whole numbers and integers are also used to **solve equations involving fractions**.

Recall that the product of a number and its reciprocal is 1. We can use this fact to solve equations like those shown below, where the **coefficient of the variable term** is a fraction.

$$\frac{2}{21}h = 18 \qquad -\frac{7}{3}d = 49$$

↑ The coefficient of h is a fraction.

↑ The coefficient of d is a negative fraction.

EXAMPLES

Solve: $\frac{2}{21}h = 18$

Recall that $\frac{2}{21}h = 18$ means $\frac{2}{21} \cdot h = 18$.

$\frac{2}{21}h = 18$ This is the equation to solve.

$\frac{21}{2} \cdot \frac{2}{21}h = \frac{21}{2} \cdot 18$ To isolate h, undo the multiplication by $\frac{2}{21}$ by multiplying both sides by the reciprocal of $\frac{2}{21}$, which is $\frac{21}{2}$.

$\left(\frac{21}{2} \cdot \frac{2}{21}\right)h = \frac{21}{2} \cdot \frac{18}{1}$ On the left side, use the associative property of multiplication to group $\frac{21}{2}$ and $\frac{2}{21}$. On the right side, write 18 as $\frac{18}{1}$.

$1h = \frac{21 \cdot 18}{2 \cdot 1}$ On the left side, the product of a number and its reciprocal is 1: $\frac{21}{2} \cdot \frac{2}{21} = 1$. On the right side, multiply the numerators and multiply the denominators.

$h = \frac{21 \cdot \overset{1}{\cancel{2}} \cdot 9}{\underset{1}{\cancel{2}} \cdot 1}$ On the left side, the coefficient of 1 need not be written: $1h = h$. To simplify on the right side, factor 18 as $2 \cdot 9$, and remove the common factor of 2.

$h = 189$ Multiply on the right side: $21 \cdot 9 = 189$.

The solution is 189. Check this result into the *original* equation.

Equations that involve integers are usually easier to solve than equations that involve fractions. We can **clear an equation of fractions** by multiplying both sides by their LCD.

Strategy for Solving Equations

1. Clear the equation of fractions.
2. Simplify each side of the equation.
3. Isolate the variable term on one side.
4. Isolate the variable.
5. Check the result in the *original* equation.

You won't always have to use all five steps to solve a given equation.

Solve: $\frac{6}{5}x = \frac{1}{10}x - \frac{5}{2}$

$10\left(\frac{6}{5}x\right) = 10\left(\frac{1}{10}x - \frac{5}{2}\right)$ Multiply both sides by the LCD of $\frac{6}{5}, \frac{1}{10}$, and $\frac{5}{2}$, which is 10.

$10\left(\frac{6}{5}x\right) = 10\left(\frac{1}{10}x\right) - 10\left(\frac{5}{2}\right)$ On the right side, distribute the multiplication by 10.

$\frac{10}{1}\left(\frac{6}{5}x\right) = \frac{10}{1}\left(\frac{1}{10}x\right) - \frac{10}{1}\left(\frac{5}{2}\right)$ Write 10 as $\frac{10}{1}$. This makes the numerators and denominators in the fraction multiplication process clearer.

$\frac{2 \cdot \overset{1}{\cancel{5}}}{1}\left(\frac{6}{\underset{1}{\cancel{5}}}x\right) = \frac{\overset{1}{\cancel{10}}}{1}\left(\frac{1}{\underset{1}{\cancel{10}}}x\right) - \frac{\overset{1}{\cancel{2}} \cdot 5}{1}\left(\frac{5}{\underset{1}{\cancel{2}}}\right)$ Factor 10 as $2 \cdot 5$. Remove the common factors of 5, 10, and 2 in the numerator and denominator.

$12x = x - 25$ Complete each multiplication.

$12x - x = x - 25 - x$ To eliminate x on the right side, subtract x from both sides.

$11x = -25$ Combine like terms on each side.

$\frac{11x}{11} = \frac{-25}{11}$ To isolate x, undo the multiplication by 11 by dividing both sides by 11.

$x = -\frac{25}{11}$

The solution is $-\frac{25}{11}$. Check by substituting into the *original* equation.

We can use the **six-step problem-solving strategy** discussed in earlier chapters to solve application problems that involve fractions.

Review Example 11 and Example 12 on pages 408 through 410.

REVIEW EXERCISES

Solve each equation and check the result.

119. $x - \dfrac{1}{40} = \dfrac{11}{40}$ $\dfrac{3}{10}$

120. $\dfrac{5}{4} = n + \dfrac{2}{3}$ $\dfrac{7}{12}$

121. $\dfrac{2}{9}x = 22$ 99

122. $\dfrac{4}{25}y = 4$ 25

123. $-\dfrac{7}{3}y = 5$ $-\dfrac{15}{7}$

124. $\dfrac{1}{8}x = \dfrac{27}{32}$ $\dfrac{27}{4}$

125. $37x = -\dfrac{19}{3}$ $-\dfrac{19}{111}$

126. $\dfrac{7}{8}x + 21 = 7$ -16

127. $\dfrac{d}{4} = \dfrac{d}{7} - 6$ -56

128. $\dfrac{1}{6}x - \dfrac{1}{2} = \dfrac{1}{3}x$ -3

In Exercises 129 and 130, let a variable represent the unknown quantity. Then write and solve an equation to answer the question.

129. TEXTBOOKS In writing a history text, the author decided to devote two-thirds of the book to events prior to World War II. The remainder of the book deals with history after the war. If pre–World War II history is covered in 220 pages, how many pages does the textbook have? 330 pages

130. SEMINARS A real estate investment seminar has three parts. In the first part, a film is shown that takes one-fourth of the class time. In the second part, the instructor lectures for 35 minutes. In the final part, successful investors take two-fifths of the class time to give their testimonials. How many minutes long is the seminar? 100 minutes

CHAPTER 4 TEST

1. Fill in the blanks.
   a. For the fraction $\frac{6}{7}$, the __numerator__ is 6 and the __denominator__ is 7.
   b. Two fractions are __equivalent__ if they represent the same number.
   c. A fraction is in __simplest__ form when the numerator and denominator have no common factors other than 1.
   d. To __simplify__ a fraction, we remove common factors of the numerator and denominator.
   e. The __reciprocal__ of $\frac{4}{5}$ is $\frac{5}{4}$.
   f. A __mixed__ number, such as $1\frac{9}{16}$, is the sum of a whole number and a proper fraction.
   g. $\dfrac{\frac{1}{8}}{\frac{7}{12}}$ and $\dfrac{\frac{3}{4}+\frac{1}{3}}{\frac{5}{12}-\frac{1}{4}}$ are examples of __complex__ fractions.

2. See the illustration to the right.
   a. What fractional part of the plant is above ground? $\frac{4}{5}$
   b. What fractional part of the plant is below ground? $\frac{1}{5}$

3. Each region outlined in black represents one whole. Write an improper fraction and a mixed number to represent the shaded portion. $\frac{13}{6}=2\frac{1}{6}$

4. Graph $2\frac{4}{5}$, $-\frac{2}{5}$, $-1\frac{1}{7}$, and $\frac{7}{6}$ on a number line.

5. Are $\frac{1}{3}$ and $\frac{5}{15}$ equivalent? yes

6. a. Express $\frac{4}{5}$ as an equivalent fraction with denominator 45. $\frac{36}{45}$
   b. Express $\frac{7}{8}$ as an equivalent fraction with denominator $24x$. $\frac{21x}{24x}$

7. Simplify each fraction, if possible.
   a. $\frac{0}{15}$ 0
   b. $\frac{9}{0}$ undefined

8. Simplify each fraction.
   a. $\frac{27}{36}$ $\frac{3}{4}$
   b. $\frac{72n^3}{180n}$ $\frac{2n^2}{5}$

9. Add and simplify, if possible: $\frac{3}{16}+\frac{7}{16}$ $\frac{5}{8}$

10. Multiply and simplify, if possible: $-\frac{3}{4}\left(\frac{1}{5}\right)$ $-\frac{3}{20}$

11. Divide and simplify, if possible: $\frac{4a}{3b} \div \frac{a^2}{9b^5}$ $\frac{12b^4}{a}$

12. Subtract and simplify, if possible: $\frac{11}{12}-\frac{11}{30}$ $\frac{11}{20}$

13. Add and simplify, if possible: $-\frac{3}{7}+2\frac{11}{7}$

14. Multiply and simplify, if possible: $\frac{9}{10}\left(-\frac{4}{15}\right)\left(-\frac{25}{18}\right)$ $\frac{1}{3}$

15. Which fraction is larger: $\frac{8}{9}$ or $\frac{9}{10}$? $\frac{9}{10}$

16. Find the reciprocal of:
    a. $-\frac{17}{53}$ $-\frac{53}{17}$
    b. $\frac{7}{3a^2}$ $\frac{3a^2}{7}$

17. Subtract: $\frac{x}{6}-\frac{4}{5}$ $\frac{5x-24}{30}$

18. COFFEE DRINKERS Two-fifths of 100 adults surveyed said they started their morning with a cup of coffee. Of the 100, how many would this be? 40

19. THE INTERNET The graph shows the fraction of the total number of Internet searches that were made using various sites in December 2012. What fraction of all the searches were done using Google, Yahoo, or Microsoft sites? $\frac{23}{24}$

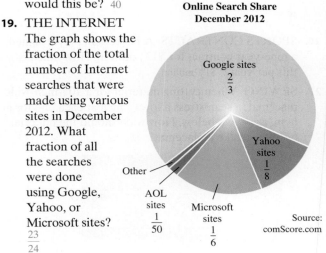

20. a. Write $\dfrac{55}{6}$ as a mixed number. $9\dfrac{1}{6}$

  b. Write $1\dfrac{18}{21}$ as an improper fraction. $\dfrac{39}{21}$

21. Find the sum of $157\dfrac{3}{10}$ and $103\dfrac{13}{15}$. Simplify the result. $261\dfrac{1}{6}$

22. Subtract and simplify, if possible: $67\dfrac{1}{4} - 29\dfrac{5}{6}$ $37\dfrac{5}{12}$

23. Divide and simplify, if possible: $6\dfrac{1}{4} \div 3\dfrac{3}{4}$ $1\dfrac{2}{3}$

24. BOXING Two of the greatest heavyweight boxers of all time are Muhammad Ali and George Foreman. Refer to the "Tale of the Tape" comparison shown below.

  a. Which fighter weighed more? How much more? Foreman, $39\dfrac{1}{2}$ lb

  b. Which fighter had the larger waist measurement? How much larger? Foreman, $5\dfrac{1}{2}$ in.

  c. Which fighter had the larger forearm measurement? How much larger? Ali, $\dfrac{1}{4}$ in.

Tale of the Tape

Muhammad Ali		George Foreman
6-3	Height	6-4
210½ lb	Weight	250 lb
82 in.	Reach	79 in.
43 in.	Chest (Normal)	48 in.
45½ in.	Chest (Expanded)	50 in.
34 in.	Waist	39½ in.
12½ in.	Fist	13½ in.
15 in.	Forearm	14¾ in.

Source: The International Boxing Hall of Fame

25. Evaluate the formula $P = 2\ell + 2w$ for $\ell = \dfrac{1}{3}$ and $w = \dfrac{1}{9}$. $\dfrac{8}{9}$

26. SPORTS CONTRACTS A basketball player signed a nine-year contract for $13\dfrac{1}{2}$ million. How much is this per year? $\$1\dfrac{1}{2}$ million

27. SEWING When cutting material for a $10\dfrac{1}{2}$-inch-wide placemat, a seamstress allows $\dfrac{5}{8}$ inch at each end for a hem, as shown below. How wide should the material be cut to make a placemat? $11\dfrac{3}{4}$ in.

28. Find the perimeter and the area of the triangle shown to the right.
perimeter: $53\dfrac{1}{3}$ in., area: $106\dfrac{2}{3}$ in.²

29. NUTRITION A box of Tic Tacs contains 40 of the $1\dfrac{1}{2}$-calorie breath mints. How many calories are there in a box of Tic Tacs? 60 calories

30. COOKING How many servings are there in an 8-pound roast, if the suggested serving size is $\dfrac{2}{3}$ pound? 12 servings

31. Evaluate:

$$\left(\dfrac{2}{3} \cdot \dfrac{5}{16}\right) - \left(-1\dfrac{3}{5} \div 4\dfrac{4}{5}\right) \quad \dfrac{13}{24}$$

32. Evaluate: $\left(\dfrac{1}{2}\right)^3 \div \left(\dfrac{3}{4} - \dfrac{1}{3}\right)$ $\dfrac{3}{10}$

33. Simplify:

$$\dfrac{\dfrac{5}{6}}{\dfrac{7}{8}} \quad \dfrac{20}{21}$$

34. Simplify:

$$\dfrac{\dfrac{1}{2} + \dfrac{1}{3}}{-\dfrac{1}{6} - \dfrac{1}{3}} - \dfrac{5}{3} \quad -\dfrac{5}{3}$$

Solve each equation and check the result.

35. $\dfrac{5}{16}y = \dfrac{5}{6}$ $\dfrac{8}{3}$

36. $-\dfrac{3}{8} = a + \dfrac{1}{2}$ $-\dfrac{7}{8}$

37. $\dfrac{6}{11}n + 35 = -7$ -77

38. $\dfrac{7}{8}x - \dfrac{5}{16} = \dfrac{1}{4}x$ $\dfrac{1}{2}$

Let a variable represent the unknown quantity. Then write and solve an equation to answer the question.

39. PETS A dog obedience class has four parts. The class begins with a 10-minute demonstration. In the second part, which takes one-half of the class time, an expert explains how to train a dog. Then, a veterinarian speaks about nutrition and health. This takes one-third of the class time. The class ends with a 25-minute question-and-answer session. How many minutes long is the dog obedience class? 210 minutes

40. Explain each mathematical concept that is shown below.

a. $\dfrac{6}{8} = \dfrac{\overset{1}{\cancel{2}} \cdot 3}{\underset{1}{\cancel{2}} \cdot 4} = \dfrac{3}{4}$ removing a common factor from the numerator and denominator (simplifying a fraction)

b.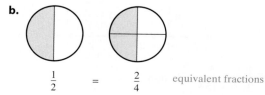

$\dfrac{1}{2} = \dfrac{2}{4}$ equivalent fractions

c. $\dfrac{3}{5} = \dfrac{3}{5} \cdot \dfrac{4}{4} = \dfrac{12}{20}$ multiplying a fraction by a form of 1 (building an equivalent fraction)

Unless otherwise noted, all art on this page is © Cengage Learning.

CHAPTERS 1–4 CUMULATIVE REVIEW

1. Consider the number 5,896,619. [Section 1.1]
   a. What digit is in the millions column? 5
   b. What is the place value of the digit 8? 8 hundred thousands
   c. Round to the nearest hundred. 5,896,600
   d. Round to the nearest ten thousand. 5,900,000

2. BANKS In 2012, the world's largest bank, with total assets of $2,799,977,000,000, was the Deutsche Bank of Germany. In what place-value column is the digit 2? (Source: gfmag.com) [Section 1.1] trillions

3. POPULATION Rank the following counties in order, from greatest to least population. [Section 1.1]
   Miami-Dade, Kings, Dallas, Queens, Riverside, San Bernardino

County	2011 Population
Dallas County, TX	2,416,014
Kings County, NY	2,532,645
Miami-Dade County, FL	2,554,766
San Bernardino County, CA	2,065,377
Queens County, NY	2,247,848
Riverside, CA	2,239,620

   (Source: *The World Almanac and Book of Facts*, 2013)

4. POOL CONSTRUCTION Refer to the rectangular swimming pool shown below.
   a. Find the perimeter of the pool. [Section 1.2] 450 ft
   b. Find the area of the pool's surface. [Section 1.3] 11,250 ft^2

5. Add: 7,897 [Section 1.2]
   6,909
   1,812
   +14,378
   30,996

6. Subtract 3,456 from 20,000. Check the result.
   [Section 1.2] 16,544, 16,544 + 3,456 = 20,000

7. SHEETS OF STICKERS There are twenty rows of twelve gold stars on one sheet of stickers. If a packet contains ten sheets, how many stars are there in one packet? [Section 1.3] 2,400 stickers

8. Multiply: 5,345 [Section 1.3]
   × 56
   299,320

9. Divide: $35\overline{)34{,}685}$. Check the result.
   [Section 1.4] 991, 991 · 35 = 34,685

10. a. List factors of 24, from least to greatest.
    [Section 1.5] 1, 2, 3, 4, 6, 8, 12, 24
    b. Find the prime factorization of 450.
    [Section 1.5] $2 \cdot 3^2 \cdot 5^2$

11. Find the LCM of 16 and 20. [Section 1.6] 80

12. Find the GCF of 63 and 84. [Section 1.6] 21

13. Evaluate: $15 + 5[12 - (2^2 + 4)]$ [Section 1.7] 35

14. REAL ESTATE A homeowner, wishing to sell his house, had it appraised by three different real estate agents. The appraisals were: $158,000, $163,000, and $147,000. He decided to use the mean of the appraisals as the listing price. For what amount was the home listed? [Section 1.7] $156,000

Solve each equation and check the result.

15. $18 = n - 47$ [Section 1.8] 65

16. $23x = 483$ [Section 1.9] 21

17. a. Write the set of integers. [Section 2.1]
    $\{\ldots, -5, -4, -3, -2, -1, 0, 1, 2, 3, 4, 5, \ldots\}$
    b. Is the statement $-9 \leq -8$ true or false?
    [Section 2.1] true

18. Find the sum of -20, 6, and -1. [Section 2.2]
    -15

19. Subtract 453 from 129. [Section 2.3]
    -324

20. Subtract: $-50 - (-60)$ [Section 2.3]
    10

21. GOLD MINING An elevator lowers gold miners from the ground level entrance to different depths in the mine. The elevator stops every 25 vertical feet to let off miners. At what depth do the miners work if they get off the elevator at the 8th stop?
    [Section 2.4] -200 ft

22. TEMPERATURE DROP During a five-hour period, the temperature steadily dropped 55°F. By how many degrees did the temperature change each hour? [Section 2.5] −11°F per hour

Evaluate each expression. [Section 2.6]

23. $7(8) + (-3)(-6)(-2)$ 20

24. $(-2)^3 - 3^3$ −35

25. $-5 + 3|-4 - (-6)|$ 1

26. $\dfrac{2(3^2 - 4^2)}{-2(3) - 1}$ 2

Solve each equation and check the result. [Section 2.9]

27. $-12y - 9 = 39$ −4

28. $-9 = 4 + \dfrac{m}{-4}$ 52

Form an equation and solve it to answer the following questions. [Section 2.9]

29. SHARKS During a research project, a diver inside a shark cage made observations at a depth of 132 feet. For a second set of observations, the cage was raised to a depth of 64 feet. How many feet was the cage raised between observations? The cage was raised 68 ft between observations.

30. PROFITS AND LOSSES In its first year of business, a pet store suffered a loss, ending the year $4,028 in the red. In the second year, it made a sizable profit. If the total profit for the first two years in business was $33,611, how much profit was made the second year? The pet store made a profit of $37,639 the second year.

31. Translate each expression into an algebraic expression involving the variable x. [Section 3.1]

  a. The sum of a number and 15 $x + 15$

  b. Eight less than a number $x - 8$

  c. The product of a number and 4 $4x$

  d. The quotient obtained when a number is divided by 10 $\dfrac{x}{10}$

32. Evaluate $3x - x^3$ for $x = -4$. [Section 3.2] 52

33. Simplify each expression. [Section 3.3]

  a. $-8k(-6)$ $48k$ **b.** $-4x(-7y)$ $28xy$

34. Multiply. [Section 3.3]

  a. $2(3x - 4)$ $6x - 8$

  b. $-5(3x - 2y + 4)$ $-15x + 10y - 20$

Simplify each expression. [Section 3.4]

35. $-3x + 8x$ $5x$

36. $4a^2 - (-3a^2)$ $7a^2$

37. $4x - 3y - 5x + 2y$ $-x - y$

38. $-2(3x - 4) + 2x$ $-4x + 8$

Solve each equation and check the result. [Section 3.5]

39. $6x - 12 = 2x + 4$ 4

40. $3(2y - 8) = -2(y - 4)$ 4

41. $-(19 - 3s) - (8s + 1) = 35$ −11

42. $14m = m + 5m + 3m - 40$ −8

Form an equation and solve it to answer the following questions. [Section 3.6]

43. OBSERVATION HOURS To earn a master's degree in educational psychology, a student must have 100 hours of observation time at a clinic. If the student has already observed for 37 hours, how many 3-hour shifts must she observe to complete the requirement? She must complete 21 more shifts.

44. GEOMETRY A rectangle is four times as long as it is wide. If its perimeter is 210 feet, find its width and its length. The width is 21 ft and the length is 84 ft.

Simplify each fraction. [Section 4.1]

45. $\dfrac{21}{28}$ $\dfrac{3}{4}$

46. $\dfrac{40x^6y^4}{16x^3y^5}$ $\dfrac{5x^3}{2y}$

Perform each operation.

47. $\dfrac{6}{5}\left(-\dfrac{2}{3}\right)$ $-\dfrac{4}{5}$ [Section 4.2]

48. $\dfrac{14p^2}{8} \div \dfrac{7p^3}{2}$ $\dfrac{1}{2p}$ [Section 4.3]

49. $\dfrac{2}{3} + \dfrac{3}{4}$ $1\dfrac{5}{12}$ [Section 4.4]

50. $\dfrac{4}{m} - \dfrac{3}{5}$ $\dfrac{20 - 3m}{5m}$ [Section 4.4]

51. FIRE HAZARDS Two terminals in an electrical switch were so close that electricity could jump the gap and start a fire. The illustration below shows a newly designed switch that will keep this from happening. By how much was the distance between the ground terminal and the hot terminal increased? [Section 4.4] $\dfrac{11}{16}$ in.

Unless otherwise noted, all art on this page is © Cengage Learning.

52. a. Write $\frac{75}{7}$ as a mixed number. [Section 4.5] $10\frac{5}{7}$

b. Write $-6\frac{5}{8}$ as an improper fraction. [Section 4.5] $-\frac{53}{8}$

Perform each operation. Simplify, if possible.

53. $2\frac{2}{5}\left(3\frac{1}{12}\right)$ [Section 4.5] $7\frac{2}{5}$

54. $15\frac{1}{3} \div 2\frac{2}{9}$ [Section 4.5] $6\frac{9}{10}$

55. $4\frac{2}{3} + 5\frac{1}{4}$ [Section 4.6] $9\frac{11}{12}$

56. $14\frac{2}{5} - 8\frac{2}{3}$ [Section 4.6] $5\frac{11}{15}$

57. LUMBER As shown below, 2-by-4's from the lumber yard do not really have dimensions of 2 inches by 4 inches. How wide and how high is the stack of 2-by-4's in the illustration? [Section 4.5] width: 28 in., height: 6 in.

58. GAS STATIONS How much gasoline is left in a 500-gallon storage tank if $225\frac{3}{4}$ gallons have been pumped out of it? [Section 4.6] $274\frac{1}{4}$ gal

59. Find the perimeter of the triangle shown below. [Section 4.6] $3\frac{5}{12}$ ft

60. Evaluate: $\left(-\frac{3}{4} \cdot \frac{9}{16}\right) + \left(\frac{1}{2} - \frac{1}{8}\right)$ [Section 4.7] $-\frac{3}{64}$

61. Simplify: $\dfrac{\frac{2}{3}}{\frac{4}{5}}$ [Section 4.7] $\frac{5}{6}$

62. Simplify: $\dfrac{\frac{3}{7} + \left(-\frac{1}{2}\right)}{1\frac{3}{4}}$ [Section 4.7] $-\frac{2}{49}$

Solve each equation and check the result. [Section 4.8]

63. $x + \frac{1}{5} = -\frac{11}{15}$ $-\frac{14}{15}$

64. $\frac{3}{4}x = \frac{5}{8}x + \frac{1}{2}$ 4

65. $\frac{2}{3}x = -10$ -15

66. $3y - 8 = 0$ $\frac{8}{3}$

Form an equation and solve it to answer the following questions. [Section 4.8]

67. SHAVING An advertisement for a new, improved model of an electric razor claims that men can shave in just two-thirds of the time it took them with the older model. Using the new model, it took a man 90 seconds to shave. If the advertising claim is true, how long would it have taken him to shave using the older model? It would have taken 135 seconds to shave with the older model.

68. GRADING In an economics class, a student's grade is based on the number of points he or she earns on tests, homework, and a term paper. Tests account for three-fourths of the total possible points, homework assigments account for one-fifth of the total possible points, and the term paper is worth 50 points. What is the maximum number of points that a student can earn in the class? The maximum number of points that a student can earn is 1,000.

5

Decimals

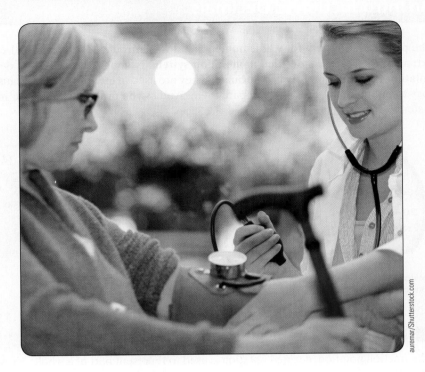

5.1 An Introduction to Decimals
5.2 Adding and Subtracting Decimals
5.3 Multiplying Decimals
5.4 Dividing Decimals
5.5 Fractions and Decimals
5.6 Square Roots
5.7 Solving Equations That Involve Decimals
Chapter Summary and Review
Chapter Test
Cumulative Review

from Campus to Careers
Home Health Aide

Home health aides provide personalized care to the elderly and the disabled in the patient's own home. They help their patients take medicine, eat, dress, and bathe. Home health aides need to have a good number sense. They must accurately take the patient's temperature, pulse, and blood pressure, and monitor the patient's calorie intake and sleeping schedule.

In **Problem 101** of **Study Set 5.2,** you will see how a home health aide uses decimal addition and subtraction to chart a patient's temperature.

JOB TITLE: Home Health Aide
EDUCATION: Successful completion of a home health aide training program as required by state law or federal regulation.
JOB OUTLOOK: Excellent due to rapid employment growth and high replacement needs.
ANNUAL EARNINGS: The average (median) salary in 2011 was $21,820.
FOR MORE INFORMATION: www.bls.gov/oes/current/oes311011.htm

443

Objectives

1. Identify the place value of a digit in a decimal number.
2. Write decimals in expanded form.
3. Read decimals and write them in standard form.
4. Compare decimals using inequality symbols.
5. Graph decimals on a number line.
6. Round decimals.
7. Read tables and graphs involving decimals.

SECTION 5.1
An Introduction to Decimals

The place-value system for whole numbers that was introduced in Section 1.1 can be extended to create the **decimal numeration system**. Numbers written using **decimal notation** are often simply called **decimals**. They are used in measurement, because it is easy to put them in order and compare them. And as you probably know, our money system is based on decimals.

The decimal 1,537.6 on the odometer represents the distance, in miles, that the car has traveled.

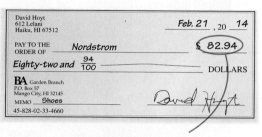

The decimal 82.94 repesents the amount of the check, in dollars.

1 Identify the place value of a digit in a decimal number.

Like fraction notation, decimal notation is used to represent part of a whole. However, when writing a number in decimal notation, we don't use a fraction bar, nor is a denominator shown. For example, consider the rectangular region below that has 1 of 10 equal parts colored red. We can use the fraction $\frac{1}{10}$ or the decimal 0.1 to describe the amount of the figure that is shaded. Both are read as "one-tenth," and we can write:

$$\frac{1}{10} = 0.1$$

Fraction: $\frac{1}{10}$ Decimal: 0.1

The square region below has 1 of 100 equal parts colored red. We can use the fraction $\frac{1}{100}$ or the decimal 0.01 to describe the amount of the figure that is shaded. Both are read as "one one-hundredth," and we can write:

$$\frac{1}{100} = 0.01$$

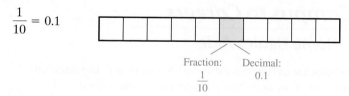

Fraction: $\frac{1}{100}$
Decimal: 0.01

Decimals are written by entering the digits 0, 1, 2, 3, 4, 5, 6, 7, 8, and 9 into place-value columns that are separated by a **decimal point**. The following **place-value chart** shows the names of the place-value columns. Those to the left of the decimal point form the **whole-number part** of the decimal number; they have the familiar names ones, tens, hundreds, and so on. The columns to the right of the decimal point form the **fractional part**. Their place value names are similar to those in the whole-number part, but they end in "*ths*." Notice that there is no one*ths* place in the chart.

Unless otherwise noted, all art on this page is © Cengage Learning.

5.1 An Introduction to Decimals

The decimal 365.24219, entered in the place-value chart above, represents the number of days it takes Earth to make one full orbit around the sun. We say that the decimal is written in **standard form** (also called **standard notation**). Each of the 2's in 365.24219 has a different place value because of its position. The place value of the red 2 is two tenths. The place value of the blue 2 is two thousandths.

EXAMPLE 1
Consider the decimal number: 2,864.709531

a. What is the place value of the digit 5?
b. Which digit tells the number of millionths?

Strategy We will locate the decimal point in 2,864.709531. Then, moving to the right, we will name each column (tenths, hundredths, and so on) until we reach 5.

WHY It's easier to remember the names of the columns if you begin at the decimal point and move to the right.

Solution

a. 2,864.709531   Say "Tenths, hundredths, thousandths, ten-thousandths" as you move from column to column.

5 ten-thousandths is the place value of the digit 5.

b. 2,864.709531   Say "Tenths, hundredths, thousandths, ten-thousandths, hundred thousandths, millionths" as you move from column to column.

The digit 1 is in the millionths column.

> **Caution!** We *do not* separate groups of three digits on the right side of the decimal point with commas as we do on the left side. For example, it would be incorrect to write:
>
> 2,864.709,531

Self Check 1

Consider the decimal number: 56,081.639724
a. What is the place value of the digit 9? 9 thousandths
b. Which digit tells the number of hundred-thousandths? 2

Now Try Problem 17

Teaching Example 1 Consider the decimal number: 8,420.167953
a. What is the place value of the digit 5?
b. Which digit tells the number of hundredths?
Answers:
a. 5 hundred-thousandths
b. 6

We can write a whole number in decimal notation by placing a decimal point immediately to its right and then entering a zero, or zeros, to the right of the decimal point. For example,

$$99 \;=\; 99.0 \;=\; 99.00 \quad \text{Because } 99 = 99\tfrac{0}{10} = 99\tfrac{00}{100}.$$

↑ A whole number   ↑ Place a decimal point here and enter a zero, or zeros, to the right of it.

When there is no whole-number part of a decimal, we can show that by entering a zero directly to the left of the decimal point. For example,

$$.83 \;=\; 0.83 \quad \text{Because } \tfrac{83}{100} = 0\tfrac{83}{100}.$$

↑ No whole-number part   ↑ Enter a zero here, if desired.

Negative decimals are used to describe many situations that arise in everyday life, such as temperatures below zero and the balance in a checking account that is overdrawn. For example, the coldest natural temperature ever recorded on Earth was −128.6°F at the Russian Vostok Station in Antarctica on July 21, 1983.

2 Write decimals in expanded form.

The decimal 4.458, entered in the place-value chart below, represents the time (in seconds) that it took women's record holder Melanie Troxel to cover a quarter mile in her top-fuel dragster. Notice that the place values of the columns for the whole-number part are 1, 10, 100, 1,000, and so on. We learned in Section 1.1 that the value of each of those columns is 10 times greater than the column directly to its right.

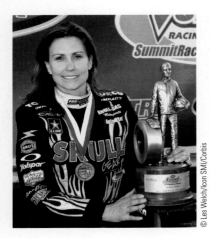

The place values of the columns for the fractional part of a decimal are $\frac{1}{10}$, $\frac{1}{100}$, $\frac{1}{1,000}$, and so on. Each of those columns has a value that is $\frac{1}{10}$ of the value of the place directly to its left. For example,

- The value of the tenths column is $\frac{1}{10}$ of the value of the ones column: $1 \cdot \frac{1}{10} = \frac{1}{10}$.
- The value of the hundredths column is $\frac{1}{10}$ of the value of the tenths column: $\frac{1}{10} \cdot \frac{1}{10} = \frac{1}{100}$.
- The value of the thousandths column is $\frac{1}{10}$ of the value of the hundredths column: $\frac{1}{100} \cdot \frac{1}{10} = \frac{1}{1,000}$.

The meaning of the decimal 4.458 becomes clear when we write it in **expanded form** (also called **expanded notation**).

$$4.458 = 4 \text{ ones} + 4 \text{ tenths} + 5 \text{ hundredths} + 8 \text{ thousandths}$$

which can be written as:

$$4.458 = 4 + \frac{4}{10} + \frac{5}{100} + \frac{8}{1,000}$$

> **The Language of Algebra** The word *decimal* comes from the Latin word *decima*, meaning a tenth part.

Self Check 2

Write the decimal number 1,277.9465 in expanded form.

Now Try Problems 23 and 27

Self Check 2 Answer
$1{,}000 + 200 + 70 + 7 + \frac{9}{10} + \frac{4}{100} + \frac{6}{1{,}000} + \frac{5}{10{,}000}$

Teaching Example 2 Write the decimal number 549.2136 in expanded form.
Answer:
$500 + 40 + 9 + \frac{2}{10} + \frac{1}{100} + \frac{3}{1{,}000} + \frac{6}{10{,}000}$

EXAMPLE 2
Write the decimal number 592.8674 in expanded form.

Strategy Working from left to right, we will give the place value of each digit and combine them with + symbols.

WHY The term *expanded form* means to write the number as an addition of the place values of each of its digits.

Solution The expanded form of 592.8674 is:

5 hundreds + 9 tens + 2 ones + 8 tenths + 6 hundredths + 7 thousandths + 4 ten-thousandths

which can be written as

$$500 + 90 + 2 + \frac{8}{10} + \frac{6}{100} + \frac{7}{1,000} + \frac{4}{10,000}$$

3 Read decimals and write them in standard form.

To understand how to read a decimal, we will examine the expanded form of 4.458 in more detail. Recall that

$$4.458 = 4 + \frac{4}{10} + \frac{5}{100} + \frac{8}{1,000}$$

To add the fractions, we need to build $\frac{4}{10}$ and $\frac{5}{100}$ so that each has a denominator that is the LCD, 1,000.

$$4.458 = 4 + \frac{4}{10} \cdot \frac{100}{100} + \frac{5}{100} \cdot \frac{10}{10} + \frac{8}{1,000}$$

$$= 4 + \frac{400}{1,000} + \frac{50}{1,000} + \frac{8}{1,000}$$

$$= 4 + \frac{458}{1,000}$$

$$= 4\frac{458}{1,000}$$

We have found that $4.458 = 4\frac{458}{1,000}$, where 4 is the whole-number part and $\frac{458}{1,000}$ is the fractional part.

We read 4.458 as "four and four hundred fifty-eight thousandths" because 4.458 is the same as $4\frac{458}{1,000}$. Notice that the last digit in 4.458 is in the thousandths place. This observation suggests the following method for reading decimals.

Reading a Decimal

To read a decimal:

1. Look to the left of the decimal point and say the name of the whole number.
2. The decimal point is read as "and."
3. Say the fractional part of the decimal as a whole number followed by the name of the last place-value column of the digit that is the farthest to the right.

We can use the steps for reading a decimal to write it in words.

EXAMPLE 3 Write each decimal in words and then as a fraction or mixed number. **You do not have to simplify the fraction.**

a. *Sputnik*, the first satellite launched into space, weighed 184.3 pounds.
b. Usain Bolt of Jamaica holds the men's world record in the 100-meter dash: 9.58 seconds.
c. A one-dollar bill is 0.0043 inch thick.
d. Liquid mercury freezes solid at −37.7°F.

Strategy We will identify the whole number to the left of the decimal point, the fractional part to its right, and the name of the place-value column of the digit the farthest to the right.

WHY We need to know those three pieces of information to read a decimal or write it in words.

Self Check 3

Write each decimal in words and then as a fraction or mixed number. **You do not have to simplify the fraction.**
a. The average normal body temperature is 98.6°F.
b. The planet Venus makes one full orbit around the sun every 224.7007 Earth days.
c. One gram is about 0.035274 ounce.
d. Liquid nitrogen freezes solid at −345.748°F.

Now Try Problems 31, 35, and 39

Self Check 3 Answers
a. ninety-eight and six tenths, $98\frac{6}{10}$
b. two hundred twenty-four and seven thousand seven ten-thousandths, $224\frac{7,007}{10,000}$
c. thirty-five thousand, two hundred seventy-four millionths, $\frac{35,274}{1,000,000}$
d. negative three hundred forty-five and seven hundred forty-eight thousandths, $-345\frac{748}{1,000}$

Teaching Example 3 Write each decimal in words and then as a fraction or mixed number. **You do not have to simplify the fraction.**
a. One gallon of milk weighs 8.6 pounds.
b. One meter is about 39.3701 inches.
c. The smallest freshwater fish is the dwarf pygmy goby. An adult male weighs 0.00014 ounce.
d. Ocean salt water freezes at $-1.94°$ Celsius.

Answers:
a. eight and six tenths, $8\frac{6}{10}$
b. thirty-nine and three thousand, seven hundred one ten-thousandths, $39\frac{3,701}{10,000}$
c. fourteen hundred-thousandths, $\frac{14}{100,000}$
d. negative one and ninety-four hundredths, $-1\frac{94}{100}$

Self Check 4

Write each number in standard form:
a. *Eight hundred six and ninety-two hundredths* 806.92
b. *Twelve and sixty-seven ten-thousandths* 12.0067

Now Try Problems 41, 45, and 47

Teaching Example 4 Write each number in standard form:
a. *Sixty-seven and thirty-eight hundredths*
b. *Nineteen and twenty-five thousandths*
Answers:
a. 67.38
b. 19.025

Solution

a. $\boxed{184}\,.\,\boxed{3}$ The whole-number part is 184. The fractional part is 3.
               The digit the farthest to the right, 3, is in the tenths place.

One hundred eighty-four and three tenths
Written as a mixed number, 184.3 is $184\frac{3}{10}$.

b. $\boxed{9}\,.\,\boxed{58}$ The whole-number part is 9. The fractional part is 58.
           The digit the farthest to the right, 8, is in the hundredths place.

Nine and fifty-eight hundredths
Written as a mixed number, 9.58 is $9\frac{58}{100}$.

c. $\boxed{0}\,.\,\boxed{0043}$ The whole-number part is 0. The fractional part is 43.
             The digit the farthest to the right, 3, is in the ten-thousandths place.

Forty-three ten-thousandths   Since the whole-number part is 0, we need not write it nor the word *and*.

Written as a fraction, 0.0043 is $\frac{43}{10,000}$.

d. $\boxed{-37}\,.\,\boxed{7}$ This is a negative decimal.

Negative thirty-seven and seven tenths.
Written as a negative mixed number, -37.7 is $-37\frac{7}{10}$.

> **The Language of Mathematics** Decimals are often read in an informal way. For example, we can read 184.3 as "one hundred eighty-four point three" and 9.69 as "nine point six nine."

The procedure for reading a decimal can be applied in reverse to convert from written-word form to standard form.

EXAMPLE 4   Write each number in standard form:
a. *One hundred seventy-two and forty-three hundredths*
b. *Eleven and fifty-one thousandths*

Strategy We will locate the word *and* in the written-word form and translate the phrase that appears before it and the phrase that appears after it separately.

WHY The whole-number part of the decimal is described by the phrase that appears before the word *and*. The fractional part of the decimal is described by the phrase that follows the word *and*.

Solution

a. $\boxed{\text{One hundred seventy-two}}$ and $\boxed{\text{forty-three hundredths}}$
                         ↓
                      172.43
                         ↑
                          This is the hundredths place-value column.

b. Sometimes, when changing from written-word form to standard form, we must insert placeholder 0's in the fractional part of a decimal so that the last digit appears in the proper place-value column.

Caution! If a placeholder 0 is not written in 11.051, an incorrect answer of 11.51 (eleven and fifty-one *hundredths,* not *thousandths*) results.

4 Compare decimals using inequality symbols.

To develop a way to compare decimals, let's consider 0.3 and 0.271. Since 0.271 contains more digits, it may appear that 0.271 is greater than 0.3. However, the opposite is true. To show this, we write 0.3 and 0.271 in fraction form:

$$0.3 = \frac{3}{10} \qquad 0.271 = \frac{271}{1,000}$$

Now we build $\frac{3}{10}$ into an equivalent fraction so that it has the same denominator as $\frac{271}{1,000}$.

$$0.3 = \frac{3}{10} \cdot \frac{100}{100} = \frac{300}{1,000}$$

Since $\frac{300}{1,000} > \frac{271}{1,000}$, it follows that $0.3 > 0.271$. This observation suggests a quicker method for comparing decimals.

Comparing Decimals

To compare two decimals:

1. Make sure both numbers have the same number of decimal places to the right of the decimal point. Write any additional zeros necessary to achieve this.
2. Compare the digits of each decimal, column by column, working from left to right.
3. *If the decimals are positive:* When two digits differ, the decimal with the greater digit is the greater number. *If the decimals are negative:* When two digits differ, the decimal with the smaller digit is the greater number.

EXAMPLE 5
Place an $<$ or $>$ symbol in the box to make a true statement:

a. 1.2679 ☐ 1.2658   b. 54.9 ☐ 54.929   c. -10.419 ☐ -10.45

Strategy We will stack the decimals and then, working from left to right, we will scan their place-value columns looking for a difference in their digits.

WHY We need only look in that column to determine which digit is the greater.

Solution

a. Since both decimals have the same number of places to the right of the decimal point, we can immediately compare the digits, column by column.

Same digit
Same digit
Same digit ——— These digits are different: Since 7 is greater than 5, it follows that the first decimal is greater than the second.

Thus, 1.2679 is greater than 1.2658 and we can write $1.2679 > 1.2658$.

b. We can write two zeros after the 9 in 54.9 so that the decimals have the same number of digits to the right of the decimal point. This makes the comparison easier.

```
54.900
54.929
   ↑
```

As we work from left to right, this is the first column in which the digits differ. Since $2 > 0$, it follows that 54.929 is greater than 54.9 (or 54.9 is less than 54.929) and we can write $54.9 < 54.929$.

Self Check 5

Place an $<$ or $>$ symbol in the box to make a true statement:

a. $3.4308 \ < \ 3.4312$

b. $678.3409 \ > \ 678.34$

c. $-703.8 \ < \ -703.78$

Now Try Problems 49, 55, and 59

Teaching Example 5 Place an $<$ or $>$ symbol in the box to make a true statement:
a. 2.7446 ☐ 2.7439
b. 39.2 ☐ 39.203
c. -10.45 ☐ -10.419
Answers:
a. $>$
b. $<$
c. $<$

Success Tip Writing additional zeros *after the last digit to the right of the decimal point does not change the value of the decimal*. Also, deleting additional zeros after the last digit to the right of the decimal point does not change the value of the decimal. For example,

54.9 = 54.90 = 54.900   Because $54\frac{90}{100}$ and $54\frac{900}{1,000}$ in simplest form are equal to $54\frac{9}{10}$.

These additional zeros do not change the value of the decimal.

c. We are comparing two negative decimals. In this case, when two digits differ, the decimal with the smaller digit is the greater number.

−10.4**1**9
−10.4**5**0   Write a zero after 5 to help in the comparison.

As we work from left to right, this is the first column in which the digits differ. Since $1 < 5$, it follows that -10.419 is greater than -10.45 and we can write $-10.419 > -10.45$.

5 Graph decimals on a number line.

Decimals can be shown by drawing points on a number line.

Self Check 6

Graph -1.1, -1.64, -0.8, and 1.9 on a number line.

Now Try Problem 61

Teaching Example 6 Graph -1.37, 0.9, -1.7, and -0.2 on a number line.
Answer:

EXAMPLE 6 Graph -1.8, -1.23, -0.3, and 1.89 on a number line.

Strategy We will locate the position of each decimal on the number line and draw a bold dot.

WHY To *graph a number* means to make a drawing that represents the number.

Solution The graph of each negative decimal is to the left of 0 and the graph of each positive decimal is to the right of 0. Since $-1.8 < -1.23$, the graph of -1.8 is to the left of -1.23.

6 Round decimals.

When we don't need exact results, we can approximate decimal numbers by **rounding**. To round the decimal part of a decimal number, we use a method similar to that used to round whole numbers.

Rounding a Decimal

1. To round a decimal to a certain decimal place value, locate the **rounding digit** in that place.
2. Look at the **test digit** directly to the right of the rounding digit.
3. If the test digit is 5 or greater, round up by adding 1 to the rounding digit and dropping all the digits to its right. If the test digit is less than 5, round down by keeping the rounding digit and dropping all the digits to its right.

EXAMPLE 7
Chemistry A student in a chemistry class uses a digital balance to weigh a compound in grams. Round the reading shown on the balance to the nearest thousandth of a gram.

Strategy We will identify the digit in the thousandths column and the digit in the ten-thousandths column.

WHY To round to the nearest thousandth, the digit in the thousandths column is the rounding digit and the digit in the ten-thousandths column is the test digit.

Solution The rounding digit in the thousandths column is 8. Since the test digit 7 is 5 or greater, we round up.

The reading on the balance is approximately 15.239 grams.

Self Check 7
Round 24.41658 to the nearest ten-thousandth. 24.4166

Now Try Problems 65 and 69

Teaching Example 7 Round 47.0676 to the nearest thousandth.
Answer:
47.068

EXAMPLE 8
Round each decimal to the indicated place value:
a. −645.1358 to the nearest tenth
b. 33.096 to the nearest hundredth

Strategy In each case, we will first identify the rounding digit. Then we will identify the test digit and determine whether it is less than 5 or greater than or equal to 5.

WHY If the test digit is less than 5, we round down; if it is greater than or equal to 5, we round up.

Solution

a. Negative decimals are rounded in the same ways as positive decimals. The rounding digit in the tenths column is 1. Since the test digit 3 is less than 5, we round down.

Thus, −645.1358 rounded to the nearest tenth is −645.1.

b. The rounding digit in the hundredths column is 9. Since the test digit 6 is 5 or greater, we round up.

Thus, 33.096 rounded to the nearest hundredth is 33.10.

> **Caution!** It would be incorrect to drop the 0 in the answer 33.10. If asked to round to a certain place value (in this case, thousandths), that place must have a digit, even if the digit is 0.

Self Check 8
Round each decimal to the indicated place value:
a. −708.522 to the nearest tenth
b. 9.1198 to the nearest thousandth

Now Try Problems 73 and 77

Self Check 8 Answers
a. −708.5
b. 9.120

Teaching Example 8 Round each decimal to the indicated place value:
a. −420.139 to the nearest tenth
b. 56.095 to the nearest hundredth
Answers:
a. −420.1
b. 56.10

There are many situations in our daily lives that call for rounding amounts of money. For example, a grocery shopper might round the unit cost of an item to the nearest cent or a taxpayer might round his or her income to the nearest dollar when filling out an income tax return.

Self Check 9
a. Round $0.076601 to the nearest cent. $0.08
b. Round $24,908.53 to the nearest dollar. $24,909

Now Try Problems 85 and 87

EXAMPLE 9

a. **Utility Bills** A utility company calculates a homeowner's monthly electric bill by multiplying the unit cost of $0.06421 by the number of kilowatt hours used that month. Round the unit cost to the nearest cent.

b. **Annual Income** A secretary earned $36,500.91 in one year. Round her income to the nearest dollar.

Strategy In part a, we will round the decimal to the nearest hundredth. In part b, we will round the decimal to the ones column.

WHY Since there are 100 cents in a dollar, each cent is $\frac{1}{100}$ of a dollar. To round to the *nearest cent* is the same as rounding to the *nearest hundredth* of a dollar. To round to the *nearest dollar* is the same as rounding to the *ones place*.

Solution

a. The rounding digit in the hundredths column is 6. Since the test digit 4 is less than 5, we round down.

Thus, $0.06421 rounded to the nearest cent is $0.06.

b. The rounding digit in the ones column is 0. Since the test digit 9 is 5 or greater, we round up.

Thus, $36,500.91 rounded to the nearest dollar is $36,501.

7 Read tables and graphs involving decimals.

Year	Pounds
1960	2.68
1970	3.25
1980	3.66
1990	4.57
2000	4.72
2010	4.43

Source: U.S. Environmental Protection Agency

The table on the left is an example of the use of decimals. It shows the number of pounds of trash generated daily per person in the United States for selected years from 1960 through 2010.

When the data in the table are presented in the form of a **bar graph**, a trend is apparent. The amount of trash generated daily per person increased steadily until the year 2000. Since then, it appears to have dropped slightly.

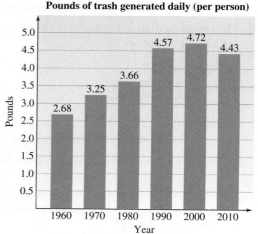

Unless otherwise noted, all art on this page is © Cengage Learning.

ANSWERS TO SELF CHECKS

1. a. 9 thousandths b. 2   2. $1,000 + 200 + 70 + 7 + \frac{9}{10} + \frac{4}{100} + \frac{6}{1,000} + \frac{5}{10,000}$
3. a. ninety-eight and six tenths, $98\frac{6}{10}$ b. two hundred twenty-four and seven thousand seven ten-thousandths, $224\frac{7,007}{10,000}$ c. thirty-five thousand, two hundred seventy-four millionths, $\frac{35,274}{1,000,000}$ d. negative three hundred forty-five and seven hundred forty-eight thousandths, $-345\frac{748}{1,000}$ 4. a. 806.92   b. 12.0067   5. a. <   b. >   c. <
6. -1.64 -1.1 -0.8 1.9 7. 24.4166   8. a. -708.5 b. 9.120
9. a. $0.08   b. $24,909

SECTION 5.1 STUDY SET

VOCABULARY

Fill in the blanks.

1. Decimals are written by entering the digits 0, 1, 2, 3, 4, 5, 6, 7, 8, and 9 into place-value columns that are separated by a decimal point .

2. The place-value columns to the left of the decimal point form the whole-number part of a decimal number and the place-value columns to the right of the decimal point form the fractional part.

3. We can show the value represented by each digit of the decimal 98.6213 by using expanded form:

   $98.6213 = 90 + 8 + \frac{6}{10} + \frac{2}{100} + \frac{1}{1,000} + \frac{3}{10,000}$

4. When we don't need exact results, we can approximate decimal numbers by rounding .

CONCEPTS

5. Write the name of each column in the following place-value chart.

6. Write the value of each column in the following place-value chart.

7	2	.	3	1	9	5	8
10	1		$\frac{1}{10}$	$\frac{1}{100}$	$\frac{1}{1,000}$	$\frac{1}{10,000}$	$\frac{1}{100,000}$

7. Fill in the blanks.
   a. The value of each place in the whole-number part of a decimal number is 10 times greater than the column directly to its right.
   b. The value of each place in the fractional part of a decimal number is $\frac{1}{10}$ of the value of the place directly to its left.

8. Represent each situation using a signed number.
   a. A checking account overdrawn by $33.45 -33.45
   b. A river 6.25 feet above flood stage   6.25
   c. 3.9 degrees below zero -3.9
   d. 17.5 seconds after liftoff   17.5

9. a. Represent the shaded part of the rectangular region as a fraction and a decimal. $\frac{7}{10}$, 0.7

   b. Represent the shaded part of the square region as a fraction and a decimal. $\frac{47}{100}$, 0.47

10. Write $400 + 20 + 8 + \frac{9}{10} + \frac{1}{100}$ as a decimal.   428.91

11. Fill in the blanks in the following illustration to label the *whole-number part* and the *fractional part*.

12. Fill in the blanks.
    a. To round $0.13506 to the *nearest cent*, the rounding digit is 3 and the test digit is 5 .
    b. To round $1,906.47 to the *nearest dollar*, the rounding digit is 6 and the test digit is 4 .

NOTATION

Fill in the blanks.

13. The columns to the right of the decimal point in a decimal number form its fractional part. Their place-value names are similar to those in the whole-number part, but they end in the letters " _ths_ ."

14. When reading a decimal, such as 2.37, we can read the decimal point as " _and_ " or as " _point_ ."

15. Write a decimal number that has . . .

    6 in the ones column,
    1 in the tens column,
    0 in the tenths column,
    8 in the hundreds column,
    2 in the hundredths column,
    9 in the thousands column,
    4 in the thousandths column,
    7 in the ten thousands column, and
    5 in the ten-thousandths column. 79,816.0245

16. Determine whether each statement is true or false.

    a. $0.9 = 0.90$ true
    b. $1.260 = 1.206$ false
    c. $-1.2800 = -1.280$ true
    d. $0.001 = .0010$ true

GUIDED PRACTICE

Answer the following questions about place value. See Example 1.

17. Consider the decimal number: 145.926
    a. What is the place value of the digit 9? 9 tenths
    b. Which digit tells the number of thousandths? 6
    c. Which digit tells the number of tens? 4
    d. What is the place value of the digit 5? 5 ones

18. Consider the decimal number: 304.817
    a. What is the place value of the digit 1? 1 hundredth
    b. Which digit tells the number of thousandths? 7
    c. Which digit tells the number of hundreds? 3
    d. What is the place value of the digit 7? 7 thousandths

19. Consider the decimal number: 6.204538
    a. What is the place value of the digit 8? 8 millionths
    b. Which digit tells the number of hundredths? 0
    c. Which digit tells the number of ten-thousandths? 5
    d. What is the place value of the digit 6? 6 ones

20. Consider the decimal number: 4.390762
    a. What is the place value of the digit 6? 6 hundred-thousandths
    b. Which digit tells the number of thousandths? 0
    c. Which digit tells the number of ten-thousandths? 7
    d. What is the place value of the digit 4? 4 ones

Write each decimal number in expanded form. See Example 2.

21. 37.89 $30 + 7 + \frac{8}{10} + \frac{9}{100}$
22. 26.93 $20 + 6 + \frac{9}{10} + \frac{3}{100}$
23. 124.575 $100 + 20 + 4 + \frac{5}{10} + \frac{7}{100} + \frac{5}{1,000}$
24. 231.973 $200 + 30 + 1 + \frac{9}{10} + \frac{7}{100} + \frac{3}{1,000}$
25. 7,498.6468 $7,000 + 400 + 90 + 8 + \frac{6}{10} + \frac{4}{100} + \frac{6}{1,000} + \frac{8}{10,000}$
26. 1,946.7221 $1,000 + 900 + 40 + 6 + \frac{7}{10} + \frac{2}{100} + \frac{2}{1,000} + \frac{1}{10,000}$
27. 6.40941 $6 + \frac{4}{10} + \frac{9}{1,000} + \frac{4}{10,000} + \frac{1}{100,000}$
28. 8.70214 $8 + \frac{7}{10} + \frac{2}{1,000} + \frac{1}{10,000} + \frac{4}{100,000}$

Write each decimal in words and then as a fraction or mixed number. See Example 3.

29. 0.3   three tenths, $\frac{3}{10}$
30. 0.9   nine tenths, $\frac{9}{10}$
31. 50.41   fifty and forty-one hundredths, $50\frac{41}{100}$
32. 60.61   sixty and sixty-one hundredths, $60\frac{61}{100}$
33. 19.529   nineteen and five hundred twenty-nine thousandths, $19\frac{529}{1,000}$
34. 12.841   twelve and eight hundred forty-one thousandths, $12\frac{841}{1,000}$
35. 304.0003   three hundred four and three ten-thousandths, $304\frac{3}{10,000}$
36. 405.0007   four hundred five and seven ten-thousandths, $405\frac{7}{10,000}$
37. -0.00137 negative one hundred thirty-seven hundred-thousandths, $-\frac{137}{100,000}$
38. -0.00613 negative six hundred thirteen hundred-thousandths, $-\frac{613}{100,000}$
39. $-1,072.499$ negative one thousand seventy-two and four hundred ninety-nine thousandths, $-1,072\frac{499}{1,000}$
40. $-3,076.177$ negative three thousand seventy-six and one hundred seventy-seven thousandths, $-3,076\frac{177}{1,000}$

Write each number in standard form. See Example 4.

41. Six and one hundred eighty-seven thousandths 6.187
42. Four and three hundred ninety-two thousandths 4.392
43. Ten and fifty-six ten-thousandths 10.0056
44. Eleven and eighty-six ten-thousandths 11.0086
45. Negative sixteen and thirty-nine hundredths -16.39
46. Negative twenty-seven and forty-four hundredths -27.44
47. One hundred four and four millionths 104.000004
48. Two hundred three and three millionths 203.000003

Place an $<$ or an $>$ symbol in the box to make a true statement. See Example 5.

49. $2.59 > 2.55$
50. $5.17 > 5.14$
51. $45.103 < 45.108$
52. $13.874 < 13.879$
53. $3.28724 > 3.2871$
54. $8.91335 > 8.9132$
55. $379.67 > 379.6088$
56. $446.166 < 446.2$
57. $-23.45 < -23.1$
58. $-301.98 > -302.45$
59. $-0.065 > -0.066$
60. $-3.99 < -3.9888$

Graph each number on a number line. See Example 6.

61. 0.8, −0.7, −3.1, 4.5, −3.9

62. 0.6, −0.3, −2.7, 3.5, −2.2

63. −1.21, −3.29, −4.25, 2.75, −1.84

64. −3.19, −0.27, −3.95, 4.15, −1.66

Round each decimal number to the indicated place value. See Example 7.

65. 506.198 nearest tenth   506.2
66. 51.451 nearest tenth   51.5
67. 33.0832 nearest hundredth   33.08
68. 64.0059 nearest hundredth   64.01
69. 4.2341 nearest thousandth   4.234
70. 8.9114 nearest thousandth   8.911
71. 0.36563 nearest ten-thousandth   0.3656
72. 0.77623 nearest ten-thousandth   0.7762

Round each decimal number to the indicated place value. See Example 8.

73. −0.137 nearest hundredth   −0.14
74. −808.0897 nearest hundredth   −808.09
75. −2.718218 nearest tenth   −2.7
76. −3,987.8911 nearest tenth   −3,987.9
77. 3.14959 nearest thousandth   3.150
78. 9.50966 nearest thousandth   9.510
79. 1.4142134 nearest millionth   1.414213
80. 3.9998472 nearest millionth   3.999847
81. 16.0995 nearest thousandth   16.100
82. 67.0998 nearest thousandth   67.100
83. 290.303496 nearest hundred-thousandth   290.30350
84. 970.457297 nearest hundred-thousandth   970.45730

Round each given dollar amount. See Example 9.

85. $0.284521 nearest cent $0.28
86. $0.312906 nearest cent $0.31
87. $27,841.52 nearest dollar $27,842
88. $44,633.78 nearest dollar $44,634

APPLICATIONS

89. READING METERS To what decimal is the arrow pointing?   −0.7

90. MEASUREMENT Estimate a length of 0.3 inch on the 1-inch-long line segment below.

91. CHECKING ACCOUNTS Complete the check shown by writing in the amount, using a decimal. $1,025.78

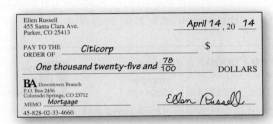

92. MONEY We use a decimal point when working with dollars, but the decimal point is not necessary when working with cents. For each dollar amount in the table, give the equivalent amount expressed as cents.

Dollars	Cents
$0.50	50¢
$0.05	5¢
$0.55	55¢
$5.00	500¢
$0.01	1¢

93. INJECTIONS A syringe is shown below. Use an arrow to show to what point the syringe should be filled if a 0.38-cc dose of medication is to be given. ("cc" stands for "cubic centimeters.")

94. LASERS The laser used in laser vision correction is so precise that each pulse can remove 39 millionths of an inch of tissue in 12 billionths of a second. Write each of these numbers as a decimal.
0.000039 in., 0.000000012 sec

95. NASCAR On April 17, 2011, in the Aaron's 499 Sprint Cup Series race at Talladega Superspeedway, Jimmie Johnson beat Clint Bowyer by a mere 0.002 second. This margin of victory tied for the closest finish since NASCAR switched to electronic timing. Write the decimal in words and then as a fraction in simplest form. (Source: deadspin.com)
two-thousandths, $\frac{2}{1,000} = \frac{1}{500}$

96. THE METRIC SYSTEM The metric system is widely used in science to measure length (meters), weight (grams), and capacity (liters). Round each decimal to the nearest hundredth.

  a. 1 ft is 0.3048 meter. 0.30
  b. 1 mi is 1,609.344 meters. 1,609.34
  c. 1 lb is 453.59237 grams. 453.59
  d. 1 gal is 3.785306 liters. 3.79

97. UTILITY BILLS A portion of a homeowner's electric bill is shown below. Round each decimal dollar amount to the nearest cent. $0.16, $1.02, $1.20, $0.00, $0.10

Billing Period		Meter		The Gas Company
From 06/05/14	To 07/05/14	Number 10694435		

Next Meter Reading Date on or about Aug 03 2014

Summary of Charges

Customer Charge	30 Days	× $0.16438
Baseline	14 Therms	× $1.01857
Over Baseline	11 Therms	× $1.20091
State Regulatory Fee	25 Therms	× $0.00074
Public Purpose Surcharge	25 Therms	× $0.09910

98. INCOME TAX A portion of a W-2 tax form is shown below. Round each dollar amount to the nearest dollar. $35,674, $7,134, $38,204, $2,369, $38,204, $550

Form **W-2** Wage and Tax Statement		**2014**
1 Wages, tips, other comp $35,673.79	2 Fed inc tax withheld $7,134.28	3 Social security wages $38,204.16
4 SS tax withheld $2,368.65	5 Medicare wages & tips $38,204.16	6 Medicare tax withheld $550.13
7 Social security tips	8 Allocated tips	9 Advance EIC payment
10 Depdnt care benefits	11 Nonqualified plans	12a

99. THE DEWEY DECIMAL SYSTEM When stacked on the shelves, the library books shown in the next column are to be in numerical order, least to greatest, *from left to right*. How should the titles be rearranged to be in the proper order?
candlemaking, crafts, hobbies, folk dolls, modern art

Crafts 745.51, Modern art 745.601, Hobbies 745.58, Folk dolls 745.6, Candlemaking 745.49

100. 2012 OLYMPICS The top six finishers in the women's individual all-around gymnastics competition at the London Olympic Games are shown below in alphabetical order.

  a. If the highest score wins, which gymnasts won the gold (1st place), silver (2nd place), and bronze (3rd place) medals? gold: Douglas, silver: Komova, bronze: tie, Mustafina and Raisman

  b. In part a, did you notice there was a tie for the bronze medal? In the case of a tie, the sum of the three highest apparatus scores is used. These sums were 45.933 for Mustafina and 45.366 for Raisman (both dropped their beam score). If the higher score wins the tie-breaker, who was awarded the bronze medal? Mustafina

Name	Nation	Score
Linlin Deng	China	58.399
Gabrielle Douglas	U.S.A.	62.232
Sandra Izbasa	Romania	58.883
Victoria Komova	Russia	61.973
Aliya Mustafina	Russia	59.566
Alexandra Raisman	U.S.A.	59.566

(Source: london2012.com)

101. TUNEUPS The six spark plugs from the engine of a Nissan Quest were removed, and the spark plug gap was checked. If vehicle specifications call for the gap to be from 0.031 to 0.035 inch, which of the plugs should be replaced? Cylinder 2, Cylinder 4

Cylinder 1: 0.035 in.
Cylinder 2: 0.029 in.
Cylinder 3: 0.033 in.
Cylinder 4: 0.039 in.
Cylinder 5: 0.031 in.
Cylinder 6: 0.032 in.

102. **GEOLOGY** Geologists classify types of soil according to the grain size of the particles that make up the soil. The four major classifications of soil are shown below. Classify each of the samples (A, B, C, and D) in the table as clay, silt, sand, or granule.

Clay 0.00 in.   Silt 0.00008 in.   Sand 0.002 in.   0.08 in.   Granule 0.15 in.

Sample	Location found	Grain size (in.)	Classification
A	Riverbank	0.009	sand
B	Pond	0.0007	silt
C	NE corner	0.095	granule
D	Dry lake	0.00003	clay

103. **MICROSCOPES** A microscope used in a lab is capable of viewing structures that range in size from 0.1 to as small as 0.0001 centimeter. Which of the structures listed in the table would be visible through this microscope?
bacterium, plant cell, animal cell, asbestos fiber

Structure	Size (cm)
Bacterium	0.00011
Plant cell	0.015
Virus	0.000017
Animal cell	0.00093
Asbestos fiber	0.0002

104. **ACCELERATION** Refer to the graph below. Find the time it takes each midsized SUV to accelerate from 0 to 60 mph. (Source: consumerreports.org)
4 Runner: 7.7 sec, Infiniti: 8.3 sec, Pilot: 9.1 sec, Lexus: 7.3 sec, Santa Fe: 9.7 sec, Edge: 8.9 sec

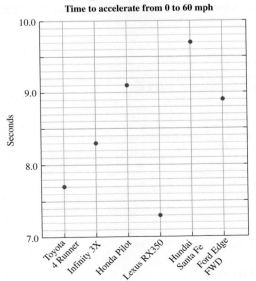

Source: consumerreports.org

105. **DIGGING FOR CLAMS** The following graph shows the time each day that the lowest tide occurs on a beach in Alaska. The negative numbers indicate the water level will be below the normal low water mark. If a tide of −4.0 feet or less is needed to expose the clam beds, on what days should there be good clamming? June 3, 4, 5, 6, 7

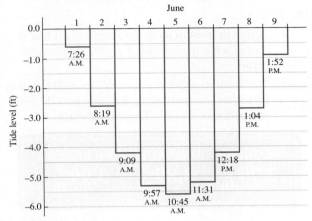

Source: ycharts.com

106. **GASOLINE PRICES** Refer to the graph below.
a. In what month of 2011 was the retail price of a gallon of gasoline the least and in what month was it the greatest? Estimate each price.
Jan: $3.09, May: $3.91
b. In what month of 2012 was the retail price of a gallon of gasoline the least and in what month was it the greatest? Estimate each price.
Dec: $3.31, April: $3.89

*Retail price includes state and federal taxes
Source: EPA Short-Term Energy Outlook, January 2013

WRITING

107. Explain the difference between ten and one-tenth.

108. "The more digits a number contains, the larger it is." Is this statement true? Explain.

109. Explain why is it wrong to read 2.103 as *"two and one hundred and three thousandths."*

110. SIGNS

  a. A sign in front of a fast-food restaurant had the cost of a hamburger listed as .99¢. Explain the error.

  b. The illustration below shows the unusual notation that some service stations use to express the price of a gallon of gasoline. Explain the error.

  REGULAR UNLEADED UNLEADED +
  $3.79\frac{9}{10}$ $3.89\frac{9}{10}$ $3.99\frac{9}{10}$

111. Write a definition for each of these words.

  decade   decathlon   decimal

112. Show that in the decimal numeration system, each place-value column for the fractional part of a decimal is $\frac{1}{10}$ of the value of the place directly to its left.

REVIEW

113. a. Find the perimeter of the rectangle shown below. $12\frac{1}{2}$ ft.

  b. Find the area of the rectangle. $9\frac{5}{8}$ ft²

114. a. Find the perimeter of the triangle shown below. $3\frac{3}{5}$ in.

  b. Find the area of the triangle. $\frac{27}{50}$ in.²

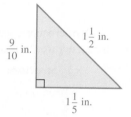

SECTION 5.2
Adding and Subtracting Decimals

Objectives
1. Add decimals.
2. Subtract decimals.
3. Add and subtract signed decimals.
4. Estimate sums and differences of decimals.
5. Solve application problems by adding and subtracting decimals.

To add or subtract objects, they must be similar. The federal income tax form shown below has a vertical line to make sure that dollars are added to dollars and cents added to cents. In this section, we show how decimal numbers are added and subtracted using this type of vertical form.

Form **1040EZ**	Department of the Treasury—Internal Revenue Service **Income Tax Return for Single and Joint Filers With No Dependents** **2014**		
Income Attach Form(s) W-2 here. Enclose, but do not attach, any payment.	1 Wages, salaries, and tips. This should be shown in box 1 of your Form(s) W-2. Attach your Form(s) W-2.	**1**	21,056 | 89
	2 Taxable interest. If the total is over $1,500, you cannot use Form 1040EZ.	**2**	42 | 06
	3 Unemployment compensation and Alaska Permanent Fund dividends (see page 11).	**3**	200 | 00
	4 Add lines 1, 2, and 3. This is your **adjusted gross income**.	**4**	21,298 | 95

1 Add decimals.

Adding decimals is similar to adding whole numbers. We use **vertical form** and stack the decimals with their corresponding place values and decimal points lined up. Then we add the digits in each column, working from right to left, making sure that hundredths are added to hundredths, tenths are added to tenths, ones are added to ones, and so on. We write the decimal point in the **sum** so that it lines up with the decimal points in the **addends**. For example, to find 4.21 + 1.23 + 2.45, we proceed as follows:

The sum is 7.89.

In this example, each addend had two decimal places, tenths and hundredths. If the number of decimal places in the addends are different, we can insert additional zeros so that the number of decimal places match.

Adding Decimals

To add decimal numbers:

1. Write the numbers in vertical form with the decimal points lined up.
2. Add the numbers as you would add whole numbers, from right to left.
3. Write the decimal point in the result from Step 2 directly below the decimal points in the addends.

As with whole number addition, if the sum of the digits in any place-value column is greater than 9, we must **carry**.

EXAMPLE 1   Add: 31.913 + 5.6 + 68 + 16.78

Strategy We will write the addition in vertical form so that the corresponding place values and decimal points of the addends are lined up. Then we will add the digits, column by column, working from right to left.

WHY We can only add digits with the same place value.

Solution To make the column additions easier, we will write two zeros after the 6 in the addend 5.6 and one zero after the 8 in the addend 16.78. Since whole numbers have an "understood" decimal point immediately to the right of their ones digit, we can write the addend 68 as 68.000 to help line up the columns.

```
   31.913
    5.600    Insert two zeros after the 6.
   68.000    Insert a decimal point and three zeros: 68 = 68.000.
 + 16.780    Insert a zero after the 8.
       ↑
       Line up the decimal points.
```

Now we add, right to left, as we would whole numbers, writing the sum from each column below the horizontal bar.

Self Check 1
Add: 41.07 + 35 + 67.888 + 4.1
Now Try Problem 19
Self Check 1 Answer
148.058

Teaching Example 1 Add:
36.824 + 8.5 + 79 + 12.65
Answer:
136.974

```
  2 2
 31.913   Carry a 2 (shown in blue) to the ones column.
  5.600   Carry a 2 (shown in red) to the tens column.
 68.000
+16.780
────────
122.293
```
Write the decimal point in the result directly below the decimal points in the addends.

The sum is 122.293.

Success Tip In Example 1, the digits in each place-value column were added from *top to bottom*. To check the answer, we can instead add from *bottom to top*. Adding down or adding up should give the same result. If it does not, an error has been made and you should re-add.

Using Your CALCULATOR   Adding Decimals

The bar graph on the right shows the number of grams of fiber in a standard serving of each of several foods. It is believed that men can significantly cut their risk of heart attack by eating at least 25 grams of fiber a day. Does this diet meet or exceed the 25-gram requirement?

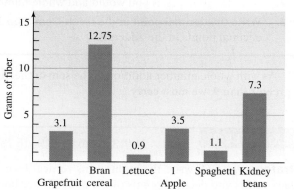

To find the total fiber intake, we add the fiber content of each of the foods. We can use a calculator to add the decimals.

3.1 [+] 12.75 [+] .9 [+] 3.5 [+] 1.1 [+] 7.3 [=]   ⎡ 28.65 ⎤

On some calculators, the [ENTER] key is pressed to find the sum.

Since 28.65 > 25, this diet exceeds the daily fiber requirement of 25 grams.

2 Subtract decimals.

Subtracting decimals is similar to subtracting whole numbers. We use **vertical form** and stack the decimals with their corresponding place values and decimal points lined up so that we subtract similar objects—hundredths from hundredths, tenths from tenths, ones from ones, and so on. We write the decimal point in the **difference** so that it lines up with the decimal points in the **minuend** and **subtrahend**. For example, to find 8.59 − 1.27, we proceed as follows:

The difference is 7.32.

Subtracting Decimals

To subtract decimal numbers:

1. Write the numbers in vertical form with the decimal points lined up.
2. Subtract the numbers as you would subtract whole numbers from right to left.
3. Write the decimal point in the result from Step 2 directly below the decimal points in the minuend and the subtrahend.

As with whole numbers, if the subtraction of the digits in any place-value column requires that we subtract a larger digit from a smaller digit, we must **borrow** or **regroup**.

EXAMPLE 2   Subtract: 279.6 − 138.7

Strategy As we prepare to subtract in each column, we will compare the digit in the subtrahend (bottom number) to the digit directly above it in the minuend (top number).

WHY If a digit in the subtrahend is greater than the digit directly above it in the minuend, we must borrow (regroup) to subtract in that column.

Solution Since 7 in the tenths column of 138.**7** is greater than 6 in the tenths column of 279.**6**, we cannot immediately subtract in that column because 6 − 7 is *not* a whole number. To subtract in the tenths column, we must regroup by borrowing as shown below.

$$\begin{array}{r} \overset{8\;\;16}{279.\cancel{6}} \\ -\;138.7 \\ \hline 140.9 \end{array}$$

To subtract in the tenths column, borrow 1 one in the form of 10 tenths from the ones column. Add 10 to the 6 in the tenths column to get 16 (shown in blue).

Recall from Section 1.3 that subtraction can be checked by addition. If a subtraction is done correctly, the sum of the difference and the subtrahend will equal the minuend: **Difference + subtrahend = minuend**.

Check:

$$\begin{array}{r} \overset{1}{140.9} \quad \text{Difference} \\ +\;138.7 \quad \text{Subtrahend} \\ \hline 279.6 \quad \text{Minuend} \end{array}$$

Since the sum of the difference and the subtrahend is the minuend, the subtraction is correct.

Some subtractions require borrowing from two (or more) place-value columns.

Self Check 2
Subtract: 382.5 − 227.1   155.4
Now Try Problem 27

Teaching Example 2
Subtract: 356.4 − 123.6
Answer:
232.8

Self Check 3

Subtract 27.122 from 29.7. 2.578

Now Try Problem 31

Teaching Example 3 Subtract 42.318 from 45.6.
Answer:
3.282

EXAMPLE 3   Subtract 13.059 from 15.4.

Strategy We will translate the sentence to mathematical symbols and then perform the subtraction. As we prepare to subtract in each column, we will compare the digit in the subtrahend (bottom number) to the digit directly above it in the minuend (top number).

WHY If a digit in the subtrahend is greater than the digit directly above it in the minuend, we must borrow (regroup) to subtract in that column.

Solution Since 13.059 is the number to be subtracted, it is the subtrahend.

Subtract 13.059 from 15.4

$$15.4 - 13.059$$

To find the difference, we write the subtraction in vertical form. To help with the column subtractions, we write two zeros to the right of 15.4 so that both numbers have three decimal places.

$$\begin{array}{r} 15.400 \\ -13.059 \end{array}$$ Insert two zeros after the 4 so that the decimal places match.

Line up the decimal points.

Since 9 in the thousandths column of 13.059 is greater than 0 in the thousandths column of 15.400, we cannot immediately subtract. It is not possible to borrow from the digit **0** in the hundredths column of 15.400. We can, however, borrow from the digit **4** in the tenths column of 15.400.

$$\begin{array}{r} \overset{3\,10}{15\,.\,\cancel{4}\,0\,0} \\ -13\,.\,0\,5\,9 \end{array}$$ Borrow 1 tenth in the form of 10 hundredths from 4 in the tenths column. Add 10 to 0 in the hundredths column to get 10 (shown in blue).

Now we complete the two-column borrowing process by borrowing from the **10** in the hundredths column. Then we subtract, column-by-column, from the right to the left to find the difference.

$$\begin{array}{r} \overset{9}{\overset{3\,\cancel{10}\,10}{15\,.\,\cancel{4}\,\cancel{0}\,0}} \\ -13\,.\,0\,5\,9 \\ \hline 2\,.\,3\,4\,1 \end{array}$$ Borrow 1 hundredth in the form of 10 thousandths from 10 in the hundredths column. Add 10 to 0 in the thousandths column to get 10 (shown in green).

When 13.059 is subtracted from 15.4, the difference is 2.341.

Check:

$$\begin{array}{r} \overset{1\,1}{2.341} \\ +13.059 \\ \hline 15.400 \end{array}$$ Since the sum of the difference and the subtrahend is the minuend, the subtraction is correct.

Using Your CALCULATOR Subtracting Decimals

A giant weather balloon is made of a flexible rubberized material that has an uninflated thickness of 0.011 inch. When the balloon is inflated with helium, the thickness becomes 0.0018 inch. To find the change in thickness, we need to subtract. We can use a calculator to subtract the decimals.

.011 $\boxed{-}$ .0018 $\boxed{=}$ $\boxed{0.0092}$

On some calculators, the $\boxed{\text{ENTER}}$ key is pressed to find the difference.

After the balloon is inflated, the rubberized material loses 0.0092 inch in thickness.

3 Add and subtract signed decimals.

To add signed decimals, we use the same rules that we used for adding integers.

> **Adding Two Decimals That Have the Same (Like) Signs**
>
> 1. To add two positive decimals, add them as usual. The final answer is positive.
> 2. To add two negative decimals, add their absolute values and make the final answer negative.

> **Adding Two Decimals That Have Different (Unlike) Signs**
>
> To add a positive decimal and a negative decimal, subtract the smaller absolute value from the larger.
>
> 1. If the positive decimal has the larger absolute value, the final answer is positive.
> 2. If the negative decimal has the larger absolute value, make the final answer negative.

EXAMPLE 4 Add: $-6.1 + (-4.7)$

Strategy We will use the rule for adding two decimals that have the same sign.

WHY Both addends, -6.1 and -4.7, are negative.

Solution Find the absolute values: $|-6.1| = 6.1$ and $|-4.7| = 4.7$.

$$-6.1 + (-4.7) = -10.8$$

Add the absolute values, 6.1 and 4.7, to get 10.8. Then make the final answer negative.

$$\begin{array}{r} 6.1 \\ +\ 4.7 \\ \hline 10.8 \end{array}$$

Self Check 4
Add: $-5.04 + (-2.32)$ -7.36
Now Try Problem 35

Teaching Example 4
Add: $-3.4 + (-8.3)$
Answer:
-11.7

EXAMPLE 5 Add: $5.35 + (-12.9)$

Strategy We will use the rule for adding two decimals that have different signs.

WHY One addend is positive and the other is negative.

Solution Find the absolute values: $|5.35| = 5.35$ and $|-12.9| = 12.9$.

$$5.35 + (-12.9) = -7.55$$

Subtract the smaller absolute value from the larger: $12.9 - 5.35 = 7.55$. Since the negative number, -12.9, has the larger absolute value, make the final answer negative.

$$\begin{array}{r} \overset{8\,10}{12.9\rlap{/}0} \\ -\ 5.35 \\ \hline 7.55 \end{array}$$

Self Check 5
Add: $-21.4 + 16.75$ -4.65
Now Try Problem 39

Teaching Example 5
Add: $15.21 + (-29.8)$
Answer:
-14.59

The rule for subtraction that was introduced in Section 2.3 can be used with signed decimals: *To subtract two decimals, add the first decimal to the opposite of the decimal to be subtracted.*

EXAMPLE 6 Subtract: $-35.6 - 5.9$

Strategy We will apply the rule for subtraction: Add the first decimal to the opposite of the decimal to be subtracted.

WHY It is easy to make an error when subtracting signed decimals. We will probably be more accurate if we write the subtraction as addition of the opposite.

Self Check 6
Subtract: $-1.18 - 2.88$ -4.06
Now Try Problem 43

Teaching Example 6
Subtract: $-46.8 - 7.6$
Answer:
-54.4

Solution The number to be subtracted is 5.9. Subtracting 5.9 is the same as adding its opposite, −5.9.

Change the subtraction to addition.

$-35.6 - 5.9 = -35.6 + (-5.9) = -41.5$ Use the rule for adding two decimals with the same sign. Make the final answer negative.

Change the number being subtracted to its opposite.

$$\begin{array}{r} \overset{1\,1}{35.6} \\ +\ 5.9 \\ \hline 41.5 \end{array}$$

Self Check 7
Subtract: $-2.56 - (-4.4)$ 1.84
Now Try Problem 47

Teaching Example 7
Subtract: $-2.98 - (-27.3)$
Answer:
24.32

EXAMPLE 7 Subtract: $-8.37 - (-16.2)$

Strategy We will apply the rule for subtraction: Add the first decimal to the opposite of the decimal to be subtracted.

WHY It is easy to make an error when subtracting signed decimals. We will probably be more accurate if we write the subtraction as addition of the opposite.

Solution The number to be subtracted is −16.2. Subtracting −16.2 is the same as adding its opposite, 16.2.

Add . . .

$-8.37 - (-16.2) = -8.37 + 16.2 = 7.83$ Use the rule for adding two decimals with different signs. Since 16.2 has the larger absolute value, the final answer is positive.

. . . the opposite

$$\begin{array}{r} \overset{5\ \overset{11}{\cancel{1}}10}{16.2\cancel{0}} \\ -\ 8.3\,7 \\ \hline 7.8\,3 \end{array}$$

Self Check 8
Evaluate: $-4.9 - (-1.2 + 5.6)$
Now Try Problem 51
Self Check 8 Answer
−9.3

Teaching Example 8
Evaluate: $-33.5 - (-16.3 + 7.9)$
Answer:
−25.1

EXAMPLE 8 Evaluate: $-12.2 - (-14.5 + 3.8)$

Strategy We will perform the operation within the parentheses first.

WHY This is the first step of the order of operations rule.

Solution We perform the addition within the grouping symbols first.

$-12.2 - (-14.5 + 3.8) = -12.2 - (-10.7)$ Perform the addition.
$= -12.2 + 10.7$ Add the opposite of −10.7.
$= -1.5$ Perform the addition.

$$\begin{array}{r} \overset{3\ 15}{14.\cancel{5}} \\ -\ 3.8 \\ \hline 10.\,7 \end{array}$$

$$\begin{array}{r} \overset{1\ 12}{12.\cancel{2}} \\ -10.\,7 \\ \hline 1.\,5 \end{array}$$

4 Estimate sums and differences of decimals.

Estimation can be used to check the reasonableness of an answer to a decimal addition or subtraction. There are several ways to estimate, but the objective is the same: Simplify the numbers in the problem so that the calculations can be made easily and quickly.

Self Check 9
a. Estimate by rounding the addends to the nearest ten: $526.93 + 284.03$ 810
b. Estimate using front-end rounding: $512.33 - 36.47$ 460
Now Try Problems 55 and 57

EXAMPLE 9
a. Estimate by rounding the addends to the nearest ten: $261.76 + 432.94$
b. Estimate using front-end rounding: $381.77 - 57.01$

Strategy We will use rounding to approximate each addend, minuend, and subtrahend. Then we will find the sum or difference of the approximations.

WHY Rounding produces numbers that contain many 0's. Such numbers are easier to add or subtract.

Solution

a. $261.76 \rightarrow 260$ Round to the nearest ten.
$+432.94 \rightarrow +430$ Round to the nearest ten.
$\overline{690}$

The estimate is 690. If we compute $261.76 + 432.94$, the sum is 694.7. We can see that the estimate is close; it's just 4.7 less than 694.7.

b. We use front-end rounding. Each number is rounded to its largest place value.

$381.77 \rightarrow 400$ Round to the nearest hundred.
$-57.01 \rightarrow -60$ Round to the nearest ten.
$\overline{340}$

The estimate is 340. If we compute $381.77 - 57.01$, the difference is 324.76. We can see that the estimate is close; it's 15.24 more than 324.76.

5 Solve application problems by adding and subtracting decimals.

To make a profit, a merchant must sell an item for more than she paid for it. The price at which the merchant sells the product, called the **retail price**, is the *sum* of the item's **cost** to the merchant plus the **markup**.

Retail price = cost + markup

EXAMPLE 10   Pricing
Find the retail price of a Rubik's Cube if a game store owner buys them for $8.95 each and then marks them up $4.25 to sell in her store.

Solution
- Rubik's Cubes cost the store owner $8.95 each.   Given
- She marks up the price $4.25.   Given
- What is the retail price of a Rubik's Cube?   Find

We translate the words of the problem to numbers and symbols.

The retail price	is equal to	the cost	plus	the markup.
The retail price	=	8.95	+	4.25

Use vertical form to perform decimal addition:

$\overset{1\,1}{8.95}$
$+4.25$
$\overline{13.20}$

The retail price of a Rubik's Cube is $13.20.

Check: We can estimate to check the result. If we use $9 to approximate the cost of a Rubik's Cube to the store owner and $4 to be the approximate markup, then the retail price is about $9 + $4 = $13. The result, $13.20, seems reasonable.

EXAMPLE 11   Kitchen Sinks
One model of kitchen sink is made of 18-gauge stainless steel that is 0.0500 inch thick. Another, less expensive, model is made from 20-gauge stainless steel that is 0.0375 inch thick. How much thicker is the 18-gauge?

Teaching Example 9
a. Estimate by rounding the addends to the nearest ten: $659.59 + 223.75$
b. Estimate using front-end rounding: $734.01 - 48.22$

Answers:
a. 880
b. 650

Self Check 10
PRICING Find the retail price of a wool coat if a clothing outlet buys them for $109.95 each and then marks them up $99.95 to sell in its stores. $209.90

Now Try Problem 91

Teaching Example 10 PRICING Find the retail price of an electric stapler if an office supply company buys them for $29.50 each and then marks them up $12.95 to sell in its stores.
Answer:
$42.45

Self Check 11

ALUMINUM How much thicker is 16-gauge aluminum that is 0.0508 inch thick than 22-gauge aluminum that is 0.0253 inch thick? 0.0255 in.

Now Try Problem 97

Teaching Example 11 TABLES A kitchen table has a 20-gauge stainless steel top that is 0.0375 inch thick and legs made of 16-gauge stainless steel that is 0.0625 inch thick. How much thicker is 16-gauge stainless steel?
Answer:
0.025 inch

Solution

- The 18-gauge stainless steel is 0.0500 inch thick. *Given*
- The 20-gauge stainless steel is 0.0375 inch thick. *Given*
- How much thicker is the 18-gauge stainless steel? *Find*

Phrases such as *how much older*, *how much longer*, and, in this case, *how much thicker*, indicate subtraction. We translate the words of the problem to numbers and symbols.

How much thicker	is equal to	the thickness of the 18-gauge stainless steel	minus	the thickness of the 20-gauge stainless steel.
How much thicker	=	0.0500	−	0.0375

Use vertical form to perform subtraction:

$$\begin{array}{r}0.05\overset{4\,\overset{9}{\cancel{10}}10}{\cancel{00}}\\-\,0.0375\\\hline 0.0125\end{array}$$

The 18-gauge stainless steel is 0.0125 inch thicker than the 20-gauge.

Check: We can add to check the subtraction:

$$\begin{array}{rl}\overset{1\,1}{}0.0125 & \text{Difference}\\ +\,0.0375 & \text{Subtrahend}\\ \hline 0.0500 & \text{Minuend}\end{array}$$

The result checks.

Sometimes more than one operation is needed to solve a problem involving decimals.

Self Check 12

WRESTLING A 195.5-pound wrestler had to lose 6.5 pounds to make his weight class. After the weigh-in, he gained back 3.7 pounds. What did he weigh then? 192.7 lb

Now Try Problem 103

EXAMPLE 12
Conditioning Programs A 350-pound football player lost 15.7 pounds during the first week of practice. During the second week, he gained 4.9 pounds. Find his weight after the first two weeks of practice.

Solution

- The football player's beginning weight was 350 pounds. *Given*
- The first week he lost 15.7 pounds. *Given*
- The second week he gained 4.9 pounds. *Given*
- What was his weight after two weeks of practice? *Find*

The word *lost* indicates subtraction. The word *gained* indicates addition. We translate the words of the problem to numbers and symbols.

The player's weight after two weeks of practice	is equal to	his beginning weight	minus	the first-week weight loss	plus	the second-week weight gain.
The player's weight after two weeks of practice	=	350	−	15.7	+	4.9

To evaluate 350 − 15.7 + 4.9, we work from left to right and perform the subtraction first, then the addition.

$$\begin{array}{r} \overset{9}{3\cancel{5}\cancel{0}.\cancel{0}} \\ -\ 15.7 \\ \hline 334.3 \end{array}$$ Write the whole number 350 as 350.0 and use a two-column borrowing process to subtract in the tenths column.

This is the player's weight after one week of practice.

Next, we add the 4.9-pound gain to the previous result to find the player's weight after two weeks of practice.

$$\begin{array}{r} \overset{1}{334.3} \\ +\ 4.9 \\ \hline 339.2 \end{array}$$

The player's weight was 339.2 pounds after two weeks of practice.

Check: We can estimate to check the result. The player lost about 16 pounds the first week and then gained back about 5 pounds the second week, for a net loss of 11 pounds. If we subtract the approximate 11-pound loss from his beginning weight, we get 350 − 11 = 339 pounds. The result, 339.2 pounds, seems reasonable.

> **Teaching Example 12 CONDITIONING PROGRAMS** A 180-pound baseball player lost 4.2 pounds during the first week of spring training. He gained back 1.9 pounds the second week. Find his weight after two weeks of spring training.
> *Answer:*
> 177.7 lb

ANSWERS TO SELF CHECKS
1. 148.058   2. 155.4   3. 2.578   4. −7.36   5. −4.65   6. −4.06   7. 1.84   8. −9.3
9. a. 810   b. 460   10. $209.90   11. 0.0255 in.   12. 192.7 lb

---

SECTION 5.2 STUDY SET

VOCABULARY

Fill in the blanks.

1. In the addition problem shown below, label each *addend* and the *sum*.

   $$\begin{array}{r} 1.72 \leftarrow \text{addend} \\ 4.68 \leftarrow \text{addend} \\ +\ 2.02 \leftarrow \text{addend} \\ \hline 8.42 \leftarrow \text{sum} \end{array}$$

2. When using the vertical form to add decimals, if the addition of the digits in any one column produces a sum greater than 9, we must **carry**.

3. In the subtraction problem shown below, label the *minuend*, *subtrahend*, and the *difference*.

   $$\begin{array}{r} 12.9 \leftarrow \text{minuend} \\ -\ 4.3 \leftarrow \text{subtrahend} \\ \hline 8.6 \leftarrow \text{difference} \end{array}$$

4. If the subtraction of the digits in any place-value column requires that we subtract a larger digit from a smaller digit, we must **borrow** or *regroup*.

5. To see whether the result of an addition is reasonable, we can round the addends and **estimate** the sum.

6. In application problems, phrases such as *how much older*, *how much longer*, and *how much thicker* indicate the operation of **subtraction**.

CONCEPTS

7. Check the following result. Use addition to determine if 15.2 is the correct difference. It is not correct: 15.2 + 12.5 ≠ 28.7

   $$\begin{array}{r} 28.7 \\ -\ 12.5 \\ \hline 15.2 \end{array}$$

8. Determine whether the *sign* of each result is positive or negative. *You do not have to find the sum.*
   a. −7.6 + (−1.8)   negative
   b. −24.99 + 29.08   positive
   c. 133.2 + (−400.43)   negative

9. Fill in the blank: To subtract signed decimals, add the **opposite** of the decimal that is being subtracted.

10. Apply the rule for subtraction and fill in the three blanks.

    $$3.6 - (-2.1) = 3.6\ +\ 2.1 = 5.7$$

11. Fill in the blanks to rewrite each subtraction as addition of the opposite of the number being subtracted.
    a. $6.8 - 1.2 = 6.8 + (\ -1.2\ )$
    b. $29.03 - (-13.55) = 29.03 + \ 13.55$
    c. $-5.1 - 7.4 = -5.1 + (\ -7.4\ )$

12. Fill in the blanks to complete the estimation.

    $567.7 \to\ 570$ Round to the nearest ten.
    $+ 214.3 \to +\ 210$ Round to the nearest ten.
    $782.0 \qquad\ 780$

NOTATION

13. Copy the following addition problem. Insert a decimal point and additional zeros so that the number of decimal places in the addends match.

    46.6 00
    11. 000
    + 15.702

14. Refer to the subtraction problem below. Fill in the blanks: To subtract in the hundredths column, we borrow 1 tenth in the form of 10 hundredths from the 3 in the tenths column.

    $\overset{211}{29.3\rlap{/}1}$
    $- 25.16$

GUIDED PRACTICE

Add. See Objective 1.

15. 32.5
    + 7.4
    39.9

16. 16.3
    + 3.5
    19.8

17. 3.04
    4.12
    + 1.43
    8.59

18. 2.11
    5.04
    + 2.72
    9.87

Add. See Example 1.

19. $36.821 + 7.3 + 42 + 15.44$ 101.561
20. $46.228 + 5.6 + 39 + 19.37$ 110.198
21. $27.471 + 6.4 + 157 + 12.12$ 202.991
22. $52.763 + 9.1 + 128 + 11.84$ 201.703

Subtract. See Objective 2.

23. 6.83
    − 3.52
    3.31

24. 9.47
    − 5.06
    4.41

25. 8.97
    − 6.22
    2.75

26. 7.56
    − 2.33
    5.23

Subtract. See Example 2.

27. 495.4
    − 153.7
    341.7

28. 977.6
    − 345.8
    631.8

29. 878.1
    − 174.6
    703.5

30. 767.2
    − 614.7
    152.5

Perform the indicated operation. See Example 3.

31. Subtract 11.065 from 18.3. 7.235
32. Subtract 15.041 from 17.8. 2.759
33. Subtract 23.037 from 66.9. 43.863
34. Subtract 31.089 from 75.6. 44.511

Add. See Example 4.

35. $-6.3 + (-8.4)$ −14.7
36. $-9.2 + (-6.7)$ −15.9
37. $-5.6 + (-5.6)$ −11.2
38. $-7.3 + (-5.4)$ −12.7

Add. See Example 5.

39. $4.12 + (-18.8)$ −14.68
40. $9.95 + (-9.95)$ 0
41. $6.45 + (-12.6)$ −6.15
42. $8.81 + (-14.9)$ −6.09

Subtract. See Example 6.

43. $-1.9 - 0.2$ −2.1
44. $-56.1 - 8.6$ −64.7
45. $-42.5 - 2.8$ −45.3
46. $-93.2 - 3.9$ −97.1

Subtract. See Example 7.

47. $-4.49 - (-11.3)$ 6.81
48. $-1.35 - (-0.5)$ −0.85
49. $-6.78 - (-24.6)$ 17.82
50. $-8.51 - (-27.4)$ 18.89

Evaluate each expression. See Example 8.

51. $-11.1 - (-14.4 + 7.8)$ −4.5
52. $-12.3 - (-13.6 + 7.9)$ −6.6
53. $-16.4 - (-18.9 + 5.9)$ −3.4
54. $-15.5 - (-19.8 + 5.7)$ −1.4

Estimate each sum by rounding the addends to the nearest ten. See Example 9.

55. $510.65 + 279.19$ 790
56. $424.08 + 169.04$ 590

Estimate each difference by using front-end rounding. See Example 9.

57. $671.01 - 88.35$ 610
58. $447.23 - 36.16$ 360

TRY IT YOURSELF

Perform the indicated operations.

59. $-45.6 + 34.7$ −10.9
60. $-6.95 + 8.2$ 1.25
61. $-9.5 - 7.1$ −16.6
62. $-1.5 - 0.35$ −1.85
63. $15.95 + (-15.95)$ 0
64. $34.7 + (-30.1)$ 4.6

65. 21.88
    + 33.12
    55.00

66. 19.05
    + 31.95
    51.00

67. $2.55 - (-1.1)$ 3.65
68. $143.3 - (-64.01)$ 207.31
69. $645 + 9.90005 + 0.12 + 3.02002$ 658.04007
70. $505.0103 + 23 + 0.989 + 12.0704$ 541.0697
71. Subtract 47.5 from 0. −47.5

72. Subtract 5.9 from 7.001. 1.101
73. (3.4 − 6.6) + 7.3 4.1
74. 3.4 − (6.6 + 7.3) −10.5
75. 247.9 + 40 + 0.56 288.46
76. 0.0053 + 1.78 + 6 7.7853

77. 78.1
    − 7.81
    ─────
     70.29

78. 202.234
    − 19.34
    ──────
    182.894

79. −7.8 + (−6.5) −14.3
80. −5.78 + (−33.1) −38.88
81. 19.93 + (−20.21) + 1.37 1.09
82. −43 − (0.032 − 0.045) −42.987
83. Find the sum of *two and forty-three hundredths* and *five and six tenths*. 8.03
84. Find the difference of *nineteen hundredths* and *six thousandths*. 0.184
85. |−14.1 + 6.9| + 8 15.2
86. 15 − |−2.3 + (−2.4)| 10.3
87. 5 − 0.023 4.977
88. 30 − 11.98 18.02
89. −2.002 − (−4.6) 2.598
90. −0.005 − (−8) 7.995

APPLICATIONS

91. **RETAILING** Find the retail price of each appliance listed in the following table if a department store purchases them for the given costs and then marks them up as shown.

Appliance	Cost	Markup	Retail price
Refrigerator	$610.80	$205.00	$815.80
Washing machine	$389.50	$155.50	$545.00
Dryer	$363.99	$167.50	$531.49

92. **PRICING** Find the retail price of a Kenneth Cole two-button suit if a men's clothing outlet buys them for $210.95 each and then marks them up $144.95 to sell in its stores. $355.90

93. **OFFSHORE DRILLING** A company needs to construct a pipeline from an offshore oil well to a refinery located on the coast. Company engineers have come up with two plans for consideration, as shown. Use the information in the illustration to complete the table that is shown in the next column.

	Pipe underwater (mi)	Pipe underground (mi)	Total pipe (mi)
Design 1	1.74	2.32	4.06
Design 2	2.90	0	2.90

94. **DRIVING DIRECTIONS** Find the total distance of the trip using the information in the MapQuest printout shown below. 28.8 mi

1: Start out going EAST on SUNKIST AVE. — 0.0 mi
2: Turn LEFT onto MERCED AVE. — 0.4 mi
3: Turn Right onto PUENTE AVE. — 0.3 mi
4: Merge onto I-10 W toward LOS ANGELES. — 2.2 mi
5: Merge onto I-605 S. — 10.6 mi
6: Merge onto I-5 S toward SANTA ANA. — 14.9 mi
7: Take the HARBOR BLVD exit, EXIT 110A. — 0.3 mi
8: Turn RIGHT onto S HARBOR BLVD. — 0.1 mi
9: End at 1313 S Harbor Blvd Anaheim, CA.

Total Distance: ___?___ miles   MAPQUEST®

95. **PIPE (PVC)** Find the *outside* diameter of the plastic sprinkler pipe shown below if the thickness of the pipe wall is 0.218 inch and the inside diameter is 1.939 inches. 2.375 in.

96. **pH SCALE** The pH scale shown below is used to measure the strength of acids and bases in chemistry. Find the difference in pH readings between
   a. bleach and stomach acid. 10.95
   b. ammonia and coffee. 7.02
   c. blood and coffee. 2.37

97. **RECORD HOLDERS** The late Florence Griffith-Joyner of the United States holds the women's world record in the 100-meter sprint: 10.49 seconds. Britta Steffen of Germany holds the women's world record in the 100-meter freestyle swim: 52.07 seconds. How much faster did Griffith-Joyner run the 100 meters than Steffen swam it? (Source: *The World Almanac and Book of Facts,* 2013) 41.58 sec

98. **WEATHER REPORTS** Barometric pressure readings are recorded on the weather map below. In a low-pressure area (L on the map), the weather is often stormy. The weather is usually fair in a high-pressure area (H). What is the difference in readings between the areas of highest and lowest pressure? 1.8

99. **BANKING** A businesswoman deposited several checks in her company's bank account, as shown on the deposit slip below. Find the *Subtotal* line on the slip by adding the amounts of the checks and total from the reverse side. If the woman wanted to get $25 in cash back from the teller, what should she write as the *Total deposit* on the slip? $523.19, $498.19

Deposit slip

Cash		
Checks (properly endorsed)	116	10
	47	93
Total from reverse side	359	16
Subtotal		
Less cash	25	00
Total deposit		

100. **SPORTS PAGES** Decimals are often used in the sports pages of newspapers. Two examples are given below.

   a. "German bobsledders set a world record today with a final run of 53.03 seconds, finishing ahead of the Italian team by only fourteen thousandths of a second." What was the time for the Italian bobsled team? 53.044 sec

   b. "The women's figure skating title was decided by only thirty-three hundredths of a point." If the winner's point total was 102.71, what was the second-place finisher's total? (*Hint:* The highest score wins in a figure skating contest.) 102.38

101. Suppose certain portions of a patient's morning (A.M.) temperature chart were not filled in. Use the given information to complete the chart below. (*Hint:* 98.6°F is considered normal.)

    from Campus to Careers
    Home Health Aide

Day of week	Patient's A.M. temperature	Amount above normal
Monday	99.7°	1.1°
Tuesday	101.1°	2.5°
Wednesday	98.6°	0°
Thursday	100.0°	1.4°
Friday	99.5°	0.9°

102. **QUALITY CONTROL** An electronics company has strict specifications for the silicon chips it uses in its computers. The company only installs chips that are within 0.05 centimeter of the indicated thickness. The table below gives the specifications for two types of chips. Fill in the blanks to complete the chart.

Chip type	Thickness specification	Acceptable range	
		Low	High
A	0.78 cm	0.73 cm	0.83 cm
B	0.643 cm	0.593 cm	0.693 cm

103. **FLIGHT PATHS** Find the added distance a plane must travel to avoid flying through the storm. 20.01 mi

104. TELEVISION The following illustration shows the six most-watched television series finales of all time.

  a. What was the combined total audience of all six shows? 447.8 million

  b. How many more people watched the last episode of *MASH* than watched the last episode of *Friends*? 53.5 million

  c. How many more people would have had to watch the last *Magnum, P.I.* to move it into a tie for fifth place? 1.8 million

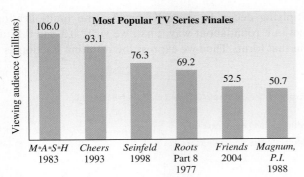

Source: businessinsider.com

105. THE HOME SHOPPING NETWORK The illustration shows a description of a cookware set that was sold on television.

  a. Find the difference between the manufacturer's suggested retail price (MSRP) and the sale price. $101.94

  b. Including shipping and handling (S & H), how much will the cookware set cost? $55.80

Item 229-442	
Continental 9-piece Cookware Set	
Stainless steel	
MSRP	$149.79
HSN Price	$59.85
On Sale	$47.85
S & H	$7.95

106. VEHICLE SPECIFICATIONS Certain dimensions of a compact car are shown. Find the wheelbase of the car. 103.4 in.

WRITING

107. Explain why we line up the decimal points and corresponding place-value columns when adding decimals.

108. Explain why we can write additional zeros to the right of a decimal such as 7.89 without affecting its value.

109. Explain what is wrong with the work shown below.

$$\begin{array}{r} 203.56 \\ 37 \\ +0.43 \\ \hline 204.36 \end{array}$$

110. Consider the following addition:

$$\begin{array}{r} \overset{2}{23.7} \\ 41.9 \\ + 12.8 \\ \hline 78.4 \end{array}$$

Explain the meaning of the small red 2 written above the ones column.

111. Write a set of instructions that explains the two-column borrowing process shown below.

$$\begin{array}{r} 2.6\,\overset{4\,\overset{9}{\cancel{10}}\,\cancel{10}}{\cancel{5}\cancel{0}\cancel{0}} \\ -1.3246 \\ \hline 1.3254 \end{array}$$

112. Explain why it is easier to add the decimals 0.3 and 0.17 than the fractions $\frac{3}{10}$ and $\frac{17}{100}$.

REVIEW

Perform the indicated operations.

113. a. $\frac{4}{5} + \frac{5}{12}$ $\frac{73}{60} = 1\frac{13}{60}$

  b. $\frac{4}{5} - \frac{5}{12}$ $\frac{23}{60}$

  c. $\frac{4}{5} \cdot \frac{5}{12}$ $\frac{1}{3}$

  d. $\frac{4}{5} \div \frac{5}{12}$ $\frac{48}{25} = 1\frac{23}{25}$

114. a. $\frac{3}{8} + \frac{1}{6}$ $\frac{13}{24}$

  b. $\frac{3}{8} - \frac{1}{6}$ $\frac{5}{24}$

  c. $\frac{3}{8} \cdot \frac{1}{6}$ $\frac{1}{16}$

  d. $\frac{3}{8} \div \frac{1}{6}$ $\frac{9}{4} = 2\frac{1}{4}$

Objectives

1. Multiply decimals.
2. Multiply decimals by powers of 10.
3. Multiply signed decimals.
4. Evaluate exponential expressions that have decimal bases.
5. Use the order of operations rule.
6. Evaluate formulas.
7. Estimate products of decimals.
8. Solve application problems by multiplying decimals.

SECTION 5.3
Multiplying Decimals

Since decimal numbers are *base-ten* numbers, multiplication of decimals is similar to multiplication of whole numbers. However, when multiplying decimals, there is one additional step—we must determine where to write the decimal point in the product.

1 Multiply decimals.

To develop a rule for multiplying decimals, we will consider the multiplication $0.3 \cdot 0.17$ and find the product in a roundabout way. First, we write 0.3 and 0.17 as fractions and multiply them in that form. Then we express the resulting fraction as a decimal.

$$0.3 \cdot 0.17 = \frac{3}{10} \cdot \frac{17}{100} \quad \text{Express the decimals 0.3 and 0.17 as fractions.}$$

$$= \frac{3 \cdot 17}{10 \cdot 100} \quad \begin{array}{l}\text{Multiply the numerators.} \\ \text{Multiply the denominators.}\end{array}$$

$$= \frac{51}{1,000}$$

$$= 0.051 \quad \text{Write the resulting fraction } \tfrac{51}{1,000} \text{ as a decimal.}$$

From this example, we can make observations about multiplying decimals.

- The digits in the answer are found by multiplying 3 and 17.

    $0.3 \cdot 0.17 = 0.051$
    $3 \cdot 17 = 51$

- The answer has 3 decimal places. The *sum* of the number of decimal places in the factors 0.3 and 0.17 is also 3.

    $0.3 \cdot 0.17 = 0.051$
    1 decimal place   2 decimal places   3 decimal places

These observations illustrate the following rule for multiplying decimals.

Multiplying Decimals

To multiply two decimals:

1. Multiply the decimals as if they were whole numbers.
2. Find the total number of decimal places in both factors.
3. Insert a decimal point in the result from step 1 so that the answer has the same number of decimal places as the total found in step 2.

Self Check 1

Multiply: $2.7 \cdot 4.3$ 11.61

Now Try Problem 9

EXAMPLE 1   Multiply: $5.9 \cdot 3.4$

Strategy We will ignore the decimal points and multiply 5.9 and 3.4 as if they were whole numbers. Then we will write a decimal point in that result so that the final answer has two decimal places.

WHY Since the factor 5.9 has 1 decimal place, and the factor 3.4 has 1 decimal place, the product should have 1 + 1 = 2 decimal places.

Solution We write the multiplication in vertical form and proceed as follows:

Vertical form
$$\begin{array}{r} 5.9 \leftarrow \text{1 decimal place} \\ \times \ 3.4 \leftarrow \text{1 decimal place} \\ \hline 236 \\ 1770 \\ \hline 20.06 \end{array}$$

The answer will have 1 + 1 = 2 decimal places.

Move 2 places from the right to the left and insert a decimal point in the answer.

Thus, $5.9 \cdot 3.4 = 20.06$.

The Language of Algebra Recall the vocabulary of multiplication.

$$\begin{array}{r} 5.9 \leftarrow \text{Factor} \\ \times \ 3.4 \leftarrow \text{Factor} \\ \hline 236 \\ 1770 \\ \hline 20.06 \leftarrow \text{Product} \end{array}$$

Partial products

Success Tip When multiplying decimals, we do not need to line up the decimal points, as the next example illustrates.

EXAMPLE 2 Multiply: $1.3(0.005)$

Strategy We will ignore the decimal points and multiply 1.3 and 0.005 as if they were whole numbers. Then we will write a decimal point in that result so that the final answer has four decimal places.

WHY Since the factor 1.3 has 1 decimal place, and the factor 0.005 has 3 decimal places, the product should have 1 + 3 = 4 decimal places.

Solution Since many students find vertical form multiplication of decimals easier if the decimal with the smaller number of nonzero digits is written on the bottom, we will write 0.005 under 1.3.

$$\begin{array}{r} 1.3 \leftarrow \text{1 decimal place} \\ \times \ 0.005 \leftarrow \text{3 decimal places} \\ \hline 0.0065 \end{array}$$

The answer will have 1 + 3 = 4 decimal places.

Write 2 placeholder zeros in front of 6. Then move 4 places from the right to the left and insert a decimal point in the answer.

Thus, $1.3(0.005) = 0.0065$.

EXAMPLE 3 Multiply: $234(5.1)$

Strategy We will ignore the decimal point and multiply 234 and 5.1 as if they were whole numbers. Then we will write a decimal point in that result so that the final answer has one decimal place.

WHY Since the factor 234 has 0 decimal places, and the factor 5.1 has 1 decimal place, the product should have 0 + 1 = 1 decimal place.

Teaching Example 1
Multiply: $6.4 \cdot 3.7$
Answer:
23.68

Self Check 2
Multiply: $(0.0002)7.2$ 0.00144
Now Try Problem 13

Teaching Example 2
Multiply: $0.004(2.3)$
Answer:
0.0092

Self Check 3
Multiply: $178(4.7)$ 836.6
Now Try Problem 17

Teaching Example 3
Multiply: $286(3.9)$
Answer:
1,115.4

Solution We write the multiplication in vertical form, with 5.1 under 234.

$$
\begin{array}{r}
234 \leftarrow \text{No decimal places} \\
\times 5.1 \leftarrow \text{1 decimal place} \\
\hline
23\,4 \\
1170\,0 \\
\hline
1193.4
\end{array}
$$

The answer will have $0 + 1 = 1$ decimal place.

Move 1 place from the right to the left and insert a decimal point in the answer.

Thus, $234(5.1) = 1{,}193.4$.

Using Your CALCULATOR Multiplying Decimals

When billing a household, a gas company converts the amount of natural gas used to units of heat energy called *therms*. The number of therms used by a household in one month and the cost per therm are shown below.

Customer charge . . . . 39 therms @ $0.72264

To find the total charges for the month, we multiply the number of therms by the cost per therm: $39 \cdot 0.72264$.

39 ⨯ .72264 = 28.18296   ` 28.18296 `

On some calculator models, the ENTER key is pressed to display the product. Rounding to the nearest cent, we see that the total charge is $28.18.

THINK IT THROUGH Overtime

"Employees covered by the Fair Labor Standards Act must receive overtime pay for hours worked in excess of 40 in a workweek of at least 1.5 times their regular rates of pay."

United States Department of Labor

The map of the United States shown below is divided into nine regions. The average hourly wage for private industry workers in each region is also listed in the legend below the map. Find the average hourly wage for the region where you live. Then calculate the corresponding average hourly overtime wage for that region.

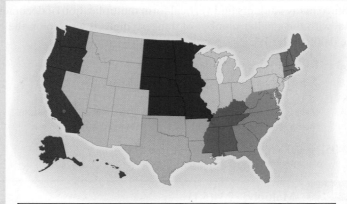

Legend
- West North Central: $20.32
- Mountain: $19.99
- Pacific: $21.86
- East South Central: $19.25
- East North Central: $17.31
- West South Central: $19.06
- New England: $23.56
- Middle Atlantic: $22.68
- South Atlantic: $19.10

Source: Bureau of Labor Statistics

2 Multiply decimals by powers of 10.

The numbers 10, 100, and 1,000 are called **powers of 10**, because they are the results when we evaluate 10^1, 10^2, and 10^3. To develop a rule to find the product when multiplying a decimal by a power of 10, we multiply 8.675 by three different powers of 10.

Multiply: 8.675 · **10**

```
    8.675
  ×    10
    0000
   86750
   86.750
```

Multiply: 8.675 · **100**

```
    8.675
  ×   100
    0000
   00000
  867500
  867.500
```

Multiply: 8.675 · **1,000**

```
     8.675
  ×   1000
     0000
    00000
   000000
  8675000
  8675.000
```

When we inspect the answers, the decimal point in the first factor 8.675 appears to be moved to the right by the multiplication process. The number of decimal places it moves depends on the power of 10 by which 8.675 is multiplied.

One zero in 10
$8.675 \cdot 10 = 86.75$
It moves 1 place to the right.

Two zeros in 100
$8.675 \cdot 100 = 867.5$
It moves 2 places to the right.

Three zeros in 1,000
$8.675 \cdot 1,000 = 8675$
It moves 3 places to the right.

These observations illustrate the following rule.

Multiplying a Decimal by 10, 100, 1,000, and So On

To find the product of a decimal and 10, 100, 1,000, and so on, move the decimal point to the right the same number of places as there are zeros in the power of 10.

EXAMPLE 4 Multiply: **a.** 2.81 · 10 **b.** 0.076(10,000)

Strategy For each multiplication, we will identify the factor that is a power of 10, and count the number of zeros that it has.

WHY To find the product of a decimal and a power of 10 that is greater than 1, we move the decimal point to the right the same number of places as there are zeros in the power of 10.

Solution
a. $2.81 \cdot 10 = 28.1$ Since 10 has one zero, move the decimal point in 2.81 one place to the right.

b. $0.076(10,000) = 00760.$ Since 10,000 has four zeros, move the decimal point in 0.076 four places to the right. Write a placeholder zero (shown in blue).
$= 760$

Numbers such as 10, 100, and 1,000 are powers of 10 that are *greater than 1*. There are also powers of 10 that are *less than 1*, such as 0.1, 0.01, and 0.001. To develop a rule to find the product when multiplying a decimal by one tenth, one hundredth, one thousandth, and so on, we will consider three examples:

Multiply: 5.19 · **0.1**

```
    5.19
  ×  0.1
   0.519
```

Multiply: 5.19 · **0.01**

```
    5.19
  × 0.01
  0.0519
```

Multiply: 5.19 · **0.001**

```
     5.19
  × 0.001
  0.00519
```

Self Check 4

Multiply:
a. 0.721 · 100   72.1
b. 6.08(1,000)   6,080

Now Try Problems **21** and **23**

Teaching Example 4
Multiply:
a. 3.725 · 100
b. 46.3(100,000)
Answers:
a. 372.5
b. 4,630,000

When we inspect the answers, the decimal point in the first factor 5.19 appears to be moved to the left by the multiplication process. The number of places that it moves depends on the power of 10 by which it is multiplied.

These observations illustrate the following rule.

> **Multiplying a Decimal by 0.1, 0.01, 0.001, and So On**
>
> To find the product of a decimal and 0.1, 0.01, 0.001, and so on, move the decimal point to the left the same number of decimal places as there are in the power of 10.

Self Check 5

Multiply:
a. 0.1(129.9) 12.99
b. 0.002 · 0.00001 0.00000002

Now Try Problems 25 and 27

Teaching Example 5 Multiply:
a. 542.6 · 0.0001
b. 0.83(0.01)
Answers:
a. 0.05426
b. 0.0083

EXAMPLE 5 Multiply: a. 145.8 · 0.01   b. 9.76(0.0001)

Strategy For each multiplication, we will identify the factor of the form 0.1, 0.01, and 0.001, and count the number of decimal places that it has.

WHY To find the product of a decimal and a power of 10 that is less than 1, we move the decimal point to the left the same number of decimal places as there are in the power of 10.

Solution

a. 145.8 · 0.01 = 1.458   Since 0.01 has *two* decimal places, move the decimal point in 145.8 two places to the left.

b. 9.76(0.0001) = 0.000976   Since 0.0001 has *four* decimal places, move the decimal point in 9.76 four places to the left. This requires that three placeholder zeros (shown in blue) be inserted in front of the 9.

Quite often, newspapers, websites, and television programs present large numbers in a shorthand notation that involves a decimal in combination with a place-value column name. For example,

- In 2011, Microsoft sold *14.6 million* Xbox 360 units worldwide. (Source: geekwire.com)
- Boston's Big Dig was the most expensive single highway project in U.S. history. It cost about *$14.63 billion.* (Source: Roadtraffic-technology.com)
- The distance that light travels in one year is about *5.878 trillion* miles. (Source: Encyclopaedia Britannica)

We can use the rule for multiplying a decimal by a power of 10 to write these large numbers in standard form.

Self Check 6

Write each number in standard notation:
a. 567.1 million 567,100,000
b. 50.82 billion 50,820,000,000
c. 4.133 trillion 4,133,000,000,000

Now Try Problems 29, 31, and 33

EXAMPLE 6 Write each number in standard notation:
a. 14.6 million   b. 14.63 billion   c. 5.878 trillion

Strategy We will express each of the large numbers as the product of a decimal and a power of 10.

WHY Then we can use the rule for multiplying a decimal by a power of 10 to find their product. The result will be in the required standard form.

Solution

a. 14.6 million = 14.6 · **1 million**

    = 14.6 · **1,000,000**   Write 1 million in standard form.

    = 14,600,000   Since 1,000,000 has six zeros, move the decimal point in 14.6 six places to the right.

b. 14.63 billion = 14.63 · **1 billion**

= 14.63 · 1,000,000,000   Write 1 billion in standard form.

= 14,630,000,000   Since 1,000,000,000 has *nine* zeros, move the decimal point in 14.63 *nine* places to the right.

c. 5.878 trillion = 5.878 · **1 trillion**

= 5.878 · 1,000,000,000,000   Write 1 trillion in standard form.

= 5,878,000,000,000   Since 1,000,000,000,000 has *twelve* zeros, move the decimal point in 5.878 *twelve* places to the right.

3 Multiply signed decimals.

The rules for multiplying integers also hold for multiplying signed decimals. The product of two decimals with like signs is positive, and the product of two decimals with unlike signs is negative.

EXAMPLE 7   Multiply: **a.** $-1.8(4.5)$ **b.** $(-1,000)(-59.08)$

Strategy In part a, we will use the rule for multiplying signed decimals that have different (unlike) signs. In part b, we will use the rule for multiplying signed decimals that have the same (like) signs.

WHY In part a, one factor is negative and one is positive. In part b, both factors are negative.

Solution

a. Find the absolute values: $|-1.8| = 1.8$ and $|4.5| = 4.5$. Since the decimals have unlike signs, their product is negative.

$-1.8(4.5) = -8.1$ Multiply the absolute values, 1.8 and 4.5, to get 8.1. Then make the final answer negative.

$$\begin{array}{r} 1.8 \\ \times\ 4.5 \\ \hline 90 \\ 720 \\ \hline 8.10 \end{array}$$

b. Find the absolute values: $|-1,000| = 1,000$ and $|-59.08| = 59.08$. Since the decimals have like signs, their product is positive.

$(-1,000)(-59.08) = 1,000(59.08)$

$= 59080.$ Multiply the absolute values, 1,000 and 59.08. Since 1,000 has 3 zeros, move the decimal point in 59.08 3 places to the right. Write a placeholder zero. The answer is positive.

4 Evaluate exponential expressions that have decimal bases.

We have evaluated exponential expressions that have whole number bases, integer bases, and fractional bases. The base of an exponential expression can also be a positive or a negative decimal.

EXAMPLE 8   Evaluate: **a.** $(2.4)^2$ **b.** $(-0.05)^2$

Strategy We will write each exponential expression as a product of repeated factors, and then perform the multiplication. This requires that we identify the base and the exponent.

WHY The exponent tells the number of times the base is to be written as a factor.

Teaching Example 6 Write each number in standard notation:
a. In 2011, the U.S had *2.2 million* farms. (Source: *The World Almanac and Book of Facts*, 2013)
b. Americans took *10.4 billion* trips on public transportation in 2011. (Source: American Public Transit Association)
c. It would take about *1.818 trillion* pennies to fill the Empire State Building. (Source: The Mega Penny Project)
Answers:
a. 2,200,000
b. 10,400,000,000
c. 1,818,000,000,000

Self Check 7
Multiply:
a. $6.6(-5.5)$ -36.3
b. $-44.968(-100)$ $4,496.8$
Now Try Problems 37 and 41

Teaching Example 7 Multiply:
a. $-3.6(2.8)$
b. $(-10,000)(-63.11)$
Answers:
a. -10.08
b. 631,100

Self Check 8
Evaluate:
a. $(-1.3)^2$ 1.69
b. $(0.09)^2$ 0.0081
Now Try Problems 45 and 47

Teaching Example 8 Evaluate:
a. $(4.2)^2$
b. $(-0.07)^2$
Answers:
a. 17.64
b. 0.0049

Solution

a. $(2.4)^2 = 2.4 \cdot 2.4$ The base is 2.4 and the exponent is 2. Write the base as a factor 2 times.

$ = 5.76$ Multiply the decimals.

$$\begin{array}{r} 2.4 \\ \times\ 2.4 \\ \hline 96 \\ 480 \\ \hline 5.76 \end{array}$$

b. $(-0.05)^2 = (-0.05)(-0.05)$ The base is -0.05 and the exponent is 2. Write the base as a factor 2 times.

$ = 0.0025$ Multiply the decimals. The product of two decimals with like signs is positive.

$$\begin{array}{r} 0.05 \\ \times\ 0.05 \\ \hline 0.0025 \end{array}$$

5 Use the order of operations rule.

Recall that the order of operations rule is used to evaluate expressions that involve more than one operation.

Self Check 9

Evaluate:
$-2|-4.4 + 5.6| + (-0.8)^2$ -1.76

Now Try Problem 49

Teaching Example 9
Evaluate: $(-0.9)^2 + 4|-7.1 + 6.8|$
Answer:
2.01

EXAMPLE 9 Evaluate: $-(0.6)^2 + 5|-3.6 + 1.9|$

Strategy The absolute value bars are grouping symbols. We will perform the addition within them first.

WHY By the order of operations rule, we must perform all calculations within parentheses and other grouping symbols (such as absolute value bars) first.

Solution

$-(0.6)^2 + 5|-3.6 + 1.9|$

$= -(0.6)^2 + 5|-1.7|$ Do the addition within the absolute value symbols. Use the rule for adding two decimals with different signs.

$= -(0.6)^2 + 5(1.7)$ Simplify: $|-1.7| = 1.7$.

$= -0.36 + 5(1.7)$ Evaluate: $(0.6)^2 = 0.36$.

$= -0.36 + 8.5$ Do the multiplication: $5(1.7) = 8.5$.

$= 8.14$ Use the rule for adding two decimals with different signs.

$$\begin{array}{r} {}^{2\ 16}\\ \cancel{3}.\cancel{6} \\ -\ 1.9 \\ \hline 1.7 \end{array}$$

$$\begin{array}{r} {}^{3}\\ 1.7 \\ \times\ 5 \\ \hline 8.5 \end{array}$$

$$\begin{array}{r} {}^{4\ 10}\\ 8.\cancel{5}\cancel{0} \\ -0.36 \\ \hline 8.14 \end{array}$$

6 Evaluate formulas.

Recall that to evaluate a formula, we replace the letters (called **variables**) with specific numbers and then use the order of operations rule.

Self Check 10

Evaluate $V = 1.3\pi r^3$ for $\pi \approx 3.14$ and $r = 3$. 110.214

Now Try Problem 53

Teaching Example 10
Evaluate: $A = P + Prt$ for $P = 75.50$, $r = 0.12$, and $t = 2$.
Answer:
93.62

EXAMPLE 10 Evaluate the formula $S = 6.28r(h + r)$ for $h = 3.1$ and $r = 6$.

Strategy In the given formula, we will replace the letter r with 6 and h with 3.1.

WHY Then we can use the order of operations rule to find the value of the expression on the right side of the $=$ symbol.

Solution

$S = 6.28r(h + r)$ $6.28r(h + r)$ means $6.28 \cdot r \cdot (h + r)$.

$S = 6.28(6)(3.1 + 6)$ Replace r with 6 and h with 3.1.

$S = 6.28(6)(9.1)$ Do the addition within the parentheses.

$S = 37.68(9.1)$ Do the multiplication: $6.28(6) = 37.68$.

$S = 342.888$ Do the multiplication.

$$\begin{array}{r} 37.68 \\ \times\ 9.1 \\ \hline 3768 \\ 339120 \\ \hline 342.888 \end{array}$$

7 Estimate products of decimals.

Estimation can be used to check the reasonableness of an answer to a decimal multiplication. There are several ways to estimate, but the objective is the same: Simplify the numbers in the problem so that the calculations can be made easily and quickly.

EXAMPLE 11

a. Estimate using front-end rounding: 27 · 6.41
b. Estimate by rounding each factor to the nearest tenth: 13.91 · 5.27
c. Estimate by rounding: 0.1245(101.4)

Strategy We will use rounding to approximate the factors. Then we will find the product of the approximations.

WHY Rounding produces factors that contain fewer digits. Such numbers are easier to multiply.

Solution

a. To estimate 27 · 6.41 by front-end rounding, we begin by rounding both factors to their *largest* place value.

$$
\begin{array}{rcl}
27 & \longrightarrow & 30 \quad \text{Round to the nearest ten.} \\
\times\ 6.41 & \longrightarrow & \times\ 6 \quad \text{Round to the nearest one.} \\
\hline
& & 180
\end{array}
$$

The estimate is 180. If we calculate 27 · 6.41, the product is exactly 173.07. The estimate is close: It's about 7 more than 173.07.

b. To estimate 13.91 · 5.27, we will round both decimals to the nearest tenth.

$$
\begin{array}{rcl}
13.91 & \longrightarrow & 13.9 \quad \text{Round to the nearest tenth.} \\
\times\ 5.27 & \longrightarrow & \times\ 5.3 \quad \text{Round to the nearest tenth.} \\
\hline
& & 417 \\
& & 6950 \\
\hline
& & 73.67
\end{array}
$$

The estimate is 73.67. If we calculate 13.91 · 5.27, the product is exactly 73.3057. The estimate is close: It's just slightly more than 73.3057.

c. Since 101.4 is approximately 100, we can estimate 0.1245(**101.4**) using 0.1245(**100**).

0.1245(100) = 12.45   Since 100 has two zeros, move the decimal point in 0.1245 two places to the right.

The estimate is 12.45. If we calculate 0.1245(101.4), the product is exactly 12.6243. Note that the estimate is close: It's slightly less than 12.6243.

Self Check 11

a. Estimate using front-end rounding: 4.337 · 65   280
b. Estimate by rounding the factors to the nearest tenth: 3.092 · 11.642   35.96
c. Estimate by rounding: 0.7899(985.34)   789.9

Now Try Problems 61 and 63

Teaching Example 11
a. Estimate using front-end rounding: 2.371 · 76
b. Estimate by rounding the factors to the nearest tenth: 9.1682 · 7.344
c. Estimate by rounding: 0.5301(97.89)

Answers:
a. 160
b. 67.16
c. 53.01

8 Solve application problems by multiplying decimals.

Application problems that involve repeated addition are often more easily solved using multiplication.

EXAMPLE 12   *Coins* Banks wrap pennies in rolls of 50 coins. If a penny is 1.55 millimeters thick, how tall is a stack of 50 pennies?

Solution

- There are 50 pennies in a stack.   Given
- A penny is 1.55 millimeters thick.   Given
- How tall is a stack of 50 pennies?   Find

Self Check 12

COINS Banks wrap nickels in rolls of 40 coins. If a nickel is 1.95 millimeters thick, how tall is a stack of 40 nickels? 78 mm

Now Try Problem 97

Teaching Example 12 COINS Banks wrap dimes in rolls of 50 coins. If a dime is 1.35 millimeters thick, how tall is a stack of 50 dimes?
Answer:
67.5 mm

The height (in millimeters) of a stack of 50 pennies, each of which is 1.55 thick, is the sum of fifty 1.55's. This repeated addition can be calculated more simply by multiplication.

The height of a stack of pennies	is equal to	the thickness of one penny	times	the number of pennies in the stack.
The height of stack of pennies	=	1.55	·	50

Use vertical form to perform the multiplication:

```
    1.55
  ×   50
    000
   7750
   77.50
```

A stack of 50 pennies is 77.5 millimeters tall.

Check: We can estimate to check the result. If we use 2 millimeters to approximate the thickness of one penny, then the height of a stack of 50 pennies is about 2 · 50 millimeters = 100 millimeters. The result, 77.5 mm, seems reasonable.

Sometimes more than one operation is needed to solve a problem involving decimals.

Self Check 13

WEEKLY EARNINGS A pharmacy assistant's basic workweek is 40 hours. After her daily shift is over, she can work overtime at a rate of 1.5 times her regular rate of $15.90 per hour. How much money will she earn in a week if she works 4 hours of overtime? $731.40

Now Try Problem 113

Teaching Example 13 WEEKLY EARNINGS An ironworker's basic workweek is 40 hours. After his daily shift is over, he can work overtime at a rate of 1.5 times his regular rate of $18.08 per hour. How much money will he earn in a week if he works 2 hours of overtime?
Answer:
$777.44

EXAMPLE 13 *Weekly Earnings* A cashier's basic workweek is 40 hours. After his daily shift is over, he can work overtime at a rate 1.5 times his regular rate of $13.10 per hour. How much money will he earn in a week if he works 6 hours of overtime?

Solution

- A cashier's basic workweek is 40 hours. *Given*
- His overtime pay rate is 1.5 times his regular rate of $13.10 per hour. *Given*
- How much money will he earn in a week if he works his regular shift and 6 hours overtime? *Find*

To find the cashier's overtime pay rate, we multiply 1.5 times his regular pay rate, $13.10.

```
   13.10
 ×  1.5
   6550
  13100
  19.650
```

The cashier's overtime pay rate is $19.65 per hour.
    We now translate the words of the problem to numbers and symbols.

The total amount the cashier earns in a week	is equal to	40 hours	times	his regular pay rate	plus	the number of overtime hours	times	his overtime rate.
The total amount the cashier earns in a week	=	40	·	$13.10	+	6	·	$19.65

We will use the rule for the order of operations to evaluate the expression:

40 · 13.10 + 6 · 19.65 = 524.00 + 117.90   Do the multiplication first.

= 641.90   Do the addition.

```
      13.10      5 3 3
              19.65
    ×   40    ×    6
    0000      117.90
    52400
    524.00
```

```
        1
     524.00
    +117.90
     641.90
```

The cashier will earn a total of $641.90 for the week.

Check: We can use estimation to check. The cashier works 40 hours per week for approximately $13 per hour to earn about 40 · $13 = $520. His 6 hours of overtime at approximately $20 per hour earns him about 6 · $20 = $120. His total earnings that week are about $520 + $120 = $640. The result, $641.90, seems reasonable.

ANSWERS TO SELF CHECKS

1. 11.61 **2.** 0.00144 **3.** 836.6 **4. a.** 72.1 **b.** 6,080 **5. a.** 12.99 **b.** 0.00000002
6. a. 567,100,000 **b.** 50,820,000,000 **c.** 4,133,000,000,000 **7. a.** −36.3 **b.** 4,496.8
8. a. 1.69 **b.** 0.0081 **9.** −1.76 **10.** 110.214 **11. a.** 280 **b.** 35.96 **c.** 789.9
12. 78 mm **13.** $731.40

SECTION 5.3   STUDY SET

VOCABULARY

Fill in the blanks.

1. In the multiplication problem shown below, label each *factor*, the *partial products*, and the *product*.

```
   3.4   ← factor
 × 2.6   ← factor
   204   ← partial product
   680   ← partial product
  8.84   ← product
```

2. Numbers such as 10, 100, and 1,000 are called _powers_ of 10.

CONCEPTS

Fill in the blanks.

3. Insert a decimal point in the correct place for each product shown below. Write placeholder zeros, if necessary.

a. 3.8   2.28
     × 0.6
       228

b. 1.79   14.499
     × 8.1
        179
      14320
      14499

c.   2.0   14.0
      ×   7
        140

d.   0.013   0.00026
      × 0.02
         0026

4. Fill in the blanks.
  a. To find the product of a decimal and 10, 100, 1,000, and so on, move the decimal point to the _right_ the same number of places as there are zeros in the power of 10.
  b. To find the product of a decimal and 0.1, 0.01, 0.001, and so on, move the decimal point to the _left_ the same number of places as there are in the power of 10.

5. Determine whether the *sign* of each result is positive or negative. **You do not have to find the product.**
  a. −7.6(−1.8) positive
  b. −4.09 · 2.274 negative

6. a. When we move its decimal point to the right, does a decimal number get larger or smaller? larger
  b. When we move its decimal point to the left, does a decimal number get larger or smaller? smaller

NOTATION

7. a. List the first five powers of 10 that are greater than 1. 10, 100, 1,000, 10,000, 100,000
  b. List the first five powers of 10 that are less than 1.
  0.1, 0.01, 0.001, 0.0001, 0.00001

8. Write each number in standard notation.
   a. one million 1,000,000
   b. one billion 1,000,000,000
   c. one trillion 1,000,000,000,000

GUIDED PRACTICE

Multiply. See Example 1.

9. $4.8 \cdot 6.2$ 29.76
10. $3.5 \cdot 9.3$ 32.55
11. $5.6(8.9)$ 49.84
12. $7.2(8.4)$ 60.48

Multiply. See Example 2.

13. $0.003(2.7)$ 0.0081
14. $0.002(2.6)$ 0.0052
15. $\quad 5.8$
    $\times 0.009$
    $\overline{0.0522}$
16. $\quad 8.7$
    $\times 0.004$
    $\overline{0.0348}$

Multiply. See Example 3.

17. $179(6.3)$ 1,127.7
18. $225(4.9)$ 1,102.5
19. $\quad 316$
    $\times 7.4$
    $\overline{2,338.4}$
20. $\quad 527$
    $\times 3.7$
    $\overline{1,949.9}$

Multiply. See Example 4.

21. $6.84 \cdot 100$ 684
22. $2.09 \cdot 100$ 209
23. $0.041(10,000)$ 410
24. $0.034(10,000)$ 340

Multiply. See Example 5.

25. $647.59 \cdot 0.01$ 6.4759
26. $317.09 \cdot 0.01$ 3.1709
27. $1.15(0.001)$ 0.00115
28. $2.83(0.001)$ 0.00283

Write each number in standard notation. See Example 6.

29. 14.2 million 14,200,000
30. 33.9 million 33,900,000
31. 98.2 billion 98,200,000,000
32. 80.4 billion 80,400,000,000
33. 1.421 trillion 1,421,000,000,000
34. 3.056 trillion 3,056,000,000,000
35. 657.1 billion 657,100,000,000
36. 422.7 billion 422,700,000,000

Multiply. See Example 7.

37. $-1.9(7.2)$ −13.68
38. $-5.8(3.9)$ −22.62
39. $-3.3(-1.6)$ 5.28
40. $-4.7(-2.2)$ 10.34
41. $(-10,000)(-44.83)$ 448,300
42. $(-10,000)(-13.19)$ 131,900
43. $678.231(-1,000)$ −678,231
44. $491.565(-1,000)$ −491,565

Evaluate each expression. See Example 8.

45. $(3.4)^2$ 11.56
46. $(5.1)^2$ 26.01
47. $(-0.03)^2$ 0.0009
48. $(-0.06)^2$ 0.0036

Evaluate each expression. See Example 9.

49. $-(-0.2)^2 + 4|-2.3 + 1.5|$ 3.16
50. $-(-0.3)^2 + 6|-6.4 + 1.7|$ 28.11
51. $-(-0.8)^2 + 7|-5.1 - 4.8|$ 68.66
52. $-(-0.4)^2 + 6|-6.2 - 3.5|$ 58.04

Evaluate each formula. See Example 10.

53. $A = P + Prt$ for $P = 85.50$, $r = 0.08$, and $t = 5$ 119.70
54. $A = P + Prt$ for $P = 99.95$, $r = 0.05$, and $t = 10$ 149.925
55. $A = lw$ for $l = 5.3$ and $w = 7.2$ 38.16
56. $A = 0.5bh$ for $b = 7.5$ and $h = 6.8$ 25.5
57. $P = 2l + 2w$ for $l = 3.7$ and $w = 3.6$ 14.6
58. $P = a + b + c$ for $a = 12.91$, $b = 19$, and $c = 23.6$ 55.51
59. $C = 2\pi r$ for $\pi \approx 3.14$ and $r = 2.5$ 15.7
60. $A = \pi r^2$ for $\pi \approx 3.14$ and $r = 4.2$ 55.3896

Estimate each product using front-end rounding. See Example 11.

61. $46 \cdot 5.3$ 250
62. $37 \cdot 4.29$ 160

Estimate each product by rounding the factors to the nearest tenth. See Example 11.

63. $17.11 \cdot 3.85$ 66.69
64. $18.33 \cdot 6.46$ 118.95

TRY IT YOURSELF

Perform the indicated operations.

65. $-0.56 \cdot 0.33$ −0.1848
66. $(-3.7)(0.4)$ −1.48
67. $(-0.1)^2$ 0.01
68. $(-0.6)^2$ 0.36
69. $(-0.7 - 0.5)(2.4 - 3.1)$ 0.84
70. $(-8.1 - 7.8)(0.3 + 0.7)$ −15.9
71. $\quad 0.008$
    $\times 0.09$
    $\overline{0.00072}$
72. $\quad 0.003$
    $\times 0.09$
    $\overline{0.00027}$
73. $-0.2 \cdot 1,000,000$ −200,000
74. $-1,000,000 \cdot 1.9$ −1,900,000
75. $(-5.6)(-2.2)$ 12.32
76. $(-7.1)(-4.1)$ 29.11
77. $-4.6(23.4 - 19.6)$ −17.48
78. $6.9(9.8 - 8.9)$ 6.21
79. $(-4.9)(-0.001)$ 0.0049
80. $(-0.001)(-7.09)$ 0.00709
81. $(-0.2)^2 + 2(7.1)$ 14.24
82. $(-6.3)(3) - (1.2)^2$ −20.34
83. $\quad 2.13$
    $\times 4.05$
    $\overline{8.6265}$
84. $\quad 3.06$
    $\times 1.82$
    $\overline{5.5692}$
85. $-7(8.1781)$ −57.2467
86. $-5(4.7199)$ −23.5995
87. $-1,000(0.02239)$ −22.39
88. $-100(0.0897)$ −8.97
89. $(0.5 + 0.6)^2(-3.2)$ −3.872
90. $(-5.1)(4.9 - 3.4)^2$ −11.475
91. $-0.2(306)(-0.4)$ 24.48
92. $(-2.5)(-4)(-9)$ −90
93. $-0.01(|-2.6 - 6.7|)^2$ −0.8649
94. $-0.01(|-8.16 + 9.9|)^2$ −0.030276

Complete each table.

95.

Decimal	Its square
0.1	0.01
0.2	0.04
0.3	0.09
0.4	0.16
0.5	0.25
0.6	0.36
0.7	0.49
0.8	0.64
0.9	0.81

96.

Decimal	Its cube
0.1	0.001
0.2	0.008
0.3	0.027
0.4	0.064
0.5	0.125
0.6	0.216
0.7	0.343
0.8	0.512
0.9	0.729

APPLICATIONS

97. REAMS OF PAPER Find the thickness of a 500-sheet ream of copier paper if each sheet is 0.0038 inch thick. 1.9 in.

98. MILEAGE CLAIMS Each month, a salesman is reimbursed by his company for any work-related travel that he does in his own car at the rate of $0.565 per mile. How much will the salesman receive if he traveled a total of 120 miles in his car on business in the month of June? $67.80

99. SALARIES Use the following formula to determine the annual salary of a recording engineer who works 38 hours per week at a rate of $37.35 per hour. Round the result to the nearest hundred dollars. $74,100

Annual salary = hourly rate · hours per week · 52.2 weeks

100. PAYCHECKS If you are paid every other week, your monthly gross income is your gross income from one paycheck times 2.17. Find the monthly gross income of a supermarket clerk who earns $1,095.70 every two weeks. Round the result to the nearest cent. $2377.67

101. BAKERY SUPPLIES A bakery buys various types of nuts as ingredients for cookies. Complete the table by filling in the cost of each purchase.

Type of nut	Price per pound	Pounds	Cost
Almonds	$5.95	16	$95.20
Walnuts	$4.95	25	$123.75

102. NEW HOMES Find the cost to build the home shown below if construction costs are $92.55 per square foot. $212,032.05

House Plan #DP-2203
Square Feet: 2,291 Sq Ft.
Stories: Single Story
Bedrooms: 3
Bathrooms: 3
Garage Bays: 2

103. BIOLOGY Cells contain DNA. In humans, it determines such traits as eye color, hair color, and height. A model of DNA appears below. If 1 Å (angstrom) = 0.000000004 inch, find the dimensions of 34 Å, 3.4 Å, and 10 Å, shown in the illustration. 0.000000136 in., 0.0000000136 in., 0.00000004 in.

104. TACHOMETERS

a. Estimate the decimal number to which the tachometer needle points in the illustration below. 4.5

b. What engine speed (in rpm) does the tachometer indicate? 4,500 rpm

105. CITY PLANNING The streets shown in blue on the city map below are 0.35 mile apart. Find the distance of each trip between the two given locations.

a. The airport to the Convention Center 2.1 mi

b. City Hall to the Convention Center 3.5 mi

c. The airport to City Hall 5.6 mi

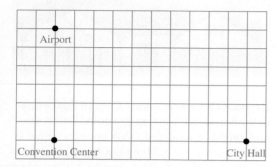

106. RETROFITS The illustration below shows the current widths of the three columns of a freeway overpass. A computer analysis indicated that the width of each column should actually be 1.4 times what it currently is to withstand the stresses of an earthquake. According to the analysis, how wide should each of the columns be? 6.3 ft, 4.9 ft, 3.5 ft

107. ELECTRIC BILLS When billing a household, a utility company charges for the number of kilowatt-hours used. A kilowatt-hour (kwh) is a standard measure of electricity. If the cost of 1 kwh is $0.14277, what is the electric bill for a household that uses 719 kwh in a month? Round the answer to the nearest cent. $102.65

108. UTILITY TAXES Some gas companies are required to tax the number of therms used each month by the customer. What are the taxes collected on a monthly usage of 31 therms if the tax rate is $0.00566 per therm? Round the answer to the nearest cent. 18¢

109. Write each highlighted number in standard form.

a. **CONSERVATION** The *19.6-million acre* Arctic National Wildlife Refuge is located in the northeast corner of Alaska. (Source: National Wildlife Federation) 19,600,000 acres

b. **POPULATION** According to the website census.gov, at 11:55 P.M. eastern time on Sunday, January 20, 2013, the population of the Earth was *7.061 billion* people. 7,061,000,000

c. **DRIVING** The U.S. Federal Highway Administration estimated that Americans drove a total of *2.9629 trillion miles* in 2011. 2,962,900,000,000 miles

110. Write each highlighted number in standard form.

a. **MILEAGE** As of July 2012, Irv Gordon of Long Island, New York, had driven a record *2.967 million miles* in his 1966 Volvo P-1800. (Source: newyork.cbslocal) 2,967,000 miles

b. **E-COMMERCE** Online spending during the 2012 holiday season (November 1 through December 23) was about *$37.2 billion*. (Source: comscore.com) $37,200,000,000

c. **FEDERAL DEBT** On January 30, 2012, the U.S. national debt limit was raised to *$16.394 trillion* by Congress. $16,394,000,000,000

111. SOCCER A soccer goal is rectangular and measures 24 feet wide by 8 feet high. Major league soccer officials are proposing to increase its width by 1.5 feet and increase its height by 0.75 foot.

a. What is the area of the goal opening now? 192 ft²

b. What would the area be if the proposal were adopted? 223.125 ft²

c. How much area would be added? 31.125 ft²

112. SALT INTAKE Studies done by the Centers for Disease Control and Prevention found that the average American eats 3.436 grams of salt each day. The recommended amount is 1.5 grams per day. How many more grams of salt does the average American eat in one week compared with what the Center recommends? 13.552 g

113. CONCERT SEATING Two types of tickets were sold for a concert. Floor seating costs $12.50 a ticket, and balcony seats cost $15.75.

a. Complete the following table and find the receipts from each type of ticket.

b. Find the total receipts from the sale of both types of tickets. $14,075

Ticket type	Price	Number sold	Receipts
Floor	$12.50	1,000	$12,500
Balcony	$15.75	100	$1,575

114. PLUMBING BILLS A corner of the invoice for plumbing work is torn. What is the labor charge for the 4 hours of work? What is the total charge (standard service charge, parts, labor)? $162.20, $226.70

Carter Plumbing 100 W. Dalton Ave.		Invoice #210
Standard service charge	$	25.75
Parts	$	38.75
Labor: 4 hr @ $40.55/hr	$	
Total charges	$	

115. WEIGHTLIFTING The barbell is evenly loaded with iron plates. How much plate weight is loaded on the barbell? 136.4 lb

45.5 lb
20.5 lb
2.2 lb

116. SWIMMING POOLS Long bricks, called *coping*, can be used to outline the edge of a swimming pool. How many meters of coping will be needed in the construction of the swimming pool shown? 160.6 m

50 m
30.3 m

117. STORM DAMAGE After a rainstorm, the saturated ground under a hilltop house began to give way. A survey team noted that the house dropped 0.57 inch initially. In the next three weeks, the house fell 0.09 inch per week. How far did the house fall during this three-week period? 0.84 in.

118. WATER USAGE In May, the water level of a reservoir reached its high mark for the year. During the summer months, as water usage increased, the level dropped. In the months of May and June, it fell 4.3 feet each month. In August and September, because of high temperatures, it fell another 8.7 feet each month. By the beginning of October, how far below the year's high mark had the water level fallen? 26 ft

WRITING

119. Explain how to determine where to place the decimal point in the answer when multiplying two decimals.

120. List the similarities and differences between whole-number multiplication and decimal multiplication.

121. Explain how to multiply a decimal by a power of 10 that is greater than 1, and by a power of 10 that is less than 1.

122. Is it easier to multiply the decimals 0.4 and 0.16 or the fractions $\frac{4}{10}$ and $\frac{16}{100}$? Explain why.

123. Why do we have to line up the decimal points when adding but not when multiplying?

124. Which vertical form for the following multiplication do you like better? Explain why.

$$\begin{array}{r} 0.000003 \\ \times \quad 2.7 \\ \hline \end{array} \qquad \begin{array}{r} 2.8 \\ \times \ 0.000003 \\ \hline \end{array}$$

REVIEW

Find the prime factorization of each number. Use exponents in your answer, when helpful.

125. 220 $2^2 \cdot 5 \cdot 11$

126. 400 $2^4 \cdot 5^2$

127. 162 $2 \cdot 3^4$

128. 735 $3 \cdot 5 \cdot 7^2$

Objectives

1. Divide a decimal by a whole number.
2. Divide a decimal by a decimal.
3. Round a decimal quotient.
4. Estimate quotients of decimals.
5. Divide decimals by powers of 10.
6. Divide signed decimals.
7. Use the order of operations rule.
8. Evaluate formulas.
9. Solve application problems by dividing decimals.

SECTION 5.4
Dividing Decimals

In Chapter 1, we used a process called long division to divide whole numbers.

Long division form

$$\text{Divisor} \rightarrow 5\overline{)10} \leftarrow \text{Dividend}$$
$$\phantom{\text{Divisor} \rightarrow 5)}2 \leftarrow \text{Quotient}$$
$$\phantom{\text{Divisor} \rightarrow 5)}\underline{10}$$
$$\phantom{\text{Divisor} \rightarrow 5)}0 \leftarrow \text{Remainder}$$

In this section, we consider division problems in which the divisor, the dividend, or both are decimals.

1 Divide a decimal by a whole number.

To develop a rule for decimal division, let's consider the problem $47 \div 10$. If we rewrite the division as $\frac{47}{10}$, we can use the long division method from Chapter 4 for changing an improper fraction to a mixed number to find the answer:

$$10\overline{)47} = 4\frac{7}{10}$$
$$\underline{-40}$$
$$7$$

Here the result is written in quotient $+ \dfrac{\text{remainder}}{\text{divisor}}$ form.

To perform this same division using decimals, we write 47 as 47.0 and divide as we would divide whole numbers.

$$10\overline{)47.0}$$
$$4.7$$
$$\underline{-40}\downarrow$$
$$70$$
$$\underline{-70}$$
$$0$$

Note that the decimal point in the quotient (answer) is placed directly above the decimal point in the dividend.

After subtracting 40 from 47, bring down the 0 and continue to divide.

The remainder is 0.

Since $4\frac{7}{10} = 4.7$, either method gives the same answer. This result suggests the following method for dividing a decimal by a whole number.

Dividing a Decimal by a Whole Number

To divide a decimal by a whole number:

1. Write the problem in long division form and place a decimal point in the quotient (answer) directly above the decimal point in the dividend.
2. Divide as if working with whole numbers.
3. If necessary, additional zeros can be written to the right of the last digit of the dividend to continue the division.

Self Check 1

Divide: $20.8 \div 4$. Check the result. 5.2

Now Try Problem 15

EXAMPLE 1   Divide: $42.6 \div 6$. Check the result.

Strategy Since the divisor, 6, is a whole number, we will write the problem in long division form and place a decimal point directly above the decimal point in 42.6. Then we will divide as if the problem were $426 \div 6$.

WHY To divide a decimal by a whole number, we divide as if working with whole numbers.

Solution
Step 1

$$6\overline{)42.6}$$

Place a decimal point in the quotient that lines up with the decimal point in the dividend.

Step 2 Now divide using the four-step division process: **estimate**, **multiply**, **subtract**, and **bring down**.

```
      7.1
  6)42.6    Ignore the decimal points and divide as if working with whole numbers.
  −42 ↓
    0 6    After subtracting 42 from 42, bring down the 6 and continue to divide.
    − 6
      0    The remainder is 0.
```

In Section 1.5, we checked whole-number division using multiplication. Decimal division is checked in the same way: *The product of the quotient and the divisor should be the dividend.*

```
    7.1  ← Quotient        7.1
  ×   6  ← Divisor      6)42.6
   42.6  — Dividend
```

The check confirms that 42.6 ÷ 6 = 7.1.

EXAMPLE 2 Divide: 71.68 ÷ 28

Strategy Since the divisor is a whole number, 28, we will write the problem in long division form and place a decimal point directly above the decimal point in 71.68. Then we will divide as if the problem were 7,168 ÷ 28.

WHY To divide a decimal by a whole number, we divide as if working with whole numbers.

Solution

Write the decimal point in the quotient (answer) directly above the decimal point in the dividend.

```
       2.56
  28)71.68    Ignore the decimal points and divide as if working
   − 56 ↓     with whole numbers.
     15 6     After subtracting 56 from 71, bring down the 6
   − 14 0↓    and continue to divide.
      1 68    After subtracting 140 from 156, bring down the 8
    − 1 68    and continue to divide.
         0    The remainder is 0.
```

We can use multiplication to check this result.

```
    2.56
  ×   28                2.56
    2048          28)71.68
    5120
   71.68
```

The check confirms that 71.68 ÷ 28 = 2.56.

Teaching Example 1 Divide: 21.9 ÷ 3. Check the result.
Answer:
7.3

Self Check 2
Divide: 101.44 ÷ 32   3.17
Now Try Problem 19

Teaching Example 2 Divide: 88.32 ÷ 24. Check the result.
Answer:
3.68

Self Check 3

Divide: 42.8 ÷ 8   5.35

Now Try Problem 23

Teaching Example 3
Divide: 37.4 ÷ 4. Check the result.
Answer:
9.35

EXAMPLE 3   Divide: 19.2 ÷ 5

Strategy We will write the problem in long division form, place a decimal point directly above the decimal point in 19.2, and divide. If necessary, we will write additional zeros to the right of the 2 in 19.2.

WHY Writing additional zeros to the right of the 2 allows us to continue the division process until we obtain a remainder of 0 or the digits in the quotient repeat in a pattern.

Solution

```
      3.8
   5)19.2
    −15 ↓
      4 2     After subtracting 15 from 19, bring down the 2 and continue to divide.
    − 4 0
        2     All the digits in the dividend have been used, but the remainder is not 0.
```

We can write a zero to the right of 2 in the dividend and continue the division process. Recall that writing additional zeros to the right of the decimal point does not change the value of the decimal. That is, 19.2 = 19.20.

```
      3.84
   5)19.20    Write a zero to the right of the 2 and bring it down.
    −15
      4 2
    − 4 0↓
        20    Continue to divide.
       −20
         0    The remainder is 0.
```

Check:

```
    3.84
  ×    5
   19.20  ← Since this is the dividend, the result checks.
```

2 Divide a decimal by a decimal.

To develop a rule for division involving a decimal divisor, let's consider the problem $0.36\overline{)0.2592}$, where the divisor is the decimal 0.36. First, we express the division in fraction form.

$0.36\overline{)0.2592}$ can be represented by $\dfrac{0.2592}{0.36}$

⎯⎯⎯⎯⎯ Divisor ⎯⎯⎯⎯⎯

To be able to use the rule for dividing decimals by a *whole number* discussed earlier, we need to move the decimal point in the divisor 0.36 two places to the right. This can be accomplished by multiplying it by 100. However, if the denominator of the fraction is multiplied by 100, the numerator must also be multiplied by 100 so that the fraction maintains the same value. It follows that $\frac{100}{100}$ is the form of 1 that we should use to build $\frac{0.2592}{0.36}$.

$$\dfrac{0.2592}{0.36} = \dfrac{0.2592}{0.36} \cdot \dfrac{100}{100} \quad \text{Multiply by a form of 1.}$$

$$= \dfrac{0.2592 \cdot 100}{0.36 \cdot 100} \quad \begin{array}{l}\text{Multiply the numerators.}\\ \text{Multiply the denominators.}\end{array}$$

$$= \dfrac{25.92}{36} \quad \begin{array}{l}\text{Multiplying both decimals by 100 moves}\\ \text{their decimal points two places to the right.}\end{array}$$

This fraction represents the division problem $36\overline{)25.92}$. From this result, we have the following observations.

- The division problem $0.36\overline{)0.2592}$ is equivalent to $36\overline{)25.92}$; that is, they have the same answer.
- The decimal points in *both* the divisor and the dividend of the first division problem have been moved two decimal places to the right to create the second division problem.

$$0.36\overline{)0.2592} \quad \text{becomes} \quad 36\overline{)25.92}$$

These observations illustrate the following rule for division with a decimal divisor.

Division with a Decimal Divisor

To divide with a decimal divisor:
1. Write the problem in long division form.
2. Move the decimal point of the divisor so that it becomes a whole number.
3. Move the decimal point of the dividend the same number of places to the right.
4. Write the decimal point in the quotient (answer) directly above the decimal point in the dividend. Divide as if working with whole numbers.
5. If necessary, additional zeros can be written to the right of the last digit of the dividend to continue the division.

EXAMPLE 4 Divide: $\dfrac{0.2592}{0.36}$

Strategy We will move the decimal point of the divisor, 0.36, two places to the right and we will move the decimal point of the dividend, 0.2592, the same number of places to the right.

WHY We can then use the rule for dividing a decimal by a *whole number*.

Solution We begin by writing the problem in long division form.

$0.36\overline{)0\,25\,.\,92}$ Move the decimal point two places to the right in the divisor and the dividend. Write the decimal point in the quotient (answer) directly above the decimal point in the dividend.

Since the divisor is now a whole number, we can use the rule for dividing a decimal by a whole number to find the quotient.

```
        0.72
   36)25.92    Now divide as with whole numbers.
     -25 2↓
        72
       -72
         0
```

Check:

```
    0.72
  ×   36
    432
   2160
   25.92    Since this is the dividend, the result checks.
```

Self Check 4

Divide: $\dfrac{0.6045}{0.65}$ 0.93

Now Try Problem 27

Teaching Example 4 Divide: $0.1596 \div 0.38$. Check the result.
Answer:
0.42

Chapter 5 Decimals

> **Success Tip** When dividing decimals, moving the decimal points the same number of places to the right in *both* the divisor and the dividend does not change the answer.

3 Round a decimal quotient.

In Example 4, the division process stopped after we obtained a 0 from the second subtraction. Sometimes when we divide, the subtractions never give a zero remainder, and the division process continues forever. In such cases, we can round the result.

EXAMPLE 5 Divide: $\dfrac{9.35}{0.7}$. Round the quotient to the nearest hundredth.

Strategy We will use the methods of this section to divide to the thousandths column.

WHY To round to the hundredths column, we need to continue the division process for one more decimal place, which is the thousandths column.

Solution We begin by writing the problem in long division form.

$0.7\overline{)93.5}$ To write the divisor as a whole number, move the decimal point one place to the right. Do the same for the dividend. Place the decimal point in the quotient (answer) directly above the decimal point in the dividend.

We need to write two zeros to the right of the last digit of the dividend so that we can divide to the thousandths column.

$7\overline{)93.500}$

After dividing to the thousandths column, we round to the hundredths column.

The rounding digit in the hundredths column is 5.
The test digit in the thousandths column is 7.

```
         13.357
      7)93.500
       - 7
         23
       - 21
          2 5
        - 2 1
            40
          - 35
             50
           - 49
              1    The division process can stop. We have divided to the thousandths column.
```

Since the test digit 7 is 5 or greater, we will round 13.357 up to approximate the quotient to the nearest hundredth.

$\dfrac{9.35}{0.7} \approx 13.36$ Read ≈ as "is approximately equal to."

Check:

13.36 ← The approximation of the quotient
$\underline{\times\ \ \ 0.7}$ ← The original divisor
9.352 ← Since this is close to the original dividend, 9.35, the result seems reasonable.

Self Check 5
Divide: 12.82 ÷ 0.9. Round the quotient to the nearest hundredth. 14.24

Now Try Problem 33

Teaching Example 5 Divide: 11.07 ÷ 0.7. Round the quotient to the nearest hundredth.
Answer:
15.81

> **Success Tip** To round a quotient to a certain decimal place value, continue the division process one more column to its right to find the *test digit*.

Using Your CALCULATOR Dividing Decimals

The nucleus of a cell contains vital information about the cell in the form of DNA. The nucleus is very small: A typical animal cell has a nucleus that is only 0.00023622 inch across. How many nuclei (plural of *nucleus*) would have to be laid end to end to extend to a length of 1 inch?

To find how many 0.00023622-inch lengths there are in 1 inch, we must use division: $1 \div 0.00023622$.

$$1 \boxed{\div} .00023622 \boxed{=} \qquad \boxed{4233.3418}$$

On some calculators, we press the $\boxed{\text{ENTER}}$ key to display the quotient.

It would take approximately 4,233 nuclei laid end to end to extend to a length of 1 inch.

4 Estimate quotients of decimals.

There are many ways to make an error when dividing decimals. Estimation is a helpful tool that can be used to determine whether or not an answer seems reasonable.

To estimate quotients, we use a method that approximates both the dividend and the divisor so that they divide easily. There is one rule of thumb for this method: If possible, round both numbers up or both numbers down.

EXAMPLE 6 Estimate the quotient: $248.687 \div 43.1$

Strategy We will round the dividend and the divisor down and find $240 \div 40$.

WHY The division can be made easier if the dividend and the divisor end with zeros. Also, 40 divides 240 exactly.

Solution

$248.687 \div 43.1 \qquad 240 \div 40 = 6 \qquad$ To divide, drop one zero from 240 and from 40, and find $24 \div 4$.

(The dividend is approximately; The divisor is approximately)

The estimate is 6.

If we calculate $248.687 \div 43.1$, the quotient is exactly 5.77. Note that the estimate is close: It's just 0.23 more than 5.77.

5 Divide decimals by powers of 10.

To develop a set of rules for division of decimals by a power of 10, we consider the problems $8.13 \div 10$ and $8.13 \div 0.1$.

```
    0.813
10)8.130      Write a zero to the
   -8 0↓      right of the 3.
     13
    -10↓
     30
    -30
      0
```

```
       813
0 1)81.3      Move the decimal points in the divisor
   -8↓        and dividend one place to the right.
    1
   -1↓
    3
   -3
    0
```

Self Check 6

Estimate the quotient:
$6,229.249 \div 68.9$

Now Try Problems 35 and 39
Self Check 6 Answer
$6,300 \div 70 = 630 \div 7 = 90$

Teaching Example 6 Estimate the quotient: $481.797 \div 62.98$
Answer:
$480 \div 60 = 48 \div 6 = 8$

Note that the quotients, 0.813 and 81.3, and the dividend, 8.13, are the same except for the location of the decimal points. The first quotient, 0.813, can be easily obtained by moving the decimal point of the dividend one place to the left. The second quotient, 81.3, is easily obtained by moving the decimal point of the dividend one place to the right. These observations illustrate the following rules for dividing a decimal by a power of 10.

> **Dividing a Decimal by 10, 100, 1,000, and So On**
>
> To find the quotient of a decimal and 10, 100, 1,000, and so on, move the decimal point to the left the same number of places as there are zeros in the power of 10.

> **Dividing a Decimal by 0.1, 0.01, 0.001, and So On**
>
> To find the quotient of a decimal and 0.1, 0.01, 0.001, and so on, move the decimal point to the right the same number of decimal places as there are in the power of 10.

Self Check 7

Find each quotient:

a. $721.3 \div 100$ 7.213

b. $\dfrac{1.07}{1,000}$ 0.00107

c. $19.4407 \div 0.0001$ 194,407

Now Try Problems 43 and 49

Teaching Example 7 Find each quotient:
a. $624.87 \div 100$
b. $\dfrac{1.67}{1,000}$
c. $49.87513 \div 0.0001$
Answers:
a. 6.2487
b. 0.00167
c. 498,751.3

EXAMPLE 7

Find each quotient:

a. $16.74 \div 10$ b. $8.6 \div 10,000$ c. $\dfrac{290.623}{0.01}$

Strategy We will identify the divisor in each division. If it is a power of 10 greater than 1, we will count the number of zeros that it has. If it is a power of 10 less than 1, we will count the number of decimal places that it has.

WHY Then we will know how many places to the right or left to move the decimal point in the dividend to find the quotient.

Solution

a. $16.74 \div 10 = 1.674$ Since the divisor 10 has one zero, move the decimal point one place to the left.

b. $8.6 \div 10,000 = .00086$ Since the divisor 10,000 has four zeros, move the decimal point four places to the left. Write three placeholder zeros (shown in blue).

$= 0.00086$

c. $\dfrac{290.623}{0.01} = 29062.3$ Since the divisor 0.01 has two decimal places, move the decimal point in 290.623 two places to the right.

6 Divide signed decimals.

The rules for dividing integers also hold for dividing signed decimals. The quotient of two decimals with *like signs* is positive, and the quotient of two decimals with *unlike signs* is negative.

Self Check 8

Divide:

a. $-100.624 \div 15.2$ -6.62

b. $\dfrac{-23.9}{-0.1}$ 239

Now Try Problems 51 and 55

EXAMPLE 8

Divide: a. $-104.483 \div 16.3$ b. $\dfrac{-38.677}{-0.1}$

Strategy In part a, we will use the rule for dividing signed decimals that have different (unlike) signs. In part b, we will use the rule for dividing signed decimals that have the same (like) signs.

WHY In part a, the divisor is positive and the dividend is negative. In part b, both the dividend and divisor are negative.

Solution

a. First, we find the absolute values: $|-104.483| = 104.483$ and $|16.3| = 16.3$. Then we divide the absolute values, 104.483 by 16.3, using the methods of this section.

$$
\begin{array}{r}
6.41 \\
163\overline{)1044.83} \\
-978 \\
\hline
66\ 8 \\
-65\ 2 \\
\hline
1\ 63 \\
-1\ 63 \\
\hline
0
\end{array}
$$

Move the decimal point in the divisor and the dividend one place to the right.

Write the decimal point in the quotient (answer) directly above the decimal point in the dividend.

Divide as if working with whole numbers.

Since the signs of the original dividend and divisor are unlike, we make the final answer negative. Thus,

$$-104.483 \div 16.3 = -6.41$$

Check the result using multiplication.

b. We can use the rule for dividing a decimal by a power of 10 to find the quotient.

$$\frac{-38.677}{-0.1} = 386.77$$

Since the divisor 0.1 has one decimal place, move the decimal point in 38.677 one place to the right. Since the dividend and divisor have like signs, the quotient is positive.

Teaching Example 8 Divide:
a. $-113.016 \div 13.6$
b. $\frac{-31.93}{-0.1}$

Answers:
a. -8.31
b. 319.3

7 Use the order of operations rule.

Recall that the order of operations rule is used to evaluate expressions that involve more than one operation.

EXAMPLE 9 Evaluate: $\dfrac{2(0.351) + 0.5592}{0.2 - 0.6}$

Strategy We will evaluate the expression above and the expression below the fraction bar separately. Then we will do the indicated division, if possible.

WHY Fraction bars are grouping symbols. They group the numerator and denominator.

Solution

$$
\begin{aligned}
\frac{2(0.351) + 0.5592}{0.2 - 0.6} &= \frac{0.702 + 0.5592}{-0.4} \\
&= \frac{1.2612}{-0.4} \\
&= -3.153
\end{aligned}
$$

In the numerator, do the multiplication. In the denominator, do the subtraction.

In the numerator, do the addition.

Do the division indicated by the fraction bar. The quotient of two numbers with unlike signs is negative.

$$
\begin{array}{r}
\overset{1}{0.351} \\
\times\ \ 2 \\
\hline
0.702
\end{array}
\qquad
\begin{array}{r}
\overset{1}{0.7}\overset{1}{0}20 \\
+\ 0.5592 \\
\hline
1.2612
\end{array}
$$

$$
\begin{array}{r}
3.153 \\
4\overline{)12.612} \\
-12 \\
\hline
0\ 6 \\
-\ 4 \\
\hline
2\ 1 \\
-2\ 0 \\
\hline
12 \\
-12 \\
\hline
0
\end{array}
$$

Self Check 9

Evaluate: $\dfrac{2.7756 + 3(-0.63)}{0.4 - 1.2}$ -1.107

Now Try Problem 59

Teaching Example 9

Evaluate: $\dfrac{2(0.932) + 0.6764}{0.3 - 0.9}$

Answer: -4.234

8 Evaluate formulas.

Self Check 10

Evaluate the formula $l = \frac{A}{w}$ for $A = 5.511$ and $w = 1.002$. 5.5

Now Try Problem 63

Teaching Example 10 Evaluate the formula $h = \frac{3V}{B}$ for $V = 3.144$ and $B = 2.4$.
Answer:
3.93

EXAMPLE 10 Evaluate the formula $b = \frac{2A}{h}$ for $A = 15.36$ and $h = 6.4$.

Strategy In the given formula, we will replace the letter A with 15.36 and h with 6.4.

WHY Then we can use the order of operations rule to find the value of the expression on the right side of the $=$ symbol.

Solution

$b = \frac{2A}{h}$ This is the given formula.

$b = \frac{2(15.36)}{6.4}$ Replace A with 15.36 and h with 6.4.

$b = \frac{30.72}{6.4}$ In the numerator, do the multiplication.

$b = 4.8$ Do the division indicated by the fraction bar.

```
   1 1
  15.36
 ×    2
  30.72

      4.8
 64)307.2
    −256
     51 2
    −51 2
        0
```

9 Solve application problems by dividing decimals.

Recall that application problems that involve forming equal-sized groups can be solved by division.

Self Check 11

FRUITCAKES A 9-inch-long fruitcake loaf is cut into 0.25-inch-thick slices. How many slices are there in one fruitcake? 36 slices

Now Try Problem 95

Teaching Example 11 **PEPPERONI** A meat slicing machine cuts 12-inch-long pepperoni sausages into 0.125-inch-thick slices. How many slices are there in one sausage?
Answer:
96 slices

EXAMPLE 11 **French Bread** A bread slicing machine cuts 25-inch-long loaves of French bread into 0.625-inch-thick slices. How many slices are there in one loaf?

Solution

- 25-inch-long loaves of French bread are cut into slices. Given
- Each slice is 0.625-inch thick. Given
- How many slices are there in one loaf? Find

Cutting a loaf of French bread into equally thick slices indicates division. We translate the words of the problem to numbers and symbols.

The number of slices in a loaf of French bread	is equal to	the length of the loaf of French bread	divided by	the thickness of one slice.
The number of slices in a loaf of French bread	=	25	÷	0.625

When we write $25 \div 0.625$ in long division form, we see that the divisor is a decimal.

$0.625\overline{)25.000}$ To write the divisor as a whole number, move the decimal point three places to the right. To move the decimal point three places to the right in the dividend, three placeholder zeros must be inserted (shown in blue).

Unless otherwise noted, all art on this page is © Cengage Learning.

5.4 Dividing Decimals

Now that the divisor is a whole number, we can perform the division.

$$\begin{array}{r} 40 \\ 625\overline{)25000} \\ -2500\downarrow \\ \hline 00 \\ -0 \\ \hline 0 \end{array}$$

There are 40 slices in one loaf of French bread.

Check: The multiplication below verifies that 40 slices, each 0.625-inch thick, makes a 25-inch-long loaf. The result checks.

$$\begin{array}{r} 0.625 \leftarrow \text{The thickness of one slice of bread (in inches)} \\ \times \quad 40 \leftarrow \text{The number of slices in one loaf} \\ \hline 0000 \\ 25000 \\ \hline 25.000 \leftarrow \text{The length of one loaf of bread (in inches)} \end{array}$$

Recall that the **arithmetic mean**, or **average**, of several numbers is a value around which the numbers are grouped. We use addition and division to find the mean (average).

EXAMPLE 12 Comparison Shopping
An online shopping website, Shopping.com, listed the four best prices for an automobile GPS receiver as shown below. What is the mean (average) price of the GPS?

Strategy We will add 169.99, 182.65, 194.84, and 204.48 and divide the sum by 4.

WHY To find the mean (average) of a set of values, we divide the sum of the values by the number of values.

Solution

$$\text{Mean} = \frac{169.99 + 182.65 + 194.84 + 204.48}{4} \quad \text{Since there are 4 prices, divide the sum by 4.}$$

$$\text{Mean} = \frac{751.96}{4} \quad \text{In the numerator, do the addition.}$$

$$\text{Mean} = 187.99 \quad \text{Do the indicated division.}$$

$$\begin{array}{r} 2\ 2\ 2\ \ 2 \\ 169.99 \\ 182.65 \\ 194.84 \\ +204.48 \\ \hline 751.96 \end{array} \qquad \begin{array}{r} 187.99 \\ 4\overline{)751.96} \\ -4 \\ \hline 35 \\ -32 \\ \hline 31 \\ -28 \\ \hline 39 \\ -36 \\ \hline 36 \\ -36 \\ \hline 0 \end{array}$$

The mean (average) price of the GPS receiver is $187.99.

Self Check 12
U.S. NATIONAL PARKS Use the following data to determine the average number of visitors per year to the national parks for the years 2007 through 2011. (Source: National Park Service)
279.2 million visitors

Year	Visitors (millions)
2011	278.8
2010	281.2
2009	285.5
2008	274.9
2007	275.6

Now Try Problem 103

Teaching Example 12 BROADWAY SHOWS Use the following data to determine the average annual paid attendance for Broadway shows in New York City for the years 2007 through 2011. (Source: The Broadway League)

Season	Attendance (millions)
2011	12.34
2010	12.53
2009	11.89
2008	12.16
2007	12.28

Answer:
12.24 million annual attendance

THINK IT THROUGH — GPA

"In considering all of the factors that are important to employers as they recruit students in colleges and universities nationwide, college major, grade point average, and work-related experience usually rise to the top of the list."

Mary D. Feduccia, Ph.D., Career Services Director, Louisiana State University

A grade point average (GPA) is a weighted average based on the grades received and the number of units (credit hours) taken. A GPA for one semester (or term) is defined as

> the quotient of the sum of the grade points earned for each class and the sum of the number of units taken. The number of grade points earned for a class is the product of the number of units assigned to the class and the value of the grade received in the class.

1. Use the table of grade values below to compute the GPA for the student whose semester grade report is shown. Round to the nearest hundredth. 2.86

Grade	Value
A	4
B	3
C	2
D	1
F	0

Class	Units	Grade
Geology	4	C
Algebra	5	A
Psychology	3	C
Spanish	2	B

2. If you were enrolled in school last semester (or term), list the classes taken, units assigned, and grades received like those shown in the grade report above. Then calculate your GPA.

ANSWERS TO SELF CHECK

1. 5.2 **2.** 3.17 **3.** 5.35 **4.** 0.93 **5.** 14.24 **6.** $6{,}300 \div 70 = 630 \div 7 = 90$
7. a. 7.213 **b.** 0.00107 **c.** 194,407 **8. a.** −6.62 **b.** 239 **9.** −1.107 **10.** 5.5
11. 36 slices **12.** 279.2 million visitors

SECTION 5.4 STUDY SET

VOCABULARY

Fill in the blanks.

1. In the division problem shown below, label the *dividend*, the *divisor*, and the *quotient*.

$$\text{divisor} \to 5\overline{)15.85} \leftarrow \text{dividend}$$
$$3.17 \leftarrow \text{quotient}$$

2. To perform the division $2.7\overline{)9.45}$, we move the decimal point of the divisor so that it becomes the __whole__ number 27. Then we move the decimal point of the __dividend__ the same number of places to the right.

CONCEPTS

3. A decimal point is missing in each of the following quotients. Write a decimal point in the proper position.

   a. $4\overline{)21.04}$ with quotient 526 → 5.26

   b. $3\overline{)0.024}$ with quotient 0008 → 0.008

4. a. How many places to the right must we move the decimal point in 6.14 so that it becomes a whole number? two places

   b. When the decimal point in 49.8 is moved three places to the right, what is the resulting number? 49,800

5. Move the decimal point in the divisor and the dividend the same number of places so that the divisor becomes a whole number. *You do not have to find the quotient.*
   a. $1.3\overline{)10.66}$ $13\overline{)106.6}$
   b. $3.71\overline{)16.695}$ $371\overline{)1669.5}$

6. Fill in the blanks: To divide with a decimal divisor, write the problem in long division form. Move the decimal point of the divisor so that it becomes a whole number. Then move the decimal point of the dividend the same number of places to the right . Write the decimal point in the quotient directly above the decimal point in the dividend and divide as working with whole numbers .

7. To perform the division $7.8\overline{)14.562}$, the decimal points in the divisor and dividend are moved 1 place to the right. This is equivalent to multiplying $\frac{14.562}{7.8}$ by what form of 1? $\frac{10}{10}$

8. Use multiplication to check the following division. Is the result correct?

   $$\frac{1.917}{0.9} = 2.13 \quad \text{yes, } 2.13 \cdot 0.9 = 1.917$$

9. When rounding a decimal to the hundredths column, to what other column must we look at first? thousandths

10. a. When 9.545 is divided by 10, is the answer smaller or larger than 9.545? smaller
    b. When 9.545 is divided by 0.1, is the answer smaller or larger than 9.545? larger

11. Fill in the blanks.
    a. To find the quotient of a decimal and 10, 100, 1,000, and so on, move the decimal point to the left the same number of places as there are zeros in the power of 10.
    b. To find the quotient of a decimal and 0.1, 0.01, 0.001, and so on, move the decimal point to the right the same number of decimal places as there are in the power of 10.

12. Determine whether the *sign* of each result is positive or negative. *You do not have to find the quotient.*
    a. $-15.25 \div (-0.5)$ positive
    b. $\frac{-25.92}{3.2}$ negative

NOTATION

13. Explain what the red arrows are illustrating in the division problem below.

    $467\overline{)3208.7}$ moving the decimal points in the divisor and dividend 2 places to the right

14. The division shown below is not finished. Why was the red 0 written after the 7 in the dividend?

    $$\begin{array}{r} 2.3 \\ 2\overline{)4.70} \\ -4 \\ \hline 07 \\ -6 \\ \hline 1 \end{array}$$

    The red 0 was written after the 7 so that the division process could continue.

GUIDED PRACTICE

Divide. Check the result. See Example 1.

15. $12.6 \div 6$ 2.1
16. $40.8 \div 8$ 5.1
17. $3\overline{)27.6}$ 9.2
18. $4\overline{)28.8}$ 7.2

Divide. Check the result. See Example 2.

19. $98.21 \div 23$ 4.27
20. $190.96 \div 28$ 6.82
21. $37\overline{)320.05}$ 8.65
22. $32\overline{)125.12}$ 3.91

Divide. Check the result. See Example 3.

23. $13.4 \div 4$ 3.35
24. $38.3 \div 5$ 7.66
25. $5\overline{)22.8}$ 4.56
26. $6\overline{)28.5}$ 4.75

Divide. Check the result. See Example 4.

27. $\frac{0.1932}{0.42}$ 0.46
28. $\frac{0.2436}{0.29}$ 0.84
29. $0.29\overline{)0.1131}$ 0.39
30. $0.58\overline{)0.1566}$ 0.27

Divide. Round the quotient to the nearest hundredth. Check the result. See Example 5.

31. $\frac{11.83}{0.6}$ 19.72
32. $\frac{16.43}{0.9}$ 18.26
33. $\frac{17.09}{0.7}$ 24.41
34. $\frac{13.07}{0.6}$ 21.78

Estimate each quotient. See Example 6.

35. $289.842 \div 72.1$ $280 \div 70 = 28 \div 7 = 4$
36. $284.254 \div 91.4$ $270 \div 90 = 27 \div 9 = 3$
37. $383.76 \div 7.8$ $400 \div 8 = 50$
38. $348.84 \div 5.7$ $360 \div 6 = 60$
39. $3{,}883.284 \div 48.12$ $4{,}000 \div 50 = 400 \div 5 = 80$
40. $5{,}556.521 \div 67.89$ $5{,}600 \div 70 = 560 \div 7 = 80$
41. $6.1\overline{)15{,}819.74}$ $15{,}000 \div 5 = 3{,}000$
42. $9.2\overline{)19{,}460.76}$ $18{,}000 \div 9 = 2{,}000$

Find each quotient. See Example 7.

43. $451.78 \div 100$ 4.5178
44. $991.02 \div 100$ 9.9102
45. $\frac{30.09}{10{,}000}$ 0.003009
46. $\frac{27.07}{10{,}000}$ 0.002707
47. $1.25 \div 0.1$ 12.5
48. $8.62 \div 0.01$ 862
49. $\frac{545.2}{0.001}$ 545,200
50. $\frac{67.4}{0.001}$ 67,400

498 Chapter 5 Decimals

Divide. See Example 8.

51. $-110.336 \div 12.8$ -8.62
52. $-121.584 \div 14.9$ -8.16
53. $-91.304 \div (-22.6)$ 4.04
54. $-66.126 \div (-32.1)$ 2.06
55. $\dfrac{-20.3257}{-0.001}$ $20,325.7$
56. $\dfrac{-48.8933}{-0.001}$ $48,893.3$
57. $0.003 \div (-100)$ -0.00003
58. $0.008 \div (-100)$ -0.00008

Evaluate each expression. See Example 9.

59. $\dfrac{2(0.614) + 2.3854}{0.2 - 0.9}$ -5.162
60. $\dfrac{2(1.242) + 0.8932}{0.4 - 0.8}$ -8.443
61. $\dfrac{5.409 - 3(1.8)}{(0.3)^2}$ 0.1
62. $\dfrac{1.674 - 5(0.222)}{(0.1)^2}$ 56.4

Evaluate each formula. See Example 10.

63. $t = \dfrac{d}{r}$ for $d = 211.75$ and $r = 60.5$ 3.5
64. $h = \dfrac{2A}{b}$ for $A = 9.62$ and $b = 3.7$ 5.2
65. $r = \dfrac{d}{t}$ for $d = 219.375$ and $t = 3.75$ 58.5
66. $\pi = \dfrac{C}{d}$ for $C = 14.4513$ and $d = 4.6$ (Round to the nearest hundredth.) 3.14

TRY IT YOURSELF

Perform the indicated operations. Round the result to the specified decimal place, when indicated.

67. $4.5\overline{)11.97}$ 2.66
68. $4.1\overline{)14.637}$ 3.57
69. $\dfrac{75.04}{10}$ 7.504
70. $\dfrac{-1.3}{10}$ -0.13
71. $8\overline{)0.036}$ 0.0045
72. $4\overline{)0.073}$ 0.01825
73. $9\overline{)2.889}$ 0.321
74. $6\overline{)3.378}$ 0.563
75. $\dfrac{-3(0.2) - 2(3.3)}{30(0.4)^2}$ -1.5
76. $\dfrac{(-1.3)^2 + 9.2}{-2(0.2) - 0.5}$ -12.1
77. Divide 1.2202 by -0.01. -122.02
78. Divide -0.4531 by -0.001. 453.1
79. $-5.714 \div 2.4$ (nearest tenth) -2.4
80. $-21.21 \div 3.8$ (nearest tenth) -5.6
81. $-39 \div (-4)$ 9.75
82. $-26 \div (-8)$ 3.25
83. $7.8915 \div .00001$ $789,150$
84. $23.025 \div 0.0001$ $230,250$
85. $\dfrac{-30.6}{5.1}$ -6
86. $\dfrac{0.0092}{0.023}$ 0.4
87. $12.243 \div 0.9$ (nearest hundredth) 13.60
88. $13.441 \div 0.6$ (nearest hundredth) 22.40
89. $1,000\overline{)34.8}$ 0.0348
90. $10,000\overline{)678.9}$ 0.06789

91. $\dfrac{40.7(3 - 8.3)}{0.4 - 0.61}$ (nearest hundredth) $1,027.19$
92. $\dfrac{(0.5)^2 - (0.3)^2}{0.005 + 0.1}$ (nearest hundredth) 1.52
93. Divide 0.25 by 1.6 0.15625
94. Divide 1.2 by 0.64 1.875

APPLICATIONS

95. BUTCHER SHOPS A meat slicer trims 0.05-inch-thick pieces from a sausage. If the sausage is 14 inches long, how many slices are there in one sausage? 280 slices

96. ELECTRONICS The volume control on a computer is shown to the right. If the distance between the Low and High settings is 21 cm, how far apart are the equally spaced volume settings? 1.75 cm

97. COMPUTERS A computer can do an arithmetic calculation in 0.00003 second. How many of these calculations could it do in 60 seconds? 2,000,000 calculations

98. THE LOTTERY In December 2008, fifteen city employees of Piqua, Ohio, who had played the Mega Millions Lottery as a group, won the jackpot. They were awarded a total of $94.5 million. If the money was split equally, how much did each person receive? (Source: pal-item.com) $6.3 million

99. SPRAY BOTTLES Each squeeze of the trigger of a spray bottle emits 0.017 ounce of liquid. How many squeezes are there in an 8.5-ounce bottle? 500 squeezes

100. CAR LOANS See the loan statement below. How many more monthly payments must be made to pay off the loan? 15 payments

American Finance Company		June
Monthly payment:	Paid to date: $547.30	
$42.10	Loan balance: $631.50	

101. HIKING Refer to the illustration below to determine how long it will take the person shown to complete the hike. Then determine at what time of the day she will complete the hike. 11 hr, 6 P.M.

The hiker walks 2.5 miles each hour.

Departure A.M. — Start — 27.5-mile hike — Finish — Arrival

Unless otherwise noted, all art on this page is © Cengage Learning.

102. **HOURLY PAY** The graph below shows the average hours worked and the average weekly earnings of U.S. production workers for the years 2000 and 2011. What did the average production worker earn per hour
    a. in 2000? $14.10
    b. in 2011? $19.50

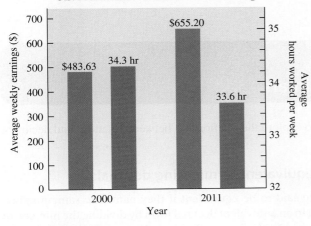

Source: U.S. Department of Labor Statistic

103. **TRAVEL** The illustration shows the annual number of person-trips of 50 miles or more (one way) for the years 2007–2011, as estimated by the Travel Industry Association of America. Find the average number of trips per year for this period of time. 1,518.4 million

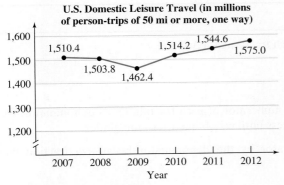

Source: U.S. Travel Association

104. **OIL WELLS** Geologists have mapped out the types of soil through which engineers must drill to reach an oil deposit. See the illustration below.

a. How far below the surface is the oil deposit? 1.48 mi
b. What is the average depth that must be drilled each week if the drilling is to be a four-week project? 0.37 mi

105. **REFLEXES** An online reaction time test is shown below. When the stoplight changes from red to green, the participant is to immediately click on the large green button. The program then displays the participant's reaction time in the table. After the participant takes the test five times, the *average* reaction time is found. Determine the average reaction time for the results shown below. 0.231 sec

Test Number	Reaction Time (in seconds)	The stoplight to watch.	The button to click.
1	0.219		
2	0.233		Click here on green light
3	0.204		
4	0.297		
5	0.202		
AVG.	?		

106. **INDY 500** Driver Ryan Briscoe, of Sydney, Australia, had the fastest average qualifying speed for the 2012 Indianapolis 500-mile race. This earned him the *pole position* to begin the race. The speeds for each of his four qualifying laps are shown below. What was his average qualifying speed? 225.078 mph

| Lap 1: 225.297 mph |
| Lap 2: 224.959 mph |
| Lap 3: 225.177 mph |
| Lap 4: 224.879 mph |

Source: indianapolismotorspeedway.com

WRITING

107. Explain the process used to divide two numbers when both the divisor and the dividend are decimals. Give an example.

108. Explain why we must sometimes use rounding when we write the answer to a division problem.

109. The division $0.5\overline{)2.005}$ is equivalent to $5\overline{)20.05}$. Explain what equivalent means in this case.

110. In $3\overline{)0.7}$, why can additional zeros be placed to the right of 0.7 without affecting the result?

111. Explain how to estimate the following quotient: $0.75\overline{)2.415}$

112. Explain why multiplying $\dfrac{4.86}{0.2}$ by the form of 1 shown below moves the decimal points in the dividend, 4.86, and the divisor, 0.2, one place to the right.

$$\dfrac{4.86}{0.2} = \dfrac{4.86}{0.2} \cdot \dfrac{10}{10}$$

REVIEW

113. a. Find the GCF of 10 and 25. 5
  b. Find the LCM of 10 and 25. 50
114. a. Find the GCF of 8, 12, and 16. 4
  b. Find the LCM of 8, 12, and 16. 48

SECTION 5.5
Fractions and Decimals

Objectives
1. Write fractions as equivalent terminating decimals.
2. Write fractions as equivalent repeating decimals.
3. Round repeating decimals.
4. Graph fractions and decimals on a number line.
5. Compare fractions and decimals.
6. Evaluate expressions containing fractions and decimals.
7. Solve application problems involving fractions and decimals.

In this section, we continue to explore the relationship between fractions and decimals.

1 Write fractions as equivalent terminating decimals.

A fraction and a decimal are said to be **equivalent** if they name the same number. Every fraction can be written in an equivalent decimal form by dividing the numerator by the denominator, as indicated by the fraction bar.

Writing a Fraction as a Decimal

To write a fraction as a decimal, divide the numerator of the fraction by its denominator.

Self Check 1
Write each fraction as a decimal.
a. $\dfrac{1}{5}$ 0.2
b. $\dfrac{3}{16}$ 0.1875
c. $\dfrac{9}{2}$ 4.5

Now Try Problems 15, 17, and 21

Teaching Example 1 Write each fraction as a decimal:
a. $\dfrac{1}{4}$ **b.** $\dfrac{7}{8}$ **c.** $\dfrac{11}{2}$
Answers:
a. 0.25
b. 0.875
c. 5.5

EXAMPLE 1 Write each fraction as a decimal.
a. $\dfrac{3}{4}$ **b.** $\dfrac{5}{8}$ **c.** $\dfrac{7}{2}$

Strategy We will divide the numerator of each fraction by its denominator. We will continue the division process until we obtain a zero remainder.

WHY We divide the numerator by the denominator because a fraction bar indicates division.

Solution

a. $\dfrac{3}{4}$ means $3 \div 4$. To find $3 \div 4$, we begin by writing it in long division form as $4\overline{)3}$. To proceed with the division, we must write the dividend 3 with a decimal point and some additional zeros. Then we use the procedure from Section 5.4 for dividing a decimal by a whole number.

$$\begin{array}{r} 0.75 \\ 4\overline{)3.00} \\ -2\;8\downarrow \\ \hline 20 \\ -20 \\ \hline 0 \end{array}$$

Write a decimal point and two additional zeros to the right of 3.

← The remainder is 0.

Thus, $\dfrac{3}{4} = 0.75$. We say that the **decimal equivalent** of $\dfrac{3}{4}$ is 0.75.

We can check the result by writing 0.75 as a fraction in simplest form:

$0.75 = \dfrac{75}{100}$ 0.75 is seventy-five hundredths.

$= \dfrac{3 \cdot \overset{1}{\cancel{25}}}{4 \cdot \underset{1}{\cancel{25}}}$ To simplify the fraction, factor 75 as 3 · 25 and 100 as 4 · 25 and remove the common factor of 25.

$= \dfrac{3}{4}$ This is the original fraction.

b. $\dfrac{5}{8}$ means $5 \div 8$.

```
      0.625
   8)5.000    Write a decimal point and three additional zeros to the right of 5.
    -4 8↓
      20
     -16
       40
      -40
        0  ← The remainder is 0.
```

Thus, $\dfrac{5}{8} = 0.625$.

c. $\dfrac{7}{2}$ means $7 \div 2$.

```
      3.5
   2)7.0    Write a decimal point and one additional zero to the right of 7.
    -6↓
     10
    -10
      0  ← The remainder is 0.
```

Thus, $\dfrac{7}{2} = 3.5$.

Caution! A common error when finding a decimal equivalent for a fraction is to *incorrectly divide the denominator by the numerator*. An example of this is shown on the right, where the decimal equivalent of $\dfrac{5}{8}$ (a number less than 1) is incorrectly found to be 1.6 (a number greater than 1).

```
    1.6
 5)8.0
  -5
   30
  -30
    0
```

In parts a, b, and c of Example 1, the division process ended because a remainder of 0 was obtained. When such a division *terminates* with a remainder of 0, we call the resulting decimal a **terminating decimal**. Thus, 0.75, 0.625, and 3.5 are three examples of terminating decimals.

The Language of Algebra To *terminate* means to bring to an end. In the movie *The Terminator*, actor Arnold Schwarzenegger plays a heartless machine sent to Earth to bring an end to his enemies.

---

2 Write fractions as equivalent repeating decimals.

Sometimes, when we are finding a decimal equivalent of a fraction, the division process never gives a remainder of 0. In this case, the result is a **repeating decimal**. Examples of repeating decimals are 0.4444 . . . and 1.373737 . . . . The three dots tell us that a block of digits repeats in the pattern shown. Repeating decimals can also be written using a bar over the repeating block of digits. For example, 0.4444... can be written as $0.\overline{4}$, and 1.373737... can be written as $1.\overline{37}$.

> **Caution!** When using an **overbar** to write a repeating decimal, use the least number of digits necessary to show the repeating block of digits.
>
> ~~0.333... = 0.333~~   ~~6.7454545... = 6.7454~~
>
> 0.333... = $0.\overline{3}$ 6.7454545... = $6.7\overline{45}$

Some fractions can be written as decimals using an alternate approach. If the denominator of a fraction in simplified form has factors of only 2's or 5's, or a combination of both, it can be written as a decimal by multiplying it by a form of 1. The objective is to write the fraction in an equivalent form with a denominator that is a power of 10, such as 10, 100, 1,000, and so on.

Self Check 2
Write each fraction as a decimal using multiplication by a form of 1:

a. $\dfrac{2}{5}$ 0.4

b. $\dfrac{8}{25}$ 0.32

Now Try Problems 27 and 29

Teaching Example 2 Write each fraction as a decimal using multiplication by a form of 1:

a. $\dfrac{1}{5}$ **b.** $\dfrac{9}{20}$

Answers:
a. 0.2
b. 0.45

EXAMPLE 2
Write each fraction as a decimal using multiplication by a form of 1:   **a.** $\dfrac{4}{5}$ **b.** $\dfrac{11}{40}$

Strategy We will multiply $\dfrac{4}{5}$ by $\dfrac{2}{2}$ and we will multiply $\dfrac{11}{40}$ by $\dfrac{25}{25}$.

WHY The result of each multiplication will be an equivalent fraction with a denominator that is a power of 10. Such fractions are then easy to write in decimal form.

Solution

a. Since we need to multiply the denominator of $\dfrac{4}{5}$ by 2 to obtain a denominator of 10, it follows that $\dfrac{2}{2}$ should be the form of 1 that is used to build $\dfrac{4}{5}$.

$\dfrac{4}{5} = \dfrac{4}{5} \cdot \dfrac{2}{2}$ Multiply $\dfrac{4}{5}$ by 1 in the form of $\dfrac{2}{2}$.

$= \dfrac{8}{10}$ Multiply the numerators. Multiply the denominators.

$= 0.8$ Write the fraction as a decimal.

b. Since we need to multiply the denominator of $\dfrac{11}{40}$ by 25 to obtain a denominator of 1,000, it follows that $\dfrac{25}{25}$ should be the form of 1 that is used to build $\dfrac{11}{40}$.

$\dfrac{11}{40} = \dfrac{11}{40} \cdot \dfrac{25}{25}$ Multiply $\dfrac{11}{40}$ by 1 in the form of $\dfrac{25}{25}$.

$= \dfrac{275}{1,000}$ Multiply the numerators. Multiply the denominators.

$= 0.275$ Write the fraction as a decimal.

Mixed numbers can also be written in decimal form.

Self Check 3
Write the mixed number $3\dfrac{17}{20}$ in decimal form.   3.85

Now Try Problem 37

EXAMPLE 3
Write the mixed number $5\dfrac{7}{16}$ in decimal form.

Strategy We need only find the decimal equivalent for the fractional part of the mixed number.

WHY The whole-number part in the decimal form is the same as the whole-number part in the mixed number form.

Solution To write $\frac{7}{16}$ as a fraction, we find $7 \div 16$.

$$
\begin{array}{r}
0.4375 \\
16\overline{)7.0000} \\
-64 \\
\hline
60 \\
-48 \\
\hline
120 \\
-112 \\
\hline
80 \\
-80 \\
\hline
0
\end{array}
$$

Write a decimal point and four additional zeros to the right of 7.

← The remainder is 0.

Since the whole-number part of the decimal must be the same as the whole-number part of the mixed number, we have:

$$5\frac{7}{16} = 5.4375$$

We would have obtained the same result if we changed $5\frac{7}{16}$ to the improper fraction $\frac{87}{16}$ and divided 87 by 16.

EXAMPLE 4 Write $\frac{5}{12}$ as a decimal.

Strategy We will divide the numerator of the fraction by its denominator and watch for a repeating pattern of nonzero remainders.

WHY Once we detect a repeating pattern of remainders, the division process can stop.

Solution $\frac{5}{12}$ means $5 \div 12$.

$$
\begin{array}{r}
0.4166 \\
12\overline{)5.0000} \\
-48 \\
\hline
20 \\
-12 \\
\hline
80 \\
-72 \\
\hline
80 \\
-72 \\
\hline
8
\end{array}
$$

Write a decimal point and four additional zeros to the right of 5.

It is apparent that 8 will continue to reappear as the remainder. Therefore, 6 will continue to reappear in the quotient. Since the repeating pattern is now clear, we can stop the division.

We can use three dots to show that a repeating pattern of 6's appears in the quotient:

$$\frac{5}{12} = 0.416666\ldots$$

Or, we can use an overbar to indicate the repeating part (in this case, only the 6), and write the decimal equivalent in more compact form:

$$\frac{5}{12} = 0.41\overline{6}$$

EXAMPLE 5 Write $-\frac{6}{11}$ as a decimal.

Strategy To find the decimal equivalent for $-\frac{6}{11}$, we will first find the decimal equivalent for $\frac{6}{11}$. To do this, we will divide the numerator of $\frac{6}{11}$ by its denominator and watch for a repeating pattern of nonzero remainders.

Teaching Example 3 Write the mixed number $1\frac{15}{16}$ in decimal form.
Answer:
1.9375

Self Check 4
Write $\frac{1}{12}$ as a decimal. $0.08\overline{3}$
Now Try Problem 41

Teaching Example 4 Write $\frac{11}{12}$ as a decimal.
Answer:
$0.91\overline{6}$

Self Check 5
Write $-\frac{13}{33}$ as a decimal. $-0.\overline{39}$
Now Try Problem 47

Teaching Example 5 Write $-\frac{9}{11}$ as a decimal.
Answer:
$-0.\overline{81}$

WHY Once we detect a repeating pattern of remainders, the division process can stop.

Solution $\frac{6}{11}$ means $6 \div 11$.

```
       0.54545
   11)6.00000     Write a decimal point and five additional zeros to the right of 6.
     -5 5
       ─────
        50
       -44
       ─────
         60
        -55
        ─────
          50
         -44
         ─────
           60
          -55
          ─────
            5
```

It is apparent that 6 and 5 will continue to reappear as remainders. Therefore, 5 and 4 will continue to reappear in the quotient. Since the repeating pattern is now clear, we can stop the division process.

We can use three dots to show that a repeating pattern of 5 and 4 appears in the quotient:

$$\frac{6}{11} = 0.545454\ldots \text{ and therefore, } -\frac{6}{11} = -0.545454\ldots$$

Or, we can use an overbar to indicate the repeating part (in this case, 54), and write the decimal equivalent in more compact form:

$$\frac{6}{11} = 0.\overline{54} \text{ and therefore, } -\frac{6}{11} = -0.\overline{54}$$

The repeating part of the decimal equivalent of some fractions is quite long. Here are some examples:

$\frac{9}{37} = 0.\overline{243}$ A block of three digits repeats.

$\frac{13}{101} = 0.\overline{1287}$ A block of four digits repeats.

$\frac{6}{7} = 0.\overline{857142}$ A block of six digits repeats.

Every fraction can be written as either a terminating decimal or a repeating decimal. For this reason, the set of fractions (**rational numbers**) form a subset of the set of decimals called the set of **real numbers**. The set of real numbers corresponds to all points on a number line.

Not all decimals are terminating or repeating decimals. For example,

$$0.2020020002\ldots$$

does not terminate, and it has no repeating block of digits. This decimal cannot be written as a fraction with an integer numerator and a nonzero integer denominator. Thus, it is not a rational number. It is an example from the set of **irrational numbers**.

3 Round repeating decimals.

When a fraction is written in decimal form, the result is either a terminating or a repeating decimal. Repeating decimals are often rounded to a specified place value.

5.5 Fractions and Decimals

EXAMPLE 6 Write $\frac{1}{3}$ as a decimal and round to the nearest hundredth.

Strategy We will use the methods of this section to divide to the thousandths column.

WHY To round to the hundredths column, we need to continue the division process for one more decimal place, which is the thousandths column.

Solution $\frac{1}{3}$ means $1 \div 3$.

```
       0.333
    3)1.000    Write a decimal point and three additional zeros to the right of 1.
     - 9↓
       10
     -  9↓
        10
      -  9
         1    The division process can stop. We have divided to the thousandths column.
```

After dividing to the thousandths column, we round to the hundredths column.

The rounding digit in the hundredths column is 3.
The test digit in the thousandths column is 3.

0.333 ...

Since 3 is less than 5, we round down, and we have

$\frac{1}{3} \approx 0.33$ Read ≈ as "is approximately equal to."

Self Check 6
Write $\frac{4}{9}$ as a decimal and round to the nearest hundredth.   0.44

Now Try Problem 51

Teaching Example 6 Write $\frac{1}{9}$ as a decimal and round to the nearest hundredth.
Answer:
0.11

EXAMPLE 7 Write $\frac{2}{7}$ as a decimal and round to the nearest thousandth.

Strategy We will use the methods of this section to divide to the ten-thousandths column.

WHY To round to the thousandths column, we need to continue the division process for one more decimal place, which is the ten-thousandths column.

Solution $\frac{2}{7}$ means $2 \div 7$.

```
       0.2857
    7)2.0000    Write a decimal point and four additional zeros to the right of 2.
    - 1 4↓
        60
      - 56↓
         40
       - 35↓
          50
        - 49
           1    The division process can stop.
                We have divided to the ten-thousandths column.
```

After dividing to the ten-thousandths column, we round to the thousandths column.

The rounding digit in the thousandths column is 5.
The test digit in the ten-thousandths column is 7.

0.2857

Since 7 is greater than 5, we round up, and $\frac{2}{7} \approx 0.286$.

Self Check 7
Write $\frac{7}{24}$ as a decimal and round to the nearest thousandth.   0.292

Now Try Problem 61

Teaching Example 7 Write $\frac{11}{12}$ as a decimal and round to the nearest thousandth.
Answer:
0.917

Chapter 5 Decimals

> **Using Your CALCULATOR** **The Fixed-Point Key**
>
> After performing a calculation, a scientific calculator can round the result to a given decimal place. This is done using the *fixed-point key*. As we did in Example 7, let's find the decimal equivalent of $\frac{2}{7}$ and round to the nearest thousandth. This time, we will use a calculator.
>
> First, we set the calculator to round to the third decimal place (thousandths) by pressing 2nd FIX 3. Then we press 2 ÷ 7 =   `0.286`
>
> Thus, $\frac{2}{7} \approx 0.286$. To round to the nearest tenth, we would fix 1; to round to the nearest hundredth, we would fix 2; and so on. After using the FIX feature, don't forget to remove it and return the calculator to the normal mode.
>
> Graphing calculators can also round to a given decimal place. See the owner's manual for the required keystrokes.

4 Graph fractions and decimals on a number line.

A number line can be used to show the relationship between fractions and their decimal equivalents. On the number line below, sixteen equally spaced marks are used to scale from 0 to 1. Some commonly used fractions that have terminating decimal equivalents are shown. For example, we see that $\frac{1}{8} = 0.125$ and $\frac{13}{16} = 0.8125$.

On the next number line, six equally spaced marks are used to scale from 0 to 1. Some commonly used fractions and their repeating decimal equivalents are shown.

5 Compare fractions and decimals.

To compare the size of a fraction and a decimal, it is helpful to write the fraction in its equivalent decimal form.

Self Check 8

Place an <, >, or = symbol in the box to make a true statement:

a. $\frac{3}{8}\;>\;0.305$

b. $0.7\overline{6}\;<\;\frac{7}{9}$

c. $\frac{11}{4}\;=\;2.75$

Now Try Problems 67, 69, and 71

EXAMPLE 8 Place an <, >, or = symbol in the box to make a true statement: a. $\frac{4}{5}$ ☐ 0.91 b. $0.3\overline{5}$ ☐ $\frac{1}{3}$ c. $\frac{9}{4}$ ☐ 2.25

Strategy In each case, we will write the given fraction as a decimal.

WHY Then we can use the procedure for comparing two decimals to determine which number is the larger and which is the smaller.

Solution

a. To write $\frac{4}{5}$ as a decimal, we divide 4 by 5.

$$\begin{array}{r} 0.8 \\ 5\overline{)4.0} \\ -4\,0 \\ \hline 0 \end{array}$$ Write a decimal point and one additional zero to the right of 4.

Thus, $\frac{4}{5} = 0.8$.

To make the comparison of the decimals easier, we can write one zero after 8 so that they have the same number of digits to the right of the decimal point.

0.8**0**   This is the decimal equivalent for $\frac{4}{5}$.
0.9**1**
↑

As we work from left to right, this is the first column in which the digits differ. Since $8 < 9$, it follows that $0.80 = \frac{4}{5}$ is less than 0.91, and we can write $\frac{4}{5} < 0.91$.

b. In Example 6, we saw that $\frac{1}{3} = 0.3333\ldots$. To make the comparison of these repeating decimals easier, we write them so that they have the same number of digits to the right of the decimal point.

0.3**5**55...   This is $0.3\overline{5}$.
0.3**3**33...   This is the decimal equivalent of $\frac{1}{3}$.
↑

As we work from left to right, this is the first column in which the digits differ. Since $5 > 3$, it follows that $0.3555\ldots = 0.3\overline{5}$ is greater than $0.3333\ldots = \frac{1}{3}$, and we can write $0.3\overline{5} > \frac{1}{3}$.

c. To write $\frac{9}{4}$ as a decimal, we divide 9 by 4.

```
  2.25
4)9.00     Write a decimal point and two additional zeros to the right of 9.
 −8
  1 0
  −8
    20
   −20
     0
```

From the division, we see that $\frac{9}{4} = 2.25$.

Teaching Example 8 Place an $<$, $>$, or an $=$ symbol in the box to make a true statement:

a. $\frac{3}{5}$ ☐ 0.66

b. $0.\overline{67}$ ☐ $\frac{2}{3}$

c. $\frac{9}{8}$ ☐ 1.125

Answers:
a. $<$
b. $>$
c. $=$

EXAMPLE 9   Write the numbers in order from least to greatest:

$$2.168,\ 2\frac{1}{6},\ \frac{20}{9}$$

Strategy We will write $2\frac{1}{6}$ and $\frac{20}{9}$ in decimal form.

WHY Then we can do a column-by-column comparison of the numbers to determine the largest and smallest.

Solution From the number line on page 506, we see that $\frac{1}{6} = 0.1\overline{6}$. Thus, $2\frac{1}{6} = 2.1\overline{6}$. To write $\frac{20}{9}$ as a decimal, we divide 20 by 9.

```
   2.222
9)20.000    Write a decimal point and three additional zeros to the right of 20.
 −18
   20
  −18
    20
   −18
     20
    −18
      2
```

Thus, $\frac{20}{9} = 2.222\ldots$.

Self Check 9

Write the numbers in order from least to greatest: $1.832,\ \frac{9}{5},\ 1\frac{5}{6}$

Now Try Problem 75

Self Check 9 Answers

$\frac{9}{5}, 1.832, 1\frac{5}{6}$

Teaching Example 9 Write the numbers in order from least to greatest: $3.584,\ 3\frac{7}{12},\ \frac{18}{5}$

Answer:
$3\frac{7}{12}, 3.584, \frac{18}{5}$

To make the comparison of the three decimals easier, we stack them as shown below.

$$2.1\underline{6}80 \quad \text{This is 2.168 with an additional 0.}$$
$$2.1\underline{6}66\ldots \quad \text{This is } 2\tfrac{1}{6} = 2.1\overline{6}.$$
$$2.\underline{2}222\ldots \quad \text{This is } \tfrac{20}{9}.$$

Working from left to right, this is the first column in which the digits differ. Since $2 > 1$, it follows that $2.222\ldots = \frac{20}{9}$ is the largest of the three numbers.

Working from left to right, this is the first column in which the top two numbers differ. Since $8 > 6$, it follows that 2.168 is the next largest number and that $2.1\overline{6} = 2\tfrac{1}{6}$ is the smallest.

Written in order from least to greatest, we have:

$$2\tfrac{1}{6},\; 2.168,\; \tfrac{20}{9}$$

6 Evaluate expressions containing fractions and decimals.

Expressions can contain both fractions and decimals. In the following examples, we show two methods that can be used to evaluate expressions of this type. With the first method we find the answer by working in terms of fractions.

Self Check 10

Evaluate by working in terms of fractions: $0.53 + \tfrac{1}{6}$ $\tfrac{209}{300}$

Now Try Problem 79

Teaching Example 10 Evaluate $\tfrac{2}{3} + 0.11$ by working in terms of fractions.
Answer: $\tfrac{233}{300}$

EXAMPLE 10 Evaluate $\tfrac{1}{3} + 0.27$ by working in terms of fractions.

Strategy We will begin by writing 0.27 as a fraction.

WHY Then we can use the methods of Chapter 3 for adding fractions with unlike denominators to find the sum.

Solution To write 0.27 as a fraction, it is helpful to read it aloud as "twenty-seven hundredths."

$$\tfrac{1}{3} + 0.27 = \tfrac{1}{3} + \tfrac{27}{100} \quad \text{Replace 0.27 with } \tfrac{27}{100}.$$

$$= \tfrac{1}{3} \cdot \tfrac{100}{100} + \tfrac{27}{100} \cdot \tfrac{3}{3} \quad \begin{array}{l}\text{The LCD for } \tfrac{1}{3} \text{ and } \tfrac{27}{100} \text{ is 300. To build each} \\ \text{fraction so that its denominator is 300,} \\ \text{multiply by a form of 1.}\end{array}$$

$$= \tfrac{100}{300} + \tfrac{81}{300} \quad \begin{array}{l}\text{Multiply the numerators.}\\\text{Multiply the denominators.}\end{array}$$

$$= \tfrac{181}{300} \quad \begin{array}{l}\text{Add the numerators and write the sum over}\\\text{the common denominator 300.}\end{array}$$

Now we will evaluate the expression from Example 10 by working in terms of decimals.

Self Check 11

Estimate the result by working in terms of decimals: $0.53 - \tfrac{1}{6}$

Now Try Problem 87

Self Check 11 Answer approximately 0.36

EXAMPLE 11 Estimate $\tfrac{1}{3} + 0.27$ by working in terms of decimals.

Strategy Since 0.27 has two decimal places, we will begin by finding a decimal approximation for $\tfrac{1}{3}$ to two decimal places.

WHY Then we can use the methods of this chapter for adding decimals to find the sum.

5.5 Fractions and Decimals

Solution We have seen that the decimal equivalent of $\frac{1}{3}$ is the repeating decimal 0.333 . . . . Rounded to the nearest hundredth: $\frac{1}{3} \approx 0.33$.

$\frac{1}{3} + 0.27 \approx 0.33 + 0.27$ Approximate $\frac{1}{3}$ with the decimal 0.33.

≈ 0.60 Do the addition.

$$\begin{array}{r} \overset{1}{0.33} \\ +0.27 \\ \hline 0.60 \end{array}$$

Teaching Example 11 Evaluate $\frac{2}{3} + 0.11$ by working in terms of decimals.
Answer:
approximately 0.78

In Examples 10 and 11, we evaluated $\frac{1}{3} + 0.27$ in different ways. In Example 10, we obtained the exact answer, $\frac{181}{300}$. In Example 11, we obtained an approximation, 0.6. The results seem reasonable when we write $\frac{181}{300}$ in decimal form: $\frac{181}{300} = 0.60333\ldots$.

EXAMPLE 12

Evaluate: $\left(\frac{4}{5}\right)(1.35) + (0.5)^2$

Strategy We will find the decimal equivalent of $\frac{4}{5}$ and then evaluate the expression in terms of decimals.

WHY It is easier to perform multiplication and addition with the given decimals than it would be converting them to fractions.

Solution We use division to find the decimal equivalent of $\frac{4}{5}$.

$$\begin{array}{r} 0.8 \\ 5\overline{)4.0} \\ -4\,0 \\ \hline 0 \end{array}$$

Write a decimal point and one additional zero to the right of the 4.

Now we use the order of operation rule to evaluate the expression.

$\left(\frac{4}{5}\right)(1.35) + (0.5)^2$

$= (\mathbf{0.8})(1.35) + (0.5)^2$ Replace $\frac{4}{5}$ with its decimal equivalent, 0.8.

$= (0.8)(1.35) + 0.25$ Evaluate: $(0.5)^2 = 0.25$.

$= 1.08 + 0.25$ Do the multiplication: $(0.8)(1.35) = 1.08$.

$= 1.33$ Do the addition.

Self Check 12

Evaluate: $(-0.6)^2 + (2.3)\left(\frac{1}{8}\right)$
0.6475
Now Try Problem 99

Teaching Example 12
Evaluate: $\frac{3}{5}(1.75) + (0.6)^2$
Answer:
1.41

$$\begin{array}{r} 0.5 \\ \times\; 0.5 \\ \hline 0.25 \end{array}$$

$$\begin{array}{r} \overset{2\;4}{1.35} \\ \times\;\; 0.8 \\ \hline 1.080 \end{array}$$

$$\begin{array}{r} \overset{1}{1.08} \\ +0.25 \\ \hline 1.33 \end{array}$$

7 Solve application problems involving fractions and decimals.

EXAMPLE 13

Shopping A shopper purchased $\frac{3}{4}$ pound of fruit, priced at $0.88 a pound, and $\frac{1}{3}$ pound of fresh-ground coffee, selling for $6.60 a pound. Find the total cost of these items.

Solution

- $\frac{3}{4}$ pound of fruit was purchased at $0.88 per pound.   *Given*
- $\frac{1}{3}$ pound of coffee was purchased at $6.60 per pound.   *Given*
- What was the total cost of the items?   *Find*

To find the total cost of each item, multiply the number of pounds purchased by the price per pound.

Self Check 13

DELICATESSENS A shopper purchased $\frac{2}{3}$ pound of Swiss cheese, priced at $2.19 per pound, and $\frac{3}{4}$ pound of sliced turkey, selling for $6.40 per pound. Find the total cost of these items. $6.26
Now Try Problem 111

		the number of pounds of fruit	times	the price per pound	plus	the number of pounds of coffee	times	the price per pound.
The total cost of the items	is equal to							

$$\text{The total cost of the items} = \frac{3}{4} \cdot \$0.88 + \frac{1}{3} \cdot \$6.60$$

Teaching Example 13 SHOPPING
A shopper purchased $\frac{2}{3}$ pound of fruit, priced at $0.99 per pound, and $\frac{1}{2}$ pound of fresh-ground coffee, selling for $7.60 per pound. Find the total cost of these items.
Answer:
$4.46

Because 0.88 is divisible by 4 and 6.60 is divisible by 3, we can work with the decimals and fractions in this form; no conversion is necessary.

$$\frac{3}{4} \cdot 0.88 + \frac{1}{3} \cdot 6.60$$

$$= \frac{3}{4} \cdot \frac{0.88}{1} + \frac{1}{3} \cdot \frac{6.60}{1} \quad \text{Express 0.88 as } \frac{0.88}{1} \text{ and 6.60 as } \frac{6.60}{1}.$$

$$= \frac{2.64}{4} + \frac{6.60}{3} \quad \begin{array}{l}\text{Multiply the numerators.}\\ \text{Multiply the denominators.}\end{array}$$

$$= 0.66 + 2.20 \quad \text{Do each division.}$$

$$= 2.86 \quad \text{Do the addition.}$$

```
    2
  0.88
 ×  3
  2.64

   0.66          2.20
 4)2.64        3)6.60
  −2 4           −6
    24            06
   −24            −6
     0            00
                  −0
                   0

  0.66
 +2.20
  2.86
```

The total cost of the items is $2.86.

Check: If approximately 1 pound of fruit, priced at approximately $1 per pound, was purchased, then about $1 was spent on fruit. If exactly $\frac{1}{3}$ of a pound of coffee, priced at approximately $6 per pound, was purchased, then about $\frac{1}{3} \cdot \$6$, or $2, was spent on coffee. Since the approximate cost of the items $1 + $2 = $3, is close to the result, $2.86, the result seems reasonable.

ANSWERS TO SELF CHECKS

1. a. 0.2   b. 0.1875   c. 4.5   2. a. 0.4   b. 0.32   3. 3.85   4. $0.08\overline{3}$ 5. $-0.\overline{39}$ 6. 0.44
7. 0.292   8. a. >   b. <   c. =   9. $\frac{9}{5}, 1.832, 1\frac{5}{6}$ 10. $\frac{209}{300}$ 11. approximately 0.36
12. 0.6475   13. $6.26

SECTION 5.5   STUDY SET

VOCABULARY

Fill in the blanks.

1. A fraction and a decimal are said to be __equivalent__ if they name the same number.
2. The __decimal__ equivalent of $\frac{3}{4}$ is 0.75.
3. 0.75, 0.625, and 3.5 are examples of __terminating__ decimals.
4. 0.3333 … and 1.666 … are examples of __repeating__ decimals.

CONCEPTS

Fill in the blanks.

5. $\frac{7}{8}$ means $7 \div 8$.
6. To write a fraction as a decimal, divide the __numerator__ of the fraction by its denominator.
7. To perform the division shown below, a decimal point and two additional __zeros__ were written to the right of 3.

$$4\overline{)3.00}$$

8. Sometimes, when finding the decimal equivalent of a fraction, the division process ends because a remainder of 0 is obtained. We call the resulting decimal a _terminating_ decimal.

9. Sometimes, when we are finding the decimal equivalent of a fraction, the division process never gives a remainder of 0. We call the resulting decimal a _repeating_ decimal.

10. If the denominator of a fraction in simplified form has factors of only 2's or 5's, or a combination of both, it can be written as a decimal by multiplying it by a form of _1_.

11. a. Round 0.3777... to the nearest hundredth. 0.38
    b. Round 0.212121... to the nearest thousandth. 0.212

12. a. When evaluating the expression $0.25 + \left(2.3 + \frac{2}{5}\right)^2$, would it be easier to work in terms of fractions or decimals? decimals
    b. What is the first step that should be performed to evaluate the expression? write $\frac{2}{5}$ as a decimal

NOTATION

13. Write each decimal in fraction form.
    a. 0.7 $\frac{7}{10}$
    b. 0.77 $\frac{77}{100}$

14. Write each repeating decimal in simplest form using an overbar.
    a. 0.888... $0.\overline{8}$
    b. 0.323232... $0.\overline{32}$
    c. 0.56333... $0.56\overline{3}$
    d. 0.8898989... $0.8\overline{89}$

GUIDED PRACTICE

Write each fraction as a decimal. See Example 1.

15. $\frac{1}{2}$ 0.5
16. $\frac{1}{4}$ 0.25
17. $\frac{7}{8}$ 0.875
18. $\frac{3}{8}$ 0.375
19. $\frac{11}{20}$ 0.55
20. $\frac{17}{20}$ 0.85
21. $\frac{13}{5}$ 2.6
22. $\frac{15}{2}$ 7.5
23. $\frac{9}{16}$ 0.5625
24. $\frac{3}{32}$ 0.09375
25. $-\frac{17}{32}$ -0.53125
26. $-\frac{15}{16}$ -0.9375

Write each fraction as a decimal using multiplication by a form of 1. See Example 2.

27. $\frac{3}{5}$ 0.6
28. $\frac{13}{25}$ 0.52
29. $\frac{9}{40}$ 0.225
30. $\frac{7}{40}$ 0.175
31. $\frac{19}{25}$ 0.76
32. $\frac{21}{50}$ 0.42
33. $\frac{1}{500}$ 0.002
34. $\frac{1}{250}$ 0.004

Write each mixed number in decimal form. See Example 3.

35. $3\frac{3}{4}$ 3.75
36. $5\frac{4}{5}$ 5.8
37. $12\frac{11}{16}$ 12.6875
38. $32\frac{9}{16}$ 32.5625

Write each fraction as a decimal. Use an overbar in your answer. See Example 4.

39. $\frac{1}{9}$ $0.\overline{1}$
40. $\frac{8}{9}$ $0.\overline{8}$
41. $\frac{7}{12}$ $0.58\overline{3}$
42. $\frac{11}{12}$ $0.91\overline{6}$
43. $\frac{7}{90}$ $0.0\overline{7}$
44. $\frac{1}{99}$ $0.\overline{01}$
45. $\frac{1}{60}$ $0.01\overline{6}$
46. $\frac{1}{66}$ $0.0\overline{15}$

Write each fraction as a decimal. Use an overbar in your answer. See Example 5.

47. $-\frac{5}{11}$ $-0.\overline{45}$
48. $-\frac{7}{11}$ $-0.\overline{63}$
49. $-\frac{20}{33}$ $-0.\overline{60}$
50. $-\frac{16}{33}$ $-0.\overline{48}$

Write each fraction in decimal form. Round to the nearest hundredth. See Example 6.

51. $\frac{7}{30}$ 0.23
52. $\frac{8}{9}$ 0.89
53. $\frac{22}{45}$ 0.49
54. $\frac{17}{45}$ 0.38
55. $\frac{24}{13}$ 1.85
56. $\frac{34}{11}$ 3.09
57. $-\frac{13}{12}$ -1.08
58. $-\frac{25}{12}$ -2.08

Write each fraction in decimal form. Round to the nearest thousandth. See Example 7.

59. $\frac{5}{33}$ 0.152
60. $\frac{5}{24}$ 0.208
61. $\frac{10}{27}$ 0.370
62. $\frac{17}{21}$ 0.810

Graph the given numbers on a number line. See Objective 4.

63. $1\frac{3}{4}$, -0.75, $0.\overline{6}$, $-3.8\overline{3}$

64. $2\frac{7}{8}$, -2.375, $0.\overline{3}$, $4.1\overline{6}$

65. 3.875, $-3.\overline{5}$, $0.\overline{2}$, $-1\frac{4}{5}$

66. 1.375, $-4\frac{1}{7}$, $0.\overline{1}$, $-2.\overline{7}$

Place an <, >, or = symbol in the box to make a true statement. See Example 8.

67. $\frac{7}{8}$ < 0.895

68. $\frac{3}{8}$ < 0.381

69. $0.\overline{7}$ > $\frac{17}{22}$

70. $0.\overline{45}$ > $\frac{7}{16}$

71. $\frac{52}{25}$ = 2.08

72. 4.4 = $\frac{22}{5}$

73. $-\frac{11}{20}$ < $-0.\overline{48}$

74. $-0.0\overline{9}$ < $-\frac{1}{11}$

Write the numbers in order from least to greatest. See Example 9.

75. $6\frac{1}{2}$, 6.25, $\frac{19}{3}$ $6.25, \frac{19}{3}, 6\frac{1}{2}$

76. $7\frac{3}{8}$, 7.08, $\frac{43}{6}$ $7.08, \frac{43}{6}, 7\frac{3}{8}$

77. $-0.\overline{81}$, $-\frac{8}{9}$, $-\frac{6}{7}$ $-\frac{8}{9}, -\frac{6}{7}, -0.\overline{81}$

78. $-0.\overline{19}$, $-\frac{1}{11}$, -0.1 $-0.\overline{19}, -0.1, -\frac{1}{11}$

Evaluate each expression. Work in terms of fractions. See Example 10.

79. $\frac{1}{9} + 0.3$ $\frac{37}{90}$

80. $\frac{2}{3} + 0.1$ $\frac{23}{30}$

81. $0.9 - \frac{7}{12}$ $\frac{19}{60}$

82. $0.99 - \frac{5}{6}$ $\frac{47}{300}$

83. $\frac{5}{11}(0.3)$ $\frac{3}{22}$

84. $(0.9)\left(\frac{1}{27}\right)$ $\frac{1}{30}$

85. $\frac{1}{4}(0.25) + \frac{15}{16}$ 1

86. $\frac{2}{5}(0.02) - (0.04)$ $-\frac{4}{125}$

Estimate the value of each expression. Work in terms of decimals. See Example 11.

87. $0.24 + \frac{1}{3}$ 0.57

88. $0.02 + \frac{5}{6}$ 0.85

89. $5.69 - \frac{5}{12}$ 5.27

90. $3.19 - \frac{2}{3}$ 2.52

91. $0.43 - \frac{1}{12}$ 0.35

92. $0.27 + \frac{5}{12}$ 0.69

93. $\frac{1}{15} - 0.55$ -0.48

94. $\frac{7}{30} - 0.84$ -0.61

Evaluate each expression. Work in terms of decimals. See Example 12.

95. $(3.5 + 6.7)\left(-\frac{1}{4}\right)$ -2.55

96. $\left(-\frac{5}{8}\right)\left(5.3 - 3\frac{9}{10}\right)$ -0.875

97. $\left(\frac{1}{5}\right)^2 (1.7)$ 0.068

98. $(2.35)\left(\frac{2}{5}\right)^2$ 0.376

99. $7.5 - (0.78)\left(\frac{1}{2}\right)^2$ 7.305

100. $8.1 - \left(\frac{3}{4}\right)^2 (0.12)$ 8.0325

101. $\frac{3}{8}(3.2) + \left(4\frac{1}{2}\right)\left(-\frac{1}{4}\right)$ 0.075

102. $(-0.8)\left(\frac{1}{4}\right) + \left(\frac{1}{5}\right)(0.39)$ -0.122

APPLICATIONS

103. DRAFTING The architect's scale shown below has several measuring edges. The edge marked 16 divides each inch into 16 equal parts. Find the decimal form for each fractional part of 1 inch that is highlighted with a red arrow. $0.0625, 0.375, 0.5625, 0.9375$

104. MILEAGE SIGNS The freeway sign shown below gives the number of miles to the next three exits. Convert the mileages to decimal notation. 0.75 mi, 2.25 mi, 3.5 mi

105. **GARDENING** Two brands of replacement line for a lawn trimmer shown below are labeled in different ways. On one package, the line's thickness is expressed as a decimal; on the other, as a fraction. Which line is thicker? $\frac{3}{40}$ in.

106. **AUTO MECHANICS** While doing a tune-up, a mechanic checks the gap on one of the spark plugs of a car to be sure it is firing correctly. The owner's manual states that the gap should be $\frac{2}{125}$ inch. The gauge the mechanic uses to check the gap is in decimal notation; it registers 0.025 inch. Is the spark plug gap too large or too small? too large

107. **HORSE RACING** In thoroughbred racing, the time a horse takes to run a given distance is measured using fifths of a second. For example, :23² (read "twenty-three and two") means $23\frac{2}{5}$ seconds. The illustration below lists four split times for a horse named *Speedy Flight* in a $1\frac{1}{16}$-mile race. Express each split time in decimal form. 23.4 sec, 23.8 sec, 24.2 sec, 32.6 sec

Speedy Flight	Turfway Park, Ky	3-year-old
	17 May 2014	$1\frac{1}{16}$ mile
Splits	:23² :23⁴ :24¹ :32³	

108. **GEOLOGY** A geologist weighed a rock sample at the site where it was discovered and found it to weigh $17\frac{7}{8}$ lb. Later, a more accurate digital scale in the laboratory gave the weight as 17.671 lb. What is the difference in the two measurements? 0.204 lb.

109. **WINDOW REPLACEMENTS** The amount of sunlight that comes into a room depends on the area of the windows in the room. What is the area of the window shown below? (*Hint:* Use the formula $A = \frac{1}{2}bh$.) 93.6 in.²

110. **FORESTRY** A command post asked each of three fire crews to estimate the length of the fire line they were fighting. Their reports came back in different forms, as shown. Find the perimeter of the fire. Round to the nearest tenth. 4.7 mi

111. **DELICATESSENS** A shopper purchased $\frac{2}{3}$ pound of green olives, priced at $4.14 per pound, and $\frac{3}{4}$ pound of smoked ham, selling for $5.68 per pound. Find the total cost of these items. $7.02

112. **CHOCOLATE** A shopper purchased $\frac{3}{4}$ pound of dark chocolate, priced at $8.60 per pound, and $\frac{1}{3}$ pound of milk chocolate, selling for $5.25 per pound. Find the total cost of these items. $8.20

WRITING

113. Explain the procedure used to write a fraction in decimal form.

114. How does the terminating decimal 0.5 differ from the repeating decimal $0.\overline{5}$?

115. A student represented the repeating decimal 0.1333 . . . as $0.1\overline{333}$. Is this the best form? Explain why or why not.

116. Is 0.10100100010000 . . . a repeating decimal? Explain why or why not.

117. A student divided 19 by 25 to find the decimal equivalent of $\frac{19}{25}$ to be 0.76. Explain how she can check this result.

118. Explain the error in the following work to find the decimal equivalent for $\frac{5}{6}$.

$$\begin{array}{r}1.2\\5\overline{)6.0}\\-5\\\hline 1\,0\\-1\,0\\\hline 0\end{array}$$ Thus, $\frac{5}{6} = 1.2$. ✗

REVIEW

119. Write each set of numbers.
    a. the first ten whole numbers {0, 1, 2, 3, 4, 5, 6, 7, 8, 9}
    b. the first ten prime numbers {2, 3, 5, 7, 11, 13, 17, 19, 23, 29}
    c. the integers {. . . , −3, −2, −1, 0, 1, 2, 3, . . .}

120. Evaluate: $(0.9)^2$ 0.81

121. Evaluate: $(-0.1)^2$ 0.01

122. Evaluate: $(0.8)^2$ 0.64

Objectives

1. Find the square root of a perfect square.
2. Find the square root of fractions and decimals.
3. Evaluate expressions that contain square roots.
4. Evaluate formulas involving square roots.
5. Approximate square roots.

SECTION 5.6
Square Roots

We have discussed the relationships between addition and subtraction and between multiplication and division. In this section, we explore the relationship between raising a number to a power and finding a root. Decimals play an important role in this discussion.

1 Find the square root of a perfect square.

When we raise a number to the second power, we are squaring it, or finding its **square**.

The square of 6 is 36, because $6^2 = 36$.

The square of -6 is 36, because $(-6)^2 = 36$.

The **square root** of a given number is a number whose square is the given number. For example, the square roots of 36 are 6 and -6, because either number, when squared, is 36.

Every positive number has two square roots. The number 0 has only one square root. In fact, it is its own square root, because $0^2 = 0$.

> **Square Root**
>
> A number is a **square root** of a second number if the square of the first number equals the second number.

Self Check 1

Find the two square roots of 64. **8 and -8**

Now Try Problem 21

Teaching Example 1 Find the two square roots of 4.
Answers:
2 and -2

EXAMPLE 1 Find the two square roots of 49.

Strategy We will ask "What positive number and what negative number, when squared, is 49?"

WHY The square root of 49 is a number whose square is 49.

Solution

7 is a square root of 49 because $7^2 = 49$

and

-7 is a square root of 49 because $(-7)^2 = 49$.

In Example 1, we saw that 49 has two square roots—one positive and one negative. The symbol $\sqrt{}$ is called a **radical symbol** and is used to indicate a positive square root of a nonnegative number. When reading this symbol, we usually drop the word *positive* and simply say *square root*. Since 7 is the positive square root of 49, we can write

$\sqrt{49} = 7$ $\sqrt{49}$ represents the positive number whose square is 49.
Read as "the square root of 49 is 7."

When a number, called the **radicand**, is written under a radical symbol, we have a radical expression.

Radical symbol ↘
$\sqrt{49}$ ← Radicand
Radical expression

Some other examples of radical expressions are:

$$\sqrt{36} \quad \sqrt{100} \quad \sqrt{144} \quad \sqrt{81}$$

To evaluate (or simplify) a radical expression like those shown above, we need to find the positive square root of the radicand. For example, if we evaluate $\sqrt{36}$ (read as "the square root of 36"), the result is

$$\sqrt{36} = 6$$

because $6^2 = 36$.

> **Caution!** Remember that the radical symbol asks you to find only the *positive* square root of the radicand. It is incorrect, for example, to say that
>
> $$\sqrt{36} \text{ is } 6 \text{ and } -6$$

The symbol $-\sqrt{}$ is used to indicate the **negative square root** of a positive number. It is the opposite of the positive square root. Since −6 is the negative square root of 36, we can write

$$-\sqrt{36} = -6 \quad \text{Read as "the negative square root of 36 is −6" or "the opposite of the square root of 36 is −6." } -\sqrt{36} \text{ represents the negative number whose square is 36.}$$

If the number under the radical symbol is 0, we have $\sqrt{0} = 0$.

Numbers, such as 36 and 49, that are squares of whole numbers, are called **perfect squares**. To evaluate square root radical expressions, it is helpful to be able to identify perfect square radicands. You need to memorize the following list of perfect squares, shown in red.

Perfect Squares

$0 = 0^2$	$16 = 4^2$	$64 = 8^2$	$144 = 12^2$
$1 = 1^2$	$25 = 5^2$	$81 = 9^2$	$169 = 13^2$
$4 = 2^2$	$36 = 6^2$	$100 = 10^2$	$196 = 14^2$
$9 = 3^2$	$49 = 7^2$	$121 = 11^2$	$225 = 15^2$

A calculator is helpful in finding the square root of a perfect square that is larger than 225.

EXAMPLE 2 Evaluate each square root: **a.** $\sqrt{81}$ **b.** $-\sqrt{100}$

Strategy In each case, we will determine what positive number, when squared, produces the radicand.

WHY The radical symbol $\sqrt{}$ indicates that the positive square root of the number written under it should be found.

Solution
a. $\sqrt{81} = 9$ Ask: What positive number, when squared, is 81?
The answer is 9 because $9^2 = 81$.

b. $-\sqrt{100}$ is the opposite (or negative) of the square root of 100. Since $\sqrt{100} = 10$, we have

$$-\sqrt{100} = -10$$

Self Check 2
Evaluate each square root:
a. $\sqrt{144}$ 12
b. $-\sqrt{81}$ −9

Now Try Problems 25 and 29

Teaching Example 2 Evaluate each square root:
a. $\sqrt{25}$
b. $-\sqrt{121}$
Answers:
a. 5
b. −11

> **Caution!** Radical expressions such as
>
> $$\sqrt{-36} \quad \sqrt{-100} \quad \sqrt{-144} \quad \sqrt{-81}$$
>
> do not represent real numbers, because there are no real numbers that when squared give a negative number.
>
> Be careful to note the difference between expressions such as $-\sqrt{36}$ and $\sqrt{-36}$. We have seen that $-\sqrt{36}$ is a real number: $-\sqrt{36} = -6$. In contrast, $\sqrt{-36}$ is not a real number.

> **Using Your CALCULATOR** **Finding a square root**
>
> We use the square root key $\boxed{\sqrt{}}$ on a scientific calculator to find square roots. For example, to find $\sqrt{729}$, we enter the number and press the square root key.
>
> 729 $\boxed{\sqrt{}}$ $\boxed{27}$
>
> We have found that $\sqrt{729} = 27$. To check this result, we need to square 27. This can be done by entering 27 and pressing the $\boxed{x^2}$ key. We obtain 729. Thus, 27 is the square root of 729.
>
> Some calculator models require keystrokes of $\boxed{2nd}$ and then $\boxed{\sqrt{}}$ followed by the radicand to find a square root.

2 Find the square root of fractions and decimals.

So far, we have found square roots of whole numbers. We can also find square roots of fractions and decimals.

Self Check 3

Evaluate:

a. $\sqrt{\dfrac{16}{49}}$ $\dfrac{4}{7}$

b. $\sqrt{0.04}$ 0.2

Now Try Problems 37 and 43

Teaching Example 3 Evaluate each square root:

a. $\sqrt{\dfrac{81}{4}}$

b. $\sqrt{0.25}$

Answers:

a. $\dfrac{9}{2}$

b. 0.5

> **EXAMPLE 3** Evaluate each square root: a. $\sqrt{\dfrac{25}{64}}$ b. $\sqrt{0.81}$
>
> **Strategy** In each case, we will determine what positive number, when squared, produces the radicand.
>
> **WHY** The radical symbol $\sqrt{}$ indicates that the positive square root of the number written under it should be found.
>
> **Solution**
>
> a. $\sqrt{\dfrac{25}{64}} = \dfrac{5}{8}$ Ask: What positive fraction, when squared, is $\dfrac{25}{64}$? The answer is $\dfrac{5}{8}$ because $\left(\dfrac{5}{8}\right)^2 = \dfrac{25}{64}$.
>
> b. $\sqrt{0.81} = 0.9$ Ask: What positive decimal, when squared, is 0.81? The answer is 0.9 because $(0.9)^2 = 0.81$.

3 Evaluate expressions that contain square roots.

In Chapters 1, 2, 3, and 4, we used the order of operations rule to evaluate expressions that involve more than one operation. If an expression contains any square roots, they are to be evaluated at the same stage in your solution as exponential expressions. (See Step 2 in the familiar order of operations rule on the next page.)

Order of Operations

1. Perform all calculations within parentheses and other grouping symbols following the order listed in Steps 2–4 below, working from the innermost pair of grouping symbols to the outermost pair.
2. Evaluate all exponential expressions and **square roots**.
3. Perform all multiplications and divisions as they occur from left to right.
4. Perform all additions and subtractions as they occur from left to right.

EXAMPLE 4
Evaluate: **a.** $\sqrt{64} + \sqrt{9}$ **b.** $-\sqrt{25} - \sqrt{225}$

Strategy We will scan the expression to determine what operations need to be performed. Then we will perform those operations, one at a time, following the order of operations rule.

WHY Every numerical expression has only one correct value. If we don't follow the correct order of operations, the expression can have more than one value.

Solution Since the expression does not contain any parentheses, we begin with Step 2 of the rules for the order of operations: Evaluate all exponential expressions and any square roots.

a. $\sqrt{64} + \sqrt{9} = 8 + 3$ Evaluate each square root first.
$= 11$ Do the addition.

b. $-\sqrt{25} - \sqrt{225} = -5 - 15$ Evaluate each square root first.
$= -20$ Do the subtraction.

Self Check 4
Evaluate:
a. $\sqrt{121} + \sqrt{1}$ 12
b. $-\sqrt{9} - \sqrt{196}$ −17

Now Try Problems 49 and 53

Teaching Example 4 Evaluate:
a. $\sqrt{81} + \sqrt{144}$
b. $-\sqrt{169} - \sqrt{100}$

Answers:
a. 21
b. −23

EXAMPLE 5
Evaluate: **a.** $6\sqrt{100}$ **b.** $-5\sqrt{16} + 3\sqrt{9}$

Strategy We will scan the expression to determine what operations need to be performed. Then we will perform those operations, one at a time, following the order of operations rule.

WHY Every numerical expression has only one correct value. If we don't follow the correct order of operations, the expression can have more than one value.

Solution Since the expression does not contain any parentheses, we begin with Step 2 of the rules for the order of operations: Evaluate all exponential expressions and any square roots.

a. We note that $6\sqrt{100}$ means $6 \cdot \sqrt{100}$.

$6\sqrt{100} = 6(10)$ Evaluate the square root first.
$= 60$ Do the multiplication.

b. $-5\sqrt{16} + 3\sqrt{9} = -5(4) + 3(3)$ Evaluate each square root first.
$= -20 + 9$ Do the multiplication.
$= -11$ Do the addition.

Self Check 5
Evaluate:
a. $8\sqrt{121}$ 88
b. $-6\sqrt{25} + 2\sqrt{36}$ −18

Now Try Problems 57 and 61

Teaching Example 5 Evaluate:
a. $5\sqrt{49}$
b. $-6\sqrt{1} + \sqrt{64}$

Answers:
a. 35
b. 2

EXAMPLE 6
Evaluate: $12 + 3[3^2 - (4 - 1)\sqrt{36}]$

Strategy We will work within the parentheses first and then within the brackets. Within each set of grouping symbols, we will follow the order of operations rule.

WHY By the order of operations rule, we must work from the *innermost* pair of grouping symbols to the *outermost*.

Self Check 6
Evaluate:
$10 - 4[2^2 - (3 + 2)\sqrt{4}]$ 34

Now Try Problems 65 and 69

Teaching Example 6 Evaluate:
$15 + 2[4^2 - (4 + 1)\sqrt{25}]$
Answer:
-3

Solution

$$12 + 3[3^2 - (4 - 1)\sqrt{36}] = 12 + 3[3^2 - 3\sqrt{36}] \quad \text{Do the subtraction within the parentheses.}$$

$$= 12 + 3[9 - 3(6)] \quad \text{Within the brackets, evaluate the exponential expression and the square root.}$$

$$= 12 + 3[9 - 18] \quad \text{Do the multiplication within the brackets.}$$

$$= 12 + 3[-9] \quad \text{Do the subtraction within the brackets.}$$

$$= 12 + (-27) \quad \text{Do the multiplication.}$$

$$= -15 \quad \text{Do the addition.}$$

4 Evaluate formulas involving square roots.

To evaluate formulas that involve square roots, we replace the letters with specific numbers and then use the order of operations rule.

Self Check 7

Evaluate $a = \sqrt{c^2 - b^2}$ for $c = 17$ and $b = 15$. 8

Now Try Problem 81

Teaching Example 7 Evaluate $c = \sqrt{a^2 + b^2}$ for $a = 5$ and $b = 12$.
Answer:
13

EXAMPLE 7 Evaluate $c = \sqrt{a^2 + b^2}$ for $a = 3$ and $b = 4$.

Strategy In the given formula, we will replace the letter a with 3 and b with 4. Then we will use the order of operations rule to find the value of the radicand.

WHY We need to know the value of the radicand before we can find its square root.

Solution

$$c = \sqrt{a^2 + b^2} \quad \text{This is the formula to evaluate.}$$
$$c = \sqrt{3^2 + 4^2} \quad \text{Replace } a \text{ with 3 and } b \text{ with 4.}$$
$$c = \sqrt{9 + 16} \quad \text{Evaluate the exponential expressions.}$$
$$c = \sqrt{25} \quad \text{Do the addition.}$$
$$c = 5 \quad \text{Evaluate the square root.}$$

5 Approximate square roots.

In Examples 2–7, we have found square roots of perfect squares. If a number is not a perfect square, we can use the $\boxed{\sqrt{}}$ key on a calculator or a table of square roots to find its *approximate* square root. For example, to find $\sqrt{17}$ using a scientific calculator, we enter 17 and press the square root key:

17 $\boxed{\sqrt{}}$

The display reads

4.123105626

This result is an approximation, because the exact value of $\sqrt{17}$ is a **nonterminating decimal** that never repeats. If we round to the nearest thousandth, we have

$\sqrt{17} \approx 4.123$ Read $\approx$ as "is approximately equal to."

To check this approximation, we square 4.123.

$(4.123)^2 = 16.999129$

n	$\sqrt{n}$
11	3.317
12	3.464
13	3.606
14	3.742
15	3.873
16	4.000
17	4.123
18	4.243
19	4.359
20	4.472

Since the result is close to 17, we know that $\sqrt{17} \approx 4.123$.

A portion of the table of square roots from Appendix 3 on page A-15 is shown in the margin on the previous page. The table gives decimal approximations of square roots of whole numbers that are not perfect squares. To find an approximation of $\sqrt{17}$ to the nearest thousandth, we locate 17 in the n column of the table and scan directly right, to the $\sqrt{n}$ column, to find that $\sqrt{17} \approx 4.123$.

EXAMPLE 8
Use a calculator to approximate each square root. Round to the nearest hundredth. **a.** $\sqrt{373}$ **b.** $\sqrt{56.2}$ **c.** $\sqrt{0.0045}$

Strategy We will identify the radicand and find the square root using the $\sqrt{}$ key. Then we will identify the digit in the thousandths column of the display.

WHY To round to the hundredths column, we must determine whether the digit in the thousandths column is less than 5, or greater than or equal to 5.

Solution
a. From the calculator, we get $\sqrt{373} \approx 19.31320792$. Rounded to the nearest hundredth, $\sqrt{373} \approx 19.31$.
b. From the calculator, we get $\sqrt{56.2} \approx 7.496665926$. Rounded to the nearest hundredth, $\sqrt{56.2} \approx 7.50$.
c. From the calculator, we get $\sqrt{0.0045} \approx 0.067082039$. Rounded to the nearest hundredth, $\sqrt{0.0045} \approx 0.07$.

Self Check 8
Use a scientific calculator to approximate each square root. Round to the nearest hundredth.
a. $\sqrt{153}$ 12.37
b. $\sqrt{607.8}$ 24.65
c. $\sqrt{0.076}$ 0.28

Now Try Problems 87 and 91

Teaching Example 8 Use a calculator to approximate each square root. Round to the nearest hundredth.
a. $\sqrt{217}$
b. $\sqrt{89.1}$
c. $\sqrt{0.032}$
Answers:
a. 14.73
b. 9.44
c. 0.18

ANSWERS TO SELF CHECKS
1. 8 and −8   **2. a.** 12 **b.** −9   **3. a.** $\frac{4}{7}$ **b.** 0.2   **4. a.** 12 **b.** −17   **5. a.** 88 **b.** −18
6. 34   **7.** 8   **8. a.** 12.37 **b.** 24.65 **c.** 0.28

SECTION 5.6   STUDY SET

VOCABULARY
Fill in the blanks.

1. When we raise a number to the second power, we are squaring it, or finding its __square__.
2. The square __root__ of a given number is a number whose square is the given number.
3. The symbol $\sqrt{}$ is called a __radical__ symbol.
4. Label the *radicand*, the *radical expression*, and the *radical symbol* in the illustration below.

    Radical symbol
    $\sqrt{64}$ ← Radicand
    Radical expression

5. Whole numbers, such as 36 and 49, that are squares of whole numbers are called __perfect__ squares.
6. The exact value of $\sqrt{17}$ is a __nonterminating__ decimal that never repeats.

CONCEPTS
Fill in the blanks.

7. **a.** The square of 5 is 25, because $5^2 =$ 25.
   b. The square of $\frac{1}{4}$ is $\frac{1}{16}$, because $\left(\frac{1}{4}\right)^2 = \frac{1}{16}$.
8. Complete the list of perfect squares: 1, 4, 9, 16, 25, 36, 49, 64, 81, 100, 121, 144, 169, 196, 225.
9. **a.** $\sqrt{49} = 7$, because $7^2 = 49$.
   b. $\sqrt{4} = 2$, because $2^2 = 4$.
10. **a.** $\sqrt{\frac{9}{16}} = \frac{3}{4}$, because $\left(\frac{3}{4}\right)^2 = \frac{9}{16}$.
    b. $\sqrt{0.16} = 0.4$, because $(0.4)^2 = 0.16$.
11. Evaluate each square root.
    a. $\sqrt{1}$ 1
    b. $\sqrt{0}$ 0
12. Evaluate each square root.
    a. $\sqrt{121}$ 11
    b. $\sqrt{144}$ 12
    c. $\sqrt{169}$ 13
    d. $\sqrt{196}$ 14
    e. $\sqrt{225}$ 15

13. In what step of the order of operations rule are square roots to be evaluated? Step 2: Evaluate all exponential expressions and any square roots.

14. Graph $\sqrt{9}$ and $-\sqrt{4}$ on a number line.

15. Graph $-\sqrt{3}$ and $\sqrt{7}$ on a number line. (*Hint:* Use a calculator or square root table to approximate each square root first.)

16. a. Between what two whole numbers would $\sqrt{19}$ be located when graphed on a number line? 4, 5
    b. Between what two whole numbers would $\sqrt{50}$ be located when graphed on a number line? 7, 8

NOTATIONS
Fill in the blanks.

17. a. The symbol $\sqrt{}$ is used to indicate a positive __square__ __root__.
    b. The symbol $-\sqrt{}$ is used to indicate the __negative__ square root of a positive number.

18. $4\sqrt{9}$ means 4 __·__ $\sqrt{9}$.

Complete each step to evaluate the expression.

19. $-\sqrt{49} + \sqrt{64} = $ __−7__ $+$ __8__
    $= 1$

20. $2\sqrt{100} - 5\sqrt{25} = 2(\underline{10}) - 5(\underline{5})$
    $= \underline{20} - 25$
    $= -5$

GUIDED PRACTICE
Find the two square roots of each number. See Example 1.

21. 25 5 and −5
22. 1 1 and −1
23. 16 4 and −4
24. 144 12 and −12

Evaluate each square root without using a calculator. See Example 2.

25. $\sqrt{16}$ 4
26. $\sqrt{64}$ 8
27. $\sqrt{9}$ 3
28. $\sqrt{25}$ 5
29. $-\sqrt{144}$ −12
30. $-\sqrt{121}$ −11
31. $-\sqrt{49}$ −7
32. $-\sqrt{81}$ −9

Use a calculator to evaluate each square root. See Objective 1, Using Your Calculator.

33. $\sqrt{961}$ 31
34. $\sqrt{841}$ 29
35. $\sqrt{3,969}$ 63
36. $\sqrt{5,625}$ 75

Evaluate each square root without using a calculator. See Example 3.

37. $\sqrt{\dfrac{4}{25}}$ $\dfrac{2}{5}$
38. $\sqrt{\dfrac{36}{121}}$ $\dfrac{6}{11}$
39. $-\sqrt{\dfrac{16}{9}}$ $-\dfrac{4}{3}$
40. $-\sqrt{\dfrac{64}{25}}$ $-\dfrac{8}{5}$
41. $-\sqrt{\dfrac{1}{81}}$ $-\dfrac{1}{9}$
42. $-\sqrt{\dfrac{1}{4}}$ $-\dfrac{1}{2}$
43. $\sqrt{0.64}$ 0.8
44. $\sqrt{0.36}$ 0.6
45. $-\sqrt{0.81}$ −0.9
46. $-\sqrt{0.49}$ −0.7
47. $\sqrt{0.09}$ 0.3
48. $\sqrt{0.01}$ 0.1

Evaluate each expression without using a calculator. See Example 4.

49. $\sqrt{36} + \sqrt{1}$ 7
50. $\sqrt{100} + \sqrt{16}$ 14
51. $\sqrt{81} + \sqrt{49}$ 16
52. $\sqrt{4} + \sqrt{36}$ 8
53. $-\sqrt{144} - \sqrt{16}$ −16
54. $-\sqrt{1} - \sqrt{196}$ −15
55. $-\sqrt{225} + \sqrt{144}$ −3
56. $-\sqrt{169} + \sqrt{16}$ −9

Evaluate each expression without using a calculator. See Example 5.

57. $4\sqrt{25}$ 20
58. $2\sqrt{81}$ 18
59. $-10\sqrt{196}$ −140
60. $-40\sqrt{4}$ −80
61. $-4\sqrt{169} + 2\sqrt{4}$ −48
62. $-6\sqrt{81} + 5\sqrt{1}$ −49
63. $-8\sqrt{16} + 5\sqrt{225}$ 43
64. $-3\sqrt{169} + 2\sqrt{225}$ −9

Evaluate each expression without using a calculator. See Example 6.

65. $15 + 4[5^2 - (6-1)\sqrt{4}]$ 75
66. $18 + 2[4^2 - (7-3)\sqrt{9}]$ 26
67. $50 - [(6^2 - 24) + 9\sqrt{25}]$ −7
68. $40 - [(7^2 - 40) + 7\sqrt{64}]$ −25
69. $\sqrt{196} + 3(5^2 - 2\sqrt{225})$ −1
70. $\sqrt{169} + 2(7^2 - 3\sqrt{144})$ 39
71. $\dfrac{\sqrt{16} - 6(2^2)}{\sqrt{4}}$ −10
72. $\dfrac{\sqrt{49} - 3(1^6)}{\sqrt{16} - \sqrt{64}}$ −1
73. $\sqrt{\dfrac{1}{16}} - \sqrt{\dfrac{9}{25}}$ $-\dfrac{7}{20}$
74. $\sqrt{\dfrac{25}{9}} - \sqrt{\dfrac{64}{81}}$ $\dfrac{7}{9}$
75. $5(-\sqrt{49})(-2)^2$ −140
76. $(-\sqrt{64})(-2)(3)^3$ 432
77. $(6^2)\sqrt{0.04} + 2.36$ 9.56
78. $(5^2)\sqrt{0.25} + 4.7$ 17.2
79. $-(-3\sqrt{1.44} + 5)$ −1.4
80. $-(-2\sqrt{1.21} - 6)$ 8.2

Evaluate each formula without using a calculator. See Example 7.

81. Evaluate $c = \sqrt{a^2 + b^2}$ for $a = 9$ and $b = 12$. 15
82. Evaluate $c = \sqrt{a^2 + b^2}$ for $a = 6$ and $b = 8$. 10

83. Evaluate $a = \sqrt{c^2 - b^2}$ for $c = 25$ and $b = 24$. 7
84. Evaluate $b = \sqrt{c^2 - a^2}$ for $c = 17$ and $a = 8$. 15

Use a calculator (or the square root table in Appendix 3) to complete each square root table. Round to the nearest thousandth when an answer is not exact. See Example 8.

85.

Number	Square Root
1	1
2	1.414
3	1.732
4	2
5	2.236
6	2.449
7	2.646
8	2.828
9	3
10	3.162

86.

Number	Square Root
10	3.162
20	4.472
30	5.477
40	6.325
50	7.071
60	7.746
70	8.367
80	8.944
90	9.487
100	10

Use a calculator (or a square root table) to approximate each of the following to the nearest hundredth. See Example 8.

87. $\sqrt{5}$ 2.24
88. $\sqrt{51}$ 7.14
89. $\sqrt{66}$ 8.12
90. $\sqrt{204}$ 14.28

Use a calculator to approximate each of the following to the nearest thousandth. See Example 8.

91. $\sqrt{24.05}$ 4.904
92. $\sqrt{70.69}$ 8.408
93. $-\sqrt{11.1}$ −3.332
94. $\sqrt{0.145}$ 0.381

APPLICATIONS

In the following problems, some lengths are expressed as square roots. Solve each problem by evaluating any square roots. You may need to use a calculator. If so, round to the nearest tenth when an answer is not exact.

95. CARPENTRY Find the length of the slanted side of each roof truss shown below.

a.

b.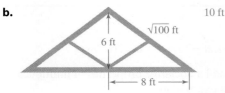

96. RADIO ANTENNAS Refer to the illustration below. How far from the base of the antenna is each guy wire anchored to the ground? (The measurements are in feet.) 12 ft, 4 ft, 6 ft

97. BASEBALL The illustration below shows some dimensions of a major league baseball field. How far is it from home plate to second base? 127.3 ft

98. SURVEYING Refer to the illustration below. Use the imaginary triangles set up by a surveyor to find the length of each lake. (The measurements are in meters.)

a. 564 m

b. 305 m

99. FLATSCREEN TELEVISIONS The picture screen on a television set is measured diagonally. What size screen is shown below? 42-inch screen

100. LADDERS A painter's ladder is shown below. How long are the legs of the ladder? 15 ft, 13 ft

WRITING

101. When asked to find $\sqrt{16}$, a student answered 8. Explain his misunderstanding of the concept of square root.

102. Explain the difference between the *square* and the *square root* of a number. Give an example.

103. What is a *nonterminating* decimal? Use an example in your explanation.

104. a. How would you check whether $\sqrt{389} = 17$?
  b. How would you check whether $\sqrt{7} \approx 2.65$?

105. Explain why $\sqrt{-4}$ does not represent a real number.

106. Is there a difference between $-\sqrt{25}$ and $\sqrt{-25}$? Explain.

107. $\sqrt{6} \approx 2.449$. Explain why an $\approx$ symbol is used and not an $=$ symbol.

108. Without evaluating the following square roots, determine which is the largest and which is the smallest. Explain how you decided.

$$\sqrt{23}, \sqrt{27}, \sqrt{11}, \sqrt{6}, \sqrt{20}$$

REVIEW

109. Simplify: $5(-2q)$ $-10q$

110. Simplify: $-4(5t)$ $-20t$

111. Solve: $-7y + 2 = 44$ -6

112. Solve: $8q - 3 = 29$ 4

Objectives

1. Simplify products.
2. Use the distributive property.
3. Simplify expressions by combining like terms.
4. Use one property of equality to solve equations that involve decimals.
5. Use more than one property of equality to solve equations that involve decimals.
6. Use equations to solve application problems that involve decimals.

SECTION 5.7
Solving Equations That Involve Decimals

In this section, we revisit the topic of solving equations. The equations that we will solve involve decimals, and some of the solutions are decimals as well. But first, we need to practice simplifying algebraic expressions that involve decimals.

1 Simplify products.

In algebra, we often replace one algebraic expression with another that is equivalent and simpler in form. That process is called *simplifying an expression*.

EXAMPLE 1 Simplify:

a. $5.4 \cdot 6x$ **b.** $9.91y(-8)$ **c.** $-1.2(-7.1t)$ **d.** $2(6.78s)4$

Strategy We will use the commutative and associative properties of multiplication to reorder and regroup the factors in each expression.

WHY We want to group all of the numerical factors of an expression together so that we can find their product.

Solution

a. $5.4 \cdot 6x = (5.4 \cdot 6)x$ — Use the associative property of multiplication to regroup the factors.

$= 32.4x$ — Do the multiplication within the parentheses.

$$\begin{array}{r} \overset{2}{5.4} \\ \times 6 \\ \hline 32.4 \end{array}$$

b. $9.91y(-8) = 9.91(-8)y$ — Use the commutative property of multiplication to reorder the factors.

$= -79.28y$ — Do the multiplication, working from left to right. Since the signs of 9.91 and -8 are unlike, make the final answer negative.

$$\begin{array}{r} \overset{7}{9.91} \\ \times 8 \\ \hline 79.28 \end{array}$$

c. $-1.2(-7.1t) = [-1.2(-7.1)]t$ — Use the associative property of multiplication to regroup the factors.

$= 8.52t$ — Do the multiplication within the brackets. Since the signs are like, make the final answer positive.

$$\begin{array}{r} 1.2 \\ \times 7.1 \\ \hline 12 \\ 840 \\ \hline 8.52 \end{array}$$

d. $2(6.78s)4 = (2 \cdot 6.78 \cdot 4)s$ — Use the commutative and associative properties of multiplication to reorder and regroup the factors.

$= (8 \cdot 6.78)s$ — Within the parentheses, use the commutative property of multiplication: multiply 2 and 4 to get 8.

$= 54.24s$ — Complete the multiplication within the parentheses.

$$\begin{array}{r} \overset{66}{6.78} \\ \times 8 \\ \hline 54.24 \end{array}$$

Self Check 1

Simplify:
a. $1.6 \cdot 9a$ $14.4a$
b. $4.23d(-3)$ $-12.69d$
c. $-8.9(-1.5x)$ $13.35x$
d. $3(7.7g)2$ $46.2g$

Now Try Problems 13, 15, 17, and 19

Teaching Example 1
Simplify:
a. $7.3 \cdot 4y$
b. $6.52b(-3)$
c. $-2.8(-7.7r)$
d. $3(5.09n)2$
Answers:
a. $29.2y$
b. $-19.56b$
c. $21.56r$
d. $30.54n$

2   Use the distributive property.

Another property that is often used to simplify algebraic expressions is the *distributive property*.

EXAMPLE 2   Multiply:

a. $3.2(3b + 4)$ **b.** $-6(5.9y - 8.2)$ **c.** $(t - 9.1)0.07$

Strategy In each case, we will distribute the multiplication by the factor outside the parentheses over each term within the parentheses.

WHY The expressions within the parentheses involve more than one term. To multiply, we must use the distributive property.

Solution

a. $3.2(3b + 4) = 3.2(3b) + 3.2(4)$ — Distribute the multiplication by 3.2.

$= (3.2 \cdot 3)b + 3.2(4)$ — In the first term, use the associative property of multiplication to regroup the factors.

$= 9.6b + 12.8$ — Do the multiplication. Try to go directly to this step.

$$\begin{array}{r} 3.2 \\ \times 3 \\ \hline 9.6 \end{array}$$

$$\begin{array}{r} 3.2 \\ \times 4 \\ \hline 12.8 \end{array}$$

Self Check 2

Multiply:
a. $4.7(2x + 8)$ $9.4x + 37.6$
b. $-9(5.3y - 3.9)$ $-47.7y + 35.1$
c. $(c - 2.2)0.08$ $0.08c - 0.176$

Now Try Problems 21, 23, and 25

Teaching Example 2
Multiply:
a. $7.3(8t + 9)$
b. $-5(8.5n - 6.9)$
c. $(m - 2.4)0.09$
Answers:
a. $58.4t + 65.7$
b. $-42.5n + 34.5$
c. $0.09m - 0.216$

b. $-6(5.9y - 8.2) = -6(5.9y) - (-6)(8.2)$ Distribute the multiplication by -6.

$= (-6 \cdot 5.9)y - (-6)(8.2)$ In the first term, use the associative property of multiplication to regroup the factors.

$= -35.4y - (-49.2)$ Do the multiplication.

$= -35.4y + 49.2$ Write the result in simpler form. Add the opposite of -49.2. Try to go directly to this step.

```
   5
  5.9
×  6
 35.4

   1
  8.2
×  6
 49.2
```

c. $(t - 9.1)\,0.07 = (t)0.07 - (9.1)0.07$ Distribute the multiplication by 0.07.

$= 0.07t - 0.637$ Do the multiplication. Try to go directly to this step.

```
  9.1
×0.07
0.637
```

3 Simplify expressions by combining like terms.

Recall that **like terms** are terms containing exactly the same variables raised to exactly the same powers. Here are several examples.

Like terms	*Unlike terms*	
$4.5x$ and $7.3x$	$4.5x$ and $7.3y$	The variables are not the same.
$-1.4p^2$ and $2.8p^2$	$-1.4p$ and $2.8p^2$	Same variable, but different powers.
$0.005c^3d$ and $1.22c^3d$	$0.005c^3d$ and $1.22a^3b$	The variables are not the same.

Like terms are combined by adding or subtracting the coefficients of the terms and keeping the same variables with the same exponents.

Self Check 3

Simplify by combining like terms:

a. $5.2y + 1.6y$ $6.8y$

b. $-3.9s + (-8.5s) + 2.8s$ $-9.6s$

c. $0.55n^2 - 0.93n^2$ $-0.38n^2$

d. $6.72x + 7 + 3.04x - 2.81$
   $9.76x + 4.19$

Now Try Problems 29, 33, and 35

Teaching Example 3
Simplify by combining like terms:
a. $4.5a + 3.3a$
b. $-5.1b + (-9.7b) + 4.9b$
c. $0.33h^2 - 0.87h^2$
d. $5.02y + 6 + 2.97y - 1.31$

Answers:
a. $7.8a$
b. $-9.9b$
c. $-0.54h^2$
d. $7.99y + 4.69$

EXAMPLE 3 Simplify by combining like terms: **a.** $2.6x + 9.3x$
b. $-8.4p + (-6.7p) + 4.5p$ **c.** $0.38s^2 - 0.52s^2$ **d.** $4.16w + 9 + 3.85w - 6.32$

Strategy We will use the distributive property in reverse to add (or subtract) the coefficients of the like terms. We will keep the same variables raised to the same powers.

WHY To combine like terms means to add or subtract the like terms in an expression.

Solution

a. Since $2.6x$ and $9.3x$ are like terms with the common variable x, we can combine them.

$2.6x + 9.3x = 11.9x$ Think: $(2.6 + 9.3)x = 11.9x$.

```
  2.6
+ 9.3
 11.9
```

b. Since $-8.4p$, $-6.7p$, and $4.5p$ are like terms with the common variable p, we can combine them.

$-8.4p + (-6.7p) + 4.5p$

$= -15.1p + 4.5p$ Work left to right. Think: $[-8.4 + (-6.7)]p = -15.1p$.

$= -10.6p$ Think: $(-15.1 + 4.5)p = -10.6p$.

```
   1
  8.4
+ 6.7
 15.1

  4 11
 1̶5̶.1̶
-  4.5
  10.6
```

c. Since $0.38s^2$ and $0.52s^2$ are like terms with the same variable s raised to the same power, 2, we can combine them.

$0.38s^2 - 0.52s^2 = -0.14s^2$ Think: $(0.38 - 0.52)s^2 = -0.14s^2$.

d. We can combine the like terms that involve w. Since the constant terms in the expression, 9 and -6.32, are like terms, we can combine them, as well.

Think: $(4.16 + 3.85)w = 8.01w$.

$4.16w + 9 + 3.85w - 6.32 = 8.01w + 2.68$

Think: $9 - 6.32 = 2.68$.

EXAMPLE 4
Simplify: $4.9(x + 5) - 3.7 - (2.5x - 6.3)$

Strategy First, we will use the distributive property to remove the parentheses. Then we will identify any like terms and combine them.

WHY Any multiplication should be performed before the addition or subtraction.

Solution

$4.9(x + 5) - 3.7 - (2.5x - 6.3)$

$= 4.9(x + 5) - 3.7 - 1(2.5x - 6.3)$ Replace the $-$ symbol in front of $(2.5x - 6.3)$ with -1.

$= 4.9x + 24.5 - 3.7 - 2.5x + 6.3$ Distribute the multiplication by 4.9 and -1.

$= 2.4x + 20.8 + 6.3$ Combine like terms.
Think: $(4.9 - 2.5)x = 2.4x$.
Think: $24.5 - 3.7 = 20.8$.

$= 2.4x + 27.1$ Combine like terms.
Think: $20.8 + 6.3 = 27.1$.

Self Check 4
Simplify:
$6.2(3y - 4) + 2.3 - (-5.8y + 4.7)$
$24.4y - 27.2$

Now Try Problem 37

Teaching Example 4
Simplify: $3.8(y + 2) - 4.9 - (1.4y + 7.9)$
Answer:
$2.4y - 5.2$

4 Use one property of equality to solve equations that involve decimals.

Recall that to **solve an equation**, we find all the values of the variable that make the equation true. The properties of equality that we used to solve equations involving whole numbers, integers, and fractions are also used to solve equations that involve decimals.

EXAMPLE 5
Solve each equation and check the result:

a. $x + 3.5 = 7.8$ **b.** $x - 1.23 = -4.52$

Strategy We will use a property of equality to isolate the variable on one side of the equation.

WHY To solve the original equation, we want to find a simpler equivalent equation of the form $x = $ **a number**, whose solution is obvious.

Self Check 5
Solve each equation and check the result:

a. $4.6 + x = 15.7$ 11.1

b. $-1.24 = r - 0.04$ -1.2

Now Try Problems 45 and 49

Teaching Example 5
Solve each equation and check the result:
a. $x + 3.7 = 10.9$
b. $x - 2.31 = -5.68$

Answers:
a. 7.2
b. −3.37

Solution

a. We will use the subtraction property of equality to isolate x on the left side of the equation. We can undo the addition of 3.5 by subtracting 3.5 from both sides.

$$x + 3.5 = 7.8 \quad \text{This is the equation to solve.}$$
$$x + 3.5 - 3.5 = 7.8 - 3.5 \quad \text{Subtract 3.5 from both sides.}$$
$$x = 4.3 \quad \text{On the left side, subtract: } 3.5 - 3.5 = 0.$$
$$\text{On the right side, subtract: } 7.8 - 3.5 = 4.3.$$

$$\begin{array}{r} 7.8 \\ -3.5 \\ \hline 4.3 \end{array}$$

To check, we substitute 4.3 for x in the original equation.

$$x + 3.5 = 7.8 \quad \text{This is the original equation.}$$
$$4.3 + 3.5 \stackrel{?}{=} 7.8 \quad \text{Substitute 4.3 for } x.$$
$$7.8 = 7.8 \quad \text{On the left side, do the addition.}$$

$$\begin{array}{r} 4.3 \\ +3.5 \\ \hline 7.8 \end{array}$$

Since the resulting statement $7.8 = 7.8$ is true, 4.3 is the solution.

b. We will use the addition property of equality to isolate x on the left side. We can undo the subtraction of 1.23 by adding 1.23 to both sides.

$$x - 1.23 = -4.52 \quad \text{This is the equation to solve.}$$
$$x - 1.23 + 1.23 = -4.52 + 1.23 \quad \text{Add 1.23 to both sides.}$$
$$x = -3.29 \quad \text{On the left side, } -1.23 + 1.23 = 0. \text{ On the right side, do the addition: } -4.52 + 1.23 = -3.29.$$

$$\begin{array}{r} \overset{4\,12}{4.\!\not{5}\!\not{2}} \\ -1.23 \\ \hline 3.29 \end{array}$$

Check:

$$x - 1.23 = -4.52 \quad \text{This is the original equation.}$$
$$-3.29 - 1.23 \stackrel{?}{=} -4.52 \quad \text{Substitute } -3.29 \text{ for } x.$$
$$-4.52 = -4.52 \quad \text{On the left side, do the subtraction: } -3.29 - 1.23 = -3.29 + (-1.23) = -4.52.$$

$$\begin{array}{r} \overset{1}{3.29} \\ +1.23 \\ \hline 4.52 \end{array}$$

Since the resulting statement $-4.52 = -4.52$ is true, -3.29 is the solution.

Self Check 6
Solve: $\dfrac{y}{3} = -13.11$ −39.33

Now Try Problem 53

Teaching Example 6
Solve:
$\dfrac{x}{4} = -9.9$
Answer: −39.6

EXAMPLE 6 Solve: $\dfrac{m}{2} = -24.8$

Strategy We will use a property of equality to isolate the variable on one side of the equation.

WHY To solve the original equation, we want to find a simpler equivalent equation of the form $m = \textbf{a number}$, whose solution is obvious.

Solution We will use the multiplication property of equality to isolate m on the left side. We can undo the division by 2 by multiplying both sides by 2.

$$\dfrac{m}{2} = -24.8 \quad \text{This is the equation to solve.}$$
$$2\left(\dfrac{m}{2}\right) = 2(-24.8) \quad \text{Multiply both sides by 2.}$$
$$m = -49.6 \quad \text{Do the multiplication.}$$

$$\begin{array}{r} \overset{1}{24.8} \\ \times \;\;\; 2 \\ \hline 49.6 \end{array}$$

Check the result to verify that -49.6 is the solution.

EXAMPLE 7 Solve: $-9.66 = -4.6x$

Strategy We will use a property of equality to isolate the variable on one side of the equation.

WHY To solve the original equation, we need to find a simpler equivalent equation of the form **a number** $= x$, whose solution is obvious.

Solution We will use the division property of equality to isolate x on the right side. We can undo multiplication by -4.6 by dividing both sides by -4.6.

$-9.66 = -4.6x$ This is the equation to solve.

$\dfrac{-9.66}{-4.6} = \dfrac{-4.6x}{-4.6}$ Divide both sides by -4.6.

$2.1 = x$ Do the division.

$$\begin{array}{r} 2.1 \\ 4.6\overline{)9.66} \\ \underline{-92} \\ 46 \\ \underline{-46} \\ 0 \end{array}$$

Check the result to verify that 2.1 is the solution.

Self Check 7
Solve: $-22.32 = -3.1m$ 7.2
Now Try Problem 57

Teaching Example 7
Solve: $-24.08 = -5.6x$
Answer:
4.3

5 Use more than one property of equality to solve equations that involve decimals.

Sometimes, more than one property must be used to solve an equation that involves decimals.

EXAMPLE 8 Solve: $8.1y - 6.04 = -13.33$

Strategy First we will use a property of equality to isolate the *variable term* on one side of the equation. Then we will use a second property of equality to isolate the *variable* itself.

WHY To solve the original equation, we want to find a simpler equivalent equation of the form $y =$ **a number**, whose solution is obvious.

Solution On the left side of the equation, y is multiplied by 8.1 and then 6.04 is subtracted from that product. To solve the equation, we undo the operations in the opposite order.

- To isolate the variable term $8.1y$, we add 6.04 to both sides to undo the subtraction of 6.04.

- To isolate the variable y, we divide both sides by 8.1 to undo the multiplication by 8.1.

$8.1y - 6.04 = -13.33$ This is the equation to solve.

$8.1y - 6.04 + 6.04 = -13.33 + 6.04$ Use the addition property of equality: Add 6.04 to both sides to isolate $8.1y$.

$8.1y = -7.29$ Do the addition: $-6.04 + 6.04 = 0$ and $-13.33 + 6.04 = -7.29$. Now isolate y.

$\dfrac{8.1y}{8.1} = \dfrac{-7.29}{8.1}$ Use the division property of equality: Divide both sides by 8.1 to isolate y.

$y = -0.9$ Do the division.

$$\begin{array}{r} {\scriptstyle 213} \\ 13.\cancel{3}\cancel{3} \\ \underline{-\ 6.04} \\ 7.29 \end{array}$$

$$\begin{array}{r} 0.9 \\ 8.1\overline{)7.29} \\ \underline{-7\ 29} \\ 0 \end{array}$$

Self Check 8
Solve: $-4.2h + 3.14 = 1.88$ 0.3
Now Try Problem 61

Teaching Example 8
Solve: $7.1m - 1.03 = -6.71$
Answer:
-0.8

Check:

$$8.1y - 6.04 = -13.33 \quad \text{This is the original equation.}$$
$$8.1(-0.9) - 6.04 \stackrel{?}{=} -13.33 \quad \text{Substitute } -0.9 \text{ for } y.$$
$$-7.29 - 6.04 \stackrel{?}{=} -13.33 \quad \text{Do the multiplication:}$$
$$8.1(-0.9) = -7.29.$$
$$-13.33 = -13.33 \quad \text{Do the subtraction by adding the opposite: } -7.29 - 6.04 = -7.29 + (-6.04) = -13.33.$$

$$\begin{array}{r} 8.1 \\ \times\, 0.9 \\ \hline 7.29 \end{array}$$

$$\begin{array}{r} \overset{1}{7}.29 \\ +\, 6.04 \\ \hline 13.33 \end{array}$$

Since the resulting statement $-13.33 = -13.33$ is true, -0.9 is the solution.

Self Check 9

Solve the equation and check the result:
$6.1b - 5.5 = 5.2b + 5.3$ 12

Now Try Problem 65

Teaching Example 9
Solve the equation and check the result: $0.1m - 4 = 0.9m + 1.6$
Answer:
-7

EXAMPLE 9 Solve: $0.2s - 3 = 0.7s + 1.5$

Strategy There are variable terms ($0.2s$ and $0.7s$) on both sides of the equation. We will eliminate $0.2s$ on the left side by subtracting $0.2s$ from both sides.

WHY To solve for s, all the terms containing s must be on the same side of the equation.

Solution We isolate the variable terms on the right side and isolate the constant terms on the left side of the equation.

$$0.2s - 3 = 0.7s + 1.5 \quad \text{This is the equation to solve.}$$
$$0.2s - 3 - \boxed{0.2s} = 0.7s + 1.5 - \boxed{0.2s} \quad \text{Subtract } 0.2s \text{ from both sides to isolate the variable term on the right.}$$
$$-3 = 0.5s + 1.5 \quad \text{Combine like terms: } 0.2s - 0.2s = 0 \text{ and } 0.7s - 0.2s = 0.5s.$$
$$-3 - \boxed{1.5} = 0.5s + 1.5 - \boxed{1.5} \quad \text{To isolate the variable term } 0.5s, \text{ undo the addition of } 1.5 \text{ by subtracting } 1.5 \text{ from both sides.}$$
$$-3 + (-1.5) = 0.5s \quad \text{On the left, write the subtraction as addition of the opposite. On the right, subtract.}$$
$$-4.5 = 0.5s \quad \text{Do the addition: } -3 + (-1.5) = -4.5. \text{ Now isolate } s.$$
$$\frac{-4.5}{0.5} = \frac{0.5s}{0.5} \quad \text{To isolate } s, \text{ undo the multiplication by } 0.5 \text{ by dividing both sides by } 0.5.$$
$$-9 = s \quad \text{Do the division.}$$

$$\begin{array}{r} 9 \\ 0.5\overline{)4.5} \\ -4\,5 \\ \hline 0 \end{array}$$

Check the result in the original equation to verify that -9 is the solution.

Success Tip In Example 9, we could have eliminated $0.7s$ from the right side of the equation by subtracting $0.7s$ from both sides.

$$0.2s - 3 - \boxed{0.7s} = 0.7s + 1.5 - \boxed{0.7s}$$
$$-0.5s - 3 = 1.5$$

However, it is usually easier to isolate the variable term on the side that will result in a *positive* coefficient. With this approach, the resulting coefficient of s is negative.

5.7 Solving Equations That Involve Decimals

EXAMPLE 10 Solve: $5(x + 1.3) + 1.1 = -8.8$

Strategy We will use the distributive property along with the process of combining like terms to simplify the left side of the equation.

WHY It is best to simplify each side of the equation before using a property of equality.

Solution First, we remove the parentheses by applying the distributive property.

$5(x + 1.3) + 1.1 = -8.8$	This is the equation to solve.
$5x + 6.5 + 1.1 = -8.8$	Distribute the multiplication by 5.
$5x + 7.6 = -8.8$	Combine like terms: $6.5 + 1.1 = 7.6$.
$5x + 7.6 - 7.6 = -8.8 - 7.6$	To isolate the variable term $5x$, undo the addition of 7.6 by subtracting 7.6 from both sides.
$5x = -8.8 + (-7.6)$	On the right side, add the opposite of 7.6.
$5x = -16.4$	Do the addition: $-8.8 + (-7.6) = -16.4$. Now isolate x.
$\dfrac{5x}{5} = \dfrac{-16.4}{5}$	To isolate the variable x, undo the multiplication by 5 by dividing both sides by 5.
$x = -3.28$	Do the division.

Check:

$5(x + 1.3) + 1.1 = -8.8$	This is the original equation.
$5(-3.28 + 1.3) + 1.1 \stackrel{?}{=} -8.8$	Substitute -3.28 for x.
$5(-1.98) + 1.1 \stackrel{?}{=} -8.8$	Do the addition within the parentheses.
$-9.9 + 1.1 \stackrel{?}{=} -8.8$	Do the multiplication: $5(-1.98) = -9.9$.
$-8.8 = -8.8$	Do the addition: $-9.9 + 1.1 = -8.8$.

Since the resulting statement $-8.8 = -8.8$ is true, -3.28 is the solution.

Self Check 10

Solve:
$2(c + 4.1) + 3.2 = -16.2$ -13.8

Now Try Problem 75

Teaching Example 10
Solve: $5(m + 1.9) + 6.3 = -11.8$
Answer:
-5.52

6 Use equations to solve application problems that involve decimals.

We can use the concepts of variable and equation to solve application problems involving decimals. Once again, we will follow the six-step problem-solving strategy of analyze, assign, form, solve, state, and check.

EXAMPLE 11 **Business Expenses** A business decides to rent a copy machine instead of buying one. Under the rental agreement, the company is charged a basic rental fee of $65 per month plus 2¢ for every copy made. If the business has budgeted $125 for copier expenses each month, how many copies can be made before exceeding the budget?

Self Check 11

FIELD TRIPS It costs a college $65 a day plus 25¢ per mile to rent a 15-passenger van. If the van is rented for one day, how many miles can be driven on a $275 budget?

Now Try Problem 111

Self Check 11 Answer
The van can be driven 840 miles on a $275 budget.

Teaching Example 11
GRADUATION A printer charges a setup fee of $16 and then 85 cents for each graduation announcement printed (tax included). If a student wants to spend $50 on her announcements, how many can she have printed?
Answer:
The student can have 40 graduation announcements printed for $50.

Analyze

- The basic rental charge is $65 a month.   Given
- There is a 2¢ charge for each copy made.   Given
- $125 is budgeted for copier expenses each month.   Given
- What is the maximum number of copies that can
  be made each month?   Find

Assign
We will let x = the maximum number of copies that can be made each month.

Form
To begin to translate the words of the problem to numbers and symbols, we first write:

| The basic fee | plus | the total cost of the copies | must equal | the amount budgeted each month. |

We can find the total cost of the copies by multiplying the cost per copy by the maximum number of copies that can be made. Notice that the monthly budgeted cost is $125 and the cost per copy is 2¢. We need to work in terms of one unit, so we write 2¢ as $0.02 and work in terms of dollars.

| The basic fee | plus | the cost per copy | times | the maximum number of copies made | must equal | the amount budgeted each month. |
| 65 | + | 0.02 | · | x | = | 125 |

Caution! Don't forget to write the 2¢ per copy cost in terms of dollars: $0.02 per copy. If you incorrectly form the equation as $65 + 2x = 125$, you are saying that the copies cost $2 each!

Solve

$$65 + 0.02x = 125$$
$$65 + 0.02x - 65 = 125 - 65$$ To isolate the variable term 0.02x, undo the addition of 65 by subtracting 65 from both sides.
$$0.02x = 60$$ Do the subtraction. Now isolate x.
$$\frac{0.02x}{0.02} = \frac{60}{0.02}$$ To isolate x, undo the multiplication by 0.02 by dividing both sides by 0.02.
$$x = 3{,}000$$ Do the division.

State
The business can make up to 3,000 copies each month without exceeding its budget.

Check
If we multiply the cost per copy and the maximum number of copies, we get $0.02 · 3,000 = $60. Then we add the $65 monthly fee: $60 + $65 = $125. The result checks.

ANSWERS TO SELF CHECKS

1. a. $14.4a$ b. $-12.69d$ c. $13.35x$ d. $46.2g$ 2. a. $9.4x + 37.6$ b. $-47.7y + 35.1$
c. $0.08c - 0.176$ 3. a. $6.8y$ b. $-9.6s$ c. $-0.38n^2$ d. $9.76x + 4.19$ 4. $24.4y - 27.2$
5. a. 11.1 b. -1.2 6. -39.33 7. 7.2 8. 0.3 9. 12 10. -13.8 11. The van can be driven 840 miles on a $275 budget.

SECTION 5.7 STUDY SET

VOCABULARY

Fill in the blanks.

1. In algebra, we often replace one algebraic expression with another that is equivalent and simpler in form. That process is called __simplifying__ an expression.

2. $4.1(x + 3) = 4.1x + 4.1(3)$ is an example of the use of the __distributive__ property.

3. When we write $6.2x + 2.7x$ as $8.9x$, we say we have __combined__ like terms.

4. To __solve__ an equation means to find all the values of the variable that make the equation true.

CONCEPTS

5. a. Fill in the blanks to simplify the expression.

   $4(3.2t) = (\boxed{4} \cdot \boxed{3.2})t = \boxed{12.8}\, t$

   b. What property did you use in part a? __associative property of multiplication__

   c. Fill in the blanks to simplify the expression.

   $-6.1y \cdot 2 = \boxed{-6.1} \cdot \boxed{2} \cdot y = \boxed{-12.2}\, y$

   d. What property did you use in part c? __commutative property of multiplication__

6. Fill in the blanks.
   a. $2.9(x + 4) = 2.9x\ \boxed{+}\ 11.6$
   b. $2.9(x - 4) = 2.9x\ \boxed{-}\ 11.6$
   c. $-2.9(x + 4) = -2.9x\ \boxed{-}\ 11.6$
   d. $-2.9(-x - 4) = 2.9x\ \boxed{+}\ 11.6$

7. Fill in the blanks to combine like terms.
   a. $4.2m + 6.3m = (\boxed{4.2} + \boxed{6.3})m = \boxed{10.5}\, m$
   b. $3.6n^2 - 5.8n^2 = (\boxed{3.6} - \boxed{5.8})n^2 = \boxed{-2.2}\, n^2$
   c. $1.2 + 3.2d + 1.5 = 3.2d + \boxed{2.7}$
   d. Like terms can be combined by adding or subtracting the __coefficients__ of the terms and keeping the same variables with the same exponents.

8. Simplify each expression, if possible.
   a. $5(2x)$ and $5 + 2x$ $10x$, can't be simplified
   b. $6(-7x)$ and $6 - 7x$ $-42x$, can't be simplified
   c. $2(3x)(3)$ and $2 + 3x + 3$ $18x$, $3x + 5$

9. Fill in the blanks.

   It takes two steps to solve the equation $2.1m - 9.1 = 5.6$.

   - To isolate the variable term $2.1m$, we undo the subtraction of 9.1 by __adding__ 9.1 to both sides.
   - To isolate the variable m, we undo the multiplication by 2.1 by __dividing__ both sides by 2.1.

10. Write each amount of money as a dollar amount.
    a. 25 cents $0.25
    b. 250 cents $2.50
    c. 1 penny $0.01
    d. 99 cents $0.99

NOTATION

Complete each step to solve the equation.

11. $$0.6a - 2.3 = -1.82$$
    $$0.6a - 2.3 + \boxed{2.3} = -1.82 + \boxed{2.3}$$
    $$0.6a = 0.48$$
    $$\frac{0.6a}{0.6} = \frac{0.48}{0.6}$$
    $$a = \boxed{0.8}$$

    Check:
    $$0.6a - 2.3 = -1.82$$
    $$0.6(\boxed{0.8}) - 2.3 \stackrel{?}{=} -1.82$$
    $$0.48 - 2.3 \stackrel{?}{=} \boxed{-1.82}$$
    $$-1.82 = -1.82 \quad \text{True}$$

    The solution is $\boxed{0.8}$.

12. $\dfrac{x}{2} + 1 = -5.2$

$\dfrac{x}{2} + 1 - \boxed{1} = -5.2 - \boxed{1}$

$\boxed{\dfrac{x}{2}} = -6.2$

$\boxed{2}\left(\dfrac{x}{2}\right) = \boxed{2}(-6.2)$

$x = \boxed{-12.4}$

Check: $\dfrac{x}{2} + 1 = -5.2$

$\dfrac{-12.4}{2} + 1 \stackrel{?}{=} -5.2$

$\boxed{-6.2} + 1 \stackrel{?}{=} \boxed{-5.2}$

$\boxed{-5.2} = -5.2$ True

The solution is $\boxed{-12.4}$.

GUIDED PRACTICE

Simplify each expression. See Example 1.

13. $3.2 \cdot 4t$ $12.8t$
14. $9.7 \cdot 3s$ $29.1s$
15. $8.06m(-7)$ $-56.42m$
16. $7.16n(-4)$ $-28.64n$
17. $-5(-6.7t)$ $33.5t$
18. $-8(-1.3a)$ $10.4a$
19. $2(4.4c)(3)$ $26.4c$
20. $3(2.7h)(2)$ $16.2h$

Multiply. See Example 2.

21. $3.7(4x + 3)$ $14.8x + 11.1$
22. $5.9(7a + 6)$ $41.3a + 35.4$
23. $-6(1.9m - 2.8)$ $-11.4m + 16.8$
24. $-8(2.8r - 1.4)$ $-22.4r + 11.2$
25. $(y - 9.4)0.06$ $0.06y - 0.564$
26. $(h - 3.9)0.07$ $0.07h - 0.273$
27. $-2.5(-1.2t + 11)$ $3t - 27.5$
28. $-2.5(-5.2y + 13)$ $13y - 32.5$

Simplify each expression by combining like terms. See Example 3.

29. $3.2x + 4.7x$ $7.9x$
30. $-9.8c + 6.5c$ $-3.3c$
31. $-2.4v + (-9.8v) + 3.4v$ $-8.8v$
32. $-9.5t + (-4.6t) + 6.4t$ $-7.7t$
33. $0.32b^2 - 0.59b^2$ $-0.27b^2$
34. $0.26d^2 - 0.89d^2$ $-0.63d^2$
35. $7.55a + 3 + 1.12a - 1.56$ $8.67a + 1.44$
36. $4.05y + 5 + 4.39y - 3.67$ $8.44y + 1.33$

Simplify each expression. See Example 4.

37. $5.2(m + 3) - 1.2 - (3.7m - 4.1)$ $1.5m + 18.5$
38. $6.3(n + 4) - 4.3 - (1.5n - 2.9)$ $4.8n + 23.8$
39. $3.8(d + 2) - 4.5 - (1.8d - 3.7)$ $2d + 6.8$
40. $7.6(s + 2) - 8.1 - (5.2s - 6.8)$ $2.4s + 13.9$
41. $7(3.4y - 1.1) - (6.7y + 4.9) + y$ $18.1y - 12.6$
42. $4(6.2r - 2.2) - (4.9r + 7.4) + r$ $20.9r - 16.2$
43. $9(2.7b - 4.2) - (14.2b + 37.8) - b$ $9.1b - 75.6$
44. $3(8.4y - 6.3) - (9.1y + 18.9) - y$ $15.1y - 37.8$

Solve each equation and check the result. See Example 5.

45. $x + 8.1 = 9.8$ 1.7
46. $x + 4.3 = 8.9$ 4.6
47. $6.75 + y = 8.99$ 2.24
48. $3.8 + f = 1.2$ -2.6
49. $m - 1.4 = -5.8$ -4.4
50. $h - 6.3 = -10.8$ -4.5
51. $7.08 = c - 0.03$ 7.11
52. $14.1 = k - 13.1$ 27.2

Solve each equation and check the result. See Example 6.

53. $\dfrac{x}{6} = -4.7$ -28.2
54. $\dfrac{x}{8} = -8.4$ -67.2
55. $\dfrac{y}{7} = 0.06$ 0.42
56. $\dfrac{r}{3} = 0.23$ 0.69

Solve each equation and check the result. See Example 7.

57. $-3.51 = -2.7x$ 1.3
58. $-10.15 = -3.5m$ 2.9
59. $-1.95 = 0.5f$ -3.9
60. $-4.92 = 0.6m$ -8.2

Solve each equation and check the result. See Example 8.

61. $3.2x - 3.01 = -5.25$ -0.7
62. $6.4a - 1.29 = -6.41$ -0.8
63. $-1.5b + 2.7 = 1.2$ 1
64. $-2.1x - 3.1 = 5.3$ -4

Solve each equation and check the result. See Example 9.

65. $0.4y - 1 = 0.8y + 3.4$ -11
66. $0.6w - 2 = 0.9w + 1.6$ -12
67. $53.7r + 2.6 = 46.3r + 17.4$ 2
68. $37.1w + 12.2 = 16.8w + 93.4$ 4
69. $1.2m - 1.2 = 1.1m + 1.4$ 26
70. $1.4s - 1.9 = 1.9s + 3.6$ -11
71. $2.36x + 0.09 = 2.47x - 0.35$ 4
72. $43.21a + 0.13 = 43.26a - 0.12$ 5

Solve each equation and check the result. See Example 10.

73. $3(y - 1.1) + 0.5 = -0.4$ 0.8
74. $4(b - 2.7) + 1.6 = -6.4$ 0.7
75. $8(m - 1.9) + 17.1 = -14.5$ -2.05
76. $2(w - 6.2) + 5.4 = -13.1$ -3.05

TRY IT YOURSELF

Simplify each expression or solve each equation.

77. $2x = -8.72$ -4.36
78. $3y = -12.63$ -4.21
79. $2.3 = 3(a - 1.1) + 3.2$ 0.8
80. $-6.2 = 2(m - 4.3) + 1.2$ 0.6
81. $5.6x - 8.3 - 6.1x + 12.2$ $-0.5x + 3.9$
82. $-17.3y - 8.01 + 12.2y - 4.4$ $-5.1y - 12.41$

5.7 Solving Equations That Involve Decimals

83. $1.5x - 0.3 = 1.3x + 2.7$ 15
84. $2.1a - 0.2 = 2.3a + 1.9$ −10.5
85. $5.3m - 1.2 = 4.9m - 1.6$ −1
86. $7.6r - 0.15 = 2.4r - 11.07$ −2.1
87. $1.2x - 1.3 = 2.4x + 0.02$ −1.1
88. $-4.4y - 1.3 = -5.1y - 5.08$ −5.4
89. $-6.7 = \dfrac{x}{2.04} - 2.7$ −8.16
90. $-7.5 = \dfrac{y}{2.22} - 1.5$ −13.32
91. $0.06x + 0.09(100 - x) = 8.85$ 5
92. $0.08(1{,}000 - x) + 0.6x = 72.72$ −14
93. $3.1r - 5.5r - 1.3r$ −3.7r
94. $3.8x - 6.5x - 2.4x$ −5.1x
95. $\dfrac{1}{3}x = -7.06$ −21.18
96. $\dfrac{1}{5}x = -3.02$ −15.1
97. $2(x + 3.9) = 3.4$ −2.2
98. $3(x - 0.4) = -4.8$ −1.2
99. $7.75 = n - (-7.85)$ −0.1
100. $3.33 = y - (-5.55)$ −2.22
101. $-5.6 - h = -17.1$ 11.5
102. $-0.05 - x = -1.25$ 1.2
103. $\dfrac{x}{100} = 0.004$ 0.4
104. $\dfrac{y}{1{,}000} = 0.0606$ 60.6

APPLICATIONS

Complete each solution.

105. PETITIONS On weekends, a college student works for a political organization collecting signatures for a petition drive. Her pay is $48 a day plus 95¢ for each signature she obtains. How many signatures does she have to collect to make $200 a day?

Analyze
- Her base pay is $ 48 a day.   Given
- She makes 95 ¢ for each signature.   Given
- She wants to make $ 200 a day.   Given
- How many signatures does she need to collect?   Find

Assign Let $x =$ the number of signatures she needs to collect.

Form We need to work in terms of the same units, so we write 95¢ as $ 0.95 . We now translate the words of the problem into an equation.

Her base pay	plus	the amount per signature	times	the number of signatures	should equal	$200.
48	+	0.95	·	x	=	200

Solve

$48 + \boxed{0.95x} = 200$

$48 + 0.95x - \boxed{48} = 200 - \boxed{48}$

$0.95x = \boxed{152}$

$\dfrac{0.95x}{\boxed{0.95}} = \dfrac{152}{\boxed{0.95}}$

$x = \boxed{160}$

State
She needs to collect 160 signatures to make $200 a day.

Check
If she collects 160 signatures, she will make
$0.95 · 160 = $152 from signatures. If we add this to $48, we get $ 200 . The result checks.

106. CONSTRUCTION A 12.78-mile-long highway is in its third and final year of construction. In the first year, 2.31 miles of the highway were completed. In the second year, 4.93 miles were finished. How many more miles of the highway need to be completed?

Analyze
- The planned highway is 12.78 miles long.   Given
- The first year, 2.31 miles were completed.   Given
- The second year, 4.93 miles were completed.   Given
- How many more miles of highway need to be completed?   Find

Assign We will let $x =$ the number of miles of highway that need to be completed.

Form

The miles completed in the first year	plus	the miles completed in the second year	plus	the miles that need to be completed	is	the length of the highway.
2.31	+	4.93	+	x	=	12.78

Solve

$2.31 + 4.93 + \boxed{x} = 12.78$

$\boxed{7.24} + x = 12.78$

$7.24 + x - \boxed{7.24} = 12.78 - \boxed{7.24}$

$x = \boxed{5.54}$

State
The number of miles of highway that need to be completed is 5.54 .

Check
Add:
$$\begin{array}{r} 2.31 \\ 4.93 \\ + \;\; 5.54 \\ \hline 12.78 \end{array}$$
The result checks.

In Exercises 107–118, let a variable represent the unknown quantity. Then write and solve an equation to answer the question.

107. **DISASTER RELIEF** After hurricane damage estimated at $27.9 million, a county looked to three sources for relief. Local agencies contributed $6.8 million toward the cleanup. A state emergency fund offered another $12.5 million. When applying for federal government help, how much money should the county ask for? $8.6 million

108. **TELETHONS** Midway through a telethon, the donations had reached $16.7 million. How much more was donated in the second half of the program if the total pledged during the program was $30 million? $13.3 million

109. **GRADE POINT AVERAGES** After receiving her grades for the fall semester, a college student noticed that her overall GPA had dropped by 0.18. If her new GPA was 3.09, what was her GPA at the beginning of the fall semester? 3.27

110. **MONTHLY PAYMENTS** A food dehydrator offered on a home shopping channel can be purchased by making three equal monthly payments. If the price is $113.25, how much is each monthly payment? $37.75

111. **AWARDS** A city honors its citizen of the year with a framed certificate. A calligrapher charges $20 for the frame and then 15 cents a word for writing out the proclamation. If the city charter prohibits gifts in excess of $50, what is the maximum number of words that can be printed on the award? 200 words

112. **GIVEAWAYS** A promotions company sells key chains with a personalized message printed on them. The key chains cost 78¢ each and there is a $32 shipping and handling fee on every order. If a car repair shop plans to spend $500 for this type of advertising, how many key chains can they order? 600 key chains

113. **HELIUM BALLOONS** The organizer of a jog-a-thon wants an archway of balloons constructed at the finish line of the race. A company charges a $100 setup fee and then 8 cents for every balloon. How many balloons will be used if $300 is spent for the decoration? 2,500 balloons

114. **UTILITY BILLS** An electric company charges customers a $17.50 basic monthly service fee plus 18¢ for every kilowatt of energy used. One resident's bill was $43.96. How many kilowatt-hours did the resident use that month? 147 kilowatt-hours

115. **TUTORING** Elementary school students enrolling in a reading improvement program must first take a placement test (cost $60) before receiving tutoring (cost $9.75 per hour). If a family has set aside $450 to get their child extra help, how many hours of tutoring can they afford? 40 hours

116. **DATA CONVERSION** The *Books2Bytes* service converts old print books to Microsoft Word electronic files for $40 per book plus $2.15 per page. If it cost $900 to convert a novel, how many pages did the novel have? 400 pages

117. **VIDEOCASSETTES** A family has a large number of VHS cassettes that they want to transfer to DVDs. The conversion costs $14.50 per tape and there is a $31 shipping and handling charge on any order. If they have $350 to spend on the project, how many VHS cassettes can they have converted to DVDs? 22 VHS cassettes

118. **LETTERMAN JACKETS** At one high school, athletes pay $185 for a varsity letterman jacket. It also costs them $4.25 per letter if they want their last name embroidered on the back. If a soccer player spent $223.25 on his letterman jacket, how many letters are in his last name? 9 letters

WRITING

119. In the following case, why is it rather easy to apply the distributive property?
    $$100(0.07x + 5.16)$$

120. Explain why the expression $5.6A + 3.4a$ cannot be simplified.

REVIEW

121. **a.** Add: $\dfrac{3}{4} + \dfrac{2}{3}$ $\dfrac{17}{12} = 1\dfrac{5}{12}$

   b. Add: $\dfrac{1}{x} + \dfrac{2}{3}$ $\dfrac{3 + 2x}{3x}$

122. **a.** Multiply: $\dfrac{3}{4} \cdot \dfrac{2}{3}$ $\dfrac{1}{2}$

   b. Multiply: $\dfrac{3}{4x} \cdot \dfrac{2x}{3}$ $\dfrac{1}{2}$

123. **a.** Subtract: $\dfrac{4}{5} - \dfrac{1}{7}$ $\dfrac{23}{35}$

   b. Subtract: $\dfrac{4}{5} - \dfrac{n}{7}$ $\dfrac{28 - 5n}{35}$

124. **a.** Divide: $\dfrac{5}{8} \div \dfrac{15}{16}$ $\dfrac{2}{3}$

   b. Divide: $\dfrac{5b}{8} \div \dfrac{15b}{16}$ $\dfrac{2}{3}$

STUDY SKILLS CHECKLIST

Do You Know the Basics?

The key to mastering the material in Chapter 5 is to know the basics. Put a checkmark in the box if you can answer "yes" to the statement.

☐ I have memorized the *place-value chart* on page 445.

☐ I know the rules for rounding a decimal to a certain decimal place value by identifying the *rounding digit* and the *test digit*.

☐ I know how to add decimals using *carrying* and how to subtract decimals using *borrowing*.

$$\begin{array}{r} \overset{1}{7}.18 \\ \overset{1\,1}{154}.20 \\ +\ \ 46.03 \\ \hline 207.41 \end{array} \qquad \begin{array}{r} \overset{9}{} \\ 5\overset{6}{3}7.\overset{10}{9}\overset{14}{4} \\ -\ \ 23.98 \\ \hline 513.06 \end{array}$$

☐ I know how to *multiply* and *divide* decimals and locate the decimal point in the answer.

$$\begin{array}{r} 1.84 \\ \times\ 7.6 \\ \hline 1104 \\ 12880 \\ \hline 13.984 \end{array} \qquad \begin{array}{r} 2.8 \\ 3.4\overline{)9.52} \\ -68\downarrow \\ \hline 272 \\ -272 \\ \hline 0 \end{array}$$

☐ I know how to use division to write a *fraction as a decimal*.

$$\frac{3}{5} = 0.6 \qquad \begin{array}{r} 0.6 \\ 5\overline{)3.0} \\ -30 \\ \hline 0 \end{array}$$

☐ I have memorized the list of *perfect squares* on page 515 and can find their *square roots*.

$$\sqrt{16} = 4 \qquad \sqrt{121} = 11$$

Teaching Guide: Refer to the Instructor's Resource Binder to find activities, worksheets on key concepts, more examples, instruction tips, overheads, and assessments.

CHAPTER 5 — SUMMARY AND REVIEW

SECTION 5.1 An Introduction to Decimals

DEFINITIONS AND CONCEPTS	EXAMPLES
The place-value system for whole numbers can be extended to create the **decimal numeration system**. The place-value columns to the left of the decimal point form the **whole-number part** of the decimal number. The value of each of those columns is 10 times greater than the column directly to its right. The columns to the right of the decimal point form the **fractional part**. Each of those columns has a value that is $\frac{1}{10}$ of the value of the place directly to its left.	The place value of the digit 3 is *3 hundredths*. The digit that tells the number of *ten-thousandths* is 1.

To write a decimal number in **expanded form** (**expanded notation**) means to write it as an addition of the place values of each of its digits.	Write 28.9341 in expanded notation: $28.9341 = 20 + 8 + \dfrac{9}{10} + \dfrac{3}{100} + \dfrac{4}{1,000} + \dfrac{1}{10,000}$
To read a decimal: 1. Look to the left of the decimal point and say the name of the whole number. 2. The decimal point is read as "and." 3. Say the fractional part of the decimal as a whole number followed by the name of the last place-value column of the digit that is the farthest to the right. We can use the steps for reading a decimal to **write it in words**.	Write the decimal in words and then as a fraction or mixed number: 28.9341 The whole-number part is 28. The fractional part is 9341. The digit the farthest to the right, 1, is in the ten-thousandths place. *Twenty-eight and nine thousand three hundred forty-one ten-thousandths* Written as a mixed number, 28.9341 is $28\dfrac{9,341}{10,000}$. Write the decimal in words and then as a fraction or mixed number: 0.079 The whole-number part is 0. The fractional part is 79. The digit the farthest to the right, 9, is in the thousandths place. *Seventy-nine thousandths* Written as a fraction, 0.079 is $\dfrac{79}{1,000}$.
The procedure for **reading a decimal** can be applied in reverse to convert from written-word form to standard form.	Write the decimal number in standard form: **Negative twelve and sixty-five ten-thousandths** -12.0065 This is the ten-thousandths place-value column. Two placeholder 0's must be inserted here so that the last digit in 65 is in the ten-thousandths column.
To compare two decimals: 1. Make sure both numbers have the same number of decimal places to the right of the decimal point. Write any additional zeros necessary to achieve this. 2. Compare the digits of each decimal, column by column, working from left to right. 3. If the decimals are *positive*: When two digits differ, the decimal with the greater digit is the greater number. If the decimals are *negative*: When two digits differ, the decimal with the smaller digit is the greater number.	Compare 47.31572 and 47.31569. 47.315 7 2 47.315 6 9 As we work from left to right, this is the first column in which the digits differ. Since $7 > 6$, it follows that 47.31572 is greater than 47.31569. Thus, $47.31572 > 47.31569$. Compare -6.418 and -6.41. $-6.41\,8$ These decimals are negative. $-6.41\,0$ Write a zero after 1 to help in the comparison. As we work from left to right, this is the first column in which the digits differ. Since $0 < 8$, it follows that -6.410 is greater than -6.418. Thus, $-6.41 > -6.418$.
To **graph a decimal number** means to make a drawing that represents the number.	Graph -2.17, 0.6, -2.89, 3.99, and -0.5 on a number line. $-2.89\ -2.17\ -0.5\ 0.6\qquad\qquad 3.99$ $\leftarrow\!\!\begin{array}{ccccccccccc}|&|&|&\bullet\,\bullet&|&\bullet\,\bullet&|&|&|&\bullet&|\\ -5&-4&-3&-2&-1&0&1&2&3&4&5\end{array}\!\!\rightarrow$

1. To **round a decimal** to a certain decimal place value, locate the **rounding digit** in that place.
2. Look at the **test digit** directly to the right of the rounding digit.
3. If the test digit is 5 or greater, round up by adding 1 to the rounding digit and dropping all the digits to its right. If the test digit is less than 5, round down by keeping the rounding digit and dropping all the digits to its right.

Round 33.41632 to the nearest thousandth.

Thus, 33.41632 rounded to the nearest thousandth is 33.416.

Round 2.798 to the nearest hundredth.

Thus, 2.798 rounded to the nearest hundredth is 2.80.

There are many situations in our daily lives that call for **rounding amounts of money**.

Rounded to the *nearest cent*, $0.14672 is $0.15.
Rounded to the *nearest dollar*, $142.39 is $142.

REVIEW EXERCISES

1. **a.** Represent the amount of the square region that is shaded, using a decimal and a fraction. 0.67, $\frac{67}{100}$

   b. Shade 0.8 of the region shown below.

   0.8

2. Consider the decimal number 2,809.6735.

   a. What is the place value of the digit 7? 7 hundredths

   b. Which digit tells the number of thousandths? 3

   c. Which digit tells the number of hundreds? 8

   d. What is the place value of the digit 5? 5 ten-thousandths

3. Write 16.4523 in expanded notation.
   $10 + 6 + \frac{4}{10} + \frac{5}{100} + \frac{2}{1,000} + \frac{3}{10,000}$

Write each decimal in words and then as a fraction or mixed number.

4. 2.3   two and three tenths, $2\frac{3}{10}$

5. −615.59   negative six hundred fifteen and fifty-nine hundredths, $-615\frac{59}{100}$

6. 0.0601   six hundred one ten-thousandths, $\frac{601}{10,000}$

7. 0.00001   one hundred-thousandth, $\frac{1}{100,000}$

Write each number in standard form.

8. One hundred and sixty-one hundredths   100.61
9. Eleven and nine hundred ninety-seven thousandths   11.997
10. Three hundred one and sixteen millionths   301.000016

Place an < or an > symbol in the box to make a true statement.

11. 5.68 < 5.75
12. 106.8199 < 106.82
13. −78.23 > −78.303
14. −555.098 > −555.0991
15. Graph: 1.55, −0.8, −2.1, and −2.7.

16. Determine whether each statement is true or false.

    a. 78 = 78.0   true   **b.** 6.910 = 6.901   false
    c. −3.4700 = −3.470   true   **d.** 0.008 = .00800   true

Round each decimal to the indicated place value.

17. 16.0998 nearest thousandth   16.100
18. −0.0614 nearest tenth   −0.1
19. 11.314964 nearest ten-thousandth   11.3150
20. 0.2222282 nearest millionth   0.222228

Round each given dollar amount.

21. $0.671456 to the nearest cent $0.67

22. $12.82 to the nearest dollar $13

23. VALEDICTORIANS At the end of the school year, the five students listed below were in the running to be class valedictorian (the student with the highest grade point average). Rank the students in order by GPA, beginning with the valedictorian. Washington, Diaz, Chou, Singh, Gerbac

Name	GPA
Diaz, Cielo	3.9809
Chou, Wendy	3.9808
Washington, Shelly	3.9865
Gerbac, Lance	3.899
Singh, Amani	3.9713

24. ALLERGY FORECAST The graph below shows a four-day forecast of pollen levels for Las Vegas, Nevada. Determine the decimal-number forecast for each day. Sun: 1.8, Mon: 0.6, Tues: 2.4, Wed: 3.8

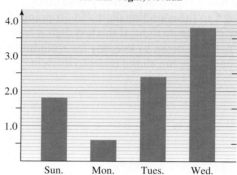

SECTION 5.2 Adding and Subtracting Decimals

DEFINITIONS AND CONCEPTS	EXAMPLES
To add or subtract decimals: 1. Write the numbers in **vertical form** with the decimal points lined up. 2. Add (or subtract) as you would whole numbers. 3. Write the decimal point in the result from Step 2 below the decimal points in the problem. If the number of decimal places in the problem are different, insert additional zeros so that the number of decimal places match. If the sum of the digits in any place-value column is greater than 9, we must **carry**. If the subtraction of the digits in any place-value column requires that we subtract a larger digit from a smaller digit, we must **borrow** or **regroup**.	Add: 15.82 + 19 + 32.995 Write the problem in vertical form and add, column by column, working right to left. 15.820 *Insert an extra zero.* 19.000 *Insert a decimal point and extra zeros.* $+32.995$ 67.815 ↑ Line up the decimal points. To **check** the result, add *bottom to top*. Subtract: 8.4 − 3.029 Write the problem in **vertical form** and subtract, column by column, working right to left. 8.400 *Insert extra zeros.* -3.029 *First, borrow from the tenths column, then* 5.371 *borrow from the hundredths column.* To **check**: The sum of the difference and the subtrahend should equal the minuend. 5.371 ← Difference $+3.029$ ← Subtrahend 8.400 ← Minuend

To **add signed decimals**, we use the same rules that are used for adding integers. **With like signs:** Add their absolute values and attach their common sign to the sum. **With unlike signs:** Subtract their absolute values (the smaller from the larger). If the positive decimal has the larger absolute value, the final answer is positive. If the negative decimal has the larger absolute value, make the final answer negative.	Add: $-21.35 + (-64.52)$ Find the absolute values: $\|-21.35\| = 21.35$ and $\|-64.52\| = 64.52$ $-21.35 + (-64.52) = -85.87$ Add the absolute values, 21.35 and 64.52, to get 85.87. Since both decimals are negative, make the final result negative. Add: $-7.4 + 9.8$ Find the absolute values: $\|-7.4\| = 7.4$ and $\|9.8\| = 9.8$ $-7.4 + 9.8 = 2.4$ Subtract the smaller absolute value from the larger: $9.8 - 7.4 = 2.4$. Since the positive number, 9.8, has the larger absolute value, the final answer is positive.
To **subtract two signed decimals**, add the first decimal to the opposite of the decimal to be subtracted.	Subtract: $-8.62 - (-1.4)$ The number to be subtracted is -1.4. Subtracting -1.4 is the same as adding its opposite, 1.4. $-8.62 - (-1.4) = -8.62 + 1.4 = -7.22$ Use the rule for adding two decimals with different signs.
Estimation can be used to check the reasonableness of an answer to a decimal addition or subtraction.	Estimate the sum by rounding the addends to the nearest ten: $328.99 + 459.02$ $328.99 \longrightarrow 330$ Round to the nearest ten. $+459.02 \longrightarrow +460$ Round to the nearest ten. $788.01 \qquad\quad 790$ This is the estimate. Estimate the difference by using **front-end rounding**: $302.47 - 36.9$ Each number is rounded to its largest place value. $302.47 \longrightarrow 300$ Round to the nearest hundred. $-\ 36.9 \longrightarrow -\ 40$ Round to the nearest ten. $265.57 \qquad\quad 260$ This is the estimate.
We can use addition and subtraction to solve many application problems that involve decimals.	See Examples 10–12 that begin on page 465 to review how to solve application problems by adding and subtracting decimals.

REVIEW EXERCISES

Perform each indicated operation.

25. $19.5 + 34.4 + 12.8$ 66.7
26. $45.8 - 17.372$ 28.428
27. $345.214 - 27.39$ 317.824
28. 8.61
    5.97
    $\underline{+9.72}$
    24.30
29. $-16.1 + 8.4$ -7.7
30. $-4.8 - (-7.9)$ 3.1
31. $-3.55 + (-1.25)$ -4.8
32. $-15.1 - 13.99$ -29.09

Evaluate each expression.

33. $-8.8 + (-7.3 - 9.5)$ -25.6
34. $(5 - 0.096) - (-0.035)$ 4.939
35. **a.** Estimate the sum by rounding the addends to the nearest ten: $612.05 + 145.006$ 760
    b. Estimate the difference by using front-end rounding: $289.43 - 21.86$ 280

36. **COINS** The thicknesses of a penny, nickel, dime, quarter, half-dollar, and presidential $1 coin are 1.55 millimeters, 1.95 millimeters, 1.35 millimeters, 1.75 millimeters, 2.15 millimeters, and 2.00 millimeters, in that order. Find the height of a stack made from one of each type of coin. 10.75 mm

37. **SALE PRICES** A calculator normally sells for $52.20. If it is being discounted $3.99, what is the sale price? $48.21

38. **MICROWAVE OVENS** A microwave oven is shown below. How tall is the window? 8.15 in.

SECTION 5.3 Multiplying Decimals

DEFINITIONS AND CONCEPTS	EXAMPLES
To multiply two decimals: 1. Multiply the decimals as if they were whole numbers. 2. Find the total number of decimal places in both factors. 3. Insert a decimal point in the result from Step 1 so that the answer has the same number of decimal places as the total found in Step 2. When multiplying decimals, we *do not* need to line up the decimal points.	Multiply: 2.76 · 4.3 Write the problem in vertical form and multiply 2.76 and 4.3 as if they were whole numbers. 　　2.76　　2 decimal places　　The answer will have 　× 4.3　　1 decimal place　　　2 + 1 = 3 decimal places. 　　828 　11040　　Move 3 places from right to left and insert 　11.868　a decimal point in the answer. Thus, 2.76 · 4.3 = 11.868.
Multiplying a decimal by 10, 100, 1,000, and so on To find the product of a decimal and 10, 100, 1,000, and so on, move the decimal point to the right the same number of places as there are zeros in the power of 10. **Multiplying a decimal by 0.1, 0.01, 0.001, and so on** To find the product of a decimal and 0.1, 0.01, 0.001, and so on, move the decimal point to the left the same number of places as there are in the power of 10.	Multiply: 　　84.561 · 10,000 = 845,610　　Since 10,000 has four zeros, move the decimal point in 84.561 four places to the right. Write a placeholder zero (shown in blue). Multiply: 　　32.67 · 0.01 = 0.3267　　Since 0.01 has two decimal places, move the decimal point in 32.67 two places to the left.

The rules for multiplying integers also hold for **multiplying signed decimals**: The product of two decimals with **like signs** is positive, and the product of two decimals with **unlike signs** is negative.	Multiply: $(-0.03)(-4.1)$ Find the absolute values: $\|-0.03\| = 0.03$ and $\|4.1\| = 4.1$ Since the decimals have like signs, the product is positive. $(-0.03)(-4.1) = 0.123$ Multiply the absolute values, 0.03 and 4.1, to get 0.123. Multiply: $-5.7(0.4)$ Find the absolute values: $\|-5.7\| = 5.7$ and $\|0.4\| = 0.4$ Since the decimals have unlike signs, the product is negative. $-5.7(0.4) = -2.28$ Multiply the absolute values, 5.7 and 0.4, to get 2.28. Make the final answer negative.
We can use the rule for multiplying a decimal by a power of 10 to **write large numbers in standard form**.	Write *4.16 billion* in standard notation: 4.16 billion $= 4.16 \cdot$ **1 billion** $\phantom{4.16 \text{ billion }} = 4.16 \cdot \mathbf{1{,}000{,}000{,}000}$ Write 1 billion in standard form. $\phantom{4.16 \text{ billion }} = 4{,}160{,}000{,}000$ Since 1,000,000,000 has nine zeros, move the decimal point in 4.16 nine places to the right.
The base of an **exponential expression** can be a positive or a negative decimal.	Evaluate: $(1.5)^2$ $(1.5)^2 = 1.5 \cdot 1.5$ The base is 1.5 and the exponent is 2. Write the base as a factor 2 times. $ = 2.25$ Multiply the decimals. Evaluate: $(-0.02)^2$ $(-0.02)^2 = (-0.02)(-0.02)$ The base is -0.02 and the exponent is 2. Write the base as a factor 2 times. $ = 0.0004$ Multiply the decimals. The product of two decimals with like signs is positive.
To **evaluate a formula**, we replace the letters with specific numbers and then use the order of operations rule.	Evaluate $P = 2l + 2w$ for $l = 4.9$ and $w = 3.4$. $P = 2l + 2w$ $ = 2(4.9) + 2(3.4)$ Replace l with 4.9 and w with 3.4. $ = 9.8 + 6.8$ Do the multiplication. $ = 16.6$ Do the addition.
Estimation can be used to check the reasonableness of an answer to a decimal multiplication.	Estimate $37 \cdot 8.49$ by **front-end rounding**. $37 \longrightarrow 40$ Round to the nearest ten. $\times 8.49 \longrightarrow \times 8$ Round to the nearest one. 320 This is the estimate. The estimate is 320. If we calculate $37 \cdot 8.49$, the product is exactly 314.13.
We can use multiplication to solve many application problems that involve decimals.	See Examples 12 and 13 that begin on page 479 to review how to solve application problems by multiplying decimals.

REVIEW EXERCISES

Multiply.

39. $2.3 \cdot 6.9$ 15.87

40. 1.7
$\underline{\times 0.004}$
0.0068

41. $1.5(-0.9)$ -1.35

42. $(-0.003)(-0.02)$ 0.00006

43. $1{,}000(90.1452)$ 90,145.2

44. $0.001(2.897)$ 0.002897

Evaluate each expression.

45. $(0.2)^2$ 0.04

46. $(0.6 + 0.7)^2 - (-3)(-4.1)$ -10.61

47. $(-3.3)^2(0.00001)$ 0.0001089

48. $(0.1)^3 + 2|-45.63 - 12.24|$ 115.741

Write each number in standard notation.

49. a. GEOGRAPHY China is the third largest country in land area with territory that extends over *9.6 million* square kilometers. (Source: china.org) 9,600,000 km²

b. PLANTING TREES Over the past three decades, the Chinese people have planted about *61.4 billion* trees in mountains, city parks, and along highways to increase the number of forests in their country. (Source: xinhuanet.com) 61,400,000,000

50. a. Estimate the product using front-end rounding: $193.28 \cdot 7.63$ 1,600

b. Estimate the product by rounding the factors to the nearest tenth: $12.42 \cdot 7.38$ 91.76

51. Evaluate the formula $A = P + Prt$ for $P = 70.05$, $r = 0.08$, and $t = 5$. 98.07

52. SHOPPING If crab meat sells for $12.95 per pound, what would 1.5 pounds of crab meat cost? Round to the nearest cent. $19.43

53. AUTO PAINTING A manufacturer uses a three-part process to finish the exterior of the cars it produces.

  Step 1: A 0.03-inch-thick rust-prevention undercoat is applied.

  Step 2: Three layers of color coat, each 0.015 inch thick, are sprayed on.

  Step 3: The finish is then buffed down, losing 0.005 inch of its thickness.

What is the resulting thickness of the automobile's finish? 0.07 in.

54. WORD PROCESSORS The Page Setup screen for a word processor is shown. Find the area that can be filled with text on an 8.5 in. × 11 in. piece of paper if the margins are set as shown. 68.62 in.²

SECTION 5.4 Dividing Decimals

DEFINITIONS AND CONCEPTS	EXAMPLES
To divide a decimal by a whole number: 1. Write the problem in long division form and place a decimal point in the quotient (answer) directly above the decimal point in the dividend. 2. Divide as if working with whole numbers. 3. If necessary, additional zeros can be written to the right of the last digit of the dividend to continue the division.	Divide: $6.2 \div 4$ Place a decimal point in the quotient that lines up with the decimal point in the dividend. 1.55 $4)\overline{6.20}$ Ignore the decimal points and divide as if $\underline{-4}\downarrow$ working with whole numbers. $2\,2$ $\underline{-2\,0}\downarrow$ Write a zero to the right of the 2 and bring 20 it down. Continue to divide. $\underline{-20}$ 0 The remainder is 0.

Chapter 5 Summary and Review   543

To **check** the result, we multiply the divisor by the quotient. The result should be the dividend.	**Check:** $$\begin{array}{r} 1.55 \leftarrow \text{Quotient} \\ \times 4 \leftarrow \text{Divisor} \\ \hline 6.20 \leftarrow \text{Dividend} \end{array}$$ The check confirms that $6.2 \div 4 = 1.55$.
To divide with a decimal divisor: 1. Write the problem in long division form. 2. Move the decimal point of the divisor so that it becomes a whole number. 3. Move the decimal point of the dividend the same number of places to the right. 4. Write a decimal point in the quotient (answer) directly above the decimal point in the dividend. Divide as if working with whole numbers. 5. If necessary, additional zeros can be written to the right of the last digit of the dividend to continue the division.	Divide: $\dfrac{1.462}{3.4}$ $3.4\overline{)1.462}$ Write the problem in long division form. Move the decimal point of the divisor, 3.4, one place to the right to make it a whole number. Move the decimal point of the dividend, 1.462, the same number of places to the right. Now use the rule for dividing a decimal by a *whole number*. $$\begin{array}{r} 0.4\,3 \\ 34\overline{)14.6\,2} \\ -13\,6\downarrow \\ \hline 1\,0\,2 \\ -1\,0\,2 \\ \hline 0 \end{array}$$ Write a decimal point in the quotient (answer) directly above the decimal point in the dividend. Divide as with whole numbers.
Sometimes when we divide decimals, the subtractions never give a zero remainder, and the division process continues forever. In such cases, we can **round the result**.	Divide: $0.77 \div 6$. Round the quotient to the nearest hundredth. To round to the hundredths column, we need to continue the division process for one more decimal place, which is the thousandths column. Rounding digit: hundredths column Test digit: Since 8 is 5 or greater, add 1 to the rounding digit and drop the test digit. $$\begin{array}{r} 0.128 \\ 6\overline{)0.770} \\ -6 \\ \hline 17 \\ -12 \\ \hline 50 \\ -48 \\ \hline 2 \end{array}$$ The remainder is still not 0. Thus, $0.77 \div 6 \approx 0.13$.
To **estimate quotients**, we use a method that approximates both the dividend and the divisor so that they divide easily. There is one rule of thumb for this method: If possible, round both numbers up or both numbers down.	Estimate the quotient: $337.96 \div 23.8$ The dividend is approximately $337.96 \div 23.8 \qquad 320 \div 20 = 16$ To divide, drop one zero from 320 and one zero from 20, and then find $32 \div 2$. The divisor is approximately The estimate is 16. (The exact answer is 14.2.)
Dividing a decimal by 10, 100, 1,000, and so on To find the quotient of a decimal and 10, 100, 1,000, and so on, move the decimal point to the left the same number of places as there are zeros in the power of 10.	Divide: $79.36 \div 10{,}000$ $79.36 \div 10{,}000 = 0.007936$ Since the divisor 10,000 has four zeros, move the decimal point four places to the left. Insert two placeholder zeros (shown in blue).

Dividing a decimal by 0.1, 0.01, 0.001, and so on To find the quotient of a decimal and 0.1, 0.01, 0.001, and so on, move the decimal point to the right the same number of decimal places as there are in the power of 10.	Divide: $\dfrac{1.6402}{0.001}$ $\dfrac{1.6402}{0.001} = 1,640.2$	Since the divisor 0.001 has three decimal places, move the decimal point in 1.6402 three places to the right.
The rules for dividing integers also hold for **dividing signed decimals**. The quotient of two decimals with *like signs* is positive, and the quotient of two decimals with *unlike signs* is negative.	Divide: $-1.53 \div 0.3 = -5.1$ Divide: $\dfrac{-0.84}{-4.2} = 0.2$	Since the signs of the dividend and divisor are unlike, the final answer is negative. Since the dividend and divisor have like signs, the quotient is positive.
We use the order of operations rule to **evaluate expressions** and **formulas**.	Evaluate: $\dfrac{37.8 - (1.2)^2}{0.1 + 0.3}$ $\dfrac{37.8 - (1.2)^2}{0.1 + 0.3} = \dfrac{37.8 - 1.44}{0.4}$ $= \dfrac{36.36}{0.4}$ $= 90.9$	In the numerator, evaluate $(1.2)^2$. In the denominator, do the addition. In the numerator, do the subtraction. Do the division indicated by the fraction bar.
We can use division to solve many application problems that involve decimals.	See Examples 11 and 12 that begin on page 494 to review how to solve application problems by dividing decimals.	

REVIEW EXERCISES

Divide. Check the result.

55. $4.4\overline{)10.516}$ 2.39

56. $\dfrac{-29.67}{-23}$ 1.29

57. $-80.625 \div 12.9$ -6.25

58. $\dfrac{0.0742}{1.4}$ 0.053

59. $\dfrac{15.75}{0.25}$ 63

60. $\dfrac{-0.003726}{-0.0046}$ 0.81

61. $89.76 \div 1,000$ 0.08976

62. $\dfrac{0.0112}{-10}$ -0.00112

63. Divide -0.8765 by -0.001. 876.5

64. $77.021 \div 0.0001$ 770,210

Estimate each quotient.

65. $4,983.01 \div 41.33$ $4,800 \div 40 = 480 \div 4 = 120$

66. $8.8\overline{)25,904.39}$ $27,000 \div 9 = 3,000$

Divide and round each result to the specified decimal place.

67. $78.98 \div 6.1$ (nearest tenth) 12.9

68. $\dfrac{-5.438}{0.007}$ (nearest hundredth) -776.86

69. Evaluate: $\dfrac{(1.4)^2 - 2(-4.6)}{0.5 + 0.3}$ 13.95

70. Evaluate the formula $C = \dfrac{5}{9}(F - 32)$ for $F = 68.9$. 20.5

71. **THANKSGIVING DINNER** The cost of purchasing the food for a Thanksgiving dinner for a family of five was $41.70. What was the cost of the dinner per person? $8.34

72. **DRINKING WATER** Water samples from five wells were taken and tested for PCBs (polychlorinated biphenyls). The number of parts per billion (ppb) found in each sample is given below. Find the average number of parts per billion for these samples. 0.51 ppb

Sample #1:	0.44 ppb
Sample #2:	0.50 ppb
Sample #3:	0.46 ppb
Sample #4:	0.52 ppb
Sample #5:	0.63 ppb

73. **SERVING SIZE** The illustration below shows the package labeling on a box of children's cereal. Use the information given to find the number of servings. 14 servings

Nutrition Facts	
Serving size	1.1 ounce
Servings per container	?
Package weight	15.4 ounces

74. **TELESCOPES** To change the position of a focusing mirror on a telescope, an adjustment knob is used. The mirror moves 0.025 inch with each revolution of the knob. The mirror needs to be moved 0.2375 inch to improve the sharpness of the image. How many revolutions of the adjustment knob does this require? 9.5 revolutions

SECTION 5.5 Fractions and Decimals

DEFINITIONS AND CONCEPTS	EXAMPLES
A fraction and a decimal are said to be **equivalent** if they name the same number. To **write a fraction as a decimal**, divide the numerator of the fraction by its denominator. Sometimes, when finding the decimal equivalent of a fraction, the division process ends because a remainder of 0 is obtained. We call the resulting decimal a **terminating decimal**.	Write $\frac{3}{5}$ as a decimal. We divide the numerator by the denominator because a fraction bar indicates division: $\frac{3}{5}$ means $3 \div 5$. $\begin{array}{r} 0.6 \\ 5\overline{)3.0} \\ -30 \\ \hline 0 \end{array}$ Write a decimal point and one additional zero to the right of 3. Since a zero remainder is obtained, the result is a terminating decimal. Thus, $\frac{3}{5} = 0.6$. We say that 0.6 is the **decimal equivalent** of $\frac{3}{5}$.
If the denominator of a fraction in simplified form has factors of only 2's or 5's, or a combination of both, it can be written as a decimal by **multiplying it by a form of 1**. The objective is to write the fraction in an equivalent form with a denominator that is a power of 10, such as 10, 100, 1,000, and so on.	Write $\frac{3}{25}$ as a decimal. Since we need to multiply the denominator of $\frac{3}{25}$ by 4 to obtain a denominator of 100, it follows that $\frac{4}{4}$ should be the form of 1 that is used to build $\frac{3}{25}$. $\frac{3}{25} = \frac{3}{25} \cdot \frac{4}{4}$ Multiply $\frac{3}{25}$ by 1 in the form of $\frac{4}{4}$. $= \frac{12}{100}$ Multiply the numerators. Multiply the denominators. $= 0.12$ Write the fraction as a decimal.
Sometimes, when we are finding the decimal equivalent of a fraction, the division process never gives a remainder of 0. We call the resulting decimal a **repeating decimal**. An **overbar** can be used instead of the three dots ... to represent the repeating pattern in a **repeating decimal**.	Write $\frac{5}{6}$ as a decimal. $\begin{array}{r} 0.833 \\ 6\overline{)5.000} \\ -48 \\ \hline 20 \\ -18 \\ \hline 20 \\ -18 \\ \hline 2 \end{array}$ Write a decimal point and three additional zeros to the right of 5. It is apparent that 2 will continue to reappear as the remainder. Therefore, 3 will continue to reappear in the quotient. Since the repeating pattern is now clear, we can stop the division. Thus, $\frac{5}{6} = 0.8333\ldots$, or, using an overbar, we have $\frac{5}{6} = 0.8\overline{3}$.
When a fraction is written in decimal form, the result is either a terminating or repeating decimal. Repeating decimals are often **rounded** to a specified place value.	The decimal equivalent for $\frac{5}{11}$ is $0.454545\ldots$. Round it to the nearest hundredth. Rounding digit: hundredths column. Test digit: Since 4 is less than 5, round down. $\frac{5}{11} = 0.454545\ldots$ Thus, $\frac{5}{11} \approx 0.45$.

To write a mixed number in decimal form, we need only find the decimal equivalent for the fractional part of the mixed number. The whole-number part in the decimal form is the same as the whole-number part in the mixed-number form.	Whole-number part $$4\frac{7}{8} = 4.875$$ Write the fraction as a decimal.
A number line can be used to show the relationship between fractions and decimals.	Graph 3.125, $-4\frac{5}{7}$, $0.\overline{6}$, $-1.\overline{09}$ on a number line.
To **compare the size of a fraction and a decimal**, it is helpful to write the fraction in its equivalent decimal form.	Place an $<$, $>$, or an $=$ symbol in the box to make a true statement: $\frac{3}{50} \quad \boxed{} \quad 0.07$ To write $\frac{3}{50}$ as a decimal, divide 50 by 3: $\frac{3}{50} = 0.06$. Since 0.06 is less than 0.07, we have: $\frac{3}{50} < 0.07$.
To evaluate **expressions that can contain both fractions and decimals**, we can work in terms of decimals or in terms of fractions.	Evaluate: $\frac{1}{6} + 0.31$ If we work in terms of fractions, we have: $\frac{1}{6} + 0.31 = \frac{1}{6} + \frac{31}{100}$ — Write 0.31 in fraction form. $= \frac{1}{6} \cdot \frac{50}{50} + \frac{31}{100} \cdot \frac{3}{3}$ — The LCD is 300. Build each fraction by multiplying by a form of 1. $= \frac{50}{300} + \frac{93}{300}$ — Multiply the numerators. Multiply the denominators. $= \frac{143}{300}$ — Add the numerators and write the sum over the common denominator 300. If we work in terms of decimals, we have: $\frac{1}{6} + 0.31 \approx 0.17 + 0.31$ — Approximate $\frac{1}{6}$ with the decimal 0.17. ≈ 0.48 — Do the addition.
We can use the six-step **problem-solving strategy** to solve application problems that involve fractions and decimals.	See Example 13 on page 509 to review how to solve application problems involving fractions and decimals.

REVIEW EXERCISES

Write each fraction or mixed number as a decimal. Use an overbar when necessary.

75. $\frac{7}{8}$ 0.875

76. $-\frac{2}{5}$ -0.4

77. $\frac{9}{16}$ 0.5625

78. $\frac{3}{50}$ 0.06

79. $\frac{6}{11}$ $0.\overline{54}$

80. $-\frac{4}{3}$ $-1.\overline{3}$

81. $3\frac{7}{125}$ 3.056

82. $\frac{26}{45}$ $0.5\overline{7}$

Write each fraction as a decimal. Round to the nearest hundredth.

83. $\frac{19}{33}$ 0.58

84. $\frac{31}{30}$ 1.03

Place an $<$, $>$, or an $=$ symbol in the box to make a true statement.

85. $\frac{13}{25}$ $>$ 0.499

86. $-\frac{4}{15}$ $=$ $-0.2\overline{6}$

87. Write the numbers in order from least to greatest: $\frac{10}{33}$, $0.\overline{3}$, 0.3 $0.3, \frac{10}{33}, 0.\overline{3}$

88. Graph $1.125, -3.\overline{3}, 2\frac{3}{4}$, and $-\frac{9}{10}$ on a number line.

Evaluate each expression. Work in terms of fractions.

89. $\frac{1}{3} + 0.4$ $\frac{11}{15}$

90. $\frac{5}{6} + 0.19$ $\frac{307}{300} = 1\frac{7}{300}$

Evaluate each expression. Work in terms of decimals.

91. $\frac{1}{2}(9.7 + 8.9)(10)$ 93

92. $7.5 - (0.78)\left(\frac{1}{2}\right)^2$
    7.305

93. ROADSIDE EMERGENCY What is the area of the reflector shown below? 34.88 in.²

94. SEAFOOD A shopper purchased $\frac{3}{4}$ pound of crab meat, priced at $13.80 per pound, and $\frac{1}{3}$ pound of lobster meat, selling for $35.70 per pound. Find the total cost of these items. $22.25

SECTION 5.6 Square Roots

DEFINITIONS AND CONCEPTS	EXAMPLES
The **square root** of a given number is a number whose square is the given number. Every positive number has two square roots. The number 0 has only one square root.	Find the two square roots of 81. 9 is a square root of 81 because $9^2 = 81$ and -9 is a square root of 81 because $(-9)^2 = 81$.
A **radical symbol** $\sqrt{}$ is used to indicate a positive square root. To **evaluate a radical expression** such as $\sqrt{4}$, find the positive square root of the radicand. Radical symbol $\underbrace{\sqrt{4}}_{\text{Radical expression}} \leftarrow$ Radicand Read as "the square root of 4." Numbers such as 4, 64, and 225, that are squares of whole numbers, are called **perfect squares**. To evaluate square root radical expressions, it is helpful to be able to identify **perfect square radicands**. Review the list of perfect squares on page 515.	Evaluate each square root: $\sqrt{4} = 2$ Ask: What positive number, when squared, is 4? The answer is 2 because $2^2 = 4$. $\sqrt{64} = 8$ Ask: What positive number, when squared, is 64? The answer is 8 because $8^2 = 64$. $\sqrt{225} = 15$ Ask: What positive number, when squared, is 225? The answer is 15 because $15^2 = 225$.
The symbol $-\sqrt{}$ is used to indicate the **negative square root** of a positive number. It is the opposite of the positive square root.	Evaluate: $-\sqrt{36}$ $-\sqrt{36}$ is the opposite (or negative) of the positive square root of 36. Since $\sqrt{36} = 6$, we have: $-\sqrt{36} = -6$
We can find the square root of fractions and decimals.	Evaluate each square root: $\sqrt{\frac{49}{100}}$ Ask: What positive fraction, when squared, is $\frac{49}{100}$? The answer is $\frac{7}{10}$ because $\left(\frac{7}{10}\right)^2 = \frac{49}{100}$. $\sqrt{0.25}$ Ask: What positive decimal, when squared, is 0.25? The answer is 0.5 because $(0.5)^2 = 0.25$.

Unless otherwise noted, all art on this page is © Cengage Learning.

When **evaluating an expression containing square roots**, evaluate square roots at the same stage in your solution as exponential expressions.	Evaluate: $20 + 6(2^3 - 4\sqrt{9})$ Perform the operations within the parentheses first. $20 + 6(2^3 - 4\sqrt{9}) = 20 + 6(8 - 4 \cdot 3)$ Within the parentheses, evaluate the exponential expression and the square root. $= 20 + 6(8 - 12)$ Within the parentheses, do the multiplication. $= 20 + 6(-4)$ Within the parentheses, do the subtraction. $= 20 + (-24)$ Do the multiplication. $= -4$ Do the addition.
To **evaluate formulas that involve square roots**, we replace the letters with specific numbers and then use the order of operations rule.	Evaluate $a = \sqrt{c^2 - b^2}$ for $c = 25$ and $b = 20$. $a = \sqrt{c^2 - b^2}$ This is the formula to evaluate. $a = \sqrt{25^2 - 20^2}$ Replace c with 25 and b with 20. $a = \sqrt{625 - 400}$ Evaluate the exponential expressions. $a = \sqrt{225}$ Do the subtraction. $a = 15$ Evaluate the square root.
If a number is not a perfect square, we can use the square root key $\boxed{\sqrt{}}$ on a calculator (or a table of square roots) to find its **approximate square root**.	Approximate $\sqrt{149}$. Round to the nearest hundredth. From a scientific calculator we get $\sqrt{149} \approx 12.20655562$. Rounded to the nearest hundredth, $\sqrt{149} \approx 12.21$

REVIEW EXERCISES

95. Find the two square roots of 25. 5 and −5
96. Fill in the blanks: $\sqrt{49} =$ 7 because 7$^2 = 49$.

Evaluate each square root without using a calculator.

97. $\sqrt{49}$ 7

98. $-\sqrt{16}$ −4

99. $\sqrt{\dfrac{64}{169}}$ $\dfrac{8}{13}$

100. $\sqrt{0.81}$ 0.9

101. Graph each square root: $\sqrt{9}, -\sqrt{2}, \sqrt{3}, -\sqrt{16}$
(*Hint:* Use a calculator or square root table to approximate any square roots, when necessary.)

$-\sqrt{16}$ $-\sqrt{2}$ $\sqrt{3}$ $\sqrt{9}$
-5 -4 -3 -2 -1 0 1 2 3 4 5

102. Use a calculator to approximate each square root to the nearest hundredth.

a. $\sqrt{19}$ 4.36   b. $\sqrt{598}$ 24.45   c. $\sqrt{12.75}$ 3.57

Evaluate each expression without using a calculator.

103. $-3\sqrt{49} - \sqrt{36}$ −27

104. $\sqrt{\dfrac{100}{9}} + \sqrt{225}$ $18\dfrac{1}{3}$

105. $40 + 6[5^2 - (7 - 2)\sqrt{16}]$ 70

106. $1 - 7[6^2 + (1 + 2)\sqrt{81}]$ −440

107. Evaluate $b = \sqrt{c^2 - a^2}$ for $c = 17$ and $a = 15$. 8

108. SHEET METAL Find the length of the side of the range hood shown in the illustration below. 33 in.

$\sqrt{1{,}089}$ in.

109. Between what two whole numbers would $\sqrt{83}$ be located when graphed on a number line? 9 and 10

110. $\sqrt{7} \approx 2.646$. Explain why an $\approx$ symbol is used and not an $=$ symbol. Since $(2.646)^2 = 7.001316$, we cannot use an $=$ symbol.

SECTION 5.7 Solving Equations That Involve Decimals

DEFINITIONS AND CONCEPTS	EXAMPLES
We often use the *commutative property of multiplication* to reorder factors and the *associative property of multiplication* to regroup factors when **simplifying expressions**.	Simplify: $5(3.1y) = (5 \cdot 3.1)y$ Use the associative property of multiplication to regroup the factors. $= 15.5y$ Do the multiplication within the parentheses. Simplify: $6.2m(-4) = 6.2(-4)m$ Use the commutative property of multiplication to reorder the factors. $= -24.8m$ Do the multiplication, working from left to right.
The **distributive property** can be used to remove parentheses.	Multiply: $7.7(2x + 3) = 7.7(2x) + 7.7(3)$ Distribute the multiplication by 7.7. $= 15.4x + 23.1$ Do the multiplication. Try to go directly to this step.
Like terms are terms with exactly the same variables raised to exactly the same powers.	$3.5x$ and $5.8x$ are like terms. $-4.2t^3$ and $3.9t^2$ are unlike terms because the variable t has different exponents.
Simplifying the sum or difference of like terms is called **combining like terms**. Like terms can be combined by adding or subtracting the coefficients of the terms and keeping the same variables with the same exponents.	Simplify: $4.3a + 3.6a = 7.9a$ Think: $(4.3 + 3.6)a = 7.9a$. Simplify: $5.4p + 17.8 - 4.2p = 1.2p + 17.8$ Think: $(5.4 - 4.2)p = 1.2p$. Like terms Simplify: $5.2(2y - 1) - (3.1y - 9.6)$ $= 5.2(2y - 1) - 1(3.1y - 9.6)$ Replace the $-$ symbol in front of $(3.1y - 9.6)$ with -1. $= 10.4y - 5.2 - 3.1y + 9.6$ Distribute the multiplication by 5.2 and -1. $= 7.3y + 4.4$ Combine like terms. Think: $(10.4 - 3.1)y = 7.3y$. Think: $-5.2 + 9.6 = 4.4$.
To **solve an equation** means to find all the values of the variable that make the equations true. To isolate the variable on one side of the equation, we use the: 1. Addition property of equality 2. Subtraction property of equality 3. Multiplication property of equality 4. Division property of equality	Solve: $x + 15.6 = 39.8$ We can use the subtraction property of equality to isolate x on the left side of the equation. $x + 15.6 - 15.6 = 39.8 - 15.6$ To undo the addition of 15.6, subtract 15.6 from both sides. $x = 24.2$ Do the subtraction. **Check:** $x + 15.6 = 39.8$ This is the original equation. $24.2 + 15.6 \stackrel{?}{=} 39.8$ Substitute 24.2 for x. $39.8 = 39.8$ Do the addition. Since the resulting statement $39.8 = 39.8$ is true, 24.2 is the solution.

Sometimes we must use two (or more) properties of equality to solve more complicated equations.	Solve: $9.3x - 2.6 = 6.2x + 9.8$ $9.3x - 2.6 - \mathbf{6.2x} = 6.2x + 9.8 - \mathbf{6.2x}$ To eliminate the term 6.2x on the right side, subtract 6.2x from both sides. $3.1x - 2.6 = 9.8$ Combine like terms on both sides. $\mathbf{3.1x} - 2.6 + \mathbf{2.6} = 9.8 + \mathbf{2.6}$ To isolate 3.1x, undo the subtraction of 2.6 by adding 2.6 to both sides. $3.1x = 12.4$ Do the addition. Now isolate x. $\dfrac{\mathbf{3.1x}}{\mathbf{3.1}} = \dfrac{12.4}{\mathbf{3.1}}$ To isolate x, undo the multiplication by 3.1 by dividing both sides by 3.1. $x = 4$ Do the division. Check the result in the original equation to verify that 4 is the solution.
We can use the concepts of *variable* and *equation* and the six-step problem-solving strategy to solve application problems involving decimals.	**ADVERTISING** A promotions company sells pens with a personalized marketing message printed on them. They charge a $12 setup fee and the pens cost 85¢ each. If a business has budgeted $250 for this type of promotion, how many pens can they order? *Analyze* • There is a setup fee of $12. Given • The pens cost 85¢ each. (This is $0.85 each.) Given • The company has budgeted $250 for the promotion. Given • How many pens can they order? Find *Assign* Let x = the number of pens that they can order. *Form* | The setup fee	plus	the cost per pen	times	the number of pens ordered	must equal	the amount budgeted.
---	---	---	---	---	---	---
12	+	0.85	·	x	=	250	 **Solve** $12 + 0.85x = 250$ $12 + \mathbf{0.85x} - \mathbf{12} = 250 - \mathbf{12}$ To isolate the variable term 0.85x, subtract 12 from both sides to undo the addition of 12. $\mathbf{0.85x} = 238$ Do the subtraction. Now isolate x. $\dfrac{0.85x}{\mathbf{0.85}} = \dfrac{238}{\mathbf{0.85}}$ To isolate x, divide both sides by 0.85 to undo the multiplication by 0.85. $x = 280$ Do the division.

State The business can order 280 pens.

Check If we multiply the number of pens (280) by the cost per pen ($0.85), and add the setup fee ($12) to that product, we should get the amount budgeted ($250).

$$\begin{array}{r} 280 \\ \times\ 0.85 \\ \hline 14\ 00 \\ 224\ 00 \\ \hline 238.00 \end{array} \qquad \begin{array}{r} \overset{1}{2}38 \\ +\ 12 \\ \hline 250 \end{array} \begin{array}{l} \leftarrow \text{Cost of pens} \\ \leftarrow \text{Setup fee} \\ \leftarrow \text{Budget} \end{array}$$

The result checks.

REVIEW EXERCISES

Simplify each expression.

111. $4.6(7w)$ $32.2w$

112. $3.8(-2t)(4)$ $-30.4t$

Multiply.

113. $2.7(2q - 1)$
$5.4q - 2.7$

114. $-7(2.9x - 8.4)$
$-20.3x + 58.8$

Simplify each expression by combining like terms.

115. $8.2p + 5.9p$
$14.1p$

116. $0.57m^2 - 0.69m^2$
$-0.12m^2$

117. $5.7a - 12.4 - 2.9a$ $2.8a - 12.4$

118. $2(0.3t - 0.4) + 3(0.8t - 0.2)$ $3t - 1.4$

Solve each equation and check the result.

119. $y + 12.4 = -6.01$
-18.41

120. $x - 0.23 = 5$
5.23

121. $\dfrac{x}{1.78} = -3$ -5.34

122. $-1.61b = -27.37$ 17

123. $1.7y + 1.24 = -1.4y - 0.62$ -0.6

124. $0.08(x - 100) = 32.5 - 0.07x$ 270

In Problems 125 and 126, let a variable represent the unknown quantity. Then write and solve an equation to answer the question.

125. U.S. GASOLINE PRICES The average price of regular-grade gasoline in 2012 was $3.63 per gallon. This was $0.82 more than the average price in 2007. What was the price per gallon of regular gasoline in 2007? (Source: Energy Information Administration) $2.81 per gallon

126. BOWLING If it costs $4.00 to rent shoes and $4.50 a game to use a lane, how many games can be bowled for $40? 8 games

CHAPTER 5 TEST

1. Fill in the blanks.
   a. Copy the following addition. Label each *addend* and the *sum*.

   $$\begin{array}{r} 2.67 \leftarrow \text{addend} \\ +6.01 \leftarrow \text{addend} \\ \hline 8.68 \leftarrow \text{sum} \end{array}$$

   b. Copy the following subtraction. Label the *minuend*, the *subtrahend*, and the *difference*.

   $$\begin{array}{r} 9.6 \leftarrow \text{minuend} \\ -6.2 \leftarrow \text{subtrahend} \\ \hline 3.4 \leftarrow \text{difference} \end{array}$$

   c. Copy the following multiplication. Label the *factors* and the *product*.

   $$\begin{array}{r} 1.3 \leftarrow \text{factor} \\ \times 7 \leftarrow \text{factor} \\ \hline 9.1 \leftarrow \text{product} \end{array}$$

   d. Copy the following division. Label the *dividend*, the *divisor*, and the *quotient*.

   $$\text{divisor} \rightarrow 2\overline{)6.8} \leftarrow \text{dividend},\ \ 3.4 \leftarrow \text{quotient}$$

   e. 0.6666… and 0.8333… are examples of __repeating__ decimals.
   f. The $\sqrt{}$ symbol is called a __radical__ symbol.
   g. When we write $4.1x + 2.7x$ as $6.8x$, we say we have __combined__ like terms.
   h. To __solve__ an equation means to find all the values of the variable that make the equations true.

2. Express the amount of the square region that is shaded using a fraction and a decimal. $\frac{79}{100}, 0.79$

3. Consider the decimal number 629.471.
   a. What is the place value of the digit 1? 1 thousandth
   b. Which digit tells the number of tenths? 4
   c. Which digit tells the number of hundreds? 6
   d. What is the place value of the digit 2? 2 tens

4. WATER PURITY A county health department sampled the pollution content of tap water in five cities, with the results shown. Rank the cities in order, from dirtiest tap water to cleanest.
   Selway, Monroe, Paston, Covington, Cadia

City	Pollution, parts per million
Monroe	0.0909
Covington	0.0899
Paston	0.0901
Cadia	0.0890
Selway	0.1001

5. Write *four thousand five hundred nineteen and twenty-seven ten-thousandths* in standard form. 4,519.0027

6. Write each decimal in:
   • expanded form
   • words
   • as a fraction or mixed number. (You do not have to simplify the fraction.)
   a. SKATEBOARDING Mischo Erban of Canada, set the skateboard speed record of 80.74 mph in 2012. (Source: skateboardballbearings.com)
   $80 + \frac{7}{10} + \frac{4}{100}$, eighty and seventy-four hundredths, $80\frac{74}{100}$
   b. MONEY A dime weighs 0.08013 ounce.
   $\frac{8}{100} + \frac{1}{10,000} + \frac{3}{100,000}$, eight thousand thirteen hundred-thousandths, $\frac{8,013}{100,000}$

7. Round each decimal number to the indicated place value.
   a. 461.728, nearest tenth 461.7
   b. 2,733.0495, nearest thousandth 2,733.050
   c. -1.9833732, nearest millionth -1.983373

8. Round $0.648209 to the nearest cent. $0.65

Perform each operation.

9. $4.56 + 2 + 0.896 + 3.3$ 10.756
10. Subtract 39.079 from 45.2 6.121
11. $(0.32)^2$ 0.1024
12. $\dfrac{0.1368}{0.24}$ 0.57
13. $(-3.2)(0.6)$ -1.92
14.   0.0348
15. $11\overline{)13}$ $1.\overline{18}$
16. $-2.4 - (-1.6)$ -0.8

17. Divide. Round the quotient to the nearest hundredth:
    $\dfrac{12.146}{-5.3}$ -2.29

18. a. Estimate the product using front-end rounding: $34 \cdot 6.83$ 210
    b. Estimate the quotient: $3,907.2 \div 19.3$
    $4,000 \div 20 = 400 \div 2 = 200$

19. Perform each operation in your head.
 a. $567.909 \div 1{,}000$ 0.567909
 b. $0.00458 \cdot 100$ 0.458

20. Write 61.4 billion in standard notation. 61,400,000,000

21. NEW YORK CITY Refer to the illustration on the right. Central Park, which lies in the middle of Manhattan, is the city's best-known park. If it is 2.5 miles long and 0.5 mile wide, what is its area? 1.25 mi²

22. TELEPHONE BOOKS To print a telephone book, 565 sheets of paper were used. If the book is 2.26 inches thick, what is the thickness of each sheet of paper? 0.004 in.

23. ACCOUNTING At an ice-skating complex, receipts on Friday were $130.25 for indoor skating and $162.25 for outdoor skating. On Saturday, the corresponding amounts were $140.50 and $175.75. On which day, Friday or Saturday, were the receipts higher? How much higher? Saturday, $23.75

24. WEIGHT OF WATER One gallon of water weighs 8.33 pounds. How much does the water in a $2\frac{1}{2}$-gallon jug weigh? 20.825 lb

25. Evaluate the formula $C = 2\pi r$ for $\pi = 3.14$ and $r = 1.7$. 10.676

26. Write each fraction as a decimal.
 a. $\frac{17}{50}$ 0.34
 b. $\frac{5}{12}$ $0.41\overline{6}$

Evaluate each expression.

27. $4.1 - (3.2)(0.4)^2$ 3.588

28. $\left(\frac{2}{5}\right)^2 + 6\left|-6.2 - 3\frac{1}{4}\right|$ 56.86

29. $8 - 2\left(2^4 - 60 + 6\sqrt{81}\right)$ -12

30. $\frac{2}{3} + 0.7$ (Work in terms of fractions.) $\frac{41}{30}$

31. a. Graph $\frac{3}{8}$, $\frac{2}{3}$, and $-\frac{4}{5}$ on the number line. Label each point using the decimal equivalent of the fraction.

 b. Graph $\sqrt{16}$, $\sqrt{2}$, $-\sqrt{9}$, and $-\sqrt{5}$ on the number line below. (*Hint:* When necessary, use a calculator or square root table to approximate a square root.)

32. SALADS A shopper purchased $\frac{3}{4}$ pound of potato salad, priced at $5.60 per pound, and $\frac{1}{3}$ pound of coleslaw, selling for $4.35 per pound. Find the total cost of these items. $5.65

33. Use a calculator to evaluate $c = \sqrt{a^2 + b^2}$ for $a = 12$ and $b = 35$. 37

34. Write each number as a decimal.
 a. $-\frac{27}{25}$ -1.08
 b. $2\frac{9}{16}$ 2.5625

35. Fill in the blank: $\sqrt{144} = $ 12 because 12² = 144.

36. Place an $<$, $>$, or an $=$ symbol in the box to make a true statement.
 a. $-6.78 > -6.79$
 b. $0.3 < \frac{3}{8}$
 c. $\sqrt{\frac{16}{81}} = 0.\overline{4}$
 d. $0.45 < 0.\overline{45}$

Evaluate each expression without using a calculator.

37. $-2\sqrt{25} + 3\sqrt{49}$ 11

38. $\sqrt{\frac{1}{36}} - \sqrt{\frac{1}{25}}$ $-\frac{1}{30}$

39. Evaluate each square root without using a calculator.
 a. $-\sqrt{0.04}$ -0.2
 b. $\sqrt{1.69}$ 1.3
 c. $\sqrt{225}$ 15
 d. $-\sqrt{121}$ -11

40. Although the decimal 3.2999 contains more digits than 3.3, it is smaller than 3.3. Explain why this is so.

Simplify each expression.

41. a. $9(1.6t)$ $14.4t$
 b. $-3(4.7a)(2)$ $-28.2a$

42. a. $-5.7c + 4.5c$ $-1.2c$
 b. $2.1(x - 3) + 3.1(x - 4)$ $5.2x - 18.7$

Solve each equation and check the result.

43. $-2.4d = 16.8$ -7

44. $\frac{x}{2.04} + 2.2 = 6.7$ 9.18

45. $2(y + 4.3) - 0.3 = 7.1$ -0.6

46. $0.3x + 6.9 = 1.6x - 8.7$ 12

In Problems 47 and 48, let a variable represent the unknown quantity. Then write and solve an equation to answer the question.

47. **CHEMISTRY** In a lab experiment, a chemist mixed three compounds together to form a mixture weighing 4.37 grams. Later, she discovered that she had forgotten to record the weight of compound C in her notes. Find the weight of compound C used in the experiment. 0.42 gram

	Weight
Compound A	1.86 grams
Compound B	2.09 grams
Compound C	?
Mixture total	4.37 grams

48. **WEDDING COSTS** A printer charges a setup fee of $24 and then 95 cents for each wedding announcement printed (tax included). If a couple has budgeted $100 for printing costs, how many announcements can they have printed? 80 announcements

CHAPTERS 1–5 CUMULATIVE REVIEW

1. Write 154,302
   a. in words [Section 1.1] one hundred fifty-four thousand, three hundred two
   b. in expanded form [Section 1.1]
      $100{,}000 + 50{,}000 + 4{,}000 + 300 + 2$

2. Use 3, 4, and 5 to express the associative property of addition. [Section 1.2] $(3 + 4) + 5 = 3 + (4 + 5)$

3. Add: $9{,}339 + 471 + 6{,}883$ [Section 1.2] 16,693

4. Subtract 199 from 301. [Section 1.2] 102

5. SUDOKU The world's largest Sudoku puzzle was carved into a hillside near Bristol, England. It measured 275 ft by 275 ft. Find the area covered by the square-shaped puzzle. (Source: joe-ks.com) [Section 1.3] 75,625 ft²

6. Divide: $43\overline{)1{,}203}$ [Section 1.4] 27 R 42

7. List the factors of 20, from smallest to largest. [Section 1.5] 1, 2, 4, 5, 10, 20

8. Find the prime factorization of 220. [Section 1.5] $2^2 \cdot 5 \cdot 11$

9. Find the LCM and the GCF of 100 and 120. [Section 1.6] 600, 20

10. Find the mean (average) of 7, 1, 8, 2, and 2. [Section 1.7] 4

11. Solve each equation and check the result.
    a. $x + 19 = 285$ [Section 1.8] 266
    b. $19x = 285$ [Section 1.9] 15

12. Place an $<$ or an $>$ symbol in the box to make a true statement: $|-50|\ \boxed{>}\ -(-40)$ [Section 2.1]

13. Add: $-8 + (-5)$ [Section 2.2] -13

14. Fill in the blank: Subtraction is the same as __adding__ the opposite. [Section 2.3]

15. WEATHER Marsha flew from her Minneapolis home to Hawaii for a week of vacation. She left blizzard conditions and a temperature of $-11°F$, and stepped off the airplane into $72°F$ weather. What temperature change did she experience? [Section 2.3] 83°F increase

16. Multiply: $-3(-5)(2)(-9)$ [Section 2.4] -270

17. Evaluate: $(-1)^5$ [Section 2.4] -1

18. SUBMARINES As part of a training exercise, the captain of a submarine ordered it to descend 350 feet, level off for 10 minutes, and then repeat the process several times. If the sub was on the ocean's surface at the beginning of the exercise, find its depth after the 6th dive. [Section 2.4] $-2{,}100$ ft.

19. Consider the division statement $\frac{-15}{-5} = 3$. What is its related multiplication statement? [Section 2.5] $3(-5) = -15$

20. Divide: $-420{,}000 \div (-7{,}000)$ [Section 2.5] 60

21. Evaluate: $(-6)^2 - 2(5 - 4 \cdot 2)$ [Section 2.6] 42

22. Solve $2y + 8 = -6$ and check the result. [Section 2.7] -7

Write an equation and solve it to answer the following question. [Section 2.7]

23. MERCURY The freezing point of mercury is $-38°F$. By how many degrees must it be heated to reach its boiling point, which is $674°F$? It must be heated 712°F.

24. Translate to mathematical symbols: The cost c split five equal ways. [Section 3.1] $\frac{c}{5}$

25. Write an algebraic expression that represents the number of minutes in h hours. [Section 3.1] $60h$ min

26. BOWLING Find the average for a bowler who rolled scores of 233, 218, and 206. [Section 3.2] 219

27. Simplify: $-5(-6x)$ [Section 3.3] $30x$

28. Combine like terms: $5x + 5x + 5x$ [Section 3.4] $15x$

29. Solve $9a + 3(a - 7) = 18 - a$ and check the result. [Section 3.5] 3

Write an equation and solve it to answer the following question. [Section 3.6]

30. AIRLINE SEATING A 132-seat passenger plane has ten times as many economy seats as first-class seats. Find the number of first-class seats. There are 12 first-class seats.

31. FLAGS What fraction of the stripes on a U.S. flag are white? [Section 4.1] $\frac{6}{13}$

32. Simplify: $\frac{90}{126}$ [Section 4.1] $\frac{5}{7}$

Perform the operations. Simplify the result.

33. $\dfrac{3}{8} \cdot \dfrac{7}{16}$ [Section 4.2] $\dfrac{21}{128}$

34. $-\dfrac{15}{8a^4} \div \dfrac{10}{a^5}$ [Section 4.3] $-\dfrac{3a}{16}$

35. $\dfrac{a}{9} + \dfrac{5}{7}$ [Section 4.4] $\dfrac{7a+45}{63}$

36. $-4\dfrac{1}{4}\left(-4\dfrac{1}{2}\right)$ [Section 4.5] $19\dfrac{1}{8}$

37. $76\dfrac{1}{6} - 49\dfrac{7}{8}$ [Section 4.6] $26\dfrac{7}{24}$

38. $\dfrac{\frac{5}{27}}{-\frac{5}{9}}$ [Section 4.7] $-\dfrac{1}{3}$

39. What is $\dfrac{1}{4}$ of $\dfrac{7}{16}$? [Section 4.2] $\dfrac{7}{64}$

40. **TAPE MEASURES** Use the information shown in the illustration below to determine the inside length of the drawer. [Section 4.6] $11\dfrac{1}{8}$ in.

41. Evaluate: $\left(\dfrac{9}{20} \div 2\dfrac{2}{5}\right) + \left(\dfrac{3}{4}\right)^2$ [Section 4.7] $\dfrac{3}{4}$

42. Solve $\dfrac{3}{4}m = -27$ and check the result. [Section 4.8] -36

43. **GLASS** Some electronic and medical equipment uses glass that is only 0.00098 inch thick. Round this number to the nearest thousandth. [Section 5.1] 0.001 in.

44. Graph $-3\dfrac{1}{4}$, 0.75, -1.5, $-\dfrac{9}{8}$, 3.8, and $\sqrt{4}$ on a number line. [Section 5.1]

Perform the operations.

45. $7.001 - 5.9$ [Section 5.2] 1.101

46. $-1.8(4.52)$ [Section 5.3] -8.136

47. $56.012(0.001)$ [Section 5.3] 0.056012

48. $\dfrac{-21.28}{-3.8}$ [Section 5.4] 5.6

49. **KITES** Find the area of the front of the kite shown below. [Section 5.3] 157.5 in.²

50. Evaluate the formula $C = \dfrac{5}{9}(F - 32)$ for $F = 451$. Round to the nearest tenth. [Section 5.4] 232.8

51. Write the fraction $\dfrac{5}{12}$ as a decimal. [Section 5.5] $0.41\overline{6}$

52. Evaluate: $\dfrac{3}{8}(-3.2) + \left(4\dfrac{1}{2}\right)\left(-\dfrac{1}{4}\right)$ [Section 5.5] -2.325

53. Evaluate: $-4\sqrt{36} + 2\sqrt{81}$ [Section 5.6] -6

54. Solve $-5.6 - 2h = -17.1 - h$ and check the result. [Section 5.7] 11.5

6

Ratio, Proportion, and Measurement

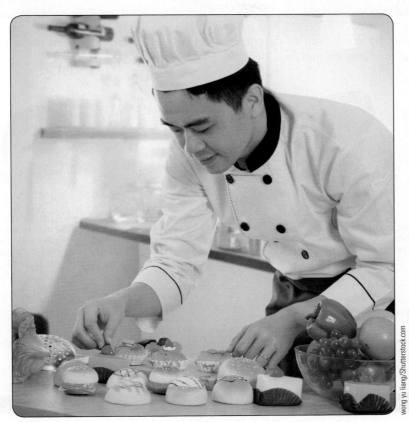

6.1 Ratios
6.2 Proportions
6.3 American Units of Measurement
6.4 Metric Units of Measurement; Medical Applications
6.5 Converting between American and Metric Units
Chapter Summary and Review
Chapter Test
Cumulative Review

from Campus to Careers
Chef

Chefs prepare and cook a wide range of foods—from soups, snacks, and salads to main dishes, side dishes, and desserts. They work in a variety of restaurants and food service kitchens. They measure, mix, and cook ingredients according to recipes, using a variety of equipment and tools. They are also responsible for directing the tasks of other kitchen workers, estimating food requirements, and ordering food supplies.

In **Problem 90** of **Study Set 6.2**, you will see how a chef can use proportions to determine the correct amounts of each ingredient needed to make a large batch of brownies.

JOB TITLE: Chef
EDUCATION: Training programs are available through culinary schools, two- or four-year college degree programs, and the armed forces.
JOB OUTLOOK: Job opportunities are expected to be good; however, competition is going to be high.
ANNUAL EARNINGS: The average (median) salary in 2011 was $46,600.
FOR MORE INFORMATION:
www.bls.gov/oes/current/oes351011.htm

557

Objectives

1. Write ratios as fractions.
2. Simplify ratios involving decimals and mixed numbers.
3. Convert units to write ratios.
4. Write rates as fractions.
5. Find unit rates.
6. Find the best buy based on unit price.

SECTION 6.1
Ratios

Ratios are often used to describe important relationships between two quantities. Here are three examples:

To prepare fuel for an outboard marine engine, gasoline must be mixed with oil in the ratio of 50 to 1.

To make 14-karat jewelry, gold is combined with other metals in the ratio of 14 to 10.

In this drawing, the eyes-to-nose distance and the nose-to-chin distance are drawn using a ratio of 2 to 3.

1 Write ratios as fractions.

Ratios give us a way to compare two numbers or two quantities measured in the same units.

> **Ratios**
>
> A **ratio** is the quotient of two numbers or the quotient of two quantities that have the same units.

There are three ways to write a ratio. The most common way is as a fraction. Ratios can also be written as two numbers separated by the word *to*, or as two numbers separated by a colon. For example, the ratios described in the illustrations above can be expressed as:

$$\frac{50}{1}, \quad 14 \text{ to } 10, \quad \text{and} \quad 2:3$$

- The fraction $\frac{50}{1}$ is read as "the ratio of 50 to 1."   *A fraction bar separates the numbers being compared.*

- 14 **to** 10 is read as "the ratio of 14 to 10."   *The word "to" separates the numbers being compared.*

- 2**:**3 is read as "the ratio of 2 to 3."   *A colon separates the numbers being compared.*

> **Writing a Ratio as a Fraction**
>
> To **write a ratio as a fraction**, write the first number (or quantity) mentioned as the numerator and the second number (or quantity) mentioned as the denominator. Then simplify the fraction, if possible.

EXAMPLE 1
Write each ratio as a fraction: **a.** 3 to 7   **b.** 10:11

Strategy We will identify the numbers before and after the word *to* and the numbers before and after the colon.

WHY The word *to* and the colon separate the numbers to be compared in a ratio.

Solution
To write the ratio as a fraction, the first number mentioned is the numerator and the second number mentioned is the denominator.

a. The ratio 3 **to** 7 can be written as $\frac{3}{7}$.   The fraction $\frac{3}{7}$ is in simplest form.

b. The ratio 10 **:** 11 can be written as $\frac{10}{11}$.   The fraction $\frac{10}{11}$ is in simplest form.

> **Caution!** When a ratio is written as a fraction, the fraction should be in simplest form. (Recall from Chapter 4 that a fraction is in **simplest form**, or **lowest terms**, when the numerator and denominator have no common factors other than 1.)

Self Check 1
Write each ratio as a fraction:
a. 4 to 9 $\frac{4}{9}$ **b.** 8:15 $\frac{8}{15}$

Now Try Problem 13

Teaching Example 1
Write each ratio as a fraction:
a. 2 to 5 **b.** 8:27
Answers:
a. $\frac{2}{5}$ **b.** $\frac{8}{27}$

EXAMPLE 2
Write the ratio 35 to 10 as a fraction in simplest form.

Strategy We will translate the ratio from its given form in words to fractional form. Then we will look for any factors common to the numerator and denominator and remove them.

WHY We need to make sure that the numerator and denominator have no common factors other than 1. If that is the case, the ratio will be in *simplest form*.

Solution

The ratio 35 **to** 10 can be written as $\frac{35}{10}$.   The fraction $\frac{35}{10}$ is not in simplest form.

Now we simplify the fraction, using the method discussed in Section 4.1.

$$\frac{35}{10} = \frac{\overset{1}{\cancel{5}} \cdot 7}{2 \cdot \underset{1}{\cancel{5}}}$$ Factor 35 as 5 · 7 and 10 as 2 · 5. Then remove the common factor of 5 in the numerator and denominator.

$$= \frac{7}{2}$$

The ratio 35 to 10 can be written as the fraction $\frac{35}{10}$, which simplifies to $\frac{7}{2}$ (read as "7 to 2"). Because the fractions $\frac{35}{10}$ and $\frac{7}{2}$ represent equal numbers, they are called **equal ratios**.

Self Check 2
Write the ratio 12 to 9 as a fraction in simplest form. $\frac{4}{3}$

Now Try Problems 17 and 23

Teaching Example 2 Write the ratio 30 to 18 as a fraction in simplest form.
Answer:
$\frac{5}{3}$

Caution! Since ratios are comparisons of two numbers, it would be *incorrect* in Example 2 to write the ratio $\frac{7}{2}$ as the mixed number $3\frac{1}{2}$. Ratios written as improper fractions are perfectly acceptable—just make sure the numerator and denominator have no common factors other than 1.

To write a ratio in simplest form, we remove any common factors of the numerator and denominator as well as any common units.

Self Check 3

CARRY-ON LUGGAGE
a. Write the ratio of the height to the length of the carry-on space shown in the illustration in Example 3 as a fraction in simplest form. $\frac{2}{3}$
b. Write the ratio of the length of the carry-on space to its height in simplest form. $\frac{3}{2}$

Now Try Problem 27

Teaching Example 3
BASKETBALL NBA basketball courts are 50 feet wide and 94 feet long.
a. Write the ratio of the width of the court to the length as a fraction in simplest form.
b. Write the ratio of the length of the court to the width as a fraction in simplest form.
Answers:
a. $\frac{25}{47}$ b. $\frac{47}{25}$

EXAMPLE 3 *Carry-on Luggage* An airline allows its passengers to carry a piece of luggage onto an airplane only if it will fit in the space shown below.

a. Write the ratio of the width of the space to its length as a fraction in simplest form.
b. Write the ratio of the length of the space to its width as a fraction in simplest form.

Strategy To write each ratio as a fraction, we will identify the quantity before the word *to* and the quantity after it.

WHY The first quantity mentioned is the numerator of the fraction and the second quantity mentioned is the denominator.

Solution

a. The ratio of the width of the space to its length is $\frac{10 \text{ inches}}{24 \text{ inches}}$.

To write a ratio in simplest form, we remove the common factors *and* the common units of the numerator and denominator.

$$\frac{10 \text{ inches}}{24 \text{ inches}} = \frac{\overset{1}{\cancel{2} \cdot 5} \cancel{\text{inches}}}{\underset{1}{\cancel{2} \cdot 12} \cancel{\text{inches}}}$$

Factor 10 as 2 · 5 and 24 as 2 · 12. Then remove the common factor of 2 and the common units of inches from the numerator and denominator.

$$= \frac{5}{12}$$

The width-to-length ratio of the carry-on space is $\frac{5}{12}$ (read as "5 to 12").

b. The ratio of the length of the space to its width is $\frac{24 \text{ inches}}{10 \text{ inches}}$.

$$\frac{24 \text{ inches}}{10 \text{ inches}} = \frac{\overset{1}{\cancel{2} \cdot 12} \cancel{\text{inches}}}{\underset{1}{\cancel{2} \cdot 5} \cancel{\text{inches}}}$$

Factor 24 and 10. Then remove the common factor of 2 and the common units of inches from the numerator and denominator.

$$= \frac{12}{5}$$

The length-to-width ratio of the carry-on space is $\frac{12}{5}$ (read as "12 to 5").

Unless otherwise noted, all art on this page is © Cengage Learning.

Caution! Example 3 shows that order is important when writing a ratio. The width-to-length ratio is $\frac{5}{12}$ while the length-to-width ratio is $\frac{12}{5}$.

2 Simplify ratios involving decimals and mixed numbers.

EXAMPLE 4 Write the ratio 0.3 to 1.8 as a fraction in simplest form.

Strategy After writing the ratio as a fraction, we will multiply it by a form of 1 to obtain an equivalent ratio of whole numbers.

WHY A ratio of whole numbers is easier to understand than a ratio of decimals.

Solution

The ratio 0.3 to 1.8 can be written as $\frac{0.3}{1.8}$.

To write this as a ratio of *whole numbers*, we need to move the decimal points in the numerator and denominator one place to the right. Recall that to find the product of a decimal and 10, we simply move the decimal point one place to the right. Therefore, it follows that $\frac{10}{10}$ is the form of 1 that we should use to build $\frac{0.3}{1.8}$ into an equivalent ratio.

$$\frac{0.3}{1.8} = \frac{0.3}{1.8} \cdot \frac{10}{10} \quad \text{Multiply the ratio by a form of 1.}$$

$$= \frac{0.3 \cdot 10}{1.8 \cdot 10} \quad \begin{array}{l}\text{Multiply the numerators.}\\ \text{Multiply the denominators.}\end{array}$$

$$= \frac{3}{18} \quad \text{Do the multiplications by moving each decimal point one place to the right. } 0.3 \cdot 10 = 3 \text{ and } 1.8 \cdot 10 = 18.$$

$$= \frac{1}{6} \quad \text{Simplify the fraction: } \frac{3}{18} = \frac{\cancel{3}}{\cancel{3} \cdot 6} = \frac{1}{6}.$$

Self Check 4

Write the ratio 0.8 to 2.4 as a fraction in simplest form. $\frac{1}{3}$

Now Try Problems 29 and 33

Teaching Example 4 Write the ratio 0.6 to 4.8 as a fraction in simplest form.
Answer:
$\frac{1}{8}$

THINK IT THROUGH Student-to-Instructor Ratio

"A more personal classroom atmosphere can sometimes be an easier adjustment for college freshmen. They are less likely to feel like a number, a feeling that can sometimes impact students' first semester grades."

From *The Importance of Class Size* by Stephen Pemberton

The data below come from a nationwide study of mathematics programs at two-year colleges. Determine which course has the lowest student-to-instructor ratio. (Assume that there is one instructor per section.)
21:1, 24:1, 25:1; Prealgebra has the lowest student-to-instructor ratio.

	Prealgebra	Elementary Algebra	Intermediate Algebra
Students enrolled	226,002	427,992	344,000
Number of sections	10,762	17,833	13,760

Source: Conference Board of the Mathematical Science, 2010 CBMS Survey of Undergraduate Programs (The data have been rounded to yield ratios involving whole numbers.)

Chapter 6 Ratio, Proportion, and Measurement

Self Check 5

Write the ratio $3\frac{1}{3}$ to $1\frac{1}{9}$ as a fraction in simplest form. $\frac{3}{1}$

Now Try Problem 37

Teaching Example 5 Write the ratio $6\frac{3}{4}$ to $1\frac{1}{8}$ as a fraction in simplest form.
Answer:
$\frac{6}{1}$

EXAMPLE 5 Write the ratio $4\frac{2}{3}$ to $1\frac{1}{6}$ as a fraction in simplest form.

Strategy After writing the ratio as a fraction, we will use the method for simplifying a complex fraction from Section 4.7 to obtain an equivalent ratio of whole numbers.

WHY A ratio of whole numbers is easier to understand than a ratio of mixed numbers.

Solution

The ratio of $4\frac{2}{3}$ to $1\frac{1}{6}$ can be written as $\dfrac{4\frac{2}{3}}{1\frac{1}{6}}$.

The resulting ratio is a complex fraction. To write the ratio in simplest form, we perform the division indicated by the main fraction bar (shown in red).

$$\dfrac{4\frac{2}{3}}{1\frac{1}{6}} = \dfrac{\frac{14}{3}}{\frac{7}{6}} \qquad \text{Write } 4\frac{2}{3} \text{ and } 1\frac{1}{6} \text{ as improper fractions.}$$

$$= \frac{14}{3} \div \frac{7}{6} \qquad \text{Write the division indicated by the main fraction bar using a } \div \text{ symbol.}$$

$$= \frac{14}{3} \cdot \frac{6}{7} \qquad \text{Use the rule for dividing fractions: Multiply the first fraction by the reciprocal of } \frac{7}{6}, \text{ which is } \frac{6}{7}.$$

$$= \frac{14 \cdot 6}{3 \cdot 7} \qquad \text{Multiply the numerators. Multiply the denominators.}$$

$$= \frac{2 \cdot \overset{1}{\cancel{7}} \cdot 2 \cdot \overset{1}{\cancel{3}}}{\underset{1}{\cancel{3}} \cdot \underset{1}{\cancel{7}}} \qquad \text{To simplify the fraction, factor 14 as } 2 \cdot 7 \text{ and 6 as } 2 \cdot 3. \text{ Then remove the common factors 3 and 7.}$$

$$= \frac{4}{1} \qquad \text{Multiply the remaining factors in the numerator. Multiply the remaining factors in the denominator.}$$

We would normally simplify the result $\frac{4}{1}$ and write it as 4. But since a ratio compares two numbers, we leave the result in fractional form.

3 Convert units to write ratios.

When a ratio compares two quantities, both quantities must be measured in the same units. For example, inches must be compared to inches, pounds to pounds, and seconds to seconds.

Self Check 6

Write the ratio *6 feet to 3 yards* as a fraction in simplest form. (*Hint:* 3 feet = 1 yard.) $\frac{2}{3}$

Now Try Problem 41

EXAMPLE 6 Write the ratio *12 ounces to 2 pounds* as a fraction in simplest form.

Strategy We will convert 2 pounds to ounces and write a ratio that compares ounces to ounces. Then we will simplify the ratio.

WHY A ratio compares two quantities that have the *same* units. When the units are different, it's usually easier to write the ratio using the smaller unit of measurement. Since ounces are smaller than pounds, we will compare in ounces.

Solution

To express 2 pounds in ounces, we use the fact that there are 16 ounces in one pound.

$$2 \cdot 16 \text{ ounces} = 32 \text{ ounces}$$

We can now express the ratio *12 ounces to 2 pounds* using the same units:

12 **ounces** to 32 **ounces**

Next, we write the ratio in fraction form and simplify.

$$\frac{12 \text{ ounces}}{32 \text{ ounces}} = \frac{\overset{1}{\cancel{3 \cdot 4}} \text{ ounces}}{\underset{1}{\cancel{4} \cdot 8} \text{ ounces}}$$

To simplify, factor 12 as $3 \cdot 4$ and 32 as $4 \cdot 8$. Then remove the common factor of 4 and the common units of ounces from the numerator and denominator.

$$= \frac{3}{8}$$

The ratio in simplest form is $\frac{3}{8}$.

Teaching Example 6
Write the ratio *18 seconds to 2 minutes* as a fraction in simplest form.
Answer:
$\frac{3}{20}$

4 Write rates as fractions.

When we compare two quantities that have different units (and neither unit can be converted to the other), we call the comparison a **rate**, and we can write it as a fraction. For example, on the label of the can of paint shown on the right, we see that 1 quart of paint is needed for every 200 square feet to be painted. Writing this as a rate in fractional form, we have

$$\frac{1 \text{ quart}}{200 \text{ square feet}}$$ Read as "1 quart per 200 square feet."

ANTIQUE RED
LATEX
SEMI-GLOSS
ENAMEL
Dries in one hour
COVERAGE: one quart covers 200 square feet

> **The Language of Algebra** The word **per** is associated with the operation of division, and it means "for each" or "for every." For example, when we say 1 quart of paint *per* 200 square feet, we mean 1 quart of paint *for every* 200 square feet.

Rates

A **rate** is a quotient of two quantities that have different units.

When writing a rate, always include the units. Some other examples of rates are:

- 16 computers **for** 75 students
- 1,550 feet **in** 4.5 seconds
- 88 tomatoes **from** 3 plants
- 250 miles **on** 2 gallons of gasoline

> **The Language of Algebra** As seen above, words such as *per, for, in, from,* and *on* are used to separate the two quantities that are compared in a rate.

Unless otherwise noted, all art on this page is © Cengage Learning.

Writing a Rate as a Fraction

To **write a rate as a fraction**, write the first quantity mentioned as the numerator and the second quantity mentioned as the denominator, and then simplify, if possible. Write the units as part of the fraction.

Self Check 7

GROWTH RATES The fastest-growing flowering plant on record grew 12 feet in 14 days. Write the rate of growth as a fraction in simplest form. $\frac{6 \text{ feet}}{7 \text{ days}}$

Now Try Problems 49 and 53

Teaching Example 7 **HOSPITALS** In a 6-hour span, 21 patients were treated in a hospital emergency room. Write the rate of patients treated in the emergency room as a fraction in simplest form.
Answer: $\frac{7 \text{ patients}}{2 \text{ hours}}$

EXAMPLE 7 *Snowfall* According to the *Guinness Book of World Records*, a total of 78 inches of snow fell at Mile 47 Camp, Cooper River Division, Arkansas, in a 24-hour period in 1963. Write the rate of snowfall as a fraction in simplest form.

Strategy We will use a fraction to compare the amount of snow that fell (in inches) to the amount of time in which it fell (in hours). Then we will simplify it.

WHY A rate is a quotient of two quantities with different units.

Solution

78 inches **in** 24 hours can be written as $\frac{78 \text{ inches}}{24 \text{ hours}}$.

Now, we simplify the fraction.

$$\frac{78 \text{ inches}}{24 \text{ hours}} = \frac{\overset{1}{\cancel{6}} \cdot 13 \text{ inches}}{4 \cdot \underset{1}{\cancel{6}} \text{ hours}}$$

To simplify, factor 78 as 6 · 13 and 24 as 4 · 6. Then remove the common factor of 6 from the numerator and denominator.

$$= \frac{13 \text{ inches}}{4 \text{ hours}}$$

Since the units are different, they cannot be removed.

The snow fell at a rate of 13 inches per 4 hours.

5 Find unit rates.

Unit Rate

A **unit rate** is a rate in which the denominator is 1.

To illustrate the concept of a unit rate, suppose a driver makes the 354-mile trip from Pittsburgh to Indianapolis in 6 hours. Then the motorist's rate (or more specifically, rate of speed) is given by

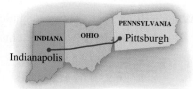

$$\frac{354 \text{ miles}}{6 \text{ hours}} = \frac{\overset{1}{\cancel{6}} \cdot 59 \text{ miles}}{\underset{1}{\cancel{6}} \cdot \text{ hours}}$$

Factor 354 as 6 · 59 and remove the common factor of 6 from the numerator and denominator.

$$= \frac{59 \text{ miles}}{1 \text{ hour}}$$

Since the units are different, they cannot be removed. Note that the denominator is 1.

Unless otherwise noted, all art on this page is © Cengage Learning.

We can also find the unit rate by dividing 354 by 6.

Rate:
$$\frac{354 \text{ miles}}{6 \text{ hours}}$$

$$\begin{array}{r} 59 \\ 6\overline{)354} \\ -\,30 \\ \hline 54 \\ -\,54 \\ \hline 0 \end{array}$$

← This quotient is the numerical part of the unit rate, written as a fraction.

The numerical part of the denominator is always 1.

Unit rate:
$$\frac{59 \text{ miles}}{1 \text{ hour}}$$

The unit rate $\frac{59 \text{ miles}}{1 \text{ hour}}$ can be expressed in any of the following forms:

$59 \frac{\text{miles}}{\text{hour}}$, 59 miles per hour, 59 miles/hour, or 59 mph

> **The Language of Algebra** A slash mark / is often used to write a unit rate. In such cases, we read the slash mark as "per." For example, 33 pounds/gallon is read as 33 pounds *per* gallon.

Writing a Rate as a Unit Rate

To **write a rate as a unit rate**, divide the numerator of the rate by the denominator.

EXAMPLE 8 **Coffee** There are 384 calories in a 16-ounce cup of caramel Frappuccino blended coffee with whip cream. Write this rate as a unit rate. (*Hint:* Find the number of calories in 1 ounce.)

Strategy We will translate the rate from its given form in words to fractional form. Then we will perform the indicated division.

WHY To write a rate as a unit rate, we divide the numerator of the rate by the denominator.

Solution

384 calories **in** 16 ounces can be written as $\frac{384 \text{ calories}}{16 \text{ ounces}}$.

To find the number of calories in 1 ounce of the coffee (the unit rate), we perform the division as indicated by the fraction bar:

$$\begin{array}{r} 24 \\ 16\overline{)384} \\ -\,32 \\ \hline 64 \\ -\,64 \\ \hline 0 \end{array}$$ Divide the numerator of the rate by the denominator.

For the caramel Frappuccino blended coffee with whip cream, the unit rate is $\frac{24 \text{ calories}}{1 \text{ ounce}}$, which can be written as 24 calories per ounce or 24 calories/ounce.

Self Check 8

NUTRITION There are 204 calories in a 12-ounce can of cranberry juice. Write this rate as a unit rate. (*Hint:* Find the number of calories in 1 ounce.) 17 calories/oz

Now Try Problem 57

Teaching Example 8 NUTRITION
There are 180 calories in a 12-ounce serving of apple juice. Write this rate as a unit rate. (*Hint:* Find the number of calories in 1 ounce.)
Answer:
15 calories/oz

Self Check 9

FULL-TIME JOBS Joan earns $436 per 40-hour week managing a dress shop. Write this rate as a unit rate. (*Hint:* Find her hourly rate of pay.) $10.90 per hour

Now Try Problem 61

Teaching Example 9 PART-TIME JOBS A student earns $90 for working 8 hours in a school cafeteria. Write this rate as a unit rate. (*Hint:* Find her hourly rate of pay.)
Answer:
$11.25/hr

EXAMPLE 9 *Part-time Jobs* A student earns $74 for working 8 hours in a bookstore. Write this rate as a unit rate. (*Hint:* Find his hourly rate of pay.)

Strategy We will translate the rate from its given form in words to fractional form. Then we will perform the indicated division.

WHY To write a rate as a unit rate, we divide the numerator of the rate by the denominator.

Solution

$74 **for** working 8 hours can be written as $\dfrac{\$74}{8 \text{ hours}}$.

To find the rate of pay for 1 hour of work (the unit rate), we divide 74 by 8.

$$\begin{array}{r} 9.25 \\ 8\overline{)74.00} \\ -72 \\ \hline 2\,0 \\ -1\,6 \\ \hline 40 \\ -40 \\ \hline 0 \end{array}$$ Write a decimal point and two additional zeros to the right of 4.

The unit rate of pay is $\dfrac{\$9.25}{1 \text{ hour}}$, which can be written as $9.25 per hour or $9.25/hr. ∎

6 Find the best buy based on unit price.

If a grocery store sells a 5-pound package of hamburger for $18.75, a consumer might want to know what the hamburger costs per pound. When we find the cost of 1 pound of the hamburger, we are finding a **unit price**. To find the unit price of an item, we begin by comparing its price to the number of units.

$\dfrac{\$18.75}{5 \text{ pounds}}$ ← Price
← Number of units

Then we divide the price by the number of units.

$$5\overline{)18.75} \quad \text{quotient } 3.75$$

The unit price of the hamburger is $3.75 per pound.
Other examples of unit prices are:

- $8.15 per ounce
- $200 per day
- $0.75 per foot

Unit Price

A **unit price** is a rate that tells how much is paid for *one* unit (or *one* item). It is the quotient of price to the number of units.

$$\text{Unit price} = \dfrac{\text{price}}{\text{number of units}}$$

When shopping, it is often difficult to determine the best buys because the items that we purchase come in so many different sizes and brands. Comparison shopping can be made easier by finding unit prices. *The best buy is the item that has the lowest unit price.*

EXAMPLE 10 Comparison Shopping
Olives come packaged in a 10-ounce jar, which sells for $2.49, or in a 6-ounce jar, which sells for $1.53. Which is the better buy?

Strategy We will find the unit price for each jar of olives. Then we will identify which jar has the lower unit price.

WHY The better buy is the jar of olives that has the lower unit price.

Solution
To find the unit price of each jar of olives, we write the quotient of its price and its weight, and then perform the indicated division. Before dividing, we convert each price from dollars to cents so that the unit price can be expressed in cents per ounce.

The 10-ounce jar:

$$\frac{\$2.49}{10 \text{ oz}} = \frac{249¢}{10 \text{ oz}}$$ Write the rate: $\frac{\text{price}}{\text{number of units}}$.
Then change $2.49 to 249 cents.

$$= 24.9¢ \text{ per oz}$$ Divide 249 by 10 by moving the decimal point 1 place to the left.

The 6-ounce jar:

$$\frac{\$1.53}{6 \text{ oz}} = \frac{153¢}{6 \text{ oz}}$$ Write the rate: $\frac{\text{price}}{\text{number of units}}$.
Then change $1.53 to 153 cents.

$$= 25.5¢ \text{ per oz}$$ Do the division.

$$\begin{array}{r} 25.5 \\ 6\overline{)153.0} \\ \underline{-12} \\ 33 \\ \underline{-30} \\ 3\ 0 \\ \underline{-3\ 0} \\ 0 \end{array}$$

One ounce for 24.9¢ is a better buy than one ounce for 25.5¢. The unit price is less when olives are packaged in 10-ounce jars, so that is the better buy.

Self Check 10

COMPARISON SHOPPING A fast-food restaurant sells a 12-ounce cola for 72¢ and a 16-ounce cola for 99¢. Which is the better buy? the 12-oz cola

Now Try Problems 65 and 103

Teaching Example 10
GUITAR LESSONS A music store charges $27 for a 30-minute guitar lesson and $36 for a 45-minute lesson. Which is the better buy?
Answer:
The 45-min lesson

ANSWERS TO SELF CHECKS

1. a. $\frac{4}{9}$ b. $\frac{8}{15}$ 2. $\frac{4}{3}$ 3. a. $\frac{2}{3}$ b. $\frac{3}{2}$ 4. $\frac{1}{3}$ 5. $\frac{3}{1}$ 6. $\frac{2}{3}$ 7. $\frac{6 \text{ feet}}{7 \text{ days}}$
8. 17 calories/oz   9. $10.90 per hour   10. the 12-oz cola

SECTION 6.1   STUDY SET

VOCABULARY

Fill in the blanks.

1. A __ratio__ is the quotient of two numbers or the quotient of two quantities that have the same units.
2. A __rate__ is the quotient of two quantities that have different units.
3. A __unit__ rate is a rate in which the denominator is 1.
4. A unit __price__ is a rate that tells how much is paid for one unit or one item.

CONCEPT

5. To write the ratio $\frac{15}{24}$ in simplest form, we remove any common factors of the numerator and denominator. What common factor do they have? 3

6. Complete the steps used to write the ratio $\frac{14}{21}$ in simplest form.

$$\frac{14}{21} = \frac{2 \cdot 7}{3 \cdot 7} = \frac{2 \cdot \overset{1}{\cancel{7}}}{3 \cdot \underset{1}{\cancel{7}}} = \frac{2}{3}$$

7. Consider the ratio $\frac{0.5}{0.6}$. By what number should we multiply numerator and denominator to make this a ratio of whole numbers? 10

8. What should be done to write the ratio $\frac{15 \text{ inches}}{22 \text{ inches}}$ in simplest form? Remove the common units of inches in the numerator and denominator.

9. Write $\frac{11 \text{ minutes}}{1 \text{ hour}}$ so that it compares the same units and then simplify, if possible. $\frac{11 \text{ minutes}}{60 \text{ minutes}} = \frac{11}{60}$

10. a. Consider the rate $\frac{\$248}{16 \text{ hours}}$. What division should be performed to find the unit rate in dollars per hour? $248 \div 16$

    b. Suppose 3 pairs of socks sell for $7.95: $\frac{\$7.95}{3 \text{ pairs}}$. What division should be performed to find the unit price of one pair of socks? $7.95 \div 3$

NOTATION

11. Write the ratio of the flag's length to its width using a fraction, using the word *to*, and using a colon. $\frac{13}{9}$, 13 to 9, 13:9

9 inches
13 inches

12. The rate $\frac{55 \text{ miles}}{1 \text{ hour}}$ can be expressed as
    - 55 __miles__ __per__ __hour__ (in three words)
    - 55 __miles__ / __hour__ (in two words with a slash)
    - 55 __m__ __p__ __h__ (in three letters)

GUIDED PRACTICE

Write each ratio as a fraction. See Example 1.

13. 5 to 8 $\frac{5}{8}$
14. 3 to 23 $\frac{3}{23}$
15. 11:16 $\frac{11}{16}$
16. 9:25 $\frac{9}{25}$

Write each ratio as a fraction in simplest form. See Example 2.

17. 25 to 15 $\frac{5}{3}$
18. 45 to 35 $\frac{9}{7}$
19. 63:36 $\frac{7}{4}$
20. 54:24 $\frac{9}{4}$
21. 22:33 $\frac{2}{3}$
22. 14:21 $\frac{2}{3}$
23. 17 to 34 $\frac{1}{2}$
24. 19 to 38 $\frac{1}{2}$

Write each ratio as a fraction in simplest form. See Example 3.

25. 4 ounces to 12 ounces $\frac{1}{3}$
26. 3 inches to 15 inches $\frac{1}{5}$
27. 24 miles to 32 miles $\frac{3}{4}$
28. 56 yards to 64 yards $\frac{7}{8}$

Write each ratio as a fraction in simplest form. See Example 4.

29. 0.3 to 0.9 $\frac{1}{3}$
30. 0.2 to 0.6 $\frac{1}{3}$
31. 0.65 to 0.15 $\frac{13}{3}$
32. 2.4 to 1.5 $\frac{8}{5}$
33. 3.8:7.8 $\frac{19}{39}$
34. 4.2:8.2 $\frac{21}{41}$
35. 7:24.5 $\frac{2}{7}$
36. 5:22.5 $\frac{2}{9}$

Write each ratio as a fraction in simplest form. See Example 5.

37. $2\frac{1}{3}$ to $4\frac{2}{3}$ $\frac{1}{2}$
38. $1\frac{1}{4}$ to $1\frac{1}{2}$ $\frac{5}{6}$
39. $10\frac{1}{2}$ to $1\frac{3}{4}$ $\frac{6}{1}$
40. $12\frac{3}{4}$ to $2\frac{1}{8}$ $\frac{6}{1}$

Write each ratio as a fraction in simplest form. See Example 6.

41. 12 minutes to 1 hour $\frac{1}{5}$
42. 8 ounces to 1 pound $\frac{1}{2}$
43. 3 days to 1 week $\frac{3}{7}$
44. 4 inches to 1 yard $\frac{1}{9}$
45. 18 months to 2 years $\frac{3}{4}$
46. 8 feet to 4 yards $\frac{2}{3}$
47. 21 inches to 3 feet $\frac{7}{12}$
48. 32 seconds to 2 minutes $\frac{4}{15}$

Write each rate as a fraction in simplest form. See Example 7.

49. 64 feet in 6 seconds $\frac{32 \text{ ft}}{3 \text{ sec}}$
50. 45 applications for 18 openings $\frac{5 \text{ applications}}{2 \text{ openings}}$
51. 75 days on 20 gallons of water $\frac{15 \text{ days}}{4 \text{ gal}}$
52. 3,000 students over a 16-year career $\frac{375 \text{ students}}{2 \text{ yr}}$
53. 84 made out of 100 attempts $\frac{21 \text{ made}}{25 \text{ attempts}}$
54. 16 right compared to 34 wrong $\frac{8 \text{ right}}{17 \text{ wrong}}$
55. 18 beats every 12 measures $\frac{3 \text{ beats}}{2 \text{ measures}}$
56. 10 inches as a result of 30 turns $\frac{1 \text{ in.}}{3 \text{ turns}}$

Write each rate as a unit rate. See Example 8.

57. 60 revolutions in 5 minutes   12 revolutions per min
58. 14 trips every 2 months   7 trips per month
59. $50,000 paid over 10 years $5,000 per year
60. 245 presents for 35 children   7 presents per child

Unless otherwise noted, all art on this page is © Cengage Learning.

Write each rate as a unit rate. **See Example 9.**

61. 12 errors in 8 hours 1.5 errors per hr
62. 114 times in a 12-month period 9.5 times per month
63. 4,007,500 people living in 12,500 square miles 320.6 people per square mi
64. 117.6 pounds of pressure on 8 square inches 14.7 pounds per square in.

Find the unit price of each item. **See Example 10.**

65. They charged $48 for 12 minutes. $4 per min
66. 150 barrels cost $4,950. $33 per barrel
67. Four sold for $272. $68 per person
68. 7,020 pesos will buy six tickets. 1,170 pesos per ticket
69. 65 ounces sell for 78 cents. 1.2 cents per ounce
70. For 7 dozen, you will pay $10.15. $1.45 per dozen
71. $3.50 for 50 feet $0.07 per ft
72. $4 billion over a 5-month span $0.8 billion per month

APPLICATIONS

73. **GEAR RATIOS** Refer to the illustration below.
    a. Write the ratio of the number of teeth of the smaller gear to the number of teeth of the larger gear in simplest form. $\frac{2}{3}$
    b. Write the ratio of the number of teeth of the larger gear to the number of teeth of the smaller gear in simplest form. $\frac{3}{2}$

74. **CARDS** The suit of hearts from a deck of playing cards is shown below. What is the ratio of the number of face cards to the total number of cards in the suit? (*Hint:* A face card is a Jack, Queen, or King.) $\frac{3}{13}$

75. **BAKING** A recipe for sourdough bread calls for $5\frac{1}{4}$ cups of all-purpose flour and $1\frac{3}{4}$ cups of water. Write the ratio of flour to water in simplest form. $\frac{3}{1}$

76. **DESSERTS** Refer to the recipe card shown below. Write the ratio of milk to sugar in simplest form. $\frac{7}{3}$

Frozen Chocolate Slush
(Serves 8)
Once frozen, this chocolate can be cut into cubes and stored in sealed plastic bags for a spur-of-the-moment dessert.

$\frac{1}{2}$ cup Dutch cocoa powder, sifted
$1\frac{1}{2}$ cups sugar
$3\frac{1}{2}$ cups skim milk

77. **SKIN** Refer to the cross-section of human skin shown below. Write the ratio of the thickness of the stratum corneum to the thickness of the dermis in simplest form. (Source: Philips Research Laboratories) $\frac{1}{55}$

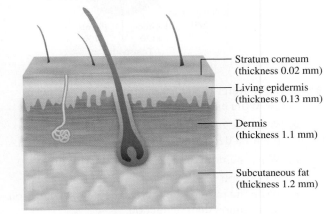

Stratum corneum (thickness 0.02 mm)
Living epidermis (thickness 0.13 mm)
Dermis (thickness 1.1 mm)
Subcutaneous fat (thickness 1.2 mm)

78. **PAINTING** A 9.5-mil thick coat of fireproof paint is applied with a roller to a wall. (A *mil* is a unit of measure equal to 1/1,000 of an inch.) The coating dries to a thickness of 5.7 mils. Write the ratio of the thickness of the coating when wet to the thickness when dry in simplest form. $\frac{5}{3}$

79. **BANKRUPTCY** After declaring bankruptcy, a company could pay its creditors only 5¢ on the dollar. Write this as a ratio in simplest form. $\frac{1}{20}$

80. **EGGS** An average-sized ostrich egg weighs 3 pounds and an average-sized chicken egg weighs 2 ounces. Write the ratio of the weight of an ostrich egg to the weight of a chicken egg in simplest form. $\frac{24}{1}$

81. **CPR** A paramedic performed 155 compressions to 25 breaths on an adult with no pulse. What compressions-to-breaths rate did the paramedic use? $\frac{31 \text{ compressions}}{5 \text{ breaths}}$

82. **FACULTY–STUDENT RATIOS** At a college, there are 125 faculty members and 2,000 students. Find the rate of faculty to students. (This is often referred to as the faculty–student *ratio*, even though the units are different.) $\frac{1 \text{ faculty member}}{16 \text{ students}}$

83. **BUDGETS** Refer to the circle graph below that shows a monthly budget for a family. Write each ratio in simplest form.
   a. Find the total amount for the monthly budget. $1,800
   b. Write the ratio of the amount budgeted for rent to the total budget. $\frac{4}{9}$
   c. Write the ratio of the amount budgeted for food to the total budget. $\frac{1}{3}$
   d. Write the ratio of the amount budgeted for the phone to the total budget. $\frac{1}{18}$

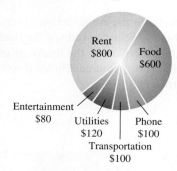

84. **TAXES** Refer to the list of tax deductions shown below. Write each ratio in simplest form.
   a. Write the ratio of the real estate tax deduction to the total deductions. $\frac{1}{7}$
   b. Write the ratio of the charitable contributions to the total deductions. $\frac{1}{5}$
   c. Write the ratio of the mortgage interest deduction to the union dues deduction. $\frac{35}{4}$

Item	Amount
Medical expenses	$875
Real estate taxes	$1,250
Charitable contributions	$1,750
Mortgage interest	$4,375
Union dues	$500
Total deductions	$8,750

85. **ART HISTORY** Leonardo da Vinci drew the human figure shown within a square. Write the ratio of the length of the man's outstretched arms to his height. (*Hint:* All four sides of a square are the same length.) $\frac{1}{1}$

86. **FLAGS** The checkered flag is composed of 24 equal-sized squares. What is the ratio of the width of the flag to its length? (*Hint:* All four sides of a square are the same length.) $\frac{2}{3}$

87. **AIRLINE COMPLAINTS** An airline had 3.29 complaints for every 1,000 passengers. Write this as a rate of whole numbers. $\frac{329 \text{ complaints}}{100,000 \text{ passengers}}$

88. **FINGERNAILS** On average, fingernails grow 0.02 inch per week. Write this rate using whole numbers. $\frac{1 \text{ inch}}{50 \text{ weeks}}$

89. **INTERNET SALES** A website determined that it had 112,500 hits in one month. Of those visiting the site, 4,500 made purchases.
   a. Those that visited the site, but did not make a purchase, are called *browsers*. How many browsers visited the website that month? 108,000
   b. What was the browsers-to-buyers unit rate for the website that month? 24 browsers per buyer

90. **TYPING** A secretary typed a document containing 330 words in 5 minutes. Write this rate as a unit rate. 66 words per minute, or 66 wpm

91. **GOLD MINING** In 2009, a geologist estimated that on a plot of land near Smith Creek, Alaska, there were about 35,500 cubic feet of gravel, from which approximately 1,065 ounces of gold could be mined. Find the predicted number of ounces of gold per cubic foot of gravel on the land. (Source: mrdata.usgs.gov) 0.03 ounce gold/cubic foot gravel

92. **BUSINESS** *Sales per square foot* is often used to measure the financial performance of a store. For example, in 2012, Apple Stores sold goods at an average rate of $6,050 per square foot of retail space—the best for all retailers in the United States. Refer to the table on the next page.
   a. About how many times greater is Apple's rate of sales per square foot compared to Tiffany's? about two times greater
   b. About how many times greater is Apple's rate of sales per square foot compared to lululemon's? about three times greater

c. The U.S. national average for regional mall stores is $341 per square foot. About how many times greater is Apple's rate of sales per square foot than that? almost 18 times greater

Top Five U.S. Retailers' Sales per Square Foot, 2012

Company	Sales per square foot
Apple Stores	$6,050
Tiffany & Co.	$3,017
lululemon athletics	$1,936
Coach	$1,871
Michael Kors	$1,431

Source: retailsails.com

93. **UNIT PRICES** A 12-ounce can of cola sells for 84¢. Find the unit price in cents per ounce. 7¢ per oz

94. **DAY CARE** A day-care center charges $32 for 8 hours of supervised care. Find the unit price in dollars per hour for the day care. $4 per hour

95. **PARKING** A parking meter requires 25¢ for 20 minutes of parking. Find the unit price to park. 1.25¢ per min

96. **GASOLINE COST** A driver pumped 17 gallons of gasoline into the tank of his pickup truck at a cost of $66.13. Find the unit price of the gasoline. $3.89 per gal

97. **LANDSCAPING** A 50-pound bag of grass seed sells for $222.50. Find the unit price of grass seed. $4.45 per lb

98. **UNIT COSTS** A 24-ounce package of green beans sells for $1.29. Find the unit price in cents per ounce. 5.375¢ per oz

99. **DRAINING TANKS** An 11,880-gallon tank of water can be emptied in 27 minutes. Find the unit rate of flow of water out of the tank. 440 gal per min

100. **PAY RATE** Ricardo worked for 27 hours to help insulate a hockey arena. For his work, he received $337.50. Find his hourly rate of pay. $12.50 per hr

101. **AUTO TRAVEL** A car's odometer reads 34,746 at the beginning of a trip. Five hours later, it reads 35,071.
  a. How far did the car travel? 325 mi
  b. What was its rate of speed? 65 mph

102. **RATES OF SPEED** An airplane travels from Chicago to San Francisco, a distance of 1,883 miles, in 3.5 hours. Find the rate of speed of the plane. 538 mph

103. **COMPARISON SHOPPING** A 6-ounce can of orange juice sells for 89¢, and an 8-ounce can sells for $1.19. Which is the better buy? the 6-oz can

104. **COMPARISON SHOPPING** A 30-pound bag of planting mix costs $12.25, and an 80-pound bag costs $30.25. Which is the better buy? the 80-lb bag

105. **COMPARISON SHOPPING** A certain brand of cold and sinus medication is sold in 20-tablet boxes for $4.29 and in 50-tablet boxes for $9.59. Which is the better buy? the 50-tablet boxes

106. **COMPARISON SHOPPING** Which tire shown is the better buy? the economy model

107. **COMPARING SPEEDS** A car travels 345 miles in 6 hours, and a truck travels 376 miles in 6.2 hours. Which vehicle is going faster? the truck

108. **READING** One seventh-grader read a 54-page book in 40 minutes. Another read an 80-page book in 62 minutes. If the books were equally difficult, which student read faster? the first student

109. **GAS MILEAGE** One car went 1,235 miles on 51.3 gallons of gasoline, and another went 1,456 miles on 55.78 gallons. Which car got the better gas mileage? the second car

110. **ELECTRICITY RATES** In one community, a bill for 575 kilowatt-hours of electricity is $38.81. In a second community, a bill for 831 kwh is $58.10. In which community is electricity cheaper? the first community

WRITING

111. Are the ratios 3 to 1 and 1 to 3 the same? Explain why or why not.

112. Give three examples of ratios (or rates) that you have encountered in the past week.

113. How will the topics studied in this section make you a better shopper?

114. What is a unit rate? Give some examples.

REVIEW

115. Simplify: $\dfrac{48}{54}$ $\dfrac{8}{9}$

116. Simplify: $\dfrac{34}{153}$ $\dfrac{2}{9}$

117. Solve: $18x = 63$ 3.5

118. Solve: $63 = 9c$ 7

Objectives

1. Write proportions.
2. Determine whether proportions are true or false.
3. Solve a proportion to find an unknown term.
4. Write proportions to solve application problems.

SECTION 6.2
Proportions

One of the most useful concepts in mathematics is the *equation*. Recall that an **equation** is a statement indicating that two expressions are equal. All equations contain an = symbol. Some examples of equations are:

$$4 + 4 = 8, \quad 15.6 - 4.3 = 11.3, \quad \frac{1}{2} \cdot 10 = 5, \quad \text{and} \quad -16 \div 8 = -2$$

Each of the equations shown above is true. Equations can also be false. For example,

$$3 + 2 = 6 \quad \text{and} \quad -40 \div (-5) = -8$$

are false equations.

In this section, we will work with equations that state that two ratios (or rates) are equal.

1 Write proportions.

Like any tool, a ladder can be dangerous if used improperly. When setting up an extension ladder, users should follow the *4-to-1 rule:* For every 4 feet of ladder height, position the legs of the ladder 1 foot away from the base of the wall. The 4-to-1 rule for ladders can be expressed using a ratio.

$$\frac{4 \text{ feet}}{1 \text{ foot}} = \frac{4 \text{ feet}}{1 \text{ foot}} = \frac{4}{1} \quad \text{Remove the common units of feet.}$$

The figure on the right shows how the 4-to-1 rule was used to properly position the legs of a ladder 3 feet from the base of a 12-foot-high wall. We can write a ratio comparing the ladder's height to its distance from the wall.

$$\frac{12 \text{ feet}}{3 \text{ feet}} = \frac{12 \text{ feet}}{3 \text{ feet}} = \frac{12}{3} \quad \text{Remove the common units of feet.}$$

Since this ratio satisfies the 4-to-1 rule, the two ratios $\frac{4}{1}$ and $\frac{12}{3}$ must be equal. Therefore, we have

$$\frac{4}{1} = \frac{12}{3}$$

Equations like this, which show that two ratios are equal, are called *proportions*.

Proportion

A **proportion** is a statement that two ratios (or rates) are equal.

Some examples of proportions are

- $\dfrac{1}{2} = \dfrac{3}{6}$ Read as "1 is to 2 as 3 is to 6."

- $\dfrac{3 \text{ waiters}}{7 \text{ tables}} = \dfrac{9 \text{ waiters}}{21 \text{ tables}}$ Read as "3 waiters are to 7 tables as 9 waiters are to 21 tables."

EXAMPLE 1
Write each statement as a proportion.

a. 22 is to 6 as 11 is to 3.
b. 1,000 administrators is to 8,000 teachers as 1 administrator is to 8 teachers.

Strategy We will locate the word *as* in each statement and identify the ratios (or rates) before and after it.

WHY The word *as* translates to the = symbol that is needed to write the statement as a proportion (equation).

Solution

a. This proportion states that two ratios are equal.

$\underbrace{22 \text{ is to } 6}$ as $\underbrace{11 \text{ is to } 3}$. Recall that the word "to" is used to separate the numbers being compared.

$$\frac{22}{6} = \frac{11}{3}$$

b. This proportion states that two rates are equal.

$\underbrace{1{,}000 \text{ administrators is to } 8{,}000 \text{ teachers}}$ as $\underbrace{1 \text{ administrator is to } 8 \text{ teachers}}$

$$\frac{1{,}000 \text{ administrators}}{8{,}000 \text{ teachers}} = \frac{1 \text{ administrator}}{8 \text{ teachers}}$$

When proportions involve rates, the units are often written outside of the proportion, as shown below:

Administrators $\longrightarrow \dfrac{1{,}000}{8{,}000} = \dfrac{1}{8} \longleftarrow$ Administrators
Teachers $\longrightarrow$ $\longleftarrow$ Teachers

Self Check 1
Write each statement as a proportion.
a. 16 is to 28 as 4 is to 7. $\frac{16}{28} = \frac{4}{7}$
b. 300 children is to 500 adults as 3 children is to 5 adults. $\frac{300 \text{ children}}{500 \text{ adults}} = \frac{3 \text{ children}}{5 \text{ adults}}$

Now Try Problems 17 and 19

Teaching Example 1 Write each statement as a proportion.
a. 45 is to 50 as 9 is to 10.
b. 900 applicants is to 600 jobs as 3 applicants is to 2 jobs.
Answers:
a. $\dfrac{45}{50} = \dfrac{9}{10}$
b. $\dfrac{900 \text{ applicants}}{600 \text{ jobs}} = \dfrac{3 \text{ applicants}}{2 \text{ jobs}}$

2 Determine whether proportions are true or false.

Since a proportion is an equation, a proportion can be true or false. A proportion is true if its ratios (or rates) are equal and false if it its ratios (or rates) are not equal. One way to determine whether a proportion is true is to use the fraction simplifying skills of Chapter 4.

EXAMPLE 2
Determine whether each proportion is true or false by simplifying.

a. $\dfrac{3}{8} = \dfrac{21}{56}$ **b.** $\dfrac{30}{4} = \dfrac{45}{12}$

Strategy We will simplify any ratios in the proportion that are not in simplest form. Then we will compare them to determine whether they are equal.

WHY If the ratios are equal, the proportion is true. If they are not equal, the proportion is false.

Solution

a. On the left side of the proportion $\frac{3}{8} = \frac{21}{56}$, the ratio $\frac{3}{8}$ is in simplest form. On the right side, the ratio $\frac{21}{56}$ can be simplified.

$$\frac{21}{56} = \frac{3 \cdot \overset{1}{7}}{\underset{1}{7} \cdot 8} = \frac{3}{8}$$ Factor 21 and 56 and then remove the common factor of 7 in the numerator and denominator.

Since the ratios on the left and right sides of the proportion are equal, the proportion is true.

Self Check 2
Determine whether each proportion is true or false by simplifying.
a. $\dfrac{4}{5} = \dfrac{16}{20}$ true   **b.** $\dfrac{30}{24} = \dfrac{28}{16}$ false

Now Try Problem 23

Teaching Example 2 Determine whether each proportion is true or false by simplifying.
a. $\dfrac{7}{9} = \dfrac{72}{81}$ **b.** $\dfrac{16}{12} = \dfrac{44}{33}$
Answers:
a. false   **b.** true

b. Neither ratio in the proportion $\frac{30}{4} = \frac{45}{12}$ is in simplest form. To simplify each ratio, we proceed as follows:

$$\frac{30}{4} = \frac{\overset{1}{\cancel{2}} \cdot 15}{\underset{1}{\cancel{2}} \cdot 2} = \frac{15}{2} \qquad \frac{45}{12} = \frac{\overset{1}{\cancel{3}} \cdot 15}{\underset{1}{\cancel{3}} \cdot 4} = \frac{15}{4}$$

Since the ratios on the left and right sides of the proportion are not equal $\left(\frac{15}{2} \neq \frac{15}{4}\right)$, the proportion is false.

There is another way to determine whether a proportion is true or false. Before we can discuss it, we need to introduce some more vocabulary of proportions.

Each of the four numbers in a proportion is called a **term**. The first and fourth terms are called the **extremes**, and the second and third terms are called the **means**.

First term (extreme) ⟶ $\frac{1}{2}$ = $\frac{3}{6}$ ⟵ Third term (mean)
Second term (mean) ⟶   ⟵ Fourth term (extreme)

In the proportion shown above, the *product of the extremes is equal to the product of the means*.

$$1 \cdot 6 = 6 \qquad \text{and} \qquad 2 \cdot 3 = 6$$

These products can be found by multiplying diagonally in the proportion in such a way that one numerator is multiplied by the other denominator. We call $1 \cdot 6$ and $2 \cdot 3$ **cross products**.

$$1 \cdot 6 = 6 \qquad 2 \cdot 3 = 6$$
$$\frac{1}{2} \times \frac{3}{6}$$

Multiplication along one diagonal is shown in red.
Multiplication along the other diagonal is shown in blue.

Note that the cross products are equal. To see why this is true in general, consider the proportion $\frac{a}{b} = \frac{c}{d}$. If the ratios $\frac{a}{b}$ and $\frac{c}{d}$ are equal, we can show that the cross products are equal by multiplying both sides of the proportion by the LCD, bd.

$$\frac{a}{b} = \frac{c}{d} \qquad \text{Assume that this proportion is true and that neither denominator is 0.}$$

$$bd \cdot \frac{a}{b} = bd \cdot \frac{c}{d} \qquad \text{To clear the equation of fractions, multiply both sides by the LCD of } \frac{a}{b} \text{ and } \frac{c}{d}, \text{ which is } bd.$$

$$\frac{bd}{1} \cdot \frac{a}{b} = \frac{bd}{1} \cdot \frac{c}{d} \qquad \text{Write } bd \text{ as a fraction: } bd = \frac{bd}{1}.$$

$$\frac{abd}{b} = \frac{bcd}{d} \qquad \text{Multiply the numerators. Write the factors in alphabetical order. Multiply the denominators.}$$

$$\frac{a\cancel{b}d}{\underset{1}{\cancel{b}}} = \frac{bc\cancel{d}}{\underset{1}{\cancel{d}}} \qquad \text{Simplify each fraction. On the left, remove the common factor of } b \text{ in the numerator and denominator. On the right, remove the common factor of } d.$$

$$ad = bc \qquad \text{The cross products are equal.}$$

If neither b nor d is 0, then the steps shown above are reversible, and it is also true that if $ad = bc$, then $\frac{a}{b} = \frac{c}{d}$. This observation leads to the following property of proportions.

Cross-Products Property (Means-Extremes Property)

To determine whether a proportion is true or false, first multiply along one diagonal, and then multiply along the other diagonal.

- If the cross products are *equal*, the proportion is true.
- If the cross products are *not equal*, the proportion is false.

(If the product of the extremes is *equal* to the product of the means, the proportion is true. If the product of the extremes is *not equal* to the product of the means, the proportion is false.)

EXAMPLE 3 Determine whether each proportion is true or false.

a. $\dfrac{3}{7} = \dfrac{9}{21}$ **b.** $\dfrac{8}{3} = \dfrac{13}{5}$

Strategy We will check to see whether the cross products are equal (the product of the extremes is equal to the product of the means).

WHY If the cross products are equal, the proportion is true. If the cross products are not equal, the proportion is false.

Solution

a. $3 \cdot 21 = 63 \qquad 7 \cdot 9 = 63$

$$\dfrac{3}{7} = \dfrac{9}{21}$$

Each cross product is 63.

Since the cross products are equal, the proportion is true.

b. $8 \cdot 5 = 40 \qquad 3 \cdot 13 = 39$

$$\dfrac{8}{3} = \dfrac{13}{5}$$

One cross product is 40 and the other is 39.

Since the cross products are not equal, the proportion is false.

Caution! We cannot remove common factors "across" an = symbol. When this is done, the true proportion from Example 3 part a, $\dfrac{3}{7} = \dfrac{9}{21}$, is changed into the false proportion $\dfrac{1}{7} = \dfrac{9}{7}$.

$$\dfrac{\overset{1}{3}}{7} = \dfrac{9}{\underset{7}{21}}$$

Self Check 3
Determine whether the proportion
$$\dfrac{6}{13} = \dfrac{18}{39}$$
is true or false. true

Now Try Problem 25

Teaching Example 3 Determine whether each proportion is true or false.

a. $\dfrac{7}{24} = \dfrac{2}{7}$ **b.** $\dfrac{12}{8} = \dfrac{9}{6}$

Answers:
a. false **b.** true

EXAMPLE 4 Determine whether each proportion is true or false.

a. $\dfrac{0.9}{0.6} = \dfrac{2.4}{1.5}$ **b.** $\dfrac{2\frac{1}{3}}{3\frac{1}{2}} = \dfrac{4\frac{2}{3}}{7}$

Strategy We will check to see whether the cross products are equal (the product of the extremes is equal to the product of the means).

WHY If the cross products are equal, the proportion is true. If the cross products are not equal, the proportion is false.

Self Check 4
Determine whether each proportion is true or false.

a. $\dfrac{9.9}{13.2} = \dfrac{1.125}{1.5}$ true

b. $\dfrac{3\frac{3}{16}}{2\frac{1}{2}} = \dfrac{4\frac{1}{4}}{3\frac{1}{3}}$ true

Now Try Problems 31 and 35

Chapter 6 Ratio, Proportion, and Measurement

Teaching Example 4 Determine whether each proportion is true or false.

a. $\dfrac{0.7}{2.4} = \dfrac{0.9}{4.1}$ b. $\dfrac{3\frac{1}{3}}{2\frac{1}{12}} = \dfrac{1\frac{1}{5}}{\frac{3}{4}}$

Answers:
a. false b. true

Solution

a.
$$\begin{array}{r} 1.5 \\ \times\, 0.9 \\ \hline 1.35 \end{array} \qquad \begin{array}{r} 2.4 \\ \times\, 0.6 \\ \hline 1.44 \end{array}$$

$$\dfrac{0.9}{0.6} = \dfrac{2.4}{1.5}$$ One cross product is 1.35 and the other is 1.44.

Since the cross products are not equal, the proportion is not true.

b.
$$2\tfrac{1}{3} \cdot 7 = \tfrac{7}{3} \cdot \tfrac{7}{1} \qquad 3\tfrac{1}{2} \cdot 4\tfrac{2}{3} = \tfrac{7}{2} \cdot \tfrac{14}{3}$$

$$= \tfrac{49}{3} \qquad\qquad\qquad = \dfrac{7 \cdot \overset{1}{\cancel{2}} \cdot 7}{\cancel{2} \cdot 3}$$

$$\qquad\qquad\qquad\qquad\qquad = \tfrac{49}{3}$$

$$\dfrac{2\frac{1}{3}}{3\frac{1}{2}} = \dfrac{4\frac{2}{3}}{7}$$ Each cross product is $\tfrac{49}{3}$.

Since the cross products are equal, the proportion is true.

When two pairs of numbers such as 2, 3 and 8, 12 form a true proportion, we say that they are **proportional**. To show that 2, 3 and 8, 12 are proportional, we check to see whether the equation

$$\dfrac{2}{3} = \dfrac{8}{12}$$

is a true proportion. To do so, we find the cross products.

$$2 \cdot 12 = 24 \qquad 3 \cdot 8 = 24$$

Since the cross products are equal, the proportion is true, and the numbers are proportional.

Self Check 5

Determine whether 6, 11 and 54, 99 are proportional. yes

Now Try Problem 37

Teaching Example 5 Determine whether 6, 8 and 81, 99 are proportional.
Answer:
The numbers are not proportional.

EXAMPLE 5
Determine whether 3, 7 and 36, 91 are proportional.

Strategy We will use the given pairs of numbers to write two ratios and form a proportion. Then we will find the cross products.

WHY If the cross products are equal, the proportion is true, and the numbers are proportional. If the cross products are not equal, the proportion is false, and the numbers are not proportional.

Solution
The pair of numbers 3 and 7 form one ratio and the pair of numbers 36 and 91 form a second ratio. To write a proportion, we set the ratios equal. Then we find the cross products.

$$3 \cdot 91 = 273 \qquad 7 \cdot 36 = 252$$

$$\dfrac{3}{7} = \dfrac{36}{91}$$ One cross product is 273 and the other is 252.

Since the cross products are not equal, the numbers are not proportional.

3 Solve a proportion to find an unknown term.

Suppose that we know three of the four terms in the following proportion.

$$\dfrac{?}{5} = \dfrac{24}{20}$$

If we let the variable x represent the unknown term, we can write:

$$\frac{x}{5} = \frac{24}{20}$$

If the proportion is to be true, the cross products must be equal.

$x \cdot 20 = 5 \cdot 24$ Find the cross products for $\frac{x}{5} = \frac{24}{20}$ and set them equal.

$20x = 120$ On the left side, rewrite $x \cdot 20$ as $20x$. On the right side, do the multiplication: $5 \cdot 24 = 120$.

$$\begin{array}{r} \overset{2}{24} \\ \times\ 5 \\ \hline 120 \end{array}$$

On the left side of the equation, the unknown number x is multiplied by 20. To undo the multiplication by 20 and isolate x, we divide both sides of the equation by 20.

$\dfrac{20x}{20} = \dfrac{120}{20}$ Use the division property of equality.

$x = 6$ Do the division: $120 \div 20 = 6$.

$$\begin{array}{r} 6 \\ 20\overline{)120} \\ -120 \\ \hline 0 \end{array}$$

Thus, x is 6. We have found that the unknown term in the proportion is 6 and we can write:

$$\frac{6}{5} = \frac{24}{20}$$

To check this result, we find the cross products.

Check:

$\dfrac{6}{5} \overset{?}{=} \dfrac{24}{20}$ $6 \cdot 20 = 120$
              $5 \cdot 24 = 120$

Since the cross products are equal, the result, 6, checks.

In the previous example, when we find the value of the variable x that makes the given proportion true, we say that we have *solved the proportion* to find the unknown term.

> **The Language of Algebra** We solve proportions by writing a series of steps that result in an equation of the form $x =$ **a number** or **a number** $= x$. We say that the variable x is *isolated* on one side of the equation. *Isolated* means alone or by itself.

Solving a Proportion to Find an Unknown Term

1. Set the cross products equal to each other to form an equation.
2. Isolate the variable on one side of the equation by dividing both sides by the number that is multiplied by that variable.
3. Check by substituting the result into the original proportion and finding the cross products.

EXAMPLE 6   Solve the proportion: $\dfrac{12}{20} = \dfrac{3}{x}$

Strategy We will set the cross products equal to each other to form an equation.

WHY Then we can isolate the variable x on one side of the equation to find the unknown term in the proportion that it represents.

Self Check 6

Solve the proportion: $\dfrac{15}{x} = \dfrac{20}{32}$

Now Try Problem 41

Chapter 6 Ratio, Proportion, and Measurement

Teaching Example 6

Solve the proportion: $\dfrac{8}{30} = \dfrac{4}{x}$

Answer:
15

Solution

$$\dfrac{12}{20} = \dfrac{3}{x} \quad \text{This is the proportion to solve.}$$

$12 \cdot x = 20 \cdot 3$ Set the cross products equal to each other to form an equation.

$12x = 60$ On the right side, do the multiplication: $20 \cdot 3 = 60$.

$\dfrac{12x}{12} = \dfrac{60}{12}$ To isolate x, undo the multiplication by 12 by dividing both sides by 12.

$x = 5$ Do the division: $60 \div 12 = 5$.

$$\begin{array}{r} 5 \\ 12\overline{)60} \\ -\,60 \\ \hline 0 \end{array}$$

Thus, x is 5. To check this result, we substitute 5 for x in the original proportion.

Check:

$$\dfrac{12}{20} \overset{?}{=} \dfrac{3}{5} \qquad \begin{array}{l} 12 \cdot 5 = 60 \\ 20 \cdot 3 = 60 \end{array}$$

Since the cross products are equal, the result, 5, checks.

Self Check 7

Solve the proportion:

$\dfrac{6.7}{x} = \dfrac{33.5}{38}$ 7.6

Now Try Problem 45

Teaching Example 7

Solve the proportion: $\dfrac{3.6}{8.1} = \dfrac{x}{12.15}$

Answer:
5.4

EXAMPLE 7

Solve the proportion: $\dfrac{3.5}{7.2} = \dfrac{x}{15.84}$

Strategy We will set the cross products equal to each other to form an equation.

WHY Then we can isolate the variable x on one side of the equation to find the unknown term in the proportion that it represents.

Solution

$$\dfrac{3.5}{7.2} = \dfrac{x}{15.84} \quad \text{This is the proportion to solve.}$$

$3.5 \cdot 15.84 = 7.2 \cdot x$ Set the cross products equal to each other to form an equation.

$55.44 = 7.2x$ On the left side, do the multiplication: $3.5 \cdot 15.84 = 55.44$.

$\dfrac{55.44}{7.2} = \dfrac{7.2x}{7.2}$ To isolate x, undo the multiplication by 7.2 by dividing both sides by 7.2.

$7.7 = x$ Do the division: $55.44 \div 7.2 = 7.7$.

$$\begin{array}{r} 15.84 \\ \times\; 3.5 \\ \hline 7920 \\ 47520 \\ \hline 55.440 \end{array}$$

$$\begin{array}{r} 7.7 \\ 7.2\overline{)55.44} \\ -\,50\,4 \\ \hline 5\,04 \\ -\,5\,04 \\ \hline 0 \end{array}$$

Thus, x is 7.7. Check the result in the original proportion.

Self Check 8

Solve the proportion:

$\dfrac{x}{2\tfrac{1}{3}} = \dfrac{2\tfrac{1}{4}}{1\tfrac{1}{2}}$

Write the result as a mixed number. $3\tfrac{1}{2}$

Now Try Problem 49

EXAMPLE 8

Solve the proportion $\dfrac{x}{4\tfrac{1}{5}} = \dfrac{5\tfrac{1}{2}}{16\tfrac{1}{2}}$. Write the result as a mixed number.

Strategy We will set the cross products equal to each other to form an equation.

WHY Then we can isolate the variable x on one side of the equation to find the unknown term in the proportion that it represents.

Solution

$$\dfrac{x}{4\tfrac{1}{5}} = \dfrac{5\tfrac{1}{2}}{16\tfrac{1}{2}} \quad \text{This is the proportion to solve.}$$

$x \cdot 16\tfrac{1}{2} = 4\tfrac{1}{5} \cdot 5\tfrac{1}{2}$ Set the cross products equal to each other to form an equation.

$x \cdot \dfrac{33}{2} = \dfrac{21}{5} \cdot \dfrac{11}{2}$ Write each mixed number as an improper fraction.

$\dfrac{33}{2}x = \dfrac{21}{5} \cdot \dfrac{11}{2}$ On the left side, write $x \cdot \dfrac{33}{2}$ as $\dfrac{33}{2}x$.

$\dfrac{2}{33} \cdot \dfrac{33}{2}x = \dfrac{2}{33} \cdot \dfrac{21}{5} \cdot \dfrac{11}{2}$ To isolate x, multiply both sides by the reciprocal of the coefficient of the variable term, $\dfrac{33}{2}x$. The reciprocal of $\dfrac{33}{2}$ is $\dfrac{2}{33}$.

$1x = \dfrac{2 \cdot 21 \cdot 11}{33 \cdot 5 \cdot 2}$ On the left side, the product of a number and its reciprocal is 1: $\dfrac{2}{33} \cdot \dfrac{33}{2} = 1$. On the right side, multiply the numerators and multiply the denominators.

$x = \dfrac{\cancel{2} \cdot \cancel{3} \cdot 7 \cdot \cancel{11}}{\cancel{3} \cdot \cancel{11} \cdot 5 \cdot \cancel{2}}$ On the left side, the coefficient of 1 need not be written since $1x = x$. To simplify the right side, factor 21 and 33. Then remove the common factors 2, 3, and 11.

$x = \dfrac{7}{5}$ Multiply the remaining factors in the numerator: $1 \cdot 1 \cdot 7 \cdot 1 = 7$. Multiply the remaining factors in the denominator: $1 \cdot 1 \cdot 5 \cdot 1 = 5$.

$x = 1\dfrac{2}{5}$ Write the improper fraction as a mixed number.

Thus, x is $1\dfrac{2}{5}$. Check this result in the original proportion.

Using Your CALCULATOR Solving Proportions with a Calculator

To solve the proportion in Example 7, we set the cross products equal and divided both sides by 7.2 to isolate the variable x.

$$\dfrac{3.5 \cdot 15.84}{7.2} = x$$

We can find x by entering these numbers and pressing these keys on a calculator.

3.5 [×] 15.84 [÷] 7.2 [=]   **7.7**

Thus, x is 7.7.

---

4 Write proportions to solve application problems.

Proportions can be used to solve application problems from a wide variety of fields such as medicine, accounting, construction, and business. It is easy to spot problems that can be solved using a proportion. You will be given a ratio (or rate) and asked to find the missing part of another ratio (or rate). It is helpful to follow the six-step problem-solving strategy seen earlier in the text to solve proportion problems.

EXAMPLE 9 **Shopping** If 5 apples cost $1.15, find the cost of 16 apples.

Analyze

- We can express the fact that 5 apples cost $1.15 using the rate: $\dfrac{5 \text{ apples}}{\$1.15}$.
- What is the cost of 16 apples?

Assign We will let the variable c represent the unknown cost of 16 apples.

Form If we compare the number of apples to their cost, we know that the two rates must be equal and we can write a proportion.

5 apples **is to** $1.15 **as** 16 apples **is to** $c.

$$\dfrac{5 \leftarrow \text{5 apples}}{1.15 \leftarrow \text{Cost of 5 apples}} = \dfrac{16 \leftarrow \text{16 apples}}{c \leftarrow \text{Cost of 16 apples}}$$ The units can be written outside of the proportion.

Teaching Example 8
Solve the proportion:
$$\dfrac{x}{1\dfrac{4}{7}} = \dfrac{3\dfrac{2}{5}}{4\dfrac{2}{5}}$$
Write the result as a mixed number.
Answer:
$1\dfrac{3}{14}$

Self Check 9

CONCERT TICKETS If 9 tickets to a concert cost $112.50, find the cost of 15 tickets. $187.50

Now Try Problem 73

Teaching Example 9 GROCERY SHOPPING If 3 avocados cost $2.67, find the cost of 10 avocados.
Answer:
$8.90

Solve To find the cost of 16 apples, we solve the proportion for c.

$5 \cdot c = 1.15 \cdot 16$ Set the cross products equal to each other to form an equation.

$5c = 18.4$ On the right side, do the multiplication: $1.15(16) = 18.4$.

$\dfrac{5c}{5} = \dfrac{18.4}{5}$ To isolate c, undo the multiplication by 5 by dividing both sides by 5.

$c = 3.68$ Do the division: $18.4 \div 5 = 3.68$.

```
   3.68
5)18.40
  -15
   3 4
  -3 0
    40
   -40
     0
```

State Sixteen apples will cost $3.68.

Check If 5 apples cost $1.15, then 15 apples would cost 3 times as much: $3 \cdot \$1.15 = \$3.45$. It seems reasonable that 16 apples would cost $3.68.

We could have compared the cost of the apples to the number of apples as shown below. If we solve that proportion for c, we obtain the same result: 3.68.

Cost of 5 apples → $\dfrac{1.15}{5}$ = $\dfrac{c}{16}$ ← Cost of 16 apples
5 apples →   ← 16 apples

Caution! When solving problems using proportions, make sure that the units of the numerators are the same and the units of the denominators are the same. For Example 9, it would be incorrect to write

~~Cost of 5 apples → $\dfrac{1.15}{5} = \dfrac{16}{c}$ ← 16 apples~~
~~5 apples →~~   ~~← Cost of 16 apples~~

Self Check 10

SCALE MODELS In a scale model of a city, a 300-foot-tall building is 4 inches high. An observation tower in the model is 9 inches high. How tall is the actual tower? 675 ft

Now Try Problem 83

Teaching Example 10
SCALE DRAWINGS
The distance between the wheels in the scale drawing of the airplane is 1.25 inches. What is the actual distance between the wheels?
Answer:
7.5 ft

EXAMPLE 10 *Scale Drawings* A **scale** is a ratio (or rate) that compares the size of a model, drawing, or map to the size of an actual object. The airplane shown below is drawn using a scale of 1 inch: 6 feet. This means that 1 inch on the drawing is actually 6 feet on the plane. The distance from wing tip to wing tip (the wingspan) on the drawing is 4.5 inches. What is the actual wingspan of the plane?

SCALE
1 inch: 6 feet

Analyze
- The airplane is drawn using a scale of 1 inch: 6 feet, which can be written as a ratio in fraction form as: $\dfrac{1 \text{ inch}}{6 \text{ feet}}$.
- The wingspan of the airplane on the drawing is 4.5 inches.
- What is the actual wingspan of the plane?

Assign We will let w represent the unknown actual wingspan of the plane.

Form If we compare the measurements on the drawing to their actual measurement of the plane, we know that those two rates must be equal and we can write a proportion.

1 inch corresponds **to** 6 feet **as** 4.5 inches corresponds **to** w feet.

$$\text{Measure on the drawing} \rightarrow \frac{1}{6} = \frac{4.5}{w} \leftarrow \text{Measure on the drawing}$$
$$\text{Measure on the plane} \phantom{\rightarrow \frac{1}{6} = \frac{4.5}{w}} \leftarrow \text{Measure on the plane}$$

Solve To find the actual wingspan of the airplane, we solve the proportion for w.

$1 \cdot w = 6 \cdot 4.5$	Set the cross products equal to form an equation.	$\begin{array}{r} \overset{3}{4.5} \\ \times\ 6 \\ \hline 27.0 \end{array}$
$w = 27$	Do the multiplication: $6 \cdot 4.5 = 27$.	

State The actual wingspan of the plane is 27 feet.

Check Every 1 inch on the scale drawing corresponds to an actual length of 6 feet on the plane. Therefore, a 5-inch measurement corresponds to an actual wingspan of $5 \cdot 6$ feet, or 30 feet. It seems reasonable that a 4.5-inch measurement corresponds to an actual wingspan of 27 feet.

EXAMPLE 11 **Baking** A recipe for chocolate cake calls for $1\frac{1}{2}$ cups of sugar for every $2\frac{1}{4}$ cups of flour. If a baker has only $\frac{1}{2}$ cup of sugar on hand, how much flour should he add to it to make chocolate cake batter?

Analyze

- The sugar-to-flour rate can be expressed as: $\dfrac{1\frac{1}{2} \text{ cups sugar}}{2\frac{1}{4} \text{ cups flour}}$

- How much flour should be added to $\frac{1}{2}$ cup of sugar?

Assign We will let the variable f represent the unknown cups of flour.

Form If we compare the cups of sugar to the cups of flour, we know that the two ratios must be equal and we can write a proportion.

$1\frac{1}{2}$ cups of sugar is **to** $2\frac{1}{4}$ cups of flour **as** $\frac{1}{2}$ cup of sugar is **to** f cups of flour

$$\text{Cups of sugar} \rightarrow \frac{1\frac{1}{2}}{2\frac{1}{4}} = \frac{\frac{1}{2}}{f} \leftarrow \text{Cup of sugar}$$
$$\text{Cups of flour} \phantom{\rightarrow \frac{1\frac{1}{2}}{2\frac{1}{4}} = \frac{\frac{1}{2}}{f}} \leftarrow \text{Cups of flour}$$

Solve To find the amount of flour that is needed, we solve the proportion for f.

$\dfrac{1\frac{1}{2}}{2\frac{1}{4}} = \dfrac{\frac{1}{2}}{f}$	This is the proportion to solve.
$1\frac{1}{2} \cdot f = 2\frac{1}{4} \cdot \frac{1}{2}$	Set the cross products equal to each other to form an equation.
$\dfrac{3}{2} \cdot f = \dfrac{9}{4} \cdot \dfrac{1}{2}$	Write each mixed number as an improper fraction.
$\dfrac{2}{3} \cdot \dfrac{3}{2} f = \dfrac{2}{3} \cdot \dfrac{9}{4} \cdot \dfrac{1}{2}$	To isolate f, multiply both sides by the reciprocal of the coefficient of the variable term $\frac{3}{2}f$. The reciprocal of $\frac{3}{2}$ is $\frac{2}{3}$.

Self Check 11

BAKING See Example 11. How many cups of flour will be needed to make several chocolate cakes that will require a total of $12\frac{1}{2}$ cups of sugar? $18\frac{3}{4}$ cups

Now Try Problem 89

Teaching Example 11 BAKING A recipe for coffee cake calls for $1\frac{3}{4}$ cups of sugar for every $5\frac{1}{4}$ cups of flour. If a baker has only $\frac{7}{8}$ cup sugar, how much flour should she add to make the coffee cake batter?
Answer:
$2\frac{5}{8}$ cups

$$1f = \frac{2 \cdot 9 \cdot 1}{3 \cdot 4 \cdot 2}$$

On the left side, the product of a number and its reciprocal is 1: $\frac{2}{3} \cdot \frac{3}{2} = 1$. On the right side, multiply the numerators and multiply the denominators.

$$f = \frac{\overset{1}{2} \cdot \overset{1}{3} \cdot 3 \cdot 1}{\underset{1}{3} \cdot 4 \cdot \underset{1}{2}}$$

On the left side, the coefficient of 1 need not be written since $1f = f$. To simplify the right side, factor 9. Then remove the common factors 2 and 3.

$$f = \frac{3}{4}$$

Multiply the remaining factors in the numerator: $1 \cdot 1 \cdot 3 \cdot 1 = 3$.
Multiply the remaining factors in the denominator: $1 \cdot 4 \cdot 1 = 4$.

State The baker should use $\frac{3}{4}$ cups of flour.

Check The rate of $1\frac{1}{2}$ cups of sugar for every $2\frac{1}{4}$ cups of flour is about 1 to 2. The rate of $\frac{1}{2}$ cup of sugar to $\frac{3}{4}$ cup flour is also about 1 to 2. The result, $\frac{3}{4}$, seems reasonable.

Success Tip In Example 11, an alternate approach would be to write each term of the proportion in its equivalent decimal form and then solve for f.

Fractions and mixed numbers   **Decimals**

$$\frac{1\frac{1}{2}}{2\frac{1}{4}} = \frac{\frac{1}{2}}{f} \longrightarrow \frac{1.5}{2.25} = \frac{0.5}{f}$$

ANSWERS TO SELF CHECKS
1. a. $\frac{16}{28} = \frac{4}{7}$ **b.** $\frac{300 \text{ children}}{500 \text{ adults}} = \frac{3 \text{ children}}{5 \text{ adults}}$ **2. a.** true **b.** false **3.** true **4. a.** true **b.** true
5. yes **6.** 24 **7.** 7.6 **8.** $3\frac{1}{2}$ **9.** $187.50 **10.** 675 ft **11.** $18\frac{3}{4}$ cups

---

SECTION 6.2 STUDY SET

VOCABULARY

Fill in the blanks.

1. A __proportion__ is a statement that two ratios (or rates) are equal.

2. In $\frac{1}{2} = \frac{5}{10}$, the terms 1 and 10 are called the __extremes__ of the proportion and the terms 2 and 5 are called the __means__ of the proportion.

3. The __cross__ products for the proportion $\frac{4}{7} = \frac{36}{x}$ are $4 \cdot x$ and $7 \cdot 36$.

4. When two pairs of numbers form a proportion, we say that the numbers are __proportional__.

5. A letter that is used to represent an unknown number is called a __variable__.

6. When we find the value of x that makes the proportion $\frac{3}{8} = \frac{x}{16}$ true, we say that we have __solved__ the proportion.

7. We solve proportions by writing a series of steps that result in an equation of the form $x =$ a number or a number $= x$. We say that the variable x is __isolated__ on one side of the equation.

8. A __scale__ is a ratio (or rate) that compares the size of a model, drawing, or map to the size of an actual object.

CONCEPTS

Fill in the blanks.

9. If the cross products of a proportion are equal, the proportion is **true**. If the cross products are *not equal*, the proportion is **false**.

10. The proportion $\frac{2}{5} = \frac{4}{10}$ will be true if the product **2** · 10 is equal to the product **5** · 4.

11. Complete the cross products.

$$9 \cdot 10 = \boxed{90} \qquad 2 \cdot \boxed{45} = \boxed{90}$$
$$\frac{9}{2} = \frac{45}{10}$$

12. In the equation $6 \cdot x = 2 \cdot 12$, to undo the multiplication by 6 and isolate x, **divide** both sides of the equation by 6.

13. Label the missing units in the proportion.

Teacher's aides ⟶ $\dfrac{12}{100} = \dfrac{3}{25}$ ⟵ **Teacher's aides**
Children ⟶ $\phantom{\dfrac{12}{100}}$ ⟵ **Children**

14. Consider the following problem: *For every 15 feet of chain link fencing, 4 support posts are used. How many support posts will be needed for 300 feet of chain link fencing?* Which of the proportions below could be used to solve this problem? **i, iv**

i. $\dfrac{15}{4} = \dfrac{300}{x}$ ii. $\dfrac{15}{4} = \dfrac{x}{300}$

iii. $\dfrac{4}{15} = \dfrac{300}{x}$ iv. $\dfrac{4}{15} = \dfrac{x}{300}$

NOTATION

Complete each step.

15. Solve the proportion: $\dfrac{2}{3} = \dfrac{x}{9}$

$$2 \cdot 9 = \boxed{3 \cdot x}$$
$$\boxed{18} = 3x$$
$$\frac{\boxed{18}}{\boxed{3}} = \frac{3x}{\boxed{3}}$$
$$\boxed{6} = x$$

The solution is $\boxed{6}$.

16. Solve the proportion: $\dfrac{14}{x} = \dfrac{49}{17.5}$

$$14 \cdot \boxed{17.5} = x \cdot 49$$
$$\boxed{245} = 49x$$
$$\frac{\boxed{245}}{49} = \frac{49x}{49}$$
$$\boxed{5} = x$$

The solution is $\boxed{5}$.

GUIDED PRACTICE

Write each statement as a proportion. See Example 1.

17. 20 is to 30 as 2 is to 3. $\dfrac{20}{30} = \dfrac{2}{3}$

18. 9 is to 36 as 1 is to 4. $\dfrac{9}{36} = \dfrac{1}{4}$

19. 400 sheets is to 100 beds as 4 sheets is to 1 bed. $\dfrac{400 \text{ sheets}}{100 \text{ beds}} = \dfrac{4 \text{ sheets}}{1 \text{ bed}}$

20. 50 shovels is to 125 laborers as 2 shovels is to 5 laborers. $\dfrac{50 \text{ shovels}}{125 \text{ laborers}} = \dfrac{2 \text{ shovels}}{5 \text{ laborers}}$

Determine whether each proportion is true or false by simplifying. See Example 2.

21. $\dfrac{7}{9} = \dfrac{70}{81}$ false   **22.** $\dfrac{2}{5} = \dfrac{8}{20}$ true

23. $\dfrac{21}{14} = \dfrac{18}{12}$ true   **24.** $\dfrac{42}{38} = \dfrac{95}{60}$ false

Determine whether each proportion is true or false by finding cross products. See Example 3.

25. $\dfrac{4}{32} = \dfrac{2}{16}$ true   **26.** $\dfrac{6}{27} = \dfrac{4}{18}$ true

27. $\dfrac{9}{19} = \dfrac{38}{80}$ false   **28.** $\dfrac{40}{29} = \dfrac{29}{22}$ false

Determine whether each proportion is true or false by finding cross products. See Example 4.

29. $\dfrac{0.5}{0.8} = \dfrac{1.1}{1.3}$ false   **30.** $\dfrac{0.6}{1.4} = \dfrac{0.9}{2.1}$ true

31. $\dfrac{1.2}{3.6} = \dfrac{1.8}{5.4}$ true   **32.** $\dfrac{3.2}{4.5} = \dfrac{1.6}{2.7}$ false

33. $\dfrac{1\frac{4}{5}}{3\frac{3}{7}} = \dfrac{2\frac{3}{16}}{4\frac{1}{6}}$ true   **34.** $\dfrac{2\frac{1}{2}}{1\frac{1}{5}} = \dfrac{3\frac{3}{4}}{2\frac{9}{10}}$ false

35. $\dfrac{\frac{1}{5}}{1\frac{1}{6}} = \dfrac{1\frac{1}{7}}{11\frac{2}{3}}$ false   **36.** $\dfrac{11\frac{1}{4}}{2\frac{1}{2}} = \dfrac{\frac{3}{4}}{\frac{1}{6}}$ true

Determine whether the numbers are proportional. See Example 5.

37. 18, 54 and 3, 9   yes   **38.** 4, 3 and 12, 9   yes

39. 8, 6 and 21, 16   no   **40.** 15, 7 and 13, 6   no

Solve each proportion. Check each result. See Example 6.

41. $\dfrac{5}{10} = \dfrac{3}{x}$ 6
42. $\dfrac{7}{14} = \dfrac{2}{x}$ 4
43. $\dfrac{2}{3} = \dfrac{x}{6}$ 4
44. $\dfrac{3}{6} = \dfrac{x}{8}$ 4

Solve each proportion. Check each result. See Example 7.

45. $\dfrac{0.6}{9.6} = \dfrac{x}{4.8}$ 0.3
46. $\dfrac{0.4}{3.4} = \dfrac{x}{13.6}$ 1.6
47. $\dfrac{2.75}{x} = \dfrac{1.5}{1.2}$ 2.2
48. $\dfrac{9.8}{x} = \dfrac{2.8}{5.4}$ 18.9

Solve each proportion. Check each result. Write each result as a fraction or mixed number. See Example 8.

49. $\dfrac{x}{1\frac{1}{2}} = \dfrac{10\frac{1}{2}}{4\frac{1}{2}}$ $3\frac{1}{2}$
50. $\dfrac{x}{3\frac{1}{3}} = \dfrac{1\frac{1}{2}}{1\frac{9}{11}}$ $2\frac{3}{4}$
51. $\dfrac{x}{1\frac{1}{6}} = \dfrac{2\frac{5}{8}}{3\frac{1}{2}}$ $\frac{7}{8}$
52. $\dfrac{x}{2\frac{2}{3}} = \dfrac{1\frac{1}{20}}{3\frac{1}{2}}$ $\frac{4}{5}$

TRY IT YOURSELF

Solve each proportion.

53. $\dfrac{4{,}000}{x} = \dfrac{3.2}{2.8}$ 3,500
54. $\dfrac{0.4}{1.6} = \dfrac{96.7}{x}$ 386.8
55. $\dfrac{12}{6} = \dfrac{x}{\frac{1}{4}}$ $\frac{1}{2}$
56. $\dfrac{15}{10} = \dfrac{x}{\frac{1}{3}}$ $\frac{1}{2}$
57. $\dfrac{x}{8} = \dfrac{900}{200}$ 36
58. $\dfrac{x}{200} = \dfrac{1{,}800}{600}$ 600
59. $\dfrac{x}{2.5} = \dfrac{3.7}{9.25}$ 1
60. $\dfrac{8.5}{x} = \dfrac{4.25}{1.7}$ 3.4
61. $\dfrac{0.8}{2} = \dfrac{x}{5}$ 2
62. $\dfrac{0.9}{0.3} = \dfrac{6}{x}$ 2
63. $\dfrac{x}{4\frac{1}{10}} = \dfrac{3\frac{3}{4}}{1\frac{7}{8}}$ $8\frac{1}{5}$
64. $\dfrac{x}{2\frac{1}{4}} = \dfrac{1\frac{1}{2}}{1\frac{1}{5}}$ $5\frac{5}{8}$
65. $\dfrac{340}{51} = \dfrac{x}{27}$ 180
66. $\dfrac{480}{36} = \dfrac{x}{15}$ 200
67. $\dfrac{0.4}{1.2} = \dfrac{6}{x}$ 18
68. $\dfrac{5}{x} = \dfrac{2}{4.4}$ 11
69. $\dfrac{4.65}{7.8} = \dfrac{x}{5.2}$ 3.1
70. $\dfrac{8.6}{2.4} = \dfrac{x}{6}$ 21.5
71. $\dfrac{\frac{3}{4}}{\frac{1}{2}} = \dfrac{0.25}{x}$ $\frac{1}{6}$
72. $\dfrac{\frac{7}{8}}{\frac{1}{2}} = \dfrac{0.25}{x}$ $\frac{1}{7}$

APPLICATIONS

To solve each problem, write and then solve a proportion.

73. **SCHOOL LUNCHES** A manager of a school cafeteria orders 750 pudding cups. What will the order cost if she purchases them wholesale, 6 cups for $1.75? $218.75

74. **CLOTHES SHOPPING** As part of a spring clearance, a men's store put dress shirts on sale, 2 for $25.98. How much will a businessman pay if he buys five shirts? $64.95

75. **ANNIVERSARY GIFTS** A florist sells a dozen long-stemmed red roses for $57.99. In honor of their 16th wedding anniversary, a man wants to buy 16 roses for his wife. What will the roses cost? (*Hint:* How many roses are in one dozen?) $77.32

76. **COOKING** A recipe for spaghetti sauce requires four 16-ounce bottles of ketchup to make 2 gallons of sauce. How many bottles of ketchup are needed to make 10 gallons of sauce? (*Hint:* Read the problem very carefully.) 20 bottles

77. **BUSINESS PERFORMANCE** The following bar graph shows the yearly costs and the revenue received for a business. Are the ratios of costs to revenue for 2012 and 2013 equal? yes

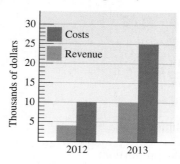

78. **RAMPS** Write a ratio of the rise to the run for each ramp shown. Set the ratios equal.
    a. Is the resulting proportion true? $\frac{12}{20} = \frac{18}{30}$, yes
    b. Is one ramp steeper than the other? no

79. **MIXING PERFUMES** A perfume is to be mixed in the ratio of 3 drops of pure essence to 7 drops of alcohol. How many drops of pure essence should be mixed with 56 drops of alcohol? 24 drops

80. **MAKING COLOGNE** A cologne can be made by mixing 2 drops of pure essence with 5 drops of distilled water. How many drops of water should be used with 15 drops of pure essence? 37.5 drops

81. **LAB WORK** In a red blood cell count, a drop of the patient's diluted blood is placed on a grid like that shown below. Instead of counting each and every red blood cell in the 25-square grid, a technician counts only the number of cells in the five highlighted squares. Then he or she uses a proportion to estimate the total red blood cell count. If there are 195 red blood cells in the blue squares, about how many red blood cells are in the entire grid? 975 red blood cells

82. **DOSAGES** The proper dosage of a certain medication for a 30-pound child is shown. At this rate, what would be the dosage for a 45-pound child? $1\frac{1}{8}$ oz

83. **DRAFTING** In a scale drawing, a 280-foot antenna tower is drawn 7 inches high. The building next to it is drawn 2 inches high. How tall is the actual building? 80 ft

84. **BLUEPRINTS** The scale for the drawing in the blueprint tells the reader that a $\frac{1}{4}$-inch length $\left(\frac{1}{4}''\right)$ on the drawing corresponds to an actual size of 1 foot (1′0″). Suppose the length of the kitchen is $2\frac{1}{2}$ inches on the blueprint. How long is the actual kitchen? 10 ft

85. **MODEL RAILROADS** An HO-scale model railroad engine is 9 inches long. If HO scale is 87 feet to 1 foot, how long is a real engine? (*Hint:* Compare feet to inches. How many inches are in one foot?) 65.25 ft = 65 ft 3 in.

86. **MODEL RAILROADS** An N-scale model railroad caboose is 4 inches long. If N scale is 169 feet to 1 foot, how long is a real caboose? (*Hint:* Compare feet to inches. How many inches are in one foot?) $56\frac{1}{3}$ ft = 56 ft 4 in.

87. **CAROUSELS** The ratio in the illustration below indicates that 1 inch on the model carousel is equivalent to 160 inches on the actual carousel. How wide should the model be if the actual carousel is 35 feet wide? (*Hint:* Convert 35 feet to inches.) 2.625 in. = $2\frac{5}{8}$ in.

88. **MIXING FUELS** The instructions on a can of oil intended to be added to lawn mower gasoline read as shown. Are these instructions correct? (*Hint*: There are 128 ounces in 1 gallon.) not exactly, but close

Recommended	Gasoline	Oil
50 to 1	6 gal	16 oz

89. **MAKING COOKIES** A recipe for chocolate chip cookies calls for $1\frac{1}{4}$ cups of flour and 1 cup of sugar. The recipe will make $3\frac{1}{2}$ dozen cookies. How many cups of flour will be needed to make 12 dozen cookies? $4\frac{2}{7}$, which is about $4\frac{1}{4}$

90. **MAKING BROWNIES** *from Campus to Careers* Chef
A recipe for brownies calls for 4 eggs and $1\frac{1}{2}$ cups of flour. If the recipe makes 15 brownies, how many cups of flour will be needed to make 130 brownies? 13

91. **COMPUTER SPEED** Using the *Mathematica 3.0* program, a Dell Dimension XPS R350 (Pentium II) computer can perform a set of 15 calculations in 2.85 seconds. How long will it take the computer to perform 100 such calculations? 19 sec

92. **QUALITY CONTROL** Out of a sample of 500 men's shirts, 17 were rejected because of crooked collars. How many crooked collars would you expect to find in a run of 15,000 shirts? 510 crooked collars

93. **DOGS** Refer to the illustration below. A Saint Bernard website lists the "ideal proportions for the *height at the withers* to *body length* as 5:6." What is the ideal height at the withers for a Saint Bernard whose body length is $37\frac{1}{2}$ inches? 31.25 in. = $31\frac{1}{4}$ in.

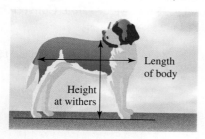

94. **MILEAGE** Under normal conditions, a Hummer can travel 325 miles on a full tank (25 gallons) of diesel. How far can it travel on its auxiliary tank, which holds 17 gallons of diesel? 221 mi

95. **PAYCHECKS** Billie earns $412 for a 40-hour week. If she missed 10 hours of work last week, how much did she get paid? $309

96. **STAFFING** A school board has determined that there should be 3 teachers for every 50 students. Complete the table by filling in the number of teachers needed at each school.

	Glenwood High	Goddard Junior High	Sellers Elementary
Enrollment	2,700	1,900	850
Teachers	162	114	51

WRITING

97. Explain the difference between a ratio and a proportion.

98. The following paragraph is from a book about dollhouses. What concept from this section is mentioned?

    Today, the internationally recognized scale for dollhouses and miniatures is 1 in. = 1 ft. This is small enough to be defined as a miniature, yet not too small for all details of decoration and furniture to be seen clearly.

99. Write a problem that could be solved using the following proportion.

    $$\text{Ounces of cashews} \longrightarrow \frac{4}{639} = \frac{10}{x} \longleftarrow \text{Ounces of cashews}$$
    $$\text{Calories} \longrightarrow \phantom{\frac{4}{639}} \phantom{\frac{10}{x}} \longleftarrow \text{Calories}$$

100. Write a problem about a situation you encounter in your daily life that could be solved by using a proportion.

REVIEW

Perform each operation.

101. $7.4 + 6.78 + 35 + 0.008$ 49.188
102. $29.5 + 34.4 + 12.8$ 76.7
103. $48.8 - 17.372$ 31.428
104. $78.47 - 53.3$ 25.17
105. $-3.8 - (-7.9)$ 4.1
106. $-17.1 + 8.4$ -8.7
107. $-35.1 - 13.99$ -49.09
108. $-5.55 + (-1.25)$ -6.8

SECTION 6.3
American Units of Measurement

Two common systems of measurement are the **American (or English) system** and the **metric system.** We will discuss American units of measurement in this section and metric units in the next. Some common American units are *inches, feet, miles, ounces, pounds, tons, cups, pints, quarts,* and *gallons.* These units are used when measuring length, weight, and capacity.

Objectives

1. Use a ruler to measure lengths in inches.
2. Define American units of length.
3. Convert from one American unit of length to another.
4. Define American units of weight.
5. Convert from one American unit of weight to another.
6. Define American units of capacity.
7. Convert from one American unit of capacity to another.
8. Define units of time.
9. Convert from one unit of time to another.

A newborn baby is 20 inches long.

First-class postage for a letter that weighs less than 1 ounce is 46¢.

Milk is sold in gallon containers.

1 Use a ruler to measure lengths in inches.

A ruler is one of the most common tools used for measuring distances or lengths. The figure below shows part of a ruler. Most rulers are 12 inches (1 foot) long. Since 12 inches = 1 foot, a ruler is divided into 12 equal lengths of 1 inch. Each inch is divided into halves of an inch, quarters of an inch, eighths of an inch, and sixteenths of an inch.

The left end of a ruler can be (but sometimes isn't) labeled with a 0. Each point on a ruler, like each point on a number line, has a number associated with it. That number is the distance between the point and 0. Several lengths on the ruler are shown below.

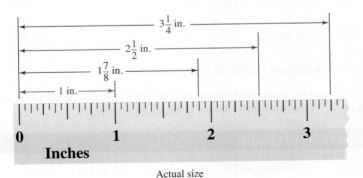
Actual size

EXAMPLE 1
Find the length of the paper clip shown here.

Strategy We will place a ruler below the paper clip, with the left end of the ruler (which could be thought of as 0) directly underneath one end of the paper clip.

WHY Then we can find the length of the paper clip by identifying where its other end lines up on the tick marks printed in black on the ruler.

Unless otherwise noted, all art on this page is © Cengage Learning.

Self Check 1

Find the length of the jumbo paper clip. $1\frac{7}{8}$ in.

Now Try Problem 28

Self Check 2

Find the width of the circle. $1\frac{1}{4}$ in.

Now Try Problem 31

Solution

Since the tick marks between 0 and 1 on the ruler create eight equal spaces, the ruler is scaled in eighths of an inch. The paper clip is $1\frac{3}{8}$ inches long.

EXAMPLE 2
Find the length of the nail shown below.

Strategy We will place a ruler below the nail, with the left end of the ruler (which could be thought of as 0) directly underneath the head of the nail.

WHY Then we can find the length of the nail by identifying where its pointed end lines up on the tick marks printed in black on the ruler.

Solution
Since the tick marks between 0 and 1 on the ruler create sixteen equal spaces, the ruler is scaled in sixteenths of an inch.

The nail is $2\frac{7}{16}$ inches long.

2 Define American units of length.

The American system of measurement uses the units of **inch**, **foot**, **yard**, and **mile** to measure length. These units are related in the following ways.

> **American Units of Length**
>
> 1 foot (ft) = 12 inches (in.)   1 yard (yd) = 36 inches
> 1 yard = 3 feet   1 mile (mi) = 5,280 feet
>
> The abbreviation for each unit is written within parentheses.

> **The Language of Algebra** According to some sources, the inch was originally defined as the length from the tip of the thumb to the first knuckle. In some languages the word for *inch* is similar to or the same as *thumb*. For example, in Spanish, *pulgada* is inch and *pulgar* is thumb. In Swedish, *tum* is inch and *tumme* is thumb. In Italian, *pollice* is both inch and thumb.

3 Convert from one American unit of length to another.

To convert from one unit of length to another, we use *unit conversion factors*. To find the unit conversion factor between yards and feet, we begin with this fact:

$$3 \text{ ft} = 1 \text{ yd}$$

If we divide both sides of this equation by 1 yard, we get

$$\frac{3 \text{ ft}}{1 \text{ yd}} = \frac{1 \text{ yd}}{1 \text{ yd}}$$

$$\frac{3 \text{ ft}}{1 \text{ yd}} = 1 \quad \text{Simplify the right side. A nonzero number divided by itself is 1: } \frac{1 \text{ yd}}{1 \text{ yd}} = 1.$$

The fraction $\frac{3 \text{ ft}}{1 \text{ yd}}$ is called a **unit conversion factor**, because its value is 1. It can be read as "3 feet per yard." Since this fraction is equal to 1, multiplying a length by this fraction does not change its measure; it changes only the *units* of measure.

To convert units of length in the American system of measurement, we use the following unit conversion factors. Each conversion factor shown below is a form of 1.

To convert from	Use the unit conversion factor	To convert from	Use the unit conversion factor
feet to inches	$\frac{12 \text{ in.}}{1 \text{ ft}}$	inches to feet	$\frac{1 \text{ ft}}{12 \text{ in.}}$
yards to feet	$\frac{3 \text{ ft}}{1 \text{ yd}}$	feet to yards	$\frac{1 \text{ yd}}{3 \text{ ft}}$
yards to inches	$\frac{36 \text{ in.}}{1 \text{ yd}}$	inches to yards	$\frac{1 \text{ yd}}{36 \text{ in.}}$
miles to feet	$\frac{5,280 \text{ ft}}{1 \text{ mi}}$	feet to miles	$\frac{1 \text{ mi}}{5,280 \text{ ft}}$

EXAMPLE 3   Convert 8 yards to feet.

Strategy We will multiply 8 yards by a carefully chosen unit conversion factor.

WHY If we multiply by the proper unit conversion factor, we can eliminate the unwanted units of yards and convert to feet.

Solution
To convert from yards to feet, we must use a unit conversion factor that relates feet to yards. Since there are 3 feet per yard, we multiply 8 yards by the unit conversion factor $\frac{3 \text{ ft}}{1 \text{ yd}}$.

$$8 \text{ yd} = \frac{8 \text{ yd}}{1} \cdot \frac{3 \text{ ft}}{1 \text{ yd}} \quad \text{Write 8 yd as a fraction: } 8 \text{ yd} = \frac{8 \text{ yd}}{1}. \text{ Then multiply by a form of 1: } \frac{3 \text{ ft}}{1 \text{ yd}}.$$

$$= \frac{8 \cancel{\text{ yd}}}{1} \cdot \frac{3 \text{ ft}}{1 \cancel{\text{ yd}}} \quad \text{Remove the common units of yards from the numerator and denominator. Notice that the units of feet remain.}$$

$$= 8 \cdot 3 \text{ ft} \quad \text{Simplify.}$$

$$= 24 \text{ ft} \quad \text{Multiply: } 8 \cdot 3 = 24.$$

8 yards is equal to 24 feet.

Success Tip Notice that in Example 3, we eliminated the units of yards and introduced the units of feet by multiplying by the appropriate unit conversion factor. In general, a unit conversion factor is a fraction with the following form:

$$\frac{\text{Unit we want to introduce}}{\text{Unit we want to eliminate}} \leftarrow \text{Numerator} \atop \leftarrow \text{Denominator}$$

Self Check 3
Convert 9 yards to feet. 27 ft
Now Try Problem 35

Teaching Example 3
Convert 5 yards to feet.
Answer:
15 ft

Self Check 4

Convert $1\frac{1}{2}$ feet to inches. 18 in.

Now Try Problem 39

Teaching Example 4
Convert $2\frac{1}{4}$ feet to inches.
Answer:
27 in.

EXAMPLE 4 Convert $1\frac{3}{4}$ feet to inches.

Strategy We will multiply $1\frac{3}{4}$ feet by a carefully chosen unit conversion factor.

WHY If we multiply by the proper unit conversion factor, we can eliminate the unwanted units of feet and convert to inches.

Solution

To convert from feet to inches, we must choose a unit conversion factor whose numerator contains the units we want to introduce (inches), and whose denominator contains the units we want to eliminate (feet). Since there are 12 inches per foot, we will use

$\frac{12 \text{ in.}}{1 \text{ ft}}$ ← This is the unit we want to introduce.
← This is the unit we want to eliminate (the original unit).

To perform the conversion, we multiply.

$1\frac{3}{4}$ ft $= \frac{7}{4}$ ft $\cdot \frac{12 \text{ in.}}{1 \text{ ft}}$ Write $1\frac{3}{4}$ as an improper fraction: $1\frac{3}{4} = \frac{7}{4}$.
Then multiply by a form of 1: $\frac{12 \text{ in.}}{1 \text{ ft}}$.

$= \frac{7}{4}$ ft $\cdot \frac{12 \text{ in.}}{1 \text{ ft}}$ Remove the common units of feet from the numerator and denominator. Notice that the units of inches remain.

$= \frac{7 \cdot 12}{4 \cdot 1}$ in.   Multiply the fractions.

$= \frac{7 \cdot 3 \cdot \overset{1}{4}}{\underset{1}{4} \cdot 1}$ in.   To simplify the fraction, factor 12. Then remove the common factor of 4 from the numerator and denominator.

$= 21$ in.   Simplify.

$1\frac{3}{4}$ feet is equal to 21 inches.

Caution! When converting lengths, if no common units appear in the numerator and denominator to remove, you have chosen the wrong conversion factor.

Sometimes we must use two (or more) unit conversion factors to eliminate the given units while introducing the desired units. The following example illustrates this concept.

Self Check 5

MARATHONS The *marathon* is a long-distance race with an official distance of 26 miles 385 yards. Convert 385 yards to miles. Give the exact answer and a decimal approximation, rounded to the nearest hundredth of a mile. $\frac{7}{32}$ mi ≈ 0.22 mi

Now Try Problem 43

EXAMPLE 5 *Football* A football field (including both end zones) is 120 yards long. Convert this length to miles. Give the exact answer and a decimal approximation, rounded to the nearest hundredth of a mile.

Strategy We will use a two-part multiplication process that converts 120 yards to feet and then converts that result to miles.

WHY We must use a two-part process because the table on page 589 does not contain a single unit conversion factor that converts from yards to miles.

Solution
Since there are 3 feet per yard, we can convert 120 yards to feet by multiplying by the unit conversion factor $\frac{3 \text{ ft}}{1 \text{ yd}}$. Since there is 1 mile for every 5,280 feet, we can convert that result to miles by multiplying by the unit conversion factor $\frac{1 \text{ mi}}{5,280 \text{ ft}}$.

$$120 \text{ yd} = \frac{120 \text{ yd}}{1} \cdot \frac{3 \text{ ft}}{1 \text{ yd}} \cdot \frac{1 \text{ mi}}{5,280 \text{ ft}}$$

Write 120 yd as a fraction: $120 \text{ yd} = \frac{120 \text{ yd}}{1}$. Then multiply by two unit conversion factors: $\frac{3 \text{ ft}}{1 \text{ yd}} = 1$ and $\frac{1 \text{ mi}}{5,280 \text{ ft}} = 1$.

$$= \frac{120 \text{ y\hspace{-0.3em}/\hspace{-0.3em}d}}{1} \cdot \frac{3 \text{ f\hspace{-0.3em}/\hspace{-0.3em}t}}{1 \text{ yd}} \cdot \frac{1 \text{ mi}}{5,280 \text{ ft}}$$

Remove the common units of yards and feet in the numerator and denominator. Notice that all the units are removed except for miles.

$$= \frac{120 \cdot 3}{5,280} \text{ mi}$$

Multiply the fractions.

$$= \frac{\overset{1}{2} \cdot \overset{1}{2} \cdot \overset{1}{2} \cdot \overset{1}{3} \cdot \overset{1}{5} \cdot 3}{\underset{1}{2} \cdot \underset{1}{2} \cdot \underset{1}{2} \cdot 2 \cdot 2 \cdot \underset{1}{3} \cdot \underset{1}{5} \cdot 11} \text{ mi}$$

To simplify the fraction, prime factor 120 and 5,280, and remove the common factors 2, 3, and 5.

$$= \frac{3}{44} \text{ mi}$$

Multiply the remaining factors in the numerator. Multiply the remaining factors in the denominator.

```
    0.068
44)3.000
   − 0
   3 00
  − 2 64
     360
   − 352
       8
```

A football field (including the end zones) is *exactly* $\frac{3}{44}$ miles long.

We can also present this conversion as a decimal. If we divide 3 by 44 (as shown on the right), and round the result to the nearest hundredth, we see that a football field (including the end zones) is *approximately* 0.07 mile long.

Teaching Example 5 THE MASTERS GOLF TOURNAMENT Hole number 8 of the Augusta National course is 570 yards long. Convert the length of the hole to miles. Give the exact answer and a decimal approximation, rounded to the nearest hundredth of a mile.
Answer:
$\frac{57}{176} \text{ mi} \approx 0.32 \text{ mi}$

4 Define American units of weight.

The American system of measurement uses the units of **ounce**, **pound**, and **ton** to measure weight. These units are related in the following ways.

American Units of Weight

1 pound (lb) = 16 ounces (oz)   1 ton (T) = 2,000 pounds

The abbreviation for each unit is written within parentheses.

---

5 Convert from one American unit of weight to another.

To convert units of weight in the American system of measurement, we use the following unit conversion factors. Each conversion factor shown below is a form of 1.

To convert from	Use the unit conversion factor	To convert from	Use the unit conversion factor
pounds to ounces	$\frac{16 \text{ oz}}{1 \text{ lb}}$	ounces to pounds	$\frac{1 \text{ lb}}{16 \text{ oz}}$
tons to pounds	$\frac{2,000 \text{ lb}}{1 \text{ ton}}$	pounds to tons	$\frac{1 \text{ ton}}{2,000 \text{ lb}}$

Self Check 6

Convert 60 ounces to pounds. $3\frac{3}{4}$ lb = 3.75 lb

Now Try Problem 47

Teaching Example 6 Convert 20 ounces to pounds.
Answer:
$1\frac{1}{4}$ lb = 1.25 lb

EXAMPLE 6 Convert 40 ounces to pounds.

Strategy We will multiply 40 ounces by a carefully chosen unit conversion factor.

WHY If we multiply by the proper unit conversion factor, we can eliminate the unwanted units of ounces and convert to pounds.

Solution

To convert from ounces to pounds, we must choose a unit conversion factor whose numerator contains the units we want to introduce (pounds), and whose denominator contains the units we want to eliminate (ounces). Since there is 1 pound for every 16 ounces, we will use

$\dfrac{1 \text{ lb}}{16 \text{ oz}}$ ← This is the unit we want to introduce.
← This is the unit we want to eliminate (the original unit).

To perform the conversion, we multiply.

$40 \text{ oz} = \dfrac{40 \text{ oz}}{1} \cdot \dfrac{1 \text{ lb}}{16 \text{ oz}}$ Write 40 oz as a fraction: 40 oz = $\frac{40 \text{ oz}}{1}$. Then multiply by a form of 1: $\frac{1 \text{ lb}}{16 \text{ oz}}$.

$= \dfrac{40 \cancel{\text{oz}}}{1} \cdot \dfrac{1 \text{ lb}}{16 \cancel{\text{oz}}}$ Remove the common units of ounces from the numerator and denominator. Notice that the units of pounds remain.

$= \dfrac{40}{16} \text{ lb}$ Multiply the fractions.

There are two ways to complete the solution. First, we can remove any common factors of the numerator and denominator to simplify the fraction. Then we can write the result as a mixed number.

$\dfrac{40}{16} \text{ lb} = \dfrac{5 \cdot \overset{1}{\cancel{8}}}{2 \cdot \underset{1}{\cancel{8}}} \text{ lb} = \dfrac{5}{2} \text{ lb} = 2\dfrac{1}{2} \text{ lb}$

A second approach is to divide the numerator by the denominator and express the result as a decimal.

$\dfrac{40}{16} \text{ lb} = 2.5 \text{ lb}$ Perform the division: 40 ÷ 16.

$$\begin{array}{r} 2.5 \\ 16\overline{)40.0} \\ \underline{-32} \\ 8\,0 \\ \underline{-8\,0} \\ 0 \end{array}$$

40 ounces is equal to $2\frac{1}{2}$ lb (or 2.5 lb).

Self Check 7

Convert 60 pounds to ounces. 960 oz

Now Try Problem 51

Teaching Example 7
Convert 90 pounds to ounces.
Answer:
1,440 oz

EXAMPLE 7 Convert 25 pounds to ounces.

Strategy We will multiply 25 pounds by a carefully chosen unit conversion factor.

WHY If we multiply by the proper unit conversion factor, we can eliminate the unwanted units of pounds and convert to ounces.

Solution

To convert from pounds to ounces, we must choose a unit conversion factor whose numerator contains the units we want to introduce (ounces), and whose denominator contains the units we want to eliminate (pounds). Since there are 16 ounces per pound, we will use

$\dfrac{16 \text{ oz}}{1 \text{ lb}}$ ← This is the unit we want to introduce.
← This is the unit we want to eliminate (the original unit).

To perform the conversion, we multiply.

$$25 \text{ lb} = \frac{25 \text{ lb}}{1} \cdot \frac{16 \text{ oz}}{1 \text{ lb}}$$ Write 25 lb as a fraction: $25 \text{ lb} = \frac{25 \text{ lb}}{1}$. Then multiply by a form of 1: $\frac{16 \text{ oz}}{1 \text{ lb}}$.

$$= \frac{25 \cancel{\text{ lb}}}{1} \cdot \frac{16 \text{ oz}}{1 \cancel{\text{ lb}}}$$ Remove the common units of pounds from the numerator and denominator. Notice that the units of ounces remain.

$$= 25 \cdot 16 \text{ oz}$$ Simplify.

$$= 400 \text{ oz}$$ Multiply: $25 \cdot 16 = 400$.

$$\begin{array}{r} 25 \\ \times\ 16 \\ \hline 150 \\ 250 \\ \hline 400 \end{array}$$

25 pounds is equal to 400 ounces.

6 Define American units of capacity.

The American system of measurement uses the units of **ounce**, **cup**, **pint**, **quart**, and **gallon** to measure capacity. These units are related as follows.

> **The Language of Algebra** The word *capacity* means the amount that can be contained. For example, a gas tank might have a *capacity* of 12 gallons.

American Units of Capacity

1 cup (c) = 8 fluid ounces (fl oz)   1 pint (pt) = 2 cups
1 quart (qt) = 2 pints   1 gallon (gal) = 4 quarts

The abbreviation for each unit is written within parentheses.

7 Convert from one American unit of capacity to another.

To convert units of capacity in the American system of measurement, we use the following unit conversion factors. Each conversion factor shown below is a form of 1.

To convert from	Use the unit conversion factor	To convert from	Use the unit conversion factor
cups to ounces	$\frac{8 \text{ fl oz}}{1 \text{ c}}$	ounces to cups	$\frac{1 \text{ c}}{8 \text{ fl oz}}$
pints to cups	$\frac{2 \text{ c}}{1 \text{ pt}}$	cups to pints	$\frac{1 \text{ pt}}{2 \text{ c}}$
quarts to pints	$\frac{2 \text{ pt}}{1 \text{ qt}}$	pints to quarts	$\frac{1 \text{ qt}}{2 \text{ pt}}$
gallons to quarts	$\frac{4 \text{ qt}}{1 \text{ gal}}$	quarts to gallons	$\frac{1 \text{ gal}}{4 \text{ qt}}$

Self Check 8

Convert 2.5 pints to fluid ounces. 40 fl oz

Now Try Problem 55

Felix Wirth/Corbis

Teaching Example 8 DESSERTS If a recipe for homemade ice cream calls for 4 pints of cream, how many fluid ounces of cream should be used?
Answer:
64 fl oz

EXAMPLE 8 Cooking
If a recipe calls for 3 pints of milk, how many fluid ounces of milk should be used?

Strategy We will use a two-part multiplication process that converts 3 pints to cups and then converts that result to fluid ounces.

WHY We must use a two-part process because the table on page 593 does not contain a single unit conversion factor that converts from pints to fluid ounces.

Solution
Since there are 2 cups per pint, we can convert 3 pints to cups by multiplying by the unit conversion factor $\frac{2\,c}{1\,pt}$. Since there are 8 fluid ounces per cup, we can convert that result to fluid ounces by multiplying by the unit conversion factor $\frac{8\,fl\,oz}{1\,c}$.

$$3\,pt = \frac{3\,pt}{1} \cdot \frac{2\,c}{1\,pt} \cdot \frac{8\,fl\,oz}{1\,c}$$ Write 3 pt as a fraction: $3\,pt = \frac{3\,pt}{1}$. Multiply by two unit conversion factors: $\frac{2\,c}{1\,pt} = 1$ and $\frac{8\,fl\,oz}{1\,c} = 1$.

$$= \frac{3\,\cancel{pt}}{1} \cdot \frac{2\,\cancel{c}}{1\,\cancel{pt}} \cdot \frac{8\,fl\,oz}{1\,\cancel{c}}$$ Remove the common units of pints and cups in the numerator and denominator. Notice that all the units are removed except for fluid ounces.

$$= 3 \cdot 2 \cdot 8 \,fl\,oz$$ Simplify.
$$= 48 \,fl\,oz$$ Multiply.

Since 3 pints is equal to 48 fluid ounces, 48 fluid ounces of milk should be used.

8 Define units of time.

The American system of measurement and the metric system both use the units of **second**, **minute**, **hour**, and **day** to measure time. These units are related as follows.

Units of Time

1 minute (min) = 60 seconds (sec)   1 hour (hr) = 60 minutes
1 day = 24 hours

The abbreviation for each unit is written within parentheses.

To convert units of time, we use the following unit conversion factors. Each conversion factor shown below is a form of 1.

To convert from	Use the unit conversion factor	To convert from	Use the unit conversion factor
minutes to seconds	$\frac{60\,sec}{1\,min}$	seconds to minutes	$\frac{1\,min}{60\,sec}$
hours to minutes	$\frac{60\,min}{1\,hr}$	minutes to hours	$\frac{1\,hr}{60\,min}$
days to hours	$\frac{24\,hr}{1\,day}$	hours to days	$\frac{1\,day}{24\,hr}$

Unless otherwise noted, all art on this page is © Cengage Learning.

9 Convert from one unit of time to another.

EXAMPLE 9 **Astronomy** A lunar eclipse occurs when the Earth is between the sun and the moon in such a way that Earth's shadow darkens the moon. (See the figure below, which is not to scale.) A total lunar eclipse can last as long as 105 minutes. Express this time in hours.

Strategy We will multiply 105 minutes by a carefully chosen unit conversion factor.

WHY If we multiply by the proper unit conversion factor, we can eliminate the unwanted units of minutes and convert to hours.

Solution
To convert from minutes to hours, we must choose a unit conversion factor whose numerator contains the units we want to introduce (hours), and whose denominator contains the units we want to eliminate (minutes). Since there is 1 hour for every 60 minutes, we will use

$\dfrac{1 \text{ hr}}{60 \text{ min}}$ ← This is the unit we want to introduce.
← This is the unit we want to eliminate (the original unit).

To perform the conversion, we multiply.

$$105 \text{ min} = \dfrac{105 \text{ min}}{1} \cdot \dfrac{1 \text{ hr}}{60 \text{ min}}$$ Write 105 min as a fraction: $105 = \dfrac{105 \text{ min}}{1}$. Then multiply by a form of 1: $\dfrac{1 \text{ hr}}{60 \text{ min}}$.

$$= \dfrac{105 \cancel{\text{ min}}}{1} \cdot \dfrac{1 \text{ hr}}{60 \cancel{\text{ min}}}$$ Remove the common units of minutes in the numerator and denominator. Notice that the units of hours remain.

$$= \dfrac{105}{60} \text{ hr}$$ Multiply the fractions.

$$= \dfrac{\overset{1}{\cancel{3}} \cdot \overset{1}{\cancel{5}} \cdot 7}{2 \cdot 2 \cdot \underset{1}{\cancel{3}} \cdot \underset{1}{\cancel{5}}} \text{ hr}$$ To simplify the fraction, prime factor 105 and 60. Then remove the common factors 3 and 5 in the numerator and denominator.

$$= \dfrac{7}{4} \text{ hr}$$ Multiply the remaining factors in the numerator. Multiply the remaining factors in the denominator.

$$= 1\dfrac{3}{4} \text{ hr}$$ Write $\dfrac{7}{4}$ as a mixed number.

A total lunar eclipse can last as long as $1\dfrac{3}{4}$ hours.

Self Check 9

THE SUN A solar eclipse (eclipse of the sun) can last as long as 450 seconds. Express this time in minutes. $7\dfrac{1}{2}$ min

Now Try Problem 59

Teaching Example 9 SPACE FLIGHT
On February 20, 1962, John Glenn was the first American astronaut to orbit the Earth. The Mercury–Atlas 6 mission in the spacecraft *Friendship 7* lasted a total of 295 minutes. Express this time in hours.
Answer:
$4\dfrac{11}{12}$ hr

ANSWERS TO SELF CHECKS

1. $1\dfrac{7}{8}$ in.   **2.** $1\dfrac{1}{4}$ in.   **3.** 27 ft   **4.** 18 in.   **5.** $\dfrac{7}{32}$ mi ≈ 0.22 mi   **6.** $3\dfrac{3}{4}$ lb = 3.75 lb
7. 960 oz   **8.** 40 fl oz   **9.** $7\dfrac{1}{2}$ min

Unless otherwise noted, all art on this page is © Cengage Learning.

SECTION 6.3 STUDY SET

VOCABULARY

Fill in the blanks.

1. A ruler is used for measuring __length__.
2. Inches, feet, and miles are examples of American units of __length__.
3. $\frac{3 \text{ ft}}{1 \text{ yd}}$, $\frac{1 \text{ ton}}{2{,}000 \text{ lb}}$, and $\frac{4 \text{ qt}}{1 \text{ gal}}$ are examples of __unit__ conversion factors.
4. Ounces, pounds, and tons are examples of American units of __weight__.
5. Some examples of American units of __capacity__ are cups, pints, quarts, and gallons.
6. Some units of __time__ are seconds, minutes, hours, and days.

CONCEPTS

Fill in the blanks.

7. a. 12 inches = __1__ foot
   b. __3__ feet = 1 yard
   c. 1 yard = __36__ inches
   d. 1 mile = __5,280__ feet
8. a. __16__ ounces = 1 pound
   b. __2,000__ pounds = 1 ton
9. a. 1 cup = __8__ fluid ounces
   b. 1 pint = __2__ cups
   c. __2__ pints = 1 quart
   d. __4__ quarts = 1 gallon
10. a. 1 day = __24__ hours
    b. 2 hours = __120__ minutes
11. The value of any unit conversion factor is __1__.
12. In general, a unit conversion factor is a fraction with the following form:

    $$\frac{\text{Unit that we want to } \underline{\text{introduce}}}{\text{Unit that we want to } \underline{\text{eliminate}}} \leftarrow \text{Numerator} \\ \leftarrow \text{Denominator}$$

13. Consider the work shown below.

    $$\frac{48 \text{ oz}}{1} \cdot \frac{1 \text{ lb}}{16 \text{ oz}}$$

    a. What units can be removed? __oz__
    b. What units remain? __lb__

14. Consider the work shown below.

    $$\frac{600 \text{ yd}}{1} \cdot \frac{3 \text{ ft}}{1 \text{ yd}} \cdot \frac{1 \text{ mi}}{5{,}280 \text{ ft}}$$

    a. What units can be removed? __yd, ft__
    b. What units remain? __mi__

15. Write a unit conversion factor to convert
    a. pounds to tons $\frac{1 \text{ ton}}{2{,}000 \text{ lb}}$
    b. quarts to pints $\frac{2 \text{ pt}}{1 \text{ qt}}$
16. Write the two unit conversion factors used to convert
    a. inches to yards $\frac{1 \text{ ft}}{12 \text{ in.}}, \frac{1 \text{ yd}}{3 \text{ ft}}$
    b. days to minutes $\frac{24 \text{ hr}}{1 \text{ day}}, \frac{60 \text{ min}}{1 \text{ hr}}$
17. Match each item with its proper measurement.
    a. Length of the U.S. coastline iv
    b. Height of a Barbie doll i
    c. Span of the Golden Gate Bridge ii
    d. Width of a football field iii

    i. $11\frac{1}{2}$ in.
    ii. 4,200 ft
    iii. 53.5 yd
    iv. 12,383 mi

18. Match each item with its proper measurement.
    a. Weight of the men's shot put used in track and field ii
    b. Weight of an African elephant iii
    c. Amount of gold that is worth $2,000 i

    i. $1\frac{1}{2}$ oz
    ii. 16 lb
    iii. 7.2 tons

19. Match each item with its proper measurement.
    a. Amount of blood in an adult iii
    b. Size of the Exxon Valdez oil spill in 1989 iv
    c. Amount of nail polish in a bottle i
    d. Amount of milk to make 2 dozen pancakes ii

    i. $\frac{1}{2}$ fluid oz
    ii. 2 cups
    iii. 6 qt
    iv. 10,080,000 gal

20. Match each item with its proper measurement.
    a. Length of first U.S. manned space flight ii
    b. A leap year iv
    c. Time difference between New York and Fairbanks, Alaska iii
    d. Length of Wright Brothers' first flight i

    i. 12 sec
    ii. 15 min
    iii. 4 hr
    iv. 366 days

NOTATION

21. What unit does each abbreviation represent?
   a. lb pound
   b. oz ounce
   c. fl oz fluid ounce

22. What unit does each abbreviation represent?
   a. qt quart
   b. c cup
   c. pt pint

Complete each step.

23. Convert 2 yards to inches.
$$2 \text{ yd} = \frac{2 \text{ yd}}{1} \cdot \frac{\boxed{36} \text{ in.}}{1 \text{ yd}}$$
$$= 2 \cdot 36 \text{ in.}$$
$$= \boxed{72} \text{ in.}$$

24. Convert 24 pints to quarts.
$$24 \text{ pt} = \frac{24 \text{ pt}}{1} \cdot \frac{1 \text{ qt}}{\boxed{2} \text{ pt}}$$
$$= \frac{24}{1} \cdot \frac{1}{2} \boxed{\text{qt}}$$
$$= \boxed{12} \text{ qt}$$

25. Convert 1 ton to ounces.
$$1 \text{ ton} = \frac{1 \text{ ton}}{1} \cdot \frac{\boxed{2{,}000 \text{ lb}}}{1 \text{ ton}} \cdot \frac{\boxed{16 \text{ oz}}}{1 \text{ lb}}$$
$$= 1 \cdot 2{,}000 \cdot 16 \boxed{\text{ oz}}$$
$$= \boxed{32{,}000} \text{ oz}$$

26. Convert 37,440 minutes to days.
$$37{,}440 \text{ min} = 37{,}440 \text{ min} \cdot \frac{1 \text{ hr}}{\boxed{60 \text{ min}}} \cdot \frac{1 \text{ day}}{\boxed{24 \text{ hr}}}$$
$$= \frac{37{,}440}{60 \cdot 24} \boxed{\text{days}}$$
$$= \boxed{26} \text{ days}$$

GUIDED PRACTICE

Refer to the given ruler to answer each question. See Example 1.

27. a. Each inch is divided into how many equal parts? 8
   b. Determine which measurements the arrows point to on the ruler. $\frac{5}{8}$ in., $1\frac{1}{4}$ in., $2\frac{7}{8}$ in.

28. Find the length of the needle. $1\frac{3}{8}$ in.

Refer to the given ruler to answer each question. See Example 2.

29. a. Each inch is divided into how many equal parts? 16
   b. Determine which measurements the arrows point to on the ruler. $\frac{9}{16}$ in., $1\frac{3}{4}$ in., $2\frac{3}{16}$ in.

30. Find the length of the bolt. $2\frac{15}{16}$ in.

Use a ruler scaled in sixteenths of an inch to measure each object. See Example 2.

31. The width of a dollar bill $2\frac{9}{16}$ in.

32. The length of a dollar bill $6\frac{1}{8}$ in.

33. The length (top to bottom) of this page $10\frac{7}{8}$ in.

34. The length of the word as printed here: supercalifragilisticexpialidocious $1\frac{7}{8}$ in.

Perform each conversion. See Example 3.

35. 4 yards to feet 12 ft
36. 6 yards to feet 18 ft
37. 35 yards to feet 105 ft
38. 33 yards to feet 99 ft

Perform each conversion. See Example 4.

39. $3\frac{1}{2}$ feet to inches 42 in.
40. $2\frac{2}{3}$ feet to inches 32 in.
41. $5\frac{1}{4}$ feet to inches 63 in.
42. $6\frac{1}{2}$ feet to inches 78 in.

Use two unit conversion factors to perform each conversion. Give the exact answer and a decimal approximation, rounded to the nearest hundredth when necessary. See Example 5.

43. 105 yards to miles $\frac{21}{352}$ mi ≈ 0.06 mi
44. 198 yards to miles $\frac{9}{80}$ mi = 0.1125 mi
45. 1,540 yards to miles $\frac{7}{8}$ mi = 0.875 mi
46. 1,512 yards to miles $\frac{189}{220}$ mi ≈ 0.86 mi

Perform each conversion. See Example 6.

47. Convert 44 ounces to pounds. $2\frac{3}{4}$ lb = 2.75 lb
48. Convert 24 ounces to pounds. $1\frac{1}{2}$ lb = 1.5 lb
49. Convert 72 ounces to pounds. $4\frac{1}{2}$ lb = 4.5 lb
50. Convert 76 ounces to pounds. $4\frac{3}{4}$ lb = 4.75 lb

Perform each conversion. See Example 7.

51. 50 pounds to ounces 800 oz
52. 30 pounds to ounces 480 oz
53. 87 pounds to ounces 1,392 oz
54. 79 pounds to ounces 1,264 oz

Perform each conversion. See Example 8.

55. 8 pints to fluid ounces 128 fl oz
56. 5 pints to fluid ounces 80 fl oz
57. 21 pints to fluid ounces 336 fl oz
58. 30 pints to fluid ounces 480 fl oz

Perform each conversion. See Example 9.

59. 165 minutes to hours $2\frac{3}{4}$ hr
60. 195 minutes to hours $3\frac{1}{4}$ hr
61. 330 minutes to hours $5\frac{1}{2}$ hr
62. 80 minutes to hours $1\frac{1}{3}$ hr

TRY IT YOURSELF

Perform each conversion.

63. 3 quarts to pints 6 pt
64. 20 quarts to gallons 5 gal
65. 7,200 minutes to days 5 days
66. 691,200 seconds to days 8 days
67. 56 inches to feet $4\frac{2}{3}$ ft
68. 44 inches to feet $3\frac{2}{3}$ ft
69. 4 feet to inches 48 in.
70. 7 feet to inches 84 in.
71. 16 pints to gallons 2 gal
72. 3 gallons to fluid ounces 384 fl oz
73. 80 ounces to pounds 5 lb
74. 8 pounds to ounces 128 oz
75. 240 minutes to hours 4 hr
76. 2,400 seconds to hours $\frac{2}{3}$ hr
77. 8 yards to inches 288 in.
78. 324 inches to yards 9 yd
79. 90 inches to yards $2\frac{1}{2}$ yd = 2.5 yd
80. 12 yards to inches 432 in.
81. 5 yards to feet 15 ft
82. 21 feet to yards 7 yd
83. 12.4 tons to pounds 24,800 lb
84. 48,000 ounces to tons $1\frac{1}{2}$ tons = 1.5 tons
85. 7 feet to yards $2\frac{1}{3}$ yd
86. $4\frac{2}{3}$ yards to feet 14 ft
87. 15,840 feet to miles 3 mi
88. 2 miles to feet 10,560 ft
89. $\frac{1}{2}$ mile to feet 2,640 ft
90. 1,320 feet to miles $\frac{1}{4}$ mi = 0.25 mi
91. 7,000 pounds to tons $3\frac{1}{2}$ tons = 3.5 tons
92. 2.5 tons to ounces 80,000 oz
93. 32 fluid ounces to pints 2 pt
94. 2 quarts to fluid ounces 64 fl oz

APPLICATIONS

95. **THE GREAT PYRAMID** The Great Pyramid in Egypt is about 450 feet high. Express this distance in yards. 150 yd

96. **THE WRIGHT BROTHERS** In 1903, Orville Wright made the world's first sustained flight. It lasted 12 seconds, and the plane traveled 120 feet. Express the length of the flight in yards. 40 yd

97. **THE GREAT SPHINX** The Great Sphinx of Egypt is 240 feet long. Express this in inches. 2,880 in.

98. **HOOVER DAM** The Hoover Dam in Nevada is 726 feet high. Express this distance in inches. 8,712 in.

99. **THE SEARS TOWER** The Sears Tower in Chicago has 110 stories and is 1,454 feet tall. To the nearest hundredth, express this height in miles. 0.28 mi

100. **NFL RECORDS** Emmit Smith, the former Dallas Cowboys and Arizona Cardinals running back, holds the National Football League record for yards rushing in a career: 18,355. How many miles is this? Round to the nearest tenth of a mile. 10.4 mi

101. **NFL RECORDS** When Dan Marino of the Miami Dolphins retired, it was noted that Marino's career passing total was nearly 35 miles! How many yards is this? 61,600 yd

102. **LEWIS AND CLARK** The trail traveled by the Lewis and Clark expedition is shown below. When the expedition reached the Pacific Ocean, Clark estimated that they had traveled 4,162 miles. (It was later determined that his guess was within 40 miles of the actual distance.) Express Clark's estimate of the distance in feet. 21,975,360 ft

103. **WEIGHT OF WATER** One gallon of water weighs about 8 pounds. Express this weight in ounces. 128 oz

104. **WEIGHT OF A BABY** A newborn baby boy weighed 136 ounces. Express this weight in pounds. $8\frac{1}{2}$ lb = 8.5 lb

105. **HIPPOS** An adult hippopotamus can weigh as much as 9,900 pounds. Express this weight in tons. $4\frac{19}{20}$ tons = 4.95 tons

106. **ELEPHANTS** An adult elephant can consume as much as 495 pounds of grass and leaves in one day. How many ounces is this? 7,920 oz

107. **BUYING PAINT** A painter estimates that he will need 17 gallons of paint for a job. To take advantage of a closeout sale on quart cans, he decides to buy the paint in quarts. How many cans will he need to buy? 68 quart cans

108. **CATERING** How many cups of apple cider are there in a 10-gallon container of cider? 160 cups

109. **SCHOOL LUNCHES** Each student attending Eagle River Elementary School receives 1 pint of milk for lunch each day. If 575 students attend the school, how many gallons of milk are used each day? $71\frac{7}{8}$ gal = 71.875 gal

110. **RADIATORS** The radiator capacity of a piece of earth-moving equipment is 39 quarts. If the radiator is drained and new coolant put in, how many gallons of new coolant will be used? $9\frac{3}{4}$ gal = 9.75 gal

111. **CAMPING** How many ounces of camping stove fuel will fit in the container shown? 320 oz

112. **HIKING** A college student walks 11 miles in 155 minutes. To the nearest tenth, how many hours does he walk? $2\frac{7}{12}$ hr ≈ 2.6 hr

113. **SPACE TRAVEL** The astronauts of the *Apollo 8* mission, which was launched on December 21, 1968, were in space for 147 hours. How many days did the mission take? $6\frac{1}{8}$ days = 6.125 days

114. **AMELIA EARHART** In 1935, Amelia Earhart became the first woman to fly across the Atlantic Ocean alone, establishing a new record for the crossing: 13 hours and 30 minutes. How many minutes is this? 810 min

WRITING

115. a. Explain how to find the unit conversion factor that will convert feet to inches.

    b. Explain how to find the unit conversion factor that will convert pints to gallons.

116. Explain why the unit conversion factor $\frac{1 \text{ lb}}{16 \text{ oz}}$ is a form of 1.

REVIEW

117. Multiply: $2.09 \cdot 100$ 209

118. Multiply: $0.0546 \cdot 1,000$ 54.6

119. Multiply: $2.83(0.001)$ 0.00283

120. Multiply: $72.361(0.01)$ 0.72361

Objectives

1. Define metric units of length.
2. Use a metric ruler to measure lengths.
3. Use unit conversion factors to convert metric units of length.
4. Use a conversion chart to convert metric units of length.
5. Define metric units of mass.
6. Convert from one metric unit of mass to another.
7. Define metric units of capacity.
8. Convert from one metric unit of capacity to another.
9. Use conversions in medical applications.

SECTION 6.4
Metric Units of Measurement; Medical Applications

The metric system is the system of measurement used by most countries in the world. All countries, including the United States, use it for scientific purposes. The metric system, like our decimal numeration system, is based on the number 10. For this reason, converting from one metric unit to another is easier than with the American system.

1 Define metric units of length.

The basic metric unit of length is the **meter** (m). One meter is approximately 39 inches, which is slightly more than 1 yard. The figure below compares the length of a yardstick to a meterstick.

1 yard: 36 inches

1 meter: about 39 inches

Longer and shorter metric units of length are created by adding **prefixes** to the front of the basic unit, *meter*.

kilo means thousands   *deci* means tenths

hecto means hundreds   *centi* means hundredths

deka means tens   *milli* means thousandths

Metric Units of Length

Prefix	kilo-meter	hecto-meter	deka-meter	meter	deci-meter	centi-meter	milli-meter
Meaning	1,000 meters	100 meters	10 meters	1 meter	$\frac{1}{10}$ or 0.1 of a meter	$\frac{1}{100}$ or 0.01 of a meter	$\frac{1}{1,000}$ or 0.001 of a meter
Abbreviation	km	hm	dam	m	dm	cm	mm

> *The Language of Algebra* It is helpful to memorize the prefixes listed above because they are also used with metric units of weight and capacity.

The most often used metric units of length are kilometers, meters, centimeters, and millimeters. It is important that you gain a practical understanding of metric lengths just as you have for the length of an inch, a foot, and a mile. Some examples of metric lengths are shown below.

1 kilometer is about the length of 60 train cars.

1 m

1 meter is about the distance from a doorknob to the floor.

1 cm

1 centimeter is about as wide as the nail on your little finger.

1 mm

1 millimeter is about the thickness of a dime.

Unless otherwise noted, all art on this page is © Cengage Learning.

6.4 Metric Units of Measurement; Medical Applications   601

2   Use a metric ruler to measure lengths.

Parts of a metric ruler, scaled in centimeters, and a ruler scaled in inches are shown below. Several lengths on the metric ruler are highlighted.

(Actual size)

EXAMPLE 1   Find the length of the nail shown below.

Strategy We will place a metric ruler below the nail, with the left end of the ruler (which could be thought of as 0) directly underneath the head of the nail.

WHY Then we can find the length of the nail by identifying where its pointed end lines up on the tick marks printed in black on the ruler.

Solution
The longest tick marks on the ruler (those labeled with numbers) mark lengths in centimeters. Since the pointed end of the nail lines up on 6, the nail is 6 centimeters long.

EXAMPLE 2   Find the length of the paper clip shown below.

Strategy We will place a metric ruler below the paper clip, with the left end of the ruler (which could be thought of as 0) directly underneath one end of the paper clip.

WHY Then we can find the length of the paper clip by identifying where its other end lines up on the tick marks printed in black on the ruler.

Self Check 1
To the nearest centimeter, find the width of the circle. 3 cm

Now Try Problem 24

Self Check 2
Find the length of the jumbo paper clip. 47 mm

Now Try Problem 26

Unless otherwise noted, all art on this page is © Cengage Learning.

Solution

On the ruler, the shorter tick marks divide each centimeter into 10 millimeters, as shown. If we begin at the left end of the ruler and count by tens as we move right to 3, and then add an additional 6 millimeters to that result, we find that the length of the paper clip is 30 + 6 = 36 millimeters.

3 Use unit conversion factors to convert metric units of length.

Metric units of length are related as shown in the following table.

Metric Units of Length

1 kilometer (km) = 1,000 meters	1 meter = 10 decimeters (dm)
1 hectometer (hm) = 100 meters	1 meter = 100 centimeters (cm)
1 dekameter (dam) = 10 meters	1 meter = 1,000 millimeters (mm)

The abbreviation for each unit is written within parentheses.

We can use the information in the table to write unit conversion factors that can be used to convert metric units of length. For example, in the table we see that

$$1 \text{ meter} = 100 \text{ centimeters}$$

From this fact, we can write two unit conversion factors.

$$\frac{1 \text{ m}}{100 \text{ cm}} = 1 \quad \text{and} \quad \frac{100 \text{ cm}}{1 \text{ m}} = 1$$

To obtain the first unit conversion factor, divide both sides of the equation 1 m = 100 cm by 100 cm. To obtain the second unit conversion factor, divide both sides by 1 m.

One advantage of the metric system is that multiplying or dividing by a unit conversion factor involves multiplying or dividing by a power of 10.

Self Check 3

Convert 860 centimeters to meters. 8.6 m

Now Try Problem 31

Teaching Example 3
Convert 410 centimeters to meters.
Answer:
4.1 m

EXAMPLE 3 Convert 350 centimeters to meters.

Strategy We will multiply 350 centimeters by a carefully chosen unit conversion factor.

WHY If we multiply by the proper unit conversion factor, we can eliminate the unwanted units of centimeters and convert to meters.

Solution

To convert from centimeters to meters, we must choose a unit conversion factor whose numerator contains the units we want to introduce (meters), and whose denominator contains the units we want to eliminate (centimeters). Since there is 1 meter for every 100 centimeters, we will use

$$\frac{1 \text{ m}}{100 \text{ cm}}$$

← This is the unit we want to introduce.
← This is the unit we want to eliminate (the original unit).

To perform the conversion, we multiply 350 centimeters by the unit conversion factor $\frac{1 \text{ m}}{100 \text{ cm}}$.

$$350 \text{ cm} = \frac{350 \text{ cm}}{1} \cdot \frac{1 \text{ m}}{100 \text{ cm}}$$
Write 350 cm as a fraction: $350 \text{ cm} = \frac{350 \text{ cm}}{1}$. Multiply by a form of 1: $\frac{1 \text{ m}}{100 \text{ cm}}$.

$$= \frac{350 \cancel{\text{cm}}}{1} \cdot \frac{1 \text{ m}}{100 \cancel{\text{cm}}}$$
Remove the common units of centimeters from the numerator and denominator. Notice that the units of meter remain.

$$= \frac{350}{100} \text{ m}$$
Multiply the fractions.

$$= \frac{350.0}{100} \text{ m}$$
Write the whole number 350 as a decimal by placing a decimal point immediately to its right and entering a zero: $350 = 350.0$

$$= 3.5 \text{ m}$$
Divide 350.0 by 100 by moving the decimal point 2 places to the left: $3.5\underset{\curvearrowleft}{00}$.

Thus, 350 centimeters = 3.5 meters.

4 Use a conversion chart to convert metric units of length.

In Example 3, we converted 350 centimeters to meters using a unit conversion factor. We can also make this conversion by recognizing that all units of length in the metric system are powers of 10 of a meter.

To see this, review the table of metric units of length on page 600. Note that each unit has a value that is $\frac{1}{10}$ of the value of the unit immediately to its left and 10 times the value of the unit immediately to its right. Converting from one unit to another is as easy as multiplying (or dividing) by the correct power of 10 or, simply moving a decimal point the correct number of places to the right (or left). For example, in the **conversion chart** below, we see that to convert from centimeters to meters, we move two places to the left.

To go from centimeters to meters, we must move 2 places to the left.

If we write 350 centimeters as 350.0 centimeters, we can convert to meters by moving the decimal point two places to the left.

350.0 centimeters = 3.500 meters = 3.5 meters

Move 2 places to the left.

With the unit conversion factor method or the conversion chart method, we get 350 cm = 3.5 m.

> **Caution!** When using a chart to help make a metric conversion, be sure to list the units from *largest to smallest* when reading from left to right.

EXAMPLE 4   Convert 2.4 meters to millimeters.

Strategy On a conversion chart, we will count the places and note the direction as we move from the original units of meters to the conversion units of millimeters.

WHY The decimal point in 2.4 must be moved the same number of places and in that same direction to find the conversion to millimeters.

Self Check 4
Convert 5.3 meters to millimeters. 5,300 mm
Now Try Problem 35

Teaching Example 4
Convert 1.9 meters to millimeters.
Answer:
1,900 mm

Solution
To construct a conversion chart, we list the metric units of length from largest (kilometers) to smallest (millimeters), working from left to right. Then we locate the original units of meters and move to the conversion units of millimeters, as shown below.

We see that the decimal point in 2.4 should be moved three places to the right to convert from meters to millimeters.

$$2.4 \text{ meters} = 2400. \text{ millimeters} = 2,400 \text{ millimeters}$$

Move 3 places to the right.

We can use the unit conversion factor method to confirm this result. Since there are 1,000 millimeters per meter, we multiply 2.4 meters by the unit conversion factor $\frac{1,000 \text{ mm}}{1 \text{ m}}$.

$$2.4 \text{ m} = \frac{2.4 \text{ m}}{1} \cdot \frac{1,000 \text{ mm}}{1 \text{ m}} \quad \text{Write 2.4 m as a fraction: } 2.4 \text{ m} = \frac{2.4 \text{ m}}{1}.$$
Multiply by a form 1: $\frac{1,000 \text{ mm}}{1 \text{ m}}$.

$$= \frac{2.4 \text{ m}}{1} \cdot \frac{1,000 \text{ mm}}{1 \text{ m}} \quad \text{Remove the common units of meters from the numerator and denominator. Notice that the units of millimeters remain.}$$

$$= 2.4 \cdot 1,000 \text{ mm} \quad \text{Multiply the fractions and simplify.}$$

$$= 2,400 \text{ mm} \quad \text{Multiply 2.4 by 1,000 by moving the decimal point 3 places to the right: 2,400.}$$

Self Check 5
Convert 5.15 centimeters to kilometers. 0.0000515 km
Now Try Problem 39

Teaching Example 5
Convert 9.1 centimeters to kilometers.
Answer:
0.000091 km

EXAMPLE 5 Convert 3.2 centimeters to kilometers.

Strategy On a conversion chart, we will count the places and note the direction as we move from the original units of centimeters to the conversion units of kilometers.

WHY The decimal point in 3.2 must be moved the same number of places and in that same direction to find the conversion to kilometers.

Solution
We locate the original units of centimeters on a conversion chart, and then move to the conversion units of kilometers, as shown below.

We see that the decimal point in 3.2 should be moved five places to the left to convert centimeters to kilometers.

$$3.2 \text{ centimeters} = 0.000032 \text{ kilometers} = 0.000032 \text{ kilometers}$$

Move 5 places to the left.

We can use the unit conversion factor method to confirm this result. To convert to kilometers, we must use two unit conversion factors so that the units of centimeters drop out and the units of kilometers remain. Since there is 1 meter for every 100 centimeters and 1 kilometer for every 1,000 meters, we multiply by $\frac{1\,m}{100\,cm}$ and $\frac{1\,km}{1{,}000\,m}$.

$$3.2\ cm = \frac{3.2\ cm}{1} \cdot \frac{1\ m}{100\ cm} \cdot \frac{1\ km}{1{,}000\ m}$$

Remove the common units of centimeters and meters. The units of km remain.

$$= \frac{3.2}{100 \cdot 1{,}000}\ km$$

Multiply the fractions.

$$= 0.000032\ km$$

Divide 3.2 by 1,000 and 100 by moving the decimal point 5 places to the left.

5 Define metric units of mass.

The **mass** of an object is a measure of the amount of material in the object. When an object is moved about in space, its mass does not change. The basic unit of mass in the metric system is the **gram** (g). A gram is defined to be the mass of water contained in a cube having sides 1 centimeter long. (See the figure below.)

Other units of mass are created by adding prefixes to the front of the basic unit, *gram*.

Metric Units of Mass							
Prefix	kilo-gram	hecto-gram	deka-gram	gram	deci-gram	centi-gram	milli-gram
Meaning	1,000 grams	100 grams	10 grams	1 gram	$\frac{1}{10}$ or 0.1 of a gram	$\frac{1}{100}$ or 0.01 of a gram	$\frac{1}{1{,}000}$ or 0.001 of a gram
Abbreviation	kg	hg	dag	g	dg	cg	mg

The most often used metric units of mass are kilograms, grams, and milligrams. Some examples are shown below.

An average bowling ball weighs about 6 kilograms.

A raisin weighs about 1 gram.

A certain vitamin tablet contains 450 milligrams of calcium.

The **weight** of an object is determined by the Earth's gravitational pull on the object. Since gravitational pull on an object decreases as the object gets farther from Earth, the object weighs less as it gets farther from Earth's surface. This is why astronauts experience weightlessness in space. However, since most of us remain near Earth's surface, we will use the words *mass* and *weight* interchangeably. Thus, a mass of 30 grams is said to weigh 30 grams.

Metric units of mass are related as shown in the following table.

Metric Units of Mass

1 kilogram (kg) = 1,000 grams	1 gram = 10 decigrams (dg)
1 hectogram (hg) = 100 grams	1 gram = 100 centigrams (cg)
1 dekagram (dag) = 10 grams	1 gram = 1,000 milligrams (mg)

The abbreviation for each unit is written within parentheses.

We can use the information in the table to write unit conversion factors that can be used to convert metric units of mass. For example, in the table we see that

$$1 \text{ kilogram} = 1,000 \text{ grams}$$

From this fact, we can write two unit conversion factors.

$$\frac{1 \text{ kg}}{1,000 \text{ g}} = 1 \quad \text{and} \quad \frac{1,000 \text{ g}}{1 \text{ kg}} = 1$$

To obtain the first unit conversion factor, divide both sides of the equation 1 kg = 1,000 g by 1,000 g. To obtain the second unit conversion factor, divide both sides by 1 kg.

Another metric unit of mass is the **microgram**, which is represented by the notation mcg. One microgram is related to the gram by

$$1 \text{ gram} = 1,000,000 \text{ micrograms}$$

This fact gives us the following conversion factors

$$\frac{1 \text{ g}}{1,000,000 \text{ mcg}} = 1 \quad \text{and} \quad \frac{1,000,000 \text{ mcg}}{1 \text{ g}} = 1$$

6 **Convert from one metric unit of mass to another.**

Self Check 6
Convert 5.83 kilograms to grams. 5,830 g
Now Try Problem 43

Teaching Example 6
Convert 1.21 kilograms to grams.
Answer:
1,210 g

EXAMPLE 6 Convert 7.86 kilograms to grams.

Strategy On a conversion chart, we will count the places and note the direction as we move from the original units of kilograms to the conversion units of grams.

WHY The decimal point in 7.86 must be moved the same number of places and in that same direction to find the conversion to grams.

Solution
To construct a conversion chart, we list the metric units of mass from largest (kilograms) to smallest (milligrams), from left to right. Then we locate the original units of kilograms and move to the conversion units of grams, as shown below.

largest unit | kg | hg | dag | g | dg | cg | mg | smallest unit

3 places to the right

We see that the decimal point in 7.86 should be moved three places to the right to change kilograms to grams.

7.86 kilograms = 7860. grams = 7,860 grams

Move 3 places to the right.

We can use the unit conversion factor method to confirm this result. To convert to grams, we must choose a unit conversion factor such that the units of kilograms drop out and the units of grams remain. Since there are 1,000 grams per 1 kilogram, we multiply 7.86 kilograms by $\frac{1{,}000 \text{ g}}{1 \text{ kg}}$.

$$7.86 \text{ kg} = \frac{7.86 \text{ kg}}{1} \cdot \frac{1{,}000 \text{ g}}{1 \text{ kg}}$$

Remove the common units of kilograms in the numerator and denominator. The units of g remain.

$$= 7.86 \cdot 1{,}000 \text{ g}$$

Simplify.

$$= 7{,}860 \text{ g}$$

Multiply 7.86 by 1,000 by moving the decimal point 3 places to the right.

EXAMPLE 7 Medications
A bottle of Verapamil, a drug taken for high blood pressure, contains 30 tablets. If each tablet has 180 mg of active ingredient, how many grams of active ingredient are in the bottle?

Strategy We will multiply the number of tablets in one bottle by the number of milligrams of active ingredient in each tablet.

WHY We need to know the total number of milligrams of active ingredient in one bottle before we can convert that number to grams.

Solution

Since there are 30 tablets, and each one contains 180 mg of active ingredient, there are

$$30 \cdot 180 \text{ mg} = 5{,}400 \text{ mg}$$

```
   180
 × 30
   000
 5400
 5,400
```

of active ingredient in the bottle. To use a conversion chart to solve this problem, we locate the original units of milligrams and then move to the conversion units of grams, as shown below.

| kg | hg | dag | g | dg | cg | mg |

3 places to the left

We see that the understood decimal point in 5,400 should be moved three places to the left to convert from milligrams to grams.

5,400 milligrams = 5.400 grams   Think: 5,400 = 5,400.0.

Move 3 places to the left.

There are 5.4 grams of active ingredient in the bottle.

We can use the unit conversion factor method to confirm this result. To convert milligrams to grams, we multiply 5,400 milligrams by $\frac{1 \text{ g}}{1{,}000 \text{ mg}}$.

$$5{,}400 \text{ mg} = \frac{5{,}400 \text{ mg}}{1} \cdot \frac{1 \text{ g}}{1{,}000 \text{ mg}}$$

Remove the common units of milligrams from the numerator and denominator. The units of g remain.

$$= \frac{5{,}400}{1{,}000} \text{ g}$$

Multiply the fractions.

$$= 5.4 \text{ g}$$

Divide 5,400 by 1,000 by moving the understood decimal point in 5,400 three places to the left.

Self Check 7
MEDICATIONS A bottle of Isoptin (a drug taken for high blood pressure) contains 90 tablets, and each has 200 mg of active ingredient. How many grams of active ingredient are in the bottle? 18 g

Now Try Problems 47 and 95

Teaching Example 7 MEDICATIONS A drug taken for high cholesterol contains 65 tablets. If each tablet contains 120 mg of active ingredient, how many grams of active ingredient are there in the bottle?
Answer:
7.8 g

7 Define metric units of capacity.

In the metric system, the basic unit of capacity is the **liter** (L), which is defined to be the capacity of a cube with sides 10 centimeters long. Other units of capacity are created by adding prefixes to the front of the basic unit, liter.

Metric Units of Capacity

Prefix	kilo- liter	hecto- liter	deka- liter	liter	deci- liter	centi- liter	milli- liter
Meaning	1,000 liters	100 liters	10 liters	1 liter	$\frac{1}{10}$ or 0.1 of a liter	$\frac{1}{100}$ or 0.01 of a liter	$\frac{1}{1,000}$ or 0.001 of a liter
Abbreviation	kL	hL	daL	L	dL	cL	mL

The most often used metric units of capacity are liters and milliliters. Here are some examples.

Soft drinks are sold in 2-liter plastic bottles.

The fuel tank of a minivan can hold about 75 liters of gasoline.

A teaspoon holds about 5 milliliters.

Metric units of capacity are related as shown in the following table.

Metric Units of Capacity

1 kiloliter (kL) = 1,000 liters   1 liter = 10 deciliters (dL)
1 hectoliter (hL) = 100 liters   1 liter = 100 centiliters (cL)
1 dekaliter (daL) = 10 liters   1 liter = 1,000 milliliters (mL)

The abbreviation for each unit is written within parentheses.

We can use the information in the table to write unit conversion factors that can be used to convert metric units of capacity. For example, in the table we see that

1 liter = 1,000 milliliters

From this fact, we can write two unit conversion factors.

$$\frac{1 \text{ L}}{1,000 \text{ mL}} = 1 \quad \text{and} \quad \frac{1,000 \text{ mL}}{1 \text{ L}} = 1$$

8 Convert from one metric unit of capacity to another.

EXAMPLE 8 **Soft Drinks** How many milliliters are in three 2-liter bottles of cola?

Strategy We will multiply the number of bottles of cola by the number of liters of cola in each bottle.

WHY We need to know the total number of liters of cola before we can convert that number to milliliters.

Solution
Since there are three bottles, and each contains 2 liters of cola, there are

$$3 \cdot 2 \text{ L} = 6 \text{ L}$$

of cola in the bottles. To construct a conversion chart, we list the metric units of capacity from largest (kiloliters) to smallest (milliliters), working from left to right. Then we locate the original units of liters and move to the conversion units of milliliters, as shown below.

largest unit | kL | hL | daL | L | dL | cL | mL | smallest unit

3 places to the right

We see that the understood decimal point in 6 should be moved three places to the right to convert from liters to milliliters.

6 liters = 6000. milliliters = 6,000 milliliters
Move 3 places to the right.

Thus, there are 6,000 milliliters in three 2-liter bottles of cola.

We can use the unit conversion factor method to confirm this result. To convert to milliliters, we must choose a unit conversion factor such that liters drop out and the units of milliliters remain. Since there are 1,000 milliliters per 1 liter, we multiply 6 liters by the unit conversion factor $\frac{1,000 \text{ mL}}{1 \text{ L}}$.

$6 \text{ L} = \frac{6 \text{ L}}{1} \cdot \frac{1,000 \text{ mL}}{1 \text{ L}}$ Remove the common units of liters in the numerator and denominator. The units of mL remain.

$= 6 \cdot 1,000 \text{ mL}$ Simplify.

$= 6,000 \text{ mL}$ Multiply 6 by 1,000 by moving the understood decimal point in 6 three places to the right.

Self Check 8
SOFT DRINKS How many milliliters are in a case of twelve 2-liter bottles of cola? 24,000 mL

Now Try Problems 51 and 97

Teaching Example 8 ORANGE JUICE How many milliliters are in six 2-liter cartons of orange juice?
Answer:
12,000 mL

9 Use conversions in medical applications.

Another metric unit of capacity is the **cubic centimeter**, which is represented by the notation cm³ or, more simply, cc. One milliliter and one cubic centimeter represent the same capacity.

$$1 \text{ mL} = 1 \text{ cm}^3 = 1 \text{ cc}$$

The units of cubic centimeters are used frequently in medicine. For example, when a nurse administers an injection containing 5 cc of medication, the dosage can also be expressed using milliliters.

$$5 \text{ cc} = 5 \text{ mL}$$

Self Check 9

MEDICAL DOSAGE A doctor orders that a patient be given 2.5 liters of a 5% dextrose solution. How many cubic centimeters were ordered? 2,500 cc

Now Try Problem 102

Teaching Example 9 MEDICAL DOSAGE A doctor orders that a patient be given 2,200 cc of 0.9% saline solution. How many liters were ordered?
Answer: 2.2 L

EXAMPLE 9

Medical Dosage A doctor orders that a patient be given 1,000 cc of dextrose solution. How many liters were ordered?

Strategy On a conversion chart, we will count the places and note the direction as we move from the original units of cubic centimeters, which is the same as milliliters, to the conversion unit of liters.

WHY The decimal point in 1,000 must be moved the same number of places and in the same direction to find the conversion to liters.

Solution
The conversion chart for capacity is

3 places to the right

We see the decimal point in 1,000 should be moved three places to the left to convert from milliliters to liters.

1,000 cc = 1,000 mL = 1,000 L = 1 L
Move 3 places to the left.

ANSWERS TO SELF CHECKS

1. 3 cm   2. 47 mm   3. 8.6 m   4. 5,300 mm   5. 0.0000515 km   6. 5,830 g   7. 18 g
8. 24,000 mL   9. 2,500 cc

SECTION 6.4 STUDY SET

VOCABULARY

Fill in the blanks.

1. The meter, the gram, and the liter are basic units of measurement in the **metric** system.
2. **a.** The basic unit of length in the metric system is the **meter**.
   b. The basic unit of mass in the metric system is the **gram**.
   c. The basic unit of capacity in the metric system is the **liter**.
3. **a.** *Deka* means **tens**.
   b. *Hecto* means **hundreds**.
   c. *Kilo* means **thousands**.
4. **a.** *Deci* means **tenths**.
   b. *Centi* means **hundredths**.
   c. *Milli* means **thousandths**.
5. We can convert from one unit to another in the metric system using **unit** conversion factors or a conversion **chart** like that shown below.

   | km | hm | dam | m | dm | cm | mm |

6. The **mass** of an object is a measure of the amount of material in the object.
7. The **weight** of an object is determined by the Earth's gravitational pull on the object.
8. Another metric unit of capacity is the cubic **centimeter**, which is represented by the notation cm³, or, more simply, cc.

CONCEPTS

Fill in the blanks.

9. **a.** 1 kilometer = **1,000** meters
   b. **100** centimeters = 1 meter
   c. **1,000** millimeters = 1 meter
10. **a.** 1 gram = **1,000** milligrams
    b. 1 kilogram = **1,000** grams
11. **a.** **1,000** milliliters = 1 liter
    b. 1 dekaliter = **10** liters
12. **a.** 1 milliliter = **1** cubic centimeter
    b. 1 liter = **1,000** cubic centimeters
13. Write a unit conversion factor to convert
    a. meters to kilometers $\frac{1 \text{ km}}{1,000 \text{ m}}$
    b. grams to centigrams $\frac{100 \text{ cg}}{1 \text{ g}}$
    c. liters to milliliters $\frac{1,000 \text{ mL}}{1 \text{ L}}$

14. Use the chart to determine how many decimal places and in which direction to move the decimal point when converting the following.
    a. Kilometers to centimeters 5 places to the right

km	hm	dam	m	dm	cm	mm

    b. Milligrams to grams 3 places to the left

kg	hg	dag	g	dg	cg	mg

    c. Hectoliters to centiliters 4 places to the right

kL	hL	daL	L	dL	cL	mL

15. Match each item with its proper measurement.
    a. Thickness of a phone book iii
    b. Length of the Amazon River i
    c. Height of a soccer goal ii

    i. 6,275 km
    ii. 2 m
    iii. 6 cm

16. Match each item with its proper measurement.
    a. Weight of a giraffe i
    b. Weight of a paper clip ii
    c. Active ingredient in an aspirin tablet iii

    i. 800 kg
    ii. 1 g
    iii. 325 mg

17. Match each item with its proper measurement.
    a. Amount of blood in an adult ii
    b. Cola in an aluminum can iii
    c. Kuwait's daily production of crude oil i

    i. 290,000 kL
    ii. 5 L
    iii. 355 mL

18. Of the objects shown below, which can be used to measure the following?
    a. Millimeters micrometer
    b. Milligrams balance
    c. Milliliters beaker

    Balance

    Beaker

    Micrometer

Unless otherwise noted, all art on this page is © Cengage Learning.

NOTATION

Complete each step.

19. Convert 20 centimeters to meters.
    $$20 \text{ cm} = \frac{20 \text{ cm}}{1} \cdot \frac{1 \text{ m}}{100 \text{ cm}}$$
    $$= \frac{20}{100} \text{ m}$$
    $$= 0.2 \text{ m}$$

20. Convert 3,000 milligrams to grams.
    $$3{,}000 \text{ mg} = \frac{3{,}000 \text{ mg}}{1} \cdot \frac{1 \text{ g}}{1{,}000 \text{ mg}}$$
    $$= \frac{3{,}000}{1{,}000} \text{ g}$$
    $$= 3 \text{ g}$$

21. Convert 0.2 kilograms to milligrams.
    $$0.2 \text{ kg} = \frac{0.2 \text{ kg}}{1} \cdot \frac{1{,}000 \text{ g}}{1 \text{ kg}} \cdot \frac{1{,}000 \text{ mg}}{1 \text{ g}}$$
    $$= 0.2 \cdot 1{,}000 \cdot 1{,}000 \text{ mg}$$
    $$= 200{,}000 \text{ mg}$$

22. Convert 400 milliliters to kiloliters.
    $$400 \text{ mL} = \frac{400 \text{ mL}}{1} \cdot \frac{1 \text{ L}}{1{,}000 \text{ mL}} \cdot \frac{1 \text{ kL}}{1{,}000 \text{ L}}$$
    $$= \frac{400}{1{,}000 \cdot 1{,}000} \text{ kL}$$
    $$= 0.0004 \text{ kL}$$

GUIDED PRACTICE

Refer to the given ruler to answer each question. See Example 1.

23. Determine which measurements the arrows point to on the metric ruler. 1 cm, 3 cm, 5 cm

24. Find the length of the birthday candle (including the wick). 6 cm

Refer to the given ruler to answer each question. See Example 2.

25. a. Each centimeter is divided into how many equal parts? What is the length of one of those parts? 10, 1 millimeter

b. Determine which measurements the arrows point to on the ruler. 27 mm, 41 mm, 55 mm

26. Find the length of the stick of gum. 74 mm

Use a metric ruler scaled in millimeters to measure each object. See Example 2.

27. The length of a dollar bill 156 mm
28. The width of a dollar bill 66 mm
29. The length (top to bottom) of this page 280 mm
30. The length of the word antidisestablishmentarianism as printed here. 42 mm

Perform each conversion. See Example 3.

31. 380 centimeters to meters 3.8 m
32. 590 centimeters to meters 5.9 m
33. 120 centimeters to meters 1.2 m
34. 640 centimeters to meters 6.4 m

Perform each conversion. See Example 4.

35. 8.7 meters to millimeters 8,700 mm
36. 1.3 meters to millimeters 1,300 mm
37. 2.89 meters to millimeters 2,890 mm
38. 4.06 meters to millimeters 4,060 mm

Perform each conversion. See Example 5.

39. 4.5 centimeters to kilometers 0.000045 km
40. 6.2 centimeters to kilometers 0.000062 km
41. 0.3 centimeters to kilometers 0.000003 km
42. 0.4 centimeters to kilometers 0.000004 km

Perform each conversion. See Example 6.

43. 1.93 kilograms to grams 1,930 g
44. 8.99 kilograms to grams 8,990 g
45. 4.531 kilograms to grams 4,531 g
46. 6.077 kilograms to grams 6,077 g

Perform each conversion. See Example 7.

47. 6,000 milligrams to grams 6 g
48. 9,000 milligrams to grams 9 g
49. 3,500 milligrams to grams 3.5 g
50. 7,500 milligrams to grams 7.5 g

Perform each conversion. See Example 8.

51. 3 liters to milliliters 3,000 mL
52. 4 liters to milliliters 4,000 mL
53. 26.3 liters to milliliters 26,300 mL
54. 35.2 liters to milliliters 35,200 mL

TRY IT YOURSELF

Perform each conversion.

55. 0.31 decimeters to centimeters 3.1 cm
56. 73.2 meters to decimeters 732 dm
57. 500 milliliters to liters 0.5 L
58. 500 centiliters to milliliters 5,000 mL
59. 2 kilograms to grams 2,000 g
60. 4,000 grams to kilograms 4 kg
61. 0.074 centimeters to millimeters 0.74 mm
62. 0.125 meters to millimeters 125 mm
63. 1,000 kilograms to grams 1,000,000 g
64. 2 kilograms to centigrams 200,000 cg
65. 658.23 liters to kiloliters 0.65823 kL
66. 0.0068 hectoliters to kiloliters 0.00068 kL
67. 4.72 cm to dm 0.472 dm
68. 0.593 cm to dam 0.000593 dam
69. 10 mL = 10 cc
70. 2,000 cc = 2 L
71. 1,250 mcg to mg 1.25 mg
72. 500 mcg to mg 0.5 mg
73. 5,689 g to kg 5.689 kg
74. 0.0579 km to mm 57,900 mm
75. 453.2 cm to m 4.532 m
76. 675.3 cm to m 6.753 m
77. 0.325 dL to L 0.0325 L
78. 0.0034 mL to L 0.0000034 L
79. 675 dam = 675,000 cm
80. 76.8 hm = 7,680,000 mm
81. 0.00777 cm = 0.0000077 dam
82. 400 liters to hL 4 hL
83. 134 m to hm 1.34 hm
84. 6.77 mm to cm 0.677 cm
85. 65.78 km to dam 6,578 dam
86. 5 g to cg 500 cg

APPLICATIONS

87. SPEED SKATING American Eric Heiden won an unprecedented five gold medals by capturing the men's 500 m, 1,000 m, 1,500 m, 5,000 m, and 10,000 m races at the 1980 Winter Olympic Games in Lake Placid, New York. Convert each race length to kilometers. 0.5 km, 1 km, 1.5 km, 5 km, 10 km

Unless otherwise noted, all art on this page is © Cengage Learning.

88. **THE SUEZ CANAL** The 163-km-long Suez Canal connects the Mediterranean Sea with the Red Sea. It provides a shortcut for ships operating between European and American ports. Convert the length of the Suez Canal to meters. 163,000 m

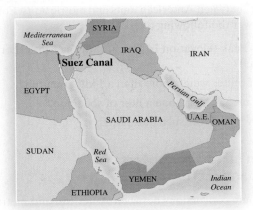

89. **SKYSCRAPERS** The John Hancock Center in Chicago has 100 stories and is 343 meters high. Give this height in hectometers. 3.43 hm

90. **WEIGHT OF A BABY** A baby weighs 4 kilograms. Give this weight in centigrams. 400,000 cg

91. **HEALTH CARE** Blood pressure is measured by a sphygmomanometer (see at right). The measurement is read at two points and is expressed, for example, as 120/80. This indicates a *systolic* pressure of 120 millimeters of mercury and a *diastolic* pressure of 80 millimeters of mercury. Convert each measurement to centimeters of mercury. 12 cm, 8 cm

92. **JEWELRY** A gold chain weighs 1,500 milligrams. Give this weight in grams. 1.5 g

93. **EYE DROPPERS** One drop from an eye dropper is 0.05 mL. Convert the capacity of one drop to liters. 0.00005 L

94. **BOTTLING** How many liters of wine are in a 750-mL bottle? $\frac{3}{4}$ L or 0.75 L

95. **MEDICINE** A bottle of hydrochlorothiazide contains 60 tablets. If each tablet contains 50 milligrams of active ingredient, how many grams of active ingredient are in the bottle? 3 g

96. **IBUPROFEN** What is the total number of grams of ibuprofen contained in all the tablets in the box shown at the right? 33 g

97. **SIX PACKS** Some stores sell Fanta orange soda in 0.5 liter bottles. How many milliliters are there in a six pack of this size bottle? 3,000 mL

98. **CONTAINERS** How many deciliters of root beer are in two 2-liter bottles? 40 dL

99. **OLIVES** The net weight of a bottle of olives is 284 grams. Find the smallest number of bottles that must be purchased to have at least 1 kilogram of olives. 4

100. **INJECTIONS** The illustration below shows a 3cc syringe. Express its capacity using units of milliliters. 3 mL

101. **MEDICAL DOSAGE** A doctor ordered 1,500 cc of a saline solution from a pharmacy. How many liters of saline solution is this? 1.5 L

102. **MEDICAL DOSAGE** A doctor ordered 2,000 cc of a saline (salt) solution from a pharmacy. How many liters of saline solution is this? 2 liters

WRITING

103. To change 3.452 kilometers to meters, we can move the decimal point in 3.452 three places to the right to get 3,452 meters. Explain why.

104. To change 7,532 grams to kilograms, we can move the decimal point in 7,532 three places to the left to get 7.532 kilograms. Explain why.

105. A *centimeter* is one hundredth of a meter. Make a list of five other words that begin with the prefix *centi* or *cent* and write a definition for each.

106. List the advantages of the metric system of measurement as compared to the American system. There have been several attempts to bring the metric system into general use in the United States. Why do you think these efforts have been unsuccessful?

REVIEW

Write each fraction as a decimal. Use an overbar in your answer.

107. $\frac{8}{9}$ $0.\overline{8}$

108. $\frac{11}{12}$ $0.91\overline{6}$

109. $\frac{7}{90}$ $0.0\overline{7}$

110. $\frac{1}{66}$ $0.0\overline{15}$

Objectives

1. Use unit conversion factors to convert between American and metric units.
2. Convert between Fahrenheit and Celsius temperatures.

SECTION 6.5
Converting between American and Metric Units

It is often necessary to convert between American units and metric units. For example, we must convert units to answer the following questions.

- Which is higher: Pikes Peak (elevation 14,110 feet) or the Matterhorn (elevation 4,478 meters)?
- Does a 2-pound tub of butter weigh more than a 1-kilogram tub?
- Is a quart of soda pop more or less than a liter of soda pop?

In this section, we discuss how to answer such questions.

1 Use unit conversion factors to convert between American and metric units.

The following table shows some conversions between American and metric units of length. In all but one case, the conversions are rounded approximations. An ≈ symbol is used to show this. The one exact conversion in the table is 1 inch = 2.54 centimeters.

Equivalent Lengths	
American to metric	**Metric to American**
1 in. = 2.54 cm	1 cm ≈ 0.39 in.
1 ft ≈ 0.30 m	1 m ≈ 3.28 ft
1 yd ≈ 0.91 m	1 m ≈ 1.09 yd
1 mi ≈ 1.61 km	1 km ≈ 0.62 mi

Unit conversion factors can be formed from the facts in the table to make specific conversions between American and metric units of length.

Self Check 1

CLOTHING LABELS Refer to the figure in Example 1. What is the inseam length, to the nearest inch? 30 in.

Now Try Problem 13

Teaching Example 1 **CLOTHING LABELS** The length of a sleeve of a dress shirt made in Italy is 92 centimeters. Express the length of the sleeve to the nearest inch.
Answer:
36 in.

EXAMPLE 1 Clothing Labels

The figure shows a label sewn into some pants made in Mexico that are for sale in the United States. Express the waist size to the nearest inch.

Strategy We will multiply 82 centimeters by a carefully chosen unit conversion factor.

WHY If we multiply by the proper unit conversion factor, we can eliminate the unwanted units of centimeters and convert to inches.

Solution

To convert from centimeters to inches, we must choose a unit conversion factor whose numerator contains the units we want to introduce (inches), and whose denominator contains the units we want to eliminate (centimeters). From the first row of the *Metric to American* column of the table, we see that there is approximately 0.39 inch per centimeter. Thus, we will use the unit conversion factor:

$$\frac{0.39 \text{ in.}}{1 \text{ cm}}$$

← This is the unit we want to introduce.
← This is the unit we want to eliminate (the original unit).

To perform the conversion, we multiply.

$$82 \text{ cm} \approx \frac{82 \text{ cm}}{1} \cdot \frac{0.39 \text{ in.}}{1 \text{ cm}}$$ Write 82 cm as a fraction: $82 \text{ cm} = \frac{82 \text{ cm}}{1}$. Multiply by a form of 1: $\frac{0.39 \text{ in.}}{1 \text{ cm}}$.

$$\approx \frac{82 \cancel{\text{ cm}}}{1} \cdot \frac{0.39 \text{ in.}}{1 \cancel{\text{ cm}}}$$ Remove the common units of centimeters from the numerator and denominator. The units of inches remain.

$$\approx 82 \cdot 0.39 \text{ in.}$$ Simplify.

$$\approx 31.98 \text{ in.}$$ Do the multiplication.

$$\approx 32 \text{ in.}$$ Round to the nearest inch (ones column).

```
    0.39
  × 82
  ─────
    78
  3120
  ─────
  31.98
```

To the nearest inch, the waist size is 32 inches.

EXAMPLE 2 Mountain Elevations

Pikes Peak, one of the most famous peaks in the Rocky Mountains, has an elevation of 14,110 feet. The Matterhorn, in the Swiss Alps, rises to an elevation of 4,478 meters. Which mountain is higher?

Strategy We will convert the elevation of Pikes Peak, which given in feet, to meters.

WHY Then we can compare the mountain's elevations in the same units, meters.

Solution
To convert Pikes Peak elevation from feet to meters, we must choose a unit conversion factor whose numerator contains the units we want to introduce (meters) and whose denominator contains the units we want to eliminate (feet). From the second row of the *American to metric* column of the table, we see that there is approximately 0.30 meter per foot. Thus, we will use the unit conversion factor:

$$\frac{0.30 \text{ m}}{1 \text{ ft}}$$ ← This is the unit we want to introduce.
← This is the unit we want to eliminate (the original unit).

To perform the conversion, we multiply.

$$14{,}110 \text{ ft} \approx \frac{14{,}110 \text{ ft}}{1} \cdot \frac{0.30 \text{ m}}{1 \text{ ft}}$$ Write 14,110 ft as a fraction: $14{,}110 \text{ ft} = \frac{14{,}110 \text{ ft}}{1}$. Multiply by a form of 1: $\frac{0.30 \text{ m}}{1 \text{ ft}}$.

$$\approx \frac{14{,}110 \cancel{\text{ ft}}}{1} \cdot \frac{0.30 \text{ m}}{1 \cancel{\text{ ft}}}$$ Remove the common units of feet from the numerator and denominator. The units of meters remain.

$$\approx 14{,}110 \cdot 0.30 \text{ m}$$ Simplify.

$$\approx 4{,}233 \text{ m}$$ Do the multiplication.

```
      1
   14,110
  × 0.30
  ───────
   000 00
  4233 00
  ───────
  4233.00
```

Since the elevation of Pikes Peak is about 4,233 meters, we can conclude that the Matterhorn, with an elevation of 4,478 meters, is higher.

Self Check 2
TRACK AND FIELD Which is longer: a 500-meter race or a 550-yard race? the 550-yard race

Now Try Problem 17

Teaching Example 2 VOLCANOS
The elevation of the Lassen Peak volcano in California is 10,456 feet. The elevation of the Fuego volcano in Guatemala is 3,763 meters. Which volcano is higher?
Answer:
the Fuego volcano

We can convert between American units of weight and metric units of mass using the rounded approximations in the following table.

Equivalent Weights and Masses	
American to metric	**Metric to American**
1 oz ≈ 28.35 g	1 g ≈ 0.035 oz
1 lb ≈ 0.45 kg	1 kg ≈ 2.20 lb

1 pound

1 kilogram

Unless otherwise noted, all art on this page is © Cengage Learning.

Chapter 6 Ratio, Proportion, and Measurement

Self Check 3

Convert 68 pounds to grams. Round to the nearest gram. *30,845 g*

Now Try Problem 21

Teaching Example 3
Convert 40 pounds to grams.
Answer:
18,144 g

EXAMPLE 3   Convert 50 pounds to grams.

Strategy We will use a two-part multiplication process that converts 50 pounds to ounces, and then converts that result to grams.

WHY We must use a two-part process because the conversion table on page 615 does not contain a single unit conversion factor that converts from pounds to grams.

Solution

Since there are 16 ounces per pound, we can convert 50 pounds to ounces by multiplying by the unit conversion factor $\frac{16 \text{ oz}}{1 \text{ lb}}$. Since there are approximately 28.35 g per ounce, we can convert that result to grams by multiplying by the unit conversion factor $\frac{28.35 \text{ g}}{1 \text{ oz}}$.

$$50 \text{ lb} \approx \frac{50 \text{ lb}}{1} \cdot \frac{16 \text{ oz}}{1 \text{ lb}} \cdot \frac{28.35 \text{ g}}{1 \text{ oz}}$$

Write 50 lb as a fraction: $50 \text{ lb} = \frac{50 \text{ lb}}{1}$. Multiply by two forms of 1: $\frac{16 \text{ oz}}{1 \text{ lb}}$ and $\frac{28.35 \text{ g}}{1 \text{ oz}}$.

$$\approx \frac{50 \cancel{\text{ lb}}}{1} \cdot \frac{16 \cancel{\text{ oz}}}{1 \cancel{\text{ lb}}} \cdot \frac{28.35 \text{ g}}{1 \cancel{\text{ oz}}}$$

Remove the common units of pounds and ounces from the numerator and denominator. The units of grams remain.

$$\approx 50 \cdot 16 \cdot 28.35 \text{ g}$$ Simplify.

$$\approx 800 \cdot 28.35 \text{ g}$$ Multiply: $50 \cdot 16 = 800$.

$$\approx 22,680 \text{ g}$$ Do the multiplication.

```
   3
  16          28.35
× 50        ×   800
 800       22680.00
```

Thus, 50 pounds ≈ 22,680 grams.

Self Check 4

BODY WEIGHT Who weighs more, a person who weighs 165 pounds or one who weighs 76 kilograms? *the person who weighs 76 kg*

Now Try Problem 25

Teaching Example 4
WEIGHTLIFTING Which weighs more, a 45-pound iron plate or a 15-kilogram dumbell?
Answer:
the 45-pound plate

EXAMPLE 4   Packaging   Does a 2.5-pound tub of butter weigh more than a 1.5-kilogram tub?

Strategy We will convert the weight of the 1.5-kilogram tub of butter to pounds.

WHY Then we can compare the weights of the tubs of butter in the same units, pounds.

Solution

To convert 1.5 kilograms to pounds we must choose a unit conversion factor whose numerator contains the units we want to introduce (pounds), and whose denominator contains the units we want to eliminate (kilograms). From the second row of the *Metric to American* column of the table, we see that there are approximately 2.20 pounds per kilogram. Thus, we will use the unit conversion factor:

$$\frac{2.20 \text{ lb}}{1 \text{ kg}}$$

← This is the unit we want to introduce.
← This is the unit we want to eliminate (the original unit).

To perform the conversion, we multiply.

$$1.5 \text{ kg} \approx \frac{1.5 \text{ kg}}{1} \cdot \frac{2.20 \text{ lb}}{1 \text{ kg}}$$

Write 1.5 kg as a fraction: $1.5 \text{ kg} = \frac{1.5 \text{ kg}}{1}$. Multiply by a form of 1: $\frac{2.20 \text{ lb}}{1 \text{ kg}}$.

$$\approx \frac{1.5 \cancel{\text{ kg}}}{1} \cdot \frac{2.20 \text{ lb}}{1 \cancel{\text{ kg}}}$$

Remove the common units of kilograms from the numerator and denominator. The units of pounds remain.

$$\approx 1.5 \cdot 2.20 \text{ lb}$$ Simplify.

$$\approx 3.3 \text{ lb}$$ Do the multiplication.

```
  2.20
× 1.5
 1100
 2200
 3.300
```

Since a 1.5-kilogram tub of butter weighs about 3.3 pounds, the 1.5-kilogram tub weighs more.

We can convert between American and metric units of capacity using the rounded approximations in the following table.

Equivalent Capacities	
American to metric	**Metric to American**
1 fl oz ≈ 29.57 mL	1 L ≈ 33.81 fl oz
1 pt ≈ 0.47 L	1 L ≈ 2.11 pt
1 qt ≈ 0.95 L	1 L ≈ 1.06 qt
1 gal ≈ 3.79 L	1 L ≈ 0.264 gal

1 liter   1 quart

THINK IT THROUGH   *Studying in Other Countries*

"The number of American college students studying abroad is now at an all-time high."

From the Institute of International Education

In 2010/2011, a record number of 273,996 college students received credit for study abroad. Since students traveling to other countries are almost certain to come into contact with the metric system of measurement, they need to have a basic understanding of metric units.

Suppose a student studying overseas needs to purchase the following school supplies. For each item in red, choose the appropriate metric units.

1. $8\frac{1}{2}$ in. × 11 in. **notebook paper:** 216 mm × 279 mm

   216 meters × 279 meters   216 centimeters × 279 centimeters

   216 millimeters × 279 millimeters

2. A backpack that can hold 20 pounds of books: 9 kilograms

   9 kilograms   9 grams   9 milligrams

3. $\frac{3}{4}$ fluid ounce bottle of Liquid Paper correction fluid: 22.2 milliliters

   22.5 hectoliters   2.5 liters   22.2 milliliters

EXAMPLE 5   *Cleaning Supplies* A bottle of window cleaner contains 750 milliliters of solution. Convert this measure to quarts. Round to the nearest tenth.

Strategy We will use a two-part multiplication process that converts 750 milliliters to liters, and then converts that result to quarts.

WHY We must use a two-part process because the conversion table at the top of this page does not contain a single unit conversion factor that converts from milliliters to quarts.

Solution
Since there is 1 liter for every 1,000 milliliters, we can convert 750 milliliters to liters by multiplying by the unit conversion factor $\frac{1\ L}{1{,}000\ mL}$. Since there

Self Check 5
DRINKING WATER A student bought a 360 mL bottle of water. Convert this measure to quarts. Round to the nearest tenth. 0.4 qt

Now Try Problem 29

Teaching Example 5 MOTOR OIL A can contains 850 milliliters of motor oil. Convert this measure to quarts. Round to the nearest tenth.
Answer:
0.9 qt

are approximately 1.06 quarts per liter, we can convert that result to quarts by multiplying by the unit conversion factor $\frac{1.06 \text{ qt}}{1 \text{ L}}$.

$$750 \text{ mL} \approx \frac{750 \text{ mL}}{1} \cdot \frac{1 \text{ L}}{1{,}000 \text{ mL}} \cdot \frac{1.06 \text{ qt}}{1 \text{ L}}$$

Write 750 mL as a fraction: $750 \text{ mL} = \frac{750 \text{ mL}}{1}$. Multiply by two forms of 1: $\frac{1 \text{ L}}{1{,}000 \text{ mL}}$ and $\frac{1.06 \text{ qt}}{1 \text{ L}}$.

$$\approx \frac{750 \text{ mL}}{1} \cdot \frac{1 \text{ L}}{1{,}000 \text{ mL}} \cdot \frac{1.06 \text{ qt}}{1 \text{ L}}$$

Remove the common units of milliliters and liters from the numerator and denominator. The units of quarts remain.

$$\approx \frac{750 \cdot 1.06}{1{,}000} \text{ qt}$$

Multiply the fractions.

$$\approx \frac{795}{1{,}000} \text{ qt}$$

Multiply: $750 \cdot 1.06 = 795$.

$$\approx 0.795 \text{ qt}$$

Divide 795 by 1,000 by moving the decimal point 3 places to the left.

$$\approx 0.8 \text{ qt}$$

Round to the nearest tenth.

```
   750
 × 1.06
  4500
  0000
 75000
 795.00
```

The bottle contains approximately 0.8 qt of cleaning solution.

2 Convert between Fahrenheit and Celsius temperatures.

In the American system, we measure temperature using **degrees Fahrenheit** (°F). In the metric system, we measure temperature using **degrees Celsius** (°C). These two scales are shown on the thermometers on the right. From the figures, we can see that

- 212°F = 100°C Water boils
- 32°F = 0°C Water freezes
- 5°F = −15°C A cold winter day
- 95°F = 35°C A hot summer day

There are formulas that enable us to convert from degrees Fahrenheit to degrees Celsius and from degrees Celsius to degrees Fahrenheit.

Conversion Formulas for Temperature

If F is the temperature in degrees Fahrenheit and C is the corresponding temperature in degrees Celsius, then

$$C = \frac{5}{9}(F - 32) \quad \text{and} \quad F = \frac{9}{5}C + 32$$

6.5 Converting between American and Metric Units

EXAMPLE 6 **Bathing** Warm bath water is 90°F. Express this temperature in degrees Celsius. Round to the nearest tenth of a degree.

Strategy We will substitute 90 for F in the formula $C = \frac{5}{9}(F - 32)$.

WHY Then we can use the rule for the order of operations to evaluate the right side of the equation and find the value of C, the temperature in degrees Celsius of the bath water.

Solution

$C = \frac{5}{9}(F - 32)$ This is the formula to find degrees Celsius.

$= \frac{5}{9}(90 - 32)$ Substitute 90 for F.

$= \frac{5}{9}(58)$ Do the subtraction within the parentheses first: 90 − 32 = 58.

$= \frac{5}{9}\left(\frac{58}{1}\right)$ Write 58 as a fraction: $58 = \frac{58}{1}$.

$= \frac{290}{9}$ Multiply the numerators. Multiply the denominators.

$= 32.222\ldots$ Do the division.

≈ 32.2 Round to the nearest tenth.

To the nearest tenth of a degree, the temperature of the bath water is 32.2°C.

```
   4
  58
 × 5
 290
```

```
  32.22
9)290.00
 −27
   20
  −18
   20
  −18
   20
  −18
    2
```

Self Check 6

COFFEE Hot coffee is 110°F. Express this temperature in degrees Celsius. Round to the nearest tenth of a degree.   43.3°C

Now Try Problem 33

Teaching Example 6 HAMBURGERS The recommended safe minimal internal temperature when cooking ground beef is 160°F. Express this temperature in degrees Celsius. Round to the nearest tenth of a degree.
Answer:
71.1°C

EXAMPLE 7 **Dishwashers** A dishwasher manufacturer recommends that dishes be rinsed in hot water with a temperature of 60°C. Express this temperature in degrees Fahrenheit.

Strategy We will substitute 60 for C in the formula $F = \frac{9}{5}C + 32$.

WHY Then we can use the rule for the order of operations to evaluate the right side of the equation and find the value of F, the temperature in degrees Fahrenheit of the water.

Solution

$F = \frac{9}{5}C + 32$ This is the formula to find degrees Fahrenheit.

$= \frac{9}{5}(60) + 32$ Substitute 60 for C.

$= \frac{540}{5} + 32$ Multiply: $\frac{9}{5}(60) = \frac{9}{5}\left(\frac{60}{1}\right) = \frac{540}{5}$.

$= 108 + 32$ Do the division.

$= 140$ Do the addition.

The manufacturer recommends that dishes be rinsed in 140°F water.

```
  60
 × 9
 540
```

```
  108
5)540
 −5
   4
  −0
   40
  −40
    0
```

Self Check 7

FEVERS To determine whether a baby has a fever, her mother takes her temperature with a Celsius thermometer. If the reading is 38.8°C, does the baby have a fever? (*Hint:* Normal body temperature is 98.6°F.)   yes

Now Try Problem 37

Teaching Example 7 OLD FAITHFUL Just before it erupts, the temperature of the water in the Old Faithful geyser in Yellowstone National Park reaches 95.6°C. Express this temperature in degrees Fahrenheit. Round to the nearest degree.
Answer:
204°F

ANSWERS TO SELF CHECKS

1. 30 in.   **2.** the 550-yard race   **3.** 30,845 g   **4.** the person who weighs 76 kg
5. 0.4 qt   **6.** 43.3°C   **7.** yes

SECTION 6.5 STUDY SET

VOCABULARY

Fill in the blanks.

1. In the American system, temperatures are measured in degrees Fahrenheit . In the metric system, temperatures are measured in degrees Celsius .
2. a. Inches and centimeters are units used to measure length .
   b. Pounds and grams are units used to measure mass (weight).
   c. Gallons and liters are units used to measure capacity .

CONCEPTS

3. Which is longer:
   a. A yard or a meter? meter
   b. A foot or a meter? meter
   c. An inch or a centimeter? inch
   d. A mile or a kilometer? mile
4. Which is heavier:
   a. An ounce or a gram? ounce
   b. A pound or a kilogram? kilogram
5. Which is the greater unit of capacity:
   a. A pint or a liter? liter
   b. A quart or a liter? liter
   c. A gallon or a liter? gallon
6. a. What formula is used for changing degrees Celsius to degrees Fahrenheit? $F = \frac{9}{5}C + 32$
   b. What formula is used for changing degrees Fahrenheit to degrees Celsius? $C = \frac{5}{9}(F - 32)$
7. Write a unit conversion factor to convert
   a. feet to meters $\frac{0.30 \text{ m}}{1 \text{ ft}}$
   b. pounds to kilograms $\frac{0.45 \text{ kg}}{1 \text{ lb}}$
   c. gallons to liters $\frac{3.79 \text{ L}}{1 \text{ gal}}$
8. Write a unit conversion factor to convert
   a. centimeters to inches $\frac{0.39 \text{ in.}}{1 \text{ cm}}$
   b. grams to ounces $\frac{0.035 \text{ oz}}{1 \text{ g}}$
   c. liters to fluid ounces $\frac{33.81 \text{ fl oz}}{1 \text{ L}}$

NOTATION

Complete each step.

9. Convert 4,500 feet to meters.

   $4{,}500 \text{ ft} \approx \frac{4{,}500 \text{ ft}}{1} \cdot \frac{0.30 \text{ m}}{1 \text{ ft}}$

   $\approx 1{,}350$ m

10. Convert 8 liters to gallons.

    $8 \text{ L} \approx \frac{8 \text{ L}}{1} \cdot \frac{0.264 \text{ gal}}{1 \text{ L}}$

    ≈ 2.112 gal

11. Convert 3 kilograms to ounces.

    $3 \text{ kg} \approx \frac{3 \text{ kg}}{1} \cdot \frac{1{,}000 \text{ g}}{1 \text{ kg}} \cdot \frac{0.035 \text{ oz}}{1 \text{ g}}$

    $\approx 3 \cdot$ 1,000 $\cdot 0.035 \text{ oz}$

    ≈ 105 oz

12. Convert 70°C to degrees Fahrenheit.

    $F = \frac{9}{5}C + 32$

    $= \frac{9}{5}($70$) + 32$

    $=$ 126 $+ 32$

    $= 158$

    Thus, 70°C = 158 °F

GUIDED PRACTICE

Perform each conversion. Round to the nearest inch. See Example 1. Since most conversions are approximate, answers may vary slightly, depending on the method used.

13. 25 centimeters to inches 10 in.
14. 35 centimeters to inches 14 in.
15. 88 centimeters to inches 34 in.
16. 91 centimeters to inches 35 in.

Perform each conversion. See Example 2.

17. 8,400 feet to meters 2,520 m
18. 7,300 feet to meters 2,190 m
19. 25,115 feet to meters 7,534.5 m
20. 36,242 feet to meters 10,872.6 m

Perform each conversion. See Example 3.

21. 20 pounds to grams 9,072 g
22. 30 pounds to grams 13,608 g
23. 75 pounds to grams 34,020 g
24. 95 pounds to grams 43,092 g

Perform each conversion. See Example 4.

25. 6.5 kilograms to pounds 14.3 lb
26. 7.5 kilograms to pounds 16.5 lb
27. 300 kilograms to pounds 660 lb
28. 800 kilograms to pounds 1,760 lb

Perform each conversion. Round to the nearest tenth. See Example 5.

29. 650 milliliters to quarts 0.7 qt
30. 450 milliliters to quarts 0.5 qt
31. 1,200 milliliters to quarts 1.3 qt
32. 1,500 milliliters to quarts 1.6 qt

Express each temperature in degrees Celsius. Round to the nearest tenth of a degree. See Example 6.

33. 120°F 48.9°C
34. 110°F 43.3°C
35. 35°F 1.7°C
36. 45°F 7.2°C

Express each temperature in degrees Fahrenheit. See Example 7.

37. 75°C 167°F
38. 85°C 185°F
39. 10°C 50°F
40. 20°C 68°F

TRY IT YOURSELF

Perform each conversion. If necessary, round answers to the nearest tenth. Since most conversions are approximate, answers may vary slightly depending on the method used.

41. 25 pounds to grams 11,340 g
42. 7.5 ounces to grams 212.6 g
43. 50°C to degrees Fahrenheit 122°F
44. 36.2°C to degrees Fahrenheit 97.2°F
45. 0.75 quart to milliliters 712.5 mL
46. 3 pints to milliliters 1,410 mL
47. 0.5 kilogram to ounces 17.6 oz
48. 35 grams to pounds 0.1 lb
49. 3.75 meters to inches 147.6 in.
50. 2.4 kilometers to miles 1.5 mi
51. 3 fluid ounces to liters 0.1 L
52. 2.5 pints to liters 1.2 L
53. 12 kilometers to feet 39,283 ft
54. 3,212 centimeters to feet 104.4 ft
55. 37 ounces to kilograms 1.0 kg
56. 10 pounds to kilograms 4.5 kg
57. −10°C to degrees Fahrenheit 14°F
58. −22.5°C to degrees Fahrenheit −8.5°F
59. 17 grams to ounces 0.6 oz
60. 100 kilograms to pounds 220 lb
61. 7.2 liters to fluid ounces 243.4 fl oz
62. 5 liters to quarts 5.3 qt
63. 3 feet to centimeters 91.4 cm
64. 7.5 yards to meters 6.8 m
65. 500 milliliters to quarts 0.5 qt
66. 2,000 milliliters to gallons 0.5 gal
67. 50°F to degrees Celsius 10°C
68. 67.7°F to degrees Celsius 19.8°C
69. 5,000 inches to meters 127 m
70. 25 miles to kilometers 40.3 km
71. −5°F to degrees Celsius −20.6°C
72. −10°F to degrees Celsius −23.3°C

APPLICATIONS

Since most conversions are approximate, answers may vary slightly depending on the method used.

73. **THE MIDDLE EAST** The distance between Jerusalem and Bethlehem is 8 kilometers. To the nearest mile, give this distance in miles. 5 mi
74. **THE DEAD SEA** The Dead Sea is 80 kilometers long. To the nearest mile, give this distance in miles. 50 mi
75. **CHEETAHS** A cheetah can run 112 kilometers per hour. Express this speed in mph. 69 mph
76. **LIONS** A lion can run 50 mph. Express this speed in kilometers per hour. 80.5 kph
77. **MOUNT WASHINGTON** The highest peak of the White Mountains of New Hampshire is Mount Washington, at 6,288 feet. Give this height in kilometers. Round to the nearest tenth. 1.9 km
78. **TRACK AND FIELD** Track meets are held on an oval track. One lap around the track is usually 400 meters. However, some older tracks in the United States are 440-yard ovals. Are these two types of tracks the same length? If not, which is longer? The 440-yard track is longer.

79. **HAIR GROWTH** When hair is short, its rate of growth averages about $\frac{3}{4}$ inch per month. How many centimeters is this a month? Round to the nearest tenth of a centimeter. 1.9 cm
80. **WHALES** An adult male killer whale can weigh as much as 12,000 pounds and be as long as 25 feet. Change these measurements to kilograms and meters. 5,400 kg, 7.5 m
81. **WEIGHTLIFTING** The table lists the personal best bench press records for two of the world's best powerlifters. Change each metric weight to pounds. Round to the nearest pound. 402 lb, 799 lb

Name	Hometown	Bench press
Liz Willet	Ferndale, Washington	182.5 kg
Brian Siders	Charleston, W. Virginia	363 kg

Source: goheavy.net, t-nation.com

82. **WORDS OF WISDOM** Refer to the wall hanging. Convert the first metric weight to ounces and the second to pounds. What famous saying results? An ounce of prevention is worth a pound of cure.

28.35 grams of prevention is worth 0.45 kilogram of cure

83. **OUNCES AND FLUID OUNCES**
    a. There are 310 calories in 8 ounces of broiled chicken. Convert 8 ounces to grams. 226.8 g
    b. There are 112 calories in a glass of fresh Valencia orange juice that holds 8 fluid ounces. Convert 8 fluid ounces to liters. Round to the nearest hundredth. 0.24 L

84. **TRACK AND FIELD** A shot-put weighs 7.264 kilograms. Convert this weight to pounds. Round to the nearest pound. 16 lb

85. **POSTAL REGULATIONS** You can mail a package weighing up to 70 pounds via priority mail. Can you mail a package that weighs 32 kilograms by priority mail? no

86. **NUTRITION** Refer to the nutrition label shown below for a packet of oatmeal. Change each circled weight to ounces. 1.61 oz, 0.0175 oz, 0.00875 oz, 0.14 oz

Nutrition Facts
Serving Size: 1 Packet (46g)
Servings Per Container: 10

Amount Per Serving
Calories 170   Calories from Fat 20
                        % Daily Value
Total fat 2g   3%
  Saturated fat (0.5g)   2%
  Polyunsaturated Fat 0.5g
  Monounsaturated Fat 1g
Cholesterol 0mg   0%
Sodium (250mg)   10%
Total carbohydrate 35g   12%
  Dietary fiber 3g   12%
  Soluble Fiber 1g
  Sugars 16g
Protein (4g)

87. **HOT SPRINGS** The thermal springs in Hot Springs National Park in central Arkansas emit water as warm as 143°F. Change this temperature to degrees Celsius. about 62°C

88. **COOKING MEAT** Meats must be cooked at temperatures high enough to kill harmful bacteria. According to the USDA and the FDA, the internal temperature for cooked roasts and steaks should be at least 145°F, and whole poultry should be 180°F. Convert these temperatures to degrees Celsius. Round up to the next degree. 63°C, 83°C

89. **TAKING A SHOWER** When you take a shower, which water temperature would you choose: 15°C, 38°C, or 50°C? 38°C

90. **DRINKING WATER** To get a cold drink of water, which temperature would you choose: −2°C, 10°C, or 25°C? 10°C

91. **SNOWY WEATHER** At which temperatures might it snow: −5°C, 0°C, or 10°C? −5°C and 0°C

92. **AIR CONDITIONING** At which outside temperature would you be likely to run the air conditioner: 15°, 20°C, or 30°C? 30°C

93. **COMPARISON SHOPPING** Which is the better buy: 3 quarts of root beer for $4.50 or 2 liters of root beer for $3.60? the 3 quarts

94. **COMPARISON SHOPPING** Which is the better buy: 3 gallons of antifreeze for $10.35 or 12 liters of antifreeze for $10.50? the 12 liters

WRITING

95. Explain how to change kilometers to miles.
96. Explain how to change 50°C to degrees Fahrenheit.
97. The United States is the only industrialized country in the world that does not officially use the metric system. Some people claim this is costing American businesses money. Do you think so? Why?
98. What is meant by the phrase *a table of equivalent measures*?

REVIEW

Perform each operation.

99. $\frac{3}{5} + \frac{4}{3}$ $\frac{29}{15}$
100. $\frac{3}{5} - \frac{4}{3}$ $-\frac{11}{15}$
101. $\frac{3}{5} \cdot \frac{4}{3}$ $\frac{4}{5}$
102. $\frac{3}{5} \div \frac{4}{3}$ $\frac{9}{20}$
103. $3.25 + 4.8$ 8.05
104. $3.25 - 4.8$ −1.55
105. $3.25 \cdot 4.8$ 15.6
106. $4.8\overline{)15.6}$ 3.25

STUDY SKILLS CHECKLIST

Proportions and Unit Conversion Factors

Before taking the test on Chapter 6, make sure that you have a solid understanding of how to write proportions and how to choose unit conversion factors. Put a checkmark in the box if you can answer "yes" to the statement.

☐ When writing a proportion, I know that the units of the numerators must be the same and the units of the denominators must be the same.

This proportion is correctly written:

$$\text{Ounces} \rightarrow \frac{150}{x} = \frac{3}{2.75} \leftarrow \text{Ounces}$$
$$\text{Cost} \rightarrow \qquad \qquad \leftarrow \text{Cost}$$

This proportion is incorrectly written:

$$\text{Ounces} \rightarrow \frac{50}{x} = \frac{2.75}{3} \leftarrow \text{Cost}$$
$$\text{Cost} \rightarrow \qquad \qquad \leftarrow \text{Ounces}$$

☐ When converting from one unit to another, I know that I must choose a unit conversion factor with the following form:

$$\frac{\text{Unit I want to introduce}}{\text{Unit I want to eliminate}}$$

For example, in the following conversion of 15 pints to cups, the units of pints are eliminated and the units of cups are introduced by choosing the unit conversion factor $\frac{2\text{ c}}{1\text{ pt}}$.

$$15\text{ pt} = \frac{15\text{ pt}}{1} \cdot \frac{2\text{ c}}{1\text{ pt}} = 30\text{ c}$$

Teaching Guide: Refer to the Instructor's Resource Binder to find activities, worksheets on key concepts, more examples, instruction tips, overheads, and assessments.

CHAPTER 6   SUMMARY AND REVIEW

SECTION 6.1   Ratios and Rates

DEFINITIONS AND CONCEPTS	EXAMPLES
Ratios are often used to describe important relationships between two quantities. A **ratio** is the quotient of two numbers or the quotient of two quantities that have the same units. Ratios are written in three ways: as fractions, in words separated by the word *to*, and using a colon.	The ratio 4 **to** 5 can be written as $\frac{4}{5}$. The ratio 5 **:** 12 can be written as $\frac{5}{12}$.
To **write a ratio as a fraction**, write the first number (or quantity) mentioned as the numerator and the second number (or quantity) mentioned as the denominator. Then simplify the fraction, if possible.	Write the ratio 30 to 36 as a fraction in simplest form. The word *to* separates the numbers to be compared. $\frac{30}{36} = \frac{5 \cdot \overset{1}{\cancel{6}}}{6 \cdot \underset{1}{\cancel{6}}}$ To simplify, factor 30 and 36. Then remove the common factor of 6 from the numerator and denominator. $= \frac{5}{6}$

To write a **ratio in simplest form**, remove any common factors of the numerator and denominator as well as any common units.	Write the ratio *14 feet: 2 feet* as a fraction in simplest form. A colon separates the quantities to be compared. $$\frac{14 \text{ feet}}{2 \text{ feet}} = \frac{\overset{1}{\cancel{2}} \cdot 7 \text{ feet}}{\underset{1}{\cancel{2} \text{ feet}}}$$ To simplify, factor 14. Then remove the common factor of 2 and the common units of feet from the numerator and denominator. $$= \frac{7}{1}$$ Since a ratio compares two numbers, we leave the result in fractional form. Do not simplify further.
To **simplify ratios involving decimals**, multiply the ratio by a form of 1 so that the numerator and denominator become whole numbers. Then simplify, if possible.	Write the ratio 0.23 to 0.71 as a fraction in simplest form. To write this as a ratio of *whole numbers*, we need to move the decimal points in the numerator and denominator two places to the right. This will occur if they are both multiplied by 100. $$\frac{0.23}{0.71} = \frac{0.23}{0.71} \cdot \frac{100}{100}$$ Multiply the ratio by a form of 1. $$= \frac{0.23 \cdot 100}{0.71 \cdot 100}$$ Multiply the numerators. Multiply the denominators. $$= \frac{23}{71}$$ To find the product of each decimal and 100, simply move the decimal point two places to the right. The resulting fraction is in simplest form.
To **simplify ratios involving mixed numbers**, use the method for simplifying complex fractions from Section 4.7. Perform the division indicated by the main fraction bar.	Write the ratio $3\frac{1}{3}$ to $4\frac{1}{6}$ as a fraction in simplest form. $$\frac{3\frac{1}{3}}{4\frac{1}{6}} = \frac{\frac{10}{3}}{\frac{25}{6}}$$ Write $3\frac{1}{3}$ and $4\frac{1}{6}$ as improper fractions. $$= \frac{10}{3} \div \frac{25}{6}$$ Write the division indicated by the main fraction bar using a ÷ symbol. $$= \frac{10}{3} \cdot \frac{6}{25}$$ Use the rule for dividing fractions: Multiply the first fraction by the reciprocal of $\frac{25}{6}$, which is $\frac{6}{25}$. $$= \frac{10 \cdot 6}{3 \cdot 25}$$ Multiply the numerators. Multiply the denominators. $$= \frac{2 \cdot \overset{1}{\cancel{5}} \cdot 2 \cdot \overset{1}{\cancel{3}}}{\underset{1}{\cancel{3}} \cdot \underset{1}{\cancel{5}} \cdot 5}$$ To simplify the fraction, factor 10, 6, and 25. Then remove the common factors 3 and 5. $$= \frac{4}{5}$$ Multiply the remaining factors in the numerator. Multiply the remaining factors in the denominator.
When a ratio compares two quantities, both quantities must be measured in the **same units**. When the units are different, it's usually easier to write the ratio using the smaller unit of measurement.	Write the ratio *5 inches to 2 feet* as a fraction in simplest form. Since inches are smaller than feet, compare in inches: 5 **inches** to 24 **inches** Because 2 feet = 24 inches. Next, write the ratio in fraction form and simplify. $$\frac{5 \text{ inches}}{24 \text{ inches}} = \frac{5}{24}$$ Remove the common units of inches.

Chapter 6 Summary and Review

When we compare two quantities that have different units (and neither unit can be converted to the other), we call the comparison a **rate**. To **write a rate as a fraction**, write the first quantity mentioned as the numerator and the second quantity mentioned as the denominator, and then simplify, if possible. Write the units as part of the fraction. Words such as *per*, *for*, *in*, *from*, and *on* are used to separate the two quantities that are compared in a rate.	Write the rate *33 miles in 6 hours* as a fraction in simplest form. 33 miles **in** 6 hours can be written as $\dfrac{33 \text{ miles}}{6 \text{ hours}}$ $\dfrac{33 \text{ miles}}{6 \text{ hours}} = \dfrac{\overset{1}{\cancel{3}} \cdot 11 \text{ miles}}{2 \cdot \underset{1}{\cancel{3}} \text{ hours}}$ To simplify, factor 33 and 6. Then remove the common factor of 3 from the numerator and denominator. $= \dfrac{11 \text{ miles}}{2 \text{ hours}}$ Write the units as part of the rate. The rate can be written as 11 miles per 2 hours.
A **unit rate** is a rate in which the denominator is 1. To **write a rate as a unit rate**, divide the numerator of the rate by the denominator. A **slash mark** / is often used to write a unit rate.	Write as a unit rate: 2,490 apples from 6 trees. To find the unit rate, divide 2,490 by 6. $\begin{array}{r} 415 \\ 6\overline{)2{,}490} \end{array}$ The unit rate is $\dfrac{415 \text{ apples}}{1 \text{ tree}}$. This rate can also be expressed as: $415 \dfrac{\text{apples}}{\text{tree}}$, 415 apples per tree, or 415 apples/tree.
A **unit price** is a rate that tells how much is paid for *one* unit (or *one* item). It is the quotient of price to the number of units. Unit price = $\dfrac{\text{price}}{\text{number of units}}$ Comparison shopping can be made easier by finding **unit prices**. The best buy is the item that has the lowest unit price.	Which is the better buy for shampoo? 12 ounces for $3.84 or 16 ounces for $4.64 To find the unit price of a bottle of shampoo, write the quotient of its price and its capacity, and then perform the indicated division. Before dividing, convert each price from dollars to cents so that the unit price can be expressed in cents per ounce. $\dfrac{\$3.84}{12 \text{ oz}} = \dfrac{384¢}{12 \text{ oz}}$ | $\dfrac{\$4.64}{16 \text{ oz}} = \dfrac{464¢}{16 \text{ oz}}$ $= 32¢ \text{ per oz}$ | $= 29¢ \text{ per oz}$ One ounce of shampoo for 29¢ is better than one ounce for 32¢. Thus, the 16-ounce bottle is the better buy.

REVIEW EXERCISES

Write each ratio as a fraction in simplest form.

1. 7 to 25 $\dfrac{7}{25}$
2. 15:16 $\dfrac{15}{16}$
3. 24 to 36 $\dfrac{2}{3}$
4. 21:14 $\dfrac{3}{2}$
5. 4 inches to 12 inches $\dfrac{1}{3}$
6. 63 meters to 72 meters $\dfrac{7}{8}$
7. 0.28 to 0.35 $\dfrac{4}{5}$
8. 5.1:1.7 $\dfrac{3}{1}$
9. $2\dfrac{1}{3}$ to $2\dfrac{2}{3}$ $\dfrac{7}{8}$
10. $4\dfrac{1}{6}:3\dfrac{1}{3}$ $\dfrac{5}{4}$
11. 15 minutes : 3 hours $\dfrac{1}{12}$
12. 8 ounces to 2 pounds $\dfrac{1}{4}$

Write each rate as a fraction in simplest form.

13. 64 centimeters in 12 years $\dfrac{16 \text{ cm}}{3 \text{ yr}}$
14. $15 for 25 minutes $\dfrac{\$3}{5 \text{ min}}$

Write each rate as a unit rate.

15. 600 tickets in 20 minutes   30 tickets per min
16. 45 inches every 3 turns   15 inches per turn
17. 195 feet in 6 rolls   32.5 feet per roll
18. 48 calories in 15 pieces   3.2 calories per piece

Find the unit price of each item.

19. 5 pairs cost $11.45. $2.29 per pair
20. $3 billion in a 12-month span $0.25 billion per month
21. AIRCRAFT Specifications for a Boeing B-52 Stratofortress are shown below. What is the ratio of the airplane's wingspan to its length? $\frac{37}{32}$

Crew: 6
Length: 160 ft
Wingspan: 185 ft
Maximum takeoff weight: 488,000 lb
Maximum speed: 595 mph
Maximum altitude: more than 50,000 ft
Range: 7,500 mi

22. PAY RATES Find the hourly rate of pay for a student who earned $333.25 for working 43 hours. $7.75
23. CROWD CONTROL After a concert is over, it takes 48 minutes for a crowd of 54,000 people to exit a stadium. Find the unit rate of people exiting the stadium. 1,125 people per min
24. COMPARISON SHOPPING Mixed nuts come packaged in a 12-ounce can, which sells for $4.95, or an 8-ounce can, which sells for $3.25. Which is the better buy? the 8-oz can

SECTION 6.2 Proportions

DEFINITIONS AND CONCEPTS	EXAMPLES
A **proportion** is a statement that two ratios or two rates are equal.	Write each statement as a proportion. 6 is to 10 as 3 is to 5 $\frac{6}{10} = \frac{3}{5}$ The word "to" is used to separate the numbers to be compared in a ratio (or rate). $300 is to 500 minutes as $3 is to 5 minutes $\frac{\$300}{500 \text{ minutes}} = \frac{\$3}{5 \text{ minutes}}$
Each of the four numbers in a proportion is called a **term**. The first and fourth terms are called the **extremes**, and the second and third terms are called the **means**.	First term (extreme) Third term (mean) $\frac{1}{2} = \frac{3}{6}$ Second term (mean) Fourth term (extreme)
Since a proportion is an equation, **a proportion can be true or false**. A proportion is true if its ratios (or rates) are equivalent and false if its ratios (or rates) are not equivalent. One way to determine whether a proportion is true or false is to use the fraction simplifying skills of Chapter 4. The two products found by multiplying diagonally in a proportion are called **cross products**. Another way to determine whether a proportion is true or false involves the cross products. If the **cross products are equal**, the proportion is true. If the cross products are *not equal*, the proportion is false.	Determine whether the proportion $\frac{3}{5} = \frac{15}{27}$ is true or false. **Method 1** Simplify any ratios in the proportion that are not in simplest form. Then compare them to determine whether they are equal. $\frac{15}{27} = \frac{\overset{1}{3} \cdot 5}{\underset{1}{3} \cdot 9} = \frac{5}{9}$ Simplify the ratio on the right side. Since the ratios on the left and right sides of the proportion are not equal, the proportion is false. **Method 2** Check to see whether the cross products are equal. Cross products $3 \cdot 27 = 81$ $5 \cdot 15 = 75$ $\frac{3}{5} \bowtie \frac{15}{27}$ Since the cross products are not equal, the proportion is not true.

When two pairs of numbers form a proportion, we say that they are **proportional**.	Determine whether 0.7, 0.3 and 2.1, 0.9 are proportional. Write two ratios and form a proportion. Then find the cross products. $$\frac{0.7}{0.3} = \frac{2.1}{0.9} \qquad 0.7 \cdot 0.9 = 0.63 \qquad 0.3 \cdot 2.1 = 0.63$$ Since the cross products are equal, the numbers are proportional.
Solving a proportion to find an unknown term: 1. Set the cross products equal to each other to form an equation. 2. Isolate the variable on one side of the equation by dividing both sides by the number that is multiplied by that variable. 3. Check by substituting the result into the original proportion and finding the cross products.	Solve the proportion: $\frac{5}{37.5} = \frac{2}{x}$ $\frac{5}{37.5} = \frac{2}{x}$ This is the proportion to solve. $5 \cdot x = 37.5 \cdot 2$ Set the cross products equal to each other to form an equation. $5x = 75$ On the right side, do the multiplication: $37.5 \cdot 2 = 75$. $\frac{5x}{5} = \frac{75}{5}$ To isolate x, undo the multiplication by 5 by dividing both sides by 5. $x = 15$ Do the division: $75 \div 5 = 15$. Thus, x is 15. Check this result in the original proportion by finding the cross products.
Proportions can be used to **solve application problems**. It is easy to spot problems that can be solved using a proportion. You will be given a ratio (or rate) and asked to find the missing part of another ratio (or rate). It is helpful to follow the **six-step problem-solving strategy** seen earlier in the text to solve proportion problems.	**PEANUT BUTTER** It takes 360 peanuts to make 8 ounces of peanut butter. How many peanuts does it take to make 12 ounces? (Source: National Peanut Board) **Analyze** - We can express the fact that it takes 360 peanuts to make 8 ounces of peanut butter as a rate: $\frac{360 \text{ peanuts}}{8 \text{ ounces}}$. - How many peanuts does it take to make 12 ounces? **Assign** We will let the variable p represent the unknown number of peanuts. **Form** 360 peanuts is **to** 8 ounces **as** p peanuts is **to** 12 ounces. Number of peanuts → $\frac{360}{8} = \frac{p}{12}$ ← Number of peanuts Ounces of peanut butter → ← Ounces of peanut butter **Solve** To find the number of peanuts needed, solve the proportion for p. $360 \cdot 12 = 8 \cdot p$ Set the cross products equal to each other to form an equation. $4{,}320 = 8p$ On the left side, do the multiplication: $360 \cdot 12 = 4{,}320$. $\frac{4{,}320}{8} = \frac{8p}{8}$ To isolate p, undo the multiplication by 8 by dividing both sides by 8. $540 = p$ Do the division: $4{,}320 \div 8 = 540$. **State** It takes 540 peanuts to make 12 ounces of peanut butter. **Check** 16 ounces of peanut butter would require twice as many peanuts as 8 ounces: $2 \cdot 360$ peanuts $= 720$ peanuts. It seems reasonable that 12 ounces would require 540 peanuts.

REVIEW EXERCISES

25. Write each statement as a proportion.
  a. 20 is to 30 as 2 is to 3. $\frac{20}{30}=\frac{2}{3}$
  b. 6 buses replace 100 cars as 36 buses replace 600 cars. $\frac{6\text{ buses}}{100\text{ cars}}=\frac{36\text{ buses}}{600\text{ cars}}$

26. Complete the cross products.
$2 \cdot 27 = 54 \qquad 9 \cdot 6 = 54$
$$\frac{2}{9} = \frac{6}{27}$$

Determine whether each proportion is true or false by simplifying.

27. $\frac{8}{12}=\frac{3}{7}$ false

28. $\frac{4}{18}=\frac{10}{45}$ true

Determine whether each proportion is true or false by finding cross products.

29. $\frac{9}{27}=\frac{2}{6}$ true

30. $\frac{17}{7}=\frac{51}{21}$ true

31. $\frac{3.5}{9.3}=\frac{1.2}{3}$ false

32. $\frac{1\frac{1}{2}}{3\frac{1}{3}}=\frac{\frac{1}{4}}{1\frac{1}{7}}$ false

Determine whether the numbers are proportional.

33. 5, 9 and 20, 36 yes

34. 7, 13 and 29, 54 no

Solve each proportion.

35. $\frac{12}{18}=\frac{3}{x}$ 4.5

36. $\frac{4}{x}=\frac{2}{8}$ 16

37. $\frac{4.8}{6.6}=\frac{x}{9.9}$ 7.2

38. $\frac{0.08}{x}=\frac{0.04}{0.06}$ 0.12

39. $\frac{1\frac{9}{11}}{x}=\frac{3\frac{1}{3}}{2\frac{3}{4}}$ $1\frac{1}{2}$

40. $\frac{\frac{4}{5}}{1\frac{1}{20}}=\frac{2\frac{2}{3}}{x}$ $3\frac{1}{2}$

41. $\frac{\frac{2}{3}}{\frac{1}{2}}=\frac{x}{0.25}$ $\frac{1}{3}$

42. $\frac{x}{300}=\frac{5,000}{1,500}$ 1,000

43. TRUCKS A Dodge Ram pickup truck can go 35 miles on 2 gallons of gas. How far can it go on 11 gallons? 192.5 mi

44. QUALITY CONTROL In a manufacturing process, 12 parts out of 66 were found to be defective. How many defective parts will be expected in a run of 1,650 parts? 300

45. SCALE DRAWINGS The illustration below shows an architect's drawing of a kitchen using a scale of $\frac{1}{8}$ inch to 1 foot $\left(\frac{1}{8}'' : 1'0''\right)$. On the drawing, the length of the kitchen is $1\frac{1}{2}$ inches. How long is the actual kitchen? (The symbol ″ means inch and ′ means foot.) 12 ft

ELEVATION B-B
SCALE: $\frac{1''}{8}$ to 1'0"

46. DOGS The American Kennel Club website gives the ideal length-to-height proportions for a German Shepherd as $10 : 8\frac{1}{2}$. What is the ideal length of a German Shepherd that is $25\frac{1}{2}$ inches high at the shoulder? 30 in.

Chapter 6 Summary and Review

SECTION 6.3 American Units of Measurement

DEFINITIONS AND CONCEPTS	EXAMPLES
The **American system of measurement** uses the units of **inch**, **foot**, **yard**, and **mile** to measure **length**. A **ruler** is one of the most common tools for measuring lengths. Most rulers are 12 inches long. Each inch is divided into halves of an inch, quarters of an inch, eighths of an inch, and sixteenths of an inch.	1 ft = 12 in. 1 yd = 3 ft 1 yd = 36 in. 1 mi = 5,280 ft Since the black tick marks between 0 and 1 on the ruler create sixteen equal spaces, the ruler is scaled in sixteenths.
To convert from one unit of length to another, we use **unit conversion factors**. They are called unit conversion factors because their value is 1. Multiplying a measurement by a unit conversion factor does not change the measure; it only changes the units of the measure. A list of unit conversion factors for American units of length is given on page 589.	Convert 4 yards to inches. To convert from yards to inches, we select a unit conversion factor that introduces the units of inches and eliminates the units of yards. Since there are 36 inches per yard, we will use: $\dfrac{36 \text{ in.}}{1 \text{ yd}}$ ← This is the unit we want to introduce. ← This is the unit we want to eliminate (the original unit). To perform the conversion, we multiply. $4 \text{ yd} = \dfrac{4 \text{ yd}}{1} \cdot \dfrac{36 \text{ in.}}{1 \text{ yd}}$ Write 4 yd as a fraction. Then multiply by a form of 1: $\dfrac{36 \text{ in.}}{1 \text{ yd}}$. $= \dfrac{4 \cancel{\text{ yd}}}{1} \cdot \dfrac{36 \text{ in.}}{1 \cancel{\text{ yd}}}$ Remove the common units of yards from the numerator and denominator. The units of inches remain. $= 4 \cdot 36 \text{ in.}$ Simplify. $= 144 \text{ in.}$ Do the multiplication. Thus, 4 yards = 144 inches.
The American system of measurement uses the units of **ounce**, **pound**, and **ton** to measure **weight**.	1 lb = 16 oz 1 ton = 2,000 lb
A list of unit conversion factors for American units of weight is given on page 591.	Convert 9,000 pounds to tons. To convert from pounds to tons, we select a unit conversion factor that introduces the units of tons and eliminates the units of pounds. Since there is 1 ton for every 2,000 pounds, we will use: $\dfrac{1 \text{ ton}}{2,000 \text{ lb}}$ ← This is the unit we want to introduce. ← This is the unit we want to eliminate (the original unit).

Unless otherwise noted, all art on this page is © Cengage Learning.

To perform the conversion, we multiply.

$$9{,}000 \text{ lb} = \frac{9{,}000 \text{ lb}}{1} \cdot \frac{1 \text{ ton}}{2{,}000 \text{ lb}}$$

Write 9,000 lb as a fraction. Then multiply by a form of 1: $\frac{1 \text{ ton}}{2{,}000 \text{ lb}}$.

$$= \frac{9{,}000 \cancel{\text{lb}}}{1} \cdot \frac{1 \text{ ton}}{2{,}000 \cancel{\text{lb}}}$$

Remove the common units of pounds from the numerator and denominator. The units of tons remains.

$$= \frac{9{,}000}{2{,}000} \text{ ton}$$

Multiply the fractions.

There are two ways to complete the solution. First, we can remove any common factors of the numerator and denominator to simplify the fraction. Then we can write the result as a mixed number.

$$\frac{9{,}000}{2{,}000} \text{ tons} = \frac{9 \cdot \cancel{1{,}000}}{2 \cdot \cancel{1{,}000}} \text{ tons} = \frac{9}{2} \text{ tons} = 4\frac{1}{2} \text{ tons}$$

A second approach is to divide the numerator by the denominator and express the result as a decimal.

$$\frac{9{,}000}{2{,}000} \text{ tons} = 4.5 \text{ tons}$$

Thus, 9,000 pounds is equal to $4\frac{1}{2}$ tons (or 4.5 tons).

The American system of measurement uses the units of **ounce**, **cup**, **pint**, **quart**, and **gallon** to measure **capacity**.	1 c = 8 fl oz 1 qt = 2 pt	1 pt = 2 c 1 gal = 4 qt
A list of unit conversion factors for American units of capacity is given on page 593. Some conversions require the use of **two** (or more) **unit conversion factors**.	Convert 5 gallons to pints. There is not a single unit conversion factor that converts from gallons to pints. We must use two unit conversion factors. Since there are 4 quarts per gallon, we can convert 5 gallons to quarts by multiplying by the unit conversion factor $\frac{4 \text{ qt}}{1 \text{ gal}}$. Since there are 2 pints per quart, we can convert that result to pints by multiplying by the unit conversion factor $\frac{2 \text{ pt}}{1 \text{ qt}}$. $5 \text{ gal} = \frac{5 \text{ gal}}{1} \cdot \frac{4 \text{ qt}}{1 \text{ gal}} \cdot \frac{2 \text{ pt}}{1 \text{ qt}}$ $= \frac{5 \cancel{\text{gal}}}{1} \cdot \frac{4 \cancel{\text{qt}}}{1 \cancel{\text{gal}}} \cdot \frac{2 \text{ pt}}{1 \cancel{\text{qt}}}$ Remove the common units of gallons and quarts in the numerator and denominator. The units of pints remain. $= 40 \text{ pt}$ Do the multiplication: $5 \cdot 4 \cdot 2 = 40$. Thus, 5 gallons = 40 pints.	
The American and metric systems of measurement both use the units of **seconds**, **minutes**, **hours**, and **days** to measure time.	1 min = 60 sec 1 hr = 60 min 1 day = 24 hr	

A list of unit conversion factors for units of time is given on page 594.

Convert 240 minutes to hours.

To convert from minutes to hours, we select a unit conversion factor that introduces the units of hours and eliminates the units of minutes. Since there is 1 hour for every 60 minutes, we will use:

$\dfrac{1 \text{ hr}}{60 \text{ min}}$ ← This is the unit we want to introduce.
← This is the unit we want to eliminate (the original unit).

To perform the conversion, we multiply.

$240 \text{ min} = \dfrac{240 \text{ min}}{1} \cdot \dfrac{1 \text{ hr}}{60 \text{ min}}$ Write 240 min as a fraction. Then multiply by a form of 1: $\dfrac{1 \text{ hr}}{60 \text{ min}}$.

$= \dfrac{240 \cancel{\text{ min}}}{1} \cdot \dfrac{1 \text{ hr}}{60 \cancel{\text{ min}}}$ Remove the common units of minutes from the numerator and denominator. The units of hours remain.

$= \dfrac{240}{60} \text{ hr}$ Multiply the fractions.

$= 4 \text{ hr}$ Do the division.

Thus, 240 minutes = 4 hours.

REVIEW EXERCISES

47. a. Refer to the ruler below. Each inch is divided into how many equal parts? 16
b. Determine which measurements the arrows point to on the ruler. $\frac{7}{16}$ in., $1\frac{1}{2}$ in., $1\frac{3}{4}$ in., $2\frac{5}{8}$ in.

48. Use a ruler to measure the length of the computer mouse. $1\frac{1}{2}$ in.

49. Write two unit conversion factors using the fact that 1 mile = 5,280 ft. $\frac{1 \text{ mi}}{5{,}280 \text{ ft}} = 1$, $\frac{5{,}280 \text{ ft}}{1 \text{ mi}} = 1$

50. Consider the work shown below.

$\dfrac{100 \text{ min}}{1} \cdot \dfrac{60 \text{ sec}}{1 \text{ min}}$

a. What units can be removed? min
b. What units remain? sec

Perform each conversion.

51. 5 yards to feet   15 ft
52. 6 yards to inches   216 in.
53. 66 inches to feet $5\frac{1}{2}$ ft = 5.5 ft
54. 9,240 feet to miles $1\frac{3}{4}$ mi = 1.75 mi
55. $4\frac{1}{2}$ feet to inches   54 in.
56. 1 mile to yards   1,760 yd
57. 32 ounces to pounds   2 lb
58. 17.2 pounds to ounces   275.2 oz
59. 3 tons to ounces   96,000 oz
60. 4,500 pounds to tons $2\frac{1}{4}$ tons = 2.25 tons
61. 5 pints to fluid ounces   80 fl oz
62. 8 cups to gallons $\frac{1}{2}$ gal = 0.5 gal
63. 17 quarts to cups   68 c
64. 176 fluid ounces to quarts   5.5 qt
65. 5 gallons to pints   40 pt
66. 3.5 gallons to cups   56 c
67. 20 minutes to seconds   1,200 sec
68. 900 seconds to minutes   15 min
69. 200 hours to days $8\frac{1}{3}$ days
70. 6 hours to minutes   360 min
71. 4.5 days to hours   108 hr
72. 1 day to seconds   86,400 sec

73. Convert 210 yards to miles. Give the exact answer and a decimal approximation, rounded to the nearest hundredth. $\frac{21}{176}$ mi ≈ 0.12 mi

74. TRUCKING Large concrete trucks can carry roughly 40,500 pounds of concrete. Express this weight in tons. $20\frac{1}{4}$ tons = 20.25 tons

75. SKYSCRAPERS The Willis Tower in Chicago is 1,454 feet high. Express this height in yards. $484\frac{2}{3}$ yd

76. BOTTLING A farmer sells apple cider in 2-quart bottles. How many bottles will he need to hold 50 gallons of cider? 100

SECTION 6.4 Metric Units of Measurement

DEFINITIONS AND CONCEPTS	EXAMPLES
The basic metric unit of measurement is the **meter**, which is abbreviated **m**. Longer and shorter metric units are created by adding **prefixes** to the front of the basic unit, meter.	*kilo* means thousands *deci* means tenths *hecto* means hundreds *centi* means hundredths *deka* means tens *milli* means thousandths *micro* means millionths
Common metric units of length are the **kilometer, hectometer, dekameter, decimeter, centimeter,** and **millimeter.** Abbreviations are often used when writing these units. See the table on page 600.	1 km = 1,000 m 1 m = 10 dm 1 hm = 100 m 1 m = 100 cm 1 dam = 10 m 1 m = 1,000 mm
A **metric ruler** can be used for measuring lengths. On most metric rulers, each centimeter is divided into 10 millimeters.	(metric ruler showing 2 cm, 43 mm, 65 mm)
To convert from one metric unit of length to another, we use **unit conversion factors**.	Convert 4 meters to centimeters. To convert from meters to centimeters, we select a unit conversion factor that introduces the units of centimeters and eliminates the units of meters. Since there are 100 centimeters per meter, we will use: $\frac{100 \text{ cm}}{1 \text{ m}}$ ← This is the unit we want to introduce. ← This is the unit we want to eliminate (the original unit). To perform the conversion, we multiply. $4 \text{ m} = \frac{4 \text{ m}}{1} \cdot \frac{100 \text{ cm}}{1 \text{ m}}$ Write 4 m as a fraction. Then multiply by a form of 1: $\frac{100 \text{ cm}}{1 \text{ m}}$. $= \frac{4 \cancel{\text{m}}}{1} \cdot \frac{100 \text{ cm}}{1 \cancel{\text{m}}}$ Remove the common units of meters from the numerator and denominator. The units of cm remain. $= 400 \text{ cm}$ Multiply the fractions and simplify. Thus, 4 meters = 400 centimeters.

The **mass** of an object is a measure of the amount of material in the object. Common metric units of mass are the **kilogram**, **hectogram**, **dekagram**, **decigram**, **centigram**, and **milligram**. Abbreviations are often used when writing these units. See the table on page 605.	1 kg = 1,000 g 1 g = 10 dg 1 hg = 100 g 1 g = 100 cg 1 dag = 10 g 1 g = 1,000 mg 1 g = 1,000,000 mcg
Converting from one metric unit to another can be done using **unit conversion factors** or a **conversion chart**. In a conversion chart, the units are listed from largest to smallest, reading left to right. We **count the places** and note the **direction** as we move from the original units to the conversion units.	Convert 820 grams to kilograms. To use a conversion chart, locate the original units of grams and move to the conversion units of kilograms. To go from grams to kilograms, we must move 3 places to the left. If we write 820 grams as 820.0 grams, we can convert to kilograms by moving the decimal point three places to the left. 820.0 grams = 0.8200 kilograms = 0.82 kilograms The unit conversion factor method gives the same result: $$820 \text{ g} = \frac{820 \text{ g}}{1} \cdot \frac{1 \text{ kg}}{1,000 \text{ g}}$$ $$= \frac{820}{1,000} \text{ kg}$$ $$= 0.82 \text{ kg}$$ Thus, 820 grams = 0.82 kilograms.
Common metric units of capacity are the **kiloliter**, **hectoliter**, **dekaliter**, **deciliter**, **centiliter**, and **milliliter**. Abbreviations are often used when writing these units. See the table on page 608.	1 kL = 1,000 L 1 L = 10 dL 1 hL = 100 L 1 L = 100 cL 1 daL = 10 L 1 L = 1,000 mL
Converting from one metric unit to another can be done using **unit conversion factors** or a **conversion chart**.	Convert 0.7 kiloliters to milliliters. To use a conversion chart, locate the original units of kiloliters and move to the conversion units of milliliters. To go from kiloliters to milliliters, we must move 6 places to the right. We can convert to milliliters by moving the decimal point six places to the right. 0.7 kiloliters = 0700000. milliliters = 700,000 milliliters

The unit conversion factor method gives the same result:

$$0.7 \text{ kL} = \frac{0.7 \text{ kL}}{1} \cdot \frac{1{,}000 \text{ L}}{1 \text{ kL}} \cdot \frac{1{,}000 \text{ mL}}{1 \text{ L}}$$
$$= 0.7 \cdot 1{,}000 \cdot 1{,}000 \text{ mL}$$
$$= 700{,}000 \text{ mL}$$

Thus, 0.7 kiloliters = 700,000 milliliters.

Another metric unit of capacity is the **cubic centimeter**, written cm^3, or, more simply, cc. One cc represents the same capacity as one milliliter. The units of cubic centimeters are used frequently in medicine.

1 milliliter = 1 cm^3 = 1 cc
5 milliliters = 5 cm^3 = 5 cc
0.6 milliliters = 0.6 cm^3 = 0.6 cc

REVIEW EXERCISES

77. a. Refer to the metric ruler below. Each centimeter is divided into how many equal parts? What is the length of one of those parts? 10, 1 millimeter

b. Determine which measurements the arrows point to on the ruler. 19 mm, 3 cm, 45 mm, 62 mm

78. Use a metric ruler to measure the length of the computer mouse to the nearest centimeter. 4 cm

79. Write two unit conversion factors using the given fact.
a. 1 km = 1,000 m $\frac{1 \text{ km}}{1{,}000 \text{ m}} = 1, \frac{1{,}000 \text{ m}}{1 \text{ km}} = 1$
b. 1 g = 100 cg $\frac{1 \text{ g}}{100 \text{ cg}} = 1, \frac{100 \text{ cg}}{1 \text{ g}} = 1$

80. Use the chart to determine how many decimal places and in which direction to move the decimal point when converting from centimeters to kilometers. 5 places to the left

km	hm	dam	m	dm	cm	mm

Perform each conversion.

81. 475 centimeters to meters   4.75 m
82. 8 meters to millimeters   8,000 mm
83. 165.7 kilometers to meters   165,700 m
84. 6,789 centimeters to decimeters   678.9 dm
85. 25 micrograms to milligrams   0.025 mg
86. 800 centigrams to grams   8 g
87. 5,425 grams to kilograms   5.425 kg
88. 5,425 grams to milligrams   5,425,000 mg
89. 150 centiliters to liters   1.5 L
90. 3,250 liters to kiloliters   3.25 kL
91. 400 milliliters to centiliters   40 cL
92. 1 hectoliter to deciliters   1,000 dL
93. THE BRAIN The adult human brain weighs about 1,350 g. Convert the weight to kilograms.   1.35 kg
94. TEST TUBES A rack holds one dozen 20 mL test tubes. Find the total capacity of the test tubes in the rack in liters.   0.24 L
95. TYLENOL A bottle of Extra-Strength Tylenol contains 100 caplets of 500 milligrams each. How many grams of Tylenol are in the bottle?   50 g
96. SURGERY A dextrose solution is being administered to a patient intravenously as shown to the right. How many milliliters of solution does the IV bag hold?   1,000 mL

SECTION 6.5 Converting between American and Metric Units

DEFINITIONS AND CONCEPTS	EXAMPLES
We **convert between American and metric units** of length using the facts on the right. In all but one case, the conversions are rounded approximations.	**American to metric** 1 in. = 2.54 cm 1 ft ≈ 0.30 m 1 yd ≈ 0.91 m 1 mi ≈ 1.61 km **Metric to American** 1 cm ≈ 0.39 in. 1 m ≈ 3.28 ft 1 m ≈ 1.09 yd 1 km ≈ 0.62 mi
Unit conversion factors can be formed from the facts in the table above right to make specific conversions between American and metric units of length.	Convert 15 inches to centimeters. To convert from inches to centimeters, we select a unit conversion factor that introduces the units of centimeters and eliminates the units of inches. Since there are 2.54 centimeters for every inch, we will use: $\dfrac{2.54 \text{ cm}}{1 \text{ in.}}$ ← This is the unit we want to introduce. ← This is the unit we want to eliminate (the original unit). To perform the conversion, we multiply. $15 \text{ in.} = \dfrac{15 \text{ in.}}{1} \cdot \dfrac{2.54 \text{ cm}}{1 \text{ in.}}$ Write 15 in. as a fraction. Then multiply by a form of 1: $\dfrac{2.54 \text{ cm}}{1 \text{ in.}}$. $= \dfrac{15 \text{ in.}}{1} \cdot \dfrac{2.54 \text{ cm}}{1 \text{ in.}}$ Remove the common units of inches from the numerator and denominator. The units of cm remain. $= 15 \cdot 2.54 \text{ cm}$ Simplify. $= 38.1 \text{ cm}$ Do the multiplication. Thus, 15 inches = 38.1 centimeters.
We **convert between American and metric units** of mass (weight) using the facts on the right. The conversions are rounded approximations.	**American to metric** 1 oz ≈ 28.35 g 1 lb ≈ 0.45 kg **Metric to American** 1 g ≈ 0.035 oz 1 kg ≈ 2.20 lb
Unit conversion factors can be formed from the facts in the table above right to make specific conversions between American and metric units of mass (weight).	Convert 6 kilograms to ounces. There is not a single unit conversion factor that converts from kilograms to ounces. We must use **two unit conversion factors**—one to convert kilograms to grams, and another to convert that result to ounces. $6 \text{ kg} \approx \dfrac{6 \text{ kg}}{1} \cdot \dfrac{1{,}000 \text{ g}}{1 \text{ kg}} \cdot \dfrac{0.035 \text{ oz}}{1 \text{ g}}$ $\approx \dfrac{6 \text{ kg}}{1} \cdot \dfrac{1{,}000 \text{ g}}{1 \text{ kg}} \cdot \dfrac{0.035 \text{ oz}}{1 \text{ g}}$ Remove the common units of kilograms and grams in the numerator and denominator. The units of oz remain. $\approx 6 \cdot 1{,}000 \cdot 0.035 \text{ oz}$ Simplify. $\approx 6 \cdot 35 \text{ oz}$ Multiply the last two factors: $1{,}000 \cdot 0.035 = 35$. $\approx 210 \text{ oz}$ Do the multiplication. Thus, 6 kilograms ≈ 210 ounces.

	American to metric	Metric to American
We **convert between American and metric units** of capacity using the facts on the right. The conversions are rounded approximations.	1 fl oz ≈ 29.57 mL	1 L ≈ 33.81 fl oz
	1 pt ≈ 0.47 L	1 L ≈ 2.11 pt
	1 qt ≈ 0.95 L	1 L ≈ 1.06 qt
	1 gal ≈ 3.79 L	1 L ≈ 0.264 gal

Unit conversion factors can be formed from the facts in the table above right to make specific conversions between American and metric units of capacity.

Convert 5 fluid ounces to milliliters. Round to the nearest tenth.

To convert from fluid ounces to milliliters, we select a unit conversion factor that introduces the units of milliliters and eliminates the units of fluid ounces. Since there are 29.57 milliliters for every fluid ounce, we will use:

$$\frac{29.57 \text{ mL}}{1 \text{ fl oz}}$$ ← This is the unit we want to introduce.
← This is the unit we want to eliminate (the original unit).

To perform the conversion, we multiply.

$5 \text{ fl oz} \approx \frac{5 \text{ fl oz}}{1} \cdot \frac{29.57 \text{ mL}}{1 \text{ fl oz}}$ Write 5 fl oz as a fraction. Then multiply by a form of 1: $\frac{29.57 \text{ mL}}{1 \text{ fl oz}}$.

$\approx \frac{5 \cancel{\text{ fl oz}}}{1} \cdot \frac{29.57 \text{ mL}}{1 \cancel{\text{ fl oz}}}$ Remove the common units of fluid ounces from the numerator and denominator. The units of mL remain.

$\approx 5 \cdot 29.57 \text{ mL}$ Simplify.

$\approx 147.85 \text{ mL}$ Do the multiplication.

$\approx 147.9 \text{ mL}$ Round to the nearest tenth.

Thus, 5 fluid ounces ≈ 147.9 milliliters.

In the American system, we measure temperature using **degrees Fahrenheit** (°F). In the metric system, we measure temperature using **degrees Celsius** (°C).

If F is the temperature in degrees Fahrenheit and C is the corresponding temperature in degrees Celsius, then

$$C = \frac{5}{9}(F - 32) \quad \text{and} \quad F = \frac{9}{5}C + 32$$

Convert 92°F to degrees Celsius. Round to the nearest tenth of a degree.

$C = \frac{5}{9}(F - 32)$ This is the formula to find degrees Celsius.

$= \frac{5}{9}(92 - 32)$ Substitute 92 for F.

$= \frac{5}{9}(60)$ Do the subtraction within the parentheses first.

$= \frac{5}{9}\left(\frac{60}{1}\right)$ Write 60 as a fraction.

$= \frac{300}{9}$ Multiply the numerators: 5 · 60 = 300. Multiply the denominators.

$= 33.333\ldots$ Do the division.

≈ 33.3 Round to the nearest tenth.

Thus, 92°F ≈ 33.3°C.

REVIEW EXERCISES

97. SWIMMING Olympic-size swimming pools are 50 meters long. Express this distance in feet. 164 ft

98. HIGH-RISE BUILDINGS The Willis Tower (formerly the Sears Tower) is 443 meters high, and the Empire State Building is 1,250 feet high. Which building is taller? Willis Tower

99. WESTERN SETTLERS The Oregon Trail was an overland route pioneers used from the 1840s through the 1870s to reach the Oregon Territory. It stretched 1,930 miles from Independence, Missouri, to Oregon City, Oregon. Find this distance to the nearest kilometer. 3,107 km

100. U.S. PRESIDENTS Abraham Lincoln was the tallest of all presidents: 6 feet, 4 inches. Express his height in centimeters. Round to the nearest centimeter. 193 cm

Perform each conversion. Since most conversions are approximate, answers may vary slightly depending on the method used.

101. 30 ounces to grams 850.5 g

102. 15 kilograms to pounds 33 lb

103. 50 pounds to grams 22,680 g

104. 2,000 pounds to kilograms about 909 kg

105. POLAR BEARS At birth, polar bear cubs weigh less than human babies—about 910 grams. Convert this to pounds. about 2.0 lb

andamanec/Shutterstock.com

106. BOTTLED WATER LaCroix bottled water can be purchased in bottles containing 17 fluid ounces. Mountain Valley water can be purchased in half-liter bottles. Which bottle contains more water? LaCroix

107. CRUDE OIL There are 42 gallons in a barrel of crude oil. How many liters of crude oil is that? about 159.2 L

108. Convert 105°C to degrees Fahrenheit. 221°F

109. Convert 77°F to degrees Celsius. 25°C

110. RECREATION Which water temperature is appropriate for swimming: 10°C, 30°C, 50°C, or 70°C? 30°C

CHAPTER 6 TEST

1. Fill in the blanks.
   a. A __ratio__ is the quotient of two numbers or the quotient of two quantities that have the same units.
   b. A __rate__ is the quotient of two quantities that have different units.
   c. A __proportion__ is a statement that two ratios (or rates) are equal.
   d. The __cross__ products for the proportion $\frac{3}{8} = \frac{6}{16}$ are $3 \cdot 16$ and $8 \cdot 6$.
   e. *Deci* means __tenths__, *centi* means __hundredths__, and *milli* means __thousandths__.
   f. The meter, the gram, and the liter are basic units of measurement in the __metric__ system.
   g. In the American system, temperatures are measured in degrees __Fahrenheit__. In the metric system, temperatures are measured in degrees __Celsius__.

2. PIANOS A piano keyboard is made up of a total of eighty-eight keys, as shown below. What is the ratio of the number of black keys to white keys? $\frac{9}{13}$, 9:13, 9 to 13

Middle C

Write each ratio as a fraction in simplest form.

3. 6 feet to 8 feet $\frac{3}{4}$
4. 8 ounces to 3 pounds $\frac{1}{6}$
5. $0.26 : 0.65$ $\frac{2}{5}$
6. $3\frac{1}{3}$ to $3\frac{8}{9}$ $\frac{6}{7}$
7. Write the rate 54 feet in 36 seconds as a fraction in simplest form. $\frac{3\text{ feet}}{2\text{ seconds}}$
8. COMPARISON SHOPPING A 2-pound can of coffee sells for $3.38, and a 5-pound can of the same brand of coffee sells for $8.50. Which is the better buy? the 2-pound can
9. UTILITY COSTS A household used 675 kilowatt-hours of electricity during a 30-day month. Find the rate of electric usage in kilowatt-hours per day. 22.5 kwh per day
10. Write the following statement as a proportion: 15 billboards to 50 miles as 3 billboards to 10 miles. $\frac{15\text{ billboards}}{50\text{ miles}} = \frac{3\text{ billboards}}{10\text{ miles}}$

11. Determine whether each proportion is true.
    a. $\frac{25}{33} = \frac{2}{3}$ no
    b. $\frac{2.2}{3.5} = \frac{1.76}{2.8}$ yes

12. Are the numbers 7, 15 and 35, 75 proportional? yes

Solve each proportion.

13. $\frac{x}{3} = \frac{35}{7}$ 15
14. $\frac{15.3}{x} = \frac{3}{12.4}$ 63.24
15. $\frac{2\frac{2}{9}}{\frac{4}{3}} = \frac{x}{1\frac{1}{2}}$ $2\frac{1}{2}$
16. $\frac{25}{\frac{1}{10}} = \frac{50}{x}$ 0.2

17. SHOPPING If 13 ounces of tea costs $2.79, how much would you expect to pay for 16 ounces of tea? $3.43

18. BAKING A recipe calls for $1\frac{2}{3}$ cup of sugar and 5 cups of flour. How much sugar should be used with 6 cups of flour? 2 c

19. a. Refer to the ruler below. Each inch is divided into how many equal parts? 16
    b. Determine which measurements the arrows point to on the ruler. $\frac{5}{16}$ in., $1\frac{3}{8}$ in., $2\frac{3}{4}$ in.

20. Fill in the blanks. In general, a unit conversion factor is a fraction with the following form:

    Unit that we want to __introduce__ ← Numerator
    Unit that we want to __eliminate__ ← Denominator

21. Convert 180 inches to feet. 15 ft
22. TOOLS If a 25-foot tape measure is completely extended, how many yards does it stretch? Write your answer as a mixed number. $8\frac{1}{3}$ yd
23. Convert $10\frac{3}{4}$ pounds to ounces. 172 oz
24. AUTOMOBILES A car weighs 1.6 tons. Find its weight in pounds. 3,200 lb
25. CONTAINERS How many fluid ounces are in a 1-gallon carton of milk? 128 fl oz

26. **LITERATURE** An excellent work of early science fiction is the book *Around the World in 80 Days* by Jules Verne (1828–1905). Convert 80 days to minutes. 115,200 min

27. **a.** A quart and a liter of fruit punch are shown below. Which is the 1-liter carton: the larger one or the smaller one? the larger one

b. The figures below show the relative lengths of a yardstick and a meterstick. Which one represents the meterstick: the longer one or the shorter one? the longer one

c. One ounce and one gram weights are placed on a balance, as shown below. On which side is the gram: the left side or the right side? the right side

28. Determine which measurements the arrows point to on the metric ruler shown below. 12 mm, 5 cm, 65 mm

29. **SPEED SKATING** American Bonnie Blair won gold medals in the women's 500-meter speed skating competitions at the 1988, 1992, and 1994 Winter Olympic Games. Convert the race length to kilometers. 0.5 km

30. How many centimeters are in 5 meters? 500 cm

31. Convert 5,000 micrograms to milligrams. 5 mg

32. Convert 70 liters to milliliters. 70,000 mL

33. **PRESCRIPTIONS** A bottle contains 50 tablets, each containing 150 mg of medicine. How many grams of medicine does the bottle contain? 7.5 g

34. **TRACK** Which is the longer distance: a 100-yard race or an 80-meter race? the 100-yard race

35. **BODY WEIGHT** Which person is heavier: Jim, who weighs 160 pounds, or Ricardo, who weighs 71 kilograms? Jim

36. Convert 810 milliliters to quarts. Round to the nearest tenth. 0.9 qt

37. Convert 16.5 inches to centimeters. Round to the nearest centimeter. 42 cm

38. **COOKING MEAT** The USDA recommends that turkey be cooked to a temperature of 83°C. Change this to degrees Fahrenheit. To be safe, *round up* to the next degree. (*Hint:* $F = \frac{9}{5}C + 32$.) 182°F

39. What is a scale drawing? Give an example. A scale is a ratio (or rate) comparing the size of a drawing and the size of an actual object. For example, 1 inch to 6 feet (1 in.:6 ft).

40. Explain the benefits of the metric system of measurement as compared to the American system. It is easier to convert from one unit to another in the metric system because it is based on the number 10.

CHAPTERS 1–6 CUMULATIVE REVIEW

1. Write 5,764,502:
   a. in words five million, seven hundred sixty-four thousand, five hundred two
   b. in expanded notation [Section 1.1]
      $5{,}000{,}000 + 700{,}000 + 60{,}000 + 4{,}000 + 500 + 2$

2. BASKETBALL RECORDS On December 13, 1983, the Detroit Pistons and the Denver Nuggets played in the highest-scoring game in NBA history. See the game summary below. [Section 1.2]
   a. What was the final score? 186 to 184
   b. Which team won? Detroit
   c. What was the total number of points scored in the game? 370 points

	Quarter				Overtime			
	1	2	3	4	1	2	3	Total
Detroit	38	36	34	37	14	12	15	?
Denver	34	40	39	32	14	12	13	?

Source: ESPN.com

3. Subtract: $70{,}006 - 348$ [Section 1.2] 69,658
4. Multiply: $504 \cdot 729$ [Section 1.3] 367,416
5. Divide: $37\overline{)743}$ [Section 1.4] 20 R3
6. List the factors of 30, from smallest to largest. [Section 1.5] 1, 2, 3, 5, 6, 10, 15, 30
7. Find the prime factorization of 360. [Section 1.5] $2^3 \cdot 3^2 \cdot 5$
8. Find the LCM and the GCF of 20 and 28. [Section 1.6] 140, 4
9. Evaluate: $81 + 9\,[7^2 - 7(11 - 4)]$ [Section 1.7] 81
10. Solve each equation and check the result.
    a. $m + 158 = 203$ [Section 1.8] 45
    b. $\dfrac{n}{57} = 300$ [Section 1.9] 17,100
11. Place an $<$ or an $>$ symbol in the box to make a true statement: $-(-10)\ \boxed{<}\ |-11|$ [Section 2.1]
12. Evaluate: $(-12 + 6) + (-6 + 8)$ [Section 2.2] -4
13. GOLF Tiger Woods won the 100th U.S. Open in June of 2000 by the largest margin in the history of that tournament. If he shot 12 under par (-12) and the second-place finisher, Miguel Angel Jimenez, shot 3 over par ($+3$), what was Tiger's margin of victory? [Section 2.3] 15 shots

14. Evaluate: -3^2 and $(-3)^2$ [Section 2.4] $-9, 9$
15. Evaluate each expression, if possible. [Sections 2.2–2.5]
    a. $0 + (-8)$ -8
    b. $\dfrac{-8}{0}$ undefined
    c. $0 - |-8|$ -8
    d. $\dfrac{0}{-8}$ 0
    e. $0 - (-8)$ 8
    f. $0(-8)$ 0
16. Evaluate: $\dfrac{3 + 3\,[5(-6) - (1 - 10)]}{-1 + (-1)}$ [Section 2.6] 30
17. Estimate the value of the following expression by rounding each number to the nearest hundred. [Section 2.6]
    $-3{,}887 + (-5{,}806) + 4{,}701$ $-5{,}000$

Write and then solve an equation to answer the following question. [Section 2.7]

18. POLLS Three months before an election, a political candidate was 18 points behind in the polls. The election was held and the candidate won by 3 points. How much support did the candidate gain over the last three months? The candidate gained 21 points over the last three months.

Solve each equation and check the result. [Section 2.7]

19. $t - 4 = -8 - (-2)$ -2
20. $-3a + (-2) = 16$ -6
21. Translate each phrase to an algebraic expression. [Section 3.1]
    a. Twenty-nine less than the weight w $w - 29$
    b. The quotient of 10 and the cube of m $\dfrac{10}{m^3}$
22. Evaluate $\dfrac{-x + 3(1 + x)}{x^2 - 3}$ for $x = -2$ [Section 3.2] -1
23. a. Simplify: $-6(15a)$ [Section 3.3] $-90a$
    b. Multiply: $5(4x - 2y + 7)$ [Section 3.3] $20x - 10y + 35$
24. Combine like terms. [Section 3.4]
    a. $5x - 11x$ $-6x$
    b. $-4(x - 3y) + 5x - 2y$ $x + 10y$

Chapters 1–6 Cumulative Review

Solve each equation and check the result. [Section 3.5]

25. $z + 12 + z = 8z - 4 - 2z$ 4

26. $2y + 7 = 2 - (4y + 7)$ -2

Write and then solve an equation to answer the following question. [Section 3.6]

27. **SPRINKLERS** A landscaper buried a water line around a rectangular lawn to serve as a supply line for a sprinkler system. The length of the lawn is 3 times its width. If 144 feet of pipe was used to do the job, what are the width and the length of the lawn? The width of the lawn is 18 ft and the length is 54 ft.

28. Simplify: $-\dfrac{16}{20}$ [Section 4.1] $-\dfrac{4}{5}$

29. Express $\dfrac{9}{10}$ as an equivalent fraction with a denominator of $60a$. [Section 4.1] $\dfrac{54a}{60a}$

30. **GEOGRAPHY** Earth has a surface area of about 197,000,000 square miles. Use the information in the circle graph below to determine the number of square miles of Earth's surface covered by land. (Source: scienceclarified.com) [Section 4.2] 59,100,000 sq mi

Land covers about $\dfrac{3}{10}$ of the Earth's surface

Water covers about $\dfrac{7}{10}$ of the Earth's surface

31. What is the formula for the area of a triangle? [Section 4.2] $A = \tfrac{1}{2}bh$

32. Divide: $\dfrac{5}{10b^3} \div \dfrac{1}{2b^2}$ [Section 4.3] $\dfrac{1}{b}$

33. Subtract: $\dfrac{11}{12} - \dfrac{7}{15}$ [Section 4.4] $\dfrac{9}{20}$

34. Determine which fraction is larger: $\dfrac{19}{15}$ or $\dfrac{5}{4}$ [Section 4.4] $\dfrac{19}{15}$

35. Add: $\dfrac{5}{m} + \dfrac{7}{8}$ [Section 4.4] $\dfrac{40 + 7m}{8m}$

36. **HARDWARE** Find the length of the wood screw shown below. [Section 4.4] $\dfrac{31}{32}$ in.

Head: $\dfrac{5}{32}$ in.
Shank: $\dfrac{5}{16}$ in.
Thread: $\dfrac{1}{2}$ in.

37. Multiply: $-15\dfrac{1}{3}\left(-\dfrac{9}{20}\right)$ [Section 4.5] $6\dfrac{9}{10}$

38. **PAPER SHREDDERS** A paper shredder cuts paper into $\tfrac{1}{4}$-inch-wide strips. If an $8\tfrac{1}{2}$ in. by 11 in. piece of notebook paper is fed into the shredder lengthwise, as shown, into how many strips will it be shredded? [Section 4.5] 34 strips

39. **MOTORS** What is the difference in horsepower (hp) between the two motors shown? [Section 4.6] $\tfrac{3}{4}$ hp

Keyed shaft
$1\tfrac{1}{2}$ hp

Thru bolt mount
$\tfrac{3}{4}$ hp

40. Simplify: $\dfrac{4 - \dfrac{3}{4}}{-1\dfrac{7}{8}}$ [Section 4.7] $-\dfrac{26}{15} = -1\dfrac{11}{15}$

Solve each equation and check the result. [Section 4.8]

41. $\dfrac{7}{8}t = -28$ -32

42. $\dfrac{4}{5}x = \dfrac{3}{4}x + \dfrac{1}{2}$ 10

Write and then solve an equation to answer the following question. [Section 4.8]

43. **CONCERT TICKETS** Seven-eighths of the total number of tickets sold for a concert were ordered by mail. The remaining 200 tickets were purchased at the concert hall box office. How many tickets were sold for the concert? 1,600 tickets were sold for the concert.

44. Place an < or > symbol in the box to make a true statement. [Section 5.1]

    $-64.22 \;>\; -64.238$

45. Graph $-1\frac{3}{4}$, 2.25, -0.5, $\frac{11}{8}$, -3.2, and $\sqrt{9}$ on a number line. [Section 5.1]

46. Add: $-20.04 + 2.4$ [Section 5.2] -17.64

47. Subtract: $-8.08 - 15.3$ [Section 5.2] -23.38

48. Multiply: $2.5 \cdot 100$ [Section 5.3] 250

49. **AQUARIUMS** One gallon of water weighs 8.33 pounds. What is the weight of the water in an aquarium that holds 55 gallons of water? [Section 5.3] 458.15 lb

50. Divide: $2.5 \div 100$ [Section 5.4] 0.025

51. Evaluate the formula $t = \frac{d}{r}$ for $d = 107.95$ and $r = 8.5$. [Section 5.4] 12.7

52. Write $\frac{1}{12}$ as a decimal. [Section 5.5] $0.08\overline{3}$

53. **LUNCH MEATS** A shopper purchased $\frac{3}{4}$ pound of barbequed beef, priced at $8.60 per pound, and $\frac{2}{3}$ pound of ham, selling for $5.25 per pound. Find the total cost of these items. [Section 5.5] $9.95

54. Evaluate: $3\sqrt{25} + 4\sqrt{4}$ [Section 5.6] 23

Solve each equation and check the result. [Section 5.7]

55. $3.2x = 74.46 - 1.9x$ 14.6

56. $-5.2x = 108 - 6.1x$ 120

57. $-2(x - 2.1) = -2.4$ 3.3

58. $\frac{1}{5}x - 2.5 = -17.2$ -73.5

Write and then solve an equation to answer the following question. [Section 5.7]

59. **PETITION DRIVES** A worker for a political organization is to collect signatures for a petition drive. Her pay is $20 plus 5¢ per signature. How many signatures must she get to earn $60? She must get 800 signatures to earn $60.

60. Express the phrase "3 inches to 15 inches" as a ratio in simplest form. [Section 6.1] $\frac{1}{5}$

61. **BUILDING MATERIALS** Which is the better buy: a 94-pound bag of cement for $4.48 or a 100-pound bag of cement for $4.80? [Section 6.1] the 94-pound bag

62. Determine whether the proportion $\frac{25}{33} = \frac{12}{17}$ is true or false. [Section 6.2] false

63. **CAFFEINE** There are 55 milligrams of caffeine in 12 ounces of Mountain Dew. How many milligrams of caffeine are there in a super-size 44-ounce cup of Mountain Dew? Round to the nearest milligram. [Section 6.2] 202 mg

64. Solve the proportion: $\frac{x}{3} = \frac{35}{7}$ [Section 6.2] 15

65. **SURVIVAL GUIDE** [Section 6.3]
    a. A person can go without food for about 40 days. How many hours is this? 960 hr
    b. A person can go without water for about 3 days. How many minutes is that? 4,320 min
    c. A person can go without breathing oxygen for about 8 minutes. How many seconds is that? 480 sec

66. Convert 40 ounces to pounds. [Section 6.3] 2.5 lb

67. Convert 2.4 meters to millimeters. [Section 6.4] 2,400 mm

68. Convert 320 grams to kilograms. [Section 6.4] 0.32 kg

69. a. Which holds more: a 2-liter bottle or a 1-gallon bottle? [Section 6.5] 1 gal
    b. Which is longer: a meterstick or a yardstick? a meterstick

70. **BELTS** A leather belt made in Mexico is 92 centimeters long. Express the length of the belt to the nearest inch. [Section 6.5] 36 in.

7

Percent

7.1 Percents, Decimals, and Fractions
7.2 Solving Percent Problems Using Percent Equations and Proportions
7.3 Applications of Percent
7.4 Estimation with Percent
7.5 Interest
  Chapter Summary and Review
  Chapter Test
  Cumulative Review

from Campus to Careers
Loan Officer

Loan officers help people apply for loans. *Commercial loan officers* work with businesses, *mortgage loan officers* work with people who want to buy a house or other real estate, and *consumer loan officers* work with people who want to buy a boat, a car, or need a loan for college. Loan officers analyze the applicant's financial history and often use banking formulas to determine the possibility of granting a loan.

In **Problem 43** of **Study Set 7.5,** you will see how a credit union loan officer calculates the interest to be charged on a loan.

JOB TITLE: Loan Officer
EDUCATION: Most have a degree in finance, economics, or a similar field. Mathematics and computer classes are good preparation for this job.
JOB OUTLOOK: Employment of loan officers is expected to grow about as fast as the average (14%) for all jobs through 2020.
ANNUAL EARNINGS: In 2011, the average salary for a loan officer was $58,030.
FOR MORE INFORMATION:
www.bls.gov/oes/current/oes132072.htm

Objectives

1. Explain the meaning of percent.
2. Write percents as fractions.
3. Write percents as decimals.
4. Write decimals as percents.
5. Write fractions as percents.

SECTION 7.1
Percents, Decimals, and Fractions

We see percents everywhere, every day. Stores use them to advertise discounts, manufacturers use them to describe the contents of their products, and banks use them to list interest rates for loans and savings accounts. Newspapers are full of information presented in percent form. In this section, we introduce percents and show how fractions, decimals, and percents are related.

1 Explain the meaning of percent.

A percent tells us the number of parts per one hundred. You can think of a percent as the *numerator* of a fraction (or ratio) that has a denominator of 100.

Percent

Percent means parts per one hundred.

> **The Language of Algebra** The word *percent* is formed from the prefix *per*, which means ratio, and the suffix *cent*, which comes from the Latin word *centum*, meaning 100.
>
> $$\underset{\text{ratio}}{\text{per}} \bullet \underset{100}{\text{cent}}$$

In the figure below, there are 100 equal-sized square regions, and 93 of them are shaded. Thus, $\frac{93}{100}$ or 93 percent of the figure is shaded. The word *percent* can be written using the symbol %, so we say that 93% of the figure is shaded.

If the entire figure had been shaded, we would say that 100 out of the 100 square regions, or 100%, was shaded. Using this fact, we can determine what percent of the

Unless otherwise noted, all art on this page is © Cengage Learning.

figure is *not* shaded by subtracting the percent of the figure that is shaded from 100%.

$$100\% - 93\% = 7\%$$

So 7% of the figure is *not* shaded.

To illustrate a percent greater than 100%, say 121%, we would shade one entire figure and 21 of the 100 square regions in a second, equal-sized grid.

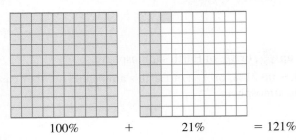

100% + 21% = 121%

EXAMPLE 1 Tossing a Coin
A coin was tossed 100 times and it landed heads up 51 times.

a. What percent of the time did the coin land heads up?
b. What percent of the time did it land tails up?

Strategy We will write a fraction that compares the number of times that the coin landed heads up (or tails up) to the total number of tosses.

WHY Since the denominator in each case will be 100, the numerator of the fraction will give the percent.

Solution
a. If a coin landed heads up 51 times after being tossed 100 times, then

$$\frac{51}{100} = 51\%$$

of the time it landed heads up.

b. The number of times the coin landed tails up is $100 - 51 = 49$ times. If a coin landed tails up 49 times after being tossed 100 times, then

$$\frac{49}{100} = 49\%$$

of the time it landed tails up.

Self Check 1
BOARD GAMES A standard Scrabble game contains 100 tiles. There are 42 vowel tiles, 2 blank tiles, and the rest are consonant tiles.

a. What percent of the tiles are vowels? 42%
b. What percent of the tiles are consonants? 56%

Now Try Problem 13

Teaching Example 1
GOVERNMENT In 2013, 20 of the 100 U.S. senators were women.
a. What percent of the senators were women?
b. What percent of the senators were men?
Answers:
a. 20% b. 80%

2 Write percents as fractions.

We can use the definition of percent to write any percent in an equivalent fraction form.

Writing Percents as Fractions

To write a percent as a fraction, drop the % symbol and write the given number over 100. Then simplify the fraction, if possible.

EXAMPLE 2 Earth
The chemical makeup of Earth's atmosphere is 78% nitrogen, 21% oxygen, and 1% other gases. Write each percent as a fraction in simplest form.

Strategy We will drop the % symbol and write the given number over 100. Then we will simplify the resulting fraction, if possible.

WHY *Percent* means parts per one hundred, and the word *per* indicates a ratio (fraction).

Self Check 2
WATERMELONS An average watermelon is 92% water. Write this percent as a fraction in simplest form. $\frac{23}{25}$

Now Try Problems 17 and 23

Unless otherwise noted, all art on this page is © Cengage Learning.

Teaching Example 2
ANATOMY The human body is about 65% water. Write this percent as a fraction in simplest form. (Source: Data from Guyton, Arthur C. [1976], *Textbook of Medical Physiology* (5th ed.), Philadelphia: W. B. Saunders, p. 424; Jackson, Sheila [1985], *Anatomy & Physiology for Nurses*, Nurses' Aids Series (9th ed.), London: Bailliere Tindall.)
Answer:
$\frac{13}{20}$

Solution We begin with nitrogen.

$$78\% = \frac{78}{100} \quad \text{Drop the \% symbol and write 78 over 100.}$$

$$= \frac{\overset{1}{\cancel{2}} \cdot 39}{\underset{1}{\cancel{2}} \cdot 50} \quad \text{To simplify the fraction, factor 78 as } 2 \cdot 39 \text{ and 100 as } 2 \cdot 50.\text{ Then remove the common factor of 2 from the numerator and denominator.}$$

$$= \frac{39}{50}$$

Nitrogen makes up $\frac{78}{100}$, or $\frac{39}{50}$, of Earth's atmosphere.

Oxygen makes up 21%, or $\frac{21}{100}$, of Earth's atmosphere. Other gases make up 1%, or $\frac{1}{100}$, of the atmosphere.

Self Check 3
UNIONS In 2002, 13.3% of the U.S. labor force belonged to a union. Write this percent as a fraction in simplest form. (Source: Bureau of Labor Statistics) $\frac{133}{1,000}$

Now Try Problems 27 and 31

Teaching Example 3
UNIONS In 1983, 20.1% of the U.S. labor force belonged to a union. Write this percent as a fraction in simplest form. (Source: Bureau of Labor Statistics)
Answer:
$\frac{201}{1,000}$

EXAMPLE 3 **Unions** In 2012, 11.3% of the U.S. labor force belonged to a union. Write this percent as a fraction in simplest form. (Source: Bureau of Labor Statistics)

Strategy We will drop the % symbol and write the given number over 100. Then we will multiply the resulting fraction by a form of 1 and simplify, if possible.

WHY When writing a percent as a fraction, the numerator and denominator of the fraction should be whole numbers that have no common factors (other than 1).

Solution
$$11.3\% = \frac{11.3}{100} \quad \text{Drop the \% symbol and write 11.3 over 100.}$$

To write this as an equivalent fraction of *whole numbers*, we need to move the decimal point in the numerator one place to the right. (Recall that to find the product of a decimal and 10, we simply move the decimal point one place to the right.) Therefore, it follows that $\frac{10}{10}$ is the form of 1 that we should use to build $\frac{11.3}{100}$.

$$\frac{11.3}{100} = \frac{11.3}{100} \cdot \frac{10}{10} \quad \text{Multiply the fraction by a form of 1.}$$

$$= \frac{11.3 \cdot 10}{100 \cdot 10} \quad \begin{array}{l}\text{Multiply the numerators.}\\ \text{Multiply the denominators.}\end{array}$$

$$= \frac{113}{1,000} \quad \begin{array}{l}\text{Since 113 and 1,000 do not have any common factors}\\ \text{(other than 1), the fraction is in simplest form.}\end{array}$$

Thus, $11.3\% = \frac{113}{1,000}$. This means that 113 out of every 1,000 workers in the U.S. labor force belonged to a union in 2012.

Self Check 4
Write $83\frac{1}{3}\%$ as a fraction in simplest form. $\frac{5}{6}$

Now Try Problem 35

Teaching Example 4 Write $16\frac{2}{3}\%$ as a fraction in simplest form.
Answer:
$\frac{1}{6}$

EXAMPLE 4 Write $66\frac{2}{3}\%$ as a fraction in simplest form.

Strategy We will drop the % symbol and write the given number over 100. Then we will perform the division indicated by the fraction bar and simplify, if possible.

WHY When writing a percent as a fraction, the numerator and denominator of the fraction should be whole numbers that have no common factors (other than 1).

Solution

$$66\frac{2}{3}\% = \frac{66\frac{2}{3}}{100} \quad \text{Drop the \% symbol and write } 66\frac{2}{3} \text{ over 100.}$$

To write this as a fraction of whole numbers, we will perform the division indicated by the fraction bar.

$$\frac{66\frac{2}{3}}{100} = 66\frac{2}{3} \div 100 \quad \text{The fraction bar indicates division.}$$

$$= \frac{200}{3} \cdot \frac{1}{100} \quad \text{Write } 66\frac{2}{3} \text{ as an improper fraction and then multiply by the reciprocal of 100.}$$

$$= \frac{200 \cdot 1}{3 \cdot 100} \quad \text{Multiply the numerators. Multiply the denominators.}$$

$$= \frac{2 \cdot \overset{1}{\cancel{100}} \cdot 1}{3 \cdot \underset{1}{\cancel{100}}} \quad \text{To simplify the fraction, factor 200 as } 2 \cdot 100. \text{ Then remove the common factor of 100 from the numerator and denominator.}$$

$$= \frac{2}{3}$$

EXAMPLE 5
a. Write 175% as a fraction in simplest form.
b. Write 0.22% as a fraction in simplest form.

Strategy We will drop the % symbol and write each given number over 100. Then we will simplify the resulting fraction, if possible.

WHY *Percent* means parts per one hundred and the word *per* indicates a ratio (fraction).

Solution

a. $175\% = \frac{175}{100}$ Drop the % symbol and write 175 over 100.

$$= \frac{\overset{1}{\cancel{5}} \cdot \overset{1}{\cancel{5}} \cdot 7}{2 \cdot 2 \cdot \underset{1}{\cancel{5}} \cdot \underset{1}{\cancel{5}}} \quad \text{To simplify the fraction, prime factor 175 and 100. Remove the common factors of 5 from the numerator and denominator.}$$

$$= \frac{7}{4}$$

5	175		2	100
5	35		2	50
	7		5	25
				5

Thus, $175\% = \frac{7}{4}$.

b. $0.22\% = \frac{0.22}{100}$ Drop the % symbol and write 0.22 over 100.

To write this as an equivalent fraction of *whole numbers*, we need to move the decimal point in the numerator two places to the right. (Recall that to find the product of a decimal and 100, we simply move the decimal point two places to the right.) Therefore, it follows that $\frac{100}{100}$ is the form of 1 that we should use to build $\frac{0.22}{100}$.

$$\frac{0.22}{100} = \frac{0.22}{100} \cdot \frac{100}{100} \quad \text{Multiply the fraction by a form of 1.}$$

$$= \frac{0.22 \cdot 100}{100 \cdot 100} \quad \text{Multiply the numerators. Multiply the denominators.}$$

$$= \frac{22}{10,000}$$

$$= \frac{\overset{1}{\cancel{2}} \cdot 11}{\underset{1}{\cancel{2}} \cdot 5,000} \quad \text{To simplify the fraction, factor 22 and 10,000. Remove the common factor of 2 from the numerator and denominator.}$$

$$= \frac{11}{5,000}$$

Thus, $0.22\% = \frac{11}{5,000}$.

Self Check 5
a. Write 210% as a fraction in simplest form. $\frac{21}{10}$
b. Write 0.54% as a fraction in simplest form. $\frac{27}{5,000}$

Now Try Problems 39 and 43

Teaching Example 5
a. Write 150% as a fraction in simplest form.
b. Write 0.42% as a fraction in simplest form.
Answers:
a. $\frac{3}{2}$
b. $\frac{21}{5,000}$

Success Tip When percents that are greater than 100% are written as fractions, the fractions are greater than 1. When percents that are less than 1% are written as fractions, the fractions are less than $\frac{1}{100}$.

3 Write percents as decimals.

To write a percent as a decimal, recall that a percent can be written as a fraction with denominator 100 and that a denominator of 100 indicates division by 100.

For example, consider 14%, which means 14 parts per 100.

$14\% = \frac{14}{100}$ Use the definition of percent: write 14 over 100.

     $= 14 \div 100$ The fraction bar indicates division.

     $= 14.0 \div 100$ Write the whole number 14 in decimal notation by placing a decimal point immediately to its right and entering a zero to the right of the decimal point.

     $= .140$ Since the divisor 100 has two zeros, move the decimal point two places to the left.

     $= 0.14$ Write a zero to the left of the decimal point.

We have found that 14% = 0.14. This example suggests the following procedure.

Writing Percents as Decimals

To write a percent as a decimal, drop the % symbol and divide the given number by 100 by moving the decimal point two places to the left.

Self Check 6
a. Write 16.43% as a decimal.
b. Write 2.06% as a decimal.

Now Try Problems 51 and 55

Self Check 6 Answers
a. 0.1643
b. 0.0206

Teaching Example 6
SOCIAL NETWORKS Refer to the graph on the right.
a. Write the percent of market share of visits for YouTube as a decimal.
b. Write the percent of market share of visits for Google+ as a decimal.
Answers:
a. 0.1980 = 0.198   b. 0.0135

EXAMPLE 6   *Social Networks*   The graph below shows the percent of market share of visits for the top five social networking sites.

a. Write the percent of market share for the Facebook site as a decimal.
b. Write the percent of market share for the Twitter site as a decimal.

Strategy We will drop the % symbol and divide each given number by 100 by moving the decimal point two places to the left.

WHY Recall from Section 5.4 that to find the quotient of a decimal and 10, 100, 1,000, and so on, move the decimal point to the left the same number of places as there are zeros in the power of 10.

Top Five Social Network Sites
by Market Share of Visits (%)
(for week ended October 13, 2012)

Site	%
Facebook	62.89%
YouTube	19.80%
Twitter	1.85%
Google+	1.35%
Pinterest	1.27%

Source: dreamgrow.com

Solution
a. From the graph, we see that the percent market share of visits for Facebook is 62.89%. To write this percent as a decimal, we proceed as follows.

$62.89\% = .6289$ Drop the % symbol and divide 62.89 by 100 by moving the decimal point two places to the left.

      $= 0.6289$ Write a zero to the left of the decimal point.

62.89%, written as a decimal, is 0.6289.

Unless otherwise noted, all art on this page is © Cengage Learning.

b. From the graph, we see that the percent market share of visits for Twitter is 1.85%. To write this percent as a decimal, we proceed as follows.

$1.85\% = .0185$ Drop the % symbol and divide 1.85 by 100 by moving the decimal point two places to the left. This requires that a placeholder zero (shown in blue) be inserted in front of the 1.

$= 0.0185$ Write a zero to the left of the decimal point.

1.85%, written as a decimal, is 0.0185.

EXAMPLE 7 *Population*

The population of the state of Oregon is approximately $1\frac{1}{4}\%$ of the population of the United States. Write this percent as a decimal. (Source: U.S. Census Bureau)

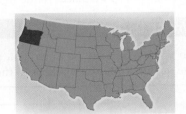

Strategy We will write the mixed number $1\frac{1}{4}$ in decimal notation.

WHY With $1\frac{1}{4}$ in mixed-number form, we cannot apply the rule for writing a percent as a decimal; there is no decimal point to move two places to the left.

Solution To change a percent to a decimal, we drop the percent symbol and divide by 100 by moving the decimal point two places to the left. In this case, however, there is no decimal point to move in $1\frac{1}{4}\%$. Since $1\frac{1}{4} = 1 + \frac{1}{4}$, and since the decimal equivalent of $\frac{1}{4}$ is 0.25, we can write $1\frac{1}{4}\%$ in an equivalent form as 1.25%.

$1\frac{1}{4}\% = 1.25\%$ Write $1\frac{1}{4}$ as 1.25.

$= .0125$ Drop the % symbol and divide 1.25 by 100 by moving the decimal point two places to the left. This requires that a placeholder zero (shown in blue) be inserted in front of the 1.

$= 0.0125$ Write a zero to the left of the decimal point.

$1\frac{1}{4}\%$, written as a decimal, is 0.0125.

Self Check 7

POPULATION The population of the state of Ohio is approximately $3\frac{3}{4}\%$ of the population of the United States. Write this percent as a decimal. (Source: U.S. Census Bureau) 0.0375

Now Try Problem 57

Teaching Example 7
POPULATION The population of the state of Virginia is approximately $2\frac{1}{2}\%$ of the population of the United States. Write this percent as a decimal. (Source: U.S. Census Bureau)
Answer:
0.025

EXAMPLE 8

a. Write 310% as a decimal.   **b.** Write 0.9% as a decimal.

Strategy We will drop the % symbol and divide each given number by 100 by moving the decimal point two places to the left.

WHY Recall that to find the quotient of a decimal and 100, we move the decimal point to the left the same number of places as there are zeros in 100.

Solution
a. $310\% = 310.0\%$ Write the whole number 310 in decimal notation: $310 = 310.0$.

$= 3.100$ Drop the % symbol and divide 310 by 100 by moving the decimal point two places to the left.

$= 3.1$ Drop the unnecessary zeros to the right of the 1.

310%, written as a decimal, is 3.1.

b. $0.9\% = .009$ Drop the % symbol and divide 0.9 by 100 by moving the decimal point two places to the left. This requires that a placeholder zero (shown in blue) be inserted in front of the 0.

$= 0.009$ Write a zero to the left of the decimal point.

0.9%, written as a decimal, is 0.009.

Self Check 8

a. Write 600% as a decimal. 6
b. Write 0.8% as a decimal. 0.008

Now Try Problems 63 and 67

Teaching Example 8
a. Write 540% as a decimal.
b. Write 0.1% as a decimal.
Answers:
a. 5.4 **b.** 0.001

Success Tip When percents that are greater than 100% are written as decimals, the decimals are greater than 1.0. When percents that are less than 1% are written as decimals, the decimals are less than 0.01.

4 Write decimals as percents.

To write a percent as a decimal, we drop the % symbol and move the decimal point two places to the left. To write a decimal as a percent, we do the opposite: we move the decimal point two places to the right and insert a % symbol.

Writing Decimals as Percents

To write a decimal as a percent, multiply the decimal by 100 by moving the decimal point two places to the right, and then insert a % symbol.

Self Check 9
Write 0.5343 as a percent. 53.43%

Now Try Problems 71 and 75

Teaching Example 9 Write 0.457 as a percent.
Answer:
45.7%

EXAMPLE 9 *Geography* Land areas make up 0.291 of Earth's surface. Write this decimal as a percent.

Strategy We will multiply the decimal by 100 by moving the decimal point two places to the right, and then insert a % symbol.

WHY To write a *decimal as a percent*, we reverse the steps used to write a *percent as a decimal*.

Solution

$0.291 = 029.1\%$ Multiply 0.291 by 100 by moving the decimal point two places to the right, and then insert a % symbol.

$= 29.1\%$

0.291, written as a percent, is 29.1%

5 Write fractions as percents.

We use a two-step process to write a fraction as a percent. First, we write the fraction as a decimal. Then we write that decimal as a percent.

Fraction ⟶ decimal ⟶ percent

Writing Fractions as Percents

To write a fraction as a percent:

1. Write the fraction as a decimal by dividing its numerator by its denominator.
2. Multiply the decimal by 100 by moving the decimal point two places to the right, and then insert a % symbol.

Self Check 10
Write 7 out of 8 as a percent.

Now Try Problem 79

Self Check 10 Answer
87.5%

EXAMPLE 10 *Television* The highest-rated television show of all time was a special episode of *M*A*S*H* that aired February 28, 1983. Surveys found that three out of every five American households watched this show. Express the rating as a percent. (Source: *The World Almanac and Book of Facts*, 2013)

Strategy First, we will translate the phrase *three out of every five* to fraction form and write that fraction as a decimal. Then we will write that decimal as a percent.

WHY A fraction-to-decimal-to-percent approach must be used to write a fraction as a percent.

Solution
Step 1 The phrase *three out of every five* can be expressed as $\frac{3}{5}$. To write this fraction as a decimal, we divide the numerator, 3, by the denominator, 5.

$$\begin{array}{r} 0.6 \\ 5\overline{)3.0} \\ -3\ 0 \\ \hline 0 \end{array}$$ Write a decimal point and one additional zero to the right of 3.

← The remainder is 0.

The result is a terminating decimal.

Step 2 To write 0.6 as a percent, we proceed as follows.

$$\frac{3}{5} = 0.6$$

0.6 = 0̰6̰0̰.%   Write a placeholder 0 to the right of the 6 (shown in blue). Multiply 0.60 by 100 by moving the decimal point two places to the right, and then insert a % symbol.

$$= 60\%$$

60% of American households watched the special episode of *M*A*S*H*.

TM and Copyright © 20th Century Fox Film Corp. All rights reserved. Courtesy: Everett Collection.

M*A*S*H was a highly popular situation comedy that aired on CBS from 1972 through 1983. It was set during the Korean War and focused on the relationships between members of a Mobile Army Surgical Hospital unit.

Teaching Example 10
ADVERTISING There is a commercial that says, "Four out of five dentists recommend sugarless gum for their patients who chew gum." Express this recommendation as a percent.
Answer:
80%

EXAMPLE 11   Write $\frac{13}{4}$ as a percent.

Strategy We will write the fraction $\frac{13}{4}$ as a decimal. Then we will write that decimal as a percent.

WHY A fraction-to-decimal-to-percent approach must be used to write a fraction as a percent.

Solution
Step 1 To write $\frac{13}{4}$ as a decimal, we divide the numerator, 13, by the denominator, 4.

$$\begin{array}{r} 3.25 \\ 4\overline{)13.00} \\ -12 \\ \hline 1\ 0 \\ -\ 8 \\ \hline 20 \\ -20 \\ \hline 0 \end{array}$$ Write a decimal point and two additional zeros to the right of 3.

← The remainder is 0.

The result is a terminating decimal.

Step 2 To write 3.25 as a percent, we proceed as follows.

3.25 = 325̰.%   Multiply 3.25 by 100 by moving the decimal point two places to the right, and then insert a % symbol.

$$= 325\%$$

The fraction $\frac{13}{4}$, written as a percent, is 325%.

Self Check 11
Write $\frac{5}{2}$ as a percent.   250%
Now Try Problem 85

Teaching Example 11
Write $\frac{15}{4}$ as a percent.
Answer:
375%

> **Success Tip** When fractions that are greater than 1 are written as percents, the percents are greater than 100%.

In Examples 10 and 11, the result of the division was a terminating decimal. Sometimes when we write a fraction as a decimal, the result of the division is a repeating decimal.

Self Check 12

Write $\frac{2}{3}$ as a percent. Give the exact answer and an approximation to the nearest tenth of one percent. $66\frac{2}{3}\% \approx 66.7\%$

Now Try Problem 91

Teaching Example 12 Write $\frac{1}{3}$ as a percent. Give the exact answer and an approximation to the nearest tenth of one percent.
Answer:
$33\frac{1}{3}\% \approx 33.3\%$

EXAMPLE 12

Write $\frac{5}{6}$ as a percent. Give the exact answer and an approximation to the nearest tenth of one percent.

Strategy We will write the fraction $\frac{5}{6}$ as a decimal. Then we will write that decimal as a percent.

WHY A fraction-to-decimal-to-percent approach must be used to write a fraction as a percent.

Solution

Step 1 To write $\frac{5}{6}$ as a decimal, we divide the numerator, 5, by the denominator, 6.

$$
\begin{array}{r}
0.8333 \\
6\overline{)5.0000} \\
-4\,8 \\
\hline
20 \\
-18 \\
\hline
20 \\
-18 \\
\hline
20 \\
-18 \\
\hline
2
\end{array}
$$

Write a decimal point and several zeros to the right of 5.

← The repeating pattern is now clear. We can stop the division.

The result is a repeating decimal.

Step 2 To write the decimal as a percent, we proceed as follows.

$$\frac{5}{6} = 0.8333\ldots$$

$0.833\ldots = 0\,83.33\ldots\%$ Multiply $0.8333\ldots$ by 100 by moving the decimal point
$ = 83.33\ldots\%$ two places to the right, and then insert a % symbol.

We must now decide whether we want an exact answer or an approximation. For an exact answer, we can represent *the repeating part of the decimal using an equivalent fraction.* For an approximation, we can round $83.333\ldots\%$ to a specific place value.

Exact answer:

$\frac{5}{6} = 83.\underline{3333\ldots}\%$

$= 83\frac{1}{3}\%$ Use the fraction $\frac{1}{3}$ to represent $.3333\ldots$.

Thus,

$\frac{5}{6} = 83\frac{1}{3}\%$

Approximation:

$\frac{5}{6} = 83.33\ldots\%$

$\approx 83.3\%$ Round to the nearest tenth.

Thus,

$\frac{5}{6} \approx 83.3\%$

Some percents occur so frequently that it is useful to memorize their fractional and decimal equivalents.

Percent	Decimal	Fraction
1%	0.01	$\frac{1}{100}$
10%	0.1	$\frac{1}{10}$
$16\frac{2}{3}\%$	$0.1666\ldots$	$\frac{1}{6}$
20%	0.2	$\frac{1}{5}$
25%	0.25	$\frac{1}{4}$

Percent	Decimal	Fraction
$33\frac{1}{3}\%$	$0.3333\ldots$	$\frac{1}{3}$
50%	0.5	$\frac{1}{2}$
$66\frac{2}{3}\%$	$0.6666\ldots$	$\frac{2}{3}$
75%	0.75	$\frac{3}{4}$
$83\frac{1}{3}\%$	$0.8333\ldots$	$\frac{5}{6}$

Unless otherwise noted, all art on this page is © Cengage Learning.

ANSWERS TO SELF CHECKS

1. a. 42% **b.** 56% **2.** $\frac{23}{25}$ **3.** $\frac{133}{1,000}$ **4.** $\frac{5}{6}$ **5. a.** $\frac{21}{10}$ **b.** $\frac{27}{5,000}$ **6. a.** 0.1643 **b.** 0.0206
7. 0.0375 **8. a.** 6 **b.** 0.008 **9.** 53.43% **10.** 87.5% **11.** 250% **12.** $66\frac{2}{3}\% \approx 66.7\%$

SECTION 7.1 STUDY SET

VOCABULARY

Fill in the blanks.

1. __Percent__ means parts per one hundred.
2. The word *percent* is formed from the prefix *per*, which means __ratio__, and the suffix *cent*, which comes from the Latin word *centum*, meaning __100__.

CONCEPTS

Fill in the blanks.

3. To write a percent as a fraction, drop the % symbol and write the given number over __100__. Then __simplify__ the fraction, if possible.
4. To write a percent as a decimal, drop the % symbol and divide the given number by 100 by moving the decimal point two places to the __left__.
5. To write a decimal as a percent, multiply the decimal by 100 by moving the decimal point two places to the __right__, and then insert a % symbol.
6. To write a fraction as a percent, first write the fraction as a __decimal__. Then multiply the decimal by 100 by moving the decimal point two places to the right, and then insert a __%__ symbol.

NOTATION

7. What does the symbol % mean? percent
8. Write the whole number 45 as a decimal. 45.0

GUIDED PRACTICE

What percent of the figure is shaded? What percent of the figure is not shaded? See Objective 1.

9. 84%, 16%
10. 7%, 93%

In the following illustrations, each set of 100 square regions represents 100%. What percent is shaded?

11.   107%

12.   249%

For Problems 13–16, see Example 1.

13. **THE INTERNET** The following sentence appeared on a technology blog: "Ask Internet users what they want from their service and 99 times out of 100 the answer will be the same: more speed." According to the blog, what percent of the time do Internet users give that answer? 99%

14. **BASKETBALL RECORDS** In 1962, Wilt Chamberlain of the Philadelphia Warriors scored a total of 100 points in an NBA game. If 28 of his points came from made free throws, what percent of his point total came from free throws? 28%

15. **QUILTS** A quilt is made from 100 squares of colored cloth.
    a. If 15 of the squares are blue, what percent of the squares in the quilt are blue? 15%
    b. What percent of the squares are not blue? 85%

16. **DIVISIBILITY** Of the natural numbers from 1 through 100, only 14 of them are divisible by 7.
    a. What percent of the numbers are divisible by 7? 14%
    b. What percent of the numbers are not divisible by 7? 86%

Write each percent as a fraction. Simplify, if possible. See Example 2.

17. 17% $\frac{17}{100}$
18. 31% $\frac{31}{100}$
19. 91% $\frac{91}{100}$
20. 89% $\frac{89}{100}$
21. 4% $\frac{1}{25}$
22. 5% $\frac{1}{20}$
23. 60% $\frac{3}{5}$
24. 40% $\frac{2}{5}$

Write each percent as a fraction. Simplify, if possible. See Example 3.

25. 1.9% $\frac{19}{1,000}$
26. 2.3% $\frac{23}{1,000}$
27. 54.7% $\frac{547}{1,000}$
28. 97.1% $\frac{971}{1,000}$
29. 12.5% $\frac{1}{8}$
30. 62.5% $\frac{5}{8}$
31. 6.8% $\frac{17}{250}$
32. 4.2% $\frac{21}{500}$

Write each percent as a fraction. Simplify, if possible. See Example 4.

33. $1\frac{1}{3}$% $\frac{1}{75}$
34. $3\frac{1}{3}$% $\frac{1}{30}$
35. $14\frac{1}{6}$% $\frac{17}{120}$
36. $10\frac{5}{6}$% $\frac{13}{120}$

Write each percent as a fraction. Simplify, if possible. See Example 5.

37. 130% $\frac{13}{10}$
38. 160% $\frac{8}{5}$
39. 220% $\frac{11}{5}$
40. 240% $\frac{12}{5}$
41. 0.35% $\frac{7}{2,000}$
42. 0.45% $\frac{9}{2,000}$
43. 0.25% $\frac{1}{400}$
44. 0.75% $\frac{3}{400}$

Write each percent as a decimal. See Objective 3.

45. 16% 0.16
46. 11% 0.11
47. 81% 0.81
48. 93% 0.93

Write each percent as a decimal. See Example 6.

49. 34.12% 0.3412
50. 27.21% 0.2721
51. 50.033% 0.50033
52. 40.083% 0.40083
53. 6.99% 0.0699
54. 4.77% 0.0477
55. 1.3% 0.013
56. 8.6% 0.086

Write each percent as a decimal. See Example 7.

57. $7\frac{1}{4}$% 0.0725
58. $9\frac{3}{4}$% 0.0975
59. $18\frac{1}{2}$% 0.185
60. $25\frac{1}{2}$% 0.255

Write each percent as a decimal. See Example 8.

61. 460% 4.6
62. 230% 2.3
63. 316% 3.16
64. 178% 1.78
65. 0.5% 0.005
66. 0.9% 0.009
67. 0.03% 0.0003
68. 0.06% 0.0006

Write each decimal or whole number as a percent. See Example 9.

69. 0.362 36.2%
70. 0.245 24.5%
71. 0.98 98%
72. 0.57 57%
73. 1.71 171%
74. 4.33 433%
75. 4 400%
76. 9 900%

Write each fraction as a percent. See Example 10.

77. $\frac{2}{5}$ 40%
78. $\frac{1}{5}$ 20%
79. $\frac{4}{25}$ 16%
80. $\frac{9}{25}$ 36%
81. $\frac{5}{8}$ 62.5%
82. $\frac{3}{8}$ 37.5%
83. $\frac{7}{16}$ 43.75%
84. $\frac{9}{16}$ 56.25%

Write each fraction as a percent. See Example 11.

85. $\frac{9}{4}$ 225%
86. $\frac{11}{4}$ 275%
87. $\frac{21}{20}$ 105%
88. $\frac{33}{20}$ 165%

Write each fraction as a percent. Give the exact answer and an approximation to the nearest tenth of one percent. See Example 12.

89. $\frac{1}{6}$ $16\frac{2}{3}\% \approx 16.7\%$
90. $\frac{2}{9}$ $22\frac{2}{9}\% \approx 22.2\%$
91. $\frac{5}{3}$ $166\frac{2}{3}\% \approx 166.7\%$
92. $\frac{4}{3}$ $133\frac{1}{3}\% \approx 133.3\%$

TRY IT YOURSELF

Complete the table. Give an exact answer and an approximation to the nearest tenth of one percent when necessary. Round decimals to the nearest hundredth when necessary.

	Fraction	Decimal	Percent
93.	$\frac{157}{5,000}$	0.0314	3.14%
94.	$\frac{21}{10,000}$	0.0021	0.21%
95.	$\frac{51}{125}$	0.408	40.8%
96.	$\frac{171}{500}$	0.342	34.2%
97.	$\frac{21}{400}$	0.0525	$5\frac{1}{4}$%
98.	$\frac{27}{400}$	0.0675	$6\frac{3}{4}$%
99.	$\frac{7}{3}$	2.33	$233\frac{1}{3}\% \approx 233.3\%$
100.	$\frac{7}{9}$	0.78	$77\frac{7}{9}\% \approx 77.8\%$

APPLICATIONS

101. **THE RED CROSS** A fact sheet released by the American Red Cross in 2012 stated, "An average of 91 cents of every dollar donated to the Red Cross is spent on services and programs." What percent of the money donated to the Red Cross went to services and programs? 91%

102. BIOLOGY Blood is 92% water. Write the percent as a fraction in simplest form. (Source: waterinfo.org) $\frac{92}{100} = \frac{23}{25}$

103. REGIONS OF THE COUNTRY The continental United States is divided into seven regions as shown below.

  a. What percent of the 50 states are in the Rocky Mountain region? 12%

  b. What percent of the 50 states are in the Southwestern region? 28%

  c. What percent of the 50 states are not located in any of the seven regions shown here? 4% (Alaska, Hawaii)

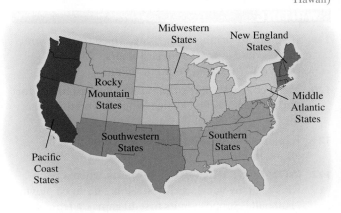

104. ROAD SIGNS Sometimes, signs like that shown below are posted to warn truckers when they are approaching a steep grade on the highway.

  a. Write the grade shown on the sign as a fraction. $\frac{5}{100} = \frac{1}{20}$

  b. Write the grade shown on the sign as a decimal. 0.05

105. INTEREST RATES Write each interest rate for the following accounts as a decimal.

  a. Home loan: 3.75% 0.0375

  b. Savings account: 2% 0.02

  c. Credit card: 14.25% 0.1425

106. DRUNK DRIVING In most states, it is illegal to drive with a blood alcohol concentration of 0.08% or higher.

  a. Write this percent as a fraction. Do not simplify. $\frac{8}{10,000}$

  b. Use your answer to part a to fill in the blanks: A blood alcohol concentration of 0.08% means 8 parts alcohol to 10,000 parts blood.

107. HUMAN SKIN The illustration below shows what percent of the total skin area that each section of the body covers. Find the missing percent for the torso, and then complete the bar graph. (Source: Burn Center at Sherman Oaks Hospital, American Medical Assn. Encyclopedia of Medicine) torso: 27.5%

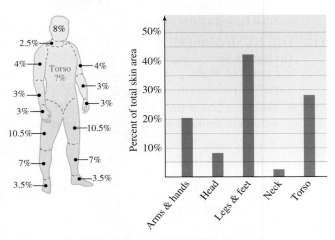

108. RAP MUSIC The table below shows what percent rap album sales were of total U.S. album sales of recorded music for the years 2006–2012.

2006	2007	2008	2009	2010	2011	2012
10.1%	8.3%	7.8%	7.1%	8.4%	8.2%	7.7%

Use the data to construct a line graph.

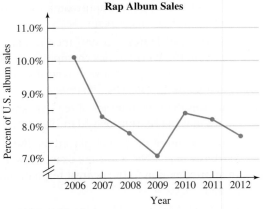

Source: businesswire.com

109. THE UN SECURITY COUNCIL The United Nations has 193 members. The United States, Russia, the United Kingdom, France, and China, along with ten other nations, make up the Security Council. (Source: *The World Almanac and Book of Facts*, 2013)

  a. What fraction of the members of the United Nations belong to the Security Council? $\frac{15}{193}$

  b. Write your answer to part a as a decimal. Round to the nearest ten-thousandth. 0.0777

  c. Write your answer to part b as a percent. 7.77%

110. SOAP Ivory soap claims to be $99\frac{44}{100}$% pure. Write this percent as a decimal. 0.9944

111. LOGOS In the illustration, what part of the company's logo is shaded red? Express your answer as a percent (exact), a fraction, and a decimal (using an overbar). $33\frac{1}{3}$%, $\frac{1}{3}$, $0.\overline{3}$

Recycling Industries Inc.

112. THE HUMAN SPINE The human spine consists of a group of bones (vertebrae) as shown.
   - 7 Cervical vertebrae
   - 12 Thoracic vertebrae
   - 5 Lumbar vertebrae
   - 1 Sacral vertebra
   - 4 Coccygeal vertebrae

   a. What fraction of the vertebrae are lumbar? $\frac{5}{29}$

   b. What percent of the vertebrae are lumbar? (Round to the nearest one percent.) 17%

   c. What percent of the vertebrae are cervical? (Round to the nearest one percent.) 24%

113. BOXING Oscar De La Hoya, also known as "the Golden Boy," generated over $600 million in pay-per-view revenue for his fights. He won 39 out of 45 of his professional fights.

   a. What fraction of his fights did he win? $\frac{13}{15}$

   b. What percent of his fights did he win? Give the exact answer and an approximation to the nearest tenth of one percent. $86\frac{2}{3}$% ≈ 86.7%

114. BASEBALL HISTORY In 1997, the Florida Marlins were the World Series champions. In 1998, they had the worst regular season record in major league baseball: 54 wins and 108 losses.

   a. What was the total number of regular season games that the Marlins played in 1998? 162 games

   b. What percent of the games played did the Marlins win in 1998? Give the exact answer and an approximation to the nearest tenth of one percent. $33\frac{1}{3}$% ≈ 33.3%

115. ECONOMIC FORECASTS One economic indicator of the national economy is the number of orders placed by manufacturers. One month, the number of orders rose *one-fourth of 1 percent*.

   a. Write this using a % symbol. $\frac{1}{4}$%

   b. Express it as a fraction. $\frac{1}{400}$

   c. Express it as a decimal. 0.0025

116. TAXES Several years ago, Springfield, Missouri, voters approved a *one-eighth of one percent* sales tax to fund transportation projects in the city.

   a. Write the percent as a decimal. 0.00125

   b. Write the percent as a fraction. $\frac{1}{800}$

117. BIRTHDAYS If the day of your birthday represents $\frac{1}{365}$ of a year, what percent of the year is it? Round to the nearest hundredth of a percent. 0.27%

118. POPULATION As a fraction, each resident of the United States represents approximately $\frac{1}{305{,}000{,}000}$ of the U.S. population. Express this as a percent. Round to one nonzero digit. 0.0000003%

WRITING

119. If you were writing advertising, which form do you think would attract more customers: "25% off" or "$\frac{1}{4}$ off"? Explain your reasoning.

120. Many coaches ask their players to give a 110% effort during practices and games. What do you think this means? Is it possible?

121. Explain how an amusement park could have an attendance that is 103% of capacity.

122. WON-LOST RECORDS In sports, when a team wins as many games as it loses, it is said to be playing "500 ball." Suppose in its first 40 games, a team wins 20 games and loses 20 games. Use the concepts in this section to explain why such a record could be called "500 ball."

REVIEW

123. The width of a rectangle is 6.5 centimeters and its length is 10.5 centimeters.
   a. Find its perimeter. 34 cm
   b. Find its area. 68.25 cm²

124. The length of a side of a square is 9.8 meters.
   a. Find its perimeter. 39.2 m
   b. Find its area. 96.04 m²

SECTION 7.2
Solving Percent Problems Using Percent Equations and Proportions

Objectives
PERCENT EQUATIONS (begins below)
1. Translate percent sentences to percent equations.
2. Solve percent equations to find the amount.
3. Solve percent equations to find the percent.
4. Solve percent equations to find the base.

PERCENT PROPORTIONS (see p. 664)
1. Write percent proportions.
2. Solve percent proportions to find the amount.
3. Solve percent proportions to find the percent.
4. Solve percent proportions to find the base.
5. Read circle graphs.

The articles on the front page of the newspaper below illustrate three types of percent problems.

Type 1 In the labor article, if we want to know how many union members voted to accept the new offer, we would ask:

  What number is 84% of 500?

Type 2 In the article on drinking water, if we want to know what percent of the wells are safe, we would ask:

  38 is what percent of 40?

Type 3 In the article on new appointees, if we want to know how many members are on the State Board of Examiners, we would ask:

  6 is 75% of what number?

DAILY NEWS
Circulation   Monday, March 23   50 cents

Transit Strike Averted!
Labor: 84% of 500-member union votes to accept new offer

Drinking Water
38 of 40 Wells Declared Safe

New Appointees
These six area residents now make up 75% of the State Board of Examiners

This section introduces two methods that can be used to solve the percent problems shown above. The first method involves writing and solving *percent equations*. The second method involves writing and solving *percent proportions*. If your instructor only requires you to learn the proportion method, then turn to page 664 and begin reading Objective 1.

To the instructor: Note that the two sets of objectives for this section are basically the same. Determine which method you want your students to use to solve percent problems: the percent equation method or the proportion method. You need not cover both. The Study Set problems for this section can be solved using either approach.

METHOD 1: PERCENT EQUATIONS

1 Translate percent sentences to percent equations.

The **percent sentences** highlighted in blue in the introduction above have three things in common.

- Each contains the word *is*. Here, *is* can be translated as an = symbol.
- Each contains the word *of*. In this case, *of* means multiply.
- Each contains a phrase such as *what number* or *what percent*. In other words, there is an unknown number that can be represented by a variable.

These observations suggest that each percent sentence contains key words that can be translated to form an equation. The equation, called a **percent equation**, will contain three numbers (two known and one unknown represented by a variable), the operation of multiplication, and, of course, an = symbol.

> **The Language of Algebra** The key words in a percent sentence translate as follows:
> - *is* translates to an equal symbol =
> - *of* translates to multiplication that is shown with a raised dot ·
> - *what number* or *what percent* translates to an unknown number that is represented by a variable

Self Check 1

Translate each percent sentence to a percent equation.

a. What number is 33% of 80?
b. What percent of 55 is 6?
c. 172% of what number is 4?

Now Try Problem 17

Self Check 1 Answers
a. $x = 33\% \cdot 80$
b. $x \cdot 55 = 6$
c. $172\% \cdot x = 4$

Teaching Example 1 Translate each percent sentence to a percent equation.
a. What number is 25% of 37?
b. What percent of 300 is 12?
c. 110% of what number is 341?
Answers:
a. $x = 25\% \cdot 37$
b. $x \cdot 300 = 12$
c. $110\% \cdot x = 341$

EXAMPLE 1
Translate each percent sentence to a percent equation.
a. What number is 12% of 64?
b. What percent of 88 is 11?
c. 165% of what number is 366?

Strategy We will look for the key words *is*, *of*, and *what number* (or *what percent*) in each percent sentence.

WHY These key words translate to mathematical symbols that form the percent equation.

Solution In each case, we will let the variable x represent the unknown number. However, any letter can be used.

a. What number is 12% of 64?   *This is the given percent sentence.*
   $x = 12\% \cdot 64$ *This is the percent equation.*

b. What percent of 88 is 11?   *This is the given percent sentence.*
   $x \cdot 88 = 11$ *This is the percent equation.*

c. 165% of what number is 366?   *This is the given percent sentence.*
   $165\% \cdot x = 366$ *This is the percent equation.*

2 Solve percent equations to find the amount.

To solve the labor union percent problem (Type 1 from the newspaper), we translate the percent sentence into a percent equation and then find the unknown number.

Self Check 2

What number is 36% of 400?   144

Now Try Problems 19 and 71

Teaching Example 2 What number is 28% of 300?
Answer:
84

EXAMPLE 2
What number is 84% of 500?

Strategy We will look for the key words *is*, *of*, and *what number* in the percent sentence and translate them to mathematical symbols to form a percent equation.

WHY Then it will be clear what operation should be performed to find the unknown number.

Solution First, we translate.

What number is 84% of 500?
$x = 84\% \cdot 500$ Translate to a percent equation.

Now we perform the multiplication on the right side of the equation.

$x = 0.84 \cdot 500$ Write 84% as a decimal: 84% = 0.84.
$x = 420$ Do the multiplication.

We have found that 420 is 84% of 500. That is, 420 union members mentioned in the newspaper article voted to accept the new offer.

> **The Language of Algebra**   When we find the value of the variable that makes a percent equation true, we say that we have **solved the equation**. In Example 2, we solved $x = 84\% \cdot 500$ to find that x is 420.

Caution! When solving percent equations, always write the percent as a decimal (or a fraction) before performing any calculations. In Example 2, we wrote 84% as 0.84 before multiplying by 500.

Percent sentences involve a comparison of numbers. In the statement "420 is 84% of 500," the number 420 is called the **amount**, 84% is the **percent**, and 500 is called the **base**. Think of the base as the standard of comparison—it represents the **whole** of some quantity. The amount is a **part** of the base, but it can exceed the base when the percent is more than 100%. The percent, of course, has the % symbol.

In any percent problem, the relationship between the amount, the percent, and the base is as follows: *Amount is percent of base*. This relationship is shown below as the **percent equation** (also called the **percent formula**).

Percent Equation (Formula)

Any percent sentence can be translated to a percent equation that has the form:

$$\text{Amount} = \text{percent} \cdot \text{base} \quad \text{or} \quad \text{Part} = \text{percent} \cdot \text{whole}$$

EXAMPLE 3   What number is 160% of 15.8?

Strategy We will look for the key words *is*, *of*, and *what number* in the percent sentence and translate them to mathematical symbols to form a percent equation.

WHY Then it will be clear what operation needs to be performed to find the unknown number.

Solution First, we translate.

What number is 160% of 15.8?
$$x = 160\% \cdot 15.8$$

x is the amount, 160% is the percent, and 15.8 is the base.

Now we solve the equation by performing the multiplication on the right side.

$x = 1.6 \cdot 15.8$ Write 160% as a decimal: 160% = 1.6.
$x = 25.28$ Do the multiplication.

```
    15.8
  × 1.6
   948
  1580
  25.28
```

Thus, 25.28 is 160% of 15.8. In this case, the amount exceeds the base because the percent is more than 100%.

Self Check 3
What number is 240% of 80.3?
Now Try Problem 23
Self Check 3 Answer
192.72

Teaching Example 3 What number is 340% of 28.4?
Answer:
96.56

3 Solve percent equations to find the percent.

In the drinking water problem (Type 2 from the newspaper), we must find the percent. Once again, we translate the words of the problem into a percent equation and solve it.

Self Check 4

4 is what percent of 80? 5%

Now Try Problems 27 and 79

Teaching Example 4 34 is what percent of 40?
Answer:
85%

EXAMPLE 4   38 is what percent of 40?

Strategy We will look for the key words *is*, *of*, and *what percent* in the percent sentence and translate them to mathematical symbols to form a percent equation.

WHY Then we can solve the equation to find the unknown percent.

Solution First, we translate.

38	is	what percent	of	40?
↓	↓	↓	↓	↓
38	=	x	·	40

38 is the amount, x is the percent, and 40 is the base.

$38 = 40x$ Use the commutative property of multiplication to write $x \cdot 40$ as $40x$.

$\dfrac{38}{40} = \dfrac{40x}{40}$ To isolate x, undo the multiplication by 40 by dividing both sides by 40.

$0.95 = x$ Do the division: $38 \div 40 = 0.95$.

$$\begin{array}{r} 0.95 \\ 40\overline{)38.00} \\ -360 \\ 200 \\ -200 \\ 0 \end{array}$$

Since we want to find the percent, we need to write the decimal 0.95 as a percent.

$095\%\kern-1em\underset{\curvearrowright}{} = x$ To write 0.95 as a percent, multiply it by 100 by moving the decimal point two places to the right, and then insert a % symbol.

$95\% = x$

We have found that 38 is 95% of 40. That is, 95% of the wells mentioned in the newspaper article were declared safe.

Self Check 5

9 is what percent of 16? 56.25%

Now Try Problem 31

Teaching Example 5 5 is what percent of 16?
Answer:
31.25%

EXAMPLE 5   14 is what percent of 32?

Strategy We will look for the key words *is*, *of*, and *what percent* in the percent sentence and translate them to mathematical symbols to form a percent equation.

WHY Then we can solve the equation to find the unknown percent.

Solution First, we translate.

14	is	what percent	of	32?
↓	↓	↓	↓	↓
14	=	x	·	32

14 is the amount, x is the percent, and 32 is the base.

$14 = 32x$ Use the commutative property of multiplication to write $x \cdot 32$ as $32x$.

$\dfrac{14}{32} = \dfrac{32x}{32}$ To isolate x, undo the multiplication by 32 by dividing both sides by 32.

$0.4375 = x$ Do the division: $14 \div 32 = 0.4375$.

$043.75\% = x$ To write the decimal 0.4375 as a percent, multiply it by 100 by moving the decimal point two places to the right, and then insert a % symbol.

$43.75\% = x$

Thus, 14 is 43.75% of 32.

```
     0.4375
32)14.0000
   -12 8
     1 20
    - 96
      240
     -224
      160
     -160
        0
```

Using Your CALCULATOR Cost of an Air Bag

An air bag is estimated to add $500 to the cost of a car. What percent of the $16,295 sticker price is the cost of the air bag?

First, we translate the words of the problem into a percent equation.

What percent	of	the $16,295 sticker price	is	the cost of the air bag?
↓	↓	↓	↓	↓
x	$\cdot$	16,295	=	500

500 is the amount, x is the percent, and 16,295 is the base.

Then we solve the equation.

$16{,}295x = 500$ Write $x \cdot 16{,}295$ as $16{,}295x$.

$\dfrac{16{,}295x}{16{,}295} = \dfrac{500}{16{,}295}$ To undo the multiplication by 16,295 and isolate x on the left side, divide both sides of the equation by 16,295.

$x = \dfrac{500}{16{,}295}$

To perform the division on the right side using a scientific calculator, enter the following:

500 ÷ 16295 = $\boxed{0.03068425\mathsf{9}}$

This display gives the answer in decimal form. To change it to a percent, we multiply the result by 100. This moves the decimal point two places to the right. (See the display.) Then we insert a % symbol. If we round to the nearest tenth of a percent, the cost of the air bag is about 3.1% of the sticker price.

$\boxed{3.068425898}$

EXAMPLE 6 What percent of 6 is 7.5?

Strategy We will look for the key words *is*, *of*, and *what percent* in the percent sentence and translate them to mathematical symbols to form a percent equation.

WHY Then we can solve the equation to find the unknown percent.

Self Check 6
What percent of 5 is 8.5? 170%
Now Try Problem 35

Teaching Example 6 What percent of 7 is 9.1?
Answer:
130%

Solution First, we translate.

What percent	of	6	is	7.5
↓	↓	↓	↓	↓
x	·	6	=	7.5

$6x = 7.5$ Use the commutative property of multiplication to write $x \cdot 6$ as $6x$.

$\dfrac{6x}{6} = \dfrac{7.5}{6}$ To isolate x, undo the multiplication by 6 by dividing both sides by 6.

$x = 1.25$ Do the division: $7.5 \div 6 = 1.25$.

$x = 125\%$ To write the decimal 1.25 as a percent, multiply it by 100 by moving the decimal point two places to the right, and then insert a % symbol.

$x = 125\%$

Thus, 7.5 is 125% of 6.

$$\begin{array}{r}1.25\\6\overline{)7.50}\\-6\\\hline 1\,5\\-1\,2\\\hline 30\\-30\\\hline 0\end{array}$$

4 Solve percent equations to find the base.

In the percent problem about the State Board of Examiners (Type 3 from the newspaper), we must find the base. As before, we translate the percent sentence into a percent equation and then find the unknown number.

Self Check 7

3 is 5% of what number?   60

Now Try Problem 39

Teaching Example 7 7 is 35% of what number?
Answer:
20

EXAMPLE 7   6 is 75% of what number?

Strategy We will look for the key words *is*, *of*, and *what number* in the percent sentence and translate them to mathematical symbols to form a percent equation.

WHY Then we can solve the equation to find the unknown number.

Solution First, we translate.

6	is	75%	of	what number?
↓	↓	↓	↓	↓
6	=	75%	·	x

6 is the amount, 75% is the percent, and x is the base.

Now we solve the equation.

$6 = 0.75x$ Write 75% as a decimal: $75\% = 0.75$. It is not necessary to write the multiplication raised dot.

$\dfrac{6}{0.75} = \dfrac{0.75x}{0.75}$ To isolate x, undo the multiplication by 0.75 by dividing both sides by 0.75.

$8 = x$ Do the division: $6 \div 0.75 = 8$.

$$\begin{array}{r}8\\75\overline{)600}\\-600\\\hline 0\end{array}$$

Thus, 6 is 75% of 8. That is, there are 8 members on the State Board of Examiners mentioned in the newspaper article.

Success Tip Sometimes the calculations to solve a percent problem are made easier if we write the percent as a fraction instead of a decimal. This is the case with percents that have *repeating* decimal equivalents such as $33\frac{1}{3}\%$, $66\frac{2}{3}\%$, and $16\frac{2}{3}\%$. You may want to review the table of percents and their fractional equivalents on page 652.

EXAMPLE 8 31.5 is $33\frac{1}{3}\%$ of what number?

Self Check 8
150 is $66\frac{2}{3}\%$ of what number? 225
Now Try Problems 43 and 83

Teaching Example 8 22.6 is $33\frac{1}{3}\%$ of what number?
Answer:
67.8

Strategy We will look for the key words *is*, *of*, and *what number* in the percent sentence and translate them to mathematical symbols to form a percent equation.

WHY Then we can solve the equation to find the unknown number.

Solution First, we translate.

$$31.5 \text{ is } 33\tfrac{1}{3}\% \text{ of what number?}$$

$$31.5 = 33\tfrac{1}{3}\% \cdot x$$

31.5 is the amount, $33\frac{1}{3}\%$ is the percent, and x is the base.

In this case, the calculations can be made easier by writing $33\frac{1}{3}\%$ as a fraction instead of as a repeating decimal.

$31.5 = \dfrac{1}{3}x$ Recall from Section 7.1 that $33\tfrac{1}{3}\% = \tfrac{1}{3}$. It is not necessary to write the multiplication raised dot.

$3(31.5) = 3\left(\dfrac{1}{3}x\right)$ The coefficient of x is the fraction $\tfrac{1}{3}$. To isolate x, multiply both sides by the reciprocal of $\tfrac{1}{3}$, which is 3.

$94.5 = \left(3 \cdot \dfrac{1}{3}\right)x$ On the left side, do the multiplication: $3(31.5) = 94.5$. On the right side, use the associative property of multiplication to group 3 and $\tfrac{1}{3}$.

$\begin{array}{r}\overset{1}{3}1.5\\ \times\ \ 3\\\hline 94.5\end{array}$

$94.5 = 1x$ On the right side, the product of a number and its reciprocal is 1.

$94.5 = x$ On the right side, the coefficient of 1 need not be written since $1x = x$.

Thus, 31.5 is $33\frac{1}{3}\%$ of 94.5.

To solve percent application problems, we often have to rewrite the facts of the problem in percent sentence form before we can translate to an equation.

EXAMPLE 9 **Rentals** In an apartment complex, 198 of the units are currently occupied. If this represents an 88% occupancy rate, how many units are in the complex?

Strategy We will carefully read the problem and use the given facts to write them in the form of a percent sentence.

Self Check 9
CAPACITY OF A GYM A total of 784 people attended a graduation in a high school gymnasium. If this was 98% of capacity, what is the total capacity of the gym? 800 people

Now Try Problem 81

Teaching Example 9
CLOSED CLASSES On the first day of classes, 36 of the algebra sections offered by a college mathematics department were closed. If this represented 72% of all the sections of algebra, what is the total number of algebra sections offered that semester?
Answer:
50 sections

WHY Then we can translate the sentence into a percent equation and solve it to find the unknown number of units in the complex.

Solution An occupancy rate of 88% means that 88% of the units are occupied. Thus, the 198 units that are currently occupied are 88% of some unknown number of units in the complex, and we can write:

198	is	88%	of	what number?
↓	↓	↓	↓	↓
198	=	88%	·	x

198 is the amount, 88% is the percent, and x is the base.

Now we solve the equation.

$198 = 0.88x$ Write 88% as a decimal: 88% = 0.88. It is not necessary to write the multiplication raised dot.

$\dfrac{198}{0.88} = \dfrac{0.88x}{0.88}$ To isolate x, undo the multiplication by 0.88 by dividing both sides by 0.88.

$225 = x$ Do the division: 198 ÷ 0.88 = 225.

$$\begin{array}{r} 225 \\ 88\overline{)19800} \\ \underline{-176} \\ 220 \\ \underline{-176} \\ 440 \\ \underline{-440} \\ 0 \end{array}$$

The apartment complex has 225 units, of which 198, or 88%, are occupied.

If you are only learning the percent equation method for solving percent problems, turn to page 671 and pick up your reading at Objective 5.

METHOD 2: PERCENT PROPORTIONS

1 Write percent proportions.

Another method to solve percent problems involves writing and then solving a proportion. To introduce this method, consider the figure on the right. The vertical line down its middle divides the figure into two equal-sized parts. Since 1 of the 2 parts is shaded red, the shaded portion of the figure can be described by the ratio $\frac{1}{2}$. We call this an **amount-to-base** (or **part-to-whole**) **ratio**.

50 of the 100 parts shaded: 50% shaded

Now consider the 100 equal-sized square regions within the figure. Since 50 of them are shaded red, we say that $\frac{50}{100}$, or 50% of the figure is shaded. The ratio $\frac{50}{100}$ is called a **percent ratio**.

Since the amount-to-base ratio, $\frac{1}{2}$, and the percent ratio, $\frac{50}{100}$, represent the same shaded portion of the figure, they must be equal, and we can write

The amount-to-base ratio → $\dfrac{1}{2} = \dfrac{50}{100}$ ← The percent ratio

Recall from Section 6.2 that statements of this type stating that two ratios are equal are called *proportions*. We call $\frac{1}{2} = \frac{50}{100}$ a **percent proportion**. The four terms of a percent proportion are shown on the following page.

Unless otherwise noted, all art on this page is © Cengage Learning.

7.2 Solving Percent Problems Using Percent Equations and Proportions

Percent Proportion

To translate a percent sentence to a **percent proportion**, use the following form:

Amount is to base as percent is to 100.   *Part is to whole as percent is to 100.*

$$\frac{\text{amount}}{\text{base}} = \frac{\text{percent}}{100} \quad \text{or} \quad \frac{\text{part}}{\text{whole}} = \frac{\text{percent}}{100}$$

This is always 100 because percent means parts per one hundred.

To write a percent proportion, you must identify three of the terms as you read the problem. (Remember, the fourth term of the proportion is always 100.) Here are some ways to identify those terms.

- The **percent** is easy to find. Look for the % symbol or the words *what percent*.
- The **base** (or **whole**) usually follows the word *of*.
- The **amount** (or **part**) is compared to the base (or whole).

EXAMPLE 1
Translate each percent sentence to a percent proportion.

a. What number is 12% of 64?
b. What percent of 88 is 11?
c. 165% of what number is 366?

Strategy A percent proportion has the form $\frac{\text{amount}}{\text{base}} = \frac{\text{percent}}{100}$. Since one of the terms of the percent proportion is always 100, we only need to identify three terms to write the proportion. We will begin by identifying the percent and the base in the given sentence.

WHY The remaining number (or unknown) must be the amount.

Solution
a. We will identify the terms in this order:
- *First:* the percent (next to the % symbol)
- *Second:* the base (usually after the word *of*)
- *Last:* the amount (the number that remains)

b.

Self Check 1
Translate each percent sentence to a percent proportion.

a. What number is 33% of 80?
b. What percent of 55 is 6?
c. 172% of what number is 4?

Now Try Problem 17

Self Check 1 Answers

a. $\frac{x}{80} = \frac{33}{100}$

b. $\frac{6}{55} = \frac{x}{100}$

c. $\frac{4}{x} = \frac{172}{100}$

Teaching Example 1 Translate each percent sentence to a percent proportion.
a. What number is 25% of 37?
b. What percent of 300 is 12?
c. 110% of what number is 341?

Answers:

a. $\frac{x}{37} = \frac{25}{100}$

b. $\frac{12}{300} = \frac{x}{100}$

c. $\frac{341}{x} = \frac{110}{100}$

c. 165% of what number is 366?

percent   base   amount

$$\frac{366}{x} = \frac{165}{100}$$

2 Solve percent proportions to find the amount.

Recall the labor union problem from the newspaper example in the introduction to this section. We can write and solve a percent proportion to find the unknown amount.

Self Check 2

What number is 36% of 400? 144

Now Try Problems 19 and 71

Teaching Example 2 What number is 28% of 300?
Answer:
84

EXAMPLE 2   What number is 84% of 500?

Strategy We will identify the percent, the base, and the amount and write a percent proportion of the form $\frac{\text{amount}}{\text{base}} = \frac{\text{percent}}{100}$.

WHY Then we can solve the proportion to find the unknown number.

Solution First, we write the percent proportion.

What number is 84% of 500?

amount   percent   base

$$\frac{x}{500} = \frac{84}{100}$$ This is the proportion to solve.

To make the calculations easier, it is helpful to simplify the ratio $\frac{84}{100}$ at this time.

$$\frac{x}{500} = \frac{21}{25}$$ On the right side, simplify: $\frac{84}{100} = \frac{\overset{1}{\cancel{4}} \cdot 21}{\underset{1}{\cancel{4}} \cdot 25} = \frac{21}{25}$.

Recall from Section 6.2 that to solve a proportion we use the cross products.

$x \cdot 25 = 500 \cdot 21$ Find the cross products: $\frac{x}{500} \diagdown \frac{21}{25}$. Then set them equal.

$25x = 10{,}500$ On the left side, use the commutative property of multiplication to write $x \cdot 25$ as $25x$. On the right side, do the multiplication: $500 \cdot 21 = 10{,}500$.

$$\frac{25x}{25} = \frac{10{,}500}{25}$$ To isolate x, undo the multiplication by 25 by dividing both sides by 25.

$x = 420$ Do the division: $10{,}500 \div 25 = 420$.

```
      500
    ×  21
      500
    10 000
    10,500

         420
    25)10,500
       -10 0
         50
        -50
         00
         -0
          0
```

We have found that 420 is 84% of 500. That is, 420 union members mentioned in the newspaper article voted to accept the new offer.

> **The Language of Algebra** When we find the value of the variable that makes a percent proportion true, we say that we have **solved the proportion**. In Example 2, we solved $\frac{x}{500} = \frac{84}{100}$ to find that x is 420.

7.2 Solving Percent Problems Using Percent Equations and Proportions

EXAMPLE 3   What number is 160% of 15.8?

Strategy We will identify the percent, the base, and the amount and write a percent proportion of the form $\frac{\text{amount}}{\text{base}} = \frac{\text{percent}}{100}$.

WHY Then we can solve the proportion to find the unknown number.

Solution First, we write the percent proportion.

To make the calculations easier, it is helpful to simplify the ratio $\frac{160}{100}$ at this time.

$\dfrac{x}{15.8} = \dfrac{8}{5}$ On the right side, simplify: $\frac{160}{100} = \frac{8 \cdot \overset{1}{\cancel{20}}}{5 \cdot \cancel{20}} = \frac{8}{5}$.

$x \cdot 5 = 15.8 \cdot 8$ Find the cross products: $\frac{x}{15.8} \bowtie \frac{8}{5}$. Then set them equal.

$5x = 126.4$ On the left side, use the commutative property of multiplication to write $x \cdot 5$ as $5x$. On the right side, do the multiplication: $15.8 \cdot 8 = 126.4$.

$\dfrac{5x}{5} = \dfrac{126.4}{5}$ To isolate x, undo the multiplication by 5 by dividing both sides by 5.

$x = 25.28$ Do the division: $126.4 \div 5 = 25.28$.

Thus, 25.28 is 160% of 15.8.

$\begin{array}{r} \overset{4\ 6}{15.8} \\ \times\ \ 8 \\ \hline 126.4 \end{array}$

$\begin{array}{r} 25.28 \\ 5\overline{)126.40} \\ -10 \\ \hline 26 \\ -25 \\ \hline 1\,4 \\ -1\,0 \\ \hline 40 \\ -40 \\ \hline 0 \end{array}$

Self Check 3
What number is 240% of 80.3?
Now Try Problem 23
Self Check 3 Answer
192.72

Teaching Example 3 What number is 340% of 28.4?
Answer:
96.56

3 Solve percent proportions to find the percent.

Recall the drinking water problem from the newspaper example in the introduction to this section. We can write and solve a percent proportion to find the unknown percent.

EXAMPLE 4   38 is what percent of 40?

Strategy We will identify the percent, the base, and the amount and write a percent proportion of the form $\frac{\text{amount}}{\text{base}} = \frac{\text{percent}}{100}$.

WHY Then we can solve the proportion to find the unknown percent.

Solution First, we write the percent proportion.

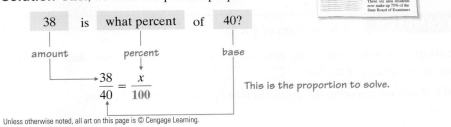

Self Check 4
4 is what percent of 80? 5%
Now Try Problems 27 and 79

Teaching Example 4 34 is what percent of 40?
Answer:
85%

To make the calculations easier, it is helpful to simplify the ratio $\frac{38}{40}$ at this time.

$$\frac{19}{20} = \frac{x}{100}$$ On the left side, simplify: $\frac{38}{40} = \frac{\overset{1}{\cancel{2}} \cdot 19}{\cancel{2} \cdot 20} = \frac{19}{20}$.

$19 \cdot 100 = 20 \cdot x$ To solve the proportion, find the cross products: $\frac{19}{20} \times \frac{x}{100}$. Then set them equal.

$1{,}900 = 20x$ On the left side, do the multiplication: $19 \cdot 100 = 1{,}900$. On the right side, write $20 \cdot x$ as $20x$.

$$\frac{1{,}900}{20} = \frac{20x}{20}$$ To isolate x, undo the multiplication by 20 by dividing both sides by 20.

$95 = x$ Do the division: $1{,}900 \div 20 = 95$.

```
      95
20)1,900
    -1 80
      100
     -100
        0
```

We have found that 38 is 95% of 40. That is, 95% of the wells mentioned in the newspaper article were declared safe.

Self Check 5

9 is what percent of 16?   56.25%

Now Try Problem 31

Teaching Example 5 5 is what percent of 16?
Answer:
31.25%

EXAMPLE 5   14 is what percent of 32?

Strategy We will identify the percent, the base, and the amount and write a percent proportion of the form $\frac{\text{amount}}{\text{base}} = \frac{\text{percent}}{100}$.

WHY Then we can solve the proportion to find the unknown percent.

Solution First, we write the percent proportion.

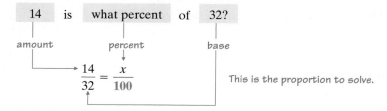

$\frac{14}{32} = \frac{x}{100}$ This is the proportion to solve.

To make the calculations easier, it is helpful to simplify the ratio $\frac{14}{32}$ at this time.

$$\frac{7}{16} = \frac{x}{100}$$ On the left side, simplify: $\frac{14}{32} = \frac{\overset{1}{\cancel{2}} \cdot 7}{\cancel{2} \cdot 16} = \frac{7}{16}$.

$7 \cdot 100 = 16 \cdot x$ To solve the proportion, find the cross products: $\frac{7}{16} \times \frac{x}{100}$. Then set them equal.

$700 = 16x$ On the left side, do the multiplication: $7 \cdot 100 = 700$. On the right side, write $16 \cdot x$ as $16x$.

$$\frac{700}{16} = \frac{16x}{16}$$ To isolate x, undo the multiplication by 16 by dividing both sides by 16.

$43.75 = x$ Do the division: $700 \div 16 = 43.75$.

```
       43.75
16)700.00
   -64
    60
   -48
    12 0
   -11 2
       80
      -80
        0
```

Thus, 14 is 43.75% of 32.

Self Check 6

What percent of 5 is 8.5?   170%

Now Try Problem 35

EXAMPLE 6   What percent of 6 is 7.5?

Strategy We will identify the percent, the base, and the amount and write a percent proportion of the form $\frac{\text{amount}}{\text{base}} = \frac{\text{percent}}{100}$.

WHY Then we can solve the proportion to find the unknown percent.

7.2 Solving Percent Problems Using Percent Equations and Proportions 669

Solution First, we write the percent proportion.

Thus, 7.5 is 125% of 6.

Teaching Example 6 What percent of 7 is 9.1?
Answer:
130%

4 Solve percent proportions to find the base.

Recall the State Board of Examiners problem from the newspaper example in the introduction to this section. We can write and solve a percent proportion to find the unknown base.

EXAMPLE 7 6 is 75% of what number?

Strategy We will identify the percent, the base, and the amount and write a percent proportion of the form $\frac{\text{amount}}{\text{base}} = \frac{\text{percent}}{100}$.

WHY Then we can solve the proportion to find the unknown number.

Solution First, we write the percent proportion.

Self Check 7

3 is 5% of what number? 60

Now Try Problem 39

Teaching Example 7 7 is 35% of what number?
Answer:
20

To make the calculations easier, it is helpful to simplify the ratio $\frac{75}{100}$ at this time.

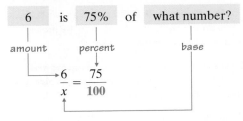

$6 \cdot 4 = x \cdot 3$ To solve the proportion, find the cross products: $\frac{6}{x} \bowtie \frac{3}{4}$. Then set them equal.

$24 = 3x$ On the left side, do the multiplication: $6 \cdot 4 = 24$. On the right side, use the commutative property of multiplication to write $x \cdot 3$ as $3x$.

Chapter 7 Percent

$$\frac{24}{3} = \frac{3x}{3}$$ To isolate x, undo the multiplication by 3 by dividing both sides by 3.

$$8 = x$$ Do the division: $24 \div 3 = 8$.

Thus, 6 is 75% of 8. That is, there are 8 members on the State Board of Examiners mentioned in the newspaper article.

Self Check 8

150 is $66\frac{2}{3}$% of what number? 225

Now Try Problems 43 and 83

Teaching Example 8 22.6 is $33\frac{1}{3}$% of what number?
Answer:
67.8

EXAMPLE 8 31.5 is $33\frac{1}{3}$% of what number?

Strategy We will identify the percent, the base, and the amount and write a percent proportion of the form $\frac{\text{amount}}{\text{base}} = \frac{\text{percent}}{100}$.

WHY Then we can solve the proportion to find the unknown number.

Solution First, we write the percent proportion.

31.5 is $33\frac{1}{3}$% of what number?
amount — percent — base

$$\frac{31.5}{x} = \frac{33\frac{1}{3}}{100}$$

To make the calculations easier, it is helpful to write the mixed number $33\frac{1}{3}$ as the improper fraction $\frac{100}{3}$.

$$\frac{31.5}{x} = \frac{\frac{100}{3}}{100}$$ Write $33\frac{1}{3}$ as $\frac{100}{3}$.

$$31.5 \cdot 100 = x \cdot \frac{100}{3}$$ To solve the proportion, find the cross products: $\frac{31.5}{x} \bowtie \frac{\frac{100}{3}}{100}$. Then set them equal.

$$3{,}150 = \frac{100}{3}x$$ On the left side, do the multiplication: $31.5 \cdot 100 = 3{,}150$. On the right side, use the commutative property of multiplication to write $x \cdot \frac{100}{3}$ as $\frac{100}{3}x$.

$$\frac{3}{100}(3{,}150) = \frac{3}{100}\left(\frac{100}{3}x\right)$$ The coefficient of x is the fraction $\frac{100}{3}$. To isolate x, multiply both sides by the reciprocal of $\frac{100}{3}$, which is $\frac{3}{100}$.

$$\frac{3}{100}\left(\frac{3{,}150}{1}\right) = \left(\frac{3}{100} \cdot \frac{100}{3}\right)x$$ On the left side, write 3,150 as a fraction: $3{,}150 = \frac{3{,}150}{1}$. On the right, use the associative property of multiplication to group $\frac{3}{100}$ and $\frac{100}{3}$.

$$\frac{3 \cdot 3{,}150}{100 \cdot 1} = 1x$$ On the left side, multiply the numerators and multiply the denominators. On the right side, the product of a number and its reciprocal is 1.

$$\frac{9{,}450}{100} = x$$ On the left side, $3 \cdot 3{,}150 = 9{,}450$. On the right side, the coefficient of 1 need not be written since $1x = x$.

$$94.50 = x$$ Divide 9,450 by 100 by moving the understood decimal point in 9,450 two places to the left.

$$\begin{array}{r} \overset{1}{3{,}150} \\ \times 3 \\ \hline 9{,}450 \end{array}$$

Thus, 31.5 is $33\frac{1}{3}$% of 94.5.

To solve percent application problems, we often have to rewrite the facts of the problem in percent sentence form before we can translate to an equation.

EXAMPLE 9

Rentals In an apartment complex, 198 of the units are currently occupied. If this represents an 88% occupancy rate, how many units are in the complex?

Strategy We will carefully read the problem and use the given facts to write them in the form of a percent sentence.

WHY Then we can write and solve a percent proportion to find the unknown number of units in the complex.

Solution An occupancy rate of 88% means that 88% of the units are occupied. Thus, the 198 units that are currently occupied are 88% of some unknown number of units in the complex, and we can write:

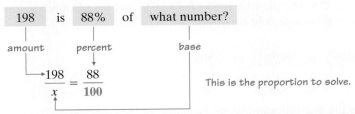

198 is 88% of what number?
amount | percent | base

$$\frac{198}{x} = \frac{88}{100}$$ This is the proportion to solve.

To make the calculations easier, it is helpful to simplify the ratio $\frac{88}{100}$ at this time.

$\frac{198}{x} = \frac{22}{25}$ On the right side, simplify: $\frac{88}{100} = \frac{\overset{1}{\cancel{4}} \cdot 22}{\underset{1}{\cancel{4}} \cdot 25} = \frac{22}{25}$.

$198 \cdot 25 = x \cdot 22$ Find the cross products. Then set them equal.

$4{,}950 = 22x$ On the left side, do the multiplication: $198 \cdot 25 = 4{,}950$. On the right side, use the commutative property of multiplication to write $x \cdot 22$ as $22x$.

$\frac{4{,}950}{22} = \frac{22x}{22}$ To isolate x, undo the multiplication by 22 by dividing both sides by 22.

$225 = x$ On the left side, do the division: $4{,}950 \div 22 = 225$.

```
  198
 × 25
  990
 3960
 4,950
```

```
      225
 22)4,950
    −44
     55
    −44
    110
   −110
      0
```

The apartment complex has 225 units, of which 198, or 88%, are occupied.

Self Check 9

CAPACITY OF A GYM A total of 784 people attended a graduation in a high school gymnasium. If this was 98% of capacity, what is the total capacity of the gym? 800 people

Now Try Problem 81

Teaching Example 9
CLOSED CLASSES On the first day of classes, 36 of the algebra sections offered by a college mathematics department were closed. If this represented 72% of all the sections of algebra, what is the total number of algebra sections offered that semester?
Answer:
50 classes

5 Read circle graphs.

Percents are used with **circle graphs**, or **pie charts**, as a way of presenting data for comparison. In the figure below, the entire circle represents the total amount of electricity generated in the United States in 2011. The pie-shaped pieces of the graph show the relative sizes of the energy sources used to generate the electricity. For example, we see that the greatest amount of electricity (42%) was generated from coal. Note that if we add the percents from all categories (42% + 1% + 8% + 25% + 19% + 5%), the sum is 100%.

The 100 tick marks equally spaced around the circle serve as a visual aid when constructing a circle graph. For example, to represent hydropower as 8%, a line was drawn from the center of the circle to a tick mark. Then we counted off 8 ticks and drew a second line from the center to that tick to complete the pie-shaped wedge.

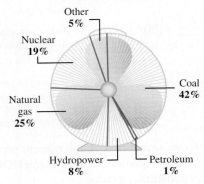

Sources of Electricity in the United States, 2011

Other 5%
Nuclear 19%
Natural gas 25%
Hydropower 8%
Petroleum 1%
Coal 42%

Source: Energy Information Administration

Unless otherwise noted, all art on this page is © Cengage Learning.

Self Check 10

PRESIDENTIAL ELECTIONS Results from the 2004 U.S. presidential election are shown in the circle graph below. Find the number of states won by President Bush.

2004 Presidential Election
States won by each candidate

Now Try Problem 85

Teaching Example 10
HOUSING At the end of 2010, approximately 112,000,000 housing units in the United States were occupied. Use the information in the following circle graph to find the number of units that were owner occupied in 2010.

Source: U.S. Census Bureau

Answer:
75,040,000 housing units

EXAMPLE 10 Presidential Elections

Results from the 2012 U.S. presidential election are shown in the circle graph to the right. Find the number of states won by Barack Obama.

2012 Presidential Election
States won by each candidate

Strategy We will rewrite the facts of the problem in percent sentence form.

WHY Then we can translate the sentence to a percent equation (or percent proportion) to find the number of states won by Barack Obama.

Solution The circle graph shows that Barack Obama won 52% of the 50 states. Thus, the percent is 52% and the base is 50. One way to find the unknown amount is to write and then solve a percent equation.

$$\text{What number is } 52\% \text{ of } 50?$$
$$x = 52\% \cdot 50 \quad \text{Translate to a percent equation.}$$

Now we perform the multiplication on the right side of the equation.

$$x = 0.52 \cdot 50 \quad \text{Write 52\% as a decimal: 52\% = 0.52.}$$
$$x = 26 \quad \text{Do the multiplication.}$$

$$\begin{array}{r} 50 \\ \times 0.52 \\ \hline 1\,00 \\ 25\,00 \\ \hline 26.00 \end{array}$$

Thus, Barack Obama won 26 of the 50 states in the 2012 U.S. presidential election.

Another way to find the unknown amount is to write and then solve a percent proportion.

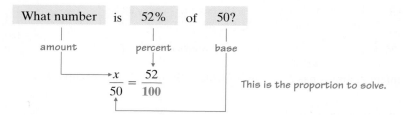

$$\frac{x}{50} = \frac{52}{100} \quad \text{This is the proportion to solve.}$$

To make the calculations easier, it is helpful to simplify the ratio $\frac{52}{100}$ at this time.

$$\frac{x}{50} = \frac{13}{25} \quad \text{On the right side, simplify: } \frac{52}{100} = \frac{\overset{1}{4} \cdot 13}{\underset{1}{4} \cdot 25} = \frac{13}{25}.$$

$$x \cdot 25 = 50 \cdot 13 \quad \text{Find the cross products: } \frac{x}{50} = \frac{13}{25}. \text{ Then set them equal.}$$

$$25x = 650 \quad \text{On the left side, use the commutative property of multiplication to write } x \cdot 25 \text{ as } 25x. \text{ On the right side, do the multiplication: } 50 \cdot 13 = 650.$$

$$\frac{25x}{25} = \frac{650}{25} \quad \text{To isolate } x, \text{ undo the multiplication by 25 by dividing both sides by 25.}$$

$$x = 26 \quad \text{On the right side, do the division: } 650 \div 25 = 26.$$

$$\begin{array}{r} 50 \\ \times 13 \\ \hline 150 \\ 500 \\ \hline 650 \end{array}$$

$$\begin{array}{r} 26 \\ 25\overline{)650} \\ -50 \\ \hline 150 \\ -150 \\ \hline 0 \end{array}$$

As we would expect, the percent proportion method gives the same answer as the percent equation method. Barack Obama won 26 of the 50 states in the 2012 U.S. presidential election.

7.2 Solving Percent Problems Using Percent Equations and Proportions

THINK IT THROUGH Community College Students

"When the history of American higher education is updated years from now, the story of our current times will highlight the pivotal role community colleges played in developing human capital and bolstering the nation's educational system."

Community College Survey of Student Engagement, 2007

More than 453,000 students responded to the 2012 Community College Survey of Student Engagement. Some results are shown below. Study each circle graph and then complete its legend.

Enrollment in Community Colleges

- 55% are enrolled in college part time.
- ? 45% are enrolled in college full time

Community College Students Who Work More Than 20 Hours per Week

- 47% of the students work more than 20 hours per week.
- ? 53% of the students work less than 20 hours per week

Community College Students Who Discussed Their Grades or Assignments with an Instructor

- 49% often or very often
- 42% sometimes
- ? 9% never

ANSWERS TO SELF CHECKS

1. a. $x = 33\% \cdot 80$ or $\frac{x}{80} = \frac{33}{100}$ b. $x \cdot 55 = 6$ or $\frac{6}{55} = \frac{x}{100}$ c. $172\% \cdot x = 4$ or $\frac{4}{x} = \frac{172}{100}$
2. 144 3. 192.72 4. 5% 5. 56.25% 6. 170% 7. 60 8. 225 9. 800 people
10. 31 states

SECTION 7.2 STUDY SET

VOCABULARY

Fill in the blanks.

1. We call "What number is 15% of 25?" a percent __sentence__. It translates to the percent __equation__ $x = 15\% \cdot 25$.

2. The key words in a percent sentence translate as follows:
   - __is__ translates to an equal symbol =
   - __of__ translates to multiplication that is shown with a raised dot ·
   - __what__ number or __what__ percent translates to an unknown number that is represented by a variable

3. When we find the value of the variable that makes a percent equation true, we say that we have __solved__ the equation.

4. In the percent sentence "45 is 90% of 50," 45 is the __amount__, 90% is the percent, and 50 is the __base__.

5. The amount is __part__ of the base. The base is the standard of comparison—it represents the __whole__ of some quantity.

6. a. *Amount is to base as percent is to 100:*

   $$\frac{\text{amount}}{\text{base}} = \frac{\text{percent}}{100}$$

   b. *Part is to whole as percent is to 100:*

   $$\frac{\text{part}}{\text{whole}} = \frac{\text{percent}}{100}$$

7. The __cross__ products for the proportion $\frac{24}{x} = \frac{36}{100}$ are $24 \cdot 100$ and $x \cdot 36$.

8. In a __circle__ graph, pie-shaped wedges are used to show the division of a whole quantity into its component parts.

CONCEPTS

9. Fill in the blanks to complete the percent equation (formula):

   Amount = percent · base

   or

   Part = percent · whole

10. a. Without doing the calculation, tell whether 12% of 55 is more than 55 or less than 55. less
    b. Without doing the calculation, tell whether 120% of 55 is more than 55 or less than 55. more

11. PIZZA SALES The circle graph shows what percent of total U.S. pizza sales the leading companies had in 2010. What percent of the sales did Pizza Hut have? 14%

12. SMARTPHONES The circle graph shows the percent U.S. market share for the leading smartphone companies in 2012. What was the market share for Android? 50.1%

NOTATION

13. When computing with percents, we must change the percent to a decimal or a fraction. Change each percent to a decimal.
    a. 12%   0.12
    b. 5.6%   0.056
    c. 125%   1.25
    d. $\frac{1}{4}$%   0.0025

14. When computing with percents, we must change the percent to a decimal or a fraction. Change each percent to a fraction.
    a. $33\frac{1}{3}$% $\frac{1}{3}$
    b. $66\frac{2}{3}$% $\frac{2}{3}$
    c. $16\frac{2}{3}$% $\frac{1}{6}$
    d. $83\frac{1}{3}$% $\frac{5}{6}$

GUIDED PRACTICE

Translate each percent sentence to a percent equation or percent proportion. Do not solve. See Example 1.

15. a. What number is 7% of 16? $x = 7\% \cdot 16$
       $\frac{x}{16} = \frac{7}{100}$
    b. 125 is what percent of 800? $125 = x \cdot 800$
       $\frac{125}{800} = \frac{x}{100}$
    c. 1 is 94% of what number? $1 = 94\% \cdot x$
       $\frac{1}{x} = \frac{94}{100}$

16. a. What number is 28% of 372? $x = 28\% \cdot 372$
       $\frac{x}{372} = \frac{28}{100}$
    b. 9 is what percent of 21? $9 = x \cdot 21$
       $\frac{9}{21} = \frac{x}{100}$
    c. 4 is 17% of what number? $4 = 17\% \cdot x$
       $\frac{4}{x} = \frac{17}{100}$

17. a. 5.4% of 99 is what number? $5.4\% \cdot 99 = x$
       $\frac{x}{99} = \frac{5.4}{100}$
    b. 75.1% of what number is 15? $75.1\% \cdot x = 15$
       $\frac{15}{x} = \frac{75.1}{100}$
    c. What percent of 33.8 is 3.8? $x \cdot 33.8 = 3.8$
       $\frac{3.8}{33.8} = \frac{x}{100}$

18. a. 1.5% of 3 is what number? $1.5\% \cdot 3 = x$
       $\frac{x}{3} = \frac{1.5}{100}$
    b. 49.2% of what number is 100? $49.2\% \cdot x = 100$
       $\frac{100}{x} = \frac{49.2}{100}$
    c. What percent of 100.4 is 50.2? $x \cdot 100.4 = 50.2$
       $\frac{50.2}{100.4} = \frac{x}{100}$

Translate to a percent equation or percent proportion and then solve to find the unknown number. See Example 2.

19. What is 34% of 200?   68
20. What is 48% of 600?   288
21. What is 88% of 150?   132
22. What number is 52% of 350?   182

Translate to a percent equation or percent proportion and then solve to find the unknown number. See Example 3.

23. What number is 224% of 7.9?   17.696
24. What number is 197% of 6.3?   12.411
25. What number is 105% of 23.2?   24.36
26. What number is 228% of 34.5?   78.66

Translate to a percent equation or percent proportion and then solve to find the unknown number. See Example 4.

27. 8 is what percent of 32? 25%
28. 9 is what percent of 18? 50%
29. 51 is what percent of 60? 85%
30. 52 is what percent of 80? 65%

Translate to a percent equation or percent proportion and then solve to find the unknown number. See Example 5.

31. 5 is what percent of 8? 62.5%
32. 7 is what percent of 8? 87.5%
33. 7 is what percent of 16? 43.75%
34. 11 is what percent of 16? 68.75%

Translate to a percent equation or percent proportion and then solve to find the unknown number. See Example 6.

35. What percent of 60 is 66? 110%
36. What percent of 50 is 56? 112%
37. What percent of 24 is 84? 350%
38. What percent of 14 is 63? 450%

Translate to a percent equation or percent proportion and then solve to find the unknown number. See Example 7.

39. 9 is 30% of what number? 30
40. 8 is 40% of what number? 20
41. 36 is 24% of what number? 150
42. 24 is 16% of what number? 150

Translate to a percent equation or percent proportion and then solve to find the unknown number. See Example 8.

43. 19.2 is $33\frac{1}{3}$% of what number? 57.6
44. 32.8 is $33\frac{1}{3}$% of what number? 98.4
45. 48.4 is $66\frac{2}{3}$% of what number? 72.6
46. 56.2 is $16\frac{2}{3}$% of what number? 337.2

TRY IT YOURSELF

Translate to a percent equation or percent proportion and then solve to find the unknown number.

47. What percent of 40 is 0.5? 1.25%
48. What percent of 15 is 0.3? 2%
49. 7.8 is 12% of what number? 65
50. 39.6 is 44% of what number? 90
51. $33\frac{1}{3}$% of what number is 33? 99
52. $66\frac{2}{3}$% of what number is 28? 42
53. What number is 36% of 250? 90
54. What number is 82% of 300? 246
55. 16 is what percent of 20? 80%
56. 13 is what percent of 25? 52%
57. What number is 0.8% of 12? 0.096
58. What number is 5.6% of 40? 2.24
59. 3.3 is 7.5% of what number? 44
60. 8.4 is 20% of what number? 42
61. What percent of 0.05 is 1.25? 2,500%
62. What percent of 0.06 is 2.46? 4,100%
63. 102% of 105 is what number? 107.1
64. 210% of 66 is what number? 138.6
65. $9\frac{1}{2}$% of what number is 5.7? 60
66. $\frac{1}{2}$% of what number is 5,000? 1,000,000
67. What percent of 8,000 is 2,500? 31.25%
68. What percent of 3,200 is 1,400? 43.75%
69. Find $7\frac{1}{4}$% of 600. 43.5
70. Find $1\frac{3}{4}$% of 800. 14

APPLICATIONS

71. **DOWNLOADING** The message on the computer monitor screen shown below indicates that 24% of the 50K bytes of information that the user has decided to view have been downloaded to her computer at that time. Find the number of bytes of information that have been downloaded. (50K stands for 50,000.) 12K bytes = 12,000 bytes

72. **LUMBER** The rate of tree growth for walnut trees is about 3% per year. If a walnut tree has 400 board feet of lumber that can be cut from it, how many more board feet will it produce in a year? (Source: Iowa Department of Natural Resources) 12 board ft

73. **REBATES** A telephone company offered its customers a rebate of 20% of the cost of all long-distance calls made in the month of July. One customer's long-distance calls for July are shown below.
   a. Find the total amount of the customer's long-distance charges for July. $20.75
   b. How much will this customer receive in the form of a rebate for these calls? $4.15

Date	Time	Place called	Min.	Amount
Jul 4	3:48 P.M.	Denver	47	$3.80
Jul 9	12:00 P.M.	Detroit	68	$7.50
Jul 20	8:59 A.M.	San Diego	70	$9.45
July Totals			185	?

74. **PRICE GUARANTEES** To assure its customers of low prices, the Home Club offers a "10% Plus" guarantee. If the customer finds the same item selling for less somewhere else, he or she receives the difference in price, plus 10% of the difference. A woman bought miniblinds at the Home Club for $120 but later saw the same blinds on sale for $98 at another store.
   a. What is the difference in the prices of the miniblinds? $22
   b. What is 10% of the difference in price? $2.20
   c. How much money can the woman expect to receive if she takes advantage of the "10% Plus" guarantee from the Home Club? $24.20

75. **ENLARGEMENTS** The enlarge feature on a copier is set at 180%, and a 1.5-inch wide picture is to be copied. What will be the width of the enlarged picture? 2.7 in.

76. **COPY MACHINES** The reduce feature on a copier is set at 98%, and a 2-inch wide picture is to be copied. What will be the width of the reduced picture? 1.96 in.

77. **DRIVER'S LICENSE** On the written part of his driving test, a man answered 28 out of 40 questions correctly. If 70% correct is passing, did he pass the test? yes

78. **HOUSING** A general budget rule of thumb is that your rent or mortgage payment should be less than 30% of your income. Together, a couple earns $4,500 per month and they pay $1,260 in rent. Are they following the budget rule of thumb for housing? yes (28%)

79. **INSURANCE** The cost to repair a car after a collision was $4,000. The automobile insurance policy paid the entire bill except for a $200 deductible, which the driver paid. What percent of the cost did he pay? 5%

80. **FLOOR SPACE** A house has 1,200 square feet on the first floor and 800 square feet on the second floor.
   a. What is the total square footage of the house? 2,000 sq ft
   b. What percent of the square footage of the house is on the first floor? 60%

81. **CHILD CARE** After the first day of registration, 84 children had been enrolled in a new day care center. That represented 70% of the available slots. What was the maximum number of children the center could enroll? 120

82. **RACING PROGRAMS** One month before a stock car race, the sale of ads for the official race program was slow. Only 12 pages, or 60% of the available pages, had been sold. What was the total number of pages devoted to advertising in the program? 20

83. **WATER POLLUTION** A recent study found that about 4,500 kilometers, or $33\frac{1}{3}$% of China's Yellow River and its tributaries were not fit for any use. What is the combined length of the river and its tributaries? (Source: Discovermagazine.com) 13,500 km

84. **SOLAR POWER** In 2011, two-thirds of the solar panel capacity installed worldwide was in Europe. If the European panels' capacity totaled 18,500 megawatts, what was the total capacity of the panels installed wordwide? (Source: huffingtonpost.com) 27,750 megawatts

85. **GOVERNMENT SPENDING** The circle graph below shows the breakdown of federal spending for fiscal year 2011. If the total spending was approximately $3.603 trillion, how many dollars were spent on social programs, Social Security, Medicare, and other retirement programs? $2.1618 trillion

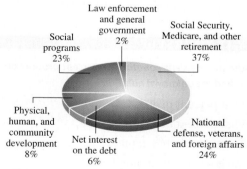

Source: 2012 Federal Income Tax Form 1040 Instructions

86. WASTE The circle graph below shows the types of trash U.S. residents, businesses, and institutions generated in 2010. If the total amount of trash produced that year was about 250 million tons, how many million tons of yard trimmings were there? 33.5 million tons

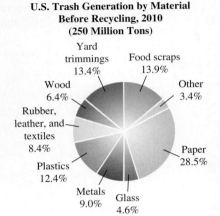

U.S. Trash Generation by Material
Before Recycling, 2010
(250 Million Tons)

Yard trimmings 13.4%
Food scraps 13.9%
Wood 6.4%
Other 3.4%
Rubber, leather, and textiles 8.4%
Paper 28.5%
Plastics 12.4%
Metals 9.0%
Glass 4.6%

Source: Environmental Protection Agency

87. PRODUCT PROMOTION To promote sales, a free 6-ounce bottle of shampoo is packaged with every large bottle. Use the information on the package to find how many ounces of shampoo the large bottle contains. 24 oz

88. NUTRITION FACTS The nutrition label on a package of corn chips is shown.

a. How many milligrams of sodium are in one serving of chips? 240 mg

b. According to the label, what percent of the daily value of sodium is this? 12%

c. What daily value of sodium intake is considered healthy? 2,000 mg

Nutrition Facts
Serving Size: 1 oz. (28g/About 29 chips)
Servings Per Container: About 11

Amount Per Serving
Calories 160   Calories from Fat 90

	% Daily Value
Total fat 10g	15%
Saturated fat 1.5 g	7%
Cholesterol 0mg	0%
Sodium 240mg	12%
Total carbohydrate 15g	5%
Dietary fiber 1g	4%
Sugars less than 1g	
Protein 2g	

89. MIXTURES Complete the table to find the number of gallons of sulfuric acid in each of two storage tanks.

	Gallons of solution in tank	% Sulfuric acid	Gallons of sulfuric acid in tank
Tank 1	60	50%	30
Tank 2	40	30%	12

90. THE ALPHABET What percent of the English alphabet do the vowels a, e, i, o, and u make up? (Round to the nearest one percent.) 19%

91. TIPS In August of 2006, a customer left Applebee's employee Cindy Kienow of Hutchinson, Kansas, a $10,000 tip for a bill that was approximately $25. What percent tip is this? (Source: cbsnews.com) 40,000%

92. ELECTIONS In Los Angeles City Council races, if no candidate receives more than 50% of the vote, a runoff election is held between the first- and second-place finishers.

a. How many total votes were cast? 17,177

b. Determine whether there must be a runoff election for District 10. yes

City council	District 10
Nate Holden	8,501
Madison T. Shockley	3,614
Scott Suh	2,630
Marsha Brown	2,432

Use a circle graph to illustrate the given data. A circle divided into 100 sections is provided to help in the graphing process.

93. ENERGY Draw a circle graph to show what percent of the total U.S. energy produced in 2010 was provided by each source.

Petroleum 1%   Renewable 11%
Natural gas 24%
Nuclear 19%
Coal 45%

Renewable	11%
Nuclear	19%
Coal	45%
Natural gas	24%
Petroleum	1%

Source: Energy Information Administration

94. GREENHOUSE GASES Draw a circle graph to show what percent of the total U.S. greenhouse gas emissions in 2010 came from each economic sector.

Electric power	34%
Transportation	27%
Industry	21%
Agriculture	7%
Commercial	6%
Residential	5%

Source: Environmental Protection Agency

95. GOVERNMENT INCOME Complete the following table by finding what percent of total federal government income in 2011 each source provided. Then draw a circle graph for the data.

Total Income, Fiscal Year 2011: $2,300 Billion

Source of income	Amount	Percent of total
Social Security, Medicare, unemployment taxes	$529 billion	23%
Personal income taxes	$690 billion	30%
Corporate income taxes	$115 billion	5%
Excise, estate, customs taxes	$138 billion	6%
Borrowing to cover deficit	$828 billion	36%

Source: 2012 Federal Income Tax Form 1040 Instructions

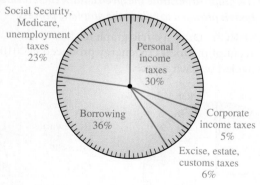

96. WATER USAGE The per-person indoor water use in the typical single family home is about 70 gallons per day. Complete the following table. Then draw a circle graph for the data.

Use	Gallons per person per day	Percent of total daily use
Showers	11.9	17%
Clothes washer	15.4	22%
Dishwasher	0.7	1%
Toilets	18.9	27%
Baths	1.4	2%
Leaks	9.8	14%
Faucets	10.5	15%
Other	1.4	2%

Source: American Water Works Association

WRITING

97. Write a real-life situation that can be described by "9 is what percent of 20?"

98. Write a real-life situation that can be translated to $15 = 25\% \cdot x$.

99. Explain why 150% of a number is more than the number.

100. Explain why each of the following problems is easy to solve.
   a. What is 9% of 100?
   b. 16 is 100% of what number?
   c. 27 is what percent of 27?
101. When solving percent problems, when is it best to write a given percent as a fraction instead of as a decimal?
102. Explain how to identify the amount, the percent, and the base in a percent problem.

REVIEW

103. Add: $2.78 + 6 + 9.09 + 0.3$ 18.17
104. Evaluate: $\sqrt{64} + 3\sqrt{9}$ 17
105. On the number line, which is closer to 5: the number 4.9 or the number 5.001?   5.001
106. Multiply: $34.5464 \cdot 1,000$ 34,546.4
107. Evaluate: $(0.2)^3$ 0.008
108. Evaluate the formula $d = 4t$ for $t = 25$.   100

SECTION 7.3
Applications of Percent

Objectives

1. Calculate sales taxes, total cost, and tax rates.
2. Calculate commissions and commission rates.
3. Find the percent of increase or decrease.
4. Calculate the amount of discount, the sale price, and the discount rate.

In this section, we discuss applications of percent. Three of them (taxes, commissions, and discounts) are directly related to purchasing. A solid understanding of these concepts will make you a better shopper and consumer. The fourth uses percent to describe increases or decreases of such things as population and unemployment.

1 Calculate sales taxes, total cost, and tax rates.

The department store sales receipt shown below gives a detailed account of what items were purchased, how many of each were purchased, and the price of each item.

```
        Bradshaw's
      Department Store #612

  4  @   1.05   GIFTS       $ 4.20
  1  @   1.39   BATTERIES   $ 1.39
  1  @  24.85   TOASTER     $24.85
  3  @   2.25   SOCKS       $ 6.75
  2  @   9.58   PILLOWS     $19.16

  SUBTOTAL                  $56.35
  SALES TAX @ 5.00%         $ 2.82
  TOTAL                     $59.17
```

The purchase price of the items bought → $56.35
The sales tax on the items purchased → $2.82
The total cost → $59.17
The sales tax rate ← 5.00%

The receipt shows that the $56.35 purchase price (labeled *subtotal*) was taxed at a rate of 5%. Sales tax of $2.82 was charged.

This example illustrates the following sales tax formula. Notice that the formula is based on the percent equation discussed in Section 7.2.

Finding the Sales Tax

The sales tax on an item is a percent of the purchase price of the item.

Sales tax = sales tax rate · purchase price
↑   ↑   ↑
amount   =   percent   ·   base

Sales tax rates are usually expressed as a percent. When necessary, sales tax dollar amounts are rounded to the nearest cent.

Self Check 1

SALES TAX What would the sales tax be if the $56.35 purchase were made in a state that has a 6.25% state sales tax? $3.52

Now Try Problem 13

Teaching Example 1
SALES TAX Find the sales tax on a purchase of $64.50 if the sales tax rate is 7%.
Answer:
$4.52

EXAMPLE 1   *Sales Tax*

Find the sales tax on a purchase of $56.35 if the sales tax rate is 5%. (This is the purchase on the sales receipt shown on the previous page.)

Strategy We will identify the sales tax rate and the purchase price.

WHY Then we can use the sales tax formula to find the unknown sales tax.

Solution The sales tax rate is 5% and the purchase price is $56.35.

Sales tax = sales tax rate · purchase price   This is the sales tax formula.

Sales tax =   5%   · $56.35   Substitute 5% for the sales tax rate and $56.35 for the purchase price.

Sales tax = 0.05 · $56.35   Write 5% as a decimal: 5% = 0.05.

Sales tax = $2.8175   Do the multiplication.

$$\begin{array}{r} \overset{3\,1\,2}{56.35} \\ \times\, 0.05 \\ \hline 2.8175 \end{array}$$

Sales tax = $2.8175   The rounding digit in the hundredths column is 1. Prepare to round the sales tax to the nearest cent (hundredth) by identifying the rounding digit and test digit. The test digit is 7.

Sales tax ≈ $2.82   Since the test digit is 5 or greater, round up.

The sales tax on the $56.35 purchase is $2.82. The sales receipt shown on the previous page is correct.

Success Tip It is helpful to see the sales tax problem in Example 1 as a type of percent problem from Section 7.2.

What number   is   5%   of $56.35?
↓   ↓   ↓   ↓   ↓
x =   5%   · $56.35

Look at the department store sales receipt once again. Note that the sales tax was added to the purchase price to get the total cost. This example illustrates the following formula for total cost.

Finding the Total Cost

The total cost of an item is the sum of its purchase price and the sales tax on the item.

Total cost = purchase price + sales tax

EXAMPLE 2 *Total Cost* Find the total cost of the child's car seat shown on the right if the sales tax rate is 7.2%.

Saftey-T First
Child's Car Seat
$249.50
Buy today!
Ships next business day

Strategy First, we will find the sales tax on the child's car seat.

WHY Then we can add the purchase price and the sales tax to find the total of the car seat.

Solution The sales tax rate is 7.2% and the purchase price is $249.50.

Sales tax = **sales tax rate · purchase price**	This is the sales tax formula.
Sales tax = 7.2% · $249.50	Substitute 7.2% for the sales tax rate and $249.50 for the purchase price.
Sales tax = 0.072 · $249.50	Write 7.2% as a decimal: 7.2% = 0.072.
Sales tax = $17.964	Do the multiplication.

```
   249.50
 ×  0.072
   49900
 1746500
 17.96400
```

Sales tax = $17.96̱4   Prepare to round the sales tax to the nearest cent (hundredth) by identifying the rounding digit and test digit.

The rounding digit in the hundredths column is 6.
The test digit is 4.

Sales tax ≈ $17.96   Since the test digit is less than 5, round down.

Thus, the sales tax on the $249.50 purchase is $17.96. The total cost of the car seat is the sum of its purchase price and the sales tax.

Total cost = **purchase price + sales tax**	This is the formula for the total cost.
Total cost = $249.50 + $17.96	Substitute $249.50 for the purchase price and $17.96 for the sales tax.

```
  ¹¹
  249.50
 + 17.96
  267.46
```

Total cost = $267.46   Do the addition.

In addition to sales tax, we pay many other taxes in our daily lives. Income tax, gasoline tax, and Social Security tax are just a few. To find such tax rates, we can use an approach like that discussed in Section 7.2.

EXAMPLE 3 *Withholding Tax* A waitress found that $11.04 was deducted from her weekly gross earnings of $240 for federal income tax. What withholding tax rate was used?

Strategy We will carefully read the problem and use the given facts to write them in the form of a percent sentence.

WHY Then we can translate the sentence into a percent equation (or percent proportion) and solve it to find the unknown withholding tax rate.

Solution There are two methods that can be used to solve this problem.

The percent equation method: Since the withholding tax of $11.04 is some unknown percent of her weekly gross earnings of $240, the percent sentence is:

$11.04   is   what percent   of $240?

11.04 = x · 240   This is the percent equation to solve.

Self Check 2

TOTAL COST Find the total cost of a $179.95 baby stroller if the sales tax rate on the purchase is 3.2%. $185.71

Now Try Problem 17

Teaching Example 2
TOTAL COST Find the total cost of a $619.50 refrigerator if the sales tax rate on the purchase is 2.8%.
Answer:
$636.85

Self Check 3

INHERITANCE TAX A tax of $5,250 was paid on an inheritance of $15,000. What was the inheritance tax rate? 35%

Now Try Problem 21

Teaching Example 3
CITY INCOME TAX For the year 2013, a resident of Detroit, Michigan, paid $528 in city income tax on taxable earnings of $22,000. What is the Detroit city income tax rate?
Answer:
2.4%

$11.04 = 240x$ — On the right side, use the commutative property of multiplication to write $x \cdot 240$ as $240x$.

$\dfrac{11.04}{240} = \dfrac{240x}{240}$ — To isolate x, undo the multiplication by 240 by dividing both sides by 240.

$0.046 = x$ — Do the division: $11.04 \div 240 = 0.046$.

$0\,04.6\% = x$ — To write the decimal 0.046 as a percent, multiply it by 100 by moving the decimal point two places to the right, and then insert a % symbol.

$4.6\% = x$

```
     0.046
240) 11.0400
    −  0
     11 04
    − 9 60
      1 440
    − 1 440
          0
```

The withholding tax rate was 4.6%.

The percent proportion method: Since the withholding tax of $11.04 is some unknown percent of her weekly gross earnings of $240, the percent sentence is:

$11.04 is what percent of $240?

amount — percent — base

$\dfrac{11.04}{240} = \dfrac{x}{100}$ This is the percent proportion to solve.

$11.04 \cdot 100 = 240 \cdot x$ — To solve the proportion, find the cross products and set them equal.

$1{,}104 = 240x$ — On the left side, do the multiplication: $11.04 \cdot 100 = 1{,}104$. On the right side, write $240 \cdot x$ as $240x$.

$\dfrac{1{,}104}{240} = \dfrac{240x}{240}$ — To isolate x, undo the multiplication by 240 by dividing both sides by 240.

$4.6 = x$ — Do the division: $1{,}104 \div 240 = 4.6$.

```
       4.6
240) 1,104.0
    − 960
      144 0
    − 144 0
          0
```

The withholding tax rate was 4.6%.

2 Calculate commissions and commission rates.

Instead of working for a salary or getting paid at an hourly rate, many salespeople are paid on **commission**. They earn a certain percent of the total dollar amount of the goods or services that they sell. The following formula to calculate a commission is based on the percent equation discussed in Section 7.2.

Finding the Commission

The amount of commission paid is a percent of the total dollar sales of goods or services.

Commission = commission rate · sales

amount = percent · base

EXAMPLE 4 Appliance Sales
The commission rate for a salesperson at an appliance store is 16.5%. Find his commission from the sale of a refrigerator that costs $500.

Strategy We will identify the commission rate and the dollar amount of the sale.

WHY Then we can use the commission formula to find the unknown amount of the commission.

Solution The commission rate is 16.5% and the dollar amount of the sale is $500.

Commission = commission rate · sales	This is the commission formula.
Commission = 16.5% · $500	Substitute 16.5% for the commission rate and $500 for the sales.
Commission = 0.165 · $500	Write 16.5% as a decimal: 16.5% = 0.165.
Commission = $82.50	Do the multiplication.

$$\begin{array}{r} \overset{3\,2}{0.165} \\ \times\ 500 \\ \hline 82.500 \end{array}$$

The commission earned on the sale of the $500 refrigerator is $82.50.

Self Check 4
SELLING INSURANCE An insurance salesperson receives a 4.1% commission on each $120 premium paid by a client. What is the amount of the commission on this premium? $4.92

Now Try Problem 25

Teaching Example 4
SELLING REAL ESTATE A real estate agent earns a 3% commission on all sales. Find her commission if she sells a house for the listing price of $212,000.
Answer: $6,360

EXAMPLE 5 Jewelry Sales
A jewelry salesperson earned a commission of $448 for selling a diamond ring priced at $5,600. Find the commission rate.

Strategy We will identify the commission and the dollar amount of the sale.

WHY Then we can use the commission formula to find the unknown commission rate.

Solution The commission is $448 and the dollar amount of the sale is $5,600.

Commission = commission rate · sales	This is the commission formula.
$448 = x · $5,600	Substitute $448 for the commission and $5,600 for the sales. Let x represent the unknown commission rate.
$448 = 5,600x$	We can drop the dollar signs. On the right side, use the commutative property of multiplication to write $x \cdot 5{,}600$ as $5{,}600x$.
$\dfrac{448}{5{,}600} = \dfrac{5{,}600x}{5{,}600}$	To isolate x, undo the multiplication by 5,600 by dividing both sides by 5,600.
$0.08 = x$	Do the division: $448 \div 5{,}600 = 0.08$.
$008\% = x$	To write the decimal 0.08 as a percent, multiply it by 100 by moving the decimal point two places to the right, and then insert a % symbol.
$8\% = x$	

$$\begin{array}{r} 0.08 \\ 5{,}600\overline{)448.00} \\ -448\ 00 \\ \hline 0 \end{array}$$

The commission rate paid the salesperson on the sale of the diamond ring was 8%.

Self Check 5
SELLING ELECTRONICS If the commission on a $430 digital camcorder is $21.50, what is the commission rate? 5%

Now Try Problem 29

Teaching Example 5
JEWELRY SALES A jewelry salesperson earned a commission of $78 for selling a pearl necklace priced at $1,300. Find the commission rate.
Answer:
6%

3 Find the percent of increase or decrease.

Percents can be used to describe how a quantity has changed. For example, consider the table on the right, which shows the number of television channels that the average U.S. home received in 2007 and 2010.

Year	Number of television channels that the average U.S. home received
2007	107
2010	135

Source: videomind.ooyala.com

From the table, we see that the number of television channels received increased considerably from 2007 to 2010. To describe this increase using a percent, we first subtract to find the **amount of increase**.

$135 - 107 = 28$ Subtract the number of TV channels received in 2007 from the number received in 2010.

Thus, the number of channels received increased by 28 from 2007 to 2010.

Next, we find what percent of the *original* 107 channels received in 2007 that the 28 channel increase represents. To do this, we translate the problem into a percent equation (or percent proportion) and solve it.

The percent equation method:

$28 = 107x$ On the right side, use the commutative property of multiplication to write $x \cdot 107$ as $107x$.

$\dfrac{28}{107} = \dfrac{107x}{107}$ To isolate x, undo the multiplication by 107 by dividing both sides by 107.

$0.2616 \approx x$ On the left side, divide 28 by 107. The division does not terminate.

$26.16\% \approx x$ To write the decimal 0.2616 as a percent, multiply it by 100 by moving the decimal point two places to the right, and then insert a % symbol.

$26\% \approx x$ Round to the nearest one percent.

The percent proportion method:

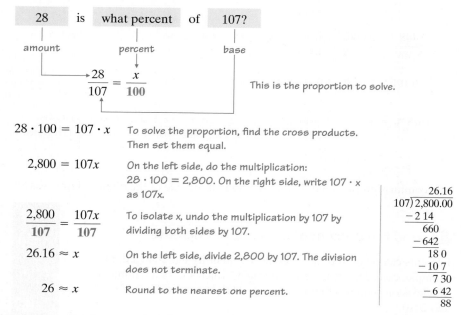

$28 \cdot 100 = 107 \cdot x$ To solve the proportion, find the cross products. Then set them equal.

$2{,}800 = 107x$ On the left side, do the multiplication: $28 \cdot 100 = 2{,}800$. On the right side, write $107 \cdot x$ as $107x$.

$\dfrac{2{,}800}{107} = \dfrac{107x}{107}$ To isolate x, undo the multiplication by 107 by dividing both sides by 107.

$26.16 \approx x$ On the left side, divide 2,800 by 107. The division does not terminate.

$26 \approx x$ Round to the nearest one percent.

With either method, we see that there was a 26% increase in the number of television channels received by the average American home from 2007 to 2010.

EXAMPLE 6

JFK A 1996 auction included an oak rocking chair used by President John F. Kennedy in the Oval Office. The chair, originally valued at $5,000, sold for $453,500. Find the percent of increase in the value of the rocking chair.

Strategy We will begin by finding the amount of increase in the value of the rocking chair.

WHY Then we can calculate what percent of the original $5,000 value of the chair that the increase represents.

Solution First, we find the amount of increase in the value of the rocking chair.

$453{,}500 - 5{,}000 = 448{,}500$ Subtract the original value from the price paid at auction.

The rocking chair increased in value by $448,500. Next, we find what percent of the original $5,000 value of the rocking chair the $448,500 increase represents by translating the problem into a percent equation (or percent proportion) and solving it.

The percent equation method:

$448{,}500$ is what percent of $5{,}000$?

$448{,}500 = x \cdot 5{,}000$ This is the percent equation to solve.

$448{,}500 = 5{,}000x$ On the right side, use the commutative property of multiplication to write $x \cdot 5{,}000$ as $5{,}000x$.

$\dfrac{448{,}500}{5{,}000} = \dfrac{5{,}000x}{5{,}000}$ To isolate x, undo the multiplication by 5,000 by dividing both sides by 5,000.

On the left side, recall that there is a shortcut for dividing a dividend by a divisor when both end with zeros. Remove two of the ending zeros in the divisor 5,000 and remove the same number of ending zeros in the dividend 448,500.

$\dfrac{4{,}485}{50} = x$

$89.7 = x$ Do the division: $4{,}485 \div 50 = 89.7$.

$89.7\% = x$ To write the decimal 89.7 as a percent, multiply it by 100 by moving the decimal point two places to the right, and then insert a % symbol.

$8{,}970\% = x$

```
        89.7
50)4,485.0
   -4 00
     485
    -450
      35 0
     -35 0
         0
```

The percent proportion method:

$\$448{,}500$ is what percent of $\$5{,}000$?

amount — percent — base

$\dfrac{448{,}500}{5{,}000} = \dfrac{x}{100}$ This is the proportion to solve.

$448{,}500 \cdot 100 = 5{,}000 \cdot x$ To solve the proportion, find the cross products. Then set them equal.

$44{,}850{,}000 = 5{,}000x$ On the left side, do the multiplication: $448{,}500 \cdot 100 = 44{,}850{,}000$. On the right side, write $5{,}000 \cdot x$ as $5{,}000x$.

Self Check 6

HOME SCHOOLING In one school district, the number of home-schooled children increased from 15 to 150 in four years. Find the percent of increase. 900%

Now Try Problem 33

Teaching Example 6
TALK RADIO Refer to the table below and find the percent increase in the number of news/talk radio stations from 2004 to 2012. Round to the nearest one percent.

Year	Number of U.S. News/ Radio Stations
2004	1,282
2012	1,503

Source: *The World Almanac and Book of Facts*, 2013

Answer:
17%

$$\frac{44{,}850{,}000}{5{,}000} = \frac{5{,}000x}{5{,}000}$$ To isolate x, undo the multiplication by 5,000 by dividing both sides by 5,000.

On the left side, recall that there is a shortcut for dividing a dividend by a divisor when both end with zeros. Remove the three ending zeros in the divisor 5,000 and remove the same number of ending zeros in the dividend 44,850,000.

$$\frac{44{,}850}{5} = x$$

$8{,}970 = x$ Divide 44,850 by 5.

With either method, we see that there was an amazing 8,970% increase in the value of the Kennedy rocking chair.

Caution! The percent of increase (or decrease) is a percent of the *original number*—that is, the number before the change occurred. Thus, in Example 6, it would be incorrect to write a percent sentence that compares the increase to the *new value* of the Kennedy rocking chair.

$448,500 is what percent of $453,500?

Finding the Percent of Increase or Decrease

To find the percent of increase or decrease:

1. Subtract the smaller number from the larger to find the amount of increase or decrease.
2. Find what percent the amount of increase or decrease is of the original amount.

Self Check 7

REDUCING FAT INTAKE One serving of the original Jif peanut butter has 16 grams of fat per serving. The new Jif Reduced Fat product contains 12 grams of fat per serving. What is the percent decrease in the number of grams of fat per serving? 25%

Now Try Problem 37

Teaching Example 7
INSURANCE COSTS A college student's good grades earned her a student discount on her car insurance premium. Find the percent of decrease if her annual premium was lowered from $1,050 to $924.
Answer:
12%

EXAMPLE 7 *Commercials* Jared Fogle credits his tremendous weight loss to exercise and a diet of low-fat Subway sandwiches. His maximum weight (reached in March 1998) was 425 pounds. His current weight is about 187 pounds. Find the percent of decrease in his weight.

Strategy We will begin by finding the amount of decrease in Jared Fogle's weight.

WHY Then we can calculate what percent of his original 425-pound weight the decrease represents.

Solution First, we find the amount of decrease in his weight.

$425 - 187 = 238$ Subtract his new weight from his weight before going on the weight-loss program.

His weight decreased by 238 pounds.

Next, we find what percent of his original 425 weight the 238-pound decrease represents by translating the problem into a percent equation (or percent proportion) and solving it.

Andrew Kent/Getty Images

The percent equation method:

$$\underbrace{238}_{} \text{ is } \underbrace{\text{what percent}}_{} \text{ of } \underbrace{425}_{}?$$

$238 = x \cdot 425$ This is the percent equation to solve.

$238 = 425x$ On the right side, use the commutative property of multiplication to write $x \cdot 425$ as $425x$.

$\dfrac{238}{425} = \dfrac{425x}{425}$ To isolate x, undo the multiplication by 425 by dividing both sides by 425.

$0.56 = x$ Do the division: $238 \div 425 = 0.56$.

$0.56\% = x$ To write the decimal 0.56 as a percent, multiply it by 100 by moving the decimal point two places to the right, and then insert a % symbol.

$56\% = x$

```
     0.56
425)238.00
    -212 5
     25 50
    -25 50
         0
```

The percent proportion method:

$$\underbrace{238}_{\text{amount}} \text{ is } \underbrace{\text{what percent}}_{\text{percent}} \text{ of } \underbrace{425}_{\text{base}}?$$

$\dfrac{238}{425} = \dfrac{x}{100}$ This is the proportion to solve.

$238 \cdot 100 = 425 \cdot x$ To solve the proportion, find the cross products. Then set them equal.

$23{,}800 = 425x$ On the left side, do the multiplication: $238 \cdot 100 = 23{,}800$. On the right side, write $425 \cdot x$ as $425x$.

$\dfrac{23{,}800}{425} = \dfrac{425x}{425}$ To isolate x, undo the multiplication by 425 by dividing both sides by 425.

$56 = x$ Do the division: $23{,}800 \div 425 = 56$.

```
       56
425)23,800
   -21 25
    2 550
   -2 550
        0
```

With either method, we see that there was a 56% decrease in Jared Fogle's weight.

THINK IT THROUGH *Studying Mathematics*

"All students, regardless of their personal characteristics, backgrounds, or physical challenges, must have opportunities to study—and support to learn—mathematics."
National Council of Teachers of Mathematics

The table below shows the number of students enrolled in Basic Mathematics classes at two-year colleges.

Year	1975	1980	1985	1990	1995	2000	2005	2010
Enrollment	100,000	146,000	142,000	147,000	134,000	122,000	104,000	146,000

Source: CBMS Survey of Undergraduate Programs, 2005, 2012

1. Over what five-year span was there the greatest percent increase in enrollment in Basic Mathematics classes? What was the percent increase?
   1975–1980, a 46% increase
2. Over what five-year span was there the greatest percent decrease in enrollment in Basic Mathematics classes? What was the percent decrease?
   2000–2005, about a 15% decrease

4 Calculate the amount of discount, the sale price, and the discount rate.

While shopping, you have probably noticed that many stores display signs advertising sales. Store managers have found that offering discounts attracts more customers. To be a smart shopper, it is important to know the vocabulary of discount sales.

The difference between the **original price** and the **sale price** of an item is called the **amount of discount**, or simply the **discount**. If the discount is expressed as a percent of the selling price, it is called the **discount rate**.

If we know the original price and the sale price of an item, we can use the following formula to find the amount of discount.

Finding the Discount

The amount of discount is the difference between the original price and the sale price.

Amount of discount = original price − sale price

If we know the original price of an item and the discount rate, we can use the following formula to find the amount of discount. Like several other formulas in this section, it is based on the percent equation discussed in Section 7.2.

Finding the Discount

The amount of discount is a percent of the original price.

Amount of discount = discount rate · original price

amount = percent · base

We can use the following formula to find the sale price of an item that is being discounted.

Finding the Sale Price

To find the sale price of an item, subtract the discount from the original price.

Sale price = original price − discount

Unless otherwise noted, all art on this page is © Cengage Learning.

EXAMPLE 8

Shoe Sales Use the information in the advertisement shown on the previous page to find the amount of the discount on the pair of men's basketball shoes. Then find the sale price.

Strategy We will identify the discount rate and the original price of the shoes and use a formula to find the amount of the discount.

WHY Then we can subtract the discount from the original price to find the sale price of the shoes.

Solution From the advertisement, we see that the discount rate on the men's shoes is 25% and the original price is $89.80.

Amount of discount =	**discount rate · original price**		This is the amount of discount formula.
Amount of discount =	25% ·	$89.80	Substitute 25% for the discount rate and $89.80 for the original price.
Amount of discount =	0.25 · $89.80		Write 25% as a decimal: 25% = 0.25.
Amount of discount =	$22.45		Do the multiplication.

$$\begin{array}{r} 89.80 \\ \times\, 0.25 \\ \hline 44900 \\ 179600 \\ \hline 22.4500 \end{array}$$

The discount on the men's shoes is $22.45. To find the sale price, we use subtraction.

Sale price =	**original price − discount**		This is the sale price formula.
Sale price =	$89.80 − $22.45		Substitute $89.80 for the original price and $22.45 for the discount.
Sale price =	$67.35		Do the subtraction.

$$\begin{array}{r} \overset{7\,10}{89.8\!\!\not{0}} \\ -\,22.45 \\ \hline 67.35 \end{array}$$

The sale price of the men's basketball shoes is $67.35.

Self Check 8

SUNGLASSES SALES Sunglasses, regularly selling for $15.40, are discounted 15%. Find the amount of the discount. Then find the sale price. $2.31, $13.09

Now Try Problem 41

Teaching Example 8
BOOT SALES Find the amount of the discount on a pair of hiking boots if they are normally priced at $124.60, but are currently on sale for 15% off. Then find the sale price.
Answer:
$18.69, $105.91

EXAMPLE 9

Discounts Find the discount rate on the ladies' cross trainer shoes shown in the advertisement on the previous page. Round to the nearest one percent.

Strategy We will think of this as a percent-of-decrease problem.

WHY We want to find what percent of the $59.99 original price the amount of discount represents.

Solution From the advertisement, we see that the original price of the women's shoes is $59.99 and the sale price is $33.99. The discount (decrease in price) is found using subtraction.

$59.99 − $33.99 = $26 Use the formula:
Amount of discount = original price − sale price.

The shoes are discounted $26. Now we find what percent of the original price the $26 discount represents.

Amount of discount =	discount rate · original price		This is the amount of discount formula.
26 =	x ·	$59.99	Substitute 26 for the amount of discount and $59.99 for the original price. Let x represent the unknown discount rate.
26 = 59.99x			We can drop the dollar signs. On the right side, use the commutative property of multiplication to write x · 59.99 as 59.99x.

Self Check 9

DINING OUT An early-bird special at a restaurant offers a $10.99 prime rib dinner for only $7.95 if it is ordered before 6 P.M. Find the rate of discount. Round to the nearest one percent. 28%

Now Try Problem 45

Teaching Example 9
FURNITURE SALES Find the discount rate on a desk lamp regularly priced at $89.95 that is on sale for $79.95. Round to the nearest one percent.
Answer:
11%

$$\frac{26}{59.99} = \frac{59.99x}{59.99}$$ To isolate x, undo the multiplication by 59.99 by dividing both sides by 59.99.

$0.433 \approx x$ On the left side, divide 26 by 59.99. The division does not terminate.

$043.3\% \approx x$ To write the decimal 0.433 as a percent, multiply it by 100 by moving the decimal point two places to the right, and then insert a % symbol.

$43\% \approx x$ Round to the nearest one percent.

```
        0.433
59.99)26 00.000
     −23 99 6
       2 00 40
      −1 79 97
         20 430
        −17 997
          2 433
```

To the nearest one percent, the discount rate on the women's shoes is 43%.

ANSWERS TO SELF CHECKS
1. $3.52 2. $185.71 3. 35% 4. $4.92 5. 5% 6. 900% 7. 25%
8. $2.31, $13.09 9. 28%

SECTION 7.3 STUDY SET

VOCABULARY

Fill in the blanks.

1. Instead of working for a salary or getting paid at an hourly rate, some salespeople are paid on __commission__. They earn a certain percent of the total dollar amount of the goods or services they sell.
2. Sales tax __rates__ are usually expressed as a percent.
3. a. When we use percent to describe how a quantity has increased compared to its original value, we are finding the percent of __increase__.
   b. When we use percent to describe how a quantity has decreased compared to its __original__ value, we are finding the percent of decrease.
4. Refer to the advertisement below for a ceiling fan on sale.
   a. The __original__ price of the ceiling fan was $199.99.
   b. The amount of the __discount__ is $40.00.
   c. The discount __rate__ is 20%.
   d. The __sale__ price of the ceiling fan is $159.99.

CONCEPTS

Fill in the blanks in each of the following formulas.

5. Sales tax = sales tax rate · __purchase price__
6. Total cost = __purchase price__ + sales tax
7. Commission = commission rate · __sales__
8. a. Amount of discount = original price − __sale price__
   b. Amount of discount = __discount rate__ · original price
   c. Sale price = __original price__ − discount
9. a. The sales tax on an item priced at $59.32 is $4.75. What is the total cost of the item? $64.07
   b. The original price of an item is $150.99. The amount of discount is $15.99. What is the sale price of the item? $135.00
10. Round each dollar amount to the nearest cent.
    a. $168.257 $168.26
    b. $57.234 $57.23
    c. $3.396 $3.40
11. Fill in the blanks: To find the percent decrease, __subtract__ the smaller number from the larger number to find the amount of decrease. Then find what percent that difference is of the __original__ amount.

12. **NEWSPAPERS** The table below shows how the circulations of two daily newspapers changed from 2011 to 2012.

Daily Circulation

	New York Times	USA Today
2011	1,150,589	1,784,242
2012	1,613,865	1,713,833

Source: poynter.org

a. What was the *amount of decrease* of *USA Today's* circulation? 70,409

b. What was the *amount of increase* of the *New York Times's* circulation? 463,276

GUIDED PRACTICE

Solve each problem to find the sales tax. See Example 1.

13. Find the sales tax on a purchase of $92.70 if the sales tax rate is 4%. $3.71

14. Find the sales tax on a purchase of $33.60 if the sales tax rate is 8%. $2.69

15. Find the sales tax on a purchase of $83.90 if the sales tax rate is 5%. $4.20

16. Find the sales tax on a purchase of $234.80 if the sales tax rate is 2%. $4.70

Solve each problem to find the total cost. See Example 2.

17. Find the total cost of a $68.24 purchase if the sales tax rate is 3.8%. $70.83

18. Find the total cost of an $86.56 purchase if the sales tax rate is 4.3%. $90.28

19. Find the total cost of a $60.18 purchase if the sales tax rate is 6.4%. $64.03

20. Find the total cost of a $70.73 purchase if the sales tax rate is 5.9%. $74.90

Solve each problem to find the tax rate. See Example 3.

21. **SALES TAX** The purchase price for a blender is $140. If the sales tax is $7.28, what is the sales tax rate? 5.2%

22. **SALES TAX** The purchase price for a camping tent is $180. If the sales tax is $8.64, what is the sales tax rate? 4.8%

23. **SELF-EMPLOYMENT TAX** A business owner paid self-employment tax of $4,590 on a taxable income of $30,000. What is the self-employment tax rate? 15.3%

24. **CAPITAL GAINS TAX** A couple paid $3,000 in capital gains tax on a profit of $20,000 made from the sale of some shares of stock. What is the capital gains tax rate? 15%

Solve each problem to find the commission. See Example 4.

25. **SELLING SHOES** A shoe salesperson earns a 12% commission on all sales. Find her commission if she sells a pair of casual shoes for $95. $11.40

26. **SELLING CARS** A used car salesperson earns an 11% commission on all sales. Find his commission if he sells a 2001 Chevy Malibu for $4,800. $528

27. **EMPLOYMENT AGENCIES** An employment counselor receives a 35% commission on the first week's salary of anyone that she places in a new job. Find her commission if one of her clients is hired as a secretary at $480 per week. $168

28. **PHARMACEUTICAL SALES** A medical sales representative is paid an 18% commission on all sales. Find her commission if she sells $75,000 of Coumadin, a blood-thinning drug, to a pharmacy chain. $13,500

Solve each problem to find the commission rate. See Example 5.

29. **AUCTIONS** An auctioneer earned a $15 commission on the sale of an antique chair for $750. What is the commission rate? 2%

30. **SELLING TIRES** A tire salesman was paid a $28 commission after one of his customers purchased a set of new tires for $560. What is the commission rate? 5%

31. **SELLING ELECTRONICS** If the commission on a $500 laptop computer is $20, what is the commission rate? 4%

32. **SELLING CLOCKS** If the commission on a $600 grandfather clock is $54, what is the commission rate? 9%

Solve each problem to find the percent of increase. See Example 6.

33. **CLUBS** The number of members of a service club increased from 80 to 88. What was the percent of increase in club membership? 10%

34. **SAVINGS ACCOUNTS** The amount of money in a savings account increased from $2,500 to $3,000. What was the percent of increase in the amount of money saved? 20%

35. **RAISES** After receiving a raise, the salary of a secretary increased from $300 to $345 dollars per week. What was the percent of increase in her salary? 15%

36. **TUITION** The tuition at a community college increased from $2,500 to $2,650 per semester. What was the percent of increase in the tuition? 6%

Solve each problem to find the percent of decrease. **See Example 7.**

37. **TRAVEL TIME** After a new freeway was completed, a commuter's travel time to work decreased from 30 minutes to 24 minutes. What was the percent of decrease in travel time? 20%

38. **LAYOFFS** A printing company reduced the number of employees from 300 to 246. What was the percent of decrease in the number of employees? 18%

39. **ENROLLMENT** Thirty-six of the 40 students originally enrolled in an algebra class completed the course. What was the percent of decrease in the number of students in the class? 10%

40. **DECLINING SALES** One year, a pumpkin patch sold 1,200 pumpkins. The next year, they only sold 900 pumpkins. What was the percent of decrease in the number of pumpkins sold? 25%

Eye for Africa/Shutterstock.com

Solve each problem to find the amount of the discount and the sale price. **See Example 8.**

41. **DINNERWARE SALES** Find the amount of the discount on a six-place dinnerware set if it regularly sells for $90, but is on sale for 33% off. Then find the sale price of the dinnerware set. $29.70, $60.30

42. **BEDDING SALES** Find the amount of the discount on a $130 bedspread that is now selling for 20% off. Then find the sale price of the bedspread. $26, $104

43. **MEN'S CLOTHING SALES** 501 Levi jeans that regularly sell for $58 are now discounted 15%. Find the amount of the discount. Then find the sale price of the jeans. $8.70, $49.30

44. **BOOK SALES** At a bookstore, the list price of $23.50 for the *Merriam-Webster's Collegiate Dictionary* is crossed out, and a 30% discount sticker is pasted on the cover. Find the amount of the discount. Then find the sale price of the dictionary. $7.05, $16.45

Solve each problem to find the discount rate. **See Example 9.**

45. **LADDER SALES** Find the discount rate on an aluminum ladder regularly priced at $79.95 that is on sale for $64.95. Round to the nearest one percent. 19%

46. **OFFICE SUPPLIES SALES** Find the discount rate on an electric pencil sharpener regularly priced at $49.99 that is on sale for $45.99. Round to the nearest one percent. 8%

47. **DISCOUNT TICKETS** The price of a one-way airline ticket from Atlanta to New York City was reduced from $209 to $179. Find the discount rate. Round to the nearest one percent. 14%

48. **DISCOUNT HOTELS** The cost of a one-night stay at a hotel was reduced from $245 to $200. Find the discount rate. Round to the nearest one percent. 18%

APPLICATIONS

49. **SALES TAX** The Massachusetts state sales tax rate is 6.25%. Find the sales tax on a dining room set that sells for $900. $56.25

50. **SALES TAX** Find the sales tax on a pair of jeans costing $40 if they are purchased in Missouri, which has a state sales tax rate of 4.225%. $1.69

51. **SALES RECEIPTS** Complete the sales receipt below by finding the subtotal, the sales tax, and the total cost of the purchase. $47.34, $2.84, $50.18

```
         NURSERY CENTER
         Your one-stop garden supply

    3 @  2.99    PLANTING MIX    $ 8.97
    1 @  9.87    GROUND COVER    $ 9.87
    2 @ 14.25    SHRUBS          $28.50

    SUBTOTAL                     $
    SALES TAX @ 6.00%            $
    TOTAL                        $
```

52. **SALES RECEIPTS** Complete the sales receipt below by finding all three prices, the subtotal, the sales tax, and the total cost of the purchase.
$450.00, $180.00, $350.00, $980.00, $41.16, $1,021.16

```
         McCOY'S FURNITURE

    1 @ 450.00   SOFA            $
    2 @  90.00   END TABLES      $
    1 @ 350.00   LOVE SEAT       $

    SUBTOTAL                     $
    SALES TAX @ 4.20%            $
    TOTAL                        $
```

53. **ROOM TAX** After checking out of a hotel, a man noticed that the hotel bill included an additional charge labeled *room tax*. If the price of the room was $129 plus a room tax of $10.32, find the room tax rate. 8%

54. **TANNING** As part of the Affordable Care Act of 2009, indoor tanning services are subject to a federal excise tax. If a customer pays $25 for a 10-minute tanning session, and the tax is $2.50, what is the excise tax rate? 10%

55. **GAMBLING** For state authorized wagers (bets) placed with legal bookmakers and lottery operators, there is a federal excise tax on the wager. What is the excise tax rate if there is an excise tax of $5 on a $2,000 bet? 0.25%

56. BUYING FISHING EQUIPMENT There are federal excise taxes on the retail price when purchasing fishing equipment. The taxes are intended to help pay for parks and conservation. What is the federal excise tax rate if there is an excise tax of $17.50 on a fishing rod and reel that has a retail price of $175? 10%

57. TAX HIKES In order to raise more revenue, some states raise the sales tax rate. How much additional money will be collected on the sale of a $24,950 car if the sales tax rate is raised 1%? $249.50

58. FOREIGN TRAVEL Value-added tax (VAT) is a consumer tax on goods and services. Currently, VAT systems are in place all around the world. (The United States is one of the few nations not using a value-added tax system.) Complete the table by determining the VAT a traveler would pay in each country on a dinner that cost $25. Round to the nearest cent.

Country	VAT rate	Tax on a $25 dinner
Mexico	16%	$4.00
Germany	19%	$4.75
France	19.6%	$4.90
Sweden	25%	$6.25

Source: www.worldwide-tax.com

59. PAYCHECKS Use the information on the paycheck stub to find the tax rate for the federal withholding, worker's compensation, Medicare, and Social Security taxes that were deducted from the gross pay.
8%, 3.75%, 1.45%, 6.2%

```
                6286244
Issue date: 03-27-14
GROSS PAY           $360.00
TAXES
   FED. TAX         $ 28.80
   WORK. COMP.      $ 13.50
   MEDICARE         $  5.22
   SOCIAL SECURITY  $ 22.32

NET PAY             $290.16
```

60. GASOLINE TAX In one state, a gallon of unleaded gasoline sells for $3.61. This price includes federal and state taxes that total approximately $0.69. Therefore, the price of a gallon of gasoline, before taxes, is $2.92. What is the tax rate on gasoline? Round to the nearest one percent. 24%

61. POLICE FORCE A police department plans to increase its 80-person force to 84 persons. Find the percent increase in the size of the police force. 5%

62. COST-OF-LIVING INCREASES A woman making $32,000 a year receives a cost-of-living increase that raises her salary to $32,768 per year. Find the percent of increase in her yearly salary. 2.4%

63. LAKE SHORELINES Because of a heavy spring runoff, the shoreline of a lake increased from 5.8 miles to 7.6 miles. What was the percent of increase in the length of the shoreline? Round to the nearest one percent. 31%

64. CROP DAMAGE After flooding damaged much of the crop, the cost of a head of lettuce jumped from $0.99 to $2.20. What percent of increase is this? Round to the nearest one percent. 122%

65. OVERTIME From May to June, the number of overtime hours for employees at a printing company increased from 42 to 106. What is the percent of increase in the number of overtime hours? Round to the nearest percent. 152%

66. TOURISM The graph below shows the number of international visitors (travelers) to the United States each year from 2005 to 2011.

a. The greatest percent of increase in the number of travelers was between 2006 and 2007. Find the percent increase. Round to the nearest one percent. 10%

b. The only decrease in the number of travelers was between 2008 and 2009. Find the percent decrease. Round to the nearest one percent. 5%

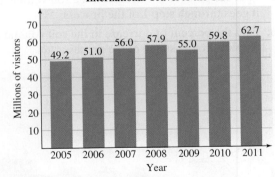

Source: U.S. Department of Commerce

67. REDUCED CALORIES A company advertised its new, improved chips as having 96 calories per serving. The original style contained 150 calories. What percent of decrease in the number of calories per serving is this? 36%

68. **CAR INSURANCE** A student paid a car insurance premium of $420 every three months. Then the premium dropped to $370, because she qualified for a good-student discount. What was the percent of decrease in the premium? Round to the nearest percent. 12%

69. **BUS PASSES** To increase the number of riders, a bus company reduced the price of a monthly pass from $112 to $98. What was the percent of decrease in the cost of a bus pass? 12.5%

70. **BASEBALL** The illustration below shows the path of a baseball hit 110 mph, with a launch angle of 35 degrees, at sea level and at Coors Field, home of the Colorado Rockies. What is the percent of increase in the distance the ball travels at Coors Field? 10%

Source: *Los Angeles Times*, September 16, 1996

71. **EARTH MOVING** The illustration below shows the typical soil volume change during earth moving. (One cubic yard of soil fits in a cube that is 1 yard long, 1 yard wide, and 1 yard high.)

   a. Find the percent of increase in the soil volume as it goes through step 1 of the process. 25%
   b. Find the percent of decrease in the soil volume as it goes through step 2 of the process. 36%

Source: U.S. Department of the Army

72. **PARKING** The management of a mall has decided to increase the parking area. The plans are shown in the next column. What will be the percent of increase in the parking area when the project is completed? 30%

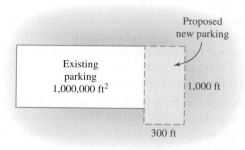

73. **REAL ESTATE** After selling a house for $198,500, a real estate agent split the 6% commission with another agent. How much did each person receive? $5,955

74. **COMMISSIONS** A salesperson for a medical supplies company is paid a regular commission of 9%. For orders exceeding $8,000, she receives an additional 2% in commission on the total amount. What is her commission on a sale of $14,600? $1,606

75. **SPORTS AGENTS** A sports agent charges her clients a fee to represent them during contract negotiations. The fee is based on a percent of the contract amount. If the agent earned $37,500 when her client signed a $2,500,000 professional football contract, what rate did she charge for her services? 1.5%

76. **ART GALLERIES** An art gallery displays paintings for artists and receives a commission from the artist when a painting is sold. What is the commission rate if a gallery received $135.30 when a painting was sold for $820? 16.5%

77. **WHOLE LIFE INSURANCE** For the first 12 months, insurance agents earn a very large commission on the monthly premium of any whole life policy that they sell. After that, the commission rate is lowered significantly. Suppose on a new policy with monthly premiums of $160, an agent is paid monthly commissions of $144. Find the commission rate. 90%

78. **TERM INSURANCE** For the first 12 months, insurance agents earn a large commission on the monthly premium of any term life policy that they sell. After that, the commission rate is lowered significantly. Suppose on a new policy with monthly premiums of $180, an agent is paid monthly commissions of $81. Find the commission rate. 45%

79. **CONCERT PARKING** A concert promoter gets a commission of $33\frac{1}{3}$% of the revenue an arena receives from parking the night of the performance. How much can the promoter make if 6,000 cars are expected and parking costs $6 a car? $12,000

80. **PARTIES** A homemaker invited her neighbors to a kitchenware party to show off cookware and utensils. As party hostess, she received 12% of the total sales. How much was purchased if she received $41.76 for hosting the party? $348

81. **WATCH SALE** Refer to the advertisement below.
   a. Find the amount of the discount on the watch. $7.99
   b. Find the sale price of the watch. $31.96

82. **SCOOTER SALE** Refer to the advertisement below.
   a. Find the amount of the discount on the scooter. $108
   b. Find the sale price of the scooter. $492

83. **SEGWAYS** Find the discount rate on a Segway PT shown in the advertisement. Round to the nearest one percent. 6%

84. **FAX MACHINES** An HP 3180 fax machine, regularly priced at $160, is on sale for $116. What is the discount rate? 27.5%

85. **DISC PLAYERS** What are the sale price and the discount rate for a Blu-ray disc player that regularly sells for $399.97 and is being discounted $50? Round to the nearest one percent. $349.97, 13%

86. **CAMCORDER SALE** What are the sale price and the discount rate for a camcorder that regularly sells for $559.97 and is being discounted $80? Round to the nearest one percent. $479.97, 14%

87. **REBATES** Find the discount rate and the new price for a case of motor oil if a shopper receives the manufacturer's rebate mentioned in the advertisement. Round to the nearest one percent. 23%, $11.88

88. **DOUBLE COUPONS** Find the discount, the discount rate, and the reduced price for a box of cereal that normally sells for $3.29 if a shopper presents the coupon at a store that doubles the value of the coupon. $0.70, 21%, $2.59

89. **TV SHOPPING** Determine the Home Shopping Network (HSN) price of the ring described in the illustration if it sells it for 55% off of the retail price. Ignore shipping and handling costs. $76.50

90. **INFOMERCIALS** The host of a TV infomercial says that the suggested retail price of a rotisserie grill is $249.95 and that it is now offered "for just 4 easy payments of only $39.95." What is the discount, and what is the discount rate? $90.15, 36%

91. **RING SALE** What does a ring regularly sell for if it has been discounted 20% and is on sale for $149.99? (*Hint:* The ring is selling for 80% of its regular price.) $187.49

92. **BLINDS SALE** What do vinyl blinds regularly sell for if they have been discounted 55% and are on sale for $49.50? (*Hint:* The blinds are selling for 45% of their regular price.) $110

WRITING

93. Explain the difference between a sales tax and a sales tax rate.

94. List the pros and cons of working on commission.

95. Suppose the price of an item increases $25 from $75 to $100. Explain why the following percent sentence *cannot* be used to find the percent of increase in the price of the item.

    25 is what percent of 100?

96. Explain how to find the sale price of an item if you know the regular price and the discount rate.

REVIEW

97. Multiply: $-5(-5)(-2)$ -50

98. Divide: $\dfrac{-320}{40}$ -8

99. Subtract: $-4 - (-7)$ 3

100. Add: $-17 + 6 + (-12)$ -23

101. Evaluate: $|-5 - 8|$ 13

102. Evaluate: $\sqrt{25} - \sqrt{16}$ 1

Objectives

1. Estimate answers to percent problems involving 1% and 10%.
2. Estimate answers to percent problems involving 50%, 25%, 5%, and 15%.
3. Estimate answers to percent problems involving 200%.
4. Use estimation to solve percent application problems.

SECTION 7.4
Estimation with Percent

Estimation can be used to find approximations when exact answers aren't necessary. For example, when dining at a restaurant, it's helpful to be able to estimate the amount of the tip. When shopping, the ability to estimate a discount or the sale price of an item also comes in handy. In this section, we will discuss some estimation methods that can be used to make quick calculations involving percents.

1 Estimate answers to percent problems involving 1% and 10%.

There is an easy way to find 1% of a number that does not require any calculations. First, recall that $1\% = \frac{1}{100} = 0.01$. Thus, to find 1% of a number, we multiply it by 0.01, and a quick way to multiply the number by 0.01 is to move its decimal point *two places to the left*.

> **Finding 1% of a Number**
>
> To find 1% of a number, move the decimal point in the number two places to the left.

Self Check 1

What is 1% of 519.3? Find the exact answer and an estimate using front-end rounding. 5.193, 5

Now Try Problem 11

Teaching Example 1 What is 1% of 722.6? Find the exact answer and an estimate using front-end rounding.
Answer:
7.226, 7

EXAMPLE 1 What is 1% of 423.1? Find the exact answer and an estimate using front-end rounding.

Strategy To find the exact answer, we will move the decimal point in 423.1 two places to the left. To find an estimate, we will move the decimal point in an approximation of 423.1 two places to the left.

WHY We move the decimal point *two* places to the left because 1% of a number means 0.01 of (times) the number.

Solution
Exact answer:

$$1\% \text{ of } 423.1 = 4.231 \quad \text{Move the decimal point in 423.1 two places to the left.}$$

Estimate: Recall from Chapter 1 that with **front-end rounding**, a number is rounded to its largest place value so that all but its first digit is zero. To estimate 1% of 423.1, we can front-end round 423.1 to 400 and find 1% of 400. If we move the understood decimal point in 400 two places to the left, we get 4. Thus,

$$1\% \text{ of } 423.1 \approx 4 \quad \text{Because 1\% of 400 = 4.}$$

> **Success Tip** To quickly find 2% of a number, find 1% of the number by moving the decimal point two places to the left, and then double (multiply by 2) the result. In Example 1, we found that 1% of 423.1 is **4.231**. Thus, 2% of 423.1 is $2 \cdot 4.231 = 8.462$. A similar approach can be used to find 3% of a number, 4% of a number, and so on.

There is also an easy way to find 10% of a number that doesn't require any calculations. First, recall that $10\% = \frac{10}{100} = \frac{1}{10}$. Thus, to find 10% of a number, we multiply the number by 0.1, and a quick way to multiply the number by 0.1 is to move its decimal point *one place to the left*.

7.4 Estimation with Percent

Finding 10% of a Number

To find 10% of a number, move the decimal point in the number one place to the left.

EXAMPLE 2 What is 10% of 6,872 feet? Find the exact answer and an estimate using front-end rounding.

Strategy To find the exact answer, we will move the decimal point in 6,872 one place to the left. To find an estimate, we will move the decimal point in an approximation of 6,872 one place to the left.

WHY We move the decimal point *one* place to the left because 10% of a number means 0.10 of (times) the number.

Solution
Exact answer:

10% of 6,872 feet = 687.2 feet   *Move the understood decimal point in 6,872 one place to the left.*

Estimate: To estimate 10% of 6,872 feet, we can front-end round 6,872 to 7,000 and find 10% of 7,000 feet. If we move the understood decimal point in 7,000 one place to the left, we get 700. Thus,

10% of 6,872 feet ≈ 700 feet   *Because 10% of 7,000 = 700.*

Caution! In Examples 1 and 2, *front-end rounding* was used to find estimates of answers to percent problems. Since there are other ways to approximate (round) the numbers involved in a percent problem, the answers to estimation problems may vary.

Self Check 2
What is 10% of 3,536 pounds? Find the exact answer and an estimate using front-end rounding. 353.6 lb, 400 lb

Now Try Problem 15

Teaching Example 2 What is 10% of 5,844 miles? Find the exact answer and an estimate using front-end rounding.
Answer:
584.4 mi, 600 mi

The rule for finding 10% of a number can be extended to help us quickly find multiples of 10% of a number.

Finding 20%, 30%, 40%, ... of a Number

To find 20% of a number, find 10% of the number by moving the decimal point one place to the left, and then double (multiply by 2) the result. A similar approach can be used to find 30% of a number, 40% of a number, and so on.

EXAMPLE 3 Estimate the answer: What is 20% of 416?

Strategy We will estimate 10% of 416, and double (multiply by 2) the result.

WHY 20% of a number is twice as much as 10% of a number.

Solution Since 10% of 416 is 41.6 (or about **42**), it follows that 20% of 416 is about 2 · **42**, which is 84. Thus,

20% of 416 ≈ 84   *Because 10% of 416 = 41.6 ≈ 42 and 2 · 42 = 84.*

Self Check 3
Estimate the answer: What is 20% of 129? 26

Now Try Problem 19

Teaching Example 3 Estimate the answer: What is 20% of 208?
Answer:
42

2 Estimate answers to percent problems involving 50%, 25%, 5%, and 15%.

There is an easy way to find 50% of a number. First, recall that $50\% = \frac{50}{100} = \frac{1}{2}$. Thus, to find 50% of a number means to find $\frac{1}{2}$ of that number, and to find $\frac{1}{2}$ of a number we simply divide it by 2.

Finding 50% of a Number

To find 50% of a number, divide the number by 2.

EXAMPLE 4 Estimate the answer: What is 50% of 2,595,603?

Strategy We will divide an approximation of 2,595,603 by 2.

WHY To find 50% of a number, we divide the number by 2.

Solution To estimate 50% of 2,595,603, we will find 50% of 2,600,000. We use 2,600,000 as an approximation because it is close to 2,595,603, because it is even, and, therefore, divisible by 2, and because it ends with many zeros.

50% of $2,595,603 \approx 1,300,000$ Because 50% of 2,600,000 = $\frac{2,600,000}{2}$ = 1,300,000.

Self Check 4
Estimate the answer: What is 50% of 14,272,549? 7,000,000
Now Try Problem 23

Teaching Example 4 Estimate the answer: What is 50% of 5,811,249?
Answer: 3,000,000

There is also an easy way to find 25% of a number. First, find 50% of the number by dividing the number by 2. Then, since 25% is one-half of 50%, divide that result by 2. Or, to save time, simply divide the original number by 4.

Finding 25% of a Number

To find 25% of a number, divide the number by 4.

EXAMPLE 5 Estimate the answer: What is 25% of 43.02?

Strategy We will divide an approximation of 43.02 by 4.

WHY To find 25% of a number, divide the number by 4.

Solution To estimate 25% of 43.02, we will find 25% of 44. We use 44 as an approximation because it is close to 43.02 and because it is divisible by 4.

25% of $43.02 \approx 11$ Because 25% of 44 = $\frac{44}{4}$ = 11.

Self Check 5
Estimate the answer: What is 25% of 27.16? 7
Now Try Problem 27

Teaching Example 5 Estimate the answer: What is 25% of 11.47?
Answer: 3

There is a quick way to find 5% of a number. First, find 10% of the number by moving the decimal point in the number one place to the left. Then, since 5% is one-half of 10%, divide that result by 2.

Finding 5% of a Number

To find 5% of a number, find 10% of the number by moving the decimal point in the number one place to the left. Then, divide that result by 2.

EXAMPLE 6 Electricity Usage
The average U.S. household uses 11,496 kilowatt-hours of electricity each year. Several energy conservation groups would like each household to take steps to reduce its electricity usage by 5%. Estimate 5% of 11,496 kilowatt-hours. (Source: Energy Information Administration)

Strategy We will find 10% of 11,496. Then, we will divide an approximation of that result by 2.

WHY 5% of a number is one-half of 10% of a number.

Solution First, we find 10% of 11,496.

10% of 11,496 = 1,149.6   Move the understood decimal point in 11,496 one place to the left.

We will use 1,150 as an approximation of this result because it is close to 1,149.6 and because it is even, and, therefore, divisible by 2. Next, we divide the approximation by 2 to estimate 5% of 11,496.

$$\frac{1{,}150}{2} = 575$$ Divide the approximation of 10% of 11,496 by 2.

Thus, 5% of 11,496 ≈ 575. A 5% reduction in electricity usage by the average U.S. household is about 575 kilowatt-hours.

Self Check 6
Estimate the answer: What is 5% of 24,198? 1,210
Now Try Problem 31

Teaching Example 6
LAYOFFS A company is going to reduce its 8,350-person work force by 5%. Estimate 5% of 8,350.
Answer:
420

We can use the shortcuts for finding 10% and 5% of a number to find 15% of a number.

Finding 15% of a Number
To find 15% of a number, find the sum of 10% of the number and 5% of the number.

EXAMPLE 7 Tipping
As a general rule, if the service in a restaurant is acceptable, a tip of 15% of the total bill should be left for the server. Estimate the 15% tip on a $77.55 dinner bill.

Strategy We will find 10% and 5% of an approximation of $77.55. Then we will add those results.

WHY To find 15% of a number, find the sum of 10% of the number and 5% of the number.

Solution To simplify the calculations, we will estimate the cost of the $77.55 dinner to be $80. Then, to estimate the tip, we find 10% of $80 and 5% of $80, and add.

10% of $80 is $8 ⟶ $8
5% of $80 (half as much as 10% of $80) ⟶ + $4

$12   Add to get the estimated tip.

The tip should be $12.

Self Check 7
TIPPING Estimate the 15% tip on a $29.55 breakfast bill. $4.50
Now Try Problems 35 and 75

Teaching Example 7
TIPPING Estimate the 15% tip on a lunch that costs $38.60.
Answer:
$6

3 Estimate answers to percent problems involving 200%.

Since 100% of a number is the number itself, it follows that 200% of a number would be twice the number. We can extend this rule to quickly find multiples of 100% of a number.

> **Finding 200%, 300%, 400%, ... of a Number**
>
> To find 200% of a number, multiply the number by 2. A similar approach can be used to find 300% of a number, 400% of a number, and so on.

Self Check 8

Estimate the answer: What is 200% of 12.437? 24

Now Try Problem 43

Teaching Example 8 Estimate the answer: What is 200% of 8.159?
Answer:
16

EXAMPLE 8 Estimate the answer: What is 200% of 5.673?

Strategy We will multiply an approximation of 5.673 by 2.

WHY To find 200% of a number, multiply the number by 2.

Solution To estimate 200% of 5.673, we will find 200% of 6. We use 6 as an approximation because it is close to 5.673 and it makes the multiplication by 2 easy.

$$200\% \text{ of } 5.673 \approx 12 \quad \text{Because 200\% of 6} = 2 \cdot 6 = 12.$$

4 Use estimation to solve percent application problems.

In the previous examples of this section, we were given the percent (1%, 10%, 50%, 25%, 5%, 15%, or 200%), we approximated the base, and then we estimated the amount. Sometimes we must approximate the percent, as well, to estimate an answer.

Self Check 9

STUDENT DRIVERS Of the 1,550 students attending a high school, 26% of them drive to school. Estimate the number of students that drive to school. 400 students

Now Try Problem 85

Teaching Example 9
DAY CARE Of the 250 children attending an elementary school, 26% of them are enrolled in after-school day care. Estimate the number of children in day care.
Answer:
60 children

EXAMPLE 9 *Music Education* Of the 350 children attending an elementary school, 24% of them are enrolled in the instrumental music program. Estimate the number of children taking instrumental music.

Strategy We will use the rule from this section for finding 25% of a number.

WHY 24% is approximately 25%, and there is a quick way to find 25% of a number.

Solution 24% of the 350 children in the school are taking instrumental music. To estimate 24% of 350, we will find 25% of 360. We use 360 as an approximation because it is close to 350 and it is divisible by 4.

$$24\% \text{ of } 350 \approx 90 \quad \text{Because 25\% of 360} = \frac{360}{4} = 90.$$

There are approximately 90 children in the school taking instrumental music.

> **ANSWERS TO SELF CHECKS**
>
> **1.** 5.193, 5 **2.** 353.6 lb, 400 lb **3.** 26 **4.** 7,000,000 **5.** 7 **6.** 1,210 **7.** $4.50
> **8.** 24 **9.** 400 students

SECTION 7.4 STUDY SET

VOCABULARY

Fill in the blanks.

1. **Estimation** can be used to find approximations when exact answers aren't necessary.
2. With **front**-end rounding, a number is rounded to its largest place value so that all but its first digit is zero.

CONCEPTS

Fill in the blanks.

3. To find 1% of a number, move the decimal point in the number **two** places to the left.
4. To find 10% of a number, move the decimal point in the number **one** place to the left.
5. To find 20% of a number, find 10% of the number by moving the decimal point one place to the left, and then double (multiply by **2**) the result.
6. To find 50% of a number, divide the number by **2**.
7. To find 25% of a number, divide the number by **4**.
8. To find 5% of a number, find 10% of the number by moving the decimal point one place to the left. Then, divide that result by **2**.
9. To find 15% of a number, find the sum of **10**% of the number and **5**% of the number.
10. To find 200% of a number, multiply the number by **2**.

GUIDED PRACTICE

What is 1% of the given number? Find the exact answer and an estimate using front-end rounding. See Example 1.

11. 275.1 2.751, 3
12. 460.9 4.609, 5
13. 12.67 0.1267, 0.1
14. 92.11 0.9211, 0.9

What is 10% of the given number? Find the exact answer and an estimate using front-end rounding. See Example 2.

15. 4,059 pounds 405.9 lb, 400 lb
16. 7,435 hours 743.5 hr, 700 hr
17. 691.4 minutes 69.14 min, 70 min
18. 881.2 kilometers 88.12 km, 90 km

Estimate each answer. (Answers may vary.) See Example 3.

19. What is 20% of 346? 70
20. What is 20% of 409? 82
21. What is 20% of 67? 14
22. What is 20% of 32? 6

Estimate each answer. (Answers may vary.) See Example 4.

23. What is 50% of 4,195,898? 2,100,000
24. What is 50% of 6,802,117? 3,400,000
25. What is 50% of 397,020? 200,000
26. What is 50% of 793,288? 400,000

Estimate each answer. (Answers may vary.) See Example 5.

27. What is 25% of 15.49? 4
28. What is 25% of 7.02? 2
29. What is 25% of 49.33? 12
30. What is 25% of 39.74? 10

Estimate each answer. (Answers may vary because of the approximation used.) See Example 6.

31. What is 5% of 16,359? 820
32. What is 5% of 44,191? 2,210
33. What is 5% of 394.182? 20
34. What is 5% of 176.001? 9

Estimate a 15% tip on each dollar amount. (Answers may vary.) See Example 7.

35. $58.99 $9
36. $38.60 $6
37. $27.16 $4.50
38. $49.05 $7.50
39. $115.75 $18
40. $135.88 $21
41. $9.74 $1.50
42. $11.75 $1.80

Estimate each answer. (Answers may vary.) See Example 8.

43. What is 200% of 4.212? 8
44. What is 200% of 5.189? 10
45. What is 200% of 35.77? 72
46. What is 200% of 80.32? 160

TRY IT YOURSELF

Find the exact answer using methods from this section.

47. What is 2% of 600? 12
48. What is 3% of 700? 21
49. What is 30% of 18? 5.4
50. What is 40% of 45? 18

Estimate each answer. (Answers may vary.)

51. What is 300% of 59.2? 180
52. What is 400% of 203.77? 800
53. What is 5% of 4,605? 230
54. What is 5% of 8,401? 420

55. What is 1% of 628.21? 6
56. What is 1% of 12,847.9? 130
57. What is 15% of 119? 18
58. What is 15% of 237? 36
59. What is 10% of 67.0056? 7
60. What is 10% of 94.2424? 9
61. What is 25% of 275? 70
62. What is 25% of 313? 80
63. What is 50% of 23,898? 12,000
64. What is 25% of 56,716? 14,000
65. What is 200% of 0.9123? 1.8
66. What is 200% of 0.4189? 0.8

Find the exact answer.

67. What is 1% of 50% of 98? 0.49
68. What is 10% of 25% of 20? 0.5
69. What is 15% of 20% of 400? 12
70. What is 5% of 10% of 30? 0.15

APPLICATIONS

Estimate each answer unless stated otherwise. (Answers may vary.)

71. **COLLEGE COURSES** 20% of the 815 students attending a small college were enrolled in a science course. How many students is this? 164 students

72. **SPECIAL OFFERS** In the grocery store, a 65-ounce bottle of window cleaner was marked "25% free." How many ounces are free? 16 oz

73. **DISCOUNTS** By how much is the price of a coat discounted if the regular price of $196.88 is reduced by 30%? $60

74. **SIGNS** The nation's largest electronic billboard is at the south intersection of Times Square in New York City. It has 12,000,000 LED lights. If just 1% of these lights burnt out, how many lights would have to be replaced? Give the exact answer. 120,000 lights

75. **TIPPING** A restaurant tip is normally 15% of the cost of the meal. Find the tip on a dinner costing $38.64. $6

76. **VISA RECEIPTS** Refer to the receipt to the right. Estimate the 15% gratuity (tip) and then find the total. $9, $67.47

CLARK'S SEAFOOD
OKLAHOMA CITY, OK

Date:
Card Type: VISA
Acct Num: ************0241
Exp Date: **/**
Customer: WONG/TOM
Server: 209 Colleen

Amount: $58.47
Gratuity: ?
Total: ?

77. **DINING OUT** A couple went out to eat at a restaurant. The food they ordered cost $28.55 and the drinks they ordered cost $19.75. Estimate a 15% tip on the total bill. $7.50

78. **SPLITTING THE TIP** The total bill for three businessmen who went out to eat at a Chinese restaurant was $121.10. If they split the tip equally, estimate each person's share. $6

79. **FIRE DAMAGE** An insurance company paid 25% of the $118,000 it cost to rebuild a home that was destroyed by fire. How much did the insurance company pay? $30,000

80. **SAFETY INSPECTIONS** Of the 2,513 vehicles inspected at a safety checkpoint, 10% had code violations. How many cars had code violations? 250 cars

81. **WEIGHTLIFTING** A 158-pound weightlifter can bench press 200% of his body weight. How many pounds can he bench press? 320 lb

82. **TESTING** On a 60-question true/false test, 5% of a student's answers were wrong. How many questions did she miss? 3 questions

83. **TRAFFIC STUDIES** According to an electronic traffic monitor, 30% of the 690 motorists who passed it were speeding. How many of these motorists were speeding? 210 motorists

84. **SELLING A HOME** A homeowner has been told she will get back 50% of her $6,125 investment if she paints her home before selling it. How much will she get back if she paints her home? $3,000

Approximate the percent and then estimate each answer. (Answers may vary.)

85. **NO-SHOWS** The attendance at a seminar was only 24% of what the organizers had anticipated. If 875 people were expected, how many actually attended the seminar? 220 people

86. **HONOR ROLL** Of the 900 students in a school, 16% were on the principal's honor roll. How many students were on the honor roll? 135 students

87. **INTERNET SURVEYS** The illustration shows an online survey question. How many people voted yes? 18,000 people

88. **SALES TAX** The state sales tax rate in Nebraska is 5.5%. Estimate the sales tax on a purchase of $596. $30

89. **VOTING** On election day, 48% of the 6,200 workers at the polls were volunteers. How many volunteers helped with the election? 3,100 volunteers

90. **BUDGETS** Each department at a college was asked to cut its budget by 21%. By how much money should the mathematics department budget be reduced if it is currently $4,715? $1,000

WRITING

91. Explain why 200% of a number is twice the number.

92. If you know 10% of a number, explain how you can find 30% of the same number.

93. If you know 10% of a number, explain how you can find 5% of the same number.

94. Explain why 25% of a number is the same as $\frac{1}{4}$ of the number.

REVIEW

Perform each operation and simplify, if possible.

95. a. $\frac{5}{6} + \frac{1}{2} \quad \frac{4}{3} = 1\frac{1}{3}$ b. $\frac{5}{6} - \frac{1}{2} \quad \frac{1}{3}$

    c. $\frac{5}{6} \cdot \frac{1}{2} \quad \frac{5}{12}$ d. $\frac{5}{6} \div \frac{1}{2} \quad \frac{5}{3} = 1\frac{2}{3}$

96. a. $\frac{7}{15} + \frac{7}{18} \quad \frac{77}{90}$ b. $\frac{7}{15} - \frac{7}{18} \quad \frac{7}{90}$

    c. $\frac{7}{15} \cdot \frac{7}{18} \quad \frac{49}{270}$ d. $\frac{7}{15} \div \frac{7}{18} \quad \frac{6}{5} = 1\frac{1}{5}$

SECTION 7.5
Interest

Objectives
1. Calculate simple interest.
2. Calculate compound interest.

When money is borrowed, the lender expects to be paid back the amount of the loan plus an additional charge for the use of the money. The additional charge is called **interest**. When money is deposited in a bank, the depositor is paid for the use of the money. The money the deposit earns is also called interest. In general, *interest is money that is paid for the use of money.*

1   Calculate simple interest.

Interest is calculated in one of two ways: either as **simple interest** or as **compound interest**. We begin by discussing simple interest. First, we need to introduce some key terms associated with borrowing or lending money.

- **Principal:** the amount of money that is invested, deposited, loaned, or borrowed.
- **Interest rate:** a percent that is used to calculate the amount of interest to be paid. The interest rate is assumed to be per year (annual interest) unless otherwise stated.
- **Time:** the length of time that the money is invested, deposited, or borrowed.

The amount of interest to be paid depends on the principal, the rate, and the time. That is why all three are usually mentioned in advertisements for bank accounts, investments, and loans, as shown below.

Chapter 7 Percent

Simple interest is interest earned only on the original principal. It is found using the following formula.

> **Simple Interest Formula**
>
> Interest = principal · rate · time   or $I = P \cdot r \cdot t$
>
> where the rate r is expressed as an annual (yearly) rate and the time t is expressed in years. This formula can be written more simply without the multiplication raised dots as
>
> $I = Prt$

Self Check 1
If $4,200 is invested for 2 years at a rate of 4%, how much simple interest is earned? $336

Now Try Problem 17

Teaching Example 1 If $5,000 is invested for 1 year at a rate of 6%, how much simple interest is earned?
Answer:
$300

EXAMPLE 1 If $3,000 is invested for 1 year at a rate of 5%, how much simple interest is earned?

Strategy We will identify the principal, rate, and time for the investment.

WHY Then we can use the formula $I = Prt$ to find the unknown amount of simple interest earned.

Solution The principal is $3,000, the interest rate is 5%, and the time is 1 year.

$P = \$3,000 \qquad r = 5\% = 0.05 \qquad t = 1$

$I = Prt$	This is the simple interest formula.
$I = \$3,000 \cdot 0.05 \cdot 1$	Substitute the values for P, r, and t. Remember to write the rate r as a decimal.
$I = \$3,000 \cdot 0.05$	Multiply: $0.05 \cdot 1 = 0.05$.
$I = \$150$	Complete the multiplication.

$$\begin{array}{r} 3,000 \\ \times 0.05 \\ \hline 150.00 \end{array}$$

The simple interest earned in 1 year is $150.

The information given in this problem and the result can be presented in a table.

Principal	Rate	Time	Interest earned
$3,000	5%	1 year	$150

If no money is withdrawn from an investment, the investor receives the principal *and* the interest at the end of the time period. Similarly, a borrower must repay the principal *and* the interest when taking out a loan. In each case, the **total amount** of money involved is given by the following formula.

> **Finding the Total Amount**
>
> The total amount in an investment account or the total amount to be repaid on a loan is the sum of the principal and the interest.
>
> Total amount = principal + interest

Self Check 2
If $600 is invested at 2.5% simple interest for 4 years, what will be the total amount of money in the investment account at the end of the 4 years? $660

EXAMPLE 2 If $800 is invested at 4.5% simple interest for 3 years, what will be the total amount of money in the investment account at the end of the 3 years?

Strategy We will find the simple interest earned on the investment and add it to the principal.

WHY At the end of 3 years, the total amount of money in the account is the sum of the principal and the interest earned.

Solution The principal is $800, the interest rate is 4.5%, and the time is 3 years. To find the interest the investment earns, we use multiplication.

$P = \$800 \qquad r = 4.5\% = 0.045 \qquad t = 3$

$I = Prt$	This is the simple interest formula.
$I = \$800 \cdot 0.045 \cdot 3$	Substitute the values for P, r, and t. Remember to write the rate r as a decimal.
$I = \$36 \cdot 3$	Multiply: $\$800 \cdot 0.045 = \$36$.
$I = \$108$	Complete the multiplication.

$$\begin{array}{r} \overset{4}{0.045} \\ \times\ 800 \\ \hline 36.000 \end{array} \qquad \begin{array}{r} \overset{1}{36} \\ \times 3 \\ \hline 108 \end{array}$$

The simple interest earned in 3 years is $108. To find the total amount of money in the account, we add.

Total amount = principal + interest	This is the total amount formula.
Total amount = $800 + $108	Substitute $800 for the principal and $108 for the interest.
Total amount = $908	Do the addition.

At the end of 3 years, the total amount of money in the account will be $908.

> **Caution!** When we use the formula $I = Prt$, the time must be expressed in years. If the time is given in days or months, we rewrite it as a fractional part of a year. For example, a 30-day investment lasts $\frac{30}{365}$ of a year, since there are 365 days in a year. For a 6-month loan, we express the time as $\frac{6}{12}$ or $\frac{1}{2}$ of a year, since there are 12 months in a year.

EXAMPLE 3 **Education Costs** A student borrowed $920 at 3% for 9 months to pay some college tuition expenses. Find the simple interest that must be paid on the loan.

Strategy We will rewrite 9 months as a fractional part of a year, and then we will use the formula $I = Prt$ to find the unknown amount of simple interest to be paid on the loan.

WHY To use the formula $I = Prt$, the time must be expressed in years, or as a fractional part of a year.

Solution Since there are 12 months in a year, we have

$$9 \text{ months} = \frac{9}{12} \text{ year} = \frac{\overset{1}{\cancel{3}} \cdot 3}{\underset{1}{\cancel{3}} \cdot 4} \text{ year} = \frac{3}{4} \text{ year}$$

Simplify the fraction $\frac{9}{12}$ by removing a common factor of 3 from the numerator and denominator.

The time of the loan is $\frac{3}{4}$ year. To find the amount of interest, we multiply.

$P = \$920 \qquad r = 3\% = 0.03 \qquad t = \frac{3}{4}$

$I = Prt$	This is the simple interest formula.
$I = \$920 \cdot 0.03 \cdot \frac{3}{4}$	Substitute the values for P, r, and t. Remember to write the rate r as a decimal.
$I = \frac{\$920}{1} \cdot \frac{0.03}{1} \cdot \frac{3}{4}$	Write $920 and 0.03 as fractions.
$I = \frac{\$82.80}{4}$	Multiply the numerators. Multiply the denominators.
$I = \$20.70$	Do the division: $82.80 \div 4 = 20.70$.

$$\begin{array}{r} 920 \\ \times 0.03 \\ \hline 27.60 \end{array} \qquad \begin{array}{r} \overset{2\ 1}{27.60} \\ \times\ \ 3 \\ \hline 82.80 \end{array}$$

$$\begin{array}{r} 20.70 \\ 4\overline{)82.80} \\ \underline{-8} \\ 02 \\ \underline{-0} \\ 2\,8 \\ \underline{-2\,8} \\ 00 \\ \underline{-0} \\ 00 \end{array}$$

The simple interest to be paid on the loan is $20.70.

Now Try Problem 21

Teaching Example 2 If $500 is invested at 3.5% simple interest for 6 years, what will be the total amount of money in the investment account at the end of the 6 years?
Answer:
$605

Self Check 3
SHORT-TERM LOANS Find the simple interest on a loan of $810 at 9% for 8 months. $48.60

Now Try Problem 25

Teaching Example 3
NAVIGATION SYSTEMS A trucker borrowed $480 at 8% for 3 months to purchase a GPS system. Find the simple interest that must be paid on the loan.
Answer:
$9.60

Self Check 4

ACCOUNTING To cover payroll expenses, a small business owner borrowed $3,200 at a simple interest rate of 15%. Find the total amount he must repay at the end of 120 days. $3,357.81

Now Try Problem 29

Teaching Example 4
ADVERTISING EXPENSES A business borrowed $15,500 for 90 days to pay for advertising on the Internet. If the simple interest rate is 16%, find the total amount that must be repaid at the end of the 90-day loan.
Answer:
$16,111.51

EXAMPLE 4 *Short-term Business Loans* To start a business, a couple borrowed $5,500 for 90 days to purchase equipment and supplies. If the loan has a 14% simple interest rate, find the total amount they must repay at the end of the 90-day period.

Strategy We will rewrite 90 days as a fractional part of a year, and then we will use the formula $I = Prt$ to find the unknown amount of simple interest to be paid on the loan.

WHY To use the formula $I = Prt$, the time must be expressed in years, or as a fractional part of a year.

Solution Since there are 365 days in a year, we have

$$90 \text{ days} = \frac{90}{365} \text{ year} = \frac{\overset{1}{\cancel{5}} \cdot 18}{\underset{1}{\cancel{5}} \cdot 73} \text{ year} = \frac{18}{73} \text{ year}$$

Simplify the fraction $\frac{90}{365}$ by removing a common factor of 5 from the numerator and denominator.

The time of the loan is $\frac{18}{73}$ year. To find the amount of interest, we multiply.

$$P = \$5,500 \qquad r = 14\% = 0.14 \qquad t = \frac{90}{365} = \frac{18}{73}$$

$I = Prt$ This is the simple interest formula.

$I = \$5,500 \cdot 0.14 \cdot \frac{18}{73}$ Substitute the values for P, r, and t.

$I = \frac{\$5,500}{1} \cdot \frac{0.14}{1} \cdot \frac{18}{73}$ Write $5,500 and 0.14 as fractions.

$I = \frac{\$13,860}{73}$ Multiply the numerators. Multiply the denominators.

$I \approx \$189.86$ Divide 13,860 by 73. The division does not terminate. Round to the nearest cent.

```
  5,500        770
 ×0.14        × 18
 22000        6160
 55000        7700
770.00       13,860
```

The interest on the loan is $189.86. To find how much they must pay back, we add.

Total amount = **principal + interest**   This is the total amount formula.

$\qquad = \$5,500 + \$189.86$ Substitute $5,500 for the principal and $189.86 for the interest.

$\qquad = \$5,689.86$ Do the addition.

The couple must pay back $5,689.86 at the end of 90 days.

2 Calculate compound interest.

Most savings accounts and investments pay *compound interest* rather than simple interest. We have seen that simple interest is paid only on the original principal. **Compound interest** is paid on the principal and *previously earned interest*. To illustrate this concept, suppose that $2,000 is deposited in a savings account at a rate of 5% for 1 year. We can use the formula $I = Prt$ to calculate the interest earned at the end of 1 year.

$I = Prt$ This is the simple interest formula.
$I = \$2,000 \cdot 0.05 \cdot 1$ Substitute for P, r, and t.
$I = \$100$ Do the multiplication.

Interest of $100 was earned. At the end of the first year, the account contains the interest ($100) plus the original principal ($2,000), for a balance of $2,100.

Suppose that the money remains in the savings account for another year at the same interest rate. For the second year, interest will be paid on a principal of $2,100.

That is, during the second year, we earn *interest on the interest* as well as on the original $2,000 principal. Using $I = Prt$, we can find the interest earned in the second year.

$I = Prt$ This is the simple interest formula.

$I = \$2{,}100 \cdot 0.05 \cdot 1$ Substitute for P, r, and t.

$I = \$105$ Do the multiplication.

In the second year, $105 of interest is earned. The account now contains that interest plus the $2,100 principal, for a total of $2,205.

As the figure below shows, we calculated the simple interest two times to find the compound interest.

If we compute only the *simple interest* on $2,000, at 5% for 2 years, the interest earned is $I = \$2{,}000 \cdot 0.05 \cdot 2 = \$200$. Thus, the account balance would be $2,200. Comparing the balances, we find that the account earning compound interest will contain $5 more than the account earning simple interest.

In the previous example, the interest was calculated at the end of each year, or **annually**. When compounding, we can compute the interest in other time spans, such as **semiannually** (twice a year), **quarterly** (four times a year), or even **daily**.

EXAMPLE 5   *Compound Interest*

As a special gift for her newborn granddaughter, a grandmother opens a $1,000 savings account in the baby's name. The interest rate is 4.2%, compounded quarterly. Find the amount of money the child will have in the bank on her first birthday.

Strategy We will use the simple interest formula $I = Prt$ four times in a series of steps to find the amount of money in the account after 1 year. Each time, the time t is $\frac{1}{4}$.

WHY The interest is compounded *quarterly*.

Solution If the interest is compounded quarterly, the interest will be computed four times in one year. To find the amount of interest $1,000 will earn in the first quarter of the year, we use the simple interest formula, where t is $\frac{1}{4}$ of a year.

Interest earned in the first quarter:

$P_{\text{1st Qtr}} = \$1{,}000 \qquad r = 4.2\% = 0.042 \qquad t = \frac{1}{4}$

$I = Prt$ This is the simple interest formula.

$I = \$1{,}000 \cdot 0.042 \cdot \frac{1}{4}$ Substitute for P, r, and t.

$I = \$42 \cdot \frac{1}{4}$ Multiply: $1,000 · 0.042 = $42.

$I = \frac{\$42}{4}$ Do the multiplication. Think of 42 as $\frac{42}{1}$.

$I = \$10.50$ Do the division: 42 ÷ 4 = 10.5.

The interest earned in the first quarter is $10.50. This now becomes part of the principal for the second quarter.

$P_{\text{2nd Qtr}} = \$1{,}000 + \$10.50 = \$1{,}010.50$ Add the original principal and the interest that it earned to find the second-quarter principal.

Self Check 5

COMPOUND INTEREST Suppose $8,000 is deposited in an account that earns 2.3% compounded quarterly. Find the amount of money in an account at the end of the first year. $8,185.59

Now Try Problem 33

Teaching Example 5
COMPOUND INTEREST As a gift for their newborn son, the parents open a $5,000 savings account in the baby's name. The interest rate is 3.8%, compounded quarterly. Find the amount of money in the account on the child's first birthday.
Answer:
$5,192.73

To find the amount of interest $1,010.50 will earn in the second quarter of the year, we use the simple interest formula, where t is again $\frac{1}{4}$ of a year.

Interest earned in the second quarter:

$P_{\text{2nd Qtr}} = \$1,010.50 \qquad r = 0.042 \qquad t = \dfrac{1}{4}$

$I = Prt$ This is the simple interest formula.

$I = \$1,010.50 \cdot 0.042 \cdot \dfrac{1}{4}$ Substitute for P, r, and t.

$I = \dfrac{\$1,010.50 \cdot 0.042 \cdot 1}{4}$ Multiply.

$I \approx \$10.61$ Use a calculator. Round to the nearest cent (hundredth).

The interest earned in the second quarter is $10.61. This becomes part of the principal for the third quarter.

$P_{\text{3rd Qtr}} = \$1,010.50 + \$10.61 = \$1,021.11$ Add the second-quarter principal and the interest that it earned to find the third-quarter principal.

To find the interest $1,021.11 will earn in the third quarter of the year, we proceed as follows.

Interest earned in the third quarter:

$P_{\text{3rd Qtr}} = \$1,021.11 \qquad r = 0.042 \qquad t = \dfrac{1}{4}$

$I = Prt$ This is the simple interest formula.

$I = \$1,021.11 \cdot 0.042 \cdot \dfrac{1}{4}$ Substitute for P, r, and t.

$I = \dfrac{\$1,021.11 \cdot 0.042 \cdot 1}{4}$ Multiply.

$I \approx \$10.72$ Use a calculator. Round to the nearest cent (hundredth).

The interest earned in the third quarter is $10.72. This now becomes part of the principal for the fourth quarter.

$P_{\text{4th Qtr}} = \$1,021.11 + \$10.72 = \$1,031.83$ Add the third-quarter principal and the interest that it earned to find the fourth-quarter principal.

To find the interest $1,031.83 will earn in the fourth quarter, we again use the simple interest formula.

Interest earned in the fourth quarter:

$P_{\text{4th Qtr}} = \$1,031.83 \qquad r = 0.042 \qquad t = \dfrac{1}{4}$

$I = Prt$ This is the simple interest formula.

$I = \$1,031.83 \cdot 0.042 \cdot \dfrac{1}{4}$ Substitute for P, r, and t.

$I = \dfrac{\$1,031.83 \cdot 0.042 \cdot 1}{4}$ Multiply.

$I \approx \$10.83$ Use a calculator. Round to the nearest cent (hundredth).

The interest earned in the fourth quarter is $10.83. Adding this to the existing principal, we get

Total amount $= \$1,031.83 + \$10.83 = \$1,042.66$ Add the fourth-quarter principal and the interest that it earned.

The total amount in the account after four quarters, or 1 year, is $1,042.66.

Calculating compound interest by hand can take a long time. The **compound interest formula** can be used to find the total amount of money that an account will contain at the end of the term quickly.

> **Compound Interest Formula**
>
> The total amount A in an account can be found using the formula
>
> $$A = P\left(1 + \frac{r}{n}\right)^{nt}$$
>
> where P is the principal, r is the annual interest rate expressed as a decimal, t is the length of time in years, and n is the number of compoundings in one year.

A calculator is very helpful in performing the operations on the right side of the compound interest formula.

> **Using Your CALCULATOR** Compound Interest
>
> A businessperson invests $9,250 at 7.6% interest, to be compounded monthly. To find what the investment will be worth in 3 years, we use the compound interest formula with the following values.
>
> $P = \$9{,}250 \quad r = 7.6\% = 0.076 \quad t = 3 \text{ years} \quad n = 12 \text{ times a year (monthly)}$
>
> $A = P\left(1 + \dfrac{r}{n}\right)^{nt}$ This is the compound interest formula.
>
> $A = 9{,}250\left(1 + \dfrac{0.076}{12}\right)^{12\,(3)}$ Substitute the values of P, r, t, and n. In the exponent, nt means $n \cdot t$.
>
> $A = 9{,}250\left(1 + \dfrac{0.076}{12}\right)^{36}$ Evaluate the exponent: $12(3) = 36$.
>
> To evaluate the expression on the right-hand side of the equation using a calculator, we enter these numbers and press these keys.
>
> 9250 [×] [(] 1 [+] .076 [÷] 12 [)] [y^x] 36 [=]   `11610.43875`
>
> On some calculator models, the [^] key is used in place of the [y^x] key. Also, the [ENTER] key is pressed instead of the [=] key for the result to be displayed.
>
> Rounded to the nearest cent, the amount in the account after 3 years will be $11,610.44.
>
> If your calculator does not have parenthesis keys, calculate the sum within the parentheses first. Then find the power. Finally, multiply by 9,250.

EXAMPLE 6 *Compounding Daily* An investor deposited $50,000 in a long-term account at 6.8% interest, compounded daily. How much money will he be able to withdraw in 7 years if the principal is to remain in the bank?

Strategy We will use the compound interest formula to find the *total amount* in the account after 7 years. Then we will subtract the original principal from that result.

WHY When the investor withdraws money, he does not want to touch the original $50,000 principal in the account.

Self Check 6

COMPOUNDING DAILY Find the amount of interest $25,000 will earn in 10 years if it is deposited in an account at 5.99% interest, compounded daily. $20,505.20

Now Try Problem 37

Teaching Example 6
COMPOUNDING DAILY An investor deposited $60,000 in a long-term account at 5.8% interest, compounded daily. How much money will she be able to withdraw in 10 years if the principal is to remain in the bank?
Answer:
$47,157.37

Solution "Compounded daily" means that compounding will be done 365 times in a year for 7 years.

$P = \$50{,}000 \quad r = 6.8\% = 0.068 \quad t = 7 \quad n = 365$

$A = P\left(1 + \dfrac{r}{n}\right)^{nt}$ This is the compound interest formula.

$A = 50{,}000\left(1 + \dfrac{0.068}{365}\right)^{365(7)}$ Substitute the values of P, r, t, and n. In the exponent, nt means n · t.

$\begin{array}{r} \overset{4\;3}{365} \\ \times\quad 7 \\ \hline 2{,}555 \end{array}$

$A = 50{,}000\left(1 + \dfrac{0.068}{365}\right)^{2{,}555}$ Evaluate the exponent: 365 · 7 = 2,555.

$A \approx 80{,}477.58$ Use a calculator. Round to the nearest cent.

The account will contain $80,477.58 at the end of 7 years. To find how much money the man can withdraw, we must subtract the original principal of $50,000 from the total amount in the account.

$80{,}477.58 - 50{,}000 = 30{,}477.58$

The man can withdraw $30,477.58 without having to touch the $50,000 principal.

ANSWERS TO SELF CHECKS
1. $336 **2.** $660 **3.** $48.60 **4.** $3,357.81 **5.** $8,185.59 **6.** $20,505.20

SECTION 7.5 STUDY SET

VOCABULARY

Fill in the blanks.

1. In general, _interest_ is money that is paid for the use of money.
2. In banking, the original amount of money invested, deposited, loaned, or borrowed is known as the _principal_.
3. The percent that is used to calculate the amount of interest to be paid is called the interest _rate_.
4. _Simple_ interest is interest earned only on the original principal.
5. The _total_ amount in an investment account is the sum of the principal and the interest.
6. _Compound_ interest is interest paid on the principal and previously earned interest.

CONCEPTS

7. Refer to the home loan advertisement below.

a. What is the principal? $125,000
b. What is the interest rate? 5%
c. What is the time? 30 years

8. Refer to the investment advertisement below.

a. What is the minimum principal? $10,000
b. What is the interest rate? 1.55%
c. What is the time? 12 months

9. When making calculations involving percents, they must be written as decimals or fractions. Change each percent to a decimal.

a. 7% 0.07   b. 9.8% 0.098   c. $6\tfrac{1}{4}\%$ 0.0625

10. Express each of the following as a fraction of a year. Simplify the fraction.

a. 6 months $\tfrac{1}{2}$ year   b. 90 days $\tfrac{18}{73}$ year
c. 120 days $\tfrac{24}{73}$ year   d. 1 month $\tfrac{1}{12}$ year

11. Complete the table by finding the simple interest earned.

Principal	Rate	Time	Interest earned
$10,000	6%	3 years	$1,800

12. Determine how many times a year the interest on a savings account is calculated if the interest is compounded
   a. annually 1
   b. semiannually 2
   c. quarterly 4
   d. daily 365
   e. monthly 12

13. a. What concept studied in this section is illustrated by the diagram below? compound interest
   b. What was the original principal? $1,000
   c. How many times was the interest found? 4
   d. How much interest was earned on the first compounding? $50
   e. For how long was the money invested? 1 year

   ┌1st qtr┐┌2nd qtr┐┌3rd qtr┐┌4th qtr┐
   $1,000 $1,050 $1,102.50 $1,157.63 $1,215.51

14. $3,000 is deposited in a savings account that earns 10% interest compounded annually. Complete the series of calculations in the illustration below to find how much money will be in the account at the end of 2 years.

NOTATION

15. Write the simple interest formula $I = P \cdot r \cdot t$ without the multiplication raised dots. $I = Prt$

16. In the formula $A = P\left(1 + \dfrac{r}{n}\right)^{nt}$, how many operations must be performed to find A? 5

GUIDED PRACTICE

Calculate the simple interest earned. **See Example 1.**

17. If $2,000 is invested for 1 year at a rate of 5%, how much simple interest is earned? $100

18. If $6,000 is invested for 1 year at a rate of 7%, how much simple interest is earned? $420

19. If $700 is invested for 4 years at a rate of 9%, how much simple interest is earned? $252

20. If $800 is invested for 5 years at a rate of 8%, how much simple interest is earned? $320

Calculate the total amount in each account. **See Example 2.**

21. If $500 is invested at 2.5% simple interest for 2 years, what will be the total amount of money in the investment account at the end of the 2 years? $525

22. If $400 is invested at 6.5% simple interest for 6 years, what will be the total amount of money in the investment account at the end of the 6 years? $556

23. If $1,500 is invested at 1.2% simple interest for 5 years, what will be the total amount of money in the investment account at the end of the 5 years? $1,590

24. If $2,500 is invested at 4.5% simple interest for 8 years, what will be the total amount of money in the investment account at the end of the 8 years? $3,400

Calculate the simple interest. **See Example 3.**

25. Find the simple interest on a loan of $550 borrowed at 4% for 9 months. $16.50

26. Find the simple interest on a loan of $460 borrowed at 9% for 9 months. $31.05

27. Find the simple interest on a loan of $1,320 borrowed at 7% for 4 months. $30.80

28. Find the simple interest on a loan of $1,250 borrowed at 10% for 3 months. $31.25

Calculate the total amount that must be repaid at the end of each short-term loan. **See Example 4.**

29. $12,600 is loaned at a simple interest rate of 18% for 90 days. Find the total amount that must be repaid at the end of the 90-day period. $13,159.23

30. $45,000 is loaned at a simple interest rate of 12% for 90 days. Find the total amount that must be repaid at the end of the 90-day period. $46,331.51

31. $40,000 is loaned at 10% simple interest for 45 days. Find the total amount that must be repaid at the end of the 45-day period. $40,493.15

32. $30,000 is loaned at 20% simple interest for 60 days. Find the total amount that must be repaid at the end of the 60-day period. $30,986.30

Calculate the total amount in each account. **See Example 5.**

33. Suppose $2,000 is deposited in a savings account that pays 3% interest, compounded quarterly. How much money will be in the account in one year? $2,060.68

34. Suppose $3,000 is deposited in a savings account that pays 2% interest, compounded quarterly. How much money will be in the account in one year? $3,060.46

35. If $5,400 earns 4% interest, compounded quarterly, how much money will be in the account at the end of one year? $5,619.27

36. If $10,500 earns 8% interest, compounded quarterly, how much money will be in the account at the end of one year? $11,365.53

Use a calculator to solve the following problems. **See Example 6.**

37. A deposit of $30,000 is placed in a savings account that pays 4.8% interest, compounded daily. How much money can be withdrawn at the end of 6 years if the principal is to remain in the bank? $10,011.96

38. A deposit of $12,000 is placed in a savings account that pays 5.6% interest, compounded daily. How much money can be withdrawn at the end of 8 years if the principal is to remain in the bank? $6,781.50

39. If 8.55% interest, compounded daily, is paid on a deposit of $55,250, how much money will be in the account at the end of 4 years? $77,775.64

40. If 4.09% interest, compounded daily, is paid on a deposit of $39,500, how much money will be in the account at the end of 9 years? $57,075.79

APPLICATIONS

41. **RETIREMENT INCOME** A retiree invests $5,000 in a savings plan that pays a simple interest rate of 6%. What will the account balance be at the end of the first year? $5,300

42. **INVESTMENTS** A developer promised a return of 8% simple interest on an investment of $15,000 in her company. How much could an investor expect to make in the first year? $1,200

43. A member of a credit union was loaned $1,200 to pay for car repairs. The loan was made for 3 years at a simple interest rate of 5.5%. Find the interest due on the loan. $198

from Campus to Careers
Loan Officer

44. **REMODELING** A homeowner borrows $8,000 to pay for a kitchen remodeling project. The terms of the loan are 9.2% simple interest and repayment in 2 years. How much interest will be paid on the loan? $1,472

45. **SMOKE DAMAGE** The owner of a café borrowed $4,500 for 2 years at 12% simple interest to pay for the cleanup after a kitchen fire. Find the total amount due on the loan. $5,580

46. **ALTERNATIVE FUELS** To finance the purchase of a fleet of natural-gas–powered vehicles, a city borrowed $200,000 for 4 years at a simple interest rate of 3.5%. Find the total amount due on the loan. $228,000

47. **SHORT-TERM LOANS** A loan of $1,500 at 12.5% simple interest is paid off in 3 months. What is the interest charged? $46.88

48. **FARM LOANS** An apple orchard owner borrowed $7,000 from a farmers' co-op bank. The money was loaned at 8.8% simple interest for 18 months. How much money did the co-op charge him for the use of the money? $924

49. **MEETING PAYROLLS** In order to meet end-of-the-month payroll obligations, a small business had to borrow $4,200 for 30 days. How much did the business have to repay if the simple interest rate was 18%? $4,262.14

50. **CAR LOANS** To purchase a car, a man takes out a loan for $2,000 for 120 days. If the simple interest rate is 9% per year, how much interest will he have to pay at the end of the 120-day loan period? $59.18

51. **SAVINGS ACCOUNTS** Find the interest earned on $10,000 at $7\frac{1}{4}$% for 2 years. Use the table to organize your work.

P	r	t	I
$10,000	$7\frac{1}{4}$% = 0.0725	2 yr	$1,450

52. **TUITION** A student borrows $300 from an educational fund to pay for books for spring semester. If the loan is for 45 days at $3\frac{1}{2}$% annual interest, what will the student owe at the end of the loan period? $301.29

53. **LOAN APPLICATIONS** Complete the following loan application.

Loan Application Worksheet

1. Amount of loan (principal) ___$1,200.00___
2. Length of loan (time) ___2 YEARS___
3. Annual percentage rate ___8%___
   (simple interest)
4. Interest charged ___$192___
5. Total amount to be repaid ___$1,392___
6. Check method of repayment:
   ☐ 1 lump sum ☑ monthly payments

Borrower agrees to pay ___24___ equal payments of ___$58___ to repay loan.

54. LOAN APPLICATIONS Complete the following loan application.

Loan Application Worksheet
1. Amount of loan (principal) __$810.00__
2. Length of loan (time) __9 mos.__
3. Annual percentage rate (simple interest) __12%__
4. Interest charged __$72.90__
5. Total amount to be repaid __$882.90__
6. Check method of repayment:
   ☐ 1 lump sum   ☑ monthly payments
   Borrower agrees to pay __9__ equal payments of __$98.10__ to repay loan.

55. LOW-INTEREST LOANS An underdeveloped country receives a low-interest loan from a bank to finance the construction of a water treatment plant. What must the country pay back at the end of $3\frac{1}{2}$ years if the loan is for $18 million at 2.3% simple interest? $19.449 million

56. REDEVELOPMENT A city is awarded a low-interest loan to help renovate the downtown business district. The $40-million loan, at 1.75% simple interest, must be repaid in $2\frac{1}{2}$ years. How much interest will the city have to pay? $1.75 million

A calculator will be helpful in solving the following problems.

57. COMPOUNDING ANNUALLY If $600 is invested in an account that earns 8%, compounded annually, what will the account balance be after 3 years? $755.83

58. COMPOUNDING SEMIANNUALLY If $600 is invested in an account that earns annual interest of 8%, compounded semiannually, what will the account balance be at the end of 3 years? $759.19

59. COLLEGE FUNDS A ninth-grade student opens a savings account that locks her money in for 4 years at an annual rate of 6%, compounded daily. If the initial deposit is $1,000, how much money will be in the account when she begins college in 4 years? $1,271.22

60. CERTIFICATE OF DEPOSITS A 3-year certificate of deposit pays an annual rate of 5%, compounded daily. The maximum allowable deposit is $90,000. What is the most interest a depositor can earn from the CD? $14,564.01

61. TAX REFUNDS A couple deposits an income tax refund check of $545 in an account paying an annual rate of 4.6%, compounded daily. What will the size of the account be at the end of 1 year? $570.65

62. INHERITANCE After receiving an inheritance of $11,000, a man deposits the money in an account paying an annual rate of 7.2%, compounded daily. How much money will be in the account at the end of 1 year? $11,821.12

63. LOTTERY Suppose you won $500,000 in the lottery and deposited the money in a savings account that paid an annual rate of 6% interest, compounded daily. How much interest would you earn in a year? $30,915.66

64. CASH GIFTS After receiving a $250,000 cash gift, a university decides to deposit the money in an account paying an annual rate of 5.88%, compounded quarterly. How much money will the account contain in 5 years? $334,728.90

65. WITHDRAWING ONLY INTEREST A financial adviser invested $90,000 in a long-term account at 5.1% interest, compounded daily. How much money will she be able to withdraw in 20 years if the principal is to remain in the account? $159,569.75

66. LIVING ON THE INTEREST A couple sold their home and invested the profit of $490,000 in an account at 6.3% interest, compounded daily. How much money will they be able to withdraw in 2 years if they don't want to touch the principal? $65,792.22

WRITING

67. What is the difference between simple and compound interest?

68. Explain this statement: *Interest is the amount of money paid for the use of money.*

69. On some accounts, banks charge a penalty if the depositor withdraws the money before the end of the term. Why would a bank do this?

70. Explain why it is better for a depositor to open a savings account that pays 5% interest compounded daily than one that pays 5% interest compounded monthly.

REVIEW

71. Evaluate: $\sqrt{\frac{1}{4}}$ $\frac{1}{2}$

72. Evaluate: $\left(\frac{1}{4}\right)^2$ $\frac{1}{16}$

73. Add: $\frac{3}{7} + \frac{2}{5}$ $\frac{29}{35}$

74. Subtract: $\frac{3}{7} - \frac{2}{5}$ $\frac{1}{35}$

75. Multiply: $2\frac{1}{2} \cdot 3\frac{1}{3}$ $8\frac{1}{3}$

76. Divide: $-12\frac{1}{2} \div 5$ $-2\frac{1}{2}$

77. Evaluate: -6^2 -36

78. Evaluate: $(0.2)^2 - (0.3)^2$ -0.05

STUDY SKILLS CHECKLIST

Percents, Decimals, and Fractions

Before taking the test on Chapter 7, read the following checklist. These skills are sometimes misunderstood by students. Put a checkmark in the box if you can answer "yes" to the statement.

☐ I know that to write a decimal as a percent, the decimal point is moved two places *to the right* and a % symbol is inserted.

Decimal	Percent
0.23	23%
0.768	76.8%
1.50	150%
0.9	90%

☐ I know that to write a percent as a decimal, the % symbol is dropped and the decimal point is moved two places *to the left*.

Percent	Decimal
44%	0.44
98.7%	0.987
0.5%	0.005
178.3%	1.783

☐ I know that to write a fraction as a percent, *a two-step process* is used:

Fraction ⟶ decimal ⟶ percent

Divide the numerator by the denominator. Move the decimal point two places to the right and insert a % symbol.

$$\frac{3}{4} \longrightarrow \begin{array}{r} 0.75 \\ 4\overline{)3.00} \\ \underline{-2\,8} \\ 20 \\ \underline{-20} \\ 0 \end{array} \longrightarrow 75\%$$

☐ I know that to find the percent increase (or decrease), we find what percent the amount of increase (or decrease) is of the *original amount*.

The number of phone calls increased from 10 to 18 per day.

Original amount   Amount of increase: $18 - 10 = 8$

Teaching Guide: Refer to the Instructor's Resource Binder to find activities, worksheets on key concepts, more examples, instruction tips, overheads, and assignments.

CHAPTER 7   SUMMARY AND REVIEW

SECTION 7.1   Percents, Decimals, and Fractions

DEFINITIONS AND CONCEPTS	EXAMPLES
Percent means parts per one hundred. The word *percent* can be written using the symbol %.	In the figure below, there are 100 equal-sized square regions, and 37 of them are shaded. We say that $\frac{37}{100}$, or 37%, of the figure is shaded. Numerator $\frac{37}{100} = 37\%$ Per 100

Unless otherwise noted, all art on this page is © Cengage Learning.

To **write a percent as a fraction**, drop the % symbol and write the given number over 100. Then simplify the fraction, if possible.	Write 22% as a fraction. $22\% = \dfrac{22}{100}$ — Drop the % symbol and write 22 over 100. $= \dfrac{\overset{1}{\cancel{2}} \cdot 11}{\underset{1}{\cancel{2}} \cdot 50}$ — To simplify the fraction, factor 22 and 100. Then remove the common factor of 2 from the numerator and denominator. Thus, $22\% = \dfrac{11}{50}$.
Percents such as 9.1% and 36.23% can be written as fractions of whole numbers by multiplying the numerator and denominator by a power of 10.	Write 9.1% as a fraction. $9.1\% = \dfrac{9.1}{100}$ — Drop the % symbol and write 9.1 over 100. $= \dfrac{9.1}{100} \cdot \dfrac{10}{10}$ — To obtain an equivalent fraction of whole numbers, we need to move the decimal point in the numerator one place to the right. Choose $\dfrac{10}{10}$ as the form of 1 to build the fraction. $= \dfrac{91}{1,000}$ — Multiply the numerators. Multiply the denominators. Thus, $9.1\% = \dfrac{91}{1,000}$.
Mixed number percents, such as $2\tfrac{1}{3}\%$ and $23\tfrac{5}{6}\%$, can be written as fractions of whole numbers by performing the indicated division.	Write $2\tfrac{1}{3}\%$ as a fraction. $2\tfrac{1}{3}\% = \dfrac{2\tfrac{1}{3}}{100}$ — Drop the % symbol and write $2\tfrac{1}{3}$ over 100. $= 2\tfrac{1}{3} \div 100$ — The fraction bar indicates division. $= \dfrac{7}{3} \cdot \dfrac{1}{100}$ — Write $2\tfrac{1}{3}$ as an improper fraction and then multiply by the reciprocal of 100. $= \dfrac{7}{300}$ — Multiply the numerators. Multiply the denominators. Thus, $2\tfrac{1}{3}\% = \dfrac{7}{300}$.
When **percents that are greater than 100%** are written as fractions, the fractions are greater than 1.	Write 170% as a fraction. $170\% = \dfrac{170}{100}$ — Drop the % symbol and write 170 over 100. $= \dfrac{\overset{1}{\cancel{10}} \cdot 17}{\underset{1}{\cancel{10}} \cdot 10}$ — To simplify the fraction, factor 170 and 100. Then remove the common factor of 10 from the numerator and denominator. Thus, $170\% = \dfrac{17}{10}$.
When **percents that are less than 1%** are written as fractions, the fractions are less than $\dfrac{1}{100}$.	Write 0.03% as a fraction. $0.03\% = \dfrac{0.03}{100}$ — Drop the % symbol and write 0.03 over 100. $= \dfrac{0.03}{100} \cdot \dfrac{100}{100}$ — To obtain an equivalent fraction of whole numbers, we need to move the decimal point in the numerator two places to the right. Choose $\dfrac{100}{100}$ as the form of 1 to build the fraction. $= \dfrac{3}{10,000}$ — Multiply the numerators and multiply the denominators. Since the numerator and denominator do not have any common factors (other than 1), the fraction is in simplified form. Thus, $0.03\% = \dfrac{3}{10,000}$.

To **write a percent as a decimal**, drop the % symbol and divide the given number by 100 by moving the decimal point two places to the left.	Write each percent as a decimal. $14\% = 14.0\% = 0.14$ — Write a decimal point and 0 to the right of the 4 in 14%. $9.35\% = 0.0935$ — Write a placeholder 0 (shown in blue) to the left of the 9. $198\% = 198.0\% = 1.98$ — Write a decimal point and 0 to the right of the 8 in 198%. $0.75\% = 0.0075$ — Write two placeholder zeros.
Mixed number percents, such as $1\frac{3}{4}\%$ and $10\frac{1}{2}\%$, can be written as decimals by writing the fractional part of the mixed number in its equivalent decimal form.	Write $1\frac{3}{4}\%$ as a decimal. There is no decimal point to move in $1\frac{3}{4}\%$. Since $1\frac{3}{4} = 1 + \frac{3}{4}$ and since the decimal equivalent of $\frac{3}{4}$ is 0.75, we can write $1\frac{3}{4}\%$ as 1.75% $1\frac{3}{4}\% = 1.75\% = 0.0175$ — Write a placeholder 0 (shown in blue) to the left of the 1.
To **write a decimal as a percent**, multiply the decimal by 100 by moving the decimal point two places to the right, and then insert a % symbol.	Write each decimal as a percent. $0.501 = 50.1\%$ $3.66 = 366\%$ $0.002 = 000.2\% = 0.2\%$
To **write a fraction as a percent**, 1. Write the fraction as a decimal by dividing its numerator by its denominator. 2. Multiply the decimal by 100 by moving the decimal point two places to the right, and then insert a % symbol. Fraction ⟶ decimal ⟶ percent	Write $\frac{3}{4}$ as a percent. **Step 1** Divide the numerator by the denominator. $$\begin{array}{r} 0.75 \\ 4\overline{)3.00} \\ -2\,8 \\ \hline 20 \\ -20 \\ \hline 0 \end{array}$$ Write a decimal point and some additional zeros to the right of 3. The remainder is 0. **Step 2** Write the decimal 0.75 as a percent. $\frac{3}{4} = 0.75 = 75\%$
Sometimes, when we want to write a fraction as a percent, the result of the division is a **repeating decimal**. In such cases, we can give an **exact answer** or an **approximate answer**.	Write $\frac{2}{3}$ as a percent. **Step 1** Divide the numerator by the denominator. $$\begin{array}{r} 0.666 \\ 3\overline{)2.000} \\ -1\,8 \\ \hline 20 \\ -18 \\ \hline 20 \\ -18 \\ \hline 2 \end{array}$$ Write a decimal point and some additional zeros to the right of 2. The repeating pattern is now clear. We can stop the division. **Step 2** Write the decimal 0.6666... as a percent. $0.6666 = 66.66\ldots\%$ **Exact Answer:** | **Approximation:** Use $\frac{2}{3}$ to represent 0.666.... | Round to the nearest tenth. $\frac{2}{3} = 66.66\ldots\%$ | $\frac{2}{3} = 66.66\ldots\%$ $\phantom{\frac{2}{3}} = 66\frac{2}{3}\%$ | $\phantom{\frac{2}{3}} \approx 66.7\%$

REVIEW EXERCISES

Express the amount of each figure that is shaded as a percent, as a decimal, and as a fraction. Each set of squares represents 100%.

1. 39%, 0.39, $\frac{39}{100}$

2. 111%, 1.11, $\frac{111}{100}$

3. In Problem 1, what percent of the figure is not shaded? 61%

4. THE INTERNET The following sentence appeared on a technology blog: "54 out of the top 100 websites failed Yahoo's performance test."
   a. What percent of the websites failed the test? 54%
   b. What percent of the websites passed the test? 46%

Write each percent as a fraction.

5. 15% $\frac{3}{20}$
6. 120% $\frac{6}{5}$
7. $9\frac{1}{4}$% $\frac{37}{400}$
8. 0.2% $\frac{1}{500}$

Write each percent as a decimal.

9. 27% 0.27
10. 8% 0.08
11. 655% 6.55
12. $1\frac{4}{5}$% 0.018
13. 0.75% 0.0075
14. 0.23% 0.0023

Write each decimal or whole number as a percent.

15. 0.83 83%
16. 1.625 162.5%
17. 0.051 5.1%
18. 6 600%

Write each fraction as a percent.

19. $\frac{1}{2}$ 50%
20. $\frac{4}{5}$ 80%
21. $\frac{7}{8}$ 87.5%
22. $\frac{1}{16}$ 6.25%

Write each fraction as a percent. Give the exact answer and an approximation to the nearest tenth of a percent.

23. $\frac{1}{3}$ $33\frac{1}{3}\% \approx 33.3\%$
24. $\frac{5}{6}$ $83\frac{1}{3}\% \approx 83.3\%$
25. $\frac{11}{12}$ $91\frac{2}{3}\% \approx 91.7\%$
26. $\frac{15}{9}$ $166\frac{2}{3}\% \approx 166.7\%$

27. WATER DISTRIBUTION The oceans contain 97.2% of all of the water on Earth. (Source: National Ground Water Association)
    a. Write this percent as a decimal. 0.972
    b. Write this percent as a fraction in simplest form. $\frac{243}{250}$

28. BILL OF RIGHTS There are 27 amendments to the Constitution of the United States. The first 10 are known as the Bill of Rights. What percent of the amendments were adopted after the Bill of Rights? (Round to the nearest one percent.) 63%

29. TAXES The city of Grand Prairie, Texas, uses a *one-fourth of one percent* sales tax to help fund park improvements.
    a. Write this percent as a decimal. 0.0025
    b. Write this percent as a fraction. $\frac{1}{400}$

30. SOCIAL SECURITY If your full retirement age is 66, your Social Security benefits are reduced by $\frac{1}{15}$ if you retire at age 65. Write this fraction as a percent. Give the exact answer and an approximation to the nearest tenth of a percent. (Source: Social Security Administration) $6\frac{2}{3}\% \approx 6.7\%$

SECTION 7.2 Solving Percent Problems Using Percent Equations and Proportions

DEFINITIONS AND CONCEPTS

The key words in a **percent sentence** can be translated to a percent equation.

- Each *is* translates to an equal symbol =
- *of* translates to multiplication that is shown with a raised dot ·
- *what* number or *what* percent translates to an unknown number that is represented by a variable.

EXAMPLES

Translate the percent sentence to a percent equation.

What number is 26% of 180?

$x = 26\% \cdot 180$ This is the percent equation.

Percent sentences involve a comparison of numbers. The relationship between the **base** (the standard of comparison, the whole), the **amount** (a part of the base), and the **percent** is:
 	
    Amount = percent · base
 	
or
 	
    Part = percent · whole
	$\underset{\substack{\uparrow \\ \text{Amount} \\ \text{(part)}}}{8}$ is $\underset{\substack{\uparrow \\ \text{percent}}}{12.5\%}$ of $\underset{\substack{\uparrow \\ \text{base} \\ \text{(whole)}}}{64.}$

The percent equation method:
 	
We can translate percent sentences to percent equations and solve to **find the amount**.
 	
Caution! When solving percent equations, always write the percent as a decimal (or fraction) before performing any calculations.
	What number is 45% of 120?
	    $\downarrow$ $\downarrow$ $\downarrow$ $\downarrow$ $\downarrow$
	    x $=$ 45% · 120 Translate.
	
	Now, solve the percent equation.
	
	    $x = 0.45 \cdot 120$ Write 45% as a decimal.
	    $x = 54$ Do the multiplication.
	
	Thus, 54 is 45% of 120.

We can translate percent sentences to percent equations and solve to **find the percent**.	12 is what percent of 192?
	$\downarrow$ $\downarrow$ $\downarrow$ $\downarrow$ $\downarrow$
	12 $=$ x · 192 Translate.
	
	Now, solve the percent equation.
	
	    $12 = 192x$ Write $x \cdot 192$ as $192x$.
	    $\dfrac{12}{192} = \dfrac{192x}{192}$ To isolate x, undo the multiplication by 192 by dividing both sides by 192.
	    $0.0625 = x$ Do the division: $12 \div 192 = 0.0625$.
	    $06.25\% = x$ To write 0.0625 as a percent, multiply it by 100 by moving the decimal point two places to the right, and then insert a % symbol.
	
	Thus, 12 is 6.25% of 192.

We can translate percent sentences to percent equations and solve to **find the base**.
 	
Caution! Sometimes the calculations to solve a percent problem are made easier if we **write the percent as a fraction instead of a decimal**. This is the case with percents that have *repeating* decimal equivalents such as $33\frac{1}{3}\%$, $66\frac{2}{3}\%$, and $16\frac{2}{3}\%$.
	8.2 is $33\frac{1}{3}\%$ of what number?
	$\downarrow$ $\downarrow$ $\downarrow$ $\downarrow$ $\downarrow$
	8.2 $=$ $33\frac{1}{3}\%$ · x Translate.
	
	Now, solve the percent equation.
	
	    $8.2 = \dfrac{1}{3}x$ Write the percent as a fraction: $33\frac{1}{3}\% = \frac{1}{3}$.
	    $3(8.2) = 3\left(\dfrac{1}{3}x\right)$ To isolate x, multiply both sides by the reciprocal of $\frac{1}{3}$, which is 3.
	    $24.6 = 1x$ On the left side, do the multiplication: $3(8.2) = 24.6$. On the right side, the product of a number and its reciprocal is 1: $3\left(\frac{1}{3}\right) = 1$.
	    $24.6 = x$ On the right side, the coefficient of 1 need not be written: $1x = x$.
	
	Thus, 8.2 is $33\frac{1}{3}\%$ of 24.6.

Chapter 7 Summary and Review

The percent proportion method: We can translate percent sentences to percent proportions and solve to **find the amount**. To translate a percent sentence to a **percent proportion**, use the following form: *Amount is to base as percent is to 100:* $$\frac{\text{amount}}{\text{base}} = \frac{\text{percent}}{100}$$ or *Part is to whole as percent is to 100:* $$\frac{\text{part}}{\text{whole}} = \frac{\text{percent}}{100}$$	What number is 45% of 120? (amount) (percent) (base) $\dfrac{x}{120} = \dfrac{45}{100}$ This is the proportion to solve. To make the calculations easier, simplify the ratio $\frac{45}{100}$. $\dfrac{x}{120} = \dfrac{9}{20}$ Simplify: $\frac{45}{100} = \frac{\overset{1}{\cancel{5}} \cdot 9}{\underset{1}{\cancel{5}} \cdot 20} = \frac{9}{20}$. To solve the proportion we use the cross products. $x \cdot 20 = 120 \cdot 9$ Find the cross products and set them equal. $20x = 1{,}080$ On the left side, write $x \cdot 20$ as $20x$. On the right side, do the multiplication: $120 \cdot 9 = 1{,}080$. $\dfrac{20x}{20} = \dfrac{1{,}080}{20}$ To isolate x, undo the multiplication by 20 by dividing both sides by 20. $x = 54$ Do the division: $1{,}080 \div 20 = 54$. Thus, 54 is 45% of 120.
We can translate percent sentences to percent proportions and solve to **find the percent**.	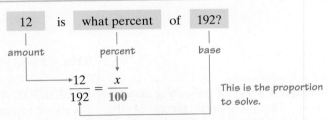 12 is what percent of 192? (amount) (percent) (base) $\dfrac{12}{192} = \dfrac{x}{100}$ This is the proportion to solve. To make the calculations easier, simplify the ratio $\frac{12}{192}$ first. $\dfrac{1}{16} = \dfrac{x}{100}$ Simplify: $\frac{12}{192} = \frac{\overset{1}{\cancel{2}} \cdot \overset{1}{\cancel{2}} \cdot \overset{1}{\cancel{3}}}{\underset{1}{\cancel{2}} \cdot \underset{1}{\cancel{2}} \cdot 2 \cdot 2 \cdot 2 \cdot \cancel{3}} = \frac{1}{16}$. $1 \cdot 100 = 16 \cdot x$ Find the cross products and set them equal. $100 = 16x$ On the left side, do the multiplication: $1 \cdot 100 = 100$. On the right side, write $16 \cdot x$ as $16x$. $\dfrac{100}{16} = \dfrac{16x}{16}$ To isolate x, undo the multiplication by 16 by dividing both sides by 16. $6.25 = x$ Do the division: $100 \div 16 = 6.25$. Thus, 12 is 6.25% of 192.
We can translate percent sentences to percent proportions and solve to **find the base**.	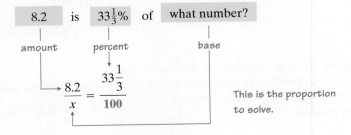 8.2 is $33\frac{1}{3}\%$ of what number? (amount) (percent) (base) $\dfrac{8.2}{x} = \dfrac{33\frac{1}{3}}{100}$ This is the proportion to solve.

To make the calculations easier, write the mixed number $33\frac{1}{3}$ as the improper fraction $\frac{100}{3}$.

$$\frac{8.2}{x} = \frac{\frac{100}{3}}{100} \qquad \text{Write } 33\frac{1}{3} \text{ as } \frac{100}{3}.$$

$$8.2 \cdot 100 = x \cdot \frac{100}{3} \qquad \text{To solve the proportion, find the cross products and set them equal.}$$

$$820 = \frac{100}{3}x \qquad \text{On the left side, do the multiplication: } 8.2 \cdot 100 = 820. \text{ On the right side, write } x \cdot \frac{100}{3} \text{ as } \frac{100}{3}x.$$

$$\frac{3}{100}(820) = \frac{3}{100}\left(\frac{100}{3}x\right) \qquad \text{To isolate } x, \text{ multiply both sides by the reciprocal of } \frac{100}{3}, \text{ which is } \frac{3}{100}.$$

$$\frac{3}{100}\left(\frac{820}{1}\right) = 1x \qquad \text{On the left side, write 820 as a fraction: } 820 = \frac{820}{1}. \text{ On the right side, the product of a number and its reciprocal is 1: } \frac{3}{100}\left(\frac{100}{3}\right) = 1.$$

$$\frac{2{,}460}{100} = x \qquad \text{On the left side, multiply the numerators and multiply the denominators. On the right side, the coefficient of 1 is not needed: } 1x = x.$$

$$24.6 = x \qquad \text{Divide 2,460 by 100 by moving the understood decimal point in 2,460 two places to the left.}$$

Thus, 8.2 is $33\frac{1}{3}\%$ of 24.6.

A **circle graph** is a way of presenting data for comparison. The pie-shaped pieces of the graph show the relative sizes of each category.

The 100 tick marks equally spaced around the circle serve as a visual aid when constructing a circle graph.

FACEBOOK As of the spring of 2012, Facebook had approximately 900 million users worldwide. Use the information in the circle graph on the right to find how many of them were male.

The circle graph shows that 40% of the 900 million users of Facebook were male.

**Facebook Users Worldwide
900 Million**

Male 40%
Female 60%

Source: tweetsmarter.com

Method 1: To find the unknown amount write and then solve a **percent equation**.

What number	is	40%	of	900 million?
↓	↓	↓	↓	↓
x	$=$	40%	$\cdot$	900

Translate.

Now, solve the percent equation.

$x = 0.40 \cdot 900 \qquad$ Write 40% as a decimal: 40% = 0.40.

$x = 360 \qquad$ Do the multiplication. The answer is in millions.

In the spring of 2012, there were approximately 360 million male users of Facebook worldwide.

To solve percent application problems, we often have to **rewrite the facts** of the problem in percent sentence form before we can translate to an equation.

Unless otherwise noted, all art on this page is © Cengage Learning.

Method 2: To find the unknown amount write and then solve a **percent proportion**.

$$\frac{x}{900} = \frac{2}{5}$$ Simplify the ratio: $\frac{40}{100} = \frac{2 \cdot 20}{5 \cdot 20} = \frac{2}{5}$.

$x \cdot 5 = 900 \cdot 2$ Find the cross products and set them equal.

$5x = 1{,}800$ On the left side, write $x \cdot 5$ as $5x$. On the right side, do the multiplication.

$\dfrac{5x}{5} = \dfrac{1{,}800}{5}$ To isolate x, undo the multiplication by 5 by dividing both sides by 5.

$x = 360$ Do the division: $1{,}800 \div 5 = 360$. The answer is in millions.

In the spring of 2012, there were approximately 360 million male users of Facebook worldwide.

REVIEW EXERCISES

31. a. Identify the amount, the base, and the percent in the statement "15 is $33\frac{1}{3}\%$ of 45."
    amount: 15, base: 45, percent: $33\frac{1}{3}\%$
  b. Fill in the blanks to complete the percent equation (formula):

      Amount = percent · base

      or

      Part = percent · whole

32. When computing with percents, we must change the percent to a decimal or a fraction. Change each percent to a decimal.
  a. 13% 0.13   **b.** 7.1% 0.071
  c. 195% 1.95   **d.** $\frac{1}{4}\%$ 0.0025

  Change each percent to a fraction.
  e. $33\frac{1}{3}\%$ $\frac{1}{3}$ **f.** $66\frac{2}{3}\%$ $\frac{2}{3}$
  g. $16\frac{2}{3}\%$ $\frac{1}{6}$

33. Translate each percent sentence into a *percent equation*. **Do not solve.**
  a. What number is 32% of 96? $x = 32\% \cdot 96$
  b. 64 is what percent of 135? $64 = x \cdot 135$
  c. 9 is 47.2% of what number? $9 = 47.2\% \cdot x$

34. Translate each percent sentence into a *percent proportion*. **Do not solve.**
  a. What number is 32% of 96? $\frac{x}{96} = \frac{32}{100}$
  b. 64 is what percent of 135? $\frac{64}{135} = \frac{x}{100}$
  c. 9 is 47.2% of what number? $\frac{9}{x} = \frac{47.2}{100}$

Translate to a percent equation or percent proportion and then solve to find the unknown number.

35. What number is 40% of 500?   200
36. 16% of what number is 20?   125
37. 1.4 is what percent of 80?   1.75%
38. $66\frac{2}{3}\%$ of 3,150 is what number?   2,100
39. Find 220% of 55.   121
40. What is 0.05% of 60,000?   30

41. 3.88 is 2.5% of what number? 155.2
42. What percent of 0.08 is 4.24? 5,300%
43. **RACING** The nitro–methane fuel mixture used to power some experimental cars is 96% nitro and 4% methane. How many gallons of methane are needed when filling a 15-gallon fuel tank? 0.6 gal methane
44. **HOME SALES** After the first day on the market, 51 homes in a new subdivision had already sold. This was 75% of the total number of homes available. How many homes were originally for sale? 68
45. **HURRICANE DAMAGE** In a mobile home park, 96 of the 110 trailers were either damaged or destroyed by hurricane winds. What percent is this? (Round to the nearest 1 percent.) 87%
46. **TIPPING** The cost of dinner for a family of five at a restaurant was $36.20. Find the amount of the tip if it should be 15% of the cost of dinner. $5.43
47. **COLLEGE EXPENSES** The table in the next column shows how the typical family pays for college, according to a 2012 Sallie Mae survey. Draw a circle graph for the data.

Percent of total cost of college paid from each source

Parent income/savings	28%	Student income/savings	12%
Grants/scholarships	29%	Parent borrowing	9%
Student borrowing	18%	Relatives/friends	4%

48. **EARTH'S SURFACE** The surface of Earth is approximately 196,800,000 square miles. Use the information in the circle graph to determine the number of square miles of Earth's surface that are covered with water. 139,531,200 mi²

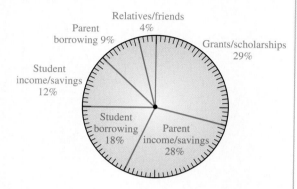

SECTION 7.3 Applications of Percent

DEFINITIONS AND CONCEPTS	EXAMPLES
The **sales tax** on an item is a percent of the purchase price of the item. Sales tax = sales tax rate · purchase price ↑ ↑ ↑ Amount = percent · base Notice that the formula is based on the percent equation discussed in Section 7.2. **Sales tax dollar amounts** are rounded to the nearest cent (hundredth). The **total cost** of an item is the sum of its purchase price and the sales tax on the item. Total cost = purchase price + sales tax	**SHOPPING** Find the sales tax and total cost of a $50.95 purchase if the sales tax rate is 8%. Sales tax = **sales tax rate · purchase price** Sales tax = 8% · $50.95 Sales tax = 0.08 · $50.95 Write 8% as a decimal: 8% = 0.08. Sales tax = $4.076 Do the multiplication. Sales tax ≈ $4.08 Round the sales tax to the nearest cent (hundredth). Thus, the sales tax is $4.08. The total cost is the sum of its purchase price and the sales tax. Total cost = **purchase price + sales tax rate** Total cost = $50.95 + $4.08 Total cost = $55.03 Do the addition. The total cost of the purchase is $55.03.
Sales tax rates are usually expressed as a percent.	**APPLIANCES** The purchase price of a toaster is $82. If the sales tax is $5.33, what is the sales tax rate? The sales tax of $5.33 is some unknown percent of the purchase price of $82. Two methods can be used to solve this problem.

Two methods can be used to find the unknown sales tax rate: • The percent equation method • The percent proportion method	**The percent equation method:** 5.33 is what percent of 82? $5.33 = x \cdot 82$ Translate. Now, solve the percent equation. $5.33 = 82x$ Write $x \cdot 82$ as $82x$. $\dfrac{5.33}{82} = \dfrac{82x}{82}$ To isolate x, undo the multiplication by 82 by dividing both sides by 82. $0.065 = x$ Do the division: $5.33 \div 82 = 0.065$. $006.5\% = x$ Write the decimal 0.065 as a percent. $6.5\% = x$ The sales tax rate is 6.5%. **The percent proportion method:** 5.33 is what percent of 82? amount percent base $\dfrac{5.33}{82} = \dfrac{x}{100}$ This is the percent proportion to solve. $5.33 \cdot 100 = 82 \cdot x$ To solve the proportion, find the cross products and set them equal. $533 = 82x$ On the left side, do the multiplication: $5.33 \cdot 100 = 533$. On the right side, write $82 \cdot x$ as $82x$. $\dfrac{533}{82} = \dfrac{82x}{82}$ To isolate x, undo the multiplication by 82 by dividing both sides by 82. $6.5 = x$ Do the division: $533 \div 82 = 6.5$. The sales tax rate is 6.5%.
Instead of working for a salary or getting paid at an hourly rate, many salespeople are paid on **commission**. The **amount of commission** paid is a percent of the total dollar sales of goods or services. Commission = commission rate · sales	**COMMISSIONS** A salesperson earns an 11% commission on all appliances that she sells. If she sells a $450 dishwasher, what is her commission? Commission = commission rate · sales Commission = 11% · $450 Commission = 0.11 · $450 Write 11% as a decimal. Commission = $49.50 Do the multiplication. The commission earned on the sale of the $450 dishwasher is $49.50.
The **commission rate** is usually expressed as a percent.	**TELEMARKETING** A telemarketer made a commission of $600 in one week on sales of $4,000. What is his commission rate? Commission = commission rate · sales $600 = x · $4,000 Let x represent the unknown commission rate. $600 = 4,000x$ Drop the dollar signs. Write $x \cdot 4{,}000$ as $4{,}000x$. $\dfrac{600}{4{,}000} = \dfrac{4{,}000x}{4{,}000}$ To isolate x, undo the multiplication by 4,000 by dividing both sides by 4,000.

	$0.15 = x$ Do the division: $600 \div 4{,}000 = 0.15$.
	$0\underset{\curvearrowright}{15}\% = x$ Write the decimal 0.15 as a percent.
	The commission rate is 15%.

To find **percent of increase or decrease**:
1. Subtract the smaller number from the larger to find the amount of increase or decrease.
2. Find what percent the amount of increase or decrease is of the original amount.
 	
Two methods can be used to find the unknown percent of increase (or decrease):
• The percent equation method
• The percent proportion method
 	
Caution! The percent of increase (or decrease) is a percent of the *original number*—that is, the number before the change occurred.
	WATCHING TELEVISION According to the Nielsen Company, the average American watched 145 hours of TV a month in 2007. That increased to 153 hours per month in 2011. Find the percent of increase. Round to the nearest one percent.
	
	First, subtract to find the amount of increase.
	
	$153 - 145 = 8$ Subtract the smaller number from the larger number.
	
	The number of hours watched per month increased by 8.
	
	Next, find what percent of the *original* 145 hours the 8-hour increase represents.
	
	The percent equation method:
	
	    8 is what percent of 145?
	    8 = x · 145 Translate.
	
	Now, solve the percent equation.
	
	$8 = 145x$ Write $x \cdot 145$ as $145x$.
	
	$\dfrac{8}{145} = \dfrac{145x}{145}$ To isolate x, undo the multiplication by 145 by dividing both sides by 145.
	
	$0.055 \approx x$ On the left side, divide 8 by 145. The division does not terminate.
	
	$0\underset{\curvearrowright}{05}.5\% \approx x$ Write the decimal 0.055 as a percent.
	
	$6\% \approx x$ Round to the nearest one percent.
	
	Between 2007 and 2011, the number of hours of television watched by the average American each month increased by 6%.
	
	If the **percent proportion method** is used, solve the following proportion for x to find the percent of increase.
	
	    8 is what percent of 145?
	  amount percent base
	
	$\dfrac{8}{145} = \dfrac{x}{100}$ This is the proportion to solve.

The **amount of discount** is a percent of the original price. $\underset{\uparrow}{\text{Amount of}\atop\text{discount}} = \underset{\uparrow}{\text{discount}\atop\text{rate}} \cdot \underset{\uparrow}{\text{original}\atop\text{price}}$ amount = percent · base Notice that the formula is based on the percent equation discussed in Section 7.2.	TOOL SALES Find the amount of the discount on a tool kit if it is normally priced at $89.95 but is currently on sale for 35% off. Then find the sale price. Amount of discount = **discount rate · original price** Amount of discount = 35% · $89.95 Amount of discount = 0.35 · $89.95 Write 35% as a decimal. Amount of discount = $31.4825 Do the multiplication. Amount of discount ≈ $31.48 Round to the neaerst cent (hundredth).

To find the **sale price** of an item, subtract the discount from the original price. Sale price = original price − discount	The discount on the tool kit is $31.48. To find the sale price, we use subtraction. Sale price = **original price** − **discount** Sale price = $89.95 − $31.48 Sale price = $58.47 *Do the subtraction.* The sale price of the tool kit is $58.47.
The difference between the original price and the sale price is the **amount of discount**. Amount of discount = original price − sale price	**FURNITURE SALES** Find the discount rate on a living room set regularly priced at $2,500 that is on sale for $1,870. Round to the nearest one percent. We will think of this as a *percent-of-decrease problem*. The discount (decrease in price) is found using subtraction. $2,500 − $1,870 = $630 Discount = original price − sale price The living room set is discounted $630. Now we find what percent of the original price the $630 discount represents. Amount of discount = discount rate · original price $630 = x · $2,500 $630 = 2,500x$ *Drop the dollar signs. Write x · 2,500 as 2,500x.* $\dfrac{630}{2,500} = \dfrac{2,500x}{2,500}$ *To isolate x, undo the multiplication by 2,500 by dividing both sides by 2,500.* $0.252 = x$ *Do the division: 630 ÷ 2,500 = 0.252.* $025.2\% = x$ *Write the decimal 0.252 as a percent.* $25\% \approx x$ *Round to the nearest one percent.* To the nearest one percent, the discount rate on the living room set is 25%.

REVIEW EXERCISES

49. SALES RECEIPTS Complete the sales receipt shown below by finding the sales tax and total cost of the camera. $3.30, $63.29

CAMERA CENTER

35mm Canon Camera	$59.99
SUBTOTAL	$59.99
SALES TAX @ 5.5%	?
TOTAL	?

50. SALES TAX RATES Find the sales tax rate if the sales tax is $492 on the purchase of an automobile priced at $12,300. 4%

51. COMMISSIONS If the commission rate is 6%, find the commission earned by an appliance salesperson who sells a washing machine for $369.97 and a dryer for $299.97. $40.20

52. SELLING MEDICAL SUPPLIES A salesperson made a commission of $646 on a $15,200 order of antibiotics. What is her commission rate? 4.25%

53. T-SHIRT SALES A stadium owner earns a commission of $33\frac{1}{3}\%$ of the T-shirt sales from any concert or sporting event. How much can the owner make if 12,000 T-shirts are sold for $25 each at a soccer match? $100,000

54. Fill in the blank: The percent of increase (or decrease) is a percent of the original number—that is, the number before the change occurred.

55. **THE UNITED NATIONS** In 2013, the United Nations expanded its peacekeeping force in the Democratic Republic of Congo from 17,000 to 19,500 troops. Find the percent of increase in the number of troops. Round to the nearest one percent. (Source: aljazeera.com) 15%

56. **GAS MILEAGE** A woman found that the gas mileage fell from 18.8 to 17.0 miles per gallon when she experimented with a new brand of gasoline in her truck. Find the percent of decrease in her mileage. Round to the nearest tenth of one percent. 9.6%

57. Fill in the blanks.
    a. Sales tax = sales tax rate · purchase price
    b. Total cost = purchase price + sales tax
    c. Commission = commission rate · sales

58. Fill in the blanks.
    a. Amount of discount = original price − sale price
    b. Amount of discount = discount rate · original price
    c. Sale price = original price − discount

59. **TOOL CHESTS** Use the information in the advertisement below to find the discount, the original price, and the discount rate on the tool chest. $180, $2,500, 7.2%

Sale price $2,320
Save $180!
Tool Chest
Professional quality
7 drawers

60. **RENTS** Find the discount rate if the monthly rent for an apartment is reduced from $980 to $931 per month. 5%

SECTION 7.4 Estimation with Percent

DEFINITIONS AND CONCEPTS	EXAMPLES

Estimation can be used to find approximations when exact answers aren't necessary.
 	
To find **1% of a number**, move the decimal point in the number two places to the left.
	What is 1% of 291.4? Find the exact answer and an estimate using front-end rounding.
	
	Exact answer:
	
	  1% of 291.4 = 2.914 *Move the decimal point two places to the left.*
	
	Estimate: 291.4 front-end rounds to 300. If we move the understood decimal point in 300 two places to the left, we get 3. Thus
	
	  1% of 291.4 ≈ 3 *Because 1% of 300 = 3.*

To find **10% of a number**, move the decimal point in the number one place to the left.	What is 10% of 40,735 pounds? Find the exact answer and an estimate using front-end rounding.
	
	Exact answer:
	
	  10% of 40,735 = 4,073.5 *Move the decimal point one place to the left.*
	
	Estimate: 40,735 front-end rounds to 40,000. If we move the understood decimal point in 40,000 one place to the left, we get 4,000. Thus
	
	  10% of 40,735 ≈ 4,000 *Because 10% of 40,000 = 4,000.*

To find **20% of a number**, find 10% of the number by moving the decimal point one place to the left, and then double (multiply by 2) the result. A similar approach can be used to find 30% of a number, 40% of a number, and so on.	Estimate the answer: What is 20% of 809? Since 10% of 809 is 80.9 (or about **81**), it follows that 20% of 809 is about 2 · **81**, which is 162. Thus, 20% of 809 ≈ 162 *Because 10% of 809 ≈ 81.*

To find **50% of a number**, divide the number by 2.	Estimate the answer: What is 50% of 1,442,957? We use 1,400,000 as an approximation of 1,442,957 because it is even, divisible by 2, and ends with many zeros. $$50\% \text{ of } 1{,}442{,}957 \approx 700{,}000 \quad \text{Because 50\% of 1,400,000} = \frac{1{,}400{,}000}{2} = 700{,}000.$$
To find **25% of a number**, divide the number by 4.	Estimate the answer: What is 25% of 21.004? We use 20 as an approximation because it is close to 21.004 and because it is divisible by 4. $$25\% \text{ of } 21.004 \approx 5 \quad \text{Because 25\% of 20} = \frac{20}{4} = 5.$$
To find **5% of a number**, find 10% of the number by moving the decimal point in the number one place to the left. Then, divide that result by 2.	Estimate the answer: What is 5% of 36,150? First, we find 10% of 36,150: $$10\% \text{ of } 36{,}150 = 3{,}615$$ We use 3,600 as an approximation of this result because it is close to 3,615 and because it is even, and therefore divisible by 2. Next, we divide the approximation by 2 to estimate 5% of 36,150. $$\frac{3{,}600}{2} = 1{,}800$$ Thus, 5% of 36,150 ≈ 1,800.
To find **15% of a number**, find the sum of 10% of the number and 5% of the number.	TIPPING Estimate the 15% tip on a dinner costing $88.55. To simplify the calculations, we will estimate the cost of the $88.55 dinner to be $90. Then, to estimate the tip, we find 10% of $90 and 5% of $90, and add. 10% of $90 is $9 $9 5% of $90 (half as much as 10% of $90) + $4.50 $13.50 The tip should be $13.50.
To find **200% of a number**, multiply the number by 2. A similar approach can be used to find 300% of a number, 400% of a number, and so on.	Estimate the answer: What is 200% of 3.509? To estimate 200% of 3.509, we will find 200% of 4. We use 4 as an approximation because it is close to 3.509 and it makes the multiplication by 2 easy. $$200\% \text{ of } 3.509 \approx 8 \quad \text{Because 200\% of 4} = 2 \cdot 4 = 8.$$
Sometimes we must **approximate the percent** to estimate an answer.	QUALITY CONTROL In a production run of 145,350 ceramic tiles, 3% were found to be defective. Estimate the number of defective tiles. To estimate 3% of 145,350, we will find 1% of 150,000, and multiply the result by 3. We use 150,000 as the approximation because it is close to 145,350 and it ends with several zeros. $$3\% \text{ of } 145{,}350 \approx 4{,}500 \quad \text{Because 1\% of 150,000} = 1{,}500 \text{ and } 3 \cdot 1{,}500 = 4{,}500.$$ There were about 4,500 defective tiles in the production run.

REVIEW EXERCISES

What is 1% of the given number? Find the exact answer and an estimate using front-end rounding.

61. 342.03 3.4203, 3

62. 8,687 86.87, 90

What is 10% of the given number? Find the exact answer and an estimate using front-end rounding.

63. 43.4 seconds
4.34 sec, 4 sec

64. 10,900 liters
1,090 L, 1,000 L

Estimate each answer. (Answers may vary.)

65. What is 20% of 63? 12

66. What is 20% of 612? 120

67. What is 50% of 279,985? 140,000

68. What is 50% of 327? 150

69. What is 25% of 13.02? 3

70. What is 25% of 39.9? 10

71. What is 5% of 7,150? 350

72. What is 5% of 19,359? 1,000

73. What is 200% of 29.78? 60

74. What is 200% of 1.125? 2

Estimate a 15% tip on each dollar amount. (Answers may vary.)

75. $243.55 $36

76. $46.99 $7.50

Estimate each answer. (Answers may vary.)

77. SPECIAL OFFERS A home improvement store sells a 50-fluid-ounce pail of asphalt driveway sealant that is labeled "25% free." How many ounces are free? about 12 fluid oz

78. JOB TRAINING 15% of the 785 people attending a job training program had a college degree. How many people is this? about 120 people

Approximate the percent and then estimate each answer. (Answers may vary.)

79. SEAT BELTS A state trooper survey on an interstate highway found that of the 3,850 cars that passed the inspection point, 6% of the drivers were not wearing a seat belt. Estimate the number not wearing a seat belt. 200

80. DOWN PAYMENTS Estimate the amount of an 11% down payment on a house that is selling for $279,950. $30,000

SECTION 7.5 Interest

DEFINITIONS AND CONCEPTS

Interest is money that is paid for the use of money.

Simple interest is interest earned on the original principal and is found using the formula

$$I = Prt$$

where P is the principal, r is the annual (yearly) interest rate, and t is the length of time in years.

The **total amount** in an investment account or the total amount to be repaid on a loan is the sum of the principal and the interest.

Total amount = principal + interest

EXAMPLES

If $4,000 is invested for 3 years at a rate of 7.2%, how much simple interest is earned?

$P = \$4,000 \qquad r = 7.2\% = 0.072 \qquad t = 3$

$I = Prt$ This is the simple interest formula.

$I = \$4,000 \cdot 0.072 \cdot 3$ Substitute the values for P, r, and t. Remember to write the rate r as a decimal.

$I = \$288 \cdot 3$ Multiply: $\$4,000 \cdot 0.072 = \$288$.

$I = \$864$ Complete the multiplication.

The simple interest earned in 3 years is $864.

HOME REPAIRS A homeowner borrowed $5,600 for 2 years at 10% simple interest to pay for a new concrete driveway. Find the total amount due on the loan.

$P = \$5,600 \qquad r = 10\% = 0.10 \qquad t = 2$

$I = Prt$ This is the simple interest formula.

$I = \$5,600 \cdot 0.10 \cdot 2$ Write the rate r as a decimal.

$I = \$560 \cdot 2$ Multiply: $\$5,600 \cdot 0.10 = \$560$.

$I = \$1,120$ Complete the multiplication.

The interest due in 2 years is $1,120. To find the total amount of money due on the loan, we add.

Total amount = principal + interest
            = $5,600 + $1,120
            = $6,720   Do the addition.

At the end of 2 years, the total amount of money due on the loan is $6,720.

When using the formula $I = Prt$, the **time must be expressed in years**. If the time is given in days or months, rewrite it as a fractional part of a year. Here are two examples: • Since there are 365 days in a year, $$60 \text{ days} = \frac{60}{365} \text{ year} = \frac{\overset{1}{\cancel{5}} \cdot 12}{\underset{1}{\cancel{5}} \cdot 73} \text{ year} = \frac{12}{73} \text{ year}$$ • Since there are 12 months in a year, $$4 \text{ months} = \frac{4}{12} \text{ year} = \frac{\overset{1}{\cancel{4}}}{3 \cdot \underset{1}{\cancel{4}}} \text{ year} = \frac{1}{3} \text{ year}$$	**FINES** A man borrowed $300 at 15% for 45 days to get his car out of an impound parking garage. Find the simple interest that must be paid on the loan. Since there are 365 days in a year, we have $$45 \text{ days} = \frac{45}{365} \text{ year} = \frac{\overset{1}{\cancel{5}} \cdot 9}{\underset{1}{\cancel{5}} \cdot 73} \text{ year} = \frac{9}{73} \text{ year} \quad \text{Simplify the fraction.}$$ The time of the loan is $\frac{9}{73}$ year. To find the amount of interest, we multiply. $P = \$300 \qquad r = 15\% = 0.15 \qquad t = \frac{9}{73}$ $I = Prt$ This is the simple interest formula. $I = \$300 \cdot 0.15 \cdot \frac{9}{73}$ Write the rate r as a decimal. $I = \frac{\$300}{1} \cdot \frac{0.15}{1} \cdot \frac{9}{73}$ Write $300 and 0.15 as fractions. $I = \frac{\$405}{73}$ Multiply the numerators. Multiply the denominators. $I \approx \$5.55$ Do the division. Round to the nearest cent. The simple interest that must be paid on the loan is $5.55.
Compound interest is interest earned on the original principal and previously earned interest. When compounding, we can calculate interest: • **annually:** once a year • **semiannually:** twice a year • **quarterly:** four times a year • **daily:** 365 times a year	**COMPOUND INTEREST** Suppose $10,000 is deposited in an account that earns 6.5% compounded semiannually. Find the amount of money in an account at the end of the first year. The word *semiannually* means that the interest will be compounded two times in one year. To find the amount of interest $10,000 will earn in the first half of the year, use the simple interest formula, where t is $\frac{1}{2}$ of a year. **Interest earned in the first half of the year:** $P = \$10,000 \qquad r = 6.5\% = 0.065 \qquad t = \frac{1}{2}$ $I = Prt$ This is the simple interest formula. $I = \$10,000 \cdot 0.065 \cdot \frac{1}{2}$ Write the rate r as a decimal. $I = \frac{\$10,000}{1} \cdot \frac{0.065}{1} \cdot \frac{1}{2}$ Write $10,000 and 0.065 as fractions. $I = \frac{\$650}{2}$ Multiply the numerators. Multiply the denominators. $I = \$325$ Do the division. The interest earned in the first half of the year is $325. The original principal and this interest now become the principal for the second half of the year. $$\$10,000 + \$325 = \$10,325$$ To find the amount of interest $10,325 will earn in the second half of the year, use the simple interest formula, where t is again $\frac{1}{2}$ of a year.

Interest earned in the second half of the year:

$$P = \$10{,}325 \qquad r = 6.5\% = 0.065 \qquad t = \frac{1}{2}$$

$I = Prt$ This is the simple interest formula.

$I = \$10{,}325 \cdot 0.065 \cdot \dfrac{1}{2}$ Write the rate r as a decimal.

$I = \dfrac{\$10{,}325}{1} \cdot \dfrac{0.065}{1} \cdot \dfrac{1}{2}$ Write $10,325 and 0.065 as fractions.

$I = \dfrac{\$671.125}{2}$ Multiply the numerators. Multiply the denominators.

$I \approx \$335.56$ Do the division. Round to the nearest cent.

The interest earned in the second half of the year is $335.56. Adding this to the principal for the second half of the year, we get

$$\$10{,}325 + \$335.56 = \$10{,}660.56$$

The total amount in the account after one year is $10,660.56

Computing compound interest by hand can take a long time. The **compound interest formula** can be used to find the amount of money that an account will contain at the end of the term.

$$A = P\left(1 + \frac{r}{n}\right)^{nt}$$

where A is the amount in the account, P is the principal, r is the annual interest rate, n is the number of compoundings in one year, and t is the length of time in years.

A **calculator** is helpful in performing the operations on the right side of the compound interest formula.

COMPOUNDING DAILY A mini-mall developer promises investors in his company $3\frac{1}{4}\%$ interest, compounded daily. If a businessman decides to invest $80,000 with the developer, how much money will be in his account in 8 years?

Compounding daily means the compounding will be done 365 times a year.

$$P = \$80{,}000 \qquad r = 3\frac{1}{4}\% = 0.0325 \qquad t = 8 \qquad n = 365$$

$A = P\left(1 + \dfrac{r}{n}\right)^{nt}$ This is the compound interest formula.

$A = 80{,}000\left(1 + \dfrac{0.0325}{365}\right)^{365\,(8)}$ Substitute for P, r, n, and t.

$A = 80{,}000\left(1 + \dfrac{0.0325}{365}\right)^{2{,}920}$ Evaluate the exponent: $365 \cdot 8 = 2{,}920$.

$A \approx 103{,}753.21$ Use a calculator. Round to the nearest cent.

There will be $103,753.21 in the account in 8 years.

REVIEW EXERCISES

81. **INVESTMENTS** Find the simple interest earned on $6,000 invested at 8% for 2 years. Use the following table to organize your work.

P	r	t	I
$6,000	8%	2 years	$960

82. **INVESTMENT ACCOUNTS** If $24,000 is invested at a simple interest rate of 4.5% for 3 years, what will be the total amount of money in the investment account at the end of the term? $27,240

83. **EMERGENCY LOANS** A teacher's credit union loaned a client $2,750 at a simple interest rate of 11% so that he could pay an overdue medical bill. How much interest does the client pay if the loan must be paid back in 3 months? $75.63

84. **CODE VIOLATIONS** A business was ordered to correct safety code violations in a production plant. To pay for the needed corrections, the company borrowed $10,000 at 12.5% simple interest for 90 days. Find the total amount that had to be paid after 90 days. $10,308.22

85. **MONTHLY PAYMENTS** A couple borrows $1,500 for 1 year at a simple interest rate of $7\frac{3}{4}\%$.
    a. How much interest will they pay on the loan? $116.25
    b. What is the total amount they must repay on the loan? $1,616.25
    c. If the couple decides to repay the loan by making 12 equal monthly payments, how much will each monthly payment be? $134.69

86. **SAVINGS ACCOUNTS** Find the amount of money that will be in a savings account at the end of 1 year if $2,000 is the initial deposit and the interest rate of 7% is compounded semi-annually. $2,142.45

87. **SAVINGS ACCOUNTS** Find the amount that will be in a savings account at the end of 3 years if a deposit of $5,000 earns interest at a rate of $6\frac{1}{2}\%$, compounded daily. $6,076.45

88. **CASH GRANTS** Each year a cash grant is given to a deserving college student. The grant consists of the interest earned that year on a $500,000 savings account. What is the cash award for the year if the money is invested at a rate of 8.3%, compounded daily? $43,265.78

CHAPTER 7 TEST

1. Fill in the blanks.
   a. __Percent__ means parts per one hundred.
   b. The key words in a percent sentence translate as follows:
      • __is__ translates to an equal symbol =
      • __of__ translates to multiplication that is shown with a raised dot ·
      • __what__ number or __what__ percent translates to an unknown number that is represented by a variable
   c. In the percent sentence "5 is 25% of 20," 5 is the __amount__, 25% is the percent, and 20 is the __base__.
   d. When we use percent to describe how a quantity has increased compared to its original value, we are finding the percent of __increase__.
   e. __Simple__ interest is interest earned only on the original principal. __Compound__ interest is interest paid on the principal and previously earned interest.

2. a. Express the amount of the figure that is shaded as a percent, as a fraction, and as a decimal. $61\%, \frac{61}{100}, 0.61$
   b. What percent of the figure is not shaded? 39%

3. In the illustration below, each set of 100 square regions represents 100%. Express as a percent the amount of the figure that is shaded. Then express that percent as a fraction and as a decimal. $199\%, \frac{199}{100}, 1.99$

4. Write each percent as a decimal.
   a. 67% 0.67   b. 12.3% 0.123   c. $9\frac{3}{4}\%$ 0.0975

5. Write each percent as a decimal.
   a. 0.06% 0.0006   b. 210% 2.1   c. 55.375% 0.55375

6. Write each fraction as a percent.
   a. $\frac{1}{4}$ 25%   b. $\frac{5}{8}$ 62.5%   c. $\frac{28}{25}$ 112%

7. Write each decimal as a percent.
   a. 0.19 19%   b. 3.47 347%   c. 0.005 0.5%

8. Write each decimal or whole number as a percent.
   a. 0.667 66.7%   b. 2 200%   c. 0.9 90%

9. Write each percent as a fraction. Simplify, if possible.
   a. 55% $\frac{11}{20}$ b. 0.01% $\frac{1}{10,000}$ c. 125% $\frac{5}{4}$

10. Write each percent as a fraction. Simplify, if possible.
    a. $6\frac{2}{3}\%$ $\frac{1}{15}$ b. 37.5% $\frac{3}{8}$ c. 8% $\frac{2}{25}$

11. Write each fraction as a percent. Give the exact answer and an approximation to the nearest tenth of a percent.
    a. $\frac{1}{30}$ $3\frac{1}{3}\% \approx 3.3\%$ b. $\frac{16}{9}$ $177\frac{7}{9}\% \approx 177.8\%$

Unless otherwise noted, all art on this page is © Cengage Learning.

12. 65 is what percent of 1,000? 6.5%

13. What percent of 14 is 35? 250%

14. **FUGITIVES** As of January 2013, exactly 467 of the 497 fugitives who have appeared on the FBI's Ten Most Wanted list have been captured or located. What percent is this? Round to the nearest hundredth of one percent. (Source: www.fbi.gov/wanted) 93.96%

15. **SWIMMING WORKOUTS** A swimmer was able to complete 18 laps before a shoulder injury forced him to stop. This was only 20% of a typical workout. How many laps does he normally complete during a workout? 90

16. **COLLEGE EMPLOYEES** The 700 employees at a community college fall into three major categories, as shown in the circle graph. How many employees are in administration? 21

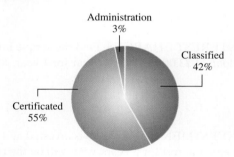

17. What number is 224% of 60? 134.4

18. 2.6 is $33\frac{1}{3}\%$ of what number? 7.8

19. **SHRINKAGE** See the following label from a new pair of jeans. The measurements are in inches. (*Inseam* is a measure of the length of the jeans.)

   WAIST 33   INSEAM 34
   Expect shrinkage of approximately **3%** in length after the jeans are washed.

   a. How much length will be lost due to shrinkage? 1.02 in.
   b. What will be the length of the jeans after being washed? 32.98 in.

20. **TOTAL COST** Find the total cost of a $25.50 purchase if the sales tax rate is 2.9%. $26.24

21. **SALES TAX** The purchase price for a watch is $90. If the sales tax is $2.70, what is the sales tax rate? 3%

22. **POPULATION INCREASES** After a new freeway was completed, the population of a city it passed through increased from 2,800 to 3,444 in two years. Find the percent of increase. 23%

23. **INSURANCE** An automobile insurance salesperson receives a 4% commission on the annual premium of any policy she sells. Find her commission on a policy if the annual premium is $898. $35.92

24. **TELEMARKETING** A telemarketer earned a commission of $528 on $4,800 worth of new business that she obtained over the telephone. Find her rate of commission. 11%

25. **COST-OF-LIVING** A teacher earning $40,000 just received a cost-of-living increase of 3.6%. What is the teacher's new salary? $41,440

26. **AUTO CARE** Refer to the advertisement below. Find the discount, the sale price, and the discount rate on the car waxing kit. $9, $66, 12%

27. TOWEL SALES Find the amount of the discount on a beach towel if it regularly sells for $20 but is on sale for 33% off. Then find the sale price of the towel. $6.60, $13.40

28. Fill in the blanks.
  a. To find 1% of a number, move the decimal point in the number two places to the left .
  b. To find 10% of a number, move the decimal point in the number one place to the left .

29. Estimate each answer. (Answers may vary slightly.)
  a. What is 20% of 396? 80
  b. What is 50% of 6,189,034? 3,000,000
  c. What is 200% of 21.2? 42

30. CREDIT Credit scores range from a low of 309 to a high of 839. The graph below shows the percent of consumers with scores in a particular range. What percent of consumers have a credit score from 750 to 799? 26%

Source: JP Morgan Chase Bank

31. Shade Figure b so that Figure a and Figure b have the same percent shaded.

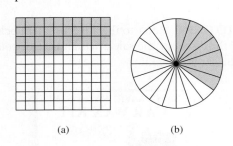

(a)   (b)

32. BRAKE INSPECTIONS Of the 1,920 trucks inspected at a safety checkpoint, 5% had problems with their brakes. Estimate the number of trucks that had brake problems. 100 trucks

33. TIPPING Estimate the amount of a 15% tip on a lunch costing $28.40. $4.50

34. CAR SHOWS 24% of 63,400 people that attended a five-day car show were female. Estimate the number of females that attended the car show. 16,000 females

35. INTEREST CHARGES Find the simple interest on a loan of $3,000 at 5% per year for 1 year. $150

36. INVESTMENTS If $23,000 is invested at $4\frac{1}{2}$% simple interest for 5 years, what will be the total amount of money in the investment account at the end of the 5 years? $28,175

37. SHORT-TERM LOANS Find the simple interest on a loan of $2,000 borrowed at 8% for 90 days. $39.45

38. Use the formula $A = P\left(1 + \dfrac{r}{n}\right)^{nt}$ to find the amount of interest earned on an investment of $24,000 paying an annual rate of 6.4% interest, compounded daily for 3 years. $5,079.60

CHAPTERS 1–7 CUMULATIVE REVIEW

1. Write 6,054,346 [Section 1.1]

   a. in words

   six million, fifty-four thousand, three hundred forty-six

   b. in expanded notation

   6,000,000 + 50,000 + 4,000 + 300 + 40 + 6

2. WEATHER The cloudiest major city in the United States is Juneau, Alaska. The tables below show the average number of cloudy days in Juneau each month. Find the total number of cloudy days in a year. (Source: Western Regional Climate Center) [Section 1.2] 279 days

Jan	Feb	Mar	Apr	May	June
23	21	23	22	23	22

July	Aug	Sept	Oct	Nov	Dec
23	22	24	27	24	25

3. Subtract: 50,055 − 7,899 [Section 1.2] 42,156

4. Multiply: 308 · 75 [Section 1.3] 23,100

5. PAINTING A square tarp has sides 8 feet long. When it is laid out on a floor, how much area will it cover? [Section 1.3] 64 ft^2

6. Divide: $37\overline{)561}$ [Section 1.4] 15 R6

7. a. List the factors of 40, from smallest to largest. [Section 1.5] 1, 2, 4, 5, 8, 10, 20, 40

   b. Find the prime factorization of 294. [Section 1.5] $2 \cdot 3 \cdot 7^2$

8. Find the LCM and the GCF of 24 and 30. [Section 1.6] 120, 6

9. Evaluate: $\dfrac{39 + 3[4^3 - 2(2^2 - 3)]}{4 \cdot 2^2 - 1}$ [Section 1.7] 15

10. AUTO INSURANCE See the premium comparison below. What is the average insurance premium for the companies listed? $2,106

Allstate	$2,672	Geico	$1,370
Auto Club	$1,680	State Farm	$2,737
Farmers	$2,485	20th Century	$1,692

11. Solve each equation and check the result.

    a. 23 + a = 207 [Section 1.8] 184

    b. 23a = 207 [Section 1.9] 9

12. Place an < or an > symbol in the box to make a true statement: $|-8|$ > $-(-5)$ [Section 2.1]

13. Evaluate: $(-20 + 9) + (-13 + 24)$ [Section 2.2] 0

14. OVERDRAFT PROTECTION A student forgot that she had only $55 in her bank account and wrote a check for $75, used an ATM to get $60 cash, and used her debit card to buy $25 worth of groceries. On each of the three transactions, the bank charged her a $10 overdraft protection fee. Find the new account balance. [Section 2.3] −$135

15. Evaluate: -6^2 and $(-6)^2$ [Section 2.4] −36, 36

16. Evaluate each expression, if possible. [Sections 2.2–2.5]

    a. $\dfrac{-14}{0}$ undefined

    b. $\dfrac{0}{-12}$ 0

    c. $-3(-4)(-5)(0)$ 0

    d. $0 - (-14)$ 14

17. Evaluate: $24 \div 2(3) - 3^3$ [Section 2.6] 9

18. Estimate the following sum by rounding each number to the nearest hundred. [Section 2.6]

    $-5,684 + (-2,270) + 3,404 + 2,689$ −1,900

19. Solve $6 = 2 - 2x$ and check the result. [Section 2.7] −2

Write and solve an equation to answer the following question. [Section 2.7]

20. MARKET SHARE After its first year of business, the manufacturer of an energy drink found its market share 51 points behind the industry leader. Five years later, it trailed the leader by only 15 points. How many points of market share did the company gain over this five-year span? The company gained 36 points of market share in five years.

21. Translate into mathematical symbols: 16 less than twice the total. [Section 3.1] $2t - 16$

22. FRUIT STORAGE Use the formula $C = \frac{5(F - 32)}{9}$ to complete the label on the box of bananas shown below. [Section 3.2] 15°C

23. a. Simplify: $4(8m)$ [Section 3.3] $32m$
b. Multiply: $(4d + 8)2$ [Section 3.3] $8d + 16$

24. Combine like terms. [Section 3.4]
a. $5x - 4x$ x
b. $l + w + l + w$ $2l + 2w$

25. Solve $-(3x - 3) = 6(2x - 7)$ and check the result. [Section 3.5] 3

Write and then solve an equation to answer the following question. [Section 3.6]

26. BUSINESS After beginning a new position with 34 established accounts, a salesperson made it her objective to add 6 new accounts every month. At this rate, how many months will it take her to reach the goal of 100 accounts? It will take her 11 months to reach the goal.

27. SPELLING What fraction of the letters in the word *Mississippi* are vowels? [Section 4.1] $\frac{4}{11}$

28. Simplify: $\frac{10y}{15y}$ [Section 4.1] $\frac{2}{3}$

29. Express $\frac{4}{5}$ as an equivalent fraction with a denominator of 45. [Section 4.1] $\frac{36}{45}$

30. What is $\frac{1}{4}$ of -240? [Section 4.2] -60

31. KITES Find the number of square inches of nylon cloth used to make the kite shown below. (*Hint:* Find the area.) [Section 4.2] 650 in.²

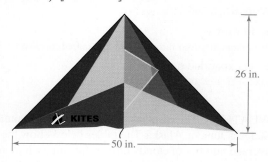

32. Multiply: $-\frac{16a}{35} \cdot \frac{25}{48a^2}$ [Section 4.2] $-\frac{5}{21a}$

33. Divide: $\frac{4}{9} \div \left(-\frac{16}{27}\right)$ [Section 4.3] $-\frac{3}{4}$

34. Subtract: $\frac{9}{10} - \frac{3}{14}$ [Section 4.4] $\frac{24}{35}$

35. Add: $\frac{4}{m} + \frac{2}{7}$ [Section 4.4] $\frac{28 + 2m}{7m}$

36. Determine which fraction is larger: $\frac{23}{20}$ or $\frac{7}{6}$ [Section 4.4] $\frac{7}{6}$

37. HAMBURGERS What is the difference in weight between a $\frac{1}{4}$-pound and a $\frac{1}{3}$-pound hamburger? [Section 4.4] $\frac{1}{12}$ lb

38. Multiply: $-3\frac{3}{4}(8)$ [Section 4.5] -30

39. BELTS Refer to the belt shown below. What is the maximum waist size that the belt will fit if it is fastened using the last hole? [Section 4.6] $35\frac{3}{4}$ in.

40. Subtract: $34\frac{1}{9} - 13\frac{5}{6}$ [Section 4.6] $20\frac{5}{18}$

41. Simplify: $\dfrac{\frac{1}{3} - \frac{3}{4}}{\frac{1}{6} + \frac{1}{3}}$ [Section 4.7] $-\frac{5}{6}$

Write and then solve an equation to answer the following question. [Section 4.8]

42. TELEPHONE BOOKS A telephone book consists of white pages and yellow pages. Find the total number of pages if the 350 white pages make up two-thirds of the telephone book. The total number of pages in the telephone book is 525.

Solve each equation and check the result. [Section 4.8]

43. $\frac{5}{6}y = -25$ -30

44. $\frac{y}{6} = \frac{y}{12} + \frac{2}{3}$ 8

45. Round each decimal. [Section 5.1]
    a. Round 452.0298 to the nearest hundredth. 452.03
    b. Round 452.0298 to the nearest thousandth. 452.030

46. Evaluate: $3.4 - (6.6 + 7.3) + 5$ [Section 5.2] -5.5

47. WEEKLY EARNINGS A welder's basic work week is 40 hours. After his daily shift is over, he can work overtime at a rate of 1.5 times his regular rate of $15.90 per hour. How much money will he earn in a week if he works 4 hours of overtime? [Section 5.3] $731.40

48. Divide: $0.58\overline{)0.1566}$ [Section 5.4] 0.27

49. Write $\frac{11}{15}$ as a decimal. Use an overbar. [Section 5.5] $0.7\overline{3}$

50. Evaluate: $3\sqrt{81} - 8\sqrt{49}$ [Section 5.6] -29

Solve each equation and check the result. [Section 5.7]

51. $37.1n + 12.2 = 12.4n + 93.4 + 4.4n$ 4

52. $2(a - 4.3) + 1.2 = -6.2$ 0.6

Write and then solve an equation to answer the following question. [Section 5.7]

53. LABOR COSTS On the repair bill shown below, one line cannot be read. How many hours of labor did it take to repair the car? It took 3.5 hr of labor to repair the car.

Brian Wood Auto Repair
Parts..$175.00
Total labor (at $35 an hour)..............
Total..$297.50

54. Write the ratio $1\frac{1}{4}$ to $1\frac{1}{2}$ as a fraction in simplest form. [Section 6.1] $\frac{5}{6}$

55. Solve the proportion: $\frac{7}{14} = \frac{2}{x}$ [Section 6.2] 4

56. TYPING A secretary typed a document containing 385 words in 7 minutes. How long will it take her to type a document containing 495 words? [Section 6.2] It will take her 9 minutes.

57. How many days are in 960 hours? [Section 6.3] 40 days

58. Convert 2,400 millimeters to meters. [Section 6.4] 2.4 m

59. Convert 6.5 kilograms to pounds. [Section 6.5] 14.3 lb

60. Complete the table. [Section 7.1]

Percent	Decimal	Fraction
29%	0.29	$\frac{29}{100}$
47.3%	0.473	$\frac{473}{1,000}$
87.5%	0.875	$\frac{7}{8}$

61. 16% of what number is 20? [Section 7.2] 125

62. GENEALOGY Through an extensive computer search, a genealogist determined that worldwide, 180 out of every 10 million people had his last name. What percent is this? [Section 7.2] 0.0018%

63. HEALTH CLUBS The number of members of a health club increased from 300 to 534. What was the percent of increase in club membership? [Section 7.3] 78%

64. GUITAR SALE What are the regular price, the sale price, the discount, and discount rate for the guitar shown in the advertisement below? [Section 7.3] $428, $321, $107, 25%

Save on the Standard Strat
Fender
Now Only
$321.00
Save $107

65. **TIPPING** Refer to the sales receipt below. [Section 7.4]
    a. Estimate the 15% tip. $12
    b. Find the total. $90.18

    STEAK STAMPEDE
    Bloomington, MN
    Server #12\ AT

VISA	67463777288
NAME	DALTON/ LIZ
AMOUNT	$78.18
GRATUITY $	12.00
TOTAL $	90.18

66. **INVESTMENTS** Find the simple interest earned on $10,000 invested for 2 years at 7.25%. [Section 7.5]
    $1,450

8

Graphs and Statistics

8.1 Reading Graphs and Tables

8.2 Mean, Median, and Mode

8.3 Equations in Two Variables; The Rectangular Coordinate System

8.4 Graphing Linear Equations

Chapter Summary and Review

Chapter Test

Cumulative Review

from Campus to Careers

Postal Service Mail Carrier

Mail carriers follow schedules as they collect and deliver mail to homes and businesses. They must have the ability to quickly and accurately compare similarities and differences among sets of letters, numbers, objects, pictures, and patterns. They also need to have strong problem-solving skills to redirect mislabeled letters and packages. Mail carriers weigh items on postal scales and make calculations with money as they read postage rate tables.

In **Problem 19** of **Study Set 8.1,** you will see how a mail carrier must be able to read a postal rate table and know American units of weight to determine the cost to send a package using priority mail.

JOB TITLE: Postal Service Mail Carrier

EDUCATION: A high school diploma (or equivalent) and a passing score on a written exam are required.

JOB OUTLOOK: Competition for jobs is high since positions usually come open only upon retirement of current mail carriers.

ANNUAL EARNINGS: Average (mean) salary $51,390

FOR MORE INFORMATION: http://stats.bls.gov/oes/current/oes/435052.htm

Objectives

1. Read tables.
2. Read bar graphs.
3. Read pictographs.
4. Read circle graphs.
5. Read line graphs.
6. Read histograms and frequency polygons.

SECTION 8.1
Reading Graphs and Tables

We live in an information age. Never before have so many facts and figures been right at our fingertips. Since information is often presented in the form of tables or graphs, we need to be able to read and make sense of data displayed in that way.

The following **table**, **bar graph**, and **circle graph** (or **pie chart**) show the results of a shopper survey. A large sample of adults were asked how far in advance they typically shop for a gift. In the bar graph, the length of a bar represents the percent of responses for a given shopping method. In the circle graph, the size of a colored region represents the percent of responses for a given shopping method.

Shopper Survey
How far in advance gift givers typically shop

A Table
Survey responses

Time in advance	Percent
A month or longer	8%
Within a month	12%
Within 3 weeks	12%
Within 2 weeks	23%
Within a week	41%
The same day as giving it	4%

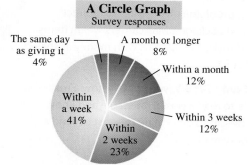

Source: Harris interactive online study via QuickQuery for Gifts.com

It is often said that a picture is worth a thousand words. That is the case here, where the graphs display the results of the survey more clearly than the table. It's easy to see from the graphs that most people shop within a week of when they need to purchase a gift. It is also apparent that same-day shopping for a gift was the least popular response. That information also appears in the table, but it is just not as obvious.

Unless otherwise noted, all art on this page is © Cengage Learning.

1 Read tables.

Data are often presented in tables, with information organized in **rows** and **columns**. To read a table, we must find the *intersection* of the row and column that contains the desired information.

EXAMPLE 1 **Postal Rates** Refer to the table of priority mail postal rates (from 2013) below. Find the cost of mailing an $8\frac{1}{2}$-pound package by priority mail to postal zone 4.

Postage Rate for Priority Mail 2013
Delivery within 2 to 3 days in most cases.

Weight Not Over (pounds)	Zones						
	Local, 1 & 2	3	4	5	6	7	8
1	$5.60	$5.70	$5.85	$6.00	$6.20	$6.45	$6.95
2	5.80	6.15	6.85	8.75	9.50	10.25	11.25
3	6.60	7.60	8.75	10.60	12.00	12.90	15.25
4	7.45	8.90	10.05	14.05	15.40	16.45	18.35
5	8.85	10.20	11.50	15.95	17.55	18.90	21.20
6	9.75	11.25	12.90	17.75	19.70	21.20	23.95
7	10.35	12.20	13.90	19.75	21.80	23.85	26.90
8	11.10	13.30	15.55	21.45	23.95	26.30	30.15
9	11.85	14.35	**16.85**	23.25	26.05	28.45	33.55
10	12.65	15.40	18.35	25.20	28.15	31.30	36.50
11	13.50	16.45	19.80	27.15	30.20	34.55	40.10
12	14.45	17.65	21.20	29.15	32.85	37.35	43.05

Strategy We will read the number at the intersection of the 9th row and the column labeled Zone 4.

WHY Since $8\frac{1}{2}$ pounds is more than 8 pounds, we cannot use the 8th row. Since $8\frac{1}{2}$ pounds does not exceed 9 pounds, we use the 9th row of the table.

Solution
The number at the intersection of the 9th row (in red) and the column labeled Zone 4 (in blue) is 16.85 (in purple). This means it would cost $16.85 to mail the $8\frac{1}{2}$-pound package by priority mail.

> **Self Check 1**
> **POSTAL RATES** Refer to the table of priority mail postal rates. Find the cost of mailing a 3.75-pound package by priority mail to postal zone 8. $18.35
> **Now Try** Problem 17
>
> Teaching Example 1 POSTAL RATES Refer to the table of priority mail postal rates. Find the cost of mailing a $5\frac{3}{4}$-pound package by priority mail to postal zone 6.
> *Answer:*
> $19.70

2 Read bar graphs.

Another popular way to display data is to use a **bar graph** with bars drawn vertically or horizontally. The relative heights (or lengths) of the bars make for easy comparisons of values. A horizontal or vertical line used for reference in a bar graph is called an **axis**. The **horizontal axis** and the **vertical axis** of a bar graph serve to frame the graph, and they are scaled in units such as years, dollars, minutes, pounds, and percent.

Unless otherwise noted, all art on this page is © Cengage Learning.

Chapter 8 Graphs and Statistics

Self Check 2

SPEED OF ANIMALS Refer to the bar graph for Example 2.
a. What is the maximum speed of a giraffe? 32 mph
b. How much greater is the maximum speed of a coyote compared to that of a reindeer? 11 mph
c. Which animals listed in the graph have a maximum speed that is slower than that of a domestic cat? a chicken and an elephant

Now Try Problem 21

Teaching Example 2 SPEED OF ANIMALS Refer to the bar graph for Example 2.
a. What is the maximum speed of a chicken?
b. How much greater is the maximum speed of a zebra compared to that of an elephant?
c. What two animals have the same maximum speed?
Answers:
a. 9 mph
b. 15 mph
c. a giraffe and a reindeer

EXAMPLE 2 Speed of Animals

The following bar graph shows the maximum speeds for several animals over a given distance.
a. What animal in the graph has the fastest maximum speed?
b. What animal in the graph has the slowest maximum speed?
c. How much greater is the maximum speed of a lion compared to that of a coyote?

Source: Infoplease.com

Strategy We will locate the name of each desired animal on the vertical axis and move right to the end of its corresponding bar.

WHY Then we can extend downward and read the animal's maximum speed on the horizontal axis scale.

Solution

a. The longest bar in the graph has a length of 70 units and corresponds to a cheetah. Of all the animals listed in the graph, the cheetah has the fastest maximum speed at 70 mph.

b. The shortest bar in the graph has a length of approximately 9 units and corresponds to a chicken. Of all the animals listed in the graph, the chicken has the slowest maximum speed at 9 mph.

c. The length of the bar that represents a lion's maximum speed is 50 units long and the length of the bar that represents a coyote's maximum speed appears to be 43 units long. To find how much greater is the maximum speed of a lion compared to that of a coyote, we subtract

$$50 \text{ mph} - 43 \text{ mph} = 7 \text{ mph}$$ Subtract the coyote's maximum speed from the lion's maximum speed.

The maximum speed of a lion is about 7 mph faster than the maximum speed of a coyote.

To compare sets of related data, groups of two (or three) bars can be shown. For **double-bar** or **triple-bar graphs**, a **key** is used to explain the meaning of each type of bar in a group.

EXAMPLE 3 The U.S. Economy

The following bar graph shows the total income generated by three sectors of the U.S. economy in each of three years.
a. What income was generated by retail sales in 2000?
b. Which sector of the economy consistently generated the most income?
c. By what amount did the income from the wholesale sector increase from 2000 to 2011?

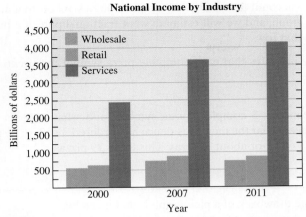

Source: *The World Almanac and Book of Facts*, 2013

Self Check 3

THE U.S. ECONOMY Refer to the bar graph for Example 3.
a. What income was generated by retail sales in 2011? about $875 billion
b. What income was generated by the wholesale sector in 2011? about $750 billion
c. In 2000, by what amount did the income from the services sector exceed the income from the retail sector? about $1,800 billion

Now Try Problems 25 and 31

Teaching Example 3 **THE U.S. ECONOMY** Refer to the bar graph for Example 3.
a. What income was generated by wholesale sales in 2007?
b. What income was generated by the services sector in 2000?
c. By what amount did the income from the services sector increase from 2000 through 2011?
Answers:
a. about $750 billion
b. about $2,500 billion
c. about $1,700 billion

Strategy To answer questions about years, we will locate the correct colored bar and look at the *horizontal axis* of the graph. To answer questions about the income, we will locate the correct colored bar and extend to the left to look at the *vertical axis* of the graph.

WHY The years appear on the horizontal axis. The height of each bar, representing income in billions of dollars, is measured on the scale on the vertical axis.

Solution

a. The first group of bars indicates income in the year 2000. According to the color key, the blue bar of that group shows the retail sales. Since the vertical axis is scaled in units of $\frac{\$500 \text{ billion}}{2} = \$250$ billion, the height of that bar is approximately 500 plus one-half of 250, or 125. Thus, the height of the blue bar is approximately $500 + 125 = 625$, which represents $625 billion in retail sales in 2000.

b. In each group, the green bar is the tallest. That bar, according to the color key, represents the income from the services sector of the economy. Thus, services consistently generated the most income.

c. According to the color key, the orange bar in each group shows income from the wholesale sector. That sector generated about $600 billion of income in 2000 and $750 billion in income in 2011. The amount of increase is the difference of these two quantities.

$$\$750 \text{ billion} - \$600 \text{ billion} = \$150 \text{ billion}$$ Subtract the 2000 wholesale income from the 2011 wholesale income.

Wholesale income increased by about $150 billion between 2000 and 2011.

3 Read pictographs.

A **pictograph** is like a bar graph, but the bars are made from pictures or symbols. A **key** tells the meaning (or value) of each symbol.

EXAMPLE 4 *Pizza Deliveries*

The pictograph on the right shows the number of pizzas delivered to the three residence halls on a college campus during final exam week. In the graph, what information does the top row of pizzas give?

Pizzas ordered during final exam week

Men's residence hall
Women's residence hall
Co-ed residence hall

= 12 pizzas

Self Check 4

PIZZA DELIVERIES In the pictograph for Example 4, what information does the last row of pizzas give? 33 pizzas were delivered to the co-ed residence hall.

Now Try Problems 33 and 35

Teaching Example 4 PIZZA DELIVERIES In the pictograph for Example 4, what information does the middle row of pizzas give?
Answer:
54 pizzas were delivered to the women's residence hall.

Strategy We will count the number of complete pizza symbols that appear in the top row of the graph, and we will estimate what fractional part of a pizza symbol also appears in that row.

WHY The key indicates that each complete pizza symbol represents one dozen (12) pizzas.

Solution

The top row contains 3 complete pizza symbols and what appears to be $\frac{1}{4}$ of another. This means that the men's residence hall ordered $3 \cdot 12$, or 36 pizzas, plus approximately $\frac{1}{4}$ of 12, or about 3 pizzas. This totals 39 pizzas.

> **Caution!** One drawback of a pictograph is that it can be difficult to determine what fractional amount is represented by a portion of a symbol. For example, if the CD shown to the right represents 1,000 units sold, we can only estimate that the partial CD symbol represents about 600 units sold.

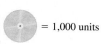

4 Read circle graphs.

In **a circle graph**, regions called **sectors** are used to show what part of the whole each quantity represents.

> **The Language of Algebra** A *sector* has the shape of a slice of pizza or a slice of pie. Thus, circle graphs are also called **pie charts**.

Self Check 5

GOLD PRODUCTION Refer to the circle graph for Example 5. To the nearest tenth of a million, how many troy ounces of gold did Russia produce in 2012? 6.9 million troy ounces

Now Try Problems 37, 41, and 43

Teaching Example 5 GOLD PRODUCTION Refer to the circle graph for Example 5.
a. What percent of the total was the combined production of Australia, the United States, and Canada in 2012?
b. How many troy ounces of gold did South Africa produce in 2012? Round to the nearest tenth of one million.
Answers:
a. 21%   b. 5.2 million troy ounces

EXAMPLE 5 *Gold Production* The circle graph to the right gives information about world gold production. The entire circle represents the world's total production of 86 million troy ounces in 2012. Use the graph to answer the following questions.
a. What percent of the total was the combined production of the United States and Canada?
b. What percent of the total production came from sources other than those listed?
c. To the nearest tenth of a million, how many troy ounces of gold did China produce in 2012?

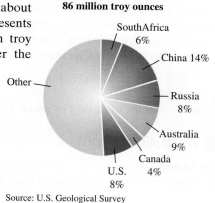

Source: U.S. Geological Survey

Strategy We will look for the key words in each problem.

WHY Key words tell us what operation (addition, subtraction, multiplication, or division) must be performed to answer each question.

Solution

a. The key word *combined* indicates addition. According to the graph, the United States produced 8% and Canada produced 4% of the total amount of gold in 2012. Together, they produced 8% + 4%, or 12% of the total.

b. The phrase *from sources other than those listed* indicates subtraction. To find the percent of gold produced by countries that are not listed, we add the contributions of all the listed sources and subtract that total from 100%.

$$100\% - (6\% + 14\% + 8\% + 9\% + 4\% + 8\%) = 100\% - 49\% = 51\%$$

Countries that are not listed in the graph produced 51% of the world's total production of gold in 2012.

c. From the graph we see that China produced 14% of the world's gold in 2012. To find the number of troy ounces produced by China (the amount), we use the method for solving percent problems from Section 7.2.

This is the percent sentence. The units are millions of troy ounces.

Translate to a percent equation.

Now we perform the multiplication on the right side of the equation.

$x = 0.14 \cdot 86$ Write 14% as a decimal: 14% = 0.14.
$x = 12.04$ Do the multiplication.

Rounded to the nearest tenth of a million, China produced 12.0 million troy ounces of gold in 2012.

```
    86
  × 0.14
  ─────
   344
   860
  ─────
  12.04
```

5 Read line graphs.

A **line graph** is used to show how quantities change with time. From such a graph, we can determine when a quantity is increasing and when it is decreasing.

> **The Language of Algebra** The symbol ⌇ is used to show a break in the scale on an axis. It enables us to omit large portions of empty space on a graph.

EXAMPLE 6 **ATMs** The line graph below shows the number of automated teller machines (ATMs) in the United States for the years 2002 through 2012. Use the graph to answer the following questions.

a. How many ATMs were there in the United States in 2005?
b. Between which two years was there the greatest increase in the number of ATMs?
c. When did the number of ATMs decrease for the first time?
d. Between which two years did the number of ATMs remain about the same?

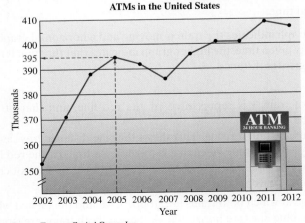

Source: Tremont Capital Group, Inc.
Unless otherwise noted, all art on this page is © Cengage Learning.

Self Check 6

ATMs Refer to the line graph for Example 6.
a. Find the increase in the number of ATMs between 2007 and 2008. about 10,000
b. How many more ATMs were there in the United States in 2012 as compared to 2002? about 55,000

Now Try Problems 45, 47, and 51

Teaching Example 6 ATMs Refer to the line graph for Example 6.
a. How many ATMs were there in the United States in 2009?
b. Find the increase in the number of ATMs between 2010 and 2011.
c. In which year did the number of ATMs first exceed 390,000?

Answers:
a. about 400,000
b. about 8,000
c. 2005

Strategy We will determine whether the graph is rising, falling, or horizontal.

WHY When the graph rises as we read from left to right, the number of ATMs is increasing. When the graph falls as we read from left to right, the number of ATMs is decreasing. If the graph is horizontal, there is no change in the number of ATMs.

Solution

a. To find the number of ATMs in 2005, we follow the dashed blue line from the label 2005 on the horizontal axis straight up to the line graph. Then we extend directly over to the scale on the vertical axis, where the arrowhead points to 395. Since the vertical scale is in thousands of ATMs, there were about 395,000 ATMs in 2005 in the United States.

b. This line graph is composed of ten line segments that connect pairs of consecutive years. The steepest of those seven segments represents the greatest increase in the number of ATMs. Since that segment is between 2002 and 2003, the greatest increase in the number of ATMs occurred between 2002 and 2003.

c. The first line segment of the graph that falls as we read from left to right is the segment connecting the data points for the years 2005 and 2006. Thus, the number of ATMs decreased from 2005 to 2006.

d. The line segment connecting the data points for the years 2009 and 2010 appears to be horizontal. Since there is little or no change in the number of ATMs for those years, the number of ATMs remained about the same from 2009 to 2010.

Two quantities that are changing with time can be compared by drawing both lines on the same graph.

Self Check 7

TRAINS In the graph for Exercise 7, what is train 1 doing at time D?

Now Try Problems 53, 55, and 59

Self Check 7 Answer
Train 1, which had been stopped, is beginning to move.

Teaching Example 7 TRAINS Refer to the line graph for Example 7.
a. Which train had traveled farther at time A?
b. Which train had traveled farther at time D?
c. Was train 2 moving or was it stopped at time B?
d. Immediately after time E, which train was traveling faster?
Answers:
a. train 1   b. train 2
c. stopped   d. train 1

EXAMPLE 7

Trains The line graph below shows the movements of two trains. The horizontal axis represents time, and the vertical axis represents the distance that the trains have traveled.

a. How are the trains moving at time A?
b. At what time (A, B, C, D, or E) are both trains stopped?
c. At what times have both trains traveled the same distance?

Strategy We will determine whether the graphs are rising or are horizontal. We will also consider the relative positions of the graphs for a given time.

WHY A rising graph indicates the train is moving and a horizontal graph means it is stopped. For any given time, the higher graph indicates that the train it represents has traveled the greater distance.

Solution
The movement of train 1 is represented by the red line, and that of train 2 is represented by the blue line.

a. At time A, the blue line is rising. This shows that the distance traveled by train 2 is increasing. Thus, at time A, train 2 is moving. At time A, the red line is horizontal. This indicates that the distance traveled by train 1 is not changing: At time A, train 1 is stopped.

b. To find the time at which both trains are stopped, we find the time at which both the red and the blue lines are horizontal. At time B, both trains are stopped.

c. At any time, the height of a line gives the distance a train has traveled. Both trains have traveled the same distance whenever the two lines are the same height—that is, at any time when the lines intersect. This occurs at times C and E.

6 Read histograms and frequency polygons.

A company that makes vitamins is sponsoring a program on a cable TV channel. The marketing department must choose from three advertisements to show during the program.

1. Children talking about a chewable vitamin that the company makes.
2. A college student talking about an active-life vitamin that the company makes.
3. A grandmother talking about a multivitamin that the company makes.

A survey of the viewing audience records the age of each viewer, counting the number in the 6-to-15-year-old age group, the 16-to-25-year-old age group, and so on. The graph of the data is displayed in a special type of bar graph called a **histogram**, as shown on the right. The vertical axis, labeled **Frequency**, indicates the number of viewers in each age group. For example, the histogram shows that 105 viewers are in the 36-to-45-year-old age group.

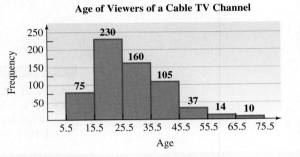

A histogram is a bar graph with three important features.

1. The bars of a histogram touch.
2. Data values never fall at the edge of a bar.
3. The widths of the bars are equal and represent a range of values.

The width of each bar of a histogram represents a range of numbers called a **class interval**. The histogram above has seven class intervals, each representing an age span of 10 years. Since most viewers are in the 16-to-25-year-old age group, the marketing department decides to advertise the active-life vitamins in commercials that appeal to young adults.

EXAMPLE 8 Carry-on Luggage
An airline weighed the carry-on luggage of 2,260 passengers. The data are displayed in the histogram below.

a. How many passengers carried luggage in the 8-to-11-pound range?

b. How many carried luggage in the 12-to-19-pound range?

Strategy We will examine the scale on the horizontal axis of the histogram and identify the interval that contains the given range of weight for the carry-on luggage.

WHY Then we can read the height of the corresponding bar to answer the question.

Solution

a. The second bar, with edges at 7.5 and 11.5 pounds, corresponds to the 8-to-11-pound range. Use the height of the bar (or the number written there) to determine that 430 passengers carried such luggage.

b. The 12-to-19-pound range is covered by two bars. The total number of passengers with luggage in this range is 970 + 540, or 1,510.

Self Check 8

CARRY-ON LUGGAGE Refer to the histogram for Example 8. How many passengers carried luggage in the 20-to-23-pound range? 200

Now Try Problem 61

Teaching Example 8 CARRY-ON LUGGAGE Refer to the histogram for Example 8.
a. How many passengers carried luggage in the 4-to-7-pound range?
b. How many passengers carried luggage in the 8-to-19-pound range?

Answers:
a. 120   b. 1,940

A special line graph, called a **frequency polygon**, can be constructed from the carry-on luggage histogram by joining the center points at the top of each bar. (See the graphs below.) On the horizontal axis, we write the middle value of each bar. After erasing the bars, we get the frequency polygon shown on the right below.

Histogram

Frequency polygon

ANSWERS TO SELF CHECKS

1. $18.35 **2. a.** 32 mph **b.** 11 mph **c.** a chicken and an elephant **3. a.** about $875 billion **b.** about $750 billion **c.** about $1,800 billion **4.** 33 pizzas were delivered to the co-ed residence hall. **5.** 6.9 million troy ounces **6. a.** about 10,000 **b.** about 55,000 **7.** Train 1, which had been stopped, is beginning to move. **8.** 200

SECTION 8.1 STUDY SET

VOCABULARY

For problems 1-6, refer to graphs a through f below. Fill in the blanks with the correct letter. Each letter is used only once.

1. Graph __(a)__ is a bar graph.
2. Graph __(e)__ is a circle graph.
3. Graph __(c)__ is a pictograph.
4. Graph __(f)__ is a line graph.
5. Graph __(d)__ is a histogram.
6. Graph __(b)__ is a frequency polygon.

(a)

(b)

(c)

(d)

(e)

(f)

Unless otherwise noted, all art on this page is © Cengage Learning.

7. A horizontal or vertical line used for reference in a bar graph is called an __axis__.

8. A circle graph uses slice-of-pie-shaped regions called __sectors__ to show what part of the whole each quantity represents.

CONCEPTS

Fill in the blanks.

9. To read a table, we must find the __intersection__ of the row and column that contains the desired information.

10. The __horizontal__ axis and the vertical axis of a bar graph serve to frame the graph; they are scaled in units such as years, dollars, minutes, pounds, and percent.

11. A pictograph is like a bar graph, but the bars are made from __pictures__ or symbols.

12. Line graphs are often used to show how a quantity changes with __time__. On such graphs, we can easily see when a quantity is increasing and when it is __decreasing__.

13. A histogram is a bar graph with three important features.
    - The __bars__ of a histogram touch.
    - Data values never fall at the __edge__ of a bar.
    - The widths of the bars are __equal__ and represent a range of values.

14. A frequency polygon can be constructed from a histogram by joining the __center__ points at the top of each bar.

NOTATION

15. If the symbol 🚌 = 1,000 buses, estimate what the symbol 🚌 represents. about 500 buses

16. Fill in the blank: The symbol ⚡ is used when graphing to show a __break__ in the scale on an axis.

GUIDED PRACTICE

Refer to the postal rate table on page 741 to answer the following questions. See Example 1.

17. Find the cost of using priority mail to send a package weighing $7\frac{1}{4}$ pounds to zone 3. $13.30

18. Find the cost of sending a package weighing $2\frac{1}{4}$ pounds to zone 5 by priority mail. $10.60

19. A woman wants to send a birthday gift and an anniversary gift to her brother, who lives in zone 6, using priority mail. One package weighs 2 pounds 9 ounces, and the other weighs 3 pounds 8 ounces. Suppose you are the woman's mail carrier and she asks you how much money will be saved by sending both gifts as one package instead of two. Make the necessary calculations to answer her question. (Hint: 16 ounces = 1 pound.) $5.60 ($27.40 − $21.80)

from Campus to Careers
Postal Service Mail Carrier

20. Juan wants to send a package weighing 11 pounds 15 ounces to a friend living in zone 2. Express (overnight) mail service would cost $39.50. How much more would it cost to send the package by express mail instead of priority mail? $25.05

Refer to the bar graph below to answer the following questions. See Example 2.

21. List the top three most commonly owned pets in the United States. fish, cat, dog

22. There are three types of pets that are owned in approximately equal numbers. What are they? birds, reptiles, small animals

23. The number of fish is approximately equal to the combined total of which two types of pets? dogs and cats

24. How many more pet cats are there than pet dogs? about 10 million

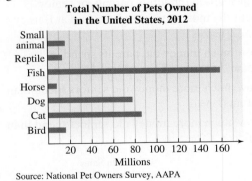

Refer to the bar graph on the next page to answer the following questions. See Example 3.

Zinc and lead are two commonly used metals. A zinc coating stops steel and iron from rusting, and zinc is combined with copper to form brass. Lead is used in batteries and in aprons to shield patients during X-rays.

25. For the years shown in the graph, has the production of zinc always exceeded the production of lead? yes

26. Estimate how many times greater the amount of zinc produced in 2000 was compared to the amount of lead produced that year? about 3 times greater

27. What is the sum of the amounts of lead produced in 1990, 2000, and 2011? about 11,000,000 metric tons
28. For which metal, lead or zinc, and for which two years, did the production remain about the same? lead: 1990, 2000
29. In what years was the amount of zinc produced at least twice that of lead? 1990, 2000, 2011
30. Find the difference between the amount of zinc produced in 2011 and the amount produced in 2000. about 4,000,000 metric tons
31. By how many metric tons did the amount of zinc produced increase between 1990 and 2011? 6,000,000 metric tons
32. Between which two years did the production of lead decrease? 1990 and 2000

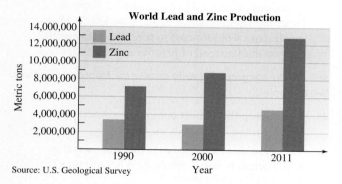

Source: U.S. Geological Survey

Refer to the pictograph below to answer the following questions. See Example 4.

33. Which group (children, parents, or seniors) spent the most money on ice cream at Barney's Café? seniors
34. How much money did parents spend on ice cream? $700
35. How much more money did seniors spend than parents? $50
36. How much more money did seniors spend than children? $200

Ice Cream Sales at Barney's Café

Children 🍦🍦🍦🍦🍦
Parents 🍦🍦🍦🍦🍦🍦🍦
Seniors 🍦🍦🍦🍦🍦🍦🍦🍦

🍦 = $100

Refer to the circle graph below to answer the following questions. See Example 5.

37. Of the languages in the graph, which is spoken by the greatest number of people? Chinese
38. Do more people speak Spanish or French? Spanish
39. Together, do more people speak English, French, Spanish, Russian, and German combined than Chinese? no
40. Three pairs of languages shown in the graph are spoken by groups of the same size. Which pairs of languages are they? French and German (1%), Hindi and Arabic (3%), English and Spanish (5%)
41. What percent of the world's population speak a language other than the eight shown in the graph? 62%
42. What percent of the world's population speak Russian or English? 7%
43. To the nearest one million, how many people in the world speak Chinese? 1,272,000,000
44. To the nearest one million, how many people in the world speak Arabic? 212,000,000

Estimated world population 2013: 7,064,000,000
Source: ethnologue.com, U.S. Census Bureau

Refer to the line graph on the next page to answer the following questions. See Example 6.

45. How many U.S. ski resorts were in operation in 2004? 494
46. How many U.S. ski resorts were in operation in 2008? 481
47. Between which two years was there a decrease in the number of ski resorts in operation? (*Hint:* there is more than one answer.) 2002 to 2003; 2004 to 2005; 2005 to 2006; 2007 to 2008; 2008 to 2009; 2009 to 2010
48. Between which two years was there an increase in the number of ski resorts in operation? (*Hint:* there is more than one answer.) 2001 to 2002; 2003 to 2004; 2006 to 2007; 2010 to 2011

49. For which two years was the number of ski resorts in operation exactly the same? 2001 and 2003
50. Find the difference in the number of ski resorts in operation in 2001 and 2008. 9 fewer resorts in 2008
51. Between which two years was there the greatest decrease in the number of ski resorts in operation? What was the decrease? 2005 to 2006; a decrease of 14 resorts
52. Between which two years was there the greatest increase in the number of ski resorts in operation? What was the increase? 2010 to 2011; an increase of 15 resorts

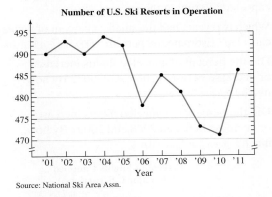

Source: National Ski Area Assn.

Refer to the line graph below to answer the following questions. See Example 7.

53. Which runner ran faster at the start of the race? 1
54. At time A, which runner was ahead in the race? 1
55. At what time during the race were the runners tied for the lead? B
56. Which runner stopped to rest first? 1
57. Which runner dropped his watch and had to go back to get it? 1
58. At which of these times (A, B, C, D, E) was runner 1 stopped and runner 2 running? A and D
59. Describe what was happening at time E. Who was running? Who was stopped? Runner 1 was running; runner 2 was stopped.
60. Which runner won the race? 2

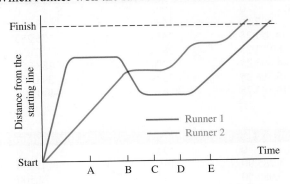

Refer to the histogram and frequency polygon below to answer the following questions. See Example 8.

61. COMMUTING MILES An insurance company collected data on the number of miles its employees drive to and from work. The data are presented in the histogram below.
    a. How many employees have a commute that is in the range of 15 to 19 miles per week? 27
    b. How many employees commute 14 miles or less per week? 22

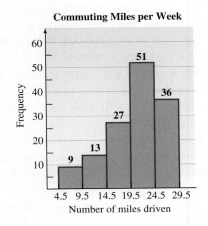

62. NIGHT SHIFT STAFFING A hospital administrator surveyed the medical staff to determine the number of room calls during the night. She constructed the frequency polygon below.
    a. On how many nights were there about 30 room calls? 90
    b. On how many nights were there about 60 room calls? 10

TRY IT YOURSELF

Refer to the 2013 federal income tax tables below.

63. **FILING A SINGLE RETURN** Herb is single and has taxable income of $79,250. Compute his federal income tax. $15,741.25

64. **MILLIONAIRES** Compute the federal income tax on a taxable income of $1,000,000 for:
   a. a single adult $353,763.75
   b. a married couple filing jointly $343,646

65. **LOTTO WINNINGS** In 2012, a married couple from Illinois won the Mega Millions jackpot and took it as a lump sum of $157,000,000. What is the federal income tax on their winnings? $62,119,646

66. **FILING A JOINT RETURN** Raul and his wife have a combined taxable income of $57,100. Compute their federal income tax if they file jointly. $7,672.50

67. **TAX-SAVING STRATEGY** Angelina is single and has a taxable income of $53,000. If she marries someone with no income, she will gain other deductions that will reduce her income by $3,900, and she can file a joint return.
   a. Compute her federal income tax if she remains single. $9,178.75
   b. Compute her federal income tax if she gets married. $6,472.50
   c. How much will she save in federal income tax by getting married? $2,706.25

68. **THE MARRIAGE PENALTY** A single man with a taxable income of $80,000 is dating a single woman with a taxable income of $75,000.
   a. Find the amount of federal income tax each of them must pay. man: $15,928.75; woman: $14,678.75
   b. Add the results from part a. $30,607.50
   c. If they get married and file a joint return, how much federal income tax will they have to pay on their combined incomes? $30,865.50
   d. Will they owe more federal income taxes if they get married or if they stay single? Find the amount of the "marriage penalty." married; $258

Refer to the bar graph below.

69. In which year was the largest percent of flights canceled? Estimate the percent. 2007; about 2.1%

70. Which two years had the smallest percent of flights canceled? Estimate the percent. 2009 and 2012; about 1.3%

71. Did the percent of canceled flights increase or decrease between 2006 and 2007? By how much? increase; about 0.5%

72. Did the percent of canceled flights increase or decrease between 2008 and 2009? By how much? decrease; about 0.6%

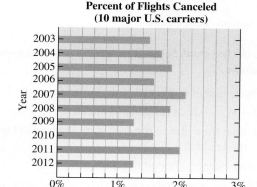

Source: Bureau of Transportation Statistics

Unmarried Individuals–Schedule X 2013

If Taxable Income Is Over	But Not Over	The Tax Is	Plus	Of the Amount Over
$ 0	$ 8,925	$ 0	10%	$ 0
8,925	36,250	892.50	15%	8,925
36,250	87,850	4,991.25	25%	36,250
87,850	183,250	17,891.25	28%	87,850
183,250	398,350	44,603.25	33%	183,250
398,350	400,000	115,586.25	35%	398,350
400,000	----------	116,163.75	39.6%	400,000

Married Filing Jointly–Schedule Y-1 2013

If Taxable Income Is Over	But Not Over	The Tax Is	Plus	Of the Amount Over
$ 0	$ 17,850	$ 0	10%	$ 0
17,850	72,500	1,785.00	15%	17,850
72,500	146,400	9,982.50	25%	72,500
146,400	223,050	28,457.50	28%	146,400
223,050	398,350	49,919.50	33%	223,050
398,350	450,000	107,768.00	35%	398,350
450,000	----------	125,846.00	39.6%	450,000

Refer to the line graph below, which shows the altitude of a small private airplane.

73. How did the plane's altitude change between times B and C? it increased
74. At what time did the pilot first level off the airplane? A
75. When did the pilot first begin his descent to land the airplane? D
76. How did the plane's altitude change between times D and E? it decreased

Refer to the double-bar graph below.

77. In which categories of moving violations have violations decreased since last month?
reckless driving and failure to yield
78. Last month, which violation occurred most often?
following too closely
79. This month, which violation occurred least often?
reckless driving
80. Which violation has shown the greatest decrease in number since last month? failure to yield

Refer to the line graph in the next column.

81. What were the average weekly earnings in mining for the year 1980? about $440
82. What were the average weekly earnings in construction for the year 1980? about $390
83. Were the average weekly earnings in mining and construction ever the same? no
84. What was the difference in a miner's and a construction worker's weekly earnings in 1995? about $75
85. In the period between 2005 and 2010, which occupation's weekly earnings were increasing more rapidly, the miner's or the construction worker's? the miner's

86. Did the weekly earnings of a miner or a construction worker ever decrease over a five-year span? no
87. In the period from 1980 to 2010, which group of workers received the greater increase in weekly earnings? the miners
88. In what five-year span was the miner's increase in weekly earnings the smallest? 1990 to 1995

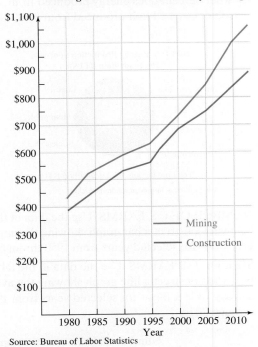

Refer to the pictograph below.

89. What is the daily parking rate for Midtown New York? about $42
90. What is the daily parking rate for Boston? about $33
91. How much more would it cost to park a car for five days in Boston compared to five days in San Francisco? about $30
92. How much more would it cost to park a car for five days in Midtown New York compared to five days in Boston? about $45

Source: Colliers International

Refer to the following circle graph below.

93. What percent of U.S. energy production comes from nuclear energy? Round to the nearest percent. 9%
94. What percent of U.S. energy production comes from natural gas? Round to the nearest percent. 26%
95. What percent of the total energy production comes from renewable and nuclear combined? 18%
96. By what percent does energy produced from petroleum exceed that produced from coal? about 16%

2011 U.S. Energy Production Sources (in quadrillion BTUs)

Natural gas: 24.8
Coal: 19.7
Renewable: 9.1
Nuclear: 8.3
Petroleum: 35.3

Total production: 97.2 quadrillion BTUs
Source: Energy Information Administration

97. **NUMBER OF U.S. FARMS** Use the data in the table below to construct a bar graph showing the number of U.S. farms for selected years from 1950 through 2010.

98. **SIZE OF U.S. FARMS** Use the data in the table below to construct a line graph showing the average acreage of U.S. farms for selected years from 1950 through 2010.

Year	Number of U.S. farms (in millions)	Average size of U.S. farms (acres)
1950	5.6	213
1960	4.0	297
1970	2.9	374
1980	2.4	426
1990	2.1	460
2000	2.2	436
2010	2.2	418

Source: U.S. Dept. of Agriculture

99. **COUPONS** Each coupon value shown in the table below provides savings for shoppers. Construct a line graph that relates the original price (in dollars, on the horizontal axis) to the sale price (in dollars, on the vertical axis).

Coupon value: amount saved	Original price of the item
$10	$100, but less than $250
$25	$250, but less than $500
$50	$500 or more

100. **DENTISTRY** To study the effect of fluoride in preventing tooth decay, researchers counted the number of fillings in the teeth of 28 patients and recorded these results:

3, 7, 11, 21, 16, 22, 18, 8, 12, 3, 7, 2, 8, 19, 12, 19, 12, 10, 13, 10, 14, 15, 14, 14, 9, 10, 12, 13

Tally the results by completing the table. Then construct a histogram. The first bar extends from 0.5 to 5.5, the second bar from 5.5 to 10.5, and so on.

Number of fillings	Frequency
1–5	3
6–10	8
11–15	11
16–20	4
21–25	2

WRITING

101. What kind of presentation (table, bar graph, line graph, circle graph, pictograph, or histogram) is most appropriate for displaying each type of information? Explain your choices.
    - The percent of students at a college, classified by major
    - The percent of biology majors at a college each year since 1970
    - The number of hours a group of students spent studying for final exams
    - The ethnic populations of the ten largest cities
    - The average annual salary of corporate executives for ten major industries

102. Explain why a histogram is a special type of bar graph.

REVIEW

103. Write the prime numbers between 10 and 30.
    11, 13, 17, 19, 23, 29
104. Write the first ten composite numbers.
    4, 6, 8, 9, 10, 12, 14, 15, 16, 18
105. Write the even whole numbers less than 6 that are not prime. 0, 4
106. Write the odd whole numbers less than 20 that are not prime. 1, 9, 15

SECTION 8.2
Mean, Median, and Mode

Objectives
1. Find the mean (average) of a set of values.
2. Find the weighted mean of a set of values.
3. Find the median of a set of values.
4. Find the mode of a set of values.
5. Use the mean, median, mode, and range to describe a set of values.

Graphs are not the only way of describing sets of numbers in a compact form. Another way to describe a set of numbers is to find *one* value around which the numbers in the set are grouped. We call such a value a **measure of central tendency**. In Section 1.7, we studied the most popular measure of central tendency, the *mean* or *average*. In this section we will examine two other measures of central tendency, called the *median* and the *mode*.

1 Find the mean (average) of a set of values.

Recall that the *mean* or *average* of a set of values gives an indication of the "center" of the set of values. To review this concept, let's consider the case of a student who has taken five tests this semester in a history class scoring 87, 73, 89, 92, and 84. To find out how well she is doing, she calculates the mean, or the average, of these scores, by finding their sum and then dividing it by 5.

$$\text{Mean} = \frac{87 + 73 + 89 + 92 + 84}{5} \quad \begin{array}{l}\leftarrow \text{The sum of the test scores} \\ \leftarrow \text{The number of test scores}\end{array}$$

$$= \frac{425}{5} \quad \text{In the numerator, do the addition.}$$

$$= 85 \quad \text{Do the division.}$$

$$\begin{array}{r} \overset{2}{8}7 \\ 73 \\ 89 \\ 92 \\ +\;84 \\ \hline 425 \end{array} \qquad \begin{array}{r} 85 \\ 5\overline{)425} \\ -\,40 \\ \hline 25 \\ -\,25 \\ \hline 0 \end{array}$$

The mean is 85. Some scores were better and some were worse, but 85 is a good indication of her performance in the class.

> **Success Tip** The mean (average) is a single value that is "typical" of a set of values. It can be, but is not necessarily, one of the values in the set. In the previous example, note that the student's mean score was 85; however, she did not score 85 on any of the tests.

> **Finding the Mean (Arithmetic Average)**
>
> The **mean**, or the **average**, of a set of values is given by the formula:
>
> $$\text{Mean (average)} = \frac{\text{the sum of the values}}{\text{the number of values}}$$

> **The Language of Algebra** The *mean* (average) of a set of values is more formally called the **arithmetic mean** (pronounced air-rith-MET-tick).

EXAMPLE 1 **Store Sales** One week's sales in men's, women's, and children's departments of the Clothes Shoppe are given in the table on the next page. Find the mean of the daily sales in the women's department for the week.

Strategy We will add $3,135, $2,310, $3,206, $2,115, $1,570, and $2,100 and divide the sum by 6.

Self Check 1
STORE SALES Find the mean of the daily sales in the men's department of the Clothes Shoppe for the week. $1,540

Now Try Problems 9 and 41

Unless otherwise noted, all art on this page is © Cengage Learning.

756 Chapter 8 Graphs and Statistics

Teaching Example 1 STORE SALES
Refer to the table for Example 1.
Find the mean of the daily sales in the children's department of the Clothes Shoppe for the week.
Answer:
$985

Total Daily Sales by Department — Clothes Shoppe			
Day	Men's department	Women's department	Children's department
Monday	$2,315	$3,135	$1,110
Tuesday	2,020	2,310	890
Wednesday	1,100	3,206	1,020
Thursday	2,000	2,115	880
Friday	955	1,570	1,010
Saturday	850	2,100	1,000

WHY To find the mean (average) of a set of values, we divide the sum of the values by the number of values. In this case, there are 6 days of sales (Monday through Saturday).

Solution
Since there are 6 days of sales, divide the sum by 6.

$$\text{Mean} = \frac{\$3,135 + \$2,310 + \$3,206 + \$2,115 + \$1,570 + \$2,100}{6}$$

$$= \frac{\$14,436}{6} \quad \text{In the numerator, do the addition.}$$

$$= \$2,406 \quad \text{Do the division.}$$

```
  1 11
  3,135
  2,310
  3,206
  2,115
  1,570
+ 2,100
 14,436
```

```
     2406
6)14,436
  -12
    24
   -24
     03
     -0
     36
    -36
      0
```

The mean of the week's daily sales in the women's department is $2,406.

Using Your CALCULATOR **Finding the Mean**

Most scientific calculators do statistical calculations and can easily find the mean of a set of numbers. To use a scientific calculator in statistical mode to find the mean in Example 1, try these keystrokes:

- Set the calculator to statistical mode.
- Reset the calculator to clear the *statistical registers*.
- Enter each number, followed by the $\boxed{\Sigma+}$ key instead of the $\boxed{+}$ key. That is, enter 3,135, press $\boxed{\Sigma+}$, enter 2,310, press $\boxed{\Sigma+}$, and so on.
- When all data are entered, find the mean by pressing the $\boxed{\bar{x}}$ key. You may need to press $\boxed{2^{\text{nd}}}$ first. The mean is 2,406.

Because keystrokes vary among calculator brands, you might have to check the owner's manual if these instructions don't work.

Self Check 2

TRUCKING If a trucker drove 3,360 miles in February, how many miles did he drive per day, on average? (Assume it is not a leap year.) 120 miles per day

Now Try Problem 43

EXAMPLE 2 *Driving* In the month of January, a trucker drove a total of 4,805 miles. On the average, how many miles did he drive per day?

Strategy We will divide 4,805 by 31 (the number of days in the month of January).

January						
S	M	T	W	T	F	S
				1	2	3	4
5	6	7	8	9	10	11
12	13	14	15	16	17	18
19	20	21	22	23	24	25
26	27	28	29	30	31	

Unless otherwise noted, all art on this page is © Cengage Learning.

WHY We do not have to find the sum of the miles driven each day in January. That total is given in the problem as 4,805 miles.

Solution

$$\text{Average number of miles driven per day} = \frac{\text{the total miles driven}}{\text{the number of days}}$$

$$= \frac{4{,}805}{31} \quad \begin{array}{l}\leftarrow \text{This is given.} \\ \leftarrow \text{January has 31 days.}\end{array}$$

$$= 155 \quad \text{Do the division.}$$

```
       155
    ┌──────
 31)4,805
     -31
     ───
     170
    -155
    ────
     155
    -155
    ────
       0
```

On average, the trucker drove 155 miles per day.

Teaching Example 2 DELIVERING PIZZAS In the month of March, delivery persons at a Pizza Hut store drove 2,945 miles in their pizza delivery car. On average, how many miles were driven per day in the delivery car?
Answer:
95 mi per day

2 Find the weighted mean of a set of values.

When a value in a set appears more than once, that value has a greater "influence" on the mean than another value that only occurs a single time. To simplify the process of finding a mean, any value that appears more than once can be "weighted" by multiplying it by the number of times it occurs. A mean that is found in this way is called a **weighted mean**.

EXAMPLE 3   Hotel Reservations

A hotel electronically recorded the number of times the reservation desk telephone rang before it was answered by a receptionist. The results of the weeklong survey are shown in the table on the right. Find the average number of times the phone rang before a receptionist answered.

Number of rings	Number of calls
1	11
2	46
3	45
4	28
5	20

Strategy First, we will determine the total number of times the reservation desk telephone rang during the week before it was answered. Then we will divide that result by the total number of calls received.

WHY To find the average of a set of values, we divide the sum of the values by the number of values.

Solution To find the total number of times the reservation desk telephone rang during the week before it was answered, we multiply each number of rings (1, 2, 3, 4, and 5) by the number of times it occurred and add those results to get 450. The calculations are highlighted in blue in the "Weighted number of rings" column.

Number of rings	Number of calls	Weighted number of rings	
1	11	1 · 11 →	11
2	46	2 · 46 →	92
3	45	3 · 45 →	135
4	28	4 · 28 →	112
5	+ 20	5 · 20 →	+ 100
Totals	**150**		**450**

Self Check 3

QUIZ RESULTS The class results on a five-question true-or-false Spanish quiz are shown in the table below. Find the average number of incorrect answers on the quiz. 2 incorrect answers

Total number of incorrect answers on the quiz	Number of students
0	8
1	8
2	5
3	15
4	3
5	1

Now Try Problem 45

Teaching Example 3 CLASS SIZE An elementary school principal created the table below that shows how many children are in each classroom. Find the average class size for the school.

Students per class	Number of classrooms
17	1
18	5
22	3
26	4
34	2

Answer:
23 children per class

To find the total number of calls received, we add the values in the "Number of calls" column of the table and get 150, as highlighted in red. To find the average, we divide.

$$\text{Average} = \frac{450}{150} \begin{array}{l} \leftarrow \text{The total number of rings} \\ \leftarrow \text{The total number of calls} \end{array}$$

$$= 3 \quad \text{Do the division.}$$

$$\begin{array}{r} 3 \\ 150{\overline{\smash{\big)}\,450}} \\ \underline{-450} \\ 0 \end{array}$$

The average number of times the phone rang before it was answered was 3.

Finding the Weighted Mean

To find the weighted mean of a set of values:

1. Multiply each value by the number of times it occurs.
2. Find the sum of products from step 1.
3. Divide the sum from step 2 by the total number of individual values.

Another example of a weighted mean is a **grade point average (GPA)**. To find a GPA, we divide:

$$\text{GPA} = \frac{\text{total number of grade points}}{\text{total number of credit hours}}$$

The Language of Algebra Some schools assign a certain number of **credit hours (credits)** to a course while others assign a certain number of **units**. For example, at San Antonio College, the Basic Mathematics course is 3 credit hours while the same course at Los Angeles City College is 3 units.

Self Check 4

FINDING GPAs Find the semester grade point average for a student that received the following grades. 2.75

Course	Grade	Credits
MATH 130	A	4
ENG 101	D	3
PHY 080	B	4
SWIM 100	C	1

Now Try Problem 51

EXAMPLE 4 *Finding GPAs* Find the semester grade point average for a student that received the following grades. Round to the nearest hundredth.

Course	Grade	Credits
Speech	C	2
Basic Mathematics	A	4
French	B	4
Business Law	D	3
Study Skills	A	1

Strategy First, we will determine the total number of grade points earned by the student. Then we will divide that result by the total number of credits.

WHY To find the mean of a set of values, we divide the sum of the values by the number of values.

Solution

The point values of grades that are used at most colleges and universities are:

A: 4 pts   B: 3 pts   C: 2 pts   D: 1 pt   F: 0 pt

To find the total number of grade points that the student earned, we multiply the number of credits for each course by the point value of the grade received. Then we add those results to find that the total number of grade points is 39. The calculations are highlighted in blue in the "Weighted grade points" column on the next page.

To find the total number of credits, we add the values in that column (highlighted in red), to get 14.

Course	Grade	Credits	Weighted grade points
Speech	C	2	$2 \cdot 2 \rightarrow$ 4
Basic Mathematics	A	4	$4 \cdot 4 \rightarrow$ 16
French	B	4	$3 \cdot 4 \rightarrow$ 12
Business Law	D	3	$1 \cdot 3 \rightarrow$ 3
Study Skills	A	+1	$4 \cdot 1 \rightarrow$ + 4
Totals		14	39

To find the GPA, we divide.

$$\text{GPA} = \frac{39}{14} \quad \begin{array}{l}\leftarrow \text{The total number of grade points} \\ \leftarrow \text{The total number of credits}\end{array}$$

≈ 2.785 Do the division.

$= 2.79$ Round 2.785 to the nearest hundredth.

The student's semester GPA is 2.79.

$$\begin{array}{r} 2.785 \\ 14{\overline{\smash{\big)}\,39.000}} \\ \underline{-28} \\ 11\,0 \\ \underline{-9\,8} \\ 1\,20 \\ \underline{-1\,12} \\ 80 \\ \underline{-70} \\ 10 \end{array}$$

Teaching Example 4 FINDING GPAs Find the semester grade point average for a student that received the following grades. Round to the nearest hundredth.

Course	Grade	Credits
COMP SCI 101	D	3
MATH 150	B	4
ART HIST 175	A	2
HEALTH 090	C	1
SPAN 130	B	3

Answer:
2.62 GPA

3 Find the median of a set of values.

The mean is not always the best measure of central tendency. It can be affected by very high or very low values. For example, suppose the weekly earnings of four workers in a small company are $280, $300, $380, and $240, and the owner pays himself $5,000 a week. At that company, the mean salary per week is

Mean $= \dfrac{\$280 + \$300 + \$380 + \$240 + \$5{,}000}{5}$ There are 4 employees plus the owner: $4 + 1 = 5$.

$= \dfrac{\$6{,}200}{5}$ In the numerator, do the addition.

$= \$1{,}240$ Do the division.

The owner could say, "Our employees earn an average of $1,240 per week." Clearly, the mean does not fairly represent the typical worker's salary there.

A better measure of the company's typical salary is the *median*: the salary in the middle when all of them are arranged by size.

Smallest $240 $280   **$300** $380 $5,000   Largest

         Two salaries   The middle salary   Two salaries

The typical worker earns $300 per week, far less than the mean salary.

The Median

The **median** of a set of values is the middle value. To find the median:

1. Arrange the values in increasing order.
2. If there is an odd number of values, the median is the middle value.
3. If there is an even number of values, the median is the mean (average) of the middle two values.

Unless otherwise noted, all art on this page is © Cengage Learning.

Self Check 5

Find the median of the following set of values:

$1\frac{7}{8}$ $2\frac{1}{2}$ $3\frac{3}{5}$ $\frac{1}{2}$ $2\frac{3}{4}$ $2\frac{1}{2}$

Now Try Problems 17 and 21

Teaching Example 5 Find the median of the following set of values:

3.8 6.3 7.4 5.4 8.8 6.3 3.8 7.1 4.9

Answer: 6.3

Self Check 6

GRADE DISTRIBUTIONS On a mathematics exam, there were four scores of 68, five scores of 83, and scores of 72, 78, and 90. Find the median score. 80.5

Now Try Problems 25 and 29

Teaching Example 6 GRADE DISTRIBUTIONS On an English exam, there were three scores of 78, two scores of 81, five scores of 85, and scores of 60 and 99. Find the median score.
Answer: 83

EXAMPLE 5 Find the median of the following set of values:

7.5 20.9 9.9 4.4 9.8 5.3 6.2 7.5 4.9

Strategy We will arrange the nine values in increasing order.

WHY It is easier to find the middle value when they are written in that way.

Solution
Since there is an odd number of values, the median is the middle value.

Smallest 4.4 4.9 5.3 6.2 **7.5** 7.5 9.8 9.9 20.9 Largest
              └─ Four values ─┘   ↑   └─ Four values ─┘
                         The middle value

The median is 7.5.

If there is an even number of values in a set, there is no middle value. In that case, the median is the mean (average) of the two values closest to the middle.

EXAMPLE 6 *Grade Distributions* On an exam, there were three scores of 59, four scores of 77, and scores of 43, 47, 53, 60, 68, 82, and 97. Find the median score.

Strategy We will arrange the fourteen exam scores in increasing order.

WHY It is easier to find the two middle scores when they are written in that way.

Solution
Since there is an even number of exam scores, we need to identify the two middle scores.

Smallest 43 47 53 59 59 59 **60 68** 77 77 77 77 82 97 Largest
              └── Six scores ──┘   ↑   └── Six scores ──┘
                           Two middle scores

Since there is an even number of scores, the median is the average (mean) of the two scores closest to the middle: the 60 and the 68.

$$\text{Median} = \frac{60 + 68}{2} = \frac{128}{2} = 64$$

The median is 64.

Success Tip The median is a single value that is "typical" of a set of values. It can be, but is not necessarily, one of the values in the set. In Example 5, the median, 7.5, was one of the given values. In Example 6, the median exam score, 64, was not in the given set of exam scores.

4 Find the mode of a set of values.

The mean and the median are not always the best measure of central tendency. For example, suppose a hardware store displays 20 outdoor thermometers. Ten of them read 80°, and the other ten all have different readings.

To choose an accurate thermometer, should we choose one with a reading that is closest to the *mean* of all 20, or to their *median*? Neither. Instead, we should choose one of the 10 that all read the same, figuring that any of those that agree will likely be correct.

By choosing that temperature that appears most often, we have chosen the *mode* of the 20 values.

Unless otherwise noted, all art on this page is © Cengage Learning.

The Mode

The **mode** of a set of values is the single value that occurs most often. The mode of several values is also called the **modal value**.

EXAMPLE 7
Find the mode of these values:

3 6 5 7 3 7 2 4 3 5 3 7 8 7 3 7 6 3 4

Strategy We will determine how many times each of the values, 2, 3, 4, 5, 6, 7, and 8 occurs.

WHY We need to know which values occur most often.

Solution
It is not necessary to list the values in increasing order. Instead, we can make a chart and use **tally marks** to keep track of the number of times that the values 2, 3, 4, 5, 6, 7, and 8 occur.

2	3	4	5	6	7	8	← These values appear in the list.
/	### /	//	//	//	###	/	← Tally marks

Because 3 occurs more times than any other value, it is the mode.

The Language of Algebra In Example 7, the given set of values has one mode. If a set of values has two modes (exactly two values that occur an equal number of times and more often than any other value) it is said to be **bimodal**. If no value in a set occurs more often than another, then there is *no mode*.

Self Check 7
Find the mode of these values:
2 3 4 6 2 4
3 4 3 4 2 5 4

Now Try Problems 33 and 37

Teaching Example 7
Find the mode of these values:
67 90 42 53 42 68
42 53 17 42 90 89
Answer:
42

5 Use the mean, median, mode, and range to describe a set of values.

Another measure that is used to describe a set of values is the **range**. It indicates the spread of the values.

The Range

The **range** of a set of values is the difference between the largest value and the smallest value.

EXAMPLE 8
Machinist's Tools The diameters (distances across) of eight stainless steel bearings were found using the calipers shown below. Find **a.** the mean, **b.** the median, **c.** the mode, and **d.** the range of the set of measurements listed below.

3.43 cm   3.25 cm   3.48 cm   3.39 cm   3.54 cm   3.48 cm   3.23 cm   3.24 cm

Calipers

Stainless steel bearing

Strategy We will determine the sum of the measurements, the number of measurements, the middle measurement(s), the most often occurring measurement, and the difference between the largest and smallest measurement.

WHY We need to know that information to find the mean, median, mode, and range.

Self Check 8
MOBILE PHONES The weights of eight different makes of mobile phones are: 4.37 oz, 5.98 oz, 4.36 oz, 4.95 oz, 5.05 oz, 5.95 oz, 4.95 oz, and 5.27 oz. Find the mean, median, and mode weight. Then find the range of the weights.

Now Try Problem 47

Self Check 8 Answers
mean: 5.11 oz; median: 5.00 oz; mode: 4.95 oz, range: 1.62 oz

Teaching Example 8 MICROSCOPES
The weights of six different glass lenses used in microscopes are: 5.28 g, 5.44 g, 5.36 g, 5.32 g, 5.44 g, and 5.50 g. Find the mean, median, and mode weight. Then find the range of the weights.
Answers:
mean: 5.39 g, median: 5.40 g, mode: 5.44 g, range: 0.22 g

Solution

a. To find the mean, we add the measurements and divide by the number of values, which is 8.

$$\text{Mean} = \frac{3.43 + 3.25 + 3.48 + 3.39 + 3.54 + 3.48 + 3.23 + 3.24}{8}$$

$$= \frac{27.04}{8} \quad \text{In the numerator, do the addition.}$$

$$= 3.38 \quad \text{Do the division.}$$

The mean is 3.38 cm.

b. To find the median, we first arrange the eight measurements in increasing order.

Smallest   3.23   3.24   3.25   **3.39**   **3.43**   3.48   3.48   3.54   Largest

↑ Two middle measurements

Because there is an even number of measurements, the median is the average of the two middle values.

$$\text{Median} = \frac{3.39 + 3.43}{2} = \frac{6.82}{2} = 3.41 \text{ cm}$$

c. Since the measurement 3.48 cm occurs most often (twice), it is the mode.

d. In part b, we see that the smallest value is 3.23 and the largest value is 3.54. To find the range, we subtract the smallest value from the largest value.

Range = 3.54 − 3.23 = 0.31 cm

THINK IT THROUGH   *The Value of an Education*

"If we want America to lead in the 21st century, nothing is more important than giving everyone the best education possible."
President Barack Obama, 2012

As college costs increase, some people wonder if it is worth it to spend years working toward a degree when that same time could be spent earning money. The following median income data makes it clear that, over time, additional education is well worth the investment. Use the given facts to complete the bar graph.

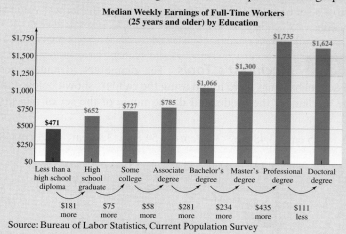

Median Weekly Earnings of Full-Time Workers (25 years and older) by Education

Source: Bureau of Labor Statistics, Current Population Survey

ANSWERS TO SELF CHECKS

1. $1,540   2. 120 miles per day   3. 2 incorrect answers   4. 2.75   5. $2\frac{1}{2}$ 6. 80.5
7. 4   8. mean: 5.11 oz; median: 5.00 oz; mode: 4.95 oz; range: 1.62 oz

SECTION 8.2 STUDY SET

VOCABULARY

Fill in the blanks.

1. The __mean__ (average) of a set of values is the sum of the values divided by the number of values in the set.
2. The __median__ of a set of values written in increasing order is the middle value.
3. The __mode__ of a set of values is the single value that occurs most often.
4. The __range__ of a set of values is the difference between the largest value and the smallest value.

CONCEPTS

5. Fill in the blank. The mean of a set of values is given by the formula

$$\text{Mean} = \frac{\text{the sum of the values}}{\text{the number of values}}$$

6. Consider the following set of values written in increasing order:

    3   6   8   10   11   15   16

    a. Is there an even or an odd number of values? an odd number
    b. What is the middle number of the list? 10
    c. What is the median of the set of values? 10
    d. What is the largest value? What is the smallest value? What is the range of the values? 16, 3, 13

7. Consider the following set of values written in increasing order:

    4   5   5   6   8   9   9   15

    a. Is there an even or odd number of values? an even number
    b. What are the middle numbers of the set of values? 6 and 8
    c. Fill in the blanks:

    $$\text{Median} = \frac{6+8}{2} = \frac{14}{2} = 7$$

    d. What is the largest value? What is the smallest value? What is the range of the values? 15, 4, 11

8. Consider the following set of values:

    1   6   8   6   10   9   10   2   6

    a. What value occurs the most often? How many times does it occur? 6; 3
    b. What is the mode of the set of values? 6
    c. What is the largest value? What is the smallest value? What is the range of the values? 10, 1, 9

GUIDED PRACTICE

Find the mean of each set of values. See Example 1.

9. 3   4   7   7   8   11   16   8
10. 13   15   17   17   15   13   15
11. 5   9   12   35   37   45   60   77   35
12. 0   0   3   4   7   9   12   5
13. 15   7   12   19   27   17   19   35   20   19
14. 45   67   42   35   86   52   91   102   65
15. 4.2   3.6   7.1   5.9   8.2   5.8
16. 19.1   12.8   16.5   20.0   17.1

Find the median of each set of values. See Example 5.

17. 29   5   1   9   11   17   2   9
18. 20   4   3   2   9   8   1   4
19. 7   5   4   7   3   6   7   4   1   5
20. 0   0   3   4   0   0   3   4   5   3
21. 15.1   44.9   19.7   13.6   17.2   17.2
22. 22.4   22.1   50.5   22.3   22.2   22.3
23. $\frac{1}{100}$ $\frac{999}{1,000}$ $\frac{16}{15}$ $\frac{1}{3}$ $\frac{5}{8}$ $\frac{5}{8}$
24. $\frac{1}{30}$ $\frac{17}{30}$ $\frac{7}{30}$ $\frac{29}{30}$ $\frac{11}{30}$ $\frac{11}{30}$

Find the median of each set of values. See Example 6.

25. 8   10   16   63   6   7   9
26. 7   2   11   5   4   17   6
27. 39   1   50   41   51   47   44
28. 47   18   35   29   27   16   28
29. 1.8   1.7   2.0   9.0   2.1   2.3   2.1   2.0   2.05
30. 5.0   1.3   5.0   2.3   4.3   5.6   3.2   4.5   4.4
31. $\frac{1}{5}$ $\frac{11}{5}$ $\frac{13}{5}$ $\frac{2}{5}$ $\frac{3}{5}$ $\frac{7}{5}$ 1
32. $\frac{1}{9}$ $\frac{2}{9}$ $\frac{7}{9}$ $\frac{11}{9}$ $\frac{13}{9}$ $\frac{29}{9}$ 1

Find the mode (if any) of each set of values. See Example 7.

33. 3 5 7 3 5 4 6 7 2 3 1 4 3

34. 12 12 17 17 12 13 17 12 12

35. −6 −7 −6 −4 −3 −6 −7 −6

36. 0 3 0 2 7 0 6 0 3 4 2 0 0

37. 23.1 22.7 23.5 22.7 34.2 22.7 22.7

38. 21.6 19.3 1.3 19.3 1.6 9.3 2.6 19.3

39. $\frac{1}{2}$ $\frac{1}{3}$ $\frac{1}{3}$ 2 $\frac{1}{2}$ 2 $\frac{1}{5}$ $\frac{1}{2}$ 5 $\frac{1}{3}$ bimodal: $\frac{1}{3}, \frac{1}{2}$

40. 5 9 12 35 37 45 60 none

APPLICATION

41. **SEMESTER GRADES** Frank's algebra grade is based on the average of four exams, which count equally. His grades are 75, 80, 90, and 85.

    a. Find his average exam score. 82.5

    b. If Frank's professor decided to count the fourth exam double, what would Frank's average be? 83

42. **HURRICANES** The table lists the number of major hurricanes to strike the mainland of the United States by decade. Find the average (mean) number per decade. Round to the nearest one. 6 hurricanes

Decade	Number	Decade	Number
1891–1900	8	1951–1960	9
1901–1910	4	1961–1970	6
1911–1920	7	1971–1980	4
1921–1930	5	1981–1990	4
1931–1940	8	1991–2000	5
1941–1950	10	2001–2010	7

Source: National Hurricane Center NOAA Technical Memorandum, 2011

43. **FLEET MILEAGE** An insurance company's sales force uses 37 cars. Last June, those cars logged a total of 98,790 miles.

    a. On average, how many miles did each car travel that month? 2,670 mi

    b. Find the average number of miles driven daily for each car. 89 mi

44. **BUDGETS** The Hinrichs family spent $519 on groceries last April.

    a. On average, how much did they spend on groceries each day? $17.30

    b. The Hinrichs family has five members. What is the average spent for groceries for one family member for one day? $3.46

45. **CASH AWARDS** A contest is to be part of a promotional kickoff for a new children's cereal. The prizes to be awarded are shown.

    a. How much money will be awarded in the promotion? $11,875

    b. How many cash prizes will be awarded? 125

    c. What is the average cash prize? $95

    > **Coloring Contest**
    > Grand prize: Disney World vacation plus $2,500
    > Four 1st place prizes of $500
    > Thirty-five 2nd place prizes of $150
    > Eighty-five 3rd place prizes of $25

46. **SURVEYS** Some students were asked to rate their college cafeteria food on a scale from 1 to 5. The responses are shown on the tally sheet. Find the average rating. 4

Poor		Fair		Excellent
1	2	3	4	5
	III	III	ШН	ШН IIII

47. **CANDY BARS** The prices (in cents) of the different types of candy bars sold in a drug store are: 50, 60, 50, 50, 70, 75, 50, 45, 50, 50, 60, 75, 60, 75, 100, 50, 80, 75, 100, 75.

    a. Find the mean price of a candy bar. 65¢

    b. Find the median price for a candy bar. 60¢

    c. Find the mode of the prices of the candy bars. 50¢

    d. Find the range of the candy bar prices. 55¢

48. **COMPUTER SUPPLIES** Several computer stores reported differing prices for toner cartridges for a laser printer (in dollars): 51, 55, 73, 75, 72, 70, 53, 59, 75.

    a. Find the mean price of a toner cartridge. $64.78

    b. Find the median price for a toner cartridge. $70

    c. Find the mode of the prices for a toner cartridge. $75

    d. Find the range of the toner cartridge prices. $24

49. TEMPERATURE CHANGES Temperatures were recorded at hourly intervals and listed in the table below. Find the average temperature of the period from midnight to 11:00 A.M. 61°

Time	Temperature	Time	Temperature
12:00 A.M.	53	12:00 noon	71
1:00	54	1:00 P.M.	73
2:00	57	2:00	76
3:00	58	3:00	77
4:00	59	4:00	78
5:00	59	5:00	71
6:00	61	6:00	70
7:00	62	7:00	64
8:00	64	8:00	61
9:00	66	9:00	59
10:00	68	10:00	53
11:00	71	11:00	51

50. AVERAGE TEMPERATURES Find the average temperature for the 24-hour period shown in the table in Exercise 49. 64°

For Exercises 51–54, find the semester grade point average for a student that received the following grades. Round to the nearest hundredth, when necessary.

51. 2.23 GPA

Course	Grade	Credits
MATH 210	C	5
ACCOUNTING 175	A	3
HEALTH 090	B	1
JAPANESE 010	D	4

52. 2.42 GPA

Course	Grade	Credits
NURSING 101	D	3
READING 150	B	4
PAINTING 175	A	2
LATINO STUDIES 090	C	3

53. 2.5 GPA

Course	Grade	Credits
PHOTOGRAPHY	D	3
MATH 020	B	4
CERAMICS 175	A	1
ELECTRONICS 090	C	3
SPANISH 130	B	5

54. 2.71 GPA

Course	Grade	Credits
ANTHROPOLOGY 050	D	3
STATISTICS 100	A	4
ASTRONOMY 100	C	1
FORESTRY 130	B	5
CHOIR 130	C	1

55. EXAM AVERAGES Roberto received the same score on each of five exams, and his mean score is 85. Find his median score and the mode of his scores. median and mode are 85

56. EXAM SCORES The scores on the first exam of the students in a history class were 57, 59, 61, 63, 63, 63, 70, 87, 89, 95, 99, and 100. Kia, who scored the 70, claims that "70 is better than average." Which of the three measures of central tendency (mean, median, or mode) is Kia's score better than? better than the mode (63) and the median (66.5), but worse than the mean (75.5)

57. COMPARING GRADES A student received scores of 37, 53, and 78 on three quizzes. His sister received scores of 53, 57, and 58. Who had the better average? Whose grades were more consistent? same average (56); sister's scores are more consistent

58. What is the average of all of the integers from −100 to 100, inclusive? 0

59. OCTUPLETS In December 1998, Nkem Chukwu gave birth to eight babies in Texas Children's Hospital. Find the mean, the median, and the range of their birth weights listed below. 22.525 oz, 25 oz, 17.3 oz

Ebuka (girl)	24 oz	Odera (girl)	11.2 oz
Chidi (girl)	27 oz	Ikem (boy)	17.5 oz
Echerem (girl)	28 oz	Jioke (boy)	28.5 oz
Chima (girl)	26 oz	Gorom (girl)	18 oz

60. COMPARISON SHOPPING A survey of grocery stores found the price of a 14-ounce box of Cheerios cereal ranging from $3.89 to $4.39, as shown below. What are the mean, median, mode, and range of the prices listed? $4.15, $4.19, $4.29, $0.50

$4.29 $3.89 $4.29 $4.09 $4.24 $3.99
$3.98 $4.19 $4.19 $4.39 $3.97 $4.29

61. EARTHQUAKES The magnitudes of 2011's major earthquakes are listed below. Find the mean (round to the nearest tenth), the median, and the range. 7.4, 7.2, 2.0

Date	Location	Magnitude
Jan. 1	Santiago del Estero, Argentina	7.0
Jan. 13	Loyalty Islands	7.0
Jan. 18	Pakistan	7.2
March 9	Honshu, Japan	7.2
March 11	Tohoku, Japan	9.0
March 11	Honshu, Japan	7.9
March 11	Honshu, Japan	7.7
April 7	Honshu, Japan	7.1
June 23	Aleutian Islands, Alaska	7.2
July 6	Kermadec Islands, New Zealand	7.6
July 10	Honshu, Japan	7.0
Aug. 20	Port Vila, Vanuatu	7.1
Sept. 3	Isangel, Vanuatu	7.0
Sept. 15	Ndoi Islands, Fiji	7.3
Oct. 21	Kermadec Islands, New Zealand	7.4
Oct. 23	Van, Turkey	7.2
Dec. 14	Lae, Papua New Guinea	7.1

Source: Infoplease.com

62. FUEL EFFICIENCY The ten most fuel-efficient cars in 2012, based on manufacturer's estimated city and highway average miles per gallon (mpg), are shown in the table below.

  a. Find the mean, median, and mode of the city mileage. mean: 61.8 mpg, median: 47.5 mpg, bimodal: 44 mpg and 43 mpg

  b. Find the mean, median, and mode of the highway mileage. mean: 57 mpg, median: 46 mpg, trimodal: 62 mpg, 44 mpg, 40 mpg

Model	mpg city/hwy
Mitsubishi i	126/99
Nissan Leaf	106/92
Ford Transit	62/62
Chevy Volt	58/62
Toyota Prius	51/48
Honda Civic Hybrid	44/44
Toyota Prius V	44/40
Lexus CT200h	43/40
Honda Insight	41/44
Toyota Camry Hybrid	43/39

Source: autoguide.com

63. SPORT FISHING The report shown below lists the fishing conditions at Pyramid Lake for a Saturday in January. Find the median, the mode, and the range of the weights of the striped bass caught at the lake. 5 lb, 4 lb, 10 lb

Pyramid Lake—Some striped bass are biting but are on the small side. Striking jigs and plastic worms. Water is cold: 38°. Weights of fish caught (lb): 6, 9, 4, 7, 4, 3, 3, 5, 6, 9, 4, 5, 8, 13, 4, 5, 4, 6, 9

64. NUTRITION Refer to the table below.

  a. Find the mean number of calories in one serving of the meats shown. 207.5 calories
  b. Find the median. 206.5 calories
  c. Find the mode. none
  d. Find the range of the number of calories. 140 calories

NUTRITIONAL COMPARISONS
Per 3.5 oz. serving of cooked meat

Species	Calories
Bison	143
Beef (Choice)	283
Beef (Select)	201
Pork	212
Chicken (Skinless)	190
Sockeye Salmon	216

Source: The National Bison Association

WRITING

65. Explain how to find the mean, the median, the mode, and the range of a set of values.

66. The mean, median, and mode are used to measure the central tendency of a set of values. What is meant by central tendency?

67. Which measure of central tendency—mean, median, or mode—do you think is the best for describing the salaries at a large company? Explain your reasoning.

68. When is the mode a better measure of central tendency than the mean or the median? Give an example and explain why.

REVIEW

Translate to a percent equation (or percent proportion) and then solve to find the unknown number.

69. 52 is what percent of 80? 65%

70. What percent of 50 is 56? 112%

71. $66\frac{2}{3}$% of what number is 28? 42

72. 56.2 is $16\frac{2}{3}$% of what number? 337.2

73. 5 is what percent of 8? 62.5%

74. What number is 52% of 350? 182

75. Find $7\frac{1}{4}$% of 600. 43.5

76. $\frac{1}{2}$% of what number is 5,000? 1,000,000

Unless otherwise noted, all art on this page is © Cengage Learning.

SECTION 8.3
Equations in Two Variables; The Rectangular Coordinate System

Objectives
1. Determine whether an ordered pair is a solution of an equation.
2. Complete ordered-pair solutions of equations.
3. Construct a rectangular coordinate system.
4. Plot ordered pairs and determine the coordinates of a point.

We have seen that information is often presented in the form of tables or graphs. In algebra, we also present information that way. For example, the following table and graph are related to the equation $d = 4t$. This formula gives the distance d (in miles) that a hiker can walk in a time t (in hours) at a rate of 4 miles per hour.

To find the distance the hiker can walk in 3 hours, we substitute 3 for t in the formula and evaluate the right side.

$d = 4t$ This is the given formula.
$ = 4(3)$ Substitute 3 for t, the time.
$ = 12$ Do the multiplication.

In 3 hours, the hiker can walk 12 miles. This result and others are shown in the table and graph below.

$d = 4t$

Both the table and the graph show the time-distance relationship for the hiker as paired data.

- In 1 hour, the hiker can walk a distance of 4 miles.
- In 2 hours, the hiker can walk a distance of 8 miles.
- In 3 hours, the hiker can walk a distance of 12 miles.
- In 4 hours, the hiker can walk a distance of 16 miles.
- In 5 hours, the hiker can walk a distance of 20 miles.

In the next two sections, we will discuss how to construct tables and graphs like those shown above.

1 Determine whether an ordered pair is a solution of an equation.

So far, we have worked with **equations in one variable**. For example, $x + 3 = 9$ is an equation in x. If we subtract 3 from both sides, we see that 6 is the solution. To check, we replace x with 6 and note that the result is a true statement: $9 = 9$.

We will now extend our equation-solving skills to find solutions of **equations in two variables**. To begin, let's consider $2x + y = 12$, an equation in x and y.

A solution of $2x + y = 12$ is a pair of values, one for x and one for y, that make the equation true. To illustrate, suppose x is 3 and y is 6. Then we have:

$2x + y = 12$ This is the given equation.
$2(3) + 6 \stackrel{?}{=} 12$ Substitute 3 for x and 6 for y.
$6 + 6 \stackrel{?}{=} 12$ Do the multiplication: $2(3) = 6$.
$12 = 12$ Do the addition: $6 + 6 = 12$.

Since the result 12 = 12 is a true statement, $x = 3$ and $y = 6$ is a solution of $2x + y = 12$. We write the solution as the **ordered pair** (3, 6), with the value of x listed first. We say that (3, 6) *satisfies* the equation.

An ordered pair: (3, 6)
$\quad\quad\quad\quad\quad\quad\quad x \uparrow\ \uparrow y$

In general, a **solution of an equation in two variables** is an ordered pair of numbers that makes the equation a true statement.

> *Caution!* Don't be confused by this new use of parentheses. (3, 6) represents an ordered pair, whereas 3(6) indicates multiplication.

Self Check 1
Is $(-4, 2)$ a solution of $2x - y = -10$? yes

Now Try Problem 27

Teaching Example 1 Is $(1, -5)$ a solution of $3x + y = -2$?
Answer:
yes

EXAMPLE 1
Is $(-2, 4)$ a solution of $3x - 4y = -22$?

Strategy We will substitute -2 for x and 4 for y and see whether the resulting equation is true.

WHY An ordered pair is a *solution* of $3x - 4y = -22$ if replacing the variables with the values of the ordered pair results in a true statement.

Solution

$$3x - 4y = -22 \quad \text{This is the given equation.}$$
$$3(-2) - 4(4) \stackrel{?}{=} -22 \quad \text{Substitute } -2 \text{ for } x \text{ and } 4 \text{ for } y.$$
$$-6 - 16 \stackrel{?}{=} -22 \quad \text{On the left side, do the multiplication.}$$
$$-6 + (-16) \stackrel{?}{=} -22 \quad \text{Write the subtraction as addition of the opposite.}$$
$$-22 = -22 \quad \text{Do the addition: } -6 + (-16) = -22.$$

Since $-22 = -22$ is a true statement, $(-2, 4)$ is a solution of $3x - 4y = -22$.

Self Check 2
Is (8, 9) a solution of $y = x - 1$? no

Now Try Problem 35

Teaching Example 2 Is $(-5, -2)$ a solution of $y = x + 7$?
Answer:
no

EXAMPLE 2
Is $(-1, -3)$ a solution of $y = x - 1$?

Strategy We will substitute -1 for x and -3 for y in $y = x - 1$ and see whether the resulting equation is true.

WHY An ordered pair is a solution of $y = x - 1$ if replacing the variables with the values of the ordered pair results in a true statement.

Solution

$$y = x - 1 \quad \text{This is the given equation.}$$
$$-3 \stackrel{?}{=} -1 - 1 \quad \text{Substitute } -1 \text{ for } x \text{ and } -3 \text{ for } y.$$
$$-3 \stackrel{?}{=} -1 + (-1) \quad \text{Write the subtraction as addition of the opposite.}$$
$$-3 = -2 \quad \text{Do the addition.}$$

Since $-3 = -2$ is false, $(-1, -3)$ is not a solution of $y = x - 1$.

2 Complete ordered-pair solutions of equations.

If only one of the values of an ordered-pair solution is known, we can substitute it into the equation to find the other value.

EXAMPLE 3

Complete the following ordered pairs so that each one is a solution of the equation $4x + 2y = 2$.

a. (0,   )   **b.** (, 2)

Strategy In each case, we will substitute the known value of the solution into $4x + 2y = 2$.

WHY Then we can solve the resulting equation in one variable to find the unknown value of the solution.

Solution

a. For (0,   ), we are given the x-value of the solution is 0. To find the corresponding y-value, we substitute 0 for x in $4x + 2y = 2$ and solve for y.

$4x + 2y = 2$	This is the given equation.
$4(0) + 2y = 2$	Substitute 0 for x.
$0 + 2y = 2$	On the left side, do the multiplication: $4(0) = 0$.
$2y = 2$	On the left side, do the addition: $0 + 2y = 2y$.
$\dfrac{2y}{2} = \dfrac{2}{2}$	To isolate y, undo the multiplication by 2 by dividing both sides by 2.
$y = 1$	Do the division. This is the missing y-value of the solution.

When $x = 0$, $y = 1$. The completed ordered pair is $(0, 1)$.

b. For (, 2), we are given the y-value of the solution is 2. To find the corresponding x-value, we substitute 2 for y in $4x + 2y = 2$ and solve for x.

$4x + 2y = 2$	This is the given equation.
$4x + 2(2) = 2$	Substitute 2 for y.
$4x + 4 = 2$	On the left side, do the multiplication: $2(2) = 4$.
$4x + 4 - 4 = 2 - 4$	To isolate the variable term $4x$, undo the addition of 4 by subtracting 4 from both sides.
$4x = -2$	Do the subtraction: $2 - 4 = 2 + (-4) = -2$.
$\dfrac{4x}{4} = \dfrac{-2}{4}$	To isolate x, undo the multiplication by 4 by dividing both sides by 4.
$x = -\dfrac{2}{4}$	Write the $-$ sign in front of the fraction.
$x = -\dfrac{1}{2}$	Simplify the fraction. This is the missing x-value of the solution.

When $y = 2$, $x = -\dfrac{1}{2}$. The completed ordered pair is $\left(-\dfrac{1}{2}, 2\right)$.

Self Check 3

Complete the following ordered pairs so that each one is a solution of the equation $2x - 7y = 14$.

a. (7,   ) (7, 0)   **b.** (, 3) $\left(\dfrac{35}{2}, 3\right)$

Now Try Problems 43 and 49

Teaching Example 3 Complete the following ordered pairs so that each one is a solution of the equation $5x + 3y = 15$.

a. (0,   )   **b.** (4,   )

Answers:

a. (0, 5)   **b.** $\left(4, -\dfrac{5}{3}\right)$

Solutions of equations in two variables are often listed in a **table of solutions** (or a **table of values**). The solutions of $4x + 2y = 2$ that we found in Example 3 are shown in the table below.

x	y	(x, y)
0	1	$(0, 1)$
$-\dfrac{1}{2}$	2	$\left(-\dfrac{1}{2}, 2\right)$

A table of solutions

Self Check 4

Complete the table of solutions for $3x + 2y = 5$.

x	y	(x, y)
3	−2	(3, −2)
5	−5	(5, −5)

Now Try Problem 51

Teaching Example 4 Complete the table of solutions for $5x + 2y = -9$.

x	y	(x, y)
3	−12	(3, −12)
−1	−2	(−1, −2)

x	y	(x, y)
7	−8	(7, −8)

x	y	(x, y)
7	−8	(7, −8)
−1	4	(−1, 4)

EXAMPLE 4
Complete the table of solutions for $3x + 2y = 5$.

x	y	(x, y)
7		(7,)
	4	(, 4)

Strategy In each case we will substitute the known value of the solution into the equation $3x + 2y = 5$.

WHY Then we can solve the resulting equation in one variable to find the unknown value of the solution.

Solution In the first row of the table, we are given an *x*-value of 7. To find the corresponding *y*-value, we substitute 7 for *x* and solve for *y*.

$3x + 2y = 5$ This is the given equation.
$3(7) + 2y = 5$ Substitute 7 for x.
$21 + 2y = 5$ On the left side, do the multiplication: 3(7) = 21.
$21 + 2y - 21 = 5 - 21$ To isolate the variable term 2y, subtract 21 from both sides.
$2y = -16$ Do the subtraction: 5 − 21 = 5 + (−21) = −16.
$\frac{2y}{2} = \frac{-16}{2}$ To isolate y, undo the multiplication by 2 by dividing both sides by 2.
$y = -8$ Do the division. This is the missing y-value of the solution.

When $x = 7$, $y = -8$. The completed ordered pair $(7, -8)$ is entered in the first row in the table on the left.

In the second row of the table, we are given a *y*-value of 4. To find the corresponding *x*-value, we substitute 4 for *y* and solve for *x*.

$3x + 2y = 5$ This is the given equation.
$3x + 2(4) = 5$ Substitute 4 for y.
$3x + 8 = 5$ On the left side, do the multiplication: 2(4) = 8.
$3x + 8 - 8 = 5 - 8$ To isolate the variable term 3x, subtract 8 from both sides.
$3x = -3$ Do the subtraction: 5 − 8 = 5 + (−8) = −3.
$\frac{3x}{3} = \frac{-3}{3}$ To isolate x, undo the multiplication by 3 by dividing both sides by 3.
$x = -1$ Do the division. This is the missing x-value of the solution.

When $y = 4$, $x = -1$. The completed ordered pair $(-1, 4)$ is entered in the second row of the table on the left.

We have seen that solutions of an equation containing two variables are ordered pairs and that the ordered pairs can be listed in a table. We will now introduce a way to represent ordered pairs as points on a graph.

3 Construct a rectangular coordinate system.

Ordered pairs of numbers can be displayed on a grid called a **rectangular coordinate system**. This system is also called the *Cartesian coordinate system* after its developer, René Descartes, a 17th-century French mathematician.

> **The Language of Algebra** A rectangular coordinate system is a *grid*—a network of uniformly spaced lines. At times, some large U.S. cities have such horrible traffic congestion that vehicles can barely move, if at all. The condition is called *gridlock*.

A rectangular coordinate system is formed by two perpendicular number lines. The horizontal number line is usually called the **x-axis**, and the vertical number line is usually called the **y-axis**. On the x-axis, the positive direction is to the right. On the y-axis, the positive direction is upward.

> **The Language of Algebra** The word *axis* is used in mathematics and science. For example, Earth rotates on its *axis* once every 24 hours. The plural of *axis* is **axes**.

The point where the axes intersect is called the **origin**. This is the zero point on each axis. The axes form a **coordinate plane**, and they divide it into four regions called **quadrants**, which are numbered counterclockwise using the Roman numerals I, II, III, and IV.

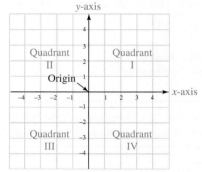

4 Plot ordered pairs and determine the coordinates of a point.

Each point in a coordinate plane can be identified by an **ordered pair** of numbers x and y written in the form (x, y). The first number in the pair is called the **x-coordinate** and the second number is called the **y-coordinate**. Some examples of such pairs are $(3, 2)$, $(-4, 4)$, $(-2, -3)$, and $(4, -4)$.

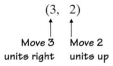

The process of locating a point on a rectangular coordinate system is called **graphing** or **plotting** the point. Here are four examples:

- Red arrows are used to show how to graph (plot) the point $(3, 2)$ on a rectangular coordinate system. Since the x-coordinate is positive, we start at the origin and move 3 units to the *right* along the x-axis. Since the y-coordinate is positive, we then move *up* 2 units and draw a dot. This locates the point $(3, 2)$.

- Blue arrows are used to show how to graph (plot) the point $(-4, 4)$. Since the x-coordinate is negative, we start at the origin and move 4 units to the *left* along the x-axis. Since the y-coordinate is positive, we then move *up* 4 units and draw a dot. This locates the point $(-4, 4)$.

$$(-4, \quad 4)$$
$$\uparrow \quad \uparrow$$
Move 4   Move 4
units left   units up

Unless otherwise noted, all art on this page is © Cengage Learning.

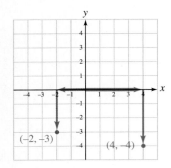

- Purple arrows are used to show how to graph (plot) the point $(-2, -3)$ on a rectangular coordinate system. Since the x-coordinate is negative, we start at the origin and move 2 units to the *left* along the x-axis. Since the y-coordinate is negative, we then move *down* 3 units and draw a dot. This locates the point $(-2, -3)$.

- Green arrows are used to show how to graph (plot) the point $(4, -4)$. Since the x-coordinate is positive, we start at the origin and move 4 units to the *right* along the x-axis. Since the y-coordinate is negative, we then move *down* 4 units and draw a dot. This locates the point $(4, -4)$.

Caution! The order of the coordinates of a point is important. The point with coordinates $(-4, 4)$ is not the same as the point with coordinates $(4, -4)$.

Success Tip Points with an x-coordinate that is 0 lie on the y-axis. Points with a y-coordinate that is 0 lie on the x-axis. Points that lie on an axis are not considered to be in any quadrant.

Self Check 5

Graph (plot) the points $(2, -2)$, $(-4, 0)$, $\left(1.5, \frac{5}{2}\right)$, and $(0, 5)$.

Now Try Problem 55

Teaching Example 5
Graph (plot) the points $(5, 4)$, $(6, -3)$, $(0, 4.5)$, $(0, -1)$, $(-6, 0)$ and $\left(-2, -\frac{5}{2}\right)$.

EXAMPLE 5 Graph (plot) each point. Then state the quadrant in which it lies or the axis on which it lies. **a.** $(4, 4)$ **b.** $(2, -3)$ **c.** $(0, 2.5)$ **d.** $(-3, 0)$ **e.** $(0, 0)$ **f.** $\left(-1, -\frac{7}{2}\right)$

Strategy After identifying the x- and y-coordinates of the ordered pair, we will move the corresponding number of units left or right and up or down to locate the point.

WHY The coordinates of a point determine its location on the coordinate plane.

Solution

a. Since the x-coordinate, 4, is positive, we start at the origin and move 4 units to the *right* along the x-axis. Since the y-coordinate, 4, is positive, we then move *up* 4 units and draw a dot. This locates the point $(4, 4)$. The point lies in quadrant I.

b. To plot $(2, -3)$, we begin at the origin and move 2 units to the *right*, because the x-coordinate is 2. Then, since the y-coordinate is negative, we move *down* 3 units. The point lies in quadrant IV.

c. To plot $(0, 2.5)$, we begin at the origin and do not move right or left, because the x-coordinate is 0. Since the y-coordinate is positive, we move 2.5 units *up*. The point lies on the y-axis.

d. To plot $(-3, 0)$, we begin at the origin and move 3 units to the *left*, because the x-coordinate is -3. Since the y-coordinate is 0, we do not move up or down. The point lies on the x-axis.

e. To plot $(0, 0)$, we begin at the origin, and we remain there because both coordinates are 0. The point with coordinates $(0, 0)$ is the origin.

f. To plot $\left(-1, -\frac{7}{2}\right)$, we begin at the origin and move 1 unit to the *left*, because the x-coordinate is -1. The y-coordinate of the given point is negative. To better understand how many units to move *down*, we note that $-\frac{7}{2} = -3\frac{1}{2}$. After moving down $3\frac{1}{2}$ units, we draw a dot. The point lies in quadrant III.

$$\begin{array}{r} 3 \\ 2\overline{\smash{)}7} \\ \underline{-6} \\ 1 \end{array}$$

> **Success Tip** To graph the point $\left(-1, -\frac{7}{2}\right)$ in Example 5, part f, we expressed the y-coordinate as $-3\frac{1}{2}$. When graphing, if the x- or y-coordinate of a point is an improper fraction, it is helpful to express such a coordinate in equivalent mixed-number or decimal form. Thus, $\left(-1, -\frac{7}{2}\right), \left(-1, -3\frac{1}{2}\right)$, and $(-1, -3.5)$ all name the same point.

To find the coordinates of a point on the rectangular coordinate system, we use the numbering (scaling) on the x- and y-axes.

> **The Language of Algebra** Points are often labeled with capital letters. For example, the notation $A(2, 3)$ indicates that point A has coordinates $(2, 3)$.

EXAMPLE 6 Find the coordinates of points $A, B, C, D, E,$ and F plotted in figure (a) below.

Self Check 6
Find the coordinates of each point in figure (b) of Example 6.
Now Try Problem 59
Self Check 6 Answer
$A(4, 0), B(0, 1), C(-3.5, -2.5)$ or $C\left(-3\frac{1}{2}, -2\frac{1}{2}\right), D(2, -4)$

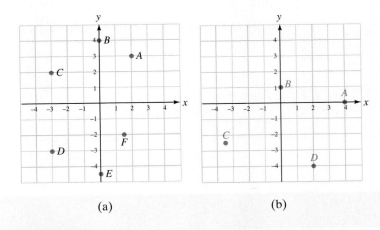

(a)   (b)

Strategy We will start at the origin and count to the left or right on the x-axis, and then up or down to reach each point.

WHY The movement left or right gives the x-coordinate of the ordered pair and the movement up or down gives the y-coordinate.

Solution To locate point A, we start at the origin, move 2 units to the right on the x-axis, and then 3 units up. Its coordinates are $(2, 3)$. The coordinates of the other points are found in the same manner.

$B(0, 4), \quad C(-3, 2), \quad D(-3, -3), \quad E(0, -4.5) \quad \text{or} \quad E\left(0, -4\frac{1}{2}\right), \quad F(1.5, -2) \quad \text{or}$

$F\left(1\frac{1}{2}, -2\right)$

THINK IT THROUGH   Population Shift

"Since 1950, the median center of the U.S. population has moved south and west at every census."

U.S. Census Bureau

In the illustration below, data from the 2010 census were used to draw a north–south line and an east–west line, so equal numbers of the nation's population lived in each "quadrant." The point of intersection of the lines occurs in Pike County, Indiana. It could be thought of as the "center" of the U.S. population in 2010. If the 2010 census recorded the population of the United States to be about 308,745,500, how many people lived in each quadrant created by the lines in the illustration? 77,186,375

Median U. S. population center, southwest of Petersburg, Pike County, Indiana

Source: U.S. Census Bureau

ANSWERS TO SELF CHECKS

1. yes **2.** no **3. a.** (7, 0) **b.** $\left(\dfrac{35}{2}, 3\right)$ **4.** 3, 3, −5, −5

5.

6. $A(4, 0), B(0, 1), C(-3.5, -2.5)$ or $C\left(-3\tfrac{1}{2}, -2\tfrac{1}{2}\right), D(2, -4)$

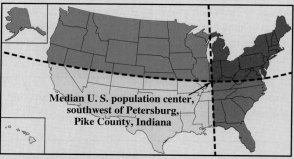

SECTION 8.3   STUDY SET

VOCABULARY

Fill in the blanks.

1. $x + y = 4$ is an equation in two ___variables___.
2. A ___solution___ of an equation in two variables is an ordered pair of numbers that makes the equation a true statement.
3. When we substitute 1 for x and 3 for y in the equation $x + y = 4$, the result is the true statement $4 = 4$. We say that $(1, 3)$ ___satisfies___ the equation.
4. $x = 2$ and $y = 3$ is a solution of the equation $x + y = 5$. We can write the solution as the ___ordered___ pair $(2, 3)$.

5. Solutions of equations in two variables can be listed in a ___table___ of solutions like the one shown below.

x	y	(x, y)
1	3	(1, 3)
4	7	(4, 7)

6. A rectangular coordinate ___system___ is shown below. Label the x-axis and the y-axis.

7. The point with coordinates $(0, 0)$ is called the ___origin___.
8. The x- and y-axes divide the rectangular coordinate system into four regions called ___quadrants___.
9. In the ordered pair $(-2, 4)$, -2 is the ___x___-coordinate and 4 is the ___y___-coordinate.
10. The process of locating a point on a rectangular coordinate system is called graphing or ___plotting___ the point.

CONCEPTS

11. BURNING CALORIES The table below shows the number of calories a 140-pound woman would burn doing light activities such as office work, cleaning house, or playing golf. Create a graph of the paired data.

Minutes of activity	Calories burned
1	4
2	8
3	12
4	16

12. CONCRETE The graph in the next column shows the number of parts of sand that should be used for a given number of parts of cement when mixing concrete for a walkway. Create a table of the paired data.

Parts cement	Parts sand
2	5
4	10
6	15
8	20

Fill in the blanks.

13. a. $2x + 5 = 10$ is an equation in ___one___ variable.
    b. $2x + 5y = 10$ is an equation in ___two___ variables.
14. If only one of the values of an ordered-pair solution is known, we can ___substitute___ it into the equation to find the other value.
15. a. On the x-axis, the positive direction is to the ___right___.
    b. On the y-axis, the positive direction is ___upward___.
16. To plot the point $\left(\frac{9}{2}, 6\right)$, it is helpful to write the x-coordinate as the mixed number $4\frac{1}{2}$ or as the decimal ___4.5___.
17. To plot the point $(3, -4)$, we start at the ___origin___ and move 3 units to the ___right___ on the x-axis and then move 4 units ___down___.
18. To plot the point $(-2, 3)$, we start at the ___origin___ and move 2 units to the ___left___ on the x-axis and then move 3 units ___up___.
19. Refer to the graph below. In which quadrant does each point lie? If a point does not lie in a quadrant, tell on what axis (or axes) it lies.

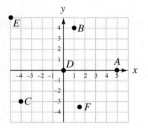

A: x-axis, B: quadrant I, C: quadrant III, D: x-axis and y-axis, E: quadrant II, F: quadrant IV

20. In which quadrant does each point lie?
    a. $(-1, 2.5)$ quadrant II b. $\left(6, -\frac{5}{2}\right)$ quadrant IV
    c. $(8, 10)$ quadrant I d. $(-2, -3)$ quadrant III

NOTATION

21. Label each coordinate with the correct letter, x or y.

(3, 6)
x y

22. Label the top row of the table of solutions shown below with the correct letters.

x	y	(x, y)
2	5	(2, 5)
6	1	(6, 1)

23. Does the ordered pair $\left(1\frac{1}{3}, -\frac{5}{2}\right)$ name the same point as $\left(\frac{4}{3}, -2.5\right)$? yes

24. List the Roman numerals from 1 to 4. What are they used to label? I, II, III, IV; the quadrants

Complete each step.

25. For the equation $4x + 3y = 14$, find the value of y when $x = 2$.

$$4x + 3y = 14$$
$$4(\;2\;) + 3y = 14$$
$$\;8\; + 3y = 14$$
$$8 + 3y - \;8\; = 14 - \;8\;$$
$$3y = \;6\;$$
$$\frac{3y}{3} = \frac{6}{3}$$
$$y = \;2\;$$

26. For the equation $2x - 5y = 20$, find the value of x when $y = 2$.

$$2x - 5y = 20$$
$$2x - 5(\;2\;) = 20$$
$$2x - \;10\; = 20$$
$$2x - 10 + \;10\; = 20 + \;10\;$$
$$2x = \;30\;$$
$$\frac{2x}{2} = \frac{30}{2}$$
$$x = \;15\;$$

GUIDED PRACTICE

Determine whether the given ordered pair is a solution of the equation. See Example 1.

27. Is $(2, 3)$ a solution of $2x + 3y = 13$? yes
28. Is $(4, 1)$ a solution of $3x - 2y = 10$? yes
29. Is $(9, -3)$ a solution of $x + 5y = -7$? no
30. Is $(3, -1)$ a solution of $x + 6y = -1$? no
31. Is $(-9, -9)$ a solution of $-3x + 4y = -11$? no
32. Is $(-6, -4)$ a solution of $-11x + 6y = 40$? no
33. Is $(0, 0)$ a solution of $10x - y = 0$? yes
34. Is $(0, 0)$ a solution of $6x + 7y = 0$? yes

Determine whether the given ordered pair is a solution of the equation. See Example 2.

35. Is $(2, 1)$ a solution of $y = 5x - 4$? no
36. Is $(5, 8)$ a solution of $y = 2x - 4$? no
37. Is $(-2, 6)$ a solution of $y = -x + 4$? yes
38. Is $(-3, 7)$ a solution of $y = -x + 4$? yes
39. Is $(-4, 25)$ a solution of $y = -3x + 13$? yes
40. Is $(-6, 20)$ a solution of $y = -3x + 2$? yes
41. Is $(0, -9)$ a solution of $y = 7x - 9$? yes
42. Is $(9, 0)$ a solution of $y = 5x - 9$? no

Complete the following ordered pairs so that each one is a solution of the given equation. See Example 3.

43. $2x + y = 8$
  a. $(0, \;8\;)$
  b. $(\;3\;, 2)$

44. $3x + y = 14$
  a. $(0, \;14\;)$
  b. $(\;4\;, 2)$

45. $x - 3y = 5$
  a. $(2, \;-1\;)$
  b. $(\;14\;, 3)$

46. $x - 7y = 10$
  a. $(3, \;-1\;)$
  b. $(\;38\;, 4)$

47. $5x + 3y = 15$
  a. $(-6, \;15\;)$
  b. $\left(\frac{3}{5}, 4\right)$

48. $9x + 4y = 36$
  a. $(-4, \;18\;)$
  b. $\left(\frac{4}{9}, 8\right)$

49. $8x - 15y = -4$
  a. $\left(1, \frac{4}{5}\right)$
  b. $(\;-8\;, -4)$

50. $7x - 4y = 9$
  a. $\left(1, -\frac{1}{2}\right)$
  b. $(\;3\;, 3)$

Complete each table of solutions for the given equation. See Example 4.

51. $4x + 3y = 24$

x	y	(x, y)
0	8	(0, 8)
6	0	(6, 0)
-3	12	(-3, 12)

52. $3x + y = 12$

x	y	(x, y)
0	12	(0, 12)
4	0	(4, 0)
−6	30	(−6, 30)

53. $5x − 4y = 20$

x	y	(x, y)
0	−5	(0, −5)
4	0	(4, 0)
−4	−10	(−4, −10)

54. $7x − 3y = 21$

x	y	(x, y)
0	−7	(0, −7)
3	0	(3, 0)
−3	−14	(−3, −14)

Graph (plot) each point on a rectangular coordinate system. See Example 5.

55. $(1, 3), (−2, 4), (4, 0), (0, 1), (−3, −2), (0, −5), (3, −2), (0, 0)$

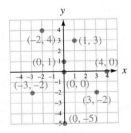

56. $\left(-\frac{3}{2}, 2\right), \left(3, -\frac{5}{2}\right), (0, 0), (0, 2), (0, −3), (3, 0), (−4, −1), (4.5, 4)$

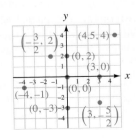

57. $(−4, −3), (1.5, 1.5), \left(\frac{7}{2}, 4\right), (0, −1), (−3.5, 0), (0, 3.5), \left(0, -\frac{9}{2}\right), (5, −2)$

58. $(0, 0), \left(-\frac{1}{2}, \frac{5}{2}\right), (3.5, 0), \left(-\frac{7}{2}, −5\right), (5, −5), (−5, 5), \left(5, -\frac{3}{2}\right), (2, 4)$

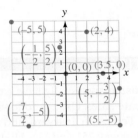

Find the coordinates of each point shown in the graph. See Example 6.

59.

$A(2, 4), B(−3, 3), C(−2, −3), D(4, −3), E(3, 0), F(0, −1)$

60.

$A(−2.5, 3)$ or $A\left(-2\frac{1}{2}, 3\right)$,
$B(2, −3.5)$ or $B\left(2, -3\frac{1}{2}\right)$,
$C(0, −4), D(4, 3), E(−4, 0), F(4, −1)$

61.

$A(−3, −4), B(2.5, 3.5)$ or $B\left(2\frac{1}{2}, 3\frac{1}{2}\right), C(−2.5, 0)$ or $C\left(-2\frac{1}{2}, 0\right), D(2.5, 0)$ or $D\left(2\frac{1}{2}, 0\right), E(3, −3), F(0, 3)$

62.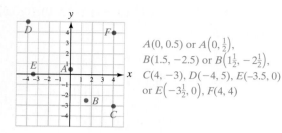

$A(0, 0.5)$ or $A\left(0, \frac{1}{2}\right)$,
$B(1.5, -2.5)$ or $B\left(1\frac{1}{2}, -2\frac{1}{2}\right)$,
$C(4, -3)$, $D(-4, 5)$, $E(-3.5, 0)$ or $E\left(-3\frac{1}{2}, 0\right)$, $F(4, 4)$

TRY IT YOURSELF

Use a check to determine whether the given ordered pair is a solution of the equation.

63. $3x - 6y = 12; (-3.6, -3.8)$ yes
64. $8x + 4y = 10; (-0.5, 3.5)$ yes
65. $y - 6x = 12; \left(\frac{5}{6}, 7\right)$ no
66. $y + 8x = 4; \left(\frac{3}{4}, 2\right)$ no

APPLICATIONS

67. **MAPS** Road maps usually have a coordinate system to help locate cities. Use the following map to locate Rockford, Forreston, Harvard, and the intersection of State Highway 251 and U.S. Highway 30. Express each answer in the form (number, letter).
Rockford (5, B), Forreston (2, C), Harvard (7, A), intersection (5, E)

68. **GAMES** In the game Battleship, coordinates are used to locate ships. What are the coordinates of three ordered pairs that locate the ship shown below? Express each answer in the form (letter, number).
(G, 2), (G, 3), (G, 4)

69. **EARTHQUAKES** On the map below, the circular area that is shaded blue shows where damage was caused by an earthquake. Important roads and freeways are also labeled.
   a. Find the coordinates of the epicenter (the source of the quake). $(2, -1)$
   b. Was damage done at the point $(4, 5)$? no
   c. Was damage done at the point $(-1, -4)$? yes
   d. Did Highway 220 suffer any damage? no

70. **AUTOMATION** A robot can be programmed to make welds on a car frame. To do this, an imaginary coordinate system is superimposed on the side of the car. Using the commands Left, Right, Up, and Down write a set of instructions for the robot arm to move from its beginning position to weld the points A, B, C, and D, in that order. right 2, up 3; right 4; left 2, down 3; right 5, up 2 (answers may vary)

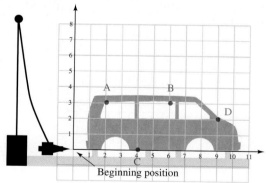

71. **THE GLOBE** A coordinate system that is used to locate places on the surface of Earth uses a series of curved lines running north and south and east and west, as shown on the next page. List the cities in order, beginning with the one that is farthest east on this map.
New Delhi, Kampala, Coats Land, Reykjavik, Buenos Aires, Havana

8.3 Equations in Two Variables; The Rectangular Coordinate System

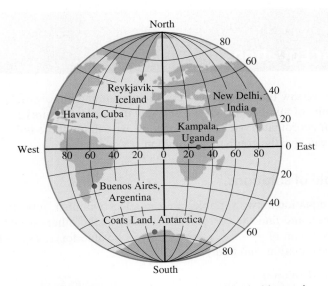

72. BLOOD TRANSFUSIONS The red shaded boxes in the illustration below show which pairs of the major blood groups (AB, A, B, and O) can be mixed without clumping occurring. List all of the ordered pairs of blood types that do not clump when combined. Express your answers in the form (donor blood type, recipient blood type). (AB, AB), (A, AB), (B, AB), (O, AB), (A, A), (O, A), (B, B), (O, B), (O, O)

73. COOKING Use the information from the table below to complete the graph for 2, 4, 6, 8, and 10 servings of instant mashed potatoes.

Number of servings	2	4	6	8	10
Flakes (cups)	$\frac{2}{3}$	$1\frac{1}{3}$	2	$2\frac{2}{3}$	$3\frac{1}{3}$

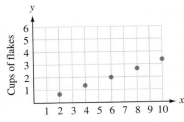

74. DICE The red point in figure (a) represents one of the 36 possible outcomes when two fair dice are rolled a single time. Draw the correct number of dots on the top face of each die in figure (b) to illustrate this outcome. 1st die: 5 dots, 2nd die: 2 dots

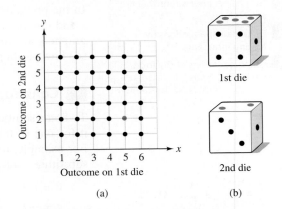

(a)   (b)

WRITING

75. Explain why the point with coordinates $(-4, 4)$ is not the same as the point with coordinates $(4, -4)$.

76. Explain the difference between $2(4)$ and $(2, 4)$.

77. Explain how to plot the point with coordinates of $(1, -7)$.

78. Explain how to plot the point with coordinates of $\left(-\frac{9}{2}, 5\right)$.

79. Explain why the coordinates of the origin are $(0, 0)$.

80. Explain the diagram shown below.

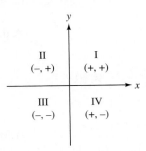

REVIEW

Evaluate each expression.

81. $(-8 - 5) - 3$ -16

82. $-1 - [5 - (-3)]$ -9

83. $-4^2 - 3^2$ -25

84. $-5 - \frac{24}{6} - 8(-3)$ 15

Solve each equation and check the result.

85. $\frac{x}{3} + 3 = 10$ 21

86. $-3x - 4 = 8$ -4

87. $5 - (7 - x) = -5$ -3

88. $2(y + 6) - 4 = 2$ -3

Objectives

1. Construct a table of solutions.
2. Graph linear equations that are solved for y.
3. Graph linear equations by finding intercepts.
4. Graph equations of the form $y = b$ and $x = a$.

SECTION 8.4
Graphing Linear Equations

In the previous section, we saw that solutions of equations containing the variables x and y were ordered pairs of numbers (x, y). We also saw that ordered pairs can be graphed on a rectangular coordinate system. In this section, we will use these skills to see how plotting points can give the graph of an equation.

1 Construct a table of solutions.

To find a solution of an equation in two variables, we can select a number, substitute it for one of the variables, and find the corresponding value of the other variable. For example, to find a solution of $y = x - 1$, we can select a value for x, say, -4, substitute -4 for x in the equation, and find y.

x	y	(x, y)
-4	-5	$(-4, -5)$

$y = x - 1$ This is the given equation.
$y = -4 - 1$ Substitute -4 for x.
$y = -5$ Do the subtraction: $-4 - 1 = -4 + (-1) = -5$.

The ordered pair $(-4, -5)$ is a solution of $y = x - 1$. We list it in the table on the left.
To find another solution of $y = x - 1$, we select another value for x, say, -2, and find the corresponding y-value.

x	y	(x, y)
-4	-5	$(-4, -5)$
-2	-3	$(-2, -3)$

$y = x - 1$ This is the given equation.
$y = -2 - 1$ Substitute -2 for x.
$y = -3$ Do the subtraction: $-2 - 1 = -2 + (-1) = -3$.

A second solution is $(-2, -3)$, and we list it in the table of solutions.
If we let $x = 0$, we can find a third ordered pair that satisfies $y = x - 1$.

x	y	(x, y)
-4	-5	$(-4, -5)$
-2	-3	$(-2, -3)$
0	-1	$(0, -1)$

$y = x - 1$ This is the given equation.
$y = 0 - 1$ Substitute 0 for x.
$y = -1$ Do the subtraction: $0 - 1 = 0 + (-1) = -1$.

A third solution is $(0, -1)$, which we also add to the table of solutions.
We can find a fourth solution by letting $x = 2$, and a fifth solution by letting $x = 4$.

x	y	(x, y)
-4	-5	$(-4, -5)$
-2	-3	$(-2, -3)$
0	-1	$(0, -1)$
2	1	$(2, 1)$
4	3	$(4, 3)$

$y = x - 1$ $y = x - 1$
$y = 2 - 1$ Substitute 2 for x. $y = 4 - 1$ Substitute 4 for x.
$y = 1$ Subtract. $y = 3$ Subtract.

A fourth solution is $(2, 1)$ and a fifth solution is $(4, 3)$. We add them to the table.
Since we can choose any number for x, and since any choice of x will give a corresponding value of y, it is apparent that the equation $y = x - 1$ has *infinitely many solutions*. We have found five of them: $(-4, -5), (-2, -3), (0, -1), (2, 1)$, and $(4, 3)$.

2 Graph linear equations that are solved for y.

It is impossible to list the infinitely many solutions of the equation $y = x - 1$ in a table. However, to show all of its solutions, we can draw a mathematical "picture" of them. We call this picture the *graph of the equation*.
To graph $y = x - 1$, we plot the ordered pairs shown in the table above on a rectangular coordinate system. Then we draw a straight line through the points,

because the graph of any solution of $y = x - 1$ will lie on this line. Furthermore, every point on this line represents a solution. We call the line the **graph of the equation**. It represents all of the solutions of $y = x - 1$. The graph below only shows a part of the line. The arrowheads indicate that it extends indefinitely in both directions.

$y = x - 1$

x	y	(x, y)
-4	-5	$(-4, -5)$
-2	-3	$(-2, -3)$
0	-1	$(0, -1)$
2	1	$(2, 1)$
4	3	$(4, 3)$

↑ Select x ↑ Find y ↑ Plot (x, y)

Construct a table of solutions.

Plot the ordered pairs.

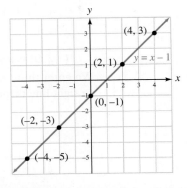

Draw a straight line through the points. This is the *graph of the equation*.

The equation $y = x - 1$ is said to be *linear* and its graph is a straight line. Some more examples of linear equations are

$$y = 2x + 4, \quad x - 3y = 6, \quad \text{and} \quad 40x + 3y = -120$$

When we graphed $y = x - 1$, we did more work than necessary. Since *two points determine a line*, only two points are needed to graph the line. However, it is always a good idea to plot a third point as a check. If the three points do not lie on a straight line, then at least one of them is incorrect.

Linear equations can be graphed in several ways. Generally, the form in which an equation is written determines the method that we use to graph it. To graph linear equations solved for y, such as $y = x - 1$ and $y = 2x + 4$, we can use the following method.

Graphing Linear Equations Solved for y by Plotting Points

1. Find three ordered pairs that are solutions of the equation by selecting three values for x and calculating the corresponding values of y.
2. Plot the solutions on a rectangular coordinate system.
3. Draw a straight line passing through the points. If the points do not lie on a line, check your calculations.

EXAMPLE 1 Graph: $y = 2x + 4$

Strategy We will find three solutions of the given equation, plot them on a rectangular coordinate system, and then draw a straight line passing through the points.

WHY To *graph* a linear equation in two variables means to make a drawing that represents all of its solutions.

Self Check 1

Graph: $y = 2x - 2$

Now Try Problem 23

Teaching Example 1 Graph:
$y = 3x - 1$
Answer:

Solution To find three solutions of $y = 2x + 4$, we select three values for x that will make the calculations easy. Then we find each corresponding value of y.

If $x = -2$:	If $x = 0$:	If $x = 2$:
$y = 2x + 4$	$y = 2x + 4$	$y = 2x + 4$
$y = 2(-2) + 4$	$y = 2(0) + 4$	$y = 2(2) + 4$
$y = -4 + 4$	$y = 0 + 4$	$y = 4 + 4$
$y = 0$	$y = 4$	$y = 8$
$(-2, 0)$ is a solution.	$(0, 4)$ is a solution.	$(2, 8)$ is a solution.

We enter these results in a table of solutions and plot the points. Then we draw a straight line through the points and label it $y = 2x + 4$.

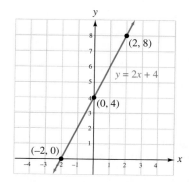

As a check, we can pick two points that the line appears to pass through, such as $(1, 6)$ and $(-1, 2)$. When we substitute their coordinates into the given equation, the two true statements that result indicate that $(1, 6)$ and $(-1, 2)$ are solutions and that the graph of the line is correctly drawn.

Check $(1, 6)$: $\quad y = 2x + 4 \qquad$ **Check $(-1, 2)$:** $\quad y = 2x + 4$

$\qquad\qquad\qquad 6 \stackrel{?}{=} 2(1) + 4 \qquad\qquad\qquad\qquad\quad 2 \stackrel{?}{=} 2(-1) + 4$

$\qquad\qquad\qquad 6 \stackrel{?}{=} 2 + 4 \qquad\qquad\qquad\qquad\qquad 2 \stackrel{?}{=} -2 + 4$

$\qquad\qquad\qquad 6 = 6 \qquad$ True $\qquad\qquad\qquad\quad 2 = 2 \qquad$ True

> **Success Tip** When selecting x-values for a table of solutions, a rule of thumb is to choose a negative number, a positive number, and 0. When $x = 0$, the calculations to find y are usually quite simple.

EXAMPLE 2
Graph: $y = -\dfrac{1}{3}x - 2$

Strategy We will find three solutions of the given equation, plot them on a rectangular coordinate system, and then draw a straight line passing through the points.

WHY To *graph* a linear equation in two variables means to make a drawing that represents all of its solutions.

Solution To find three solutions of $y = -\frac{1}{3}x - 2$, each value of x must be multiplied by $-\frac{1}{3}$. This calculation is made easier if we select x-values that are multiples of the denominator 3, such as $-3, 0,$ and 3.

If $x = -3$:
$$y = -\frac{1}{3}x - 2$$
$$y = -\frac{1}{3}(-3) - 2$$
$$y = -\frac{1}{3}\left(-\frac{3}{1}\right) - 2$$
$$y = \frac{3}{3} - 2$$
$$y = 1 - 2$$
$$y = -1$$
$(-3, -1)$ is a solution.

If $x = 0$:
$$y = -\frac{1}{3}x - 2$$
$$y = -\frac{1}{3}(0) - 2$$
$$y = 0 - 2$$
$$y = -2$$
$(0, -2)$ is a solution.

If $x = 3$:
$$y = -\frac{1}{3}x - 2$$
$$y = -\frac{1}{3}(3) - 2$$
$$y = -\frac{1}{3}\left(\frac{3}{1}\right) - 2$$
$$y = -\frac{3}{3} - 2$$
$$y = -1 - 2$$
$$y = -3$$
$(3, -3)$ is a solution.

We enter these results in a table of solutions and plot the points. Then we draw a straight line through the points and label it $y = -\frac{1}{3}x - 2$.

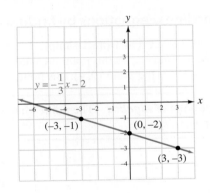

Self Check 2
Graph: $y = -\frac{1}{4}x + 3$
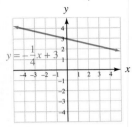

Now Try Problem 27

Teaching Example 2 Graph: $y = -\frac{1}{3}x + 4$
Answer:

Success Tip In Example 2, when we select x-values that are multiples of the denominator 3, the corresponding y-values are integers, and not difficult-to-plot fractions.

EXAMPLE 3 Graph: $y = 20x$

Strategy We will find three solutions of the given equation, plot them on a rectangular coordinate system, and then draw a straight line passing through the points.

WHY To *graph* a linear equation in two variables means to make a drawing that represents all of its solutions.

Solution We begin by selecting three values for x: $-2, 0,$ and 2. If $x = -2$, we can calculate the corresponding value of y by substituting -2 for x in $y = 20x$.

$$y = 20x \quad \text{This is the equation to graph.}$$
$$y = 20(-2) \quad \text{Substitute } -2 \text{ for } x.$$
$$y = -40 \quad \text{Do the multiplication.}$$

Self Check 3
Graph: $y = 25x$

Now Try Problem 31

Teaching Example 3 Graph: $y = 30x$
Answer:

We see that $x = -2$ and $y = -40$ is a solution of $y = 20x$. In a similar manner, we find the corresponding values for y when x is 0 and 2 and enter them in the table below.

Because of the sizes of the y-coordinates of the points $(-2, -40)$ and $(2, 40)$, we must adjust the scale on the y-axis. (If we used grid lines 1 unit apart, the graph would be very large.)

One way to make these points fit is to scale the y-axis in units of 5, 10, or 20. If we choose divisions of 20 units, plot the three solutions from the table, and draw a line through them, we get the graph shown below.

$y = 20x$

x	y	(x, y)
-2	-40	$(-2, -40)$
0	0	$(0, 0)$
2	40	$(2, 40)$

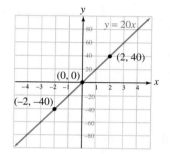

3 Graph linear equations by finding intercepts.

The graph of $y = 2x + 4$ from Example 1 is shown below. We see that the graph crosses the x-axis at $(-2, 0)$; this point is called the **x-intercept** of the graph. The graph crosses the y-axis at the point $(0, 4)$; this point is called the **y-intercept** of the graph.

> **The Language of Algebra** The point where a line intersects the x- or y-axis is called an *intercept*.

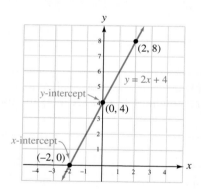

We see that the x-intercept has a y-coordinate of 0, and the y-intercept has an x-coordinate of 0. These observations suggest the following procedures for finding the intercepts of a graph from its equation.

> **Finding Intercepts**
>
> To find the y-intercept, substitute 0 for x in the given equation and solve for y.
> To find the x-intercept, substitute 0 for y in the given equation and solve for x.

Plotting the x- and y-intercepts of a graph and drawing a line through them is called the **intercept method of graphing a line**. This method is useful when graphing linear equations that have x- and y-terms on one side and a constant on the other side, such as $x - 3y = 6$ and $40x + 3y = -120$.

EXAMPLE 4   Graph $x - 3y = 6$ by finding the x- and y-intercepts.

Strategy We will let $y = 0$ to find the x-intercept. We will then let $x = 0$ to find the y-intercept of the graph.

WHY Since two points determine a line, the x-intercept and y-intercept are enough information to graph this linear equation.

Solution

x-intercept: let $y = 0$

$x - 3y = 6$
$x - 3(0) = 6$ Substitute 0 for y.
$x - 0 = 6$
$x = 6$

The x-intercept is $(6, 0)$.

y-intercept: let $x = 0$

$x - 3y = 6$
$0 - 3y = 6$ Substitute 0 for x.
$-3y = 6$
$y = -2$ To isolate y, divide both sides by -3.

The y-intercept is $(0, -2)$.

Since each intercept of the graph is a solution of the equation, we enter the intercepts in the table of solutions below.

As a check, we find one more point on the line. We select a convenient value for x, say, 3, and find the corresponding value of y. The check point should lie on the same line as the x- and y-intercepts. If it does not, check your work to find the incorrect coordinate or coordinates.

$x - 3y = 6$
$3 - 3y = 6$ Substitute 3 for x.
$-3y = 3$ To isolate the variable term, $-3y$, subtract 3 from both sides.
$y = -1$ To isolate y, divide both sides by -3.

Therefore, $(3, -1)$ is a solution. It is also entered in the table.

We plot the intercepts and the check point, draw a straight line through them, and label the line as $x - 3y = 6$.

$x - 3y = 6$

x	y	(x, y)	
6	0	$(6, 0)$	← x-intercept
0	-2	$(0, -2)$	← y-intercept
3	-1	$(3, -1)$	← Check point

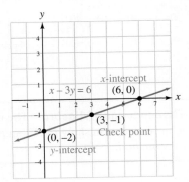

Self Check 4
Graph $x - 2y = 2$ by finding the x- and y-intercepts.

Now Try Problem 35

Teaching Example 4 Graph $2x - 3y = 6$ by finding the x- and y-intercepts.
Answer:

The calculations for finding intercepts can be simplified if we realize what occurs when we substitute 0 for y or 0 for x in a linear equation.

EXAMPLE 5   Graph $40x + 3y = -120$ by finding the x- and y-intercepts.

Strategy We will let $y = 0$ to find the x-intercept. We will then let $x = 0$ to find the y-intercept of the graph.

WHY Since two points determine a line, the x-intercept and y-intercept are enough information to graph this linear equation.

Self Check 5

Graph $32x + 5y = -160$ by finding the x- and y-intercepts.

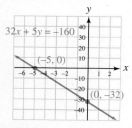

Now Try Problem 39

Teaching Example 5 Graph $40x + 7y = -280$ by finding the x- and y-intercepts.
Answer:

Solution When we substitute 0 for y, it follows that the term $3y$ will be equal to 0. Therefore, to find the x-intercept, we can cover the $3y$ and solve the remaining equation for x.

$40x + 3y = -120$ If $y = 0$, then $3y = 3(0) = 0$. Cover the $3y$ term.
$x = -3$ To solve $40x = -120$, divide both sides by 40.

The x-intercept is $(-3, 0)$.

When we substitute 0 for x, it follows that the term $40x$ will be equal to 0. Therefore, to find the y-intercept, we can cover the $40x$ and solve the remaining equation for y.

$40x + 3y = -120$ If $x = 0$, then $40x = 40(0) = 0$. Cover the $40x$ term.
$y = -40$ To solve $3y = -120$, divide both sides by 3.

The y-intercept is $(0, -40)$.

> **Caution!** When using the cover-over method to find the y-intercept, be careful not to cover the sign in front of the y-term.

We can find a third solution by selecting a convenient value for x and finding the corresponding value for y. If we choose $x = -6$, we find that $y = 40$. The solution $(-6, 40)$ is entered in the table, and the equation is graphed as shown.

> **Success Tip** To fit y-values of 40 and -40 on the graph, the y-axis was scaled in units of 10.

$40x + 3y = -120$

x	y	(x, y)	
-3	0	$(-3, 0)$	← x-intercept
0	-40	$(0, -40)$	← y-intercept
-6	40	$(-6, 40)$	← Check point

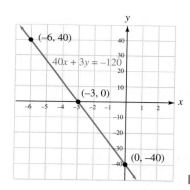

> **The Language of Algebra** The method to find the intercepts of the graph of a linear equation shown in Example 5 is commonly referred to as the *cover-over method*.

4 Graph equations of the form $y = b$ and $x = a$.

When a linear equation contains only one variable, such as $y = 4$ or $x = -2$, its graph is either a horizontal or a vertical line.

EXAMPLE 6 Graph: $y = 4$

Strategy To find three ordered-pair solutions of this equation to plot, we will select three values for x and use 4 for y each time.

WHY The given equation requires that $y = 4$.

Solution We can write the equation in the form $0x + y = 4$. Since the coefficient of x is 0, the numbers chosen for x have no effect on y. The value of y is always 4. For example, if $x = 2$, we have

$$0x + y = 4$$
$$0(2) + y = 4 \quad \text{Substitute 2 for } x.$$
$$y = 4 \quad \text{Simplify the left side.}$$

One solution is $(2, 4)$. To find two more solutions, we select $x = 0$ and $x = -3$. For any x-value, the y-value is always 4, so we enter $(0, 4)$ and $(-3, 4)$ in the table. If we plot the ordered pairs and draw a straight line through the points, the result is a horizontal line. The y-intercept is $(0, 4)$ and there is no x-intercept.

Self Check 6
Graph: $y = -2$

Now Try Problem 43

Teaching Example 6 Graph: $y = -3$
Answer:

$y = 4$

x	y	(x, y)
2	4	$(2, 4)$
0	4	$(0, 4)$
-3	4	$(-3, 4)$

↑ Select any number for x.
↑ Each value of y must be 4.

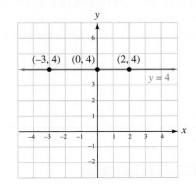

EXAMPLE 7   Graph: $x = -3$

Strategy To find three ordered-pair solutions of this equation to plot, we must select -3 for x each time.

WHY The given equation requires that $x = -3$.

Solution We can write the equation in the form $x + 0y = -3$. Since the coefficient of y is 0, the numbers chosen for y have no effect on x. The value of x is always -3. For example, if $y = -2$, we have

$$x + 0y = -3$$
$$x + 0(-2) = -3 \quad \text{Substitute } -2 \text{ for } y.$$
$$x = -3 \quad \text{Simplify the left side.}$$

One solution is $(-3, -2)$. To find two more solutions, we select $y = 0$ and $y = 3$. For any y-value, the x-value is always -3, so we enter $(-3, 0)$ and $(-3, 3)$ in the table. If we plot the ordered pairs and draw a straight line through the points, the result is a vertical line. The x-intercept is $(-3, 0)$ and there is no y-intercept.

Self Check 7
Graph: $x = -4$

Now Try Problem 47

Teaching Example 7 Graph: $x = 2$
Answer:

$x = -3$

x	y	(x, y)
-3	-2	$(-3, -2)$
-3	0	$(-3, 0)$
-3	3	$(-3, 3)$

↑ Each value of x must be -3.
↑ Select any number for y.

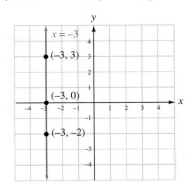

Unless otherwise noted, all art on this page is © Cengage Learning.

Equations of Horizontal and Vertical Lines

The equation $y = b$ represents the horizontal line that intersects the y-axis at $(0, b)$.

The equation $x = a$ represents the vertical line that intersects the x-axis at $(a, 0)$.

Success Tip The graph of the equation $y = 0$ has special importance; it is the x-axis. Similarly, the graph of the equation $x = 0$ is the y-axis.

ANSWERS TO SELF CHECKS

1.

2.

3.

4.

5.

6.

7.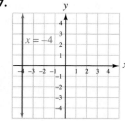

SECTION 8.4 STUDY SET

VOCABULARY

Fill in the blanks.

1. $y = 2x + 3$ is an equation in __two__ variables, x and y.
2. A __solution__ of an equation in two variables is an ordered pair of numbers that makes the equation a true statement.
3. The __graph__ of a linear equation is a mathematical "picture" of all of its solutions.
4. The graph of a linear equation is a straight __line__.
5. The point where the graph of a linear equation crosses the x-axis is called the __x-intercept__.
6. The y-intercept of the graph of a linear equation is the point where it crosses the __y__-axis.

CONCEPTS

Fill in the blanks.

7. To graph $y = 2x + 3$, we can use the following steps:
   Step 1. Find three ordered pairs that are solutions of $y = 2x + 3$ by selecting three values for __x__ and calculating the corresponding values of __y__.
   Step 2. __Plot/graph__ the solutions on a rectangular coordinate system.
   Step 3. Draw a straight __line__ passing through the points. If the points do not lie on a line, __check__ your calculations.

8. The graph of a linear equation is shown here. What three points were plotted to obtain the graph? Enter them in the table of solutions.

x	y	(x, y)
-2	-1	$(-2, -1)$
0	1	$(0, 1)$
2	3	$(2, 3)$

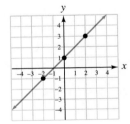

9. **a.** Find the y-intercept of the line graphed on the right. $(0, 1)$
   b. What is its x-intercept? $(-2, 0)$
   c. Does the line pass through the point $(4, 3)$? yes

10. **a.** Find the y-intercept of the line graphed on the right. $(0, -3)$
    b. What is its x-intercept? $(2, 0)$
    c. Does the line pass through the point $(1, -1)$? no

11. Fill in the blanks. To graph $3x + 4y = 12$, we can use the following two-step process:
    Step 1. To find the x-intercept, substitute __0__ for y in $3x + 4y = 12$ and solve for x.
    Step 2. To find the y-intercept, substitute 0 for __x__ in $3x + 4y = 12$ and solve for y.

12. Fill in the blanks in the following table of solutions.

$3x + 2y = 6$

x	y	(x, y)	
2	0	$(2, 0)$	← x-intercept
0	3	$(0, 3)$	← y-intercept

13. **a.** Fill in the blanks. The graph of the equation $y = 3$ is a __horizontal__ line.
    b. The graph of the equation $x = -2$ is a __vertical__ line.

14. **a.** What is the equation of the x-axis? $y = 0$
    b. What is the equation of the y-axis? $x = 0$

15. **a.** Name three points on the line graphed in figure (a) below. $(-2, 4), (0, 4), (2, 4)$ (answers may vary)
    b. Name three points on the line graphed in figure (b) below. $(-1, 2), (-1, 0), (-1, -2)$ (answers may vary)

(a)

(b)

16. The graph of a linear equation is shown below. Name six ordered-pair solutions of the equation from the graph. $(-2, -2), (-1, -1), (0, 0), (1, 1), (2, 2), (3, 3)$ (answers may vary)

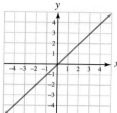

NOTATION

Complete each step.

17. Find the x-intercept of the graph of $2x - 4y = 8$.

$$2x - 4y = 8$$
$$2x - 4(\boxed{0}) = 8$$
$$2x - \boxed{0} = 8$$
$$2x = \boxed{8}$$
$$\frac{2x}{2} = \frac{8}{2}$$
$$x = \boxed{4}$$

The x-intercept of the graph is ($\boxed{4}$, 0).

18. Find the y-intercept of the graph of $2x - 4y = 8$.

$$2x - 4y = 8$$
$$2(\boxed{0}) - 4y = 8$$
$$\boxed{0} - 4y = 8$$
$$-4y = \boxed{8}$$
$$\frac{-4y}{-4} = \frac{8}{-4}$$
$$y = \boxed{-2}$$

The y-intercept of the graph is (0, $\boxed{-2}$).

GUIDED PRACTICE

Complete each table of solutions. See Objective 1.

19. $y = 3x + 3$

x	y	(x, y)
-2	-3	$(-2, -3)$
0	3	$(0, 3)$
2	9	$(2, 9)$

20. $y = 4x + 2$

x	y	(x, y)
-2	-6	$(-2, -6)$
0	2	$(0, 2)$
2	10	$(2, 10)$

21. $y = -2x - 3$

x	y	(x, y)
-2	1	$(-2, 1)$
0	-3	$(0, -3)$
2	-7	$(2, -7)$

22. $y = -3x + 4$

x	y	(x, y)
-2	10	$(-2, 10)$
0	4	$(0, 4)$
2	-2	$(2, -2)$

Graph each equation. See Example 1.

23. $y = 2x - 5$

24. $y = 3x + 1$

25. $y = -3x + 2$

26. $y = -4x - 2$

Graph each equation. See Example 2.

27. $y = -\dfrac{1}{3}x + 1$

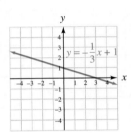

28. $y = -\dfrac{1}{3}x - 1$

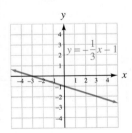

29. $y = -\dfrac{1}{2}x + 1$

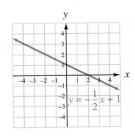

30. $y = -\dfrac{1}{2}x - 2$

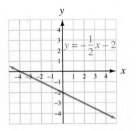

Graph each equation. See Example 3.

31. $y = 100x$

32. $y = 50x$

33. $y = -30x$

34. $y = -20x$

Complete the table and graph the equation using the intercept method. See Example 4.

35. $x - 2y = -4$

x	y
0	2
-4	0
4	4

36. $3x + y = -3$

x	y
0	-3
-1	0
1	-6

37. $4x + 5y = 20$

x	y
0	4
5	0
1	$\frac{16}{5} = 3\frac{1}{5}$

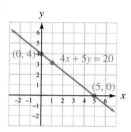

38. $3x - 5y = 15$

x	y
0	-3
5	0
3	$-\frac{6}{5} = -1\frac{1}{5}$

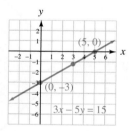

Use the intercept method to graph each equation. See Example 5.

39. $30x + y = -30$

40. $20x - y = -20$

41. $4x - 20y = 60$

42. $6x - 30y = 30$

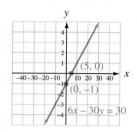

Graph each equation. See Example 6.

43. $y = 5$

44. $y = 1$

45. $y = -4$

46. $y = -3$

Graph each equation. See Example 7.

47. $x = 4$

48. $x = 5$

49. $x = -2$

50. $x = -1$

TRY IT YOURSELF

Find the coordinates of the x- and y-intercepts of the graph of each equation. You do not have to graph the equation.

51. $x + y = 8$ x-intercept is (8, 0); y-intercept is (0, 8)
52. $x - y = 9$ x-intercept is (9, 0); y-intercept is (0, -9)
53. $4x + 5y = 100$ x-intercept is (25, 0); y-intercept is (0, 20)
54. $3x - 5y = 75$ x-intercept is (25, 0); y-intercept is (0, -15)

Graph each equation.

55. $y = \dfrac{2}{3}x - 2$

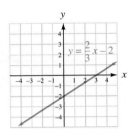

56. $y = \dfrac{5}{6}x - 5$

57. $x + y = 5$

58. $x - y = 2$

59. $y = x$

60. $y = -2x$

61. $x = 0$

62. $y = 0$

63. $3x + 5y = -150$

64. $x + 5y = 50$

65. $y = \dfrac{x}{3}$

66. $y = \dfrac{3}{4}x$

67. $y = -50x - 25$

68. $y = 200x - 400$

69. $4x - 3y = 12$

70. $5x - 10y = 20$

71. $y = \dfrac{5}{2}$

72. $x = \dfrac{4}{3}$

APPLICATIONS

73. HOURLY WAGES The following table gives the amount y (in dollars) that a student can earn by working x hours. Plot the ordered pairs in the table and draw a straight line through the points. Then estimate how much the student will earn in 3 hours. $22.50

x	y
2	15
4	30
6	45

Hours worked

74. VALUE OF A CAR The following table shows the value y (in thousands of dollars) of a car that is x years old. Plot the ordered pairs in the table and draw a straight line through the points. Then estimate the value of the car when it is 7 years old. $1,000

x	y
3	7
4	5.5
5	4

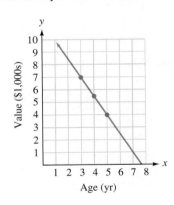

Age (yr)

75. DISTANCE, RATE, AND TIME The formula $d = 2t$ gives the distance d (in miles) that a child can walk in a time t (in hours) at the rate of 2 mph. Complete the table of solutions in the next column, and then graph the equation to get a picture of the relationship between distance and time. (*Hint:* Plot t on the horizontal axis and d on the vertical axis.)

$d = 2t$

t	d
1	2
2	4
3	6
4	8
5	10

76. INVESTMENTS If $100 is invested in a savings account paying 6% per year simple interest, the amount A in the account over a period of time t is given by the formula $A = 6t + 100$. Complete the table of solutions, and then graph this equation to get a picture of how the account grows over a period of time. (*Hint:* Plot t on the horizontal axis and A on the vertical axis.)

$A = 6t + 100$

t	A
1	106
2	112
3	118
4	124
5	130
6	136

The symbol ⌇ is used to indicate a break in the labeling of the vertical axis.

77. BILLIARDS Refer to the billiard table shown below. The path traveled by the black 8-ball is described by the equations $y = 2x - 4$ and $y = -2x + 12$. Construct a table of solutions for the equation $y = 2x - 4$ using the x-values 1, 2, and 4. Do the same for the equation $y = -2x + 12$, using the x-values 4, 6, and 8. Then graph the path of the 8-ball.

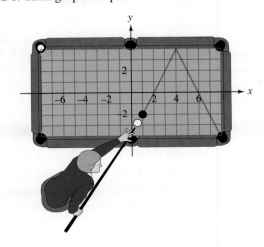

78. **AIR TRAFFIC CONTROL** The equations describing the paths of two airplanes are $y = -\frac{1}{2}x + 3$ and $2x - 3y = -2$. Each equation is graphed on the radar screen shown below. If the planes are flying at the same altitude, is there a possibility of a midair collision? If so, where? yes, (2, 2)

79. **SPRAY BOTTLES** Sometimes, the variables used to label the axes are changed to better model a real-life situation. The following graph shows the amount A of sore throat medication (in ounces) that remains in the bottle after the spray trigger has been pushed down a total of n times. What information can be obtained from the n-intercept? The A-intercept? There are 300 sprays of throat medication in a bottle. The bottle holds 6 ounces of throat spray.

80. **DEPRECIATION** The graph on the right shows how the value of a computer decreased over the age of the computer. What information can be obtained from the x-intercept? The y-intercept? In 8 years ($x = 8$), the computer will have no value. When new ($x = 0$), the computer was worth $3,000.

WRITING

81. When we say that $(-2, -1)$ is a solution of $5x - 6y = -4$, what do we mean?

82. What does it mean when we say that the equation $y = 2x + 4$ has *infinitely many* solutions?

83. On a quiz, students were asked to graph $y = 3x - 1$. One student made the table of solutions on the left below. Another student made the table of solutions on the right. Which table is incorrect? Or could they both be correct? Explain.

x	y	(x, y)
0	−1	(0, −1)
2	5	(2, 5)
3	8	(3, 8)

x	y	(x, y)
−2	−7	(−2, −7)
−1	−4	(−1, −4)
1	2	(1, 2)

84. Explain how the *cover-over method* can be used to quickly find the x- and y-intercepts of the graph of $3x - 2y = 12$.

85. To graph $y = -x + 1$, a student constructed a table of solutions and plotted the ordered pairs as shown. Instead of drawing a curve through the points, what should he have done?

$y = -x + 1$

x	y	(x, y)
−3	−2	(−3, −2)
0	1	(0, 1)
2	−1	(2, −1)

86. What is wrong with the graph of $x - y = 3$ below?

x	y	(x, y)
0	−3	(0, −3)
3	0	(3, 0)
1	−2	(1, −2)

REVIEW

Find the prime factorization of each number.

87. 180 $2^2 \cdot 3^2 \cdot 5$

88. 270 $2 \cdot 3^3 \cdot 5$

Evaluate each expression for $a = -2$ and $b = 3$.

89. $\dfrac{3(b - a)}{5a + 7}$ -5

90. $\dfrac{-2b^2 - b}{b}$ -7

91. **LIGHTNING** The average flash of lightning lasts 0.25 second. Write the decimal as a fraction in simplest form. $\frac{1}{4}$ sec

92. Simplify: $4(a + 1) - 5(6 - a)$ $9a - 26$

STUDY SKILLS CHECKLIST

Know the Definitions

Before taking the test on Chapter 8, make sure that you have memorized the definitions of *mean*, *median*, *mode*, and *range*. Put a checkmark in the box if you can answer "yes" to the statement.

☐ I know that the *mean* of a set of values is often referred to as the *average*.

☐ I know that the *mean* of a set of values is given by the formula:

$$\text{Mean} = \frac{\text{sum of the values}}{\text{number of values}}$$

☐ I know that the *median* of a set of values is the middle value when they are arranged in increasing order.

☐ I know that the *range* of a set of values is the difference of the largest value and the smallest value.

☐ I know how to find the *median* of a set of values if there is an odd number of values.

2 4 5 8 10 13 14   7 values
         ↑
   Median = Middle value

☐ I know how to find the *median* of a set of values if there is an even number of values.

2 4 5 8 10 13 14 16   8 values

$$\text{Median} = \frac{8 + 10}{2} = 9$$

☐ I know that the *mode* of a set of values is the value that occurs most often.

☐ I know that a set of values may have one *mode*, or more than one *mode*.

| 2 | 8 | 5 | 8 | 10 | 8 | 14 | mode: 8

| 2 | 8 | 5 | 8 | 2 | 8 | 2 | two modes: 2, 8

Teaching Guide: Refer to the Instructor's Resource Binder to find activities, worksheets on key concepts, more examples, instruction tips, overheads, and assessments.

CHAPTER 8 SUMMARY AND REVIEW

SECTION 8.1 Reading Graphs and Tables

DEFINITIONS AND CONCEPTS	EXAMPLES
To read a **table** and locate a specific fact in it, we find the *intersection* of the correct row and column that contains the desired information.	SALARY SCHEDULES Find the annual salary for a teacher with a master's degree plus 15 additional units of study who is beginning her 4th year of teaching.

Teacher Salary Schedule

Step	BA	BA+15	BA+30	BA+45	MA	MA+15	MA+30
1	37,295	38,362	39,416	40,480	41,556	42,612	43,669
2	38,504	39,581	40,652	41,728	42,812	43,879	44,952
3	39,716	40,802	41,885	42,973	44,066	45,147	46,234
4	40,926	42,021	43,120	44,220	45,321	46,417	47,514
5	42,135	43,240	44,356	45,465	46,577	47,682	48,795
6	44,458	45,567	46,683	47,782	48,897	50,010	51,113
7	46,780	47,891	49,003	50,115	51,226	52,330	53,438

The annual salary is $46,417. It can be found by looking on the fourth row (labeled Step 4) in the 6th column (labeled MA + 15).

A **bar graph** presents data using vertical or horizontal bars. A **horizontal axis** and **vertical axis** serve to frame the graph, and they are scaled in units such as years, dollars, minutes, pounds, and percent.

CANCER DEATHS Refer to the bar graph below. Estimate how many more deaths were caused by lung cancer than by colon cancer in the United States in 2012.

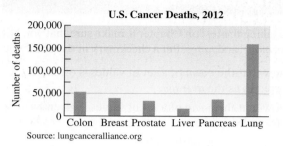

Source: lungcanceralliance.org

From the graph, we see that there were about 160,000 deaths caused by lung cancer and about 50,000 deaths from colon cancer. To find the difference, we subtract:

160,000 − 50,000 = 110,000

There were about 110,000 more deaths caused by lung cancer than deaths caused by colon cancer in the United States in 2012.

To compare sets of related data, groups of two (or three) bars can be shown. For **double-bar** or **triple-bar graphs**, a key is used to explain the meaning of each type of bar in a group.

SEAT BELTS Refer to the double-bar graph below. How did the percent of male high school students that rarely or never wore seat belts change from 2001 to 2011?

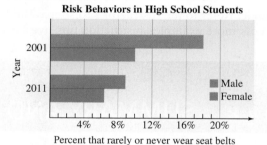

Source: *The World Almanac and Book of Facts*, 2003, 2013

From the graph, we see that in 2001 about 18% of male high school students rarely or never wore seat belts. By 2011, the percent was about 9%, a decrease of 18% − 9%, or 9%.

A **pictograph** is like a bar graph, but the bars are made from pictures or symbols. A **key** tells the meaning (or value) of each symbol.

MEDICAL SCHOOLS Refer to the pictograph below. In 2012, how many students were enrolled in California medical schools?

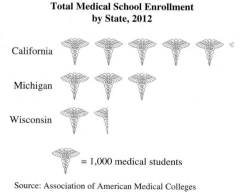

Source: Association of American Medical Colleges

	The California row contains five complete symbols and a small part of another. This means that there were 5 · 1,000, or 5,000 medical students, plus approximately 100 more. In 2012, about 5,100 students were enrolled in California medical schools.
In a **circle graph**, regions called *sectors* (they look like slices of pizza) are used to show what part of the whole each quantity represents.	**CHECKING EMAIL** The circle graph to the right shows the results of a survey of adults who were asked how many personal email addresses they regularly check. What percent of the adults surveyed check 4 or more email addresses regularly? 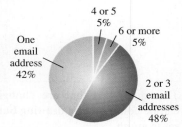 Source: Ipsos for Habeas We add the percent of the responses for 4 or 5 email addresses and the percent of the responses for 6 or more email addresses: 5% + 5% = 10% Thus, 10% of the adults surveyed check 4 or more email addresses regularly. Use the survey results to predict the number of adults in a group of 5,000 that would check only one email address regularly. In the survey, 42% said they check only one email address. We need to find: What number is 42% of 5,000? $x = 42\% \cdot 5{,}000$ Translate. $x = 0.42 \cdot 5{,}000$ Write 42% as a decimal. $x = 2{,}100$ Do the multiplication. According to the survey, about 2,100 of the 5,000 adults would check only one email address regularly.
A **line graph** is used to show how quantities change with time. From such a graph, we can determine when a quantity is increasing and when it is decreasing.	**SNOWBOARDING** The line graph below shows the number of people who participated in snowboarding in the United States for the years 1997–2012. Source: NSAA Demographic Study

When did the popularity of snowboarding seem to peak?

The year with the highest participation was 2010.

Between which two years was there the greatest decrease in the number of snowboarding participants?

The line segment with the greatest "fall" as we read left to right is the segment connecting the data points for the years 2011 and 2012. Thus, the greatest decrease in the number of snowboarding participants occurred between 2011 and 2012.

Two quantities that are changing with time can be compared by **drawing both lines on the same graph**.

SKATEBOARDING Refer to the line graphs below that show the results of a skateboarding race.

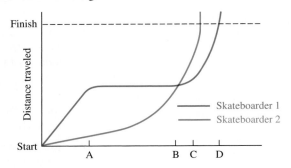

Observations:

- Since the red graph is well above the blue graph at time A, skateboarder 1 was well ahead of skateboarder 2 at that stage of the race.
- Since the red graph is horizontal from time A to time B, skateboarder 1 had stopped.
- Since the blue graph crosses the red graph at time B, at that instant, the skateboarders are tied for the lead.
- Since the blue graph crosses the dashed finish line at time C, which is sooner than time D, skateboarder 2 won the race.

A **histogram** is a bar graph with these features:

1. The bars of the histogram touch.
2. Data values never fall at the edge of a bar.
3. The widths of the bars are equal and represent a range of values.

SLEEP A group of parents of junior high students were surveyed and asked to estimate the number of hours that their children slept each night. The results are displayed in the histogram to the right. How many children sleep 6 to 9 hours a night?

The bar with edges 5.5 and 7.5 corresponds to the 6- to 7-hour range. The height of that bar indicates that 42 children sleep 6 to 7 hours. The bar with edges 7.5 and 9.5 corresponds to the 8- to 9-hour range. The height of that bar indicates that 93 children sleep 8 to 9 hours. The total number of children sleeping 6 to 9 hours is found using addition:

42 + 93 = 135

135 of the junior high children sleep 6 to 9 hours a night.

A **frequency polygon** is a special line graph formed from a histogram by joining the center points at the top of each bar. On the horizontal axis, we write the coordinate of the middle value of each **class interval**. Then we erase the bars.

Frequency polygon

REVIEW EXERCISES

Refer to the table below to answer the following questions.

1. **WINDCHILL TEMPERATURES**
   a. Find the windchill temperature on a 10°F day when a 15-mph wind is blowing. $-18°$
   b. Find the windchill temperature on a –15°F day when a 30-mph wind is blowing. $-71°$

2. **WIND SPEEDS**
   a. The windchill temperature is −25°F, and the actual outdoor temperature is 15°F. How fast is the wind blowing? 30 mph
   b. The windchill temperature is −38°F, and the actual outdoor temperature is –5°F. How fast is the wind blowing? 15 mph

As of 2013, the United States had the most nuclear power plants in operation worldwide, with 104. The bar graph below shows the rest of the top ten countries and the number of nuclear power plants they have in operation.

3. How many nuclear power plants does Korea have in operation? 23
4. How many nuclear power plants does France have in operation? 58
5. Which countries have fewer than 20 nuclear power plants in operation? Ukraine, United Kingdom, China, and Canada
6. How many more nuclear power plants in operation does Japan have than Canada? 31

Determining the Windchill Temperature

Wind speed	Actual temperature							
	20°F	15°F	10°F	5°F	0°F	−5°F	−10°F	−15°F
5 mph	16°	12°	7°	0°	−5°	−10°	−15°	−21°
10 mph	3°	−3°	−9°	−15°	−22°	−27°	−34°	−40°
15 mph	−5°	−11°	−18°	−25°	−31°	−38°	−45°	−51°
20 mph	−10°	−17°	−24°	−31°	−39°	−46°	−53°	−60°
25 mph	−15°	−22°	−29°	−36°	−44°	−51°	−59°	−66°
30 mph	−18°	−25°	−33°	−41°	−49°	−56°	−64°	−71°
35 mph	−20°	−27°	−35°	−43°	−52°	−58°	−67°	−74°

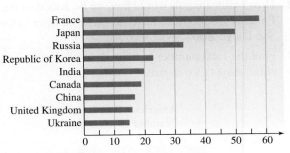

Number of Nuclear Power Plants in Operation, 2013

Source: euronuclear.org

In a workplace survey, employed adults were asked if they would date a coworker. The results of the survey are shown below. Use the double-bar graph to answer the following questions.

7. What percent of the women said they would not date a coworker? 29%

8. Did more men or women say that they would date a coworker? What percent more? men; 15% more

9. When asked, were more men or more women unsure if they would date a coworker? women

10. Which of the three responses to the survey was given by approximately the same percent of men and women? No, I would not date a coworker (31% to 29%)

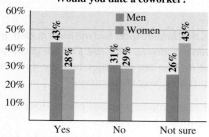

Source: Spherion Workplace Survey

Refer to the pictograph below to answer the following questions.

11. How many animals are there at the San Diego Zoo? about 4,100 animals

12. Which of the zoos listed has the most animals? How many? the Columbus Zoo; about 6,250 animals

13. How many animals would have to be added to the Phoenix Zoo for it to have the same number as the San Diego Zoo? about 3,000 animals

14. Find the total number of animals in all three zoos. about 11,500 animals

Source: zoo.findthebest.com

Refer to the circle graph below to answer the following questions.

15. What element makes up the largest percent of the body weight of a human? oxygen

16. Elements other than oxygen, carbon, hydrogen, and nitrogen account for what percent of the weight of a human body? 4%

17. Hydrogen accounts for how much of the body weight of a 135-pound woman? 13.5 lb

18. Oxygen and carbon account for how much of the body weight of a 200-pound man? 166 lb

Source: General Chemistry Online

Refer to the line graph on the next page to answer the following questions.

19. How many eggs were produced in Nebraska in 2001? about 3,000 million eggs

20. How many eggs were produced in North Carolina in 2008? about 3,050 million eggs

21. In what year was the egg production of Nebraska equal to that of North Carolina? How many eggs? 2007; about 2,950 million eggs

22. What was the total egg production of Nebraska and North Carolina in 2005? about 5,750 million eggs

23. Between what two years did the egg production in North Carolina increase dramatically? between 2006 and 2007

24. Between what two years did the egg production in Nebraska decrease dramatically? between 2007 and 2008

25. How many more eggs did North Carolina produce in 2009 compared to Nebraska? about 400 million more eggs

26. How many more eggs did North Carolina produce in 2011 compared to Nebraska? about 500 million more eggs

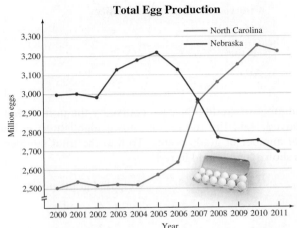

29. How many households watch 11 hours or more each week? 160

30. Create a frequency polygon from the histogram shown above.

A survey of the weekly television viewing habits of 320 households produced the histogram in the next column. Use the graph to answer the following questions.

27. How many households watch between 1 and 5 hours of TV each week? 60

28. How many households watch between 6 and 15 hours of TV each week? 180

SECTION 8.2 Mean, Median, and Mode

DEFINITIONS AND CONCEPTS	EXAMPLES
It is often useful to find one number to represent the "center" of all the numbers in a set of data. There are three measures of **central tendency**: mean, median, mode. The **mean** of a set of values is given by the formula $$\text{Mean} = \frac{\text{sum of the values}}{\text{number of values}}$$	Find the mean of the following set of values: 6 8 3 5 9 8 10 7 8 5 To find the mean, we divide the sum of the values by the number of values, which is 10. $$\frac{6+8+3+5+9+8+10+7+8+5}{10} = \frac{69}{10}$$ $$= 6.9$$ Thus, 6.9 is the mean.

When a value in a set appears more than once, that value has a greater "influence" on the mean than another value that only occurs a single time. To simplify the process of finding the mean, any value that appears more than once can be "weighted" by multiplying it by the number of times it occurs.

To find the **weighted mean** of a set of values:

1. Multiply each value by the number of times it occurs.
2. Find the sum of the products from step 1.
3. Divide the sum from step 2 by the total number of individual values.

A student's **grade point average (GPA)** can be found using a weighted mean.

Some schools assign a certain number of **credit hours** to a course while others assign a certain number of **units**.

GPAs Find the semester grade point average for a student that received the following grades. (The point values are A = 4, B = 3, C = 2, D = 1, and F = 0.)

Course	Grade	Credits
Algebra	A	5
History	C	3
Art	D	4

Multiply the number of credits for each course by the point value of the grade received. Add the results (as shown highlighted in blue) to get the total number of grade points. To find the total number of credits, add as shown highlighted in red.

Course	Grade	Credits	Weighted grade points
Algebra	A	5	$4 \cdot 5 \rightarrow$ 20
History	C	3	$2 \cdot 3 \rightarrow$ 6
Art	D	+ 4	$1 \cdot 4 \rightarrow$ + 4
Totals		12	30

To find the GPA, we divide.

$$\text{GPA} = \frac{30}{12} \quad \begin{array}{l}\leftarrow \text{The total number of grade points} \\ \leftarrow \text{The total number of credits}\end{array}$$

$$= 2.5 \quad \text{Do the division.}$$

The student's semester GPA is 2.5.

To find the **median** of a set of values:

1. Arrange the values in increasing order.
2. If there is an odd number of values, the median is the middle value.
3. If there is an even number of values, the median is the mean (average) of the middle two values.

To find the median of

6   8   3   5   9   8   10   7   8   5

arrange them in increasing order:

Smallest   Largest

3   5   5   6   **7   8**   8   8   9   10   There are 10 values.

↑ Middle two values

Since there is an even number of values, the median is the mean (average) of the two middle values:

$$\frac{7 + 8}{2} = \frac{15}{2} = 7.5$$

Thus, 7.5 is the median.

The **range** of a set of values is the difference between the largest value and the smallest value.

The range of the data listed above is:

$$\text{range} = 10 - 3 = 7 \quad \text{Subtract the smallest value, 3, from the largest value, 10.}$$

| The **mode** of a set of values is the single value that occurs most often. | To find the mode of

6 8 3 5 9 8 10 7 8 5

we find the value that occurs most often.

6 **8** 3 5 9 **8** 10 7 **8** 5

     3 times

Since 8 occurs the most times, it is the mode. |
| When a collection of values has two modes, it is called **bimodal**. | The collection of values

1 2 **3 3** 4 5 **6 6** 7 8

has two modes: 3 and 6. |

REVIEW EXERCISES

31. GRADES Jose worked hard this semester, earning grades of 87, 92, 97, 100, 100, 98, 90, and 98. If he needs a 95 average to earn an A in the class, did he make it? yes

32. GRADE SUMMARIES The students in a mathematics class had final averages of 43, 83, 40, 100, 40, 36, 75, 39, and 100. When asked how well her students did, their teacher answered, "43 was typical." What measure was the teacher using: mean, median, or mode? median

33. PRETZEL PACKAGING Samples of SnacPak pretzels were weighed to find out whether the package claim "Net weight 1.2 ounces" was accurate. The tally appears in the table. Find the mode of the weights. 1.2 oz

| Weights of SnacPak Pretzels ||
Ounces	Number
0.9	1
1.0	6
1.1	18
1.2	23
1.3	2

34. Find the mean weight and the range of the weights of the samples in Exercise 33. 1.138 oz, 0.4 oz

35. BLOOD SAMPLES A medical laboratory technician examined a blood sample under a microscope and measured the sizes (in microns) of the white blood cells. The data are listed below. Find the mean, median, mode, and range. 7.3 microns, 7.2 microns, 6.9 microns, 1.3 microns

 7.8 6.9 7.9 6.7 6.8 8.0 7.2 6.9 7.5

36. SUMMER READING A paperback version of the classic *Gone with the Wind* is 960 pages long. If a student wants to read the entire book during the month of June, how many pages must she average per day? 32 pages per day

37. WALK-A-THONS Use the data in the table to find the mean (average) donation to a charity walk-a-thon. $20

Donation amount	$5	$10	$20	$50	$100
Number received	20	65	25	5	10

38. GPAs Find the semester grade point average for a student who received the grades shown below. Round to the nearest hundredth. (Assume the following standard point values for the letter grades: A = 4, B = 3, C = 2, D = 1, and F = 0.) 2.62 GPA

Course	Grade	Credits
Chemistry	A	5
Sociology	C	3
Economics	D	4
Archery	A	1

SECTION 8.3 Equations in Two Variables; The Rectangular Coordinate System

DEFINITIONS AND CONCEPTS

A **solution of an equation in two variables** is an **ordered pair** of numbers that makes the equation a true statement when the numbers are substituted for the variables.

An ordered pair:

$(2, -3)$
 $x\uparrow$ $\uparrow y$

EXAMPLES

Is $(2, -3)$ a solution of $2x - y = 7$?

$2x - y = 7$ This is the given equation.
$2(2) - (-3) \stackrel{?}{=} 7$ Substitute 2 for x and -3 for y.
$4 + 3 \stackrel{?}{=} 7$ Evaluate the left side.
$7 = 7$ Do the addition.

Since $7 = 7$ is a true statement, $(2, -3)$ is a solution of $2x - y = 7$.

If only one coordinate of an ordered-pair solution is known:

1. Substitute it into the equation for the proper variable.
2. Solve the resulting equation to find the unknown coordinate.

To complete the solution (, 8) for $3x + y = -1$, we substitute 8 for y and solve the resulting equation for x.

$3x + y = -1$ This is the given equation.
$3x + 8 = -1$ Substitute 8 for y.
$3x + 8 - 8 = -1 - 8$ To isolate the variable term 3x, subtract 8 from both sides.
$3x = -9$ Do the subtraction.
$\dfrac{3x}{3} = \dfrac{-9}{3}$ To isolate x, undo the multiplication by 3 by dividing both sides by 3.
$x = -3$ Do the division. This is the missing x-value of the solution.

When $y = 8$, $x = -3$. The completed ordered pair is $(-3, 8)$.

Solutions of an equation in two variables can be listed in a **table of solutions**.

Two solutions of $y = x + 4$ are shown below.

x	y	(x, y)
0	4	(0, 4)
6	10	(6, 10)

A **rectangular coordinate system** is composed of a horizontal number line, called the **x-axis**, and a vertical number line, called the **y-axis**. The two axes intersect at the **origin**.

To **plot** or **graph** ordered pairs means to locate their position on a rectangular coordinate system.

The x- and y-axes divide the coordinate plane into four regions called **quadrants**.

Plot the points: $(2, 3)$, $(-4, 2)$, $(-3, -1)$, $(0, -2.5)$, and $(4, -2)$

To graph each point, start at the origin and move the given number of units to the right or left on the x-axis and then the given number of units up or down. Then draw a dot and label it.

REVIEW EXERCISES

39. Is $(2, -3)$ a solution of $2x + 5y = -11$? yes

40. Is $(-3, 2)$ a solution of $y = 8x + 15$? no

41. Complete the solutions of the equation $3x - 4y = 12$: $(0, -3)$ and $(4, 0)$

42. Complete the table of solutions for $y = -3x - 2$.

x	y	(x, y)
1	-5	(1, -5)
3	-11	(3, -11)
-2	4	(-2, 4)

43. Graph the points $(2, 3), (-3, 4), (-5, 0), (0, -4),$ $(-1.5, -3),$ and $\left(\frac{7}{2}, -1\right)$.

44. Give the coordinates of each point graphed below.

$A(4,3)$, $B(-3, 3)$, $C(-4, 0)$, $D(-1.5, -3.5)$ or $D(-1\frac{1}{2}, -3\frac{1}{2})$, $E(2\frac{1}{2}, -1\frac{1}{2})$ or $E(2.5, -1.5)$, $F(0, 0)$

45. In what quadrant does the point $(-3, -4)$ lie? III

46. THEATER SEATING Your ticket at the theater is for seat B-10. Locate your seat on the diagram.

SECTION 8.4 Graphing Linear Equations

DEFINITIONS AND CONCEPTS

The **graph of an equation** in two variables is a mathematical "picture" of all of its solutions.

To **graph a linear equation** solved for y:

1. Find three solutions by selecting three values of x and finding the corresponding values of y.
2. Plot each ordered-pair solution.
3. Draw a straight line through the points.

EXAMPLES

Graph: $y = -2x + 1$

We construct a table of solutions, plot the points, and draw the line.

$y = -2x + 1$

x	y	(x, y)
-2	5	(-2, 5)
0	1	(0, 1)
2	-3	(2, -3)

↑ Select x ↑ Find y ↑ Plot (x, y)

The point where a line intersects the x-axis is called the **x-intercept**. The point where a line intersects the y-axis is called the **y-intercept**.

To **find the x-intercept**, substitute 0 for y and solve for x. To **find the y-intercept**, substitute 0 for x in the given equation and solve for y.

Plotting the x- and y-intercepts of a graph and drawing a line through them is called the **intercept method for graphing a line**.

Use the x- and y-intercepts to graph $3x + 2y = -6$.

x-intercept: $y = 0$	y-intercept: $x = 0$
$3x + 2y = -6$	$3x + 2y = -6$
$3x + 2(0) = -6$	$3(0) + 2y = -6$
$3x = -6$	$2y = -6$
$x = -2$	$y = -3$

The x-intercept is $(-2, 0)$ and the y-intercept is $(0, -3)$.

As a check, a third point on the line can be found by selecting a convenient value for x and finding the corresponding value for y.

$3x + 2y = -6$

x	y	(x, y)	
-2	0	$(-2, 0)$	← x-intercept
0	-3	$(0, -3)$	← y-intercept
-4	3	$(-4, 3)$	← Check point

The equation $y = b$ represents the **horizontal line** that intersects the y-axis at $(0, b)$.

The equation $x = a$ represents the **vertical line** that intersects the x-axis at $(a, 0)$.

Graph: $y = 3$

x	y	(x, y)
-2	3	$(-2, 3)$
0	3	$(0, 3)$
2	3	$(2, 3)$

↑ Each value of y must be 3.

Graph: $x = -3$

x	y	(x, y)
-3	-2	$(-3, -2)$
-3	0	$(-3, 0)$
-3	2	$(-3, 2)$

↑ Each value of x must be -3.

REVIEW EXERCISES

Graph each equation.

47. $y = 2x - 3$

48. $y = \dfrac{1}{2}x - 1$

Unless otherwise noted, all art on this page is © Cengage Learning.

49. $y = -3x + 2$ **50.** $y = 4x$

Graph each equation.

55. $y = 2$ **56.** $x = 1$

51. The graph of a linear equation is shown on the right. Name six ordered-pair solutions of the equations from the graph. $(-3, -2)$, $(-1, 0)$, $(0, 1)$, $(1, 2)$, $(2, 3)$, $(3, 4)$ (answers may vary)

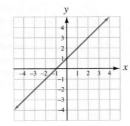

57. MACHINERY Refer to the graph below.
  a. Which intercept tells the purchase price of the machinery? What was that price?
  y-intercept: $(0, 80{,}000)$; $80,000
  b. Which intercept indicates when the machinery will have lost all of its value? When is that?
  x-intercept: $(30, 0)$; 30 years after purchase

52. Identify the x- and y-intercepts of the graph shown on the right. $(-3, 0)$, $(0, 2.5)$

Use the intercept method to graph each equation.

53. $8x + 4y = -24$ **54.** $30x - y = 30$

58. BOTTLED WATER DISPENSER The graph below shows the number of gallons g of water remaining in a bottle after c six-ounce cups have been served from it. Find the intercepts of the graph. What information do they give? The g-intercept is $(0, 5)$: Before any cups of water have been served from the bottle, it contains 5 gallons of water. The c-intercept is $(106\frac{2}{3}, 0)$: The bottle will be empty after $106\frac{2}{3}$ six-ounce cups of water have been served from it.

CHAPTER 8 TEST

Fill in the blanks.

1. a. A horizontal or vertical line used for reference in a bar graph is called an __axis__.
    b. The __mean__ (average) of a set of values is the sum of the values divided by the number of values in the set.
    c. The __median__ of a set of values written in increasing order is the middle value.
    d. The __mode__ of a set of values is the single value that occurs most often.
    e. The mean, median, and mode are three measures of __central__ tendency.

2. WORKOUTS Refer to the table below to answer the following questions.
    a. How many calories will a 155-pound person burn if she runs for one hour at a rate of 5 mph? 563 calories
    b. In one hour, how many more calories will a 190-pound person burn if he runs at a rate of 7 mph instead of 6 mph? 129 calories
    c. At what rate does a 130-pound person have to run for one hour to burn approximately 800 calories? about 8 mph

Number of Calories Burned While Running for One Hour

Running speed (mph)	Body Weight		
	130 lb	155 lb	190 lb
5	472	563	690
6	590	704	863
7	679	809	992
8	797	950	1,165
9	885	1,056	1,294

Source: nutristrategy.com

3. MOVING Refer to the bar graph in the next column to answer the following questions.
    a. Which piece of furniture shown in the graph requires the greatest number of feet of bubble wrap? How much? love seat; 150 ft
    b. How many more feet of bubble wrap is needed to wrap a desk than a coffee table? 50 feet more
    c. How many feet of bubble wrap is needed to cover a bedroom set that has a headboard, a dresser, and two end tables? 340 ft

Source: transitsystems.com

4. CANCER SURVIVAL RATES Refer to the graph below to answer the following questions.
    a. What was the survival rate (in percent) from breast cancer in 1976? 75%
    b. By how many percent did the cancer survival rate for breast cancer increase by 2008? about 15%
    c. Which type of cancer shown in the graph has the lowest survival rate? lung cancer
    d. Which type of cancer has had the greatest increase in survival rate from 1976 to 2008? How much of an increase? prostate cancer; about 35%

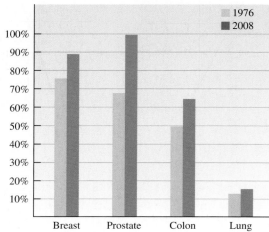
Source: lungcanceralliance.org

5. **ENERGY DRINKS** Refer to the pictograph below to answer the following questions.
   a. How many grams of sugar are there in 12 ounces of Big Red? about 38 g
   b. For a 12-ounce serving, how many more grams of sugar are there in Monster Energy Drink than in Starbucks Tall Caffè Mocha? about 15 g

Sugar Content in Energy Drinks and Coffee (12-ounce serving)

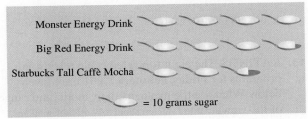

Source: energyfiend.com

6. **FIRES** Refer to the graph below to answer the following questions.
   a. In 2011, what percent of the fires in the United States were vehicle fires? 16%
   b. In 2011, there were a total of 1,389,500 fires in the United States. How many were structure fires? 486,325

Where Fires Occurred, 2011

Source: U.S. Fire Administration

7. **NYPD** Refer to the graph in the next column to answer the following questions.
   a. How many uniformed police officers did the NYPD have in 1999? about 39,000 police officers
   b. When was the number of uniformed police officers the least? How many officers were there at that time? 2011; about 33,800 police officers
   c. When was the number of uniformed police officers the greatest? How many officers were there at that time? 2000; about 40,300 police officers
   d. Find the decrease in the number of uniformed police officers from 2003 to 2012. about 1,500 police officers

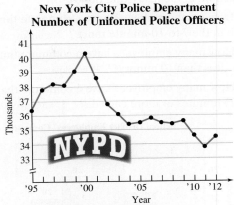

Source: Mayor's Management Reports

8. **BICYCLE RACE** Refer to the graph below to answer each of following questions about a two-man bicycle race.
   a. Which bicyclist had traveled farther at time A? bicyclist 1
   b. Explain what was happening in the race at time B. Bicyclist 1 is stopped, but is ahead in the race. Bicyclist 2 is beginning to catch up.
   c. When was the first time that bicyclist 2 stopped to rest? time C
   d. Did bicyclist 2 ever lead the race? If so, at what time? Bicyclist 2 never led.
   e. Which bicyclist won the race? bicyclist 1

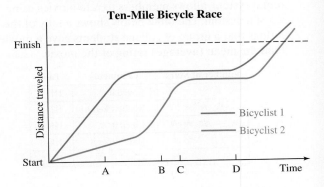

9. **COMMUTING TIME** A school district collected data on the number of minutes it took its employees to drive to work in the morning. The results are presented in the histogram below.

   a. How many employees have a commute time that is in the 7-to-10-minute range? 22 employees

   b. How many employees have a commute time that is less than 10 minutes? 30 employees

   c. How many employees have a commute that takes 15 minutes or more each day? 57 employees

School District Employees' Commute

10. **VOLUNTEER SERVICE** The number of hours served last month by each of the volunteers at a homeless shelter are listed below:

    4 6 8 2 8 10 11 9 5 12 5 18 7 5 1 9

    a. Find the mean (average) of the hours of volunteer service. 7.5 hr

    b. Find the median of the hours of volunteer service. 7.5 hr

    c. Find the mode of the hours of volunteer service. 5 hr

    d. Find the range of the number of hours of volunteer service. 17 hr

11. **RATING MOVIES** Netflix, a popular online DVD rental system, allows members to rate movies using a 5-star system. The table below shows a tally of the ratings that a group of college students gave a movie. Find the mean (average) rating of the movie. 3 stars

Number of Stars	Comments	Tally
★★★★★	Loved it	III
★★★★	Really liked it	IIII
★★★	Liked it	℔
★★	Didn't like it	℔ I
★	Hated it	II

12. **GPAs** Find the semester grade point average for a student who received the following grades. Round to the nearest hundredth. 3.36 GPA

Course	Grade	Credits
WEIGHT TRAINING	C	1
TRIGONOMETRY	A	3
GOVERNMENT	B	2
PHYSICS	A	4
PHYSICS LAB	D	1

13. **RATINGS** The eight top-rated cable television programs for Friday, February 1, 2013, are given below. What are the mean, median, mode, and range of the viewer data? mean: about 3.27 million; median: 2.99 million; mode: 2.57 million; range: 2.13 million

Show/day/time/network	Millions of viewers
Gold Rush, 9 P.M., DISC	4.70
Jessie, 8:30 P.M., DSNY	3.84
Phineas and Ferb, 9 P.M., DSNY	3.77
SpongeBob, 6 P.M., NICK	3.01
WWE Smackdown, 8 P.M., SYFY	2.97
SpongeBob, 6:30 P.M., NICK	2.76
Austin & Ally, 7:30 P.M., DSNY	2.57
Bering Sea Gold, 10 P.M., DISC	2.57

Source: The Nielsen Company

14. **REAL ESTATE** In December 2012, the median sales price of an existing single-family home in the United States was $180,300. Explain what is meant by the median sales price. (Source: National Association of Realtors) Of all the existing single-family homes sold in December 2012, half of them sold for less than $180,300 and half sold for more than $180,300.

15. Is $(-1, 2)$ a solution of $4x + 5y = 6$? yes

16. Is $(3, -2)$ a solution of $y = 12x - 34$? no

17. Complete the ordered pairs so that each one is a solution of the equation $x - 2y = 4$.

    $(0, -2)$, $(4, 0)$ and $(2, -1)$

18. Complete the table of solutions for $x - 3y = -3$.

x	y	(x, y)
0	1	$(0, 1)$
3	2	$(3, 2)$
-3	0	$(-3, 0)$

19. **PANTS SALE** In the illustration below, an X indicates the sizes of jeans that a store has in stock. List the jeans sizes that are *not* available as ordered pairs of the form (waist, length). (30, 32), (30, 34), (31, 34), (38, 30)

25. a. What is the x-intercept of the graph on the right?
(−3, 0)
   b. What is the y-intercept of the graph on the right?
(0, −1)

26. The graph of a linear equation is shown on the right. Name three ordered-pair solutions of the equation from the graph. (−3, −1), (0, 0), (3, 1) (answers may vary)

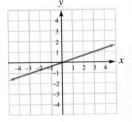

20. Graph the points: $(0, -4.5)$, $\left(\frac{3}{2}, 1\right)$, $(4, 2)$, $(-1, 3)$, $(-2, 0)$, and $(4, -3)$

Use the intercept method to graph each equation.

27. $3x + 4y = 12$

28. $20x + y = -20$

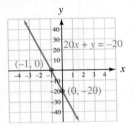

21. Give the coordinates of each point on the graph.
$A(0, 0)$, $B(2.5, 3.5)$ or $B(2\frac{1}{2}, 3\frac{1}{2})$, $C(-3, -2)$, $D(0, -2)$, $E(4, 0)$, $F(-5, 5)$

Graph each equation.

29. $y = -2$

30. $x = 3$

22. In what quadrant does the point $(-7, -1)$ lie?
quadrant III

Graph each equation.

23. $y = 4x - 2$

24. $y = -\frac{3}{2}x - 1$

CHAPTERS 1–8 CUMULATIVE REVIEW

1. **AUTOMOBILES** In 2011, a total of 59,929,016 cars were produced in the world. Write this number in words and in expanded notation. (Source: worldometers.info) [Section 1.1]
Fifty-nine million, nine hundred twenty-nine thousand, sixteen; $50{,}000{,}000 + 9{,}000{,}000 + 900{,}000 + 20{,}000 + 9{,}000 + 10 + 6$

2. Round 59,999 to the nearest hundred. [Section 1.1] 60,000

Perform each operation.

3. $\begin{array}{r} 48{,}908 \\ + \ 5{,}696 \\ \hline 54{,}604 \end{array}$ [Section 1.2]

4. $\begin{array}{r} 8{,}700 \\ - \ 5{,}491 \\ \hline 3{,}209 \end{array}$ [Section 1.2]

5. $\begin{array}{r} 408 \\ \times \ 67 \\ \hline 27{,}336 \end{array}$ [Section 1.3]

6. $87\overline{)2{,}001}$ 23 [Section 1.4]

7. Explain how to check the following result using addition. [Section 1.2]
$\begin{array}{r} 2{,}142 \\ - \ 459 \\ \hline 1{,}683 \end{array}$ $1{,}683 + 459 = 2{,}142$

8. **GEOMETRY** Find the perimeter and the area of the rectangle shown below. [Section 1.3] 40 in., 84 in.²

6 in.
14 in.

9. a. Find the factors of 36. [Section 1.5] 1, 2, 3, 4, 6, 9, 12, 18, 36
   b. Write the first ten prime numbers. [Section 1.5] 2, 3, 5, 7, 11, 13, 17, 19, 23, 29
   c. Find the prime factorization of 36. [Section 1.5] $2^2 \cdot 3^2$

10. a. Find the LCM of 8 and 12. [Section 1.6] 24
    b. Find the GCF of 8 and 12. 4

Evaluate each expression. [Section 1.7]

11. $15 + 5[12 - (2^2 + 4)]$ 35

12. $\dfrac{12 + 5 \cdot 3}{3^2 - 2 \cdot 3}$ 9

Solve each equation and check the result.

13. $96 = m - 83$ [Section 1.8] 179
14. $73n = 219$ [Section 1.9] 3

15. Graph the integers greater than -3 but less than 4. [Section 2.1]

16. a. Simplify: $-(-6)$ [Section 2.1] 6
    b. Find the absolute value: $|-5|$ 5
    c. Is the statement $-12 > -10$ true or false? false

17. *Perform each operation.*
    a. $-35 + 5$ [Section 2.2] -30
    b. $-35 - (-5)$ [Section 2.3] -30
    c. $-35(5)$ [Section 2.4] -175
    d. $\dfrac{-35}{-5}$ [Section 2.5] 7

18. **PLANETS** Mercury orbits closer to the sun than does any other planet. Temperatures on Mercury can get as high as 810°F and as low as $-290°$F. What is the temperature range? [Section 2.3] 1,100°F

Evaluate each expression. [Section 2.6]

19. $\dfrac{(-6)^2 - 1^5}{-4 - 3}$ -5

20. $-3 + 3(-4 - 4 \cdot 2)^2$ 429

21. $-\left|\dfrac{45}{-9} - (-9)\right|$ -4

22. $-10^2 - (-10)^2$ -200

23. Solve $\dfrac{x}{-4} + 4 = -5$ and check the result. [Section 2.7] 36

Write and then solve an equation to answer the following question.

24. **WEATHER FORECASTS** The weather forecast for Barrow, Alaska, warned listeners that the daytime high temperature of 3° below zero would drop to a nighttime low of 31° below. By how many degrees did the temperature fall overnight? [Section 2.8] The temperature fell 28° overnight.

25. Translate to mathematical symbols: 4 less than the square of x. [Section 3.1] $x^2 - 4$

26. **ROAD TRIPS** Find the distance covered by a car traveling 45 miles per hour for 5 hours. [Section 3.2] 225 mi

27. a. Simplify: $(-9t)(-8)$ [Section 3.3] $72t$
    b. Multiply: $-5(3x - 2y - 10)$ [Section 3.3] $-15x + 10y + 50$

28. Combine like terms. [Section 3.4]
   a. $x + x + x + x$ $4x$
   b. $5(3 - 2x) + 4(2 - 3x) + 19x$ $-3x + 23$

29. Solve $8 + 4(2a - 2) = -16 + 4a$ and check the result. [Section 3.5] -4

Write and then solve an equation to answer the following question. [Section 3.6]

30. CIVIL SERVICE A candidate for a position with the Department of Homeland Security scored 4 points higher on the written part of the civil service exam than he did on his interview. If his combined score was 98, what were his scores on the interview and on the written part of the exam? His score on the interview was 47 and his score on the written part was 51.

31. Simplify each fraction. [Section 4.1]
   a. $\frac{60}{108}$ $\frac{5}{9}$
   b. $\frac{24a^3}{16a}$ $\frac{3a^2}{2}$

32. Simplify, if possible. [Section 4.1]
   a. $\frac{0}{64}$ 0
   b. $\frac{27}{0}$ undefined

Perform each operation. Simplify, if possible.

33. $\frac{4}{5} \cdot \frac{2}{7}$ [Section 4.2] $\frac{8}{35}$

34. $\frac{8m^2}{63n^5} \div \frac{2m^2}{n^4}$ [Section 4.3] $\frac{4}{63n}$

35. Subtract $\frac{2}{3}$ from $\frac{1}{2}$. [Section 4.4] $-\frac{1}{6}$

36. $\frac{11}{12} + \frac{1}{30}$ [Section 4.4] $\frac{19}{20}$

37. $\frac{x}{9} - \frac{1}{8}$ [Section 4.4] $\frac{8x - 9}{72}$

38. CLASS TIME In a chemistry course, students spend a total of 300 minutes in lab and lecture each week. If $\frac{7}{15}$ of the time is spent in lab each week, how many minutes are spent in lecture each week? [Section 4.3] 160 min

39. Divide: $2\frac{4}{5} \div \left(-2\frac{2}{3}\right)$ [Section 4.5] $-\frac{21}{20} = -1\frac{1}{20}$

40. TENNIS Find the length of the handle on the tennis racquet shown below. [Section 4.6] $6\frac{3}{4}$ in.

41. Evaluate the formula $A = \frac{1}{2}h(a + b)$ for $a = 4\frac{1}{2}$, $b = 5\frac{1}{2}$, and $h = 2\frac{1}{8}$. [Section 4.7] $10\frac{5}{8}$

42. Simplify the complex fraction: $\dfrac{-\frac{1}{5}}{\frac{8}{15}}$ [Section 4.7] $-\frac{3}{8}$

Solve each equation and check the result. [Section 4.8]

43. $-\frac{5}{6}y = -2$ $\frac{12}{5}$

44. $\frac{a}{5} - \frac{5}{6} = \frac{a}{3}$ $-\frac{25}{4}$

Write and then solve an equation to answer the following question. [Section 4.8]

45. TEAM ROSTERS One-third of the players on a women's basketball team are forwards and one-fifth are centers. The remaining 7 players are guards. How many players are on the team? There are 15 players on the team.

46. Write $400 + 20 + 8 + \frac{9}{10} + \frac{1}{100}$ as a decimal. [Section 5.1] 428.91

47. CHECKBOOKS Find the total dollar amount of the checks written in the register shown below. [Section 5.2] $1,815.19

DATE	CHECK NUMBER	TRANSACTION DESCRIPTION	✓ T	(•) AMOUNT OF PAYMENT OR DEBIT
3 17	703	TO: Albertsons FOR: Groceries		$ 213 16
3 19	704	TO: Brian Auto FOR: Car Repair		$1,504 80
3 19	705	TO: Nordstrom FOR: Sweater		$ 89 73
3 21	706	TO: Girl Scouts FOR: Cookies		$ 7 50

48. Perform each operation in your head. [Section 5.3]
   a. Multiply: $3.45 \cdot 100$ 345
   b. Divide: $3.45 \div 10{,}000$ 0.000345

Perform each operation.

49. Subtract: $\begin{array}{r} 760.2 \\ -614.7 \\ \hline 145.5 \end{array}$ [Section 5.2]

50. Multiply: $(-0.31)(2.4)$ [Section 5.3] -0.744

51. Divide: $0.72 \overline{)536.4}$ [Section 5.4] 745

52. Divide: $4 \overline{)0.073}$ [Section 5.4] 0.01825

53. Write $\frac{8}{11}$ as a decimal. [Section 5.5] $0.\overline{72}$

54. Evaluate: $15 + \sqrt{16}[5^2 - (\sqrt{9} + 2)\sqrt{4}]$
    [Section 5.6] 75

55. Solve $2(3.6t - 4.1) + 0.9t = 16.1$ and check the result. [Section 5.7] 3

Write and then solve an equation to answer the following question. [Section 5.7]

56. BUSINESS EXPENSES A business decides to rent a copy machine instead of buying one. Under the rental agreement, the company is charged a basic rental fee of $35 per month plus 3¢ for every copy made. If the business has budgeted $110 for copier expenses each month, how many copies can be made before exceeding the budget? The business can make up to 2,500 copies each month without exceeding the budget.

57. Express the phrase "8 feet to 4 yards" as a ratio in simplest form. [Section 6.1] $\frac{2}{3}$

58. CLOTHES SHOPPING As part of a summer clearance, a women's store put turtleneck sweaters on sale, 3 for $35.97. How much will five turtleneck sweaters cost? [Section 6.2] $59.95

59. Solve the proportion: $\dfrac{\frac{7}{8}}{\frac{1}{2}} = \dfrac{\frac{1}{4}}{x}$ [Section 6.2] $\frac{1}{7}$

60. Convert 8 pints to fluid ounces. [Section 6.3] 128 fl oz
61. Convert 640 centimeters to meters. [Section 6.4] 6.4 m
62. Convert 67.7°F to degrees Celsius. Round to the nearest tenth. [Section 6.5] 19.8°C
63. Complete the table below. [Section 7.1]

Fraction	Decimal	Percent
$\frac{3}{100}$	0.03	3%
$\frac{9}{4}$	2.25	225%
$\frac{41}{1,000}$	0.041	4.1%

64. 90 is what percent of 525? Round to the nearest one percent. [Section 7.2] 17%
65. What number is 105% of 23.2? [Section 7.2] 24.36
66. 19.2 is $33\frac{1}{3}$% of what number? [Section 7.2] 57.6
67. SALES TAX Find the sales tax on a purchase of $98.95 if the sales tax rate is 8%. [Section 7.3] $7.92
68. SELLING ELECTRONICS If the commission on a $1,500 laptop computer is $240, what is the commission rate? [Section 7.3] 16%

69. TIPPING Estimate the 15% tip on a $77.55 dinner bill. [Section 7.4] $12
70. REMODELING A homeowner borrows $18,000 to pay for a kitchen remodeling project. The terms of the loan are 9.2% annual simple interest and repayment in 2 years. How much interest will be paid on the loan? [Section 7.5] $3,312
71. LOANS $12,600 is loaned at an annual simple interest rate of 18%. Find the total amount that must be repaid at the end of a 90-day period. [Section 7.5] $13,159.23
72. SPINAL CORD INJURIES Refer to the circle graph below. [Section 8.1]
    a. What percent of spinal cord injuries are caused by sports accidents? 8%
    b. If there are approximately 12,000 new cases of spinal cord injury each year, according to the graph, how many of them were caused by vehicle crashes? 4,680 cases

Causes of Spinal Cord Injury in the United States

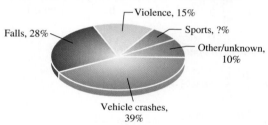

Source: National Spinal Cord Injury Statistical Center

73. AVALANCHES The bar graph on the next page shows the number of deaths from avalanches in the United States for the winter seasons ending in the years 2000 to 2011. Use the graph to answer the following questions. [Section 8.1]
    a. In which year(s) were there the most deaths from avalanches? How many deaths were there?
       2008 and 2010; 36
    b. Between what two years was there the greatest increase in the number of deaths from avalanches? What was the increase? 2007 to 2008; an increase of 16 deaths
    c. Between what two years was there the greatest decrease in the number of deaths from avalanches? What was the decrease? 2010 to 2011; a decrease of 11 deaths

Source: Northwest Weather and Avalanche Center

74. TEAM GPA The grade point averages of the players on a badminton team are listed below. Find the mean, median, mode, and range of the team's GPAs. [Section 8.2] mean: 3.02; median: 3.00; mode: 2.75; range: 1.79

3.04 4.00 2.75 3.23 3.87 2.21
3.02 2.25 2.98 2.56 3.58 2.75

75. Is $(-2, 3)$ a solution of $4x - 5y = -23$? [Section 8.3]
yes

76. Complete the solution (, 4) for the equation $4x - 5y = -4$. [Section 8.3] 4

77. Graph the points: $(-1, 3)$, $(0, 1.5)$, $(-4, -4)$, $\left(2, \frac{7}{2}\right)$, and $(4, 0)$ [Section 8.3]

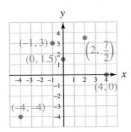

78. Graph: $x = 4$ [Section 8.4]

79. Graph: $y = -x - 1$ [Section 8.4]

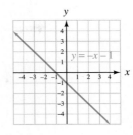

80. Graph: $3x - 3y = 9$ [Section 8.4]

81. Refer to the graph below that gives mileage data for a Toyota Prius.

a. What is the g-intercept of the graph? $(0, 12)$

b. What information does it give? When the car has been driven 0 miles, the fuel tank contains 12 gallons of gasoline.

c. What is the m-intercept of the graph? $(600, 0)$

d. What information does it give? After 600 miles of driving, the fuel tank contains 0 gallons of gasoline.

An Introduction to Geometry

9.1 Basic Geometric Figures; Angles
9.2 Parallel and Perpendicular Lines
9.3 Triangles
9.4 The Pythagorean Theorem
9.5 Congruent Triangles and Similar Triangles
9.6 Quadrilaterals and Other Polygons
9.7 Perimeters and Areas of Polygons
9.8 Circles
9.9 Volume
Chapter Summary and Review
Chapter Test
Cumulative Review

from Campus to Careers

Surveyor

Surveyors measure distances, directions, elevations (heights), contours (curves), and angles between lines on Earth's surface. Surveys are also done in the air and underground. Surveyors often work in teams. They use a variety of instruments and electronics, including the Global Positioning System (GPS). In general, people who like surveying also like math—primarily geometry and trigonometry. The field attracts people with geology, forestry, history, engineering, computer science, and astronomy backgrounds, too.

In **Problem 83** of **Study Set 9.5**, you will see how a surveyor, using geometry, can stay on dry land and yet measure the width of a river.

JOB TITLE: Surveyor
EDUCATION: Courses in algebra, geometry, trigonometry, and computer science are required.
JOB OUTLOOK: Job growth is expected to be 25% through 2020—much faster than the average for all occupations.
ANNUAL EARNINGS: In 2011, the median annual income was $58,740.
FOR MORE INFORMATION:
http://www.bls.gov/oes/current/oes171022.htm

Objectives

1. Identify and name points, lines, and planes.
2. Identify and name line segments and rays.
3. Identify and name angles.
4. Use a protractor to measure angles.
5. Solve problems involving adjacent angles.
6. Use the property of vertical angles to solve problems.
7. Solve problems involving complementary and supplementary angles.

SECTION 9.1
Basic Geometric Figures; Angles

Geometry is a branch of mathematics that studies the properties of two- and three-dimensional figures such as triangles, circles, cylinders, and spheres. More than 5,000 years ago, Egyptian surveyors used geometry to measure areas of land in the flooded plains of the Nile River after heavy spring rains. Even today, engineers marvel at the Egyptians' use of geometry in the design and construction of the pyramids. History records many other practical applications of geometry made by Babylonians, Chinese, Indians, and Romans.

> **The Language of Algebra** The word *geometry* comes from the Greek words *geo* (meaning earth) and *metron* (meaning measure).

Many scholars consider **Euclid** (330?–275? BCE) to be the greatest of the Greek mathematicians. His book *The Elements* is an impressive study of geometry and number theory. It presents geometry in a highly structured form that begins with several simple assumptions and then expands on them using logical reasoning. For more than 2,000 years, *The Elements* was the textbook that students all over the world used to learn geometry.

Euclid

1 Identify and name points, lines, and planes.

Geometry is based on three undefined words: *point*, *line*, and *plane*. Although we will make no attempt to define these words formally, we can think of a **point** as a geometric figure that has position but no length, width, or depth. Points can be represented on paper by drawing small dots, and they are labeled with capital letters. For example, point A is shown in figure (a) below.

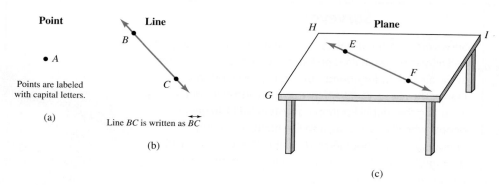

Lines are made up of points. A line extends infinitely far in both directions, but has no width or depth. A line can be represented on paper by drawing a straight line with arrowheads at either end. We can name a line using any two points on the line. In figure (b) above, the line that passes through points B and C is written as $\overleftrightarrow{BC}$.

Planes are also made up of points. A plane is a flat surface, extending infinitely far in every direction, that has length and width but no depth. The top of a table, a floor, or a wall is part of a plane. We can name a plane using any three points that lie in the plane. In figure (c) above, $\overleftrightarrow{EF}$ lies in plane GHI.

As figure (b) illustrates, points B and C determine exactly one line, the line $\overleftrightarrow{BC}$. In figure (c), the points E and F determine exactly one line, the line $\overleftrightarrow{EF}$. In general, any two different points determine exactly one line.

As figure (c) illustrates, points G, H, and I determine exactly one plane. In general, any three different points that do not lie on a line determine exactly one plane.

Other geometric figures can be created by using parts or combinations of points, lines, and planes.

2 Identify and name line segments and rays.

Line Segment

The **line segment** AB, written as $\overline{AB}$, is the part of a line that consists of points A and B and all points in between (see the figure below). Points A and B are the **endpoints** of the segment.

Line segment

Line segment AB is written as $\overline{AB}$.

Every line segment has a **midpoint**, which divides the segment into two parts of equal length. In the figure below, M is the midpoint of segment AB, because the measure of $\overline{AM}$, which is written as m($\overline{AM}$), is equal to the measure of $\overline{MB}$ which is written as m($\overline{MB}$).

$$m(\overline{AM}) = 4 - 1$$
$$m(\overline{AM}) = 3$$

and

$$m(\overline{MB}) = 7 - 4$$
$$m(\overline{MB}) = 3$$

Since the measure of both segments is 3 units, we can write m($\overline{AM}$) = m($\overline{MB}$).

When two line segments have the same measure, we say that they are **congruent**. Since m($\overline{AM}$) = m($\overline{MB}$), we can write

$$\overline{AM} \cong \overline{MB}$$ Read the symbol $\cong$ as "is congruent to."

Another geometric figure is the ray, as shown below.

Ray

A **ray** is the part of a line that begins at some point (say, A) and continues forever in one direction. Point A is the **endpoint** of the ray.

Ray

Ray AB is written as $\overrightarrow{AB}$. The endpoint of the ray is always listed first.

To name a ray, we list its endpoint and then one other point on the ray. Sometimes it is possible to name a ray in more than one way. For example, in the figure on the right, $\overrightarrow{DE}$ and $\overrightarrow{DF}$ name the same ray. This is because both have point D as their endpoint and extend forever in the same direction. In contrast, $\overrightarrow{DE}$ and $\overrightarrow{ED}$ are not the same ray. They have different endpoints and point in opposite directions.

3 Identify and name angles.

> **Angle**
>
> An **angle** is a figure formed by two rays with a common endpoint. The common endpoint is called the **vertex**, and the rays are called **sides**.

The angle shown below can be written as $\angle BAC$, $\angle CAB$, $\angle A$, or $\angle 1$. The symbol $\angle$ means angle.

Angle

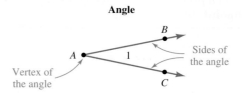

> *Caution!* When using three letters to name an angle, be sure the letter name of the vertex is the middle letter. Furthermore, we can only name an angle using a single vertex letter when there is no possibility of confusion. For example, in the figure on the right, we cannot refer to any of the angles as simply $\angle X$, because we would not know if that meant $\angle WXY$, $\angle WXZ$, or $\angle YXZ$.

4 Use a protractor to measure angles.

One unit of measurement of an angle is the **degree**. The symbol for degree is a small raised circle, °. An angle measure of 1° (read as "one degree") means that one side of an angle is rotated $\frac{1}{360}$ of a complete revolution about the vertex from the other side of the angle. The measure of $\angle ABC$, shown below, is 1°. We can write this in symbols as m($\angle ABC$) = 1°.

The following figures show the measures of several other angles. An angle measure of 90° is equivalent to $\frac{90}{360} = \frac{1}{4}$ of a complete revolution. An angle measure of 180° is equivalent to $\frac{180}{360} = \frac{1}{2}$ of a complete revolution, and an angle measure of 60° is equivalent to $\frac{60}{360} = \frac{1}{6}$ of a complete revolution.

m($\angle FED$) = 90°   m($\angle IHG$) = 180°   m($\angle JKL$) = 60°

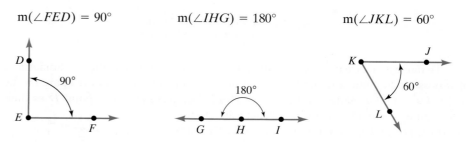

Unless otherwise noted, all art on this page is © Cengage Learning.

We can use a **protractor** to measure angles. To begin, we place the center of the protractor at the vertex of the angle, with the edge of the protractor aligned with one side of the angle, as shown below. The angle measure is found by determining where the other side of the angle crosses the scale. Be careful to use the appropriate scale, inner or outer, when reading an angle measure.

If we read the protractor from right to left, using the outer scale, we see that m($\angle ABC$) = 30°. If we read the protractor from left to right, using the inner scale, we can see that m($\angle GBF$) = 30°.

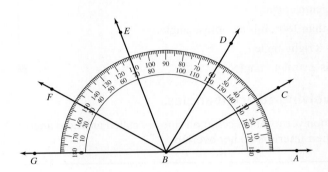

Angle	Measure in degrees
$\angle ABC$	30°
$\angle ABD$	60°
$\angle ABE$	110°
$\angle ABF$	150°
$\angle ABG$	180°
$\angle GBF$	30°
$\angle GBC$	150°

When two angles have the same measure, we say that they are **congruent**. Since m($\angle ABC$) = 30° and m($\angle GBF$) = 30°, we can write

$\angle ABC \cong \angle GBF$ Read the symbol $\cong$ as "is congruent to."

We classify angles according to their measure.

Classifying Angles

Acute angles: Angles whose measures are greater than 0° but less than 90°.
Right angles: Angles whose measures are 90°.
Obtuse angles: Angles whose measures are greater than 90° but less than 180°.
Straight angles: Angles whose measures are 180°.

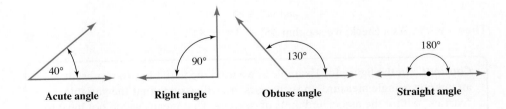

The Language of Algebra A ⌐ symbol is often used to label a right angle. For example, in the figure on the right, the ⌐ symbol drawn near the vertex of $\angle ABC$ indicates that m($\angle ABC$) = 90°.

Unless otherwise noted, all art on this page is © Cengage Learning.

Self Check 1

Classify ∠EFG, ∠DEF, ∠1, and ∠GED in the figure as an acute angle, a right angle, an obtuse angle, or a straight angle.

Now Try Problems 57, 59, and 61
Self Check 1 Answer
right angle, obtuse angle, acute angle, straight angle

EXAMPLE 1
Classify each angle in the figure as an acute angle, a right angle, an obtuse angle, or a straight angle.

Strategy We will determine how each angle's measure compares to 90° or to 180°.

WHY Acute, right, obtuse, and straight angles are defined with respect to 90° and 180° angle measures.

Solution
Since m(∠1) < 90°, it is an acute angle.
Since m(∠2) > 90° but less than 180°, it is an obtuse angle.
Since m(∠BDE) = 90°, it is a right angle.
Since m(∠ABC) = 180°, it is a straight angle.

5 Solve problems involving adjacent angles.

Two angles that have a common vertex and a common side are called **adjacent angles** if they are side by side and their interiors do not overlap.

> **Success Tip** We can use the algebra concepts of variable and equation that were introduced in Chapter 3 to solve many types of geometry problems.

Self Check 2

Use the information in the figure to find x.   35°

Now Try Problem 65

Teaching Example 2 Use the information in the figure to find x.

Answer: 25°

EXAMPLE 2
Two angles with degree measures of x and 35° are adjacent angles, as shown. Use the information in the figure to find x.

Strategy We will write an equation involving x that mathematically models the situation.

WHY We can then solve the equation to find the unknown angle measure.

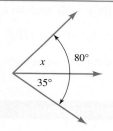

Adjacent angles

Solution
Since the sum of the measures of the two adjacent angles is 80°, we have

$x + 35° = 80°$ The word sum indicates addition.
$x + 35° - 35° = 80° - 35°$ To isolate x, undo the addition of 35° by subtracting 35° from both sides.
$x = 45°$ Do the subtractions: 35° − 35° = 0° and 80° − 35° = 45°.

$$\begin{array}{r} \overset{7\ 10}{8\cancel{0}} \\ -35 \\ \hline 45 \end{array}$$

Thus, x is 45°. As a check, we see that 45° + 35° = 80°.

> **Caution!** In the figure for Example 2, we used the variable x to represent an unknown angle measure. In such cases, we will assume that the variable "carries" with it the associated units of degrees. That means we do not have to write a ° symbol next to the variable. Furthermore, if x represents an unknown number of degrees, then expressions such as 3x, x + 15°, and 4x − 20° also have units of degrees.

6 Use the property of vertical angles to solve problems.

When two lines intersect, pairs of nonadjacent angles are called **vertical angles**. In the following figure, ∠1 and ∠3 are vertical angles and ∠2 and ∠4 are vertical angles.

Vertical angles

- ∠1 and ∠3
- ∠2 and ∠4

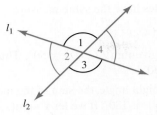

The Language of Algebra When we work with two (or more) lines at one time, we can use **subscripts** to name the lines. The prefix *sub* means below or beneath, as in *sub*marine or *sub*way. To name the first line in the figure above, we use l_1, which is read as "l sub one." To name the second line, we use l_2, which is read as "l sub two."

To illustrate that vertical angles always have the same measure, refer to the figure below, with angles having measures of x, y, and 30°. Since the measure of any straight angle is 180°, we have

$30° + x = 180°$ and $30° + y = 180°$
$x = 150°$ $y = 150°$ To undo the addition of 30°, subtract 30° from both sides.

Since x and y are both 150°, we conclude that $x = y$.

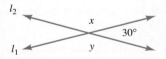

Note that the angles having measures x and y are vertical angles.

The previous example illustrates that vertical angles have the same measure. Recall that when two angles have the same measure, we say that they are *congruent*. Therefore, we have the following important fact.

Property of Vertical Angles

Vertical angles are congruent (have the same measure).

EXAMPLE 3 Refer to the figure. Find:

a. m(∠1)   **b.** m(∠ABF)

Strategy To answer part a, we will use the property of vertical angles. To answer part b, we will write an equation involving m(∠ABF) that mathematically models the situation.

WHY For part a, we note that $\overleftrightarrow{AD}$ and $\overleftrightarrow{BC}$ intersect to form vertical angles. For part b, we can solve the equation to find the unknown, m(∠ABF).

Solution

a. If we ignore $\overleftrightarrow{FE}$ for the moment, we see that $\overleftrightarrow{AD}$ and $\overleftrightarrow{BC}$ intersect to form the pair of vertical angles ∠CBD and ∠1. By the property of vertical angles,

∠CBD ≅ ∠1   Read as "angle CBD is congruent to angle one."

Self Check 3

Refer to the figure for Example 3. Find:

a. m(∠2) 100°
b. m(∠DBE) 30°

Now Try Problems 69 and 71

Teaching Example 3 For the given figure, find:
a. m(∠2)
b. m(∠PTR)
Answers:
a. 40°
b. 110°

Since congruent angles have the same measure,

$$m(\angle CBD) = m(\angle 1)$$

In the figure, we are given $m(\angle CBD) = 50°$. Thus, $m(\angle 1)$ is also 50°, and we can write $m(\angle 1) = 50°$.

b. Since $\angle ABD$ is a straight angle, the sum of the measures of $\angle ABF$, the 100° angle, and the 50° angle is 180°. If we let $x = m(\angle ABF)$, we have

$x + 100° + 50° = 180°$	The word *sum* indicates addition.
$x + 150° = 180°$	On the left side, combine like terms: $100° + 50° = 150°$.
$x = 30°$	To isolate x, undo the addition of 150° by subtracting 150° from both sides: $180° - 150° = 30°$.

Thus, $m(\angle ABF) = 30°$

Self Check 4

In the figure below, find:

a. y 15°
b. $m(\angle XYZ)$ 50°
c. $m(\angle MYX)$ 130°

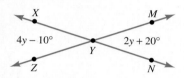

Now Try Problem 75

Teaching Example 4 In the following figure, find:
a. x
b. $m(\angle MNP)$
c. $m(\angle PNO)$
Answers:
a. 30°
b. 140°
c. 40°

EXAMPLE 4
In the figure on the right, find:

a. x **b.** $m(\angle ABC)$ **c.** $m(\angle CBE)$

Strategy We will use the property of vertical angles to write an equation that mathematically models the situation.

WHY $\overleftrightarrow{AE}$ and $\overleftrightarrow{DC}$ intersect to form two pairs of vertical angles.

Solution

a. In the figure, two vertical angles have degree measures that are represented by the algebraic expressions $4x - 20°$ and $3x + 15°$. Since the angles are vertical angles, they have equal measures.

$4x - 20° = 3x + 15°$	Set the algebraic expressions equal.
$4x - 20° - 3x = 3x + 15° - 3x$	To eliminate $3x$ from the right side, subtract $3x$ from both sides.
$x - 20° = 15°$	Combine like terms: $4x - 3x = x$ and $3x - 3x = 0$.
$x = 35°$	To isolate x, undo the subtraction of 20° by adding 20° to both sides.

Thus, x is 35°.

b. To find $m(\angle ABC)$, we evaluate the expression $3x + 15°$ for $x = 35°$.

$3x + 15° = 3(35°) + 15°$	Substitute 35° for x.
$= 105° + 15°$	Do the multiplication.
$= 120°$	Do the addition.

$$\begin{array}{r} \overset{1}{3}5 \\ \times\,3 \\ \hline 105 \end{array}$$

Thus, $m(\angle ABC) = 120°$.

c. $\angle ABE$ is a straight angle. Since the measure of a straight angle is 180° and $m(\angle ABC) = 120°$, $m(\angle CBE)$ must be $180° - 120°$, or 60°.

7 Solve problems involving complementary and supplementary angles.

> **Complementary and Supplementary Angles**
>
> Two angles are **complementary angles** when the sum of their measures is 90°.
> Two angles are **supplementary angles** when the sum of their measures is 180°.

In figure (a) below, ∠ABC and ∠CBD are complementary angles because the sum of their measures is 90°. Each angle is said to be the **complement** of the other. In figure (b) below, ∠X and ∠Y are also complementary angles, because m(∠X) + m(∠Y) = 90°. Figure (b) illustrates an important fact: Complementary angles need not be adjacent angles.

Complementary angles

60° + 30° = 90°
(a)

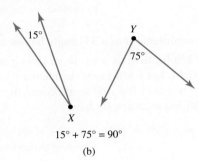

15° + 75° = 90°
(b)

In figure (a) below, ∠MNO and ∠ONP are supplementary angles because the sum of their measures is 180°. Each angle is said to be the **supplement** of the other. Supplementary angles need not be adjacent angles. For example, in figure (b) below, ∠G and ∠H are supplementary angles, because m(∠G) + m(∠H) = 180°.

Supplementary angles

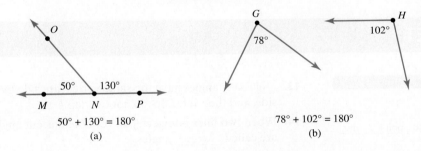

50° + 130° = 180°   78° + 102° = 180°
(a)   (b)

Caution! The definition of supplementary angles requires that the sum of *two* angles be 180°. Three angles of 40°, 60°, and 80° are not supplementary even though their sum is 180°.

EXAMPLE 5

a. Find the complement of a 35° angle.
b. Find the supplement of a 105° angle.

Strategy We will use the definitions of complementary and supplementary angles to write equations that mathematically model each situation.

WHY We can then solve each equation to find the unknown angle measure.

Self Check 5

a. Find the complement of a 50° angle. 40°
b. Find the supplement of a 50° angle. 130°

Now Try Problems 77 and 79

Teaching Example 5
a. Find the complement of a 47° angle.
b. Find the supplement of a 118° angle.
Answers:
a. 43°
b. 62°

Solution

a. It is helpful to draw a figure, as shown to the right. Let x represent the measure of the complement of the 35° angle. Since the angles are complementary, we have

$x + 35° = 90°$ The sum of the angles' measures must be 90°.

$x = 55°$ To isolate x, undo the addition of 35° by subtracting 35° from both sides: $90° - 35° = 55°$.

The complement of a 35° angle has measure 55°.

b. It is helpful to draw a figure, as shown on the right. Let y represent the measure of the supplement of the 105° angle. Since the angles are supplementary, we have

$y + 105° = 180°$ The sum of the angles' measures must be 180°.

$y = 75°$ To isolate y, undo the addition of 105° by subtracting 105° from both sides: $180° - 105° = 75°$.

The supplement of a 105° angle has measure 75°.

ANSWERS TO SELF CHECKS

1. right angle, obtuse angle, acute angle, straight angle   2. 35°   3. a. 100°   b. 30°
4. a. 15°   b. 50°   c. 130°   5. a. 40°   b. 130°

SECTION 9.1 STUDY SET

VOCABULARY

Fill in the blanks.

1. Three undefined words in geometry are point , line , and plane .
2. A line segment has two endpoints.
3. A midpoint divides a line segment into two parts of equal length.
4. A ray is the part of a line that begins at some point and continues forever in one direction.
5. An angle is formed by two rays with a common endpoint.
6. An angle is measured in degrees .
7. A protractor is used to measure angles.
8. The measure of an acute angle is less than 90°.
9. The measure of a right angle is 90°.
10. The measure of an obtuse angle is greater than 90° but less than 180°.
11. The measure of a straight angle is 180° .
12. When two segments have the same length, we say that they are congruent .
13. Adjacent angles have the same vertex, are side by side, and their interiors do not overlap.
14. When two lines intersect, pairs of nonadjacent angles are called vertical angles.
15. When two angles have the same measure, we say that they are congruent .
16. The word *sum* indicates the operation of addition .
17. The sum of two complementary angles is 90° .
18. The sum of two supplementary angles is 180°.

CONCEPTS

19. **a.** Given two points (say, M and N), how many different lines pass through these two points? one
    b. Fill in the blank: In general, two different points determine exactly one line .
20. Refer to the figure.
    a. Name $\overrightarrow{NM}$ in another way. $\overrightarrow{NC}$
    b. Do $\overrightarrow{MN}$ and $\overrightarrow{NM}$ name the same ray? no

21. Consider the acute angle shown below.
    a. What two rays are the sides of the angle? $\overrightarrow{SR}, \overrightarrow{ST}$
    b. What point is the vertex of the angle? S
    c. Name the angle in four ways. $\angle RST, \angle TSR, \angle S, \angle 1$

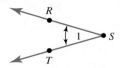

22. Estimate the measure of each angle. Do not use a protractor.

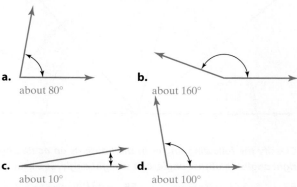

    a. about 80°   b. about 160°
    c. about 10°   d. about 100°

23. Draw an example of each type of angle.
    a. an acute angle   b. an obtuse angle

    c. a right angle   d. a straight angle

24. Fill in the blanks with the correct symbol.
    a. If $m(\overline{AB}) = m(\overline{CD})$, then $\overline{AB} \cong \overline{CD}$.
    b. If $\angle ABC \cong \angle DEF$, then $m(\angle ABC) = m(\angle DEF)$.

25. a. Draw a pair of adjacent angles. Label them $\angle ABC$ and $\angle CBD$.

   b. Draw two intersecting lines. Label them lines l_1 and l_2. Label one pair of vertical angles that are formed as $\angle 1$ and $\angle 2$.

   c. Draw two adjacent complementary angles.

   d. Draw two adjacent supplementary angles.

26. Fill in the blank:
    If $\angle MNO \cong \angle BFG$, then $m(\angle MNO) = m(\angle BFG)$.

27. Fill in the blank:
    The vertical angle property: Vertical angles are congruent.

28. Refer to the figure below. Fill in the blanks.
    a. $\angle XYZ$ and $\angle$ VYW are vertical angles.
    b. $\angle XYZ$ and $\angle ZYW$ are supplementary angles.
    c. $\angle ZYW$ and $\angle XYV$ are vertical angles.

29. Refer to the figure below and tell whether each statement is true.
    a. $\angle AGF$ and $\angle BGC$ are vertical angles. false
    b. $\angle FGE$ and $\angle BGA$ are adjacent angles. false
    c. $m(\angle AGB) = m(\angle BGC)$. false
    d. $\angle AGC \cong \angle DGF$. true

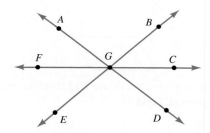

30. Refer to the figure below and tell whether the angles are congruent.
    a. $\angle 1$ and $\angle 2$ yes   b. $\angle FGB$ and $\angle CGE$ yes
    c. $\angle AGF$ and $\angle FGE$ d. $\angle CGD$ and $\angle CGB$
       no   no

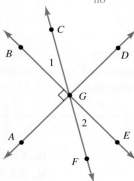

Refer to the figure above and tell whether each statement is true.

31. $\angle 1$ and $\angle CGD$ are adjacent angles. true
32. $\angle FGA$ and $\angle AGC$ are supplementary. true
33. $\angle AGB$ and $\angle BGC$ are complementary. false
34. $\angle AGF$ and $\angle 2$ are complementary. true

NOTATION

Fill in the blanks.

35. The symbol $\overleftrightarrow{AB}$ is read as " _line_ AB."
36. The symbol $\overline{AB}$ is read as " _segment_ AB."
37. The symbol $\overrightarrow{AB}$ is read as " _ray_ AB."
38. We read m($\overline{AB}$) as "the _measure_ of segment AB."
39. We read $\angle ABC$ as " _angle_ ABC."
40. We read m($\angle ABC$) as "the _measure_ of angle ABC."
41. The symbol for _degree_ is a small raised circle, °.
42. The symbol ⌐ indicates a _right_ angle.
43. The symbol ≅ is read as "is _congruent_ to."
44. The symbol l_1 can be used to name a line. It is read as "line l _sub_ one."

GUIDED PRACTICE

45. Draw each geometric figure and label it completely. **See Objective 1.**

   a. Point T T

   b. $\overleftrightarrow{JK}$

   c. Plane ABC

46. Draw each geometric figure and label it completely. **See Objectives 2 and 3.**

   a. $\overline{RS}$

   b. $\overrightarrow{PQ}$

   c. $\angle XYZ$

   d. $\angle L$

47. Refer to the figure and find the length of each segment. **See Objective 2.**

   a. $\overline{AB}$ 2
   b. $\overline{CE}$ 3
   c. $\overline{DC}$ 1
   d. $\overline{EA}$ 6

48. Refer to the figure above and find each midpoint. **See Objective 2.**

   a. Find the midpoint of $\overline{AD}$. B
   b. Find the midpoint of $\overline{BE}$. D
   c. Find the midpoint of $\overline{EA}$. C

Use the protractor to find each angle measure listed below. **See Objective 4.**

49. m($\angle GDE$) 50°
50. m($\angle ADE$) 140°
51. m($\angle EDS$) 25°
52. m($\angle EDR$) 105°
53. m($\angle CDR$) 75°
54. m($\angle CDA$) 40°
55. m($\angle CDG$) 130°
56. m($\angle CDS$) 155°

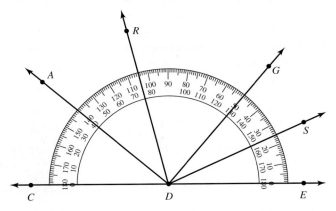

Classify the following angles in the figure as an acute angle, a right angle, an obtuse angle, or a straight angle. **See Example 1.**

57. $\angle MNO$ right
58. $\angle OPN$ acute
59. $\angle NOP$ acute
60. $\angle POS$ obtuse
61. $\angle MPQ$ straight
62. $\angle PNO$ right
63. $\angle QPO$ obtuse
64. $\angle MNQ$ straight

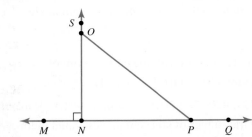

Find x. **See Example 2.**

65. 10°, 55°, 45°, x

66. 112°, x, 56°, 168°

67. 27.5°, x, 50°, 22.5°

68. 90°, 130°, x, 40°

Refer to the figure below. Find the measure of each angle. See Example 3.

69. ∠1   70°

70. ∠MYX   65°

71. ∠NYZ   65°

72. ∠2   45°

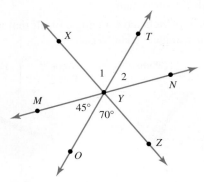

First find x. Then find m(∠ABD) and m(∠DBE). See Example 4.

73.
30°, 60°, 120°

74.
12°, 80°, 100°

First find x. Then find m(∠ZYQ) and m(∠PYQ). See Example 4.

75.
25°, 115°, 65°

76.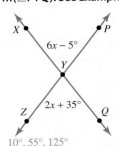
10°, 55°, 125°

Let x represent the unknown angle measure. Write an equation and solve it to find x. See Example 5.

77. Find the complement of a 30° angle.   60°

78. Find the supplement of a 30° angle.   150°

79. Find the supplement of a 105° angle.   75°

80. Find the complement of a 75° angle.   15°

TRY IT YOURSELF

81. Refer to the figure in the next column and tell whether each statement is true. If a statement is false, explain why.

  a. $\overrightarrow{GF}$ has point G as its endpoint.   true

  b. $\overline{AG}$ has no endpoints.   false, a segment has two endpoints

  c. $\overleftrightarrow{CD}$ has three endpoints.   false, a line does not have an endpoint

  d. Point D is the vertex of ∠DGB.   false, point G is the vertex of the angle

  e. m(∠AGC) = m(∠BGD)   true

  f. ∠AGF ≅ ∠BGE   true

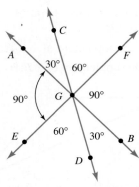

82. Refer to the figure for Problem 81 and tell whether each angle is an acute angle, a right angle, an obtuse angle, or a straight angle.

  a. ∠AGC   acute

  b. ∠EGA   right

  c. ∠FGD   obtuse

  d. ∠BGA   straight

Use a protractor to measure each angle.

83.

84.

85.

86.

87. Refer to the figure below, in which m(∠1) = 50°. Find the measure of each angle or sum of angles.

  a. ∠3   50°

  b. ∠4   130°

  c. m(∠1) + m(∠2) + m(∠3)   230°

  d. m(∠2) + m(∠4)   260°

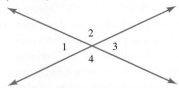

88. Refer to the figure below, in which m(∠1) + m(∠3) + m(∠4) = 180°, ∠3 ≅ ∠4, and ∠4 ≅ ∠5. Find the measure of each angle.

  a. ∠1   100°

  b. ∠2   80°

  c. ∠3   40°

  d. ∠6   140°

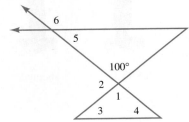

89. Refer to the figure below where ∠1 ≅ ∠ACD, ∠1 ≅ ∠2, and ∠BAC ≅ ∠2.
    a. What is the complement of ∠BAC? 66°
    b. What is the supplement of ∠BAC? 156°

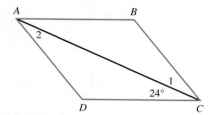

90. Refer to the figure below where ∠EBS ≅ ∠BES.
    a. What is the measure of ∠AEF? 128°
    b. What is the supplement of ∠AET? 142°

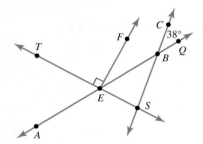

91. Find the supplement of the complement of a 51° angle. 141°
92. Find the complement of the supplement of a 173° angle. 83°
93. Find the complement of the complement of a 1° angle. 1°
94. Find the supplement of the supplement of a 6° angle. 6°

APPLICATIONS

95. MUSICAL INSTRUMENTS Suppose that you are a beginning band teacher describing the correct posture needed to play various instruments. Using the diagrams shown below, approximate the angle measure (in degrees) at which each instrument should be held in relation to the student's body.

   a. flute   b. clarinet   c. trumpet

   about 80°   about 30°   about 65°

96. PLANETS The figures below show the direction of rotation of several planets in our solar system. They also show the angle of tilt of each planet.
    a. Which planets have an angle of tilt that is an acute angle? Earth, Saturn
    b. Which planets have an angle of tilt that is an obtuse angle? Pluto, Venus

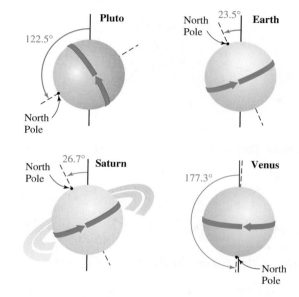

97. a. AVIATION How many degrees from the horizontal position are the wings of the airplane? 27°

   b. GARDENING What angle does the handle of the lawn mower make with the ground? 30°

98. SYNTHESIZER Find x and y. 65°, 115°

WRITING

99. PHRASES Explain what you think each of these phrases means. How is geometry involved?
  a. The president did a complete 180-degree flip on the subject of a tax cut.
  b. The rollerblader did a "360" as she jumped off the ramp.

100. In the statements below, the ° symbol is used in two different ways. Explain the difference.

$$m(\angle A) = 85° \quad \text{and} \quad 85°F$$

101. Can two angles that are complementary be equal? Explain.

102. Explain why the angles highlighted below are not vertical angles.

REVIEW

103. Solve: $x + 3x + 20 = 180$ 40

104. Solve: $3x + 4 = 5x - 40$ 22

105. Solve: $5x = 6x - 15$ 15

106. Solve: $50 + x + 45 = 180$ 85

SECTION 9.2
Parallel and Perpendicular Lines

Objectives
1. Identify and define parallel and perpendicular lines.
2. Identify corresponding angles, interior angles, and alternate interior angles.
3. Use properties of parallel lines cut by a transversal to find unknown angle measures.

In this section, we will consider *parallel* and *perpendicular* lines. Since parallel lines are always the same distance apart, the railroad tracks shown in figure (a) illustrate one application of parallel lines. Figure (b) shows one of the events of men's gymnastics, the parallel bars. Since perpendicular lines meet and form right angles, the monument and the ground shown in figure (c) illustrate one application of perpendicular lines.

(a)   (b)   (c)

The symbol ⌐ indicates a right angle.

1 Identify and define parallel and perpendicular lines.

If two lines lie in the same plane, they are called **coplanar**. Two coplanar lines that do not intersect are called **parallel lines**. See figure (a) on the next page. If two lines do not lie in the same plane, they are called noncoplanar. Two noncoplanar lines that do not intersect are called **skew lines**.

Parallel lines **Perpendicular lines**

(a) (b)

Parallel Lines

Parallel lines are coplanar lines that do not intersect.

Some lines that intersect are perpendicular. See figure (b) above.

Perpendicular Lines

Perpendicular lines are lines that intersect and form right angles.

The Language of Algebra If lines l_1 (read as "l sub 1") and l_2 (read as "l sub 2") are parallel, we can write $l_1 \parallel l_2$, where the symbol $\parallel$ is read as "is parallel to."

If lines l_1 and l_2 are perpendicular, we can write $l_1 \perp l_2$, where the symbol $\perp$ is read as "is perpendicular to."

2 Identify corresponding angles, interior angles, and alternate interior angles.

A line that intersects two coplanar lines in two distinct (different) points is called a **transversal**. For example, line l_1 in the figure to the right is a transversal intersecting lines l_2 and l_3.

When two lines are cut by a transversal, all eight angles that are formed are important in the study of parallel lines. Descriptive names are given to several pairs of these angles.

In the figure below, four pairs of **corresponding angles** are formed.

Corresponding angles

- $\angle 1$ and $\angle 5$
- $\angle 3$ and $\angle 7$
- $\angle 2$ and $\angle 6$
- $\angle 4$ and $\angle 8$

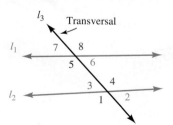

Corresponding Angles

If two lines are cut by a transversal, then the angles on the same side of the transversal and in corresponding positions with respect to the lines are called corresponding angles.

In the figure below, four **interior angles** are formed.

Interior angles

- ∠3, ∠4, ∠5, and ∠6

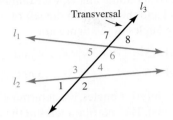

In the figure below, two pairs of **alternate interior angles** are formed.

Alternate interior angles

- ∠4 and ∠5
- ∠3 and ∠6

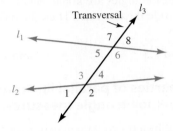

Alternate Interior Angles

If two lines are cut by a transversal, then the nonadjacent angles on opposite sides of the transversal and on the interior of the two lines are called alternate interior angles.

Success Tip Alternate interior angles are easily spotted because they form a Z-shape or a backward Z-shape, as shown below.

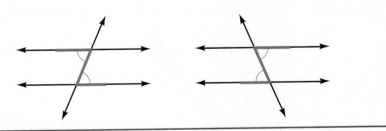

EXAMPLE 1 Refer to the figure. Identify:

a. all pairs of corresponding angles
b. all interior angles
c. all pairs of alternate interior angles

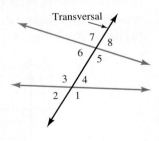

Strategy When two lines are cut by a transversal, eight angles are formed. We will consider the relative position of the angles with respect to the two lines and the transversal.

WHY There are four pairs of corresponding angles, four interior angles, and two pairs of alternate interior angles.

Self Check 1

Refer to the figure below. Identify:
a. all pairs of corresponding angles
b. all interior angles
c. all pairs of alternate interior angles

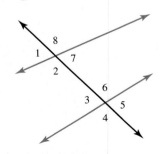

Now Try Problem 21

Self Check 1 Answers
a. ∠1 and ∠3, ∠2 and ∠4, ∠8 and ∠6, ∠7 and ∠5
b. ∠2, ∠7, ∠3, and ∠6
c. ∠2 and ∠6, ∠7 and ∠3

Solution

a. To identify corresponding angles, we examine the angles to the right of the transversal and the angles to the left of the transversal. The pairs of corresponding angles in the figure are

- ∠1 and ∠5
- ∠2 and ∠6
- ∠4 and ∠8
- ∠3 and ∠7

b. To identify the interior angles, we determine the angles inside the two lines cut by the transversal. The interior angles in the figure are

∠3, ∠4, ∠5, and ∠6

c. Alternate interior angles are nonadjacent angles on opposite sides of the transversal inside the two lines. Thus, the pairs of alternate interior angles in the figure are

- ∠3 and ∠5
- ∠4 and ∠6

3 Use properties of parallel lines cut by a transversal to find unknown angle measures.

Lines that are cut by a transversal may or may not be parallel. When a pair of parallel lines are cut by a transversal, we can make several important observations about the angles that are formed.

1. **Corresponding angles property:** If two parallel lines are cut by a transversal, each pair of corresponding angles are congruent. In the figure below, if $l_1 \parallel l_2$, then $\angle 1 \cong \angle 5$, $\angle 3 \cong \angle 7$, $\angle 2 \cong \angle 6$, and $\angle 4 \cong \angle 8$.

2. **Alternate interior angles property:** If two parallel lines are cut by a transversal, alternate interior angles are congruent. In the figure below, if $l_1 \parallel l_2$, then $\angle 3 \cong \angle 6$ and $\angle 4 \cong \angle 5$.

3. **Interior angles property:** If two parallel lines are cut by a transversal, interior angles on the same side of the transversal are supplementary. In the figure below, if $l_1 \parallel l_2$, then $\angle 3$ is supplementary to $\angle 5$ and $\angle 4$ is supplementary to $\angle 6$.

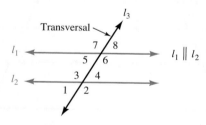

4. If a transversal is perpendicular to one of two parallel lines, it is also perpendicular to the other line. In figure (a) below, if $l_1 \parallel l_2$ and $l_3 \perp l_1$, then $l_3 \perp l_2$.

5. If two lines are parallel to a third line, they are parallel to each other. In figure (b) below, if $l_1 \parallel l_2$ and $l_1 \parallel l_3$, then $l_2 \parallel l_3$.

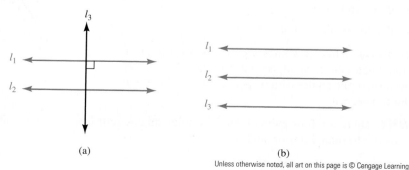

(a)   (b)

9.2 Parallel and Perpendicular Lines

EXAMPLE 2 Refer to the figure. If $l_1 \parallel l_2$ and $m(\angle 3) = 120°$, find the measures of the other seven angles that are labeled.

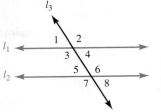

Strategy We will look for vertical angles, supplementary angles, and alternate interior angles in the figure.

WHY The facts that we have studied about vertical angles, supplementary angles, and alternate interior angles enable us to use known angle measures to find unknown angle measures.

Solution

$m(\angle 1) = 60°$	$\angle 3$ and $\angle 1$ are supplementary: $m(\angle 3) + m(\angle 1) = 180°$.
$m(\angle 2) = 120°$	Vertical angles are congruent: $m(\angle 2) = m(\angle 3)$.
$m(\angle 4) = 60°$	Vertical angles are congruent: $m(\angle 4) = m(\angle 1)$.
$m(\angle 5) = 60°$	If two parallel lines are cut by a transversal, alternate interior angles are congruent: $m(\angle 5) = m(\angle 4)$.
$m(\angle 6) = 120°$	If two parallel lines are cut by a transversal, alternate interior angles are congruent: $m(\angle 6) = m(\angle 3)$.
$m(\angle 7) = 120°$	Vertical angles are congruent: $m(\angle 7) = m(\angle 6)$.
$m(\angle 8) = 60°$	Vertical angles are congruent: $m(\angle 8) = m(\angle 5)$.

Self Check 2
Refer to the figure for Example 2. If $l_1 \parallel l_2$ and $m(\angle 8) = 50°$, find the measures of the other seven angles that are labeled.
Now Try Problem 23
Self Check 2 Answers
$m(\angle 5) = 50°, m(\angle 7) = 130°$,
$m(\angle 6) = 130°, m(\angle 3) = 130°$,
$m(\angle 4) = 50°, m(\angle 1) = 50°$,
$m(\angle 2) = 130°$

Teaching Example 2 Refer to the figure for Example 2. If $l_1 \parallel l_2$ and $m(\angle 1) = 38°$, find the measures of the other seven angles.
Answers:
$m(\angle 2) = 142°, m(\angle 3) = 142°$,
$m(\angle 4) = 38°, m(\angle 5) = 38°$,
$m(\angle 6) = 142°, m(\angle 7) = 142°$,
$m(\angle 8) = 38°$

Some geometric figures contain two transversals.

EXAMPLE 3 Refer to the figure. If $\overline{AB} \parallel \overline{DE}$, which pairs of angles are congruent?

Strategy We will use the corresponding angles property twice to find two pairs of congruent angles.

WHY Both $\overleftrightarrow{AC}$ and $\overleftrightarrow{BC}$ are transversals cutting the parallel line segments $\overline{AB}$ and $\overline{DE}$.

Solution Since $\overline{AB} \parallel \overline{DE}$, and $\overleftrightarrow{AC}$ is a transversal cutting them, corresponding angles are congruent. So we have

$\angle A \cong \angle 1$

Since $\overline{AB} \parallel \overline{DE}$ and $\overleftrightarrow{BC}$ is a transversal cutting them, corresponding angles must be congruent. So we have

$\angle B \cong \angle 2$

Self Check 3
See the figure below. If $\overline{YZ} \parallel \overline{MN}$, which pairs of angles are congruent?

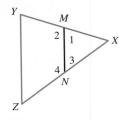

Now Try Problem 25
Self Check 3 Answer
$\angle 1 \cong \angle Y, \angle 3 \cong \angle Z$

EXAMPLE 4 In the figure, $l_1 \parallel l_2$. Find x.

Strategy We will use the corresponding angles property to write an equation that mathematically models the situation.

WHY We can then solve the equation to find x.

Solution In the figure, two corresponding angles have degree measures that are represented by the algebraic expressions $9x - 15°$ and $6x + 30°$. Since $l_1 \parallel l_2$, this pair of corresponding angles are congruent.

Unless otherwise noted, all art on this page is © Cengage Learning.

Self Check 4
In the figure below, $l_1 \parallel l_2$. Find y.
8°

Now Try Problem 27

Teaching Example 4 In the figure, $l_1 \parallel l_2$. Find x.

Answer:
9°

Self Check 5

In the figure below, $l_1 \parallel l_2$.

a. Find x. 30°
b. Find the measures of both angles labeled in the figure.
110°, 70°

Now Try Problem 29

Teaching Example 5 In the figure, $l_1 \parallel l_2$.
a. Find x.
b. Find the measures of both angles labeled in the figure.

Answers:
a. 25°
b. 74°, 106°

$9x - 15° = 6x + 30°$ Since the angles are congruent, their measures are equal.

$3x - 15° = 30°$ To eliminate 6x from the right side, subtract 6x from both sides.

$3x = 45°$ To isolate the variable term 3x, undo the subtraction of 15° by adding 15° to both sides: 30° + 15° = 45°.

$x = 15°$ To isolate x, undo the multiplication by 3 by dividing both sides by 3.

Thus, x is 15°.

EXAMPLE 5   In the figure, $l_1 \parallel l_2$.

a. Find x.

b. Find the measures of both angles labeled in the figure.

Strategy We will use the interior angles property to write an equation that mathematically models the situation.

WHY We can then solve the equation to find x.

Solution

a. Because the angles are interior angles on the same side of the transversal, they are supplementary.

$3x - 80° + 3x + 20° = 180°$ The sum of the measures of two supplementary angles is 180°.

$6x - 60° = 180°$ Combine like terms: 3x + 3x = 6x.

$6x = 240°$ To undo the subtraction of 60°, add 60° to both sides: 180° + 60° = 240°.

$x = 40°$ To isolate x, undo the multiplication by 6 by dividing both sides by 6.

Thus, x is 40°.

This problem may be solved using a different approach. In the figure below, we see that $\angle 1$ and the angle with measure $3x - 80°$ are corresponding angles.

Because l_1 and l_2 are parallel, all pairs of corresponding angles are congruent. Therefore,

$m(\angle 1) = 3x - 80°$

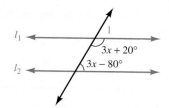

In the figure, we also see that $\angle 1$ and the angle with measure $3x + 20°$ are supplementary. That means that the sum of their measures must be 180°. We have

$m(\angle 1) + 3x + 20° = 180°$

$3x - 80° + 3x + 20° = 180°$ Replace m(∠1) with 3x − 80°.

This is the same equation that we obtained in the previous solution. When it is solved, we find that x is 40°.

b. To find the measures of the angles in the figure, we evaluate the expressions $3x + 20°$ and $3x - 80°$ for $x = 40°$.

$$3x + 20° = 3(40°) + 20° \qquad 3x - 80° = 3(40°) - 80°$$
$$= 120° + 20° \qquad\qquad\quad = 120° - 80°$$
$$= 140° \qquad\qquad\qquad\quad = 40°$$

The measures of the angles labeled in the figure are 140° and 40°.

ANSWERS TO SELF CHECKS

1. a. $\angle 1$ and $\angle 3$, $\angle 2$ and $\angle 4$, $\angle 8$ and $\angle 6$, $\angle 7$ and $\angle 5$ **b.** $\angle 2, \angle 7, \angle 3,$ and $\angle 6$ **c.** $\angle 2$ and $\angle 6$, $\angle 7$ and $\angle 3$ **2.** $m(\angle 5) = 50°, m(\angle 7) = 130°, m(\angle 6) = 130°, m(\angle 3) = 130°, m(\angle 4) = 50°, m(\angle 1) = 50°,$ and $m(\angle 2) = 130°$ **3.** $\angle 1 \cong \angle Y, \angle 3 \cong \angle Z$ **4.** 8° **5. a.** 30° **b.** 110°, 70°

SECTION 9.2 STUDY SET

VOCABULARY

Fill in the blanks.

1. Two lines that lie in the same plane are called __coplanar__. Two lines that lie in different planes are called __noncoplanar__.

2. Two coplanar lines that do not intersect are called __parallel__ lines. Two noncoplanar lines that do not intersect are called __skew__ lines.

3. __Perpendicular__ lines are lines that intersect and form right angles.

4. A line that intersects two coplanar lines in two distinct (different) points is called a __transversal__.

5. In the figure below, $\angle 4$ and $\angle 6$ are __alternate__ interior angles.

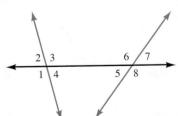

6. In the figure above, $\angle 2$ and $\angle 6$ are __corresponding__ angles.

CONCEPTS

7. **a.** Draw two parallel lines. Label them l_1 and l_2.

   b. Draw two lines that are not parallel. Label them l_1 and l_2.

8. **a.** Draw two perpendicular lines. Label them l_1 and l_2.

   b. Draw two lines that are not perpendicular. Label them l_1 and l_2.

9. **a.** Draw two parallel lines cut by a transversal. Label the lines l_1 and l_2 and label the transversal l_3.

   b. Draw two lines that are not parallel and cut by a transversal. Label the lines l_1 and l_2 and label the transversal l_3.

10. Draw three parallel lines. Label them $l_1, l_2,$ and l_3.

In Problems 11–14, two parallel lines are cut by a transversal. Fill in the blanks.

11. In the figure below, on the left, $\angle ABC \cong \angle BEF$. When two parallel lines are cut by a transversal, __corresponding__ angles are congruent.

12. In the figure below, on the right, $\angle 1 \cong \angle 2$. When two parallel lines are cut by a transversal, __alternate interior__ angles are congruent.

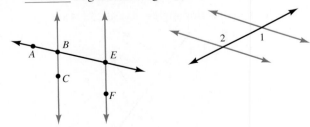

Unless otherwise noted, all art on this page is © Cengage Learning.

13. In the figure below, on the left, m(∠ABC) + m(∠BCD) = 180°. When two parallel lines are cut by a transversal, __interior__ angles on the same side of the transversal are supplementary.

14. In the figure below, on the right, ∠8 ≅ ∠6. When two parallel lines are cut by a transversal, __corresponding__ angles are congruent.

15. In the figure below, on the left, $l_1 \parallel l_2$. What can you conclude about l_1 and l_3? They are perpendicular.

16. In the figure below, on the right, $l_1 \parallel l_2$ and $l_2 \parallel l_3$. What can you conclude about l_1 and l_3? They are parallel.

NOTATION

Fill in the blanks.

17. The symbol ⌐ indicates a __right__ angle.
18. The symbol ∥ is read as "is __parallel__ to."
19. The symbol ⊥ is read as "is __perpendicular__ to."
20. The symbol l_1 is read as "line l __sub__ one."

GUIDED PRACTICE

21. Refer to the figure below and identify each of the following. **See Example 1.**
    a. corresponding angles ∠1 and ∠5, ∠4 and ∠8, ∠2 and ∠6, ∠3 and ∠7
    b. interior angles ∠3, ∠4, ∠5, and ∠6
    c. alternate interior angles ∠3 and ∠5, ∠4 and ∠6

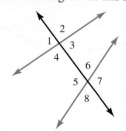

22. Refer to the figure below and identify each of the following. **See Example 1.**
    a. corresponding angles ∠ABC and ∠EFB, ∠ABF and ∠EFH, ∠CBD and ∠BFG, ∠DBF and ∠GFH
    b. interior angles ∠ABF, ∠EFB, ∠DBF, ∠GFB
    c. alternate interior angles ∠ABF and ∠GFB, ∠DBF and ∠EFB

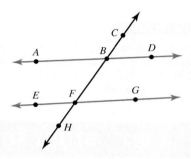

23. In the figure below, $l_1 \parallel l_2$ and m(∠4) = 130°. Find the measures of the other seven angles that are labeled. **See Example 2.** m(∠1) = 130°, m(∠2) = 50°, m(∠3) = 50°, m(∠5) = 130°, m(∠6) = 50°, m(∠7) = 50°, m(∠8) = 130°

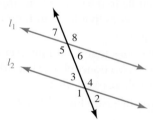

24. In the figure below, $l_1 \parallel l_2$ and m(∠2) = 40°. Find the measures of the other angles. **See Example 2.** m(∠1) = 140°, m(∠3) = 140°, m(∠4) = 40°, m(∠5) = 40°, m(∠6) = 140°, m(∠7) = 40°, m(∠8) = 140°

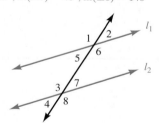

25. In the figure below, $\overline{YM} \parallel \overline{XN}$. Which pairs of angles are congruent? **See Example 3.** ∠1 ≅ ∠X, ∠2 ≅ ∠N

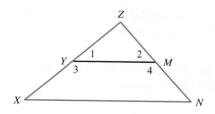

26. In the figure below, $\overline{AE} \parallel \overline{BD}$. Which pairs of angles are congruent? **See Example 3.** $\angle 1 \cong \angle A, \angle 2 \cong \angle E$

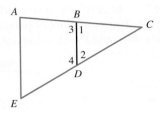

In Problems 27 and 28, $l_1 \parallel l_2$. First find x. Then determine the measure of each angle that is labeled in the figure.
See Example 4.

27. 12°, 40°, 40°

28. 10°, 30°, 30°

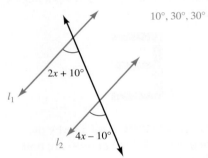

In Problems 29 and 30, $l_1 \parallel l_2$. First find x. Then determine the measure of each angle that is labeled in the figure. See Example 5.

29. 10°, 50°, 130°

30. 25°, 130°, 50°

TRY IT YOURSELF

31. In the figure below, $\overleftrightarrow{AB}$. Find:
    a. m($\angle 1$), m($\angle 2$), m($\angle 3$), and m($\angle 4$) 50°, 135°, 45°, 85°
    b. m($\angle 3$) + m($\angle 4$) + m($\angle ACD$) 180°
    c. m($\angle 1$) + m($\angle ABC$) + m($\angle 4$) 180°

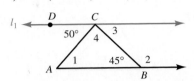

32. In the figure below, $\overline{AB} \parallel \overline{DE}$. Find m($\angle B$), m($\angle E$), and m($\angle 1$). 60°, 90°, 30°

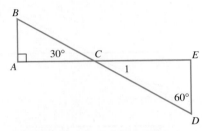

33. In the figure below, $\overline{AB} \parallel \overline{DE}$. What pairs of angles are congruent? Explain your reasoning.
vertical angles: $\angle 1 \cong \angle 2$;
alternate interior angles: $\angle B \cong \angle D$, $\angle E \cong \angle A$

34. In the figure below, $l_1 \parallel l_2$. Find x. 6.5°

In Problems 35–38, first find x. Then determine the measure of each angle that is labeled in the figure.

35. $l_1 \parallel \overline{CA}$ 40°, 40°, 140°

36. $\overline{AB} \parallel \overline{DE}$ 22°, 70°, 70°

37. $\overline{AB} \parallel \overline{DE}$ 12°, 70°, 70°

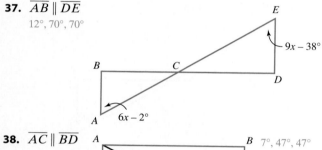

38. $\overline{AC} \parallel \overline{BD}$ 7°, 47°, 47°

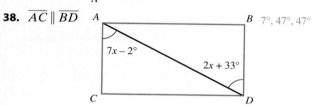

APPLICATIONS

39. CONSTRUCTING PYRAMIDS The Egyptians used a device called a **plummet** to tell whether stones were properly leveled. A plummet (shown below) is made up of an A-frame and a plumb bob suspended from the peak of the frame. How could a builder use a plummet to tell that the two stones on the left are not level and that the three stones on the right are level?
The plummet string should hang perpendicular to the top of the stones.

40. DIAGRAMMING SENTENCES English instructors have their students diagram sentences to help teach proper sentence structure. A diagram of the sentence *The cave was rather dark and damp* is shown below. Point out pairs of parallel and perpendicular lines used in the diagram.

41. BEAUTY TIPS The figure to the right shows how one website illustrated the "geometry" of the ideal eyebrow. If $l_1 \parallel l_2$ and m($\angle DCF$) = 130°, find m($\angle ABE$).   50°

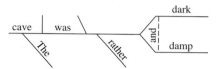

42. PAINTING SIGNS For many sign painters, the most difficult letter to paint is a capital E because of all the right angles involved. How many right angles are there?   12

43. HANGING WALLPAPER Explain why the concepts of *perpendicular* and *parallel* are both important when hanging wallpaper.
The strips of wallpaper should be hung on the wall parallel to each other, and they should be perpendicular to the floor and the ceiling.

44. TOOLS What geometric concepts are seen in the design of the rake shown here?
The teeth of the rake are parallel to each other. The teeth of the rake are perpendicular to the cross bar.

45. SEISMOLOGY The figure shows how an earthquake fault occurs when two blocks of earth move apart and one part drops down. Determine the measures of $\angle 1$, $\angle 2$, and $\angle 3$.   75°, 105°, 75°

46. CARPENTRY A carpenter cross braced three 2 × 4's as shown below and then used a tool to measure the three highlighted angles in red. Are the 2 × 4's parallel? Explain your answer. No; if the 2 × 4's were parallel, the three angles would be the same.

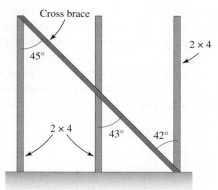

WRITING

47. PARKING DESIGN Using terms from this section, write a paragraph describing the parking layout shown below.

48. In the figure below, $l_1 \parallel l_2$. Explain why m($\angle BDE$) = 91°.

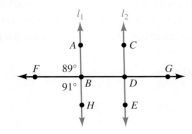

49. In the figure below, $l_1 \parallel l_2$. Explain why m($\angle FEH$) = 100°.

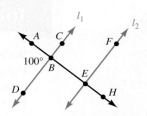

50. In the figure below, $l_1 \parallel l_2$. Explain why the figure must be mislabeled.

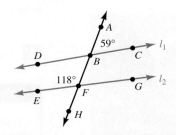

51. Are pairs of alternate interior angles always congruent? Explain.

52. Are pairs of interior angles on the same side of a transversal always supplementary? Explain.

REVIEW

53. Find 60% of 120. 72

54. 80% of what number is 400? 500

55. What percent of 500 is 225? 45%

56. Simplify: 3.45 + 7.37 · 2.98 25.4126

57. Is every whole number an integer? yes

58. Multiply: $2\frac{1}{5} \cdot 4\frac{3}{7}$ $9\frac{26}{35}$

59. Express the phrase as a ratio in lowest terms: 4 ounces to 12 ounces $\frac{1}{3}$

60. Convert 5,400 milligrams to kilograms. 0.0054 kg

Objectives

1. Classify polygons.
2. Classify triangles.
3. Identify isosceles triangles.
4. Find unknown angle measures of triangles.

SECTION 9.3
Triangles

We will now discuss geometric figures called *polygons*. We see these shapes every day. For example, the walls of most buildings are rectangular in shape. Some tile and vinyl floor patterns use the shape of a pentagon or a hexagon. Stop signs are in the shape of an octagon.

In this section, we will focus on one specific type of polygon called a *triangle*. Triangular shapes are especially important because triangles contribute strength and stability to walls and towers. The gable roofs of houses are triangular, as are the sides of many ramps.

The House of the Seven Gables, Salem, Massachusetts

1 Classify polygons.

Polygon

A **polygon** is a closed geometric figure with at least three line segments for its sides.

Polygons are formed by fitting together line segments in such a way that

- no two of the segments intersect, except at their endpoints, and
- no two line segments with a common endpoint lie on the same line.

The line segments that form a polygon are called its **sides**. The point where two sides of a polygon intersect is called a **vertex** of the polygon (plural **vertices**). The polygon shown to the right has five sides and five vertices.

Polygons are classified according to the number of sides that they have. For example, in the figure below, we see that a polygon with four sides is called a *quadrilateral*, and a polygon with eight sides is called an *octagon*. If a polygon has sides that are all the same length and angles that are the same measure, we call it a **regular polygon**.

	Triangle 3 sides	Quadrilateral 4 sides	Pentagon 5 sides	Hexagon 6 sides	Heptagon 7 sides	Octagon 8 sides	Nonagon 9 sides	Decagon 10 sides	Dodecagon 12 sides
Polygons									
Regular polygons									

Self Check 1

Give the number of vertices of:

a. a quadrilateral 4
b. a pentagon 5

Now Try Problems 25 and 27

EXAMPLE 1

Give the number of vertices of:

a. a triangle   b. a hexagon

Strategy We will determine the number of angles that each polygon has.

WHY The number of its vertices is equal to the number of its angles.

Solution

a. From the figure on the previous page, we see that a triangle has three angles and therefore three vertices.

b. From the figure on the previous page, we see that a hexagon has six angles and therefore six vertices.

> **Teaching Example 1** Give the number of vertices of:
> **a.** an octagon
> **b.** a decagon
> *Answers:*
> **a.** 8
> **b.** 10

> **Success Tip** From the results of Example 1, we see that *the number of vertices of a polygon is equal to the number of its sides.*

2 Classify triangles.

A **triangle** is a polygon with three sides (and three vertices). Recall that in geometry points are labeled with capital letters. We can use the capital letters that denote the vertices of a triangle to name the triangle. For example, when referring to the triangle in the box below, with vertices A, B, and C, we can use the notation △ABC (read as "triangle ABC").

> **The Language of Algebra** When naming a triangle, we may begin with any vertex. Then we move around the figure in a clockwise (or counterclockwise) direction as we list the remaining vertices. Other ways of naming the triangle shown here are △ACB, △BCA, △BAC, △CAB, and △CBA.

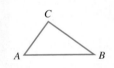

> **The Language of Algebra** The figures below show how triangles can be classified according to the lengths of their sides. The single **tick marks** drawn on each side of the equilateral triangle indicate that the sides are of equal length. The double tick marks drawn on two of the sides of the isosceles triangle indicate that they have the same length. Each side of the scalene triangle has a different number of tick marks to indicate that the sides have different lengths.

Equilateral triangle   **Isosceles triangle**   **Scalene triangle**
(all sides equal length)   (at least two sides of   (no sides of equal length)
                                       equal length)

> **The Language of Algebra** Since every angle of an equilateral triangle has the same measure, an equilateral triangle is also equiangular.

> **The Language of Algebra** Since equilateral triangles have at least two sides of equal length, they are also isosceles. However, isosceles triangles are not necessarily equilateral.

Triangles may also be classified by their angles, as shown below.

Acute triangle
(has three acute angles)

Obtuse triangle
(has an obtuse angle)

Right triangle
(has one right angle)

Right triangles have many real-life applications. For example, in figure (a) below, we see that a right triangle is formed when a ladder leans against the wall of a building.

The longest side of a right triangle is called the **hypotenuse**, and the other two sides are called **legs**. The hypotenuse of a right triangle is always opposite the 90° (right) angle. The legs of a right triangle are adjacent to (next to) the right angle, as shown in figure (b).

(a) (b)

3 Identify isosceles triangles.

In an isosceles triangle, the angles opposite the sides of equal length are called **base angles**, the sides of equal length form the **vertex angle**, and the third side is called the **base**. Two examples of isosceles triangles are shown below.

Isosceles triangles

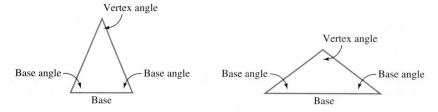

We have seen that isosceles triangles have two sides of equal length. The **isosceles triangle theorem** states that such triangles have one other important characteristic: Their base angles are congruent.

Isosceles Triangle Theorem

If two sides of a triangle are congruent, then the angles opposite those sides are congruent.

The Language of Algebra **Tick marks** can be used to denote the sides of a triangle that have the same length. They can also be used to indicate the angles of a triangle with the same measure. For example, we can show that the base angles of the isosceles triangle below are congruent by using single tick marks.

$\angle D$ is opposite $\overline{FE}$, and $\angle E$ is opposite $\overline{FD}$. By the isosceles triangle theorem, if $m(\overline{FD}) = m(\overline{FE})$, then $m(\angle D) = m(\angle E)$.

If a mathematical statement is written in the form *if p . . . , then q . . .* , we call the statement *if q . . . , then p . . .* its **converse**. The converses of some statements are true, while the converses of other statements are false. It is interesting to note that the converse of the isosceles triangle theorem is true.

Converse of the Isosceles Triangle Theorem

If two angles of a triangle are congruent, then the sides opposite the angles have the same length, and the triangle is isosceles.

EXAMPLE 2 Is the triangle shown here an isosceles triangle?

Strategy We will consider the measures of the angles of the triangle.

WHY If two angles of a triangle are congruent, then the sides opposite the angles have the same length, and the triangle is isosceles.

Solution $\angle A$ and $\angle B$ have the same measure, 50°. By the converse of the isosceles triangle theorem, if $m(\angle A) = m(\angle B)$, we know that $m(\overline{BC}) = m(\overline{AC})$ and that $\triangle ABC$ is isosceles.

Self Check 2

Is the triangle shown below an isosceles triangle? no

Now Try Problems 33 and 35

4 Find unknown angle measures of triangles.

If you draw several triangles and carefully measure each angle with a protractor, you will find that the sum of the angle measures of each triangle is 180°. Two examples are shown below.

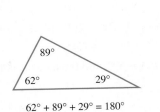

62° + 89° + 29° = 180°

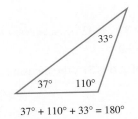

37° + 110° + 33° = 180°

Another way to show this important fact about the sum of the angle measures of a triangle is discussed in Problem 82 of the Study Set at the end of this section.

Angles of a Triangle

The sum of the angle measures of any triangle is 180°.

Self Check 3

In the figure, find y. 30°

Now Try Problem 37

Teaching Example 3 In the figure, find x.
Answer: 15°

EXAMPLE 3
In the figure, find x.

Strategy We will use the fact that the sum of the angle measures of any triangle is 180° to write an equation that models the situation.

WHY We can then solve the equation to find the unknown angle measure, x.

Solution Since the sum of the angle measures of any triangle is 180°, we have

$x + 40° + 90° = 180°$ The ⌐ symbol indicates that the measure of the angle is 90°.

$x + 130° = 180°$ Do the addition.

$x = 50°$ To isolate x, undo the addition of 130° by subtracting 130° from both sides.

$$\begin{array}{r} 90 \\ +40 \\ \hline 130 \end{array}$$

Thus, x is 50°.

Self Check 4

In $\triangle DEF$, the measure of $\angle D$ exceeds the measure of $\angle E$ by 5°, and the measure of $\angle F$ is three times the measure of $\angle E$. Find the measure of each angle of $\triangle DEF$.

Now Try Problem 41

Self Check 4 Answer
m($\angle D$) = 40°, m($\angle E$) = 35°, m($\angle F$) = 105°

Teaching Example 4 Find the measure of each angle of $\triangle ABC$.

Answers:
m($\angle A$) = 10°, m($\angle B$) = 130°, m($\angle C$) = 40°

EXAMPLE 4
In the figure, find the measure of each angle of $\triangle ABC$.

Strategy We will use the fact that the sum of the angle measures of any triangle is 180° to write an equation that models the situation.

WHY We can then solve the equation to find the unknown angle measure x, and use it to evaluate the expressions $2x$ and $x + 32°$.

Solution

$x + 32° + x + 2x = 180°$ The sum of the angle measures of any triangle is 180°.

$4x + 32° = 180°$ Combine like terms: $x + x + 2x = 4x$.

$4x + 32° - 32° = 180° - 32°$ To isolate the variable term, 4x, subtract 32° from both sides.

$4x = 148°$ Do the subtractions.

$\dfrac{4x}{4} = \dfrac{148°}{4}$ To isolate x, divide both sides by 4.

$x = 37°$ Do the divisions. This is the measure of $\angle B$.

$$\begin{array}{r} {}^{7}1\cancel{8}0 \\ -32 \\ \hline 148 \end{array} \qquad \begin{array}{r} 37 \\ 4\overline{)148} \\ -12 \\ \hline 28 \\ -28 \\ \hline 0 \end{array}$$

To find the measures of $\angle A$ and $\angle C$, we evaluate the expressions $x + 32°$ and $2x$ for $x = 37°$.

$x + 32° = 37° + 32°$ Substitute 37 for x. $\quad 2x = 2(37°)$ Substitute 37 for x.
$= 69°$ $= 74°$

The measure of $\angle B$ is 37°, the measure of $\angle A$ is 69°, and the measure of $\angle C$ is 74°.

Self Check 5

If one base angle of an isosceles triangle measures 33°, what is the measure of the vertex angle? 114°

Now Try Problem 45

EXAMPLE 5
If one base angle of an isosceles triangle measures 70°, what is the measure of the vertex angle?

Strategy We will use the isosceles triangle theorem and the fact that the sum of the angle measures of any triangle is 180° to write an equation that models the situation.

WHY We can then solve the equation to find the unknown angle measure.

Solution By the isosceles triangle theorem, if one of the base angles measures 70°, so does the other. (See the figure on the right.) If we let x represent the measure of the vertex angle, we have

$x + 70° + 70° = 180°$ The sum of the measures of the angles of a triangle is 180°.

$x + 140° = 180°$ Combine like terms: 70° + 70° = 140°.

$x = 40°$ To isolate x, undo the addition of 140° by subtracting 140° from both sides.

The vertex angle measures 40°.

Teaching Example 5 If one base angle of an isosceles triangle measures 46°, what is the measure of the vertex angle?
Answer:
88°

EXAMPLE 6
If the vertex angle of an isosceles triangle measures 99°, what are the measures of the base angles?

Strategy We will use the fact that the base angles of an isosceles triangle have the same measure and the sum of the angle measures of any triangle is 180° to write an equation that mathematically models the situation.

WHY We can then solve the equation to find the unknown angle measures.

Solution The base angles of an isosceles triangle have the same measure. If we let x represent the measure of one base angle, the measure of the other base angle is also x. (See the figure to the right.) Since the sum of the measures of the angles of any triangle is 180°, the sum of the measures of the base angles and of the vertex angle is 180°. We can use this fact to form an equation.

$x + x + 99° = 180°$

$2x + 99° = 180°$ Combine like terms: x + x = 2x.

$2x = 81°$ To isolate the variable term, 2x, undo the addition of 99° by subtracting 99° from both sides.

$\dfrac{2x}{2} = \dfrac{81°}{2}$ To isolate x, undo the multiplication by 2 by dividing both sides by 2.

$x = 40.5°$

The measure of each base angle is 40.5°.

```
   40.5
2)81.0
  -8
   01
   -0
   10
  -10
    0
```

Self Check 6
If the vertex angle of an isosceles triangle measures 57°, what are the measures of the base angles? 61.5°

Now Try Problem 49

Teaching Example 6 If the vertex angle of an isosceles triangle measures 74°, what are the measures of the base angles?
Answer:
53°

ANSWERS TO SELF CHECKS

1. a. 4 **b.** 5 **2.** no **3.** 30° **4.** m(∠D) = 40°, m(∠E) = 35°, m(∠F) = 105°
5. 114° **6.** 61.5°

SECTION 9.3   STUDY SET

VOCABULARY

Fill in the blanks.

1. A <u>polygon</u> is a closed geometric figure with at least three line segments for its sides.

2. The polygon shown to the right has seven <u>sides</u> and seven vertices.

Unless otherwise noted, all art on this page is © Cengage Learning.

3. A point where two sides of a polygon intersect is called a <u>vertex</u> of the polygon.

4. A <u>regular</u> polygon has sides that are all the same length and angles that all have the same measure.

5. A triangle with three sides of equal length is called an <u>equilateral</u> triangle. An <u>isosceles</u> triangle has at least two sides of equal length. A <u>scalene</u> triangle has no sides of equal length.

6. An acute triangle has three acute angles. An obtuse triangle has one obtuse angle. A right triangle has one right angle.

7. The longest side of a right triangle is called the hypotenuse . The other two sides of a right triangle are called legs .

8. The base angles of an isosceles triangle have the same measure. The sides of equal length of an isosceles triangle form the vertex angle.

9. In this section, we discussed the sum of the measures of the angles of a triangle. The word *sum* indicates the operation of addition .

10. Complete the table.

Number of Sides	Name of Polygon
3	triangle
4	quadrilateral
5	pentagon
6	hexagon
7	heptagon
8	octagon
9	nonagon
10	decagon
12	dodecagon

CONCEPTS

11. Draw an example of each type of regular polygon.
    a. hexagon
    b. octagon
    c. quadrilateral
    d. triangle
    e. pentagon
    f. decagon

12. Refer to the triangle below.
    a. What are the names of the vertices of the triangle? H, I, J
    b. How many sides does the triangle have? Name them. 3; $\overline{HI}, \overline{IJ}, \overline{JH}$
    c. Use the vertices to name this triangle in three ways. △HIJ, △IJH, △JHI

13. Draw an example of each type of triangle.
    a. isosceles
    b. equilateral
    c. scalene
    d. obtuse
    e. right
    f. acute

14. Classify each triangle as an acute, an obtuse, or a right triangle.
    a. obtuse
    b. 90° right
    c. acute
    d. 91° obtuse

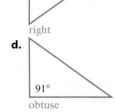

15. Refer to the triangle shown below.
    a. What is the measure of ∠B? 90°
    b. What type of triangle is it? right
    c. What two line segments form the legs? $\overline{AB}, \overline{BC}$
    d. What line segment is the hypotenuse? $\overline{AC}$
    e. Which side of the triangle is the longest? $\overline{AC}$
    f. Which side is opposite ∠B? $\overline{AC}$

16. Fill in the blanks.
    a. The sides of a right triangle that are adjacent to the right angle are called the legs .
    b. The hypotenuse of a right triangle is the side opposite the right angle.

17. Fill in the blanks.
    a. The isosceles triangle theorem states that if two sides of a triangle are congruent, then the angles opposite those sides are congruent.
    b. The converse of the isosceles triangle theorem is true and states that if two angles of a triangle are congruent, then the sides opposite the angles have the same length, and the triangle is isosceles.

18. Refer to the given triangle.
    a. What two sides are of equal length? $\overline{XY}, \overline{ZY}$
    b. What type of triangle is △XYZ? isosceles

Unless otherwise noted, all art on this page is © Cengage Learning.

c. Name the base angles. ∠X, ∠Z
d. Which side is opposite ∠X? $\overline{ZY}$
e. What is the vertex angle? ∠Y
f. Which angle is opposite side $\overline{XY}$? ∠Z
g. Which two angles are congruent? ∠X, ∠Z

19. Refer to the triangle below.
    a. What do we know about $\overline{EF}$ and $\overline{GF}$? $\overline{EF} \cong \overline{GF}$
    b. What type of triangle is △EFG? isosceles

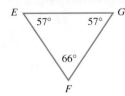

20. a. Find the sum of the measures of the angles of △JKL, shown in figure (a). 180°
    b. Find the sum of the measures of the angles of △CDE, shown in figure (b). 180°
    c. What is the sum of the measures of the angles of *any* triangle? 180°

 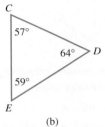
(a)   (b)

NOTATION

Fill in the blanks.

21. The symbol △ means __triangle__.
22. The symbol m(∠A) means the __measure__ of angle *A*.

Refer to the triangle below.

23. What fact about the sides of △ABC do the tick marks indicate? $\overline{AB} \cong \overline{CB}$
24. What fact about the angles of △ABC do the tick marks indicate? ∠A ≅ ∠C

GUIDED PRACTICE

For each polygon, give the number of sides it has, give its name, and then give the number of vertices that it has. See Example 1.

25. a.   b.
    4, quadrilateral, 4   6, hexagon, 6

26. a.   b.
    3, triangle, 3   8, octagon, 8

27. a.   b.
    7, heptagon, 7   9, nonagon, 9

28. a.   b.
    5, pentagon, 5   10, decagon, 10

Classify each triangle as an equilateral triangle, an isosceles triangle, or a scalene triangle. See Objective 2.

29. a.   b.
    scalene   isosceles

30. a.   b.
    isosceles   equilateral

31. a.   b.
    equilateral   scalene

32. a.   b.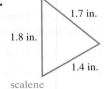
    isosceles   scalene

State whether each of the triangles is an isosceles triangle. See Example 2.

33.
yes

34.
yes

35.
no

36.
no

Find y. See Example 3.

37.

38.

39.

40.

The degree measures of the angles of a triangle are represented by algebraic expressions. First find x. Then determine the measure of each angle of the triangle. See Example 4.

41.
50°; 50°, 60°, 70°

42.
30°; 30°, 35°, 115°

43.
20°; 20°, 80°, 80°

44.
50°; 50°, 65°, 65°

Find the measure of the vertex angle of each isosceles triangle given the following information. See Example 5.

45. The measure of one base angle is 56°. 68°
46. The measure of one base angle is 68°. 44°
47. The measure of one base angle is 85.5°. 9°
48. The measure of one base angle is 4.75°. 170.5°

Find the measure of one base angle of each isosceles triangle given the following information. See Example 6.

49. The measure of the vertex angle is 102°. 39°
50. The measure of the vertex angle is 164°. 8°
51. The measure of the vertex angle is 90.5°. 44.75°
52. The measure of the vertex angle is 2.5°. 88.75°

TRY IT YOURSELF

Find the measure of each vertex angle.

53.

54.
114°

55.

56.

The measures of two angles of △ABC are given. Find the measure of the third angle.

57. m(∠A) = 30° and m(∠B) = 60°; find m(∠C). 90°
58. m(∠A) = 45° and m(∠C) = 105°; find m(∠B). 30°
59. m(∠B) = 100° and m(∠A) = 35°; find m(∠C). 45°
60. m(∠B) = 33° and m(∠C) = 77°; find m(∠A). 70°
61. m(∠A) = 25.5° and m(∠B) = 63.8°; find m(∠C). 90.7°
62. m(∠B) = 67.25° and m(∠C) = 72.5°; find m(∠A). 40.25°
63. m(∠A) = 29° and m(∠C) = 89.5°; find m(∠B). 61.5°
64. m(∠A) = 4.5° and m(∠B) = 128°; find m(∠C). 47.5°

In Problems 65–68, find x.

65. 12°

66.

67. 52.5°

68.

69. One angle of an isosceles triangle has a measure of 39°. What are the possible measures of the other angles? 39°, 39°, 102° or 70.5°, 70.5°, 39°

70. One angle of an isosceles triangle has a measure of 2°. What are the possible measures of the other angles? 2°, 2°, 176° or 89°, 89°, 2°

71. Find m(∠C). 73°

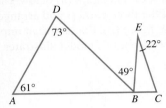

72. Find:
   a. m(∠MXZ) 24°
   b. m(∠MYN) 63°

73. Find m(∠NOQ). 75°

74. Find m(∠S). 79°

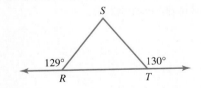

APPLICATIONS

75. POLYGONS IN NATURE As seen below, a starfish fits the shape of a pentagon. What polygon shape do you see in each of the other objects?
   a. lemon octagon
   b. chili pepper triangle
   c. apple pentagon

76. CHEMISTRY Polygons are used to represent the chemical structure of compounds. In the figure below, what types of polygons are used to represent methylprednisolone, the active ingredient in an anti-inflammatory medication? pentagon, hexagon

77. AUTOMOBILE JACK Refer to the figure below. No matter how high the jack is raised, it always forms two isosceles triangles. Explain why. As the jack is raised, the two sides of the jack remain the same length.

78. **EASELS** Refer to the figure below. What type of triangle studied in this section is used in the design of the legs of the easel? isosceles

79. **POOL** The rack shown below is used to set up the billiard balls when beginning a game of pool. Although it does not meet the strict definition of a polygon, the rack has a shape much like a type of triangle discussed in this section. Which type of triangle? equilateral

80. **DRAFTING** Among the tools used in drafting are the two clear plastic triangles shown below. Classify each according to the lengths of its sides and then according to its angle measures. first: isosceles, right triangle; second: scalene, right triangle

WRITING

81. In this section, we discussed the definition of a pentagon. What is *the* Pentagon? Why is it named that?

82. A student cut a triangular shape out of blue construction paper and labeled the angles $\angle 1$, $\angle 2$, and $\angle 3$, as shown in figure (a) below. Then she tore off each of the three corners and arranged them as shown in figure (b). Explain what important geometric concept this model illustrates.

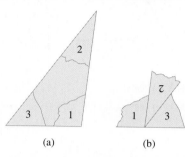

(a)   (b)

83. Explain why a triangle cannot have two right angles.

84. Explain why a triangle cannot have two obtuse angles.

REVIEW

85. Find 20% of 110. 22
86. Find 15% of 50. 7.5
87. What percent of 200 is 80? 40%
88. 20% of what number is 500? 2,500
89. Evaluate: $0.85 \div 2(0.25)$ 0.10625
90. **FIRST AID** When checking an accident victim's pulse, a paramedic counted 13 beats during a 15-second span. How many beats would be expected in 60 seconds? 52

SECTION 9.4
The Pythagorean Theorem

Objectives
1. Use the Pythagorean theorem to find the exact length of a side of a right triangle.
2. Use the Pythagorean theorem to approximate the length of a side of a right triangle.
3. Use the converse of the Pythagorean theorem.

A **theorem** is a mathematical statement that can be proven. In this section, we will discuss one of the most widely used theorems of geometry—the Pythagorean theorem. It is named after Pythagoras, a Greek mathematician who lived about 2,500 years ago. He is thought to have been the first to develop a proof of it. The Pythagorean theorem expresses the relationship between the lengths of the sides of any right triangle.

1. Use the Pythagorean theorem to find the exact length of a side of a right triangle.

Recall that a right triangle is a triangle that has a right angle (an angle with measure 90°). In a right triangle, the longest side is called the **hypotenuse**. It is the side opposite the right angle. The other two sides are called **legs**. It is common practice to let the variable c represent the length of the hypotenuse and the variables a and b represent the lengths of the legs, as shown on the right.

Pythagoras

If we know the lengths of any two sides of a right triangle, we can find the length of the third side using the **Pythagorean theorem**.

> **Pythagorean Theorem**
>
> If a and b are the lengths of two legs of a right triangle and c is the length of the hypotenuse, then
>
> $$a^2 + b^2 = c^2$$

In words, the Pythagorean theorem is expressed as follows:

In a right triangle, the sum of the squares of the lengths of the two legs is equal to the square of the length of the hypotenuse.

> **Caution!** When using the **Pythagorean equation** $a^2 + b^2 = c^2$, we can let a represent the length of either leg of the right triangle. We then let b represent the length of the other leg. The variable c must always represent the length of the hypotenuse.

EXAMPLE 1
Find the length of the hypotenuse of the right triangle shown here.

Strategy We will use the Pythagorean theorem to find the length of the hypotenuse.

WHY If we know the lengths of any two sides of a right triangle, we can find the length of the third side using the Pythagorean theorem.

Self Check 1
Find the length of the hypotenuse of the right triangle shown below.

Now Try Problem 13

Teaching Example 1 Find the length of the hypotenuse of the right triangle.

Answer:
15 yd

Solution We will let $a = 3$ and $b = 4$, and substitute into the Pythagorean equation to find c.

$a^2 + b^2 = c^2$ This is the Pythagorean equation.
$3^2 + 4^2 = c^2$ Substitute 3 for a and 4 for b.
$9 + 16 = c^2$ Evaluate each exponential expression.
$25 = c^2$ Do the addition.
$c^2 = 25$ Reverse the sides of the equation so that c^2 is on the left.

To find c, we must find a number that, when squared, is 25. There are two such numbers, one positive and one negative; they are the square roots of 25. Since c represents the length of a side of a triangle, c cannot be negative. For this reason, we need only find the positive square root of 25 to get c.

$c = \sqrt{25}$ The symbol $\sqrt{}$ is used to indicate the positive square root of a number.
$c = 5$ $\sqrt{25} = 5$ because $5^2 = 25$.

The length of the hypotenuse is 5 in.

Success Tip The Pythagorean theorem is used to find the lengths of sides of right triangles. A calculator with a square root key $\boxed{\sqrt{}}$ is often helpful in the final step of the solution process when we must find the positive square root of a number.

Self Check 2
In Example 2, can the crews communicate by radio if the distance from point B to point C remains the same but the distance from point A to point C increases to 2,520 yards? no

Now Try Problems 19 and 43

Teaching Example 2
IRON WORKS Is a 120-inch-long iron bar long enough to cross brace the rectangular gate shown in the illustration?

Answer:
yes

EXAMPLE 2 *Firefighting* To fight a forest fire, the forestry department plans to clear a rectangular fire break around the fire, as shown in the following figure. Crews are equipped with mobile communications that have a 3,000-yard range. Can crews at points A and B remain in radio contact?

Strategy We will use the Pythagorean theorem to find the distance between points A and B.

WHY If the distance is less than 3,000 yards, the crews can communicate by radio. If it is greater than 3,000 yards, they cannot.

Solution The line segments connecting points A, B, and C form a right triangle. To find the distance c from point A to point B, we can use the Pythagorean equation, substituting 2,400 for a and 1,000 for b and solving for c.

$a^2 + b^2 = c^2$ This is the Pythagorean equation.
$2,400^2 + 1,000^2 = c^2$ Substitute for a and b.
$5,760,000 + 1,000,000 = c^2$ Evaluate each exponential expression.
$6,760,000 = c^2$ Do the addition.
$c^2 = 6,760,000$ Reverse the sides of the equation so that c^2 is on the left.
$c = \sqrt{6,760,000}$ If $c^2 = 6,760,000$, then c must be a square root of 6,760,000. Because c represents a length, it must be the positive square root of 6,760,000.
$c = 2,600$ Use a calculator to find the square root.

The two crews are 2,600 yards apart. Because this distance is less than the 3,000-yard range of the radios, they can communicate by radio.

EXAMPLE 3
The lengths of two sides of a right triangle are given in the figure. Find the missing side length.

Strategy We will use the Pythagorean theorem to find the missing side length.

WHY If we know the lengths of any two sides of a right triangle, we can find the length of the third side using the Pythagorean theorem.

Solution We may substitute 11 for either a or b, but 61 must be substituted for the length c of the hypotenuse. If we substitute 11 for b, we can find the unknown side length a as follows.

$a^2 + b^2 = c^2$	This is the Pythagorean equation.
$a^2 + 11^2 = 61^2$	Substitute 11 for b and 61 for c.
$a^2 + 121 = 3{,}721$	Evaluate each exponential expression.
$a^2 + 121 - 121 = 3{,}721 - 121$	To isolate a^2 on the left side, subtract 121 from both sides.
$a^2 = 3{,}600$	Do the subtraction.
$a = \sqrt{3{,}600}$	If $a^2 = 3{,}600$, then a must be a square root of 3,600. Because a represents a length, it must be the positive square root of 3,600.
$a = 60$	Use a calculator, if necessary, to find the square root.

$$\begin{array}{r} 3{,}721 \\ -\ 121 \\ \hline 3{,}600 \end{array}$$

The missing side length is 60 ft.

Self Check 3
The lengths of two sides of a right triangle are given. Find the missing side length. 56 in.

Now Try Problem 23

Teaching Example 3 Find the missing side length of the right triangle.

Answer:
9 cm

2 Use the Pythagorean theorem to approximate the length of a side of a right triangle.

When we use the Pythagorean theorem to find the length of a side of a right triangle, the solution is sometimes the square root of a number that is not a perfect square. In that case, we can use a calculator to *approximate* the square root.

EXAMPLE 4
Refer to the right triangle shown here. Find the missing side length. Give the exact answer and an approximation to the nearest hundredth.

Strategy We will use the Pythagorean theorem to find the missing side length.

WHY If we know the lengths of any two sides of a right triangle, we can find the length of the third side using the Pythagorean theorem.

Solution We may substitute 2 for either a or b, but 6 must be substituted for the length c of the hypotenuse. If we choose to substitute 2 for a, we can find the unknown side length b as follows.

$a^2 + b^2 = c^2$	This is the Pythagorean equation.
$2^2 + b^2 = 6^2$	Substitute 2 for a and 6 for c.
$4 + b^2 = 36$	Evaluate each exponential expression.
$4 + b^2 - 4 = 36 - 4$	To isolate b^2 on the left side, undo the addition of 4 by subtracting 4 from both sides.
$b^2 = 32$	Do the subtraction.

Self Check 4
Refer to the triangle below. Find the missing side length. Give the exact answer and an approximation to the nearest hundredth. $\sqrt{24}$ m $\approx$ 4.90 m

Now Try Problem 35

Teaching Example 4 Find the missing side length. Give the exact answer and an approximation to the nearest hundredth.

Answer:
$\sqrt{19}$ m ≈ 4.36 m

We must find a number that, when squared, is 32. Since b represents the length of a side of a triangle, we consider only the positive square root.

$b = \sqrt{32}$ This is the exact length.

The missing side length is exactly $\sqrt{32}$ inches long. Since 32 is not a perfect square, its square root is not a whole number. We can use a calculator to *approximate* $\sqrt{32}$. To the nearest hundredth, the missing side length is 5.66 inches.

$\sqrt{32}$ in. ≈ 5.66 in.

Using Your CALCULATOR   Finding the Width of a TV Screen

The size of a television screen is the diagonal measure of its rectangular screen. To find the width of a 27-inch screen that is 17 inches high, we use the Pythagorean theorem with $c = 27$ and $b = 17$.

$$c^2 = a^2 + b^2$$
$$27^2 = a^2 + 17^2$$
$$27^2 - 17^2 = a^2$$

Since the variable a represents the width of the television screen, it must be positive. To find a, we find the positive square root of the result when 17^2 is subtracted from 27^2.

Using a radical symbol to indicate this, we have

$$\sqrt{27^2 - 17^2} = a$$

We can evaluate the expression on the left side by entering:

(27 x^2 − 17 x^2) √   `20.97617696`

To the nearest inch, the width of the television screen is 21 inches.

3 Use the converse of the Pythagorean theorem.

If a mathematical statement is written in the form *if p ... , then q ...* , we call the statement *if q ... , then p ...* its **converse**. The converses of some statements are true, while the converses of other statements are false. It is interesting to note that the converse of the Pythagorean theorem is true.

Converse of the Pythagorean Theorem

If a triangle has three sides of lengths a, b, and c, such that $a^2 + b^2 = c^2$, then the triangle is a right triangle.

Self Check 5
Is the triangle below a right triangle? yes

EXAMPLE 5   Is the triangle shown here a right triangle?

Strategy We will substitute the side lengths, 6, 8, and 11, into the Pythagorean equation $a^2 + b^2 = c^2$.

WHY By the converse of the Pythagorean theorem, the triangle is a right triangle if a true statement results. The triangle is not a right triangle if a false statement results.

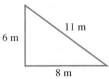

Solution We must substitute the longest side length, 11, for c, because it is the possible hypotenuse. The lengths of 6 and 8 may be substituted for either a or b.

$a^2 + b^2 = c^2$ This is the Pythagorean equation.
$6^2 + 8^2 \stackrel{?}{=} 11^2$ Substitute 6 for a, 8 for b, and 11 for c.
$36 + 64 \stackrel{?}{=} 121$ Evaluate each exponential expression.
$100 = 121$ This is a false statement.

Since $100 \neq 121$, the triangle is not a right triangle.

Now Try Problem 39

Teaching Example 5 Is the triangle a right triangle?

Answer: no

ANSWERS TO SELF CHECKS

1. 13 ft   **2.** no   **3.** 56 in.   **4.** $\sqrt{24}$ m ≈ 4.90 m   **5.** yes

---

SECTION 9.4 STUDY SET

VOCABULARY

Fill in the blanks.

1. In a right triangle, the side opposite the 90° angle is called the **hypotenuse**. The other two sides are called **legs**.

2. The Pythagorean theorem is named after the Greek mathematician **Pythagoras**, who is thought to have been the first to prove it.

3. The **Pythagorean** theorem states that in any right triangle, the square of the length of the hypotenuse is equal to the sum of the squares of the lengths of the two legs.

4. $a^2 + b^2 = c^2$ is called the Pythagorean **equation**.

CONCEPTS

Fill in the blanks.

5. If a and b are the lengths of two legs of a right triangle and c is the length of the hypotenuse, then $\boxed{a^2} + \boxed{b^2} = \boxed{c^2}$.

6. The two solutions of $c^2 = 36$ are $c = \boxed{-6}$ or $c = \boxed{6}$. If c represents the length of the hypotenuse of a right triangle, then we can discard the solution $\boxed{-6}$.

7. The converse of the Pythagorean theorem: If a triangle has three sides of lengths a, b, and c, such that $a^2 + b^2 = c^2$, then the triangle is a **right** triangle.

8. Use a protractor to draw an example of a right triangle.

9. Refer to the triangle on the right.
   a. What side is the hypotenuse? $\overline{BC}$
   b. What side is the longer leg? $\overline{AB}$
   c. What side is the shorter leg? $\overline{AC}$

10. What is the first step when solving the equation $25 + b^2 = 81$ for b? Subtract 25 from both sides.

NOTATION

Complete the steps to solve the equation, where $a > 0$ and $c > 0$.

11. $8^2 + 6^2 = c^2$
    $\boxed{64} + 36 = c^2$
    $\boxed{100} = c^2$
    $\sqrt{\boxed{100}} = c$
    $10 = c$

12. $a^2 + 15^2 = 17^2$
    $a^2 + \boxed{225} = \boxed{289}$
    $a^2 + 225 - \boxed{225} = 289 - \boxed{225}$
    $a^2 = \boxed{64}$
    $a = \sqrt{\boxed{64}}$
    $a = \boxed{8}$

GUIDED PRACTICE

Find the length of the hypotenuse of the right triangle shown below if it has the given side lengths. See Examples 1 and 2.

13. $a = 6$ ft and $b = 8$ ft   10 ft
14. $a = 12$ mm and $b = 9$ mm   15 mm
15. $a = 5$ m and $b = 12$ m   13 m
16. $a = 16$ in. and $b = 12$ in.   20 in.
17. $a = 48$ mi and $b = 55$ mi   73 mi
18. $a = 80$ ft and $b = 39$ ft   89 ft
19. $a = 88$ cm and $b = 105$ cm   137 cm
20. $a = 132$ mm and $b = 85$ mm   157 mm

Refer to the right triangle below. See Example 3.

21. Find b if $a = 10$ cm and $c = 26$ cm.   24 cm
22. Find b if $a = 14$ in. and $c = 50$ in.   48 in.
23. Find a if $b = 18$ m and $c = 82$ m.   80 m
24. Find a if $b = 9$ yd and $c = 41$ yd.   40 yd
25. Find a if $b = 21$ m and $c = 29$ m.   20 m
26. Find a if $b = 16$ yd and $c = 34$ yd.   30 yd
27. Find b if $a = 180$ m and $c = 181$ m.   19 m
28. Find b if $a = 630$ ft and $c = 650$ ft.   160 ft

The lengths of two sides of a right triangle are given. Find the missing side length. Give the exact answer and an approximation to the nearest hundredth. See Example 4.

29. $a = 5$ cm and $c = 6$ cm
    $\sqrt{11}$ cm ≈ 3.32 cm
30. $a = 4$ in. and $c = 8$ in.
    $\sqrt{48}$ in. ≈ 6.93 in.
31. $a = 12$ m and $b = 8$ m
    $\sqrt{208}$ m ≈ 14.42 m
32. $a = 10$ ft and $b = 4$ ft
    $\sqrt{116}$ ft ≈ 10.77 ft
33. $a = 9$ in. and $b = 3$ in.
    $\sqrt{90}$ in. ≈ 9.49 in.
34. $a = 5$ mi and $b = 7$ mi
    $\sqrt{74}$ mi ≈ 8.60 mi
35. $b = 4$ in. and $c = 6$ in.
    $\sqrt{20}$ in. ≈ 4.47 in.
36. $b = 9$ mm and $c = 12$ mm
    $\sqrt{63}$ mm ≈ 7.94 mm

Is a triangle with the following side lengths a right triangle? See Example 5.

37. 12, 14, 15   no
38. 15, 16, 22   no
39. 33, 56, 65   yes
40. 20, 21, 29   yes

APPLICATIONS

41. **ADJUSTING LADDERS** A 20-foot ladder reaches a window 16 feet above the ground. How far from the wall is the base of the ladder?   12 ft

42. **LENGTH OF GUY WIRES** A 30-foot tower is to be fastened by three guy wires attached to the top of the tower and to the ground at positions 20 feet from its base. How much wire is needed? Round to the nearest tenth. $3\sqrt{1,300}$ ft ≈ 108.2 ft

43. **PICTURE FRAMES** After gluing and nailing two pieces of picture frame molding together, a frame maker checks her work by making a diagonal measurement. If the sides of the frame form a right angle, what measurement should the frame maker read on the yardstick?   25 in.

44. **CARPENTRY** The gable end of the roof shown is divided in half by a vertical brace, 8 feet in height. Find the length of the roof line.   17 ft

45. **BASEBALL** A baseball diamond is a square with each side 90 feet long. How far is it from home plate to second base? Round to the nearest hundredth. $\sqrt{16,200}$ ft ≈ 127.28 ft

46. PAPER AIRPLANE The figure below gives the directions for making a paper airplane from a square piece of paper with sides 8 inches long. Find the length of the plane when it is completed in Step 3. Round to the nearest hundredth. $\sqrt{128}$ in. $\approx$ 11.31 in.

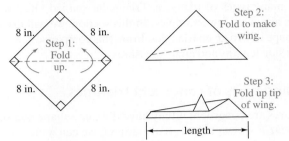

47. FIREFIGHTING The base of the 37-foot ladder shown in the figure below is 9 feet from the wall. Will the top reach a window ledge that is 35 feet above the ground? Explain how you arrived at your answer. yes, $\sqrt{1,288}$ ft $\approx$ 35.89 ft

48. WIND DAMAGE A tree was blown over in a wind storm. Find the height of the tree when it was standing vertically upright. 81 ft

WRITING

49. State the Pythagorean theorem in your own words.

50. When the lengths of the sides of the triangle shown below are substituted into the equation $a^2 + b^2 = c^2$, the result is a false statement. Explain why.

$a^2 + b^2 = c^2$
$2^2 + 4^2 = 5^2$
$4 + 16 = 25$
$20 = 25$

51. In the figure below, equal-sized squares have been drawn on the sides of right triangle $\triangle ABC$. Explain how this figure demonstrates that $3^2 + 4^2 = 5^2$.

52. In the movie *The Wizard of Oz*, the scarecrow was in search of a brain. To prove that he had found one, he recited the following:

"The sum of the square roots of any two sides of an isosceles triangle is equal to the square root of the remaining side."

Unfortunately, this statement is not true. Correct it so that it states the Pythagorean theorem.

REVIEW

Use a check to determine whether the given number in red is a solution of the equation.

53. $2b + 3 = -15$, **−8** no

54. $5t - 4 = -16$, **−2** no

55. $0.5x = 2.9$, **5** no

56. $1.2 + x = 4.7$, **3.5** yes

57. $33 - \dfrac{x}{2} = 30$, **−6** no

58. $\dfrac{x}{4} + 98 = 100$, **−8** no

59. $3x - 2 = 4x - 5$, **12** no

60. $5y + 8 = 3y - 2$, **5** no

SECTION 9.5
Congruent Triangles and Similar Triangles

Objectives
1. Identify corresponding parts of congruent triangles.
2. Use congruence properties to prove that two triangles are congruent.
3. Determine whether two triangles are similar.
4. Use similar triangles to find unknown lengths in application problems.

In our everyday lives, we see many types of triangles. Triangular-shaped kites, sails, roofs, tortilla chips, and ramps are just a few examples. In this section, we will discuss how to compare the size and shape of two given triangles. From this comparison, we can make observations about their side lengths and angle measures.

1 Identify corresponding parts of congruent triangles.

Simply put, two geometric figures are congruent if they have the same shape and size. For example, if △ABC and △DEF shown below are congruent, we can write

△ABC ≅ △DEF   Read as "Triangle ABC is congruent to triangle DEF."

One way to determine whether two triangles are congruent is to see if one triangle can be moved onto the other triangle in such a way that it fits exactly. When we write △ABC ≅ △DEF, we are showing how the vertices of one triangle are matched to the vertices of the other triangle to obtain a "perfect fit." We call this matching of points a **correspondence**.

A ↔ D	Read as "Point A corresponds to point D."
B ↔ E	Read as "Point B corresponds to point E."
C ↔ F	Read as "Point C corresponds to point F."

When we establish a correspondence between the vertices of two congruent triangles, we also establish a correspondence between the angles and the sides of the triangles. Corresponding angles and corresponding sides of congruent triangles are called **corresponding parts**. *Corresponding parts of congruent triangles are always congruent.* That is, corresponding parts of congruent triangles always have the same measure. For the congruent triangles shown above, we have

m(∠A) = m(∠D)   m(∠B) = m(∠E)   m(∠C) = m(∠F)

m($\overline{BC}$) = m($\overline{EF}$)   m($\overline{AC}$) = m($\overline{DF}$)   m($\overline{AB}$) = m($\overline{DE}$)

Congruent Triangles
Two triangles are congruent if and only if their vertices can be matched so that the corresponding sides and the corresponding angles are congruent.

EXAMPLE 1
Refer to the figure below, where △XYZ ≅ △PQR.

a. Name the six congruent corresponding parts of the triangles.
b. Find m(∠P).
c. Find m($\overline{XZ}$).

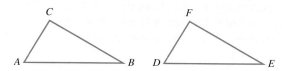

Strategy We will establish the correspondence between the vertices of △XYZ and the vertices of △PQR.

WHY This will, in turn, establish a correspondence between the congruent corresponding angles and sides of the triangles.

Solution
a. The correspondence between the vertices is

△XYZ ≅ △PQR

X ↔ P   Y ↔ Q   Z ↔ R

Corresponding parts of congruent triangles are congruent. Therefore, the congruent corresponding angles are

∠X ≅ ∠P   ∠Y ≅ ∠Q   ∠Z ≅ ∠R

The congruent corresponding sides are

$\overline{YZ} \cong \overline{QR}$ $\overline{XZ} \cong \overline{PR}$ $\overline{XY} \cong \overline{PQ}$

b. From the figure, we see that m(∠X) = 27°. Since ∠X ≅ ∠P, it follows that m(∠P) = 27°.

c. From the figure, we see that m($\overline{PR}$) = 11 inches. Since $\overline{XZ} \cong \overline{PR}$, it follows that m($\overline{XZ}$) = 11 inches.

2 Use congruence properties to prove that two triangles are congruent.

Sometimes it is possible to conclude that two triangles are congruent without having to show that three pairs of corresponding angles are congruent and three pairs of corresponding sides are congruent. To do so, we apply one of the following properties.

SSS Property

If three sides of one triangle are congruent to three sides of a second triangle, the triangles are congruent.

We can show that the triangles shown below are congruent by the SSS property:

$\overline{CD} \cong \overline{ST}$ Since m($\overline{CD}$) = 3 and m($\overline{ST}$) = 3, the segments are congruent.
$\overline{DE} \cong \overline{TR}$ Since m($\overline{DE}$) = 4 and m($\overline{TR}$) = 4, the segments are congruent.
$\overline{EC} \cong \overline{RS}$ Since m($\overline{EC}$) = 5 and m($\overline{RS}$) = 5, the segments are congruent.

Therefore, △CDE ≅ △STR.

SAS Property

If two sides and the angle between them in one triangle are congruent, respectively, to two sides and the angle between them in a second triangle, the triangles are congruent.

Self Check 1

Refer to the figure below, where △ABC ≅ △EDF.

a. Name the six congruent corresponding parts of the triangles.
b. Find m(∠C).
c. Find m($\overline{FE}$).

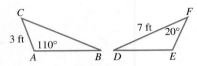

Now Try Problem 33

Self Check 1 Answers
a. ∠A ≅ ∠E, ∠B ≅ ∠D, ∠C ≅ ∠F,
$\overline{AB} \cong \overline{ED}$, $\overline{BC} \cong \overline{DF}$, $\overline{CA} \cong \overline{FE}$
b. 20°
c. 3 ft

Teaching Example 1 Refer to the figure, where △RST ≅ △NMO
a. Name the six congruent corresponding parts of the triangles.
b. Find m(∠T).
c. Find m($\overline{MN}$).

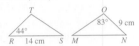

Answers:
a. ∠R ≅ ∠N, ∠S ≅ ∠M, ∠T ≅ ∠O,
$\overline{RS} \cong \overline{NM}$, $\overline{ST} \cong \overline{MO}$, $\overline{TR} \cong \overline{ON}$
b. 83°
c. 14 cm

We can show that the triangles shown below are congruent by the SAS property:

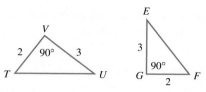

$\overline{TV} \cong \overline{FG}$ Since m($\overline{TV}$) = 2 and m($\overline{FG}$) = 2, the segments are congruent.
$\angle V \cong \angle G$ Since m($\angle V$) = 90° and m($\angle G$) = 90°, the angles are congruent.
$\overline{UV} \cong \overline{EG}$ Since m($\overline{UV}$) = 3 and m($\overline{EG}$) = 3, the segments are congruent.

Therefore, $\triangle TVU \cong \triangle FGE$.

ASA Property

If two angles and the side between them in one triangle are congruent, respectively, to two angles and the side between them in a second triangle, the triangles are congruent.

We can show that the triangles shown below are congruent by the ASA property:

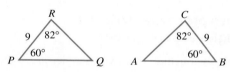

$\angle P \cong \angle B$ Since m($\angle P$) = 60° and m($\angle B$) = 60°, the angles are congruent.
$\overline{PR} \cong \overline{BC}$ Since m($\overline{PR}$) = 9 and m($\overline{BC}$) = 9, the segments are congruent.
$\angle R \cong \angle C$ Since m($\angle R$) = 82° and m($\angle C$) = 82°, the angles are congruent.

Therefore, $\triangle PQR \cong \triangle BAC$.

Caution! There is no SSA property. To illustrate this, consider the triangles shown below. Two sides and an angle of $\triangle ABC$ are congruent to two sides and an angle of $\triangle DEF$. But the congruent angle is not between the congruent sides.

We refer to this situation as SSA. Obviously, the triangles are not congruent because they are not the same shape and size.

The tick marks indicate congruent parts. That is, the sides with one tick mark are the same length, the sides with two tick marks are the same length, and the angles with one tick mark have the same measure.

Teaching Example 2 Explain why the triangles in the figure are congruent.

Answer:
$\triangle RST \cong \triangle UVT$ by the SAS property: $\overline{RT} \cong \overline{UT}$, $\angle 1 \cong \angle 2$, and $\overline{ST} \cong \overline{VT}$

EXAMPLE 2 Explain why the triangles in the figure on the following page are congruent.

Strategy We will show that two sides and the angle between them in one triangle are congruent, respectively, to two sides and the angle between them in a second triangle.

WHY Then we know that the two triangles are congruent by the SAS property.

Solution Since vertical angles are congruent,

$$\angle 1 \cong \angle 2$$

From the figure, we see that

$$\overline{AC} \cong \overline{EC} \quad \text{and} \quad \overline{BC} \cong \overline{DC}$$

Since two sides and the angle between them in one triangle are congruent, respectively, to two sides and the angle between them in a second triangle, $\triangle ABC \cong \triangle EDC$ by the SAS property.

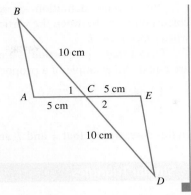

Self Check 2
Are the triangles in the figure below congruent? Explain why or why not. yes, by the SAS property

Now Try Problem 35

EXAMPLE 3 Are $\triangle RST$ and $\triangle RUT$ in the figure on the right congruent?

Strategy We will show that two angles and the side between them in one triangle are congruent, respectively, to two angles and the side between them in a second triangle.

WHY Then we know that the two triangles are congruent by the ASA property.

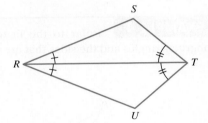

Solution From the markings on the figure, we know that two pairs of angles are congruent.

$\angle SRT \cong \angle URT$ These angles are marked with 1 tick mark, which indicates that they have the same measure.

$\angle STR \cong \angle UTR$ These angles are marked with 2 tick marks, which indicates that they have the same measure.

From the figure, we see that the triangles have side $\overline{RT}$ in common. Furthermore, $\overline{RT}$ is between each pair of congruent angles listed above. Since every segment is congruent to itself, we also have

$$\overline{RT} \cong \overline{RT}$$

Knowing that two angles and the side between them in $\triangle RST$ are congruent, respectively, to two angles and the side between them in $\triangle RUT$, we can conclude that $\triangle RST \cong \triangle RUT$ by the ASA property.

Self Check 3
Are the triangles in the following figure congruent? Explain why or why not. yes, by the SSS property

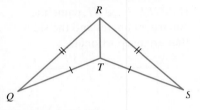

Now Try Problem 37

Teaching Example 3 Are $\triangle ABC$ and $\triangle ADC$ congruent?

Answer:
yes, by the ASA property:
$\angle BAC \cong \angle DAC$, $\overline{AC} \cong \overline{AC}$, and $\angle BCA \cong \angle DCA$

3 Determine whether two triangles are similar.

We have seen that congruent triangles have the same shape and size. **Similar triangles** have the same shape, but not necessarily the same size. That is, one triangle is an exact scale model of the other triangle. If the triangles in the figure below are similar, we can write $\triangle ABC \sim \triangle DEF$ (read the symbol $\sim$ as "is similar to").

Success Tip Note that congruent triangles are always similar, but similar triangles are not always congruent.

The formal definition of similar triangles requires that we establish a correspondence between the vertices of the triangles. The definition also involves the word *proportional*.

Recall that a **proportion** is a mathematical statement that two ratios (fractions) are equal. An example of a proportion is

$$\frac{1}{2} = \frac{4}{8}$$

In this case, we say that $\frac{1}{2}$ and $\frac{4}{8}$ are *proportional*.

Similar Triangles

Two triangles are similar if and only if their vertices can be matched so that corresponding angles are congruent and the lengths of corresponding sides are proportional.

Self Check 4

If $\triangle GEF \sim \triangle IJH$, name the congruent angles and the sides that are proportional.

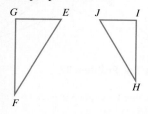

Now Try Problem 39

Self Check 4 Answer
$\angle G \cong \angle I, \angle E \cong \angle J, \angle F \cong \angle H$;
$\frac{EG}{JI} = \frac{GF}{IH} = \frac{FE}{HJ}$

Teaching Example 4 If $\triangle RST \sim \triangle WYX$, name the congruent angles and the sides that are proportional.

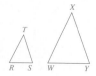

Answer:
$\angle R \cong \angle W, \angle S \cong \angle Y, \angle T \cong \angle X$;
$\frac{RS}{WY} = \frac{RT}{WX} = \frac{TS}{XY}$

EXAMPLE 4 Refer to the figure below. If $\triangle PQR \sim \triangle CDE$, name the congruent angles and the sides that are proportional.

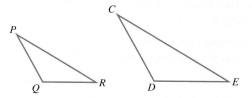

Strategy We will establish the correspondence between the vertices of $\triangle PQR$ and the vertices of $\triangle CDE$.

WHY This will, in turn, establish a correspondence between the congruent corresponding angles and proportional sides of the triangles.

Solution When we write $\triangle PQR \sim \triangle CDE$, a correspondence between the vertices of the triangles is established.

$$\triangle PQR \sim \triangle CDE$$

Since the triangles are similar, corresponding angles are congruent:

$$\angle P \cong \angle C \qquad \angle Q \cong \angle D \qquad \angle R \cong \angle E$$

The lengths of the corresponding sides are proportional. To simplify the notation, we will now let $PQ = m(\overline{PQ})$, $CD = m(\overline{CD})$, $QR = m(\overline{QR})$, and so on.

$$\frac{PQ}{CD} = \frac{QR}{DE} \qquad \frac{QR}{DE} = \frac{PR}{CE} \qquad \frac{PQ}{CD} = \frac{PR}{CE}$$

Written in a more compact way, we have

$$\frac{PQ}{CD} = \frac{QR}{DE} = \frac{PR}{CE}$$

Property of Similar Triangles

If two triangles are similar, all pairs of corresponding sides are in proportion.

It is possible to conclude that two triangles are similar without having to show that all three pairs of corresponding angles are congruent and that the lengths of all three pairs of corresponding sides are proportional.

AAA Similarity Theorem

If the angles of one triangle are congruent to corresponding angles of another triangle, the triangles are similar.

EXAMPLE 5 In the figure on the right, $\overline{PR} \parallel \overline{MN}$. Are $\triangle PQR$ and $\triangle NQM$ similar triangles?

Strategy We will show that the angles of one triangle are congruent to corresponding angles of another triangle.

WHY Then we know that the two triangles are similar by the AAA property.

Solution Since vertical angles are congruent,

$\angle PQR \cong \angle NQM$ This is one pair of congruent corresponding angles.

In the figure, we can view $\overleftrightarrow{PN}$ as a transversal cutting parallel line segments $\overline{PR}$ and $\overline{MN}$. Since alternate interior angles are then congruent, we have:

$\angle RPQ \cong \angle MNQ$ This is a second pair of congruent corresponding angles.

Furthermore, we can view $\overleftrightarrow{RM}$ as a transversal cutting parallel line segments $\overline{PR}$ and $\overline{MN}$. Since alternate interior angles are then congruent, we have:

$\angle QRP \cong \angle QMN$ This is a third pair of congruent corresponding angles.

These observations are summarized in the figure on the right. We see that corresponding angles of $\triangle PQR$ are congruent to corresponding angles of $\triangle NQM$. By the AAA similarity theorem, we can conclude that

$\triangle PQR \sim \triangle NQM$

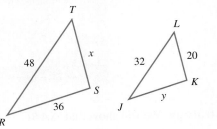

Self Check 5

In the figure below, $\overline{YA} \parallel \overline{ZB}$. Are $\triangle XYA$ and $\triangle XZB$ similar triangles?

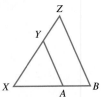

Now Try Problems 41 and 43

Self Check 5 Answer
yes, by the AAA similarity theorem: $\angle X \cong \angle X$, $\angle XYA \cong \angle XZB$, $\angle XAY \cong \angle XBZ$

Teaching Example 5 In the figure, $\overline{MN} \parallel \overline{QP}$. Are $\triangle MNO$ and $\triangle PQO$ similar triangles?

Answer:
yes, by the AAA similarity theorem: $\angle MON \cong \angle POQ$, $\angle M \cong \angle P$, $\angle N \cong \angle Q$

EXAMPLE 6 In the figure below, $\triangle RST \sim \triangle JKL$. Find: **a.** x **b.** y

Strategy To find x, we will write a proportion of corresponding sides so that x is the only unknown. Then we will solve the proportion for x. We will use a similar method to find y.

WHY Since $\triangle RST \sim \triangle JKL$, we know that the lengths of corresponding sides of $\triangle RST$ and $\triangle JKL$ are proportional.

Solution
a. When we write $\triangle RST \sim \triangle JKL$, a correspondence between the vertices of the two triangles is established.

Self Check 6

In the figure below, $\triangle DEF \sim \triangle GHI$. Find:

a. x 6 **b.** y 11.25

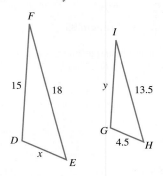

Now Try Problem 53

Unless otherwise noted, all art on this page is © Cengage Learning.

Teaching Example 6 In the figure below, △XYZ ~ △ABC. Find:
a. s   b. t

Answers:
a. 6
b. 9

The lengths of corresponding sides of these similar triangles are proportional.

$$\frac{RT}{JL} = \frac{ST}{KL}$$ Each fraction is a ratio of a side length of △RST to its corresponding side length of △JKL.

$$\frac{48}{32} = \frac{x}{20}$$ Substitute: RT = 48, JL = 32, ST = x, and KL = 20.

$48(20) = 32x$ Find each cross product and set them equal.

$960 = 32x$ Do the multiplication.

$30 = x$ To isolate x, undo the multiplication by 32 by dividing both sides by 32.

Thus, x is 30.

b. To find y, we write a proportion of corresponding side lengths in such a way that y is the only unknown.

$$\frac{RT}{JL} = \frac{RS}{JK}$$

$$\frac{48}{32} = \frac{36}{y}$$ Substitute: RT = 48, JL = 32, RS = 36, and JK = y.

$48y = 32(36)$ Find each cross product and set them equal.

$48y = 1,152$ Do the multiplication.

$y = 24$ To isolate y, undo the multiplication by 48 by dividing both sides by 48.

Thus, y is 24.

4 Use similar triangles to find unknown lengths in application problems.

Similar triangles and proportions can be used to find lengths that would normally be difficult to measure. For example, we can use the reflective properties of a mirror to calculate the height of a flagpole while standing safely on the ground.

Self Check 7

In the figure below, △ABC ~ △EDC. Find h.   500 ft

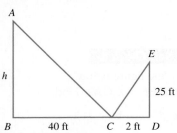

Now Try Problem 85

Teaching Example 7 In the figure, △SQR ~ △STU. Find h.

Answer:
55 in.

EXAMPLE 7   To determine the height of a flagpole, a woman walks to a point 20 feet from its base, as shown below. Then she takes a mirror from her purse, places it on the ground, and walks 2 feet farther away, where she can see the top of the pole reflected in the mirror. Find the height of the pole.

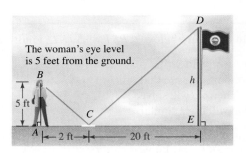

Strategy We will show that △ABC ~ △EDC.

WHY Then we can write a proportion of corresponding sides so that h is the only unknown and we can solve the proportion for h.

Solution To show that △ABC ~ △EDC, we begin by applying an important fact about mirrors. When a beam of light strikes a mirror, it is reflected at the same angle as it hits the mirror. Therefore, ∠BCA ≅ ∠DCE. Furthermore, ∠A ≅ ∠E because the woman and the flagpole are perpendicular to the ground. Finally, if two pairs of corresponding angles are congruent, it follows that the third pair of corresponding angles are also congruent: $\angle B \cong \angle D$. By the AAA similarity theorem, we conclude that $\triangle ABC \sim \triangle EDC$.

Since the triangles are similar, the lengths of their corresponding sides are in proportion. If we let h represent the height of the flagpole, we can find h by solving the following proportion.

Height of the flagpole → $\dfrac{h}{5} = \dfrac{20}{2}$ ← Distance from flagpole to mirror
Height of the woman → ← Distance from woman to mirror

$2h = 5(20)$ Find each cross product and set them equal.
$2h = 100$ Do the multiplication.
$h = 50$ To isolate h, divide both sides by 2.

The flagpole is 50 feet tall.

ANSWERS TO SELF CHECKS

1. **a.** $\angle A \cong \angle E, \angle B \cong \angle D, \angle C \cong \angle F, \overline{AB} \cong \overline{ED}, \overline{BC} \cong \overline{DF}, \overline{CA} \cong \overline{FE}$ **b.** 20° **c.** 3 ft
2. yes, by the SAS property **3.** yes, by the SSS property **4.** $\angle G \cong \angle I, \angle E \cong \angle J$, $\angle F \cong \angle H$; $\dfrac{EG}{JI} = \dfrac{GF}{IH} = \dfrac{FE}{HJ}$ **5.** yes, by the AAA similarity theorem: $\angle X \cong \angle X$, $\angle XYA \cong \angle XZB, \angle XAY \cong \angle XBZ$ **6. a.** 6 **b.** 11.25 **7.** 500 ft

SECTION 9.5 STUDY SET

VOCABULARY

Fill in the blanks.

1. _Congruent_ triangles are the same size and the same shape.
2. When we match the vertices of $\triangle ABC$ with the vertices of $\triangle DEF$, as shown below, we call this matching of points a _correspondence_.

   $A \leftrightarrow D \qquad B \leftrightarrow E \qquad C \leftrightarrow F$

3. Two angles or two line segments with the same measure are said to be _congruent_.
4. Corresponding _parts_ of congruent triangles are congruent.
5. If two triangles are _similar_, they have the same shape but not necessarily the same size.
6. A mathematical statement that two ratios (fractions) are equal, such as $\dfrac{x}{18} = \dfrac{4}{9}$, is called a _proportion_.

CONCEPTS

7. Refer to the triangles below.

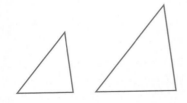

a. Do these triangles appear to be congruent? Explain why or why not. No, they are different sizes.

b. Do these triangles appear to be similar? Explain why or why not. Yes, they appear to have the same shape.

8. a. Draw a triangle that is congruent to $\triangle CDE$ shown below. Label it $\triangle ABC$.

b. Draw a triangle that is similar to, but not congruent to, $\triangle CDE$. Label it $\triangle MNO$.

Fill in the blanks.

9. $\triangle XYZ \cong \triangle$ _PRQ_

10. △ __ABC__ ≅ △DEF

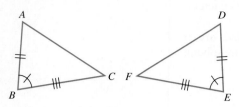

11. △RST ~ △ __MNO__

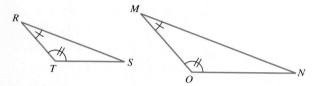

12. △ __BDE__ ~ △TAC

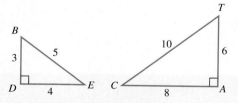

13. Name the six corresponding parts of the congruent triangles shown below. ∠A ≅ ∠B, ∠Y ≅ ∠T, ∠Z ≅ ∠R, $\overline{YZ} \cong \overline{TR}$, $\overline{AZ} \cong \overline{BR}$, $\overline{AY} \cong \overline{BT}$

14. Name the six corresponding parts of the congruent triangles shown below. ∠S ≅ ∠F, ∠T ≅ ∠G, ∠R ≅ ∠E, $\overline{TS} \cong \overline{GF}$, $\overline{TR} \cong \overline{GE}$, $\overline{RS} \cong \overline{EF}$

Fill in the blanks.

15. Two triangles are __congruent__ if and only if their vertices can be matched so that the corresponding sides and the corresponding angles are congruent.

16. SSS property: If three __sides__ of one triangle are congruent to three __sides__ of a second triangle, the triangles are congruent.

17. SAS property: If two sides and the __angle__ between them in one triangle are congruent, respectively, to two sides and the __angle__ between them in a second triangle, the triangles are congruent.

18. ASA property: If two angles and the __side__ between them in one triangle are congruent, respectively, to two angles and the __side__ between them in a second triangle, the triangles are congruent.

Solve each proportion.

19. $\dfrac{x}{15} = \dfrac{20}{3}$ 100

20. $\dfrac{5}{8} = \dfrac{35}{x}$ 56

21. $\dfrac{h}{2.6} = \dfrac{27}{13}$ 5.4

22. $\dfrac{11.2}{4} = \dfrac{h}{6}$ 16.8

Fill in the blanks.

23. Two triangles are similar if and only if their vertices can be matched so that corresponding angles are congruent and the lengths of corresponding sides are __proportional__.

24. If the angles of one triangle are congruent to corresponding angles of another triangle, the triangles are __similar__.

25. Congruent triangles are always similar, but similar triangles are not always __congruent__.

26. For certain application problems, similar triangles and __proportions__ can be used to find lengths that would normally be difficult to measure.

NOTATION

Fill in the blanks.

27. The symbol ≅ is read as " __is__ __congruent__ to ."

28. The symbol ~ is read as " __is__ __similar__ to ."

29. Use tick marks to show the congruent parts of the triangles shown below.

∠K ≅ ∠H $\overline{KR} \cong \overline{HJ}$ ∠M ≅ ∠E

30. Use tick marks to show the congruent parts of the triangles shown below.

∠P ≅ ∠T $\overline{LP} \cong \overline{RT}$ $\overline{FP} \cong \overline{ST}$

GUIDED PRACTICE

Name the six corresponding parts of the congruent triangles. See Objective 1.

31. $\overline{AC} \cong \overline{DF}$
    $\overline{DE} \cong \overline{AB}$
    $\overline{BC} \cong \overline{EF}$
    $\angle A \cong \angle D$
    $\angle E \cong \angle B$
    $\angle F \cong \angle C$

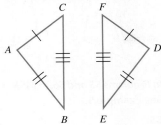

32. $\overline{AB} \cong \overline{DE}$
    $\overline{EC} \cong \overline{BC}$
    $\overline{AC} \cong \overline{DC}$
    $\angle D \cong \angle A$
    $\angle B \cong \angle E$
    $\angle 1 \cong \angle 2$

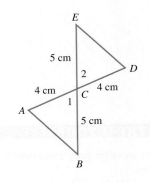

33. Refer to the figure below, where $\triangle BCD \cong \triangle MNO$.
    a. Name the six congruent corresponding parts of the triangles. See Example 1. $\angle B \cong \angle M, \angle C \cong \angle N,$ $\angle D \cong \angle O, \overline{BC} \cong \overline{MN},$ $\overline{CD} \cong \overline{NO}, \overline{BD} \cong \overline{MO}$
    b. Find m($\angle N$). 72°
    c. Find m($\overline{MO}$). 10 ft
    d. Find m($\overline{CD}$). 9 ft

34. Refer to the figure below, where $\triangle DCG \cong \triangle RST$.
    a. Name the six congruent corresponding parts of the triangles. See Example 1. $\angle D \cong \angle R, \angle C \cong \angle S,$ $\angle G \cong \angle T, \overline{DC} \cong \overline{RS},$ $\overline{CG} \cong \overline{ST}, \overline{DG} \cong \overline{RT}$
    b. Find m($\angle R$). 60°
    c. Find m($\overline{DG}$). 2 in.
    d. Find m($\overline{ST}$). 3 in.

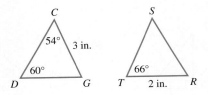

Determine whether each pair of triangles is congruent. If they are, tell why. See Examples 2 and 3.

35.
yes, SSS

36.
yes, SAS

37.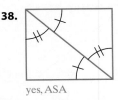
not necessarily

38.
yes, ASA

39. Refer to the similar triangles shown below. See Example 4.
    a. Name three pairs of congruent angles. $\angle L \cong \angle H,$ $\angle M \cong \angle J, \angle R \cong \angle E$
    b. Complete each proportion.
    $$\frac{LM}{HJ} = \frac{MR}{JE} \qquad \frac{MR}{JE} = \frac{LR}{HE} \qquad \frac{LM}{HJ} = \frac{LR}{HE}$$
    c. We can write the answer to part b in a more compact form:
    $$\frac{LM}{HJ} = \frac{MR}{JE} = \frac{LR}{HE}$$

40. Refer to the similar triangles shown below. See Example 4.
    a. Name three pairs of congruent angles. $\angle W \cong \angle D,$ $\angle Y \cong \angle F, \angle X \cong \angle E$
    b. Complete each proportion.
    $$\frac{WY}{DF} = \frac{YX}{FE} \qquad \frac{WX}{DE} = \frac{YX}{FE} \qquad \frac{XY}{EF} = \frac{WY}{DF}$$
    c. We can write the answer to part b in a more compact form:
    $$\frac{WY}{DF} = \frac{YX}{FE} = \frac{WX}{DE}$$

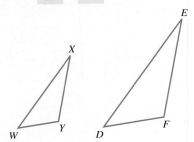

Tell whether the triangles are similar. See Example 5.

41.
yes

42.
yes

43.
not necessarily

44.
not necessarily

45.
yes

46.
yes

47.
not necessarily

48.
not necessarily

49. $\overline{XY} \parallel \overline{ZD}$ yes

50. $\overline{QR} \parallel \overline{TU}$ yes

51.
not necessarily

52.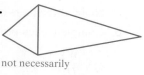
not necessarily

In Problems 53 and 54, △MSN ~ △TPR. Find x and y. See Example 6.

53.
8, 35

54.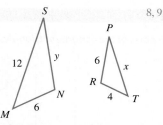
8, 9

In Problems 55 and 56, △MSN ~ △TPN. Find x and y. See Example 6.

55.
60, 38

56.
19.2, 30

TRY IT YOURSELF

Tell whether each statement is true. If a statement is false, tell why.

57. If three sides of one triangle are the same length as the corresponding three sides of a second triangle, the triangles are congruent. true

58. If two sides of one triangle are the same length as two sides of a second triangle, the triangles are congruent. false: need SSS or SAS

59. If two sides and an angle of one triangle are congruent, respectively, to two sides and an angle of a second triangle, the triangles are congruent. false: the angles must be between congruent sides

60. If two angles and the side between them in one triangle are congruent, respectively, to two angles and the side between them in a second triangle, the triangles are congruent. true

Determine whether each pair of triangles are congruent. If they are, tell why.

61.
yes, SSS

62.
not necessarily

63.
yes, SAS

64.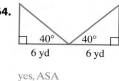
yes, ASA

9.5 Congruent Triangles and Similar Triangles

65. $\overline{AB} \parallel \overline{DE}$ yes, ASA
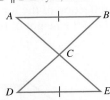

66. $\overline{XY} \parallel \overline{ZQ}$ yes, ASA

67.

not necessarily

68.

not necessarily

75.

76.

77. If $\overline{DE}$ in the figure below is parallel to $\overline{AB}$, $\triangle ABC$ will be similar to $\triangle DEC$. Find x. $\frac{25}{6} = 4\frac{1}{6}$
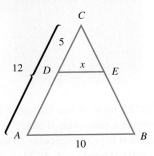

In Problems 69 and 70, $\triangle ABC \cong \triangle DEF$. Find x and y.

69. 80°, 2 yd

78. If $\overline{SU}$ in the figure below is parallel to $\overline{TV}$, $\triangle SRU$ will be similar to $\triangle TRV$. Find x. 18

70.
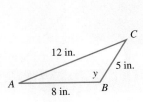
12 in., 135°

79. If $\overline{DE}$ in the figure below is parallel to $\overline{CB}$, $\triangle EAD$ will be similar to $\triangle BAC$. Find x. 16
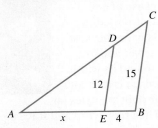

In Problems 71 and 72, find x and y.

71. $\triangle ABC \cong \triangle ABD$ 19°, 14 m
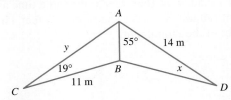

72. $\triangle ABC \cong \triangle DEC$ 8 mi, 37°
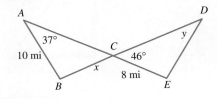

80. If $\overline{HK}$ in the figure below is parallel to $\overline{AB}$, $\triangle HCK$ will be similar to $\triangle ACB$. Find x. 4
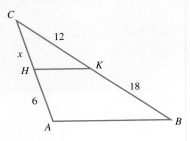

In Problems 73–76, find x.

73.

74.
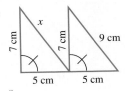

Unless otherwise noted, all art on this page is © Cengage Learning.

APPLICATIONS

81. SEWING The pattern that is sewn on the rear pocket of a pair of blue jeans is shown below. If $\triangle AOB \cong \triangle COD$, how long is the stitching from point A to point D? 17.5 cm

82. CAMPING The base of the tent pole is placed at the midpoint between the stake at point A and the stake at point B, and it is perpendicular to the ground, as shown below. Explain why $\triangle ACD \cong \triangle BCD$.
SAS property: $\overline{AD} \cong \overline{BD}$, $\angle ADC \cong \angle BDC$, $\overline{CD} \cong \overline{CD}$

83. A surveying crew needs to find the width of the river shown in the illustration below. Because of a dangerous current, they decide to stay on the west side of the river and use geometry to find its width. Their approach is to create two similar right triangles on dry land. Then they write and solve a proportion to find w. What is the width of the river? 59.2 ft

from Campus to Careers
Surveyor

84. HEIGHT OF A BUILDING A man places a mirror on the ground and sees the reflection of the top of a building, as shown below. Find the height of the building. 36 ft

85. HEIGHT OF A TREE The tree shown below casts a shadow 24 feet long when a man 6 feet tall casts a shadow 4 feet long. Find the height of the tree. 36 ft

86. WASHINGTON, D.C. The Washington Monument casts a shadow of $166\frac{1}{2}$ feet at the same time as a 5-foot-tall tourist casts a shadow of $1\frac{1}{2}$ feet. Find the height of the monument. 555 ft

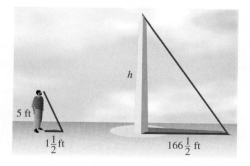

87. HEIGHT OF A TREE A tree casts a shadow of 29 feet at the same time as a vertical yardstick casts a shadow of 2.5 feet. Find the height of the tree. 34.8 ft

88. GEOGRAPHY The diagram below shows how a laser beam was pointed over the top of a pole to the top of a mountain to determine the elevation of the mountain. Find h. 1,204 ft + 5 ft = 1,209 ft

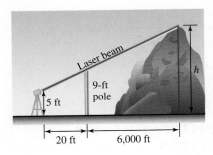

89. FLIGHT PATH An airplane ascends 200 feet as it flies a horizontal distance of 1,000 feet, as shown in the following figure. How much altitude is gained as it flies a horizontal distance of 1 mile? (*Hint:* 1 mile = 5,280 feet.) 1,056 ft

WRITING

90. Tell whether the statement is true or false. Explain your answer.
  a. Congruent triangles are always similar.
  b. Similar triangles are always congruent.

91. Explain why there is no SSA property for congruent triangles.

REVIEW

Find the LCM of the given numbers.

92. 16, 20 80

93. 21, 27 189

Find the GCF of the given numbers.

94. 18, 96 6

95. 63, 84 21

SECTION 9.6
Quadrilaterals and Other Polygons

Recall from Section 9.3 that a polygon is a closed geometric figure with at least three line segments for its sides. In this section, we will focus on polygons with four sides, called *quadrilaterals*. One type of quadrilateral is the *square*. The game boards for Monopoly and Scrabble have a square shape. Another type of quadrilateral is the *rectangle*. Most picture frames and many mirrors are rectangular. Utility knife blades and swimming fins have shapes that are examples of a third type of quadrilateral called a *trapezoid*.

Objectives

1. Classify quadrilaterals.
2. Use properties of rectangles to find unknown angle measures and side lengths.
3. Find unknown angle measures of trapezoids.
4. Use the formula for the sum of the angle measures of a polygon.

1 Classify quadrilaterals.

A **quadrilateral** is a polygon with four sides. Some common quadrilaterals are shown below.

Parallelogram
(Opposite sides parallel)

Rectangle
(Parallelogram with four right angles)

Square
(Rectangle with sides of equal length)

Rhombus
(Parallelogram with sides of equal length)

Trapezoid
(Exactly two sides parallel)

We can use the capital letters that label the vertices of a quadrilateral to name it. For example, when referring to the quadrilateral shown on the right, with vertices A, B, C, and D, we can use the notation quadrilateral $ABCD$.

Quadrilateral ABCD

> **The Language of Algebra** When naming a quadrilateral (or any other polygon), we may begin with any vertex. Then we move around the figure in a clockwise (or counterclockwise) direction as we list the remaining vertices. Some other ways of naming the quadrilateral above are quadrilateral $ADCB$, quadrilateral $CDAB$, and quadrilateral $DABC$. It would be unacceptable to name it as quadrilateral $ACDB$, because the vertices would not be listed in clockwise (or counterclockwise) order.

A segment that joins two nonconsecutive vertices of a polygon is called a **diagonal** of the polygon. Quadrilateral $ABCD$ shown below has two diagonals, $\overline{AC}$ and $\overline{BD}$.

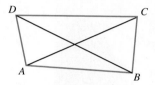

2 Use properties of rectangles to find unknown angle measures and side lengths.

Recall that a **rectangle** is a quadrilateral with four right angles. The rectangle is probably the most common and recognizable of all geometric figures. For example, most doors and windows are rectangular in shape. The boundaries of soccer fields and basketball courts are rectangles. Even our paper currency, such as the $1, $5, and $20 bills, is in the shape of a rectangle. Rectangles have several important characteristics.

Properties of Rectangles

In any rectangle:

1. All four angles are right angles.
2. Opposite sides are parallel.
3. Opposite sides have equal length.
4. The diagonals have equal length.
5. The diagonals intersect at their midpoints.

EXAMPLE 1 In the figure, quadrilateral $WXYZ$ is a rectangle. Find each measure:

a. $m(\angle YXW)$ **b.** $m(\overline{XY})$ **c.** $m(\overline{WY})$ **d.** $m(\overline{XZ})$

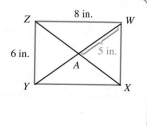

Strategy We will use properties of rectangles to find the unknown angle measure and the unknown measures of the line segments.

WHY Quadrilateral $WXYZ$ is a rectangle.

Solution

a. In any rectangle, all four angles are right angles. Therefore, $\angle YXW$ is a right angle, and m($\angle YXW$) = 90°.

b. $\overline{XY}$ and $\overline{WZ}$ are opposite sides of the rectangle, so they have equal length. Since the length of $\overline{WZ}$ is 8 inches, m($\overline{XY}$) is also 8 inches.

c. $\overline{WY}$ and $\overline{ZX}$ are diagonals of the rectangle, and they intersect at their midpoints. That means that point A is the midpoint of $\overline{WY}$. Since the length of $\overline{WA}$ is 5 inches, m($\overline{WY}$) is 2 · 5 inches, or 10 inches.

d. The diagonals of a rectangle are of equal length. In part c, we found that the length of $\overline{WY}$ is 10 inches. Therefore, m($\overline{XZ}$) is also 10 inches.

We have seen that if a quadrilateral has four right angles, it is a rectangle. The following statements establish some conditions that a parallelogram must meet to ensure that it is a rectangle.

Parallelograms That Are Rectangles

1. If a parallelogram has one right angle, then the parallelogram is a rectangle.
2. If the diagonals of a parallelogram are congruent, then the parallelogram is a rectangle.

EXAMPLE 2 **Construction** A carpenter wants to build a shed with a 9-foot-by-12-foot base. How can he make sure that the foundation has four right-angle corners?

Strategy The carpenter should find the lengths of the diagonals of the foundation.

WHY If the diagonals are congruent, then the foundation is rectangular in shape and the corners are right angles.

Solution The four-sided foundation, which we will label as parallelogram $ABCD$, has opposite sides of equal length. The carpenter can use a tape measure to find the lengths of the diagonals $\overline{AC}$ and $\overline{BD}$. If these diagonals are of equal length, the foundation will be a rectangle and have right angles at its four corners. This process is commonly referred to as "squaring a foundation." Picture framers use a similar process to make sure their frames have four 90° corners.

3 Find unknown angle measures of trapezoids.

A **trapezoid** is a quadrilateral with exactly two sides parallel. For the trapezoid shown on the next page, the parallel sides $\overline{AB}$ and $\overline{DC}$ are called **bases**. To distinguish between the two bases, we will refer to $\overline{AB}$ as the **upper base** and $\overline{DC}$ as the **lower base**. The angles on either side of the upper base are called **upper base angles**, and the angles on either side of the lower base are called **lower base angles**. The nonparallel sides are called **legs**.

Unless otherwise noted, all art on this page is © Cengage Learning.

Self Check 1

In rectangle $RSTU$ shown below, the length of $\overline{RT}$ is 13 ft. Find each measure:

a. m($\angle SRU$) 90°
b. m($\overline{ST}$) 12 ft
c. m($\overline{TG}$) 6.5 ft
d. m($\overline{SG}$) 6.5 ft

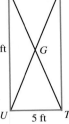

Now Try Problem 27

Teaching Example 1 In rectangle $ABCD$ shown below, the length of $\overline{AX}$ is 20 inches. Find each measure:
a. m($\angle ABC$)
b. m($\overline{AD}$)
c. m($\overline{XC}$)
d. m($\overline{BD}$)
Answers:
a. 90°
b. 32 in.
c. 20 in.
d. 40 in.

Now Try Problem 59

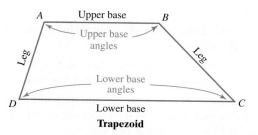

Trapezoid

In the figure above, we can view $\overleftrightarrow{AD}$ as a transversal cutting the parallel lines $\overleftrightarrow{AB}$ and $\overleftrightarrow{DC}$. Since $\angle A$ and $\angle D$ are interior angles on the same side of a transversal, they are supplementary. Similarly, $\overleftrightarrow{BC}$ is a transversal cutting the parallel lines $\overleftrightarrow{AB}$ and $\overleftrightarrow{DC}$. Since $\angle B$ and $\angle C$ are interior angles on the same side of a transversal, they are also supplementary. These observations lead us to the conclusion that *there are always two pairs of supplementary angles in any trapezoid.*

Self Check 3
Refer to trapezoid *HIJK* below, with $\overline{HI} \parallel \overline{KJ}$. Find x and y.
87°, 101°

Now Try Problem 29

Teaching Example 3 Refer to trapezoid *ABCD* below, with $\overline{AB} \parallel \overline{DC}$. Find x and y.

Answers:
79°, 96°

EXAMPLE 3
Refer to trapezoid *KLMN* below, with $\overline{KL} \parallel \overline{NM}$. Find x and y.

Strategy We will use the interior angles property twice to write two equations that mathematically model the situation.

WHY We can then solve the equations to find x and y.

Solution $\angle K$ and $\angle N$ are interior angles on the same side of transversal $\overleftrightarrow{KN}$ that cuts the parallel lines segments $\overline{KL}$ and $\overline{NM}$. Similarly, $\angle L$ and $\angle M$ are interior angles on the same side of transversal $\overleftrightarrow{LM}$ that cuts the parallel lines segments $\overline{KL}$ and $\overline{NM}$. Recall that if two parallel lines are cut by a transversal, interior angles on the same side of the transversal are supplementary. We can use this fact twice — once to find x and a second time to find y.

$$m(\angle K) + m(\angle N) = 180°$$ The sum of the measures of supplementary angles is 180°.
$$x + 82° = 180°$$ Substitute x for m($\angle$K) and 82° for m($\angle$N).
$$x = 98°$$ To isolate x, subtract 82° from both sides.

Thus, x is 98°.

$$m(\angle L) + m(\angle M) = 180°$$ The sum of the measures of supplementary angles is 180°.
$$121° + y = 180°$$ Substitute 121° for m($\angle$L) and y for m($\angle$M).
$$y = 59°$$ To isolate y, subtract 121° from both sides.

Thus, y is 59°.

If the nonparallel sides of a trapezoid are the same length, it is called an **isosceles trapezoid**. The figure on the right shows isosceles trapezoid *DEFG* with $\overline{DG} \cong \overline{EF}$. In an isosceles trapezoid, *both pairs of base angles are congruent.* In the figure, $\angle D \cong \angle E$ and $\angle G \cong \angle F$.

Isosceles trapezoid

EXAMPLE 4   Landscaping
A cross section of a drainage ditch shown below is an isosceles trapezoid with $\overline{AB} \parallel \overline{DC}$. Find x and y.

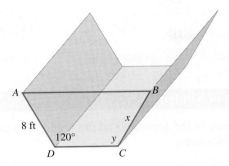

Self Check 4
Refer to the isosceles trapezoid shown below with $\overline{RS} \parallel \overline{UT}$. Find x and y.   10 in., 58°

Now Try Problem 31

Teaching Example 4 Refer to the isosceles trapezoid below, with $\overline{MN} \parallel \overline{PO}$. Find x and y.

Answers: 47°, 9 yd

Strategy We will compare the nonparallel sides and compare a pair of base angles of the trapezoid to find each unknown.

WHY The nonparallel sides of an isosceles trapezoid have the same length and both pairs of base angles are congruent.

Solution Since $\overline{AD}$ and $\overline{BC}$ are the nonparallel sides of an isosceles trapezoid, m($\overline{AD}$) and m($\overline{BC}$) are equal, and x is 8 ft.

Since $\angle D$ and $\angle C$ are a pair of base angles of an isosceles trapezoid, they are congruent and m($\angle D$) = m($\angle C$). Thus, y is 120°.

4 Use the formula for the sum of the angle measures of a polygon.

In the figure shown below, a protractor was used to find the measure of each angle of the quadrilateral. When we add the four angle measures, the result is 360°.

88° + 79° + 127° + 66° = 360°

This illustrates an important fact about quadrilaterals: The sum of the measures of the angles of *any* quadrilateral is 360°. This can be shown using the diagram in figure (a) on the following page. In the figure, the quadrilateral is divided into two triangles. Since the sum of the angle measures of any triangle is 180°, the sum of the measures of the angles of the quadrilateral is 2 · 180°, or 360°.

A similar approach can be used to find the sum of the measures of the angles of any pentagon or any hexagon. The pentagon in figure (b) is divided into three triangles. The sum of the measures of the angles of the pentagon is 3 · 180°, or 540°. The hexagon in figure (c) is divided into four triangles. The sum of the measures of the angles of the hexagon is 4 · 180°, or 720°. In general, a polygon with n sides can be divided into $n - 2$ triangles. Therefore, the sum of the angle measures of a polygon can be found by multiplying 180° by $n - 2$.

Quadrilateral   Pentagon   Hexagon

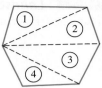

2 · 180° = 360°   3 · 180° = 540°   4 · 180° = 720°
(a)   (b)   (c)

Sum of the Angles of a Polygon

The sum S, in degrees, of the measures of the angles of a polygon with n sides is given by the formula

$$S = (n - 2)180°$$

Self Check 5
Find the sum of the angle measures of the polygon shown below. 900°

Now Try Problem 33

Teaching Example 5 Find the sum of the angle measures of a 12-sided polygon.
Answer:
1,800°

EXAMPLE 5 Find the sum of the angle measures of a 13-sided polygon.

Strategy We will substitute 13 for n in the formula $S = (n - 2)180°$ and evaluate the right side.

WHY The variable S represents the unknown sum of the measures of the angles of the polygon.

Solution

$S = (n - 2)180°$ This is the formula for the sum of the measures of the angles of a polygon.

$S = (13 - 2)180°$ Substitute 13 for n, the number of sides.

$S = (11)180°$ Do the subtraction within the parentheses.

$S = 1,980°$ Do the multiplication.

$$\begin{array}{r} 180 \\ \times\, 11 \\ \hline 180 \\ 1800 \\ \hline 1,980 \end{array}$$

The sum of the measures of the angles of a 13-sided polygon is 1,980°.

Self Check 6
The sum of the measures of the angles of a polygon is 1,620°. Find the number of sides the polygon has. 11 sides

Now Try Problem 43

Teaching Example 6 The sum of the measures of the angles of a polygon is 540°. Find the number of sides the polygon has.
Answer:
5 sides

EXAMPLE 6 The sum of the measures of the angles of a polygon is 1,080°. Find the number of sides the polygon has.

Strategy We will substitute 1,080° for S in the formula $S = (n - 2)180°$ and solve for n.

WHY The variable n represents the unknown number of sides of the polygon.

Solution

$S = (n - 2)180°$ This is the formula for the sum of the measures of the angles of a polygon.

$1,080° = (n - 2)180°$ Substitute 1,080° for S, the sum of the measures of the angles.

$1,080° = 180°n - 360°$ Distribute the multiplication by 180°.

$1,080° + 360° = 180°n - 360° + 360°$ To isolate 180°n, add 360° to both sides.

$1,440° = 180°n$ Do the additions.

$\dfrac{1,440°}{180°} = \dfrac{180°n}{180°}$ To isolate n, divide both sides by 180°.

$8 = n$ Do the division.

$$\begin{array}{r} \overset{1}{1,080} \\ +\, 360 \\ \hline 1,440 \end{array} \qquad \begin{array}{r} 8 \\ 180\overline{)1,440} \\ -1\,440 \\ \hline 0 \end{array}$$

The polygon has 8 sides. It is an octagon.

SECTION 9.6 STUDY SET

ANSWERS TO SELF CHECKS
1. a. 90° b. 12 ft c. 6.5 ft d. 6.5 ft 3. 87°, 101° 4. 10 in., 58° 5. 900°
6. 11 sides

VOCABULARY

Fill in the blanks.

1. A __quadrilateral__ is a polygon with four sides.
2. A __parallelogram__ is a quadrilateral with opposite sides parallel.
3. A __rectangle__ is a quadrilateral with four right angles.
4. A rectangle with all sides of equal length is a __square__.
5. A __rhombus__ is a parallelogram with four sides of equal length.
6. A segment that joins two nonconsecutive vertices of a polygon is called a __diagonal__ of the polygon.
7. A __trapezoid__ has two sides that are parallel and two sides that are not parallel. The parallel sides are called __bases__. The legs of an __isosceles__ trapezoid have the same length.
8. A __regular__ polygon has sides that are all the same length and angles that are all the same measure.

CONCEPTS

9. Refer to the polygon below.
   a. How many vertices does it have? List them. four; A, B, C, D
   b. How many sides does it have? List them. four; $\overline{AB}, \overline{BC}, \overline{CD}, \overline{DA}$
   c. How many diagonals does it have? List them. two; $\overline{AC}, \overline{BD}$
   d. Tell which of the following are acceptable ways of naming the polygon.

   quadrilateral $ABCD$ yes
   quadrilateral $CDBA$ no
   quadrilateral $ACBD$ no
   quadrilateral $BADC$ yes

   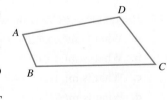

10. Draw an example of each type of quadrilateral.
    a. rhombus
    b. parallelogram
    c. trapezoid
    d. square
    e. rectangle
    f. isosceles trapezoid

11. A parallelogram is shown below. Fill in the blanks.
    a. $\overline{ST} \parallel \overline{VU}$
    b. $\overline{SV} \parallel \overline{TU}$

12. Refer to the rectangle below.
    a. How many right angles does the rectangle have? List them.
       four; $\angle M, \angle N, \angle O, \angle P$
    b. Which sides are parallel?
       $\overline{MN}$ and $\overline{PO}$, $\overline{NO}$ and $\overline{MP}$
    c. Which sides are of equal length?
       $\overline{MN}$ and $\overline{PO}$, $\overline{NO}$ and $\overline{MP}$
    d. Copy the figure and draw the diagonals. Call the point where the diagonals intersect point X. How many diagonals does the figure have? List them.
       two; $\overline{MO}$ and $\overline{NP}$

13. Fill in the blanks. In any rectangle:
    a. All four angles are __right__ angles.
    b. Opposite sides are __parallel__.
    c. Opposite sides have equal __length__.
    d. The diagonals have equal __length__.
    e. The diagonals intersect at their __midpoint__.

14. Refer to the figure below.
    a. What is m($\overline{CD}$)? 12
    b. What is m($\overline{AD}$)? 6

15. In the figure below, $\overline{TR} \parallel \overline{DF}$, $\overline{DT} \parallel \overline{FR}$, and m($\angle D$) = 90°. What type of quadrilateral is $DTRF$?
    rectangle

16. Refer to the parallelogram shown below. If m($\overline{GI}$) = 4 and m($\overline{HJ}$) = 4, what type of figure is quadrilateral GHIJ? rectangle

17. a. Is every rectangle a square? no
    b. Is every square a rectangle? yes
    c. Is every parallelogram a rectangle? no
    d. Is every rectangle a parallelogram? yes
    e. Is every rhombus a square? no
    f. Is every square a rhombus? yes

18. Trapezoid WXYZ is shown below. Which sides are parallel? $\overline{XY} \parallel \overline{WZ}$

19. Trapezoid JKLM is shown below.
    a. What type of trapezoid is this? isosceles
    b. Which angles are the lower base angles? ∠J, ∠M
    c. Which angles are the upper base angles? ∠K, ∠L
    d. Fill in the blanks:
       m(∠J) = m(∠M)
       m(∠K) = m(∠L)
       m($\overline{JK}$) = m($\overline{ML}$)

20. Find the sum of the measures of the angles of the hexagon below. 720°

NOTATION

21. What do the tick marks in the figure indicate?
    The four sides of the quadrilateral are the same length.

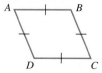

22. Rectangle ABCD is shown below. What do the tick marks indicate about point X?
    Point X is the midpoint of $\overline{AC}$ and $\overline{BD}$.

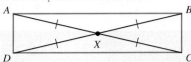

23. In the formula S = (n − 2)180°, what does S represent? What does n represent? the sum of the measures of the angles of a polygon; the number of sides of the polygon

24. Suppose n = 12. What is (n − 2)180°? 1,800°

GUIDED PRACTICE

In Problems 25 and 26, classify each quadrilateral as a rectangle, a square, a rhombus, or a trapezoid. Some figures may be correctly classified in more than one way. See Objective 1.

25. a. square   b. rhombus

    c. trapezoid   d. square

26. a. rectangle   b. rhombus

    c. trapezoid   d. square

27. Rectangle ABCD is shown below. See Example 1.
    a. What is m(∠DCB)? 90°
    b. What is m($\overline{AX}$)? 9
    c. What is m($\overline{AC}$)? 18
    d. What is m($\overline{BD}$)? 18

28. Refer to rectangle EFGH shown below. See Example 1.
    a. Find m(∠EHG). 90°   b. Find m($\overline{FH}$). 32
    c. Find m($\overline{EI}$). 16   d. Find m($\overline{EG}$). 32

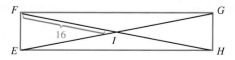

29. Refer to the trapezoid shown below. **See Example 3.**
    a. Find *x*. 42°
    b. Find *y*. 95°

30. Refer to trapezoid *MNOP* shown below. **See Example 3.**
    a. Find m(∠*O*). 60.5°
    b. Find m(∠*M*). 90°

31. Refer to the isosceles trapezoid shown below. **See Example 4.**
    a. Find m($\overline{BC}$). 9
    b. Find *x*. 70°
    c. Find *y*. 110°
    d. Find *z*. 110°

32. Refer to the trapezoid shown below. **See Example 4.**
    a. Find m(∠*T*). 47.5°
    b. Find m(∠*R*). 132.5°
    c. Find m(∠*S*). 132.5°

Find the sum of the angle measures of the polygon. See Example 5.

33. a 14-sided polygon 2,160°
34. a 15-sided polygon 2,340°
35. a 20-sided polygon 3,240°
36. a 22-sided polygon 3,600°
37. an octagon 1,080°
38. a decagon 1,440°
39. a dodecagon 1,800°
40. a nonagon 1,260°

Find the number of sides a polygon has if the sum of its angle measures is the given number. See Example 6.

41. 540° 5
42. 720° 6
43. 900° 7
44. 1,620° 11
45. 1,980° 13
46. 1,800° 12
47. 2,160° 14
48. 3,600° 22

TRY IT YOURSELF

49. Refer to rectangle *ABCD* shown below.
    a. Find m(∠1). 30°
    b. Find m(∠3). 30°
    c. Find m(∠2). 60°
    d. If m($\overline{AC}$) is 8 cm, find m($\overline{BD}$). 8 cm
    e. Find m($\overline{PD}$). 4 cm

50. The following problem appeared on a quiz. Explain why the instructor must have made an error when typing the problem.

    The sum of the measures of the angles of a polygon is 1,000°. How many sides does the polygon have?

    $n = \frac{1,360}{180} \approx 7.6$; *n* should be a whole number

For Problems 51 and 52, find *x*. Then find the measure of each angle of the polygon.

51. 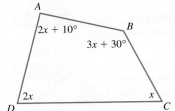 40°; m(∠*A*) = 90°, m(∠*B*) = 150°, m(∠*C*) = 40°, m(∠*D*) = 80°

52. 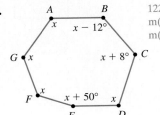 122°; m(∠*A*) = m(∠*D*) = m(∠*F*) = m(∠*G*) = 122°, m(∠*B*) = 110°, m(∠*C*) = 130°, m(∠*E*) = 172°

APPLICATIONS

53. QUADRILATERALS IN EVERYDAY LIFE
What quadrilateral shape do you see in each of the following objects?

a. podium (upper portion) *trapezoid*   b. checkerboard *square*

c. dollar bill *rectangle*   d. swimming fin *trapezoid*

e. camper shell window *parallelogram*

54. FLOWCHART A flowchart shows a sequence of steps to be performed by a computer to solve a given problem. When designing a flowchart, the programmer uses a set of standardized symbols to represent various operations to be performed by the computer. Locate a rectangle, a rhombus, and a parallelogram in the flowchart shown to the right. *rectangle: Open/Close the Files; rhombus: More Records; parallelogram: Read*

55. BASEBALL Refer to the figure to the right. Find the sum of the measures of the angles of home plate. *540°*

56. TOOLS The utility knife blade shown below has the shape of an isosceles trapezoid. Find x, y, and z. $\frac{3}{4}$ in.; 65°; 115°

WRITING

57. Explain why a square is a rectangle.

58. Explain why a trapezoid is not a parallelogram.

59. MAKING A FRAME
After gluing and nailing the pieces of a picture frame together, it didn't look right to a frame maker. (See the figure to the right.) How can she use a tape measure to make sure the corners are 90° (right) angles?

60. A decagon is a polygon with ten sides. What could you call a polygon with one hundred sides? With one thousand sides? With one million sides?

REVIEW

Write each number in words.

61. 254,309 *two hundred fifty-four thousand, three hundred nine*

62. 504,052,040 *five hundred four million, fifty-two thousand, forty*

63. 82,000,415 *eighty-two million, four hundred fifteen*

64. 51,000,201,078 *fifty-one billion, two hundred one thousand, seventy-eight*

SECTION 9.7
Perimeters and Areas of Polygons

Objectives
1. Find the perimeter of a polygon.
2. Find the area of a polygon.
3. Find the area of figures that are combinations of polygons.

In this section, we will discuss how to find perimeters and areas of polygons. Finding perimeters is important when estimating the cost of fencing a yard or installing crown molding in a room. Finding area is important when calculating the cost of carpeting, painting a room, or fertilizing a lawn.

1 Find the perimeter of a polygon.

The **perimeter** of a polygon is the distance around it. To find the perimeter P of a polygon, we simply add the lengths of its sides.

Triangle

Quadrilateral

Pentagon

$P = 6 + 7 + 8$
$ = 21$
The perimeter is 21 ft.

$P = 10 + 18 + 24 + 18$
$ = 70$
The perimeter is 70 m.

$P = 1.2 + 7.1 + 6.6 + 5.2 + 3.4$
$ = 23.5$
The perimeter is 23.5 yd.

For some polygons, such as a square and a rectangle, we can simplify the computations by using a perimeter formula. Since a square has four sides of equal length s, its perimeter P is $s + s + s + s$, or $4s$.

Perimeter of a Square

If a square has a side of length s, its perimeter P is given by the formula

$P = 4s$

EXAMPLE 1
Find the perimeter of a square whose sides are 7.5 meters long.

Strategy We will substitute 7.5 for s in the formula $P = 4s$ and evaluate the right side.

WHY The variable P represents the unknown perimeter of the square.

Solution

$P = 4s$ This is the formula for the perimeter of a square.
$P = 4(7.5)$ Substitute 7.5 for s, the length of one side of the square.
$P = 30$ Do the multiplication.

$$\begin{array}{r} \overset{2}{7.5} \\ \times\ 4 \\ \hline 30.0 \end{array}$$

The perimeter of the square is 30 meters.

Self Check 1
A Scrabble game board has a square shape with sides of length 38.5 cm. Find the perimeter of the game board. 154 cm

Now Try Problems 17 and 19

Teaching Example 1 Find the perimeter of a square whose sides are 2.5 miles long.
Answer:
10 mi

Since a rectangle has two lengths *l* and two widths *w*, its perimeter *P* is given by *l* + *w* + *l* + *w*, or 2*l* + 2*w*.

Perimeter of a Rectangle

If a rectangle has length *l* and width *w*, its perimeter *P* is given by the formula

$$P = 2l + 2w$$

Caution! When finding the perimeter of a polygon, the lengths of the sides must be expressed in the same units.

Self Check 2

Find the perimeter of the triangle shown below, in inches. 50 in.

Now Try Problem 21

Teaching Example 2 Find the perimeter (in inches) of a rectangle that has a length of 4 feet and a width of 7 inches.
Answer:
110 in.

EXAMPLE 2 Find the perimeter of the rectangle shown on the right, in inches.

Strategy We will express the length of the rectangle in inches and then use the formula $P = 2l + 2w$ to find the perimeter of the figure.

WHY We can only add quantities that are measured in the same units.

Solution Since 1 foot = 12 inches, we can convert 3 feet to inches by multiplying 3 feet by the unit conversion factor $\frac{12 \text{ in.}}{1 \text{ foot}}$.

$$3 \text{ ft} = 3 \text{ ft} \cdot \frac{12 \text{ in.}}{1 \text{ ft}} \qquad \text{Multiply by 1: } \frac{12 \text{ in.}}{1 \text{ ft}} = 1.$$

$$= \frac{3 \text{ ft}}{1} \cdot \frac{12 \text{ in.}}{1 \text{ ft}} \qquad \text{Write 3 ft as a fraction. Remove the common units of feet from the numerator and denominator. The units of inches remain.}$$

$$= 36 \text{ in.} \qquad \text{Do the multiplication.}$$

The length of the rectangle is 36 inches. We can now substitute 36 for *l*, the length, and 8 for *w*, the width, in the formula for the perimeter of a rectangle.

$P = 2l + 2w$	This is the formula for the perimeter of a rectangle.	$\overset{1}{3}6$
$P = 2(36) + 2(8)$	Substitute 36 for the length, and 8 for the width.	$\underline{\times 2}$
$P = 72 + 16$	Do the multiplication.	72
$P = 88$	Do the addition.	72
		$\underline{+16}$
		88

The perimeter of the rectangle is 88 inches.

Self Check 3

The perimeter of an isosceles triangle is 58 meters. If one of its sides of equal length is 15 meters long, how long is its base? 28 m

Now Try Problem 25

Teaching Example 3 The perimeter of an isosceles triangle is 150 yards. If one of its sides of equal length is 55 yards long, how long is its base?
Answer:
40 yd

EXAMPLE 3 **Structural Engineering** The truss shown below is made up of three parts that form an isosceles triangle. If 76 linear feet of lumber were used to make the truss, how long is the base of the truss?

Analyze

- The truss is in the shape of an isosceles triangle.   *Given*
- One of the sides of equal length is 20 feet long.   *Given*
- The perimeter of the truss is 76 feet.   *Given*
- What is the length of the base of the truss?   *Find*

Assign We can let b equal the length of the base of the truss (in feet). At this stage, it is helpful to draw a sketch. (See the figure on the right.) If one of the sides of equal length is 20 feet long, so is the other.

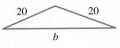

Form Because 76 linear feet of lumber were used to make the triangular-shaped truss,

The length of the base of the truss	plus	the length of one side	plus	the length of the other side	equals	the perimeter of the truss.
b	+	20	+	20	=	76

Solve

$b + 20 + 20 = 76$

$b + 40 = 76$ Combine like terms.

$b = 36$ To isolate b, subtract 40 from both sides.

$$\begin{array}{r} 76 \\ -40 \\ \hline 36 \end{array}$$

State The length of the base of the truss is 36 ft.

Check If we add the lengths of the parts of the truss, we get $36 \text{ ft} + 20 \text{ ft} + 20 \text{ ft} = 76 \text{ ft}$. The result checks.

Using Your CALCULATOR Perimeters of Figures That Are Combinations of Polygons

To find the perimeter of the figure shown below, we need to know the values of x and y. Since the figure is a combination of two rectangles, we can use a calculator to see that

$x = 20.25 - 10.17$ and $y = 12.5 - 4.75$

$x = 10.08$ cm $y = 7.75$ cm

The perimeter P of the figure is

$P = 20.25 + 12.5 + 10.17 + 4.75 + x + y$

$P = 20.25 + 12.5 + 10.17 + 4.75 + \mathbf{10.08} + \mathbf{7.75}$

We can use a scientific calculator to make this calculation.

$20.25 \boxed{+} 12.5 \boxed{+} 10.17 \boxed{+} 4.75 \boxed{+} 10.08 \boxed{+} 7.75 \boxed{=}$ $\boxed{\text{65.5}}$

The perimeter is 65.5 centimeters.

2 Find the area of a polygon.

The **area** of a polygon is the measure of the amount of surface it encloses. Area is measured in square units, such as square inches or square centimeters, as shown below.

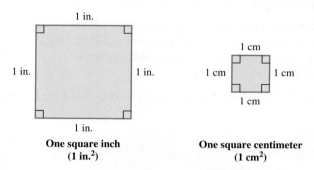

One square inch
(1 in.^{2})

One square centimeter
(1 cm^2)

In everyday life, we often use areas. For example,

- To carpet a room, we buy square yards.
- A can of paint will cover a certain number of square feet.
- To measure vast amounts of land, we often use square miles.
- We buy house roofing by the "square." One square is 100 square feet.

The rectangle shown below has a length of 10 centimeters and a width of 3 centimeters. If we divide the rectangular region into square regions as shown in the figure, each square has an area of 1 square centimeter—a surface enclosed by a square measuring 1 centimeter on each side. Because there are 3 rows with 10 squares in each row, there are 30 squares. Since the rectangle encloses a surface area of 30 squares, its area is 30 square centimeters, which can be written as 30 cm^2.

This example illustrates that to find the area of a rectangle, we multiply its length by its width.

> **Caution!** Do not confuse the concepts of perimeter and area. Perimeter is the distance around a polygon. It is measured in linear units, such as centimeters, feet, or miles. Area is a measure of the surface enclosed within a polygon. It is measured in square units, such as square centimeters, square feet, or square miles.

In practice, we do not find areas of polygons by counting squares. Instead, we use formulas to find areas of geometric figures.

9.7 Perimeters and Areas of Polygons   889

Figure	Name	Formula for Area
	Square	$A = s^2$, where s is the length of one side.
	Rectangle	$A = lw$, where l is the length and w is the width.
	Parallelogram	$A = bh$, where b is the length of the base and h is the height. (A height is always perpendicular to the base.)
	Triangle	$A = \frac{1}{2}bh$, where b is the length of the base and h is the height. The segment perpendicular to the base and representing the height (shown here using a dashed line) is called an **altitude**.
	Trapezoid	$A = \frac{1}{2}h(b_1 + b_2)$, where h is the height of the trapezoid and b_1 and b_2 represent the lengths of the bases.

EXAMPLE 4 Find the area of the square shown on the right.

Strategy We will substitute 15 for s in the formula $A = s^2$ and evaluate the right side.

WHY The variable A represents the unknown area of the square.

Solution

$A = s^2$ This is the formula for the area of a square.

$A = 15^2$ Substitute 15 for s, the length of one side of the square.

$A = 225$ Evaluate the exponential expression.

```
   15
× 15
   75
 150
 225
```

Recall that area is measured in square units. Thus, the area of the square is 225 square centimeters, which can be written as 225 cm².

Self Check 4

Find the area of the square shown below. 400 in.²

Now Try Problems 29 and 31

Teaching Example 4 Find the area of a square if the length of one of its sides is 16 feet.
Answer:
256 ft²

EXAMPLE 5 Find the number of square feet in 1 square yard.

Strategy A figure is helpful to solve this problem. We will draw a square yard and divide each of its sides into 3 equally long parts.

WHY Since a square yard is a square with each side measuring 1 yard, each side also measures 3 feet.

Self Check 5

Find the number of square centimeters in 1 square meter.
10,000 cm²

Now Try Problems 33 and 39

Unless otherwise noted, all art on this page is © Cengage Learning.

Teaching Example 5 Find the number of square inches in a square yard.
Answer:
1,296 in.²

Solution

$1 \text{ yd}^2 = (1 \text{ yd})^2$
$= (3 \text{ ft})^2$ Substitute 3 feet for 1 yard.
$= (3 \text{ ft})(3 \text{ ft})$
$= 9 \text{ ft}^2$

There are 9 square feet in 1 square yard.

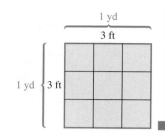

Self Check 6

PING-PONG A regulation-size Ping-Pong table is 9 feet long and 5 feet wide. Find its area in square inches. 6,480 in.²

Now Try Problem 41

Teaching Example 6
YOUTH SOCCER The recommended soccer field size for youth under 10 years old is 70 yards long by 40 yards wide. Find the area of the field in square feet.
Answer:
25,200 ft²

EXAMPLE 6   Women's Sports

Field hockey is a team sport in which players use sticks to try to hit a ball into their opponents' goal. Find the area of the rectangular field shown on the right. Give the answer in square feet.

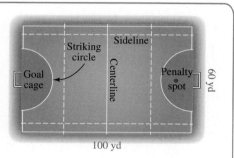

Strategy We will substitute 100 for l and 60 for w in the formula $A = lw$ and evaluate the right side.

WHY The variable A represents the unknown area of the rectangle.

Solution

$A = lw$ This is the formula for the area of a rectangle.
$A = 100(60)$ Substitute 100 for l, the length, and 60 for w, the width.
$A = 6,000$ Do the multiplication.

The area of the rectangle is 6,000 square yards. Since there are 9 square feet per square yard, we can convert this number to square feet by multiplying 6,000 square yards by $\frac{9 \text{ ft}^2}{1 \text{ yd}^2}$.

$6,000 \text{ yd}^2 = 6,000 \text{ yd}^2 \cdot \dfrac{9 \text{ ft}^2}{1 \text{ yd}^2}$ Multiply by the unit conversion factor: $\frac{9 \text{ ft}^2}{1 \text{ yd}^2} = 1$.

$= 6,000 \cdot 9 \text{ ft}^2$ Remove the common units of square yards in the numerator and denominator. The units of ft² remain.

$= 54,000 \text{ ft}^2$ Multiply: $6,000 \cdot 9 = 54,000$.

The area of the field is 54,000 ft².

THINK IT THROUGH   Dorm Rooms

"The United States has more than 4,000 colleges and universities, with 2.3 million students living in college dorms."
The New York Times, 2007

The average dormitory room in a residence hall has about 180 square feet of floor space. The rooms are usually furnished with the following items having the given dimensions:

- 2 extra-long twin beds (each is 39 in. wide × 80 in. long × 24 in. high)
- 2 dressers (each is 18 in. wide × 36 in. long × 48 in. high)
- 2 bookcases (each is 12 in. wide × 24 in. long × 40 in. high)
- 2 desks (each is 24 in. wide × 48 in. long × 28 in. high)

How many square feet of floor space are left? about 108 ft²

Unless otherwise noted, all art on this page is © Cengage Learning.

EXAMPLE 7

Find the area of the triangle shown on the right.

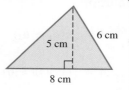

Strategy We will substitute 8 for b and 5 for h in the formula $A = \frac{1}{2}bh$ and evaluate the right side. (The side having length 6 cm is additional information that is not used to find the area.)

WHY The variable A represents the unknown area of the triangle.

Solution

$A = \dfrac{1}{2}bh$ This is the formula for the area of a triangle.

$A = \dfrac{1}{2}(8)(5)$ Substitute 8 for b, the length of the base, and 5 for h, the height.

$A = 4(5)$ Do the first multiplication: $\frac{1}{2}(8) = 4$.

$A = 20$ Complete the multiplication.

The area of the triangle is 20 cm².

Self Check 7

Find the area of the triangle shown below. 90 mm²

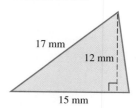

Now Try Problem 45

Teaching Example 7 Find the area of the triangle.

Answer: 18 cm²

EXAMPLE 8

Find the area of the triangle shown on the right.

Strategy We will substitute 9 for b and 13 for h in the formula $A = \frac{1}{2}bh$ and evaluate the right side. (The side having length 15 cm is additional information that is not used to find the area.)

WHY The variable A represents the unknown area of the triangle.

Solution In this case, the altitude falls outside the triangle.

$A = \dfrac{1}{2}bh$ This is the formula for the area of a triangle.

$A = \dfrac{1}{2}(9)(13)$ Substitute 9 for b, the length of the base, and 13 for h, the height.

$A = \dfrac{1}{2}\left(\dfrac{9}{1}\right)\left(\dfrac{13}{1}\right)$ Write 9 as $\frac{9}{1}$ and 13 as $\frac{13}{1}$.

$A = \dfrac{117}{2}$ Multiply the fractions.

$A = 58.5$ Do the division.

The area of the triangle is 58.5 cm².

$\begin{array}{r} \overset{2}{1}3 \\ \times\,9 \\ \hline 117 \end{array}$ $\begin{array}{r} 58.5 \\ 2\overline{)117.0} \\ -10 \\ \hline 17 \\ -16 \\ \hline 1\,0 \\ -1\,0 \\ \hline 0 \end{array}$

Self Check 8

Find the area of the triangle shown below. 10.5 ft²

Now Try Problem 49

Teaching Example 8 Find the area of the triangle.

Answer: 38.5 ft²

EXAMPLE 9

Find the area of the trapezoid shown on the right.

Strategy We will express the height of the trapezoid in inches and then use the formula $A = \frac{1}{2}h(b_1 + b_2)$ to find the area of the figure.

WHY The height of 1 foot must be expressed as 12 inches to be consistent with the units of the bases.

Self Check 9

Find the area of the trapezoid shown below. 54 m²

Unless otherwise noted, all art on this page is © Cengage Learning.

Now Try Problem 53

Teaching Example 9 Find the area of the trapezoid.

Answer: 132 in.²

Self Check 10

The area of the parallelogram below is 96 cm². Find its height.

Now Try Problem 57

Teaching Example 10 The area of the parallelogram below is 32 ft². Find the height.

Answer: 4 ft

Self Check 11

Find the area of the shaded figure below. 41 yd²

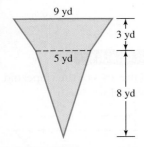

Now Try Problem 65

Solution

$A = \frac{1}{2}h(b_1 + b_2)$ This is the formula for the area of a trapezoid.

$A = \frac{1}{2}(12)(10 + 6)$ Substitute 12 for h, the height; 10 for b_1, the length of the lower base; and 6 for b_2, the length of the upper base.

$A = \frac{1}{2}(12)(16)$ Do the addition within the parentheses.

$A = 6(16)$ Do the first multiplication: $\frac{1}{2}(12) = 6$.

$A = 96$ Complete the multiplication.

$$\begin{array}{r} \overset{3}{16} \\ \times 6 \\ \hline 96 \end{array}$$

The area of the trapezoid is 96 in².

EXAMPLE 10 The area of the parallelogram shown on the right is 360 ft². Find the height.

Strategy To find the height of the parallelogram, we will substitute the given values in the formula $A = bh$ and solve for h.

WHY The variable h represents the unknown height.

Solution From the figure, we see that the length of the base of the parallelogram is

5 feet + 25 feet = 30 feet

$A = bh$ This is the formula for the area of a parallelogram.

$360 = 30h$ Substitute 360 for A, the area, and 30 for b, the length of the base.

$\frac{360}{30} = \frac{30h}{30}$ To isolate h, undo the multiplication by 30 by dividing both sides by 30.

$12 = h$ Do the division.

$$\begin{array}{r} 12 \\ 30\overline{)360} \\ -30 \\ \hline 60 \\ 60 \\ \hline 0 \end{array}$$

The height of the parallelogram is 12 feet.

3 Find the area of figures that are combinations of polygons.

> **Success Tip** To find the area of an irregular shape, break up the shape into familiar polygons. Find the area of each polygon and then add the results.

EXAMPLE 11 Find the area of one side of the tent shown below.

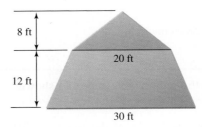

Strategy We will use the formula $A = \frac{1}{2}h(b_1 + b_2)$ to find the area of the lower portion of the tent and the formula $A = \frac{1}{2}bh$ to find the area of the upper portion of the tent. Then we will combine the results.

WHY The side of the tent is a combination of a trapezoid and a triangle.

Solution To find the area of the lower portion of the tent, we proceed as follows.

$A_{\text{trap.}} = \dfrac{1}{2}h(b_1 + b_2)$ This is the formula for the area of a trapezoid.

$A_{\text{trap.}} = \dfrac{1}{2}(12)(30 + 20)$ Substitute 30 for b_1, 20 for b_2, and 12 for h.

$A_{\text{trap.}} = \dfrac{1}{2}(12)(50)$ Do the addition within the parentheses.

$A_{\text{trap.}} = 6(50)$ Do the first multiplication: $\dfrac{1}{2}(12) = 6$.

$A_{\text{trap.}} = 300$ Complete the multiplication.

The area of the trapezoid is 300 ft².

To find the area of the upper portion of the tent, we proceed as follows.

$A_{\text{triangle}} = \dfrac{1}{2}bh$ This is the formula for the area of a triangle.

$A_{\text{triangle}} = \dfrac{1}{2}(20)(8)$ Substitute 20 for b and 8 for h.

$A_{\text{triangle}} = 80$ Do the multiplications, working from left to right: $\dfrac{1}{2}(20) = 10$ and then $10(8) = 80$.

The area of the triangle is 80 ft².

To find the total area of one side of the tent, we add:

$A_{\text{total}} = A_{\text{trap.}} + A_{\text{triangle}}$

$A_{\text{total}} = 300 \text{ ft}^2 + 80 \text{ ft}^2$

$A_{\text{total}} = 380 \text{ ft}^2$

The total area of one side of the tent is 380 ft².

EXAMPLE 12
Find the area of the shaded region shown on the right.

Strategy We will subtract the unwanted area of the square from the area of the rectangle.

Area of shaded region = Area of rectangle − Area of square

WHY The area of the rectangular-shaped shaded figure does not include the square region inside of it.

Solution

$A_{\text{shaded}} = lw - s^2$ The formula for the area of a rectangle is $A = lw$. The formula for the area of a square is $A = s^2$.

$A_{\text{shaded}} = 15(8) - 5^2$ Substitute 15 for the length l and 8 for the width w of the rectangle. Substitute 5 for the length s of a side of the square.

$A_{\text{shaded}} = 120 - 25$

$A_{\text{shaded}} = 95$

The area of the shaded region is 95 ft².

$\begin{array}{r} \overset{4}{15} \\ \times\, 8 \\ \hline 120 \end{array}$

$\begin{array}{r} \overset{11}{\cancel{1}}\overset{}{10} \\ \cancel{1}\cancel{2}\cancel{0} \\ -\,25 \\ \hline 95 \end{array}$

Teaching Example 11 Find the area of the shaded figure.

Answer:
148 in.²

Self Check 12
Find the area of the shaded region shown below. 119 ft²

Now Try Problem 69

Teaching Example 12 Find the area of the shaded region.

Answer:
140 ft²

EXAMPLE 13 **Carpeting a Room** A living room/dining room has the floor plan shown in the figure. If carpet costs $29 per square yard, including pad and installation, how much will it cost to carpet both rooms? (Assume no waste.)

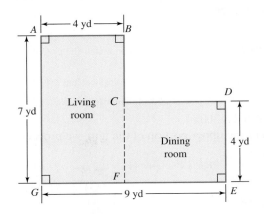

Strategy We will find the number of square yards of carpeting needed and multiply the result by $29.

WHY Each square yard costs $29.

Solution First, we must find the total area of the living room and the dining room:

$$A_{total} = A_{living\ room} + A_{dining\ room}$$

Since $\overline{CF}$ divides the space into two rectangles, the areas of the living room and the dining room are found by multiplying their respective lengths and widths. Therefore, the area of the living room is 4 yd · 7 yd = **28 yd²**.

The width of the dining room is given as 4 yd. To find its length, we subtract:

$$m(\overline{CD}) = m(\overline{GE}) - m(\overline{AB}) = 9\ yd - 4\ yd = 5\ yd$$

Thus, the area of the dining room is 5 yd · 4 yd = **20 yd²**. The total area to be carpeted is the sum of these two areas.

$$A_{total} = A_{living\ room} + A_{dining\ room}$$
$$A_{total} = 28\ yd^2 + 20\ yd^2$$
$$A_{total} = 48\ yd^2$$

```
    48
  × 29
   432
   960
 1,392
```

Now Try Problem 73

At $29 per square yard, the cost to carpet both rooms will be 48 · $29, or $1,392.

ANSWERS TO SELF CHECKS
1. 154 cm **2.** 50 in. **3.** 28 m **4.** 400 in.² **5.** 10,000 cm² **6.** 6,480 in.² **7.** 90 mm²
8. 10.5 ft² **9.** 54 m² **10.** 8 cm **11.** 41 yd² **12.** 119 ft²

SECTION 9.7 STUDY SET

VOCABULARY

Fill in the blanks.

1. The distance around a polygon is called the perimeter .
2. The perimeter of a polygon is measured in linear units such as inches, feet, and miles.
3. The measure of the surface enclosed by a polygon is called its area .
4. If each side of a square measures 1 foot, the area enclosed by the square is 1 square foot.
5. The area of a polygon is measured in square units.
6. The segment that represents the height of a triangle is called an altitude .

CONCEPTS

7. The figure below shows a kitchen floor that is covered with 1-foot-square tiles. Without counting *all* of the squares, determine the area of the floor.
8 ft · 16 ft = 128 ft²

8. Tell which concept applies, perimeter or area.
   a. The length of a walk around New York's Central Park perimeter
   b. The amount of office floor space in the White House area
   c. The amount of fence needed to enclose a playground perimeter
   d. The amount of land in Yellowstone National Park area

9. Give the formula for the perimeter of a
   a. square $P = 4s$
   b. rectangle $P = 2l + 2w$

10. Give the formula for the area of a
   a. square $A = s^2$
   b. rectangle $A = lw$
   c. triangle $A = \frac{1}{2}bh$
   d. trapezoid $A = \frac{1}{2}h(b_1 + b_2)$
   e. parallelogram $A = bh$

11. For each figure below, draw the altitude to the base *b*.

a. **b.**

c. **d.**

12. For each figure below, label the base *b* for the given altitude.

a. **b.**

c. **d.**

13. The shaded figure below is a combination of what two types of geometric figures? a rectangle and a triangle

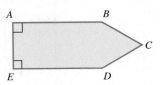

14. Explain how you would find the area of the following shaded figure. subtract the area of the triangle from the area of the parallelogram

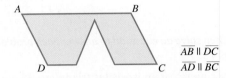

$\overline{AB} \parallel \overline{DC}$
$\overline{AD} \parallel \overline{BC}$

NOTATION

Fill in the blanks.

15. a. The symbol 1 in.² means one __square__ __inch__.
   b. One square meter is expressed as __1 m²__.

16. In the figure below, the symbol ⌐ indicates that the dashed line segment, called an *altitude*, is __perpendicular__ to the base.

GUIDED PRACTICE

Find the perimeter of each square. **See Example 1.**

17. 8 in. square (sides 8 in.)   32 in.

18. 93 in. square (sides 93 in.)   372 in.

19. A square with sides 5.75 miles long 23 mi

20. A square with sides 3.4 yards long 13.6 yd

Find the perimeter of each rectangle, in inches. **See Example 2.**

21. 2 ft by 7 in.   62 in.

22. 6 ft by 2 in.   148 in.

23. 11 in., 3 ft, 94 in.
24. 9 in., 4 ft, 114 in.

Write and then solve an equation to answer each problem.
See Example 3.

25. The perimeter of an isosceles triangle is 35 feet. One of the sides of equal length is 10 feet long. Find the length of the base of the triangle. 15 ft

26. The perimeter of an isosceles triangle is 94 feet. One of the sides of equal length is 42 feet long. Find the length of the base of the triangle. 10 ft

27. The perimeter of an isosceles trapezoid is 35 meters. The upper base is 10 meters long, and the lower base is 15 meters long. How long is each leg of the trapezoid? 5 m

28. The perimeter of an isosceles trapezoid is 46 inches. The upper base is 12 inches long, and the lower base is 16 inches long. How long is each leg of the trapezoid? 9 in.

Find the area of each square. See Example 4.

29. 4 cm, 4 cm — 16 cm²
30. 24 in., 24 in. — 576 in.²

31. A square with sides 2.5 meters long 6.25 m²
32. A square with sides 6.8 feet long 46.24 ft²

For Problems 33–40, see Example 5.

33. How many square inches are in 1 square foot? 144 in.²
34. How many square inches are in 1 square yard? 1,296 in.²
35. How many square millimeters are in 1 square meter? 1,000,000 mm²
36. How many square decimeters are in 1 square meter? 100 dm²
37. How many square feet are in 1 square mile? 27,878,400 ft²
38. How many square yards are in 1 square mile? 3,097,600 yd²
39. How many square meters are in 1 square kilometer? 1,000,000 m²
40. How many square dekameters are in 1 square kilometer? 10,000 dam²

Find the area of each rectangle. Give the answer in square feet.
See Example 6.

41. 3 yd, 5 yd   135 ft²
42. 9 yd, 10 yd   810 ft²
43. 20 yd, 62 yd   11,160 ft²
44. 7 yd, 15 yd   945 ft²

Find the area of each triangle. See Example 7.

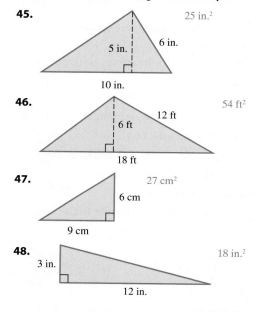

45. 5 in., 6 in., 10 in.   25 in.²
46. 6 ft, 12 ft, 18 ft   54 ft²
47. 6 cm, 9 cm   27 cm²
48. 3 in., 12 in.   18 in.²

Find the area of each triangle. See Example 8.

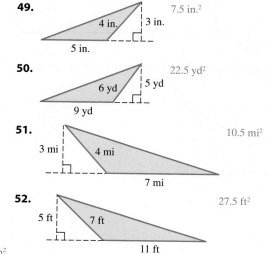

49. 4 in., 3 in., 5 in.   7.5 in.²
50. 6 yd, 5 yd, 9 yd   22.5 yd²
51. 3 mi, 4 mi, 7 mi   10.5 mi²
52. 5 ft, 7 ft, 11 ft   27.5 ft²

Unless otherwise noted, all art on this page is © Cengage Learning.

Find the area of each trapezoid. See Example 9.

53. 40 ft²

54. 496 in.²

55. 91 cm²

56. 169 mm²

Solve each problem. See Example 10.

57. The area of a parallelogram is 60 m², and its height is 15 m. Find the length of its base. 4 m

58. The area of a parallelogram is 95 in.², and its height is 5 in. Find the length of its base. 19 in.

59. The area of a rectangle is 36 cm², and its length is 3 cm. Find its width. 12 cm

60. The area of a rectangle is 144 mi², and its length is 6 mi. Find its width. 24 mi

61. The area of a triangle is 54 m², and the length of its base is 3 m. Find the height. 36 m

62. The area of a triangle is 270 ft², and the length of its base is 18 ft. Find the height. 30 ft

63. The perimeter of a rectangle is 64 mi, and its length is 21 mi. Find its width. 11 mi

64. The perimeter of a rectangle is 26 yd, and its length is 10.5 yd. Find its width. 2.5 yd

Find the area of each shaded figure. See Example 11.

65. 102 in.²

66. 80 m²

67. 360 ft²

68. 207 mm²

Find the area of each shaded figure. See Example 12.

69. 75 m²

70. 335 cm²

71. 75 yd²

72. 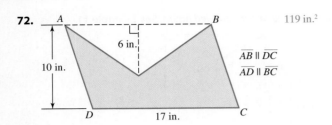 119 in.²

$\overline{AB} \parallel \overline{DC}$
$\overline{AD} \parallel \overline{BC}$

Solve each problem. See Example 13.

73. **FLOORING** A rectangular family room is 8 yards long and 5 yards wide. At $30 per square yard, how much will it cost to put down vinyl sheet flooring in the room? (Assume no waste.) $1,200

74. **CARPETING** A rectangular living room measures 10 yards by 6 yards. At $32 per square yard, how much will it cost to carpet the room? (Assume no waste.) $1,920

75. **FENCES** A man wants to enclose a rectangular yard with fencing that costs $12.50 a foot, including installation. Find the cost of enclosing the yard if its dimensions are 110 ft by 85 ft. $4,875

76. **FRAMES** Find the cost of framing a rectangular picture with dimensions of 24 inches by 30 inches if framing material costs $0.75 per inch. $81

TRY IT YOURSELF

Sketch and label each of the figures.

77. Two different rectangles, each having a perimeter of 40 in. length 15 in. and width 5 in.; length 16 in. and width 4 in. (answers may vary)
78. Two different rectangles, each having an area of 40 in.² length 10 in. and width 4 in.; length 20 in. and width 2 in. (answers may vary)
79. A square with an area of 25 m² sides of length 5 m
80. A square with a perimeter of 20 m sides of length 5 m
81. A parallelogram with an area of 15 yd² base 5 yd and height 3 yd (answers may vary)
82. A triangle with an area of 20 ft² base 5 ft and height 8 ft (answers may vary)
83. A figure consisting of a combination of two rectangles, whose total area is 80 ft² length 5 ft and width 4 ft; length 20 ft and width 3 ft (answers may vary)
84. A figure consisting of a combination of a rectangle and a square, whose total area is 164 ft² length 4 ft and width 25 ft; length 8 ft and width 8 ft (answers may vary)

Find the area of each parallelogram.

85. 60 cm²

86. 60 m²

87. The perimeter of an isosceles triangle is 80 meters. If the length of one of the congruent sides is 22 meters, how long is the base? 36 m

88. The perimeter of a square is 35 yards. How long is a side of the square? 8.75 yd

89. The perimeter of an equilateral triangle is 85 feet. Find the length of each side. $28\frac{1}{3}$ ft

90. An isosceles triangle with congruent sides of length 49.3 inches has a perimeter of 121.7 inches. Find the length of the base. 23.1 in.

Find the perimeter of the figure.

91.

92.

Find x and y. Then find the perimeter of the figure.

93. $x = 3.7$ ft, $y = 10.1$ ft; 50.8 ft

94. $x = 3.44$ in., $y = 3.27$ in.; 77.98 in.

APPLICATIONS

95. LANDSCAPING A woman wants to plant a pine-tree screen around three sides of her rectangular-shaped backyard. (See the figure below.) If she plants the trees 3 feet apart, how many trees will she need? 80 + 1 = 81 trees

120 ft

60 ft The first tree is to be planted here, even with the back of her house.

96. GARDENING A gardener wants to plant a border of marigolds around the garden shown below, to keep out rabbits. How many plants will she need if she allows 6 inches between plants? 144 plants

16 ft   20 ft

97. COMPARISON SHOPPING Which is more expensive: a ceramic-tile floor costing $3.75 per square foot or vinyl costing $34.95 per square yard? vinyl

98. COMPARISON SHOPPING Which is cheaper: a hardwood floor costing $6.95 per square foot or a carpeted floor costing $37.50 per square yard? carpeting

99. TILES A rectangular basement room measures 14 by 20 feet. Vinyl floor tiles that are 1 ft^2 cost $1.29 each. How much will the tile cost to cover the floor? (Assume no waste.) $361.20

100. PAINTING The north wall of a barn is a rectangle 23 feet high and 72 feet long. There are five windows in the wall, each 4 by 6 feet. If a gallon of paint will cover 300 ft^2, how many gallons of paint must the painter buy to paint the wall? 6 gal

101. SAILS If nylon is $12 per square yard, how much would the fabric cost to make a triangular sail with a base of 12 feet and a height of 24 feet? $192

102. REMODELING The gable end of a house is an isosceles triangle with a height of 4 yards and a base of 23 yards. It will require one coat of primer and one coat of finish to paint the triangle. Primer costs $17 per gallon, and the finish paint costs $23 per gallon. If one gallon of each type of paint covers 300 square feet, how much will it cost to paint the gable, excluding labor? $80

103. GEOGRAPHY Use the dimensions of the trapezoid that is superimposed over the state of Nevada to estimate the area of the "Silver State." 111,825 mi^2

104. SOLAR COVERS A swimming pool has the shape shown below. How many square feet of a solar blanket material will be needed to cover the pool? How much will the cover cost if it is $1.95 per square foot? (Assume no waste.) 400 ft^2; $780

105. CARPENTRY How many sheets of 4-foot-by-8-foot sheetrock are needed to drywall the inside walls on the first floor of the barn shown below? (Assume that the carpenters will cover each wall entirely before cutting out areas for the doors.) 51 sheets

106. **CARPENTRY** If it costs $90 per square foot to build a one-story home in northern Wisconsin, find the cost of building the house with the floor plan shown below. $94,320

WRITING

107. Explain the difference between perimeter and area.
108. Why is it necessary that area be measured in square units?
109. A student expressed the area of the square in the figure below as 25^2 ft. Explain his error.

110. Refer to the figure below. What must be done before we can use the formula to find the area of this rectangle?

REVIEW

Simplify each expression.

111. $8\left(\dfrac{3}{4}t\right)$ $6t$

112. $27\left(\dfrac{2}{3}m\right)$ $18m$

113. $-\dfrac{2}{3}(3w - 6)$ $-2w + 4$

114. $(2y - 8)\dfrac{1}{2}$ $y - 4$

115. $-\dfrac{7}{16}x - \dfrac{3}{16}x$ $-\tfrac{5}{8}x$

116. $-\dfrac{5}{18}x - \dfrac{7}{18}x$ $-\tfrac{2}{3}x$

117. $60\left(\dfrac{3}{20}r - \dfrac{4}{15}\right)$ $9r - 16$

118. $72\left(\dfrac{7}{8}f - \dfrac{8}{9}\right)$ $63f - 64$

Objectives

1. Define circle, radius, chord, diameter, and arc.
2. Find the circumference of a circle.
3. Find the area of a circle.

SECTION 9.8
Circles

In this section, we will discuss the circle, one of the most useful geometric figures of all. In fact, the discoveries of fire and the circular wheel are two of the most important events in the history of the human race. We will begin our study by introducing some basic vocabulary associated with circles.

1 Define circle, radius, chord, diameter, and arc.

> **Circle**
>
> A **circle** is the set of all points in a plane that lie a fixed distance from a point called its **center**.

A segment drawn from the center of a circle to a point on the circle is called a **radius**. (The plural of *radius* is *radii*.) From the definition, it follows that all radii of the same circle are the same length.

A **chord** of a circle is a line segment that connects two points on the circle. A **diameter** is a chord that passes through the center of the circle. Since a diameter D of a circle is twice as long as a radius r, we have

$$D = 2r$$

Each of the previous definitions is illustrated in figure (a) below, in which O is the center of the circle.

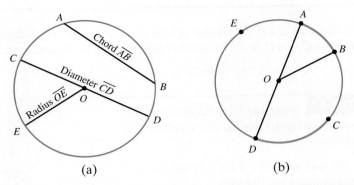

(a) (b)

Any part of a circle is called an **arc**. In figure (b) above, the part of the circle from point A to point B that is highlighted in blue is $\overset{\frown}{AB}$, read as "arc AB." $\overset{\frown}{CD}$ is the part of the circle from point C to point D that is highlighted in green. An arc that is half of a circle is a **semicircle**.

Semicircle

A **semicircle** is an arc of a circle whose endpoints are the endpoints of a diameter.

If point O is the center of the circle in figure (b), $\overline{AD}$ is a diameter and $\overset{\frown}{AED}$ is a semicircle. The middle letter E distinguishes semicircle $\overset{\frown}{AED}$ (the part of the circle from point A to point D that includes point E) from semicircle $\overset{\frown}{ABD}$ (the part of the circle from point A to point D that includes point B).

An arc that is shorter than a semicircle is a **minor arc**. An arc that is longer than a semicircle is a **major arc**. In figure (b),

$\overset{\frown}{AE}$ is a minor arc   and $\overset{\frown}{ABE}$ is a major arc.

Success Tip It is often possible to name a major arc in more than one way. For example, in figure (b), major arc $\overset{\frown}{ABE}$ is the part of the circle from point A to point E that includes point B. Two other names for the same major arc are $\overset{\frown}{ACE}$ and $\overset{\frown}{ADE}$.

2 Find the circumference of a circle.

Since early history, mathematicians have known that the ratio of the distance around a circle (the circumference) to the length of its diameter is approximately 3. First Kings, Chapter 7, of the Bible describes a round bronze tank that was 15 feet from brim to brim and 45 feet in circumference, and $\frac{45}{15} = 3$. Today, we use a more precise value for this ratio, known as π (pi). If C is the circumference of a circle and D is the length of its diameter, then

$$\pi = \frac{C}{D} \quad \text{where } \pi = 3.141592653589\ldots \quad \frac{22}{7} \text{ and } 3.14 \text{ are often used as estimates of } \pi.$$

If we multiply both sides of $\pi = \frac{C}{D}$ by D, we have the following formula.

> **Circumference of a Circle**
>
> The circumference of a circle is given by the formula
>
> $C = \pi D$ where C is the circumference and D is the length of the diameter

Since a diameter of a circle is twice as long as a radius r, we can substitute $2r$ for D in the formula $C = \pi D$ to obtain another formula for the circumference C:

$C = 2\pi r$ The notation $2\pi r$ means $2 \cdot \pi \cdot r$.

Self Check 1
Find the circumference of the circle shown below. Give the exact answer and an approximation. 24π m $\approx$ 75.4 m

Now Try Problem 25

Teaching Example 1 Find the circumference of a circle with a radius of 9 meters. Give the exact answer and an approximation.
Answer:
18π m $\approx$ 56.5 m

EXAMPLE 1 Find the circumference of the circle shown on the right. Give the exact answer and an approximation.

Strategy We will substitute 5 for r in the formula $C = 2\pi r$ and evaluate the right side.

WHY The variable C represents the unknown circumference of the circle.

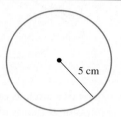

Solution

$C = 2\pi r$ This is the formula for the circumference of a circle.

$C = 2\pi(5)$ Substitute 5 for r, the radius.

$C = 2(5)\pi$ When a product involves π, we usually rewrite it so that π is the last factor.

$C = 10\pi$ Do the first multiplication: $2(5) = 10$. This is the exact answer.

The circumference of the circle is exactly 10π cm. If we replace π with 3.14, we get an approximation of the circumference.

$C = 10\pi$

$C \approx 10(3.14)$ To multiply by 10, move the decimal point in 3.14 one place to the right.

$C \approx 31.4$

The circumference of the circle is approximately 31.4 cm.

Using Your CALCULATOR   Calculating Revolutions of a Tire

When the $\boxed{\pi}$ key on a scientific calculator is pressed (on some models, the $\boxed{\text{2nd}}$ key must be pressed first), an approximation of π is displayed. To illustrate how to use this key, consider the following problem. How many times does the tire shown to the right revolve when a car makes a 25-mile trip?

We first find the circumference of the tire. From the figure, we see that the diameter of the tire is 15 inches. Since the circumference of a circle is the product of π and the length of its diameter, the tire's circumference is $\pi \cdot 15$ inches, or 15π inches. (Normally, we rewrite a product such as $\pi \cdot 15$ so that π is the second factor.)

Unless otherwise noted, all art on this page is © Cengage Learning.

We then change the 25 miles to inches using two unit conversion factors.

$$\frac{25 \text{ miles}}{1} \cdot \frac{5{,}280 \text{ feet}}{1 \text{ mile}} \cdot \frac{12 \text{ inches}}{1 \text{ foot}} = 25 \cdot 5{,}280 \cdot 12 \text{ inches}$$ The units of miles and feet can be removed.

The length of the trip is $25 \cdot 5{,}280 \cdot 12$ inches.

Finally, we divide the length of the trip by the circumference of the tire to get

$$\frac{\text{The number of revolutions of the tire}} = \frac{25 \cdot 5{,}280 \cdot 12}{15\pi}$$

We can use a scientific calculator to make this calculation.

The tire makes about 33,614 revolutions.

EXAMPLE 2 Architecture A Norman window is constructed by adding a semicircular window to the top of a rectangular window. Find the perimeter of the Norman window shown here.

Strategy We will find the perimeter of the rectangular part and the circumference of the circular part of the window and add the results.

WHY The window is a combination of a rectangle and a semicircle.

Solution The perimeter of the rectangular part is

$P_{\text{rectangular part}} = 8 + 6 + 8 = 22$ Add only 3 sides of the rectangle.

The perimeter of the semicircle is one-half of the circumference of a circle that has a 6-foot diameter.

$P_{\text{semicircle}} = \frac{1}{2}C$ This is the formula for the circumference of a semicircle.

$P_{\text{semicircle}} = \frac{1}{2}\pi D$ Since we know the diameter, replace C with πD. We could also have replaced C with $2\pi r$.

$P_{\text{semicircle}} = \frac{1}{2}\pi(6)$ Substitute 6 for D, the diameter.

$P_{\text{semicircle}} \approx 9.424777961$ Use a calculator to do the multiplication.

The total perimeter is the sum of the two parts.

$P_{\text{total}} = P_{\text{rectangular part}} + P_{\text{semicircle}}$
$P_{\text{total}} \approx 22 + 9.424777961$
$P_{\text{total}} \approx 31.424777961$

To the nearest hundredth, the perimeter of the window is 31.42 feet.

3 Find the area of a circle.

If we divide the circle shown in figure (a) on the following page into an even number of pie-shaped pieces and then rearrange them as shown in figure (b), we have a figure that looks like a parallelogram. The figure has a base b that is one-half the circumference of the circle, and its height h is about the same length as a radius of the circle.

Self Check 2
Find the perimeter of the figure shown below. Round to the nearest hundredth. (Assume the arc is a semicircle.) 39.42 m

Now Try Problem 29

Teaching Example 2 Find the perimeter of the figure shown below. Round to the nearest hundredth. (Assume the arc is a semicircle.)

Answer:
76.27 in.

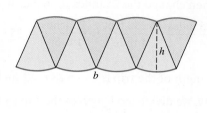

(a)   (b)

If we divide the circle into more and more pie-shaped pieces, the figure will look more and more like a parallelogram, and we can find its area by using the formula for the area of a parallelogram.

$A = bh$

$A = \frac{1}{2}Cr$ Substitute $\frac{1}{2}$ of the circumference for b, the length of the base of the "parallelogram." Substitute r for the height of the "parallelogram."

$A = \frac{1}{2}(2\pi r)r$ Substitute $2\pi r$ for C.

$A = \pi r^2$ Simplify: $\frac{1}{2} \cdot 2 = 1$ and $r \cdot r = r^2$.

This result gives the following formula.

Area of a Circle

The area of a circle with radius r is given by the formula.

$A = \pi r^2$

Self Check 3

Find the area of a circle with a diameter of 12 feet. Give the exact answer and an approximation to the nearest tenth. 36π ft² ≈ 113.1 ft²

Now Try Problem 33

Teaching Example 3 Find the area of a circle with a diameter of 16 feet. Give the exact answer and an approximation to the nearest tenth.
Answer:
64π ft² ≈ 201.1 ft²

EXAMPLE 3   Find the area of the circle shown on the right. Give the exact answer and an approximation to the nearest tenth.

Strategy We will find the radius of the circle, substitute that value for r in the formula $A = \pi r^2$, and evaluate the right side.

WHY The variable A represents the unknown area of the circle.

Solution Since the length of the diameter is 10 centimeters and the length of a diameter is twice the length of a radius, the length of the radius is 5 centimeters.

$A = \pi r^2$ This is the formula for the area of a circle.

$A = \pi(5)^2$ Substitute 5 for r, the radius of the circle. The notation πr^2 means $\pi \cdot r^2$.

$A = \pi(25)$ Evaluate the exponential expression.

$A = 25\pi$ Write the product so that π is the last factor.

The exact area of the circle is 25π cm². We can use a calculator to approximate the area.

$A \approx 78.53981634$ Use a calculator to do the multiplication: $25 \cdot \pi$.

To the nearest tenth, the area is 78.5 cm².

Using Your CALCULATOR   Painting a Helicopter Landing Pad

Orange paint is available in gallon containers at $39 each, and each gallon will cover 375 ft². To calculate how much the paint will cost to cover a circular helicopter landing pad 60 feet in diameter, we first calculate the area of the helicopter pad.

$A = \pi r^2$ This is the formula for the area of a circle.
$A = \pi(30)^2$ Substitute one-half of 60 for r, the radius of the circular pad.
$A = 30^2 \pi$ Write the product so that π is the last factor.

The area of the pad is exactly $30^2\pi$ ft². Since each gallon of paint will cover 375 ft², we can find the number of gallons of paint needed by dividing $30^2\pi$ by 375.

$$\text{Number of gallons needed} = \frac{30^2 \pi}{375}$$

We can use a scientific calculator to make this calculation.

$30 \boxed{x^2} \boxed{\times} \boxed{\pi} \boxed{=} \boxed{\div} 375 \boxed{=}$ $\boxed{7.539822369}$

Because paint comes only in full gallons, the painter will need to purchase 8 gallons. The cost of the paint will be 8 · $39, or $312.

EXAMPLE 4
Find the area of the shaded figure on the right. Round to the nearest hundredth.

Strategy We will find the area of the entire shaded figure using the following approach:

$$A_{\text{total}} = A_{\text{triangle}} + A_{\text{smaller semicircle}} + A_{\text{larger semicircle}}$$

WHY The shaded figure is a combination of a triangular region and two semicircular regions.

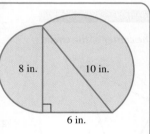

Solution The area of the triangle is

$$A_{\text{triangle}} = \frac{1}{2}bh = \frac{1}{2}(6)(8) = \frac{1}{2}(48) = 24$$

Since the formula for the area of a circle is $A = \pi r^2$, the formula for the area of a semicircle is $A = \frac{1}{2}\pi r^2$. Thus, the area enclosed by the smaller semicircle is

$$A_{\text{smaller semicircle}} = \frac{1}{2}\pi r^2 = \frac{1}{2}\pi(4)^2 = \frac{1}{2}\pi(16) = 8\pi$$

The area enclosed by the larger semicircle is

$$A_{\text{larger semicircle}} = \frac{1}{2}\pi r^2 = \frac{1}{2}\pi(5)^2 = \frac{1}{2}\pi(25) = 12.5\pi$$

The total area is the sum of the three results:

$$A_{\text{total}} = 24 + 8\pi + 12.5\pi \approx 88.4026494 \quad \text{Use a calculator to perform the operations.}$$

```
   12.5
2)25.0
  -2
   05
   -4
   1 0
   -1 0
      0
```

To the nearest hundredth, the area of the shaded figure is 88.40 in.².

Self Check 4
Find the area of the shaded figure below. Round to the nearest hundredth.   424.73 yd²

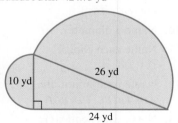

Now Try Problem 37

Teaching Example 4 Find the area of the shaded figure. Round to the nearest hundredth.

Answer: 12.28 in.²

ANSWERS TO SELF CHECKS
1. 24π m ≈ 75.4 m   2. 39.42 m   3. 36π ft² ≈ 113.1 ft²   4. 424.73 yd²

Unless otherwise noted, all art on this page is © Cengage Learning.

SECTION 9.8 STUDY SET

VOCABULARY

Fill in the blanks.

1. A segment drawn from the center of a circle to a point on the circle is called a __radius__.
2. A segment joining two points on a circle is called a __chord__.
3. A __diameter__ is a chord that passes through the center of a circle.
4. An arc that is one-half of a complete circle is a __semicircle__.
5. The distance around a circle is called its __circumference__.
6. The surface enclosed by a circle is called its __area__.
7. A diameter of a circle is __twice__ as long as a radius.
8. Suppose the exact circumference of a circle is 3π feet. When we write $C \approx 9.42$ feet, we are giving an __approximation__ of the circumference.

CONCEPTS

Refer to the figure below, where point O is the center of the circle.

9. Name each radius. $\overline{OA}, \overline{OC}, \overline{OB}$
10. Name a diameter. $\overline{AC}$
11. Name each chord. $\overline{DA}, \overline{DC}, \overline{AC}$
12. Name each minor arc. $\overset{\frown}{AD}, \overset{\frown}{AB}, \overset{\frown}{BC}, \overset{\frown}{CD}, \overset{\frown}{BCD}$
13. Name each semicircle. $\overset{\frown}{ABC}, \overset{\frown}{ADC}$
14. Name major arc $\overset{\frown}{ABD}$ in another way. $\overset{\frown}{ACD}$
15. a. If you know the radius of a circle, how can you find its diameter? Multiply the radius by 2.
    b. If you know the diameter of a circle, how can you find its radius? Divide the diameter by 2.
16. a. What are the two formulas that can be used to find the circumference of a circle? $C = \pi D$, $C = 2\pi r$
    b. What is the formula for the area of a circle? $A = \pi r^2$
17. If C is the circumference of a circle and D is its diameter, then $\frac{C}{D} = $ __π__.
18. If D is the diameter of a circle and r is its radius, then $D = $ __2__ r.
19. When evaluating $\pi(6)^2$, what operation should be performed first? square 6
20. Round $\pi = 3.141592653589\ldots$ to the nearest hundredth. 3.14

NOTATION

Fill in the blanks.

21. The symbol $\overset{\frown}{AB}$ is read as "__arc__ __AB__."
22. To the nearest hundredth, the value of π is __3.14__.
23. a. In the expression $2\pi r$, what operations are indicated? multiplication: $2 \cdot \pi \cdot r$
    b. In the expression πr^2, what operations are indicated? raising to a power and multiplication: $\pi \cdot r^2$
24. Write each expression in better form. Leave π in your answer.
    a. $\pi(8)$ 8π
    b. $2\pi(7)$ 14π
    c. $\pi \cdot \frac{25}{3}$ $\frac{25\pi}{3}$

GUIDED PRACTICE

The answers to the problems in this Study Set may vary slightly, depending on which approximation of π is used.

Find the circumference of the circle shown below. Give the exact answer and an approximation to the nearest tenth. See Example 1.

25.
8π ft ≈ 25.1 ft

26.
16π in. ≈ 50.3 in.

27.
12π m ≈ 37.7 m

28.
20π mm ≈ 62.8 mm

Find the perimeter of each figure. Assume each arc is a semicircle. Round to the nearest hundredth. See Example 2.

29.
50.85 cm   12 cm

30.
25.42 ft

31.
31.42 m   6 m

32.
67.42 in.

Find the area of each circle given the following information. Give the exact answer and an approximation to the nearest tenth. See Example 3.

33.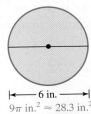
6 in.
9π in.$^2 \approx 28.3$ in.^{2}

34.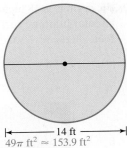
14 ft
49π ft$^2 \approx 153.9$ ft^2

35. Find the area of a circle with diameter 18 inches.
81π in.$^2 \approx 254.5$ in.^{2}

36. Find the area of a circle with diameter 20 meters.
100π m$^2 \approx 314.2$ m^2

Find the total area of each figure. Assume each arc is a semicircle. Round to the nearest tenth. See Example 4.

37.
12 cm
12 cm
128.5 cm^2

38.
6 in.
10 in.
88.3 in.^{2}

39.
8 cm
4 cm
57.1 cm^2

40.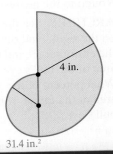
4 in.
31.4 in.^{2}

TRY IT YOURSELF

Find the area of each shaded region. Round to the nearest tenth.

41.
4 in.
10 in
27.4 in.^{2}

42.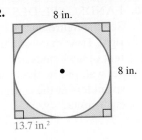
8 in.
8 in.
13.7 in.^{2}

43.
$r = 4$ in.
$h = 9$ in.
13 in.
66.7 in.^{2}

44.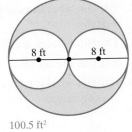
8 ft   8 ft
100.5 ft^2

45. Find the circumference of the circle shown below. Give the exact answer and an approximation to the nearest hundredth. 50π yd ≈ 157.08 yd

50 yd

46. Find the circumference of the semicircle shown below. Give the exact answer and an approximation to the nearest hundredth. $\frac{25\pi}{2}$ cm ≈ 39.27 cm

25 cm

47. Find the circumference of the circle shown below if the square has sides of length 6 inches. Give the exact answer and an approximation to the nearest tenth. 6π in. ≈ 18.8 in.

48. Find the circumference of the semicircle shown below if the length of the rectangle in which it is enclosed is 8 feet. Give the exact answer and an approximation to the nearest tenth. 4π ft ≈ 12.6 ft

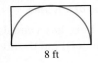
8 ft

49. Find the area of the circle shown below if the square has sides of length 9 millimeters. Give the exact answer and an approximation to the nearest tenth. 20.25π mm$^2 \approx 63.6$ mm^2

50. Find the area of the shaded semicircular region shown below. Give the exact answer and an approximation to the nearest tenth. 10.5625π mi$^2 \approx 33.2$ mi^2

6.5 mi

Unless otherwise noted, all art on this page is © Cengage Learning.

APPLICATIONS

51. Suppose the two "legs" of the compass shown below are adjusted so that the distance between the pointed ends is 1 inch. Then a circle is drawn.

a. What will the radius of the circle be? 1 in.

b. What will the diameter of the circle be? 2 in.

c. What will the circumference of the circle be? Give an exact answer and an approximation to the nearest hundredth. 2π in. ≈ 6.28 in.

d. What will the area of the circle be? Give an exact answer and an approximation to the nearest hundredth. π in.^{2} ≈ 3.14 in.^{2}

52. Suppose we find the distance around a can and the distance across the can using a measuring tape, as shown to the right. Then we make a comparison, in the form of a ratio:

$$\frac{\text{The distance around the can}}{\text{The distance across the top of the can}}$$

After we do the indicated division, the result will be close to what number? π

When appropriate, give the exact answer and an approximation to the nearest hundredth. Answers may vary slightly, depending on which approximation of π is used.

53. LAKES Round Lake has a circular shoreline that is 2 miles in diameter. Find the area of the lake.
π mi^2 ≈ 3.14 mi^2

54. HELICOPTERS Refer to the figure below. How far does a point on the tip of a rotor blade travel when it makes one complete revolution? 36π ft ≈ 113.10 ft

55. GIANT SEQUOIA The largest sequoia tree is the General Sherman Tree in Sequoia National Park in California. In fact, it is considered to be the largest living thing in the world. According to the *Guinness Book of World Records*, it has a diameter of 32.66 feet, measured $4\frac{1}{2}$ feet above the ground. What is the circumference of the tree at that height?
32.66π ft ≈ 102.60 ft

56. TRAMPOLINE See the figure below. The distance from the center of the trampoline to the edge of its steel frame is 7 feet. The protective padding covering the springs is 18 inches wide. Find the area of the circular jumping surface of the trampoline, in square feet. 30.25π ft^2 ≈ 95.03 ft^2

57. JOGGING Joan wants to jog 10 miles on a circular track $\frac{1}{4}$ mile in diameter. How many times must she circle the track? Round to the nearest lap. 13 times

58. CARPETING A state capitol building has a circular floor 100 feet in diameter. The legislature wishes to have the floor carpeted. The lowest bid is $83 per square yard, including installation. How much must the legislature spend for the carpeting project? Round to the nearest dollar. $72,431

59. ARCHERY See the figure on the right. Find the area of the entire target and the area of the bull's eye. What percent of the area of the target is the bull's eye?
4π ft^2 ≈ 12.57 ft^2;
0.25π ft^2 ≈ 0.79 ft^2; 6.25%

60. LANDSCAPE DESIGN See the figure on the right. How many square feet of lawn does not get watered by the four sprinklers at the center of each circle? 193.14 ft^2

WRITING

61. Explain what is meant by the circumference of a circle.

62. Explain what is meant by the area of a circle.

63. Explain the meaning of π.

64. Explain what it means for a car to have a small *turning radius*.

REVIEW

65. Write $\frac{9}{10}$ as a percent. 90%

66. Write $\frac{7}{8}$ as a percent. $87\frac{1}{2}\% = 87.5\%$

67. Write 0.827 as a percent. 82.7%

68. Write 0.036 as a percent. 3.6%

69. UNIT COSTS A 24-ounce package of green beans sells for $1.29. Give the unit cost in cents per ounce. 5.375¢ per oz

70. MILEAGE One car went 1,235 miles on 51.3 gallons of gasoline, and another went 1,456 on 55.78 gallons. Which car got the better gas mileage? the second car

71. How many sides does a pentagon have? five

72. What is the sum of the measures of the angles of a triangle? 180°

SECTION 9.9
Volume

Objectives

1 Find the volume of rectangular solids, prisms, and pyramids.

2 Find the volume of cylinders, cones, and spheres.

We have studied ways to calculate the perimeter and the area of two-dimensional figures that lie in a plane, such as rectangles, triangles, and circles. Now we will consider three-dimensional figures that occupy space, such as rectangular solids, cylinders, and spheres. In this section, we will introduce the vocabulary associated with these figures as well as the formulas that are used to find their volume. Volumes are measured in cubic units, such as cubic feet, cubic yards, or cubic centimeters. For example,

- We measure the capacity of a refrigerator in cubic feet.
- We buy gravel or topsoil by the cubic yard.
- We often measure amounts of medicine in cubic centimeters.

1 Find the volume of rectangular solids, prisms, and pyramids.

The **volume** of a three-dimensional figure is a measure of its capacity. The following illustration shows two common units of volume: cubic inches, written as in.³, and cubic centimeters, written as cm³.

The volume of a figure can be thought of as the number of cubic units that will fit within its boundaries. If we divide the figure shown in black below into cubes, each cube represents a volume of 1 cm³. Because there are 2 levels with 12 cubes on each level, the volume of the prism is 24 cm³.

Unless otherwise noted, all art on this page is © Cengage Learning.

Self Check 1

How many cubic centimeters are in 1 cubic meter? 1,000,000 cm³

Now Try Problem 25

Teaching Example 1 How many cubic inches are in 1 cubic yard?
Answer:
46,656 in.³

EXAMPLE 1 How many cubic inches are there in 1 cubic foot?

Strategy A figure is helpful to solve this problem. We will draw a cube and divide each of its sides into 12 equally long parts.

WHY Since a cubic foot is a cube with each side measuring 1 foot, each side also measures 12 inches.

Solution The figure on the right helps us understand the situation. Note that each level of the cubic foot contains $12 \cdot 12$ cubic inches and that the cubic foot has 12 levels. We can use multiplication to count the number of cubic inches contained in the figure. There are

$$12 \cdot 12 \cdot 12 = 1,728$$

cubic inches in 1 cubic foot. Thus, 1 ft³ = 1,728 in.³.

Cube	Rectangular Solid	Sphere
$V = s^3$	$V = lwh$	$V = \dfrac{4}{3}\pi r^3$
where s is the length of a side	where l is the length, w is the width, and h is the height	where r is the radius

Prism	Pyramid
$V = Bh$	$V = \dfrac{1}{3}Bh$
where B is the area of the base and h is the height	where B is the area of the base and h is the height

Cylinder	Cone
$V = Bh$ or $V = \pi r^2 h$	$V = \dfrac{1}{3}Bh$ or $V = \dfrac{1}{3}\pi r^2 h$
where B is the area of the base, h is the height, and r is the radius of the base	where B is the area of the base, h is the height, and r is the radius of the base

Unless otherwise noted, all art on this page is © Cengage Learning.

In practice, we do not find volumes of three-dimensional figures by counting cubes. Instead, we use the formulas shown in the table on the preceding page. Note that several of the volume formulas involve the variable B. It represents the area of the base of the figure.

Caution! The height of a geometric solid is always measured along a line perpendicular to its base.

EXAMPLE 2 Storage Tanks
An oil storage tank is in the form of a rectangular solid with dimensions 17 feet by 10 feet by 8 feet. (See the figure below.) Find its volume.

Self Check 2
Find the volume of a rectangular solid with dimensions 8 meters by 12 meters by 20 meters. 1,920 m³

Now Try Problem 29

Teaching Example 2 Find the volume of a rectangular solid that is 6 inches long, 5 inches wide, and 9 inches tall.
Answer:
270 in.³

Strategy We will substitute 17 for l, 10 for w, and 8 for h in the formula $V = lwh$ and evaluate the right side.

WHY The variable V represents the volume of a rectangular solid.

Solution

$V = lwh$ This is the formula for the volume of a rectangular solid.
$V = 17(10)(8)$ Substitute 17 for l, the length, 10 for w, the width, and 8 for h, the height of the tank.
$V = 1,360$ Do the multiplication.

$$\begin{array}{r} \overset{5}{1}70 \\ \times\ \ 8 \\ \hline 1{,}360 \end{array}$$

The volume of the tank is 1,360 ft³.

EXAMPLE 3
Find the volume of the prism shown on the right.

Strategy First, we will find the area of the base of the prism.

WHY To use the volume formula $V = Bh$, we need to know B, the area of the prism's base.

Solution The area of the triangular base of the prism is $\frac{1}{2}(6)(8) = 24$ square centimeters. To find its volume, we proceed as follows:

$V = Bh$ This is the formula for the volume of a triangular prism.
$V = 24(50)$ Substitute 24 for B, the area of the base, and 50 for h, the height.
$V = 1{,}200$ Do the multiplication.

$$\begin{array}{r} \overset{2}{2}4 \\ \times\ 50 \\ \hline 1{,}200 \end{array}$$

The volume of the triangular prism is 1,200 cm³.

Caution! Note that the 10 cm measurement was not used in the calculation of the volume.

Self Check 3
Find the volume of the prism shown below. 300 in.³

Now Try Problem 33

Teaching Example 3 Find the volume of the prism.

Answer:
12 ft³

Unless otherwise noted, all art on this page is © Cengage Learning.

Self Check 4

Find the volume of the pyramid shown below. 640 cm³

Now Try Problem 37

Teaching Example 4 Find the volume of the pyramid.

Answer: 480 mm³

Self Check 5

Find the volume of the cylinder shown below. Give the exact answer and an approximation to the nearest hundredth.

100π yd³ ≈ 314.16 yd³

Now Try Problem 45

Teaching Example 5 Find the volume of the cylinder shown below. Give the exact answer and an approximation to the nearest hundredth.

Answer: 24π ft³ ≈ 75.40 ft³

Teaching Example 6 Find the volume of the cone shown below. Give the exact answer and an approximation to the nearest hundredth.

Answer: 432π yd³ ≈ 1,357.17 yd³

EXAMPLE 4
Find the volume of the pyramid shown on the right.

Strategy First, we will find the area of the square base of the pyramid.

WHY The volume of a pyramid is $\frac{1}{3}$ of the product of the area of its base and its height.

Solution Since the base is a square with each side 6 meters long, the area of the base is (6 m)², or 36 m². To find the volume of the pyramid, we proceed as follows:

$V = \frac{1}{3}Bh$ This is the formula for the volume of a pyramid.

$V = \frac{1}{3}(36)(9)$ Substitute 36 for B, the area of the base, and 9 for h, the height.

$V = 12(9)$ Multiply: $\frac{1}{3}(36) = \frac{36}{3} = 12$.

$V = 108$ Complete the multiplication.

$$\begin{array}{r} \overset{1}{12} \\ \times 9 \\ \hline 108 \end{array}$$

The volume of the pyramid is 108 m³.

2 Find the volume of cylinders, cones, and spheres.

EXAMPLE 5
Find the volume of the cylinder shown on the right. Give the exact answer and an approximation to the nearest hundredth.

Strategy First, we will find the radius of the circular base of the cylinder.

WHY To use the formula for the volume of a cylinder, $V = \pi r^2 h$, we need to know r, the radius of the base.

Solution Since a radius is one-half of the diameter of the circular base, $r = \frac{1}{2} \cdot 6$ cm = 3 cm. From the figure, we see that the height of the cylinder is 10 cm. To find the volume of the cylinder, we proceed as follows.

$V = \pi r^2 h$ This is the formula for the volume of a cylinder.

$V = \pi(3)^2(10)$ Substitute 3 for r, the radius of the base, and 10 for h, the height.

$V = \pi(9)(10)$ Evaluate the exponential expression: (3)² = 9.

$V = 90\pi$ Multiply: (9)(10) = 90. Write the product so that π is the last factor.

$V \approx 282.7433388$ Use a calculator to do the multiplication.

The exact volume of the cylinder is 90π cm³. To the nearest hundredth, the volume is 282.74 cm³.

EXAMPLE 6
Find the volume of the cone shown on the right. Give the exact answer and an approximation to the nearest hundredth.

Strategy We will substitute 4 for r and 6 for h in the formula $V = \frac{1}{3}\pi r^2 h$ and evaluate the right side.

WHY The variable V represents the volume of a cone.

Solution

$V = \frac{1}{3}\pi r^2 h$ This is the formula for the volume of a cone.

$V = \frac{1}{3}\pi(4)^2(6)$ Substitute 4 for r, the radius of the base, and 6 for h, the height.

$V = \frac{1}{3}\pi(16)(6)$ Evaluate the exponential expression: $(4)^2 = 16$.

$V = 2\pi(16)$ Multiply: $\frac{1}{3}(6) = 2$.

$V = 32\pi$ Multiply: $2(16) = 32$. Write the product so that π is the last factor.

$V \approx 100.5309649$ Use a calculator to do the multiplication.

The exact volume of the cone is 32π ft³. To the nearest hundredth, the volume is 100.53 ft³.

Self Check 6

Find the volume of the cone shown below. Give the exact answer and an approximation to the nearest hundredth.

$\frac{20}{3}\pi$ mi³ ≈ 20.94 mi³

Now Try Problem 49

EXAMPLE 7 Water Towers

How many cubic feet of water are needed to fill the spherical water tank shown on the right? Give the exact answer and an approximation to the nearest tenth.

Strategy We will substitute 15 for r in the formula $V = \frac{4}{3}\pi r^3$ and evaluate the right side.

WHY The variable V represents the volume of a sphere.

Solution

$V = \frac{4}{3}\pi r^3$ This is the formula for the volume of a sphere.

$V = \frac{4}{3}\pi(15)^3$ Substitute 15 for r, the radius of the sphere.

$V = \frac{4}{3}\pi(3,375)$ Evaluate the exponential expression: $(15)^3 = 3,375$.

$V = \frac{13,500}{3}\pi$ Multiply: $4(3,375) = 13,500$.

$V = 4,500\pi$ Divide: $\frac{13,500}{3} = 4,500$. Write the product so that π is the last factor.

$V \approx 14,137.16694$ Use a calculator to do the multiplication.

```
  1 3 2
  3375
×    4
 13,500
```

The tank holds exactly $4,500\pi$ ft³ of water. To the nearest tenth, this is 14,137.2 ft³.

Self Check 7

Find the volume of a spherical water tank with a radius of 7 meters. Give the exact answer and an approximation to the nearest tenth. $\frac{1,372}{3}\pi$ m³ ≈ 1,436.8 m³

Now Try Problem 53

Teaching Example 7 Find the volume of a spherical tank with a radius of 30 inches. Give the exact answer and an approximation to the nearest tenth.
Answer:
$36,000\pi$ in.³ ≈ 113,097.3 in.³

Using Your CALCULATOR Volume of a Silo

A silo is a structure used for storing grain. The silo shown on the right is a cylinder 50 feet tall topped with a dome in the shape of a hemisphere. To find the volume of the silo, we add the volume of the cylinder to the volume of the dome.

$\text{Volume}_{\text{cylinder}} + \text{Volume}_{\text{dome}} = (\text{Area}_{\text{cylinder's base}})(\text{Height}_{\text{cylinder}}) + \frac{1}{2}(\text{Volume}_{\text{sphere}})$

$\text{Volume}_{\text{cylinder}} + \text{Volume}_{\text{dome}} = \pi r^2 h + \frac{1}{2}\left(\frac{4}{3}\pi r^3\right)$

$\text{Volume}_{\text{cylinder}} + \text{Volume}_{\text{dome}} = \pi r^2 h + \frac{2\pi r^3}{3}$ Multiply and simplify: $\frac{1}{2}(\frac{4}{3}\pi r^3) = \frac{4}{6}\pi r^3 = \frac{2\pi r^3}{3}$.

$\text{Volume}_{\text{cylinder}} + \text{Volume}_{\text{dome}} = \pi(10)^2(50) + \frac{2\pi(10)^3}{3}$ Substitute 10 for r and 50 for h.

914 Chapter 9 An Introduction to Geometry

We can use a scientific calculator to make this calculation.

$\pi \times 10\, x^2 \times 50 + (\, 2 \times \pi \times 10\, y^x\, 3\,) \div 3 =$

```
17802.35837
```

The volume of the silo is approximately 17,802 ft³.

ANSWERS TO SELF CHECKS

1. 1,000,000 cm³   2. 1,920 m³   3. 300 in.³   4. 640 cm³   5. 100π yd³ ≈ 314.16 yd³
6. $\frac{20}{3}\pi$ mi³ ≈ 20.94 mi³   7. $\frac{1{,}372}{3}\pi$ m³ ≈ 1,436.8 m³

SECTION 9.9 STUDY SET

VOCABULARY

Fill in the blanks.

1. The __volume__ of a three-dimensional figure is a measure of its capacity.
2. The volume of a figure can be thought of as the number of __cubic__ units that will fit within its boundaries.

Give the name of each figure.

3.
cone

4.
sphere

5.
cylinder

6.
prism

7.
pyramid

8.
rectangular solid

CONCEPTS

9. Draw a cube. Label a side s.
10. Draw a cylinder. Label the height h and radius r.
11. Draw a pyramid. Label the height h and the base.
12. Draw a cone. Label the height h and radius r.
13. Draw a sphere. Label the radius r.
14. Draw a rectangular solid. Label the length l, the width w, and the height h.
15. Which of the following are acceptable units with which to measure volume? cubic inches, mi³, m³

ft²	mi³	seconds	days
cubic inches	mm	square yards	in.
pounds	cm²	meters	m³

16. In the figure on the right, the unit of measurement of length used to draw the figure is the inch.

   a. What is the area of the base of the figure? 24 in.²
   b. What is the volume of the figure? 72 in.³

Unless otherwise noted, all art on this page is © Cengage Learning.

17. Which geometric concept (perimeter, circumference, area, or volume) should be applied when measuring each of the following?
    a. The distance around a checkerboard perimeter
    b. The size of a trunk of a car volume
    c. The amount of paper used for a postage stamp area
    d. The amount of storage in a cedar chest volume
    e. The amount of beach available for sunbathing area
    f. The distance the tip of a propeller travels circumference

18. Complete the table.

Figure	Volume formula
Cube	$V = s^3$
Rectangular solid	$V = lwh$
Prism	$V = Bh$
Cylinder	$V = \pi r^2 h$
Pyramid	$V = \frac{1}{3}Bh$
Cone	$V = \frac{1}{3}\pi r^2 h$
Sphere	$V = \frac{4}{3}\pi r^3$

19. Evaluate each expression. Leave π in the answer.
    a. $\frac{1}{3}\pi(25)6$ 50π
    b. $\frac{4}{3}\pi(125)$ $\frac{500}{3}\pi$

20. a. Evaluate $\frac{1}{3}\pi r^2 h$ for $r = 2$ and $h = 27$. Leave π in the answer. 36π
    b. Approximate your answer to part a to the nearest tenth. 113.1

NOTATION

21. a. What does "in.³" mean? cubic inch
    b. Write "one cubic centimeter" using symbols. 1 cm^3

22. In the formula $V = \frac{1}{3}Bh$, what does B represent? the area of the base

23. In a drawing, what does the symbol ⌐ indicate? a right angle

24. Redraw the figure below using dashed lines to show the hidden edges.

GUIDED PRACTICE

Convert from one unit of measurement to another. See Example 1.

25. How many cubic feet are in 1 cubic yard? 27
26. How many cubic decimeters are in 1 cubic meter? 1,000
27. How many cubic meters are in 1 cubic kilometer? 1,000,000,000
28. How many cubic inches are in 1 cubic yard? 46,656

Find the volume of each figure. See Example 2.

29.
    56 ft³

30.
    320 mm³

31.
    125 in.³

32.
    64,000 ft³

Find the volume of each figure. See Example 3.

33.
    120 cm³

34.
    2,400 cm³

35.
    1,296 in.³

36. 720 in.³

Find the volume of each figure. See Example 4.

37.
700 yd³

38.
245 m³

39.
32 ft³

40.
858 in.³

41.
69.72 ft³

42.
58.24 mm³

43.
6 yd³

44.
121 ft³

Find the volume of each cylinder. Give the exact answer and an approximation to the nearest hundredth. See Example 5.

45.
192π ft³ ≈ 603.19 ft³

46.
24π mi³ ≈ 75.40 mi³

47.
$3{,}150\pi$ cm³ ≈ 9,896.02 cm³

48.
$201{,}840\pi$ in.³ ≈ 634,099.06 in.³

Find the volume of each cone. Give the exact answer and an approximation to the nearest hundredth. See Example 6.

49.
39π m³ ≈ 122.52 m³

50.
28π mm³ ≈ 87.96 mm³

51.
189π yd³ ≈ 593.76 yd³

52.
250π ft³ ≈ 785.40 ft³

Find the volume of each sphere. Give the exact answer and an approximation to the nearest tenth. See Example 7.

53.
288π in.$^3 \approx 904.8$ in.^{3}

54.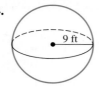
972π ft$^3 \approx 3{,}053.6$ ft^3

55.
$\frac{32}{3}\pi$ cm$^3 \approx 33.5$ cm^3

56.
$\frac{500}{3}\pi$ in.$^3 \approx 523.6$ in.^{3}

TRY IT YOURSELF

Find the volume of each figure. If an exact answer contains π, approximate to the nearest hundredth.

57. A hemisphere with a radius of 9 inches (*Hint:* a **hemisphere** is an exact half of a sphere.) 486π in.$^3 \approx 1{,}526.81$ in.^{3}

58. A hemisphere with a diameter of 22 feet (*Hint:* a **hemisphere** is an exact half of a sphere.) $\frac{2{,}662}{3}\pi$ ft$^3 \approx 2{,}787.64$ ft^3

59. A cylinder with a height of 12 meters and a circular base with a radius of 6 meters 423π m$^3 \approx 1{,}357.17$ m^3

60. A cylinder with a height of 4 meters and a circular base with a diameter of 18 meters 324π m$^3 \approx 1{,}017.88$ m^3

61. A rectangular solid with dimensions of 3 cm by 4 cm by 5 cm 60 cm^3

62. A rectangular solid with dimensions of 5 m by 8 m by 10 m 400 m^3

63. A cone with a height of 12 centimeters and a circular base with a diameter of 10 centimeters 100π cm$^3 \approx 314.16$ cm^3

64. A cone with a height of 3 inches and a circular base with a radius of 4 inches 16π in.$^3 \approx 50.27$ in.^{3}

65. A pyramid with a square base 10 meters on each side and a height of 12 meters 400 m^3

66. A pyramid with a square base 6 inches on each side and a height of 4 inches 48 in.^{3}

67. A prism whose base is a right triangle with legs 3 meters and 4 meters long and whose height is 8 meters 48 m^3

68. A prism whose base is a right triangle with legs 5 feet and 12 feet long and whose height is 25 feet 750 ft^3

Find the volume of each figure. Give the exact answer and, when needed, an approximation to the nearest hundredth.

69.
576 cm^3

70.
$\frac{320}{3}\pi$ in.$^3 \approx 335.10$ in.^{3}

71.
180π cm$^3 \approx 565.49$ cm^3

72.
52 in.^{3}

APPLICATIONS

Solve each problem. If an exact answer contains π, approximate the answer to the nearest hundredth.

73. SWEETENERS A sugar cube is $\frac{1}{2}$ inch on each edge. How much volume does it occupy? $\frac{1}{8}$ in.$^3 = 0.125$ in.^{3}

74. VENTILATION A classroom is 40 feet long, 30 feet wide, and 9 feet high. Find the number of cubic feet of air in the room. 10,800 ft^3

75. WATER HEATERS Complete the advertisement for the high-efficiency water heater shown below. 2.125

76. REFRIGERATORS The largest refrigerator advertised in a JC Penny catalog has a capacity of 25.2 cubic feet. How many cubic inches is this? 43,545.6 in.³

77. TANKS A cylindrical oil tank has a diameter of 6 feet and a length of 7 feet. Find the volume of the tank. 63π ft³ ≈ 197.92 ft³

78. DESSERTS A restaurant serves pudding in a conical dish that has a diameter of 3 inches. If the dish is 4 inches deep, how many cubic inches of pudding are in each dish? 3π in.³ ≈ 9.42 in.³

79. HOT-AIR BALLOONS The lifting power of a spherical balloon depends on its volume. How many cubic feet of gas will a balloon hold if it is 40 feet in diameter? $\frac{32,000}{3}\pi$ ft³ ≈ 33,510.32 ft³

80. CEREAL BOXES A box of cereal measures 3 inches by 8 inches by 10 inches. The manufacturer plans to market a smaller box that measures $2\frac{1}{2}$ by 7 by 8 inches. By how much will the volume be reduced? 100 in.³

81. ENGINES The *compression ratio* of an engine is the volume in one cylinder with the piston at bottom-dead-center (B.D.C.), divided by the volume with the piston at top-dead-center (T.D.C.). From the data given in the following figure, what is the compression ratio of the engine? Use a colon to express your answer. 8:1

Volume before compression: 30.4 in.³
Volume after compression: 3.8 in.³
T.D.C.
B.D.C.

82. GEOGRAPHY Earth is not a perfect sphere but is slightly pear-shaped. To estimate its volume, we will assume that it is spherical, with a diameter of about 7,926 miles. What is its volume, to the nearest one billion cubic miles? 261,000,000,000 mi³

83. BIRDBATHS
    a. The bowl of the birdbath shown on the right is in the shape of a hemisphere (half of a sphere). Find its volume. $2,250\pi$ in.³ ≈ 7,068.58 in.³
    b. If 1 gallon of water occupies 231 cubic inches of space, how many gallons of water does the birdbath hold? Round to the nearest tenth. 30.6 gal

← 30 in. →

84. CONCRETE BLOCKS Find the number of cubic inches of concrete used to make the hollow, cube-shaped, block shown below. 312 in.³

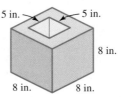
5 in.   5 in.
8 in.
8 in.   8 in.

WRITING

85. What is meant by the *volume* of a cube?

86. The stack of 3 × 5 index cards shown in figure (a) forms a right rectangular prism, with a certain volume. If the stack is pushed to lean to the right, as in figure (b), a new prism is formed. How will its volume compare to the volume of the right rectangular prism? Explain your answer.

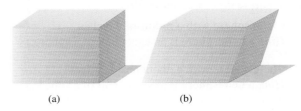

(a)   (b)

87. Are the units used to measure area different from the units used to measure volume? Explain.

88. The dimensions (length, width, and height) of one rectangular solid are entirely different numbers from the dimensions of another rectangular solid. Would it be possible for the rectangular solids to have the same volume? Explain.

REVIEW

89. Evaluate: $-5(5-2)^2 + 3$ -42

90. BUYING PENCILS Carlos bought 6 pencils at $0.60 each and a notebook for $1.25. He gave the clerk a $5 bill. How much change did he receive? $0.15

91. Solve: $-x = 4$ -4

92. 38 is what percent of 40?   95%

93. Express the phrase "3 inches to 15 inches" as a ratio in simplest form. $\frac{1}{5}$ or $1:5$

94. Convert 40 ounces to pounds.   2.5 lb

95. Convert 2.4 meters to millimeters.   2,400 mm

96. State the Pythagorean equation. $a^2 + b^2 = c^2$

STUDY SKILLS CHECKLIST
Know the Vocabulary

A large amount of vocabulary has been introduced in Chapter 9. Before taking the test, put a checkmark in the box if you can define and draw an example of each of the given terms.

- ☐ Point, line, plane
- ☐ Line segment, midpoint
- ☐ Ray, angle, vertex
- ☐ Acute angle, obtuse angle, right angle, straight angle
- ☐ Adjacent angles, vertical angles
- ☐ Complementary angles, supplementary angles
- ☐ Congruent segments, congruent angles
- ☐ Parallel lines, perpendicular lines, a transversal
- ☐ Alternate interior angles, interior angles, corresponding angles
- ☐ Polygon, triangle, quadrilateral, pentagon, hexagon, octagon
- ☐ Equilateral triangle, isosceles triangle, scalene triangle
- ☐ Acute triangle, obtuse triangle
- ☐ Right triangle, hypotenuse, legs
- ☐ Congruent triangles, similar triangles
- ☐ Parallelogram, rectangle, square, rhombus, trapezoid, isosceles trapezoid
- ☐ Circle, arc, semicircle, radius, diameter
- ☐ Rectangular solid, cube, sphere, prism, pyramid, cylinder, cone

Teaching Guide: Refer to the Instructor's Resource Binder to find activities, worksheets on key concepts, more examples, instruction tips, overheads, and assignments.

CHAPTER 9 SUMMARY AND REVIEW

SECTION 9.1 Basic Geometric Figures; Angles

DEFINITIONS AND CONCEPTS	EXAMPLES
The word **geometry** comes from the Greek words *geo* (meaning Earth) and *metron* (meaning measure). Geometry is based on three undefined words: **point**, **line**, and **plane**.	**Point** • A — **Line** $\overleftrightarrow{BC}$ (points B and C) — **Plane** EFG (points E, F, G) Points are labeled with capital letters. — We can name a line using any two points on it. — Floors, walls, and table tops are all parts of planes.

Unless otherwise noted, all art on this page is © Cengage Learning.

A **line segment** is a part of a line with two endpoints. Every line segment has a **midpoint**, which divides the segment into two parts of equal length. The notation m($\overline{AM}$) is read as "the measure of line segment $\overline{AM}$." When two line segments have the same measure, we say that they are **congruent**. Read the symbol ≅ as "is congruent to." A **ray** is a part of a line with one **endpoint**.	
An **angle** is a figure formed by two rays (called **sides**) with a common endpoint. The common endpoint is called the **vertex** of the angle. We read the symbol ∠ as "angle."	The angle below can be written as ∠BAC, ∠CAB, ∠A, or ∠1.
When two angles have the same measure, we say that they are **congruent**. A **protractor** is used to find the measure of an angle. One unit of measurement of an angle is the **degree**. The notation m(∠DEF) is read as "the measure of ∠DEF."	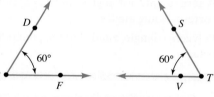 Since m(∠DEF) = m(∠STV), we say that ∠DEF ≅ ∠STV.
An **acute angle** has a measure that is greater than 0° but less than 90°. An **obtuse angle** has a measure that is greater than 90° but less than 180°. A **straight angle** measures 180°.	 Acute angle Obtuse angle Straight angle
A **right angle** measures 90°.	 Right angle A ⌐ symbol is often used to label a right angle.
Two angles that have the same vertex and are side by side are called **adjacent angles**.	Two angles with degree measures of x and 21° are adjacent angles, as shown here. Use the information in the figure to find x. **Adjacent angles**
We can use the algebra concepts of variable and equation to solve many types of geometry problems.	The sum of the measures of the two adjacent angles is 32°: $x + 21° = 32°$ The word sum indicates addition. $x + 21° - 21° = 32° - 21°$ Subtract 21° from both sides. $x = 11°$ Do the subtraction. Thus, x is 11°.

When two lines intersect, pairs of nonadjacent angles are called **vertical angles**.	**Vertical angles**
Vertical angles are congruent (have the same measure).	Refer to the figure below. Find x and $m(\angle XYZ)$. 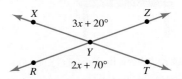 Since the angles are vertical angles, they have equal measures. $3x + 20° = 2x + 70°$ Set the expressions equal. $3x + 20° - 2x = 2x + 70° - 2x$ Eliminate $2x$ from the right side. $x + 20° = 70°$ Combine like terms. $x = 50°$ Subtract 20° from both sides. Thus, x is 50°. To find $m(\angle XYZ)$, evaluate the expression $3x + 20°$ for $x = 50°$. $3x + 20° = 3(50°) + 20°$ Substitute 50° for x. $= 150° + 20°$ Do the multiplication. $= 170°$ Do the addition. Thus, $m(\angle XYZ) = 170°$.
If the sum of two angles is 90°, the angles are **complementary**. If the sum of two angles is 180°, the angles are **supplementary**.	**Complementary angles** **Supplementary angles** $63° + 27° = 90°$ $146° + 34° = 180°$
We can use algebra to find the complement of an angle.	Find the complement of an 11° angle. Let $x =$ the measure of the complement (in degrees). $x + 11° = 90°$ The sum of the angles' measures must be 90°. $x = 79°$ To isolate x, subtract 11° from both sides. The complement of an 11° angle has measure 79°.
We can use algebra to find the supplement of an angle.	Find the supplement of a 68° angle. Let $x =$ the measure of the supplement (in degrees). $x + 68° = 180°$ The sum of the angles' measures must be 180°. $x = 112°$ To isolate x, subtract 68° from both sides. The supplement of a 68° angle has measure 112°.

REVIEW EXERCISES

1. In the illustration, give the name of a point, a line, and a plane. points C and D, line CD, plane GHI

2. a. In the figure below, find m($\overline{AG}$). 6 units
   b. Find the midpoint of $\overline{BH}$. E
   c. Is $\overline{AC} \cong \overline{GE}$? yes

3. Give four ways to name the angle shown below.
   ∠ABC, ∠CBA, ∠B, ∠1

4. a. Is the angle shown above acute or obtuse? acute
   b. What is the vertex of the angle? B
   c. What rays form the sides of the angle? $\overrightarrow{BA}$ and $\overrightarrow{BC}$
   d. Use a protractor to find the measure of the angle. 48°

5. Identify each acute angle, right angle, obtuse angle, and straight angle in the figure below. ∠1 and ∠2 are acute, ∠ABD, and ∠CBD are right angles, ∠CBE is obtuse, and ∠ABC is a straight angle.

6. In the figure above, is ∠ABD ≅ ∠CBD? yes
7. In the figure above, are $\overrightarrow{AC}$ and $\overrightarrow{AB}$ the same ray? yes
8. The measures of several angles are given below. Identify each angle as an acute angle, a right angle, an obtuse angle, or a straight angle.
   a. m(∠A) = 150° obtuse angle
   b. m(∠B) = 90° right angle
   c. m(∠C) = 180° straight angle
   d. m(∠D) = 25° acute angle

9. The two angles shown here are adjacent angles. Find x. 15°

10. Line AB is shown in the figure below. Find y. 150°

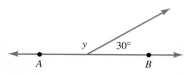

11. Refer to the figure on the right.
    a. Find m(∠1). m(∠1) = 65°
    b. Find m(∠2). m(∠2) = 115°

12. Refer to the figure below with lines $\overleftrightarrow{AE}$, $\overleftrightarrow{GD}$, and $\overleftrightarrow{CF}$.
    a. What is m(∠ABG)? 39°
    b. What is m(∠FBE)? 90°
    c. What is m(∠CBD)? 51°
    d. What is m(∠FBG)? 51°
    e. Are ∠CBD and ∠DBE complementary angles? yes

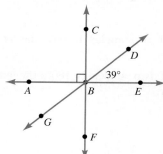

13. Refer to the figure with lines $\overleftrightarrow{EH}$ and $\overleftrightarrow{GI}$.
    a. Find x. 20°
    b. What is m(∠HFI)? 125°
    c. What is m(∠GFH)? 55°

14. Find the complement of a 71° angle. 19°
15. Find the supplement of a 143° angle. 37°
16. Are angles measuring 30°, 60°, and 90° supplementary? No, only two angles can be supplementary.

SECTION 9.2 Parallel and Perpendicular Lines

DEFINITIONS AND CONCEPTS	EXAMPLES
If two lines lie in the same plane, they are called **coplanar**. **Parallel lines** are coplanar lines that do not intersect. We read the symbol ∥ as "is parallel to." **Perpendicular lines** are lines that intersect and form right angles. We read the symbol ⊥ as "is perpendicular to."	
A line that intersects two coplanar lines in two distinct (different) points is called a **transversal**. When a transversal intersects two coplanar lines, four pairs of **corresponding angles** are formed. If two parallel lines are cut by a transversal, *corresponding angles are congruent* (have equal measures).	
When a transversal intersects two coplanar lines, two pairs of **interior angles** and two pairs of **alternate interior angles** are formed. If two parallel lines are cut by a transversal, *alternate interior angles are congruent* (have equal measures). If two parallel lines are cut by a transversal, *interior angles on the same side of the transversal are supplementary*.	

We can use algebra to find the unknown measures of corresponding angles.	In the figure, $l_1 \parallel l_2$. Find x and the measure of each angle that is labeled. 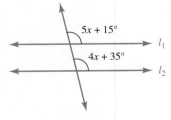
	Since the lines are parallel, and the angles are corresponding angles, the angles are congruent.
	$5x + 15° = 4x + 35°$ The angle measures are equal.
	$x + 15° = 35°$ Subtract 4x from both sides.
	$x = 20°$ To isolate x, subtract 15° from both sides.
	Thus, x is 20°. To find the measures of the angles labeled in the figure, we evaluate each expression for $x = 20°$.
	$5x + 15° = 5(\mathbf{20°}) + 15°$ $4x + 35° = 4(\mathbf{20°}) + 35°$
	             $= 100° + 15°$ $= 80° + 35°$
	             $= 115°$ $= 115°$
	The measure of each angle is 115°.

We can use algebra to find the unknown measures of interior angles.

In the figure, $l_1 \parallel l_2$. Find x and the measure of each angle that is labeled.

Since the angles are interior angles on the same side of the transversal, they are supplementary.

$4x + 17° + x - 12° = 180°$ The sum of the measures of two supplementary angles is 180°.
$5x + 5° = 180°$ Combine like terms.
$5x = 175°$ Subtract 5° from both sides.
$x = 35°$ Divide both sides by 5.

Thus, x is 35°. To find the measures of the angles in the figure, we evaluate the expressions for $x = 35°$.

$4x + 17° = 4(35°) + 17°$ | $x - 12° = 35° - 12°$
$= 140° + 17°$ | $= 23°$
$= 157°$

The measures of the angles labeled in the figure are 157° and 23°.

REVIEW EXERCISES

17. a. Lines l_1 and l_2 shown in figure (a) below do not intersect and are coplanar. What word describes the lines? parallel

b. In figure (a), line l_3 intersects lines l_1 and l_2 in two distinct (different) points. What is the name given to line l_3? transversal

c. What word describes the two lines shown in figure (b) below? perpendicular

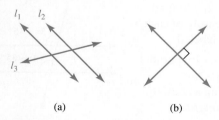

(a)   (b)

18. Identify all pairs of alternate interior angles shown in the figure below. ∠4 and ∠6, ∠3 and ∠5

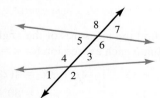

19. Refer to the figure in Problem 18. Identify all pairs of corresponding angles. ∠1 and ∠5, ∠4 and ∠8, ∠2 and ∠6, ∠3 and ∠7

20. Refer to the figure in Problem 18. Identify all pairs of vertical angles. ∠1 and ∠3, ∠2 and ∠4, ∠5 and ∠7, ∠6 and ∠8

21. In the figure below, $l_1 \parallel l_2$. Find the measure of each angle.

m(∠1) = m(∠3) = m(∠5) = m(∠7) = 70°;
m(∠2) = m(∠4) = m(∠6) = 110°

22. In the figure on the right, $\overline{DC} \parallel \overline{AB}$. Find the measure of each angle that is labeled.
m(∠1) = 60°, m(∠2) = 120°,
m(∠3) = 130°, m(∠4) = 50°

23. In the figure below, $l_1 \parallel l_2$.

a. Find x. 40°

b. Find the measure of each angle that is labeled. 50°, 50°

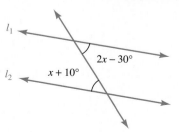

24. In the figure below, $l_1 \parallel l_2$.
  a. Find x. 20°
  b. Find the measure of each angle that is labeled. 110°, 70°

26. In the figure below, $\overline{EF} \parallel \overline{HI}$.
  a. Find x. 23°
  b. Find the measure of each angle that is labeled. 82°, 82°

25. In the figure below, $\overline{AB} \parallel \overline{DC}$.
  a. Find x. 11°
  b. Find the measure of each angle that is labeled. 31°, 31°

SECTION 9.3 Triangles

DEFINITIONS AND CONCEPTS	EXAMPLES
A **polygon** is a closed geometric figure with at least three line segments for its sides. The points at which the sides intersect are called **vertices**. A **regular polygon** has sides that are all the same length and angles that are all the same measure. The number of vertices of a polygon is equal to the number of sides it has.	Polygon Regular polygon

Classifying Polygons

Number of Sides	Name of Polygon	Number of Sides	Name of Polygon
3	triangle	8	octagon
4	quadrilateral	9	nonagon
5	pentagon	10	decagon
6	hexagon	12	dodecagon

 Quadrilateral (4 sides)   Hexagon (6 sides)   Octagon (8 sides)

A **triangle** is a polygon with three sides (and three vertices).

Triangles can be classified according to the lengths of their sides.

Tick marks indicate sides that are of equal length.

Equilateral triangle (all sides of equal length)   Isosceles triangle (at least two sides of equal length)   Scalene triangle (no sides of equal length)

Triangles can be classified by their angles.	 **Acute triangle** (has three acute angles) **Obtuse triangle** (has an obtuse angle) **Right triangle** (has one right angle)
The longest side of a right triangle is called the **hypotenuse**, and the other two sides are called **legs**. The hypotenuse of a right triangle is always opposite the 90° (right) angle. The legs of a right triangle are adjacent to (next to) the right angle.	**Right triangle** Hypotenuse (longest side) Leg Leg
In an isosceles triangle, the angles opposite the sides of equal length are called **base angles**. The third angle is called the **vertex angle**. The third side is called the **base**. **Isosceles triangle theorem:** If two sides of a triangle are congruent, then the angles opposite those sides are congruent. **Converse of the isosceles triangle theorem:** If two angles of a triangle are congruent, then the sides opposite the angles have the same length, and the triangle is isosceles.	**Isosceles triangles** Vertex angle Base angle Base angle Base
The **sum of the measures of the angles** of any triangle is 180°. We can use algebra to find unknown angle measures of a triangle.	Find the measure of each angle of $\triangle ABC$. 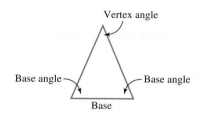 The sum of the angle measures of any triangle is 180°: $x + 3x - 25° + x - 5° = 180°$ $5x - 30° = 180°$ Combine like terms. $5x = 210°$ Add 30° to both sides. $x = 42°$ Divide both sides by 5. To find the measures of $\angle B$ and $\angle C$, we evaluate the expressions $3x - 25°$ and $x - 5°$ for $x = 42°$. $3x - 25° = 3(42°) - 25°$ | $x - 5° = 42° - 5°$ $= 126° - 25°$ | $= 37°$ $= 101°$ Thus, m($\angle A$) = 42°, m($\angle B$) = 101°, and m($\angle C$) = 37°.
We can use algebra to find unknown angle measures of an isosceles triangle.	If the vertex angle of an isosceles triangle measures 26°, what is the measure of each base angle? If we let x represent the measure of one base angle, the measure of the other base angle is also x. (See the figure.) Since the sum of the measures of the angles of any triangle is 180°, we have $x + x + 26° = 180°$ $2x + 26° = 180°$ On the left side, combine like terms. $2x = 154°$ To isolate 2x, subtract 26° from both sides. $x = 77°$ To isolate x, divide both sides by 2. The measure of each base angle is 77°.

REVIEW EXERCISES

27. For each of the following polygons, give the number of sides it has, tell its name, and then give the number of vertices that it has.

a.
8, octagon, 8

b.
5, pentagon, 5

c.
3, triangle, 3

d.
6, hexagon, 6

e.
4, quadrilateral, 4

f.
10, decagon, 10

28. Classify each of the following triangles as an equilateral triangle, an isosceles triangle, or a scalene triangle. Some figures may be correctly classified in more than one way.

a.
isosceles

b.
scalene

c.
equilateral

d.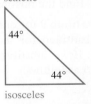
isosceles

29. Classify each of the following triangles as an acute, an obtuse, or a right triangle.

a.
acute

b.
right

c.
obtuse

d.
acute

30. Refer to the triangle shown here.
a. What is the measure of ∠X? 90°
b. What type of triangle is it? right
c. What two line segments are the legs? $\overline{XY}, \overline{XZ}$
d. What line segment is the hypotenuse? $\overline{YZ}$
e. Which side of the triangle is the longest? $\overline{YZ}$
f. Which side is opposite ∠X? $\overline{YZ}$

In each triangle shown below, find x.

31.

32.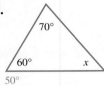

33. In △ABC, m(∠B) = 32° and m(∠C) = 77°. Find m(∠A). 71°

34. For the triangle shown below, find x. Then determine the measure of each angle of the triangle. 18°; 36°, 28°, 116°

35. One base angle of an isosceles triangle measures 65°. Find the measure of the vertex angle. 50°

36. The measure of the vertex angle of an isosceles triangle is 68°. Find the measure of each base angle.

37. Find the measure of ∠C of the triangle shown here. 67°

38. Refer to the figure shown here. Find m(∠C). 83°

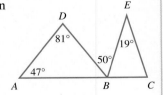

SECTION 9.4 The Pythagorean Theorem

DEFINITIONS AND CONCEPTS

Pythagorean theorem
If a and b are the lengths of the legs of a right triangle and c is the length of the hypotenuse, then

$$a^2 + b^2 = c^2$$

$a^2 + b^2 = c^2$ is called the **Pythagorean equation**.

When we use the Pythagorean theorem to find the length of a side of a right triangle, the solution is sometimes the square root of a number that is not a perfect square. In that case, we can use a calculator to *approximate* the square root.

EXAMPLES

Find the length of the hypotenuse of the right triangle shown here.

We will let $a = 6$ and $b = 8$, and substitute into the Pythagorean equation to find c.

$a^2 + b^2 = c^2$	This is the Pythagorean equation.
$6^2 + 8^2 = c^2$	Substitute 6 for a and 8 for b.
$36 + 64 = c^2$	Evaluate the exponential expressions.
$100 = c^2$	Do the addition.
$c^2 = 100$	Reverse the sides of the equation so that c^2 is on the left.

To find c, we must find a number that, when squared, is 100. There are two such numbers, one positive and one negative; they are the square roots of 100. Since c represents the length of a side of a triangle, c cannot be negative. For this reason, we need only find the positive square root of 100 to get c.

$c = \sqrt{100}$	The symbol $\sqrt{}$ is used to indicate the positive square root of a number.
$c = 10$	Because $10^2 = 100$.

The length of the hypotenuse of the triangle is 10 in.

The lengths of two sides of a right triangle are shown here. Find the missing side length.

We may substitute 9 for either a or b, but 11 must be substituted for the length c of the hypotenuse. If we substitute 9 for a, we can find the unknown side length b as follows.

$a^2 + b^2 = c^2$	This is the Pythagorean equation.
$9^2 + b^2 = 11^2$	Substitute 9 for a and 11 for c.
$81 + b^2 = 121$	Evaluate each exponential expression.
$81 + b^2 - 81 = 121 - 81$	To isolate b^2 on the left side, subtract 81 from both sides.
$b^2 = 40$	

We must find a number that, when squared, is 40. Since b represents the length of a side of a triangle, we consider only the positive square root.

$b = \sqrt{40}$	This is the exact length.

The missing side length is exactly $\sqrt{40}$ feet long. Since 40 is not a perfect square, we use a calculator to approximate $\sqrt{40}$. To the nearest hundredth, the missing side length is 6.32 ft.

The converse of the Pythagorean theorem:
If a triangle has sides of lengths a, b, and c, such that $a^2 + b^2 = c^2$, then the triangle is a right triangle.

Is the triangle shown here a right triangle?
We must substitute the longest side length, 12, for c, because it is the possible hypotenuse. The lengths of 8 and 10 may be substituted for either a or b.

$$a^2 + b^2 = c^2 \quad \text{This is the Pythagorean equation.}$$
$$8^2 + 10^2 \stackrel{?}{=} 12^2 \quad \text{Substitute 8 for a, 10 for b, and 12 for c.}$$
$$64 + 100 \stackrel{?}{=} 144 \quad \text{Evaluate each exponential expression.}$$
$$164 = 144 \quad \text{This is a false statement.}$$

Since $164 \neq 144$, the triangle is not a right triangle.

REVIEW EXERCISES

Refer to the right triangle below.

39. Find c, if $a = 5$ cm and $b = 12$ cm. 13 cm
40. Find c, if $a = 8$ ft and $b = 15$ ft. 17 ft
41. Find a, if $b = 77$ in. and $c = 85$ in. 36 in.
42. Find b, if $a = 21$ ft and $c = 29$ ft. 20 ft

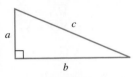

The lengths of two sides of a right triangle are given. Find the missing side length. Give the exact answer and an approximation to the nearest hundredth.

43. $\sqrt{231}$ m ≈ 15.20 m

44. $\sqrt{1{,}300}$ in. ≈ 36.06 in.

45. HIGH-ROPES ADVENTURE COURSES A builder of a high-ropes adventure course wants to secure a pole by attaching a support cable from the anchor stake 55 inches from the pole's base to a point 48 inches up the pole. See the illustration in the next column. How long should the cable be? 73 in.

46. TV SCREENS Find the height of the television screen shown. Give the exact answer and an approximation to the nearest inch. $\sqrt{1{,}023}$ in. ≈ 32 in.

Determine whether each triangle shown here is a right triangle.

47.
not a right triangle

48.
not a right triangle

SECTION 9.5 Congruent Triangles and Similar Triangles

DEFINITIONS AND CONCEPTS	EXAMPLES
If two triangles have the same size and the same shape, they are **congruent triangles**.	$\triangle ABC \cong \triangle DEF$
Corresponding parts of congruent triangles are congruent (have the same measure).	There are six pairs of congruent parts: three pairs of congruent angles and three pairs of congruent sides. • m($\angle A$) = m($\angle D$) • m($\overline{BC}$) = m($\overline{EF}$) • m($\angle B$) = m($\angle E$) • m($\overline{AC}$) = m($\overline{DF}$) • m($\angle C$) = m($\angle F$) • m($\overline{AB}$) = m($\overline{DE}$)
Three ways to show that two triangles are congruent are: **1. The SSS property:** If three sides of one triangle are congruent to three sides of a second triangle, the triangles are congruent.	$\triangle MNO \cong \triangle RST$ by the SSS property. 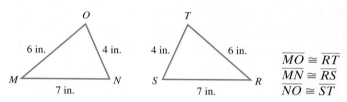 $\overline{MO} \cong \overline{RT}$ $\overline{MN} \cong \overline{RS}$ $\overline{NO} \cong \overline{ST}$
2. The SAS property: If two sides and the angle between them in one triangle are congruent, respectively, to two sides and the angle between them in a second triangle, the triangles are congruent.	$\triangle DEF \cong \triangle XYZ$ by the SAS property. 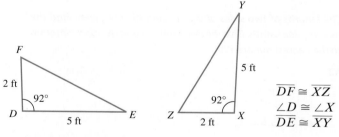 $\overline{DF} \cong \overline{XZ}$ $\angle D \cong \angle X$ $\overline{DE} \cong \overline{XY}$
3. The ASA property: If two angles and the side between them in one triangle are congruent, respectively, to two angles and the side between them in a second triangle, the triangles are congruent.	$\triangle ABC \cong \triangle TUV$ by the ASA property. $\angle A \cong \angle T$ $\overline{AB} \cong \overline{TU}$ $\angle B \cong \angle U$
Similar triangles have the same shape, but not necessarily the same size. We read the symbol ~ as "is similar to." **AAA similarity theorem** If the angles of one triangle are congruent to corresponding angles of another triangle, the triangles are similar.	$\triangle EFG \sim \triangle WXY$ by the AAA similarity theorem. 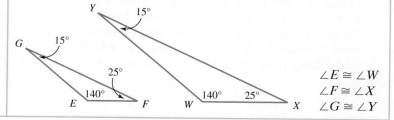 $\angle E \cong \angle W$ $\angle F \cong \angle X$ $\angle G \cong \angle Y$

Property of similar triangles
If two triangles are similar, all pairs of corresponding sides are in proportion.

Similar triangles are determined by the tree and its shadow and the man and his shadow. Since the triangles are similar, the lengths of their corresponding sides are in proportion.

LANDSCAPING A tree casts a shadow 27 feet long at the same time as a man 5 feet tall casts a shadow 3 feet long. Find the height of the tree.

If we let h = the height of the tree, we can find h by solving the following proportion.

The height of the tree → $\dfrac{h}{5} = \dfrac{27}{3}$ ← The length of the tree's shadow
The height of the man →   ← The length of the man's shadow

$3h = 5(27)$ Find each cross product and set them equal.
$3h = 135$ Do the multiplication.
$\dfrac{3h}{3} = \dfrac{135}{3}$ To isolate h, divide both sides by 3.
$h = 45$ Do the division.

The tree is 45 feet tall.

REVIEW EXERCISES

49. Two congruent triangles are shown below. Complete the list of corresponding parts.

  a. $\angle A$ corresponds to $\underline{\angle D}$.
  b. $\angle B$ corresponds to $\underline{\angle E}$.
  c. $\angle C$ corresponds to $\underline{\angle F}$.
  d. $\overline{AC}$ corresponds to $\underline{DF}$.
  e. $\overline{AB}$ corresponds to $\underline{DE}$.
  f. $\overline{BC}$ corresponds to $\underline{EF}$.

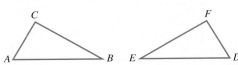

50. Refer to the figure below, where $\triangle ABC \cong \triangle XYZ$.

  a. Find m($\angle X$).   32°
  b. Find m($\angle B$).   61°
  c. Find m($\overline{YZ}$).   6 in.
  d. Find m($\overline{AB}$).   9 in.

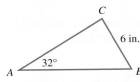

Determine whether the triangles in each pair are congruent. If they are, tell why.

51.   congruent, SSS

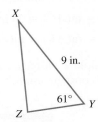

52.   congruent, SAS

53.   not necessarily congruent

54.   congruent, ASA

Determine whether the triangles are similar.

55.   yes   **56.**   yes

57. In the figure below, $\triangle RST \sim \triangle MNO$. Find x and y.   4, 28

58. HEIGHT OF A TREE A tree casts a 26-foot shadow at the same time a woman 5 feet tall casts a 2-foot shadow. What is the height of the tree? (*Hint:* Draw a diagram first and label the side lengths of the similar triangles.)   65 ft

SECTION 9.6 Quadrilaterals and Other Polygons

DEFINITIONS AND CONCEPTS	EXAMPLES
A **quadrilateral** is a polygon with four sides. Use the capital letters that label the vertices of a quadrilateral to name it. A segment that joins two nonconsecutive vertices of a polygon is called a **diagonal** of the polygon.	Quadrilateral *WXYZ* with Diagonal $\overline{WY}$ and Diagonal $\overline{XZ}$
Some special types of quadrilaterals are shown on the right.	**Parallelogram** (Opposite sides parallel) **Rectangle** (Parallelogram with four right angles) **Square** (Rectangle with sides of equal length) **Rhombus** (Parallelogram with sides of equal length) **Trapezoid** (Exactly two sides parallel)
A **rectangle** is a quadrilateral with four right angles. **Properties of rectangles:** 1. All four angles are right angles. 2. Opposite sides are parallel. 3. Opposite sides have equal length. 4. Diagonals have equal length. 5. The diagonals intersect at their midpoints.	Rectangle *ABCD* (30 in., 16 in., 17 in.) 1. m(∠*DAB*) = m(∠*ABC*) = m(∠*BCD*) = m(∠*CDA*) = 90° 2. $\overline{AD} \parallel \overline{BC}$ and $\overline{AB} \parallel \overline{DC}$ 3. m($\overline{AD}$) = 16 in. and m($\overline{DC}$) = 30 in. 4. m($\overline{DB}$) = m($\overline{AC}$) = 34 in. 5. m($\overline{DE}$) = m($\overline{AE}$) = m($\overline{EC}$) = 17 in.
Conditions that a parallelogram must meet to ensure that it is a rectangle: 1. If a parallelogram has one right angle, then the parallelogram is a rectangle. 2. If the diagonals of a parallelogram are congruent, then the parallelogram is a rectangle.	Read Example 2 on page 877 to see how these two conditions are used in construction to "square a foundation." 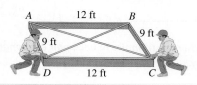

A **trapezoid** is a quadrilateral with exactly two sides parallel.

The parallel sides of a trapezoid are called **bases**. The nonparallel sides are called **legs**.

If the legs (the nonparallel sides) of a trapezoid are of equal length, it is called an **isosceles trapezoid**.

In an isosceles trapezoid, both pairs of **base angles** are congruent.

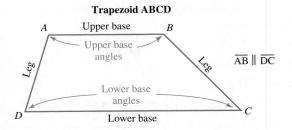

Trapezoid ABCD

$\overline{AB} \parallel \overline{DC}$

The sum S, in degrees, of the measures of the angles of a polygon with n sides is given by the formula

$$S = (n - 2)180°$$

Find the sum of the angle measures of a hexagon.

Since a hexagon has six sides, we will substitute 6 for n in the formula.

$S = (n - 2)180°$	
$S = (6 - 2)180°$	Substitute 6 for n, the number of sides.
$S = (4)180°$	Do the subtraction within the parentheses.
$S = 720°$	Do the multiplication.

The sum of the measures of the angles of a hexagon is 720°.

We can use the formula $S = (n - 2)180°$ to find the number of sides a polygon has.

The sum of the measures of the angles of a polygon is 2,340°. Find the number of sides the polygon has.

$S = (n - 2)180°$	
$2,340° = (n - 2)180°$	Substitute 2,340° for S. Now solve for n.
$2,340° = 180°n - 360°$	Distribute the multiplication by 180°.
$2,340° + 360° = 180°n - 360° + 360°$	Add 360° to both sides.
$2,700° = 180°n$	Do the addition.
$\dfrac{2,700°}{180°} = \dfrac{180°n}{180°}$	Divide both sides by 180°.
$15 = n$	Do the division.

The polygon has 15 sides.

REVIEW EXERCISES

59. Classify each of the following quadrilaterals as a parallelogram, a rectangle, a square, a rhombus, or a trapezoid. Some figures may be correctly classified in more than one way.

a.

trapezoid

b.
square

c.
parallelogram

d.
rectangle

e.
rhombus

f.
rectangle

60. The length of diagonal $\overline{AC}$ of rectangle $ABCD$ shown below is 15 centimeters. Find each measure.

a. m($\overline{BD}$) 15 cm
b. m($\angle 1$) 40°
c. m($\angle 2$) 100°
d. m($\overline{EC}$) 7.5 cm
e. m($\overline{AB}$) 14 cm

61. Refer to rectangle $WXYZ$ below. Tell whether each statement is true or false.

a. m($\overline{WX}$) = m($\overline{ZY}$) true   b. m($\overline{ZE}$) = m($\overline{EX}$) true
c. Triangle WEX is isosceles. true
d. m($\overline{WY}$) = m($\overline{WX}$) false

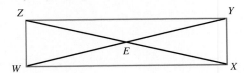

62. Refer to isosceles trapezoid *ABCD* below. Find each measure.

    a. m(∠B) 65°
    b. m(∠C) 115°
    c. m($\overline{CB}$) 4 yd

63. Find the sum of the angle measures of an octagon.
    1,080°

64. The sum of the measures of the angles of a polygon is 3,240°. Find the number of sides the polygon has. 20 sides

SECTION 9.7 Perimeters and Areas of Polygons

DEFINITIONS AND CONCEPTS	EXAMPLES
The **perimeter** of a polygon is the distance around it.	Find the perimeter of the triangle shown below.

Figure	Perimeter Formula
Square	$P = 4s$
Rectangle	$P = 2l + 2w$
Triangle	$P = a + b + c$

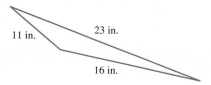

$P = a + b + c$ This is the formula for the perimeter of a triangle.
$P = 11 + 16 + 23$ Substitute 11 for a, 16 for b, and 23 for c.
$P = 50$ Do the addition.

The perimeter of the triangle is 50 inches.

The **area** of a polygon is the measure of the amount of surface it encloses.

Find the area of the triangle shown here.

Figure	Area Formulas
Square	$A = s^2$
Rectangle	$A = lw$
Parallelogram	$A = bh$
Triangle	$A = \frac{1}{2}bh$
Trapezoid	$A = \frac{1}{2}h(b_1 + b_2)$

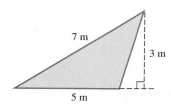

$A = \frac{1}{2}bh$ This is the formula for the area of a triangle.

$A = \frac{1}{2}(5)(3)$ Substitute 5 for b, the length of the base, and 3 for h, the height. Note that the side length 7 m is not used in the calculation.

$A = \frac{1}{2}\left(\frac{5}{1}\right)\left(\frac{3}{1}\right)$ Write 5 as $\frac{5}{1}$ and 3 as $\frac{3}{1}$.

$A = \frac{15}{2}$ Multiply the numerators. Multiply the denominators.

$A = 7.5$ Do the division.

The area of the triangle is 7.5 m².

To find the perimeter or area of a polygon, all the measurements must be in the **same units**. If they are not, use unit conversion factors to change them to the same unit.	To find the perimeter or area of the rectangle shown here, we need to express the length in inches. $$4 \text{ ft} = \frac{4 \text{ ft}}{1} \cdot \frac{12 \text{ in.}}{1 \text{ ft}}$$ Convert 4 feet to inches using a unit conversion factor. $$= 4 \cdot 12 \text{ in.}$$ Remove the common units of feet in the numerator and denominator. The unit of inches remain. $$= 48 \text{ in.}$$ Do the multiplication. The length of the rectangle is 48 inches. Now we can find the perimeter (in inches) or area (in in.²) of the rectangle.
If we know the area of a polygon, we can often use algebra to find an unknown measurement.	The area of the parallelogram shown here is 208 ft². Find the height. $A = bh$ This is the formula for the area of a parallelogram. $208 = 26h$ Substitute 208 for A, the area, and 26 for b, the length of the base. $\dfrac{208}{26} = \dfrac{26h}{26}$ To isolate h, undo the multiplication by 26 by dividing both sides by 26. $8 = h$ Do the division. The height of the parallelogram is 8 feet.
To find the area of an irregular shape, break up the shape into familiar polygons. Find the area of each polygon, and then add the results.	Find the area of the shaded figure shown here. We will find the area of the lower portion of the figure (the trapezoid) and the area of the upper portion (the square) and then add the results. $A_{\text{trapezoid}} = \dfrac{1}{2}h(b_1 + b_2)$ This is the formula for the area of a trapezoid. $A_{\text{trapezoid}} = \dfrac{1}{2}(10)(8 + 18)$ Substitute 8 for b_1, 18 for b_2, and 10 for h. $A_{\text{trapezoid}} = \dfrac{1}{2}(10)(26)$ Do the addition within the parentheses. $A_{\text{trapezoid}} = 130$ Do the multiplication. The area of the trapezoid is 130 cm². $A_{\text{square}} = s^2$ This is the formula for the area of a square. $A_{\text{square}} = 8^2$ Substitute 8 for s. $A_{\text{square}} = 64$ Evaluate the exponential expression. The area of the square is 64 cm².

The total area of the shaded figure is

$A_{\text{total}} = A_{\text{trapezoid}} + A_{\text{square}}$

$A_{\text{total}} = \mathbf{130\ cm^2 + 64\ cm^2}$

$A_{\text{total}} = 194\ cm^2$

The area of the shaded figure is 194 cm².

| To find the area of an irregular shape, we must sometimes use subtraction. | To find the area of the shaded figure below, we subtract the area of the triangle *from* the area of the rectangle. $A_{\text{shaded}} = A_{\text{rectangle}} - A_{\text{triangle}}$ |

REVIEW EXERCISES

65. Find the perimeter of a square with sides 18 inches long. 72 in.

66. Find the perimeter (in inches) of a rectangle that is 7 inches long and 3 feet wide. 86 in.

Find the perimeter of each polygon.

67.

68.

69. The perimeter of an isosceles triangle is 107 feet. If one of the congruent sides is 34 feet long, how long is the base? 39 ft

70. a. How many square feet are there in 1 square yard? 9 ft²

  b. How many square inches are in 1 square foot? 144 in.²

Find the area of each polygon.

71.   9.61 cm²

72.   7,500 ft²

73.   450 ft²

74.   200 in.²

75.   120 cm²

76.   232 ft²

77.   152 ft²

78.

79. The area of a parallelogram is 240 ft². If the length of the base is 30 feet, what is its height? 8 ft

80. The perimeter of a rectangle is 48 mm and its width is 6 mm. Find its length. 18 mm

81. **FENCES** A man wants to enclose a rectangular front yard with chain link that costs $8.50 a foot (the price includes installation). Find the cost of enclosing the yard if its dimensions are 115 ft by 78 ft. $3,281

82. **LAWNS** A family is going to have artificial turf installed in their rectangular backyard that is 36 feet long and 24 feet wide. If the turf costs $48 per *square yard*, and the installation is free, what will this project cost? (Assume no waste.) $4,608

SECTION 9.8 Circles

DEFINITIONS AND CONCEPTS	EXAMPLES
A **circle** is the set of all points in a plane that lie a fixed distance from a point called its **center**. The fixed distance is the circle's **radius**. A **chord** of a circle is a line segment connecting two points on the circle. A **diameter** is a chord that passes through the circle's center. Any part of a circle is called an **arc**. A **semicircle** is an arc of a circle whose endpoints are the endpoints of a diameter.	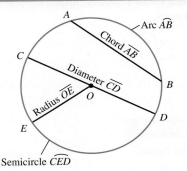
The **circumference** (perimeter) of a circle is given by the formulas $C = \pi D$ or $C = 2\pi r$ where $\pi = 3.14159\ldots$. If an exact answer contains π, we can use 3.14 as an approximation, and complete the calculations by hand. Or, we can use a calculator that has a pi key $\boxed{\pi}$ to find an approximation.	Find the circumference of the circle shown here. Give the exact answer and an approximation. $C = 2\pi r$ This is the formula for the circumference of a circle. $C = 2\pi(8)$ Substitute 8 for r, the radius. $C = 2(8)\pi$ Rewrite the product so that π is the last factor. $C = 16\pi$ Do the first multiplication: 2(8) = 16. This is the exact answer. The circumference of the circle is exactly 16π inches. If we replace π with 3.14, we get an approximation of the circumference. $C = 16\pi$ $C \approx 16(3.14)$ Substitute 3.14 for π. $C \approx 50.24$ Do the multiplication. The circumference of the circle is approximately 50.2 inches. We can also use a calculator to approximate 16π. $C \approx 50.26548246$

Unless otherwise noted, all art on this page is © Cengage Learning.

The **area** of a circle is given by the formula

$$A = \pi r^2$$

Find the area of the circle shown here. Give the exact answer and an approximation to the nearest tenth.

Since the diameter is 28 meters, the radius is half of that, or 14 meters.

$A = \pi r^2$	This is the formula for the area of a circle.
$A = \pi(14)^2$	Substitute 14 for r, the radius of the circle.
$A = \pi(196)$	Evaluate the exponential expression.
$A = 196\pi$	Write the product so that π is the last factor.

The exact area of the circle is 196π m². We can use a calculator to approximate the area.

$A \approx 615.7521601$ Use a calculator to do the multiplication.

To the nearest tenth, the area is 615.8 m².

To find the area of an irregular shape, break it up into familiar figures.

To find the area of the shaded figure shown here, find the area of the triangle and the area of the semicircle, and then add the results.

$$A_{\text{shaded figure}} = A_{\text{triangle}} + A_{\text{semicircle}}$$

REVIEW EXERCISES

83. Refer to the figure.
    a. Name each chord.
       $\overline{CD}, \overline{AB}$
    b. Name each diameter.
       $\overline{AB}$
    c. Name each radius.
       $\overline{OA}, \overline{OC}, \overline{OD}, \overline{OB}$
    d. Name the center. O

84. Find the circumference of a circle with a diameter of 21 feet. Give the exact answer and an approximation to the nearest hundredth. 21π ft ≈ 65.97 ft

85. Find the perimeter of the figure shown below. Round to the nearest tenth. 45.1 cm

86. Find the area of a circle with a diameter of 18 inches. Give the exact answer and an approximation to the nearest hundredth. 81π in.² ≈ 254.47 in.²

87. Find the area of the figure shown in Problem 85. Round to the nearest tenth. 130.3 cm²

88. Find the area of the shaded region shown on the right. Round to the nearest tenth. 6,073.0 in.²

Chapter 9 Summary and Review

SECTION 9.9 Volume

DEFINITIONS AND CONCEPTS

The volume of a figure can be thought of as the number of **cubic units** that will fit within its boundaries.

Two common units of volume are cubic inches (in.³) and cubic centimeters (cm³).

EXAMPLES

1 cubic inch: 1 in.³   1 cubic centimeter: 1 cm³

The **volume** of a solid is a measure of the space it occupies.

Figure	Volume Formula
Cube	$V = s^3$
Rectangular solid	$V = lwh$
Prism	$V = Bh$*
Pyramid	$V = \frac{1}{3}Bh$*
Cylinder	$V = \pi r^2 h$
Cone	$V = \frac{1}{3}\pi r^2 h$
Sphere	$V = \frac{4}{3}\pi r^3$

*B represents the area of the base.

CARRY-ON LUGGAGE The largest carry-on bag that Alaska Airlines allows on board a flight is shown on the right. Find the volume of space that a bag that size occupies.

Height: 17 in.
Width: 10 in.
Length: 24 in.

$V = lwh$ This is the formula for the volume of a rectangular solid.

$V = 24(10)(17)$ Substitute 24 for l, the length, 10 for w, the width, and 17 for h, the height of the bag.

$V = 4,080$ Do the multiplication.

The volume of the space that the bag occupies is 4,080 in.³.

Caution! When finding the volume of a figure, only use the measurements that are called for in the formula. Sometimes a figure may be labeled with measurements that are not used.

Find the volume of the prism shown here.

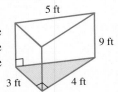

5 ft, 9 ft, 3 ft, 4 ft

The area of the triangular base of the prism is $\frac{1}{2}(3)(4) = 6$ square feet. (The 5-ft measurement is not used.) To find the volume of the prism, proceed as follows:

$V = Bh$ This is the formula for the volume of a prism.

$V = 6(9)$ Substitute 6 for B, the area of the base, and 9 for h, the height.

$V = 54$ Do the multiplication.

The volume of the triangular prism is 54 ft³.

The letter B appears in two of the volume formulas. It represents the area of the base of the figure.

Note that the volume formulas for a pyramid and a cone contain a factor of $\frac{1}{3}$.

Cone: $V = \frac{1}{3}\pi r^2 h$

Pyramid: $V = \frac{1}{3}Bh$

Find the volume of the pyramid shown here.

6 cm, 5 cm, 5 cm

Since the base is a square with each side 5 centimeters long, the area of the base is $5 \cdot 5 = 25$ cm².

$V = \frac{1}{3}Bh$ This is the formula for the volume of a pyramid.

$V = \frac{1}{3}(25)(6)$ Substitute 25 for B, the area of the base, and 6 for h, the height.

$V = 25(2)$ Multiply the first and third factors: $\frac{1}{3}(6) = 2$.

$V = 50$ Complete the multiplication by 25.

The volume of the pyramid is 50 cm³.

Unless otherwise noted, all art on this page is © Cengage Learning.

Note that the volume formulas for a cone, a cylinder, and a sphere contain a factor of π.

Cone: $V = \dfrac{1}{3}\pi r^2 h$

Cylinder: $V = \pi r^2 h$

Sphere: $V = \dfrac{4}{3}\pi r^3$

Find the volume of the cylinder shown here. Give the exact answer and an approximation to the nearest hundredth.

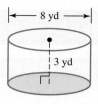

Since a radius is one-half of the diameter of the circular base, $r = \frac{1}{2} \cdot 8$ yd $= 4$ yd. To find the volume of the cylinder, proceed as follows:

$V = \pi r^2 h$ This is the formula for the volume of a cylinder.

$V = \pi(4)^2(3)$ Substitute 4 for r, the radius of the base, and 3 for h, the height.

$V = \pi(16)(3)$ Evaluate the exponential expression.

$V = 48\pi$ Write the product so that π is the last factor.

$V \approx 150.7964474$ Use a calculator to do the multiplication.

The exact volume of the cylinder is 48π yd³. To the nearest hundredth, the volume is 150.80 yd³.

If an exact answer contains π, we can use 3.14 as an approximation, and complete the calculations by hand. Or, we can use a calculator that has a pi key $\boxed{\pi}$ to find an approximation.

Find the volume of the sphere shown here. Give the exact answer and an approximation to the nearest tenth.

$V = \dfrac{4}{3}\pi r^3$ This is the formula for the volume of a sphere.

$V = \dfrac{4}{3}\pi(6)^3$ Substitute 6 for r, the radius of the sphere.

$V = \dfrac{4}{3}\pi(216)$ Evaluate the exponential expression.

$V = \dfrac{864}{3}\pi$ Multiply: $4(216) = 864$.

$V = 288\pi$ Divide: $\frac{864}{3} = 288$.

$V \approx 904.7786842$ Use a calculator to do the multiplication.

The volume of the sphere is exactly 288π ft³. To the nearest tenth, this is 904.8 ft³.

REVIEW EXERCISES

Find the volume of each figure. If an exact answer contains π, approximate to the nearest hundredth.

89.
125 cm³

90.
480 m³

91.
1,728 mm³

92.
$\frac{500}{3}\pi$ in.³ ≈ 523.60 in.³

93.
250π in.³ ≈ 785.40 in.³

94.
2,000 yd³

95.
2,940 m³

96.
$\frac{1,024}{3}\pi$ in.³ ≈ 1,072.33 in.³

97. FARMING Find the volume of the corn silo shown below. Round to the nearest one cubic foot.
1,518 ft³

98. WAFFLE CONES Find the volume of the ice cream cone shown above. Give the exact answer and an approximation to the nearest tenth.
3.125π in.³ ≈ 9.8 in.³

99. How many cubic inches are there in 1 cubic foot?
1,728 in.³

100. How many cubic feet are there in 2 cubic yards?
54 ft³

CHAPTER 9 TEST

1. Estimate each angle measure. Then tell whether it is an acute, right, obtuse, or straight angle.

   a. 135°, obtuse

   b. 90°, right

   c. 40°, acute

   d. 180°, straight

2. Fill in the blanks.
   a. If $\angle ABC \cong \angle DEF$, then the angles have the same __measure__.
   b. Two congruent segments have the same __length__.
   c. Two different points determine one __line__.
   d. Two angles are called __complementary__ if the sum of their measures is 90°.

3. Refer to the figure below. What is the midpoint of $\overline{BE}$? D

4. Refer to the figure below and tell whether each statement is true or false.
   a. $\angle AGF$ and $\angle BGC$ are vertical angles. false
   b. $\angle EGF$ and $\angle DGE$ are adjacent angles. true
   c. $m(\angle AGB) = m(\angle EGD)$. true
   d. $\angle CGD$ and $\angle DGF$ are supplementary angles. true
   e. $\angle EGD$ and $\angle AGB$ are complementary angles. false

   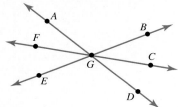

5. Find x. Then find $m(\angle ABD)$ and $m(\angle CBE)$.   20°; 60°, 60°

6. Find the supplement of a 47° angle. 133°

7. Refer to the figure below. Fill in the blanks.
   a. l_1 intersects two coplanar lines. It is called a __transversal__.
   b. $\angle 4$ and $\angle 6$ are alternate interior angles.
   c. $\angle 3$ and $\angle 7$ are corresponding angles.

8. In the figure below, $l_1 \parallel l_2$ and $m(\angle 2) = 25°$. Find the measures of the other numbered angles.

   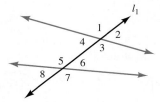

   $m(\angle 1) = 155°$, $m(\angle 3) = 155°$, $m(\angle 4) = 25°$, $m(\angle 5) = 25°$, $m(\angle 6) = 155°$, $m(\angle 7) = 25°$, $m(\angle 8) = 155°$

9. In the figure below, $l_1 \parallel l_2$. Find x. Then determine the measure of each angle that is labeled in the figure.   50°; 110°, 70°

   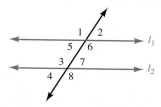

10. For each polygon, give the number of sides it has, tell its name, and then give the number of vertices it has.

a.
8, octagon, 8

b.
5, pentagon, 5

c.
6, hexagon, 6

d.
4, quadrilateral, 4

11. Classify each triangle as an equilateral triangle, an isosceles triangle, or a scalene triangle.

a.
isosceles

b.
scalene

c.
equilateral

d.
isosceles

12. Find x. 70°

13. The measure of the vertex angle of an isosceles triangle is 12°. Find the measure of each base angle. 84°

14. Refer to rectangle $EFGH$ shown below.
a. Find m($\overline{HG}$). 12
b. Find m($\overline{FH}$). 13
c. Find m($\angle FGH$). 90°
d. Find m($\overline{EH}$). 5

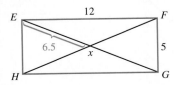

15. Refer to isosceles trapezoid $QRST$ shown below.
a. Find m($\overline{RS}$). 10
b. Find x. 65°
c. Find y. 115°
d. Find z. 115°

16. Find the sum of the measures of the angles of a decagon. 1,440°

17. Find the perimeter of the figure shown below. 188 in.

18. The perimeter of an equilateral triangle is 45.6 m. Find the length of each side. 15.2 m

19. Find the area of the shaded part of the figure shown below. 360 cm²

20. DECORATING A patio has the shape of a trapezoid, as shown on the right. If indoor/outdoor carpeting sells for $18 a *square yard* installed, how much will it cost to carpet the patio? $864

21. How many square inches are in one square foot? 144 in.²

22. Find the area of the rectangle shown below in square inches. 120 in.²

    10 ft
    1 in.

23. Refer to the figure below, where O is the center of the circle.
    a. Name each chord. $\overline{RS}, \overline{XY}$
    b. Name a diameter. $\overline{XY}$
    c. Name each radius. $\overline{OX}, \overline{OR}, \overline{OS}, \overline{OY}$

    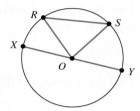

24. Fill in the blank: If C is the circumference of a circle and D is the length of its diameter, then $\frac{C}{D} = \pi$.

In Problems 25–27, when appropriate, give the exact answer and an approximation to the nearest tenth.

25. Find the circumference of a circle with a diameter of 21 feet. 21π ft ≈ 66.0 ft

26. Find the perimeter of the figure shown below. Assume that the arcs are semicircles. $(40 + 12\pi)$ ft ≈ 77.7 ft

27. HISTORY Stonehenge is a prehistoric monument in England, believed to have been built by the Druids. The site, 30 meters in diameter, consists of a circular arrangement of stones, as shown below. What area does the monument cover? 225π m² ≈ 706.9 m²

28. See the figure below, where $\triangle MNO \cong \triangle RST$. Name the six corresponding parts of the congruent triangles.

    $\angle M \cong \underline{\angle R}$ $\overline{MO} \cong \underline{\overline{RT}}$
    $\angle N \cong \underline{\angle S}$ $\overline{MN} \cong \underline{\overline{RS}}$
    $\angle O \cong \underline{\angle T}$ $\overline{NO} \cong \underline{\overline{ST}}$

29. Tell whether each pair of triangles are congruent. If they are, tell why.

    a. congruent, SSS

    b. congruent, ASA
       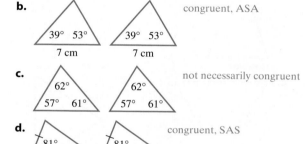

    c. not necessarily congruent

    d. congruent, SAS

30. Refer to the figure below, in which $\triangle ABC \cong \triangle DEF$.
    a. Find m($\overline{DE}$). 8 in.
    b. Find m($\angle E$). 50°

    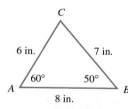

31. Tell whether the triangles in each pair are similar.

    a. yes

    b. yes

32. Refer to the triangles below. The units are meters.
   a. Find x. 6 m
   b. Find y. 12 m

33. SHADOWS If a tree casts a 7-foot shadow at the same time as a man 6 feet tall casts a 2-foot shadow, how tall is the tree? 21 ft

34. Refer to the right triangle below. Find the missing side length. Approximate any exact answers that contain a square root to the nearest tenth.
   a. Find c if $a = 10$ cm and $b = 24$ cm. 26 cm
   b. Find b if $a = 6$ in. and $c = 8$ in. $\sqrt{28}$ in. ≈ 5.3 in.

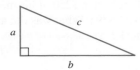

35. TELEVISIONS To the nearest tenth of an inch, what is the diagonal measurement of the television screen shown below? $\sqrt{986}$ in. ≈ 31.4 in.

36. How many cubic inches are there in 1 cubic foot? 1,728 in.³

Find the volume of each figure. Give the exact answer and an approximation to the nearest hundredth if an answer contains π.

37.
216 m³

38.
480 m³

39.
1,296π in.³ ≈ 4,071.50 in.³

40.
600 in.³

41.
1,890 ft³

42.
63π yd³ ≈ 197.92 yd³

43.
400 mi³

44.
$\frac{256}{3}\pi$ in.³ ≈ 268.08 in.³

45. FARMING A silo is used to store wheat and corn. Find the volume of the silo shown below. Give the exact answer and an approximation to the nearest cubic foot. 11,250π ft³ ≈ 35,343 ft³

46. Give a real-life example in which the concept of perimeter is used. Do the same for area and for volume. Be sure to discuss the type of units used in each case.

Unless otherwise noted, all art on this page is © Cengage Learning.

CHAPTERS 1–9 CUMULATIVE REVIEW

1. Write 104,052,005 in words. [Section 1.1] one hundred four million, fifty-two thousand, five

2. Add: $257 + 99{,}085 + 4{,}101 + 33$ [Section 1.2] 103,476

3. Multiply: $\begin{array}{r}196\\ \times\ 78\\ \hline 15{,}288\end{array}$ [Section 1.3]

4. Divide: $34\overline{)2{,}006}$ [Section 1.4] 59

5. a. Find the factors of 32. [Section 1.5] 1, 2, 4, 8, 16, 32
   b. Find the prime factorization of 140. [Section 1.5] $2^2 \cdot 5 \cdot 7$

6. Find the LCM and the GCF of 35 and 45. [Section 1.6] 315; 5

7. Evaluate: $46 + 3[5^2 - 4(9 - 5)]$ [Section 1.7] 73

8. Solve each equation and check the result.
   a. $x + 33 = 91$ [Section 1.8] 58
   b. $24m = 312$ [Section 1.9] 13

9. Is the statement $-72 > -73$ true or false? [Section 2.1] true

10. Perform each operation, if possible.
    a. $-17 + 8$ [Section 2.2] -9
    b. $-14 - (-2)$ [Section 2.3] -12
    c. $-3(72)$ [Section 2.4] -216
    d. $\dfrac{-60}{0}$ [Section 2.5] undefined

11. Evaluate: $\dfrac{2 + 3[5 - (1 - 10)]}{|2[(-2)^3 + 2] + 10|}$ [Section 2.6] 22

12. Solve $-2(5) = \dfrac{y}{-3} + 3$ and check the result. [Section 2.7] 39

Write an equation and solve it to answer the following question. [Section 2.7]

13. BANKING After she made deposits of $55 and $80, a student's account was still $82 overdrawn. What was her checking account balance before the deposits? Her checking account balance before the deposit was $-\$217$.

14. Translate to mathematical symbols: 18 less than twice the width w. [Section 3.1] $2w - 18$

15. Evaluate $b^2 - 4ac$ for $a = -1$, $b = -5$, and $c = -2$. [Section 3.2] 17

16. Simplify: $-4(6u)(-2)$ [Section 3.3] $48u$

17. PING-PONG Write an algebraic expression that represents the perimeter (in feet) of the Ping-Pong table shown here. [Section 3.4] $4x + 8$

18. Solve $6(2j + 6) + 4j = 4(j - 30)$ [Section 3.5] -13

Write an equation and solve it to answer the following question. [Section 3.6]

19. TAX REFUNDS After receiving their tax refund, a husband and wife split the refunded money equally. The husband then gave $250 of his money to charity, leaving him with $395. What was the amount of their tax refund check? The amount of the tax refund check was $1,290.

20. Simplify: $\dfrac{60x^5}{108x^3}$ [Section 4.1] $\dfrac{5x^2}{9}$

Perform each operation. Simplify, if possible.

21. $\dfrac{4}{5} \cdot \dfrac{15}{16}$ [Section 4.2] $\dfrac{3}{4}$

22. $-\dfrac{7a}{4b^2} \div \left(-\dfrac{21a}{8b}\right)$ [Section 4.3] $\dfrac{2}{3b}$

23. Subtract $\dfrac{3}{4}$ from $\dfrac{4}{n}$. [Section 4.4] $\dfrac{16 - 3n}{4n}$

24. $\dfrac{7}{10} + \dfrac{3}{14}$ [Section 4.4] $\dfrac{32}{35}$

25. TIRES The road surface "footprint" of a sport truck tire is approximately rectangular, as shown here. If the width of the tire is 6 inches, what is the area of the tire "footprint"? [Section 4.5] 45 in.²

26. Subtract: $140\dfrac{5}{6} - 129\dfrac{4}{5}$ [Section 4.6] $11\dfrac{1}{30}$

27. Simplify the complex fraction: $\dfrac{-\dfrac{1}{2} + \dfrac{7}{8}}{\dfrac{3}{4} - \dfrac{1}{2}}$ [Section 4.7] $\dfrac{3}{2}$

Unless otherwise noted, all art on this page is © Cengage Learning.

Write an equation and solve it to answer the following question. [Section 4.8]

28. CAR REPAIRS Two-fifths of the cars that an automobile repair shop serviced last year had transmission problems. If the shop repaired 140 transmissions, how many cars did they service that year? The shop serviced 350 cars last year.

Solve each equation and check the result.

29. $-\dfrac{5}{6}x = 10$ [Section 4.8] -12

30. $\dfrac{12}{5} + \dfrac{n}{2} = \dfrac{n}{10}$ [Section 4.8] -6

31. Round 9.50966 to the nearest thousandth. [Section 5.1] 9.510

32. Use a check to determine whether the subtraction shown below is correct. [Section 5.2]

$$\begin{array}{r} 451.3 \\ -89.8 \\ \hline 361.5 \end{array}$$ It is correct: $361.5 + 89.8 = 451.3$.

33. GEOMETRY The length of a rectangle is 2.72 feet and the width is 3.81 feet. Find its area. [Section 5.3] 10.3632 ft²

34. Divide: $\dfrac{14.637}{-4.1}$ [Section 5.4] -3.57

35. Write $\dfrac{11}{12}$ as a decimal. Use an overbar in your answer. [Section 5.5] $0.91\overline{6}$

36. Evaluate: $-4\sqrt{81} + 2\sqrt{4}$ [Section 5.6] -32

37. Solve $3(y - 1.1) + 3.2 = 2.3$ and check the result. [Section 5.7] 0.8

38. Write the ratio *21 inches to 3 feet* in simplest form. [Section 6.1] $\dfrac{7}{12}$

39. MAKING BROWNIES A recipe for brownies calls for 4 eggs and $1\frac{1}{2}$ cups of flour. If the recipe makes 15 brownies, how many cups of flour will be needed to make 260 brownies? [Section 6.2] 26 cups

40. Make each conversion.
  a. 50 pounds to ounces [Section 6.3] 800 oz
  b. 500 milliliters to liters [Section 6.4] 0.5 L
  c. 35 centimeters to inches [Section 6.5] about 13.7 in.

Unless otherwise noted, all art on this page is © Cengage Learning.

41. Write $\dfrac{27}{50}$ as a decimal and as a percent. [Section 7.1] 0.54, 54%

42. 36 is 24% of what number? [Section 7.2] 150

43. Find the total cost of a $196.56 purchase if the sales tax rate is 4.3%. [Section 7.3] $205.01

44. Estimate 20% of 6,700. [Section 7.4] 1,400

45. Find the simple interest on a loan of $1,250 borrowed at 10% for 3 months. [Section 7.5] $31.25

46. Find the mean, median, mode, and range of the values listed below. [Section 8.2]

73 5 38 45 11 36 27 35 45

mean: 35, median: 36, mode: 45, range: 68

47. Find the coordinates of each point shown in red. [Section 8.3]
$A(2, 3)$, $B(-3, 4)$, $C(-3, -4)$, $D(4, -4)$, $E(0, 0)$, $F(4, 0)$, $G(-4, 0)$, $H(1.5, -2.5)$ or $H\left(1\dfrac{1}{2}, -2\dfrac{1}{2}\right)$

48. Is $(9, -3)$ a solution of $x + 5y = -7$? [Section 8.3] no

Graph each equation. [Section 8.4]

49. $y = 3x + 1$

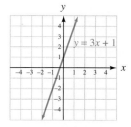

50. $5x + 15y = -15$

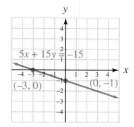

51. First find x. Then find $m(\angle ABD)$ and $m(\angle DBE)$. [Section 9.1] 12°, 80°, 100°

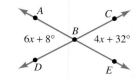

52. The measure of one base angle of an isosceles triangle is 76°. Find the measure of the vertex angle. [Section 9.3] 28°

53. The length of the hypotenuse of a right triangle is 41 yards and the length of one leg is 9 yards. Find the length of the other leg. [Section 9.4] 40 yards

54. Find the area of a circle with diameter 10 centimeters. Round to the nearest tenth. [Section 9.8] 78.5 cm²

10 Exponents and Polynomials

10.1 Multiplication Rules for Exponents
10.2 Introduction to Polynomials
10.3 Adding and Subtracting Polynomials
10.4 Multiplying Polynomials
Chapter Summary and Review
Chapter Test
Cumulative Review

from Campus to Careers
Police Officer

People depend on the police to protect their lives and property. The job can be dangerous because police officers must arrest suspects and respond to emergencies. The daily activities of police officers can vary greatly depending on their specialty, such as patrol officer, game warden, or detective. Regardless of their duties, they must write reports and maintain records that will be needed if they testify in court.

In **Problem 51** of **Study Set 10.2,** you will see how police officers can compute the stopping distance of a car.

JOB TITLE: Police Officer
EDUCATION: For many departments, two years of college or a college degree may be required. Physical education courses are helpful. Foreign language skills are desirable.
JOB OUTLOOK: Fair—employment opportunities are expected to grow 7 percent through 2020.
ANNUAL EARNINGS: Median annual salary in 2011 was $54,230. Earnings often exceed their salary because of payments for overtime.
FOR MORE INFORMATION:
www.bls.gov/oes/current/oes333051.htm

Objectives

1. Identify bases and exponents.
2. Multiply exponential expressions that have like bases.
3. Raise exponential expressions to a power.
4. Find powers of products.

SECTION 10.1
Multiplication Rules for Exponents

In this section, we will use the definition of exponent to develop some rules for simplifying expressions that contain exponents.

1 Identify bases and exponents.

Recall that an **exponent** indicates repeated multiplication. It indicates how many times the base is used as a factor. For example, 3^5 represents the product of five 3's.

$$3^5 = 3 \cdot 3 \cdot 3 \cdot 3 \cdot 3$$

where the Exponent is 5, the Base is 3, and there are 5 factors of 3.

In general, we have the following definition.

Natural-Number Exponents

A natural-number* exponent tells how many times its base is to be used as a factor.

For any number x and any natural number n,

$$x^n = \underbrace{x \cdot x \cdot x \cdot \cdots \cdot x}_{n \text{ factors of } x}$$

*The set of natural numbers is $\{1, 2, 3, 4, 5, \ldots\}$.

Expressions of the form x^n are called **exponential expressions**. The base of an exponential expression can be a number, a variable, or a combination of numbers and variables. Some examples are:

$10^5 = 10 \cdot 10 \cdot 10 \cdot 10 \cdot 10$ The base is 10. The exponent is 5. Read as "10 to the fifth power."

$y^2 = y \cdot y$ The base is y. The exponent is 2. Read as "y squared."

$(-2s)^3 = (-2s)(-2s)(-2s)$ The base is −2s. The exponent is 3. Read as "negative 2s raised to the third power" or "negative 2s cubed."

$-8^4 = -(8 \cdot 8 \cdot 8 \cdot 8)$ Since the − sign is not written within parentheses, the base is 8. The exponent is 4. Read as "the opposite (or the negative) of 8 to the fourth power."

When an exponent is 1, it is usually not written. For example, $4 = 4^1$ and $x = x^1$.

Caution! Bases that contain a − sign *must* be written within parentheses.

$$(-2s)^3 \leftarrow \text{Exponent}$$

with Base indicated as $-2s$.

EXAMPLE 1
Identify the base and the exponent in each expression:
a. 8^5 **b.** $7a^3$ **c.** $(7a)^3$

Strategy To identify the base and exponent, we will look for the form ■■.

WHY The exponent is the small raised number to the right of the base.

Solution
a. In 8^5, the base is 8 and the exponent is 5.
b. $7a^3$ means $7 \cdot a^3$. Thus, the base is a, not $7a$. The exponent is 3.
c. Because of the parentheses in $(7a)^3$, the base is $7a$ and the exponent is 3.

Self Check 1
Identify the base and the exponent:
a. $3y^4$ base: y, exponent: 4
b. $(3y)^4$ base: $3y$, exponent: 4
Now Try Problems 13 and 17

Teaching Example 1 Identify the base and the exponent:
a. 6^4 **b.** $5x^2$ **c.** $(5x)^2$
Answers:
a. base: 6, exponent: 4
b. base: x, exponent: 2
c. base: $5x$, exponent: 2

EXAMPLE 2
Write each expression in an equivalent form using an exponent: **a.** $b \cdot b \cdot b \cdot b$ **b.** $5 \cdot t \cdot t \cdot t$

Strategy We will look for repeated factors and count the number of times each appears.

WHY We can use an exponent to represent repeated multiplication.

Solution
a. Since there are four repeated factors of b in $b \cdot b \cdot b \cdot b$, the expression can be written as b^4.
b. Since there are three repeated factors of t in $5 \cdot t \cdot t \cdot t$, the expression can be written as $5t^3$.

Self Check 2
Write as an exponential expression: $(x + y)^5$
$(x + y)(x + y)(x + y)(x + y)(x + y)$
Now Try Problems 25 and 29

Teaching Example 2 Write each expression in an equivalent form using an exponent:
a. $m \cdot m \cdot m \cdot m \cdot m$
b. $2 \cdot d \cdot d \cdot d$
Answers:
a. m^5 b. $2d^3$

2 Multiply exponential expressions that have like bases.

To develop a rule for multiplying exponential expressions that have the same base, we consider the product $6^2 \cdot 6^3$. Since 6^2 means that 6 is to be used as a factor two times, and 6^3 means that 6 is to be used as a factor three times, we have

$$6^2 \cdot 6^3 = \underbrace{6 \cdot 6}_{\text{2 factors of 6}} \cdot \underbrace{6 \cdot 6 \cdot 6}_{\text{3 factors of 6}}$$

$$= \underbrace{6 \cdot 6 \cdot 6 \cdot 6 \cdot 6}_{\text{5 factors of 6}}$$

$$= 6^5$$

We can quickly find this result if we keep the common base 6 and add the exponents on 6^2 and 6^3.

$$6^2 \cdot 6^3 = 6^{2+3} = 6^5$$

This example illustrates the following rule for exponents.

Product Rule for Exponents

To multiply exponential expressions that have the same base, keep the common base and add the exponents.

For any number x and any natural numbers m and n,

$$x^m \cdot x^n = x^{m+n}$$

Read as "x to the mth power times x to the nth power equals x to the m plus nth power."

Self Check 3
Simplify:
a. $7^8(7^7)$ 7^{15}
b. $x^2 x^3 x$ x^6
c. $(y-1)^5(y-1)^5$ $(y-1)^{10}$
d. $(s^4 t^3)(s^4 t^4)$ $s^8 t^7$

Now Try Problems 33, 35, and 37

Teaching Example 3 Simplify:
a. $8^3(8^6)$
b. $x^2 \cdot x^4$
c. $y^3 y^8 y$
d. $(a^2 b^3)(a^3 b^9)$
Answers:
a. 8^9 b. x^6 c. y^{12} d. $a^5 b^{12}$

EXAMPLE 3 Simplify:
a. $9^5(9^6)$ b. $x^3 \cdot x^4$ c. $y^2 y^4 y$ d. $(c^2 d^3)(c^4 d^5)$

Strategy In each case, we want to write an equivalent expression using one base and one exponent. We will use the product rule for exponents to do this.

WHY The product rule for exponents is used to multiply exponential expressions that have the same base.

Solution

a. $9^5(9^6) = 9^{5+6} = 9^{11}$ Keep the common base, 9, and add the exponents. Since 9^{11} is a very large number, we will leave the answer in this form. We won't evaluate it.

Caution! Don't make the mistake of multiplying the bases when using the product rule. Keep the *same* base.
$$9^5(9^6) \neq 81^{11}$$

b. $x^3 \cdot x^4 = x^{3+4} = x^7$ Keep the common base, x, and add the exponents.

c. $y^2 y^4 y = y^2 y^4 y^1$ Write y as y^1.
$= y^{2+4+1}$ Keep the common base, y, and add the exponents.
$= y^7$

d. $(c^2 d^3)(c^4 d^5) = (c^2 c^4)(d^3 d^5)$ Use the commutative and associative properties of multiplication to group like bases together.
$= (c^{2+4})(d^{3+5})$ Keep the common base, c, and add the exponents. Keep the common base, d, and add the exponents.
$= c^6 d^8$

Caution! We cannot use the product rule to simplify expressions like $3^2 \cdot 2^3$, where the bases are not the same. However, we can simplify this expression by doing the arithmetic:
$$3^2 \cdot 2^3 = 9 \cdot 8 = 72 \qquad 3^2 = 3 \cdot 3 = 9 \text{ and } 2^3 = 2 \cdot 2 \cdot 2 = 8.$$

Recall that *like terms* are terms with exactly the same variables raised to exactly the same powers. To add or subtract exponential expressions, they must be like terms. To multiply exponential expressions, only the bases need to be the same.

$x^5 + x^2$ These are not like terms; the exponents are different. We cannot add.

$x^2 + x^2 = 2x^2$ These are like terms; we can add. Recall that $x^2 = 1x^2$.

$x^5 \cdot x^2 = x^7$ The bases are the same; we can multiply.

3 Raise exponential expressions to a power.

To develop another rule for exponents, we consider $(5^3)^4$. Here, an exponential expression, 5^3, is raised to a power. Since 5^3 is the base and 4 is the exponent, $(5^3)^4$ can be written as $5^3 \cdot 5^3 \cdot 5^3 \cdot 5^3$. Because each of the four factors of 5^3 contains three factors of 5, there are $4 \cdot 3$ or 12 factors of 5.

$$(5^3)^4 = 5^3 \cdot 5^3 \cdot 5^3 \cdot 5^3 = \underbrace{5 \cdot 5 \cdot 5 \cdot 5 \cdot 5 \cdot 5 \cdot 5 \cdot 5 \cdot 5 \cdot 5 \cdot 5 \cdot 5}_{\text{12 factors of 5}} = 5^{12}$$

We can quickly find this result if we keep the common base of 5 and multiply the exponents.

$$(5^3)^4 = 5^{3 \cdot 4} = 5^{12}$$

This example illustrates the following rule for exponents.

> **Power Rule for Exponents**
>
> To raise an exponential expression to a power, keep the base and multiply the exponents.
>
> For any number x and any natural numbers m and n,
>
> $$(x^m)^n = x^{m \cdot n} = x^{mn}$$ Read as "the quantity of x to the mth power raised to the nth power equals x to the mnth power."

> **The Language of Algebra** An exponential expression raised to a power, such as $(5^3)^4$, is also called a *power of a power*.

EXAMPLE 4 Simplify: **a.** $(2^3)^7$ **b.** $[(-6)^2]^5$ **c.** $(z^8)^8$

Strategy In each case, we want to write an equivalent expression using one base and one exponent. We will use the power rule for exponents to do this.

WHY Each expression is a power of a power.

Solution

a. $(2^3)^7 = 2^{3 \cdot 7} = 2^{21}$ Keep the base, 2, and multiply the exponents. Since 2^{21} is a very large number, we will leave the answer in this form.

b. $[(-6)^2]^5 = (-6)^{2 \cdot 5} = (-6)^{10}$ Keep the base, -6, and multiply the exponents. Since $(-6)^{10}$ is a very large number, we will leave the answer in this form.

c. $(z^8)^8 = z^{8 \cdot 8} = z^{64}$ Keep the base, z, and multiply the exponents.

Self Check 4

Simplify:
a. $(4^6)^5$ 4^{30}
b. $(y^5)^2$ y^{10}

Now Try Problems 49, 51, and 53

Teaching Example 4 Simplify:
a. $(7^2)^4$
b. $[(-8)^6]^5$
c. $(r^{10})^3$
Answers:
a. 7^8 **b.** $(-8)^{30}$ **c.** r^{30}

EXAMPLE 5 Simplify: **a.** $(x^2 x^5)^2$ **b.** $(z^2)^4 (z^3)^3$

Strategy In each case, we want to write an equivalent expression using one base and one exponent. We will use the product and power rules for exponents to do this.

WHY The expressions involve multiplication of exponential expressions that have the same base and they involve powers of powers.

Solution

a. $(x^2 x^5)^2 = (x^7)^2$ Within the parentheses, keep the common base, x, and add the exponents: $2 + 5 = 7$.

$ = x^{14}$ Keep the base, x, and multiply the exponents: $7 \cdot 2 = 14$.

b. $(z^2)^4 (z^3)^3 = z^8 z^9$ For each power of z raised to a power, keep the base and multiply the exponents: $2 \cdot 4 = 8$ and $3 \cdot 3 = 9$.

$ = z^{17}$ Keep the common base, z, and add the exponents: $8 + 9 = 17$.

Self Check 5

Simplify:
a. $(a^4 a^3)^3$ a^{21}
b. $(a^3)^3 (a^4)^2$ a^{17}

Now Try Problems 57 and 61

Teaching Example 5
Simplify: **a.** $(y^3 y^5)^4$ **b.** $(m^2)^5 (m^3)^2$
Answers:
a. y^{32} **b.** m^{16}

4 Find powers of products.

To develop another rule for exponents, we consider the expression $(2x)^3$, which is a *power of the product* of 2 and x.

$$(2x)^3 = 2x \cdot 2x \cdot 2x \qquad \text{Write the base 2x as a factor 3 times.}$$
$$= (2 \cdot 2 \cdot 2)(x \cdot x \cdot x) \qquad \text{Change the order of the factors and group like bases.}$$
$$= 2^3 x^3 \qquad \text{Write each product of repeated factors in exponential form.}$$
$$= 8x^3 \qquad \text{Evaluate: } 2^3 = 8.$$

This example illustrates the following rule for exponents.

> **Power of a Product**
>
> To raise a product to a power, raise each factor of the product to that power. For any numbers x and y, and any natural number n,
>
> $$(xy)^n = x^n y^n$$

Self Check 6

Simplify:
a. $(2t)^4$ $16t^4$
b. $(c^3 d^4)^6$ $c^{18} d^{24}$

Now Try Problems 65 and 69

Teaching Example 6
Simplify: a. $(2n)^5$ b. $(s^6 t^8)^9$
Answers:
a. $32n^5$ b. $s^{54} t^{72}$

EXAMPLE 6 Simplify: a. $(3c)^4$ b. $(x^2 y^3)^5$

Strategy In each case, we want to write the expression in an equivalent form in which each base is raised to a single power. We will use the power of a product rule for exponents to do this.

WHY Within each set of parentheses is a product, and each of those products is raised to a power.

Solution

a. $(3c)^4 = 3^4 c^4$ Raise each factor of the product 3c to the 4th power.
   $ = 81 c^4$ Evaluate: $3^4 = 81$.

b. $(x^2 y^3)^5 = (x^2)^5 (y^3)^5$ Raise each factor of the product $x^2 y^3$ to the 5th power.
   $ = x^{10} y^{15}$ For each power of a power, keep each base, x and y, and multiply the exponents: $2 \cdot 5 = 10$ and $3 \cdot 5 = 15$.

Self Check 7

Simplify: $(4y^3)^2 (3y^4)^3$ $432 y^{18}$

Now Try Problem 73

Teaching Example 7
Simplify: $(8m^4)^2 (2m^5)^3$
Answer:
$512 m^{23}$

EXAMPLE 7 Simplify: $(2a^2)^2 (4a^3)^3$

Strategy We want to write an equivalent expression using one base and one exponent. We will begin the process by using the power of a product rule for exponents.

WHY Within each set of parentheses is a product, and each product is raised to a power.

Solution

$$(2a^2)^2 (4a^3)^3 = 2^2 (a^2)^2 \cdot 4^3 (a^3)^3 \qquad \text{Raise each factor of the product } 2a^2 \text{ to the 2nd power. Raise each factor of the product } 4a^3 \text{ to the 3rd power.}$$
$$= 4a^4 \cdot 64 a^9 \qquad \text{Evaluate: } 2^2 = 4 \text{ and } 4^3 = 64. \text{ For each power of a power, keep each base and multiply the exponents: } 2 \cdot 2 = 4 \text{ and } 3 \cdot 3 = 9.$$
$$= (4 \cdot 64)(a^4 \cdot a^9) \qquad \text{Group the numerical factors. Group the factors that have the same base.}$$
$$= 256 a^{13} \qquad \text{Do the multiplication: } 4 \cdot 64 = 256. \text{ Keep the common base } a \text{ and add the exponents: } 4 + 9 = 13.$$

$$\begin{array}{r} 1 \\ 64 \\ \times\ 4 \\ \hline 256 \end{array}$$

The rules for natural-number exponents are summarized as follows.

> **Rules for Exponents**
>
> If m and n represent natural numbers and there are no divisions by zero, then
>
> | Exponent of 1 | Product rule | Power rule |
> |---|---|---|
> | $x^1 = x$ | $x^m x^n = x^{m+n}$ | $(x^m)^n = x^{mn}$ |
>
> **Power of a product**
> $$(xy)^n = x^n y^n$$

> **ANSWERS TO SELF CHECKS**
>
> **1. a.** base: y, exponent: 4   **b.** base: $3y$, exponent: 4   **2.** $(x+y)^5$ **3. a.** 7^{15} **b.** x^6
> **c.** $(y-1)^{10}$ **d.** $s^8 t^7$ **4. a.** 4^{30} **b.** y^{10} **5. a.** a^{21} **b.** a^{17} **6. a.** $16t^4$ **b.** $c^{18}d^{24}$
> **7.** $432y^{18}$

SECTION 10.1 STUDY SET

VOCABULARY

Fill in the blank.

1. Expressions such as x^4, 10^3, and $(5t)^2$ are called __exponential__ expressions.

2. Match each expression with the proper description.

   $(a^4 b^2)^5 \qquad (a^8)^4 \qquad a^5 \cdot a^3$

   a. Product of exponential expressions with the same base $a^5 \cdot a^3$

   b. Power of an exponential expression $(a^8)^4$

   c. Power of a product $(a^4 b^2)^5$

CONCEPTS

Fill in the blanks.

3. **a.** $(3x)^4 = $ $3x \cdot 3x \cdot 3x \cdot 3x$

   b. $(-5y)(-5y)(-5y) = $ $(-5y)^3$

4. **a.** $x = x^{\,1}$ **b.** $x^m x^n = x^{m+n}$

   c. $(xy)^n = x^n y^n$ **d.** $(a^b)^c = a^{bc}$

5. To simplify each expression, determine whether you add, subtract, multiply, or divide the exponents.

   a. $b^6 \cdot b^9$ add

   b. $(n^8)^4$ multiply

   c. $(a^4 b^2)^5$ multiply

6. To simplify $(2y^3 z^2)^4$, what factors within the parentheses must be raised to the fourth power? $2, y^3, z^2$

Simplify each expression, if possible.

7. **a.** $x^2 + x^2$ $2x^2$ **b.** $x^2 \cdot x^2$ x^4

8. **a.** $x^2 + x$ doesn't simplify   **b.** $x^2 \cdot x$ x^3

9. **a.** $x^3 - x^2$ doesn't simplify   **b.** $x^3 \cdot x^2$ x^5

10. **a.** $4^2 \cdot 2^4$ 256   **b.** $x^3 \cdot y^2$ doesn't simplify

NOTATION

Complete each step to simplify each expression.

11. $(x^4 x^2)^3 = (x^{\,6})^3$
    $= x^{\,18}$

12. $(x^4)^3 (x^2)^3$
    $= x^{\,12} \cdot x^6$
    $= x^{\,18}$

GUIDED PRACTICE

Identify the base and the exponent in each expression. See Example 1.

13. 4^3
    base 4, exponent 3

14. $(-8)^2$
    base -8, exponent 2

15. x^5
    base x, exponent 5

16. $\left(\dfrac{5}{x}\right)^3$
    base $\dfrac{5}{x}$, exponent 3

17. $(-3x)^2$
    base $-3x$, exponent 2

18. $(2xy)^{10}$
    base $2xy$, exponent 10

19. $-\dfrac{1}{3}y^6$
base y, exponent 6

20. $-x^4$
base x, exponent 4

21. $9m^{12}$
base m, exponent 12

22. $3.14r^4$
base r, exponent 4

23. $(y+9)^4$
base $y+9$, exponent 4

24. $(z-2)^3$
base $z-2$, exponent 3

Write each expression in an equivalent form using an exponent. See Example 2.

25. $m \cdot m \cdot m \cdot m \cdot m \quad m^5$
26. $r \cdot r \cdot r \cdot r \cdot r \cdot r \quad r^6$
27. $4t \cdot 4t \cdot 4t \cdot 4t \quad (4t)^4$
28. $-5u(-5u)(-5u)(-5u)(-5u) \quad (-5u)^5$
29. $4 \cdot t \cdot t \cdot t \cdot t \cdot t \quad 4t^5$
30. $5 \cdot u \cdot u \cdot u \quad 5u^3$
31. $a \cdot a \cdot b \cdot b \cdot b \quad a^2 b^3$
32. $m \cdot m \cdot m \cdot n \cdot n \quad m^3 n^2$

Use the product rule for exponents to simplify each expression. Write the results using exponents. See Example 3.

33. $5^3 \cdot 5^4 \quad 5^7$
34. $3^4 \cdot 3^6 \quad 3^{10}$
35. $a^3 \cdot a^3 \quad a^6$
36. $m^7 \cdot m^7 \quad m^{14}$
37. $bb^2 b^3 \quad b^6$
38. $aa^3 a^5 \quad a^9$
39. $(c^5)(c^8) \quad c^{13}$
40. $(d^4)(d^{20}) \quad d^{24}$
41. $(a^2 b^3)(a^3 b^3) \quad a^5 b^6$
42. $(u^3 v^5)(u^4 v^5) \quad u^7 v^{10}$
43. $cd^4 \cdot cd \quad c^2 d^5$
44. $ab^3 \cdot ab^4 \quad a^2 b^7$
45. $x^2 \cdot y \cdot x \cdot y^{10} \quad x^3 y^{11}$
46. $x^3 \cdot y \cdot x \cdot y^{12} \quad x^4 y^{13}$
47. $m^{100} \cdot m^{100} \quad m^{200}$
48. $n^{600} \cdot n^{600} \quad n^{1,200}$

Use the power rule for exponents to simplify each expression. Write the results using exponents. See Example 4.

49. $(3^2)^4 \quad 3^8$
50. $(4^3)^3 \quad 4^9$
51. $[(-4.3)^3]^8 \quad (-4.3)^{24}$
52. $[(-1.7)^9]^8 \quad (-1.7)^{72}$
53. $(m^{50})^{10} \quad m^{500}$
54. $(n^{25})^4 \quad n^{100}$
55. $(y^5)^3 \quad y^{15}$
56. $(b^3)^6 \quad b^{18}$

Use the product and power rules for exponents to simplify each expression. See Example 5.

57. $(x^2 x^3)^5 \quad x^{25}$
58. $(y^3 y^4)^4 \quad y^{28}$
59. $(p^2 p^3)^5 \quad p^{25}$
60. $(r^3 r^4)^2 \quad r^{14}$
61. $(t^3)^4 (t^2)^3 \quad t^{18}$
62. $(b^2)^5 (b^3)^2 \quad b^{16}$
63. $(u^4)^2 (u^3)^2 \quad u^{14}$
64. $(v^5)^2 (v^3)^4 \quad v^{22}$

Use the power of a product rule for exponents to simplify each expression. See Example 6.

65. $(6a)^2 \quad 36a^2$
66. $(3b)^3 \quad 27b^3$
67. $(5y)^4 \quad 625y^4$
68. $(4t)^4 \quad 256t^4$
69. $(3a^4 b^7)^3 \quad 27a^{12} b^{21}$
70. $(5m^9 n^{10})^2 \quad 25m^{18} n^{20}$
71. $(-2r^2 s^3)^3 \quad -8r^6 s^9$
72. $(-2x^2 y^4)^5 \quad -32x^{10} y^{20}$

Use the power of a product rule for exponents to simplify each expression. See Example 7.

73. $(2c^3)^3 (3c^4)^2 \quad 72c^{17}$
74. $(5b^4)^2 (3b^8)^2 \quad 225b^{24}$
75. $(10d^7)^2 (4d^9)^3 \quad 6,400d^{41}$
76. $(2x^7)^3 (4x^8)^2 \quad 128x^{37}$

TRY IT YOURSELF

Simplify each expression.

77. $(7a^9)^2 \quad 49a^{18}$
78. $(12b^6)^2 \quad 144b^{12}$
79. $t^4 \cdot t^5 \cdot t \quad t^{10}$
80. $n^4 \cdot n \cdot n^3 \quad n^8$
81. $y^3 y^2 y^4 \quad y^9$
82. $y^4 y y^6 \quad y^{11}$
83. $(-6a^3 b^2)^3 \quad -216a^9 b^6$
84. $(-10r^3 s^2)^2 \quad 100r^6 s^4$
85. $(n^4 n)^3 (n^3)^6 \quad n^{33}$
86. $(y^3 y)^2 (y^2)^2 \quad y^{12}$
87. $(b^2 b^3)^{12} \quad b^{60}$
88. $(s^3 s^3)^3 \quad s^{18}$
89. $(2b^4 b)^5 (3b)^2 \quad 288b^{27}$
90. $(2aa^7)^3 (3a)^3 \quad 216a^{27}$
91. $(c^2)^3 (c^4)^2 \quad c^{14}$
92. $(t^5)^2 (t^3)^3 \quad t^{19}$
93. $(3s^4 t^3)^3 (2st)^4 \quad 432s^{16} t^{13}$
94. $(2a^3 b^5)^2 (4ab)^3 \quad 256a^9 b^{13}$
95. $x \cdot x^2 \cdot x^3 \cdot x^4 \cdot x^5 \quad x^{15}$
96. $x^{10} \cdot x^9 \cdot x^8 \cdot x^7 \quad x^{34}$

APPLICATIONS

97. **ART HISTORY** Leonardo da Vinci's drawing relating a human figure to a square and a circle is shown below. Find an expression for the area of the square if the man's height is $5x$ feet. $25x^2$ ft^2

98. PACKAGING Find an expression for the volume of the box shown below. $216x^3$ in.^{3}

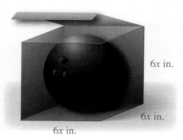

6x in.
6x in.
6x in.

WRITING

99. Explain the mistake in the following work.

$$2^3 \cdot 2^2 = 4^5 = 1{,}024$$

100. Explain why we can simplify $x^4 \cdot x^5$, but cannot simplify $x^4 + x^5$.

REVIEW

101. JEWELRY A lot of what we refer to as gold jewelry is actually made of a combination of gold and another metal. For example, 18-karat gold is $\frac{18}{24}$ gold by weight. Simplify this ratio. $\frac{3}{4}$

102. After evaluation, what is the sign of $(-13)^5$? negative

103. Divide: $\frac{-25}{-5}$ 5

104. How much did the temperature change if it went from $-4°F$ to $-17°F$? It dropped 13°F.

105. Evaluate: $2\left(\frac{12}{-3}\right) + 3(5)$ 7

106. Solve: $-10 = x + 1$ -11

107. Solve: $-x = -12$ 12

108. Divide: $\frac{0}{10}$ 0

SECTION 10.2
Introduction to Polynomials

Objectives
1. Know the vocabulary for polynomials.
2. Evaluate polynomials.

1 Know the vocabulary for polynomials.

Recall that an **algebraic term**, or simply a **term**, is a number or a product of a number and one or more variables, which may be raised to powers. Some examples of terms are

$$17, \quad 5x, \quad 6t^2, \quad \text{and} \quad -8z^3$$

The *coefficients* of these terms are 17, 5, 6, and -8, in that order.

> **Polynomials**
>
> A **polynomial** is a single term or a sum of terms in which all variables have whole-number exponents and no variable appears in the denominator.

Some examples of polynomials are

$$141, \quad 8y^2, \quad 2x + 1, \quad 4y^2 - 2y + 3, \quad \text{and} \quad 7a^3 + 2a^2 - a - 1$$

The polynomial $8y^2$ has one term. The polynomial $2x + 1$ has two terms, $2x$ and 1. Since $4y^2 - 2y + 3$ can be written as $4y^2 + (-2y) + 3$, it is the sum of three terms, $4y^2$, $-2y$, and 3.

We classify some polynomials by the number of terms they contain. A polynomial with one term is called a **monomial**. A polynomial with two terms is called a **binomial**. A polynomial with three terms is called a **trinomial**. Some examples of these polynomials are shown in the table below.

Monomials	Binomials	Trinomials
$5x^2$	$2x - 1$	$5t^2 + 4t + 3$
$-6x$	$18a^2 - 4a$	$27x^3 - 6x + 2$
29	$-27z^4 + 7z^2$	$32r^2 + 7r - 12$

Unless otherwise noted, all art on this page is © Cengage Learning.

Chapter 10 Exponents and Polynomials

Self Check 1

Classify each polynomial as a monomial, a binomial, or a trinomial:

a. $8x^2 + 7$ binomial

b. $5x$ monomial

c. $x^2 - 2x - 1$ trinomial

Now Try Problems 11, 13, and 17

Teaching Example 1 Classify each polynomial as a monomial, a binomial, or a trinomial:
a. $x^5 + x^3 - 10$
b. $t^{15} + 6$
c. $-4n$

Answers:
a. trinomial
b. binomial
c. monomial

EXAMPLE 1
Classify each polynomial as a monomial, a binomial, or a trinomial: a. $3x + 4$ b. $3x^2 + 4x - 12$ c. $25x^3$

Strategy We will count the number of terms in the polynomial.

WHY The number of terms determines the type of polynomial.

Solution
a. Since $3x + 4$ has two terms, it is a binomial.
b. Since $3x^2 + 4x - 12$ has three terms, it is a trinomial.
c. Since $25x^3$ has one term, it is a monomial.

The monomial $7x^3$ is called a **monomial of third degree**, or a **monomial of degree 3**, because the variable occurs three times as a factor.

- $5x^2$ is a monomial of degree 2.   *Because the variable occurs two times as a factor: $x^2 = x \cdot x$.*

- $-8a^4$ is a monomial of degree 4.   *Because the variable occurs four times as a factor: $a^4 = a \cdot a \cdot a \cdot a$.*

- $\frac{1}{2}m^5$ is a monomial of degree 5.   *Because the variable occurs five times as a factor: $m^5 = m \cdot m \cdot m \cdot m \cdot m$.*

- 8 is a monomial of degree 0.   *The degree of a nonzero constant is 0.*

We define the degree of a polynomial by considering the degrees of each of its terms.

Degree of a Polynomial

The **degree of a polynomial** is equal to the highest degree of any term of the polynomial.

For example,

- $x^2 + 5x$ is a binomial of degree 2, because the degree of its term with largest degree (x^2) is 2.

- $4y^3 + 2y - 7$ is a trinomial of degree 3, because the degree of its term with largest degree ($4y^3$) is 3.

- $\frac{1}{2}z + 3z^4 - 2z^2$ is a trinomial of degree 4, because the degree of its term with largest degree ($3z^4$) is 4.

Self Check 2

Find the degree of each polynomial:

a. $7p^3$ 3

b. $17r^4 + 2r^8 - r$ 8

c. $-2g^5 - 7g^6 + 12g^7$ 7

Now Try Problems 23, 25, and 29

Teaching Example 2 Find the degree of each polynomial:
a. $100 - 8x$
b. $7a^4 + 5a^3 + 4a$
c. $x - x^2 + x^3 - x^6$

Answers:
a. 1   b. 4   c. 6

EXAMPLE 2
Find the degree of each polynomial:
a. $-2x + 4$ b. $5t^3 + t^4 - 7$ c. $3 - 9z + 6z^2 - z^3$

Strategy We will determine the degree of each term of the polynomial.

WHY The degree of the term with the highest degree gives the degree of the polynomial.

Solution
a. Since $-2x$ can be written as $-2x^1$, the degree of the term with largest degree is 1. Thus, the degree of the polynomial $-2x + 4$ is 1.

b. In $5t^3 + t^4 - 7$, the degree of the term with largest degree (t^4) is 4. Thus, the degree of the polynomial is 4.

c. In $3 - 9z + 6z^2 - z^3$, the degree of the term with largest degree ($-z^3$) is 3. Thus, the degree of the polynomial is 3.

2 Evaluate polynomials.

When a number is substituted for the variable in a polynomial, the polynomial takes on a numerical value. Finding this value is called **evaluating the polynomial**.

EXAMPLE 3 Evaluate each polynomial for $x = 3$:
a. $3x - 2$ **b.** $-2x^2 + x - 3$

Strategy We will substitute the given value for each x in the polynomial and follow the order of operations rule.

WHY To *evaluate a polynomial* means to find its numerical value, once we know the value of its variable.

Solution

a. $3x - 2 = 3(3) - 2$ Substitute 3 for x.
$ = 9 - 2$ Multiply: 3(3) = 9.
$ = 7$ Subtract.

b. $-2x^2 + x - 3 = -2(3)^2 + 3 - 3$ Substitute 3 for x.
$ = -2(9) + 3 - 3$ Evaluate the exponential expression.
$ = -18 + 3 - 3$ Multiply: −2(9) = −18.
$ = -15 - 3$ Add: −18 + 3 = −15.
$ = -18$ Subtract: −15 − 3 = −15 + (−3) = −18.

Self Check 3
Evaluate each polynomial for $x = -1$:
a. $-2x^2 - 4$ −6
b. $3x^2 - 4x + 1$ 8

Now Try Problems 35 and 45

Teaching Example 3 Evaluate each polynomial for $a = -6$.
a. $-2a^2 + 2$ **b.** $a^3 - 7a + 1$
Answers:
a. −70 **b.** −173

EXAMPLE 4 **Height of an Object** The polynomial $-16t^2 + 28t + 8$ gives the height (in feet) of an object t seconds after it has been thrown into the air. Find the height of the object after 1 second.

Strategy We will substitute 1 for t and evaluate the polynomial.

WHY The variable t represents the time since the object was thrown into the air.

Solution
To find the height at 1 second, we evaluate the polynomial for $t = 1$.

$-16t^2 + 28t + 8 = -16(1)^2 + 28(1) + 8$ Substitute 1 for t.
$ = -16(1) + 28(1) + 8$ Evaluate the exponential expression.
$ = -16 + 28 + 8$ Multiply: −16(1) = −16 and 28(1) = 28.
$ = 12 + 8$ Add: −16 + 28 = 12.
$ = 20$ Add.

At 1 second, the height of the object is 20 feet.

Self Check 4
Refer to Example 4. Find the height of the object after 2 seconds. 0 ft

Now Try Problems 47 and 49

Teaching Example 4
PROJECTILES The polynomial $-16t^2 + 30t + 10$ gives the height (in feet) of an object t seconds after it has been thrown into the air. Find the height of the object after 2 seconds.
Answer:
6 ft

ANSWERS TO SELF CHECKS

1. a. binomial **b.** monomial **c.** trinomial **2. a.** 3 **b.** 8 **c.** 7
3. a. −6 **b.** 8 **4.** 0 ft

SECTION 10.2 STUDY SET

VOCABULARY

Fill in the blanks.

1. A __polynomial__ is a single term or a sum of terms in which all variables have whole-number exponents and no variable appears in the denominator.
2. A polynomial with one term is called a __monomial__.
3. A polynomial with three terms is called a __trinomial__.
4. A polynomial with two terms is called a __binomial__.

CONCEPTS

5. How many terms does each of the following polynomials have?
   a. $x^2 + 3x$ 2
   b. $3a^4$ 1
   c. $-4r^2 + 9r + 11$ 3
6. Fill in the blank so that $10c^{\,3}$ has degree 3.
7. Fill in the blank. The degree of a polynomial is equal to the __highest__ degree of any term of the polynomial.
8. What is the degree of each term of the polynomial $4x^3 + x^2 - 7x + 5$? 3, 2, 1, 0

NOTATION

Complete each step.

9. Evaluate $3a^2 + 2a - 7$ for $a = 2$.

$$3a^2 + 2a - 7 = 3(\,2\,)^2 + 2(\,2\,) - 7$$
$$= 3(\,4\,) + 2(2) - 7$$
$$= 12 + \,4\, - 7$$
$$= \,16\, - 7$$
$$= 9$$

10. Evaluate $-q^2 - 3q + 2$ for $q = -1$.

$$-q^2 - 3q + 2 = -(\,-1\,)^2 - 3(\,-1\,) + 2$$
$$= -(\,1\,) - 3(-1) + 2$$
$$= -1 + \,3\, + 2$$
$$= \,2\, + 2$$
$$= 4$$

GUIDED PRACTICE

Classify each polynomial as a monomial, a binomial, or a trinomial. See Example 1.

11. $3x^2 - 4$ binomial
12. $5t^2 - t + 1$ trinomial
13. $17e^4$ monomial
14. $x^2 + x + 7$ trinomial
15. $25u^2$ monomial
16. $x^2 - 9$ binomial
17. $q^5 + q^2 + 1$ trinomial
18. $4d^3 - 3d^2$ binomial
19. $81x^3 - 27$ binomial
20. $125m^3 - 8$ binomial
21. $4c^2 - 8c + 12$ trinomial
22. $16n^4 - 8n^2 + n$ trinomial

Find the degree of each polynomial. See Example 2.

23. $5x^3$ 3
24. $3t^5 + 3t^2$ 5
25. $2x^2 - 3x + 2$ 2
26. $\frac{1}{2}p^4 - p^2$ 4
27. $2m$ 1
28. $7q - 5$ 1
29. $25w^6 + 5w^7$ 7
30. $p^6 - p^8$ 8
31. $a^2 - 9$ 2
32. $b^2 - 25$ 2
33. $-m + 3m^4 + 5m^2$ 4
34. $-r + 6r^8 - r^7$ 8

Evaluate each polynomial for the given value. See Example 3.

35. $3x + 4$ for $x = 3$ 13
36. $5n - 10$ for $n = 6$ 20
37. $2x^2 + 4$ for $x = -1$ 6
38. $9r^2 + 12$ for $r = -3$ 93
39. $\frac{1}{2}x - 3$ for $x = -6$ -6
40. $-\frac{1}{2}x^2 - 1$ for $x = 2$ -3
41. $0.5t^3 - 1$ for $t = 4$ 31
42. $0.75a^2 + 2.5a + 2$ for $a = 0$ 2
43. $\frac{2}{3}b^2 - b + 1$ for $b = 3$ 4
44. $\frac{3}{2}n^2 - n + 2$ for $n = 2$ 6
45. $-2s^2 - 2s + 1$ for $s = -1$ 1
46. $-4r^2 - 3r - 1$ for $r = -2$ -11

APPLICATIONS

The height h (in feet) of a ball shot straight up with an initial velocity of 64 feet per second is given by the equation $h = -16t^2 + 64t$. Find the height of the ball after the given number of seconds.

47. 0 second 0 ft
48. 1 second 48 ft
49. 2 seconds 64 ft
50. 4 seconds 0 ft

51. The number of feet that a car travels before stopping depends on the driver's reaction time and the braking distance, as shown below. For one driver, the stopping distance d is given by the equation

$$d = 0.04v^2 + 0.9v$$

where v is the velocity (speed) of the car. Find the stopping distance when the driver is traveling at 30 mph. 63 ft

from Campus to Careers
Police Officer

In Problems 52–54, refer to Problem 51. Then find the stopping distance for each of the following speeds.

52. 50 mph 145 ft
53. 60 mph 198 ft
54. 70 mph 259 ft

WRITING

55. Explain how to find the degree of the polynomial $2x^3 + 5x^5 - 7x$.
56. Explain how to evaluate the polynomial $-2x^2 - 3$ for $x = 5$.

REVIEW

Perform the operations.

57. $\dfrac{2}{3} + \dfrac{4}{3}$ 2
58. $\dfrac{36}{7} - \dfrac{23}{7}$ $\dfrac{13}{7} = 1\dfrac{6}{7}$
59. $\dfrac{5}{12} \cdot \dfrac{18}{5}$ $\dfrac{3}{2} = 1\dfrac{1}{2}$
60. $\dfrac{23}{25} \div \dfrac{46}{5}$ $\dfrac{1}{10}$

Solve each equation.

61. $x - 4 = 12$ 16
62. $4z = 108$ 27
63. $2(x - 3) = 6$ 6
64. $3(a - 5) = 4(a + 9)$ -51

SECTION 10.3
Adding and Subtracting Polynomials

Objectives
1. Add polynomials.
2. Subtract polynomials.

Polynomials can be added, subtracted, and multiplied just like numbers in arithmetic. In this section, we show how to find sums and differences of polynomials.

1 Add polynomials.

Recall that like terms have exactly the same variables and the same exponents. For example, the monomials

$3z^2$ and $-2z^2$ are like terms   Both have the same variable, z, with the same exponent, 2.

However, the monomials

$7b^2$ and $8a^2$ are not like terms   They have different variables.

$32p^2$ and $25p^3$ are not like terms   The exponents of p are different.

Also recall that we use the distributive property in reverse to simplify a sum or difference of like terms. We **combine like terms** by adding their coefficients and keeping the same variables and exponents. For example,

$2y + 5y = (2 + 5)y$ and $-3x^2 + 7x^2 = (-3 + 7)x^2$
$\qquad\quad = 7y$ $\qquad\qquad\quad\,\, = 4x^2$

These examples suggest the following rule.

Chapter 10 Exponents and Polynomials

> **Adding Polynomials**
> To add polynomials, combine their like terms.

Self Check 1
Add: $7y^3 + 12y^3$ $19y^3$
Now Try Problem 15

Teaching Example 1 Add: $15x^5 + 5x^5$
Answer:
$20x^5$

EXAMPLE 1 Add: $5x^3 + 7x^3$

Strategy We will use the distributive property in reverse and add the coefficients of the terms.

WHY $5x^3$ and $7x^3$ are like terms and therefore can be added.

Solution

$5x^3 + 7x^3 = 12x^3$ Think: $(5 + 7)x^3 = 12x^3$.

Self Check 2
Add:
$\frac{1}{9}a^3 + \frac{2}{9}a^3 + \frac{5}{9}a^3$ $\frac{8}{9}a^3$
Now Try Problem 21

Teaching Example 2
Add: $\frac{1}{16}c^2 + \frac{7}{16}c^2 + \frac{3}{16}c^2$
Answer:
$\frac{11}{16}c^2$

EXAMPLE 2 Add: $\frac{3}{2}t^2 + \frac{5}{2}t^2 + \frac{7}{2}t^2$

Strategy We will use the distributive property in reverse and add the coefficients of the terms.

WHY $\frac{3}{2}t^2$, $\frac{5}{2}t^2$, and $\frac{7}{2}t^2$ are like terms and therefore can be added.

Solution
Since the three monomials are like terms, we add the coefficients and keep the variables and exponents.

$\frac{3}{2}t^2 + \frac{5}{2}t^2 + \frac{7}{2}t^2 = \left(\frac{3}{2} + \frac{5}{2} + \frac{7}{2}\right)t^2$ To add the fractions, add the numerators and keep the denominator: $3 + 5 + 7 = 15$.

$= \frac{15}{2}t^2$

To add two polynomials, we write a + sign between them and combine like terms.

Self Check 3
Add:
$5y - 2$ and $-3y + 7$ $2y + 5$
Now Try Problem 25

Teaching Example 3
Add: $11n + 4$ and $6n - 3$
Answer:
$17n + 1$

EXAMPLE 3 Add: $2x + 3$ and $7x - 1$

Strategy We will reorder and regroup to get the like terms together. Then we will combine like terms.

WHY To add polynomials means to combine their like terms.

Solution

$(2x + 3) + (7x - 1)$ Write a + sign between the binomials.

$= (2x + 7x) + (3 - 1)$ Use the associative and commutative properties to group like terms together.

$= 9x + 2$ Combine like terms.

The binomials in Example 3 can be added by writing the polynomials so that like terms are in columns.

$2x + 3$
$\underline{+\ 7x - 1}$
$9x + 2$ Add the like terms, one column at a time.

EXAMPLE 4   Add: $(5x^2 - 2x + 4) + (3x^2 - 5)$

Strategy We will combine the like terms of the trinomial and binomial.

WHY To add polynomials, we combine like terms.

Solution

$(5x^2 - 2x + 4) + (3x^2 - 5)$
$= (5x^2 + 3x^2) + (-2x) + (4 - 5)$ Use the associative and commutative properties to group like terms together.
$= 8x^2 - 2x - 1$ Combine like terms.

The polynomials in Example 4 can be added by writing the polynomials so that like terms are in columns.

$5x^2 - 2x + 4$
$+\ 3x^2 - 5$
$\overline{8x^2 - 2x - 1}$ Add the like terms, one column at a time.

Self Check 4
Add: $3b^2 - b - 1$
$(2b^2 - 4b) + (b^2 + 3b - 1)$
Now Try Problem 29

Teaching Example 4
Add: $(6t^2 - 5t) + (t^2 + 2t - 9)$
Answer:
$7t^2 - 3t - 9$

EXAMPLE 5   Add: $(3.7x^2 + 4x - 2) + (7.4x^2 - 5x + 3)$

Strategy We will combine the like terms of the two trinomials.

WHY To add polynomials, we combine like terms.

Solution

$(3.7x^2 + 4x - 2) + (7.4x^2 - 5x + 3)$
$= (3.7x^2 + 7.4x^2) + (4x - 5x) + (-2 + 3)$ Use the associative and commutative properties to group like terms together.
$= 11.1x^2 - x + 1$ Combine like terms.

The trinomials in Example 5 can be added by writing them so that like terms are in columns.

$3.7x^2 + 4x - 2$
$+\ 7.4x^2 - 5x + 3$
$\overline{11.1x^2 - x + 1}$ Add the like terms, one column at a time.

Self Check 5
Add: $4s^2 - 1.3s - 1$
$(s^2 + 1.2s - 5) + (3s^2 - 2.5s + 4)$
Now Try Problem 31

Teaching Example 5 Add:
$(m^2 + 3.7m - 7) + (8m^2 - 4.8m + 2)$
Answer:
$9m^2 - 1.1m - 5$

2 Subtract polynomials.

To subtract one monomial from another, we add the opposite of the monomial that is to be subtracted. In symbols, $x - y = x + (-y)$.

EXAMPLE 6   Subtract: $8x^2 - 3x^2$

Strategy We will add the opposite of $3x^2$ to $8x^2$.

WHY To subtract monomials, we add the opposite of the monomial that is to be subtracted.

Solution

$8x^2 - 3x^2 = 8x^2 + (-3x^2)$ Add the opposite of $3x^2$.
$= 5x^2$ Add the coefficients and keep the same variable and exponent. Think: $[8 + (-3)]x^2 = 5x^2$

Self Check 6
Subtract: $6y^3 - 9y^3$ $-3y^3$
Now Try Problem 39

Teaching Example 6
Subtract: $10b^2 - 12b^2$
Answer:
$-2b^2$

Recall from Chapter 1 that we can use the distributive property to find the opposite of several terms enclosed within parentheses. For example, we consider $-(2a^2 - a + 9)$.

$-(2a^2 - a + 9) = -\mathbf{1}(2a^2 - a + 9)$ Replace the − symbol in front of the parentheses with −1.

$= -2a^2 + a - 9$ Use the distributive property to remove parentheses.

This example illustrates the following method of subtracting polynomials.

Subtracting Polynomials

To subtract two polynomials, change the signs of the terms of the polynomial being subtracted, drop the parentheses, and combine like terms.

Self Check 7

Subtract:
$(3.3a - 5) - (7.8a + 2)$ $-4.5a - 7$

Now Try Problem 43

Teaching Example 7
Subtract: $(1.7d - 2) - (0.9d + 11)$
Answer:
$0.8d - 13$

EXAMPLE 7   Subtract: $(3x - 4.2) - (5x + 7.2)$

Strategy We will change the signs of the terms of $5x + 7.2$, drop the parentheses, and combine like terms.

WHY This is the method for subtracting two polynomials.

Solution

$(3x - 4.2) - (5x + 7.2)$

$= 3x - 4.2 - 5x - 7.2$ Change the signs of each term of $5x + 7.2$ and drop the parentheses.

$= -2x - 11.4$ Combine like terms: Think: $(3 - 5)x = -2x$ and $(-4.2 - 7.2) = -11.4$.

The binomials in Example 7 can be subtracted by writing them so that like terms are in columns.

$$\begin{array}{r} 3x - 4.2 \\ -(5x + 7.2) \end{array} \longrightarrow \begin{array}{r} 3x - 4.2 \\ + -5x - 7.2 \\ \hline -2x - 11.4 \end{array}$$ Change signs and add, column by column.

Self Check 8

Subtract: $2y^2 - 6y + 3$
$(5y^2 - 4y + 2) - (3y^2 + 2y - 1)$

Now Try Problem 47

Teaching Example 8 Subtract:
$(9x^2 - 4x + 1) - (8x^2 + x - 8)$
Answer:
$x^2 - 5x + 9$

EXAMPLE 8   Subtract: $(3x^2 - 4x - 6) - (2x^2 - 6x + 12)$

Strategy We will change the signs of the terms of $2x^2 - 6x + 12$, drop the parentheses, and combine like terms.

WHY This is the method for subtracting two polynomials.

Solution

$(3x^2 - 4x - 6) - (2x^2 - 6x + 12)$

$= 3x^2 - 4x - 6 - 2x^2 + 6x - 12$ Change the signs of each term of $2x^2 - 6x + 12$ and drop the parentheses.

$= x^2 + 2x - 18$ Combine like terms: Think: $(3 - 2)x^2 = x^2$, $(-4 + 6)x = 2x$, and $(-6 - 12) = -18$.

10.3 Adding and Subtracting Polynomials

The trinomials in Example 8 can be subtracted by writing them so that like terms are in columns.

$$3x^2 - 4x - 6$$
$$-(2x^2 - 6x + 12)$$

$\longrightarrow$

$$3x^2 - 4x - 6$$
$$+\ -2x^2 + 6x - 12$$
$$\overline{x^2 + 2x - 18}$$

Change signs and add, column by column.

ANSWERS TO SELF CHECKS

1. $19y^3$ 2. $\frac{8}{9}a^3$ 3. $2y + 5$ 4. $3b^2 - b - 1$ 5. $4s^2 - 1.3s - 1$ 6. $-3y^3$
7. $-4.5a - 7$ 8. $2y^2 - 6y + 3$

SECTION 10.3 STUDY SET

VOCABULARY

Fill in the blanks.

1. If two algebraic terms have exactly the same variables and exponents, they are called __like__ terms.
2. Because the exponents on x are different, $3x^3$ and $3x^2$ are __unlike__ terms.

CONCEPTS

Fill in the blanks.

3. To add two monomials, we add the __coefficients__ and keep the same __variables__ and exponents.
4. To subtract one monomial from another, we add the __opposite__ of the monomial that is to be subtracted.

Determine whether the monomials are like terms. If they are, combine them.

5. $3y, 4y$ yes, $7y$
6. $3x^2, 5x^2$ yes, $8x^2$
7. $3x, 3y$ no
8. $3x^2, 6x$ no
9. $3x^3, 4x^3, 6x^3$ yes, $13x^3$
10. $-2y^4, -6y^4, 10y^4$ yes, $2y^4$
11. $-5x^2, 13x^2, 7x^2$ yes, $15x^2$
12. $23, 12x, 25x$ no

NOTATION

Complete each step.

13. $(3x^2 + 2x - 5) + (2x^2 - 7x)$
  $= (3x^2 + \boxed{2x^2}) + (2x - \boxed{7x}) + (-5)$
  $= \boxed{5x^2} + (-5x) - 5$
  $= 5x^2 - 5x - 5$

14. $(3x^2 + 2x - 5) - (2x^2 - 7x)$
  $= 3x^2 + 2x - 5 \ \boxed{-}\ 2x^2 \ \boxed{+}\ 7x$
  $= (\boxed{3x^2} - \boxed{2x^2}) + (2x + 7x) - 5$
  $= x^2 + 9x - 5$

GUIDED PRACTICE

Add. See Example 1.

15. $4y + 5y$ $9y$
16. $-2x + 3x$ x
17. $8t^2 + 4t^2$ $12t^2$
18. $15x^2 + 10x^2$ $25x^2$

Add. See Example 2.

19. $\frac{1}{8}a + \frac{3}{8}a + \frac{5}{8}a$ $\frac{9}{8}a$
20. $\frac{1}{4}b + \frac{3}{4}b + \frac{1}{4}b$ $\frac{5}{4}b$
21. $\frac{2}{3}c^2 + \frac{1}{3}c^2 + \frac{2}{3}c^2$ $\frac{5}{3}c^2$
22. $\frac{4}{9}d^3 + \frac{1}{9}d^3 + \frac{3}{9}d^3$ $\frac{8}{9}d^3$

Add. See Example 3.

23. $3x + 7$ and $4x - 3$ $7x + 4$
24. $2y - 3$ and $4y + 7$ $6y + 4$
25. $2x^2 + 3$ and $5x^2 - 10$ $7x^2 - 7$
26. $-4a^2 + 1$ and $5a^2 - 1$ a^2

Add. See Example 4.

27. $(5x^3 - 42x) + (7x^3 - 107x)$ $12x^3 - 149x$
28. $(-43a^3 + 25a) + (58a^3 - 10a)$ $15a^3 + 15a$
29. $(3x^2 + 2x - 4) + (5x^2 - 17)$ $8x^2 + 2x - 21$
30. $(5a^2 - 2a) + (-2a^2 + 3a + 4)$ $3a^2 + a + 4$

Add. See Example 5.

31. $(2.5a^2 + 3a - 9) + (3.6a^2 + 7a - 10)$ $6.1a^2 + 10a - 19$
32. $(1.9b^2 - 4b + 10) + (3.7b^2 - 3b - 11)$ $5.6b^2 - 7b - 1$
33. $(3n^2 - 5.8n + 7) + (-n^2 + 5.8n - 2)$ $2n^2 + 5$
34. $(-3t^2 - t + 3.4) + (3t^2 + 2t - 1.8)$ $t + 1.6$

35. $\quad 3x^2 + 4x + 5$
    $+\ 2x^2 - 3x + 6$
    $\overline{5x^2 + x + 11}$

36. $\quad 2x^2 - 3x + 5$
    $+\ -4x^2 - x - 7$
    $\overline{-2x^2 - 4x - 2}$

37. $\quad -3x^2 \quad\quad - 7$
    $+\ -4x^2 - 5x + 6$
    $\overline{-7x^2 - 5x - 1}$

38. $\quad 4x^2 - 4x + 9$
    $+\ \quad\quad\ 9x - 3$
    $\overline{4x^2 + 5x + 6}$

Subtract. See Example 6.

39. $32u^3 - 16u^3$ $16u^3$
40. $25y^2 - 7y^2$ $18y^2$
41. $18x^5 - 11x^5$ $7x^5$
42. $17x^6 - 22x^6$ $-5x^6$

Subtract. See Example 7.

43. $(4.5a + 3.7) - (2.9a - 4.3)$ $1.6a + 8$
44. $(5.1b - 7.6) - (3.3b + 5.9)$ $1.8b - 13.5$
45. $(7.2x^2 - 3.1x) - (9.4x^2 + 6.8x)$ $-2.2x^2 - 9.9x$
46. $(3.7y^3 + 9.8y^2) - (2.4y^3 - 1.1y^2)$ $1.3y^3 + 10.9y^2$

Subtract. See Example 8.

47. $(2b^2 + 3b - 5) - (2b^2 - 4b - 9)$ $7b + 4$
48. $(3a^2 - 2a + 4) - (a^2 - 3a + 7)$ $2a^2 + a - 3$
49. $(5p^2 - p + 71) - (4p^2 + p + 71)$ $p^2 - 2p$
50. $(10m^2 - m - 19) - (6m^2 + m - 19)$ $4m^2 - 2m$

51. $\quad 3x^2 + 4x - 5$
$\quad -(-2x^2 - 2x + 3)$
$\quad \overline{5x^2 + 6x - 8}$

52. $\quad 3y^2 - 4y + 7$
$\quad -(6y^2 - 6y - 13)$
$\quad \overline{-3y^2 + 2y + 20}$

53. $\quad -2x^2 - 4x + 12$
$\quad -(10x^2 + 9x - 24)$
$\quad \overline{-12x^2 - 13x + 36}$

54. $\quad 25x^3 - 45x^2 + 31x$
$\quad -(12x^3 + 27x^2 - 17x)$
$\quad \overline{13x^3 - 72x^2 + 48x}$

TRY IT YOURSELF

Perform the operations.

55. $(30x^2 - 4) - (11x^2 + 1)$ $19x^2 - 5$
56. $(5x^3 - 8) - (2x^3 + 5)$ $3x^3 - 13$
57. $(7y^2 + 5y) + (y^2 - y - 2)$ $8y^2 + 4y - 2$
58. $(4p^2 - 4p + 5) + (6p - 2)$ $4p^2 + 2p + 3$
59. $(3x^2 - 3x - 2) + (3x^2 + 4x - 3)$ $6x^2 + x - 5$
60. $(4c^2 + 3c - 2) + (3c^2 + 4c + 2)$ $7c^2 + 7c$
61. $(m^2 - m - 5) - (m^2 + 5.5m - 75)$ $-6.5m + 70$
62. $(3.7y^2 - 5) - (2y^2 - 3.1y + 4)$ $1.7y^2 + 3.1y - 9$
63. $(t^2 - 4.5t + 5) - (2t^2 - 3.1t - 1)$ $-t^2 - 1.4t + 6$
64. $(a^4 - 5.1a^3 + 1.1a) - (3a^4 - 6.7a^3 + 0.1a)$
$-2a^4 + 1.6a^3 + a$

65. $\quad -3x^2 + 4x + 25.4$
$\quad + \underline{5x^2 - 3x - 12.5}$
$\quad 2x^2 + x + 12.9$

66. $\quad -6x^3 - 4.2x^2 + 7$
$\quad + \underline{-7x^3 + 9.7x^2 - 21}$
$\quad -13x^3 + 5.5x^2 - 14$

67. $3s^2 + 4s^2 + 7s^2$ $14s^2$
68. $-2a^3 + 7a^3 - 3a^3$ $2a^3$
69. $\frac{1}{3}b^4 + \frac{2}{3}b^4 - \frac{5}{3}b^4$ $-\frac{2}{3}b^4$
70. $\frac{4}{5}n^6 - \frac{1}{5}n^6 - \frac{2}{5}n^6$ $\frac{1}{5}n^6$

71. $\quad z^3 + 6z^2 - 7z + 16$
$\quad +\underline{9z^3 - 6z^2 + 8z - 18}$
$\quad 10z^3 \quad\quad + z - 2$

72. $\quad 3x^3 + 4x^2 - 3x + 5$
$\quad +\underline{3x^3 - 4x^2 - x - 7}$
$\quad 6x^3 \quad\quad - 4x - 2$

73. $(-4h^3 + 5h^2 + 15) - (h^3 - 15)$ $-5h^3 + 5h^2 + 30$
74. $(-c^5 + 5c^4 - 12) - (2c^5 - c^4)$ $-3c^5 + 6c^4 - 12$
75. $0.6x^3 + 0.8x^4 + 0.7x^3 + (-0.8x^4)$ $1.3x^3$
76. $1.9m^4 - 2.4m^6 - 3.7m^4 + 2.8m^6$ $0.4m^6 - 1.8m^4$

77. $(12.1h^3 - 9.9h^2 + 9.5) + (7.3h^3 - 1.2h^2 - 10.1)$
$19.4h^3 - 11.1h^2 - 0.6$
78. $(7.1a^2 + 2.2a - 5.8) - (3.4a^2 - 3.9a + 11.8)$
$3.7a^2 + 6.1a - 17.6$

79. $\quad 4x^3 - 3x + 10$
$\quad -\underline{(5x^3 - 4x - 4)}$
$\quad -x^3 + x + 14$

80. $\quad 3x^3 + 4x^2 + 12$
$\quad -\underline{(-4x^3 + 6x^2 - 3)}$
$\quad 7x^3 - 2x^2 + 15$

APPLICATIONS

81. BILLIARDS Billiard tables vary in size, but all tables are twice as long as they are wide.

  a. If the billiard table is x feet wide, write an expression that represents its length. $2x$ ft

  b. Write an expression that represents the perimeter of the table. $6x$ ft

82. GARDENING Find a polynomial that represents the length of the wooden handle of the shovel.
$(x^2 + x + 3)$ in.

83. READING BLUEPRINTS

  a. What is the difference between the length and width of the one-bedroom apartment shown below? $(6x + 5)$ ft

  b. Find the perimeter of the apartment. $(4x^2 + 26)$ ft

84. **PIÑATAS** Find the polynomial that represents the length of the rope used to hold up the piñata. $(2a^2 + 6a + 5)$ in.

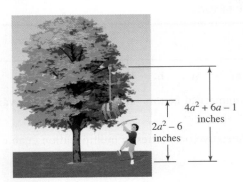

85. **GREEK ARCHITECTURE** Find a polynomial that represents the difference in the heights of the columns shown here. $(x^2 - 8x + 12)$ ft

86. **GREEK ARCHITECTURE** Find a polynomial that represents the height of the columns shown in problem 85 if they were stacked one atop the other. $(x^2 + 2x - 8)$ ft

WRITING

87. What are *like terms*?
88. Explain how to add two polynomials.
89. Explain how to subtract two polynomials.
90. When two binomials are added, is the result always a binomial? Explain.

REVIEW

91. **BASKETBALL SHOES** Use the following information to find how much lighter the Kobe Bryant shoe is than the LeBron James shoe. 7.2 oz

Kobe VIII	LeBron X
Lightest Nike basketball sneaker	Lightweight support, lockdown fit
Sizes 7–18	Sizes 6–20
Weight: 9.6 oz	Weight: 16.8 oz

92. **AEROBICS** The number of calories burned when doing step aerobics depends on the step height. How many more calories are burned during a 10-minute workout using an 8-inch step instead of a 4-inch step? 19 calories

Step height (in.)	Calories burned per minute
4	4.5
6	5.5
8	6.4
10	7.2

Source: *Reebok Instructor News* (Vol. 4, No. 3, 1991)

SECTION 10.4
Multiplying Polynomials

Objectives
1. Multiply monomials.
2. Multiply a polynomial by a monomial.
3. Multiply binomials.
4. Multiply polynomials.

We now discuss how to multiply polynomials. We will begin with the simplest case—finding the product of two monomials.

1 Multiply monomials.

To multiply $4x^2$ by $2x^3$, we use the commutative and associative properties of multiplication to reorder and regroup the factors.

$(4x^2)(2x^3) = (4 \cdot 2)(x^2 \cdot x^3)$ Group the coefficients together and the variables together.

$= 8x^5$ Simplify: $x^2 \cdot x^3 = x^{2+3} = x^5$.

This example suggests the following rule.

Multiplying Two Monomials

To multiply two monomials, multiply the numerical factors (the coefficients) and then multiply the variable factors.

Self Check 1
Multiply: $-7a^3 \cdot 2a^5$ $-14a^8$
Now Try Problems 13 and 15

Teaching Example 1
Multiply: $-9f^3 \cdot 4f^7$
Answer:
$-36f^{10}$

EXAMPLE 1 Multiply: **a.** $3y \cdot 6y$ **b.** $-3x^5(2x^5)$

Strategy We will multiply the numerical factors and then multiply the variable factors.

WHY The commutative and associative properties of multiplication enable us to reorder and regroup factors.

Solution

a. $3y \cdot 6y = (3 \cdot 6)(y \cdot y)$ Group the numerical factors and group the variables.
$\qquad = 18y^2$ Multiply: $3 \cdot 6 = 18$ and $y \cdot y = y^2$.

b. $(-3x^5)(2x^5) = (-3 \cdot 2)(x^5 \cdot x^5)$ Group the numerical factors and group the variables.
$\qquad = -6x^{10}$ Multiply: $-3 \cdot 2 = -6$ and $x^5 \cdot x^5 = x^{5+5} = x^{10}$.

2 Multiply a polynomial by a monomial.

To find the product of a polynomial and a monomial, we use the distributive property. To multiply $x + 4$ by $3x$, for example, we proceed as follows:

$3x(x + 4) = 3x(x) + 3x(4)$ Use the distributive property.
$\qquad = 3x^2 + 12x$ Multiply the monomials: $3x(x) = 3x^2$ and $3x(4) = 12x$.

The results of this example suggest the following rule.

Multiplying Polynomials by Monomials

To multiply a polynomial by a monomial, multiply each term of the polynomial by the monomial.

Self Check 2
Multiply:
a. $3y(5y^3 - 4y)$ $15y^4 - 12y^2$
b. $5x(3x^2 - 2x + 3)$ $15x^3 - 10x^2 + 15x$

Now Try Problems 17 and 19

Teaching Example 2 Multiply:
a. $4x(7x^2 - 3x)$ **b.** $6t(3t^2 - 5t + 11)$
Answers:
a. $28x^3 - 12x^2$ **b.** $18t^3 - 30t^2 + 66t$

EXAMPLE 2 Multiply: **a.** $2a^2(3a^2 - 4a)$ **b.** $8x(3x^2 + 2x - 3)$

Strategy We will multiply each term of the polynomial by the monomial.

WHY We use the distributive property to multiply a monomial and a polynomial.

Solution

a. $2a^2(3a^2 - 4a)$
$= 2a^2(3a^2) - 2a^2(4a)$ Use the distributive property.
$= 6a^4 - 8a^3$ Multiply: $2a^2(3a^2) = 6a^4$ and $2a^2(4a) = 8a^3$.

b. $8x(3x^2 + 2x - 3)$
$= 8x(3x^2) + 8x(2x) - 8x(3)$ Use the distributive property.
$= 24x^3 + 16x^2 - 24x$ Multiply: $8x(3x^2) = 24x^3$, $8x(2x) = 16x^2$, and $8x(3) = 24x$.

3 Multiply binomials.

The distributive property can also be used to multiply binomials. For example, to multiply $2a + 4$ and $3a + 5$, we think of $2a + 4$ as a single quantity and distribute it over each term of $3a + 5$.

$$
\begin{aligned}
(2a + 4)(3a + 5) &= (2a + 4)3a + (2a + 4)5 \\
&= (2a + 4)3a + (2a + 4)5 \\
&= (2a)3a + (4)3a + (2a)5 + (4)5 &&\text{Distribute the multiplication by } 3a \text{ and by } 5.\\
&= 6a^2 + 12a + 10a + 20 &&\text{Multiply the monomials.}\\
&= 6a^2 + 22a + 20 &&\text{Combine like terms.}
\end{aligned}
$$

In the third line of the solution, notice that each term of $3a + 5$ has been multiplied by each term of $2a + 4$. This example suggests the following rule.

> **Multiplying Binomials**
>
> To multiply two binomials, multiply each term of one binomial by each term of the other binomial, and then combine like terms.

We can use a shortcut method, called the **FOIL method,** to multiply binomials. FOIL is an acronym for **F**irst terms, **O**uter terms, **I**nner terms, **L**ast terms. To use the FOIL method to multiply $2a + 4$ by $3a + 5$, we

1. multiply the **F**irst terms $2a$ and $3a$ to obtain $6a^2$,
2. multiply the **O**uter terms $2a$ and 5 to obtain $10a$,
3. multiply the **I**nner terms 4 and $3a$ to obtain $12a$, and
4. multiply the **L**ast terms 4 and 5 to obtain 20.

Then we simplify the resulting polynomial, if possible.

$$
(2a + 4)(3a + 5) = 2a(3a) + 2a(5) + 4(3a) + 4(5)
$$

$$
\begin{aligned}
&= 6a^2 + 10a + 12a + 20 &&\text{Multiply the monomials.}\\
&= 6a^2 + 22a + 20 &&\text{Combine like terms.}
\end{aligned}
$$

> **The Language of Algebra** An *acronym* is an abbreviation of several words in such a way that the abbreviation itself forms a word. The *acronym* FOIL helps us remember the order to follow when multiplying two binomials: **F**irst, **O**uter, **I**nner, **L**ast.

EXAMPLE 3 Multiply: **a.** $(x + 5)(x + 7)$ **b.** $(3x + 4)(2x - 3)$

Strategy We will use the FOIL method.

WHY In each case we are to find the product of two binomials, and the FOIL method is a shortcut for multiplying two binomials.

Self Check 3

Multiply:
a. $(y + 3)(y + 1)$ $y^2 + 4y + 3$
b. $(2a - 1)(3a + 2)$ $6a^2 + a - 2$

Now Try Problems 21 and 23

Teaching Example 3 Multiply:
a. $(p + 4)(p + 3)$ b. $(6r - 1)(2r + 9)$
Answers:
a. $p^2 + 7p + 12$ b. $12r^2 + 52r - 9$

Solution

a.
$$(x + 5)(x + 7) = x(x) + x(7) + 5(x) + 5(7)$$
$$= x^2 + 7x + 5x + 35 \quad \text{Multiply the monomials.}$$
$$= x^2 + 12x + 35 \quad \text{Combine like terms.}$$

b.
$$(3x + 4)(2x - 3) = 3x(2x) + 3x(-3) + 4(2x) + 4(-3)$$
$$= 6x^2 - 9x + 8x - 12 \quad \text{Multiply the monomials.}$$
$$= 6x^2 - x - 12 \quad \text{Combine like terms.}$$

Self Check 4

Find: $(5x + 4)^2$ $25x^2 + 40x + 16$

Now Try Problem 25

Teaching Example 4
Find: $(8x + 5)^2$
Answer:
$64x^2 + 80x + 25$

EXAMPLE 4   Find: $(5x - 4)^2$

Strategy We will write the base, $5x - 4$, as a factor twice, and perform the multiplication.

WHY In the expression $(5x - 4)^2$, the binomial $5x - 4$ is the base and 2 is the exponent.

Solution

$$(5x - 4)^2 = (5x - 4)(5x - 4) \quad \text{Write the base as a factor twice.}$$
$$= 5x(5x) + 5x(-4) + (-4)(5x) + (-4)(-4)$$
$$= 25x^2 - 20x - 20x + 16 \quad \text{Multiply the monomials.}$$
$$= 25x^2 - 40x + 16 \quad \text{Combine like terms.}$$

Caution! A common error when squaring a binomial is to square only its first and second terms. For example, it is incorrect to write

$$(5x - 4)^2 = (5x)^2 - (4)^2$$
$$= 25x^2 - 16$$

The correct answer is $25x^2 - 40x + 16$.

4 Multiply polynomials.

To develop a general rule for multiplying any two polynomials, we will find the product of $2x + 3$ and $3x^2 + 3x + 5$. In the solution, the distributive property is used four times.

$$(2x + 3)(3x^2 + 3x + 5)$$
$$= (2x + 3)3x^2 + (2x + 3)3x + (2x + 3)5 \quad \text{Distribute.}$$
$$= (2x + 3)3x^2 + (2x + 3)3x + (2x + 3)5$$
$$= (2x)3x^2 + (3)3x^2 + (2x)3x + (3)3x + (2x)5 + (3)5 \quad \text{Distribute.}$$
$$= 6x^3 + 9x^2 + 6x^2 + 9x + 10x + 15 \quad \text{Multiply the monomials.}$$
$$= 6x^3 + 15x^2 + 19x + 15 \quad \text{Combine like terms.}$$

In the third line of the solution, note that each term of $3x^2 + 3x + 5$ has been multiplied by each term of $2x + 3$. This example suggests the following rule.

> **Multiplying Polynomials**
>
> To multiply two polynomials, multiply each term of one polynomial by each term of the other polynomial, and then combine like terms.

EXAMPLE 5   Multiply: $(7y + 3)(6y^2 - 8y + 1)$

Strategy We will multiply each term of the trinomial, $6y^2 - 8y + 1$, by each term of the binomial, $7y + 3$.

WHY To multiply two polynomials, we must multiply each term of one polynomial by each term of the other polynomial.

Solution

$(7y + 3)(6y^2 - 8y + 1)$

$= 7y(6y^2) + 7y(-8y) + 7y(1) + 3(6y^2) + 3(-8y) + 3(1)$
$= 42y^3 - 56y^2 + 7y + 18y^2 - 24y + 3$ Multiply the monomials.
$= 42y^3 - 38y^2 - 17y + 3$ Combine like terms.

Caution! The FOIL method cannot be applied here—only to products of two binomials.

Self Check 5
Multiply:
$(3a^2 - 1)(2a^4 - a^2 - a)$
Now Try Problem 29

Answer to Self Check 5
$6a^6 - 5a^4 - 3a^3 + a^2 + a$

Teaching Example 5 Multiply:
$(3x^2 + 1)(2x^2 - 3x + 1)$
Answer:
$6x^4 - 9x^3 + 5x^2 - 3x + 1$

It is often convenient to multiply polynomials using a **vertical form** similar to that used to multiply whole numbers.

> **Success Tip** Multiplying two polynomials in vertical form is much like multiplying two whole numbers in arithmetic.
>
> $\quad\quad 347$
> $\underline{\times\ \ 25}$
> $\quad 1\ 735$
> $\underline{+\ 6\ 940}$
> $\quad 8{,}675$

EXAMPLE 6   Multiply using vertical form:

a. $(3a^2 - 4a + 7)(2a + 5)$ **b.** $(6y^3 - 5y + 4)(-4y^2 - 3)$

Strategy First, we will write one polynomial underneath the other and draw a horizontal line beneath them. Then, we will multiply each term of the upper polynomial by each term of the lower polynomial.

WHY *Vertical form* means to use an approach similar to that used in arithmetic to multiply two whole numbers.

Self Check 6
Multiply using vertical form:
a. $(3x + 2)(2x^2 - 4x + 5)$
b. $(-2x^2 + 3)(2x^2 - 4x - 1)$
Now Try Problem 33

Answers to Self Check 6
a. $6x^3 - 8x^2 + 7x + 10$
b. $-4x^4 + 8x^3 + 8x^2 - 12x - 3$

Teaching Example 6 Multiply using vertical form:
a. $(x^2 - 2x + 1)(x + 2)$
b. $(x^3 + 5x^2 - 4x - 4)(x^2 - 8)$

Answers:
a. $x^3 - 3x + 2$
b. $x^5 + 5x^4 - 12x^3 - 44x^2 + 32x + 32$

Solution

a. Multiply:
$$\begin{array}{r} 3a^2 - 4a + 7 \\ \times \quad 2a + 5 \\ \hline 15a^2 - 20a + 35 \\ 6a^3 - 8a^2 + 14a \quad\quad \\ \hline 6a^3 + 7a^2 - 6a + 35 \end{array}$$

Multiply $3a^2 - 4a + 7$ by 5.
Multiply $3a^2 - 4a + 7$ by $2a$.
In each column, combine like terms.

b. With this method, it is often necessary to leave a space for a missing term to vertically align like terms.

Multiply:
$$\begin{array}{r} 6y^3 - 5y + 4 \\ \times \quad -4y^2 - 3 \\ \hline -18y^3 \quad + 15y - 12 \\ -24y^5 + 20y^3 - 16y^2 \quad\quad\quad \\ \hline -24y^5 + 2y^3 - 16y^2 + 15y - 12 \end{array}$$

Multiply $6y^3 - 5y + 4$ by -3.
Multiply $6y^3 - 5y + 4$ by $-4y^2$.
Leave a space for any missing powers of y.
In each column, combine like terms.

ANSWERS TO SELF CHECKS

1. $-14a^8$ 2. a. $15y^4 - 12y^2$ b. $15x^3 - 10x^2 + 15x$ 3. a. $y^2 + 4y + 3$ b. $6a^2 + a - 2$
4. $25x^2 + 40x + 16$ 5. $6a^6 - 5a^4 - 3a^3 + a^2 + a$ 6. a. $6x^3 - 8x^2 + 7x + 10$
b. $-4x^4 + 8x^3 + 8x^2 - 12x - 3$

SECTION 10.4 STUDY SET

VOCABULARY

Fill in the blanks.

1. $(2x^3)(3x^4)$ is the product of two __monomials__.
2. $(2a - 4)(3a + 5)$ is the product of two __binomials__.
3. In the acronym FOIL, F stands for __first__ terms, O for __outer__ terms, I for __inner__ terms, and L for __last__ terms.
4. $(2a - 4)(3a^2 + 5a - 1)$ is the product of a __binomial__ and a __trinomial__.

CONCEPTS

Fill in the blanks.

5. To multiply two polynomials, multiply __each__ term of one polynomial by __each__ term of the other polynomial, and then combine like terms.

6. Label each arrow using one of the letters F, O, I, or L. Then fill in the blanks.

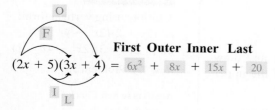

$(2x + 5)(3x + 4) = $ __$6x^2$__ + __$8x$__ + __$15x$__ + __20__

7. Simplify each polynomial by combining like terms.
   a. $6x^2 - 8x + 9x - 12$ $6x^2 + x - 12$
   b. $5x^4 + 3x^2 + 5x^2 + 3$ $5x^4 + 8x^2 + 3$

8. a. Add: $(x - 4) + (x + 8)$ $2x + 4$
   b. Subtract: $(x - 4) - (x + 8)$ -12
   c. Multiply: $(x - 4)(x + 8)$ $x^2 + 4x - 32$

NOTATION

Complete each step.

9. $(9n^3)(8n^2) = (9 \cdot \underline{8})(\underline{n^3} \cdot n^2) = \underline{72n^5}$

10. $7x(3x^2 - 2x + 5) = \underline{7x}(3x^2) - \underline{7x}(2x) + \underline{7x}(5)$
    $= \underline{21x^3} - 14x^2 + 35x$

11. $(2x + 5)(3x - 2) = 2x(3x) - \underline{2x}(2) + \underline{5}(3x) - \underline{5}(2)$
    $= 6x^2 - \underline{4x} + \underline{15x} - 10$
    $= 6x^2 + \underline{11x} - 10$

12. $$\begin{array}{r} 3x^2 + 4x - 2 \\ \times \quad 2x + 3 \\ \hline 9x^2 + 12x - 6 \\ 6x^3 + 8x^2 - 4x \quad\quad \\ \hline 6x^3 + 17x^2 + \underline{8x} - 6 \end{array}$$

GUIDED PRACTICE

Multiply. See Example 1.

13. $(3x^2)(4x^3)$ $12x^5$
14. $(-2a^3)(3a^2)$ $-6a^5$
15. $(3b^2)(-2b)$ $-6b^3$
16. $(3y)(-y^4)$ $-3y^5$

Multiply. See Example 2.

17. $2x^2(3x^2 + x)$
    $6x^4 + 2x^3$
18. $4b^3(2b^2 - 2b)$
    $8b^5 - 8b^4$
19. $2x(3x^2 + 4x - 7)$
    $6x^3 + 8x^2 - 14x$
20. $3y(2y^2 - 7y - 8)$
    $6y^3 - 21y^2 - 24y$

Multiply. See Example 3.

21. $(a + 4)(a + 5)$
    $a^2 + 9a + 20$
22. $(y - 3)(y + 5)$
    $y^2 + 2y - 15$
23. $(3x - 2)(x + 4)$
    $3x^2 + 10x - 8$
24. $(t + 4)(2t - 3)$
    $2t^2 + 5t - 12$

Square each binomial. See Example 4.

25. $(2x + 3)^2$
    $4x^2 + 12x + 9$
26. $(2y + 5)^2$
    $4y^2 + 20y + 25$
27. $(9b - 2)^2$
    $81b^2 - 36b + 4$
28. $(7m - 2)^2$
    $49m^2 - 28m + 4$

Multiply. See Example 5.

29. $(2x + 1)(3x^2 - 2x + 1)$ $6x^3 - x^2 + 1$
30. $(x + 2)(2x^2 + x - 3)$ $2x^3 + 5x^2 - x - 6$
31. $(x - 1)(x^2 + x + 1)$ $x^3 - 1$
32. $(x + 2)(x^2 - 2x + 4)$ $x^3 + 8$

Multiply. See Example 6.

33. $\quad x^2 - x + 1$
    $\times \quad x + 1$
    $\overline{\quad x^3 + 1}$
34. $\quad 4x^2 - 2x + 1$
    $\times \quad 2x + 1$
    $\overline{\quad 8x^3 + 1}$
35. $\quad 4x^2 + 3x - 4$
    $\times \quad 3x + 2$
    $\overline{12x^3 + 17x^2 - 6x - 8}$
36. $\quad 5r^2 + r + 6$
    $\times \quad 2r - 1$
    $\overline{10r^3 - 3r^2 + 11r - 6}$

TRY IT YOURSELF

Perform the operations.

37. $(2a + 4)(3a - 5)$
    $6a^2 + 2a - 20$
38. $(2b - 1)(3b + 4)$
    $6b^2 + 5b - 4$
39. $-p(2p^2 - 3p + 2)$
    $-2p^3 + 3p^2 - 2p$
40. $-2t(t^2 - t + 1)$
    $-2t^3 + 2t^2 - 2t$
41. $(-2x^2)(3x^3)$ $-6x^5$
42. $(-7x^3)(-3x^3)$ $21x^6$
43. $\quad 4x + 3$
    $\times \quad x + 2$
    $\overline{4x^2 + 11x + 6}$
44. $\quad 5r + 6$
    $\times \quad 2r - 1$
    $\overline{10r^2 + 7r - 6}$
45. $(2x - 3)^2$
    $4x^2 - 12x + 9$
46. $(2y - 5)^2$
    $4y^2 - 20y + 25$
47. $3q^2(q^2 - 2q + 7)$
    $3q^4 - 6q^3 + 21q^2$
48. $4v^3(-2v^2 + 3v - 1)$
    $-8v^5 + 12v^4 - 4v^3$
49. $\left(-\dfrac{2}{3}y^5\right)\left(\dfrac{3}{4}y^2\right)$ $-\dfrac{1}{2}y^7$
50. $\left(\dfrac{2}{5}r^4\right)\left(\dfrac{3}{5}r^2\right)$ $\dfrac{6}{25}r^6$

51. $(x + 2)(x^2 - 3x + 1)$ $x^3 - x^2 - 5x + 2$
52. $(x + 3)(x^2 + 3x + 2)$ $x^3 + 6x^2 + 11x + 6$
53. $\quad 2a^2 + 3a + 1$
    $\times \quad 3a^2 - 2a + 4$
    $\overline{6a^4 + 5a^3 + 5a^2 + 10a + 4}$
54. $\quad 3y^2 + 2y - 4$
    $\times \quad 2y^2 - 4y + 3$
    $\overline{6y^4 - 8y^3 - 7y^2 + 22y - 12}$
55. $(x + 6)(x^3 + 5x^2 - 4x - 4)$
    $x^4 + 11x^3 + 26x^2 - 28x - 24$
56. $(x - 8)(x^3 - 4x^2 - 2x - 2)$
    $x^4 - 12x^3 + 30x^2 + 14x + 16$
57. $(3n + 1)(3n - 1)$
    $9n^2 - 1$
58. $(5a + 4)(5a - 4)$
    $25a^2 - 16$
59. $(r^2 - r + 3)(r^2 - 4r - 5)$ $r^4 - 5r^3 + 2r^2 - 7r - 15$
60. $(w^2 + w - 9)(w^2 - w + 3)$ $w^4 - 7w^2 + 12w - 27$
61. $(5t - 1)^2$
    $25t^2 - 10t + 1$
62. $(6a - 3)^2$
    $36a^2 - 36a + 9$
63. $3x(x - 2)$ $3x^2 - 6x$
64. $4y(y + 5)$ $4y^2 + 20y$

APPLICATIONS

65. **GEOMETRY** Find a polynomial that represents the area of the rectangle (*Hint:* Recall that the area of a rectangle is the product of its length and width). $(x^2 - 4)$ ft²

66. **SAILING** The height h of the triangular sail is $4x$ feet, and the base b is $(3x - 2)$ feet. Find a polynomial that represents the area of the sail. (*Hint:* The area of a triangle is given by the formula $A = \tfrac{1}{2}bh$.) $(6x^2 - 4x)$ ft²

67. STAMPS Find a polynomial that represents the area of the stamp. $(6x^2 + x - 1)$ cm^2

68. PARKING Find a polynomial that represents the total area of the van-accessible parking space and its access aisle. $(2x^2 + 20x)$ ft^2

69. TOYS Find a polynomial that represents the area of the Etch-A-Sketch screen. $(35x^2 + 43x + 12)$ in.^{2}

70. PLAYPENS Find a polynomial that represents the area of the floor of the playpen. $(x^2 + 12x + 36)$ in.^{2}

71. SIGNS Find a polynomial that represents the area of the front of the sign. $\left(\frac{1}{2}h^2 - 2h\right)$ in.^{2}

72. GEOMETRY The length, width, and height of the rectangular solid shown here are consecutive integers.

a. Write a polynomial function that gives the volume of the solid.
$x(x+1)(x+2) = x^3 + 3x^2 + 2x$

b. What is the volume of the solid if the shortest dimension is 5 inches? 210 in.^{3}

WRITING

73. Explain how to multiply two binomials.

74. Explain how to find $(2x+1)^2$.

75. Explain why $(x+1)^2 \neq x^2 + 1^2$. (Read $\neq$ as "is not equal to.")

76. If two terms are to be added, they have to be like terms. If two terms are to be multiplied, must they be like terms? Explain.

REVIEW

77. THE EARTH It takes 23 hours, 56 minutes, and 4.091 seconds for the Earth to rotate on its axis once. Write 4.091 in words.
four and ninety-one thousandths

78. TAKE-OUT FOOD The sticker shows the amount and the price per pound of some spaghetti salad that was purchased at a delicatessen. Find the total price of the salad. $3.08

79. Write $\frac{7}{64}$ as a decimal. 0.109375

80. Write $-\frac{6}{10}$ as a decimal. -0.6

81. Evaluate: $56.09 + 78 + 0.567$ 134.657

82. Evaluate: $-679.4 - (-599.89)$ -79.51

83. Evaluate: $\sqrt{16} + \sqrt{36}$ 10

84. Divide: $103.6 \div 0.56$ 185

Teaching Guide: Refer to the Instructor's Resource Binder to find activities, worksheets on key concepts, more examples, instruction tips, overheads, and assessments.

CHAPTER 10 SUMMARY AND REVIEW

SECTION 10.1 Multiplication Rules for Exponents

DEFINITIONS AND CONCEPTS

An **exponent** indicates repeated multiplication. It tells how many times the **base** is to be used as a factor.

Exponent ⟶ $x^n = \underbrace{x \cdot x \cdot x \cdots \cdot x}_{n \text{ factors of } x}$
Base ⟶

EXAMPLES

Identify the base and the exponent for each given expression.

$2^6 = 2 \cdot 2 \cdot 2 \cdot 2 \cdot 2 \cdot 2$ 2 is the base and 6 is the exponent.

$(-xy)^3 = (-xy)(-xy)(-xy)$ Because of the parentheses, $-xy$ is the base and 3 is the exponent.

$5t^4 = 5 \cdot t \cdot t \cdot t \cdot t$ The base is t and 4 is the exponent.

$8^1 = 8$ The base is 8 and 1 is the exponent.

Rules for Exponents: If m and n represent integers,

Product rule: $x^m x^n = x^{m+n}$

Power rule: $(x^m)^n = x^{m \cdot n} = x^{mn}$

Power of a product rule: $(xy)^m = x^m y^m$

Simplify each expression:

$5^2 5^7 = 5^{2+7} = 5^9$ Keep the common base, 5, and add the exponents.

$(6^3)^7 = 6^{3 \cdot 7} = 6^{21}$ Keep the base, 6, and multiply the exponents.

$(2p)^5 = 2^5 p^5 = 32p^5$ Raise each factor of the product $2p$ to the 5th power.

To simplify some expressions, we must apply two (or more) rules for exponents.

Simplify: $(c^2 c^5)^4 = (c^7)^4$ Within the parentheses, keep the common base, c, and add the exponents: $2 + 5 = 7$.

$= c^{28}$ Keep the base, c, and multiply the exponents: $7 \cdot 4 = 28$.

Simplify: $(t^2)^4 (t^3)^3 = t^8 t^9$ For each power of t raised to a power, keep the base and multiply the exponents: $2 \cdot 4 = 8$ and $3 \cdot 3 = 9$.

$= t^{17}$ Keep the common base, t, and add the exponents: $8 + 9 = 17$.

REVIEW EXERCISES

1. Identify the base and the exponent in each expression.
   a. n^{12} base n, exponent 12
   b. $(2x)^6$ base $2x$, exponent 6
   c. $3r^4$ base r, exponent 4
   d. $(y - 7)^3$ base $y - 7$, exponent 3

2. Write each expression in an equivalent form using an exponent.
   a. $m \cdot m \cdot m \cdot m \cdot m$ m^5
   b. $-3 \cdot x \cdot x \cdot x \cdot x$ $-3x^4$
   c. $a \cdot a \cdot b \cdot b \cdot b \cdot b$ $a^2 b^4$
   d. $(pq)(pq)(pq)$ $(pq)^3$ or $p^3 q^3$

3. Simplify, if possible.
   a. $x^2 \cdot x^2$ x^4
   b. $x^2 + x^2$ $2x^2$
   c. $x \cdot x^2$ x^3
   d. $x + x^2$ does not simplify

4. Explain each error.
   a. $3^2 \cdot 3^4 = 9^6$ Keep the base 3, don't multiply the bases.
   b. $(3^2)^4 = 3^6$ Multiply the exponents, don't add them.

Simplify each expression.

5. $7^4 \cdot 7^8$ 7^{12}
6. $mmnn^2$ m^2n^3
7. $(y^7)^3$ y^{21}
8. $(3x)^4$ $81x^4$
9. $(6^3)^{12}$ 6^{36}
10. $-b^3b^4b^5$ $-b^{12}$
11. $(-16s^3)^2s^4$ $256s^{10}$
12. $(2.1x^2y)^2$ $4.41x^4y^2$
13. $[(-9)^3]^5$ $(-9)^{15}$
14. $(a^5)^3(a^2)^4$ a^{23}
15. $(2x^2x^3)^3$ $8x^{15}$
16. $(m^2m^3)^2(n^2n^4)^3$ $m^{10}n^{18}$
17. $(3a^4)^2(2a^3)^3$ $72a^{17}$
18. $x^{100} \cdot x^{100}$ x^{200}
19. $(4m^3)^3(2m^2)^2$ $256m^{13}$
20. $(3t^4)^3(2t^5)^2$ $108t^{22}$

SECTION 10.2 Introduction to Polynomials

DEFINITIONS AND CONCEPTS	EXAMPLES
A **polynomial** is a single term or a sum of terms in which all variables have whole-number exponents and no variable appears in a denominator.	Polynomials: 32, $-5x^2y^3$, $\frac{7}{4}p^3 - 14q^3$, $4m^2 + 5m - 12$
A polynomial with exactly one term is called a **monomial**. A polynomial with exactly two terms is called a **binomial**. A polynomial with exactly three terms is called a **trinomial**.	**Monomials** **Binomials** **Trinomials** $3x^2$ $2y^3 + 3y$ $3p^2 - 7p + 12$ $-12m^3n^2$ $87t - 25$ $4p^2q^3 - 8p^2q^2 + 12p^2q$
The **degree of a polynomial** is equal to the highest degree of any term of the polynomial.	**Polynomial** **Degree of the polynomial** $7m^3 - 4m^2 + 5m - 12$ 3 $3a^3 - 2a^2 + a^4$ 4
To **evaluate a polynomial** for a given value, substitute the value for the variable and follow the order of operations rule.	Evaluate $3x^2 - 4x + 2$ for $x = 2$. $3x^2 - 4x + 2 = 3(2)^2 - 4(2) + 2$ Substitute 2 for each x. $= 3(4) - 4(2) + 2$ Evaluate $(2)^2$ first. $= 12 - 8 + 2$ Multiply. $= 6$

REVIEW EXERCISES

Classify each polynomial as a monomial, a binomial, or a trinomial.

21. $3x^2 + 4x - 5$ trinomial
22. $3t^2$ monomial
23. $2x^2 - 1$ binomial
24. $\frac{1}{2}d^5 + \frac{3}{2}d^3 + \frac{5}{2}d$ trinomial

Give the degree of each polynomial.

25. $3x^2 + 2x^3$ 3
26. $3t^4 - 4t^2 - 3$ 4
27. $3q^2 - 4q^5$ 5
28. $0.2a - 4.5a^5 + 1.3a^3$ 5

29. Evaluate $2t^2 + t - 2$ for $t = -3$. 13

30. **WATER BALLOONS** Some college students launched water balloons from the balcony of their dormitory on unsuspecting sunbathers. The height h in feet of the balloons at a time t seconds after being launched is given by the polynomial

$$h = -16t^2 + 12t + 20$$

What was the height of the balloons 1 second after being launched? 16 ft

SECTION 10.3 Adding and Subtracting Polynomials

DEFINITIONS AND CONCEPTS	EXAMPLES	
To **add polynomials**, combine their like terms.	Add: $(4x^2 + 9x + 4) + (3x^2 - 5x - 1)$	
	$= (4x^2 + 3x^2) + (9x - 5x) + (4 - 1)$	Group like terms.
	$= 7x^2 + 4x + 3$	Combine like terms.
To **subtract two polynomials**, change the signs of the terms of the polynomial being subtracted, drop the parentheses, and combine like terms.	Subtract: $(8a^3 - 4a) - (-3a^3 + 9a)$	
	$= 8a^3 - 4a + 3a^3 - 9a$	Change the sign of each term of $-3a^3 + 9a$ and drop the parentheses.
	$= 11a^3 - 13a$	Combine like terms.

REVIEW EXERCISES

Add.

31. $3x^3 + 2x^3$ $5x^3$
32. $\frac{1}{2}p^2 + \frac{5}{2}p^2 + \frac{7}{2}p^2$ $\frac{13}{2}p^2$
33. $(3x - 1) + (6x + 5)$ $9x + 4$
34. $(3x^2 - 2x + 4) + (-x^2 - 1)$ $2x^2 - 2x + 3$
35. $5x - 2$
    $\underline{+ 3x + 5}$ $8x + 3$
36. $3x^2 - 2x + 7$
    $\underline{+ -5x^2 + 3x - 5}$ $-2x^2 + x + 2$

Subtract.

37. $16p^3 - 9p^3$ $7p^3$
38. $4y^2 - 9y^2$ $-5y^2$
39. $(2.5x + 4) - (1.4x + 12)$ $1.1x - 8$
40. $(3z^2 - z + 4) - (2z^2 + 3z - 2)$ $z^2 - 4z + 6$
41. $5x - 2$
    $\underline{-(3x + 5)}$ $2x - 7$
42. $3x^2 - 2x + 7$
    $\underline{-(-5x^2 + 3x - 5)}$ $8x^2 - 5x + 12$

SECTION 10.4 Multiplying Polynomials

DEFINITIONS AND CONCEPTS	EXAMPLES	
To **multiply two monomials**, multiply the numerical factors (the coefficients) and then multiply the variable factors.	Multiply: $(5p^6)(2p^5) = (5 \cdot 2)(p^6 \cdot p^5)$	Group the coefficients together and the variables together.
	$= 10p^{11}$	Think: $5 \cdot 2 = 10$ and $p^6 \cdot p^5 = p^{6+5} = p^{11}$.
To **multiply a monomial and a polynomial**, multiply each term of the polynomial by the monomial.	Multiply: $3r^2(2r^4 + 7r^2 - 4)$	
	$= 3r^2(2r^4) + 3r^2(7r^2) + 3r^2(-4)$	Distribute the multiplication by $3r^2$.
	$= 6r^6 + 21r^4 - 12r^2$	Multiply the monomials.

To **multiply two binomials**, use the *FOIL* method:

F: First
O: Outer
I: Inner
L: Last

Multiply: $(3m + 4)(2m - 5) = 3m(2m) + 3m(-5) + 4(2m) + 4(-5)$
$= 6m^2 - 15m + 8m - 20$ Multiply the monomials.
$= 6m^2 - 7m - 20$ Combine like terms.

To **multiply two polynomials**, multiply each term of one polynomial by each term of the other polynomial and then combine like terms.

Multiply: $(a - 3)(6a^2 - 4a + 1)$

$= a(6a^2) + a(-4a) + a(1) - 3(6a^2) - 3(-4a) - 3(1)$
$= 6a^3 - 4a^2 + a - 18a^2 + 12a - 3$ Multiply the monomials.
$= 6a^3 - 22a^2 + 13a - 3$ Combine like terms.

REVIEW EXERCISES

Multiply.

43. $3x^2 \cdot 5x^3$ $15x^5$

44. $(3z^2)(-2z^2)$ $-6z^4$

45. $2x^2(3x + 2)$ $6x^3 + 4x^2$

46. $-5t^3(7t^2 - 6t - 2)$ $-35t^5 + 30t^4 + 10t^3$

47. $(2x - 1)(3x + 2)$ $6x^2 + x - 2$

48. $(5t + 4)(7t - 6)$ $35t^2 - 2t - 24$

49. $(3x + 2)(2x^2 - x + 1)$ $6x^3 + x^2 + x + 2$

50. $(2r - 3)(3r^2 + 2r - 3)$ $6r^3 - 5r^2 - 12r + 9$

51. $\begin{array}{r} 5x^2 - 2x + 3 \\ \times \underline{ 3x + 5} \end{array}$ $15x^3 + 19x^2 - x + 15$

52. $\begin{array}{r} 3x^2 - 2x - 1 \\ \times \underline{ 5x - 2} \end{array}$ $15x^3 - 16x^2 - x + 2$

Square each binomial.

53. $(x + 2)^2$ $x^2 + 4x + 4$

54. $(8a - 3)^2$ $64a^2 - 48a + 9$

CHAPTER 10 TEST

Fill in the blanks.

1. In the exponential expression 7^5, 7 is the __base__ and 5 is the __exponent__.
2. Expressions such as x^4, 10^3, and $(5t)^2$ are called __exponential__ expressions.
3. A __polynomial__ is a term or a sum of terms in which all variables have whole-number exponents and no variable appears in a denominator.
4. A __monomial__ is a polynomial with exactly one term. A __binomial__ is a polynomial with exactly two terms. A __trinomial__ is a polynomial with exactly three terms.
5. The __degree__ of the term $3x^7$ is 7 because x appears as a factor 7 times: $3 \cdot x \cdot x \cdot x \cdot x \cdot x \cdot x \cdot x$.
6. To __evaluate__ the polynomial $x^2 - 2x + 1$ for $x = 6$, we substitute 6 for x and follow the order of operations rule.
7. $(b^3 - b^2 - 9b + 1) + (b^3 - b^2 - 9b + 1)$ is the sum of two __polynomials__.
8. $(b^2 - 9b + 11) - (4b^2 - 14b)$ is the __difference__ of a trinomial and a binomial.
9. In the acronym FOIL, F stands for __first__ terms, O for __outer__ terms, I for __inner__ terms, and L for __last__ terms.
10. $(2a - 4)(3a^2 + 5a - 1)$ is the product of a __binomial__ and a __trinomial__.
11. $(2x + 3)^2$ is the __square__ of a binomial.
12. To multiply two polynomials, multiply __each__ term of one polynomial by __each__ term of the other polynomial, and then combine like terms.

13. Identify the base and the exponent of each expression.
    a. 6^5 base: 6, exponent: 5
    b. $7b^4$ base: b, exponent: 4

14. Simplify each expression, if possible.
    a. $a^2 + a^2$ $2a^2$
    b. $a^2 \cdot a^2$ a^4
    c. $a^2 + a$ does not simplify
    d. $a^2 \cdot a$ a^3

Simplify each expression.

15. h^2h^4 h^6
16. $(m^{10})^2$ m^{20}
17. $b^2 \cdot b \cdot b^5$ b^8
18. $(x^3)^4(x^2)^3$ x^{18}
19. $(a^2b^3)(a^4b^7)$ a^6b^{10}
20. $(12a^9b)^2$ $144a^{18}b^2$
21. $(2x^2)^3(3x^3)^3$ $216x^{15}$
22. $(t^2t^3)^3$ t^{15}

Classify each polynomial as a monomial, a binomial, or a trinomial.

23. $5x^2 + 4x$ binomial
24. $-\dfrac{3}{4}t^{15}$ monomial
25. $-3x^2 - 2x + 3$ trinomial
26. $x - 8$ binomial

Give the degree of each polynomial.

27. $3t^4 - 2t^3 + 5t^6 - t$ 6
28. $10q^7 + 9q^5 - 8q^2$ 7

Evaluate each polynomial.

29. $3x^2 - 2x + 4$ for $x = 3$ 25
30. $-2r^2 - r + 3$ for $r = -1$ 2

Perform the operations.

31. $(2.1p^2 - 2p - 2) - (3.3p^2 - 5p - 2)$ $-1.2p^2 + 3p$
32. $(-2x^3)(4x^2)$ $-8x^5$
33. $(3x^2 + 2x) + (2x^2 - 5x + 4)$ $5x^2 - 3x + 4$
34. $(2x - 5)(3x + 4)$ $6x^2 - 7x - 20$
35. $\begin{array}{r} 3d^2 - 3d + 7.2 \\ -(-5d^2 + 6d - 5.3) \end{array}$ $8d^2 - 9d + 12.5$
36. $3y^2(y^2 - 2y + 3)$ $3y^4 - 6y^3 + 9y^2$
37. $\begin{array}{r} 4x^2 - 5x + 5 \\ + 3x^2 + 7x - 7 \end{array}$ $7x^2 + 2x - 2$
38. $(2x - 3)(x^2 - 2x + 4)$ $2x^3 - 7x^2 + 14x - 12$

39. **FILTERS** The length of one side of the square furnace filter shown below is $(x + 4)$ in. Find the perimeter of the filter. $(4x + 16)$ in.

$(x + 4)$ in.

40. Explain what is wrong with the following work:

    $5^4 \cdot 5^3 = 25^7$ Keep the common base 5, and add the exponents. Do not multiply the common bases to get 25.

CHAPTERS 1–10 CUMULATIVE REVIEW

1. **USED CARS** The following ad appeared in *The Car Trader*. (O.B.O. means "or best offer.") If offers of $8,750, $8,875, $8,900, $8,850, $8,800, $7,995, $8,995, and $8,925 were received, what was the selling price of the car? [Section 1.1] $8,995

   > 1969 Ford Mustang. New tires
   > Must sell!!!! $10,500 O.B.O.

2. Round 2,109,567 to the nearest thousand. [Section 1.1] 2,110,000
3. Add: $458 + 8,099 + 23,419 + 58$ [Section 1.2] 32,034
4. Subtract: $35,021 - 23,999$ [Section 1.2] 11,022
5. **PARKING** The length of a rectangular parking lot is 204 feet and its width is 97 feet. [Section 1.3]
   a. Find the perimeter of the lot. 602 ft
   b. Find the area of the lot. 19,788 ft^2
6. Divide: $1,363 \div 41$ [Section 1.4] 33 R 10
7. a. Prime factor 220. [Section 1.5] $2^2 \cdot 5 \cdot 11$
   b. Find all the factors of 12. [Section 1.5] 1, 2, 3, 4, 6, 12
8. a. Find the LCM of 16 and 24. [Section 1.6] 48
   b. Find the GCF of 16 and 24. [Section 1.6] 8
9. Evaluate: $\dfrac{(3+5)^2 + 2}{2(8-5)}$ [Section 1.7] 11
10. Solve each equation and check the result.
    a. $y - 81 = 243$ [Section 1.8] 324
    b. $81y = 243$ [Section 1.9] 3
11. a. Write the set of integers. [Section 2.1] $\{\ldots, -3, -2, -1, 0, 1, 2, 3, \ldots\}$
    b. Simplify: $-(-3)$ [Section 2.1] 3
12. Perform the operations.
    a. $-16 + 4$ [Section 2.2] -12
    b. $16 - (-4)$ [Section 2.3] 20
    c. $-16(4)$ [Section 2.4] -64
    d. $\dfrac{-16}{-4}$ [Section 2.5] 4
    e. -4^2 [Section 2.4] -16
    f. $(-4)^2$ [Section 2.4] 16
13. **OVERDRAFT PROTECTION** A student forgot that she had only $30 in her bank account and wrote a check for $55 and used her debit card to buy $75 worth of groceries. On each of the two transactions, the bank charged her a $20 overdraft protection fee. Find the new account balance. [Section 2.3] $-$140

14. Evaluate: $10 - 4|6 - (-3)^2|$ [Section 2.6] -2
15. Solve $-x + 2 = 13$ and check the result. [Section 2.7] -11
16. Solve $4 + \dfrac{x}{-5} - 6 = -1$ and check the result. [Section 2.7] -5

Write and solve an equation to answer the following question. [Section 2.7]

17. **ACCOUNTING** Because of bad economic times, Acme corporation lost $63 million in 2009. Just one year before, the corporation made a very large profit. If Acme lost a total of $17 million in this two-year span, how much profit did Acme make in 2008? In 2008, Acme made $46 million in profit.
18. See the illustration below. [Section 3.1]
    a. Let k represent the length (in inches) of the key. Write an algebraic expression that represents the length of the match (in inches). $k + 1$
    b. Let m represent the length (in inches) of the match. Write an algebraic expression that represents the length of the key (in inches). $m - 1$

19. Translate into mathematical symbols: 5 less than a number. [Section 3.1] $x - 5$ (answers may vary)
20. **AIRLINES** Find the distance traveled by a jet if it travels for 3 hours at 475 miles per hour. [Section 3.2] 1,425 mi
21. a. Simplify: $-10(-10a)$ [Section 3.3] $100a$
    b. Simplify: $-(-4y + 6)$ [Section 3.3] $4y - 6$
    c. Simplify: $-3(-6y - 8) - 12 - 4(5 - y)$ [Section 3.4] $22y - 8$
22. **SHOPPING** What is the value (in cents) of x coupons, each of which gives the shopper 50¢ off? [Section 3.6] $50x$¢

Solve each equation and check the result. [Section 3.5]

23. $5r - 24 = r + 5r + 2r$ -8
24. $2(4a + 8) = -3(2 - 3a) - 3a$ -11

Write and solve an equation to answer the following question. [Section 3.6]

25. **WISHING WELLS** A city park employee collected 600 cents in nickels, dimes, and quarters at the bottom of a wishing well. There were 10 nickels, and a combined total of 25 dimes and quarters. How many dimes and quarters were at the bottom of the well? There were 5 dimes and 20 quarters at the bottom of the wishing well.

26. Simplify: $\dfrac{35a^2}{28a}$ [Section 4.1] $\dfrac{5a}{4}$

27. a. Write $\dfrac{3}{8}$ as an equivalent fraction with denominator 48. [Section 4.1] $\dfrac{18}{48}$

    b. What is the reciprocal of $\dfrac{9}{8}$? [Section 4.3] $\dfrac{8}{9}$

    c. Write $7\dfrac{1}{2}$ as an improper fraction. [Section 4.5] $\dfrac{15}{2}$

28. **GRAVITY** Objects on the moon weigh only one-sixth as much as on Earth. If a rock weighs 54 ounces on the Earth, how much does it weigh on the moon? [Section 4.2] 9 oz

Perform the operations.

29. $-\dfrac{5}{77}\left(\dfrac{33}{50}\right)$ [Section 4.2] $-\dfrac{3}{70}$

30. $\dfrac{15b^3}{16c^5} \div \dfrac{45b}{8c}$ [Section 4.3] $\dfrac{b^2}{6c^4}$

31. $\dfrac{3}{4} - \dfrac{3}{5}$ [Section 4.4] $\dfrac{3}{20}$

32. $\dfrac{m}{9} - \dfrac{4}{5}$ [Section 4.4] $\dfrac{5m-36}{45}$

33. $-\dfrac{6}{25}\left(2\dfrac{7}{24}\right)$ [Section 4.5] $-\dfrac{11}{20}$

34. $45\dfrac{2}{3} + 96\dfrac{4}{5}$ [Section 4.6] $142\dfrac{7}{15}$

35. Simplify: $\dfrac{7 - \dfrac{2}{3}}{4\dfrac{5}{6}}$ [Section 4.7] $\dfrac{38}{29} = 1\dfrac{9}{29}$

36. **PET MEDICATION** A pet owner was told to use an eye dropper to administer medication to his sick kitten. The cup shown below contains 8 doses of the medication. Determine the size of a single dose. [Section 4.3] $\dfrac{3}{32}$ fluid oz

37. **BAKING** A bag of all-purpose flour contains $17\dfrac{1}{2}$ cups. A baker uses $3\dfrac{3}{4}$ cups. How many cups of flour are left? [Section 4.6] $13\dfrac{3}{4}$ cups

38. Evaluate: $a + b^2c$ for $a = \dfrac{3}{4}$, $b = -\dfrac{1}{3}$, and $c = \dfrac{5}{4}$. [Section 4.7] $\dfrac{8}{9}$

Solve each equation and check the result. [Section 4.8]

39. $\dfrac{x}{12} + \dfrac{2}{3} = \dfrac{x}{6}$ 8

40. $\dfrac{2}{3}q - 1 = -6$ $-\dfrac{15}{2}$

Write and solve an equation to answer the following question. [Section 4.8]

41. **COFFEE DRINKERS** Three-fifths of 275 students surveyed said they started their morning with a cup of coffee. Of the 275 students, how many would this be? 165

42. a. Round the number pi to the nearest ten thousandth: $\pi = 3.141592654\ldots$ [Section 5.1] 3.1416

    b. Place the proper symbol ($>$ or $<$) in the blank: 154.34 $>$ 154.33999. [Section 5.1]

43. a. Write 6,510,345.798 in words. [Section 5.1] six million, five hundred ten thousand, three hundred forty-five and seven hundred ninety-eight thousandths

    b. Write 7,498.6461 in expanded notation. [Section 5.1] $7,000 + 400 + 90 + 8 + \dfrac{6}{10} + \dfrac{4}{100} + \dfrac{6}{1,000} + \dfrac{1}{10,000}$

Perform the operations.

44. $3.4 + 106.78 + 35 + 0.008$ [Section 5.2] 145.188

45. $5,091.5 - 1,287.89$ [Section 5.2] 3,803.61

46. $-8.8 + (-7.3 - 9.5)$ [Section 5.2] -25.6

47. $-5.5(-3.1)$ [Section 5.3] 17.05

48. $\dfrac{0.0742}{1.4}$ [Section 5.4] 0.053

49. $\dfrac{7}{8}(9.7 + 15.8)$ [Section 5.5] 22.3125

50. **PAYCHECKS** If you are paid every other week, your *monthly gross income* is your gross income from one paycheck times 2.17. Find the monthly gross income of a secretary who earns $1,250 every two weeks. [Section 5.3] $2,712.50

51. Perform each operation in your head.

    a. $(89.9708)(10,000)$ [Section 5.3] 899,708

    b. $\dfrac{89.9708}{100}$ [Section 5.4] 0.899708

52. Estimate the quotient: $9.2\overline{)18,460.76}$ [Section 5.4] $18,000 \div 9 = 2,000$

53. Evaluate $\dfrac{(-1.3)^2 + 6.7}{-0.9}$ and round the result to the nearest hundredth. [Section 5.4] -9.32

54. Write $\dfrac{2}{15}$ as a decimal. Use an overbar. [Section 5.5] $0.1\overline{3}$

55. Evaluate each expression. [Section 5.6]

   a. $2\sqrt{121} - 3\sqrt{64}$ -2

   b. $\sqrt{\dfrac{49}{81}}$ $\dfrac{7}{9}$

56. Graph each number on the number line. [Section 5.6]
$$\left\{-4\dfrac{5}{8},\ \sqrt{17},\ 2.89,\ \dfrac{2}{3},\ -0.1,\ -\sqrt{9},\ \dfrac{3}{2}\right\}$$

57. Solve $1.7y + 1.24 = -1.4y - 0.62$ and check the result. [Section 5.7] -0.6

Write and solve an equation to answer the following question. [Section 5.8]

58. DECORATIONS A mother has budgeted $20 for decorations for her daughter's birthday party. She decides to buy a tank of helium for $15.15 and some balloons. If the balloons sell for 5 cents apiece, how many balloons can she buy? 97

59. Write each phrase as a ratio (fraction) in simplest form. [Section 6.1]

   a. 3 centimeters to 7 centimeters $\dfrac{3}{7}$

   b. 13 weeks to 1 year $\dfrac{1}{4}$

60. COMPARISON SHOPPING A dry-erase whiteboard with an area of 400 in.² sells for $24. A larger board, with an area of 600 in.², sells for $42. Which board is the better buy? [Section 6.1] the smaller board

61. Solve the proportion: $\dfrac{x}{14} = \dfrac{13}{28}$ [Section 6.2] $6\dfrac{1}{2} = 6.5$

62. INSURANCE CLAIMS In one year, an auto insurance company had 3 complaints per 1,000 policies. If a total of 375 complaints were filed that year, how many policies did the company have? [Section 6.2] 125,000

63. SCALE DRAWINGS On the scale drawing below, $\dfrac{1}{4}$-inch represents an actual length of 3 feet. The length of the house on the drawing is $6\dfrac{1}{4}$ inches. What is the actual length of the house? [Section 6.2] 75 ft

Scale $\dfrac{1}{4}$ in. : 3 ft

64. Make each conversion. [Section 6.3]

   a. Convert 168 inches to feet. 14 ft

   b. Convert 212 ounces to pounds. 13.25 lb = $13\dfrac{1}{4}$ lb

   c. Convert 30 gallons to quarts. 120 quarts

   d. Convert 12.5 hours to minutes. 750 min

65. Make each conversion. [Section 6.4]

   a. Convert 1.538 kilograms to grams. 1,538 g

   b. Convert 500 milliliters to liters. 0.5 L

   c. Convert 0.3 centimeters to kilometers. 0.000003 km

66. THE AMAZON The Amazon River enters the Atlantic Ocean through a broad estuary, roughly estimated at 240,000 m in width. Convert the width to kilometers. [Section 6.4] 240 km

67. COOKING What is the weight of a 10-pound ham in kilograms? [Section 6.5] about 4.5 kg

68. Convert 75°C to degrees Fahrenheit. [Section 6.5] 167°F

69. Complete the table. [Section 7.1]

Percent	Decimal	Fraction
57%	0.57	$\dfrac{57}{100}$
0.1%	0.001	$\dfrac{1}{1,000}$
$33\dfrac{1}{3}\%$	$0.\overline{3}$	$\dfrac{1}{3}$

70. Refer to the figure on the right. [Section 7.1]

   a. What percent of the figure is shaded? 93%

   b. What percent is not shaded? 7%

71. What number is 15% of 450? [Section 7.2] 67.5

72. 24.6 is 20.5% of what number? [Section 7.2] 120

73. 51 is what percent of 60? [Section 7.2] 85%

74. CLOTHING SALES Find the amount of the discount and the sale price of the coat shown below. [Section 7.3] $205, $615

75. SALES TAX If the sales tax rate is $6\dfrac{1}{4}\%$, how much sales tax will be added to the price of a new car selling for $18,550? [Section 7.3] $1,159.38

76. **COLLECTIBLES** A porcelain figurine, which was originally purchased for $125, was sold by a collector ten years later for $750. What was the percent increase in the value of the figurine? [Section 7.3] 500%

77. **TIPPING** Estimate a 15% tip on a dinner that cost $135.88. [Section 7.4] $21

78. **PAYING OFF LOANS** To pay for tuition, a college student borrows $1,500 for six months. If the annual interest rate is 9%, how much will the student have to repay when the loan comes due? [Section 7.5] $1,567.50

79. **FREEWAYS** Refer to the pictograph below to answer the following questions. [Section 8.1]

Freeway Traffic
Average number of vehicles daily

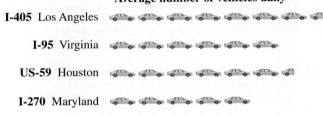

= 50,000 vehicles

Source: Office of Highway Policy Information, Federal Highway Administration

a. Estimate the number of vehicles that travel the I-405 Freeway in Los Angeles each day. 375,000 vehicles

b. Estimate the number of vehicles that travel the I-270 Freeway in Maryland each day. 250,000 vehicles

c. Estimate how many more vehicles travel US-59 in Houston than the I-95 Freeway in Virginia each day. 25,000 vehicles

80. **VEGETARIANS** The graph below gives the results of a recent study by *Vegetarian Times*. [Section 8.1]

Survey Results: Ages of Adult Vegetarians in the United States

Source: *Vegetarian Times*

a. According to the study, what percent of the adult vegetarians in the United States are over 55 years old? 18%

b. The study estimated that there were 7,300,000 adult vegetarians in the United States. How many of them are 35 to 54 years old? 2,920,000

81. **SPENDING ON PETS** Refer to the bar graph below to answer the following questions. [Section 8.1]

a. In what category was the most money spent on pets? Estimate how much. food: about $20.5 billion

b. Estimate how much money was spent on purchasing pets. about $2.2 billion

c. Estimate how much more money was spent on vet care than on grooming and boarding. about $9.5 billion

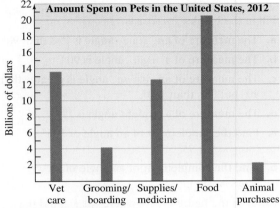

Source: americanpetproducts.org

82. **TABLE TENNIS** The weights (in ounces) of eight Ping-Pong balls that are to be used in a tournament are as follows: 0.85, 0.87, 0.88, 0.88, 0.85, 0.86, 0.84, and 0.85. Find the mean, median, and mode of the weights. [Section 8.2] mean: 0.86 oz, median: 0.855 oz, mode: 0.85 oz

83. Graph each ordered pair:
$(-4, -3)$
$(1.5, 1.5)$
$(-\frac{7}{2}, 0)$
$(0, 3.5)$ [Section 8.3]

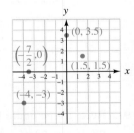

84. Is $(-2, 1)$ a solution of $3x - y = -8$? [Section 8.3] no

Graph each equation. [Section 8.4]

85. $y = -2x$

86. $2x - 3y = 12$

87. $x = 4$

88. Fill in the blanks. [Section 9.1]
  a. The measure of an acute angle is less than 90°.
  b. The measure of a right angle is 90°.
  c. The measure of an obtuse angle is greater than 90° but less than 180°.
  d. The measure of a straight angle is 180° .

89. a. Find the supplement of an angle of 105°. [Section 9.1] 75°
  b. Find the complement of an angle of 75°. [Section 9.1] 15°

90. Refer to the figure below, where $l_1 \parallel l_2$. Find the measure of each angle. [Section 9.2]
  a. m($\angle 1$) 50°
  b. m($\angle 3$) 50°
  c. m($\angle 2$) 130°
  d. m($\angle 4$) 50°

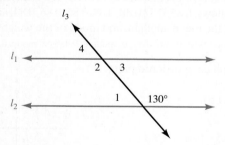

91. Refer to the figure below, where $AB \parallel DE$ and m($\overline{AC}$) = m($\overline{BC}$). Find the measure of each angle. [Section 9.3]
  a. m($\angle 1$) 75°
  b. m($\angle C$) 30°
  c. m($\angle 2$) 105°
  d. m($\angle 3$) 105°

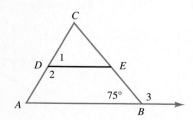

92. JAVELIN THROW Refer to the illustration below. Determine x and y. [Section 9.3] 46°, 134°

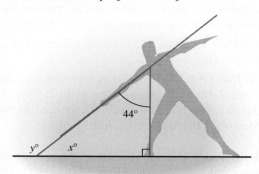

93. If the vertex angle of an isosceles triangle measures 34°, what is the measure of each base angle? [Section 9.3] 73°

94. If the legs of a right triangle measure 10 meters and 24 meters, how long is the hypotenuse? [Section 9.4] 26 m

95. Determine whether a triangle with sides of length 16 feet, 63 feet, and 65 feet is a right triangle. [Section 9.4] yes

96. SHADOWS If a tree casts a 35-foot shadow at the same time as a man 6 feet tall casts a 5-foot shadow, how tall is the tree? [Section 9.5] 42 ft

97. Find the sum of the angles of a pentagon. [Section 9.6] 540°

98. Find the perimeter and the area of a square that has sides each 12 meters long. [Section 9.7] 48 m, 144 m²

99. Find the area of a triangle with a base that is 14 feet long and a height of 18 feet. [Section 9.7] 126 ft²

100. Find the area of a trapezoid that has bases that are 12 inches and 14 inches long and a height of 7 inches. [Section 9.7] 91 in.²

101. How many square inches are in 1 square foot? [Section 9.7] 144 in.²

102. Find the circumference and the area of a circle that has a diameter of 14 centimeters. For each, give the exact answer and an approximation to the nearest hundredth. [Section 9.8]
circumference: 14π cm ≈ 43.98 cm,
area: 49π cm² ≈ 153.94 cm²

103. Find the area of the shaded region shown below, which is created using two semicircles. Round to the nearest hundredth. [Section 9.8] 98.31 yd²

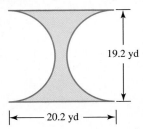

104. ICE Find the volume of a block of ice that is in the shape of a rectangular solid with dimensions 15 in. × 24 in. × 18 in. [Section 9.9] 6,480 in.³

105. Find the volume of a sphere that has a diameter of 18 inches. Give the exact answer and an approximation to the nearest hundredth. [Section 9.9] 972π in.³ ≈ 3,053.63 in.³

106. Find the volume of a cone that has a circular base with a radius of 4 meters and a height of 9 meters. Give the exact answer and an approximation to the nearest hundredth. [Section 9.9] 48π m³ ≈ 150.80 m³

107. Find the volume of a cylindrical pipe that is 20 feet long and has a radius of 1 foot. Give the exact answer and an approximation to the nearest hundredth. [Section 9.9] 20π ft³ ≈ 62.83 ft³

108. How many cubic inches are there in 1 cubic foot? [Section 9.9] 1,728 in.³

Simplify each expression. [Section 10.1]

109. $s^4 \cdot s^5$ s^9

110. $(a^5)^7$ a^{35}

111. $(y^5)^2(y^4)^3$ y^{22}

112. $(2b^3c^6)^3$ $8b^9c^{18}$

113. Classify $3x^2 - 7x + 1$ as a monomial, a binomial, or a trinomial. Then give its degree. [Section 10.2] trinomial; 2

114. Evaluate $0.5t^3 - t^2 - 4t$ for $t = 4$. [Section 10.2] 0

Perform the operations.

115. $(5x^2 - 2x + 4) - (3x^2 - 5)$ [Section 10.3] $2x^2 - 2x + 9$

116. $(6.2a^3 + 7.1a^2 - 4.1a) + (3.8a^3 - 4.1a)$ [Section 10.3] $10a^3 + 7.1a^2 - 8.2a$

117. $-3h^9(-5h)$ [Section 10.4] $15h^{10}$

118. $-3p(2p^2 + 3p - 4)$ [Section 10.4] $-6p^3 - 9p^2 + 12p$

119. $(3x + 5)(2x - 1)$ [Section 10.4] $6x^2 + 7x - 5$

120. $(2y - 7)^2$ [Section 10.4] $4y^2 - 28y + 49$

APPENDIX 1

Probability

Objectives

1. Identify the sample space for an experiment.
2. Identify an event of an experiment.
3. Find the probability of an event.

SECTION A1.1
Probability

If we flip a coin, it can land in one of two equally likely ways—either heads or tails. Because one of these two results is heads, we say the *probability* of obtaining heads in a single toss is $\frac{1}{2}$. If records show that out of 100 days with weather conditions like today's, 70 have received rain, we say that there is a $\frac{70}{100}$, or more simply, a $\frac{7}{10}$ probability of rain today. If there are two chances in three that a basketball player will make a free throw, we say that the probability that the player will make a free throw is $\frac{2}{3}$.

In this section, we will discuss the concept of probability. But before we can do that, we need to discuss some important vocabulary.

1 Identify the sample space for an experiment.

An **experiment** is any process for which the result is uncertain. Some examples of experiments are

- Flipping a coin
- Rolling dice
- Drawing a card from a standard deck of playing cards
- Selecting a marble from a bag of colored marbles

The results of an experiment are called **outcomes**.

The Language of Algebra A **die** (the singular form of the word *dice*) is a cube with six faces, each containing a number of dots from one to six.

A **standard deck** of 52 playing cards has two red **suits** (hearts and diamonds) and two black suits (clubs and spades). Each suit has 13 cards, including a king, queen, jack (called **face cards**), an ace, and cards numbered from 2 to 10. Assume that the card decks we work with in this section are well shuffled.

For any experiment, the set of all possible outcomes is called a **sample space**. When writing a sample space, we list all possible outcomes within **set braces { }**. For example, the sample space for flipping a single coin one time is

{H, T}   H represents the outcome "heads" and T represents the outcome "tails."

The sample space for flipping a coin twice is a set of four ordered pairs:

{(H, H), (H, T), (T, H), (T, T)}   The pair (H, T), for example, represents the outcome "heads on the first coin and tails on the second coin."

One way to determine all possible outcomes of a multistep experiment is to draw a **tree diagram**. For the experiment of flipping a coin twice, we have:

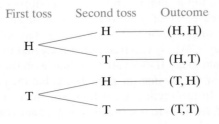

Self Check 1

Each of the letters in the word *truck* is written on a card and placed in a jar. Then one card is drawn from the jar. List the sample space for the experiment. {t, r, u, c, k}

Now Try Problem 15

Teaching Example 1 What is the sample space for an experiment in which a coin is flipped and then a die is rolled?
Answer:
{(H, 1), (H, 2), (H, 3), (H, 4), (H, 5), (H, 6), (T, 1), (T, 2), (T, 3), (T, 4), (T, 5), (T, 6)}

EXAMPLE 1

What is the sample space of the experiment "rolling two dice one time"?

Strategy We will write all the possible outcomes of this experiment as ordered pairs.

WHY The outcome from rolling the first die can be any number from 1 to 6 and the outcome from rolling the second die can be any number from 1 to 6.

Solution
We will let the first number in each ordered pair be the result on the first die and the second number in each ordered pair be the result on the second die.
The sample space contains 36 ordered pairs:

{(1, 1), (1, 2), (1, 3), (1, 4), (1, 5), (1, 6),
(2, 1), (2, 2), (2, 3), (2, 4), (2, 5), (2, 6),
(3, 1), (3, 2), (3, 3), (3, 4), (3, 5), (3, 6),
(4, 1), (4, 2), (4, 3), (4, 4), (4, 5), (4, 6),
(5, 1), (5, 2), (5, 3), (5, 4), (5, 5), (5, 6),
(6, 1), (6, 2), (6, 3), (6, 4), (6, 5), (6, 6)}

2 Identify an event of an experiment.

An **event** is one or more outcomes of an experiment. Events are subsets of the sample space and are usually labeled with the capital letter E. For the experiment of rolling a die, some events are

- Getting an even number: $E = \{2, 4, 6\}$
- Getting a number greater than 2: $E = \{3, 4, 5, 6\}$

For the experiment of tossing a coin twice, some events are

- Getting at least one heads: $E = \{(H, H), (H, T), (T, H)\}$
- Getting no tails: $E = \{(H, H)\}$

EXAMPLE 2 What is the event "getting a prime number" for the experiment of rolling a die once?

Strategy We will consider which of the numbers 1, 2, 3, 4, 5, and 6 are prime numbers.

WHY The only possible outcomes when rolling a die once are the numbers 1 through 6.

Solution
Recall that a prime number is a whole number greater than 1 that has only 1 and itself as factors. Thus:

$E = \{2, 3, 5\}$ 1, 4, and 6 are not prime numbers.

> **The Language of Algebra** When performing an experiment, we often speak of **favorable outcomes**. These are the outcomes of an experiment that satisfy the requirements of a particular event. For the experiment of rolling a die once and the event "getting a prime number," the outcomes that are considered *favorable* to the event are 2, 3, and 5.

Self Check 2
What is the event "getting at least one tails" for the experiment of tossing a coin twice?
$E = \{(T, T), (H, T), (T, H)\}$
Now Try Problem 19

Teaching Example 2 What is the event "getting a heads and an even number" for the experiment of tossing a coin and then rolling a die?
Answer:
$E = \{(H, 2), (H, 4), (H, 6)\}$

3 Find the probability of an event.

Let's again consider the experiment of tossing a coin twice. We have seen that the sample space contains four ordered pairs:

$\{(H, H), (H, T), (T, H), (T, T)\}$

We have also seen that the event E "getting at least one heads" contains three outcomes:

$E = \{(H, H), (H, T), (T, H)\}$

Because the outcome of getting at least one heads can occur in 3 ways out of a total of 4 possible outcomes, we say that the probability of event E is $\frac{3}{4}$. In symbols, we write

$P(E) = P(\text{at least one heads}) = \dfrac{3}{4}$ Read P(E) as "the probability of event E."

In general, the **probability** of an event is a measure of the likelihood (or chance) of it occurring. We define the probability of an event as follows.

> **Probability of an Event**
>
> If a sample space of an experiment has n distinct and equally likely outcomes, and E is an event that occurs in s favorable ways, the **probability of event E** is
>
> $P(E) = \dfrac{s}{n}$

Because $0 \leq s \leq n$, it follows that $0 \leq \dfrac{s}{n} \leq 1$. This implies that all probabilities have values from 0 to 1, including 0 and 1. An event that cannot happen has a probability of 0. An event that is certain to happen has a probability of 1.

EXAMPLE 3 Find the probability of drawing an ace from a standard card deck.

Strategy First, we will determine how many ways the required event (drawing an ace) can occur. Then, we will determine how many equally likely outcomes there are when drawing one card from a standard deck.

Self Check 3
Find the probability of drawing a diamond from a standard card deck. $\frac{13}{52} = \frac{1}{4}$
Now Try Problem 23

Teaching Example 3 Find the probability of drawing a face card from a well-shuffled standard card deck.
Answer:
$\frac{12}{52} = \frac{3}{13}$

WHY The probability of the event is the ratio (fraction) of those two numbers.

Solution

Since the sample space is 52 cards, $n = 52$. We let E be the set of outcomes that give an ace:

$$E = \{\text{ace of hearts, ace of diamonds, ace of spades, ace of clubs}\}$$

Since there are 4 ways to draw an ace, $s = 4$. The probability of drawing an ace is the ratio of the number of favorable outcomes, s, to the number of possible outcomes, n.

$$P(\text{ace}) = \frac{s}{n} = \frac{4}{52} = \frac{1}{13} \quad \text{Factor 52 as } 4 \cdot 13 \text{ and simplify the fraction: } \frac{\overset{1}{\cancel{4}}}{\underset{1}{\cancel{4}} \cdot 13}.$$

The probability of drawing an ace from a standard card deck is $\frac{1}{13}$.

Self Check 4

Find the probability of the event "rolling a sum of 4 on one roll of two dice." $\frac{3}{36} = \frac{1}{12}$

Now Try Problem 27

Teaching Example 4 Find the probability of the event "rolling a sum of 8 on one roll of two dice."
Answer:
$\frac{5}{36}$

EXAMPLE 4 Find the probability of the event "rolling a sum of 7 on one roll of two dice."

Strategy First, we will determine how many ways the required event (rolling a sum of 7) can occur. Then, we will determine how many equally likely outcomes there are when rolling two dice.

WHY The probability of the event is the ratio (fraction) of those two numbers.

Solution

The sample space for this experiment was listed in Example 1. Since there are 36 possible outcomes, $n = 36$.

We let E be the set of outcomes that give a sum of 7:

$$E = \{(1, 6), (2, 5), (3, 4), (4, 3), (5, 2), (6, 1)\}$$

Since there are 6 favorable outcomes, $s = 6$.

The probability of rolling a sum of 7 is the ratio of the number of favorable outcomes, s, to the number of possible outcomes, n.

$$P(\text{rolling a 7}) = \frac{s}{n} = \frac{6}{36} = \frac{1}{6} \quad \text{Always simplify the fraction, if possible.}$$

The probability of rolling a sum of 7 on one roll of two dice is $\frac{1}{6}$.

ANSWERS TO SELF CHECKS

1. $\{t, r, u, c, k\}$ **2.** $E = \{(T, T), (H, T), (T, H)\}$ **3.** $\frac{13}{52} = \frac{1}{4}$ **4.** $\frac{3}{36} = \frac{1}{12}$

APPENDIX 1 STUDY SET

VOCABULARY

Fill in the blanks.

1. An experiment is any process for which the result is uncertain.
2. The results of an experiment are called outcomes .
3. For any experiment, the set of all possible outcomes is called a sample   space .
4. An event is one or more outcomes of an experiment.
5. Favorable outcomes of an experiment satisfy the requirements of that particular event.
6. The probability of an event is a measure of the likelihood (or chance) of its occurring.

CONCEPTS

Fill in the blanks.

7. If a sample space of an experiment has n distinct and equally likely outcomes, and E is an event that occurs in s favorable ways, the probability of event E is $P(E) = \frac{s}{n}$.
8. The probability of an event is always a number from 0 to 1 , including 0 and 1 .
9. If an event cannot happen, its probability is 0 .
10. If an event is certain to happen, its probability is 1 .
11. In a standard deck of playing cards,
    a. How many cards are there? 52
    b. How many red cards are there? 26
    c. How many clubs are there? 13
    d. How many face cards are there in each suit? 3
12. How many outcomes are there when rolling a die? List them. Six; {1, 2, 3, 4, 5, 6}

NOTATION

Fill in the blanks.

13. When writing a sample space, we list all possible outcomes within braces { } .
14. We read $P(E)$ as "the probability of event E."

GUIDED PRACTICE

List the sample space of each experiment. See Example 1.

15. Rolling a die and tossing a coin {(1, H), (2, H), (3, H), (4, H), (5, H), (6, H), (1, T), (2, T), (3, T), (4, T), (5, T), (6, T)}
16. Tossing three coins {(H, H, H), (H, H, T), (H, T, H), (H, T, T), (T, T, T), (T, T, H), (T, H, T), (T, H, H)}
17. Selecting a letter of the alphabet that is a vowel {a, e, i, o, u}
18. Guessing a one-digit number {0, 1, 2, 3, 4, 5, 6, 7, 8, 9}

List each event. See Example 2.

19. What is the event "getting an even number" for the experiment of rolling a die once? $E = \{2, 4, 6\}$
20. What is the event "getting a red face card" for the experiment of drawing a card from a standard deck? {king of diamonds, queen of diamonds, jack of diamonds, king of hearts, queen of hearts, jack of hearts}
21. What is the event "getting at least one heads" for the experiment of tossing a coin twice? $E = \{(H, H), (H, T), (T, H)\}$
22. What is the event "getting a tails and an odd number" for the experiment of tossing a coin and then rolling a die? *Answer:* $E = \{(T, 1), (T, 3), (T, 5)\}$

Find the probability of each event. See Example 3.

23. Drawing a black card on one draw from a standard card deck $\frac{1}{2}$
24. Drawing a diamond on one draw from a standard card deck $\frac{1}{4}$
25. Drawing a face card on one draw from a standard card deck $\frac{3}{13}$
26. Drawing a jack or a king on one draw from a standard card deck $\frac{2}{13}$

Two dice are rolled. Find the probability of each event. See Example 4.

27. Getting a sum of 6 $\frac{5}{36}$
28. Getting a sum of 10 $\frac{1}{12}$
29. Getting a sum of 13 0
30. Getting a sum of less than 13 1

TRY IT YOURSELF

A die is rolled once. Find the probability of each event.

31. Rolling a 2 $\frac{1}{6}$
32. Rolling a number greater than 4 $\frac{1}{3}$
33. Rolling a number larger than 1 but less than 6 $\frac{2}{3}$
34. Rolling a number that is an odd number $\frac{1}{2}$

Balls numbered from 1 to 42 are placed in a jar and stirred. If one is drawn at random, find the probability of each result.

35. The number is less than 20. $\frac{19}{42}$
36. The number is less than 50. 1
37. The number is a prime number. $\frac{13}{42}$
38. The number is less than 10 or greater than 40. $\frac{11}{42}$

Find the probability of each event.

39. Drawing a red egg from a basket containing 5 red eggs and 7 blue eggs $\frac{5}{12}$
40. Drawing an orange cube from a bowl containing 5 orange cubes and 1 beige cube $\frac{5}{6}$
41. Obtaining a tails and an even number when a coin is flipped and then a die is rolled $\frac{3}{12} = \frac{1}{4}$
42. Obtaining a heads and a number greater than 2 when a coin is flipped and then a die is rolled $\frac{4}{12} = \frac{1}{3}$

A basket contains 40 red poker chips, 20 blue poker chips, and 30 black poker chips. One chip is chosen at random. Find each probability.

43. The chip chosen is not red. $\frac{50}{90} = \frac{5}{9}$
44. The chip chosen is neither red nor black. $\frac{20}{90} = \frac{2}{9}$
45. The chip chosen is not blue. $\frac{70}{90} = \frac{7}{9}$
46. The chip chosen is red or blue. $\frac{60}{90} = \frac{2}{3}$

If the spinner shown below is spun, find the probability of each event. Assume that the spinner never stops on a line.

47. The spinner stops on red. $\frac{3}{8}$
48. The spinner stops on green. $\frac{1}{4}$
49. The spinner stops on brown. 0
50. The spinner stops on yellow. $\frac{1}{8}$

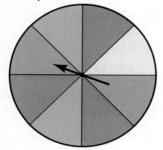

APPLICATIONS

51. An aircraft has 4 engines. Assume that an engine has an equal chance of surviving or failing during a test.
    a. Find the sample space. Let S represent an engine that survives and F represent an engine that fails. A tree diagram is helpful.
    {(S, S, S, S), (S, S, S, F), (S, S, F, S), (S, S, F, F), (S, F, S, S), (S, F, S, F), (S, F, F, S), (S, F, F, F), (F, S, S, S), (F, S, F, F), (F, S, F, S), (F, S, F, F), (F, F, S, S), (F, S, S, F), (F, F, F, S), (F, F, F, F)}
    b. Find the probability that all engines will survive the test. $\frac{1}{16}$
    c. Find the probability that exactly 1 engine will survive. $\frac{1}{4}$
    d. Find the probability that exactly 2 engines will survive. $\frac{3}{8}$
    e. Find the probability that exactly 3 engines will survive. $\frac{1}{4}$
    f. Find the probability that no engines will survive. $\frac{1}{16}$

52. Find the sum of the probabilities in problem 51. What do you discover? 1

A survey of 282 people is taken to determine the opinions of doctors, teachers, and lawyers on a proposed piece of legislation, with the results shown in the table. A person is chosen at random from those surveyed. Refer to the table to find the probability of each event.

	Number that favor	Number that oppose	Number with no opinion	Total
Doctors	70	32	17	119
Teachers	83	24	10	117
Lawyers	23	15	8	46
Total	176	71	35	282

53. The person favors the legislation. $\frac{88}{141}$
54. A doctor opposes the legislation. $\frac{32}{119}$
55. A person who opposes the legislation is a lawyer. $\frac{15}{71}$
56. The person has no opinion. $\frac{35}{282}$

WRITING

57. Explain why all probability values range from 0 to 1.
58. Explain the concept of probability.

APPENDIX 2

Inductive and Deductive Reasoning

Objectives

1. Use inductive reasoning to solve problems.
2. Use deductive reasoning to solve problems.

SECTION A2.1
Inductive and Deductive Reasoning

To reason means to think logically. The objective of this appendix is to develop your problem-solving ability by improving your reasoning skills. We will introduce two fundamental types of reasoning that can be applied in a wide variety of settings. They are known as *inductive reasoning* and *deductive reasoning*.

1 Use inductive reasoning to solve problems.

In a laboratory, scientists conduct experiments and observe outcomes. After several repetitions with similar outcomes, the scientist will generalize the results into a statement that appears to be true:

- If I heat water to 212°F, it will boil.
- If I drop a weight, it will fall.
- If I combine an acid with a base, a chemical reaction occurs.

When we draw general conclusions from specific observations, we are using **inductive reasoning**. The next examples show how inductive reasoning can be used in mathematical thinking. Given a list of numbers or symbols, called a *sequence*, we can often find a missing term of the sequence by looking for patterns and applying inductive reasoning.

Self Check 1

Find the next number in the sequence $-3, -1, 1, 3, \ldots$. 5

Now Try Problem 11

Teaching Example 1 Find the next number in the sequence $2, 8, 14, 20, \ldots$.
Answer:
26

EXAMPLE 1 Find the next number in the sequence $5, 8, 11, 14, \ldots$.

Strategy We will find the *difference* between pairs of numbers in the sequence.

WHY This process will help us discover a pattern that we can use to find the next number in the sequence.

Solution
The numbers in the sequence $5, 8, 11, 14, \ldots$ are **increasing**. We can find the difference between each pair of successive numbers as follows:

$8 - 5 = 3$ *Subtract the first number, 5, from the second number, 8.*
$11 - 8 = 3$ *Subtract the second number, 8, from the third number, 11.*
$14 - 11 = 3$ *Subtract the third number, 11, from the fourth number, 14.*

The difference between each pair of numbers is 3. This means that each number in the sequence is 3 greater than the previous one. Thus, the next number in the sequence is $14 + 3$, or 17.

Teaching Example 2 Find the next number in the sequence $-5, -8, -11, -14, \ldots$.
Answer:
-17

EXAMPLE 2 Find the next number in the sequence $-2, -4, -6, -8, \ldots$.

Strategy The terms of the sequence are **decreasing**. We will determine how each number differs from the previous number.

WHY This type of examination helps us discover a pattern that we can use to find the next number in the sequence.

A-7

Self Check 2

Find the next number in the sequence −0.1, −0.3, −0.5, −0.7, ...   −0.9

Now Try Problem 15

Solution

Since each successive number is 2 less than the previous one, the next number in the sequence is −8 − 2, or −10.

−2, −4, −6, −8, ...

(This number is 2 less than the previous number.)

Self Check 3

Find the next letter in the sequence B, G, D, I, F, K, H, ...   M

Now Try Problem 19

Teaching Example 3 Find the next letter in the sequence A, F, D, I, G, L, J, . . .
Answer: O

EXAMPLE 3   Find the next letter in the sequence A, D, B, E, C, F, D, ...

Strategy We will create a letter–number correspondence and rewrite the sequence in an equivalent numerical form.

WHY Many times, it is easier to determine the pattern if we examine a sequence of numbers instead of letters.

Solution

The letter A is the 1st letter of the alphabet, D is the 4th letter, B is the 2nd letter, and so on. We can create the following letter–number correspondence:

Letter	Number
A →	1) Add 3.
D →	4) Subtract 2.
B →	2) Add 3.
E →	5) Subtract 2.
C →	3) Add 3.
F →	6) Subtract 2.
D →	4

The numbers in the sequence 1, 4, 2, 5, 3, 6, 4, . . . **alternate** in size. They change from smaller to larger, to smaller, to larger, and so on.

We see that 3 is added to the first number to get the second number. Then 2 is subtracted from the second number to get the third number. To get successive numbers in the sequence, we alternately add 3 to one number and then subtract 2 from that result to get the next number.

Applying this pattern, the next number in the given numerical sequence would be 4 + 3, or 7. The next letter in the original sequence would be G, because it is the 7th letter of the alphabet.

Self Check 4

Find the next shape in the sequence below.

Now Try Problem 23

EXAMPLE 4   Find the next shape in the sequence below.

Strategy To find the next shape in the sequence, we will focus on the changing positions of the dots.

WHY The star does not change in any way from term to term.

Solution

We see that each of the three dots moves from one point of the star to the next, in a counterclockwise direction. This is a **circular** pattern. The next shape in the sequence will be the one shown here.

EXAMPLE 5
Find the next shape in the sequence below.

Strategy To find the next shape in the sequence, we must consider two changing patterns at the same time.

WHY The shapes are changing and the number of dots within them are changing.

Solution
The first figure has three sides and one dot, the second figure has four sides and two dots, and the third figure has five sides and three dots. Thus, we would expect the next figure to have six sides and four dots, as shown to the right.

Self Check 5
Find the next shape in the sequence below.

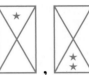

Now Try Problem 27

Self-Check 5 Answer

2 Use deductive reasoning to solve problems.

As opposed to inductive reasoning, deductive reasoning moves from the general case to the specific. For example, if we know that the sum of the angles in any triangle is 180°, we know that the sum of the angles of $\triangle ABC$ shown on the right is 180°. Whenever we apply a general principle to a particular instance, we are using deductive reasoning.

A deductive reasoning system is built on four elements:

1. **Undefined terms:** terms that we accept without giving them formal meaning
2. **Defined terms:** terms that we define in a formal way
3. **Axioms** or **postulates:** statements that we accept without proof
4. **Theorems:** statements that we can prove with formal reasoning

Many problems can be solved by deductive reasoning. For example, suppose a student knows that his college offers algebra classes in the morning, afternoon, and evening and that Professors Anderson, Medrano, and Ling are the only algebra instructors at the school. Furthermore, suppose that the student plans to enroll in a morning algebra class. After some investigating, he finds out that Professor Anderson teaches only in the afternoon and Professor Ling teaches only in the evening. Without knowing anything about Professor Medrano, he can conclude that she will be his algebra teacher, since she is the only remaining possibility.

The following examples show how to use deductive reasoning to solve problems.

EXAMPLE 6
Scheduling Classes An online college offers only one calculus course, one algebra course, one statistics course, and one trigonometry course. Each course is to be taught by a different professor. The four professors who will teach these courses have the following course preferences:

1. Professors A and B don't want to teach calculus.
2. Professor C wants to teach statistics.
3. Professor B wants to teach algebra.

Who will teach trigonometry?

Strategy We will construct a table showing all the possible teaching assignments. Then we will cross off those classes that the professors do not want to teach.

Now Try Problem 31

WHY The best way to examine this much information is to describe the situation using a table.

Solution
The following table shows each course, with each possible instructor.

Calculus	Algebra	Statistics	Trigonometry
A	A	A	A
B	B	B	B
C	C	C	C
D	D	D	D

Since Professors A and B don't want to teach calculus, we can cross them off the calculus list. Since Professor C wants to teach statistics, we can cross her off every other list. This leaves Professor D as the only person to teach calculus, so we can cross her off every other list. Since Professor B wants to teach algebra, we can cross him off every other list. Thus, the only remaining person left to teach trigonometry is Professor A.

Calculus	Algebra	Statistics	Trigonometry
A̶	A	A	A
B̶	B	B̶	B̶
C̶	C̶	C	C̶
D	D̶	D̶	D̶

Self Check 7
USED CARS Of the 50 cars on a used-car lot, 9 are red, 31 are foreign models, and 6 are red, foreign models. If a customer wants to buy an American model that is not red, how many cars does she have to choose from? 16

Now Try Problem 35

EXAMPLE 7 *State Flags*
The graph below gives the number of state flags that feature an eagle, a star, or both. How many state flags have neither an eagle nor a star?

Strategy We will use two intersecting circles to model this situation.

WHY The intersection is a way to represent the number of state flags that have both an eagle and a star.

Solution
In figure (a) on the following page, the intersection (overlap) of the circles shows that there are 5 state flags that have both an eagle and a star. If an eagle appears on a total of 10 flags, then the red circle must contain 5 more flags outside of the

intersection, as shown in figure (b). If a total of 27 flags have a star, the blue circle must contain 22 more flags outside the intersection, as shown.

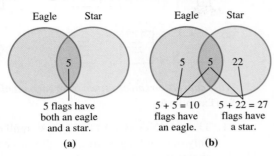

From figure (b), we see that 5 + 5 + 22, or 32 flags have an eagle, a star, or both. To find how many flags have neither an eagle nor a star, we subtract this total from the number of state flags, which is 50.

$50 - 32 = 18$

There are 18 state flags that have neither an eagle nor a star.

ANSWERS TO SELF CHECKS

1. 5   **2.** −0.9   **3.** M   **4.**   **5.** [flag image]   **7.** 16

APPENDIX 2   STUDY SET

VOCABULARY

Fill in the blanks.

1. <u>Inductive</u> reasoning draws general conclusions from specific observations.
2. <u>Deductive</u> reasoning moves from the general case to the specific.

CONCEPTS

Tell whether the pattern shown is increasing, decreasing, alternating, or circular.

3. 2, 3, 4, 2, 3, 4, 2, 3, 4, ...   circular
4. 8, 5, 2, −1, ...   decreasing
5. −2, −4, 2, 0, 6, ...   alternating
6. 0.1, 0.5, 0.9, 1.3, ...   increasing
7. a, c, b, d, c, e, ...   alternating
8. ...   circular

9. ROOM SCHEDULING From the chart, determine what time(s) on a Wednesday morning a practice room in a music building is available. The symbol X indicates that the room has already been reserved.   10 A.M.

	M	T	W	Th	F
9 A.M.	X		X		X
10 A.M.	X	X			X
11 A.M.		X	X	X	

10. COUNSELING QUESTIONNAIRE A group of college students were asked whether they were taking a mathematics course and whether they were taking an English course. The results are displayed below.

   a. How many students were taking a mathematics course and an English course?   11 students
   b. How many students were taking an English course but not a mathematics course?   18 students
   c. How many students were taking a mathematics course?   21 students

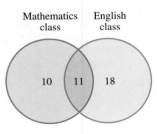

GUIDED PRACTICE

Find the number that comes next in each sequence.
See Example 1.

11. 1, 5, 9, 13, ...   17
12. 11, 20, 29, 38, ...   47
13. 5, 9, 14, 20, ...   27
14. 6, 8, 12, 18, ...   26

Find the number that comes next in each sequence.
See Example 2.

15. 15, 12, 9, 6, ...   3
16. 81, 77, 73, 69, ...   65
17. −3, −5, −8, −12, ...   −17
18. 1, −8, −16, −25, −33, ...   −42

Find the letter that comes next in each sequence.
See Example 3.

19. E, I, H, L, K, O, N, ...   R
20. C, H, D, I, E, J, F, ...   K
21. c, b, d, c, e, d, f, ...   e
22. z, w, y, v, x, u, w, ...   t

Find the figure that comes next in each sequence.
See Example 4.

23.
24.
25.
26.

Find the figure that comes next in each sequence.
See Example 5.

27.   ...
28.

29.
30.

What conclusion can be drawn from each set of information?
See Example 6.

31. **TEACHING SCHEDULES** A small college offers only one biology course, one physics course, one chemistry course, and one zoology course. Each course is to be taught by a different adjunct professor. The four professors who will teach these courses have the following course preferences:

    1. Professors B and D don't want to teach zoology.
    2. Professor A wants to teach biology.
    3. Professor B wants to teach physics.

    Who will teach chemistry?   D

32. **DISPLAYS** Four companies will be displaying their products on tables at a convention. Each company will be assigned one of the displays shown below. The companies have expressed the following preferences:

    1. Companies A and C don't want display 2.
    2. Company A wants display 3.
    3. Company D wants display 1.

    Which company will get display 4?   C

    Display 1   Display 2   Display 3   Display 4

33. **OCUPATIONS** Four people named John, Luis, Maria, and Paula have occupations as teacher, butcher, baker, and candlestick maker.

    1. John and Paula are married.
    2. The teacher plans to marry the baker in December.
    3. Luis is the baker.

    Who is the teacher?   Maria

34. **PARKING** A Ford, a Buick, a Dodge, and a Mercedes are parked side by side.

    1. The Ford is between the Mercedes and the Dodge.
    2. The Mercedes is not next to the Buick.
    3. The Buick is parked on the left end.

    Which car is parked on the right end?   the Mercedes

Use a circle diagram to solve each problem. See Example 7.

35. **EMPLOYMENT HISTORY** One hundred office managers were surveyed to determine their employment backgrounds. The survey results are shown below. How many office managers had neither sales nor manufacturing experience? 6 office managers

    Sales experience 63
    Manufacturing experience 47
    Both 16

36. **PURCHASING TEXTBOOKS** Sixty college sophomores were surveyed to determine where they purchased their textbooks during their freshman year. The survey results are shown below. How many students did not purchase a book at a bookstore or online? 8 students

    Online 23
    Bookstore 35
    Both 6

37. **SIBLINGS** When 27 children in a first-grade class were asked, "How many of you have a brother?" 11 raised their hands. When asked, "How many have a sister?" 15 raised their hands. Eight children raised their hands both times. How many children didn't raise their hands either time? 9 children

38. **PETS** When asked about their pets, a group of 35 sixth-graders responded as follows:

    - 21 said they had at least one dog.
    - 11 said they had at least one cat.
    - 5 said they had at least one dog and at least one cat.

    How many of the students do not have a dog or a cat? 8 students

TRY IT YOURSELF

Find the next letter or letters in the sequence.

39. A, c, E, g, ... I
40. R, SS, TTT, ... UUUU
41. Z, A, Y, B, X, C, ... W
42. B, N, C, N, D, ... N

Find the missing figure in each sequence.

43.

44.

Find the next letter in the sequence.

45. C, B, F, E, I, H, L, ... K
46. d, h, g, k, j, n, ... m

Find the next number in the sequence.

47. −7, 9, −6, 8, −5, 7, −4, ... 6
48. 2, 5, 3, 6, 4, 7, 5, ... 8
49. 9, 5, 7, 3, 5, 1, ... 3
50. 1.3, 1.6, 1.4, 1.7, 1.5, 1.8, ... 1.6
51. −2, −3, −5, −6, −8, −9, ... −11
52. 8, 5, 1, −4, −10, −17, ... −25
53. 6, 8, 9, 7, 9, 10, 8, 10, 11, ... 9
54. 10, 8, 7, 11, 9, 8, 12, 10, 9, ... 13

55. **ZOOS** In a zoo, a zebra, a tiger, a lion, and a monkey are to be placed in four cages numbered from 1 to 4, from left to right. The following decisions have been made:

    1. The lion and the tiger should not be side by side.
    2. The monkey should be in one of the end cages.
    3. The tiger is to be in cage 4.

    In which cage is the zebra? cage 3

56. **FARM ANIMALS** Four animals—a cow, a horse, a pig, and a sheep—are kept in a barn, each in a separate stall.

    1. The cow is in the first stall.
    2. Neither the pig nor the sheep can be next to the cow.
    3. The pig is between the horse and the sheep.

    What animal is in the last stall? the sheep

57. **OLYMPIC DIVING** Four divers at the Olympics finished first, second, third, and fourth.

    1. Diver B beat diver D.
    2. Diver A placed between divers D and C.
    3. Diver D beat diver C.

    In which order did they finish? B, D, A, C

58. **FLAGS** A green, a blue, a red, and a yellow flag are hanging on a flagpole.

    1. The only flag between the green and yellow flags is blue.
    2. The red flag is next to the yellow flag.
    3. The green flag is above the red flag.

    What is the order of the flags from top to bottom? green, blue, yellow, red

APPLICATIONS

59. JURY DUTY The results of a jury service questionnaire are shown below. Determine how many of the 20,000 respondents have served on neither a criminal court nor a civil court jury. 18,935 respondents

Jury Service Questionnaire

997	Served on a criminal court jury
103	Served on a civil court jury
35	Served on both

60. ELECTRONIC POLL For the Internet poll shown below, the first choice was clicked on 124 times, the second choice was clicked on 27 times, and both the first and second choices were clicked on 19 times. How many times was the third choice, "Neither," clicked on? 46 times

Internet Poll	You may vote for more than one.
What would you do if gasoline reached $5.50 a gallon?	◉ Cut down on driving ◉ Buy a more fuel-efficient car ◉ Neither
Number of people voting	178

61. THE SOLAR SYSTEM The graph below shows some important characteristics of the nine planets in our solar system. How many planets are neither rocky nor have moons? 0 planets

Rocky planets: 4
Planets with moons: 7
Rocky planets with moons: 2

62. WORKING TWO JOBS Andres, Barry, and Carl each have a completely different pair of jobs from the following list: jeweler, musician, painter, chauffeur, barber, and gardener. Use the facts below to find the two occupations of each man.

1. The painter bought a ring from the jeweler.
2. The chauffeur offended the musician by laughing at his mustache.
3. The chauffeur dated the painter's sister.
4. Both the musician and the gardener used to go hunting with Andres.
5. Carl beat both Barry and the painter at monopoly.
6. Barry owes the gardener $100.

Andres: painter and barber; Barry: jeweler and musician; Carl: chauffer and gardener

WRITING

63. Describe deductive reasoning and inductive reasoning.

64. Describe a real-life situation in which you might use deductive reasoning.

65. Describe a real-life situation in which you might use inductive reasoning.

66. Write a problem in such a way that the diagram below can be used to solve it.

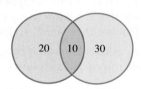

APPENDIX 3

Roots and Powers

n	n²	√n	n	n²	√n
1	1	1.000	51	2,601	7.141
2	4	1.414	52	2,704	7.211
3	9	1.732	53	2,809	7.280
4	16	2.000	54	2,916	7.348
5	25	2.236	55	3,025	7.416
6	36	2.449	56	3,136	7.483
7	49	2.646	57	3,249	7.550
8	64	2.828	58	3,364	7.616
9	81	3.000	59	3,481	7.681
10	100	3.162	60	3,600	7.746
11	121	3.317	61	3,721	7.810
12	144	3.464	62	3,844	7.874
13	169	3.606	63	3,969	7.937
14	196	3.742	64	4,096	8.000
15	225	3.873	65	4,225	8.062
16	256	4.000	66	4,356	8.124
17	289	4.123	67	4,489	8.185
18	324	4.243	68	4,624	8.246
19	361	4.359	69	4,761	8.307
20	400	4.472	70	4,900	8.367
21	441	4.583	71	5,041	8.426
22	484	4.690	72	5,184	8.485
23	529	4.796	73	5,329	8.544
24	576	4.899	74	5,476	8.602
25	625	5.000	75	5,625	8.660
26	676	5.099	76	5,776	8.718
27	729	5.196	77	5,929	8.775
28	784	5.292	78	6,084	8.832
29	841	5.385	79	6,241	8.888
30	900	5.477	80	6,400	8.944
31	961	5.568	81	6,561	9.000
32	1,024	5.657	82	6,724	9.055
33	1,089	5.745	83	6,889	9.110
34	1,156	5.831	84	7,056	9.165
35	1,225	5.916	85	7,225	9.220
36	1,296	6.000	86	7,396	9.274
37	1,369	6.083	87	7,569	9.327
38	1,444	6.164	88	7,744	9.381
39	1,521	6.245	89	7,921	9.434
40	1,600	6.325	90	8,100	9.487
41	1,681	6.403	91	8,281	9.539
42	1,764	6.481	92	8,464	9.592
43	1,849	6.557	93	8,649	9.644
44	1,936	6.633	94	8,836	9.695
45	2,025	6.708	95	9,025	9.747
46	2,116	6.782	96	9,216	9.798
47	2,209	6.856	97	9,409	9.849
48	2,304	6.928	98	9,604	9.899
49	2,401	7.000	99	9,801	9.950
50	2,500	7.071	100	10,000	10.000

INDEX

AAA similarity theorem, 867, 930
Absolute value, 140–141, 208
Absolute value symbols, 140, 188, 216
Acute angles, 823, 920
Acute triangles, 846, 926
Addends, 16, 459
Addition
   of algebraic fractions, 351–352, 425
   associative property of, 18, 116, 153, 210
   checking subtraction with, 25–26
   commutative property of, 17, 116, 153, 210
   of decimals, 458–460, 463, 465–467, 538–540
   distributive property of, 252, 253, 293
   of fractions, 343–354, 424–427
   horizontal form, 16, 376
   of integers, 148–160, 210–211
   key words and phrases that indicate, 19, 117, 226
   of mixed numbers, 374–376, 430–431
   of polynomials, 961–963, 977
   properties of, 17–18
   repeated, 120
   of signed decimals, 463–464, 539
   vertical form, 16, 17, 116, 376–378, 430, 458–459, 538
   of whole numbers, 15–33, 116–119
Addition symbol, 16
Additive inverses (opposites), 154–155, 210
Adjacent angles, 824, 920
Algebraic expressions. *See* Expressions
Algebraic fractions
   adding, 351–352, 425
   building, 312–313, 418
   definition of, 312, 418
   dividing, 337–338, 423
   multiplying, 322–323, 420
   reciprocal of, 423
   simplifying, 312–313, 418
   subtracting, 351–352, 425
Algebraic terms. *See* Terms
Alternate interior angles
   definition of, 835, 923
   identifying, 834–836
Alternate interior angles property, 836
American system of measurement, 587, 629
American units
   of capacity, 593–594, 617, 630
   converting to metric units, 614–619, 635–637
   equivalent weights and masses, 615, 617
   of length, 588, 589–591, 629
   of measurement, 587–599, 614–619, 629–632
   of weight, 591–593, 615, 629
Amount-to-base (or part-to-whole) ratios, 664
Angles, 820–833, 919–922. *See also specific types of angles*
   adjacent, 824, 920
   alternate interior, 834–836, 923
   classifying, 823
   complementary, 826, 827, 921
   congruent, 823, 920
   corresponding, 834–836, 923
   corresponding angles property, 836
   definition of, 822, 920
   finding, with properties of rectangles, 876–877
   identifying, 822
   interior, 834–836, 923
   interior angles property, 836
   measurement of
      with properties of parallel lines cut by transversals, 836–839
      with protractor, 822–824
   naming, 822
   of polygons, 880
   sides of, 822, 920
   straight, 823, 920
   supplementary, 826, 827, 878, 921
   of trapezoids, 877–879
   of triangles, 847–849, 926
   vertex, 822, 920, 926
   vertex angles, 846, 926
   vertical, 824–825, 921
Apparent magnitude scale, 147
Approximate square roots, 518–519, 548
Approximation, 7, 716, 727. *See also* Estimation
Arc of a circle, 901, 937
Area
   of circle, 904, 938
   of polygon, 890, 934
   of polygon combinations, 892–894
   of rectangle
      definition of, 42, 120
      formula for, 43, 120, 890, 934
   of triangle, 326, 390, 421, 890, 934
Arithmetic: fundamental theorem of, 66
Arithmetic mean or average, 90, 495, 755
Arrays, rectangular, 41, 120
ASA property, 864, 930
Associative property
   of addition, 18, 116, 153, 210
   of multiplication, 39, 120, 172, 213, 292, 549
Average (mean)
   on calculator, 756
   definition of, 90, 127, 291, 495, 803
   formula for, 244, 755
Axioms, A-9
Axis
   horizontal axis, 741, 798
   vertical axis, 741, 798
   x-axis, 771, 806
   y-axis, 771, 806

Bar graphs, 115, 452, 740, 798
   definition of, 9
   double-bar graphs, 742, 798
   example, 740
   horizontal axis, 741, 798
   reading, 741–743
   triple-bar graphs, 742, 798
   vertical axis, 741, 798
Base
   decimal, 477–478
   of exponent, 68, 124
   fractional, 324–325, 420
   negative, 173–175
   solving percent equations (formulas) to find, 662–664
   solving percent proportions to find, 669–671
Base angles
   of isosceles triangle, 846, 926
   of trapezoid, 877, 933
Base-10 number system, 2
Bimodal values, 761, 805
Binomials
   definition of, 957, 976
   multiplying, 969–970
   multiplying two, 978
Borrowing
   in decimal subtraction, 461, 538
   in subtraction of mixed numbers, 379, 431
   in subtraction of whole numbers, 23, 117
Brackets, 188
Business formulas, 240–242, 291

Calculators
   adding decimals on, 460
   addition key, 28
   calculating revolutions of a tire on, 902
   compound interest on, 709, 730
   direct entry on, 156
   dividing decimals on, 491
   division key, 58
   division with negative numbers on, 184
   entering negative numbers, 156
   exponential key, 69–70
   finding air bag costs on, 661
   finding mean on, 756
   finding silo volume on, 913–914
   finding square root on, 516
   finding TV screen width on, 858
   fixed-point key, 506
   multiplication key, 42
   multiplication with negative numbers on, 171
   multiplying decimals on, 474
   order of operations on, 91
   painting helicopter landing pads on, 905
   parenthesis keys, 91
   perimeters of combinations of polygons on, 887
   raising negative numbers to a power on, 175
   reverse entry on, 156
   solving proportions with, 579
   subtracting decimals on, 462
   subtraction key, 28
   subtraction with negative numbers on, 165
Capacity
   American units of, 593–594, 630
   definition of, 593
   equivalent capacities, 617
   metric units of, 608, 609, 633
Carrying
   in addition of whole numbers, 16, 116
   in decimal addition, 459, 538
   in fraction addition, 378, 430
Cartesian coordinate system, 770
Celsius, 243, 618–619, 636
Center of a circle, 900, 937
Central tendency: measures of, 755, 803
Change, 212
Chord, 901, 937
Circle graphs, 720, 740
   examples, 353, 720, 740
   reading, 671–672, 744–745
   sectors, 744, 799
Circles, 900–905, 937–938
   area of, 904, 938
   center of, 900, 937
   circumference of, 902, 937
   definition of, 900, 937
   semicircles, 901, 937
Circumference
   definition of, 902, 937
   finding, 901–903
Class interval, 747, 801
Coefficients
   definition of, 260, 294
   examples, 957
   of terms, 259–260
   of variable terms, 435
Commission rates, 682–683, 723
Commissions
   amount of, 723
   calculating, 682–683
   definition of, 723
Common denominators. *See* Least common denominator (LCD)
Common divisors. *See* Greatest common divisor (GCD)
Common factors, 78. *See also* Greatest common factor (GCF)

I-1

Common multiples. *See* Least common multiple (LCM)
Commutative property
 of addition, 17, 116, 153, 210
 of multiplication, 38, 120, 172, 213, 292, 549
Complementary angles, 826, 827, 921
Complex fractions
 definition of, 391, 433
 order of operations with, 387–389, 432–434
 simplifying, 391–393, 433
Composite numbers, 65, 123
Compound interest. *See* Interest
Cones: volume of, 910, 912–913, 939
Congruence properties, 863–865
Congruent angles, 823, 920
Congruent line segments, 821, 920
Congruent triangles, 862–875, 930–931
 corresponding parts of, 862, 930
 definition of, 862, 930
 proving congruence, 863–865
Constant terms, 289, 294
Conversion
 between Fahrenheit and Celsius temperatures, 243, 618–619, 636
 between metric units, 603–605, 606–607, 609, 632, 633
 unit conversion factors, 602–603, 632, 633
Conversion charts, 603–605, 633
Coordinate plane, 771
Coordinate systems
 Cartesian, 770
 rectangular, 767–774, 806
Coplanar lines, 833, 923
Correspondence of congruent triangles, 862–863
Corresponding angles, 834–836, 923
Cost, total. *See* Total cost
Cover-over method for graphing lines, 786
Credit hours (credits), 758, 804
Cross products, 574, 626
Cross-products property (means-extremes property), 575
Cubes, 910, 939
Cubic centimeter, 609, 634, 909
Cubic inch, 909
Cubic units, 609, 634, 909, 939
Cylinders: volume of
 finding, 912
 formula for, 910, 939

Decagons, 844, 925
Decimal bases, 477–478
Decimal notation, 444
Decimal numeration system, 444, 535
Decimal point, 444
Decimal quotients, 490–491
Decimals, 714–717
 adding, 458–460, 538–540
 adding signed, 463–464, 539
 comparing
  with fractions, 506–508, 546

 with inequality symbols, 449–450
 comparing two, 536
 definition of, 446
 dividing, 486–496, 542–545
  by 0.1, 0.01, 0.001, and so on, 492, 544
  by 10, 100, 1,000, and so on, 492, 543
  on calculator, 491
  by decimals, 488–490, 543
  by powers of 10, 491–492
  by whole numbers, 486–488, 542
 equivalents of fractions, 500, 545
 estimating products of, 479
 estimating quotients of, 491
 estimating sums and differences of, 464–465
 expanded form of, 446, 536
 fractional part of, 444, 535
 graphing, 450, 506, 536
 introduction to, 444–458, 535–538
 multiplying, 472–474, 540–542
  by 0.1, 0.01, 0.001, and so on, 476, 540
  by 10, 100, 1,000, and so on, 475, 540
  on calculator, 474
  by powers of 10, 475–477
 multiplying two, 472, 540
 negative, 445
 nonterminating, 518
 order of operations rule with, 478
 parts of, 444
 percent equivalents, 652
 place value of digits in, 444–445
 ratios that involve, 561–562, 624
 reading, 447, 448, 536
 repeating, 501–502, 545, 716
 rounding, 450–452, 504–506, 537, 545
 square root of, 516
 standard form of, 445, 447–448
 subtracting, 460–462, 538–540
 subtracting signed decimals, 463–464
 tables and graphs involving, 452–453
 terminating, 501, 545
 whole-number part of, 444, 535
 writing
  as percents, 650, 716
  in words, 536
 writing fractions as, 500, 545, 582
 writing mixed numbers as, 546, 582
 writing percents as, 648–650, 716
Deductive reasoning, A-9–A-11
Degrees
 of angles, 822, 920
 of polynomials, 958, 976
Denominators, 85, 126, 187, 216
 common. *See* Least common denominator (LCD)

 definition of, 387, 416
 identifying, 304–305
 least common denominator (LCD), 345, 348–350, 425
Diagonal of a polygon, 876, 932
Diameter, 901, 937
Differences
 of decimals, 460–461, 464–465
 of integers, 161
 of whole numbers, 22, 26
Dimensions of rectangles, 21
Discounts
 amount of, 688–690
 formula for, 724, 725
Distance: formula for, 291
Distance fallen: formula for, 244, 291
Distribution
 definition of, 252
 of factor of –1, 256
Distributive property
 extended to other forms, 255, 293
 of multiplication over addition, 253, 293
  for addition, 252, 293
  for subtraction, 253, 293
 simplifying expressions with, 252–255, 549
 solving equations with, 271, 523–524, 529
Dividend, 48
Divisibility tests, 55–56, 122
Division
 by 0, 182–183, 215
 of 0, 182–183, 215
 with 0, 50, 122
 of algebraic fractions, 337–338
 with decimal divisors, 489
 of decimals, 486–496, 542–545
  by 0.1, 0.01, 0.001, and so on, 492, 544
  by 10, 100, 1,000, and so on, 492, 543
  on calculator, 491
  by decimals, 488–490, 543
  by powers of 10, 491–492
  by whole numbers, 486–488, 542
 four-step process, 487
 of fractions, 334–337, 422–424, 423
 rule for dividing two, 335
 sign rules for, 422
 of integers, 175–182, 179–186, 214–215
 key words and phrases that indicate, 58, 122
 long, 48, 121, 486
  with no remainders, 50–53
  with remainders, 53–54
 of mixed numbers, 364–366, 427–430, 429
 with negative numbers, 184
 properties of, 49–50, 122
 related multiplication statements, 48–49, 121, 180, 182
 of signed decimals, 492–493, 544

 solving equations with, 105–113, 130–131
 translating, 227
 of whole numbers, 48–62, 121–123
 of whole numbers that end with zero, 56
Division ladders, 67, 124
Division property of equality, 107, 108, 130, 196
Division statements, 50, 180
Division symbol, 48
Divisors, 48
 decimal, 489, 543
 greatest common divisor (GCD), 79
Dodecagons, 844, 925
Double-bar graphs, 742, 798

The Elements (Euclid), 820
Endpoints, 821, 920
English system of measurement, 587
Equal ratios, 559
Equality
 addition property of, 97, 128, 196
 division property of, 107, 108, 130, 196
 multiplication property of, 106, 130, 196
 properties of, 196
 solving equations with, 107–108
 subtraction property of, 99, 128, 196
Equal-sized groups, 215, 423
Equations, 572
 clearing fractions from, 405–408, 435
 definition of, 96, 572
 equivalent, 97, 98, 128
 forming, 296
 of horizontal lines, 788
 linear, 781
  graphing, 780–797, 807–809
  that are solved for *y*, 780–784
 in one variable, 767
 parts of, 96
 percent equation (formula), 659
  definition of, 659
  solving, 658–659, 659–662, 660, 662–664
  translating percent sentences to, 657–658
 satisfying, 128, 268–269, 295
 solution of, 128, 268, 295
 solving, 252, 296, 399
  with addition and subtraction, 96–105, 128–129
  with addition property of equality, 97–99
  by combining like terms, 269–270
  definition of, 217, 549, 658
  with distributive property, 271
  with division property of equality, 107–108
  with multiplication and division, 105–113, 130–131

Index

with multiplication property of equality, 106–107
process for, 96–97, 98
with properties of equality, 196–199, 200–202
with reciprocals, 400–405
by simplifying expressions, 268–275, 295–296
steps in, 98
with subtraction property of equality, 99
that have variable terms on both sides, 270–271
that involve decimals, 522–535, 549–551
that involve fractions, 399–411, 435–436
that involve integers, 196–206, 217–218
that involve $-x$, 199
in two variables, 767–774, 806–807
of vertical lines, 788
$y = b$ and $x = a$ form, 786–788
Equilateral triangles, 845, 925
Equivalent decimals, 545
Equivalent equations, 97, 98, 128
Equivalent expressions, 251
Equivalent fractions, 307, 417, 545
Equivalent repeating decimals, 501–504
Equivalent terminating decimals, 500–501
Estimation
in addition, 19, 464–465
with decimals, 464–465, 479, 491, 539, 541, 543
in division, 56–57, 122, 491, 543
of expressions, 191–192, 216
with integers, 187–195
in long division, 50, 52, 53, 54, 487
in multiplication, 40, 120, 479, 541
with percents, 696–700, 726–728
in subtraction, 26, 464–465
with whole numbers, 19, 40, 56–57, 120, 122
Euclid, 820
Evaluation, 84, 126
Even powers of negative integers, 174, 213
Even whole numbers, 64–65, 123
Exact answers, 716
Expanded form
of decimals, 446, 536
of whole numbers, 4–5, 114
Expanded notation, 4–5, 114, 446, 536
Experiments
definition of, A-1
identifying events of, A-2–A-3
sample spaces for, A-1–A-2
Exponential expressions
with decimal bases, 477–478, 541
definition of, 68, 950
evaluating, 68–70, 124
with fractional bases, 324–325, 420
with like bases, 951–952

with negative bases, 173–175
raising to a power, 952–953
Exponential notation, 68
Exponents
base of, 68, 124
definition of, 68, 124, 975
multiplication rules for, 950–957, 975–976
natural-number, 950
power of a product rule, 954, 975
power rule for, 953, 975
product rule for, 951, 975
rules for, 955, 975
Expressions, 84
definition of, 226, 237, 289
equivalent, 251
estimating, 191–192
evaluating, 210, 213, 237–240, 290, 544
with addition, 153–154
with multiplication, 171–173
exponential
with decimal bases, 477–478, 541
definition of, 68, 950
evaluating, 68–70, 124
with fractional bases, 324–325, 420
with like bases, 951–952
with negative bases, 173–175
raising to a power, 952–953
key words and phrases that can be translated into, 289
radical, 514, 516, 547
simplifying, 292–293, 522, 549
by combining like terms, 261–263
by combining terms, 524–525
with distributive property, 252–255
with order of operations, 252
to solve equations, 268–275, 295–296
terms of, 294
that contain absolute values, 190–191
that contain fractions and decimals, 508–509, 546
that contain grouping symbols, 87–90, 188–190
that contain square roots, 516–518, 548
that involve addition and subtraction, 27–28, 163–164
translating word phrases to, 226–227
writing, to represent unknown quantities, 228–231
Extremes
means-extremes property (cross-products property), 575
in proportions, 574, 626

Factor trees, 65–66, 124
Factoring
definition of, 63, 123
of whole numbers, 63–64

Factors
of -1, 256
common, 78
definition of, 34, 63, 123
greatest common (GCF)
definition of, 78, 125
finding, by listing factors, 78–79
finding, with prime factorization, 79–81, 125
grouping, 251
listing, 78–79
order of, 251
prime, 63–72, 123–124
finding LCM with, 75–78
finding, with division ladders, 67
finding, with factor trees, 65–66
of terms, 294
Fahrenheit, 149, 243, 618–619, 636
Figures. See Geometric figures
FOIL method, 969
Foot (feet), 588
Formulas
for area of a circle, 938
for area of a parallelogram, 934
for area of a polygon, 890, 934
for area of a rectangle, 43, 120, 890, 934
for area of a square, 934
for area of a trapezoid, 934
for area of a triangle, 326, 390, 421, 890, 934
for arithmetic average (mean), 755
from business, 240–242, 291
for circumference, 937
for compound interest, 709, 730
conversion formula for temperature, 243, 618
definition of, 240, 291
for discounts, 724, 725
for distance, 291
for distance fallen, 244, 291
for mean (average), 244, 755
percent, 659
for perimeter of a rectangle, 264, 886, 934
for perimeter of a square, 264, 934
for perimeter of a triangle, 934
for profit, 241
for retail price, 241
for sale price, 240–241
from science, 242–244, 291
for simple interest, 704
for sum of angle measures of a polygon, 879–881
that involve square roots, 518, 548
for volume, 910
Fraction (word), 304
Fraction bar —, 48, 188, 416
Fractional bases, 324–325, 420
Fractions, 714–717
adding, 343–354, 424–427
algebraic
adding, 351–352, 425
building, 312–313, 418
definition of, 312, 418

dividing, 337–338, 423
multiplying, 322–323, 420
reciprocal of, 423
simplifying, 312–313, 418
subtracting, 351–352, 425
building, 308, 417
clearing from equations, 405–408, 435
comparing, 352, 426
comparing with decimals, 506–508, 546
complex, 387–398
definition of, 391, 433
order of operations with, 387–389, 432–434
simplifying, 391–393, 433
decimal equivalents, 500, 545
definition of, 304, 416
with denominator of 0, 306
with denominator of 1, 306
dividing, 334–337, 422–424
dividing two, 335, 422
equations that involve, 399–411, 435–436
equivalent, 307, 417, 545
expressing in higher terms, 308
expressions that contain, 508–509, 546
followed by the word *of*, 420
form of 1, 306, 417
fundamental property of, 308
graphing, 428
graphing on number line, 363–364, 506
improper, 305, 417, 427
writing as mixed numbers, 362–363, 428
writing mixed numbers as, 361–362, 374, 428
introduction to, 304–318, 416–419
in lowest terms, 309, 417, 559
multiplying, 307, 318–332, 419–421
multiplying two, 419
with numerator of 0, 306
parts of. See Denominators; Numerators
parts of decimals, 444, 535
percent equivalents, 652
proper, 305, 417
reciprocals of
definition of, 333
finding, 333–334, 422
sign rules for, 422
simplest form of, 309, 417, 559
simplifying, 309–312, 417
special forms of, 306–307, 417
square root of, 516
subtracting, 343–354, 424–427
undetermined, 306
writing
as decimals, 500, 545, 582
as equivalent repeating decimals, 501–504
as equivalent terminating decimals, 500–501
as percents, 650–653, 716
writing percents as, 645–648, 715
writing rates as, 563–564, 625
writing ratios as, 558–561, 623

Frequency polygons, 748, 801
Front-end rounding
  in addition, 19, 116
  in decimal addition or subtraction, 464, 539
  in decimal multiplication, 479, 541
  in multiplication, 40, 120
  in percent problems, 696, 697
  in subtraction, 26, 117, 539
Fundamental property of fractions, 308
Fundamental theorem of arithmetic, 66

GCD. *See* Greatest common divisor
GCF. *See* Greatest common factor
Geometric figures
  basic, 820–833, 919–922
  circles, 900–905, 937–938
  congruent triangles, 862–875, 930–931
  polygons, 875–884, 932–934
  quadrilaterals, 844, 875–884, 925, 932–934
  rectangles, 875, 932
  squares, 875, 932
  triangles, 845, 925
Geometry
  definition of, 820, 919
  introduction to, 819–947
Grade point average (GPA), 496, 758, 804
Gram, 605
Graphing decimals, 450, 506, 536
Graphing equations in two variables, 807
Graphing fractions, 363–364, 428, 506
Graphing integers, 137–139
Graphing linear equations, 780–797, 807–809
Graphing lines
  cover-over method for, 786
  intercept method for, 784, 807
Graphing mixed numbers, 363–364, 428
Graphing numbers, 5, 138, 450
Graphing ordered pairs, 806
Graphing points, 771–774
Graphs and tables
  bar graphs, 9, 115, 452, 741–743, 798
  circle graphs, 353, 671–672, 720, 744–745, 799
  double-bar graphs, 742, 798
  of equations, 780
  frequency polygons, 748
  histograms, 747–748
  key for, 742, 798
  line graphs, 9, 115, 745–747, 799
  pictographs, 743–744, 798
  pie charts, 353, 671, 744
  reading, 740–754, 797–803
  tables, 115, 740, 741, 797
  that involve decimals, 452–453
  that involve whole numbers, 8–10
  triple-bar graphs, 742, 798

Greatest common divisor (GCD), 79
Greatest common factor (GCF)
  definition of, 78, 125
  finding, by listing factors, 78–79
  finding, with prime factorization, 79–81, 125
Gridlock, 770
Grouping
  equal-sized groups, 215, 423
  of factors, 251
Grouping symbols, 87, 188
  innermost pair, 89
  outermost pair, 89

Height, 911
Heptagons, 844
Hexagons, 844, 925
Histograms, 747–748, 800
Horizontal axis, 741, 798
Horizontal form
  of addition, 16, 376
  of multiplication, 34
  of subtraction, 22
Horizontal lines, 788, 807
Horizontal number lines, 138
Hypotenuse, 846, 855, 926

Improper fractions
  as algebraic solutions, 401
  definition of, 305, 417
  and mixed numbers, 360, 427
  writing as mixed numbers, 362–363, 428
  writing mixed numbers as, 361–362, 374, 428
Inches, 588
Inductive reasoning, A-7–A-9
Inequality, strict, 140
Inequality symbols
  comparing decimals with, 449–450
  comparing integers with, 139–140
  comparing whole numbers with, 5–6, 114
  definition of, 6, 114, 140, 208
Infinite, 64
Integers
  with 0, 215
  absolute value of, 140–141
  adding, 148–160, 210–211
  comparing, 139–140
  dividing, 179–186, 214–215
  dividing two, 175–182, 214–215
  equations that involve, 196–206, 217–218
  estimation with, 187–195
  graphing on a number line, 137–139
  introduction to, 136–148, 208–209
  multiplying, 169–176, 213–214
  negative
    dividing, 180
    even powers of, 174, 213
    multiplying, 173, 213
    odd powers of, 174, 213
    set of, 137
  opposites of, 141, 208
  order of operations for, 187–195

positive
  dividing, 180
  set of, 137
  set of, 137, 208
  subtracting, 160–168, 211–213
Intercept method for graphing lines, 784, 807
Intercepts
  definition of, 784
  finding, 784–786
  x-intercept, 784, 807
  y-intercept, 784, 807
Interest, 703–714, 728–731
  compound
    calculating, 706–710
    on calculator, 709, 730
    definition of, 703, 729
    formula for, 709
  definition of, 703, 728
  simple
    calculating, 703–706
    definition of, 703, 704, 728
    formula for, 704
Interest rate, 703
Interior angles, 835, 923
Interior angles property, 836
Inverse operations, 25, 49
Inverses, additive (opposites), 154–155, 210
Irrational numbers, 504
Isolation of variables, 200, 577
Isosceles trapezoids, 878, 933
Isosceles triangle theorem, 846, 926
  converse of, 847, 926
Isosceles triangles, 845, 846, 925, 926

Key, of graphs and tables, 742, 798

Large numbers, 541
Least common denominator (LCD), 345, 425
  definition of, 345
  finding, 348–350
Least common multiple (LCM), 72–84, 125–126, 425
  definition of, 73, 74, 125, 348
  finding, 125
    by listing multiples, 72–75, 125
    with prime factorization, 75–78, 125
Legs
  of trapezoids, 877, 933
  of triangles, 846, 926
Like terms
  combining, 261–263, 549
  definition of, 262, 294
  simplifying expressions by, 524–525
  solving equations by, 269–270
  definition of, 269, 294, 524, 549
  examples, 524
  identifying, 261
Line graphs, 9, 115, 745–747, 799
Line segments
  congruent, 821, 920
  definition of, 821, 920

Linear equations
  definition of, 781
  graphing, 780–797, 807–809
  that are solved for y, 780–784
Lines. *See also* Number lines
  coplanar, 833, 923
  graphing
    cover-over method for, 786
    intercept method for, 784, 807
  horizontal, 788, 807
  identifying, 820–821, 919
  naming, 820–821, 825
  parallel, 833–843, 923–925
  perpendicular, 833–843, 923–925
  skew, 833
  transversal, 834, 923
  vertical, 788, 807
Long division
  with no remainders, 50–53
  process, 48, 121, 486
  with remainders, 53–54
Lower base, 877
Lower base angles, 877
Lowest common denominator. *See* Least common denominator (LCD)
Lowest terms, 309, 417, 559

Major arc, 901
Markup, 465
Mass (weight)
  American units, 591–593, 629
  converting between American and metric units, 635
  definition of, 633
  equivalent weights and masses, 615
  metric units, 605–606, 606–607, 615, 633
  unit conversion factors, 591
Mean (average)
  arithmetic mean, 90, 495, 755
  on calculator, 756
  definition of, 90, 127, 291, 495, 803
  describing a set of values with, 761–762
  finding, 90–91, 127, 244–245, 755–757
  formula for, 244, 755
  weighted, 757–759, 804
Means, 574, 626
Means-extremes property (cross-products property), 575
Measurement
  American (or English) system, 587, 629
  American units of, 587–599, 629–632
  converting between American and metric units, 614–619, 635–637
  metric system, 587
  metric units of, 600–613, 632–634
  with ruler, 587–588
  units of time, 594, 595, 630
Measure of central tendency, 755

Median
  definition of, 759
  describing a set of values with, 761–762
  finding, 759–760, 804
Metric ruler, 601–602, 632
Metric system, 587
Metric units
  of capacity, 608, 609, 633
  converting between American units and, 614–619, 635–637
  cubic units, 609, 634, 909, 939
  equivalent weights and masses, 615, 617
  of length, 600, 602–603, 603–605, 632
  of mass (weight), 605–606, 606–607, 633
  of measurement, 600–613, 632–634
  prefixes, 600, 632
Midpoints, 821
Mile, 588
Minor arc, 901
Minuend, 22, 460–461
Minus sign, 22, 142, 199, 208
Mixed number percents, 716
Mixed numbers
  adding, 374–376, 430–431
    in horizontal form, 376
    method 1, 374
    method 2, 374
    in vertical form, 376–378, 430
  definition of, 427
  dividing, 364–366, 427–430, 429
  graphing, 363–364, 428
  and improper fractions, 427
  multiplying, 364–366, 427–430, 428
  parts of, 360–361
  ratios that involve, 561–562, 624
  subtracting, 378–380, 430–431
  writing
    as decimals, 546, 582
    as improper fractions, 361–362, 374, 428
  writing improper fractions as, 362–363, 428
Modal value, 761
Mode
  bimodal values, 761, 805
  definition of, 761, 805
  describing a set of values with, 761–762
  finding, 760–761
Money, 537
Monomials
  definition of, 957, 976
  multiplying, 967–968
  multiplying polynomials by, 968, 977
  multiplying two, 968
  of third degree or of degree 3, 958
Multiples
  definition of, 72, 125
  least common multiple (LCM), 72–84, 125–126, 348, 425
  listing, 72–75

Multiplication
  of algebraic fractions, 322–323, 420
  associative property of, 39, 120, 172, 213, 292, 549
  of binomials, 969–970, 978
  commutative property of, 38, 120, 172, 213, 292, 549
  of decimals, 472–474, 540–542
    by 0.1, 0.01, 0.001, and so on, 476, 540
    by 10, 100, 1,000, and so on, 475, 540
    on calculator, 474
    estimating products, 479
    by powers of 10, 475–477
  of decimals with like signs, 541
  of decimals with unlike signs, 541
  distributive property of, over addition, 252, 253, 293
  evaluating expressions with, 171–173
  of even numbers of negative integers, 173, 213
  of exponential expressions with like bases, 951–952
  FOIL method for, 969
  of fractions, 307, 318–332, 419–421
  horizontal form of, 34
  of integers, 169–176, 171, 213–214
  of integers with different (unlike) signs, 169–170, 213
  of integers with same (like) signs, 170–171, 213
  key words and phrases that indicate, 42, 120
  in long division, 50, 52, 53, 54, 487
  of mixed numbers, 364–366, 366–368, 427–430, 428
  of monomials, 967–968, 977
  of negative integers, 171, 173, 213
  of odd numbers of negative integers, 173, 213
  of polynomials, 970–972, 977–978
  of polynomials by monomials, 968, 977
  properties of, 38–40, 172
  related multiplication statements for division, 48–49, 121, 180, 182
  rules for exponents, 950–957
  of signed decimals, 477
  of signed fractions, 419
  solving equations with, 105–113, 130–131
  in vertical form, 34, 119, 473, 971
  vocabulary of, 473
  of whole numbers, 34–47, 119–121
    by 10, 100, 1,000, and so on, 35, 119
    by numbers that end with zero, 35–36
    by one-digit numbers, 34–35
    by two- (or more) digit numbers, 36–38
Multiplication property
  of 0, 39, 119, 172
  of 1, 39, 119, 172, 307
  of equality, 106, 130, 196
    clearing equations of fractions with, 405–407, 408
    solving equations with, 106–107, 526
Multiplication symbol, 34

Natural-number exponents, 950
Negative bases, 173–175
Negative decimals, 445
Negative integers
  multiplying, 173, 213
  powers of, 174, 213
  set of, 137
Negative numbers
  definition of, 136, 137, 208
  division with, 184
  entering, 156
  multiplication with, 171
  raising to a power, 175
  subtraction with, 165
  words and phrases that indicate, 202
Negative sign, 136, 199, 208
Negative square root, 515, 547
Net gain or loss, 192
Net income, 139
Nonagons, 844, 925
Notation
  decimal, 444
  expanded, 4–5, 114, 446, 536
  exponential, 68
  standard, 2, 445
Number lines
  graphing decimals on, 450, 506
  graphing fractions on, 363–364, 506
  graphing integers on, 137–139
  graphing mixed numbers on, 363–364
  horizontal, 138
  origin on, 5
  values on, 208
  vertical, 138
  whole numbers and, 5, 114
Number system, base-10, 2
Numbers. See also Whole numbers
  composite, 65, 123
  decimal. See Decimals
  graphing, 5, 138, 450
  irrational, 504
  large, 541
  mixed
    adding, 374–376, 376–378, 430–431
    definition of, 427
    dividing, 364–366, 427–430, 429
    graphing, 363–364, 428
    and improper fractions, 427
    multiplying, 364–366, 427–430, 428
    parts of, 360–361
    ratios that involve, 561–562
    subtracting, 378–380, 430–431
    writing as decimals, 546, 582
    writing as improper fractions, 361–362, 374, 428
    writing improper fractions as, 362–363, 428
  negative
    definition of, 136, 137, 208
    division with, 184
    entering, 156
    multiplication with, 171
    raising to a power, 175
    subtraction with, 165
    words and phrases that indicate, 202
  perfect, 72
  positive, 136, 137, 208
  prime, 65, 123
  proportional, 576, 627
  rational, 307, 504
  real, 504
  signed, 136
  whole. See Whole numbers
Number-value problems, 279–282, 297
Numeration, decimal, 444, 535
Numerators, 85, 126, 187, 216
  0, 306
  definition of, 387, 416
  identifying, 304–305

Obtuse angles, 823, 920
Obtuse triangles, 846, 926
Octagons, 844, 925
Odd powers of negative integers, 174, 213
Odd whole numbers, 64–65, 123
Of (word), 420
One (1), 65
  fraction form of, 306, 417
  fractions with denominator of, 306
  implied coefficients of, 260
  multiplication property of, 39, 119, 172, 307
Operations
  inverse, 25, 49
  order of
    on calculators, 91
    with complex fractions, 387–389, 432–434
    with decimals, 493, 544
    with integers, 187, 212, 216
    with square roots, 517
    with whole numbers, 84–87, 126–127
Opposite of the opposite rule, 141, 208
Opposite symbol, 142, 208
Opposites (additive inverses), 141, 154–155, 208
  addition property of, 155, 210, 211
  distributing a factor of –1, 256
  of integers, 141–142
  solving equations that involve –x, 199
  of a sum, 256, 293
Order of factors, 251
Order of operations rule
  on calculators, 91
  with complex fractions, 387–389, 432–434

with decimals, 478, 493, 544
with integers, 187, 212, 216
with square roots, 517
with whole numbers, 84–87, 126–127
Ordered pairs
  complete ordered-pair solutions of equations, 768–770
  definition of, 768, 771
  plotting or graphing, 771–774, 806
  as solutions of equations, 767–768, 806
Origin, 5
Overbar, 502, 545

Parallel lines, 833–843, 923–925
  cut by transversals, 836–839
  definition of, 833, 834, 923
  identifying, 833–834
Parallelograms, 932
  area of, 890, 934
  definition of, 875
  rectangles, 877
Parentheses, 17, 34, 188
Partial products, 38
Part-to-whole (or amount-to-base) ratios, 664
Pentagons, 844, 885, 925
Per (word), 563, 565
Percent
  approximating, 727
  decimal equivalents, 652
  of decrease, 683–687, 724
  definition of, 644, 714
  estimation with, 696–700, 726–728
  fractional equivalents, 652
  greater than 100%, 715
  of increase, 683–687, 724
  less than 1%, 715
  mixed number, 716
  solving percent equations to find, 659–662
  solving percent proportions to find, 667–669
  writing
    as a decimal, 648–650, 716
    as a fraction, 645–648, 715
  writing decimals as, 650, 716
  writing fractions as, 650–653, 716
Percent equation (formula)
  definition of, 657, 659
  solving
    to find the amount, 658–659
    to find the base, 662–664
    to find the percent, 659–662
  solving percent problems with, 657–673, 717–722
  translating percent sentences to, 657–658, 717
Percent equation method
  finding percent of increase or decrease with, 684, 685, 687
  solving percent problems with, 717, 720, 723, 724
Percent problems
  solving
    with percent equation method, 717

with percent equations and proportions, 657–673, 717–722
  that involve 1% and 10%, 696–697
  that involve 20%, 30%, 40%, 697
  that involve 50%, 25%, 5%, and 15%, 698–699
  that involve 200%, 700
  types of, 657
Percent proportion method
  for calculating tax rates, 680–681
  finding percent of increase or decrease with, 684, 685–686, 687
  solving percent problems with, 719, 721, 723, 724
Percent proportions
  amount (or part) of, 665
  base (or whole) of, 665
  definition of, 664, 719
  solving
    to find the amount, 666–667
    to find the base, 669–671
    to find the percent, 667–669
  translating percent sentences to, 665
  ways to identify terms in, 665
  writing, 664–666
Percent ratios, 664
Percent sentences
  definition of, 657
  translating to percent equations, 657–658, 717
  translating to percent proportions, 665
Perfect numbers, 72
Perfect square radicands, 547
Perfect squares, 515, 547
Perimeters, 885–894, 934–937
  circumference of a circle, 937
  definition of, 21, 117, 934
  of polygon, 885–887
  of polygon combinations, 887
  of rectangle, 21–22, 117
    finding, 263–265
    formula for, 264, 886
  of square, 21–22, 117, 885
    finding, 263–265
    formula for, 264
Periods, 2, 114
Perpendicular lines, 833–843, 923–925
  definition of, 834, 923
  identifying, 833–834
Pictographs
  key for, 798
  reading, 743–744
Pie charts, 353, 671, 740, 744
Place value, 2–3, 114
  identifying, 444–445
Place-value chart, 114, 444
Planes
  identifying, 820–821, 919
  naming, 820–821, 919
Plotting ordered pairs, 771–774, 806
Plotting points, 771–774
Plus sign, 16

Points
  coordinates of, 771–774
  endpoints, 821, 920
  graphing or plotting, 771–774
  identifying, 820–821, 919
  midpoints, 821, 920
  naming, 820–821, 919
Polygons, 875–884, 932–934
  area of, 890, 934
  classifying, 844–845, 925
  combinations of
    area of, 892–894
    perimeter of, 887
  definition of, 844, 925
  diagonal of, 876, 932
  frequency polygons, 748, 801
  number of sides of, 933
  octagons, 844
  perimeter of, 885–887
  quadrilaterals, 844, 874–884, 925, 932–934
  regular, 844, 925
  sides of, 844
  sum of angles of, 880, 933
  vertex of, 844, 925
Polynomials
  adding, 961–963, 977
  definition of, 957, 976
  degree of, 958, 976
  evaluating, 959, 976
  introduction to, 957–961, 976
  multiplying, 970–972, 977–978
  multiplying by monomials, 968, 977
  multiplying two, 977, 978
  subtracting, 963–965, 977
  subtracting two, 977
  vocabulary for, 957–958
Positive integers
  dividing, 180
  set of, 137
Positive numbers, 136, 137, 208
Positive sign, 136
Postulates, A-9
Power of a product rule, 954, 975
Power rule for exponents, 953, 975
Powers
  even, of negative integers, 174
  odd, of negative integers, 174
  of powers, 953
  of products, 954–955, 975
  raising exponential expressions to, 952–953
  raising negative numbers to, 175
  roots and, A-15
  of ten
    dividing decimals by, 491–492
    multiplying decimals by, 475–477
  of two, 68
Prices
  retail price, 241, 465
  sale price, 240–241, 688–690, 725
  unit prices, 566, 625
Prime factorization
  with division ladders, 67
  with factor trees, 65–66
  finding, 124

Prime factors
  finding
    with division ladders, 67
    with factor trees, 65–66
  finding GCF with, 79–81
  finding LCM with, 75–78
Prime numbers, 65, 123
Principal, 703
Prisms: volume of, 910, 911, 939
Probability, A-1–A-6
Problem solving, six-step strategy for, 130–131, 100, 296–297, 436, 546, 627
Problems: analyzing, 296
Product rule for exponents, 951, 975
Products
  cross products, 574, 625
  cross-products property (means-extremes property), 575
  of decimals, 479
  definition of, 34, 63
  partial products, 38
  power of a product rule, 975
  powers of, 954–955, 975
  simplifying, 251–252, 522–523
  of whole numbers, 40
Profit, 241
Proper fractions, 305, 417
Properties
  of addition, 17–18
  congruence properties, 863–865
  of division, 49–50
  of equality, 196
  of multiplication, 38–40
  of parallel lines cut by transversal lines, 836–839
  of rectangles, 876
  of similar triangles, 931
Property of vertical angles, 825
Proportional numbers, 576, 627
Proportions, 572–586, 626–628
  definition of, 572, 626, 866
  determining whether true or false, 573–576, 626
  percent, 664–673, 719
  solving, 576–579, 627, 666
  solving percent problems with, 657–673, 717–722
  terms, 626
  writing, 572–573, 579–582
Protractors, 823, 920
Pyramids: volume of, 910, 912, 939
Pythagoras, 855
Pythagorean equation, 855, 928
Pythagorean theorem, 855, 928–929
  approximating lengths of sides of right triangles with, 857–858
  converse of, 858–859, 929
  finding lengths of sides of right triangles with, 855–857

Quadrilaterals, 875–884, 932–934. *See also specific types*
  classifying, 844, 875–876, 925
  definition of, 925, 932
  naming, 876
  perimeter of, 885
  types of, 932

Quantity (word), 18, 38, 88
Quotients
  decimal, 490–491
  definition of, 48
  estimating
    for decimals, 491, 543
    for whole numbers, 56–57, 122

Radical expressions, 514, 516, 547
Radical symbol, 514, 547
Radicands, 514, 547
Radius, 900, 937
Raised dot, 34
Range
  definition of, 212, 761, 804
  describing a set of values with, 761–762
Rates, 623–626
  definition of, 563, 625
  unit rates, 564–566, 625
  words to separate quantities compared in, 563
  writing
    as fractions, 563–564, 625
    as unit rates, 565, 625
Rational numbers, 307, 504. *See also* Fractions
Ratios, 558–571, 623–626
  amount-to-base (or part-to-whole), 664
  converting units for, 562–563
  definition of, 558, 623
  equal, 559
  percent, 664
  simplifying, 561–562, 624
  that involve decimals and mixed numbers, 561–562, 624
  writing
    as fractions, 558–561, 623
    in simplest form, 624
Rays, 821, 822, 920
Real numbers, 504
Reasoning
  deductive, A-9–A-11
  inductive, A-7–A-9
Reciprocals
  of algebraic fractions, 423
  definition of, 333, 422
  finding, 333–334, 422
Rectangles
  area of
    definition of, 42, 120
    formula for, 43, 120, 890, 934
  definition of, 875, 932
  dimensions of, 21
  parallelograms, 877
  perimeter of, 21–22, 117
    finding, 263–265
    formula for, 264, 886, 934
  properties of, 876, 932
Rectangular arrays, 41, 120
Rectangular coordinate system, 767–774, 806
  constructing, 770–771
  definition of, 770
Rectangular solids: volume of, 910, 911, 939
Regrouping
  in decimal subtraction, 461, 538
  in subtraction of whole numbers, 23

Remainders
  definition of, 53
  long division with, 53–54
Repeating decimals, 501–502, 545, 716
  rounding, 504–506
  writing fractions as, 501–504
Retail price, 241, 465
Rhombus, 875, 932
Right angles, 823, 920
Right triangles, 846, 926
  approximating lengths of sides of, 857–858
  finding lengths of sides of, 855–857
Roots, A-15
Rounding amounts of money, 537
Rounding decimal quotients, 490–491
Rounding decimals, 450–452, 537, 543, 545
Rounding digits, 7, 450, 537
Rounding down, 6
Rounding, front-end
  in addition, 19, 116
  in decimal addition or subtraction, 464, 539
  in decimal multiplication, 479, 541
  in multiplication, 40, 120
  in percent problems, 696, 697
  in subtraction, 26, 117, 539
Rounding repeating decimals, 504–506
Rounding up, 6
Rounding whole numbers, 6–8, 115
Rulers
  in inches, 587–588, 629
  metric, 601–602, 632

Sale price
  finding, 688, 725
  formula to find, 240–241
Sales per square foot, 570
Sales tax dollar amounts, 722
Sales tax rates, 722
Sales taxes
  definition of, 722
  finding, 680
SAS property, 863, 930
Scale, 580
Scalene triangles, 845, 925
Science formulas
  for converting degrees Fahrenheit to degrees Celsius, 243, 291, 618
  for distance, 291
  for distance fallen, 244, 291
Semicircles, 901, 937
Sets
  of integers, 137, 208
  of whole numbers, 2, 114
Side lengths, 876–877
  of right triangles
    approximating, 857–858
    finding, 855–857
Sides
  of angles, 822, 920
  of polygons, 844, 933
Sign rules, 422

Signed decimals
  adding, 463–464, 539
  dividing, 492–493, 544
  multiplying, 477, 541
  subtracting, 463–464, 539
Signed numbers, 136
Similar triangles, 862–875, 930–931
  definition of, 866, 930
  determining similarity of, 865–868
  property of, 866, 931
Similarity, AAA, 867
Simple interest
  definition of, 703, 704, 728
  formula for, 704
Simplifying algebraic fractions, 312–313, 418
Simplifying complex fractions, 391–393, 433
Simplifying expressions, 251–258, 252, 263, 292–293, 522, 549
  to solve equations, 268–275, 295–296
Simplifying fractions, 309–312
Simplifying products, 251–252, 321–322, 522–523
Simplifying ratios that involve decimals, 624
Simplifying ratios that involve decimals and mixed numbers, 561–562
Simplifying ratios that involve mixed numbers, 624
Simplifying special fraction forms, 306–307
Skew lines, 833
Slash mark /, 565, 625
Solids
  height of, 911
  volume of, 910, 911, 939
Solutions
  definition of, 128, 295
  determining whether numbers are, 96–97
  table of, 769
Special fraction forms, 306–307
Spheres: volume of, 910, 939
Square radicands, perfect, 547
Square roots
  approximate, 518–519, 548
  on calculator, 516
  definition of, 514, 547
  expressions that contain, 516–518, 548
  formulas that involve, 518, 548
  of fractions and decimals, 516
  negative, 515, 547
  of perfect square, 514–516
  perfect squares, 547
Squares
  area of, 890, 934
  definition of, 21, 875, 932
  perfect squares, 515, 547
  perimeter of, 21–22, 117, 885
    finding, 263–265
    formula for, 264, 934
SSS property, 863, 930
Standard form
  of whole numbers, 2, 3–4
  writing large numbers in, 541
Standard notation, 2, 445

Straight angles, 823, 920
Strict inequalities, 140
Subscripts, 825
Subsets, 137
Substitution, 96
Subtraction
  of algebraic fractions, 351–352, 425
  checking with addition, 25–26
  of decimals, 460–462, 538–540
  distributive property for, 253
  distributive property of multiplication over, 253, 293
  expressions involving addition and subtraction, 27–28, 163–164
  of fractions, 343–354, 424–427
  of fractions with different denominators, 345–348, 424
  of fractions with same denominator, 343–345, 424
  horizonal form of, 22
  of integers, 160–168, 211–213
  key words and phrases that indicate, 27, 118, 226
  in long division, 50, 52, 53, 54, 487
  of mixed numbers, 378–380, 430–431
  with negative numbers, 165
  of polynomials, 963–965, 977
  related addition statements, 25
  rule for, 211
  of signed decimals, 463–464, 539
  solving equations with, 96–105, 128–129
  translating, 227
  of two polynomials, 977
  of two signed decimals, 539
  in vertical form, 22, 117, 378–380, 430, 460–461, 538
  of whole numbers, 15–33, 116–119
Subtraction property of equality, 99, 128, 196
Subtraction rule, 160–163
Subtraction symbol, 22, 208
Subtrahend, 22, 161, 460–461
Sums
  in decimal addition, 459, 464–465
  definition of, 16
  opposite of, 256
  of whole numbers, 19
Supplementary angles
  definition of, 826, 827, 921
  in trapezoids, 878

Table of solutions, 769, 780, 806
Table of values, 769
Tables and graphs. *See* Graphs and tables
Taxes
  sales taxes, 679–682
  tax rates, 679–682
Temperature conversion, 243, 618–619, 636
Terminating decimals, 501, 545
Termination, 501

Terms
  coefficients of, 259–260, 294
  constant, 289, 294
  defined and undefined, A-9
  definition of, 253, 294, 957, A-9
  in expressions, 294
  factors of, 294
  identifying, 259–260
  like terms
    combining, 261–263, 269–270, 524–525, 549
    definition of, 269, 294, 524, 549
    examples, 524
    identifying, 261
  lowest, 309, 417, 559
  in proportions, 574, 626
  unknown, 576–579
  unlike, 524
  variable, 435
Test digits, 7, 115, 450, 537
Theorems
  AAA similarity theorem, 867, 930
  definition of, 855, A-9
  fundamental theorem of arithmetic, 66
  isosceles triangle theorem, 846, 926
    converse of, 847, 926
  Pythagorean, 855, 928–929
    approximating lengths of sides of right triangles with, 857–858
    converse of, 858–859, 929
    finding lengths of sides of right triangles with, 855–857
Tick marks, 845, 847, 925
Time
  definition of, 703
  unit conversion factors, 594
  units of, 594, 595, 630
Times symbol, 34
Total, 16
Total amount, 704, 728
Total cost
  calculating, 679–682
  definition of, 722
  finding, 680, 722
  formula for, 722
Transversal lines, 834, 923
  properties of parallel lines cut by, 836–839
Trapezoids
  base of, 878, 933
  definition of, 875, 877
  finding angle measures of, 877–879
  formula for area of, 890, 934
  isosceles, 878, 933
  legs of, 877, 933
  supplementary angles in, 878
Tree diagrams, A-2
Triangles, 844–854, 925–927
  acute, 846, 926
  angles of, 847
  area of: formula for, 326, 390, 421, 890, 934

classifying, 844, 845–846, 925
congruent, 862–875, 930–931
  corresponding parts of, 862, 930
  definition of, 862, 930
  proving congruence of, 863–865
definition of, 845, 925
equilateral, 845, 925
finding angle measures of, 847–849
formula for perimeter of, 934
isosceles, 845, 846, 925, 926
isosceles triangle theorem, 846, 926
legs of, 846, 926
naming, 845
obtuse, 846, 926
perimeter of, 885
right, 846, 926
  approximating lengths of sides of, 857–858
  finding the lengths of sides of, 855–857
scalene, 845, 925
similar, 862–875, 930–931
  definition of, 866, 930
  determining similarity of, 865–868
  property of, 866, 931
sum of the measures of the angles of, 926
Trinomials, 957, 976
Triple-bar graphs, 742, 798

Unit conversion factors
  for American and metric units, 614–618, 635, 636
  for American units of capacity, 593, 630
  for American units of length, 589
  for American units of weight, 591
  definition of, 629
  for metric units, 602–603, 632, 633
  for units of time, 594
Unit prices, 566, 625
Unit rates
  definition of, 564, 625
  finding, 564–566
  writing rates as, 565, 625
Units of measurement
  American, 587–599, 614–619, 629–632
  converting between American and metric, 614–619, 635–637
  cubic units, 609, 634, 909, 939
  metric, 600–613, 632–634
Units of time, 594
Unknowns
  definition of, 96
  expressions to represent, 228–231
  solving proportions to find, 576–579
Unlike terms, 524

Values: table of, 769
Variable terms, 435
Variables
  in algebraic expressions, 228
  assigning, 296
  definition of, 17, 43, 116, 120, 289
  in equations, 96
  equations in one variable, 767
  equations in two variables, 767–774, 806–807
  equations with variable terms on both sides, 270–271
  in formulas, 390
  isolating, 200, 577
  replacing with numerical value, 238
Vertex
  of angle, 822, 920, 926
  of polygon, 844, 925
Vertex angles, 846, 926
Vertical angles, 824–825, 921
Vertical axis, 741, 798
Vertical form
  of addition, 16, 17, 116
  of addition of mixed numbers, 376–378, 430
  of decimal addition, 458–459, 538
  of decimal subtraction, 460–461, 538
  of multiplication, 34, 119
  of multiplication of decimals, 473
  of multiplication of polynomials, 971
  of subtraction, 22, 117, 430
  of subtraction of mixed numbers, 378–380
Vertical lines, 788, 807
Vertical number lines, 138
Volume
  of cylinders, cones, and spheres, 912–914
  definition of, 909, 939
  formulas for, 910, 939
  of rectangular solids, prisms, and pyramids, 909–912, 939
  of solids, 909–912, 912–914, 939

Weight (mass)
  American units, 591–593, 629
  converting between American and metric units, 635
  definition of, 633
  equivalent weights and masses, 615
  metric units, 605–606, 606–607, 633
  unit conversion factors, 591
Weighted mean
  definition of, 757
  finding, 757–759, 804
Whole numbers
  adding, 15–33, 116–119
  comparing, 5–6
  composite numbers, 65, 123
  differences of, 22, 26
  dividing, 48–62, 121–123

dividing decimals by, 486–488, 542
estimating differences of, 26
estimating products of, 40, 120
estimating quotients of, 56–57
estimating sums of, 19
even, 64–65, 123
expanded form of, 4–5, 114
factoring, 63–64
four-digit, 4
introduction to, 2–15, 114–116
multiplying, 34–47, 119–121
  by 10, 100, 1,000, and so on, 35, 119
  by numbers that end with zero, 35–36
  by one-digit numbers, 34–35
  by two- (or more) digit numbers, 36–38
odd, 64–65, 123
order of operations for, 84–91, 126–127
part of decimals, 444, 535
prime numbers, 65, 123
prime-factored form of, 66
products of, 40
quotients of, 56–57
reading and writing, 3, 114
rounding, 6–8, 115
rounding of, 115
set of, 2, 114, 137
standard form of, 2, 3–4, 114
as subset, 137
subtracting, 15–33, 116–119
sums of, 19
tables and graphs involving, 8–10
that end with zero
  dividing, 56, 122
  multiplying, 35–36
in words, 3–4, 114
written-word form of, 114
Width, 21

x-axis, 771, 806
x-coordinate, 771
x-intercept, 784, 807

Yards, 588
y-axis, 771, 806
y-coordinate, 771
y-intercept, 784, 807

Zero (0), 65, 137, 333
  addition property of, 18, 154, 210
  division by, 182–183, 215
  division of, 182–183, 215
  division with, 50, 122, 182, 215
  fractions with denominator of, 306
  fractions with numerator of, 306
  multiplication property of, 39, 119, 172
  undetermined fractions, 306
  whole numbers that end with
    dividing, 56, 122
    multiplying whole numbers by, 35